BUSINESS/SCIENCE/TECHNOLOGY DIVISION
CHICAGO PUBLIC LIBRARY
400 SOUTH STATE STREET
CHICAGO, ILLINOIS 60605

REF
QB
1
.S775
2004

HWLCTC

Chicago Public Library

Y0-DLD-943

...guides plus : a world-wide directory

Chicago Public Library

REFERENCE

Form 178 rev. 11-00

StarGuides Plus

StarGuides Plus

A World-Wide Directory of Organizations in Astronomy and Related Space Sciences

Compiled by

André Heck
Strasbourg Astronomical Observatory, France

KLUWER ACADEMIC PUBLISHERS
DORDRECHT / BOSTON / LONDON

A C.I.P. Catalogue record for this book is available from the Library of Congress.

ISBN 1-4020-1926-2

Published by Kluwer Academic Publishers,
P.O. Box 17, 3300 AA Dordrecht, The Netherlands.

Sold and distributed in North, Central and South America
by Kluwer Academic Publishers,
101 Philip Drive, Norwell, MA 02061, U.S.A.

In all other countries, sold and distributed
by Kluwer Academic Publishers,
P.O. Box 322, 3300 AH Dordrecht, The Netherlands.

Printed on acid-free paper

All Rights Reserved
© 2004 Kluwer Academic Publishers
No part of this work may be reproduced, stored in a retrieval system, or transmitted
in any form or by any means, electronic, mechanical, photocopying, microfilming, recording
or otherwise, without written permission from the Publisher, with the exception
of any material supplied specifically for the purpose of being entered
and executed on a computer system, for exclusive use by the purchaser of the work.

Printed in the Netherlands.

Table of Contents

Table of Contents	i
Foreword	ix
How to use this directory	1
Albania	3
Algeria	4
Argentina	6
Armenia	12
Australia	14
Austria	34
Azerbaijan	43
Bangladesh	44
Barbados	45
Belarus	46
Belgium	47
Bolivia	62
Brazil	65
Bulgaria	74
Canada	78
Channel Islands	106
Chile	107
China (PRC)	112
Colombia	119
Croatia	121
Cuba	123
Czech Republic	124
Denmark	134
Ecuador	141
Egypt	142
El Salvador	143
Estonia	144
Finland	146
France	154
Georgia	206
Germany	207
Gibraltar	267
Greece	268
Guam	271
Honduras	272
Hungary	273
Iceland	279
India	281
Indonesia	290
Iran	292
Iraq	294
Ireland	295
Israel	299
Italy	302
Japan	342
Jordan	363

Kazakhstan	364
Kenya	365
Korea (DPRK)	366
Korea (ROK)	367
Latvia	371
Lebanon	373
Lithuania	374
Luxembourg	376
Macedonia	377
Malaysia	378
Malta	380
Mauritius	381
Mexico	382
Moldova	387
Mongolia	388
Netherlands	389
New Zealand	408
Nigeria	414
Norway	415
Pakistan	421
Paraguay	422
Peru	423
Philippines	425
Poland	427
Portugal	433
Puerto Rico	437
Romania	438
Russia	440
Saudi Arabia	453
Serbia-Montenegro	454
Singapore	456
Slovak Republic	458
Slovenia	462
South Africa	464
Spain	470
Sri Lanka	487
Sweden	488
Switzerland	500
Syria	515
Taiwan (ROC)	516
Tajikistan	518
Thailand	519
Trinidad & Tobago	521
Turkey	522
Ukraine	525
United Kingdom	531
Uruguay	586
USA - Alabama	589
USA - Alaska	592
USA - Arizona	593
USA - Arkansas	606
USA - California	608
USA - Colorado	652
USA - Connecticut	662
USA - Delaware	667
USA - District of Columbia	669
USA - Florida	678
USA - Georgia	687

USA - Hawaii	692
USA - Idaho	697
USA - Illinois	698
USA - Indiana	710
USA - Iowa	716
USA - Kansas	721
USA - Kentucky	725
USA - Louisiana	729
USA - Maine	732
USA - Maryland	734
USA - Massachusetts	746
USA - Michigan	761
USA - Minnesota	770
USA - Mississippi	774
USA - Missouri	776
USA - Montana	780
USA - Nebraska	782
USA - Nevada	786
USA - New Hampshire	788
USA - New Jersey	791
USA - New Mexico	799
USA - New York	805
USA - North Carolina	827
USA - North Dakota	834
USA - Ohio	835
USA - Oklahoma	846
USA - Oregon	848
USA - Pennsylvania	851
USA - Rhode Island	864
USA - South Carolina	866
USA - South Dakota	869
USA - Tennessee	870
USA - Texas	877
USA - Utah	890
USA - Vermont	893
USA - Virginia	895
USA - Washington	907
USA - West Virginia	912
USA - Wisconsin	915
USA - Wyoming	922
US Virgin Islands	924
Uzbekistan	925
Vatican	926
Venezuela	927
Zimbabwe	929
Telephone and Telefax National Codes	931
Abbreviations, Acronyms, Contractions, and Symbols	935
Index	937
Updating Form	1138

Foreword

StarGuides Plus is the enlarged and updated version of a long line of directories published since 1978 (see Heck 2003a for a history of those resources) and gathering together all practical data available on associations, societies, scientific committees, agencies, companies, institutions, universities, etc., and more generally organizations, involved in astronomy and space sciences. Many other types of entries have also been included such as academies, advisory and expert committees, bibliographic services, consultants, data and documentation centres, dealers, distributors, funding agencies and organizations, journals, manufacturers, meteorological services, norms and standards offices, planetariums, public observatories, publishers, research institutions in related fields, science museums, software producers and distributors, etc.

Besides astronomy and space sciences, related fields such as aeronautics, aeronomy, astronautics, atmospheric sciences, chemistry, communications, computer sciences, data processing, education, electronics, energetics, engineering, environment, geodesy, geophysics, information retrieval, management, mathematics, meteorology, optics, physics, remote sensing, and so on, are also covered when justified.

The basic philosophy of this directory is to provide practical data which one seeks to always have at one's disposal. Over the years they have proved to not only be valuable auxiliaries for improving national and international relationships, but also to be efficient tools for helping laypersons and public bodies to contact organizations easily.

These directories have taken advantage of the experience gained with each successive edition, especially in the development of techniques for collecting, verifying and treating the data. To compile a directory of real value is a quite different venture to just reproducing and distributing, with comments of greater or lesser interest, data collected indiscriminately from all available sources (including the World-Wide Web which already contains too many obsolete pointers and too much outdated information). If professional file construction techniques are necessary, they cannot spare the extensive background, unrewarding and very careful work which is indispensable for the compilation of a valuable directory.

StarGuides Plus gathers together about 6,000 entries from 100 countries. The information is given in an uncoded way for easy and direct use. For each entry, all practical data are listed: city, postal and electronic-mail addresses; telephone and telefax numbers; WWW sites; foundation years; numbers of members or staff; main activities; titles, frequencies, ISS-Numbers and circulations of periodicals produced; names and geographical coordinates of observing sites; names of planetariums; awards, prizes or distinctions granted; and so on. City reference coordinates are provided for each entry (based on the location of the head office or of the main centre of activities). Organizations not answering several updating requests have been deleted, and numerous new ones have been introduced since the last edition.

The entries are listed alphabetically in each country. An exhaustive index gives a breakdown not only by different designations and acronyms, but also by location and major terms in the names. A search for information in the directory would normally begin with consultation of this index.

Thematic subindices of academies, awards, bibliographical services, consultants, data centres, dealers and distributors, funding organizations, IAU observatory codes, Internet service providers, ISS-Numbers, manufacturers, meteorological offices, norms and standards institutes, observatories, periodicals, planetariums, publishers, science museums, software producers, etc. are also provided as well as a list of telephone and telefax national codes.

The quality behind the master files of *StarGuides* has been recognized by their implementation as on-line databases by the European Space Information System (ESIS) group under the label *StarWays* (Heck et al. 1992), by the European Southern Observatory (ESO) under the label *StarGates* (Albrecht & Heck 1994), and at the *Centre de Données astronomiques de Strasbourg (CDS)* as the database *StarWorlds* (Heck et al. 1994). They are now reachable on the web site of *Kluwer Academic Publishers* (http://www.wkap.nl/).

An extensive dictionary of abbreviations, acronyms, contractions, and symbols used in astronomy, space sciences, and related fields has been compiled in parallel and is available on paper as the sister publication *StarBriefs Plus* (Heck 2003b).

When compiling resources such as *StarGuides* and *StarBriefs*, one cannot but be impressed by the very broad spectrum of disciplines to which astronomy and related space sciences are linked, and by the very large variety of techniques applied in these fields.

The successive editions of the directories give fairly accurate global pictures of the active organizations in the fields covered. Their sequence testifies to the sometimes rapid evolution of scientific interests, of data collecting and handling techniques, as well as of communications in the broad sense. A few countries have also rearranged the structure of their national facilities in the course of the past years. With the rapid spreading of the WWW and the globalization of telecommunications, we now include only the headquarters of commercial organizations since the data (often changing) of their subsidiaries and branches can be obtained from their main web sites.

Studies based on data extracted from the master files have been published. They provide geographical distributions and general characteristics of various categories of astronomy-related organizations. See Heck (2000) for an exhaustive review.

Feedback from readers on possible modifications or additions to the data published here would be highly appreciated in order to ensure permanent accuracy of the master files (see updating form at the end of the volume). The information is provided in *StarGuides* 'bone fide'. The best is done to keep track of the modifications happening and to implement them as soon as they are confirmed or recognized by the international community.

Acknowledgements

Finally it is a very pleasant duty to express our gratitude to all persons and organizations who contributed over about thirty years to the very substance of the master files used here, by returning the questionnaires, by providing the relevant documentation, by participating in the various procedures of maintenance, validation and verification of the information, or otherwise. Our resources have been conceived for them and for the vast community of users.

The implementation, as databases, of our data by the European Space Agency, the European Southern Observatory, Strasbourg Astronomical Data Centre, and now Kluwer Academic Publishers have been strong incentives to continue and always improve these time-consuming compilations for the benefit of the best possible communication within the astronomical community, as well as between it and the outside world.

November 2003.

REFERENCES

Albrecht, M. & Heck, A.: 1994, StarGates – An On-line Database of Astronomy, Space Sciences, and Related Organizations of the World, *Astron. Astrophys. Suppl.* **103**, 473-474.

Heck, A.: 2000, Astronomy-Related Organizations: Geographical Distributions, Ages and Sizes, in *Organizations and Strategies in Astronomy – Vol. 1*, Ed. A. Heck, Kluwer Acad. Publ., Dordrecht, 7-66.

Heck, A.: 2003a, From Early Directories to Current Yellow-Page Services, in *Information Handling in Astronomy – Historical Vistas*, Ed. A. Heck, Kluwer Acad. Publ., Dordrecht, 183-205.

Heck, A.: 2003b, StarBriefs Plus – A Dictionary of Abbreviations, Acronyms and Symbols in Astronomy and Related Space Sciences, Kluwer Acad. Publ., Dordrecht, iv + 1114 pp. (ISBN 1-4020-1925-4).

Heck, A., Ciarlo, A. & Stokke, H.: 1992, StarWays – An On-line Database of Astronomy, Space Sciences, and Related Organizations of the World, *Astron. Astrophys. Suppl.* **96**, 565-566.

Heck, A., Egret, D. & Ochsenbein, F.: 1994, StarWorlds – StarBits (announcement of two databases), *Astron. Astrophys. Suppl.* **108**, 447-448.

Author's address: **André HECK**
Observatoire Astronomique
11, rue de l'Université
F-67000 Strasbourg
France
Telephone: (+33)(0) 390 242 420
Telefax: (+33)(0) 388 508 772
WWW: http://vizier.u-strasbg.fr/~heck

How to use this directory

A search for information in the directory would normally begin with consultation of the three-level index located at the end the volume (yellow pages). It may be tackled by the name or by any of the major terms appearing in the title of the organization. Acronyms may also be used, as well as geographical names (city, region, etc.).

Thematic subindices are available for academies, awards, bibliographic services, consultants, data centres, dealers and distributors, funding organizations, IAU observatory codes, Internet service providers, ISS-Numbers, manufacturers, meteorological offices, norms and standards institutes, observatories, periodicals, planetariums, publishers, science museums, software producers, and so on.

In the directory itself, the information is given in an uncoded way for easy and direct use. A one-column layout has been selected to avoid cutting, across two lines, important data such as addresses, World-Wide Web URLs, coordinates, and so on. The ample blank space can also be used for annotations.

Besides the astronomy-related organizations (about 6,000), listed alphabetically in each country (over a hundred), we systematically included the national representative meteorological offices (routinely approached by observers world-wide) and the national institutions responsible for norms and standards (to be consulted regarding compatibility of instrumentation and similar technical issues, for instance).

If several addresses are listed for a given entry, the first one should be given preference for all contacts by post. Eighteen entries have neither postal nor street addresses, but do have electronic contacts. About a thousand cross-references (or pointers) are provided throughout the directory for accessing information via alternate denominations.

The individual telephone and telefax numbers are preceded, whenever possible and wherever they still exist, by the national inter-area first digits (0, 9, and so on). National telephone and telefax codes for countries and territories around the world are provided in a separate section. They should be preceded by the international access number of the country or territory from which the call is placed (*e.g.* 00, 011, etc.). Unless otherwise indicated, all electronic-mail addresses are given for Internet.

This directory includes about 10,000 e-mail addresses and Uniform Resource Locators (URLs) for World-Wide Web (WWW) access. Users are however advised that such electronic addresses are still somewhat volatile and that the best has been done to keep the master files as up-to-date as possible. The latest information in our possession can always be obtained from the electronic version of the directory, called *StarWorlds*, reachable on the web site of *Kluwer Academic Publishers* (http://www.wkap.nl/).

The foundation years correspond to the current organizations and not to possible forerunners (unless clearly stated). The 'activities' listed have been homogenized and standardized as much as possible through keywords from the information delivered by the organizations themselves on the questionnaires returned and from the accompanying documentation.

As already mentioned, coding has been kept to a minimum. The annual frequencies of the periodicals and awards are indicated between parentheses. We insist that we have considered only the periodicals (not occasional books nor publications) currently published. At the request of librarians, we included sub-indices as per periodicals titles and ISS-Numbers.

The geographical coordinates of observing sites are self-explanatory. For each organization, we also entered a position based on the location of its head office or main center of activities that can be useful for mapping or selection purposes.

Finally let us recall that the information presented here is validated and authenticated (from signed and documented questionnaires). The data have been systematically compiled and presented, with a permament updating-process scheme. They give a faithful snapshot of the astronomy-related organizations world-wide at the beginning of the third millenium.

Feedback from readers on possible modifications or additions to the data published here would be highly appreciated in order to ensure permanent accuracy of the master files (see updating form at the end of the volume).

Albania

Albanian Physical Society
c/o L. Cano (Secretary General)
Physics Department
University of Tirana
Tirana
Telephone: (0)42-39479
Electronic Mail: lcano@fshn.tirana.al
Membership: 60
City Reference Coordinates: 019°50'00"E 41°20'00"N

Directorate of Standardization
(Drejtoria e Standardizimit dhe Cilesise - DSC)
Rruga Mine Peza 143/3
Tirana
Telephone: (0)42-26255
Telefax: (0)42-26255
Electronic Mail: dsc@icc.al.eu.org
Staff: 25
City Reference Coordinates: 019°50'00"E 41°20'00"N

Hydrometeorological Institute of Albania
Rruga 'Kongresi i Permetit' 26
Tirana
or :
Rruga Durresi 219
Tirana
Telephone: (0)42-37947
(0)42-23518
Telefax: (0)42-37947
(0)42-23518
Founded: 1949
Staff: 120
Activities: meteorology * hydrology * climatology * oceanography
Periodicals: "Botime Hidrometeorologike"
Observatories: 110 sites for meteorology and 120 sites for hydrometry
City Reference Coordinates: 019°50'00"E 41°20'00"N

Algeria

Association d'Astronomie El-Battani (AAB)
c/o Centre Culturel Communal Ibn-Mahrez
6, rue de Lourmel
Boîte Postale 08 RP
Oran 31000
Telephone: (0)41422657
Electronic Mail: aab.president@caramail.com
WWW: http://www.multimania.com/battani/
Founded: 1985
Membership: 300
Activities: observing * lectures * education * exhibitions * scientific expeditions
Periodicals: "El-Fellek"
City Reference Coordinates: 000°43'00"W 35°43'00"N

Association Scientifique d'Astronomie (ASA) El-Bouzdjani
c/o Maison de la Culture
Medea 26000
Telephone: (0)3-585114
Telefax: (0)3-581208
Founded: 1988
Membership: 50
Activities: observing * meetings * lectures * popularization * exhibitions * meteorology * rocketry
Periodicals: (6) "El-Marsad"
Coordinates: 002°05'00"E 36°25'00"N H1,070m (Mont Tibarine)
City Reference Coordinates: 002°50'00"E 36°12'00"N

Association Sirius d'Astronomie
Cité du 20 Août 1955
Boîte Postale 18
25000 Constantine
Telephone: (0)4-900316
Electronic Mail: sirius_astro@hotmail.com
WWW: http://www.geocities.com/CapeCanaveral/Station/9720
Founded: 1995
Membership: 40
Activities: observing * lectures * exhibitions * radio broadcast * workshops * software * advising
Periodicals: (1-4) "Sirius Voice" (in Arabic)
Coordinates: 006°37'00"E 36°22'00"N H600m
City Reference Coordinates: 006°40'00"E 36°22'00"N

Centre de Recherches en Astronomie, Astrophysique et Géophysique (CRAAG)
Observatoire d'Alger
Route de l'Observatoire
Boîte Postale 63
Bouzareah
16340 Alger
Telephone: (0)2-903160
(0)2-901424
(0)2-941572
(0)2-903267
Telefax: (0)2-903160
(0)2-901424
Electronic Mail: astro@ist.cerist.dz
WWW: http://www.craag.edu.dz/
Founded: 1980
Activities: seismology * geomagnetism * gravimetry * astronomy
Coordinates: 003°01'50"E 36°48'04"N H345m (IAU Code 008)
City Reference Coordinates: 003°08'00"E 36°42'00"N

Institut National Algérien de la Propriété Industrielle (INAPI)
Boîte Postale 403
Centre de Tri
Alger
or :
42, rue Larbi Ben M'hidi
Alger
Telephone: (0)21-732358
(0)21-736084
Telefax: (0)21-739644
Electronic Mail: info@inapi.org
WWW: http://www.inapi.org/

Founded: 1998
Staff: 65
Activities: registration of patents, trademarks and designs
Periodicals: "Bulletin Officiel de la Propriété Industrielle"
City Reference Coordinates: 003°08'00"E 36°42'00"N

Observatoire d'Alger
• See "Centre de Recherches en Astronomie, Astrophysique et Géophysique (CRAAG)"

Office National de la Météorologie
Avenue Khemisti
Boîte Postale 153
Dar El Beïda
Alger
Telephone: (0)2-507393
Telefax: (0)2-508849
WWW: http://www.meteo.dz/
City Reference Coordinates: 003°08'00"E 36°42'00"N

Argentina

Asociación Argentina Amigos de la Astronomía (AAAA)
Avenida Patricias Argentinas 550
1405 Buenos Aires
Telephone: (0)11-48633366
Telefax: (0)11-48633366
Electronic Mail: postmaster@aaaa.org.ar
WWW: http://www.asaramas.com.ar/
Founded: 1929
Membership: 650
Activities: education * library * telescope making * visits * lectures * observing (comets, minor planets, variable stars, occultations, Sun, meteors) * astrometry
Periodicals: (4) "Revista Astronómica" (ISSN 0044-9253, circ.: 1,500)
Coordinates: 058°26'04"W 34°36'19"S H26m (IAU Code 834)
City Reference Coordinates: 058°27'00"W 34°36'00"S

Asociación Argentina de Astronomía (AAA)
c/o Marta Rovira (President)
Instituto de Astronomía y Física del Espacio (IAFE)
Casilla de Correos 67 - Succ. 28
1428 Buenos Aires
or :
c/o Observatorio Astronómico de Córdoba
Laprida 854
5000 Córdoba
Telephone: (0)351-4331064
            (0)351-4331065
Telefax: (0)351-4331063
Electronic Mail: lapasset@mail.oac.uncor.edu.ar (Emilio Lapasset, Vice President)
            rovira@iafe.uba.ar
Founded: 1958
Membership: 260
Activities: organizing national scientific meetings * sponsoring international symposia and workshops in the country
Periodicals: (1) "Boletín de la AAA" (circ.: 300)
City Reference Coordinates: 058°27'00"W 34°36'00"S (Buenos Aires)
            064°11'00"W 31°25'00"S (Córdoba)

Centro Regional de Investigaciones Científicas y Tecnológicas (CRICYT)
Casilla de Correos 131
5500 Mendoza
or :
Avenida Ruiz Leal s/n
Parque General San Martin
Mendoza
Telephone: (0)61-288314
Telefax: (0)61-287370
Electronic Mail: ntcricyt@criba.edu.ar
WWW: http://www.cricyt.edu.ar/
Founded: 1973
Staff: 45
Activities: astronomy * mathematics * geology * space sciences * biology * social sciences
Periodicals: (1) "Memoria Anual"
Coordinates: 068°51'45"W 32°52'50"S
City Reference Coordinates: 068°52'00"W 32°48'00"S

Club de Astronomía Felix Aguilar
c/o Vicente López
Instituto Municipal José Hernandez
Juan B. Alberdi 1294
1636 Olivos
Electronic Mail: clubastronomia@caifa.com.ar
            clubastronomia@yahoo.com.ar
WWW: http://www.caifa.com.ar/
City Reference Coordinates: 058°27'00"W 34°36'00"S (Buenos Aires)

Comisión Nacional de Actividades Espaciales (CONAE)
Paseo Colón 751
1063 Buenos Aires
Telephone: (0)11-43310074
Telefax: (0)11-43313446
Electronic Mail: <userid>@conae.gov.ar
WWW: http://www.conae.gov.ar/

Founded: 1991
Staff: 100
Activities: peaceful uses of space
City Reference Coordinates: 058°27'00"W 34°36'00"S

Complejo Astronómico El Leoncito (CASLEO)
Casilla de Correos 467
5400 San Juan
or :
Avenida España 1512 Sur
5400 San Juan
Telephone: (0)264-4213653 (San Juan offices)
              (0)2648-441088 (Observatory)
Telefax: (0)264-4213693
Electronic Mail: <userid>@casleo.gov.ar
WWW: http://www.casleo.gov.ar/
Founded: 1986
Staff: 40
Activities: observing
Coordinates: 069°18'00"W 31°47'57"S H2,550m (IAU Codes 808 & 829)
City Reference Coordinates: 068°30'00"W 31°30'00"S

Consejo Nacional de Investigaciones Científicas y Técnicas (CONICET)
Rivadavia 1917
1033 Buenos Aires
Telephone: (0)11-49537230
              (0)11-49537239
Telefax: (0)11-49518552
            (0)11-49544955
Electronic Mail: comunica@conicet.gov.ar
WWW: http://www.conicet.gov.ar/
Founded: 1958
Activities: promoting and organizing science and technology nationally
City Reference Coordinates: 058°27'00"W 34°36'00"S

Estación Astronómica Rio Grande (EARG)
Casilla de Correo 160
9420 Rio Grande
or :
Acceso Aeropuerto
9420 Rio Grande
(Tierra del Fuego)
Telephone: (0)2964-430123
Telefax: (0)2964-430123
Electronic Mail: earg@earg.gov.ar
WWW: http://www.earg.gov.ar/
Founded: 1979
Staff: 12
Activities: astronomy * geodesy * geophysics
Coordinates: 067°45'05"W 53°47'10"S H18m
City Reference Coordinates: 067°40'00"W 53°50'00"S

Fundación José María Aragón
Avenida Felicia Moreau de Justo 1750
Piso 1 - Oficina "C"
1107 Buenos Aires
Telephone: (0)11-43120055
Telefax: (0)11-43122299
Electronic Mail: info@aragon.org.ar
WWW: http://www.aragon.org.ar/
Founded: 1962
Staff: 15
Activities: information service * post-graduate grants * education * funding
Periodicals: (12) "Becas y Cursos" (circ.: 1,400); (1/2) "Fellowship Guide"
City Reference Coordinates: 058°27'00"W 34°36'00"S

Instituto Argentino de Racionalización de Materiales (IRAM)
Perú 552/556
1068 Buenos Aires
Telephone: (0)11-43460600
Electronic Mail: iram@iram.org.ar
WWW: http://www.iram.org.ar/
Founded: 1935
Staff: 110
Activities: standardization and certification of products and quality systems
Periodicals: "Dinámica" (ISSN 3025-0733), "Guía de Licenciatorios"

City Reference Coordinates: 058°27'00"W 34°36'00"S

Instituto Argentino de Radioastronomía (IAR)
Casilla de Correos 5
1894 Villa Elisa
or :
Parque Pereyra Iraola
Camino General Belgrano Km. 40
Pdo. de Berazategui
Telephone: (0)221-4254909
Telefax: (0)221-4254909
Electronic Mail: <userid>@iar.unlp.edu.ar
WWW: http://www.iar.unlp.edu.ar/
Founded: 1963
Staff: 40
Activities: radioastronomy * instrumentation
Coordinates: 058°08'12"W 34°52'06"S H11m
City Reference Coordinates: 058°25'00"W 32°10'00"S (Villa Elisa)
058°12'00"W 34°45'00"S (Berazategui)

Instituto Copérnico
Casilla de Correos 85
5600 San Rafael
Telephone: (0)2627-441093
Telefax: (0)2627-441093
Electronic Mail: info@institutocopernico.org
WWW: http://www.institutocopernico.org/
Founded: 1973
Staff: 10
City Reference Coordinates: 068°21'00"W 34°40'00"S

Instituto de Astronomía y Física del Espacio (IAFE)
Edificio IAFE
Cuidad Universitaria
Intendente Guiraldes 2620
1428 Buenos Aires
Telephone: (0)11-47816755
(0)11-47832642
(0)11-47890179
Telefax: (0)11-47868114
Electronic Mail: <userid>@iafe.uba.ar
WWW: http://www.iafe.uba.ar/
Founded: 1969
Staff: 50
Activities: globular clusters * SNR * symbiotic stars * interstellar dust * solar flares * solar prominences * stellar atmospheres * cool stars * solar and stellar winds * astrophysical plasmas * coronal heating * cosmology * relativity * quantum field theory * string theory * atomic collisions * photon-atom collisions * high-energy astrophysics
City Reference Coordinates: 058°27'00"W 34°36'00"S

Observatorio Astronómico de Córdoba (OAC)
• See "Universidad Nacional de Córdoba, Observatorio Astronómico de Córdoba (OAC)"

Observatorio Astronómico de La Plata
• See "Universidad Nacional de La Plata (UNLP), Observatorio Astronómico"

Observatorio Astronómico Felix Aguilar (OAFA)
• See "Universidad Nacional de San Juan, Observatorio Astronómico Felix Aguilar (OAFA)"

Observatorio Astronómico Móvil (OAM)
Salvador María del Carril 2341
(C1419ZGC)
Buenos Aires
Telephone: (0)11-45734979
Telefax: (0)11-45717777
Electronic Mail: newoam@hotmail.com
oam@ciudad.com.ar
WWW: http://www.starlab.webprovider.com/index.htm
Founded: 1984
City Reference Coordinates: 058°27'00"W 34°36'00"S

Observatorio Astronómico Municipal de Mercedes (OAMM)
Calle 35 N° 876
6600 Mercedes
Telephone: (0)2324-426775
Telefax: (0)2324-422442

Electronic Mail: oamm@satlink.com
madel@iafe.edu.ar
WWW: http://oamm.50megs.com/
Founded: 1974
Staff: 3
Activities: photoelectric photometry (eclipsing binaries, variable stars)
Periodicals: (1) "Manual Astronómico"
Coordinates: 059°25'59"W 34°37'55"S H34m (IAU Code 833)
City Reference Coordinates: 059°30'00"W 34°42'00"S

Observatorio Astronómico, Planetario y Museo Experimental de Ciencias de Rosario
Parque Urquiza
Casilla de Correos 606
2000 Rosario
Telephone: (0)341-4802533
Telefax: (0)341-4802533
Electronic Mail: oamr@ifir.ifir.edu.ar
aquilano@ifir.ifir.edu.ar (Roberto Oscar Aquilano, Director)
WWW: http://www.ifir.edu.ar/~planetario
Founded: 1970
Staff: 50
Activities: astrophysics * cosmology * solar physics * planets
Periodicals: (2) "Acrux"
City Reference Coordinates: 060°40'00"W 32°57'00"S

Observatorio Nacional de Física Cosmica, San Miguel (ONFCSM)
Avenida Mitre 3100
1663 San Miguel
Telephone: (0)1-4557044
(0)1-4651225
Telefax: (0)1-4651225
Founded: 1935
Staff: 15
Activities: sunspot monitoring
Periodicals: (12) "Sunspot Data"; (1) "General Catalog"; "Observaciones Solares"
Coordinates: 058°43'54"W 34°53'24"S H37m
City Reference Coordinates: 057°36'00"W 28°01'00"S

Observatorio Naval de Buenos Aires (ONBA)
Avenida España 2099
1107 Buenos Aires
Telephone: (0)11-43611162
(0)11-43614644
Telefax: (0)11-43611162
Electronic Mail: postmast@onba.mil.ar
<userid>@onba.mil.ar
WWW: http://www.ara.mil.ar/Servicios/Observatorio_Naval/observatorio.htm
Founded: 1881
Staff: 11
Activities: national time service * ephemerides for navigation and geodesy * research on time and reference frames
Periodicals: (1) "Almanaque Nautico y Aeronautico", "Suplemento al Almanaque"
Coordinates: 058°21'18"W 34°37'18"S H6m
City Reference Coordinates: 058°27'00"W 34°36'00"S

Observatorio San José (OSJ)
Azcuenaga 158
1029 Buenos Aires
Telephone: (0)951-4303
Electronic Mail: <userid>@sanjo.edu.ar
WWW: http://members.xoom.com/observatorio
Founded: 1913
Staff: 5
Activities: Sun * occultations * variable stars * comets * meteor streams * eclipses * photography * planets
Coordinates: 058°24'02"W 34°36'30"S H61m
City Reference Coordinates: 058°27'00"W 34°36'00"S

Planetario de la Ciudad de Buenos Aires Galileo Galilei
Avenidas Sarmiento y B. Roldán
1425 Buenos Aires
Telephone: (0)11-47716629
(0)11-47729265
Telefax: (0)11-47716629
(0)11-47729265
Electronic Mail: <userid>@bibpla.bib.cyt.edu.ar
WWW: http://www.earg.gov.ar/planetario/
Founded: 1967

Staff: 25
Activities: shows * lectures * exhibitions
City Reference Coordinates: 058°27'00"W 34°36'00"S

Servicio Meteorológico Nacional (SMN)
25 de Mayo 658
1002 Buenos Aires
Telephone: (0)1-3124481
(0)1-3117176
Telefax: (0)1-3113968
Electronic Mail: dtec@udmeteor.mil.ar
WWW: http://www.meteofa.mil.ar/
Founded: 1872
Staff: 300
Activities: national meteorological office
Periodicals: (daily) "Boletín Meteorológico Diario"; (12) "Boletín Climatológico Mensual"; "Boletines Informativos"
Observatories: large network of stations
City Reference Coordinates: 058°27'00"W 34°36'00"S

Universidad Nacional de Córdoba, Observatorio Astronómico de Córdoba (OAC)
Laprida 854
5000 Córdoba
Telephone: (0)51-230491
(0)51-236876
(0)51-331064
(0)51-331065
Telefax: (0)51-210613
(0)51-331063
Electronic Mail: <userid>@uncbob.edu.ar
<userid>@oac.uncor.edu
WWW: http://www.oac.uncor.edu/
Founded: 1871
Staff: 70
Activities: open clusters * globular clusters * binary stars * intrinsic variable stars * celestial mechanics * galaxies * cosmology * astrometry
Periodicals: "Reprints"
Coordinates: 064°11'48"W 31°25'18"S H439m (IAU Code 834)
064°32'48"W 31°35'54"S H1,250m (Bosque Alegre Astr. Stn – IAU Code 821)
City Reference Coordinates: 064°11'00"W 31°25'00"S

Universidad Nacional de La Plata (UNLP), Facultad de Ciencias Astronómicas y Geofísicas (FCAG)
Paseo del Bosque s/n
1900 La Plata
Telephone: (0)221-4236593
(0)221-4236594
Telefax: (0)221-4236591
Electronic Mail: <userid>@fcaglp.unlp.edu.ar
WWW: http://www.fcaglp.unlp.edu.ar/
Founded: 1883
Staff: 53
Activities: research and education in astronomy and geophysics
Periodicals: "Universidad Nacional de La Plata Series (Astronomy, Geophysics, Circulars and Special)"
Coordinates: 069°18'06"W 31°48'00"S (El Leoncito - see separate entry)
067°45'06"W 53°47'12"S (Rio Grande - see separate entry)
057°41'24"W 35°00'01"S (Trelew)
065°22'54"W 43°16'06"S (Observatorio Magnético Las Acacias)
City Reference Coordinates: 057°55'00"W 34°52'00"S

Universidad Nacional de La Plata (UNLP), Observatorio Astronómico
Paseo del Bosque s/n
1900 La Plata
Telephone: (0)221-4236593
(0)221-4236594
Telefax: (0)221-4236591
Electronic Mail: <userid>@fcaglp.edu.ar
WWW: http://www.fcaglp.unlp.edu.ar/
Founded: 1883
Staff: 160
Activities: research and education in astronomy and geophysics * observing (IAU Code 839)
Coordinates: 069°18'06"W 31°48'00"S (El Leoncito - see separate entry)
067°45'06"W 53°47'12"S (Rio Grande - see separate entry)
065°22'54"W 43°16'06"S (Trelew)
City Reference Coordinates: 057°55'00"W 34°52'00"S

Universidad Nacional de San Juan, Observatorio Astronómico Felix Aguilar (OAFA)
Avenida Benavidez 8175 Oeste

5400 San Juan
Telephone: (0)64-231467
Telefax: (0)64-238494
Electronic Mail: oafa@iinfo.unsj.edu.ar
WWW: http://www.unsj.edu.ar/oafa/
Staff: 50
Activities: astrometry * celestial mechanics * astrophysics
Periodicals: "Contributions"
Coordinates: 068°37'15"W 31°30'38"S H700m (Estación Carlos Cesco)
City Reference Coordinates: 068°30'00"W 31°30'00"S

Armenia

Armenian Academy of Sciences, Byurakan Astrophysical Observatory
378433 Byurakan
Telephone: (0)8852-283453
  (0)8852-284142
Telefax: (0)8852-523640
Electronic Mail: byur@mav.yerphi.am
  <userid>@bao.sci.am
Periodicals: "Soobshcheniia" (ISSN 0370-8691), "Reprints"
Coordinates: 044°17'30"E 40°20'06"N H1,500m (IAU Code 123)
City Reference Coordinates: 044°18'00"E 40°20'00"N

Armenian Academy of Sciences, Garny Space Astronomy Institute (GSAI)
378534 Garny
Telephone: (0)2-647010
Telefax: (0)2-647030
Electronic Mail: gsai@arminco.com
WWW: http://www.gsai.am/
City Reference Coordinates: 044°30'00"E 40°11'00"N (Yerevan)

Armenian Physical Society
c/o R. Avakian (President)
Alikhanian Brothers Street 2
375036 Yerevan
Telephone: (0)8852-341347
  (0)2-341347
Telefax: (0)8852-350030
  (0)2-350030
Electronic Mail: r.avakian@hermes.desy.de
Membership: 200
City Reference Coordinates: 044°30'00"E 40°11'00"N

Byurakan Astrophysical Observatory
• See "Armenian Academy of Sciences, Byurakan Astrophysical Observatory"

Department for Standardization, Metrology and Certification (SARM)
Komitas Avenue 49/2
375051 Yerevan
Telephone: (0)2-235600
Telefax: (0)2-285620
Electronic Mail: sarm@sarm.com
Founded: 1931
Staff: 450
City Reference Coordinates: 044°30'00"E 40°11'00"N

Garny Space Astronomy Institute (GSAI)
• See "Armenian Academy of Sciences, Garny Space Astronomy Institute (GSAI)"

Hydrometeorological Department of Armenia
54 Leo Street
375002 Yerevan
Telephone: (0)1-532001
Telefax: (0)1-533575
Electronic Mail: armhydromet@meteo.am
WWW: http://www.meteo.am/
Founded: 1930
Staff: 850
Activities: hydrology * meteorology * radiation * climate variability * weather services
City Reference Coordinates: 044°30'00"E 40°11'00"N

Yerevan Physics Institute (YerPhi), Department of Theoretical Physics
Alikhanian Brothers Street 2
375036 Yerevan
Telephone: (0)2-350150
  (0)2-341500
Telefax: (0)2-350030
Electronic Mail: <userid>@lx2.yerphi.am
WWW: http://www.yerphi.am/
Founded: 1943
Staff: 10
Activities: stellar dynamics * evolution of galaxies and clusters * cosmology

Periodicals: "Preprints"
City Reference Coordinates: 044°30'00"E 40°11'00"N

Yerevan State University (YSU), Department of Astrophysics
Al. Manukian Street 1
375025 Yerevan
Telefax: (0)8852-597939
Electronic Mail: <userid>@ysu.am
WWW: http://www.physic.ysu.am/Astrophys/
Founded: 1945
Staff: 3
City Reference Coordinates: 044°30'00"E 40°11'00"N

Australia

Adelaide Planetarium
University of South Australia
Levels Campus
Mawson Lakes Boulevard
Mawson Lakes
Adelaide, SA
Telephone: (0)8-8302-3138
        (0)8-8353-5762
WWW: http://ching.apana.org.au/~oliri/stars.html
        http://www.unisa.edu.au/erm/planetarium/pla-home.htm
Founded: 1972
City Reference Coordinates: 138°35'00"E 34°55'00"S

Advanced Telescope Supplies
P.O. Box 447
Engadine, NSW 2233
Telephone: (0)2-9541-1676
Telefax: (0)2-9541-4449
Electronic Mail: pjward@atscope.com.au
WWW: http://www.atscope.com.au/
City Reference Coordinates: 151°01'00"E 34°04'00"S

Anglo-Australian Observatory (AAO), Coonabarabran
Siding Spring Mountain
Private Bag
Coonabarabran, NSW 2357
Telephone: (0)2-6842-6291
Telefax: (0)2-6884-2298
Electronic Mail: <userid>@aaocbn.aao.gov.au
WWW: http://www.aao.gov.au/
        http://www.aao.gov.au/ukst/ (UK Schmidt telescope)
Founded: 1974 (UK Schmidt: 1973)
Staff: 60
Activities: operating and maintaining the Anglo-Australian 3.9m Telescope (AAT) and the UK 1.2m Schmidt Telescope (UKST - see separate entry) * optical and near-IR astronomy (IAU Code 413)
Periodicals: (4) "AAO Newsletter" (ISSN 0728-5833); (1) "AAO Annual Report" (ISSN 0728-6554)
Coordinates: 149°03'58"E 31°16'37"S H1,164m (AAT)
        149°04'12"E 31°16'24"S H1,130m (UKST)
City Reference Coordinates: 149°18'00"E 31°16'00"S

Anglo-Australian Observatory (AAO), Epping Laboratory
P.O. Box 296
Epping, NSW 2121
or :
167 Vimiera Road
Eastwood, NSW 2122
Telephone: (0)2-9372-4800
Telefax: (0)2-9372-4880
Electronic Mail: <userid>@aaoepp.aao.gov.au
WWW: http://www.aao.gov.au/
Founded: 1974
Staff: 50
Activities: research in all aspects of optical and near-IR astronomy * instrumentation * photography
Periodicals: (1) "Annual Report of Anglo-Australian Telescope Board" (ISSN 0728-6554); (4) "AAO Newsletter" (ISSN 0728-5833)
Coordinates: 149°03'58"E 31°16'37"S H1,164m (AAT)
City Reference Coordinates: 145°02'00"E 37°39'00"S (Epping)
        151°05'00"E 33°48'00"S (Eastwood)

Astral Press
P.O. Box 107
Wembley, WA 6913
or :
46 Oceanic Drive
Floreat Park, WA
Telephone: (0)8-9387-4250
Telefax: (0)8-9387-3981
Electronic Mail: astral@dragon.net.au
Founded: 1985
Staff: 1
Periodicals: (2) "Journal of Astronomical History and Heritage" (ISSN 1440-2807)

Awards: "Astral Award"
• Publisher
City Reference Coordinates: 115°50'00"E 31°56'00"S (Perth)

AstroDomes
P.O. Box 52
Yandina, QLD 4561
Telephone: (0)7-5446-7449
Telefax: (0)7-5446-8544
Electronic Mail: astrodomes@astrodomes.com
WWW: http://www.astrodomes.com/
Activities: manufacturing astronomical domes
City Reference Coordinates: 152°52'00"E 26°36'00"S

Astronomical Society of Alice Springs Inc.
P.O. Box 739
Alice Springs, NT 0871
Telephone: (0)8-8952-8560 (Alan Viegas, President)
Electronic Mail: baney@topend.com.au
WWW: http://www.ozemail.com.au/~kkramer/astalice.html
Founded: 1973
Membership: 30
Activities: meetings * observing
Coordinates: 133°56'13"E 23°41'02"S H580m
City Reference Coordinates: 133°52'00"E 23°42'00"S

Astronomical Society of Australia (ASA)
c/o Department of Mathematics and Physics
University of Tasmania
G.P.O. Box 252-21
Hobart, TAS 7001
Telephone: (0)3-6226-2022
Telefax: (0)3-6223-3057
Electronic Mail: marc.duldig@utas.edu.au
WWW: http://www.atnf.csiro.au/asa_www/asa.html
http://www.atnf.csiro.au/pasa/ (Electronic PASA)
Founded: 1965
Membership: 377
Activities: annual conference
Periodicals: (3) "Publications of the Astronomical Society of Australia (PASA)" (ISSN 1323-3580); "Newsletter", "Proceedings" (ISSN 0066-9997)
Awards: (1/2) "Page Medal", "Ellery Lectureship"; (1) "Bok Prize"
City Reference Coordinates: 149°19'00"E 42°31'00"S

Astronomical Society of Frankston (ASF) Inc.
P.O. Box 596
Frankston, VIC 3199
Electronic Mail: aggro@peninsula.starway.net.au (Peter Lowe)
WWW: http://peninsula.starway.net.au/~aggro
Founded: 1969
Membership: 160
Activities: observing (minor-planet occultations, lunar grazes, variable stars, Jovian-satellite eclipses, Earth satellites, deep sky) * education
Periodicals: (6) "Scorpius" (circ.: 150)
Coordinates: 145°02'25"E 38°16'28"S H60m (The Briars, Mount Martha)
City Reference Coordinates: 145°07'00"E 38°08'00"S

Astronomical Society of Melbourne
P.O. Box 92
Bentleigh, VIC 3204
WWW: http://www.astromelb.i.net.au/
City Reference Coordinates: 144°58'00"E 37°49'00"S

Astronomical Society of New South Wales (ASNSW) Inc.
G.P.O. Box 1123
Sydney, NSW 2001
Telephone: (0)2-9337-3371
Telefax: (0)2-9337-3378
Electronic Mail: secretary@asnsw.com
WWW: http://www.asnsw.com/
Founded: 1954
Membership: 250
Activities: public outreach * education * observing * South Pacific Star Party (SPSP)
Periodicals: (6) "Universe"
Coordinates: 150°37'15"E 33°33'50"S H469m (Crago Observatory)

149°46'00"E 33°01'00"S H1020m (Wiruna)
City Reference Coordinates: 151°13'00"E 33°52'00"S

Astronomical Society of South Australia (ASSA) Inc.
G.P.O. Box 199
Adelaide, SA 5001
Telephone: (0)8-8821-1751 (Secretary)
(0)8-8338-1231 (Technical Information)
Telefax: (0)8-8379-4145
WWW: http://www.assa.org.au/
Founded: 1892
Membership: 400
Activities: observing * public nights * education * photography * CCDs * variable stars
Periodicals: (12) "Bulletin of the Astronomical Society of South Australia"
City Reference Coordinates: 138°35'00"E 34°55'00"S

Astronomical Society of Tasmania, Inc. (AST)
G.P.O. Box 1654
Hobart, TAS 7001
or :
c/o Martin George (Vice President)
Queen Victoria Museum and Art Gallery
Wellington Street
Launceston, TAS 7250
or :
c/o Norwood Avenue P.O.
Launceston, TAS 7250
Telephone: (0)3-6331-6777 (Vice President)
Electronic Mail: shevillm@southcom.com.au
WWW: http://www.southcom.com.au/~shevillm/ast/
Founded: 1934
Membership: 80
Activities: occultations * photography * public involvment * computing
Periodicals: (6) "Bulletin of the Astronomical Society of Tasmania Inc."; (1) "AST Ephemeris"
City Reference Coordinates: 149°19'00"E 42°31'00"S (Hobart)
147°08'00"E 41°26'00"S (Launceston)

Astronomical Society of the Hunter (ASH)
c/o Colin Maybury
P.O. Box 69
Kurri Kurri, NSW 2327
or :
P.O. Box 193
Wallsend, NSW 2287
Telephone: (0)2-4961-5448
(0)2-4937-4664
Founded: 1973
Membership: 20
Activities: observing (solar system, Sun, comets, deep sky, meteors) * instrumentation * photography * transient phenomena
Periodicals: (4) "Observations"
Awards: "Norther Instruments Trophy", "ASH Annual Achievement Award", "Jim Maybury Memorial Award"
Observatories: 3 (Blackhill, Wallaroo State Forest, Kurri Technical College)
City Reference Coordinates: 151°30'00"E 32°49'00"S (Kurri Kurri)
151°40'00"E 32°55'00"S (Wallsend)

Astronomical Society of the South West (ASSW) Inc.
P.O. Box 1100
Bunbury, WA 6231
Telephone: (0)8-9725-0036
(0)8-9721-1586
Founded: 1980
Membership: 46
Activities: observing * popularization * education
Periodicals: "The Celestial Onlooker"
Coordinates: 115°22'00"E 33°35'30"S
City Reference Coordinates: 115°38'00"E 33°20'00"S

Astronomical Society of Victoria (ASV) Inc.
Box 1059J - G.P.O.
Melbourne, VIC 3001
Telephone: (0)3-9877-3181
WWW: http://www.gsat.net.au/astrovic
Founded: 1922
Membership: 680
Activities: photography * aurorae * Sun * demonstrations * history of astronomy * instrumentation * activities for juniors * Moon * planets * meteors * variable stars * observing * planetarium

Periodicals: (4) "ASV Newsletter"; (1) "ASV Astronomical Yearbook" (ISSN 0067-0006); "Journal of the Astronomical Society of Victoria" (ISSN 0044-9814)
Coordinates: 144°58'24"E 37°49'54"S H28m (IAU Code 907)
City Reference Coordinates: 144°58'00"E 37°49'00"S

Astronomical Society of Western Australia (ASWA) Inc.
P.O. Box 421
Subiaco, WA 6008
Telephone: (0)8-9384-2264
Electronic Mail: aswa@cleo.murdoch.edu.au
WWW: http://cleo.murdoch.edu.au/gen/aswa/
Founded: 1950
Membership: 150
Activities: meetings * viewing nights * camps * public education
Periodicals: (6) "Sidereal Times" (circ.: 200)
City Reference Coordinates: 115°50'00"E 31°56'00"S (Perth)

Astrovisuals
6 Lind Street
Strathmore, VIC 3041
Telephone: (0)3-9379-5753
Telefax: (0)3-9379-5753
Electronic Mail: mail@astrovisuals.com.au
WWW: http://www.astrovisuals.com.au/
Founded: 1983
Staff: 1
Activities: producing and distributing astronomy posters, slide sets, videos, VCDs, maps and postcards
City Reference Coordinates: 142°32'00"E 17°48'00"S

Auspace Ltd.
50 Hoskins Street
P.O. Box 17
Mitchell, ACT 2911
Telephone: (0)6-6242-2611
Telefax: (0)6-6241-6664
Electronic Mail: admin@auspace.com.au
WWW: http://www.auspace.com.au/
Founded: 1984
Staff: 35
Activities: manufacturing and designing space surveillance and communications equipment
City Reference Coordinates: 147°58'00"E 26°29'00"S

Australasian Planetarium Society (APS)
P.O. Box 207
Dickson, ACT 2602
Telephone: (0)2-6249-7817
Telefax: (0)2-6249-7238
Electronic Mail: aps@ctuc.asn.au
WWW: http://www.ctuc.asn.au/planetarium/aps/
Founded: 1998
Membership: 9
City Reference Coordinates: 149°08'00"E 35°55'00"S

Australasian Society for General Relativity and Gravitation (ASGRG)
c/o Susan M. Scott (Treasurer)
Australian National University
Department of Physics and Theoretical Physics
Canberra, ACT 0200
Telefax: (0)2-6249-0741
Electronic Mail: asgrg@physics.adelaide.edu.au
WWW: http://www.physics.adelaide.edu.au/ASGRG/
Founded: 1994
City Reference Coordinates: 149°08'00"E 35°55'00"S

Australian Academy of Science (AAS)
G.P.O. Box 783
Canberra, ACT 2601
or :
Ian Potter House
Gordon Street
Acton, ACT 2600
Telephone: (0)2-6247-3966
         (0)2-6247-5777
Telefax: (0)2-6257-4620
Electronic Mail: eb@science.org.au

WWW: http://www.science.org.au/aashome.htm
Founded: 1954
Membership: 280
Staff: 20
Periodicals: "Academy Newsletter"
City Reference Coordinates: 149°08'00"E 35°55'00"S

Australian Defence Force Academy (AFDA), School of Physics
Canberra, ACT 2600
Electronic Mail: <userid>@afda.edu.au
WWW: http://www.adfa.edu.au/physics/astro/astron.html
Founded: 1986
Staff: 13
Activities: infrared, optical, X-ray and gamma-ray astronomy * laboratory astrophysics
• Joint undertaking between the "University of New South Wales (UNSW)" and the "Australian Defence Force"
City Reference Coordinates: 149°08'00"E 35°55'00"S

Australian Institute of Physics (AIP)
1/21 Vale Street
North Melbourne, VIC 3051
Telephone: (0)3-9326-6669
Telefax: (0)3-9328-2670
Electronic Mail: vjrjm@cc.newcastle.edu.au
physics@raci.org.au
WWW: http://www.physics.usyd.edu.au/aipaust/
http://www.physics.usyd.edu.au/aipaust/ANZPhysicist.html (The Physicist)
Periodicals: (11) "The Physicist" (ISSN 1036-3831)
City Reference Coordinates: 144°58'00"E 37°49'00"S

Australian National University (ANU), Research School of Astronomy and Astrophysics (RSAA), Mount Stromlo and Siding Spring Observatories (MSSSO)
Cotter Road
Weston Creek, ACT 2611
Telephone: (0)2-6125-0230 (Mount Stromlo)
(0)2-6842-6262 (Siding Spring)
Telefax: (0)2-6125-0233 (Mount Stromlo)
(0)2-6842-6240 (Siding Spring)
Electronic Mail: <userid>@mso.anu.edu.au
WWW: http://msowww.anu.edu.au/home.html
http://msowww.anu.edu.au/exploratory/ (Exploratory)
Founded: 1924
Staff: 104
Activities: galactic structure * stellar populations * stellar dynamics * interstellar physics * radio galaxies * galaxy clusters * QSOs * star formation * Magellanic Clouds * Hubble flow * clusters of galaxies * stellar pulsations * stellar evolution * abundances * cooling flows
Periodicals: (1) "Annual Report"
Coordinates: 149°03'42"E 31°16'24"S H1,149m (Siding Spring Obs. – IAU Code 413)
149°00'30"E 35°19'12"S H767m (Mount Stromlo Obs. – IAU Code 414)
City Reference Coordinates: 149°08'00"E 35°55'00"S (Canberra)

Australian National University (ANU), Research School of Earth Sciences (RSES)
Canberra, ACT 0200
Telephone: (0)2-6249-3406
Telefax: (0)2-6249-0738
Electronic Mail: school.secretary.rses@anu.edu.au
WWW: http://wwwrses.anu.edu.au/
Founded: 1973
Staff: 50
Activities: geochronology and isotope geochemistry * petrochemistry and experimental petrology * geophysical fluid dynamics * environmental geochemistry * geodynamics * ore genesis * petrophysics * seismology and geomagnetism
Periodicals: "ANU RSES Annual Report" (ISSN 0155-624x), "ANU RSES Research Papers" (ISSN 0084-7518), "Earth Sciences at ANU" (ISSN 1032-5999)
City Reference Coordinates: 149°08'00"E 35°55'00"S

Australian National University (ANU), Research School of Physical Sciences, Department of Theoretical Physics
G.P.O. Box 4
Canberra, ACT 2601
Telephone: (0)2-6249-2943
Telefax: (0)2-6249-1884
Electronic Mail: <userid>@rsphy2.anu.edu.au
WWW: http://rsphysse.anu.edu.au/theophys/
Founded: 1949
Staff: 17
Activities: theoretical physics * cosmology * strings * nuclear astrophysics
City Reference Coordinates: 149°08'00"E 35°55'00"S

Australian National University (ANU), Research School of Physical Sciences and Engineering, Plasma Research Laboratory (PRL)
Canberra, ACT 0200
Telephone: (0)2-6249-4680
Telefax: (0)2-6249-2575
Electronic Mail: prl@anu.edu.au
WWW: http://www.anu.edu.au/rsphyse/prl/PRL.html
Founded: 1980
Staff: 20
Activities: plasma physics * fusion * plasma processing * thin films * radio frequency heating * space plasma * helicon waves
City Reference Coordinates: 149°08'00"E 35°55'00"S

Australian Planetarium Association
• See now "Australasian Planetarium Society (APS)"

Australian Space Industry Chamber of Commerce (ASICC)
G.P.O. Box 7048
Sydney, NSW 2001
Telephone: (0)2-6228-1327
Telefax: (0)2-6233-2858
Founded: 1993
Staff: 1
Activities: biennial Australian space development conference * promotion of space development * space industry liaison
City Reference Coordinates: 151°13'00"E 33°52'00"S

Australian Space Research Institute (ASRI)
P.O. Box 20
Elizabeth, SA 5112
Telephone: (0)8-8259-5316
Electronic Mail: mark.blair@dsto.defence.gov.au (Mark Blair, Chairman)
WWW: http://www.asri.org.au/
Founded: 1993
Membership: 60
Activities: space technology education
Periodicals: (6) "ASRI News"
City Reference Coordinates: 138°40'00"E 34°43'00"S

Australia Telescope National Facility (ATNF)
• See "Commonwealth Scientific and Industrial Research Organisation (CSIRO), Australia Telescope National Facility (ATNF)"

Ballaarat Astronomical Society (BAS), Inc.
P.O. Box 284
Ballarat, VIC 3353
or :
Corner Magpie and Cobden Streets
Mount Pleasant
Ballarat, VIC
Electronic Mail: wfiddian@cbl.com.au
WWW: http://www.giant.net.au/astronomy/society.html
Founded: 1958 (Observatory: 1886)
Membership: 40
Activities: education * public viewing * maintenance of historical telescopes
Periodicals: (4) "Oddie-Baker Bulletin"
Coordinates: 143°51'30"E 37°34'59"S (Ballarat Observatory)
City Reference Coordinates: 143°52'00"E 37°34'00"S (Ballarat)

Bendigo District Astronomical Society (BDAS) Inc.
P.O. Box 123
Golden Square
Bendigo, VIC 3550
Telephone: (0)3-5448-4563
Electronic Mail: rbath@bendigo.net.au
WWW: http://www.bendigo.net.au/~rbath/
Founded: 1985
Membership: 15
Activities: monthly meetings * field nights * education
City Reference Coordinates: 144°17'00"E 36°46'00"S

Binocular and Telescope Shop
55 York Street
Sydney, NSW 2000
Telephone: (0)2-9262-1344
Telefax: (0)2-9262-1844
Electronic Mail: mike@bintel.com.au

WWW: http://www.bintel.com.au/
Founded: 1984
Staff: 2
Activities: retail sales of binoculars, telescopes, astronomy books and related material
Periodicals: (12) "The Night Sky" (circ.: 2,500)
City Reference Coordinates: 151°13'00"E 33°52'00"S

Brisbane Astronomical Society (BAS), Inc.
P.O. Box 204
Morningside, QLD 4170
Telephone: (0)7-3321-8511
Electronic Mail: basmail@bas.asn.au
WWW: http://www.bas.asn.au/
Founded: 1986
Membership: 60
Activities: public star nights * school astronomy nights * meetings * observing * education * astrocamps
Periodicals: (6) "Brisbane Astronomical Society Newsletter"
City Reference Coordinates: 153°02'00"E 27°28'00"S (Brisbane)

Brisbane Planetarium (STBP) (Sir Thomas_)
• See "Sir Thomas Brisbane Planetarium (STBP)"

Bundaberg Astronomical Society Inc.
c/o Alloway Observatory
Goodwood Road
P.O. Box 4221
Bundaberg, QLD 4670
Telephone: (0)7-4159-7231
Electronic Mail: observe@interworx.com.au
WWW: http://www.angelfire.com/al/AstronDirectory
Founded: 1964
Membership: 20
Activities: observing (variable stars, planetary phenomena) * photography * public lectures and viewing * telescope making * occultation timing * computing
Coordinates: 152°22'38"E 24°56'35"S H28m (Alloway Observatory)
City Reference Coordinates: 152°21'00"E 24°52'00"S

British Astronomical Association (BAA), New South Wales Branch
c/o Sydney Observatory
P.O. Box K346
Haymarket, NSW 1238
or :
Observatory Hill
The Rocks
Sydney, NSW 2000
Electronic Mail: wo@aaoepp.aao.gov.au (Wayne Orchiston)
Founded: 1895
Membership: 60
Activities: monthly meetings * observing * library
Periodicals: (6) "The Astronomers' Bulletin"
City Reference Coordinates: 151°13'00"E 33°52'00"S

Canberra Astronomical Society (CAS), Inc.
P.O. Box 1338
Woden, ACT 2606
Telephone: (0)2-6248-0552
Electronic Mail: dm@isd.canberra.edu.au
WWW: http://msowww.anu.edu.au/cas/
       http://www.mso.anu.edu.au/cas/
Founded: 1969
Membership: 140
Activities: observing (grazing occultations * deep sky) * photography * instrumentation * CCDs * education
Periodicals: (12) "The Southern Cross"
City Reference Coordinates: 149°08'00"E 35°55'00"S

Canberra Deep Space Communication Complex (CDSCC)
P.O. Box 638
Fyshwick, ACT 2609
or :
Discovery Drive
Tidbindilla
Telephone: (0)2-6201-7800
Telefax: (0)2-6201-7975
Electronic Mail: cdscc-prc@anbe.cdscc.nasa.gov
WWW: http://www.cdscc.nasa.gov/

Founded: 1965
Staff: 150
Activities: NASA Space Tracking Station * tracking of Earth orbiting and deep-space spacecraft * communication between the Jet Propulsion Laboratory (JPL) (see separate entry, USA-CA) and spacecraft
Coordinates: 148°59'00"E 35°24'00"S H660m
City Reference Coordinates: 149°08'00"E 35°55'00"S (Canberra)

Canberra Space Dome and Observatory (CSDO)
P.O. Box 207
Dickson, ACT 2602
or :
Hawdon Place
Dickson, ACT
Telephone: (0)2-6248-5333
Telefax: (0)2-6249-7238
Electronic Mail: planetarium@ctucu.asn.au
WWW: http://www.ctuc.asn.au/planetarium/
Founded: 1985
Staff: 14
Activities: public education in astronomy
Periodicals: (12) "Canberra Observer"
City Reference Coordinates: 149°08'00"E 35°55'00"S

Commonwealth Bureau of Meteorology, Head Office
P.O. Box 1289 K
Melbourne, VIC 3001
or :
150 Lonsdale Street
Melbourne, VIC 3000
Telephone: (0)3-9669-4000
Telefax: (0)3-9669-4699
Electronic Mail: info@bom.gov.au
WWW: http://www.bom.gov.au/
Founded: 1908
Staff: 1600
Activities: national meteorological service * network of 60 observing sites
Periodicals: (4) "Australian Meteorological Magazine"; (1) "Annual Report"
City Reference Coordinates: 144°58'00"E 37°49'00"S

Commonwealth Scientific and Industrial Research Organisation (CSIRO), Australia Telescope National Facility (ATNF), Headquarters
P.O. Box 76
Epping, NSW 2121
or :
c/o Radiophysics Laboratory
Vimiera and Pembroke Roads
Marsfield, NSW 2122
Telephone: (0)2-9372-4100
Telefax: (0)2-9372-4310
Electronic Mail: atnf-enquiries@atnf.csiro.au
WWW: http://www.atnf.csiro.au/
Founded: 1988
Staff: 75
Activities: operating the Australia Telescope as a national facility * galactic and extragalactic research
Periodicals: "ATNF News" (ISSN 1323-6326)
Coordinates: 151°05'42"E 33°46'42"S
City Reference Coordinates: 145°02'00"E 37°39'00"S (Epping)
151°07'00"E 33°47'00"S (Marsfield)

Commonwealth Scientific and Industrial Research Organisation (CSIRO), Australia Telescope National Facility (ATNF), Mopra Telescope
(Paul Wild Observatory)
Locked Bag 194
Narrabri, NSW 2390
Telephone: (0)2-6849-1801
Telefax: (0)2-6849-1888
Electronic Mail: narrabri@atnf.csiro.au
<userid>@atnf.csiro.au
WWW: http://www.narrabri.atnf.csiro.au/mopra/
Founded: 1988 (ATNF)
Activities: mm single-dish observing and VLBI
Periodicals: see main entry
Coordinates: 149°16'00"E 31°19'48"S
City Reference Coordinates: 149°47'00"E 30°19'00"S

Commonwealth Scientific and Industrial Research Organisation (CSIRO), Australia Telescope National Facility (ATNF), Narrabri
(Paul Wild Observatory)
Locked Bag 194
Narrabri, NSW 2390
Telephone: (0)2-6790-4000
Telefax: (0)2-6790-4090
Electronic Mail: narrabri@atnf.csiro.au
    <userid>@atnf.csiro.au
WWW: http://www.narrabri.atnf.csiro.au/www/public/site_info/site_info.html
    http://www.atnf.csiro.au/
Founded: 1988
Staff: 33
Activities: control centre of and site of Compact Array of Australia Telescope
Periodicals: see main entry
City Reference Coordinates: 149°47'00"E 30°19'00"S

Commonwealth Scientific and Industrial Research Organisation (CSIRO), Australia Telescope National Facility (ATNF), Parkes Radio Observatory
P.O. Box 276
Parkes, NSW 2870
Telephone: (0)2-6861-1700
Telefax: (0)2-6861-1730
Electronic Mail: parkes@atnf.csiro.au
    <userid>@atnf.csiro.au
WWW: http://www.parkes.atnf.csiro.au/
    http://www.atnf.csiro.au/
    http://wwwpks.atnf.csiro.au/visitors_centre/VCHomePage.html (Visitor Centre)
Founded: 1988 (ATNF)
Staff: 22
Activities: radio observing * VLBI
Periodicals: see main entry
Coordinates: 148°15'44"E 33°00'00"S H392m
City Reference Coordinates: 148°13'00"E 33°10'00"S

Commonwealth Scientific and Industrial Research Organisation (CSIRO), Earth Observation Centre (EOC)
P.O. Box 3023
Canberra, ACT 2600
or :
Corner North and Daley Roads
ANU Campus
Acton, ACT 2601
Telephone: (0)2-6216-7200
Telefax: (0)2-6216-7222
Electronic Mail: <userid>@cossa.csiro.au
WWW: http://www.eoc.csiro.au/
Founded: 1984
Staff: 16
City Reference Coordinates: 149°08'00"E 35°55'00"S

Commonwealth Scientific and Industrial Research Organisation (CSIRO), Editorial Services
• See now "CSIRO Publishing"

Commonwealth Scientific and Industrial Research Organisation (CSIRO), Office of Space Science and Applications (COSSA)
P.O. Box 3023
Canberra, ACT 2600
or :
Corner North and Daley Roads
ANU Campus
Acton, ACT 2601
Telephone: (0)2-6216-7200
Telefax: (0)2-6216-7222
Electronic Mail: <userid>@thor.cbr.cossa.csiro.au
WWW: http://www.eoc.csiro.au/cossa/cossa.htm
Founded: 1984
Staff: 11
Activities: international liaison on space projects * information on space programs * remote sensing instruments and aircraft * managing CSIRO space projects
Periodicals: (6) "CSIRO Space Industry News" (ISSN 1037-5759)
City Reference Coordinates: 149°08'00"E 35°55'00"S

Commonwealth Scientific and Industrial Research Organisation (CSIRO), Paul Wild Observatory
• See "Commonwealth Scientific and Industrial Research Organisation (CSIRO), Australia Telescope National Facility, Narrabri"

CSIRO Publishing
150 Oxford Street
P.O. Box 1139
Collingwood, VIC 3066
Telephone: (0)3-9662-7500
Telefax: (0)3-9662-7555
Electronic Mail: publishing@csiro.au
WWW: http://www.publish.csiro.au/
Founded: 1949
Staff: 50
Activities: publishing books, journals and multimedia
Periodicals: among others: (12) "Australian Journal of Chemistry" (ISSN 0004-9425), "Australian Journal of Physics" (ISSN 0004-9506); (3) "Publications of the Astronomical Society of Australia" (ISSN 1323-3580)
City Reference Coordinates: 145°00'00"E 37°48'00"S

Federation of Australian Scientific and Technological Societies (FASTS)
c/o Toss Gascoigne (Executive Director)
P.O. Box 218
Deakin West, ACT 2601
Telephone: (0)2-6257-2891 (Work)
Telefax: (0)2-6257-2987
Electronic Mail: fasts@anu.edu.au
WWW: http://www.usyd.edu.au/su/fasts/
Founded: 1985
Membership: 50 (societies)
Activities: representing the interests of scientists and technologists throughout Australia
City Reference Coordinates: 149°08'00"E 35°55'00"S (Canberra)

Grove Creek Observatory (GCO)
Unit 16
1-5 Stokes Street
Lane Cove, NSW 2066
Telephone: (0)2-9428-4334
          (0)2-9636-8611 (Observatory)
Telefax: (0)2-9558-2721
          (0)2-9418-6005
Electronic Mail: info@gco.org.au
WWW: http://www.gco.org.au/
Founded: 1978
Coordinates: 149°22'23"E 33°51'05"S H830m
City Reference Coordinates: 151°13'00"E 33°52'00"S (Sydney)

Hawkesbury Astronomical Association (HAA) Inc.
P.O. Box 670
Windsor, NSW 2756
Telephone: (0)2-4572-1568
WWW: http://www.rpi.net.au/~haa/
Founded: 1989
Membership: 30
City Reference Coordinates: 150°50'00"E 39°39'00"S

H.V. McKay Planetarium
• See now "Melbourne Planetarium at Scienceworks"

Institution of Engineers, Australia (IEAUST)
11 National Circuit
Barton, ACT 2600
Telephone: (0)2-6270-6552
Telefax: (0)2-6273-1488
Electronic Mail: contact_cust_service@ieaust.org.au
          memberservices@ieaust.org.au
WWW: http://www.ieaust.org.au/
Founded: 1919
Membership: 60000
Staff: 100
Activities: professional society * National Committee on Space Engineering
Periodicals: (4) "Civil Engineering Transactions" (ISSN 0020-3319), "Mechanical Engineering Transactions", "Electrical Engineering Transactions"; (2) "Multi-Disciplinary Transactions"
City Reference Coordinates: 132°39'00"E 30°31'00"S

IPS Radio & Space Services
P.O. Box 1386
Haymarket, NSW 1240
or :
Level 6 - North Wing

477 Pitt Street
Sydney, NSW 2000
Telephone: (0)2-9213-8000
Telefax: (0)2-9213-8060
Electronic Mail: office@ips.gov.au
    <userid>@ips.gov.au
WWW: http://www.ips.gov.au/
    http://www.ips.gov.au/culgoora/ (Culgoora Solar Observatory)
    http://www.ips.gov.au/learmonth/ (Learmonth Solar Observatory)
Founded: 1948
Staff: 40
Activities: providing space weather data, products and services
Periodicals: (12) "IPS Monthly Solar and Geophysical Summary"
City Reference Coordinates: 151°13'00"E 33°52'00"S (Sydney)

La Trobe University, Space Physics Group
Bundoora, VIC 3083
Telephone: (0)3-9479-2735
Telefax: (0)3-9479-1552
Electronic Mail: <userid>@latrobe.edu.au
WWW: http://www.latrobe.edu.au/www/physics/space/space.htm
Founded: 1967
Staff: 8
Activities: ionosphere * magnetosphere * relativity * atomic physics * liquid physics
Coordinates: 144°06'00"E 37°28'00"S (Beveridge)
City Reference Coordinates: 144°06'00"E 37°28'00"S

Launceston Planetarium
c/o Queen Victoria Museum and Art Gallery
Wellington Street
Launceston, TAS 7250
Telephone: (0)3-6323-3777
Telefax: (0)3-6323-3776
Electronic Mail: martin@qvmag.tased.edu.au (Martin George, Director)
WWW: http://www.vision.net.au/~peter/AST/launplan/launplan.htm
Founded: 1967
Staff: 2
Activities: public and school programs
City Reference Coordinates: 147°08'00"E 41°26'00"S

McKay Planetarium (H.V._)
• See now "Melbourne Planetarium at Scienceworks"

Melbourne Planetarium at Scienceworks
Museum of Victoria
P.O. Box 666E
Melbourne, VIC 3001
Telephone: (0)3-9392-4815
Telefax: (0)3-9399-2028
WWW: http://www.mov.vic.gov.au/planetarium/
Founded: 1965
Staff: 4
Activities: shows for general public and schools
• Formerly known as "H.V. McKay Planetarium"
City Reference Coordinates: 144°58'00"E 37°49'00"S

Molonglo Radio Observatory
• See "University of Sydney, Department of Astrophysics, Molonglo Radio Observatory"

Monash University, Department of Mathematics, Astrophysics Group
Melbourne, VIC 3168
Telephone: (0)3-9905-4431
Telefax: (0)3-9905-3867
Electronic Mail: <userid>@sci.monash.edu.au
    astro_contact@hal.maths.monash.edu.au
WWW: http://www.maths.monash.edu.au/astro/
Founded: 1961
Staff: 10
Activities: solar physics * stellar evolution * star formation * numerical hydrodynamics * planetary physics
City Reference Coordinates: 144°58'00"E 37°49'00"S

Monash University, Department of Physics
Melbourne, VIC 3168
Telephone: (0)3-9905-3639
Telefax: (0)3-9905-3637

Electronic Mail: <userid>@sci.monash.edu.au
WWW: http://www.physics.monash.edu.au/
Founded: 1970
Staff: 2
Activities: photometry * astrometry * education
Coordinates: 145°07'52"E 37°54'08"S H50m
　　　　　　 145°29'39"E 37°58'37"S H311m (Mount Burnett)
City Reference Coordinates: 144°58'00"E 37°49'00"S

Mopra Observatory
● See "Commonwealth Scientific and Industrial Research Organisation (CSIRO), Australia Telescope National Facility (ATNF), Mopra Observatory"

Morel Astrographics
c/o Mati Morel
6 Blakewell Road
Thornton, NSW 2322
Telephone: (0)2-4966-2078
Telefax: (0)2-4966-2078
Electronic Mail: morel@ozemail.com.au
Founded: 1985
Staff: 1
Activities: collecting stellar data * preparing charts * compiling indices and directories of stellar data * variable stars
Periodicals: (6) "Variable Star Memorandum"; (1) "Stellar Data"
City Reference Coordinates: 151°39'00"E 32°47'00"S

Mount Stromlo and Siding Spring Observatories (MSSSO)
● See "Australian National University (ANU), Research School of Astronomy and Astrophysics (RSAA), Mount Stromlo and Siding Spring Observatories (MSSSO)"

Murdoch Astronomical Society (MAS)
c/o Physics and Energy Studies
Murdoch University
South Street
Murdoch, WA 6150
Telephone: (0)8-9360-2433
　　　　　　(0)8-9360-2865
Telefax: (0)8-9310-1711
Electronic Mail: mas@cleo.murdoch.edu.au
WWW: http://cleo.murdoch.edu.au/clubs/mas/
Founded: 1978
Membership: 36
Activities: meetings * dark-sky nights * public viewing nights
Periodicals: "Murdoch Astronomical Society Newsletter"
City Reference Coordinates: 115°50'00"E 31°56'00"S (Perth)

National Committee for Astronomy (NCA), Australia
G.P.O. Box 783
Canberra, ACT 2601
or :
Ian Potter House
Gordon Street
Acton, ACT 2600
Telephone: (0)2-6247-3966
Telefax: (0)2-6257-4620
Electronic Mail: ns@science.org.au
WWW: http://msowww.anu.edu.au/~jrm/NCA/
　　　 http://www.science.org.au/
Membership: 11
City Reference Coordinates: 149°08'00"E 35°55'00"S

National Space Society of Australia Ltd. (NSSA)
G.P.O. Box 7048
Sydney, NSW 2001
Telephone: (0)2-9687-3301
WWW: http://www.nssa.com.au/
Founded: 1982
Membership: 300
Activities: promoting space development, space education and space industry liaison
Periodicals: (6) "Ad Astra" (circ.: 25,000)
City Reference Coordinates: 151°13'00"E 33°52'00"S

Nepean Astronomy Centre (NAC)
● See "University of Western Sydney (UWS), Nepean Astronomy Centre (NAC)"

Newcastle Astronomical Society (NAS)
c/o Department of Physics
University of Newcastle
Callaghan, NSW 2308
Telephone: (0)2-4942-6029
Electronic Mail: physpuls4@cc.newcastle.edu.au (Ian Dunlop)
　　　　　　　　phmc@cc.newcastle.edu.au (Michael Cvetanovski)
WWW: http://www.newcastle.edu.au/department/ph/plasma/NAS/nas_home.html
City Reference Coordinates: 151°56'00"E 32°56'00"S (Newcastle)

Northern Sydney Astronomical Society (NSAS)
P.O. Box 3002
North Turramurra, NSW 2074
Electronic Mail: nsas@easy.com.au
WWW: http://www.easy.com.au/nsas/
Periodicals: (4) "Reflections"
Awards: (1) "Geoff Welch Literary Award"
City Reference Coordinates: 151°13'00"E 33°52'00"S (Sydney)

Parkes Radio Observatory
• See "Commonwealth Scientific and Industrial Research Organisation (CSIRO), Australia Telescope National Facility (ATNF), Parkes Radio Observatory"

Paul Wild Observatory
• See "Commonwealth Scientific and Industrial Research Organisation (CSIRO), Australia Telescope National Facility (ATNF), Narrabri"

Perth Observatory
337 Walnut Road
Bickley, WA 6076
Telephone: (0)8-9293-8255
　　　　　　(0)8-9293-8109 (Information Line)
Telefax: (0)8-9293-8138
Electronic Mail: perthobs@calm.wa.gov.au
WWW: http://www.wa.gov.au/perthobs/
Founded: 1896
Staff: 11
Activities: photometry of solar system objects * education * SN search * minor planet and comet astrometry * stellar photometry
Periodicals: "Astronomical Handbook", "Communications of Perth Observatory" (ISSN 0079-1067)
Coordinates: 116°08'06"E 32°00'30"S H391m (IAU Code 323)
City Reference Coordinates: 115°50'00"E 31°56'00"S (Perth)

Perth Omni Theatre and Planetarium
29 City West Centre
Railway Parade
West Perth, WA 6005
Telephone: (0)8-9481-6481
　　　　　　(0)8-9481-5890
Telefax: (0)8-9481-6530
Electronic Mail: omni@wt.com.au
WWW: http://edsitewa.iinet.net.au/omni/
　　　　http://edsitewa.iinet.net.au/omni/planet.html
City Reference Coordinates: 115°50'00"E 31°56'00"S

Port Macquarie Astronomical Association
P.O. Box 1453
Port Macquarie, NSW 2444
or :
Observatory
Rotary Park
Port Macquarie, NSW
Telephone: (0)2-6583-1933
Electronic Mail: jaidal@bigpond.com.au (James K. Daniel, President)
Founded: 1962
Activities: lectures * observing
City Reference Coordinates: 152°55'00"E 31°26'00"S

Scitech Discovery Centre
P.O. Box 1155
West Perth, WA 6872
or :
City West Centre
Railway Parade
West Perth, WA 6005

Telephone: (0)8-9481-6295
Telefax: (0)8-9321-2869
Electronic Mail: <userid>@scitech.org.au
WWW: http://www.scitech.org.au/
Founded: 1987
Staff: 50
Activities: exhibitions * portable Starlab planetariums
City Reference Coordinates: 115°50'00"E 31°56'00"S

Siding Spring Observatory
• See "Australian National University (ANU), Research School of Astronomy and Astrophysics (RSAA), Mount Stromlo and Siding Spring Observatories (MSSSO)"

Sir Thomas Brisbane Planetarium (STBP)
Botanic Gardens
Mount Coot-tha
Brisbane, QLD
Telephone: (0)7-3403-2578
WWW: http://enterprise.powerup.com.au/~stbp/stbp_.html
http://www.powerup.com.au/~stbp/stbp_.html
Founded: 1978
Activities: shows * observing
City Reference Coordinates: 153°02'00"E 27°28'00"S

Sky & Space Publishing
PO Box 1690
Bondi Junction 1335
or :
80 Ebley Street
Bondi Junction, NSW 2022
Telephone: (0)2-9369-3344
Telefax: (0)2-9369-3366
Electronic Mail: 100246.2000@compuserve.com
editor@skyandspace.com.au
Founded: 1988
Periodicals: "Sky & Space" (6) (ISSN 1035-932x)
City Reference Coordinates: 151°13'00"E 33°52'00"S (Sydney)

South East Queensland Astronomical Society (SEQAS)
c/o The Secretary
P.O. Box 60
Everton Park, QLD 4053
Electronic Mail: glong@eis.net.au (Graham Long, President)
WWW: http://www.powerup.com.au/~mcerlean/
Founded: 1985
Membership: 57
Periodicals: (4) "Universal Times"
City Reference Coordinates: 152°59'00"E 27°19'00"S (Strathpine)

Southern Astronomical Society (SAS) Inc.
P.O. Box 867
Beenleigh, QLD 4207
Telephone: (0)7-3209-1741 (Sue Dreves)
Electronic Mail: renato@odyssey.com.au (Renato Langersek)
WWW: http://www.sas.org.au/
Founded: 1986
City Reference Coordinates: 153°12'00"E 27°43'00"S

Space Association of Australia (SAA) Inc.
P.O. Box 351
Mulgrave North, VIC 3170
Telephone: (0)3-9729-5538
Electronic Mail: abdilla@rmit.edu.au (Andrew Rennie, President)
WWW: http://vicnet.net.au/~saa/
Founded: 1981
Activities: promoting astronautics * monthly public meetings * weekly radio programmes
City Reference Coordinates: 145°12'00"E 37°56'00"S

Space Centre for Satellite Navigation (SCSN)
c/o Queensland University of Technology
G.P.O. Box 2434
Brisbane, QLD 4001
or :
2 George Street
Brisbane, QLD 4000

Telephone: (0)7-3864-2392
Telefax: (0)7-3864-1361
WWW: http://www.scsn.bee.qut.edu.au/
Founded: 1993
Staff: 10
Activities: satellite navigation * vehicle positioning * sensor integration
City Reference Coordinates: 153°02'00"E 27°28'00"S

Standards Australia International (SAI) Ltd.
G.P.O. Box 5420
Sydney, NSW 2001
or :
286 Sussex Street
Sydney, NSW 2000
Telephone: (0)2-8206-6000
Telefax: (0)2-8206-6001
Electronic Mail: mail@standards.com.au
WWW: http://www.standards.com.au/
Founded: 1922
City Reference Coordinates: 151°13'00"E 33°52'00"S

Sutherland Astronomical Society Inc. (SASI)
P.O. Box 31
Sutherland, NSW 1499
Telephone: (0)2-9589-1014
Telefax: (0)2-9589-1014
Electronic Mail: sasi@ozemail.com.au
WWW: http://www.ozemail.com.au/~sasi/
Founded: 1962
Membership: 110
Activities: observing * education * star nights * field nights * junior group
Periodicals: (6) "Southern Observer"
Awards: "SASI Award"
Coordinates: 151°04'18"E 34°00'13"S H54m
City Reference Coordinates: 151°04'00"E 34°02'00"S

Swinburne University of Technology, Astrophysics and Supercomputing
Mail Number 31
P.O. Box 218
Hawthorn, VIC 3122
or :
523 Burwood Road
Hawthorn, VIC 3122
Telephone: (0)3-9214-8782
Telefax: (0)3-9819-0856
Electronic Mail: mbailes@swin.edu.au (Matthew Bailes)
WWW: http://www.swin.edu.au/astronomy/
City Reference Coordinates: 144°58'00"E 37°49'00"S (Melbourne)

Sydney Observatory
c/o Museum of Applied Arts and Sciences
P.O. Box K346
Haymarket, NSW 1238
or :
Observatory Hill
The Rocks
Sydney, NSW
Telephone: (0)2-9217-0485
Telefax: (0)2-9217-0489
WWW: http://www.phm.gov.au/observe/default.htm
Founded: 1858
Staff: 8
Activities: public education * public observing * astronomical museum
Periodicals: (1) "The Sydney Sky Guide" (ISSN 1039-3048)
Coordinates: 151°12'12"E 33°51'40"S H44m (IAU Code 420)
City Reference Coordinates: 151°13'00"E 33°52'00"S (Sydney)

Sydney Outdoor Lighting Improvement Society (SOLIS) Inc.
P.O. Box 999
North Turramurra, NSW 2074
Electronic Mail: solislp@netscape.net
WWW: http://sites.netscape.net/solislp/homepage
Periodicals: "Newsletter of the Sydney Outdoor Lighting Improvement Society"
City Reference Coordinates: 151°13'00"E 33°52'00"S (Sydney)

Sydney Space Frontier Society (SSFS)
c/o Wayne Short
G.P.O. Box 7048
Sydney, NSW 2001
Telephone: (0)2-9502-3063
Electronic Mail: wshor@doh.health.nsw.gov.au
  ssfs@nssa.com.au
Founded: 1982
Membership: 100
Activities: promoting space development
Periodicals: (4) "Space Frontier News" (ISSN 1032-4739, circ.: 300)
City Reference Coordinates: 151°13'00"E 33°52'00"S

Taree Astronomical Society (TAS)
P.O. Box 111
Taree, NSW 2430
Telephone: (0)2-6550-2213
Electronic Mail: rosco@midcoast.com.au (Jim Ross)
WWW: http://www.midcoast.com.au/users/rosco/rosco.html
City Reference Coordinates: 152°26'00"E 31°54'00"S

Telescopes and Astronomy
P.O. Box 292
O'Halloran Hill, SA 5158
or :
8 Campbell Drive
Reynella, SA 5161
Telephone: (0)8-8381-3188
Electronic Mail: telescopes@bigpond.com
WWW: http://www.telescopes-astronomy.com.au/
Founded: 1994
Staff: 2
Activities: sales * manufacturing * information * viewing nights * travelling telescope
Periodicals: (12) "Newsletter"
City Reference Coordinates: 138°41'00"E 34°56'00"S (Adelaide)

Thomas Brisbane Planetarium (STBP) (Sir_)
• See "Sir Thomas Brisbane Planetarium (STBP)"

United Kingdom Schmidt Telescope (UKST)
Private Bag
Coonabarabran, NSW 2357
Telephone: (0)2-6842-1622
Telefax: (0)2-6842-2288
Electronic Mail: schmidt@aaocbn3.aao.gov.au
  <userid>@aaocbn3.aao.gov.au
WWW: http://www.aao.gov.au/schmidt.html
Founded: 1973
Activities: service facility providing deep-sky photography for systematic sky surveys and for special research projects
Coordinates: 149°04'12"E 31°16'24"S H1,130m
City Reference Coordinates: 149°18'00"E 31°16'00"S

University of Adelaide, Department of Physics, High-Energy Astrophysics Research Group
North Terrace
G.P.O. Box 498
Adelaide, SA 5001
Telephone: (0)8-8228-5996
Telefax: (0)8-8224-0464
Electronic Mail: <userid>@physics.adelaide.edu.au
WWW: http://www.physics.adelaide.edu.au/astrophysics/
  http://pilot.physics.adelaide.edu.au/astrophysics/home.html
  http://bragg.physics.adelaide.edu.au/astrophysics/home.html
Founded: 1966 (University: 1874)
Staff: 6
Activities: VHE, UHE and theoretical gamma-ray astronomy
Coordinates: 138°34'58"E 34°55'38"S H41m
City Reference Coordinates: 138°35'00"E 34°55'00"S

University of Melbourne, School of Physics, Astrophysics Group
Parkville, VIC 3010
Telephone: (0)3-9344-7670
Telefax: (0)3-9347-4783
Electronic Mail: <userid>@physics.unimelb.edu.au
WWW: http://www.ph.unimelb.edu.au/
  http://www.ph.unimelb.edu.au/astro/home.html

http://astro.ph.unimelb.edu.au/
Founded: 1853
Staff: 25
Activities: education * research
City Reference Coordinates: 144°58'00"E 37°49'00"S (Melbourne)

University of New South Wales (UNSW), Centre for Remote Sensing
P.O. Box 1
Kensington, NSW 2033
Telephone: (0)2-9697-4964
Telefax: (0)2-9313-7493
Electronic Mail: <userid>@unsw.edu.au
WWW: http://www.geog.unsw.edu.au/~web/crsgis/info.html
Founded: 1981
Staff: 10
Activities: remote sensing * education * research * consulting
Periodicals: (1) "Annual Report"
City Reference Coordinates: 151°14'00"E 33°55'00"S

University of New South Wales (UNSW), Department of Astrophysics and Optics
Sydney, NSW 2052
Telephone: (0)2-9385-5649
Telefax: (0)2-9385-6060
WWW: http://www.phys.unsw.edu.au/astro.html
http://newt.phys.unsw.edu.au/~mgb/jacara.html (JACARA)
Founded: 1962
Staff: 15
Activities: star formation * IR astronomy * PN * variable stars * telescope automation * extragalactic astronomy * Antarctic astronomy * mm astronomy * Joint Australian Centre for Astrophysical Research in Antarctica (JACARA)
City Reference Coordinates: 151°13'00"E 33°52'00"S

University of New South Wales (UNSW), University College, Department of Physics
Canberra, ACT 2600
Telephone: (0)2-6268-8801
Telefax: (0)2-6268-8786
Electronic Mail: r-sood@adfa.edu.au (Ravi Sood, Head of School)
WWW: http://www.ph.adfa.edu.au/
Founded: 1986
Staff: 13
Activities: IR astronomy * laboratory astrophysics * high-energy astrophysics * nuclear magnetic resonance
Periodicals: (1) "Annual Report"
City Reference Coordinates: 149°08'00"E 35°55'00"S

University of Queensland, Department of Physics
Saint Lucia, QLD 4072
Telephone: (0)7-3365-3424
(0)7-3365-2422 (MTO)
Telefax: (0)7-3371-5896
(0)7-3365-1242 (MTO)
Electronic Mail: <userid>@physics.uq.edu.au
WWW: http://www.physics.uq.edu.au/
http://www.physics.uq.edu.au/ap/ap.html (Astrophysics Group)
Founded: 1910 (Observatory: 1973)
Staff: 4
Activities: photoelectric photometry (chromospherically active stars, cataclysmic and eruptive variable stars, eclipsing binaries, Be stars) * transient planetary and cometary phenomena * catalogs * Nova search programme
Periodicals: "Nova Search" (ISSN 1031-508x)
Observatories: 1 (Mount Tamborine Observatory - MTO)
Coordinates: 153°12'48"E 27°58'21"S H560m
City Reference Coordinates: 153°02'00"E 27°28'00"S (Brisbane)

University of South Australia, School of Environmental and Recreation Management, Planetarium
Building P
Mawson Lakes Boulevard
Mawson Lakes, SA 5095
Telephone: (0)8-8302-3138
Telefax: (0)8-8302-5082
Electronic Mail: wayne.looker@unisa.edu.au (Wayne Looker, Manager)
WWW: http://ching.apana.org.au/~oliri/planet.html
http://www.unisa.edu.au/erm/planetarium/pla-home.htm
Founded: 1972
Staff: 3
Activities: shows * education
Coordinates: 138°30'00"E 34°52'00"S
City Reference Coordinates: 138°35'00"E 34°55'00"S (Adelaide)

University of Sydney, Chatterton Astronomy Department
School of Physics A28
Sydney, NSW 2006
Telephone: (0)2-9351-2544
Telefax: (0)2-9351-7726
Electronic Mail: <userid>@astron.physics.usyd.oz.au
WWW: http://www.physics.usyd.edu.au/astron/astron.html
http://www.physics.usyd.edu.au/astron/susi/ (Sydney University Stellar Interferometer - SUSI)
Founded: 1961
Staff: 5
Activities: high-resolution stellar interferometry * determination of fundamental stellar quantities * high-precision stellar photometry * astronomical seeing studies * Sydney University Stellar Interferometer (SUSI)
City Reference Coordinates: 151°13'00"E 33°52'00"S

University of Sydney, Chatterton Astronomy Department, Molonglo Radio Observatory
R.M.B. 107
Hoskinstown, NSW 2621
Telephone: (0)6-2238-2262
Telefax: (0)6-2238-2290
WWW: http://www.physics.usyd.edu.au/astrop/astrop.html
http://www.physics.usyd.edu.au/astrop/most/
Founded: 1964
Activities: operating the Molonglo Observatory Synthesis Telescope (MOST)
Coordinates: 149°25'25"E 35°22'19"S H732m (MOST)
City Reference Coordinates: 151°13'00"E 33°52'00"S (Sydney)

University of Sydney, Research Centre for Theoretical Astrophysics (RCfTA)
c/o School of Physics
Sydney, NSW 2006
Telephone: (0)2-9351-2542
Telefax: (0)2-9351-7726
Electronic Mail: rcfta@physics.usyd.edu.au
<userid>@physics.usyd.edu.au
WWW: http://www.physics.usyd.edu.au/rcfta/
Founded: 1991
Staff: 10
Activities: theoretical astrophysics
Periodicals: (1) "Annual Report"
City Reference Coordinates: 151°13'00"E 33°52'00"S

University of Sydney, School of Mathematics and Statistics
F07
Sydney, NSW 2006
Telephone: (0)2-9351-3039
Telefax: (0)2-9351-4534
WWW: http://www.maths.usyd.edu.au/
Founded: 1991
Staff: 70
Activities: mathematical research * solar and stellar astrophysics * geophysics
City Reference Coordinates: 151°13'00"E 33°52'00"S

University of Tasmania, Department of Physics
G.P.O. Box 252-21
Hobart, TAS 7001
Telephone: (0)3-6226-2401
Telefax: (0)3-6226-2410
Electronic Mail: <userid>@phys.utas.edu.au
<userid>@physvax.phys.utas.edu.au
WWW: http://physics01.phys.utas.edu.au/physics/
http://www-ra.phys.utas.edu.au/ (Radio Astronomy Group)
http://www.phys.utas.edu.au/physics/cosray/default.htm (Cosmic-Ray Astronomy)
http://xray03.phys.utas.edu.au/ (Optical and X-ray Astronomy)
Founded: 1960 (Mount Pleasant Radio Astronomy Observatory)
Activities: radio, X-ray, cosmic-ray and optical astronomy * pulsars and collapsed objects * molecular lines * VLBI * geodesy
Periodicals: (1) "Research Report" (ISSN 1035-4751)
Observatories: 2 (Mount Pleasant, Ceduna)
Coordinates: 147°25'00"E 42°50'00"S H300m
147°26'21"E 42°48'18"S H55m (Mount Pleasant)
City Reference Coordinates: 147°25'00"E 42°50'00"S

University of Western Australia (UWA), Department of Physics
Nedlands, WA 6907
Telephone: (0)8-9380-2738
Telefax: (0)8-9380-1014
Electronic Mail: head@physics.uwa.edu.au
WWW: http://www.pd.uwa.edu.au/Physics/Research/astro.html

Founded: 1987
Staff: 15
Activities: image processing * SN * flare stars * transient astronomical phenomena * gravitational collapse * pulsars * neutron stars * gravitational-wave detection
Observatories: 1 (Perth Observatory - see separate entry)
City Reference Coordinates: 115°49'00"E 31°59'00"S

University of Western Sydney (UWS), Nepean Astronomy Centre (NAC)
P.O. Box 10
Kingswood, NSW 2747
Telephone: (0)47-360-222
Telefax: (0)47-360-779
Electronic Mail: d.bailey@nepean.uws.edu.au (David Bailey)
WWW: http://www.nepean.uws.edu.au/astronomy/
City Reference Coordinates: 151°13'00"E 33°52'00"S

University of Wollongong (UoW), Department of Physics, Astronomy and Astrophysics Group
Northfields Avenue
Wollongong, NSW 2500
Telephone: (0)2-4221-3517
Telefax: (0)2-4221-5944
Electronic Mail: <userid>@uow.edu.au
WWW: http://www.uow.edu.au/eng/phys/astronomy.html
Founded: 1967
Staff: 4
Activities: image digitizing and analysis * radio and IR observing of galactic star forming regions * cooling flows * hydrodynamics * planetary surfaces * education * asteroid mining * Illawara Science Centre Planetarium and Observatory
City Reference Coordinates: 150°54'00"E 34°25'00"S

Warsash Pty. Ltd.
P.O. Box 1685
Strawberry Hills, NSW 2012
or :
7/1 Marian Street
Redfern, NSW 2016
Telephone: (0)2-9319-0122
Telefax: (0)2-9318-2192
Electronic Mail: warsash@ozemail.com.au
WWW: http://www.ozemail.com.au/~warsash/
Founded: 1975
Staff: 5
Activities: distributing optical components, micropositioning systems, lasers, adaptive optics, off & on-axis parabolics, acousto-optic modulators, piezo-actuators, optical benches, servo control systems, and more
City Reference Coordinates: 151°13'00"E 33°52'00"S (Sydney)

Western Sydney Amateur Astronomy Group (WSAAG)
c/o Brett White
P.O. Box 400
Kingswood NSW 2747
Telephone: (0)2-4753-1041
WWW: http://www4.tpgi.com.au/users/wsaag/
City Reference Coordinates: 151°13'00"E 33°52'00"S (Sydney)

Wild Observatory (Paul_)
• See "Commonwealth Scientific and Industrial Research Organisation (CSIRO), Australia Telescope National Facility, Narrabri"

Wollongong Amateur Astronomy Club (WAAC)
P.O. Box 398
Unanderra, NSW 2526
Telephone: (0)2-4272-4505
Electronic Mail: paul.b@bigpond.com (Paul Brown)
WWW: http://www.users.bigpond.com/paul.b/index.htm
Founded: 1999
Membership: 15
Activities: observing nights * education * visits
Periodicals: (6) "Wollongong Observer"
City Reference Coordinates: 150°52'00"E 34°25'00"S (Wollongong)

Wollongong Science Centre and Planetarium
c/o University of Wollongong
Northfields Avenue
Wollongong, NSW 2522
or :
Cowper Street

Fairy Meadow, NSW 2519
Telephone: (0)2-4286-5000
Telefax: (0)2-4283-6665
Electronic Mail: <userid>@uow.edu.au
WWW: http://www.uow.edu.au/science_centre/
   http://www.uow.edu.au/public/science_centre/planetarium.html
Founded: 1989
Staff: 35
Activities: science and astronomy education for public and schools
City Reference Coordinates: 150°54'00"E 34°23'00"S (Fairy Meadow)
   150°52'00"E 34°25'00"S (Wollongong)

Austria

Adalbert Jeszenkowitsch Gesellschaft (Dr._)
• See "Dr. Adalbert Jeszenkowitsch Gesellschaft"

Arbeitsgruppe für Astronomie Haus der Natur
Museumsplatz 5
A-5020 Salzburg
Telephone: (0)662-842653
(0)662-842322
Telefax: (0)662-847905
Electronic Mail: hausdernatur@salzburg.co.at
WWW: http://www.salzburg.co.at/hausdernatur/default.htm
http://www.salzburg.co.at/hausdernatur/sterne.htm (Volkssternwarte Voggenberg)
Founded: 1954
Membership: 140
Activities: lectures * meetings
Periodicals: "Astronomische Kurzinformationen und Eilnachrichten"
Awards: "Golden Saturn"
Coordinates: 013°02'24"E 47°52'14"N H597m (Volkssternwarte Voggenberg)
City Reference Coordinates: 013°03'00"E 47°48'00"N

Astronomischer Arbeitskreis Salzkammergut (AAS)
c/o Erwin Filimon
Sachsenstraße 2
A-4863 Seewalchen-am-Attersee
Telephone: (0)7662-6490
Electronic Mail: info@astronomie.at
WWW: http://www.astronomie.at/
Founded: 1980
Membership: 500
Activities: comets * meteor streams * photography * planets
Periodicals: (10) "Astro-Info" (circ.: 900)
Coordinates: 013°36'33"E 47°54'47"N H860m (Gahberg Obs. – IAU Code 563)
City Reference Coordinates: 013°39'00"E 48°01'00"N (Vöcklabruck)
013°35'00"E 47°57'00"N (Seewalchen)

Astronomischer Jugendclub Dingi-Vindemiatrix
Richard-Wagner-Platz 2/8
A-1160 Wien
Telephone: (0)1-4956340
Telefax: (0)1-4956340
Founded: 1969
Membership: 50
Activities: variable stars * photography
Periodicals: (4) "Die Sternenrundschau"
City Reference Coordinates: 016°20'00"E 48°13'00"N

Astronomische Vereinigung Kärntens (AVK)
Villacherstraße 239
A-9020 Klagenfurt
Telephone: (0)463-21700
Telefax: (0)463-21700
Electronic Mail: avk@uni-klu.ac.at
WWW: http://www.buk.ktn.gv.at/sterne/avk.htm
http://www.buk.ktn.gv.at/sterne/planet/planet.htm (Planetarium Europapark)
Founded: 1961
Membership: 359
Activities: education
Periodicals: (11) "Sternenwelt"
Observatories: 1 (Kreuzbergl - see separate entry)
City Reference Coordinates: 014°18'00"E 46°38'00"N

Astroteam Mariazellerland
c/o Günther Eder
Hangweg 12
A-8630 Mariazell
Telephone: (0)3882-3540
Telefax: (0)3882-217817
Electronic Mail: aastroteam.mariazell@gmx.net
WWW: http://ajax.kfunigraz.ac.at/~webers/
Founded: 1995
Membership: 57

Activities: popularization * education * observing * CCD-imaging
Observatories: 2 (Raiba Volkssternwarte Mariazellerland, Sternwarte Sankt Sebastian)
City Reference Coordinates: 015°19'00"E 47°47'00"N

Austrian Academy of Sciences
• See "Österreichische Akademie der Wissenschaften (ÖAW)"

Austrian Society for Aerospace Medicine (ASM)
Lustkandlgasse 52/3
A-1090 Wien
Telephone: (0)1-3155777
Telefax: (0)1-31557778
Electronic Mail: headoffice@asm.at
WWW: http://www.asm.at/
Founded: 1991
Activities: biomedical investigations into spaceflight medical issues
City Reference Coordinates: 016°20'00"E 48°13'00"N

Burgenländische Amateurastronomen (BAA)
Postgasse 2
A-7202 Bad Sauerbrunn
Telephone: (0)2687-54159
Electronic Mail: baa@astronomie.at
WWW: http://www.astronomie.at/burgenland/
Founded: 1992
Membership: 70
Activities: observing (meteors, occultations) * education * meetings
Periodicals: (4) "Alrukaba" (circ.: 300)
City Reference Coordinates: 016°32'00"E 47°51'00"N (Eisenstadt)

Dr. Adalbert Jeszenkowitsch Gesellschaft
c/o Norbert Hauser
Dr. Karl Rennerstraße 1
A-7000 Eisenstadt
Telephone: (0)2682-66819
Founded: 1979
Membership: 80
Activities: guided tours * library * lectures * exhibitions * photography
Coordinates: 016°31'21"E 47°50'39"N H194m (Burgenländische Landessternwarte)
City Reference Coordinates: 016°32'00"E 47°50'00"N

Eisner-Sternwarte
c/o Karl Silber
Cumberlandpark 16
A-4810 Gmunden
Telephone: (0)7612-67777
Founded: 1949
Membership: 3
Activities: lectures
Coordinates: 013°48'00"E 47°55'27"N H490m
City Reference Coordinates: 013°48'00"E 47°55'00"N

Fonds zur Förderung der Wissenschaftlichen Forschung (FWF)
(Austrian Science Fund)
Weyringergasse 35
A-1040 Wien
Telephone: (0)1-50567400
Telefax: (0)1-5056739
　　　　(0)1-505674045
WWW: http://www.fwf.ac.at/index-en.html
Founded: 1967
Staff: 40
Activities: funding research * public relations
Periodicals: "FWF-Info", "FWF-Statistics", "Annual Report"
City Reference Coordinates: 016°20'00"E 48°13'00"N

Franz-Kroller-Sternwarte (FKS)
Hauptplatz 18
A-2514 Traiskirchen
WWW: http://www.astro.or.at/astro/orgs/fks/
　　　　http://www.mycity.at/privat/98013062/fks/maksutov.htm
City Reference Coordinates: 016°18'00"E 48°01'00"N

Gnomonicae Societas Austriaca (GSA)
(Österreichischer Astronomischer Verein, Arbeitsgruppe Sonnenuhren)
Am Tigls 76A
A-6073 Sistrans
Telephone: (0)512-378868
Telefax: (0)512-378868
Electronic Mail: sundial@tirol.com
   k.schwarzinger@tirol.com (Karl Schwarzinger, Chairman)
WWW: http://www.tirol.com/sundial/
   http://www.sundials.co.uk/gsa.htm
Founded: 1990
Membership: 94
Activities: preservation of historical sundials * survey of Austrian sundials * meetings and conferences
Periodicals: (2) "Rundschreiben"
City Reference Coordinates: 011°27'00"E 47°14'00"N

Hotel und Sternwarte Heiligkreuz
Reimmichlstraße 18
A-6060 Hall in Tirol
Telephone: (0)5223-57114
Telefax: (0)5223-571145
Electronic Mail: info@heiligkreuz.at
WWW: http://www.heiligkreuz.at/
Founded: 1994
Staff: 3
Activities: observing * photography
Coordinates: 011°29'55"E 47°17'10"N H572m
City Reference Coordinates: 014°28'00"E 47°28'00"N

International Union of Geological Sciences (IUGS)
c/o Werner Janoschek (Secretary General)
Geological Survey of Austria
Rasumofskygasse 23
Postfach 127
A-1031 Wien
Telephone: (0)1-7125674180
Telefax: (0)1-712567456
Electronic Mail: wjanoschek@cc.geolba.ac.at
WWW: http://www.iugs.org/
Founded: 1961
Membership: 110 (nations)
Activities: organizing international geoscience meetings and cooperative scientific programs
Sections: (commissions) Comparative Planetology * Geological Documentation * Geology Teaching * Global Sedimentary Geology * History of Geological Sciences * Igneous and Metamorphic Petrogenesis * Marine Geology * Storage, Automatic Processing and Retrieval of Geological Data * Stratigraphy * Systematics in Petrology * Tectonics * Fossil Fuels * Geoscience for Environmental Planning
Periodicals: (4) "Episodes" (ISSN 0705-3797)
City Reference Coordinates: 016°20'00"E 48°13'00"N

Jeszenkowitsch Gesellschaft (Dr. Adalbert_)
• See "Dr. Adalbert Jeszenkowitsch Gesellschaft"

Karl-Franzens-Universität Graz
• See "Universität Graz"

Kroller-Sternwarte (Franz-_)
• See "Franz-Kroller-Sternwarte"

Kuffner-Sternwarte (KSW)
Johann Staud-Straße 10
A-1160 Wien
Telephone: (0)1-9148130
Telefax: (0)1-914813031
Electronic Mail: admin@kuffner.ac.at
WWW: http://www.kuffner.ac.at/
Founded: 1884
Staff: 6
Activities: public observatory * museum * education * history of astronomy
Periodicals: (1/3) "Tätigkeitsbericht"
Coordinates: 016°17'47"E 48°12'47"N H302m
City Reference Coordinates: 016°20'00"E 48°13'00"N

Leopold-Figl-Observatorium für Astrophysik
● See "Universität Wien, Institut für Astronomie"

Leopold-Franzens-Universität Innsbruck
● See "Universität Innsbruck"

Linzer Astronomische Gemeinschaft (LAG)
Sternwarteweg 5
A-4020 Linz
Telephone: (0)732-674042
Telefax: (0)732-673105
Electronic Mail: herbert.raab@ris.at (Herbert Raab, President)
WWW: http://www.sternwarte.at/
Founded: 1947
Membership: 200
Activities: education * amateur astronomy
Periodicals: (12) "WEGA"
Coordinates: 014°16'09"E 48°17'38"N H341m (Johannes-Kepler-Sternw. – IAU Code 540)
City Reference Coordinates: 014°18'00"E 48°18'00"N

Observatorium Lustbühel
● See "Universität Graz, Institut für Astronomie, Observatorium Lustbühel"

Österreichische Akademie der Wissenschaften (ÖAW)
(Austrian Academy of Sciences)
Dr.-Ignaz-Seipel-Platz 2
A-1010 Wien
Telephone: (0)1-515810
Telefax: (0)1-5139541
Electronic Mail: <userid>@oeaw.ac.at
WWW: http://www.oeaw.ac.at/deutsch.html
http://www.oeaw.ac.at/english.html
Founded: 1847
City Reference Coordinates: 016°20'00"E 48°13'00"N

Österreichische Akademie der Wissenschaften (ÖAW), Institut für Weltraumforschung (IWF)
Schmiedlstraße 6
A-8042 Graz
Telephone: (0)316-4120700
Telefax: (0)316-4120790
Electronic Mail: office.iwf@oeaw.ac.at
<userid>@oeaw.ac.at
WWW: http://www.iwf.oeaw.ac.at/
Founded: 1972
Staff: 60
Activities: ionosphere and magnetosphere of planets and comets * interplanetary space * satellites * rockets * balloons * time keeping and distribution
Coordinates: 015°29'40"E 47°04'05"N H494m (Lustbühel – IAU Code 580)
City Reference Coordinates: 015°27'00"E 47°05'00"N

Österreichische Gesellschaft für Astronomie und Astrophysik (ÖGAA)
c/o Institut für Astronomie
Universität Wien
Türkenschanzstraße 17
A-1180 Wien
WWW: http://www.oegaa.at/
City Reference Coordinates: 016°20'00"E 48°13'00"N

Österreichische Gesellschaft für Weltraumfragen GesmbH
(Austrian Space Agency - ASA)
Garnisongasse 7
A-1090 Wien
Telephone: (0)1-40381770
Telefax: (0)1-4058228
Electronic Mail: mgitsch@asaspace.at
WWW: http://www.asaspace.at/
Founded: 1972
Staff: 6
Activities: coordinating space activities in Austria
Periodicals: (1/2) "Report of Activities"; (5) "ASA-Information Service"
City Reference Coordinates: 016°20'00"E 48°13'00"N

Österreichische Physikalische Gesellschaft (ÖPG)
c/o Atominstitut der Österreichischen Universitäten
Schüttelstraße 115
A-1020 Wien
Telephone: (0)1-588015578
(0)1-588015575
Telefax: (0)1-5864203
Electronic Mail: oepg@kph.tuwien.ac.at
WWW: http://www.ati.ac.at/OePG/
http://www.oepg.at/
Membership: 1350
Periodicals: (4) "ÖPG-Mitteilungsblatt"
Awards: (1) "Physik-Preis", "Roman-Ulrich-Sexl-Preis", "Viktor-Heß-Preis"
City Reference Coordinates: 016°20'00"E 48°13'00"N

Österreichischer Astronomischer Verein
(Astronomisches Büro)
Hasenwartgasse 32
A-1238 Wien
Telephone: (0)1-8893541
Telefax: (0)1-8893541
Electronic Mail: astbuero@astronomisches-buero-wien.or.at
WWW: http://members.ping.at/astbuero/
Founded: 1924 (Büro: 1908)
Membership: 1500
Staff: 5
Activities: astronomical phenomenology * lunar occultations * comets * minor planets * meteors * sundials (see separate entry) * education * Sterngarten
Periodicals: (12) "Der Sternenbote" (ISSN 0039-1271); (1) "Österreichischer Himmelskalender", "Seminarpapiere"
City Reference Coordinates: 016°20'00"E 48°13'00"N

Österreichischer Astronomischer Verein, Arbeitsgruppe Sonnenuhren
• See "Gnomonicae Societas Austriaca (GSA)"

Österreichisches Normungsinstitut (ÖN)
Heinestraße 38
Postfach 130
A-1021 Wien
Telephone: (0)1-213000
Telefax: (0)1-21300650
Electronic Mail: <userid>@on-norm.at
WWW: http://www.on-norn.at/
City Reference Coordinates: 016°20'00"E 48°13'00"N

Peter-Anich-Planetarium Kufstein
• See now "Planetarium Schwaz GmbH"

Planetarium der Stadt Wien
Oswald-Thomas-Platz
A-1020 Wien
Telephone: (0)1-7295494
Telefax: (0)1-729549477
Electronic Mail: admin@planetarium-wien.at
WWW: http://www.planetarium-wien.at/
Founded: 1927
Staff: 4
Activities: popularization * shows
Periodicals: (1) "Seminars"
City Reference Coordinates: 016°20'00"E 48°13'00"N

Planetarium Schwaz GmbH
Alte Landstraße 15
Postfach 42
A-6130 Schwaz
WWW: http://www.planetarium-schwaz.at/
City Reference Coordinates: 011°44'00"E 47°21'00"N

Salzburger Sterngucker
Himmelreichstrasse 30
A-5071 Wals
Telephone: (0)662-852938
Electronic Mail: reifberger.jo@salzburg.co.at
WWW: http://www.pgliefering.asn-sbg.ac.at/herbert/sterngucker.htm
City Reference Coordinates: 013°03'00"E 47°48'00"N (Salzburg)

Sonnenobservatorium Kanzelhöhe
• See "Universität Graz, Sonnenobservatorium Kanzelhöhe"

Sport- und Kulturvereins Forschungszentrum Seibersdorf (SKVFZ), Sektion Amateurastronomie
c/o Erich Weber
Österreichisches Forschungszentrum Seibersdorf GesmbH
A-2444 Seibersdorf
Telephone: (0)2254-7803104
Telefax: (0)2254-72133
Electronic Mail: erich@zditf2.arcs.ac.at
WWW: http://www1.arcs.ac.at/skvfz/sekt-aa.html
Founded: 1994
City Reference Coordinates: 016°20'00"E 48°13'00"N (Wien)

Star Observer Verlag
Linzerstraße 100
A-3002 Gablitz
Telephone: (0)2231-63577
Telefax: (0)2231-63691
Electronic Mail: office@starobserver.com
WWW: http://www.starobserver.com/
http://www.snro.net/
Founded: 1993
Staff: 16
Periodicals: (10) "Star Observer"
City Reference Coordinates: 016°20'00"E 48°13'00"N (Wien)

Steirischer Astronomen Verein (StAV)
Postfach 8
A-8029 Graz
WWW: http://www.stav.at/
http://www.emt.tu-graz.ac.at/~stav/de/f_stw.html (Observatory)
Founded: 1981
Membership: 140
Activities: occultations by Moon and minor planets * meteors * comets * photography * public observing * CCD imaging
Sections: occultations * meteors * comets * astrophotography
Periodicals: (2) "Nova" (circ.: 140); (12) "Monthly Newsletter"
Coordinates: 015°19'55"E 47°04'08"N H554m (Johannes-Kepler-Volkssternw., Steinberg)
015°28'20"E 47°11'00"N H810m (Rinnegg - private obs.)
015°35'52"E 46°51'13"N H300m (Sankt Georgen - private obs.)
City Reference Coordinates: 015°27'00"E 47°05'00"N

Sternwarte Heiligkreuz
• See "Hotel und Sternwarte Heiligkreuz"

Sternwarte Königsleiten
Königsleiten 29
A-5742 Wald im Pinzgau
Telephone: (0)6564-877042
Telefax: (0)6564-877043
Electronic Mail: sternwarte.kgl@aon.at
WWW: http://www.sternwarte-koenigsleiten.com/
City Reference Coordinates: 012°14'00"E 47°16'00"N

Sternwarte Kreuzbergl
Giordano Bruno Weg 1
A-9020 Klagenfurt
Telephone: (0)463-21700
Telefax: (0)463-21700
WWW: http://www.buk.ktn.gv.at/sterne/kreuzb/kreuzb.htm
Founded: 1965
Membership: 10
Activities: observing * education * guided celestial tours
Coordinates: 014°17'24"E 46°37'47"N H600m
City Reference Coordinates: 014°18'00"E 46°38'00"N

Technische Universität Graz, Institut für Angewandte Geodäsie, Abteilung für Positionierung und Navigation
Steyrergasse 30
A-8010 Graz
Telephone: (0)316-8736830
Telefax: (0)316-8738888
Electronic Mail: <userid>@ftug.tu-graz.ac.at
Founded: 1930
Staff: 4

Activities: geodesy, including space geodesy
Periodicals: "Mitteilungen" (circ.: 400)
City Reference Coordinates: 015°27'00"E 47°05'00"N

United Nations Organization (UNO), Office for Outer Space Affairs (OOSA)
Room E-0953
Vienna International Centre
Postfach 500
A-1400 Wien
Telephone: (0)1-211314951
Telefax: (0)1-213455830
Electronic Mail: oosa@unov.un.or.at
WWW: http://www.un.or.at/OOSA/
Founded: 1962
Staff: 16
Activities: providing services to the Committee on the Peaceful Uses of Outer Space (COPUOS) and subsidiary bodies * administering Space Applications Programme * carrying out studies recommended by COPUOS
Periodicals: (12) "Monthly Survey of Selected Events in the Peaceful Exploration and Use of Outer Space"; (1) "Seminars of the United Nations Programme on Space Applications", "Highlights in Space: Progress in Space Science, Technology and Applications, International Cooperation and Space Law"
City Reference Coordinates: 016°20'00"E 48°13'00"N

Universität Graz, Institut für Geophysik, Astrophysik und Meteorologie (IGAM)
Universitätsplatz 5
A-8010 Graz
Telephone: (0)316-3805255
 (0)316-3805270
Telefax: (0)316-3809825
Electronic Mail: mail.igam@kfunigraz.ac.at
 <userid>@kfunigraz.ac.at
WWW: http://www.kfunigraz.ac.at/igamwww/
 http://www.kfunigraz.ac.at/igamwww/olg/start_de.html (Obs. Lustbühel)
Founded: 1999
Activities: climate * atmosphere * ionosphere * minor planets * solar physics
Coordinates: 015°26'54"E 47°04'36"N H375m
 015°29'42"E 47°03'56"N H480m (Obs. Lustbühel – IAU Code 580)
• Full university name: "Karl-Franzens-Universität Graz"
City Reference Coordinates: 015°27'00"E 47°05'00"N

Universität Graz, Sonnenobservatorium Kanzelhöhe
A-9521 Treffen
Telephone: (0)4248-27170
Telefax: (0)4248-271715
WWW: http://www.solobskh.ac.at/
Founded: 1943
Staff: 5
Activities: daily sunspot drawings * white-light integral pictures * solar flare patrol * solar instrumentation * solar activity * solar differential rotation
Periodicals: "Mitteilungen des Sonnenobservatoriums Kanzelhöhe"
Coordinates: 013°54'24"E 46°40'42"N H1,526m
• Full university name: "Karl-Franzens-Universität Graz"
City Reference Coordinates: 015°27'00"E 47°05'00"N

Universität Innsbruck, Institut für Astronomie
Technikerstraße 25
A-6020 Innsbruck
Telephone: (0)512-5076031
Telefax: (0)512-5072923
Electronic Mail: astro@uibk.ac.at
WWW: http://astro.uibk.ac.at/
Founded: 1892
Staff: 9
Activities: galactic structure * emission nebulae * reduction methods
Coordinates: 011°22'54"E 47°16'05"N (IAU Code 041)
• Full university name: "Leopold-Franzens-Universität Innsbruck"
City Reference Coordinates: 011°25'00"E 47°17'00"N

Universität Wien, Institut für Astronomie
Türkenschanzstraße 17
A-1180 Wien
Telephone: (0)1-427751801
Telefax: (0)1-42779518
Electronic Mail: <userid>@astro.univie.ac.at
WWW: http://www.astro.univie.ac.at/
Founded: 1755
Staff: 16

Activities: stellar astrophysics * dynamical astronomy * extragalactic research * astrometry * history of astronomy
Coordinates: 016°20'12"E 48°13'54"N H241m (IAU Code 045)
015°55'24"E 48°05'00"N H890m (Leopold-Figl-Obs. – IAU Code 562)
City Reference Coordinates: 016°20'00"E 48°13'00"N

Universität Wien, Institut für Geochemie
Althanstraße 14
A-1090 Wien
Telephone: (0)1-313361714
(0)1-313361725
Telefax: (0)1-31336781
Electronic Mail: geochemie@univie.ac.at
WWW: http://www.univie.ac.at/geochemistry/
Founded: 1985
Staff: 5
Activities: cosmochemistry (impact craters, meteorites, lunar rocks, cosmic dust)
City Reference Coordinates: 016°20'00"E 48°13'00"N

Universität Wien, Institut für Mathematik
Boltzmanngasse 5
A-1090 Wien
Telephone: (0)1-427750602
Telefax: (0)1-42779506
Electronic Mail: mathematik@univie.ac.at
WWW: http://www.mat.univie.ac.at/
http://gauss.mat.univie.ac.at/
Staff: 2
Activities: numerical hydrodynamics * 3D convection modelling * stellar radial pulsations * stellar atmospheres
City Reference Coordinates: 016°20'00"E 48°13'00"N

Urania-Sternwarte, Wien
Uraniastraße 1
A-1010 Wien
Telephone: (0)1-7126191
Telefax: (0)1-729549477
Electronic Mail: admin@urania-sternwarte.at
WWW: http://www.urania-sternwarte.at/
Founded: 1910
Staff: 2
Activities: popularization * phenomenology * lunar occultations * minor planets * Galilean satellites of Jupiter * Sun
Coordinates: 016°23'08"E 48°12'43"N H186m (IAU Code 602)
City Reference Coordinates: 016°20'00"E 48°13'00"N

Verein Antares
Schuhmeierstraße 1
A-3100 Sankt Pölten
Telephone: (0)2742-882492
Electronic Mail: office@teleskope.at
WWW: http://noe-sternwarte.at/
Founded: 1996
Coordinates: 015°45'00"E 48°05'00"N
City Reference Coordinates: 015°37'00"E 48°12'00"N

Vorarlberger Amateur Astronomen (VAA)
c/o Manfred Böhler
Hofsteigstraße 33
A-6890 Lustenau
Telephone: (0)5577-88647
WWW: http://www.vobs.at/astronomen/
Founded: 1989
Membership: 60
Activities: observing * lectures
City Reference Coordinates: 009°39'00"E 47°26'00"N

Wiener Arbeitsgemeinschaft für Astronomie (WAA)
c/o Alexander Pikhard
Dreyhausenstraße 11/53
A-1140 Wien
Telephone: (0)664-2561221
Electronic Mail: info@waa.at
WWW: http://www.waa.at/
Founded: 1998
Membership: 120
Activities: education * lectures * star parties
City Reference Coordinates: 016°20'00"E 48°13'00"N

Zentralanstalt für Meteorologie und Geodynamik (ZAMG)
Hohe Warte 38
A-1190 Wien
Telephone: (0)1-36026
Telefax: (0)1-3691233
Electronic Mail: <userid>@zamg.ac.at
WWW: http://www.zamg.ac.at/
Founded: 1851
Staff: 190
Activities: meteorology * climatology * environmental meteorology * geophysics
City Reference Coordinates: 016°20'00"E 48°13'00"N

Azerbaijan

Azerbaijan State Standardization and Metrology Centre
Mardanov Gardashlary 124
3700748 Baku
Telephone: (0)12-400396
Telefax: (0)12-403798
City Reference Coordinates: 049°51'00"E 40°23'00"N

Azerbaijan Academy of Sciences
Istiglaliyyat Street 10
370001 Baku
Telephone: (0)12-923529
          (0)12-925883
Telefax: (0)12-925699
WWW: http://wwwinfo.cern.ch/~mrashid/antaas.htm
Founded: 1916
City Reference Coordinates: 049°51'00"E 40°23'00"N

Azerbaijan Academy of Sciences, Shemakha Astrophysical Observatory
373243 Shemakha
or :
9 Agayev Street
370148 Baku
Telephone: (0)276-91902
          (0)8922-398248
          (0)8922-394304
WWW: http://www.ab.az/ASC/Institutes/Observation.htm
Founded: 1960
Staff: 150
Activities: non-stationary processes in stars * celestial mechanics * Sun * planets * small solar-system bodies
Periodicals: "Shemakha Circulars" (ISSN 0135-0420)
Coordinates: 048°35'04"E 40°46'20"N H1,580m
City Reference Coordinates: 048°37'00"E 40°38'00"N (Shemakha)
                    049°51'00"E 40°23'00"N (Baku)

Azerbaijanian National Aerospace Agency (ANASA)
159 Azadlyg Street
370106 Baku
Telephone: (0)12-629387
Telefax: (0)12-621738
Founded: 1975
Staff: 400
Activities: remote sensing * space image processing * developing hardware and software
City Reference Coordinates: 049°51'00"E 40°23'00"N

Shemakha Astrophysical Observatory
● See "Azerbaijan Academy of Sciences, Shemakha Astrophysical Observatory"

State Hydrometeorological Committee of Azerbaijan
3 Rasul Rza Street
370000 Baku
Telephone: (0)12-931619
Telefax: (0)12-936937
Electronic Mail: azmet@inteko.net
WWW: http://www.weather.inteko.net/
Founded: 1991
Staff: 2500
Activities: meteorology * hydrology * oceanography * environment * pollution monitoring * ozone layer monitoring
Periodicals: "Hydrometeorology and Environment Monitoring"
City Reference Coordinates: 049°51'00"E 40°22'00"N

Bangladesh

Bangladesh Meteorological Department (BMD)
Abhawa Bhaban
Agargaon
Sher-e-Bangla Nagar
Dhaka 1207
Telephone: (0)2-8119832
Telefax: (0)2-8118230
Electronic Mail: bmdswc@bdonline.com
WWW: http://www.bangladeshonline.com/gob/bmd/
Founded: 1972
Staff: 1057
Activities: observing * data handling * weather forecast and analysis * research * education
Coordinates: 090°23'00"E 23°46'00"N H9m
City Reference Coordinates: 085°10'00"E 26°41'00"N

Bangladesh Standards and Testing Institution (BSTI)
116/A Tejgaon Industrial Area
Dhaka 1208
Telephone: (0)2-881462
Telefax: (0)2-885685
Electronic Mail: bsti@bangla.net
City Reference Coordinates: 085°10'00"E 26°41'00"N

Space Research and Remote Sensing Organization (SPARRSO)
Agargaon
Sher-e-Bangla Nagar
Dhaka 1207
Telephone: (0)2-327913
         (0)2-319301
Telefax: (0)2-813080
Electronic Mail: sparrso@bangla.net
WWW: http://www.bangla.net/sparrso/
Founded: 1980
Staff: 142
Activities: R&D * applications of remote-sensing technology and GIS * resource management * environmental monitoring
Periodicals: (1) "Annual Report"; (4) "SPARRSO Newsletter"
City Reference Coordinates: 085°10'00"E 26°41'00"N

Barbados

Barbados Astronomical Society (BAS)
c/o Harry Bayley Observatory
P.O. Box 41-B
Brittons Hill
Saint-Michael
Telephone: 426-1317
          422-2394
Founded: 1956
Membership: 58
Activities: weekly meetings * education * open nights * limited research
Periodicals: (2) "Journal" (circ.: 70)
Coordinates: 059°35'10"W 13°05'07"N (Harry Bayley Observatory)
City Reference Coordinates: 059°37'00"W 13°06'00"N (Bridgetown)

Barbados National Standards Institution (BNSI)
"Flodden"
Culloden Road
Saint-Michael
Telephone: 426-3870
Telefax: 436-1495
Founded: 1973
Staff: 31
Activities: standards development and implementation * metrology * quality control * education * product development * product certification * quality management systems promotion
Periodicals: "BNSI Standards News", "National Standards and Codes of Practice"
City Reference Coordinates: 059°37'00"W 13°06'00"N (Bridgetown)

Caribbean Institute for Meteorology and Hydrology (CIMH)
P.O. Box 130
Bridgetown
or :
Husbands
Saint-James
Telephone: 425-1362
          425-1365
Telefax: 424-4733
WWW: http://inaccs.com.bb/carimet/top.htm
Founded: 1967
Staff: 40
Activities: training and research in meteorology and hydrology
City Reference Coordinates: 059°37'00"W 13°06'00"N (Bridgetown)

Belarus

Belarusian Physical Society (BPS)
c/o Institute of Molecular and Atomic Physics
National Academy of Sciences of Belarus
70 F. Skaryna Avenue
220072 Minsk
Telephone: (0)17-2840954
Telefax: (0)17-2840030
Electronic Mail: ershov@imaph.bas-net.by
WWW: http://imaph.bas-net.by/BPS/
Founded: 1990
Membership: 245
City Reference Coordinates: 027°34'00"E 53°54'00"N

Committee for Standardization, Metrology and Certification (BELST)
Starovilensky Trakt 93
220053 Minsk
Telephone: (0)0172-375213
Telefax: (0)0172-372588
Electronic Mail: belst@mcsm.belpak.minsk.by
City Reference Coordinates: 027°34'00"E 53°54'00"N

Belgium

Académie Royale des Sciences, des Lettres et des Beaux-Arts de Belgique
Palais des Académies
Rue Ducale 1
B-1000 Bruxelles
Telephone: (0)25502211
Telefax: (0)25502205
Electronic Mail: academie.royaledebelgique@cfwb.be
WWW: http://www.cfwb.be/arb/home.htm
Founded: 1772
Membership: 300
Periodicals: "Bulletin de la Classe des Sciences" (ISSN 0001-4141), "Mémoires de la Classe des Sciences in 8°" (ISSN 0365-0936), "Mémoires de la Classe des Sciences in 4°" (ISSN 0365-0952)
Awards: (1) "Prix Léon et Henri Fredericq", "Prix Jean-Servais Stas", "Prix Adolphe Wetrems"; (1/2) "Prix Professeur Louis Baes", "Prix de Boelpaepe", "Prix Théophile Gluge", "Prix Charles Lemaire", "Prix Paul et Marie Stroobant", "Prix Frédéric Swarts", "Prix Baron van Ertborn", "Prix Baron Nicolet"; (1/3) "Prix Albert Brachet", "Prix Henri Buttgenbach", "Prix Théophile De Donder", "Prix de la Fondation Jean-Marie Delwart", "Prix Jean De Meyer", "Prix Agathon De Potter", "Prix Léo Errera", "Prix Paul Fourmarier", "Prix Jean Lebrun", "Prix de l'Adjudant Hubert Lefebvre", "Prix Camille Liégeois", "Prix Joseph Schepkens", "Prix P.-J. et Édouard Van Beneden", "Prix Max Poll"; (1/4) "Prix François Deruyts", "Prix Jacques Deruyts", "Prix Dubois-Debauque", "Prix Charles Lagrange", "Prix Émile Laurent", "Prix Édouard Mailly", "Prix Louis Melsens", "Prix Pol et Christiane Swings", "Prix Georges Van der Linden"; (1/5) "Prix de la Belgica", "Prix Eugène Catalan", "Prix Edmond de Selys Longchamps", "Prix Lamarck", "Prix Fondation Jacques et Yvonne Ochs-Lefêbvre"; (1/6) "Prix Auguste Sacré"; (1/8) "Prix Octave Dupont"
City Reference Coordinates: 004°20'00"E 50°50'00"N

Advanced Mechanical and Optical Systems (AMOS)
Parc de Recherches du Sart Tilman
Rue des Chasseurs Ardennais
B-4031 Angleur
Telephone: (0)43614040
Telefax: (0)43672007
Electronic Mail: info@amos.be
WWW: http://www.amos.be/
Founded: 1983
Staff: 45
Activities: mechanical and optical engineering * mirror polishing * telescope collimators * mechanical and optical ground support equipment * space simulator * telescopes
City Reference Coordinates: 005°34'00"E 50°38'00"N

Alcatel, Études Techniques et Constructions Aérospatiales (ETCA)
Rue Chapelle Beaussart 101
B-6032 Mont-sur-Marchienne
Telephone: (0)71442853
Telefax: (0)71435778
WWW: http://www.charline.be/alcatelecom-etca/
Founded: 1963
Staff: 400
Activities: designing and developing i.a. satellite power conditioning systems, spectrum analyzers, digital processing circuits and camera control electronics
City Reference Coordinates: 004°26'00"E 50°25'00"N (Charleroi)

Astro CORAM
c/o Bertrand Seynaeve
Rue des Haies 83
B-7712 Herseaux
Telephone: (0)56346242
  (0)75953946
Electronic Mail: 101544.2100@compuserve.com
  olivierduquesne@csi.com
WWW: http://ourworld.compuserve.com/homepages/OlivierDuquesne/coram.htm
Founded: 1986
Membership: 19
Activities: observing * journeys * meetings * astronomy * space
Periodicals: (6) "Quasar/NOUT"
Coordinates: 003°12'45"E 50°43'35"N H30m
City Reference Coordinates: 003°14'00"E 50°43'00"N

Astronomie Centre Ardenne (ACA)
Chaussée de Bastogne 91
B-6840 Neufchâteau
Telephone: (0)61279316
Electronic Mail: astroaca@france-mail.com

WWW: http://www.astrosurf.org/mercure/aca/
Founded: 1988
City Reference Coordinates: 005°26'00"E 49°51'00"N

Belgian Physical Society (BPS)
c/o J. Ingels (Secretary)
Belgisch Instituut voor Ruimte-Aeronomie
Ringlaan 3
B-1180 Brussel
Telephone: (0)23730378
Telefax: (0)23748423
Electronic Mail: johan.ingels@bira-iasb.oma.be
WWW: http://inwfnu07.rug.ac.be/bps/
Membership: 450
Periodicals: "Physicalia"
City Reference Coordinates: 004°20'00"E 50°50'00"N

Belgian Space Industry Association (BELGOSPACE)
Boulevard Auguste Reyers 80
B-1030 Brussels
Telephone: (0)27067948
Telefax: (0)27067952
Electronic Mail: belgospace@fabrimetal.be
WWW: http://www.fabrimetal.be/
Founded: 1962
Membership: 12 (companies)
Activities: coordination between the members and with the public authorities * industrial forum for discussing common problems and major space options concerning Belgium
City Reference Coordinates: 004°20'00"E 50°50'00"N

Belgisch Instituut voor Ruimte-Aëronomie (BIRA)
Ringlaan 3
B-1180 Brussel
Telephone: (0)23730400
(0)23730402
(0)23730404
Telefax: (0)23748323
Electronic Mail: <userid>@bira-iasb.oma.be
WWW: http://www.oma.be/BIRA-IASB/
Founded: 1964
Staff: 90
Activities: public service * atmospheric physics and chemistry * solar-terrestrial physics * global change * space observations * data analysis
Periodicals: "Aeronomica Acta"
City Reference Coordinates: 004°20'00"E 50°50'00"N

BELGOSPACE
• See "Belgian Space Industry Association (BELGOSPACE)"

Canberra Semiconductor nv
Lammerdries 25
B-2250 Olen
Telephone: (0)14221975
Telefax: (0)14221991
Electronic Mail: csnv@canberra.com
Founded: 1981
Staff: 25
Activities: producing HPGe and Si detectors i.a. for space applications
Periodicals: "Information", "NPG-News"
City Reference Coordinates: 004°51'00"E 51°09'00"N

Centre National de Documentation Scientifique et Technique (CNDST)
c/o Bibliothèque Royale de Belgique
Boulevard de l'Empereur 4
B-1000 Bruxelles
Telephone: (0)25195640
Telefax: (0)25195679
Electronic Mail: cndst@kbr.be
WWW: http://www.belspo.be/stis/index_fr.htm
Founded: 1964
Staff: 11
Activities: on-line scientific, technical, economical, patent documentation and information * European Union National Awareness Partner * ESA-IRS National Aerospace Centre
City Reference Coordinates: 004°20'00"E 50°50'00"N

Centre Permanent d'Étude la Nature (CPEN), Observatoire
Route de Mons 52
B-6470 Sivry
Telephone: (0)60455128
Telefax: (0)60456142
WWW: http://www.vifcom.be/Local/Cpen.html
Founded: 1967
Membership: 12
Activities: meteorology * astronomy * education * planetarium
Coordinates: 004°13'45"E 50°10'50"N H220m
City Reference Coordinates: 004°11'00"E 50°10'00"N

Centre Spatial de Liège (CSL)
• See "Université de Liège (ULg), Centre Spatial de Liège (CSL)"

Cercle Astronomique de Bruxelles (CAB)
c/o Jean Schwaenen (Secretary)
Allé D 5
B-6001 Marcinelle
Telephone: (0)71434040
Telefax: (0)71434040
Electronic Mail: jean.schwaenen@worldonline.be
WWW: http://users.skynet.be/zmn/cab/cab.html
Founded: 1963
Membership: 200
Activities: monthly lectures * observing (occultations, double stars, variable stars) * camps * instruments on loan
Periodicals: (10) "OBAFGKM Informations"; (4) "OBAFGKM" (ISSN 0775-0315)
Coordinates: 004°20'39"E 50°47'44"N H78m
City Reference Coordinates: 004°27'00"E 50°24'00"N

Cercle Astronomique de Tournai (CAT)
c/o Luc Gillemain
Rue du Maréchal 6
B-7730 Estaimbourg
Telephone: (0)69557095
          (0)27372096
Electronic Mail: lgi@rtbf.be
WWW: http://www.ecoles.cfwb.be/ipestournai/caty
City Reference Coordinates: 003°24'00"E 50°36'00"N

Cercle Astronomique Mosan (CAM)
c/o Joseph Marteleur
Rue de la Barrière 9
B-5600 Romedenne
or :
Rue de Soulme 94
B-5620 Morville
Telephone: (0)496574307
          (0)82688212
Electronic Mail: cercle.cam@swing.be
WWW: http://users.swing.be/cercle.cam
Founded: 1977
Membership: 45
Activities: Sun * Moon * planets * nebulae
Periodicals: (6) "L'Univers du Namurois" (circ.: 45)
Coordinates: 004°55'39"E 50°14'56"N H215m (Herbuchenne, Dinant)
City Reference Coordinates: 004°45'00"E 50°14'00"N (Dinant)

Cercle d'Astronomie Olympus Mons
c/o Service de Physique Expérimentale et Biologique
Faculté de Médecine et de Pharmacie
Avenue du Champ de Mars 24
B-7000 Mons
Telephone: (0)65373536
          (0)65373468
Telefax: (0)65373537
Electronic Mail: moiny@umh.ac.be (Francis Moiny)
WWW: http://olympus.umh.ac.be/
Periodicals: (6) "Galactée" (in collaboration with "Cercle d'Astronomie Tycho")
City Reference Coordinates: 003°56'00"E 50°27'00"N

Cercle d'Astronomie Tycho
c/o Pierre Dehombreux
Service de Mécanique Rationnelle, Dynamique et Vibrations
Boulevard Dolez 31

B-7000 Mons
Telephone: (0)65374179
(0)65374184 (answering machine)
Telefax: (0)65374183
Electronic Mail: tycho@fpms.ac.be
WWW: http://www.fpms.ac.be/~tycho/
Founded: 1997
Membership: 50
Periodicals: (6) "Galactée" (in collaboration with "Olympus Mons")
City Reference Coordinates: 003°56'00"E 50°27'00"N

Cercle des Astronomes Amateurs de Waterloo (CAAW)
c/o Guy Alexandre Nefve (President)
Rue Bodrissart 59bis
B-1410 Waterloo
Telephone: (0)23548087
WWW: http://www.chez.com/caaw/accueil.html
City Reference Coordinates: 004°24'00"E 50°43'00"N

Club Astronomique Rochefortois (CAR)
c/o Nicolas Imbreckx
Rue des Falizes 10
B-5580 Rochefort
Electronic Mail: nimbreckx@hotmail.com
WWW: http://users.swing.be/car/
City Reference Coordinates: 005°13'00"E 50°10'00"N

Club d'Astronomie d'Amateurs "Orion"
c/o Francis Lecoyer (President)
Rive de Meuse 14 bte 13
B-5170 Profondeville
Telephone: (0)498814189
Electronic Mail: francislecoyer@swing.be
WWW: http://www.cluborion.be.tf/
Founded: 1996
Membership: 20
Activities: observing (variable stars) * photography * education
Coordinates: 004°52'42"E 50°24'23"N H89m
City Reference Coordinates: 004°52'00"E 50°23'00"N

Club d'Astronomie d'Ottignies (CAO)
c/o Pierre Ernotte
Rue de la Chapelle 42
B-1340 Ottignies
Telephone: (0)10419265
Electronic Mail: pierre.ernotte@advalvas.be
WWW: http://www.cao.be.tf/
City Reference Coordinates: 004°34'00"E 50°40'00"N

Comité Belge des Astronomes Amateurs (CBAA)
Rue de la Grange des Champs 72
B-1420 Braine-l'Alleud
Telephone: (0)2-3871642
WWW: http://www.cbaa.be.tf/
Founded: 1965
Membership: 10 (associations and observatories)
Periodicals: "Info Astro"
City Reference Coordinates: 004°22'00"E 50°41'00"N

Comité Belge pour l'Investigation Scientifique des Phénomènes Réputés Paranormaux (Comité PARA)
c/o Observatoire Royal de Belgique
Avenue Circulaire 3
B-1180 Bruxelles
Telephone: (0)23730241
WWW: http://www.comitepara.be/
Founded: 1949
Membership: 150
Activities: critical review of allegations by paranormal believers * research * experiments
Periodicals: "Nouvelles Brèves" (ISSN 0774-5834, circ.: 300), "Le Trait d'Union"
City Reference Coordinates: 004°20'00"E 50°50'00"N

École Royale Militaire (ERM), Département d'Astronomie, de Géodésie et de Topographie
Avenue de la Renaissance 30
B-1000 Bruxelles

Telephone: (0)27376120
  (0)27376123
Electronic Mail: <userid>@elec.rma.ac.be
WWW: http://www.asge.rma.ac.be/
Founded: 1834
Staff: 3
Activities: (D)GPS * remote sensing
City Reference Coordinates: 004°20'00"E 50°50'00"N

European Association of Research and Technology Organisations (EARTO)
Rue du Luxembourg 3
B-1000 Bruxelles
Telephone: (0)25028698
Telefax: (0)25028693
Electronic Mail: info@earto.org
WWW: http://www.earto.org/
Activities: trade association of Europe's specialized research and technology organisations
City Reference Coordinates: 004°20'00"E 50°50'00"N

European Committee for Electrotechnical Standardization
(Comité Européen de Normalisation Électrotechnique - CENELEC)
Rue de Stassartstraat 35
B-1050 Brussels
Telephone: (0)25196871
Telefax: (0)25196919
Electronic Mail: cenelec@cenelec.org
WWW: http://www.cenelec.org/
Founded: 1973
Activities: promoting voluntary technical harmonization in Europe
City Reference Coordinates: 004°20'00"E 50°50'00"N

European Committee for Standardization
(Comité Européen de Normalisation - CEN)
Rue de Stassart 36
B-1050 Bruxelles
Telephone: (0)25500811
Telefax: (0)25500819
Electronic Mail: infodesk@cenclcbel.be
WWW: http://www.cenorm.be/
Founded: 1961
Activities: promoting voluntary technical harmonization in Europe
City Reference Coordinates: 004°20'00"E 50°50'00"N

Europlanetarium vzw
Planetariumweg 19
B-3600 Genk
Telephone: (0)89307990
Telefax: (0)89307991
Electronic Mail: planetar@europlanetarium.com
WWW: http://www.europlanetarium.com/
Founded: 1977
Membership: 350
Staff: 11
Activities: popularization of astronomy, space sciences and meteorology
Periodicals: (12) "Info"; (1) "Jaarverslag"
Coordinates: 005°32'10"E 50°57'30"N H90m
• Earlier know as "Limburgse Volkssterrenwacht (LVS) vzw"
City Reference Coordinates: 005°30'00"E 50°58'00"N

Euro Space Center (ESC)
Rue Devant les Hêtres 1
B-6890 Transinne
Telephone: (0)61656465
Telefax: (0)61656461
Electronic Mail: info@eurospacecenter.be
WWW: http://www.eurospacecenter.be/
City Reference Coordinates: 005°12'00"E 50°00'00"N

Facultés Universitaires Notre-Dame de la Paix (FUNDP), Département de Mathématique
Rempart de la Vierge 8
B-5000 Namur
Telephone: (0)81724903
Telefax: (0)81724914
Electronic Mail: <userid>@fundp.ac.be
WWW: http://www.fundp.ac.be/sciences/math/dpt.html

Founded: 1971
Activities: celestial mechanics * dynamics of satellites and minor planets * perturbation methods
City Reference Coordinates: 004°52'00"E 50°28'00"N

Flemish Aerospace Group (FLAG)
Brouwersvliet 5 Box 4
B-2000 Antwerpen
Telephone: (0)32024452
Telefax: (0)32264334
Electronic Mail: flag@vev.be
WWW: http://www.flag-be.com/
Founded: 1980
Staff: 3
Activities: providing support for company-members active in aeronautics and aerospace
City Reference Coordinates: 004°25'00"E 51°13'00"N

Fonds National de la Recherche Scientifique (FNRS)
Rue d'Egmont 5
B-1000 Bruxelles
Telephone: (0)25049211
Telefax: (0)25049292
Electronic Mail: mjsimoen@fnrs.be (M.J. Simoen, General Secretary)
WWW: http://www.fnrs.be/
Founded: 1928
Activities: research council * funding basic research
Periodicals: (1) "Annual Report"
Awards: (1/5) "Prix Dr. A. De Leeuw-Damry-Bourlart"; "Prix Scientifique Ernest-John Solvay", "Prix Scientifique Joseph Maisin"
City Reference Coordinates: 004°20'00"E 50°50'00"N

Fonds voor Wetenschappelijk Onderzoek (FWO) - Vlaanderen
Egmontstraat 5
B-1000 Brussel
Telephone: (0)25129110
Telefax: (0)25125890
Electronic Mail: jose.traest@fwo.be (José Traest, General Secretary)
WWW: http://www.fwo.be/
Founded: 1928
Activities: funding research
Periodicals: (1) "Annual Report"
Awards: (1/5) "Prijs Dr. A. De Leeuw-Damry-Bourlart"
City Reference Coordinates: 004°20'00"E 50°50'00"N

Groupe Astronomie de Spa (GAS)
c/o Thierry Thysebaert
Rue des Petits Sapins 29
B-4900 Spa
Electronic Mail: ejehin@eso.org (Emmanuël Jehin)
WWW: http://www.ping.be/eurospace/astro.htm#GAS
Founded: 1989
Membership: 190
Activities: observing * star parties * lectures * exhibitions * photography
Periodicals: (6) "Formation Information Liaison (FIL)"
Coordinates: 005°52'00"E 50°29'00"N (Aérodrome de la Sauvenière)
City Reference Coordinates: 005°52'00"E 50°30'00"N

Groupe d'Astronomie Quasar
c/o Marc Lefort
Rue des Carrières 1
B-5353 Emptinne
Electronic Mail: quasar@trifide.com
WWW: http://www.trifide.com/quasar/
City Reference Coordinates: 005°07'00"E 50°20'00"N

Institut Belge de Normalisation (IBN)
Avenue de la Brabançonne 29
B-1000 Bruxelles
Telephone: (0)27380091
         (0)27380111
Telefax: (0)27334264
Electronic Mail: <userid>@ibn.be
WWW: http://www.ibn.be/
City Reference Coordinates: 004°20'00"E 50°50'00"N

Institut d'Aéronomie Spatiale de Belgique (IASB)
Avenue Circulaire 3
B-1180 Bruxelles
Telephone: (0)23730400
    (0)23730402
    (0)23730404
Telefax: (0)23748323
Electronic Mail: <userid>@bira-iasb.oma.be
WWW: http://www.oma.be/BIRA-IASB/
Founded: 1964
Staff: 90
Activities: public service * atmospheric physics and chemistry * solar-terrestrial physics * global change * space observations * data analysis
Periodicals: "Aeronomica Acta"
City Reference Coordinates: 004°20'00"E 50°50'00"N

Institut Royal Météorologique de Belgique (IRMB)
Avenue Circulaire 3
B-1180 Bruxelles
Telephone: (0)23730611
Telefax: (0)23751259
Electronic Mail: <userid>@oma.be
WWW: http://www.meteo.oma.be/IRM-KMI/
Founded: 1913
Staff: 190
Activities: climatology * hydrology * dynamic climatology * applied meteorology * aerology * aeronomy * internal/external geophysics * computing
Periodicals: (12) "Observations Climatologiques" (ISSN 0029-7682), "Observations Géophysiques" (ISSN 0020-2525), "Observations Ionosphériques - Rayonnement Cosmique" (ISSN 0020-2533); (4) "Observations de l'Ozone" (ISSN 0770-0164); (1) "Rayonnement Solaire" (ISSN 0524-7780), "Magnétisme Terrestre" (ISSN 0770-4569), "Rapport Annuel" (ISSN 0770-7371); "Publications Série A" (ISSN 0020-255x), "Publications Série B" (ISSN 0770-4615); "Hors Série" (ISSN 0772-4330); "Documentation Météorologique" (ISSN 0772-4349), "Miscellanea A" (ISSN 0770-0261), "Miscellanea B" (ISSN 0770-0318), "Miscellanea C" (ISSN 0772-4357), "Publications Scientifiques et Techniques"
Coordinates: 004°21'00"E 50°48'00"N H104m (Uccle)
    004°36'00"E 50°06'00"N H230m (Dourbes)
City Reference Coordinates: 004°21'00"E 50°48'00"N

International Astronomical Youth Camps (IAYC) Workshop for Astronomy (IWA)
c/o Sam Vereecke
Ter Weibroek 20
B-9880 Aalter
Telephone: (0)93742752
Electronic Mail: info@iayc.org
WWW: http://www.iayc.org/
Founded: 1977
Membership: 40
Activities: organizing international astronomical youth camps
Periodicals: (10) "Communicator"; (1) "IAYC Report"
City Reference Coordinates: 003°27'00"E 51°05'00"N

International Union of Radio Science (URSI)
c/o INTEC
Universiteit Gent
Sint-Pietersnieuwstraat 41
B-9000 Gent
Telephone: (0)92643320
Telefax: (0)92644288
Electronic Mail: ursi@intec.rug.ac.be
WWW: http://www.intec.rug.ac.be/ursi/
Founded: 1919
Membership: 48 (countries)
Activities: radiocommunications * telecommunications * research
Sections: (commissions) A. Electromagnetic Metrology * B. Fields and Waves * C. Signals and Systems * D. Electronics and Photonics * E. Electromagnetic Noise and Interference * F. Wave Propagation and Remote Sensing * G. Ionospheric Radio and Propagation * H. Waves in Plasmas * J. Radio Astronomy * K. Electromagnetics in Biology and Medicine
Periodicals: (4) "The Radio Science Bulletin" (ISSN 1024-4530); (1/3) "Proceedings of URSI General Assemblies", "Review of Radio Science", "Modern Radio Science"
Awards: (1/3) "Balthasar van der Pol Gold Medal", "John Howard Dellinger Gold Medal", "Issac Koga Gold Medal", "Booker Gold Medal", "Appleton Prize" (together with the Royal Society)
City Reference Coordinates: 003°43'00"E 51°03'00"N

International Union of the History and Philosophy of Science (IUHPS)
c/o Centre d'Histoire des Sciences
Université de Liège

Avenue des Tilleuls 15
B-4000 Liège
Telephone: (0)43669479
Telefax: (0)43669447
Electronic Mail: chst@ulg.ac.be
   chstulg@vml.ulg.ac.be
WWW: http://www.icsu.org/Membership/SUM/iuhps.html
Founded: 1956 (1947 as "International Union of History of Science")
Membership: 80 (member committees)
City Reference Coordinates: 005°34'00"E 50°38'00"N

Jongerenvereniging voor Sterrenkunde (JvS)
H. d'Ydewallestraat 46
B-8730 Beernem
Telephone: (0)50789705
Electronic Mail: info@vvs.be
   jvsstaf@listserv.cc.kuleuven.ac.be
WWW: http://www.vvs.be/
Founded: 1971
Membership: 350
Activities: meetings * camps * observing
Periodicals: (6) "Astra"
City Reference Coordinates: 003°20'00"E 51°09'00"N

Katholieke Universiteit Leuven
• See "Universiteit Leuven"

Koninklijke Vlaamse Academie van België voor Wetenschappen en Kunsten
Paleis der Academiën
Hertogsstraat 1
B-1000 Brussel
Telephone: (0)25502323
Telefax: (0)25502325
Electronic Mail: kawlsk@skynet.be
Founded: 1938
Staff: 15
Periodicals: "Mededelingen", "Academiae Analecta"
Awards: "H.L. Vanderlinden Prize", "MacLeod Prize", "P. van Oye Prize", "H. Schouteden Prize", "J. Gillis Prize", "O. Callebaut Prize", "J. Coppens Prize", "C. de Clercq Prize", "F. Van der Mueren Prize", "R. Lenaerts Prize"
City Reference Coordinates: 004°20'00"E 50°50'00"N

Koninklijk Meteorologisch Instituut van België (KMIB)
Ringlaan 3
B-1180 Brussel
Telephone: (0)23730611
Telefax: (0)23751259
Electronic Mail: <userid>@oma.be
WWW: http://www.meteo.oma.be/IRM-KMI/
Founded: 1913
Staff: 190
Activities: climatology * hydrology * dynamic climatology * applied meteorology * aerology * aeronomy * internal/external geophysics * computing
Periodicals: (12) "Klimatologische Waarnemingen" (ISSN 0029-7682), "Geofysische Waarnemingen" (ISSN 0020-2525), "Ionosferische Waarnemingen - Kosmische Straling" (ISSN 0020-2533); (4) "Ozon Waarnemingen" (ISSN 0770-0164); (1) "Zonnestraling" (ISSN 0524-7780), "Aardmagnetisme" (ISSN 0770-4569), "Jaarverslag" (ISSN 0772-3288); "Publikaties Reeks A" (ISSN 0020-255x), "Publikaties Reeks B" (ISSN 0770-4615); "Buiten Reeks" (ISSN 0772-4330); "Meteorologische Documentatie" (ISSN 0772-4349), "Miscellanea A" (ISSN 0770-0261), "Miscellanea B" (ISSN 0770-0318), "Miscellanea C" (ISSN 0772-4357), "Wetenschappelijke en Technische Publikaties"
Coordinates: 004°21'00"E 50°48'00"N H104m (Ukkel)
   004°36'00"E 50°06'00"N H230m (Dourbes)
City Reference Coordinates: 004°21'00"E 50°48'00"N

Koninklijk Sterrenkundig Genootschap van Antwerpen (KSGA)
Kapelsesteenweg 340
B-2930 Brasschaat
Telephone: (0)36642893
Founded: 1905
Membership: 110
Activities: lectures * public observing * library * excursions
Periodicals: (1) "Jaarverslag"; (12) "Astronomische Gazet"
Coordinates: 004°24'12"E 51°13'24"N H20m
City Reference Coordinates: 004°30'00"E 51°17'00"N

Koninklijke Sterrenwacht van België (KSB)
Ringlaan 3

B-1180 Brussel
Telephone: (0)23730211
Telefax: (0)23749822
Electronic Mail: rob_info@oma.be
WWW: http://www.oma.be/KSB-ORB/
Founded: 1826
Staff: 100
Activities: fundamental astronomy * time and frequency * satellite geodesy * geodynamics * seismology * gravimetry * physics of planetary interiors * astrometry * celestial mechanics * astrophysics * radioastronomy * solar physics * stellar statistics
Periodicals: (1) "Jaarboek" (ISSN 0524-7780)
Coordinates: 004°21'29"E 50°47'51"N H105m (IAU Code 012)
　　　　　　　　005°15'19"E 50°11'31"N H293m (Humain Radioastronomy Station)
City Reference Coordinates: 004°20'00"E 50°50'00"N

Kot Astro
Place de l'Escholier 1/208
B-1348 Louvain-la-Neuve
Telephone: (0)10453008
Electronic Mail: kotastro@hotmail.com
WWW: http://www.kotastro.be.tf/
Founded: 1989
Membership: 8
Activities: observing * astronomy week * lectures * popularization * exhibitions
City Reference Coordinates: 004°37'00"E 50°40'00"N

Lichtenknecker Optics (LO) sa
Grote Breemstraat 21
B-3500 Hasselt
Telephone: (0)11253052
Telefax: (0)11250090
Founded: 1973
Activities: manufacturing astronomical optics and instrumentation (including optical components up to 600mm in diameter) * designing * prototype making * distributor for Belgium of most astronomical instruments
City Reference Coordinates: 005°20'00"E 50°56'00"N

Limburgse Volkssterrenwacht (LVS) vzw
● See now "Europlanetarium vzw"

Nationaal Centrum voor Wetenschappelijke en Technische Dokumentatie (NCWTD)
Keizerslaan 4
B-1000 Brussel
Telephone: (0)25195640
Telefax: (0)25195645
　　　　　(0)25195679
Electronic Mail: cnsdt@kbr.be
WWW: http://www.belspo.be/stis/index_nl.htm
Founded: 1964
Staff: 11
Activities: on-line scientific, technical, economical, patent documentation and information * European Union National Awareness Partner * ESA-IRS National Aerospace Centre
City Reference Coordinates: 004°20'00"E 50°50'00"N

North Atlantic Treaty Organization (NATO), Division of Scientific Affairs
(Organisation du Traité de l'Atlantique Nord, OTAN, Division des Affaires Scientifiques)
Boulevard Leopold III
B-1110 Brussels
Telephone: (0)27284111
Telefax: (0)27284232
Electronic Mail: <userid>@hq.nato.int
　　　　　　　natodoc@hq.nato.int
WWW: http://www.nato.int/science/
Founded: 1958
Staff: 20
Activities: funding international collaborative scientific research and training
Periodicals: (4) "NATO Science & Society Newsletter"; "NATO Scientific Publications" (ISSN 0770-3244), "NATO Yearbook"

City Reference Coordinates: 004°20'00"E 50°50'00"N

Observatoire Royal de Belgique (ORB)
Avenue Circulaire 3
B-1180 Bruxelles
Telephone: (0)23730211
Telefax: (0)23749822
Electronic Mail: rob_info@oma.be

WWW: http://www.oma.be/KSB-ORB/
Founded: 1826
Staff: 100
Activities: fundamental astronomy * time and frequency * satellite geodesy * geodynamics * seismology * gravimetry * physics of planetary interiors * astrometry * celestial mechanics * astrophysics * radioastronomy * solar physics * stellar statistics
Periodicals: (1) "Annuaire" (ISSN 0524-7780)
Coordinates: 004°21'29"E 50°47'51"N H105m (IAU Code 012)
005°15'19"E 50°11'31"N H293m (Humain Radioastronomy Station)
City Reference Coordinates: 004°20'00"E 50°50'00"N

Organisation du Traité de l'Atlantique Nord (OTAN), Division des Affaires Scientifiques
• See "North Atlantic Treaty Organization (NATO), Division of Scientific Affairs"

Rijksuniversitair Centrum Antwerpen (RUCA)
• See "Universiteit Antwerpen"

Rijksuniversiteit Gent (RUG)
• See "Universiteit Gent"

Royal Observatory of Belgium, Planetarium
Boechoutlaan - Avenue Bouchout 10
B-1020 Brussels
Telephone: (0)24747050
(0)24747060
Telefax: (0)24783026
WWW: http://www.planetarium.be/
Founded: 1973
Activities: education * lectures * colloquia * exhibitions
City Reference Coordinates: 004°20'00"E 50°50'00"N

Société Astronomique de Liège (SAL)
c/o Institut d'Astrophysique
Avenue de Cointe 5
B-4000 Liège
Telephone: (0)42533590
Telefax: (0)42547511
Electronic Mail: sal@astro.ulg.ac.be
WWW: http://www.astro.ulg.ac.be/~sal
http://www.ulg.ac.be/musees/planetarium/ (Planetarium)
Founded: 1938
Membership: 850
Activities: education * library * observing * monthly lectures * instruments on loan * trips * planetarium
Periodicals: (10) "Le Ciel" (circ.: 1,100)
Observatories: 1 (Nandrin)
City Reference Coordinates: 005°34'00"E 50°38'00"N

Société Royale Belge d'Astronomie, de Météorologie et de Physique du Globe (SRBA)
c/o Observatoire Royal de Belgique
Avenue Circulaire 3
B-1180 Bruxelles
Telephone: (0)23730253
Telefax: (0)23749822
Electronic Mail: srba@oma.be
WWW: http://www.oma.be/BIRA-IASB/SRBA/
Founded: 1884 ("Ciel et Terre": 1880)
Membership: 800
Activities: lectures * publishing * essentially Brussels-oriented
Periodicals: (6) "Ciel et Terre" (ISSN 0009-6709, circ.: 1,200)
City Reference Coordinates: 004°20'00"E 50°50'00"N

Société Royale des Sciences de Liège (SRSL)
c/o Institut de Mathématique
Université de Liège
Bâtiment B37
Grande Traverse 12
B-4000 Liège
Telephone: (0)43669371
Telefax: (0)43669547
Electronic Mail: n.naa@ulg.ac.be
WWW: http://www.ulg.ac.be/ipne/srsl/
Founded: 1835
Membership: 220
Activities: publishing monographs, proceedings, theses and original scientific papers * organizing lectures by foreign scientists * exchange of periodicals
Periodicals: (6) "Bulletin" (circ.: 500)

Awards: (1/5) "Prix Louis D'Or" (chemistry), "Prix Lucien Godeaux" (mathematics), "Prix Pol Swings" (physics), "Prix E. van Beneden" (biology)
City Reference Coordinates: 005°34'00"E 50°38'00"N

Société Scientifique de Bruxelles
Rue de Bruxelles 61
B-5000 Namur
Telephone: (0)81724464
Telefax: (0)81724465
Electronic Mail: charles.courtoy@fundp.ac.be (Charles Courtoy, Secretary)
Founded: 1875
Membership: 140
Periodicals: "Revue des Questions Scientifiques" (ISSN 0035-2160)
City Reference Coordinates: 004°52'00"E 50°28'00"N

Solar Influences Data analysis Center (SIDC)
c/o Pierre Cugnon (Director)
Observatoire Royal de Belgique (ORB)
Avenue Circulaire 3
B-1180 Bruxelles
Telephone: (0)2-3730276
Telefax: (0)2-3730224
Electronic Mail: sidc@oma.be
              p.cugnon@oma.be (Pierre Cugnon)
WWW: http://sidc.oma.be/
Founded: 1981
Staff: 7
Activities: sunspot number * forecast * space weather * solar activity * solar cycle * solar indices
Periodicals: (12) "Sunspot Bulletin"; (4) "SIDC News"
Coordinates: 004°21'29"E 50°47'55"N H105m
• Formerly "Sunspot Index Data Center (SIDC)"
City Reference Coordinates: 004°20'00"E 50°50'00"N

Spacebel Instrumentation SA
Parc Industriel de Recherches du Sart Tilman
Rue des Chasseurs Ardennais
B-4031 Angleur
Telephone: (0)43666611
Telefax: (0)43660433
Electronic Mail: christine.mottard@spacebel.be (Public Relations)
WWW: http://www.spacebel.be/
Founded: 1988
Staff: 57
Activities: design and development of space-based optical equipment
City Reference Coordinates: 005°34'00"E 50°38'00"N

Sunspot Index Data Center (SIDC)
• See now "Solar Influences Data analysis Center (SIDC)"

Union of International Associations (UIA)
Rue Washington 40
B-1050 Bruxelles
Telephone: (0)26401808
Telefax: (0)26460525
Electronic Mail: info@uia.be
WWW: http://www.uia.org/
Founded: 1907
Membership: 195
Staff: 20
Activities: collecting, analyzing and publishing data on over 30,000 international governmental and non-governmental organizations, as well as on their meetings, publications and strategies, and on over 10,000 world problems * maintaining on-line databases * carrying out surveys and studies
Periodicals: (6) "Transnational Associations"; (4) "International Congress Calendar"; (1) "Yearbook of International Organizations"; "Encyclopedia of World Problems and Human Potential"
City Reference Coordinates: 004°20'00"E 50°50'00"N

Universitaire Instelling Antwerpen (UIA)
• See "Universiteit Antwerpen"

Université Catholique de Louvain (UCL), Institut d'Astronomie et de Géophysique G. Lemaître
Chemin du Cyclotron 2
B-1348 Louvain-la-Neuve
Telephone: (0)10473297
Telefax: (0)10474722
Electronic Mail: <userid>@astr.ucl.ac.be

WWW: http://www.astr.ucl.ac.be/
Founded: 1966
Staff: 50
Activities: education * research in climatology and meteorology
Periodicals: "Contributions", "Bulletin Climatologique de la Station de Louvain-la-Neuve", "Scientific Reports", "Progress Reports"
Coordinates: 004°37'30"E 50°39'55"N
City Reference Coordinates: 004°38'00"E 50°40'00"N

Université de Liège (ULg), Centre Spatial de Liège (CSL)
Parc Scientifique du Sart Tilman
Avenue du Pré-Aily
B-4031 Liège-Angleur
Telephone: (0)43676668
Telefax: (0)43675613
Electronic Mail: cslulg@ulg.ac.be
WWW: http://www.ulg.ac.be/
Founded: 1988
Staff: 100
Activities: space qualification of instruments for optoelectronic observing * astronomical payloads * designing optical instrumentation
City Reference Coordinates: 005°34'00"E 50°38'00"N

Université de Liège (ULg), Institut d'Astrophysique et de Géophysique (IAGL)
Allée du 6 Août 17, B5C
B-4000 Sart Tilman
Telephone: (0)43669716
Telefax: (0)43669746
Electronic Mail: <userid>@astro.ulg.ac.be
WWW: http://www.astro.ulg.ac.be/
Founded: 1880
Staff: 85
Activities: theoretical and observational astrophysics * spectroscopy * solar atmosphere * solar and atmospheric spectroscopy * planetary physics * geophysics * cosmology
Awards: "Prix Dehalu"
Coordinates: 005°33'54"E 50°37'06"N H127m (Cointe – IAU Code 623)
City Reference Coordinates: 005°34'00"E 50°38'00"N (Liège)

Université de Mons-Hainaut, Département d'Astrophysique et de Spectroscopie
Place du Parc 20
B-7000 Mons
Telephone: (0)65373727
(0)65373728
Telefax: (0)65373054
Electronic Mail: <userid>@umh.ac.be
WWW: http://www.umh.ac.be/~astro/indexgb.shtml
Founded: 1968
Staff: 5
Activities: education in astronomy, astrophysics and atomic physics * research in astrophysics and atomic spectroscopy
Periodicals: "Mons Astrophysical Papers" (circ.: 300)
Coordinates: 016°30'00"W 28°18'00"N (Izaña, Teide – IAU Code 954)
City Reference Coordinates: 003°56'00"E 50°27'00"N

Universiteit Antwerpen, Rijksuniversitair Centrum Antwerpen (RUCA), Onderzoeksgroep Astrofysica
Groenenborgerlaan 171
B-2020 Antwerpen
Telephone: (0)32180355
Telefax: (0)32180204
Electronic Mail: astro@ruca.ua.ac.be
<userid>@ruca.ua.ac.be
WWW: http://www.ruca.ua.ac.be/
Founded: 1985
Staff: 2
Activities: open clusters * associations * very young stellar systems * dynamical evolution and models * numerical N-body calculations * radial velocities * echelle spectroscopy
City Reference Coordinates: 004°25'00"E 51°13'00"N

Universiteit Antwerpen, Universitaire Instelling Antwerpen (UIA), Department of Physics
Universiteitsplein 1
B-2610 Antwerpen (Wilrijk)
Telephone: (0)38202457
Telefax: (0)38202245
WWW: http://nat-www.uia.ac.be/
Founded: 1972
Staff: 30

Activities: theoretical astrophysics * cosmology * relativity
City Reference Coordinates: 004°25'00"E 51°13'00"N

Universiteit Brussel, Astronomy Group
Pleinlaan 2
B-1050 Brussel
Telephone: (0)26293469
Telefax: (0)26293424
Electronic Mail: <userid>@vub.ac.be
WWW: http://www.vub.ac.be/STER/ster.html
Founded: 1990
Staff: 8
Activities: education * research * stellar evolution * binaries * stellar atmospheres * pulsating stars * photometry
Coordinates: 004°23'00"E 50°48'48"N H147m
• Full university name: "Vrije Universiteit Brussel (VUB)"
City Reference Coordinates: 004°20'00"E 50°50'00"N

Universiteit Gent, Sterrenkundig Observatorium
Gebouw S9
Krijgslaan 281
B-9000 Gent
Telephone: (0)92644798
Telefax: (0)92644989
Electronic Mail: <userid>@rug.ac.be
WWW: http://www.rug.ac.be/~hdejongh/astro/astro.html
Founded: 1907
Staff: 5
Activities: stellar dynamics * numerical astrophysics * radiation transfer * plasma astrophysics * popularization
Periodicals: "Mededelingen" (ISSN 0072-4432)
Coordinates: 003°04'00"E 51°02'00"N
• Full university name: "Rijksuniversiteit Gent (RUG)"
City Reference Coordinates: 003°04'00"E 51°02'00"N

Universiteit Leuven, Center for Plasma Astrophysics
Celestijnenlaan 200 B
B-3001 Heverlee
Telephone: (0)16327007
Telefax: (0)16327998
WWW: http://cpa.wis.kuleuven.ac.be/
Founded: 1992
Staff: 15
Activities: plasma astrophysics * solar physics * MHD * education
• Full university name: "Katholieke Universiteit Leuven"
City Reference Coordinates: 004°42'00"E 50°53'00"N

Universiteit Leuven, Instituut voor Sterrenkunde
Celestijnenlaan 200 B
B-3001 Heverlee
Telephone: (0)16327027
Telefax: (0)16327999
Electronic Mail: <userid>@ster.kuleuven.ac.be
WWW: http://www.ster.kuleuven.ac.be/
Staff: 14
Activities: astronomy and astrophysics
• Full university name: "Katholieke Universiteit Leuven"
City Reference Coordinates: 004°42'00"E 50°53'00"N

Université Libre de Bruxelles (ULB), Institut d'Astronomie et d'Astrophysique (IAA)
Case Postale 226
Boulevard du Triomphe 2
B-1050 Bruxelles
Telephone: (0)26502842
Telefax: (0)26504226
Electronic Mail: <userid>@astro.ulb.ac.be
WWW: http://www-astro.ulb.ac.be/
Founded: 1923
Staff: 11
Activities: theoretical astrophysics * theoretical and experimental nuclear astrophysics * stellar evolution * stellar structure * stellar abundances * CP stars * close binaries * nucleosynthesis * spectroscopy, photometry and radial velocities of red giant stars
City Reference Coordinates: 004°20'00"E 50°50'00"N

Université Libre de Bruxelles (ULB), Physique Nucléaire Théorique et Physique Mathématique (PNTPM)
Campus Plaine

Avenue F.D. Roosevelt
Case Postale 229
B-1050 Bruxelles
Telephone: (0)26505560
(0)26505562
Telefax: (0)26505045
Electronic Mail: <userid>@ulb.ac.be
WWW: http://www.ulb.ac.be/recherche/cr/cr.html
Staff: 14
Activities: theoretical physics * nuclear reactions in general and of astrophysical interest * nuclear spectroscopy * nuclear and atomic physics * mathematical physics
City Reference Coordinates: 004°20'00"E 50°50'00"N

Urania, Volkssterrenwacht van Antwerpen
J. Mattheessensstraat 60
B-2540 Hove
Telephone: (0)34552493
Telefax: (0)34542297
Electronic Mail: info@urania.be
WWW: http://www.urania.be/
Founded: 1969
Membership: 520
Activities: education * lectures * visits * observing * meteorology * working groups * exhibitions * planetarium (Antwerp Zoo) * astronomy software * radio meteors
Periodicals: (15) "Urania in de Kijker" (circ.: 150); (10) "Oberonnieuws" (circ: 100); (4) "De Sterrenwachter" (circ.: 650)
Coordinates: 004°28'02"E 51°08'45"N H18m
City Reference Coordinates: 004°29'00"E 51°09'00"N

Vereniging voor Sterrenkunde, Meteorologie, Geofysica en Aanverwante Wetenschappen (VVS)
Brieversweg 147
B-8310 Brugge
Brugge
Telephone: (0)50358872
Telefax: (0)50355007
Electronic Mail: info@vvs.be
WWW: http://www.vvs.be/
Founded: 1944
Membership: 1900
Activities: amateur astronomy and meteorology
Sections: Astrophotography * Occultations * Computing Technology * Comets * Artificial Satellites * Meteors * Planets * Space * Variable Stars * Meteorology * Sun * Light Pollution * Solar Eclipses * Deep Sky
Periodicals: (12) "Heelal" (ISSN 0772-6422, circ.: 2,100), "Halo"; (4) "Distant Targets"
Awards: (1) "Galilei Prizes"
City Reference Coordinates: 003°14'00"E 51°13'00"N

Volkssterrenwacht Beisbroek vzw
Zeeweg 96
B-8200 Brugge
Telephone: (0)50390566
Telefax: (0)50393244
Electronic Mail: beisbroek@unicall.be
WWW: http://www.urania.be/beisbroek/
Founded: 1984
Membership: 170
Staff: 1
Activities: popularization * visits of schools * education * observing (meteors, variable stars) * investigation of selected astrophysical topics * planetarium
Periodicals: (1) "Jaarverslag"
Coordinates: 003°09'00"E 51°10'00"N H20m
City Reference Coordinates: 003°09'00"E 51°10'00"N

Volkssterrenwacht Mira vzw
Abdijstraat 20
B-1850 Grimbergen
Telephone: (0)22691280
Telefax: (0)22691075
Electronic Mail: info@mira.be
WWW: http://www.mira.be/
Founded: 1967
Membership: 300
Staff: 3
Activities: popularization (astronomy, meteorology, space sciences) * lectures * exhibitions * library and documentation center * observing (Sun) * static planetarium
Periodicals: (4) "Mira Ceti"
Coordinates: 004°22'13"E 50°56'08"N
City Reference Coordinates: 004°22'00"E 50°56'00"N

Volkssterrenwacht Nysa
Pastorijstraat 1
B-9403 Neigem
Telephone: (0)54328420
(0)54323003
(0)95333242
(0)95335236
Electronic Mail: nysa.volkssterrenwacht@skynet.be
WWW: http://users.skynet.be/volkssterrenwacht.nysa/
Founded: 1990
City Reference Coordinates: 004°04'00"E 50°48'00"N

Vrije Universiteit Brussel (VUB)
• See "Universiteit Brussel"

Wega vzw
Escoriallaan 18
B-3000 Leuven
Telephone: (0)16605131
Electronic Mail: info@wega-astro.be
WWW: http://www.wega-astro.be/
Founded: 1975
Membership: 60
Activities: observing * lectures * education * camps * instrumentation * star parties * meetings
Periodicals: (12) "WIP"
City Reference Coordinates: 004°42'00"E 50°53'00"N

Bolivia

Academia Nacional de Ciencias de Bolivia (ANCB)
Avenida 16 de Julio 1732
Casilla 5829
La Paz
Telephone: (0)2-363990
Telefax: (0)2-379681
Electronic Mail: ancb@ancb.bo
Founded: 1960
Membership: 60
Activities: science policy studies * advising government * environmental sciences * astronomy
Periodicals: (4) "Revista"
City Reference Coordinates: 068°10'00"W 16°30'00"N

Asociación Boliviana de Astronomía (ABA)
Casilla 7707
La Paz
or :
Calle México 1771
La Paz
Telephone: (0)2-326922
         (0)2-352660
Telefax: (0)2-791696
Electronic Mail: lcabezas@entelsa.entelnet.bo (Luis Cabezas Tito, Vice President)
Founded: 1969
Membership: 160
Activities: observing * popularization
Sections: Meteors
Periodicals: (6) "Astromundo"; "Firmamento", "Alajpacha", "Meteoros"
Coordinates: 068°00'00"W 16°00'00"S (La Paz)
         066°00'00"W 20°00'00"S (Potosí)
         062°00'00"W 18°00'00"S (Santa Cruz)
         066°00'00"W 19°00'00"S (Sucre)
         066°00'00"W 21°00'00"S (Tarija)
City Reference Coordinates: 068°10'00"W 16°30'00"N

Astronomía Sigma Octante (ASO), Centro de Investigación y Estudio en Astronomía
Casilla 1491
Cochabamba
or :
Centro "Simón I. Patiño"
Avenida Potosí 1450
Cochabamba
Telephone: (0)4-4280602
Telefax: (0)4-4281099
Electronic Mail: aso@ucbcba.edu.bo
WWW: http://www.ucbcba.edu.bo/aso/
Founded: 1977
Membership: 20
Activities: observing (Sun, planets, satellites, minor planets, variable stars, comets, meteors) * theoretical astronomy * computational astronomy * education * astrometry
Sections: positional astronomy * meteors * Sun * solar system * stellar observing * popularization
Periodicals: (12) "Sunspots Report"
Coordinates: 066°09'48"W 17°24'06"S H2,570m
         066°09'29"W 17°22'28"S H2,633m
City Reference Coordinates: 066°10'00"W 17°26'00"S

Instituto Boliviano de Normalización y Calidad (IBNORCA)
Avenida Camacho Esq. Bueno 1488
Casilla 5034
La Paz
Telephone: (0)2-317262
Telefax: (0)2-317262
City Reference Coordinates: 068°10'00"W 16°30'00"N

Liga Ibero-Americana de Astronomía (LIADA)
c/o Rodolfo Zalles (President)
Casilla 7090
La Paz
or :
c/o Mirko Raljevic (Secretary)
Casilla 7090

La Paz
or :
c/o Raul Cagigao (Treasurer)
Casilla 387
Sucre
WWW: http://www.cnba.uba.ar/acad/astro/liada/liada.html
Founded: 1982
Membership: 300
Activities: promoting observing (meteors, comets, Sun, planets, variable stars) * grouping amateur astronomers * developing respect and appreciation of Earth as a planet * publishing works by members
Periodicals: (4) "Universo" (ISSN 0012-9820), "La Red"
City Reference Coordinates: 068°10'00"W 16°30'00"N (La Paz)
　　　　　　　　　　　　　　　071°08'00"W 08°36'00"N (Sucre)

Observatorio Astronómico Boliviano-Ruso
Santa Ana
Casilla 20
Tarija
Telephone: (0)1-860979
Telefax: (0)66-44238
Electronic Mail: obsrzb@uajms.bo
Founded: 1984
Staff: 4
Activities: astrometry * photometric catalogs
Coordinates: 064°36'00"W 21°35'00"S H2,000m (IAU Code 820)
City Reference Coordinates: 064°36'00"W 21°35'00"S

Observatorio Astronómico de Patacamaya (OAP)
• See "Universidad Mayor de San Andrés, Observatorio Astronómico de Patacamaya (OAP)"

Observatorio San Calixto (OSC)
Indaburu 944
Casilla 12656
La Paz
Telephone: (0)2-356098
Telefax: (0)2-376805
Electronic Mail: sancalixto@mail.megalink.com
Founded: 1913
Staff: 6
Activities: astronomical demonstrations to visiting school groups * seismology * meteorology
Periodicals: (12) "Lecturas de la Estación de la Paz" (circ.: 60)
Observatories: network of seismic stations
City Reference Coordinates: 068°10'00"W 16°30'00"N

Planetario Dr. Max Schreier
Calle Frederico Zuazo 1976
Casilla 3164
La Paz
Telephone: (0)2-359522
Telefax: (0)2-316738
Electronic Mail: planetar@astro.bo
WWW: http://www.umsanet.edu.bo/org/astro/
Founded: 1976
Staff: 7
Activities: astronomy * archaeoastronomy * solar activity * geomagnetism
Periodicals: "Intijiwaña"; (1) "Calendario"
Observatories: 1 (see Estación Astronómica de Patacamaya)
City Reference Coordinates: 068°10'00"W 16°30'00"S

Servicio Nacional de Meteorología e Hidrología (SENAMHI)
Avenida Arce 2579
Casilla 10993
La Paz
Telephone: (0)2-326165
　　　　　　　(0)2-355824
Telefax: (0)2-392413
Founded: 1944
Periodicals: "Boletines Meteorológicos", "Anuarios Meteorológicos"
Observatories: from 57°59'W to 69°20'W and from 11°5'S to 22°30'S
City Reference Coordinates: 068°10'00"W 16°30'00"S

Universidad Mayor de San Andrés (UMSA), Observatorio Astronómico de Patacamaya (OAP)
Calle Federico Zuazo 1976
Casilla 3164
La Paz

Telephone: (0)2-2441822
Telefax: (0)2-2441738
Electronic Mail: planetar@astro.bo
              planetar@utama.bolnet.bo
WWW: http://www.umsanet.edu.bo/org/astro/
Founded: 1974
Activities: astrometry * photometry * special events * Sun * planetarium (see Planetario Dr. Max Schreier) * ethnoastronomy
Periodicals: (4) "Boletín de Actividad Solar"; "Calendario Astronómico"
Coordinates: 067°57'07"W 17°15'57"S H3,789m
City Reference Coordinates: 068°10'00"W 16°30'00"N

Brazil

Academia Brasileira de Ciencias (ABC)
Caixa Postal 229
20001-970 Rio de Janeiro - RJ
or :
Rua Anfilófio de Carvalho 29 - 3°
20030-060 Rio de Janeiro - RJ
Telephone: (0)21-2204794
Telefax: (0)21-5325807
Electronic Mail: abc@abc.org.br
WWW: http://www.abc.org.br/
Founded: 1916
Membership: 530
Activities: publishing * meetings * advising
Periodicals: (4) "Anais da Academia Brasileira de Ciências" (ISSN 0001-3765)
Awards: (1/4) "Ruschi Award"
City Reference Coordinates: 043°15'00"W 22°54'00"S

Associação Astronômica da Poços de Caldas
Caixa Postal 340
37701-010 Poços de Caldas - MG
or :
Rua Assis Figueiredo 1.174 Conj. 34
37701-000 Poços de Caldas - MG
Telephone: (0)35-7222148
Founded: 1987
Membership: 20
Activities: education * celestial mechanics * astrophysics * Sun * Moon
Periodicals: (4) "As Plêiades" (circ.: 50)
Coordinates: 046°33'53"W 21°50'20"S H1,600m (Morro de São Domingo)
City Reference Coordinates: 046°33'00"W 21°48'00"S

Associação Brasileira de Normas Técnicas (ABNT)
Avenida Treze de Maio 13 - 28° andar
20003-900 Rio de Janeiro - RJ
Telefone (21) 3974-2300 Fax (21) 3974-2347
Telephone: (0)21-39742300
Telefax: (0)21-39742347
Electronic Mail: drl@abnt.org.br
WWW: http://www.abnt.org.br/
Founded: 1940
City Reference Coordinates: 043°15'00"W 22°54'00"S

Centro de Divulgação da Astronomia (CDA)
c/o Centro de Divulgação Cientifica e Cultural
Rua 9 de Julho 1227
13569-590 São Carlos - SP
Telephone: (0)16-2739191
         (0)16-2739772
Telefax: (0)16-2723910
Electronic Mail: cda@cdcc.sc.usp.br
WWW: http://www.cdcc.sc.usp.br/cda/observat.htm
Founded: 1986
Staff: 10
Activities: education * observing * exhibitions
Coordinates: 047°53'48"W 22°00'40"S H940m
City Reference Coordinates: 047°53'00"W 22°02'00"S

Centro de Estudios Astronômicos de Alagoas (CEAAL)
Caixa Postal 215
57.020-970 Maceió - AL
Electronic Mail: ceaal@fapeal.br
WWW: http://www.fapeal.br/ceaal/
Founded: 1978
Membership: 12
Periodicals: (12) "Apolo"
City Reference Coordinates: 035°43'00"W 09°40'00"S

Centro de Previsão de Tempo e Estudos Climáticos (CPTEC)
• See "Instituto Nacional de Pesquisas Espaciais (INPE), Centro de Previsão de Tempo e Estudos Climáticos (CPTEC)"

Centro de Radio-Astronomia e Aplicações Espaciais (CRAAE)
• See "Instituto Presbiteriano Mackenzie, Centro de Radio-Astronomia e Aplicações Espaciais (CRAAE)"

Clube de Astronomia do Rio de Janeiro (CARJ)
Caixa Postal 65.090
20072-970 Rio de Janeiro - RJ
Telephone: (0)21-33594790
Electronic Mail: carj@astronomia-carj.com.br
WWW: http://www.astronomia-carj.com.br/
Founded: 1976
Membership: 8
Activities: education * popularization
Periodicals: (12) "Guia do Amador de Astronomia", "Circular Astronômica"
City Reference Coordinates: 043°15'00"W 22°54'00"S

Clube Estudantil de Astronomia (CEA)
Caixa Postal 736
50001-000 Recife - PE
or :
Conjunto Residencial INOCOOP
Rua Azeredo Continho s/n
Várzea
50741-000 Recife - PE
WWW: http://www.cea.org.br/
Founded: 1972
Membership: 120
Activities: occultations by minor planets * comets * binaries * education
Periodicals: (4) "Boletim Astronômico" (circ.: 400)
Awards: "Messier Club Member Certificate"
Coordinates: 034°57'28"W 08°03'03"S H2m (Jorge Polman Observatory)
City Reference Coordinates: 034°53'00"W 08°06'00"S

Colégio Magno, Observatório
Rua Duque Costa 164
04671-160 São Paulo - SP
Telephone: (0)11-5486166
Electronic Mail: colmagno@eu.ansp.br
WWW: http://www.colmagno.com.br/observatorio/observatorio.htm
City Reference Coordinates: 046°37'00"W 23°32'00"S

Conselho Nacional Desenvolvimento Scientifico e Tecnologico
Edificio CNPq
SEPN 507 - Bloco "B"
Ed. Sede CNPq
70740-901 Brasília - DF
Telephone: (0)61-3489000
Telefax: (0)61-2741950
Electronic Mail: <userid>@cnpq.br
WWW: http://www.cnpq.br/
Founded: 1951
City Reference Coordinates: 044°26'00"W 16°12'00"S

Fundação de Amparo à Pesquisa do Estado de São Paulo (FAPESP)
Rua Pio XI 1500
Alto da Lapa
05468-901 São Paulo - SP
Telephone: (0)11-8384000
Telefax: (0)11-2614167
Electronic Mail: info@fapesp.br
WWW: http://www.fapesp.br/
Founded: 1960
City Reference Coordinates: 046°37'00"W 23°32'00"S

Fundação Planetário da Cidade do Rio de Janeiro
Avenida Padre Leonel Franca 240
Gávea
22451-000 Rio de Janeiro - RJ
Telephone: (0)21-2740046
        (0)21-2740096
Telefax: (0)21-2396927
Electronic Mail: planet@rdc.puc-rio.br
WWW: http://www.rio.rj.gov.br/planetario/
    http://io.rdc.puc-rio.br/planetario/
    http://io.rdc.puc-rio.br/planetario/ingles/
Founded: 1970

Staff: 40
Activities: shows * public observing * education * lectures
Coordinates: 043°10'00"W 22°53'00"S H10m
City Reference Coordinates: 043°10'00"W 22°53'00"S

Grupo de Estudos de Astronomia (GEA)
Planetário
Campus UFSC
88.000-000 Florianopolis - SC
Telephone: (0)48-3319241
Electronic Mail: geraldomattos@hotmail.com (Geraldo Mattos)
WWW: http://www.gea.org.br/
Founded: 1985
Activities: education * popularization * lectures
City Reference Coordinates: 048°31'00"W 27°35'00"S

Instituto Nacional de Meteorologia (INMET)
Eixo Monumental
Via S-1
Cruzeiro
70610-400 Brasília - DF
Telephone: (0)61-2247944
  (0)61-2251230
  (0)61-3443333
Telefax: (0)61-2266967
WWW: http://www.inmet.gov.br/
Founded: 1909
Activities: weather forecast * climatology
Periodicals: "Boletim de Radiação Solar", "Boletim Agrometeorológico"
City Reference Coordinates: 044°26'00"W 16°12'00"S

Instituto Nacional de Pesquisas Espaciais (INPE), Cachoeira Paulista
Rodovia Presidente Dutra - Km. 40
Caixa Postal 01
12630-000 Cachoeira Paulista - SP
Telephone: (0)125-609200
Telefax: (0)125-612088
WWW: http://www.dgi.inpe.br/
  http://www.inpe.br/
Founded: 1961 (INPE)
City Reference Coordinates: 045°01'00"W 22°40'00"S

Instituto Nacional de Pesquisas Espaciais (INPE), Centro de Previsão de Tempo e Estudos Climáticos (CPTEC)
Rodovia Presidente Dutra Km 40
Caixa Postal 01
12630-000 Cachoeira Paulista
Telephone: (0)12-5608400
Telefax: (0)12-5612835
Electronic Mail: clima@cptec.inpe.br
  <userid>@cptec.inpe.br
WWW: http://www.cptec.inpe.br/
Founded: 1994 (INPE: 1961)
Staff: 150
Activities: weather and climate prediction
Periodicals: (12) "Climanálise" (ISSN 0103-0019)
City Reference Coordinates: 045°01'00"W 22°40'00"S

Instituto Nacional de Pesquisas Espaciais (INPE), São José dos Campos
Caixa Postal 515
12201-970 São José dos Campos - SP
or :
Avenida dos Astronautas 1758
Jardim da Granja
12227-010 São José dos Campos - SP
Telephone: (0)123-418977
  (0)12-3456000
Telefax: (0)12-3218743
Electronic Mail: <userid>@inpe.br
WWW: http://www.inpe.br/astro/home
  http://www.inpe.br/
Founded: 1961
Staff: 1400
Activities: astrophysics * space geophysics * meteorology * remote sensing * space sciences * spacecraft engineering
Periodicals: "Espacial" (ISSN 0103-0795), "INPE Space News", "Relatório de Atividades", "Climanálise"
City Reference Coordinates: 045°13'00"W 23°11'00"S

Instituto Presbiteriano Mackenzie, Centro de Radio-Astronomia e Aplicações Espaciais (CRAAE)
Rua Consolação 896
01302-000 São Paulo - SP
Telephone: (0)11-2368331
Telefax: (0)11-2142990
Electronic Mail: <userid>@craae.mackenzie.br
WWW: http://craae.mackenzie.br/
Founded: 1989 (Mackenzie: 1870)
Staff: 15
Activities: radioastronomy * VLBI * space geodesy * solar physics * STP
Coordinates: 046°33'48"W 23°11'03"S (ROI - see separate entry)
             038°25'35"W 03°52'40"S (ROEN - see separate entry)
City Reference Coordinates: 046°37'00"W 23°32'00"S

International Mathematical Union (IMU)
c/o Jacob Palis Jr.
IMPA
Estrada Dona Castorina 110
Jardim Botânico
22460-320 Rio de Janeiro - RJ
Telephone: (0)21-2949032
           (0)21-5111749
Telefax: (0)21-5124112
Electronic Mail: imu@impa.br
WWW: http://elib.zib.de/IMU/
Founded: 1951
Membership: 60 (countries)
Periodicals: "Bulletin" (ISSN 1015-8081), "World Directory of Mathematicians"
Awards: "Fields Medal", "Rolf Nevanlinna Prize"
City Reference Coordinates: 043°15'00"W 22°54'00"S

Laboratório Nacional de Astrofísica (LNA), Itajubá
Rua Estados Unidos 154
Bairro das Nações
Caixa Postal 21
37500-000 Itajubá - MG
Telephone: (0)35-6231500
           (0)35-6212121 (Observatório do Pico dos Dias)
Telefax: (0)35-6231544
         (0)35-6212137 (Observatório do Pico dos Dias)
Electronic Mail: <userid>@lna.br
                 secret@lna.br
                 hotel@elisa.lna.br (Observatório do Pico dos Dias)
WWW: http://www.lna.br/
Founded: 1980
Staff: 58
Activities: instrumentation * stellar and extragalactic astronomy * young and active stars * PN * image processing
Coordinates: 045°34'59"W 22°32'04"S H1,864m (Obs. do Pico dos Dias – IAU Code 874)
City Reference Coordinates: 045°27'00"W 22°26'00"S

Museu de Astronomia e Ciências Afins
Rua General Bruce 586
São Cristóvão
20921-000 Rio de Janeiro - RJ
Telephone: (0)21-5809432
           (0)21-5804531
Electronic Mail: mast@omega.lncc.br
WWW: http://pub2.lncc.br:80/mast/
     http://www.info.lncc.br/mast
Founded: 1985
Staff: 90
Periodicals: "Cadernos de Astronomia", "Perspicillum" (ISSN 0102-9495)
City Reference Coordinates: 043°15'00"W 22°54'00"S

Observatório Astronômico Antares (OAA)
Rua da Barra 925
Jardim Cruzeiro
44015-430 Feira de Santana - BA
Telephone: (0)75-6241921
Telefax: (0)75-6241921
Electronic Mail: vmartin@uefs.br (Vera Aparecida Fernandes Martin)
WWW: http://www.uefs.br/antares/default_htm.htm
Founded: 1971
Staff: 10
Activities: fundamental astronomy * solar physics * planetarium
City Reference Coordinates: 038°57'00"W 12°15'00"S

Observatório Astronômico de Piedade (OAP)
• See "Universidade Federal de Minas Gerais, Departamento de Física, Observatório Astronômico de Piedade (OAP)"

Observatório Astronômico de Uberlândia (OAU)
c/o Roberto F. Silvestre
Rua das Seriemas 475
Cidade Jardim
38412-158 Uberlândia - MG
Electronic Mail: silvestre@ufu.br
WWW: http://lfc.df.ufu.br/~silvestr/
Founded: 1996
Activities: private observatory
City Reference Coordinates: 048°17'00"W 18°57'00"S

Observatório Astronômico Herschel-Einstein (OAHE)
c/o Iolanda Siqueira Pamplona (General Coordinator)
Rua General Joaquim de Andrade 68
Aldeota
60150-030 Fortaleza - CE
Telephone: (0)85-2245496
         (0)85-2241741
Founded: 1965
Membership: 12
Activities: comets * meteors * Moon * Sun * eclipses * planets * meteorology
Periodicals: "Pleyades"
Coordinates: 038°30'36"W 03°43'53"S H16m
City Reference Coordinates: 038°30'00"W 03°43'00"S

Observatório Astronômico Monoceros (OAM)
Rua Luiz Marota
Bairro Santa Marta
36660-000 Além-Paraíba - MG
Telephone: (0)32-34624582
Electronic Mail: monocero@openminds.com.br
WWW: http://www.observatoriomonoceros.hpg.com.br/
Founded: 1975
Membership: 7
Activities: solar activity * education * popularization * meteorology
Periodicals: (12) "Circular"
Coordinates: 042°40'32"W 21°52'23"S H300m
City Reference Coordinates: 042°36'00"W 21°49'00"S

Observatório Copérnico
Praça da Bandeira 28
36680-000 São João Nepomuceno - MG
Telephone: (0)32-2612463
Telefax: (0)32-2612762
Founded: 1977
Staff: 7
Activities: education * lectures * exhibitions * investigations
Coordinates: 043°01'10"W 21°32'00"S
City Reference Coordinates: 043°04'00"W 21°33'00"S

Observatório Municipal de Americana (OMA)
Avenida Brasil 2525
13465-000 Americana - SP
Telephone: (0)19-4607026
Electronic Mail: oma@dglnet.com.br
WWW: http://www.dglnet.com.br/oma/
Founded: 1985
Staff: 2
Activities: solar research * public observing
Coordinates: 047°21'10"W 22°45'20"S H582m
City Reference Coordinates: 047°19'00"W 22°44'00"S

Observatório Municipal de Campinas (OMC)
Caixa Postal 27
Distrito de Sousa
13100-000 Campinas - SP
or :
Serra das Cabras
Avenida Anchieta 200 - 6°
13100-000 Campinas - SP
Telephone: (0)192-423252
Founded: 1977

Staff: 15
Activities: education * popularization * planets * comets * Sun * astrometry * photometry
Periodicals: "Boletim Astronômico"
Coordinates: 047°04'39"W 22°53'20"S (IAU Code 870)
City Reference Coordinates: 047°06'00"W 22°54'00"S

Observatório Municipal de Piracicaba (OMP)
Rue Antonio Corrêa Barbosa 2.233 - 6° andar
13400-900 Piracicaba - SP
Telephone: (0)194-334233
Telefax: (0)194-342823
Founded: 1992
Staff: 3
Activities: education * popularization * observing (solar system, double stars)
City Reference Coordinates: 047°38'00"W 22°43'00"S

Observatório Nacional (ON)
Rua General José Cristino 77
20921-400 Rio de Janeiro - RJ
Telephone: (0)21-25806087 (Director)
   (0)21-38789100 (Switchboard)
Telefax: (0)21-25800332
   (0)21-25806041
Electronic Mail: <userid>@on.br
WWW: http://www.on.br/
Founded: 1827
Staff: 170
Activities: astronomy * geophysics * time & frequency service
Periodicals: (1) "Efemérides Astronômicas" (ISSN 0101-935x, circ.: 1,000); "Preprints", "Publicações do Observatório Nacional" (circ.: 500)
Coordinates: 043°13'22"W 22°53'42"S H33m (IAU Code 880)
City Reference Coordinates: 043°15'00"W 22°54'00"S

Observatório Young
Instituto de Aeronautica e Espaço (IAE)
Praça Marechal Eduardo Gomes 50
12228-901 São José dos Campos
Electronic Mail: naee@iconet.com.br
Founded: 1960
Staff: 3
Activities: education
City Reference Coordinates: 045°13'00"W 23°11'00"S

Planetário da Cidade do Rio de Janeiro
● See "Fundação Planetário da Cidade do Rio de Janeiro"

Planetário Prof. José Baptista Pereira
● See "Universidade Federal do Rio Grande do Sul (UFRGS), Planetário Prof. José Baptista Pereira"

Rádio Observatório do Itapetinga (ROI)
Rodovia Municipal Engenharia do Mackenzie s/n
Bairro do Itapetinga
Caixa Postal 200
12940-000 Atibaia - SP
Telephone: (0)11-78711503
Telefax: (0)11-78711784
WWW: http://www.craam.mackenzie.br/roi.htm
Founded: 1971
Coordinates: 046°33'48"W 23°11'03"S
City Reference Coordinates: 046°33'00"W 23°07'00"S

Rádio Observátorio Espacial do Nordeste (ROEN)
Rua José Hipólito s/n
Bairro Tupuiù
Caixa Postal 21
61760-000 Fortaleza - CE
Telephone: (0)85-2609266
Telefax: (0)85-260959
WWW: http://www.roen.inpe.br/english/vlbi.htm
Coordinates: 038°25'35"W 03°52'40"S
City Reference Coordinates: 038°30'00"W 03°43'00"S

Sociedade Astronômica Brasileira (SAB)
Avenida Miguel Stefano 4200
04301-904 São Paulo - SP

Telephone: (0)11-5778599
Telefax: (0)11-2763848
Electronic Mail: sab@iagusp.usp.br
　　　　　　　sab@orion.iagusp.usp.br
WWW: http://www.iagusp.usp.br/sab
Founded: 1974
Membership: 460
Activities: annual meetings
Periodicals: (3) "Boletim da Sociedade Astronômica Brasileira" (ISSN 0101-3440, circ.: 600)
City Reference Coordinates: 046°37'00"W 23°32'00"S

Sociedade Astronômica Maranhense de Amadores (SAMA)
Rua Arimatéia Cisme 234 - Apeadouro
65035-800 São Luís - MA
Telephone: (0)98-2212008
Electronic Mail: br_sana@yahoo.com.br
WWW: http://www.elo.com.br/~cportela/sama01.html
Founded: 1976
Membership: 12
Activities: observing (Sun, sunspots) * telescope gatherings * videos shows * meetings
Coordinates: 044°14'09"W 02°35'12"S
City Reference Coordinates: 044°16'00"W 02°31'00"S

Sociedade Brasileira dos Amigos da Astronomia (SBAA)
Avenida do Imperador 1330
Centro
60015-000 Fortaleza - CE
Telephone: (0)85-2525077
　　　　　(0)85-2525237
Telefax: (0)85-2525238
Electronic Mail: sbaa@sbaa.com.br
WWW: http://www.fortalnet.com.br/sbaa/
Founded: 1947
City Reference Coordinates: 038°30'00"W 03°43'00"S

Sociedade Brasileira para o Ensino da Astronomia (SBEA)
Rua Texas 1.177
Sala 04
04557-001 São Paulo - SP
Telefax: (0)11-5506-7838
Electronic Mail: sbea@mandic.com.br
WWW: http://pessoal.mandic.com.br/~sbea/
Periodicals: (12) "Astronomia Novae"
City Reference Coordinates: 046°37'00"W 23°32'00"S

Sociedade de Astronomia e Astrofísica de Diadema (SAAD)
Caixa Postal 3030
09970-990 Diadema - SP
or :
Avenida Antônio Cunha Silva Bueno 1322
09970-160 Diadema - SP
Telephone: (0)11-40435723
Telefax: (0)11-40435723
Electronic Mail: observatorio@diadema.com.br
WWW: http://www.observatorio.diadema.com.br/
Founded: 1989
Membership: 60
Activities: observing (eclipses, occultations, variable stars, Moon, Sun, planets, meteors) * instrumentation * popularization * education
Periodicals: (2) "Boletim da SAAD"
Coordinates: 046°36'38"W 23°41'10"S H820m
City Reference Coordinates: 046°37'00"W 23°42'00"S

Sociedade de Estudos Astronômicos de Ouro Preto (SEAOP)
Praça Tiradentes 20
35400-000 Ouro Preto - MG
Telephone: (0)31-5511533
Telefax: (0)31-5591528
Electronic Mail: seaop@em.ufop.br
WWW: http://www.seaop.em.ufop.br/
Founded: 1992
Membership: 6
Activities: education * public outreach
Coordinates: 043°30'16"W 20°22'59"S (Observatório Astronômico da Escuela de Minas)
City Reference Coordinates: 043°30'00"W 20°25'00"S

Universidade de São Paulo (USP), Instituto de Astronomia, Geofísica e Ciências Atmosféricas, Departamento de Astronomia
Rua do Matão 1226
Cidade Universitaria
05508-900 São Paulo - SP
Telephone: (0)11-30912710
Telefax: (0)11-30912860
Electronic Mail: secret@astro.iag.usp.br
   <userid>@astro.iag.usp.br
WWW: http://www.astro.iag.usp.br/
   http://www.iag.usp.br/
Founded: 1932
Staff: 35
Activities: astrophysics (stars, ISM, galaxies, solar system) * cosmology * fundamental astronomy * dynamical astronomy * Observatório Abrahão de Moraes (Valinhos)
Periodicals: "Anuário Astronômico"
Coordinates: 046°58'00"W 23°06'00"S H850m (São Paulo)
   046°37'22"W 23°39'07"S H800m (Valinhos – IAU Code 860)
City Reference Coordinates: 046°58'00"W 23°06'00"S

Universidade Estadual de Campinas, Instituto de Física, Departamento Raios Cosmicos e Cronologia
Caixa Postal 6165
13083-970 Campinas - SP
Telephone: (0)19-7885522
   (0)19-7885314
Telefax: (0)19-7885512
Electronic Mail: <userid>@ifi.unicamp.br
WWW: http://www.ifi.unicamp.br/ifgw.html
Founded: 1967
Staff: 13
Activities: cosmic rays * high-energy physics * hadronic interactions
City Reference Coordinates: 047°05'00"W 22°54'00"S

Universidade Estadual Paulista (UNESP), Instituto de Física Teorica
Rua Pamplona 145
01405-900 São Paulo - SP
Telephone: (0)11-2515155
Electronic Mail: <userid>@unesp.br
WWW: http://www.ift.unesp.br/
Founded: 1951
Staff: 26
Activities: field theory * elementary particles * gravitation * cosmology * nuclear physics * mathematical physics
City Reference Coordinates: 046°37'00"W 23°32'00"S

Universidade Federal de Goiás (UFG), Planetário
Caixa Postal 131
74055-140 Goiânia - GO
Telephone: (0)62-2245787
Telefax: (0)62-2245787
WWW: http://www.ac-digital.com/abplanetarios/goiania/
   http://www.fis.ufg.br/
Founded: 1970
Staff: 5
Activities: education * shows
Coordinates: 049°15'29"W 16°41'21"S H802m
City Reference Coordinates: 049°16'00"W 16°40'00"S

Universidade Federal do Espírito Santo (UFES), Observatório Astronômico
Avenida Fernando Ferrari
Goiabeiras
Caixa Postal 19015
29060-970 Vitória - ES
Telephone: (0)27-3352484
   (0)27-3251711 x267
Electronic Mail: oaufes@cce.ufes.br
WWW: http://www.cce.ufes.br/~oaufes
Founded: 1986
City Reference Coordinates: 040°21'00"W 20°19'00"S

Universidade Federal de Minas Gerais (UFMG), Departamento de Física, Observatório Astronômico de Piedade (OAP)
Caixa Postal 702
30161-000 Belo Horizonte - MG
Telephone: (0)31-4412541
Telefax: (0)31-4481372
WWW: http://www.fisica.ufmg.br/

http://www.fisica.ufmg.br/~astrof/
http://www.fisica.ufmg.br/OAP/
Founded: 1972
Staff: 10
Activities: ISM * eclipsing binaries * photometry * stellar evolution
Coordinates: 043°30'43"W 19°49'20"S H1,746m
City Reference Coordinates: 043°56'00"W 19°55'00"S

Universidade Federal de Rio de Janeiro (UFRJ), Observatório do Valongo
Ladeira do Pedro Antônio 43 - Saúde
20080-090 Rio de Janeiro - RJ
Telephone: (0)21-2630685
Telefax: (0)21-2630685
Electronic Mail: ov@ov.ufrj.br
WWW: http://www.ufrj.br/ov/
Founded: 1881
Staff: 11
Activities: education * fundamental astronomy * stellar astrophysics * ISM * extragalactic astronomy * laboratory astrophysics * PNs
Coordinates: 043°11'12"W 22°53'54"S H43m
City Reference Coordinates: 043°15'00"W 22°54'00"S

Universidade Federal do Rio Grande do Sul (UFRGS), Instituto de Física, Departamento de Astronomia
Avenida Bento Gonçalves 9500
Campus do Vale
Caixa Postal 15051
91501-970 Porto Alegre - RS
Telephone: (0)51-3167111
Telefax: (0)51-3191762
Electronic Mail: <userid>@if.ufrgs.br
WWW: http://www.if.ufrgs.br/ast
http://sofia.if.ufrgs.br/ast/
http://www.cesup.ufrgs.br/ufrgs.html
Founded: 1968
Staff: 10
Activities: education * research
Coordinates: 051°06'15"W 30°04'31"S H400m (Morro Santana)
City Reference Coordinates: 051°11'00"W 30°04'00"S

Universidade Federal do Rio Grande do Sul (UFRGS), Planetário Prof. José Baptista Pereira
Avenida Ipiranga 2000
90160-091 Porto Alegre - RS
Telephone: (0)51-3165384
Telefax: (0)51-3165387
Electronic Mail: planeta@vortex.ufrgs.br
WWW: http://www.ufrgs.br/planetario/
Founded: 1972
Staff: 7
Activities: education * shows * school visits * popularization * sky of the month
Periodicals: (1) "Agenda Astronômica", "Cadernos de Astronomia"
City Reference Coordinates: 051°11'00"W 30°04'00"S

Bulgaria

Amateur Astronomical Club Canopus
Primorski Park 4
P.O. Box 108
BG-9000 Varna
Telephone: (0)52-222890
Electronic Mail: astro@ms3.tu-varna.acad.bg
Founded: 1963
Membership: 100
Activities: observing (meteors, variable stars, minor planets, comets, Sun)
City Reference Coordinates: 027°55'00"E 43°13'00"N

Astronomical Association - Sofia (AAS)
49 Tzar Assen Street
BG-1463 Sofia
Telephone: (0)2-9811327 (Office)
(0)2-9810898
Electronic Mail: aas@bgcict.acad.bg
WWW: http://www.geocities.com/aas_andromeda
Founded: 1993
Membership: 40
Activities: popularization * observing * exhibitions
Periodicals: (6) "Andromeda" (ISSN 1310-3571); "Telescop" (ISSN 1311-3879)
City Reference Coordinates: 023°19'00"E 42°41'00"N

Astronomical Observatory "Slavey Zlatev"
P.O. Box 134
BG-6600 Kardjali
Telephone: (0)361-22595
(0)361-26457
Telefax: (0)361-26457
Electronic Mail: obsk@yahoo.com
Founded: 1972
Staff: 9
Activities: meteors * Sun * planets * education
Coordinates: 023°25'23"E 41°38'28"N H330m
City Reference Coordinates: 026°19'00"E 42°40'00"N

Astronomical Observatory, Sliven
P.O. Box 7
BG-8800 Sliven
Telephone: (0)44-28353
Electronic Mail: <userid>@sliven.osf.acad.bg
aosliven@sliven.osf.acad.bg
WWW: http://www.sliven.osf.acad.bg/
Founded: 1979
Staff: 10
Activities: variable stars * Sun * minor planets * meteors * globular clusters * photometry * instrumentation * history of astronomy * education
Periodicals: (1) "Calendars"
Coordinates: 026°15'14"E 42°40'50"N H259m (Sliven)
026°22'00"E 42°43'00"N H1,150m (Haramiata)
• Alternate name: "Dr. Petur Beron Observatory"
City Reference Coordinates: 026°19'00"E 42°40'00"N

Astronomical Observatory and Planetarium N. Copernicus
Primorski Park 4
P.O. Box 120
BG-9000 Varna
Telephone: (0)52-222890
Electronic Mail: astro@ms3.tu-varna.acad.bg
WWW: http://aol.skyarchive.org/varna/astro
Founded: 1961 (Planetarium: 1968)
Staff: 8
Activities: popularization * lectures * comets * minor planets * meteors * Sun * education
Coordinates: 027°55'26"E 43°12'12"N H10m (Varna)
027°40'14"E 43°07'13"N H398m (Avren)
City Reference Coordinates: 027°55'00"E 43°13'00"N

Beron Observatory (Dr. Petur_)
• See "Astronomical Observatory, Sliven"

Bulgarian Academy of Sciences, Central Laboratory for Geodesy
Acad. G. Bonchev Street 1
BG-1113 Sofia
or :
Ul. 15 Noemvri 1
BG-1040 Sofia
Telephone: (0)2-720841
Telefax: (0)2-720841
Electronic Mail: <userid>@bgearn.bitnet
WWW: http://www.acad.bg/BulRTD/earth/geodesy2.html
Founded: 1948
Staff: 37
Activities: geodesy * satellite geodesy * optical astrometry * geodynamics * geodetic networks
Periodicals: (1) "Geodesy" (ISSN 0324-1114)
Coordinates: 023°24'37"E 42°29'55"N H1,200m (Plana)
City Reference Coordinates: 023°19'00"E 42°41'00"N

Bulgarian Academy of Sciences, Institute of Astronomy and National Astronomical Observatory, Smolyan
P.O. Box 136
BG-4700 Smolyan
Telephone: (0)3-021356
(0)3-021357
(0)3-0128902
Telefax: (0)3-021356
Electronic Mail: rozhen@mbox.digsys.bg
WWW: http://www.astro.bas.bg/
Founded: 1980 (reestablished: 1995)
Staff: 35
Activities: comets * minor planets * stellar atmospheres * variable stars * stellar clusters * galaxies
Coordinates: 024°44'38"E 41°41'35"N H1,759m (IAU Code 071)
• Alternate name: "Rozhen Observatory"
City Reference Coordinates: 024°45'00"E 41°42'00"N

Bulgarian Academy of Sciences, Institute of Astronomy and National Astronomical Observatory, Sofia
72 Tsarigradsko Chaussee Boulevard
BG-1784 Sofia
Telephone: (0)2-758927
(0)2-7144614
Telefax: (0)2-758927
Electronic Mail: <userid>@astro.bas.bg
kpanov@astro.bas.bg (Kiril Panteleev Panov, Director)
WWW: http://www.astro.bas.bg/
Founded: 1958
Staff: 90
Activities: astronomy * astrophysics * education
Periodicals: (1) "Astronomical Calendar" (ISSN 0861-1270)
Coordinates: 024°43'00"E 41°43'00"N H1,750m (Rozhen)
022°40'30"E 43°37'35"N H630m (Belogradtchik)
City Reference Coordinates: 023°19'00"E 42°41'00"N

Bulgarian Academy of Sciences, Geophysical Institute
Acad. G. Bonchev Street
Block 3
BG-1113 Sofia
Telephone: (0)2-700128
Telefax: (0)2-700226
Electronic Mail: office@geophys.acad.bg
WWW: http://www.geophys.bas.bg/
Founded: 1961
Staff: 150
Activities: atmospheric and ionospheric physics * seismology * geomagnetism * gravimetry * network of seismic stations
Periodicals: (12) "Seismic Bulletin"; (4) "Bulgarian Geophysical Journal" (ISSN 0323-9918, circ.: 250); (1) "Ionospheric Bulletin", "Magnetic Bulletin"
City Reference Coordinates: 023°19'00"E 42°41'00"N

Bulgarian Academy of Sciences, National Institute of Meteorology and Hydrology (NIMH)
66 Tsarigradsko Chaussee Boulevard
BG-1784 Sofia
Telephone: (0)2-9733831
Telefax: (0)2-884494
Electronic Mail: office@meteo.bg
WWW: http://www.meteo.bg/
Founded: 1890
Observatories: 35 (plus 2000 observing stations)

City Reference Coordinates: 023°19'00"E 42°41'00"N

Bulgarian Academy of Sciences, Space Research Institute
6 Moskovska Street
BG-1000 Sofia
Telephone: (0)883503
(0)2-9793422
Telefax: (0)2-9813347
Founded: 1974
Staff: 310
Activities: space astronomy * space physics * space instrumentation * remote sensing * aerospace technologies
Periodicals: (12) "Aerospace Investigations in Bulgaria"
City Reference Coordinates: 023°19'00"E 42°41'00"N

Committee for Standardization and Metrology (BDS)
6th September Street 21
BG-1000 Sofia
Telephone: (0)2-8591
Telefax: (0)2-801402
Electronic Mail: csm@techno-link.com
City Reference Coordinates: 023°19'00"E 42°41'00"N

Dr. Petur Beron Observatory
• See "Astronomical Observatory, Sliven"

Petur Beron Observatory (Dr._)
• See "Astronomical Observatory, Sliven"

Rozhen Observatory
• See "Bulgarian Academy of Sciences, Department of Astronomy and National Astronomical Observatory, Smolyan"

Shoumen University, Department of Physics, Astronomical Group
BG-9700 Shoumen
Telephone: (0)54-63151 x289
Telefax: (0)54-63171
Electronic Mail: diana@uni-shoumen.bg (Diana Kjurkchieva)
<userid>@shu-bg.net
Staff: 4
Activities: spotted stars * cataclysmic stars * close binaries * imaging
• Formerly "Higher Pedagogical Institute"
• Full university name: Shoumen University "Konstantin Preslavsky"
City Reference Coordinates: 026°55'00"E 43°16'00"N

Smolyan Planetarium
20 Bulgaria Boulevard
P.O. Box 132
BG-4700 Smolyan
Telephone: (0)301-23074
Electronic Mail: planet@sm.unacs.bg
WWW: http://planetarium.bg.8m.com/planet.htm
Founded: 1975
Staff: 20
Activities: shows * observing
Coordinates: 024°41'00"E 41°35'00"N H980m
City Reference Coordinates: 024°41'00"E 41°35'00"N

Union of Physicists in Bulgaria
c/o I. Lalov (President)
Faculty of Physics
Sofia University
James Bourchier Boulevard 5
BG-1126 Sofia
Telephone: (0)2-627660
Telefax: (0)2-689085
Electronic Mail: rector@uni-sofia.bg
Founded: 1972
Membership: 100
City Reference Coordinates: 023°19'00"E 42°41'00"N

University of Sofia, Department of Astronomy
James Bourchier Boulevard 5
BG-1126 Sofia
Telephone: (0)2-688176
Telefax: (0)2-689085
WWW: http://www.phys.uni-sofia.bg/

Founded: 1891
Staff: 16
Activities: education * research
City Reference Coordinates: 023°19'00"E 42°41'00"N

Canada

ADR Spacelink & Commercialization Inc.
2426 Georgina Drive
Suite B
Ottawa, ON K2B 7M7
Telephone: 613-596-6032
Telefax: 613-596-2986
Electronic Mail: 102746.313@compuserve.com
WWW: http://adr.on.ca/
Activities: technology transfer
City Reference Coordinates: 075°42'00"W 45°25'00"N

Agence Spatiale Canadienne (ASC)
• See "Canadian Space Agency (CSA)"

Association Canadienne des Physiciens (ACP)
• See "Canadian Association of Physicists (CAP)"

Association des Astronomes Amateurs Abitibi-Témiscamingue (AAAAT)
c/o Luc Aubut
24, 9e avenue Ouest
La Sarre, QC J9Z 1N7
Telephone: 819-333-8365
Electronic Mail: lucaubut@cablevision.qc.ca
aaaat@hotmail.com
WWW: http://astro.uqat.uquebec.ca/Astro/
City Reference Coordinates: 079°41'00"W 48°48'00"N

Astro-Club du Collège de Lévis
9 Monseigneur Gosselin
Lévis, QC G6V 5K1
Telephone: 418-833-1249
Telefax: 418-833-1974
WWW: http://www.engsoc.carleton.ca/~wiswaud/acclf.htm
Founded: 1970
Membership: 30
Activities: observing * photography
Coordinates: 070°42'12"W 46°43'48"N H305m (Saint-Nérée-de-Bellechasse)
City Reference Coordinates: 071°11'00"W 46°49'00"N

Astronomical Society of Fort McMurray (ASFM)
c/o Doug Morgan
129 Sifton Avenue
Fort McMurray, AB T8J 1L3
or :
c/o Josyane Cloutier
290 Ross Haven Drive
Fort McMurray, AB T9H 3P4
Telephone: 403-791-6372
Electronic Mail: dmorgan@ccinet.ab.ca
josyane@telusplanet.net
Founded: 1986
Membership: 12
Activities: observing * public nights
City Reference Coordinates: 111°07'00"W 56°40'00"N

Atlantic Space Sciences Foundation (TASSF) Inc. (The_)
• See "The Atlantic Space Sciences Foundation (TASSF) Inc."

Barrie Astronomy Club
532 Grove Street East
Barrie, ON L4M 5Z2
Telephone: 705-739-7151
Electronic Mail: info@deepskies.com
WWW: http://www.deepskies.com/
Activities: monthly meetings * observing * education
City Reference Coordinates: 079°42'00"W 44°22'00"N

Bibliothèque Nationale du Canada (BNC)
Rue Wellington 395
Ottawa ON K1A 0N4

Telephone: 613-995-7969
Telefax: 613-991-9871
Electronic Mail: publications@nlc-bnc.ca
WWW: http://www.nlc-bnc.ca/
Founded: 1953
Staff: 500
Activities: national library
Periodicals: (10) "Nouvelles de la Bibliothèque Nationale"
• See also "National Library of Canada (NLC)"
City Reference Coordinates: 075°42'00"W 45°25'00"N

Big Sky Astronomical Society
P.O. Box 510
Vulcan, AB T0L 2B0
Electronic Mail: info@bigsky.ab.ca
                <userid>@bigsky.ab.ca
WWW: http://www.bigsky.ab.ca/
Founded: 1998
Membership: 20
Activities: education * establishing Big Sky Observatory (BSO)
Periodicals: (4) "Eye on the Sky"
Coordinates: 113°15'00"W 50°24'00"N
City Reference Coordinates: 113°12'00"W 50°27'00"N

Bondar Planetarium (Roberta_)
• See "Seneca College, Roberta Bondar Planetarium"

Brandon University, Department of Physics and Astronomy
270-18th Street
Brandon, MB R7A 6A9
Telephone: 204-727-9680
Telefax: 204-726-4573
Electronic Mail: <userid>@brandonu.ca
WWW: http://www.brandonu.ca/Physics/
Founded: 1938
Staff: 4
Activities: stellar spectra * atmospheres * Be and Ap stars
City Reference Coordinates: 099°57'00"W 49°50'00"N

Brockville Astronomical Society
c/o Devin Fetter
14 Bramshot Avenue
Brockville, ON K6V 1Y5
Telephone: 613-342-2668
Electronic Mail: kfetter@geocities.com
WWW: http://www.geocities.com/CapeCanaveral/Lab/7855/
Membership: 4
City Reference Coordinates: 075°44'00"W 44°35'00"N

Bruce County Astronomical Society
c/o Charlie Szabototh
477 Johnston Avenue
Port Elgin, ON N0H 2C1
Telephone: 519-389-3150
Electronic Mail: csz@bmts.com
WWW: http://www.geocities.com/CapeCanaveral/Hall/6380/
City Reference Coordinates: 064°08'00"W 46°03'00"N

Bureau de Normalisation du Québec (BNQ)
Parc Technologique du Québec Métropolitain
333, rue Franquet
Sainte-Foy, QC G1P 4C7
Telephone: 418-652-2238
          1-800-386-5114
Telefax: 418-652-2292
Electronic Mail: bnq@criq.qc.ca
WWW: http://www.criq.qc.ca/bnq/
Founded: 1961
Staff: 28
Activities: normalizing * certifying * recording quality systems * informing * advising
City Reference Coordinates: 071°18'00"W 46°47'00"N

Burke-Gaffney Observatory (BGO) (Rev. M.W._)
• See "Saint Mary's University, Department of Astronomy and Physics, Rev. M.W. Burke-Gaffney Observatory (BGO)"

Calgary Science Center
P.O. Box 2100
Station "M"
Location Code 73
Calgary, AB T2P 2M5
or :
701 - 11 Street SW
Calgary, AB
Telephone: 403-221-3700
Telefax: 403-237-0186
Electronic Mail: discover@calgaryscience.ca
WWW: http://www.calgaryscience.ca/
Founded: 1966
Staff: 30
Activities: education * shows * exhibitions * Centennial Planetarium
Periodicals: (6) "The Spark"
Coordinates: 114°05'21"W 51°02'51"N H1,084m (Public Obs. – IAU Code 681)
City Reference Coordinates: 114°05'00"W 51°03'00"N

Canadian Aeronautics and Space Institute (CASI)
(Institut Aéronautique et Spatial du Canada)
130 Slater
Suite 818
Ottawa, ON K1P 6E2
Telephone: 613-234-0191
Telefax: 613-234-9039
Electronic Mail: casi@asi.ca
WWW: http://www.casi.ca/
Founded: 1954
Staff: 4
Activities: technical symposia * publishing * branch programs
Periodicals: (4) "Canadian Aeronautics and Space Journal/Le Journal Aéronautique et Spatial du Canada" (ISSN 0008-2821); "Canadian Journal of Remote Sensing" (ISSN 0763-8992)
Awards: "McCurdy Award", "C.D. Howe Award", "Romeo Vachon Award", "F.W. (Casey) Baldwin Award", "Dr. Wilbur Franklin Award", "Remote Sensing Gold Medal"
City Reference Coordinates: 075°42'00"W 45°25'00"N

Canadian Association of Physicists (CAP)
(Association Canadienne des Physiciens - ACP)
c/o Gary D. Enright (Secretary/Treasurer)
Steacie Institute
100 Sussex Drive
Room 126
Ottawa, ON K1A 0R6
Telephone: 613-993-7393
Telefax: 613-954-5242
Electronic Mail: enright@ned1.sims.nrc.ca
cap@physics.uottawa.ca (General Inquiries)
WWW: http://www.cap.ca/
Founded: 1945
Staff: 3
Sections: (divisions) Aeronomy and Space Physics * Atomic and Molecular Physics * Condensed Matter Physics * Medical and Biological Physics * Nuclear Physics * Optics & Photonics * Particle Physics * Physics Education * Plasma Physics * Theoretical Physics * Industrial and Applied Physics * Surface Science
Periodicals: (1) "Annual Report" (ISSN 0068-8339); "Physics in Canada - La Physique au Canada" (ISSN 0031-9147)
Awards: (1) "CAP Medal of Achievement in Physics", "Herzberg Medal", "Lloyd G. Elliot Prize"; (12) "CAP Medal for Outstanding Achievement in Industrial and Applied Physics"
City Reference Coordinates: 075°42'00"W 45°25'00"N

Canadian Astronomical Society (CASCA)
(Société Canadienne d'Astronomie)
c/o Serge Demers (Secretary)
Département de Physique
Université de Montréal
Montréal, QC H3C 3J7
Telephone: 514-343-2364
Electronic Mail: casca@astro.umontreal.ca
WWW: http://www.astro.queensu.ca/~casca/
Founded: 1971
Membership: 360
Activities: support of astronomical research and education
Periodicals: "Cassiopeia"
Awards: "Carlyle S. Beals Award", "J.S. Plaskett Medal", "R.M. Petrie Prize", "Helen Sawyer Hogg Prize"
City Reference Coordinates: 073°34'00"W 45°31'00"N

Canadian Astronomy Data Centre (CADC)
• See "National Research Council of Canada (NRCC), Herzberg Institute of Astrophysics (HIA), Canadian Astronomy Data Centre (CADC)"

Canadian Coast Guard College Planetarium
P.O. Box 4500
Sydney, NS B1P 6L4
Telephone: 902-564-3660
Telefax: 902-564-3672
Electronic Mail: <userid>@cgc.ns.ca
WWW: http://www.cgc.ns.ca/
Founded: 1965
Staff: 105 (College)
Activities: education * astronomical navigation
City Reference Coordinates: 060°11'00"W 46°09'00"N

Canadian Council of Science Centres (CCSC)
(Conseil Canadien des Centres des Sciences - CCCS)
c/o Pacific Space Centre
1100 Chestnut Street
Vanier Park
Vancouver, BC V6J 3J9
Telephone: 604-738-STAR
          604-738-7827
Telefax: 604-736-5665 (Administration)
Electronic Mail: jdickens@pacific-space-centre.bc.ca (John Dickenson, President)
WWW: http://pacific-space-centre.bc.ca/
Founded: 1986
Membership: 26
Periodicals: "Newsletter"
City Reference Coordinates: 123°07'00"W 49°16'00"N

Canadian Institute for Advanced Research (CIAR)
180 Dundas Street West
Suite 1400
Toronto, ON M5G 1Z8
Telephone: 416-971-4251
Telefax: 416-971-6169
Electronic Mail: info@ciar.ca
WWW: http://www.ciar.ca/
Founded: 1982 (Programme in Cosmology and Gravity: 1986)
Staff: 14
City Reference Coordinates: 079°23'00"W 43°39'00"N

Canadian Institute for Scientific and Technical Information (CISTI)
• See "National Research Council of Canada (NRCC), Canadian Institute for Scientific and Technical Information (CISTI)"

Canadian Institute for Theoretical Astrophysics (CITA)
(Institut Canadien d'Astrophysique Théorique - ICAT)
c/o University of Toronto
McLennan Physical Laboratories
Room 1212A
60 Saint George Street
Toronto, ON M5S 3H8
Telephone: 416-978-6879
Telefax: 416-978-3921
Electronic Mail: office@cita.utoronto.ca
WWW: http://www.cita.utoronto.ca/
       http://www.physics.utoronto.ca/department/groups_partners/cita.html
Founded: 1984
Staff: 25
Activities: theoretical astrophysics * cosmology
City Reference Coordinates: 079°23'00"W 43°39'00"N

Canadian Meteorological Centre
4905 Dufferin Street
Downsview, ON M3H 5T4
Telephone: 416-739-4810
Telefax: 416-739-4999
Electronic Mail: <userid>@dow.on.doe.ca
WWW: http://www.tor.ec.gc.ca/cmc.html
       http://weatheroffice.ec.gc.ca/
City Reference Coordinates: 079°28'00"W 43°45'00"N

Canadian Space Agency (CSA), David Florida Laboratory (DFL)
(Agence Spatiale Canadienne - ASC, Laboratoire David Florida)
3701 Carling Avenue
Ottawa, ON K2H 8S2
Telephone: 613-998-2383
Telefax: 613-993-6103
Electronic Mail: <userid>@space.gc.ca
WWW: http://www.space.gc.ca/space_qualification/david_florida_lab/default.asp
Founded: 1972
City Reference Coordinates: 075°42'00"W 45°25'00"N

Canadian Space Agency (CSA), Space Science Program
(Agence Spatiale Canadienne - ASC, Programme des Sciences Spatiales)
6767, route de l'Aéroport
Saint-Hubert, PQ J3Y 8Y9
or :
100 Sussex Drive
P.O. Box 7275
Ottawa, ON K1L 8E3
Telephone: 450-926-4771
613-990-0790
Telefax: 450-926-4766
613-952-0970
Electronic Mail: <userid>@space.gc.ca
WWW: http://www.space.gc.ca/
Founded: 1989
Staff: 30
Activities: selection, management, development, and operation of space experiments proposed by Canadian scientists
City Reference Coordinates: 075°42'00"W 45°25'00"N (Ottawa)
073°25'00"W 45°30'00"N (Saint-Hubert)

Canadian Space Resource Centre, Atlantic
1593 Barrington Street
Halifax, NS B3J 1Z7
Telephone: 902-492-4422
1-800-511-3500 (Canada and USA only)
Telefax: 902-492-3170
Electronic Mail: space@hercules.stmarys.ca
WWW: http://apwww.stmarys.ca/space
Founded: 1995
Staff: 2
Activities: space science educational outreach and resource centre
City Reference Coordinates: 063°36'00"W 44°39'00"N

Canadian Space Resource Centre, Ontario
c/o Marc Garneau Collegiate Institute
135 Overlea Boulevard
East York, ON M3C 1B3
Telephone: 416-396-2421
Telefax: 416-396-2423
Electronic Mail: csrs@interlog.com
WWW: http://www.interlog.com/~csrc
Staff: 2
Activities: space science educational outreach and resource centre
City Reference Coordinates: 076°43'00"W 39°58'00"N

Cape Breton Astronomical Society (CBAS)
c/o John Reppa
131 Green Acres Drive
Sydney, NS B1S 1K5
Electronic Mail: ptech45@avtccb.nscc.ns.ca (John Reppa)
WWW: http://www.cbnet.ns.ca/cbnet/comucntr/astronomy/cbas.html
http://highlander.cbnet.ns.ca/~cbas/
Founded: 1986
City Reference Coordinates: 060°11'00"W 46°09'00"N

Carleton University, Centre for Research in Particle Physics (CRPP)
Ottawa, ON K1S 5B6
Telephone: 613-786-7552
Telefax: 613-786-7546
Electronic Mail: admin@crpp.carleton.ca
<userid>@physics.carleton.ca
WWW: http://www.crpp.carleton.ca/
http://www.physics.carleton.ca/
Founded: 1990
Staff: 15

Activities: research in particle physics and neutrino astrophysics
City Reference Coordinates: 075°42'00"W 45°25'00"N

Carleton University, Department of Physics, High-Energy Physics Group
Colonel By Drive
Herzberg Building
Ottawa, ON K1S 5B6
Telephone: 613-788-4377
Telefax: 613-788-4061
Electronic Mail: carnegie@physics.carleton.ca
　　　　　　<userid>@physics.carleton.ca
WWW: http://www.hepnet.carleton.ca/
　　　http://www.physics.carleton.ca/
Staff: 40
Activities: particle physics * astrophysics * medical physics
City Reference Coordinates: 075°42'00"W 45°25'00"N

Centre for Climate and Global Change Research (C2GCR)
• See "McGill University, Centre for Climate and Global Change Research (C2GCR)"

Centre for Research in Earth and Space Technology (CRESTech)
4850 Keele Street
Toronto, ON M3J 3K1
Telephone: 416-665-3311
Telefax: 416-665-2032
Electronic Mail: enquiries@admin.crestech.ca
　　　　　　<userid>@crestech.ca
　　　　　　<userid>@ists.ca
WWW: http://www.crestech.ca/
Founded: 1997
Staff: 20
Activities: space science and technology * atmospheric studies * land resources * water resources * human performance in a aerospace environment * controlled environment systems
• Results of the merger between the "Institute for Space and Terrestrial Science (ISTS)" and the "Waterloo Center for Groundwater Research (WCGR)"
City Reference Coordinates: 079°23'00"W 43°39'00"N

Club d'Astronomie de Beloeil
1020a, rue Dupré
Beloeil, PQ J4G 4A8
Telephone: 514-536-3460 (François Daoust, President)
Telefax: 450-536-5444
Electronic Mail: daoust.f@videotron.ca
WWW: http://www.quebectel.com/beloeil/
Founded: 1981
Membership: 15
Activities: meetings * observing
City Reference Coordinates: 073°12'00"W 45°34'00"N

Club d'Astronomie de Dorval
c/o Marjolaine Savoie
81 Dieppe
Pointe-Claire, QC H9R 1X2
Telephone: 514-697-8093
Electronic Mail: marjos@total.net
WWW: http://www.multimania.com/cdadfs/cdadfs1.htm
Founded: 1971
Membership: 20
Activities: meetings * observing
City Reference Coordinates: 073°44'00"W 45°26'00"N (Dorval)

Club d'Astronomie de Drummondville (CAD)
c/o Michel Dionne
2115, Saint-Nicolas
Drummondville, QC J2B 7A8
Telephone: 819-474-6716
Electronic Mail: michel.dionne2@dr.cgocable.ca
WWW: http://www.cafe.rapidus.net/albecote/index.html
Founded: 1976
Membership: 7
Activities: weekly meetings * observing
City Reference Coordinates: 072°30'00"W 45°53'00"N

Club d'Astronomie de la Péninsule Acadienne
c/o Jacques Robichaud (President)

177 Rue Bellefeuille
Shippagan, NB E8S 1G5
Telephone: 506-336-3427
Telefax: 506-336-3477
Electronic Mail: jacques@admin.cus.ca
Founded: 1994
Membership: 25
Coordinates: 064°42'42"W 47°44'42"N
City Reference Coordinates: 064°42'00"W 47°45'00"N

Club d'Astronomie de l'Observatoire "L'Étoile d'Acadie"
98 Rue Marcoux
Gr. 37 - Boîte 20
Balmoral, NB E0B 1C0
Telephone: 506-826-9893
Electronic Mail: raudet@nbnet.nb.ca
Founded: 1995
Membership: 12
Activities: observing * meetings * photography * CCDs
Coordinates: 066°27'00"W 47°58'00"N H150m
City Reference Coordinates: 066°27'00"W 47°58'00"N

Club d'Astronomie de l'Université du Québec à Rimouski
c/o Jean-Paul Caron
300 Rue des Ursulines
Local E205
Rimouski, QC G5L 3A1
Telephone: 418-723-5197
           418-723-4402
Electronic Mail: astro_uqar@uqar.uquebec.ca
                dlemay@quebectel.com
WWW: http://wwwb.uqar.uquebec.ca/astro/
       http://www.quebectel.com/astro/
City Reference Coordinates: 068°32'00"W 48°27'00"N

Club d'Astronomie de Rimouski
c/o Jean-Paul Caron
51, Place de Bretagne
Rimouski, QC G5L 5H4
Telephone: 418-723-4402
Electronic Mail: jeanpcar@cgocable.ca
WWW: http://www.quebectel.com/astro/
Founded: 1981
Membership: 60
Activities: meetings * observing
City Reference Coordinates: 068°32'00"W 48°27'00"N

Club d'Astronomie de Sept-Îles
c/o Jean-Clément Saint-Gelais
58 Ungava
Sept-Îles, QC G4R 4E3
Telephone: 418-962-8541
Electronic Mail: sirius@globetrotter.net
WWW: http://www.bbsi.net/astrosi/
Founded: 1997
Membership: 10
Activities: meetings * observing
City Reference Coordinates: 066°23'00"W 50°12'00"N

Club d'Astronomie Jupiter
c/o Michel Héroux (President)
7625, Mgr. C.E. Bourgeois
Trois-Rivières, QC G8Y 6Z8
Telephone: 819-691-0003
Electronic Mail: mike.heroux@tr.cgocable.ca
WWW: http://www.astroclub.net/mars/clubjupiter/
Founded: 1996
Membership: 25
Activities: meetings * observing
City Reference Coordinates: 072°33'00"W 46°21'00"N

Club d'Astronomie Les Almucantars
425 Boulevard du Collège
Case Postale 1500
Rouyn-Noranda, QC J9X 5E5

Telephone: 819-762-0931 x1528
Electronic Mail: glesage@lino.com (Gilles Lesage)
Activities: observing * photography
City Reference Coordinates: 079°01'00"W 48°14'00"N

Club d'Astronomie Orion
c/o Roger Leduc
88, Donald
Valleyfield, QC J6S 2Z5
Telephone: 450-371-2796
Electronic Mail: rjleduc@rocler.qc.ca
WWW: http://www.rocler.qc.ca/caori/orion/indexn.htm
Activities: meetings * observing
City Reference Coordinates: 074°08'00"W 45°15'00"N

Club d'Astronomie Pégase des Laurentides
c/o Christian Lanctot
30, Chemin Lac Morency
St.-Hyppolyte, QC J0R 1T0
Telephone: 450-563-4129
WWW: http://www.cam.org/~astrolv/pegase/
Membership: 10
Activities: meetings * observing
City Reference Coordinates: 074°08'00"W 45°15'00"N

Club d'Astronomie Sirius du Saguenay
c/o Alain Marin
3968, de la Bretagne
Jonquière, QC G7X 3W3
Telephone: 418-695-0976
Electronic Mail: alain.marin@sympatico.ca
WWW: http://www3.sympatico.ca/alain.marin/
Activities: meetings * observing
City Reference Coordinates: 071°15'00"W 48°25'00"N

Club des Astronomes Amateurs Centre Mauricie (CAACM)
c/o Jacques Pellerin
200, 14e rue
Grand-Mère, G9T 3X6
Telephone: 819- 538-5905
Electronic Mail: jacquespellerin@sympatico.ca
WWW: http://www2.cybertechs.qc.ca/org/caacm/
Founded: 1997
Membership: 10
Activities: meetings * observing
City Reference Coordinates: 072°41'00"W 46°37'00"N

Club des Astronomes Amateurs de Laval (CAAL)
c/o Jean-Marc Richard (President)
Case Postale 214 - Succ. Saint-Martin
Laval, QC H7V 3P5
Telephone: 450-661-9390
Electronic Mail: astrolv@ca.org
WWW: http://www.cam.org/~astrolv/
Founded: 1986
Membership: 120
Activities: weekly meetings * star parties * education
Periodicals: "L'Orionide"
Coordinates: 073°39'00"W 45°36'00"N
City Reference Coordinates: 073°44'00"W 45°33'00"N

Club des Astronomes Amateurs de Longueuil (CAAL)
c/o Raymond Pronovost
241, De Lorraine
Saint-Lambert, QC J4S 1R1
Telephone: 450-671-9974
Electronic Mail: rpronovo@aei.ca
WWW: http://www.aei.ca/~caal/
Founded: 1985
Membership: 60
Activities: meetings * observing * public outreach
Periodicals: "FoCAAL"
City Reference Coordinates: 073°30'00"W 45°33'00"N (Longueuil)

Club des Astronomes Amateurs de Sherbrooke (CAAS)
c/o Gisele Gilbert
Case Postale 352
Sherbrooke, QC J1H 5J1
Telephone: 819-821-1138
Electronic Mail: gisgil@videotron.ca
WWW: http://www.caas.sherbrooke.qc.ca/
Founded: 1980
Membership: 107
Activities: amateur astronomy * conferences * exhibitions * camps
Periodicals: (2) "Ciel de Nuit"
Coordinates: 072°01'53"W 45°12'10"N H345m (Les Sommets)
 071°53'53"W 45°27'17"N H240m (Beauvoir)
City Reference Coordinates: 071°54'00"W 45°24'00"N

Commission des Cadrans Solaires du Québec (CCSQ)
42 Avenue de la Brunante
Outremont, QC H3T 1R4
Telephone: 514-341-3997
Telefax: 514-341-3997*51
Electronic Mail: 600009@ican.net
WWW: http://cadrans_solaires.scg.ulaval.ca/
Founded: 1974
Membership: 85
Activities: cataloguing provincial sundials * publishing * valuation * consulting
Periodicals: (4) "Le Gnomoniste"
City Reference Coordinates: 073°34'00"W 45°31'00"N (Montréal)

Comox Valley Astronomy Club (CVAC)
c/o John C. McKee
Site 514 - RR5
1758 Greenwood Crescent
Comox, BC V9N 8B5
Telephone: 250-339-3090
Electronic Mail: dgraham@mail.comox.island.net
WWW: http://www.island.net/~dgraham/cvac.htm
Founded: 1989
Membership: 25
City Reference Coordinates: 124°55'00"W 49°40'00"N

Conseil Canadien des Centres des Sciences (CCCS)
• See "Canadian Council of Science Centres (CCSC)"

Conseil Canadien des Normes (CCN)
• See "Standards Council of Canada (SCC)"

Conseil National de Recherches du Canada (CNRC)
Montréal Road
Ottawa, ON K1A 0R6
Telephone: 613-993-9101 (general enquiry)
 613-993-2054 (NRC publications)
 613-745-1576 (time signal)
 613-745-9426 (horloge parlante)
 613-993-6549 (IAU contact)
Telefax: 613-952-6602
WWW: http://www.nrc.ca/
Founded: 1916
Periodicals: "Canadian Journal of Physics" (ISSN 0008-4204), "Canadian Journal of Earth Sciences"
• See also "National Research Council of Canada (NRCC)"
City Reference Coordinates: 075°42'00"W 45°25'00"N

Conseil National de Recherches du Canada (CNRC), Institut Canadien de l'Information Scientifique et Technique (ICIST)
• See "National Research Council of Canada (NRCC), Canadian Institute for Scientific and Technical Information (CISTI)"

Conseil National de Recherches du Canada (CNRC), Institut Herzberg d'Astrophysique, Observatoire Fédéral de Radioastrophysique (OFR)
• See "National Research Council of Canada (NRCC), Herzberg Institute of Astrophysics (HIA), Dominion Radio Astrophysical Observatory (DRAO)"

Conseil National de Recherches du Canada (CNRC), Institut Herzberg d'Astrophysique, Observatoire Fédéral d'Astrophysique (OFA)
• See "National Research Council of Canada (NRCC), Herzberg Institute of Astrophysics (HIA), Dominion Astrophysical Observatory (DAO)"

Cosmodôme
2150 Autoroute des Laurentides
Laval, QC H7T 2T8
Telephone: 514-978-3600
          514-978-3602 (Space Resource Centre)
          1-800-565-CAMP
Telefax: 514-978-3601
Electronic Mail: info@cosmodome.org
                document@cosmodome.org (Space Resource Centre)
WWW: http://www.cosmodome.org/
Founded: 1991 (Space Resource Centre: 1996)
Staff: 20
Activities: space science educational outreach and resource centre * hosting a Canadian Space Resource Centre (Québec)
City Reference Coordinates: 073°44'00"W 45°33'00"N

Cowichan Valley StarFinders Astronomy Club (CVSF)
c/o Janet Curley
#12 - 2121 Tzouhalem Road
Duncan, BC V9L 5J5
Telephone: 250-748-2919
Electronic Mail: curley@islandnet.com
WWW: http://starfinders.cvnet.net/home.htm
Founded: 1988
Membership: 70
City Reference Coordinates: 123°19'00"W 48°46'00"N

CRESTech
• See "Centre for Research in Earth and Space Technology (CRESTech)"

Cyanogen Productions Inc.
25 Conover Street
Nepean, ON K2G 4C3
Telephone: 613-225-2732
Telefax: 613-225-9688
Electronic Mail: cyanogen@cyanogen.on.ca
                cyanogen@cyanogen.com
WWW: http://www.cyanogen.on.ca/
      http://www.cyanogen.com/
Founded: 1996
Activities: CCD camera control and image processing
City Reference Coordinates: 075°42'00"W 45°25'00"N (Ottawa)

David Florida Laboratory (DFL)
• See "Canadian Space Agency (CSA), David Florida Laboratory (DFL)"

Deep River Astronomy Club
c/o Harry Adams
P.O. Box 606
Deep River, ON K0J 1P0
Telephone: 613-584-3246
Electronic Mail: adams@intranet.on.ca
WWW: http://intranet.on.ca/~rbirchal/draco.html
      http://intranet.ca/~scarlisl/draco/draco.html
Founded: 1995
City Reference Coordinates: 077°30'00"W 46°06'00"N

Dominion Astrophysical Observatory (DAO)
• See "National Research Council of Canada (NRCC), Herzberg Institute of Astrophysics (HIA), Dominion Astrophysical Observatory (DAO)"

Dominion Radio Astrophysical Observatory (DRAO)
• See "National Research Council of Canada (NRCC), Herzberg Institute of Astrophysics (HIA), Dominion Radio Astrophysical Observatory (DRAO)"

Edmonton Space and Science Centre (ESSC)
• See now "Odyssium"

Fédération des Astronomes Amateurs du Québec (FAAQ)
4545 Avenue Pierre-de-Coubertin
Casier Postal 1000
Station "M"
Montréal, QC H1V 3R2
Telephone: 514-252-3038
Telefax: 514-251-8038
Electronic Mail: ddurand@stsim.com (Danis Durand, President)
WWW: http://www.quebectel.com/faaq/
Founded: 1975
Membership: 650 (30 clubs)
Activities: gathering amateur astronomers * astronomical camps * lectures * popularization
Periodicals: (6) "Astronomie-Québec" (ISSN 1183-5362)
Awards: (1) "Prix Meritas", "Prix Les Pléiades"
City Reference Coordinates: 073°34'00"W 45°31'00"N

Fraser Valley Astronomers Society (FVAS)
2659 Valemont Crescent
Abbotsford, BC V2T 3V6
Telephone: 604-850-3931
Electronic Mail: victorp@uniserve.com (James Victor Pollock, President)
WWW: http://www.geocities.com/capecanaveral/hangar/6395/
Founded: 1987
Membership: 25
Activities: telescope making * stargazing
City Reference Coordinates: 122°17'00"W 49°03'00"N

Great Island Science and Adventure Park
P.O. Box 430
Kensington, PE C0B 1M0
or :
Route 6
Stanley Bridge, PE C0A 1E0
Telephone: 902-836-3883
Electronic Mail: scifun@auracom.com
WWW: http://www.sciencefun.com/
Founded: 1986
Staff: 10
Activities: science park * planetarium
City Reference Coordinates: 063°39'00"W 46°26'00"N

Groupe Astronomie & CCD
c/o Luc Bellavance (President)
434 Des Passereaux
Rimouski, QC G5L 8K4
Telephone: 418-722-1529 (Office)
          418-723-5197 (Home)
Electronic Mail: astroccd@globetrotter.net
WWW: http://www.quebectel.com/astroccd/
City Reference Coordinates: 068°32'00"W 48°27'00"N

Hamilton Amateur Astronomers (HAA)
P.O. Box 65578
Dundas, ON L9H 6Y6
Telephone: 905-827-9105
Telefax: 905-627-3683
Electronic Mail: haa@www.science.mcmaster.ca
WWW: http://www.science.mcmaster.ca/HAA/
Founded: 1993
Membership: 110
Activities: monthly meetings * observing * education * instrumentation
Periodicals: (10) "Event Horizon"
City Reference Coordinates: 079°58'00"W 43°16'00"N

High Res Technologies (HRT)
40 Richview Road
Suite 409
Etobicoke, ON M9A 5C1
Telephone: 416-248-4473
Telefax: 416-248-4125
Electronic Mail: hrt@planeteer.com
WWW: http://toronto.planeteer.com/~hrt/
Activities: manufacturing and distributing video frame grabbers
City Reference Coordinates: 079°34'00"W 43°39'00"N

H.R. MacMillan Planetarium and Gordon Southam Observatory
• See now "H.R. MacMillan Space Centre (HRMSC)"

H.R. MacMillan Space Centre (HRMSC)
1100 Chestnut Street
Vanier Park
Vancouver, BC V6J 3J9
Telephone: 604-738-STAR
604-738-7827
Telefax: 604-736-5665 (Administration)
604-738-5086 (Planetarium)
Electronic Mail: star@hrmacmillanspacecentre.com
WWW: http://www.hrmacmillanspacecentre.com/
Founded: 1987 (Planetarium: 1968) (Observatory: 1972)
Membership: 3000
Staff: 37
Activities: administering "H.R. MacMillan Planetarium" and "Gordon Southam Observatory" * motion simulator * space flight simulations * "Canada in Space" exhibitions
Periodicals: (6) "The Starry Messenger"
Coordinates: 123°08'31"W 49°16'32"N H5m
• Formerly know as "H.R. MacMillan Planetarium and Gordon Southam Observatory"
City Reference Coordinates: 123°07'00"W 49°16'00"N

Institut Aéronautique et Spatial du Canada
• See "Canadian Aeronautics and Space Institute (CASI)"

Institut Canadien d'Astrophysique Théorique (ICAT)
• See "Canadian Institute for Theoretical Astrophysics (CITA)"

International Federation of Institutes for Advanced Study (IFIAS)
c/o Institute for Environmental Studies
39 Willcocks Street
Suite 1016V
Toronto, ON M5R 2S9
Telephone: 416-410-4196
Telefax: 416-410-4196
Electronic Mail: info@ifias.ca
WWW: http://www.ifias.ca/
Founded: 1972
Staff: 15
Activities: international non-governmental federation of research institutes * transdisciplinary, transnational projects on development and environment issues aimed at developing specific policy alternative for decision makers
City Reference Coordinates: 079°23'00"W 43°39'00"N

Khan Scope Centre
3243 Dufferin Street
Toronto, ON M6A 2T2
Telephone: 416-783-4140
Telefax: 416-783-7697
Electronic Mail: khan@globalserve.net
WWW: http://www.khanscope.com/
Founded: 1986
City Reference Coordinates: 079°23'00"W 43°39'00"N

Laurentian University, Department of Physics and Astronomy
Ramsey Lake Road
Sudbury, ON P3E 2C6
Telephone: 705-675-1151 x2220
705-675-1151 x2227 (Doran Planetarium)
Telefax: 705-675-4868
Electronic Mail: <userid>@nickel.laurentian.ca
plegault@nickel.laurentian.ca (Paul-Émile Legault, Director, Doran Planetarium)
WWW: http://www.laurentian.ca/www/physics/
http://www.laurentian.ca/www/physics/planetarium/planetarium.html (Doran Planetarium)
Founded: 1960 (Doran Planetarium: 1968)
Staff: 17 (Doran Planetarium: 1)
Activities: research * education * shows for schools and public * Doran Planetarium
Coordinates: 081°00'00"W 46°30'00"N H900m
City Reference Coordinates: 081°00'00"W 46°30'00"N

Laurier University (WLU) (Wilfrid_)
• See "Wilfrid Laurier University (WLU)"

Les Observateurs de la Magnitude Absolue
c/o Gaetan Cormier

9070 Boivin
LaSalle, PQ H8R 2E4
Telephone: 514-623-7287
Electronic Mail: lafre@odyssee.net
WWW: http://www.microid.com/loma.htm.
City Reference Coordinates: 073°40'00"W 45°26'00"N

Lethbridge Astronomy Society (LAS)
P.O. Box 1104
Lethbridge, AB T1J 4A2
Telephone: 403-381-7827
403-381-STAR
Electronic Mail: lasa@telusplanet.net
WWW: http://www.lethbridgeastronomysociety.ca/
Founded: 1988
Membership: 40
Activities: monthly meetings * public education
Periodicals: (12) "Newsletter"
Coordinates: 112°50'00"W 49°42'00"N (Oldman River Observatory)
City Reference Coordinates: 112°48'00"W 49°43'00"N

Lire La Nature Inc. (LLN)
1198 Chemin de Chambly
Longueuil, QC J2Y 1J1
Telephone: 450-463-5072
Telefax: 450-463-3409
Electronic Mail: lirelanature@videotron.ca
info@broquet.qc.ca
WWW: http://www.stjeannet.ca/broquet/zindex.html
Founded: 1988
Staff: 2
Activities: distributing telescopes, binoculars, other optical instrumentation, books, other products
City Reference Coordinates: 073°30'00"W 45°33'00"N

London Regional Children's Museum
21 Wharncliffe Road South
London ON N6J 4G5
Telephone: 519-434-5726
Telefax: 519-434-1443
Founded: 1977
Activities: science gallery * space gallery * science experiment demonstrations * hands-on activities * planetarium
Periodicals: (12) "Thumbprint"
City Reference Coordinates: 081°14'00"W 42°59'00"N

MacMillan Planetarium (H.R._)
● See now "H.R. MacMillan Space Centre (HRMSC)"

Maison de l'Astronomie P.L. Inc.
7974 Saint-Hubert
Montréal, QC H2R 2P3
Telephone: 514-279-0063
Telefax: 514-279-9628
Electronic Mail: rlotte@interlink.net
WWW: http://www.microid.com/maison.htm
Founded: 1987
Activities: distributing and renting telescopes and accessories * education
City Reference Coordinates: 073°34'00"W 45°31'00"N

Manitoba Astronomy Club
c/o Bert Valentin
Museum of Man and Nature
190 Rupert Avenue
Winnipeg, MB R3B 0N2
Telephone: 204-988-0625
Telefax: 204-942-3679
Electronic Mail: valentin@mbnet.mb.ca
WWW: http://www.manitobamuseum.mb.ca/cla.htm
Founded: 1982
Membership: 20
Activities: meetings * instrumentation * lectures * observing
Periodicals: (12) "The Nebula"
City Reference Coordinates: 097°09'00"W 49°53'00"N

Manitoba Planetarium
c/o Manitoba Museum of Man and Nature

190 Rupert Avenue
Winnipeg, MB R3B 0N2
Telephone: 204-956-2830
Telefax: 204-942-3679
Electronic Mail: info@manitobamuseum.mb.ca
WWW: http://www.manitobamuseum.mb.ca/planet.htm
Founded: 1968
Staff: 76
Activities: shows * lectures * exhibitions * education
City Reference Coordinates: 097°09'00"W 49°53'00"N

McGill University, Department of Earth and Planetary Sciences
3450 University Street
Montréal, QC H3A 2A7
Telephone: 514-398-6767
Telefax: 514-398-4680
Electronic Mail: eps@stoner.eps.mcgill.ca
WWW: http://stoner.eps.mcgill.ca/
   http://www.eps.mcgill.ca/
   http://www.mcgill.ca/
Founded: 1992
City Reference Coordinates: 073°34'00"W 45°31'00"N

McGill University, Department of Physics
Ernest Rutherford Physics Building
3600 University Street
Montréal, QC H3A 2T8
Telephone: 514-398-6485
Telefax: 514-398-8434
Electronic Mail: <userid>@physics.mcgill.ca
WWW: http://www.physics.mcgill.ca/
   http://www.mcgill.ca/
Founded: 1891
City Reference Coordinates: 073°34'00"W 45°31'00"N

McGill University, Institute and Centre of Air and Space Law (IASL)
3661 Peel Street
Montréal, QC H3A 1X1
Telephone: 514-398-5094
Telefax: 514-398-8197
Electronic Mail: <userid>@falaw.lan.mcgill.ca
WWW: http://www.mcgill.ca/iasl/
Founded: 1951
Staff: 7
Activities: education * research * air and space law
Periodicals: "Annals of Air and Space Law - Annales de Droit Aérien et Spatial"
City Reference Coordinates: 073°34'00"W 45°31'00"N

McLaughlin Planetarium
c/o Royal Ontario Museum
100 Queens Park
Toronto, ON M5S 2C6
Telephone: 416-586-5736
Telefax: 416-586-5887
Electronic Mail: tomc@rom.on.ca
WWW: http://www.rom.on.ca/
   http://ddo.astro.utoronto.ca/planetarium.html
Founded: 1968
Staff: 1
Activities: public and school programs * astronomy gallery
• Closed on 15 Dec. 1995 - reopening undefined
City Reference Coordinates: 079°23'00"W 43°39'00"N

McMaster University, Department of Physics and Astronomy
1280 Main Street West
Hamilton, ON L8S 4M1
Telephone: 905-525-9140
Telefax: 905-546-1252
Electronic Mail: physics@mcmaster.ca
   planetarium@physics.mcmaster.ca (William J. McCallion Planetarium)
WWW: http://www.physics.mcmaster.ca/
   http://www.physics.mcmaster.ca/Planetarium/Planetarium.html (Planetarium)
Founded: 1894 (Planetarium: 1949)
Staff: 20
Activities: undergraduate and graduate education * research * William J. McCallion Planetarium
City Reference Coordinates: 079°51'00"W 43°15'00"N

Memorial University of Newfoundland (MUN), Department of Physics
Saint John's, NF A1B 3X7
Telephone: 709-737-8735
709-737-8736
709-737-8737
709-737-8738
Telefax: 709-737-8739
Electronic Mail: <userid>@kelvin.physics.mun.ca
WWW: http://www.physics.mun.ca/
Founded: 1960
Activities: education * research in molecular physics * emission spectroscopy of diatomic molecules * collision-induced absorption in hydrogen
City Reference Coordinates: 052°43'00"W 47°34'00"N

Merlan Scientific Ltd.
247 Armstrong Avenue
Georgetown, ON L7G 4X6
Telephone: 905-877-0171
1-800-387-2474
Telefax: 905-877-0929
Electronic Mail: sales@merlan.ca
WWW: http://www.merlan.ca/
Founded: 1970
Staff: 25
Activities: science and technology sales to the educational market
City Reference Coordinates: 079°55'00"W 43°39'00"N

Micromedia Ltd.
20 Victoria Street
Toronto, ON M5C 2N8
Telephone: 1-800-387-2689 (Canada only)
416-362-5211
Telefax: 416-362-6161
Electronic Mail: info@micromedia.ca
WWW: http://www.micromedia.on.ca/
Founded: 1972
City Reference Coordinates: 079°23'00"W 43°39'00"N

Mount Allison University, Department of Physics, Engineering and Geoscience
Sackville, NB E0A 3C0
Telephone: 506-564-2580
Telefax: 506-364-2580
Electronic Mail: <userid>@mta.ca
WWW: http://aci.mta.ca/peg/
Founded: 1992 (University: 1838)
Staff: 15
Activities: meteors * meteorites * fireballs * aurorae
City Reference Coordinates: 064°22'00"W 45°54'00"N

Mount Kobau Star Party (MKSP)
c/o Caroline Wallace
P.O. Box 20119 TCM
Kelowna, BC V1Y 9H2
Telephone: 250-498-3244
Telefax: 250-498-3244
Electronic Mail: loveseth@vip.net
WWW: http://www.bcinternet.com/~mksp/
City Reference Coordinates: 119°29'00"W 49°50'00"N

M.W. Burke-Gaffney Observatory (BGO) (Rev._)
• See "Saint Mary's University, Department of Astronomy and Physics, Rev. M.W. Burke-Gaffney Observatory (BGO)"

National Library of Canada (NLC)
395 Wellington Street
Ottawa ON K1A 0N4
Telephone: 613-995-7969
Telefax: 613-991-9871
Electronic Mail: publications@nlc-bnc.ca
WWW: http://www.nlc-bnc.ca/
Founded: 1953
Staff: 500
Activities: national library
Periodicals: (10) "National Library News"
• See also "Bibliothèque Nationale du Canada (BNC)"
City Reference Coordinates: 075°42'00"W 45°25'00"N

National Museum of Science and Technology Corp. (NMSTC)
P.O. Box 9754
Ottawa Terminal
Ottawa, ON K1G 5A3
Telephone: 613-991-3044
Telefax: 613-990-3636
WWW: http://www.smnst.ca/
http://www.nmstc.ca/
Founded: 1967
Staff: 200
Activities: science and technology education * public programs * related historical collections
Periodicals: (6) "Sky News"; "StarGazing", "Ciel Info"
City Reference Coordinates: 075°42'00"W 45°25'00"N

National Research Council of Canada (NRCC)
Montréal Road
Ottawa, ON K1A 0R6
Telephone: 613-993-9101 (general enquiry)
613-993-2054 (NRC publications)
613-745-1576 (time signal)
613-745-9426 (horloge parlante)
613-990-0928 (IAU contact)
613-990-6091 (IAU contact)
Telefax: 613-952-6602
613-952-9907
WWW: http://www.nrc.ca/
Founded: 1916
Periodicals: "Canadian Journal of Physics" (ISSN 0008-4204), "Canadian Journal of Earth Sciences"
• See also "Conseil National de Recherches du Canada (CNRC)"
City Reference Coordinates: 075°42'00"W 45°25'00"N

National Research Council of Canada (NRCC), Canadian Institute for Scientific and Technical Information (CISTI)
Ottawa, ON K1A 0S2
Telephone: 613-993-1600 (inquiries)
613-993-1095 (administration)
613-232-6727 (National Science Film Library)
Telefax: 613-952-9112
WWW: http://www.nrc.ca/icist/
Founded: 1974
Staff: 193
Activities: government agency providing scientific and technical information
Periodicals: "Annual Report", "CISTI News"
City Reference Coordinates: 075°42'00"W 45°25'00"N

National Research Council of Canada (NRCC), Herzberg Institute of Astrophysics (HIA)
5071 West Saanich Road
Victoria, BC V8X 4M6
Telephone: 250-363-0040 (Director General)
250-363-0567 (Administrator)
250-363-0007 (Director's office)
250-363-0001 (Dominion Astrophysical Observatory - DAO)
250-493-2277 (Dominion Radio Astrophysical Observatory - DRAO)
Telefax: 250-363-8483
Electronic Mail: don.morton@hia.nrc.ca
<userid>@hia.nrc.ca
WWW: http://www.hia.nrc.ca/
Founded: 1975
Staff: 150
Activities: optical and radio astronomy
City Reference Coordinates: 123°22'00"W 48°25'00"N

National Research Council of Canada (NRCC), Herzberg Institute of Astrophysics (HIA), Canadian Astronomy Data Centre (CADC)
c/o Dominion Astrophysical Observatory
5071 West Saanich Road
Victoria, BC V8X 4M6
Telephone: 250-363-0023
250-363-0024
250-363-0025
250-363-0052
Telefax: 250-363-0045
Electronic Mail: cadc@dao.nrc.ca
WWW: http://cadcwww.dao.nrc.ca/CADC-homepage.html (English Homepage)
http://cadcwww.dao.nrc.ca/CADC-homepage_fr.html (French Homepage)
http://www.dao.nrc.ca/

http://www.hia.nrc.ca/
Founded: 1986
Staff: 7
Activities: astronomical data archiving * data analysis
City Reference Coordinates: 123°22'00"W 48°25'00"N

National Research Council of Canada (NRCC), Herzberg Institute of Astrophysics (HIA), Dominion Astrophysical Observatory (DAO)
5071 West Saanich Road
Victoria, BC V8X 4M6
Telephone: 250-363-0001 (receptionist/office)
Telefax: 250-363-0045
Electronic Mail: <userid>@hia.nrc.ca
WWW: http://www.hia.nrc.ca/facilities/dao/dao.html
Founded: 1916
Staff: 80
Activities: cosmology * galactic and extragalactic astronomy * interstellar clouds * stellar astronomy * star clusters * magnetic fields * minor planets * optical instrumentation * radio submillimeter instrumentation * public outreach * astronomical software * Canadian Astronomy Data Centre (CADC - see separate entry)
Coordinates: 123°25'01"W 48°31'12"N H238m (IAU Code 656)
City Reference Coordinates: 123°22'00"W 48°25'00"N

National Research Council of Canada (NRCC), Herzberg Institute of Astrophysics (HIA), Dominion Radio Astrophysical Observatory (DRAO)
P.O. Box 248
Penticton, BC V2A 2S8
or :
White Lake Road
Regional District of Okanagan-Similkameen, BC
Telephone: 604-493-2277
Telefax: 604-493-7767
Electronic Mail: <userid>@drao.nrc.ca
WWW: http://www.drao.nrc.ca/
http://www.hia.nrc.ca/DRAO/DRAO-homepage.html
http://www.hia.nrc.ca/
Founded: 1959
Staff: 25
Activities: SNR * HII regions * ISM * PN * star formation * clusters of galaxies * aperture synthesis * radio instrumentation and software
Periodicals: "Preprint Series"
Coordinates: 119°37'12"W 49°19'12"N H545m
City Reference Coordinates: 119°35'00"W 49°30'00"N (Penticton)
119°43'00"W 48°06'00"N (Okanagan)
120°32'00"W 49°18'00"N (Similkameen)

Natural Resources Canada, Geodetic Survey Division
(Ressources Naturelles Canada - Division des Levés Géodésiques)
615 Booth Street
Ottawa, ON K1A 0E9
Telephone: 613-995-4410
613-992-2061 (French)
Telefax: 613-995-3215
Electronic Mail: information@geod.nrcan.ca
WWW: http://www.geod.nrcan.gc.ca/
Founded: 1909
Staff: 94
Activities: providing and maintaining Canadian spatial reference system, standards and national networks of gravity and survey control points for Canada * ensuring availability of spatial referencing information, expertise and services
City Reference Coordinates: 075°42'00"W 45°25'00"N

New Brunswick Astronomy - Astronomie Nouveau Brunswick (NBANB) Inc.
c/o Adrien Bordage (President)
390 Kingsville Road
Saint John, NB E2M 4T2
Telephone: 506-635-3004
Electronic Mail: astro-l@admin.cus.ca
abordage@sjfn.nb.ca
rachett@nbnet.nb.ca
WWW: http://www.geocities.com/beausejournb/nbanb.html
Founded: 1995
Membership: 125
Activities: annual star party * observing * talks * instrumentation
Periodicals: (4) "Pleiades"
City Reference Coordinates: 066°03'00"W 45°16'00"N

Newfoundland Science Centre
The Murray Premises
5 Beck's Cove
P.O. Box 1312
Station C
Saint John's, NF A1C 5N5
Telephone: 709-754-0807
Telefax: 709-738-3276
WWW: http://www.nsc.nfld.com/
City Reference Coordinates: 052°43'00"W 47°34'00"N

North Bay Astronomy Club
c/o Merlin Clayton
50 Van Horne Crescent
North Bay, ON P1A 3L3
Telephone: 705-472-1182
Electronic Mail: brentt@efni.com
WWW: http://www.efni.com/~brentt/nbclub/nbclub.htm
City Reference Coordinates: 079°28'00"W 46°20'00"N

North York Astronomical Association (NYAA)
26 Chryessa Avenue
Toronto, ON M6N 4T5
Electronic Mail: walmac@spanit.com (Walter MacDonald)
      tonyward@home.com
WWW: http://www.nyaa-starfest.com/
Activities: monthly meetings * photography observing (comets, meteors, planets, Sun, SN, variable stars, deep sky) * instrumentation
Periodicals: (4) "AstroTent"
City Reference Coordinates: 079°23'00"W 43°39'00"N

Nova Astronomics
P.O. Box 31013
Halifax, NS B3K 5T9
Telephone: 902-499-6196
Telefax: 902-826-7957
Electronic Mail: info@nova-astro.com
WWW: http://www.nova-astro.com/
Founded: 1991
Staff: 2
• Software producer
City Reference Coordinates: 063°36'00"W 44°39'00"N

Observateurs de la Magnitude Absolue (Les_)
• See "Les Observateurs de la Magnitude Absolue"

Observatoire Astronomique du Mont Mégantic (OAMM)
• See "Université de Montréal, Observatoire Astronomique du Mont Mégantic (OAMM)"

Observatoire Mont-Cosmos
c/o Philippe Moussette
1314, Choisy le Roi
Cap-Rouge, QC GiY 3G6
Telephone: 418-654-1577
Electronic Mail: vegacr@hotmail.com
      mont_cosmos@hotmail.com
WWW: http://www.quebectel.com/mtcosmos/
Founded: 1971
Coordinates: 071°05'10"W 46°22'28"N H495m (St.-Elzéar-de-Beauce)
City Reference Coordinates: 071°05'00"W 46°46'00"N

Odyssium
11211 - 142 Street NW
Edmonton, AB T5M 4A1
Telephone: 403-452-9100
      780-451-3344
Telefax: 780-455-5882
Electronic Mail: info@odyssium.com
WWW: http://www.odyssium.com/
Founded: 1984
Staff: 60
Activities: educational and public programs * IMAX theater * planetarium * computer laboratory * observatory * galleries * gift shop

Periodicals: (4) "Odyssium Observer"; (1) "Annual Report"
• Formerly known as "Edmonton Space and Science Centre (ESSC)"
City Reference Coordinates: 113°28'00"W 53°33'00"N

Ontario Science Centre (OSC)
770 Don Mills Road
North York, ON M3C 1T3
Telephone: 416-696-3127
          416-696-3147 (information in French)
Electronic Mail: <userid>@osc.on.ca
WWW: http://www.osc.on.ca/
Founded: 1969
City Reference Coordinates: 079°25'00"W 43°46'00"N

Ottawa Valley Astronomy and Observers Group (OAOG)
c/o Rock MAllin
P.O. Box 8285
Ottawa, ON K1K 0S1
Telephone: 613-741-1612
Electronic Mail: observers-group@home.com
WWW: http://www.members.home.net/observers-group/
City Reference Coordinates: 075°42'00"W 45°25'00"N

OZ Optics Ltd.
219 Westbrook Road
West Carleton Industrial Park
Carp, ON K0A 1L0
Telephone: 613-831-0981
Telefax: 613-836-5089
Electronic Mail: sales@ozoptics.com
WWW: http://www.ozoptics.com/
Founded: 1985
Activities: manufacturing fiber optic components
City Reference Coordinates: 076°02'00"W 45°21'00"N (Carp)

Physics and Astronomy Student Society (PASS)
c/o Department of Physics and Astronomy
University of Victoria
P.O. Box 3055 STN CSC
Victoria, BC V8W 3P6
Telephone: 250-721-7700
Telefax: 250-721-7715
Electronic Mail: uvpass@uvastro.phys.uvic.ca
WWW: http://astrowww.phys.uvic.ca/uvpass/pass.html
Founded: 1985
City Reference Coordinates: 123°22'00"W 48°25'00"N

Planetarium Association of Canada (PAC)
• See now "Canadian Council of Science Centres (CCSC)"

Planétarium de Montréal
1000 Rue Saint-Jacques Ouest
Montréal, QC H3C 1G7
Telephone: 514-872-4530
          514-872-3611
Telefax: 514-872-8102
Electronic Mail: info@planetarium.montreal.qc.ca
WWW: http://www.planetarium.montreal.qc.ca/
Founded: 1966
Staff: 25
Activities: public and school shows * lectures * exhibits * workshops * star parties
• Alternate name: "Planétarium Dow"
City Reference Coordinates: 073°34'00"W 45°31'00"N

Prince George Astronomical Society (PGAS)
c/o College of New Caledonia
3330 - 22nd Avenue
Prince George, BC V2N 1P8
Telephone: 250-964-3600
Telefax: 250-561-5848
Electronic Mail: bnelson@pgonline.com
WWW: http://www.pgweb.com/~astronomical/
Founded: 1979
Membership: 30
Activities: education * star parties * photography * observing * photometry

Periodicals: (12) "PeGASus"
Coordinates: 122°50'52"W 53°45'29"N H600m (Prince George Astronomical Observatory)
City Reference Coordinates: 122°45'00"W 53°55'00"N

Queen's University, Department of Physics, Astronomy Research Group (QUARG)
Kingston, ON K7L 3N6
Telephone: 613-533-2707
613-533-2711
Telefax: 613-533-6463
Electronic Mail: <userid>@qucdnast.queensu.cdn
<userid>@astro.queensu.ca
WWW: http://www.astro.queensu.ca/
Founded: 1960
Staff: 20
Activities: galactic structure * star formation * theoretical astrophysics
Observatories: 1 (Queen's Observatory)
City Reference Coordinates: 076°30'00"W 44°14'00"N

Resonance Ltd.
143 Ferndale Drive North
Barrie, ON L4M 4S4
Telephone: 705-733-3633
Telefax: 705-733-1388
Electronic Mail: res@resonance.on.ca
WWW: http://www.resonance.on.ca/
Founded: 1980
Staff: 11
Activities: research and development and production of light sources, remote sensors and detector technology for commercial, industrial and space applications
City Reference Coordinates: 079°40'00"W 44°24'00"N

Ressources Naturelles Canada, Division des Levés Géodésiques
• See "Natural Resources Canada, Geodetic Survey Division"

Rev. M.W. Burke-Gaffney Observatory (BGO)
• See "Saint Mary's University, Department of Astronomy and Physics, Rev. M.W. Burke-Gaffney Observatory (BGO)"

Roberta Bondar Planetarium
• See "Seneca College, Roberta Bondar Planetarium"

Rothney Astrophysical Observatory (RAO)
• See "University of Calgary, Department of Physics and Astronomy, Rothney Astrophysical Observatory (RAO)"

Royal Astronomical Society of Canada (RASC)
136 Dupont Street
Toronto, ON M5R 1V2
Telephone: 416-924-7973
Telefax: 416-924-7973
Electronic Mail: rasc@vela.astro.utoronto.ca
WWW: http://www.rasc.ca/
Founded: 1890
Membership: 3000
Activities: observing * education * telescope making
Periodicals: (6) "RASC Journal" (ISSN 0035-872x); (1) "RASC Observer's Handbook" (ISSN 0080-4193), "RASC Annual Report"
Awards: "Gold Medal", "Chant Medal", "Ken Chilton Prize", "Plaskett Medal", "Messier Certificate", "Service Award", "Simon Newcomb Award"
• See also "Société Royale d'Astronomie du Canada (SRAC)"
City Reference Coordinates: 079°23'00"W 43°39'00"N

Saint Mary's University (SMU), Department of Astronomy and Physics, Rev. M.W. Burke-Gaffney Observatory (BGO)
923 Robie Street
Halifax, NS B3H 3C3
Telephone: 902-420-5633
902-420-5828
Telefax: 920-420-5561
902-420-5141
Electronic Mail: gwelch@ap.stmarys.ca
dlane@ap.stmarys.ca
WWW: http://apwww.stmarys.ca/
http://apwww.stmarys.ca/bgo/bgo.html
Founded: 1971
Staff: 2
Activities: education * research

Coordinates: 063°34'52"W 44°37'50"N H125m (IAU Code 851)
City Reference Coordinates: 063°36'00"W 44°39'00"N

Science North Solar Observatory
100 Ramsey Lake Road
Sudbury, ON P3E 5S9
Telephone: 705-522-3701 x243
    1-800-461-4898
Telefax: 705-522-4954
Electronic Mail: yee@ramsey.cs.laurentian.ca
    <userid>@sciencenorth.on.ca
WWW: http://sciencenorth.on.ca/
Founded: 1984
Staff: 9
Activities: interactive exhibits * solar observing * theatre programs
Coordinates: 080°59'46"W 46°28'12"N H260m
City Reference Coordinates: 081°00'00"W 46°30'00"N

Sciencetech Inc.
45 Meg Drive
London, ON N6E 2V2
Telephone: 519-668-0131
Telefax: 519-668-0132
Electronic Mail: scitech@execulink.com
WWW: http://www.execulink.com/~scitech/
Founded: 1985
Staff: 15
Activities: designing and manufacturing optical spectroscopy instruments
City Reference Coordinates: 081°14'00"W 42°59'00"N

Science World British Columbia
1455 Québec Street
Vancouver, BC V6A 3Z7
Telephone: 604-443-7443 (24-hour information)
    604-443-7440 (Administration)
Telefax: 604-443-7430
Electronic Mail: question@scienceworld.bc.ca
WWW: http://www.scienceworld.bc.ca/
Founded: 1989
Staff: 100
Activities: hands-on science centre * exhibition galleries * OmniMax theater * 3D laser theater
Periodicals: (4) "Newsletter"; (1) "Annual Report"
City Reference Coordinates: 123°07'00"W 49°16'00"N

Seneca College, Roberta Bondar Planetarium
1750 Finch Avenue East
North York, ON M2J 2X5
Telephone: 416-491-5050 x2227
Electronic Mail: stars@mars.senecac.on.ca
WWW: http://www.senecac.on.ca/bondar/planet.htm
Founded: 1972
City Reference Coordinates: 079°25'00"W 43°46'00"N

Sky Optics
4031 Fairview Street
Unit 216B
Burlington, ON L7L 2A4
Telephone: 905-9631-9944
Electronic Mail: sbarnes@worldchat.com
WWW: http://www.worldchat.com/commercial/skyoptics/
City Reference Coordinates: 079°48'00"W 43°19'00"N

Société Canadienne d'Astronomie (CASCA)
• See "Canadian Astronomical Society (CASCA)"

Société d'Astronomie de Montréal (SAM)
Casier Postal 206
Succursale Saint-Michel
Montréal, QC H2A 3L9
or :
7110 8e Avenue
Montréal, QC
Telephone: 514-728-4422
WWW: http://www.cam.org/~sam/
Founded: 1968

Membership: 106
Activities: lectures * library * telescope making * shop * observing
Periodicals: (1) "Annuaire Astronomique" (ISSN 0825-9984, circ.: 1,300); (4) "Astro-Notes" (ISSN 0843-8978, circ.: 300)
Awards: "Prix Georgette-Lemoyne", "Prix Étoile d'Argent"
Observatories: 1 (Saint-Valérien de Milton)
City Reference Coordinates: 073°34'00"W 45°31'00"N

Société Royale d'Astronomie du Canada (SRAC)
● See "Royal Astronomical Society of Canada (RASC)"

Solar Terrestrial Dispatch (STD)
P.O. Box 357
Stirling, AB T0K 2E0
Telephone: 403-756-3008
Electronic Mail: coler@solar.stanford.edu (Cary Oler)
WWW: http://www.spacew.com/
Founded: 1990
Activities: providing time-critical solar and geophysical information and data * daily summaries and reports of solar and geophysical activity * alerts and warnings of impending activity
City Reference Coordinates: 112°31'00"W 49°30'00"N

Space West Science Activities (SWSA) Inc.
2115 McEown Avenue
P.O. Box 1811
Saskatoon, SK S7K 3S2
or :
2115 McEown Avenue (Porter Street Door)
Saskatoon, SK, S7J 3K8
Telephone: 306-374-1395
          1-877-IDSPACE
          1-877-437-7223
Telefax: 306-374-6270
Electronic Mail: spacewest@link.ca
WWW: http://www.spacewest.com/
Founded: 1997
Staff: 3
Activities: providing space camps for children and adults
City Reference Coordinates: 106°38'00"W 52°07'00"N

Standards Council of Canada (SCC)
(Conseil Canadien des Normes - CCN)
270 Albert Street
Suite 100
Ottawa, ON K1P 6N7
Telephone: 613-238-3222
Telefax: 613-569-7808
Electronic Mail: info@scc.ca
WWW: http://www.scc.ca/
Founded: 1977
Staff: 79
Activities: coordinating voluntary standardization activities in Canada
Periodicals: (4) "Consensus" (English edition: ISSN 0380-1314, French edition: ISSN 0380-1322); (1) "Annual Report/Rapport Annuel"
Awards: (1) "Jean P. Carrière Award"
City Reference Coordinates: 075°42'00"W 45°25'00"N

Stargazer Steve
c/o Steve Dodson
1752 Rutherglen Crescent
Sudbury, ON P3A 2K3
Telephone: 705-566-1314
Electronic Mail: stargazer@isys.ca
WWW: http://stargazer.isys.ca/
Founded: 1994
Staff: 5
Activities: manufacturing and marketing innovative telescopes
City Reference Coordinates: 081°00'00"W 46°30'00"N

Sudbury Astronomy Club
Suite M250
800 Lasalle Boulevard
Sudbury, ON P3A 4V4
WWW: http://ww2.isys.ca/astroclub
Observatories: 1 (Walden Observatory)
City Reference Coordinates: 081°00'00"W 46°30'00"N

Sudbury Neutrino Observatory Institute (SNOI)
c/o A.B. McDonald (Director)
Department of Physics
Queen's University at Kingston
Stirling Hall
Kingston, ON K7L 3N6
Telephone: 613-533-2702
Telefax: 613-533-6813
Electronic Mail: snoinfo@neutrino.phys.laurentian.ca
WWW: http://www.sno.phy.queensu.ca/ (Sudbury Neutrino Obs., SNO – IAU Code 817)
Founded: 1956
Staff: 300
Activities: condensed matter * theoretical, neutrino, nuclear and accelerator physics * analytical, physical and general chemistry * radiation applications * fusion
Periodicals: "SNOI Progress Reports" (ISSN 0067-0367)
Observatories: 1 (Sudbury Neutrino Observatory - SNO) (H-2300m)
City Reference Coordinates: 076°30'00"W 44°14'00"N

The Atlantic Space Sciences Foundation (TASSF) Inc.
P.O. Box 31011
Halifax, NS B3K 5T9
Telephone: 902-864-7256
Telefax: 902-492-3170
Electronic Mail: tassf@hercules.stmarys.ca
WWW: http://halifax.rasc.ca/tassf/tassf.html
　　　http://halifax.rasc.ca/hp/ (Halifax Planetarium)
Founded: 1990
Staff: 8
Activities: astronomy education for teachers and students * portable and fixed planetariums
City Reference Coordinates: 063°36'00"W 44°39'00"N

Toronto Sidewalk Astronomers (TSA)
Electronic Mail: burns@astro.utoronto.ca (Chris Burns)
WWW: http://www.astro.utoronto.ca/~burns/TSA.html
Membership: 4
City Reference Coordinates: 079°23'00"W 43°39'00"N

Université de Moncton, Département de Physique
Moncton, NB E1A 3E9
Telephone: 506-858-4339
Telefax: 506-858-4541
Electronic Mail: <userid>@umoncton.ca
WWW: http://139.103.16.55/phys/dphys.htm
　　　http://www.umoncton.ca/leblanfn/observ.html
Activities: stellar astrophysics * optical instrumentation * micro-gravity * thin layers * astronomical * electroluminescence * observatory projet
City Reference Coordinates: 064°50'00"W 46°04'00"N

Université de Montréal, Département de Physique, Groupe d'Astronomie
2900 Ed. Montpetit
Casier Postal 6128
Succursale Centre-Ville
Montréal, QC H3C 3J7
Telephone: 514-343-6718
Telefax: 514-343-2071
Electronic Mail: <userid>@astro.umontreal.ca
WWW: http://www.astro.umontreal.ca/
　　　http://www.astro.umontreal.ca/index_eng.html (English Homepage)
Founded: 1968
Staff: 10
Activities: stellar and extragalactic astronomy
City Reference Coordinates: 073°34'00"W 45°31'00"N

Université de Montréal, Observatoire Astronomique du Mont Mégantic (OAMM)
2900 Ed. Montpetit
Casier Postal 6128
Succursale Centre-Ville
Montréal, QC H3C 3J7
Telephone: 514-343-6718
Telefax: 514-343-2071
Electronic Mail: megantic@astro.umontreal.ca
WWW: http://www.astro.umontreal.ca/omm/
Founded: 1978
Staff: 21
Activities: research * education * optical astronomy * instrumentation
Coordinates: 071°09'11"W 45°27'20"N H1,111m (IAU Code 301)

City Reference Coordinates: 073°34'00"W 45°31'00"N

Université du Québec à Trois-Rivières (UQTR), Département de Physique
3351 Boulevard des Forges
Casier Postal 500
Trois-Rivières, QC G9A 5H7
Telephone: 819-376-5107
Telefax: 819-376-5164
Electronic Mail: <userid>@uqtr.ca
WWW: http://www.uqtr.uquebec.ca/dphy/
Founded: 1969
Staff: 7
Activities: education * research
City Reference Coordinates: 072°33'00"W 46°21'00"N

Université Laval, Département de Physique, Groupe de Recherche en Astrophysique
Sainte-Foy, QC G1K 7P4
Telephone: 418-656-2652
Telefax: 418-656-2040
Electronic Mail: <userid>@phy.ulaval.ca
WWW: http://andromede.phy.ulaval.ca/astro/
Founded: 1975
Staff: 5
Activities: evolution and dynamics of galaxies and ISM * stellar formation * stellar populations and synthesis * massive stars * starbursts * energetic phenomena (impacts on ISM) * instrumentation
Coordinates: 071°09'12"W 45°27'20"N H1,108m (Obs. du Mont Mégantic – IAU Code 301)
City Reference Coordinates: 071°18'00"W 46°47'00"N

University of Alberta, Department of Physics
Edmonton, AB T6G 2J1
Telephone: 780-492-5286 (General Office)
         780-492-0721 (Space Physics Group)
         780-987-3933 (Devon Astronomical Observatory)
Telefax: 780-492-0714
Electronic Mail: dept@phys.ualberta.ca
         <userid>@phys.ualberta.ca
         <userid>@space.ualberta.ca (Space Physics Group)
WWW: http://www.phys.ualberta.ca/
         http://www.space.ualberta.ca/~sarah/ (Space Physics Group)
Founded: 1968 (Devon Astronomical Observatory)
Staff: 2 (Astronomy) 5 (Space Physics)
Activities: photometry of binary and variable stars * CCD imagery of SNRs * aurorae
Coordinates: 113°45'31"W 53°23'27"N H708m (Devon Astronomical Observatory)
City Reference Coordinates: 113°28'00"W 53°33'00"N

University of British Columbia (UBC), Department of Physics and Astronomy
129-2219 Main Mall
Vancouver, BC V6T 1Z4
or :
6224 Agricultural Road
Vancouver, BC V6T 1Z1
Telephone: 604-822-2267
         604-822-3853
         604-822-6673
Telefax: 604-822-6047
         604-822-5324
Electronic Mail: <userid>@physics.ubc.ca
         <userid>@geop.ubc.ca
WWW: http://www.physics.ubc.ca/Homepage.html
         http://www.astro.ubc.ca/
Founded: 1996
Staff: 55
Activities: variable sources * insterstellar molecules * star formation * instrumentation * radial velocities * globular clusters * close groups of galaxies * QSOs * observational cosmology * model atmospheres * ISM
Periodicals: (1) "Annual Report" (ISSN 0068-1725)
Coordinates: 123°15'24"W 49°15'30"N
City Reference Coordinates: 123°07'00"W 49°16'00"N

University of Calgary, Department of Physics and Astronomy, Rothney Astrophysical Observatory (RAO)
2500 University Drive NW
Calgary, AB T2N 1N4
Telephone: 403-220-5385 (Department Office)
         403-931-2366 (RAO)
         403-220-5410
Telefax: 403-289-3331

Electronic Mail: <userid>@ucalgary.ca
WWW: http://www.phas.ucalgary.ca/
http://www.acs.ucalgary.ca/~milone/rao.html (RAO)
Founded: 1972
Staff: 12
Activities: variable-star photometry * analysis of eclipsing binary stars light and radial-velocity curves * solar IR spectroscopy * Be stars * IR and optical astronomy * optical imaging
Periodicals: "Publications of the Rothney Astrophysical Observatory"
Coordinates: 114°17'18"W 50°52'06"N H1,272m (Priddis – IAU Code 661)
City Reference Coordinates: 114°05'00"W 51°03'00"N

University of Calgary, Institute for Space Research
2500 University Drive NW
Calgary, AB T2N 1N4
Telephone: 403-220-7219
Telefax: 403-282-5016
Electronic Mail: isr@phys.ucalgary.ca
WWW: http://www.phys.ucalgary.ca/
Staff: 10
Activities: space and atmospheric environments * planetary science
City Reference Coordinates: 114°05'00"W 51°03'00"N

University of Guelph, Department of Physics
MacNaughton Building
Gordon Street
Guelph, ON N1G 2W1
Telephone: 519-824-4120 x2261
Telefax: 519-836-9967
Electronic Mail: contact@physics.uoguelph.ca
<userid>@physics.uoguelph.ca
WWW: http://www.physics.uoguelph.ca/physics.html
Founded: 1964
Activities: education
Coordinates: 080°13'42"W 43°31'52"N H364m
City Reference Coordinates: 080°15'00"W 43°33'00"N

University of Lethbridge, Department of Physics
4401 University Drive
Lethbridge, AB T1K 3M4
Telephone: 403-329-2280
Telefax: 403-329-2057
Electronic Mail: naylor@uleth.ca (David A. Naylor)
WWW: http://home.uleth.ca/fas/phy/
Founded: 1967
Staff: 700
Activities: education * high-resolution IR spectroscopy
Coordinates: 113°00'00"W 50°00'00"N H1,000m (Astrophysical Observatory)
City Reference Coordinates: 110°50'00"W 49°42'00"N

University of Manitoba, Department of Physics and Astronomy
Winnipeg, MB R3T 2N2
Telephone: 204-474-9817
204-883-2567 (Glenlea Astronomical Observatory)
Telefax: 204-474-7622
Electronic Mail: physics@umanitoba.ca
WWW: http://www.physics.umanitoba.ca/
http://www.umanitoba.ca/faculties/science/astronomy/
Founded: 1997 (Department of Physics: 1904)
Coordinates: 097°07'15"W 49°38'43"N (Glenlea Astron. Obs. – IAU Code 729)
City Reference Coordinates: 097°09'00"W 49°53'00"N

University of New Brunswick, Department of Physics
IUC Building
Bailey Drive
P.O. Box 4400
Fredericton, NB E3B 5A3
Telephone: 506-453-4723
Telefax: 506-453-4581
Electronic Mail: physics@unb.ca
WWW: http://www.unb.ca/physics/dept/
City Reference Coordinates: 066°39'00"W 45°58'00"N

University of Regina, Department of Physics
Regina, SK S4S 0A2
Telephone: 306-585-4149

Telefax: 306-585-4890
Electronic Mail: <userid>@meena.cc.uregina.ca
WWW: http://www.uregina.ca/
Founded: 1974
Staff: 16
Activities: subatomic physics * theoretical physics * general relativity * low-temperature metal physics * molecular biophysics * meteoroid streams
Coordinates: 104°35'26"W 50°25'08"N H575m
City Reference Coordinates: 104°39'00"W 50°25'00"N

University of Saskatchewan, Department of Physics and Engineering Physics
116 Science Place
Saskatoon, SK S7N 5E2
Telephone: 306-966-6396
Telefax: 306-966-6400
Electronic Mail: <userid>@physics.usask.ca
WWW: http://physics.usask.ca/
http://physics.usask.ca/observ.htm (Duncan Observatory)
Founded: 1911
Staff: 117
Activities: education * observing
Coordinates: 106°37'53"W 52°07'55"N H500m (Duncan Observatory)
City Reference Coordinates: 106°38'00"W 52°07'00"N

University of Saskatchewan, Institute of Space and Atmospheric Studies (ISAS)
116 Science Place
Saskatoon, SK S7N 5E2
Telephone: 306-966-6401
Telefax: 306-966-6428
Electronic Mail: isas@dansas.usask.ca
WWW: http://www.usask.ca/physics/isas/
Founded: 1957
Staff: 35
Activities: research in atmospheric, space and planetary environments * satellite and ground-based systems * education
Coordinates: 103°70'00"W 58°20'00"N (Rabbit Lake)
107°07'00"W 52°12'00"N (Park Site)
106°27'00"W 52°15'00"N (Bakker's Farm)
106°32'00"W 52°09'00"N (Kernen Farm)
City Reference Coordinates: 106°38'00"W 52°07'00"N

University of Toronto, Department of Astronomy and Astrophysics
Room 1403
60 Saint George Street
Toronto, ON M5S 3H8
Telephone: 416-978-2016
905-884-9562 (DDO)
Telefax: 416-971-2026
416-946-7287
905-884-2672 (DDO)
Electronic Mail: info@astro.utoronto.ca
<userid>@astro.utoronto.ca
WWW: http://www.astro.utoronto.ca/home.html
http://ddo.astro.utoronto.ca/ddohome (David Dunlap Observatory)
http://www.astro.utoronto.ca/~utso/ (Univ. Toronto Southern Obs.)
Founded: 1935
Staff: 30
Activities: education * research * stellar and extragalactic astrophysics * David Dunlap Observatory (DDO)
Coordinates: 079°30'00"W 43°54'00"N H244m (DDO, Richmond Hill – IAU Code 779)
070°42'00"W 29°00'00"S H2,280m (Las Campanas, Chile – IAU Code 304)
City Reference Coordinates: 079°30'00"W 43°54'00"N

University of Toronto, Department of Astronomy and Astrophysics, Erindale Campus
Mississauga Road
Mississauga, ON L5L 1C6
Telephone: 905-828-5351
905-828-3818
Telefax: 416-828-5425
Electronic Mail: <userid>@astro.erin.utoronto.ca
WWW: http://www.erin.utoronto.ca/~astro/
Founded: 1967
Staff: 3
Activities: education * studies of stellar atmospheres * variable stars * stellar spectroscopy * planetarium (Erindale Campus Planetarium)
City Reference Coordinates: 079°37'00"W 43°35'00"N

University of Toronto, Institute for Aerospace Studies (UTIAS)
4925 Dufferin Street
Downsview, ON M3H 5T6
Telephone: 416-667-7700
Telefax: 416-667-7799
Electronic Mail: info@utias.utoronto.ca
WWW: http://www.utias.utoronto.ca/
Founded: 1949
Staff: 70
Activities: aerospace engineering * applied physics * research * education
Periodicals: (1) "Annual Progress Report" (ISSN 0082-5239, circ.: 1,200); "UTIAS Reports" (ISSN 0082-5255, circ.: 350), "UTIAS Reviews" (ISSN 0082-5247, circ.: 350), "UTIAS Technical Notes" (ISSN 0082-5263, circ.: 350)
City Reference Coordinates: 079°28'00"W 43°45'00"N

University of Toronto Press Inc.
5201 Dufferin Street
Toronto ON M3H 5T8
Telephone: 1-800-565-9523 (Canada and USA only)
            416-667-7791
Telefax: 1-800-221-9985 (Canada and USA only)
            416-667-7832
Electronic Mail: info@utpress.utoronto.ca
                <userid>@utpress.utoronto.ca
WWW: http://www.utpress.utoronto.ca/
Founded: 1901
City Reference Coordinates: 079°30'00"W 43°54'00"N

University of Victoria, Department of Physics and Astronomy
P.O. Box 3055 STN CSC
Victoria, BC V8W 3P6
or :
Elliott Building
3800 Finnerty Road
Victoria, BC V8P 1A1
Telephone: 250-721-7700
            250-721-7750 (Climenhaga Observatory)
Telefax: 250-721-7715
Electronic Mail: office@phys.uvic.ca
                <userid>@phys.uvic.ca
                <userid>@uvastro.phys.uvic.ca
WWW: http://astrowww.phys.uvic.ca/
        http://info.phys.uvic.ca/
        http://info.phys.uvic.ca/uvphys_welcome.html
        http://almuhit.phys.uvic.ca/
Founded: 1963
Staff: 22
Coordinates: 123°18'30"W 48°27'48"N H74m (Climenhaga Obs. – IAU Code 657)
City Reference Coordinates: 123°22'00"W 48°25'00"N

University of Waterloo, Department of Physics, Astronomy/Gravitation Group
Waterloo, ON N2L 3G1
Telephone: 519-885-1211
Telefax: 519-746-8115
Electronic Mail: observe@astro.uwaterloo.ca
                <userid>@waterloo.ca
WWW: http://astro.uwaterloo.ca/
Periodicals: "Contributions of the University of Waterloo"
City Reference Coordinates: 080°31'00"W 43°28'00"N

University of Western Ontario (UWO), Physics and Astronomy Department
London, ON N6A 3K7
Telephone: 519-661-3183
            519-225-2620 (Elginfield Observatory)
Telefax: 519-661-2033
Electronic Mail: <userid>@uwo.ca
WWW: http://www.physics.uwo.ca/
        http://www.astro.uwo.ca/ (Astronomy Group)
Founded: 1966
Staff: 12
Activities: education * research * high-resolution spectroscopy * polarimetry * photometry * theory of Be stars * Ap stars * cool stars * stellar interiors * cosmology * ISM
Coordinates: 081°19'00"W 43°12'00"N H323m (Elginfield Observatory)
            081°16'34"W 43°00'18"N (Hume Cronyn Observatory)
City Reference Coordinates: 081°14'00"W 42°59'00"N

Western Space Education Network (WSEN) Inc.
2115 McEown Avenue
P.O. Box 1811
Saskatoon, SK S7K 3S2
Telephone: 306-374-1375
Telefax: 306-374-1395
Electronic Mail: wsen@the.link.ca
WWW: http://www.sfn.saskatoon.sk.ca/science/wsen
Founded: 1993
Staff: 3
Activities: space science educational outreach and resource centre * Canadian Space Resource Centre (Prairies)
City Reference Coordinates: 106°38'00"W 52°07'00"N

Wilfrid Laurier University (WLU), Department of Physics and Computing
75 University Avenue West
Waterloo, ON N2L 3C5
Telephone: 519-884-1970
Telefax: 519-746-0677
Electronic Mail: <userid>@mach1.wlu.ca
  jlit@mach1.wlu.ca (John Lit, Chair)
WWW: http://info.wlu.ca/~wwwphys/
Founded: 1963
Staff: 1100
Activities: academic courses in astronomy and optics
City Reference Coordinates: 080°31'00"W 43°28'00"N

York-Simcoe Amateur Astronomers
c/o Scott Gilbert
262 Patterson Street
Newmarket, ON L3Y3L8
Telephone: 905-895-6718 x305
Electronic Mail: gilbes1@investorsgroup.com
WWW: http://www.ysastronomers.com/
Membership: 40
City Reference Coordinates: 079°27'00"W 44°03'00"N

York University Astronomy Physics Club
c/o Room 329
Petrie Science Building
4700 Keele Street
North York, ON M3J 1P3
Telephone: 416-736-2100 x77763
Electronic Mail: yuac@yorku.ca
WWW: http://www.yorku.ca/yuac/
Founded: 1998 (1987 as "York University Astronomy Club - YUAC")
City Reference Coordinates: 079°25'00"W 43°46'00"N

York University, Department of Earth and Atmospheric Sciences
Room 101/102
Petrie Science Building
4700 Keele Street
North York, ON M3J 1P3
Telephone: 416-736-5245
Telefax: 416-736-5817
Electronic Mail: eas@science.yorku.ca
WWW: http://www.eas.yorku.ca/
Founded: 1983
Staff: 12
Activities: geophysics * atmospheric dynamics * Earth's core * ozone layer * boundary layer * laboratory fluid dynamics * mesoscale meteorology * aeronomy * remote sensing * space sciences * instrumentation
City Reference Coordinates: 079°25'00"W 43°46'00"N

York University, Department of Physics and Astronomy
128 Petrie Science Building
4700 Keele Street
North York, ON M3J 1P3
Telephone: 416-736-5249
Telefax: 416-736-5516
Electronic Mail: phas@yorku.ca
  <userid>@yorku.ca
WWW: http://science.yorku.ca/units/phas/home.html
  http://aries.phys.yorku.ca/~delaney/ (Observatory)
Founded: 1965
Staff: 28
Activities: education * research
City Reference Coordinates: 079°25'00"W 43°46'00"N

Channel Islands

La Société Guernesiaise, Section Astronomique
Highcliffe
Avenue Beauvais
Ville au Roi
Saint Peter Port
Guernsey
Telephone: (0)1481-724101
Electronic Mail: vs76@dial.pipex.com (David Le Conte, Newsletter Editor)
WWW: http://ds.dial.pipex.com/nightsky/astro/
Founded: 1972
Membership: 45
Activities: amateur astronomy * meetings * lectures * observing
Periodicals: (12) "Sagittarius"; (1) "Transactions"
Coordinates: 002°38'09"W 49°26'57"N H46m
City Reference Coordinates: 002°32'00"W 49°27'00"N

Chile

Asociación Chilena de Astronomía y Astronáutica (ACHAYA)
Casilla 3904
Santiago-Centro
or :
Marcoleta 485
Departamento H
Santiago
Telephone: (0)2-6327556
Telefax: (0)2-6327556
Electronic Mail: contacto@achaya.cl
WWW: http://www.achaya.cl/
Founded: 1957
City Reference Coordinates: 070°40'00"W 33°27'00"S

Carnegie Institution of Washington (CIW), Observatories, Las Campanas Observatory (LCO)
Casilla 601
La Serena
Telephone: (0)51-224680 (El Pino Office)
(0)51-211254
(0)51-211032 (Office)
(0)51-211413 (LCO Site)
Telefax: (0)51-227817
WWW: http://www.lco.cl/
Founded: 1969
Coordinates: 070°42'00"W 29°00'30"S H2,282m (IAU Code 304)
City Reference Coordinates: 071°18'00"W 29°54'00"S

Centro de Estudios Científicos de Santiago (CECS)
Casilla 16443
Santiago 9
or :
Avenida Presidente Errazuriz 3132
Las Condes
Santiago
Telephone: (0)2-2338342
(0)2-2060092
Telefax: (0)2-2338336
Electronic Mail: cecs@cecs.cl
WWW: http://www.cecs.cl/
Founded: 1984
Staff: 20
Activities: theoretical physics * biophysics
City Reference Coordinates: 070°40'00"W 33°27'00"S

Cerro Tololo Interamerican Observatory (CTIO)
• See "National Optical Astronomy Observatories (NOAO), Cerro Tololo Interamerican Observatory (CTIO)"

Comisión Nacional de Investigaciones Científicas y Técnologicas (CONICYT)
Casilla 297-V
Santiago 21
or :
Canadá 308
Providencia
Santiago
Telephone: (0)2-3654400
(0)2-2744537
Telefax: (0)2-2096729
(0)2-6551395
Electronic Mail: president@conicyt.cl
<userid>@conicyt.cl
WWW: http://www.conicyt.cl/
Founded: 1968
Staff: 120
Activities: scientific policy * financing * information * administration
City Reference Coordinates: 070°40'00"W 33°27'00"S

Dirección Meteorológica de Chile (DMC)
Aeropuerto Arturo Merino Benítez
Casilla 63
Correo Internacional
Santiago

Telephone: (0)2-6019001
(0)2-6011168
(0)2-6763437
Telefax: (0)2-6019590
(0)2-6763466
Electronic Mail: dimetchi@meteochile.cl
WWW: http://www.meteochile.cl/
Founded: 1880
Staff: 273
Activities: weather forecast * climatology * consulting * instrumentation * network of about 200 stations
Periodicals: (12) "Boletín Climatológico", "Boletín Agrometeorológico"; (1) "Anuarios Meteorológicos"
City Reference Coordinates: 070°40'00"W 33°27'00"S

European Southern Observatory (ESO), Antofagasta Office
Casilla 540
Antofagasta
or :
Balmaceda 2536
Oficina 504
Antofagasta
Telephone: (0)55-260032
(0)55-260048
Telefax: (0)55-260081
Electronic Mail: <userid>@ls.eso.org
WWW: http://www.hq.eso.org/paranal/ (Paranal Observatory)
http://www.hq.eso.org/
Founded: 1962 (ESO)
Coordinates: 070°24'10"W 24°37'30"S H2,635m (IAU Code 309)
• See main entry in Germany
City Reference Coordinates: 070°24'00"W 23°39'00"S

European Southern Observatory (ESO), Chile
Casilla 19001
Santiago 19
or :
Alonso de Córdova 3107
Vitacura
Santiago
Telephone: (0)2-4633000
Telefax: (0)2-4633001
Electronic Mail: <userid>@ls.eso.org
<userid>@sc.eso.org
WWW: http://www.ls.eso.org/
http://www.sc.eso.org/santiago/
http://www.hq.eso.org/
Founded: 1962 (ESO)
• See main entry in Germany
City Reference Coordinates: 070°40'00"W 33°27'00"S

European Southern Observatory (ESO), La Serena Office
Casilla 567
La Serena
or :
Avenida El Santo 1535
La Serena
Telephone: (0)51-215175
(0)2-4644100
Telefax: (0)51-225387
(0)2-6954263
Electronic Mail: <userid>@ls.eso.org
WWW: http://www.ls.eso.org/ (La Silla Observatory)
http://www.hq.eso.org/
Founded: 1962 (ESO)
• See main entry in Germany
City Reference Coordinates: 071°16'00"W 29°54'00"S

European Southern Observatory (ESO), La Silla Observatory
Casilla 19001
Santiago 19
Telephone: (0)2-4644100 (through Santiago)
(0)2-6988757 (through Santiago)
(0)51-224527 (through La Serena)
(0)51-224932 (through La Serena)
Telefax: (0)2-6954263 (through Santiago)
(0)51-224932 (through La Serena)
Electronic Mail: lasilla@eso.org

<userid>@ls.eso.org
WWW: http://www.ls.eso.org/
http://www.ls.eso.org/lasilla/
http://www.hq.eso.org/
Founded: 1962
Activities: astronomical research in the southern hemisphere
Coordinates: 070°43'48"W 29°15'24"S H2,347m (IAU Code 809)
• See main entry in Germany
City Reference Coordinates: 070°40'00"W 33°27'00"S (Santiago)

European Southern Observatory (ESO), Paranal Observatory
Telephone: (0)281291
Telefax: (0)285064
WWW: http://www.hq.eso.org/paranal/
Founded: 1962 (ESO)
Coordinates: 070°24'10"W 24°37'30"S H2,635m (IAU Code 309)
• Use Antofagasta office for mail
• See main entry in Germany
City Reference Coordinates: 070°24'00"W 23°39'00"S (Antofagasta)

Gemini Observatory, Southern Operations Center
c/o AURA
Casilla 603
La Serena
Telephone: (0)51-205600
Telefax: (0)51-205650
Electronic Mail: gemini@gemini.edu
<userid>@gemini.edu
WWW: http://www.gemini.edu/
http://www.conicyt.cl/gemini/
Founded: 1998
Activities: international partnership (USA, United Kingdom, Canada, Australia, Chile, Argentina, Brazil) aiming at setting up two high-performance 8m telescopes on Mauna Kea (USA-HI) and Cerro Pachón (Chile)
Periodicals: "The Gemini Newsletter"
Coordinates: 155°28'18"W 19°49'36"N (Gemini North)
070°43'24"W 30°13'42"S (Gemini South)
City Reference Coordinates: 071°18'00"W 29°54'00"S

Instituto Nacional de Normalización (INN)
Matías Cousiño 64 - 6° piso
Casilla 995
Santiago 1
Telephone: (0)2-6968144
Telefax: (0)2-6960247
Electronic Mail: inn@reuna.cl
WWW: http://www.inn.cl/
Founded: 1973
City Reference Coordinates: 070°40'00"W 33°27'00"S

Las Campanas Observatory (LCO)
• See "Carnegie Institution of Washington (CIW), Observatories, Las Campanas Observatory (LCO)"

La Silla Observatory
• See "European Southern Observatory (ESO), La Silla Observatory"

National Optical Astronomy Observatories (NOAO), Cerro Tololo Interamerican Observatory (CTIO), La Serena
Casilla 603
La Serena
Telephone: (0)51-205200
Telefax: (0)51-205342
(0)51-205212
Electronic Mail: <userid>@noao.edu
WWW: http://ctios2.ctio.noao.edu/ctio.html
http://www.ctio.noao.edu/ctio.html
http://ctiot6.ctio.noao.edu/ (Southern Columbia Millimeter Telescope - SCMT)
Founded: 1963
Staff: 117
Activities: astronomical research * observing * instrumentation * data reduction
Periodicals: (4) "NOAO Newsletter", "NOAO Quarterly Report"; (1) "NOAO Annual Report"
Coordinates: 070°48'53"W 30°09'57"S H2,215m (IAU Code 807)
City Reference Coordinates: 071°18'00"W 29°54'00"S

Observatorio Cerro Mamalluca
Calle Gabriela Mistral 260

China (PRC)

Academia Sinica
• See "Chinese Academy of Sciences"

Acta Astronomica Sinica
c/o Purple Mountain Observatory
2 Beijing Xilu
Nanjing 210008
Telephone: (0)25-3302147
Telefax: (0)25-3301459
• Journal (4) (ISSN 0001-5245)
City Reference Coordinates: 118°47'00"E 32°03'00"N

Acta Astrophysica Sinica
• See now "Chinese Journal of Astronomy and Astrophysics (CJAA)"

Beijing Ancient Observatory
Jian Guo Men Street
Beijing
Telephone: (0)542202
WWW: http://www.bjww.gov.cn/msgj/english/me13.htm
Founded: 1442
Activities: exhibitions * ancient astronomy research
City Reference Coordinates: 116°25'00"E 39°55'00"N

Beijing Astronomical Observatory (BAO)
• See "Chinese Academy of Sciences, Beijing Astronomical Observatory (BAO)"

Beijing Astronomical Data Center (BADC)
• See "Chinese Academy of Sciences, Beijing Astronomical Data Center (BADC)"

Beijing Astrophysics Center (BAC)
Peking University
5501 Yifu No. 1 Building
Beijing 100871
WWW: http://vega.bac.pku.edu.cn/
• Jointly run by the "Chinese Academy of Sciences" and "Beijing University"
City Reference Coordinates: 116°25'00"E 39°55'00"N

Beijing Normal University, Department of Astronomy
Beijing 100875
Telephone: (0)1-2012255 x2618
Telefax: (0)1-2013929
WWW: http://www.bnu.edu.cn/
Founded: 1960
Staff: 35
Activities: education * research * Chinese Visiting Lecturer Programme * planetarium
Coordinates: 116°21'37"E 39°57'27"N H70m
City Reference Coordinates: 116°25'00"E 39°55'00"N

Beijing Planetarium
138 Xi Wai Street
Beijing 100044
Telephone: (0)10-8361691
Telefax: (0)10-8353003
WWW: http://www.bao.ac.cn/cas/pic/bp.gif
Founded: 1957
Staff: 130
Activities: shows * exhibitions * education * observing * research * instrumentation
Periodicals: "Amateur Astronomer"
City Reference Coordinates: 116°25'00"E 39°55'00"N

Beijing University, Department of Geophysics, Astrophysics Division
Beijing 100871
Telephone: (0)1-2552471 x3929
Telefax: (0)1-2564095
WWW: http://www.pku.edu.cn/
City Reference Coordinates: 116°25'00"E 39°55'00"N

Changchun Observatory
• See "Chinese Academy of Sciences, Changchun Observatory"

China Meteorological Administration (CMA)
46 Baishiqiao Road
Beijing 100081
Telephone: (0)10-62174797
Telefax: (0)10-62174797
          (0)10-62173417
WWW: http://www.cma.gov.cn/
Founded: 1951
Staff: 65000 (whole country)
Activities: meteorological observing, telecommunications, data processing, analysis and prediction * research and education * training
Periodicals: "Meteorology", "Meteorological Knowledge", "Journal of Meteorology", "Journal of Tropical Meteorology", "Journal of Tropical Meteorology", "Journal of Applied Meteorology"
City Reference Coordinates: 116°25'00"E 39°55'00"N

China State Bureau of Technical Supervision (CSBTS)
P.O. Box 8010
Beijing 100088
or :
4 Zhi Chun Road
Haidian District
Beijing
Telephone: (0)1-2032424
          (0)10-62032424
Telefax: (0)1-2031010
         (0)10-62031010
Electronic Mail: intl@std.csbts.cn.net
WWW: http://www.csbts.cn.net/
City Reference Coordinates: 116°25'00"E 39°55'00"N

Chinese Academy of Sciences, Beijing Astronomical Data Center (BADC)
c/o Beijing Astronomical Observatory
Datun Road 20A
Chaoyang District
Beijing 100012
Telephone: (0)10-64888763
Telefax: (0)10-64888731
Electronic Mail: ghf@bdc.bao.ac.cn (Guo Hongfeng)
WWW: http://www.bao.ac.cn/bdc/
Founded: 1987
City Reference Coordinates: 116°25'00"E 39°55'00"N

Chinese Academy of Sciences, Beijing Astronomical Observatory (BAO)
20A Datun Road
Chaoyang District
Beijing 100012
Telephone: (0)10-64888712
          (0)10-64888708
Electronic Mail: <userid>@bao.ac.cn
              lqb@bao.ac.cn (Li Qibin, Director)
WWW: http://www.bao.ac.cn/bao/
       http://www.bao.ac.cn/bao/index-e.html
Founded: 1279
Staff: 400
Activities: solar physics * stellar physics * galaxies * AGNs * QSOs * cosmology * radio astronomy * stellar catalogs * atomic time
Periodicals: "Acta Astrophysica Sinica" (see separate entry), "Publications of the Beijing Astronomical Observatory", "Chinese Solar Geophysical Data", "Chinese Astronomy Abstracts"
Coordinates: 117°34'30"E 40°23'42"N H960m (Xinglong Station – IAU Code 327)
          116°45'54"E 40°33'24"N H160m (Miyun Station)
          116°19'42"E 40°06'06"N H40m (Shahe Station – IAU Code 324)
          116°36'00"E 40°19'00"N H62m (Huairou Station)
City Reference Coordinates: 116°25'00"E 39°55'00"N

Chinese Academy of Sciences, Changchun Observatory
Jing Yue Tan Xi Shan
Changchun
Jilin 130117
Telephone: (0)431-4513834
          (0)431-4511337

Telefax: (0)431-4513550
Electronic Mail: ccslr@public.cc.jl.cn
WWW: http://www.cho.ac.cn/
Founded: 1957
Staff: 60
Activities: artificial-satellite observing * satellite laser ranging * GPS * celestial mechanics
Coordinates: 125°26'36"E 43°47'26"N H268m
City Reference Coordinates: 126°35'00"E 43°53'00"N

Chinese Academy of Sciences, Institute of High-Energy Physics, Laboratory of Cosmic-Ray and High-Energy Astrophysics
P.O. Box 918-3
Beijing 100039
or :
19 Yuguanlu
Beijing
Telephone: (0)10-68236114
Telefax: (0)10-88211623
WWW: http://www.ihep.ac.cn/
Founded: 1970
Staff: 70
Activities: cosmic rays * high-energy astrophysics * space observing
City Reference Coordinates: 116°25'00"E 39°55'00"N

Chinese Academy of Sciences, National Astronomical Observatories
Beijing 100012
City Reference Coordinates: 116°25'00"E 39°55'00"N

Chinese Academy of Sciences, Purple Mountain Observatory
2 Beijing Xilu
Nanjing 210008
Telephone: (0)25-3303921
          (0)25-3308986
Telefax: (0)25-3300818
Founded: 1934
Staff: 250
Activities: astrophysics * celestial mechanics * astrometry * history of astronomy * instrumentation
Periodicals: (4) "Publications of the Purple Mountain Observatory" (ISSN 1000-3681), "Acta Astronomica Sinica" (see separate entry)
Coordinates: 118°49'18"E 32°04'00"N H267m (Purple Mountain – IAU Code 330)
             097°43'00"E 37°22'00"N H3,200m (Qinghai - see separate entry)
City Reference Coordinates: 118°47'00"E 32°03'00"N

Chinese Academy of Sciences, Purple Mountain Observatory, Qinghai Radio Observatory
5 Qing Xin Lu
Delingha 817000
Telephone: (0)977-221935
Telefax: (0)25-3301459
Founded: 1986
Staff: 39
Activities: mm-wave astronomy
Coordinates: 097°43'00"E 37°22'00"N H3,200m
• Formerly "Delingha Radio Astronomy Station"
City Reference Coordinates: 097°43'00"E 37°22'00"N

Chinese Academy of Sciences, Shaanxi Astronomical Observatory (CSAO)
P.O. Box 18
Lintong
Xian
Shaanxi
Telephone: (0)29-3890207
Telefax: (0)29-3890344
Electronic Mail: kyc@ms.sxso.ac.cn
                 <userid>@ms.sxso.ac.cn
Founded: 1966
Staff: 370
Activities: astrometry * astrophysics * STP * celestial mechanics * natural satellites * applied historical astronomy * time service
Periodicals: (12) "Time and Frequency Bulletin" (ISSN 1001-1811); (2) "Publications of the Shaanxi Astronomical Observatory" (ISSN 1001-1544)
Coordinates: 109°33'06"E 34°56'42"N H468m
City Reference Coordinates: 108°52'00"E 34°15'00"N

Chinese Academy of Sciences, Shanghai Astronomical Observatory (SHAO)
80 Nandan Road

Shanghai 200030
Telephone: (0)21-64386191
(0)21-57651763 (Station)
Telefax: (0)21-64384618
Electronic Mail: <userid>@center.shao.ac.cn
WWW: http://center.shao.ac.cn/
http://www.shao.ac.cn/
Founded: 1872
Staff: 300
Activities: astro-geodynamics * astrometry * stellar astronomy * astrophysics * satellite dynamics
Periodicals: "Annals of Shanghai Observatory"
Coordinates: 121°11'12"E 31°05'48"N H98m (Sheshan Station)
City Reference Coordinates: 121°28'00"E 31°14'00"N

Chinese Academy of Sciences, Urumqi Astronomical Observatory (UAO)
40 Suburb. 5
South Beijing Road
Urumqi
Xinjiang 830011
Telephone: (0)991-3838007
Telefax: (0)991-3838628
Electronic Mail: uao@public.wl.xj.cn
WWW: http://ivs.crl.go.jp/mirror/publications/ar1999/nsurumqi/
Founded: 1957
Staff: 92
Coordinates: 087°14'00"E 43°15'00"N H2000m
City Reference Coordinates: 087°38'00"E 43°43'00"N

Chinese Academy of Sciences, Yunnan Observatory
P.O. Box 110
Kunming 650011
Telephone: (0)871-3347087
(0)871-3368497
Telefax: (0)871-3358437
(0)871-3911845
Electronic Mail: ynao@public.km.yn.cn
Founded: 1972
Staff: 304
Activities: stellar theory * stellar spectra * photometry * extragalactic astrophysics * solar activity and corresponding forecast * solar radio observing * solar seismology * VLBI * astrometry * celestial mechanics * observational techniques * planetarium * popular science and educational center
Periodicals: (4) "Publications of Yunnan Observatory" (ISSN 1001-7526)
Coordinates: 102°47'18"E 25°01'32"N H2,002m (IAU Code 286)
City Reference Coordinates: 102°41'00"E 25°04'00"N

Chinese Astronomical Society (CAS)
c/o Purple Mountain Observatory
2 Beijing Xilu
Nanjing 210008
Telephone: (0)25-3302147
Telefax: (0)25-3301459
Electronic Mail: lqb@bao01.bao.ac.cn (Li Qibin, President)
WWW: http://www.bao.ac.cn/cas/
Founded: 1922
Membership: 1800
Activities: promoting study and development of astronomy * disseminating knowledge of astronomy * sponsoring scientific meetings * improving astronomical education * directing amateur astronomical activities * developing international collaborations * acting as consultant for related decision-making bodies
Sections: (committees) Organising * Teaching of Astronomy * Astronomical Terminology * Popularization of Astronomy * Publications, Library and Information
Sections: (commissions) Stellar and Planetary Physics * Galaxies and Cosmology * Radio Astronomy * Solar Physics and Solar-Terrestrial Relationships * Catalogs and Astronomical Constants * Astrogeodynamics * Time and Frequency * Celestial Mechanics * Instruments and Techniques * History of Astronomy * Satellite Dynamics * High-Energy Astrophysics * Astronomy from Space
Periodicals: (24) "Tianwen Aihaozhe" (Astronomical Amateur) (ISSN 0493-2285); (4) "Acta Astronomica Sinica" (see separate entry), "Acta Astrophysica Sinica" (see separate entry), "Progress in Astronomy" (see separate entry); (1-2) "CAS Bulletin" (ISSN 0001-5245); "Astronomical Circulars"
City Reference Coordinates: 118°47'00"E 32°03'00"N

Chinese Journal of Astronomy and Astrophysics (CJAA)
c/o Editorial Office
Beijing Astronomical Observatory (BAO)
20A Datun Road
Chaoyang District
Beijing 100012
Telephone: (0)10-64853746

Telefax: (0)10-64859720
Electronic Mail: zjz@bao.ac.cn (Zhao Jing-zhi, Editor)
WWW: http://www.chjaa.org/
• Journal (6) (ISSN 1009-9271)
• Formerly "Acta Astrophysica Sinica"
City Reference Coordinates: 116°25'00"E 39°55'00"N

Delingha Radio Astronomy Station
• See now "Chinese Academy of Sciences, Purple Mountain Observatory, Qinghai Radio Observatory"

Hong Kong Astronomical Society (HKAS)
G.P.O. Box 2872
Hong Kong
or :
5/F, 315 Des Voeux Road West
Sai Ying Pun
Hong Kong
Telephone: (0)852-25474543
Telefax: (0)852-27152345
Electronic Mail: <userid>@hkas.org
WWW: http://www.hkas.org/
Founded: 1970
• Formerly "Hong Kong Amateur Astronomical Society (HKAAS)"
City Reference Coordinates: 114°10'00"E 22°17'00"N

Hong Kong Observatory (HKO)
134A Nathan Road
Kowloon
Hong Kong
Telephone: (0)852-29268200
Telefax: (0)852-27215034
Electronic Mail: mailbox@hko.gcn.gov.hk
WWW: http://www.weather.gov.hk/
Founded: 1883
Staff: 336
Activities: meteorological and geophysical services
Coordinates: 114°10'19"E 22°18'13"N
• Formerly "Royal Hong Kong Observatory (RHKO)"
City Reference Coordinates: 114°10'00"E 22°17'00"N

Hong Kong Space Museum, Planetarium
10 Salisbury Road
Tsim Sha Tsui
Kowloon
Hong Kong
Telephone: (0)852-27342722
Telefax: (0)852-23115804
Electronic Mail: spacem@space.lcsd.gov.hk
WWW: http://www.lcsd.gov.hk/hkspm
Founded: 1980
Staff: 135 (Museum)
Activities: sky shows * omnimax shows * lectures * education * exhibitions
Periodicals: (4) "Newsletter"; "Astro Calendar"
City Reference Coordinates: 114°10'00"E 22°17'00"N

Hong Kong University (HKU), Astronomy Club
Pokfulam Road
Hong Kong
Telephone: (0)852-28583657
Electronic Mail: suastro@hkusua.hku.hk
WWW: http://www.hku.hk/suastro/home41/
City Reference Coordinates: 114°10'00"E 22°17'00"N

Nanjing Astronomical Instruments Co. Ltd.
6-10 Hua Yuan Lu
Jiang Su 210042
Telephone: (0)25-5422281
Telefax: (0)25-5422281
Electronic Mail: hzhao@mail.nairc.com
WWW: http://www.nairc.com/
Founded: 1958
Staff: 330
Activities: research and manufacturing of astronomical instruments
City Reference Coordinates: 118°47'00"E 32°03'00"N (Nanjing)

Nanjing Normal University, Department of Physics
Nanjing 210097
City Reference Coordinates: 118°47'00"E 32°03'00"N

Nanjing University, Astronomy Department
22 Hankou Road
Nanjing 210093
Telephone: (0)25-3592882
(0)25-3593510
Telefax: (0)25-3302728
Electronic Mail: <userid>@nju.edu.cn
WWW: http://www.nju.edu.cn/
Founded: 1952
Staff: 40
Activities: high-energy astrophysics * solar physics * astrometry * ERP * celestial mechanics * history of astronomy * radio astronomy * QSOs * AGNs * galactic astronomy
Periodicals: "Journal of Nanjing University (Natural Sciences Edition)" (ISSN 0465-7926)
Coordinates: 118°46'07"E 32°02'26"N H60m
118°51'00"E 32°03'00"N H36m (solar tower)
City Reference Coordinates: 118°47'00"E 32°03'00"N

National Astronomical Observatories of the Chinese Academy of Sciences
• See "Chinese Academy of Sciences, National Astronomical Observatories"

Progress in Astronomy
c/o Editorial Office
80 Nandan Road
200030 Shanghai
Telephone: (0)21-64386191
Telefax: (0)21-64384618
Electronic Mail: twxjz@center.shao.ac.cn
Founded: 1983
Staff: 5
• Journal (4) (ISSN 1000-8349)
City Reference Coordinates: 121°28'00"E 31°14'00"N

Purple Mountain Observatory
• See "Chinese Academy of Sciences, Purple Mountain Observatory"

Qinghai Radio Observatory
• See "Chinese Academy of Sciences, Purple Mountain Observatory, Qinghai Radio Observatory"

Shaanxi Astronomical Observatory
• See "Chinese Academy of Sciences, Shaanxi Astronomical Observatory (CSAO)"

Shanghai Astronomical Observatory (SHAO)
• See "Chinese Academy of Sciences, Shanghai Astronomical Observatory (SHAO)"

Shanghai Institute of Electron Physics
• See "Shanghai University, Shanghai Institute of Electron Physics"

Shanghai Jiao Tong University (SJTU), Institute for Space Astrophysics
Department of Applied Physics
Shanghai 200030
WWW: http://www.sjtu.edu.cn/
City Reference Coordinates: 121°28'00"E 31°14'00"N

Shanghai Teachers University, Department of Physics
100 Guilin Road
Shanghai 200234
Telephone: (0)21-64322726
Telefax: (0)21-64321853
Electronic Mail: jzlu@shtu.edu.cn
WWW: http://www.shtu.edu.cn/
Founded: 1954
Staff: 25
Activities: black holes * cosmology * early universe * particle physics
City Reference Coordinates: 121°28'00"E 31°14'00"N

Shanghai University, Shanghai Institute of Electron Physics
Jia Ding
Shanghai 201800
Telephone: (0)21-59528602
Telefax: (0)21-59529932

Founded: 1966
Staff: 120
Activities: electronics * microwave engineering * remote sensing * ISM * VLBI * radioastronomical instrumentation
Periodicals: (4) "Journal of Shanghai University" (ISSN 0258-7041)
Coordinates: 121°15'00"E 31°23'00"N H23m
City Reference Coordinates: 121°15'00"E 31°23'00"N

Sky Observers' Association (SOA)
155 Fuk Wing Street
4th. Floor - Room 6
Shamshuipo
Hong Kong
Telephone: (0)852-3861658
Electronic Mail: skyweb@hk.super.net
WWW: http://www.skyobserver.org/
Founded: 1972
Membership: 410
Activities: observing * lectures * instrument making * photography * education * camps
Periodicals: "The Sky Observers"
Awards: "Astronomical Observation Award"
City Reference Coordinates: 114°10'00"E 22°17'00"N

University of Science and Technology of China (USTC), Center for Astrophysics
Hefei 230026
Telephone: (0)551-3601861
Telefax: (0)551-3631760
Electronic Mail: cfaoffice@ustc.edu.cn
WWW: http://www.cfa.ustc.edu.cn/
Founded: 1973
Staff: 16
Activities: research * education
City Reference Coordinates: 117°17'00"E 31°51'00"N

Urumqi Astronomical Observatory (UAO)
• See "Chinese Academy of Sciences, Urumqi Astronomical Observatory (UAO)"

World Data Center D for Astronomy
c/o National Astronomical Observatory of China
Datun Road 20A
Chao Yang District
Beijing 100012
Telephone: (0)10-64864582
Telefax: (0)10-64888734
Electronic Mail: ghf@bao.ac.cn (Guo Hongfong)
Founded: 1988
City Reference Coordinates: 116°25'00"E 39°55'00"N

Yunnan Observatory
• See "Chinese Academy of Sciences, Yunnan Observatory"

Colombia

Asociación de Astrónomos Aficionados de Colombia (ASAFI)
Apartado Aéreo 5614
Cali
or :
Calle 5a No. 68-24
Santiago de Cali
Telephone: (0)72-3304653
(0)72-3349894
Telefax: (0)72-3304653
(0)72-6808363
Electronic Mail: asafi@caliescali.com
Founded: 1991
Membership: 5
Activities: public outreach * observing * weekly meetings
Periodicals: (6) "Boletín de Observación"
Coordinates: 076°32'37"W 03°22'58"N H1,000m
City Reference Coordinates: 076°30'00"W 03°24'00"N (Cali)

Asociación de Astrónomos Autodidactos de Colombia (ASASAC)
Calle 116 40'64
Apartado 112
Bogotá 2 DE
or :
Carretera 32 # 71-A-31
Apartado Aéreo 59534
Bogotá 2 DE
Telephone: (0)2401-721
(0)1-2151850
Telefax: (0)1-2181894
Electronic Mail: dc34930@unalcol.bitnet
Founded: 1965
Membership: 35
Activities: telescope making * observing (occultations, variable stars, comets) * ephemerides * education * planetarium
Sections: (commissions) Celestial Mechanics * Astrophotography * Instruments * Variable Stars * Comets and Meteors * Occultations * Planets * Galactic Astronomy * Ephemerides
Periodicals: "Quasar"
Coordinates: 074°05'00"W 04°39'00"N
City Reference Coordinates: 074°05'00"W 04°39'00"N

Asociación Colombiana de Estudios Astronómicos (ACDA)
Carrera 50 No. 27 -70
Módulo 5 Oficina 301
Bogotá
or :
Apartado Aéreo 47731
Bogotá 2 DE
or :
Carretera 11 # 82-60
Apartamento 202
Bogotá 8 DE
Telephone: (0)1-2563030
(0)2-225525
(0)2-563030
Electronic Mail: acda_colombia@yahoo.com
WWW: http://www.geocities.com/acda_colombia
Founded: 1986
Membership: 44
Activities: weekly meetings * education * lectures * telescope making
Periodicals: (4) "Kajuyali"
Coordinates: 073°48'00"W 05°06'00"N H2,584 (Suesca)
073°54'00"W 05°21'00"N H2,980 (Carmen de Carupa)
073°34'00"W 05°38'00"N H2,143 (Villa de Leyva)
City Reference Coordinates: 073°48'00"W 05°06'00"N

Instituto Colombiano de Normas Técnicas (ICONTEC)
Carrera 37 N° 52-95
Apartado 14237
Santafé de Bogotá
Telephone: (0)1-3150377
Telefax: (0)1-2221435
Electronic Mail: sicontec@col1.telecom.com.co

WWW: http://www.icontec.org.co/
Founded: 1963
Periodicals: (1) "Memoria Anual"; "Normas y Calidad", "Boletín Informativo"
City Reference Coordinates: 074°05'00"W 04°39'00"N

Instituto de Hidrología, Meteorología y Estudios Ambientales (IDEAM)
Apartado Aéreo 20032
Bogotá 1 DE
or :
Diagonal 97 N°17-60
Santa Fé de Bogotá DE
Telephone: (0)1-2836937
　　　　　(0)1-2860658
　　　　　(0)1-6356007
Telefax: (0)1-2842402
　　　　(0)1-2860658
Electronic Mail: <userid>@ideam.gov.co
WWW: http://www.ideam.gov.co/
City Reference Coordinates: 074°05'00"W 04°39'00"N

Observatorio Astronómico Nacional (OAN), Bogotá
Apartado Aéreo 2584
Bogotá 1 DE
or :
Carrera 8
Calle 8
Bogotá
Telephone: (0)1-3423786
　　　　　(0)1-3165222
Telefax: (0)1-3165383
Electronic Mail: observat@ciencias.ciencias.unal.edu.co
WWW: http://argos.observatorio.unal.edu.co/
Founded: 1803
Staff: 10
Activities: education * astrometry * celestial mechanics * open clusters * cosmology * stellar structures * history of astronomy
Periodicals: "Publicaciones del Observatorio Astronómico Nacional"; (1) "Anuario del Observatorio Astronómico Nacional"
Coordinates: 074°04'54"W 04°35'54"N H2,640m
City Reference Coordinates: 074°05'00"W 04°39'00"N

Croatia

Astronomical Astronautical Society Zadar (AASZ)
(Astronomsko Astronautičko Društvo Zadar)
Brne Karnarutica 2
HR-23000 Zadar
Telefax: (0)23-314052
Electronic Mail: aad@zadarnet.hr
WWW: http://pubwww.srce.hr/astro/zadaren.html
Founded: 1980
Membership: 28
City Reference Coordinates: 015°14'00"E 44°07'00"N

Astronomical Society Ivan Štefek
Školska bb
HR-44320 Kutina
Telephone: (0)44-687952
Telefax: (0)44-687952
Electronic Mail: astro-kutina@sk.tel.hr
WWW: http://www.angelfire.com/sc2/ADIvanS/
City Reference Coordinates: 016°48'00"E 45°29'00"N

Astronomical Society Leo Brenner
Zabrebačka 2/II kat
HR-51550 Mali Lošinj
Telephone: (0)51-233203
Electronic Mail: dorian.bozicevic@aad.hr (Dorian Božičevic, Vice President)
            bozicevic.dorian@excite.com
Founded: 1994
Membership: 35
Activities: observing * popularization * education
Periodicals: "foton" (ISSN 1331-5765)
Coordinates: 014°31'24"E 44°27'48"N H100m (Veli Lošinj)
         014°28'42"E 44°29'30"N H200m (Mali Lošinj)
         014°27'48"E 44°30'12"N H90m (Mali Lošinj)
City Reference Coordinates: 014°25'00"E 44°35'00"N (Lošinj)

Astronomical Society Pitomaca
Trg Kralja Tomislava 9
HR-43405 Pitomača
Telephone: (0)33-782217
Founded: 1981
Membership: 100
City Reference Coordinates: 017°14'00"E 45°57'00"N

Astronomical Society "Gea X"
Maruliceva 5
HR-55000 Slavonski Brod
Telephone: (0)35-441743
Membership: 90
City Reference Coordinates: 018°00'00"E 45°09'00"N

Croatian Physical Society
Bijenička cesta 32
P.O. Box 162
HR-10000 Zagreb
Telephone: (0)41-432480
Telefax: (0)41-432525
Electronic Mail: hfd@phy.hr
            <userid>@hfd.hr
WWW: http://www.hfd.hr/hfd/
Membership: 45
City Reference Coordinates: 015°58'00"E 45°48'00"N

Hvar Observatory
• See "University of Zagreb, Faculty of Geodesy, Hvar Observatory"

State Hydrometeorological Institute of Croatia
Grič 3
HR-41100 Zagreb
Telephone: (0)1-421222
Telefax: (0)1-278703

WWW: http://madhz.dhz.hr/
Founded: 1947
Staff: 450
Activities: official meteorological and hydrological service
City Reference Coordinates: 015°58'00"E 45°48'00"N

State Office for Standardization and Metrology (DZNM)
Ulica grada Vukovara 78
HR-10000 Zagreb
Telephone: (0)1-6133444
Telefax: (0)1-536668
Electronic Mail: <userid>@dznm.hr
WWW: http://www.dznm.hr/
City Reference Coordinates: 015°58'00"E 45°48'00"N

University of Zagreb, Faculty of Geodesy, Hvar Observatory
Kaciceva 26
HR-10000 Zagreb
or :
P.O. Box 18
HR-21450 Hvar
Telephone: (0)1-4561222
          (0)21-525966
Telefax: (0)21-741427
          (0)1-4828081
Electronic Mail: <userid>@geodet.geof.hr
WWW: http://pubwww.srce.hr/geo/hrv/gf.htm
Founded: 1972
Activities: solar physics * photoelectric photometry of variable stars * observing station of the University of Zagreb, Faculty of Geodesy
Periodicals: (1) "Hvar Observatory Bulletin" (ISSN 0351-2657, circ.: 400)
Coordinates: 016°01'18"E 45°49'30"N H146m (Zagreb)
          016°27'00"E 43°10'00"N H245m (Hvar)
City Reference Coordinates: 015°58'00"E 45°48'00"N (Zagreb)
          016°27'00"E 43°10'00"N (Hvar)

University of Zagreb, Institute of Physics
Bijenička 46
P.O.Box 304
HR-10000 Zagreb
Telephone: (0)1-4605555
Telefax: (0)1-4680336
WWW: http://www.phys.hr/
City Reference Coordinates: 015°58'00"E 45°48'00"N

Visnjan Observatory
Ul. Istarska 5
HR-52463 Visnjan
Telephone: (0)52-449212
Telefax: (0)52-449212
Electronic Mail: korado@astro.hr (Korado Korlevic)
WWW: http://www.astro.hr/
          http://www.visnjan.hr/
Founded: 1976
Membership: 32
Staff: 1
Activities: meteors * minor planets * instrumentation * education
Periodicals: (12) "Nebeske Krijesnice" (ISSN 1330-4410); (1) "Acta Astronomica Visnianiensis" (ISSN 1330-2620)
Coordinates: 013°43'26"E 45°16'38"N H243m (Visnjan – IAU Code 120)
          013°46'00"E 45°11'00"N H297m (Rusnjak)
City Reference Coordinates: 013°43'00"E 45°16'00"N

Cuba

Academia de Ciencias de Cuba (ACC)
Capitolio Nacional
La Habana 12400
Telephone: (0)6-8914
Electronic Mail: acc@ceniai.cu
WWW: http://www3.cuba.cu/ciencia/acc/index0.htm
Founded: 1861
City Reference Coordinates: 082°22'00"W 23°08'00"N

Comité Estatal de Normalización (CN)
Egido 610 entre Gloria y Apodaca
La Habana 10100
Telephone: (0)7-621503
Telefax: (0)7-338048
Electronic Mail: ncnorma@ceniai.inf.cu
Founded: 1976
Staff: 300
Activities: standardization * metrology * quality assurance
Periodicals: (2) "Normalización" (ISSN 0138-8118)
City Reference Coordinates: 082°22'00"W 23°08'00"N

Instituto de Geofísica y Astronomía (IGA)
Calle 212 No. 2906, entre 29 y 31
La Coronela
Lisa
La Habana 11600
Telephone: (0)218435
 (0)210644
 (0)214331
Telefax: (0)339497
Founded: 1964
Staff: 22
Activities: solar physics * International Sun Service daily observations * computing * astrometry
Periodicals: (1) "Datos Astronómicos para Cuba" (ISSN 0864-0645), "Ciencias de la Tierra y del Espacio" (ISSN 0138-6026)
Coordinates: 082°28'00"W 23°04'00"N H10m (La Habana Radioastronomical Station)
 082°23'00"W 22°57'00"N H160m (La Habana Optical Station)
City Reference Coordinates: 082°28'00"W 23°04'00"N

Instituto de Meteorología (INSMET)
La Habana
Electronic Mail: meteoro@ceniai.inf.cu
WWW: http://www.met.inf.cu/
Founded: 1965
City Reference Coordinates: 082°28'00"W 23°04'00"N

Sociedad Meteorológica de Cuba
Apartado Postal 4279
La Habana 11400
Telephone: 730-8996
Telefax: 733-3681
Electronic Mail: lulu@met.inf.cu
WWW: http://www.met.inf.cu/sometcub/default.htm
Membership: 500
Periodicals: "Boletín de SOMETCUBA" (ISSN 1025-921x)
City Reference Coordinates: 082°28'00"W 23°04'00"N

Czech Republic

Academy of Sciences of the Czech Republic
Národní 3
117 20 Praha 1
Telephone: (0)2-21403111
Telefax: (0)2-24240572
Electronic Mail: info@cas.cz
WWW: http://www.cas.cz/
Founded: 1992
Staff: 6000
Activities: research * education
City Reference Coordinates: 014°26'00"E 50°05'00"N

Academy of Sciences of the Czech Republic, Astronomical Institute, Ondřejov
Fričova 298
251 65 Ondřejov
Telephone: (0)204-649201
(0)204-620116
Telefax: (0)204-620110
(0)204-620117
Electronic Mail: asu@asu.cas.cz
WWW: http://www.asu.cas.cz/
Founded: 1953
Staff: 119
Activities: stellar, solar and meteor astronomy * satellite dynamics * celestial mechanics
Periodicals: "Time and Latitude" (ISSN 1210-0463), "Solar Radio Data"
Coordinates: 014°47'00"E 49°54'40"N H528m (Ondřejov Obs. – IAU Code 557)
City Reference Coordinates: 014°47'00"E 49°55'00"N

Academy of Sciences of the Czech Republic, Astronomical Institute, Praha
Boční II 1401
141 31 Praha 4
Telephone: (0)2-67103111
Telefax: (0)2-769023
Electronic Mail: asu@asu.cas.cz
WWW: http://www.asu.cas.cz/
Founded: 1953
Staff: 133
City Reference Coordinates: 014°26'00"E 50°05'00"N

Academy of Sciences of the Czech Republic, Institute of Physics
Na Slovance 2
180 40 Praha 8
Telephone: (0)2-66052660
(0)2-66051111
Telefax: (0)2-821227
(0)2-8584569
Electronic Mail: <userid>@fzu.cz
fzu@cas.cz
WWW: http://www.fzu.cz/
Founded: 1954
City Reference Coordinates: 014°26'00"E 50°05'00"N

Academy of Sciences of the Czech Republic, Institute of Plasma Physics
Za Slovankou 3
P.O. Box 17
182 21 Praha 8
Telephone: (0)2-66052052
(0)2-66051111
Telefax: (0)2-8586389
Electronic Mail: ipp@ipp.cas.cz
<userid>@ipp.cas.cz
WWW: http://www.ipp.cas.cz/
Founded: 1959
Staff: 91
Activities: basic and applied research in high-temperature and low-temperature plasma physics (experimental and theoretical)
City Reference Coordinates: 014°26'00"E 50°05'00"N

Academy of Sciences of the Czech Republic, Optical Development Workshop
Skálova 89
511 01 Turnov
Telephone: (0)436-322622

Telefax: (0)436-322913
Electronic Mail: cas-tur@telecom.cz
WWW: http://www.optikavod.cz/
Founded: 1965
Staff: 20
Activities: research * development * manufacturing * optical components * glass * crystals * X-rays * thin films
City Reference Coordinates: 015°10'00"E 50°35'00"N

Association of Observatories and Planetariums
c/o Eva Marková (Chairman)
Úpice Observatory
P.O. Box 8
542 32 Úpice
or :
Královská Obora 233
170 21 Praha 7
Telephone: (0)439-881731
Telefax: (0)439-881289
Electronic Mail: markova@obsupice.cz (Eva Marková, Chairman)
WWW: http://www.sdruzeni.hvezdarna.cz/
Founded: 1991
Membership: 24 (observatories and planetariums)
Activities: protecting common interests of observatories and planetariums * educating staff of public observatories * advising
City Reference Coordinates: 016°01'00"E 50°30'00"N (Úpice)
014°26'00"E 50°05'00"N (Praha)

Astro Klub Kostkov
Návsí 645
739 92 Jablunkov
Telephone: (0)659-932623
Telefax: (0)659-25877
Electronic Mail: honza@elanor.sci.muni.cz
WWW: http://www.sci.muni.cz/~honza
http://monoceros.physics.muni.cz/~honza/asko.htm
Founded: 1994
Membership: 41
Activities: observing * lectures * instrumentation * photography * CCDs * interplanetary matter * symbiotic stars * software
City Reference Coordinates: 018°47'00"E 49°35'00"N

Astronomers in Lelekovice
Lelekovice 393
664 31 Lelekovice
Telephone: (0)5-41232105
Electronic Mail: hornoch@astro.sci.muni.cz (Kamil Hornoch)
xplsek01@dcse.fee.vutbr.cz
WWW: http://astro.sci.muni.cz/lelek/
Founded: 1996
Membership: 2
Activities: observing (comets, novae, SNs, meteors)
Coordinates: 016°39'18"E 49°21'15"N H365m
City Reference Coordinates: 016°39'00"E 49°21'00"N

Benátky nad Jizerou Popular Observatory
c/o Radovan Kováč
Platanová 647
294 71 Benátky nad Jizerou
Telephone: (0)326-762664
Founded: 1980
Staff: 3
Activities: popularization * observing
Coordinates: 014°50'09"E 50°17'24"N H210m
City Reference Coordinates: 014°51'00"E 50°17'00"N

Boskovice Observatory
Kulturní Zařízení Města Boskovice
Kpt Jaroše 15
680 01 Boskovice
Telephone: (0)501-453544
Telefax: (0)501-453253
Electronic Mail: kopecky@fch.vutbr.cz (Pavel Kopecký)
nemecm@post.cz
Founded: 1960
Staff: 5
Activities: public observatory
Coordinates: 016°29'00"E 49°30'00"N H370m

Hradec Králové Observatory and Planetarium
Zámeček 456/30
500 08 Hradec Králové
Telephone: (0)49-5264087
Telefax: (0)49-5267952
Electronic Mail: astrohk@astrohk.cz
WWW: http://www.astrohk.cz/
http://www.astrohk.cz/indexen.html (English page)
Founded: 1947
Staff: 13
Activities: education (schools, public) * shows * lectures * observing
Coordinates: 015°50'21"E 50°10'38"N H287m (IAU Code 048)
City Reference Coordinates: 015°50'00"E 50°13'00"N

Jeseník School Observatory
Postovní 115
790 01 Jeseník
Telephone: (0)645-402408
Founded: 1965
Activities: lectures * observing * photography
Coordinates: 017°12'00"E 50°14'00"N H450m
City Reference Coordinates: 017°12'00"E 50°14'00"N

Jindřichuv Hradec Public Observatory
Gymnázium V. Nováka 333/II
377 01 Jindřichuv Hradec
Telephone: (0)331-22564
Telefax: (0)331-362248
Electronic Mail: strobl@port.troja.mff.cuni.cz
Founded: 1961
Membership: 15
Activities: popularization
Coordinates: 015°00'04"E 49°07'48"N H492m
City Reference Coordinates: 015°00'00"E 49°09'00"N

Johann Palisa Observatory and Planetarium
c/o VSB Technical University
tr. 17. Listopadu 15
708 33 Ostrava-Poruba
Telephone: (0)596-911005
Telefax: (0)596-911009
Electronic Mail: thomas.graf@vsb.cz
WWW: http://www.vsb.cz/planet
Founded: 1980
Staff: 11
Activities: education * observing
Periodicals: (6) "Planetárium"
Coordinates: 018°08'49"E 49°50'18"N H276m
City Reference Coordinates: 018°17'00"E 49°50'00"N

Karlovy Vary City Observatory
(Hvězdárna Úřadu Města Karlovy Vary)
c/o Miroslav Spurný
Moskevská 20
360 20 Karlovy Vary
Telephone: (0)17-3225772
Telefax: (0)17-3222913 (attn: Hvězdárna p. Spurný)
Electronic Mail: hvezdarna.kv@email.cz
WWW: http://astropatrola.kvary.cz/
Founded: 1963
Staff: 1
Activities: popularization * observing (variable stars, comets, AGNs)
Coordinates: 012°54'29"E 50°12'52"N H615m (Hurky)
City Reference Coordinates: 012°52'00"E 50°11'00"N

Klet Observatory
Zátkovo Nábřeží 4
370 01 České Budějovice
Telephone: (0)337-711242
Electronic Mail: klet@klet.cz
klet@asteroid.cz
WWW: http://www.klet.cz/
Founded: 1957

Staff: 4
Activities: minor planets * near-Earth objects * comets * astrometry
Coordinates: 014°17'17"E 48°51'46"N H1,070m (IAU Code 046)
City Reference Coordinates: 014°28'00"E 48°59'00"N

Liberec Astronomy Club
c/o Jirí Kapras
P.O. Box 24
463 12 Liberec 25
Telephone: (0)48-29557
(0)48-24553
(0)48-5133155
Electronic Mail: jkapras@volny.cz
Founded: 1963
Membership: 15
Activities: variable stars
Coordinates: 015°03'50"E 50°50'51"N H425m
015°04'10"E 50°42'00"N H568m
City Reference Coordinates: 015°03'00"E 50°46'00"N

Masaryk University, Department of Theoretical Physics and Astrophysics
Kotlářská 2
611 37 Brno
Telephone: (0)5-41129251
Telefax: (0)5-41211214
Electronic Mail: <userid>@physics.muni.cz
WWW: http://astro.sci.muni.cz/
http://www.physics.muni.cz/
http://www.physics.muni.cz/mb/mb.htm
Founded: 1945
Staff: 12
Activities: education * relativity * quantum field theory * gravitation * spectroscopy * CCD photometry * close binary stars * carbon stars
Coordinates: 016°35'18"E 49°12'15"N H310m (Kraví Hora)
City Reference Coordinates: 016°37'00"E 49°12'00"N

Meduza
c/o Petr Sobotka
N. Copernicus Observatory and Planetarium
Kraví Hora 2
616 00 Brno
Telephone: (0)604-126547
Electronic Mail: sobotka@meduza.org
WWW: http://www.meduza.org/
Founded: 1996
Membership: 61
Activities: variable stars * lightcurves
Periodicals: (5) "Meduza Circular"
Awards: "Meduza Certificates"
Coordinates: 016°40'00"E 49°20'00"N H300m
City Reference Coordinates: 016°37'00"E 49°12'00"N

Most Planetarium
P.O. Box 138
434 01 Most
or :
c/o Cultural Institution of Miners
Náměstí VMS c. 4
434 25 Most
Telephone: (0)35-796381
(0)35-42256
(0)35-769384
WWW: http://www.mumost.cz/english/turisti/planetarium/eplanet.htm
Founded: 1984
Activities: shows for schools and public
City Reference Coordinates: 013°39'00"E 50°32'00"N

Olomouc Observatory
c/o Eva Kobzova
Department of Theoretical Physics
Palacky University
Trida Svobody 26
771 46 Olomouc
Telephone: (0)68-411460
Electronic Mail: kobzova@prfnw.upol.cz
Activities: observing (Sun, meteors, occultations) * education * public parties

Coordinates: 017°13'06"E 49°34'14"N
City Reference Coordinates: 017°16'00"E 49°36'00"N

Ondřejov Observatory
• See "Czech Academy of Sciences, Astronomical Institute"

Ostrava Amateur Society
(Amatérská Prohlídka Oblohy - APO)
c/o Marek Kolasa
Dr. Martínka 1
700 30 Ostrava-Hrabuvka
Telephone: (0)604-740528
Electronic Mail: apo@seznam.cz
WWW: http://www.ian.cz/APO/
City Reference Coordinates: 018°17'00"E 49°50'00"N

Palacky University, Department of Theoretical Physics
Trida Svobody 26
771 46 Olomouc
Telephone: (0)68-5222451 x374
Telefax: (0)68-5225737
Electronic Mail: <userid>@risc.upol.cz
Founded: 1954
Staff: 3
Activities: solid state theory, optical activity of crystals, disequilibrium thermodynamics, quantum physics
City Reference Coordinates: 017°16'00"E 49°36'00"N

Planetarium Praha
• See "Praha Observatory and Planetarium, Planetarium Praha"

Plzeň Observatory and Planetarium
Dráhy 11
318 03 Plzeň
Telephone: (0)19-288400
Electronic Mail: hvezdarna@mmp.plzen-city.cz
Founded: 1954
Staff: 8
Activities: education * occultations
Periodicals: (12) "Observatory and Planetarium Reporter"; "Occultation Express"
Coordinates: 013°26'24"E 49°41'58"N H405m
City Reference Coordinates: 013°25'00"E 49°45'00"N

Praha Observatory and Planetarium, Planetarium Praha
Královská Obora 233
170 21 Praha 7
Telephone: (0)2-371746
         (0)2-377069
         (0)2-377746
Telefax: (0)2-375970
Electronic Mail: planetarium@planetarium.cz
WWW: http://www.planetarium.cz/
Founded: 1960
Staff: 30
Activities: school programmes * shows * education * exhibitions
City Reference Coordinates: 014°26'00"E 50°05'00"N

Praha Observatory and Planetarium
Petřín 205
118 46 Praha 1
Telephone: (0)2-57320540 (Observatory)
         (0)2-33376452 (Planetarium)
Telefax: (0)2-33376437
Electronic Mail: observat@ms.anet.cz
WWW: http://planet.infoserver.cz/
Founded: 1928
Membership: 45
Activities: popularization * education * publishing * ISM * maps * atlases * lunar occultations
Periodicals: (1) "Hvězdářská Ročenka" (ISSN 0373-8280)
Coordinates: 014°23'58"E 50°04'56"N H327m (Stefanik Observatory, Petrín)
            014°28'00"E 50°08'00"N H325m (Ďáblice - see separate entry)
            014°17'17"E 48°51'46"N H1,068m (Klet)
City Reference Coordinates: 014°26'00"E 50°05'00"N

Praha Observatory and Planetarium, Ďáblice Observatory
182 00 Praha 8

Telephone: (0)2-83910644
Electronic Mail: dabliceobs@planetarium.cz
WWW: http://www.planetarium.cz/
Founded: 1956
Staff: 5
Activities: occultations * variable stars
Coordinates: 014°28'00"E 50°08'00"N H325m
City Reference Coordinates: 014°26'00"E 50°05'00"N (Praha)

Přerov Astronomical Club
Hvězdárna
Macharova 21
750 02 Přerov
Telephone: (0)641-242361
Founded: 1961
Membership: 28
Activities: education * observing (occultations, meteors)
Periodicals: (1) "Annual Report"
Coordinates: 017°28'54"E 49°26'51"N H230m
City Reference Coordinates: 017°27'00"E 49°27'00"N

Prostějov Popular Observatory
Kolářovy Sady 3348
796 01 Prostějov
Telephone: (0)508-24130
Electronic Mail: hvezdarna.pv@mbox.vol.cz
WWW: http://www.oku-pv.cz/hvezdarna
Founded: 1962
Staff: 5
Activities: variable stars * solar activity * photography of planets
Periodicals: (12) "Zpravodaj"
Coordinates: 017°05'58"E 49°28'08"N H225m
City Reference Coordinates: 017°07'00"E 49°29'00"N

Research Institute of Geodesy, Topography and Cartography
250 66 Zdiby 98
Telephone: (0)2-6857351
Telefax: (0)2-6857056
Electronic Mail: vugtk@vugtk.cz
WWW: http://panurgos.fsv.cvut.cz/~odis/vugtk.shtml
Founded: 1954
Staff: 40
Activities: geodetic and cartographic research * astrometry
Periodicals: "Proceedings of Research Works"
Coordinates: 015°00'00"E 50°00'00"N
City Reference Coordinates: 015°00'00"E 50°00'00"N

Rokycany Astronomical Observatory
Voldusska 721/II
337 11 Rokycany
Telephone: (0)181-722622
Electronic Mail: halir@oku-ro.cz (Karel Halír, Director)
WWW: http://www.oku-ro.cz/hvezdarna/
Founded: 1947
Staff: 5
Activities: Sun * occultations
Coordinates: 013°36'16"E 49°45'07"N H400m
City Reference Coordinates: 013°36'00"E 49°45'00"N

Rtyně v Podkrkonoší Observatory
Soukromá Hvězdárna 143
542 33 Rtyně v Podkrkonoší
Telephone: (0)439-787384
Electronic Mail: modr1@volny.cz
Founded: 1980
Staff: 2
Activities: private observatory * astronomical optics * mechanical manufacturing
City Reference Coordinates: 016°01'00"E 50°30'00"N (Úpice)

Teplice Observatory and Planetarium
Koperníkova 3062
415 01 Teplice
Telephone: (0)417-28507 (Observatory)
         (0)417-576571 (Planetarium)
Electronic Mail: planettp@tep.cesnet.cz

WWW: http://www.teplice-city.cz/hap
City Reference Coordinates: 013°49'00"E 50°40'00"N

Uherský Brod Cultural House Observatory
c/o Rostislav Rajchl
Praksická 2222
688 11 Uherský Brod
Telephone: (0)633-634690
Founded: 1961
Membership: 15
Activities: solar activity * education * popularization * history of astronomy * archaeoastronomy
Coordinates: 017°38'51"E 49°02'18"N H310m
City Reference Coordinates: 017°39'00"E 49°02'00"N

Union of Czech Mathematicians and Physicists, Physics Section
Na Slovance 2
180 40 Praha 8
Telephone: (0)2-66052910
(0)2-825459
Telefax: (0)2-8586389
(0)2-8584569
Electronic Mail: krejci@ipp.cas.cs
WWW: http://www-hep.fzu.cz/~www/
http://www.ujf.cas.cz:8080/fvs/fvs.html
Founded: 1968
Membership: 763
Activities: organizing activities of physicists in the Czech Republic * seminars * conferences * promoting knowledge of physics in the society
Periodicals: (6) "Bulletin of the Physics Section" (together with the "Czechoslovak Journal of Physics" - ISSN 0009-0700)
Awards: "Gold Medal", "Silver Medal"
City Reference Coordinates: 014°26'00"E 50°05'00"N

Úpice Observatory
P.O. Box 8
542 32 Úpice
Telephone: (0)439-881731
Telefax: (0)439-881289
Electronic Mail: markova@obsupice.cz (Eva Marková, Director)
WWW: http://www.obsupice.cz/
Founded: 1959
Staff: 13
Activities: observing (solar activity, eclipses, comets, minor planets) * popularization * optical counterparts of gamma-ray bursts * meteorology * seismology * air pollution
Periodicals: (4) "Kvart" (ISSN 1211-2453); "Proceedings", "Circulars"
Coordinates: 016°00'44"E 50°30'27"N H416m (IAU Code 166)
City Reference Coordinates: 016°01'00"E 50°30'00"N

Valašské Meziříčí Astronomical Observatory
Vsetínská 78
757 01 Valašské Meziříčí
Telephone: (0)651-611928
Telefax: (0)651-611928
Electronic Mail: libor.lenza@vm.inext.cz
hvezdhvm@vm.inext.cz
WWW: http://www.inext.cz/hvezdarna/
Founded: 1955
Staff: 11
Activities: Sun * occultations * symbiotic stars * education * popularization
Periodicals: "Leaflets"
Coordinates: 017°58'32"E 49°27'51"N H338m
City Reference Coordinates: 017°58'00"E 49°28'00"N

Veselí nad Moravou Observatory
Benátky 32
698 01 Veselí nad Moravou
Telephone: (0)631-322614
Electronic Mail: micek@vmb.cz
hvezd.veseli@es-servis.cz
Founded: 1960
Activities: observing & photography (meteors, comets, occultations, sunspots, planets) * popularization of astronomy and astronautics
Coordinates: 017°22'17"E 48°57'17"N H163m
City Reference Coordinates: 017°22'00"E 48°58'00"N

Vlašim Astronomical Society (VAS)
B. Martinů 1341
258 01 Vlašim
Telephone: (0)303-42923
Telefax: (0)303-45169
Electronic Mail: jan.urban@csop.cz (Jan B. Urban, President)
              vas@vas.cz
WWW: http://www.vas.cz/
Founded: 1991
Membership: 53
Activities: solar activity * meteorology * gamma-ray burst sources * occultations * popularization * audio-visual programs
Periodicals: "Sunspot Report", "SEA Report", "Meteorological Report"
Coordinates: 014°53'47"E 49°41'37"N H423m
City Reference Coordinates: 014°54'00"E 49°42'00"N

Vsetín Observatory
Jabloňová 231
755 11 Vsetín
Telephone: (0)657-611819
Electronic Mail: hvezdarna@vs.inext.cz
WWW: http://www.inext.cz/hvezdarna.vsetin
City Reference Coordinates: 018°00'00"E 49°20'00"N

Ždánice Public Observatory
Lovecká 678
696 32 Ždánice
Telephone: (0)629-633356
Founded: 1965
Staff: 2
Activities: variable stars
Coordinates: 017°02'27"E 49°03'58"N H226m
City Reference Coordinates: 017°02'00"E 49°04'00"N

Žebrák Popular Observatory
Lidova Hvězdárna
267 53 Žebrák
Founded: 1951
Staff: 3
Activities: observing (Moon, Sun, stars) * photography
Coordinates: 013°53'56"E 49°52'28"N H370m
City Reference Coordinates: 013°54'00"E 49°53'00"N

Zlín Astronomical Observatory
Hvězdárna
760 01 Zlín
Telephone: (0)67-36045
Electronic Mail: coufal@batahospital.iqnet.cz (Zdeněk Coufal, Chairman)
WWW: http://www.zas.cz/
Founded: 1993
Membership: 47
Activities: variable stars * occultations
Periodicals: (6) "Zpravodaj Zorné Pole"
Coordinates: 017°41'53"E 49°13'08"N H333m
City Reference Coordinates: 017°40'00"E 49°14'00"N

Denmark

Ålborg Amateur Radio Astronomy Observatory (AARAO)
Kalmanparken 20
DK-9200 Frejlev
Electronic Mail: i5sp@civil.auc.dk
steffen.petersen@civil.auc.dk
WWW: http://www.protein.auc.dk/~i5sp/radioastronomy/AARAO.html
City Reference Coordinates: 009°56'00"E 57°03'00"N (Ålborg)

Amateur Astronomers of Eastern Jutland
(Østjyske Amatør Astronomer - ØAA)
c/o Elisabeth Siegel
Agertoften 2
DK-8340 Malling
Telephone: (0)86933458
Telefax: (0)86930015
Electronic Mail: wsiegel@daimi.au.dk
WWW: http://www.imf.au.dk/~kls/astro/oaa.html
Founded: 1977
Membership: 45
Activities: observing * instrumentation * meetings * computing
Periodicals: "Stjerneskuddet"
City Reference Coordinates: 010°12'00"E 56°02'00"N

Århus University, History of Science Department
Ny Munkegade
Building 521
DK-8000 Århus C
Telephone: (0)89423512
Telefax: (0)89423510
Electronic Mail: ievhlk@ifa.aau.dk
WWW: http://www.dfi.aau.dk/ievh/home.htm
Founded: 1965
Staff: 10
Activities: research in the history of science from Babylonian to modern times * museum for the history of science * archive with Hertzsprung and Strömgren documents
Periodicals: (4) "Centaurus" (ISSN 0008-8994)
City Reference Coordinates: 010°13'00"E 56°09'00"N

Århus University, Institute of Physics and Astronomy
Bygning 520
DK-8000 Århus C
Telephone: (0)86-128899
Telefax: (0)86-120740
Electronic Mail: <userid>@ifa.au.dk
WWW: http://www.ifa.au.dk/
http://www.obs.aau.dk/
Founded: 1911
Activities: minor planets * astro- and helioseismology * stellar abundances * CCD photometry and spectroscopy * formation and structure of galaxies * cosmology
Coordinates: 010°14'00"E 56°07'42"N H46m (Ole Rømer Obs. – IAU Code 155)
City Reference Coordinates: 010°13'00"E 56°09'00"N

Association Internationale de Géodésie (AIG)
• See "International Association of Geodesy (IAG)"

Astro I/S
Nøvlingvej 130
DK-9260 Gistrup
Telephone: (0)98-134396
Telefax: (0)98-164226
WWW: http://www.2astro.dk/
Activities: distributing astronomical instruments and accessories
City Reference Coordinates: 009°56'00"E 57°03'00"N (Ålborg)

Astronomical Society of South Zealand
(Astronomisk Forening for Sydsjaelland - AFS)
c/o Ove Larsen (President)
Fasanvej 2
DK-4370 Store Merløse
Telephone: (0)57801812

Telefax: (0)57801812
Electronic Mail: otlastro@post.tele.dk
Founded: 1969
Membership: 50
Activities: education * observing * CCD imaging
Coordinates: 011°45'00"E 55°13'10"N H63m
City Reference Coordinates: 011°44'00"E 55°33'00"N

Astronomisk Forening for Sydsjaelland (AFS)
• See "Astronomical Society of South Zealand"

Astronomisk Forlag
c/o Tycho Brahe Planetarium
Gl. Kongevej 10
DK-1610 København V
Telephone: (0)1-33144666
Telefax: (0)1-33142888
Electronic Mail: tycho@tycho.dk
WWW: http://www.tycho.dk/
Founded: 1978
• Publisher
City Reference Coordinates: 012°35'00"E 55°40'00"N

Astronomisk Selskab i Danmark
• See "Danish Astronomical Association"

Bohr Institute (NBI) (Niels_)
• See "Copenhagen University, Niels Bohr Institute for Astronomy, Physics and Geophysics (NBIfAFG), Niels Bohr Institute (NBI)"

Brahe Planetarium (Tycho_)
• See "Tycho Brahe Planetarium"

Center for Planetary Science
Juliane Maries Vej 30
DK-2100 København Ø
Telephone: (0)35325999
Telefax: (0)35325989
Electronic Mail: info@planetcenter.dk
WWW: http://www.planetcenter.dk/
Founded: 2001
Activities: interdisciplinary research (involving astronomy, physics, geophysics, and space instrumentation to geology and biology)
City Reference Coordinates: 012°35'00"E 55°40'00"N

Copenhagen Astronomical Society
(Københavns Astronomiske Forening - KAF)
c/o Claus Jensen (President)
Måløvvang 35 7B
DK-2760 Måløv
or :
c/o Gert Larsen (Secretary)
Auroravej 58
DK-2610 Rodovre
Telephone: (0)44-65-17-66
         (0)36-722011
Telefax: (0)36-72-20-11
WWW: http://www2.dk-online.dk/users/hpersson/kaf.htm
Founded: 1978
Membership: 175
Activities: general astronomy * all aspects of amateur astronomy
City Reference Coordinates: 012°35'00"E 55°40'00"N (Copenhagen)

Copenhagen Astronomical Observatory
• See "Copenhagen University, Niels Bohr Institute for Astronomy, Physics and Geophysics (NBIfAFG), Astronomical Observatory"

Copenhagen University, Niels Bohr Institute for Astronomy, Physics and Geophysics (NBIfAFG)
Blegdamsvej 17
DK-2100 København Ø
Telephone: (0)35325241
Telefax: (0)35431087
Electronic Mail: <userid>@nbi.dk
WWW: http://www.nbi.dk/NBIfAFG/
Founded: 1993

Staff: 145
City Reference Coordinates: 012°35'00"E 55°40'00"N

Copenhagen University, Niels Bohr Institute for Astronomy, Physics and Geophysics (NBIfAFG), Astronomical Observatory
Juliane Maries Vej 30
DK-2100 København Ø
Telephone: (0)35323999
Telefax: (0)35323989
Electronic Mail: <userid>@astro.ku.dk
WWW: http://www.astro.ku.dk/
Founded: 1993 (NBIfAFG)
Staff: 14
Activities: cosmology * extragalactic research * galactic research * stellar structure * astrometry
Coordinates: 012°34'45"E 55°41'12"N (IAU Code 035)
City Reference Coordinates: 012°35'00"E 55°40'00"N

Copenhagen University, Niels Bohr Institute for Astronomy, Physics and Geophysics (NBIfAFG), Ørsted Laboratory
Universitetsparken 5
DK-2100 København Ø
Telephone: (0)35-321818
Telefax: (0)35-320460
Electronic Mail: <userid>@fys.ku.dk
WWW: http://ntserv.fys.ku.dk/hco/
Founded: 1993 (NBIfAFG)
Staff: 20
Activities: research * education * solid-state physics * ion implantation * metastable phases * chaos * hydrodynamics * Mössbauer spectroscopy * laser spectroscopy * planetary and meteoritic sciences
City Reference Coordinates: 012°35'00"E 55°40'00"N

Copenhagen University, Niels Bohr Institute for Astronomy, Physics and Geophysics (NBIfAFG), Niels Bohr Institute (NBI)
Blegdamsvej 17
DK-2100 København Ø
Telephone: (0)35-325209
Telefax: (0)35-325016
Electronic Mail: <userid>@nbi.dk
WWW: http://www.nbi.dk/
Founded: 1993 (NBIfAFG)
Activities: extragalactic research * galactic dynamics
City Reference Coordinates: 012°35'00"E 55°40'00"N

Danish Astronautical Society
(Dansk Selskab for Rumfartsforksning - DSR)
Postboks 31
DK-1002 København K
or :
Branddamsvej 179
DK-2860 Soeborg
Telephone: (0)29677633
Telefax: (0)35362282
Electronic Mail: dsr@forening.dk (Thomas A.E. Andersen, President)
WWW: http://www.rumfart.dk/
Founded: 1949
Membership: 170
Activities: meetings * exhibitions
Periodicals: (4) "Dansk Rumfart" (ISSN 0905-2410)
City Reference Coordinates: 012°35'00"E 55°40'00"N (København)

Danish Astronomical Association
(Astronomisk Selskab i Danmark)
c/o Observatoriet
Juliane Maries Vej 30
DK-2100 København Ø
Telephone: (0)3532-5999
Electronic Mail: paldrich@inet.uni-c.dk
WWW: http://www.dsri.dk/AS/
http://www.dsri.dk/AS/as-uk.html
Founded: 1916
Membership: 700
Activities: observing * computing * public lectures and presentations
Sections: planets * meteors * satellites * occultations * variable stars * telescopes and instrumentation * history of astronomy
Periodicals: "Knudepunktet"
City Reference Coordinates: 012°35'00"E 55°40'00"N

Danish Meteorological Institute (DMI)
Lyngbyvej 100
DK-2100 København
Telephone: (0)39157500
Telefax: (0)39271080
WWW: http://www.dmi.dk/
Founded: 1872
Staff: 400
Activities: monitoring and forecasting weather and climate, including the middle and upper atmosphere * research
City Reference Coordinates: 012°35'00"E 55°40'00"N

Danish Natural Science Research Council
Randergade 60
DK-2100 København
Telephone: (0)35446200
Telefax: (0)35446201
Electronic Mail: fr@forskraad.dk
WWW: http://www.forskraad.dk/
Founded: 1968
Membership: 6
City Reference Coordinates: 012°35'00"E 55°40'00"N

Danish Physical Society
(Dansk Fysisk Selskab - DFS)
c/o Jens Olaf Pepke Pedersen
Danish Center for Earth System Science (DCESS)
Juliane Maries Vej 30
DK-2100 København Ø
Telephone: (0)35320573
Telefax: (0)35320576
Electronic Mail: jopp@dcess.ku.dk
WWW: http://dfswww.fysik.dtu.dk/dfs/
Membership: 620
City Reference Coordinates: 012°35'00"E 55°40'00"N

Danish Space Research Advisory Board
Bredgade 43
DK-1260 København K
Telephone: (0)33929734
Telefax: (0)33150205
Electronic Mail: cbl@fsk.dk
WWW: http://www.fsk.dk/
Founded: 1996
Membership: 12
Activities: advisory board on space research
City Reference Coordinates: 012°35'00"E 55°40'00"N

Danish Space Research Institute (DSRI)
Juliane Maries Vej 30
DK-2100 København Ø
Telephone: (0)35325830
Telefax: (0)35362475
Electronic Mail: <userid>@dsri.dk
WWW: http://www.dsri.dk/
Founded: 1966
Activities: developing balloon, rocket and satellite instrumentation for X- and gamma-ray astronomy * cosmic rays * electromagnetic and particle interactions in the magnetosphere
City Reference Coordinates: 012°35'00"E 55°40'00"N

Danish Standards Association (DS)
Kollegievej 6
DK-2920 Charlottenlund
Telephone: (0)39-966101
Telefax: (0)39-966102
Electronic Mail: dansk.standard@ds.dk
WWW: http://www.ds.dk/
Founded: 1926
Staff: 150
Activities: standardization
Periodicals: "Dansk Standard"
City Reference Coordinates: 012°35'00"E 55°40'00"N (København)

Dansk Fysisk Selskab (DFS)
• See "Danish Physical Society"

C.F. Møllers Allé
Building 100
University Park
DK-8000 Århus C
Telephone: (0)89423975
Telefax: (0)89423995
Electronic Mail: stenomus@au.dk
　　　　　　stenoplt@au.dk (Planetarium)
WWW: http://www.stenomuseet.dk/
　　　http://www.stenomuseet.dk/planetarium/planetarium.html
Founded: 1994
Staff: 30
Activities: exhibitions * herbal garden * planetarium * Ole Rømer Observatory (see separate entry) * lectures * guided tours * museum shop * class room
Coordinates: 010°14'00"E 56°07'42"N H46m (Ole Rømer Obs. – IAU Code 155)
City Reference Coordinates: 010°13'00"E 56°09'00"N

Stjernekammeret
Bellahøj Skole
Svenskelejren 18
DK-2700 Brønshøj
Telephone: (0)38603359
Electronic Mail: ca.str@ci.kk.dk
Founded: 1937
Staff: 1
Activities: planetarium shows * education
City Reference Coordinates: 012°35'00"E 55°40'00"N (Copenhagen)

Theoretical Astrophysics Center (TAC)
Juliane Maries Vej 30
DK-2100 København Ø
Telephone: (0)35325900
Telefax: (0)35325910
Electronic Mail: <userid>@tac.dk
WWW: http://www.tac.dk/
Founded: 1994
Staff: 25
Activities: large-scale structure * formation and evolution of galaxies * AGNs * structure and oscillations of Sun and stars
City Reference Coordinates: 012°35'00"E 55°40'00"N

Tycho Brahe Planetarium
Gl. Kongevej 10
DK-1610 København V
Telephone: (0)33-121224 (Booking)
　　　　　(0)33-144888 (Administration)
Telefax: (0)33-142888
Electronic Mail: tycho@tycho.dk
WWW: http://www.tycho.dk/
Founded: 1989
Staff: 30
Activities: shows * OmniMax films
Periodicals: (4) "Aktuel Astronomi" (ISSN 0905-8958)
City Reference Coordinates: 012°35'00"E 55°40'00"N

World Association of Industrial and Technological Research Organisations (WAITRO)
c/o Secretariat
Danish Technological Institute
P.O. Box 141
DK-2630 Tåstrup
Telephone: (0)43507045
Telefax: (0)43507050
Electronic Mail: waitro@dti.dk
WWW: http://waitro.dti.dk/
Founded: 1970
Membership: 150 (from 76 countries)
Activities: international cooperation in industrial and technological R&D
Periodicals: (4) "WAITRO News"
Awards: (1/2) "WAITRO Honorary Award"
City Reference Coordinates: 012°29'00"E 55°39'00"N

Ecuador

Fundación Mundo Juvenil, Planetario
Avenida Los Shyris y Rusia s/n
Quito
Telephone: (0)2-244314
Founded: 1966
Staff: 12
City Reference Coordinates: 078°30'00"W 00°13'00"S

Instituto Ecuatoriano de Normalización (INEN)
A. Baquerizo Moreno 454 y 6 de Diciembre
Casilla 17-01-3999
Quito
Telephone: (0)2-524499
 (0)2-565626
 (0)2-544885
Telefax: (0)2-527561
 (0)2-567815
 (0)2-222223
Electronic Mail: inen1@inen.gov.ec
WWW: http://www4.ecua.net.ec/inen/
Founded: 1970
City Reference Coordinates: 078°30'00"W 00°13'00"S

Instituto Nacional de Meteorología e Hidrología (INAMHI)
Iñaquito 700 (36-14) y Corea
Quito
Telephone: (0)2-433934
 (0)2-265331
Telefax: (0)2-433934
Electronic Mail: inamhi@ecnet.ec
WWW: http://www.ecua.net.ec/inamhi
Founded: 1961
Staff: 4
Activities: weather and climate forecast * network of stations in the whole country
Periodicals: (12) "Meteorological Monthly Bulletin"; (1) "Meteorological and Hydrological Annual Report"
City Reference Coordinates: 078°30'00"W 00°13'00"S

Observatorio Astronómico de Quito (OAQ)
Parque Alameda
Apartado 17-01-165
Quito
Telephone: (0)2-570765
 (0)2-583541
Telefax: (0)2-567848
Electronic Mail: observaq@uio.satnet.net
WWW: http://www.satnet.net/observatorio/
Founded: 1873
Staff: 10
Activities: astronomy * meteorology * astrophysics * archaeoastronomy
Periodicals: "Boletín del Observatorio Astronómico de Quito"
Coordinates: 078°29'56"W 00°12'57"S H2,818m (IAU Code 781)
City Reference Coordinates: 078°30'00"W 00°13'00"S

Egypt

Academy of Scientific Research and Technology (ASRT)
101 Kasr El-Eini Street
Cairo 11516
Telephone: (0)2-3557972
(0)2-3542714
(0)2-5940127
Telefax: (0)2-356280
(0)2-3547807
WWW: http://www.asrt.sci.eg/
Founded: 1971
Membership: 2250
Activities: enhancing contribution of Egyptian scientific capabilities to development * ensuring high-level planning and coordinating among leading research workers, technologists and users
Periodicals: "Proceedings of the Mathematical and Physical Society of Egypt" and many other scientific journals in various fields
Awards: 32 annual prizes in mathematics, physics, geology, chemistry, biology, agriculture, engineering and medicine
City Reference Coordinates: 031°15'00"E 30°03'00"N

Cairo University, Department of Astronomy and Meteorology
Gamaa Street
Geza
Telephone: (0)727022
Electronic Mail: <userid>@frcu.eun.eg
WWW: http://www.cairo.eun.eg/
Founded: 1937
Staff: 19
Activities: education * research * thesis supervision
City Reference Coordinates: 031°15'00"E 30°03'00"N

Egyptian Meteorological Authority (EMA)
P.O. Box 11784
Koubry El-Quobba
Cairo
Telephone: (0)2-830053
(0)2-830069
(0)2-2820790
(0)2-2849858
Telefax: (0)2-849857
(0)2-2849857
Electronic Mail: met@nwp.gov.eg
WWW: http://nwp.gov.eg/
Founded: 1971
City Reference Coordinates: 031°15'00"E 30°03'00"N

Egyptian Organization for Standardization and Quality Control (EOS)
16 Tadreeb El-Modarrebeen Street
El-Ameriya
Cairo
Telephone: (0)2-2566022
Telefax: (0)2-2593480
(0)2-2593481
Electronic Mail: moi@idsc.gov.eg
Founded: 1957
Staff: 1000
Activities: standardization * quality control * testing * information * metrology
City Reference Coordinates: 031°15'00"E 30°03'00"N

Helwân Observatory
• See "National Research Institute of Astronomy and Geophysics (NRIAG), Helwân Observatory"

National Research Institute of Astronomy and Geophysics (NRIAG), Helwân Observatory
Helwân
Cairo 11421
Telephone: (0)5541100
(0)5549780
Telefax: (0)5548020
Electronic Mail: astro3@frcu.eun.eg
Founded: 1903
Staff: 200
Activities: astronomy * geophysics
Periodicals: "Bulletin of the National Research Institute of Astronomy ad Geophysics"
Coordinates: 031°22'48"E 29°51'30"N H116m (IAU Code 087)
City Reference Coordinates: 031°15'00"E 30°03'00"N

El Salvador

Asociación Salvadoreña de Astronomía
c/o Museo de Ciencias
Calle La Reforma 179
Colonia San Benito
San Salvador
Telephone: (0)2233027
Telefax: (0)2638925
Electronic Mail: fegaram@es.com.sv
astro.grada@integra.com.sv
WWW: http://www.astro.org.sv/
Founded: 1991
Membership: 60
Activities: observing * meetings * lectures * public discussions * education * popularization
Periodicals: (12) "Astro"
Coordinates: 089°04'59"W 13°29'41"N H175m
City Reference Coordinates: 089°12'00"W 13°42'00"N

Consejo Nacional de Ciencia y Tecnología (CONACYT)
Colonia Médica
Avenida Dr. E. Alvarez y Pje
Dr. Guillermo Rodriguez 51
(Edificio Espinoza)
San Salvador
Telephone: (0)2262800
Telefax: (0)2256255
Electronic Mail: iso@ns.conacyt.gob.sv
City Reference Coordinates: 089°12'00"W 13°42'00"N

Servicio de Meteorología e Hidrología Nacional (SMHN)
Apartado Postal 2265
San Salvador
Telephone: (0)2940566 x43
Telefax: (0)2944750
(0)2940575
Electronic Mail: hidrometeoro@salnet.net
Founded: 1953
Staff: 120
Activities: operational forecasting * climatology * hydrology
City Reference Coordinates: 089°12'00"W 13°42'00"N

Estonia

Estonian Academy of Sciences
Kohtu 6
EE-10130 Tallinn
Telephone: (0)6442149
Telefax: (0)6451805
WWW: http://www.aca.ee/
Founded: 1938
City Reference Coordinates: 024°45'00"E 59°25'00"N

Estonian Physical Society
(Eesti Füüsika Selts - EFS)
Tähe 4
EE-51010 Tartu
Telephone: (0)7-383006
Telefax: (0)7-383033
Electronic Mail: piret@fi.tartu.ee (Piret Kuusk, Chairman)
WWW: http://www.physic.ut.ee/efs/
Founded: 1989
Membership: 200
Activities: organizing the "Estonian Days of Physics"
Periodicals: (1) "Annual Report of the Estonian Physical Society" (ISSN 1406-0574)
Awards: (1) "EFS Prize", "EFS Student Prize"
City Reference Coordinates: 026°44'00"E 58°20'00"N

National Standards Board of Estonia (EVS)
Aru 10
EE-10317 Tallinn
Telephone: (0)2-493572
Telefax: (0)2-492002
Electronic Mail: info@evs.ee
WWW: http://www.evs.ee/
Founded: 1991
Staff: 24
Activities: standardization * metrology * accreditation
Periodicals: (12) "EVS Teataja"
City Reference Coordinates: 024°45'00"E 59°25'00"N

Tartu Observatory
EE-61602 Tõravere
Telephone: (0)7-410265
Telefax: (0)7-410205
Electronic Mail: aai@aai.ee
                  <userid>@aai.ee
WWW: http://www.aai.ee/
Founded: 1804
Staff: 65
Activities: stellar atmospheres * variable stars * galactic structure * large-scale structure * remote sensing * dynamic weather forecast
Periodicals: (1) "Calendar of Tartu Observatory" (ISSN 0202-2214), "Annual Report"
Awards: (1) "Ernst Julius Öpik Fellowship"
Coordinates: 026°43'18"E 58°22'47"N H67m (IAU Code 075)
City Reference Coordinates: 026°44'00"E 58°20'00"N

Tartu Old Observatory
Tähetorn Toomel
EE-2400 Tartu
Electronic Mail: astro@obs.ee
                  <userid>@obs.ee
WWW: http://www.obs.ee/
Coordinates: 026°43'18"E 58°22'47"N
City Reference Coordinates: 026°44'00"E 58°20'00"N

Tartu University, Institute of Physics
Riia 142
EE-51014 Tartu
Telephone: (0)27-383039
Telefax: (0)27-383033
Electronic Mail: dir@fi.tartu.ee
WWW: http://www.fi.tartu.ee/
Founded: 1973
Membership: 170

Activities: materials science * matter structure * laser physics * laser optical technologies * environmental science * biophysics
Periodicals: (1) "Annual Report"
City Reference Coordinates: 026°44'00"E 58°20'00"N

Finland

Amateur Astronomers Society Seulaset
• See "Tähtitieteen Harrastajan Yhdistys Seulaset ry"

Astronomical Association Tampereen Ursa
• See "Tähtitieteellinen Yhdistys Tampereen Ursa ry"

Etelä-Karjalan Nova
(South Carelian Amateur Astronomical Association Nova)
P.O. Box 244
FIN-53101 Lappeenranta
Electronic Mail: nova@ursa.fi
WWW: http://www.ursa.fi/yhd/nova/novaeng.html
Founded: 1974
Membership: 45
Periodicals: (2) "Tähtiharrastustietoa" (ISSN 1238-0091)
City Reference Coordinates: 028°11'00"E 61°04'00"N

European Incoherent Scatter Facility (EISCAT), Sodankylä Station
c/o Geophysical Observatory
FIN-99600 Sodankylä
Telephone: (0)16-619880
Telefax: (0)16-610375
Electronic Mail: eiscat@sgo.fi
WWW: http://eiscat.sgo.fi/
Founded: 1975
Staff: 5
Activities: incoherent scatter radar receiving facility for ionospheric and magnetospheric research
Coordinates: 026°37'00"E 67°22'00"N H198m
City Reference Coordinates: 026°37'00"E 67°22'00"N

Finnish Academy of Science and Letters
Mariankatu 5
FIN-00170 Helsinki
Telephone: (0)9-636800
Telefax: (0)9-660117
WWW: http://www.acadsci.fi/
Founded: 1908
Membership: 450
Periodicals: "Annales Academiae Scientiarum Fennicae", "Finnish Academy Year Book" (ISSN 1238-9137)
Awards: (1) "Academy Award"
City Reference Coordinates: 024°58'00"E 60°10'00"N

Finnish Academy of Science and Letters, Sodankylä Geophysical Observatory (SGO)
FIN-99600 Sodankylä
Telephone: (0)16-619811
Telefax: (0)16-619875
WWW: http://www.sgo.fi/
Founded: 1913
Staff: 25
Activities: geomagnetism * ionosphere * magnetosphere * seismology * EISCAT site (see separate entry)
Periodicals: "Publications" (ISSN 0355-0826)
Coordinates: 026°37'55"E 67°22'05"N H180m
City Reference Coordinates: 026°37'00"E 67°22'00"N

Finnish Academy of Technology
Tekniikantie 12
FIN-02150 Espoo
Telephone: (0)9-4554565
Telefax: (0)9-4554626
Founded: 1957
Staff: 4
Activities: meetings * seminars * reports
Periodicals: "Acta Polytechnica Scandinavica"
Awards: "Craftsmans Award"
City Reference Coordinates: 024°40'00"E 60°13'00"N

Finnish Astronautical Society
• See "Suomen Avaruustutkimusseura"

Finnish Astronomical Society
(Suomen Tähtitieteilijäseura)
c/o University of Helsinki Observatory
P.O. Box 14
FIN-00014 Helsinki
or :
Tähtitorninmäki
FIN-00130 Helsinki 13
Telephone: (0)19122909
Telefax: (0)19122952
Electronic Mail: kimmo.lehtinen@helsinki.fi
WWW: http://cc.oulu.fi/tati/kauko/tts.html
Founded: 1969
Membership: 100
Activities: conferences * astronomy days * representing professional astronomers in Finland
Periodicals: "Proceedings of the Finnish Astronomy Days"
City Reference Coordinates: 024°58'00"E 60°10'00"N

Finnish Meteorological Institute (FMI)
Vuorikatu 24
P.O. Box 503
FIN-00101 Helsinki
Telephone: (0)9-19291
Telefax: (0)9-179581
Electronic Mail: <userid>@fmi.fi
WWW: http://www.fmi.fi/
Founded: 1838
Staff: 485
Activities: weather forecast * climatology * air pollution * geomagnetism * space physics
Periodicals: (1) "Annual Report"
Observatories: 3 (Nurmijärví, Jokioinen, Sodankylä)
City Reference Coordinates: 024°58'00"E 60°10'00"N

Finnish Physical Society
c/o S. Nokkanen (Secretary)
P.O. Box 9
FIN-00014 Helsinki
Telephone: (0)9-1918375
Telefax: (0)9-1918378
Electronic Mail: nokkanen@phcu.helsinki.fi
    finphys@pcu.helsinki.fi
WWW: http://www.physics.helsinki.fi/~sfs/
Founded: 1947
Membership: 775
City Reference Coordinates: 024°58'00"E 60°10'00"N

Finnish Standards Association (SFS)
P.O. Box 116
FIN-00241 Helsinki
or :
Maistraatinportli 2
FIN-00240 Helsinki
Telephone: (0)9-1499331
Telefax: (0)9-1464925
Electronic Mail: sfs@sfs.fi
WWW: http://www.sfs.fi/
Founded: 1924
Staff: 70
Periodicals: (6) "SFS-tiedotus" (ISSN 0356-1089, circ.: 1,900)
City Reference Coordinates: 024°58'00"E 60°10'00"N

Hämeenlinnan Tähtitieteen Harrastajlen Yhdistys Vega ry
c/o Jari Vento (President)
Viertokatu 18 A 1
FIN-13210 Hämeenlinna
Telephone: (0)3-6172696
Electronic Mail: vega@ursa.fi
WWW: http://www.helsinki.fi/~tmjlehto/engvega.htm
Founded: 1981
Membership: 100
City Reference Coordinates: 024°27'00"E 61°61'00"N

Helsinki University of Technology (HUT), Institute of Photogrammetry and Remote Sensing
P.O. Box 1200
FIN-02015 HUT
or :
Otakaari 1
FIN-02150 Espoo
Telephone: (0)9-4513900
Telefax: (0)9-465077
WWW: http://foto.hut.fi/
http://www.hut.fi/
Founded: 1957
Staff: 15
Activities: digital photogrammetry * digital image processing in remote sensing * real-time photogrammetry * projective transformations
Periodicals: "The Photogrammetric Journal of Finland" (ISSN 0557-1069), "HUT Annual Report" (ISSN 0788-5474)
City Reference Coordinates: 024°40'00"E 60°13'00"N

Helsinki University of Technology (HUT), Laboratory of Space Technology
c/o Department of Electrical Engineering
Otakaari 5 A
FIN-02150 Espoo
Telephone: (0)9-4512378
Telefax: (0)9-460224
Electronic Mail: <userid>@delta.hut.fi
WWW: http://www.space.hut.fi/
City Reference Coordinates: 024°40'00"E 60°13'00"N

Helsinki University of Technology (HUT), Metsähovi Radio Observatory
Metsähovintie 114
FIN-02540 Kylmälä
Telephone: (0)9-2564831
Telefax: (0)9-2564531
Electronic Mail: <userid>@hut.fi
vlbi@hut.fi (VLBI group)
solar@hut.fi (solar group)
WWW: http://kurp-www.hut.fi/
Founded: 1974
Staff: 18
Activities: radio astronomy * microwaves * solar research * QSOs * molecular lines * VLBI
Periodicals: "Metsähovi Publications on Radio Science" (ISSN 1455-9587) "Metsähovi Radio Observatory Report" (ISSN 1455-9579)
Coordinates: 024°23'38"E 60°13'05"N H61m
City Reference Coordinates: 024°24'00"E 60°13'00"N

Jyväskylän Sirius ry
Sepänaukion Vapaa-Aikakeskus
Kyllikinkatu 1
FIN-40100 Jyväskylä
Telephone: (0)14-3731250
Electronic Mail: sirius@ursa.fi
WWW: http://www.ursa.fi/sirius
http://www.ursa.fi/sirius/siriuseng.html
Founded: 1959
Membership: 180
Activities: meetings with lectures, slides, films * telescope making * optics * observing * photography * study trips
Periodicals: (4) "Valkoinen Kääpiö" (White Dwarf) (ISSN 0781-0466)
Awards: (1) "Amateur Astronomer of the Year"
Coordinates: 025°42'24"E 62°14'03"N H151m (Rihlaperä)
City Reference Coordinates: 025°44'00"E 62°14'00"N

Kuopion Tähtitieteellinen Seura Saturnus
• See "Saturnus Astronomical Association of Kuopio"

Lahden URSA ry
(URSA Astronomical Society)
c/o Yrjö Pullinen (Secretary)
Koulutie 8 A 25
FIN-15860 Hollola
Telephone: (0)3-7804534 (Yrjö Pullinen)
(0)3-7534002 (Observatory + answering machine)
Electronic Mail: yrjo.pullinen@pp.kolumbus.fi
lahden.ursa@ursa.fi
WWW: http://www.ursa.fi/yhd/LahdenUrsa/

Founded: 1948
Membership: 300
Activities: meetings * observing * public star shows
Coordinates: 025°35'31"E 60°59'03"N H193m (Pirttiharju)
City Reference Coordinates: 025°36'00"E 60°59'00"N

Metsähovi Radio Observatory
• See "Helsinki University of Technology (HUT), Metsähovi Radio Observatory"

Nurmijärvi Geophysical Observatory
• See "Finnish Meteorological Institute, Nurmijärvi Geophysical Observatory"

Olympos Astronomical Association
• See "Tähtitieteellinen Yhdistys Olympos ry"

Pollux
P.O. Box 69
FIN-02151 Espoo
Telephone: (0)9-4683124 (Chairman)
(0)40-5103523 (Secretary)
Telefax: (0)9-4683218
Electronic Mail: pollux@tky.hut.fi
pollux@otax.tky.hut.fi
eri@iki.fi (Eero Rinne, Chairman)
poutanen@cc.hut.fi (Matti Poutanen, Secretary)
WWW: http://www.tky.hut.fi/~pollux/
Founded: 1994
Membership: 62
Activities: observing * visits
City Reference Coordinates: 024°40'00"E 60°13'00"N

Rauma Vocational College, Maritime Department, Planetarium
Suojantie 2
FIN-26100 Rauma
Telephone: (0)2-837721
Telefax: (0)2-83772222
Electronic Mail: skarlsson@rai.rauma.fi (Sune Karlsson)
Founded: 1880
Staff: 70
Activities: education
City Reference Coordinates: 021°30'00"E 61°08'00"N

Saturnus Astronomical Association of Kuopio
(Kuopion Tähtitieteellinen Seura Saturnus)
P.O. Box 293
FIN-70101 Kuopio
Founded: 1956
Membership: 67
Activities: observing (deep sky, planets, Sun) * meetings * education * photography
Coordinates: 027°39'03"E 62°53'11"N H159m (Huuhanmäki)
City Reference Coordinates: 027°40'00"E 62°54'00"N

Seinäjoen Ursa Astronomical Association
c/o Juha-Matti Tapio
Mällikkäläntie 125
FIN-61410 Ylistaro
Telephone: (0)9-34756630
Electronic Mail: seinajoen.ursa@ursa.fi
WWW: http://www.ursa.fi/yhd/sjursa/index_en.html
Membership: 54
Activities: lectures * observing
City Reference Coordinates: 022°50'00"E 62°47'00"N (Seinäjoki)

Sodankylä Geophysical Observatory (SGO)
• See "Finnish Academy of Science and Letters, Sodankylä Geophysical Observatory (SGO)"

South Carelian Amateur Astronomical Association Nova
• See "Etelä-Karjalan Nova"

Suomen Avaruustutkimusseura (ATS)
(Finnish Astronautical Society)
P.O. Box 507
FIN-00101 Helsinki
or :

Kauppalantie 6-8
Southern-Haaga
Helsinki
Telephone: (0)9-5874433
Electronic Mail: paul.stigell@vtt.fi (Paul Stigell)
WWW: http://netlander.fmi.fi/∼sats/index_en.html
Founded: 1959
Membership: 100
Activities: space technology * rocketry * space research
Periodicals: (4) "Avaruusluotain" (ISSN 0356-021x, circ.: 150)
City Reference Coordinates: 024°58'00"E 60°10'00"N

Tähtitieteellinen Yhdistys Olympos ry
(Olympos Astronomical Association)
P.O. Box 18
FIN-33501 Tampere
Electronic Mail: markku.nyfelt@nmp.nokia.com (Markku Nyfelt)
WWW: http://www.mikrolog.fi/olympos/olympos.htm
Founded: 1993
Membership: 28
Activities: deep-sky observing * telescope making * optics
Periodicals: (4) "Equatorial Dust Lane (EDL)"
Coordinates: 023°53'48"E 61°23'65"N H158m
City Reference Coordinates: 023°45'00"E 61°32'00"N

Tähtitieteellinen Yhdistys Tampereen Ursa ry
(Astronomical Association Tampereen Ursa)
P.O. Box 18
FIN-33501 Tampere
Telephone: (0)3-2611005
Electronic Mail: ursa@sci.fi
WWW: http://www.sci.fi/∼ursa/
Founded: 1950
Membership: 170
Activities: lectures * public star shows * observing
Periodicals: (4) "Radiantti" (ISSN 0785-5672)
Coordinates: 023°47'38"E 61°30'44"N H162m
City Reference Coordinates: 023°45'00"E 61°32'00"N

Tähtitieteen Harrastajan Yhdistys Seulaset ry
(Amateur Astronomers Society Seulaset)
c/o Pertti Pääkkönen
Department of Physics
University of Joensuu
P.O. Box 111
FIN-80101 Joensuu
Telephone: (0)13-2513238
Telefax: (0)13-2513290
Electronic Mail: pertti.paakkonen@joensuu.fi
WWW: http://cc.joensuu.fi/seulaset
Founded: 1973
Membership: 70
Activities: stargazing * public observing * education
Coordinates: 029°59'49"E 62°43'39"N H155m
City Reference Coordinates: 029°45'00"E 62°36'00"N

Tampereen Särkänniemi Oy, Planetarium
Särkänniemi
FIN-33230 Tampere
Telephone: (0)3-2488111
Telefax: (0)3-2121279
Electronic Mail: timo.rahunen@sarkanniemi.fi
WWW: http://www.sarkanniemi.fi/
    http://www.sarkanniemi.fi/oppimateriaali/tahtiakatemia/
Founded: 1969
Staff: 5
City Reference Coordinates: 023°45'00"E 61°32'00"N

Technical Research Centre of Finland
• See "VTT"

Technology Development Centre
• See "TEKES"

TEKES
(Technology Development Centre)
P.O.Box 69
FIN-00101 Helsinki
or :
Kyllikinportti 2
Helsinki
Telephone: (0)105-2151
Telefax: (0)9-6949196
Electronic Mail: tekes@tekes.fi
WWW: http://www.tekes.fi/
Founded: 1983
Staff: 200
Activities: coordinating and financing technological research in Finland
Periodicals: "Views on Finnish Technology"
City Reference Coordinates: 024°58'00"E 60°10'00"N

Teknofokus
P.O. Box 47
FIN-00711 Helsinki
or :
Graniittitie 9
FIN-00710 Helsinki
Telephone: (0)9-3870471
Telefax: (0)9-3877388
Electronic Mail: hannu.m@teknofokus.fi
WWW: http://www.teknofokus.fi/
Activities: manufacturing telescope mirrors and related accessories * supplying optical components and materials, abrasives and polishing compounds * designing instrumentation * prototype services
City Reference Coordinates: 024°58'00"E 60°10'00"N

Tuorla Observatory
• See "University of Turku, Tuorla Observatory"

Turun Ursa Astronomical Association
c/o Iso-Heikkilä Observatory
FIN-20200 Turku
Telephone: (0)2-302195
Electronic Mail: ursa@ursa.utu.fi
WWW: http://org.utu.fi/(ef,hr)/yhd/ursa/index.html
Founded: 1928
Membership: 200
Activities: public shows * photography * observing * popularization
Periodicals: (4) "Ceres" (ISSN 1235-1083)
Coordinates: 022°13'46"E 60°27'09"N H28m (Iso-Heikkilä Observatory)
   022°13'46"E 60°25'14"N H28m (Kevola Obs. – IAU Code 064)
City Reference Coordinates: 022°17'00"E 60°27'00"N

University of Helsinki, Observatory
P.O. Box 14
FIN-00014 University of Helsinki
or :
Tähtitorninmäki
FIN-00130 Helsinki 13
Telephone: (0)9-19122940
Telefax: (0)9-19122952
Electronic Mail: <userid>@astro.helsinki.fi
WWW: http://www.astro.helsinki.fi/
Founded: 1834
Staff: 40
Periodicals: "Report" (ISSN 0355-9289)
Coordinates: 024°57'18"E 60°09'42"N H33m (IAU Code 569)
City Reference Coordinates: 024°58'00"E 60°10'00"N

University of Oulu, Department of Physical Sciences, Astronomy Division
P.O. Box 3000
FIN-90401 Oulu
Telephone: (0)8-5531280
Telefax: (0)8-5531934
Electronic Mail: <userid>@oulu.fi
WWW: http://physics.oulu.fi/fysiikka/
Founded: 1964
Staff: 20

Activities: astrophysics * planetology * stellar dynamics
City Reference Coordinates: 025°54'00"E 65°05'00"N

University of Turku, Space Research Laboratory
Vesilinnantie 4
FIN-20014 Turku
Telephone: (0)2-3335639
Telefax: (0)2-3335993
Electronic Mail: <userid>@utu.fi
WWW: http://www.srl.utu.fi/
Founded: 1988
Staff: 14
Activities: solar and heliospheric particle physics
City Reference Coordinates: 022°17'00"E 60°27'00"N

University of Turku, Tuorla Observatory
Väisäläntie 20
FIN-21500 Piikkiö
Telephone: (0)2-2744244
Telefax: (0)2-2433767
Electronic Mail: <userid>@astro.utu.fi
WWW: http://www.astro.utu.fi/
Founded: 1927
Staff: 30
Activities: positional astronomy * celestial mechanics * optical photo-polarimetry * radio astronomy * dynamical astronomy * cosmology
Periodicals: "Tuorla Observatory Reports", "Informo" (ISSN 0789-6719)
Coordinates: 022°26'48"E 60°25'00"N H40m (IAU Code 063)
City Reference Coordinates: 022°31'00"E 60°26'00"N

Ursa Astronomical Association
Raatimiehenkatu 3 A 2
FIN-00140 Helsinki
Telephone: (0)9-68404000
Telefax: (0)9-68404040
Electronic Mail: ursa@ursa.fi
WWW: http://www.ursa.fi/ursa/
      http://www.ursa.fi/ursa/planetaario/ (Planetarium)
Founded: 1921
Membership: 10000
Activities: publishing * meetings * lectures * courses * youth camps * topical sections * library * portable planetariums * public events
Periodicals: (6) "Tähdet ja Avaruus" (ISSN 0355-9467), "Ursa Minor" (ISSN 0780-7945); (1) "Tähdet 200x" (ISSN 0355-9459)

Awards: "Stella Arcti"
Coordinates: 024°57'31"E 60°09'20"N H20m (Kaivopuisto Observatory)
City Reference Coordinates: 024°58'00"E 60°10'00"N

URSA Astronomical Society
• See "Lahden URSA ry"

Väisälä Institute for Space Physics and Astronomy (VISPA)
Founded: 2001
• Joining of the "University of Turku, Space Research Laboratory" and of the "University of Turku, Tuorla Observatory"

Verne Theatre
Heureka - The Finnish Science Centre
Tiedepuisto 1
FIN-01301 Vantaa
Telephone: (0)9-85799
Telefax: (0)9-8734142
Electronic Mail: info@heureka.fi
WWW: http://www.heureka.fi/
Founded: 1989
Staff: 8 (100 for science center)
Activities: education * exhibitions
City Reference Coordinates: 024°59'00"E 60°30'00"N

VTT
(Technical Research Centre of Finland)
P.O. Box 1000
FIN-02044 VTT
or :
Vuorimiehentie 5
Espoo

FIN-02044 VTT
Telephone: (0)9-4561
      (0)9-4564330
Telefax: (0)9-4567011
      (0)9-4553349
Electronic Mail: <userid>@vtt.fi
WWW: http://www.vtt.fi/
Founded: 1942
Staff: 3000
Activities: technical research * space instrumentation * remote sensing * satellite communications * mechanics * electronics * computers * information handling * optics * Earth observing
City Reference Coordinates: 024°40'00"E 60°13'00"N

France

Académie des Sciences
c/o Institut de France
23, quai de Conti
F-75006 Paris
Telephone: (0)143266621
Telefax: (0)143546399
WWW: http://www.acad-sciences.institut-de-france.fr/
Founded: 1666
Membership: 126
Activities: meetings * publishing
Periodicals: "Comptes-Rendus de l'Académie des Sciences - Série I" (ISSN 0764-4442), "Comptes-Rendus de l'Académie des Sciences - Série II" (ISSN 0764-4450), "Sciences de la Terre" (ISSN 1164-5873)
City Reference Coordinates: 002°20'00"E 48°52'00"N

Aerospatiale
37, boulevard de Montmorency
F-75781 Paris Cedex 16
Telephone: (0)142242424
Telefax: (0)145245414
WWW: http://www.aerospatiale.fr/
Founded: 1970
Activities: manufacturing aerospace equipment
Periodicals: (12) "Revue Aerospatiale"
• Now part of the "European Aeronautic, Defense and Space Co. (EADS)"
City Reference Coordinates: 002°20'00"E 48°52'00"N

Agence Spatiale Européenne (ASE)
• See "European Space Agency (ESA)"

Air and Space Europe
23 rue Linois
F-75724 Paris Cedex
Telephone: (0)145589863
Telefax: (0)145589421
Electronic Mail: ase@elsevier.fr
WWW: http://www.airandspaceeurope.com/
• Periodical (6) (ISSN 1290-0958)
City Reference Coordinates: 002°20'00"E 48°52'00"N

Albiréo - Astronomes Amateurs Tarnais
c/o Jean-Luc Floutard
24, rue des Acacias
F-81000 Albi
or :
Observatoire de Saint-Caprais
F-81800 Rabastens
Telephone: (0)563607312
Telefax: (0)563405612
Electronic Mail: rieugnie@club-internet.fr
Founded: 1992
Membership: 52
Activities: observing * photography * CCDs * mirror polishing
Coordinates: 001°43'10"E 43°52'30"N H196m (Saint-Caprais, Rabastens)
City Reference Coordinates: 002°09'00"E 43°56'00"N (Albi)

Amis du Planétarium d'Aix-en-Provence (APAP)
c/o Maison des Associations
Place de l'Eglise
F-13540 Puyricard
Telephone: (0)442921546
         (0)442204366 (bookings)
WWW: http://aix.planet.free.fr/
Founded: 1989
Membership: 100
Activities: managing a Planetarium Peiresc * popularization
City Reference Coordinates: 005°26'00"E 43°32'00"N

Andromède
c/o Observatoire de Marseille
2, place Le Verrier
F-13248 Marseille Cedex 4

Telephone: (0)495044100
Telefax: (0)491621190
Electronic Mail: duval@observatoire.cnrs-mrs.fr
WWW: http://www-obs.cnrs-mrs.fr/Andromede/
Founded: 1976
Membership: 40
Activities: school visits * observing * planetarium * lectures
City Reference Coordinates: 005°24'00"E 43°18'00"N

Angénieux
F-42570 Saint-Héand
Telephone: (0)477304210
Telefax: (0)477304875
Electronic Mail: angenieux@calva.net
WWW: http://www.angenieux.com/
Founded: 1935
Staff: 200
Activities: manufacturing optical components
City Reference Coordinates: 004°22'00"E 45°31'00"N

Annales Geophysicae
Editorial Office
Centre d'Étude Spatiale des Rayonnements
9, avenue du Colonel Roche
F-31029 Toulouse Cedex
Telephone: (0)561558370
Telefax: (0)561556535
Electronic Mail: anngeo@cesr.fr
Founded: 1983
Staff: 1
• Journal (12) (ISSN 0992-7689)
City Reference Coordinates: 001°26'00"E 43°36'00"N

Arcane
3, rue du Puits d'Argent
F-02240 Itancourt
Telephone: (0)323088842
Telefax: (0)323088875
Electronic Mail: arcane.astro@wanadoo.fr
WWW: http://perso.wanadoo.fr/arcane
Founded: 1986
Activities: manufacturing telescopes and pieces * studies * designs * modifications * adjustments * software * CCD
Coordinates: 003°21'51"E 49°48'33"N H115m
City Reference Coordinates: 003°28'00"E 49°48'00"N (Ribemont)

Arianespace, Head Office
Boulevard de l'Europe
Boîte Postale 177
F-91006 Évry Cedex
Telephone: (0)160876000
Telefax: (0)160876217
          (0)160876247
WWW: http://www.arianespace.com/
Founded: 1980
Staff: 255
Activities: satellite launching services
Periodicals: (1) "Annual Report"
City Reference Coordinates: 002°27'00"E 48°38'00"N

Association À Ciel Ouvert
Moulin du Roy
F-32500 Fleurance
Telephone: (0)562060976
Telefax: (0)562062499
Electronic Mail: etoiles.fleurance@mipnet.fr
WWW: http://www.gascogne.com/Ferme/ciel.htm
Founded: 1994
Membership: 7
Activities: public education * activities in schools * small planetarium
City Reference Coordinates: 000°40'00"E 43°51'00"N

Association Aéronautique et Astronautique de France (AAAF)
6, rue Galilée
F-75782 Paris Cedex 16
Telephone: (0)147230749

Telefax: (0)147230748
WWW: http://www.aaaf.asso.fr/
City Reference Coordinates: 002°20'00"E 48°52'00"N

Association Astronomie Tycho Brahe
c/o Philippe Jacquot
Les Closets
F-74270 Clermont
WWW: http://www.astrosurf.com/tychobrahe/
Founded: 1992
City Reference Coordinates: 006°07'00"E 45°54'00"N (Annecy)

Association Astronomique d'Anjou (AAA)
15, rue Marc Sangnier
F-49000 Angers
Telephone: (0)241475294
WWW: http://perso.wanadoo.fr/aaanjou/
Founded: 1979
Membership: 75
Activities: photography * radioastronomy * telescope making * popularization * solar observing
Periodicals: (6) "Pégase" (ISSN 0981-6410)
Observatories: 2
City Reference Coordinates: 000°33'00"W 47°28'00"N

Association Astronomique de Franche-Comté (AAFC)
Parc de l'Observatoire
34, avenue de l'Observatoire
F-25000 Besançon
Telephone: (0)381888788
WWW: http://www.iap.fr/saf/claafc.htm
Founded: 1978
Membership: 80
Activities: observing (minor planets, comets) * education * lectures * workshops * planetarium
Periodicals: (3) "Le Point Astro" (ISSN 1168-1195)
Coordinates: 006°00'00"E 47°00'00"N H309m
City Reference Coordinates: 006°02'00"E 47°15'00"N

Association Astronomique de la Côte d'Or (AACO)
c/o Maison des Associations
Esplanade F. Manière
F-21850 Saint-Apollinaire
Telephone: (0)380458780 (Thierry Guillot, President)
Electronic Mail: aaco@astrosurf.com
                guth3@wanadoo.fr (President)
WWW: http://www.astrosurf.com/aaco/
Founded: 1996
Membership: 30
Activities: popularization * planetarium project * debunking astrology
Coordinates: 004°43'00"E 47°16'16"N H531m
City Reference Coordinates: 005°02'00"E 47°20'00"N (Dijon)

Association Astronomique de l'Ain (AAA)
c/o Maison des Sociétés
Boulevard Joliot-Curie
F-01000 Bourg-en-Bresse
Telephone: (0)474250447
Electronic Mail: astro.ain@free.fr
WWW: http://www.astroclub.net/astro-ain
Founded: 1966
Membership: 75
Activities: observing * photography * public events
Coordinates: 005°20'23"E 46°11'35"N
City Reference Coordinates: 005°13'00"E 46°12'00"N

Association Astronomique de La Réunion
18, rue Georges-Bizet
Les Makes
F-97421 La Rivière
WWW: http://perso.wanadoo.fr/observatoire.makes/ (Obs. des Makes)
City Reference Coordinates: 055°36'00"E 21°06'00"S

Association Astronomique de la Vallée (AAV)
c/o Maison des Associations
Mairie d'Orsay
2, place du Général Leclerc

Boîte Postale 47
F-91401 Orsay Cedex
WWW: http://membres.lycos.fr/aav/
Founded: 1978
City Reference Coordinates: 002°17'00"E 48°46'00"N

Association Astronomique de l'Indre (AAI)
21, rue Paul-Verlaine
F-36000 Chateauroux
Telephone: (0)254495016
Founded: 1983
Membership: 28
Activities: monthly meetings * observing
City Reference Coordinates: 001°42'00"E 46°49'00"N

Association Astronomique de Loir-et-Cher (AALC)
10, rue Alexandre-Dumas
F-41350 Vineuil
Telephone: (0)254421954
Founded: 1978
Membership: 200
Activities: observing * education * documentation
Periodicals: (4) "Astro 41"
City Reference Coordinates: 001°22'00"E 47°35'00"N

Association Astronomique du Soissonnais (AAS)
16, rue de la Congrégation
F-02200 Soissons
Telephone: (0)323728650
Electronic Mail: aas@astrosurf.com
WWW: http://www.astrosurf.com/aas/
Founded: 1979
Membership: 40
Activities: lectures * observing * variable stars * mirror making * sundials
City Reference Coordinates: 003°20'00"E 49°22'00"N

Association Astronomique Picarde M80
3, Le Pré du Bois
F-80260 Rubempré
or :
Le Safran
Rue Georges Guynemer
F-80080 Amiens
Telephone: (0)322934902 (answering machine)
          (0)681494492 (President)
WWW: http://www.astroclub.net/terre/m80/
Founded: 1996
Membership: 5
Periodicals: "Bulletin Interne"
City Reference Coordinates: 002°20'00"E 50°00'00"N (Villers Bocage)
                               002°18'00"E 49°54'00"N (Amiens)

Association Ciel d'Anjou (ACA)
62, rue de Villoutreys
F-49000 Angers
Telephone: (0)241475493
Electronic Mail: cielanjou@astrosurf.com
WWW: http://www.astrosurf.com/cielanjou
Founded: 1994
Membership: 52
Activities: popularization * education * exhibitions * slide shows
Periodicals: (6) "La Petite Ourse"
City Reference Coordinates: 000°33'00"W 47°28'00"N

Association Copernic (Gap Astronomie -_)
• See "Gap Astronomie - Association Copernic"

Association d'Astronomie Véga
Beffroi des Associations
50, Avenue de Saintonge
F-78450 Villepreux
or :
20, chemin de Rambouillet
F-78450 Villepreux
Telephone: (0)134623052

Coordinates: 005°45'00"E 43°17'75"N H550m (Riboux)
City Reference Coordinates: 005°24'00"E 43°18'00"N

Association Méditerranéenne des Sciences de l'Environnement et de l'Espace (AMSEE)
c/o Patrice Poyet
182, route de Gairaut
Parc Château d'Azur
Villa la Chamaille
F-06100 Nice
Telephone: (0)493090651
Electronic Mail: poyet@cstb.fr
WWW: http://www.cstb.fr/ILC/amsee/
http://cic.cstb.fr/amsee/projects/
Founded: 1990
Activities: observing (planets, double stars, deep sky) * CCDs
City Reference Coordinates: 007°15'00"E 43°42'00"N

Association Narbonnaise d'Astronomie Populaire (ANAP)
c/o Observatoire de Narbonne
31, rue de la Distillerie
F-11110 Vinassan
Telephone: (0)468453113
Electronic Mail: jacques.cazenove@wanadoo.fr (President)
WWW: http://perso.wanadoo.fr/jacques.cazenove/
Founded: 1981
Membership: 30
Activities: popularization * research * CCD imaging
Coordinates: 002°58'00"E 43°09'00"N (Domaine de Montplaisir, Narbonne)
City Reference Coordinates: 002°58'00"E 43°09'00"N

Association Nationale pour l'Amélioration de la Vue (ASNAV)
F-92038 Paris La Défense Cedex
or :
39-41, rue Louis-Blanc
F-92400 Courbevoie
Telephone: (0)147176478
Telefax: (0)147176883
Founded: 1954
City Reference Coordinates: 002°20'00"E 48°52'00"N (Paris)
002°15'00"E 48°54'00"N (Courbevoie)

Association Nationale Sciences Techniques Jeunesse (ANSTJ)
● See now "Planète Sciences"

Association Normande d'Astronomie (ASNORA)
52, rue de la Folie
F-14000 Caen
Telephone: (0)231738554
WWW: http://www.cpod.com/monoweb/asnora/
Founded: 1948
Membership: 50
Activities: meetings * observing * lectures * radioastronomy * outings
Periodicals: (4) "Capella" (ISSN 1266-7390)
Observatories: 1 (Laize-la-Ville)
City Reference Coordinates: 000°21'00"W 49°11'00"N

Association Novae
584, boulevard Jean-Ossola
F-06700 Saint-Laurent-du-Var
Telephone: (0)493075403
Electronic Mail: novae@castor.unice.fr
WWW: http://www.eurecom.fr/~roudier/Astro/Novae/
Founded: 1988
Membership: 71
Activities: education * observing * photography * electronics * spectroscopy * mirror polishing * exchanges * image processing * rocketry
Awards: (1/2) "Concours National de Poésie sur l'Astronomie"
Observatories: 1 (Observatoire de Nice)
City Reference Coordinates: 007°11'00"E 43°40'00"N

Association pour le Développement International de l'Observatoire de Nice (ADION)
c/o Observatoire de Nice
Boîte Postale 4229
F-06304 Nice Cedex 4
Telephone: (0)492003011

WWW: http://www.obs-nice.fr/adion/
Founded: 1964
Activities: international relations * relations with local authorities * awarding prizes and medals
Periodicals: (1) "ADION Bulletin" (ISSN 0249-7522)
Awards: (1) "Médaille de l'ADION", "Prix pour Bénévoles"
City Reference Coordinates: 007°15'00"E 43°42'00"N

Association Sammielloise d'Astronomie (ASA)
c/o Sylvain Jannot
28, rue de Saint-Mihiel
F-55300 Dompcevrin
Telephone: (0)329901183
Founded: 1983
Membership: 25
Activities: observing * photography * planetarium
Coordinates: 005°29'22"E 48°54'52"N H304m (Les Paroches)
City Reference Coordinates: 005°30'00"E 48°56'00"N

Association Saranaise des Astronomes Amateurs (ASAA)
41, allée des Pyrénées
F-45770 Saran
Telephone: (0)238430507
Electronic Mail: asaa_h5@yahoo.com
WWW: http://www.astrosurf.org/asaa
Founded: 1988
Membership: 14
Activities: Sun * variable stars * photography * general observing * education * popularization
Periodicals: (10) "Astromag"
Coordinates: 001°52'00"E 48°00'00"N
City Reference Coordinates: 001°52'00"E 48°00'00"N

Association Sportive et Culturelle Toussaintaise (ASCT), Section Astronomie
3, rue du 19 Mars 1962
F-76400 Toussaint
Telephone: (0)235271940
Electronic Mail: philippe.ledoux@wanadoo.fr
Founded: 1995
Membership: 55
Activities: observing * photography * popularization * education
City Reference Coordinates: 000°23'00"E 49°45'00"N (Fécamp)

Association Stéphanoise d'Astronomie M42
28, rue Pierre et Dominique Ponchardier
F-42100 Saint-Étienne
Telephone: (0)477336363
Founded: 1988
Membership: 55
Activities: observing * education * lectures * workshops * visits
Coordinates: 004°07'00"E 45°22'00"N H1,000m
City Reference Coordinates: 004°07'00"E 45°22'00"N

Association Sterenn, Groupement d'Études Astronomiques de Queven (GEAQ)
c/o Bernard Goumon
29, rue Chateaubriand
F-56530 Queven
Telephone: (0)297053040
Founded: 1983
Membership: 25
Activities: instrument making * photography * computing * introduction to astronomy * CCDs
Coordinates: 003°23'00"W 47°45'00"N
City Reference Coordinates: 003°23'00"W 47°45'00"N

Astro Club Aubagnais (ACA)
MJC l'Escale
Les Aires Saint-Michel
F-13400 Aubagne
or :
c/o Lionel Ruiz (President)
Quartier le Clos
F-13360 Roquevaire
Telephone: (0)442031532
Electronic Mail: kirstin@vulcain.u-3mrs.fr
WWW: http://pegase.unice.fr/~skylink/pages_clubs/aubagne/ACA.html
http://www.astrosurf.org/pluton/aubagne/ACA.html
Founded: 1993

Membership: 25
Activities: observing * lectures * education
Observatories: 1 (La Sainte Baume)
City Reference Coordinates: 005°34'00"E 43°17'00"N (Aubagne)
005°36'00"E 43°21'00"N (Roquevaire)

Astro-Club de Limagne-Sud
3, rue des Sources
F-63730 Les Marthes de Veyre
Telephone: (0)473392336
Electronic Mail: guy.berson@inserm.u-clermont1.fr (Guy Berson, President)
Founded: 1993
Membership: 15
Activities: observing * education * telescope making
City Reference Coordinates: 003°11'00"E 45°41'00"N

Astronomes Amateurs Tarnais
• See "Albiréo - Astronomes Amateurs Tarnais"

Astronomie en Chinonais
Le Vauroux
F-37500 Chinon
Telephone: (0)247588038
Electronic Mail: atco@wanadoo.fr
WWW: http://www.atco-fr.com/aec/aec.php3
Founded: 1984
Membership: 25
Activities: popularization * lectures * slide shows * telescope making * meetings
Coordinates: 002°30'00"E 47°10'00"N H100m
City Reference Coordinates: 000°15'00"E 47°10'00"N

Astronomie en Touraine et Centre-Ouest (ATCO)
• See now "Astronomie, Techniques et Communication (ATCO)"

Astronomie Magazine
18, boulevard Léon-Blum
F-02100 Saint-Quentin
Telephone: (0)323651919
Telefax: (0)323651909
Electronic Mail: astronomie.magazine@wanadoo.fr
WWW: http://pro.wanadoo.fr/astronomie.magazine/
• Periodical (11)
City Reference Coordinates: 003°17'00"E 47°51'00"N

Astronomie, Physique, Élaboration, Instrumentation et Observation (APHELIE)
Pavillon Colbert
35, rue Jean-Longuet
F-92290 Châtenay-Malabry
Telephone: (0)546612492 (Hervé Dole)
(0)547021584 (Pascal Audureau)
Electronic Mail: dole@mesiom.obspm.fr
audureau@ief-paris-sud.fr
WWW: http://wwwfirback.ias.fr/users/dole/aphelie
Founded: 1994
Membership: 20
Activities: meetings * lectures * instrumentation
Periodicals: "La Lettre d'Aphélie"
Observatories: 3 (Orsay, Thorigné-sur-Dué, Ablis)
City Reference Coordinates: 002°17'00"E 48°46'00"N

Astronomie, Techniques et Communication (ATCO)
Le Vauroux
F-37500 Chinon
Telephone: (0)247932744
Electronic Mail: atco@atco-fr.com
WWW: http://www.atco-fr.com/
Founded: 2001
Membership: 120
Activities: linking together amateur astronomers and associations in the area * meetings
Periodicals: (4) "Astronomie en Touraine et Centre-Ouest" (ISSN 1243-8219, circ.: 200)
• Formerly "Astronomie en Touraine et Centre-Ouest (ATCO)"
City Reference Coordinates: 000°15'00"E 47°10'00"N

Astronomy and Space Information Service (ASIS)
c/o Service RNGC-0311

5, rue des Mésanges
F-67120 Duttlenheim
Telephone: (0)388150743
Telefax: (0)388491255
Founded: 1985
Staff: 1
Activities: directories * databases * dictionaries * WWW * compilations * documentation * astronomy and space sciences * lectures * trainings * colloquia * conferences
City Reference Coordinates: 007°39'00"E 48°33'00"N

AstroQueyras (AQ)
Mairie de Saint-Véran
F-05350 Saint-Véran
Telephone: (0)160753105
Electronic Mail: secretaire@astroqueyras.com
WWW: http://www.astroqueyras.com/
Founded: 1989
Membership: 160
Activities: observing
Periodicals: (4) "Lettre AstroQueyras"
Coordinates: 006°54'24"E 44°41'56"N H2930m (IAU Code 615)
City Reference Coordinates: 002°11'00"E 48°53'00"N (Rueil-Malmaison)
006°52'00"E 44°42'00"N (Saint-Véran)

Astro-Terre
c/o Observatoire des Vallons
F-83630 Bauduen
Telephone: (0)494843919
WWW: http://www.ec-lille.fr/~astro/
http://www.astrosurf.org/pluton/bauduen/home.html
Founded: 1989
Membership: 6
Activities: observing * photography * construction of telescopes and observing domes for amateur astronomers
Coordinates: 006°11'03"E 43°43'21"N
City Reference Coordinates: 006°10'00"E 43°44'00"N

Atelier d'Helios (L'_)
• See "L'Atelier d'Helios"

Blénod Animation Loisirs, Section Astronomie
c/o Patrick Brandebourg
16, rue de la Fontaine
F-54700 Blénod-lès-Pont-à-Mousson
Telephone: (0)383823105
Founded: 1971
Membership: 15
Activities: photometry * popularization * activities in schools
City Reference Coordinates: 006°03'00"E 48°53'00"N

Bureau des Longitudes (BDL)
• See now "Observatoire de Paris, Institut de Mécanique Céleste et de Calcul des Éphémérides (IMCCE)"

Bureau Gravimétrique International (BGI)
c/o Observatoire Midi-Pyrénées
18, avenue Edouard-Belin
F-31055 Toulouse Cedex
Telephone: (0)561332889
Telefax: (0)561253098
Electronic Mail: balmino.uggi@cnes.fr (Georges Balmino, Director)
WWW: http://www.obs-mip.fr/uggi/bgi.html
Founded: 1951
Staff: 6
Activities: collecting, checking, archiving and distributing gravity measurements
Periodicals: (2) "Bulletin d'Information"
City Reference Coordinates: 001°26'00"E 43°36'00"N

Bureau International des Poids et Mesures (BIPM)
Pavillon de Breteuil
F-92312 Sèvres Cedex
Telephone: (0)145077070
Telefax: (0)145342021
Electronic Mail: info@bipm.org
<userid>@bipm.org
WWW: http://www.bipm.org/
Founded: 1875

Staff: 60
Activities: unification of the basic units for physics, with the Time Section establishing the International Atomic Time and, jointly with the International Earth Rotation Service, the Coordinated Universal Time
Periodicals: (12) "Circular T"; (1) "Annual Report of the Time Section" (ISSN 1016-6114); "Comptes-Rendus des Séances de la Conférence Générale des Poids et Mesures", "Procès-Verbaux des Séances du Comité International des Poids et Mesures", "Sessions des Comités Consultatifs" (ISSN 0588-6228), "Metrologia"
City Reference Coordinates: 002°17'00"E 48°49'00"N

Bureau National de Métrologie (BNM), Laboratoire Primaire du Temps et des Fréquences (LPTF)
c/o Observatoire de Paris
61, avenue de l'Observatoire
F-75014 Paris
Telephone: (0)140512213
Telefax: (0)143255542
Electronic Mail: lptfop@obspm.fr
WWW: http://opdaf1.obspm.fr/www/lptf.html
Founded: 1975
Staff: 30
Activities: time and frequency metrology
Periodicals: (12) "Bulletin"
City Reference Coordinates: 002°20'00"E 48°52'00"N

Cassini Astronomie
18, avenue de l'Europe
(Piscine Municipale)
F-94190 Villeneuve-Saint-Georges
Telephone: (0)143896099
Founded: 1972
Membership: 30
Activities: instrumentation * photography * CCDs * observing * popularization
Periodicals: (12) "Albiréo"
• Previously called "Club Astronomique Cassini"
City Reference Coordinates: 002°27'00"E 48°44'00"N

Centre Astronomique Vendéen (CAV)
c/o Christophe Guillou
12, impasse Viala
F-85000 La Roche-sur-Yon
Electronic Mail: christophe.guillou@wanadoo.fr
WWW: http://www.atco-fr.com/cav/cav.php3
Founded: 1995
City Reference Coordinates: 001°25'00"W 46°40'00"N

Centre Culturel de l'Astronomie (CCA)
Boîte Postale 1088
F-34007 Montpellier
Telephone: (0)467617401
Telefax: (0)467611008
Founded: 1984
Membership: 50
Activities: popularization * studies of astronomical heritage
Periodicals: (4) "Revue Montpelliéraine d'Astronomie Languedocienne" (ISSN 0246-1390)
City Reference Coordinates: 003°53'00"E 43°36'00"N

Centre d'Analyse des Images (CAI)
• See "Institut National des Sciences de l'Univers (INSU), Centre d'Analyse des Images (CAI)"

Centre d'Astronomie
Le Moulin à Vent
F-04870 Saint-Michel-l'Observatoire
Telephone: (0)492766969
Telefax: (0)492766767
Electronic Mail: centre.astro@wanadoo.fr
WWW: http://www.astrosurf.com/centre.astro/
Founded: 1998
Activities: observing * displays * education * festival * camps
City Reference Coordinates: 005°43'00"E 43°56'00"N

Centre de Culture Scientifique et Technique (CCST)
28, rue Albert 1er
F-17000 La Rochelle
Telephone: (0)546411825
Telefax: (0)546506365
Electronic Mail: jacques.vialle@wanadoo.fr (Jacques Vialle, Astronomy Advisor)
Founded: 1990

Staff: 3
Activities: public education * exhibitions * lectures * mobile planetarium
Coordinates: 001°11'00"W 46°10'12"N
City Reference Coordinates: 001°10'00"W 46°10'00"N

Centre de Données astronomiques de Strasbourg (CDS)
c/o Observatoire Astronomique
11, rue de l'Université
F-67000 Strasbourg
Telephone: (0)390242476 (Director)
(0)390242475 (Secretary)
Telefax: (0)390242417
Electronic Mail: question@simbad.u-strasbg.fr
<userid>@simbad.u-strasbg.fr
WWW: http://cdsweb.u-strasbg.fr/CDS.html (English Homepage)
http://cdsweb.u-strasbg.fr/CDS-f.html (French Homepage)
Founded: 1972
Staff: 12
Activities: astronomical data center * bibliographical service * compilation, critical evaluation and distribution of astronomical data * on-line catalog server * SIMBAD database * Aladin project * yellow-page services * research in classification, astronomical statistics, quality control
Periodicals: "Publications Spéciales du CDS" (ISSN 0764-9614)
• Operated under agreement between the "Institut National des Sciences de l'Univers (INSU)" and the "Université de Strasbourg I (Université Louis Pasteur - ULP)"
City Reference Coordinates: 007°45'00"E 48°35'00"N

Centre de Physique des Particules de Marseille (CPPM)
163, avenue de Luminy
Case 907
F-13288 Marseille Cedex 9
Telephone: (0)491827200
Telefax: (0)491827299
Electronic Mail: <userid>@cppm.in2p3.fr
WWW: http://cppm.in2p3.fr/
http://antares.in2p3.fr/ (Antares Project)
Founded: 1982
Staff: 120
Activities: particle physics * high-energy physics * astroparticle physics * astrophysics * neutrino astronomy * electroweak and strong interactions * conservation laws * standard model * supersymmetry * detectors * data acquisition
Coordinates: 005°57'00"E 42°50'00"N (Antares Neutrino Detector)
• Jointly run by the "Centre National de la Recherche Scientifique (CNRS)", the "Institut National de Physique Nucléaire et de Physique des Particules (IN2P3)" and the "Université de la Méditerranée (Aix-Marseille II)"
City Reference Coordinates: 005°24'00"E 43°18'00"N

Centre de Recherche Astronomique de Lyon (CRAL)
9, avenue Charles André
F-69561 Saint-Genis-Laval Cedex
or :
46, allée d'Italie
F-69364 Lyon Cedex 07
Telephone: (0)478868383
(0)472728000
Telefax: (0)478868386
(0)472728080
Electronic Mail: <userid>@obs.univ-lyon1.fr
WWW: http://www-obs.univ-lyon1.fr/
http://image.univ-lyon1.fr/
http://www-obs.univ-lyon1.fr/base/leda-doc.html (LEDA)
Founded: 1995 (1878 as "Observatoire de Lyon")
Staff: 70
Activities: research * cosmology * nuclei of galaxies * stellar astrophysics * instrumentation * education * Lyon-Meudon Extragalactic Database (LEDA)
Periodicals: "Reprints"
Coordinates: 004°47'06"E 45°41'42"N H299m (IAU Code 513)
• Merging of the "Observatoire de Lyon" and of the "Groupe d'Astrophysique" from the "École Normale Supérieure de Lyon (ENSL)"
City Reference Coordinates: 004°48'00"E 45°41'00"N (Saint-Genis-Laval)
004°51'00"E 45°45'00"N (Lyon)

Centre de Recherches en Physique de l'Environnement Terrestre et Planétaire (CRPE)
• See now "Centre d'Étude des Environnements Terrestre et Planétaires (CETP)"

Centre de Spectrométrie Nucléaire et de Spectrométrie de Masse (CSNSM)
• See "Université de Paris XI, Centre de Spectrométrie Nucléaire et de Spectrométrie de Masse (CSNSM)"

WWW: http://www.inist.fr/
Founded: 1988
Staff: 330
Activities: producing bibliographical databases PASCAL, FRANCIS and article@inist * providing primary documents
Periodicals: "Special'IST" (ISSN 1244-9148)
City Reference Coordinates: 006°12'00"E 48°41'00"N

Centre National de la Recherche Scientifique (CNRS), Institut des Sciences de la Terre, de l'Eau et de l'Espace de Montpellier (ISTEEM)
• See "Institut des Sciences de la Terre, de l'Eau et de l'Espace de Montpellier (ISTEEM)"

Centre National de la Recherche Scientifique (CNRS), Laboratoire de l'Accélérateur Linéaire (LAL)
• See "Laboratoire de l'Accélérateur Linéaire (LAL)"

Centre National de la Recherche Scientifique (CNRS), Observatoire de Haute Provence (OHP)
• See "Observatoire de Haute Provence (OHP)"

Centre National de la Recherche Scientifique (CNRS), Service d'Aéronomie
Boîte Postale 3
F-91371 Verrières-le-Buisson Cedex
or :
Réduit de Verrières
Route des Gatines
F-91370 Verrières-le-Buisson
Telephone: (0)164474245
Telefax: (0)169202999
Electronic Mail: <userid>@aero.jussieu.fr
WWW: http://www.aero.jussieu.fr/
Founded: 1958
Staff: 140
Activities: optical studies of Earth and planetary atmospheres (UV, visible, IR, lidars)
Observatories: 1 (Station Géophysique de l'Observatoire de Haute Provence - see separate entry)
City Reference Coordinates: 002°16'00"E 48°45'00"N

Centre National d'Enseignement à Distance (CNED)
Avenue du Téléport
Boîte Postale 200
F-86960 Futuroscope Cedex
Telephone: (0)549499494
Telefax: (0)549499696
Electronic Mail: Teletel/Minitel: 3615 Code CNED
accueil@cned.fr
WWW: http://www.cned.fr/
Founded: 1939
City Reference Coordinates: 000°20'00"E 46°35'00"N

Centre National d'Études Spatiales (CNES)
(French Space Agency)
2, place Maurice-Quentin
F-75039 Paris Cedex 01
Telephone: (0)144767500
Telefax: (0)144767676
Electronic Mail: <userid>@cnes.fr
WWW: http://www.cnes.fr/
Founded: 1962
Staff: 2500
Activities: developing French space activities (national projects, cooperative ventures, participation in ESA's programmes)
Periodicals: (6) "Calendrier des Manifestations Spatiales"; (4) "CNES Magazine" (ISSN 1283-9817, circ.: 9000); (2) "Qualité Espace" (ISSN 1556-6558, circ.: 3000); (1) "Rapport Annuel d'Activités"; (1/2) "Rapport au COSPAR" (circ.: 2,500)
City Reference Coordinates: 002°20'00"E 48°52'00"N

Centre National d'Études Spatiales (CNES), Centre Spatial Guyanais (CSG)
Boîte Postale 726
F-97387 Kourou Cedex
(Guyane Française)
Telephone: (0)594335111
(0)594325111
Telefax: (0)594334766
WWW: http://www.csg-spatial.tm.fr/
Founded: 1964
Staff: 1800
Activities: operational launch site
Periodicals: (4) "Latitude 5" (ISSN 0293-0072, circ.: 2,500)
City Reference Coordinates: 052°39'00"W 05°09'00"N

Centre National d'Études Spatiales (CNES), Délégation à la Communication et à l'Éducation
c/o Centre Spatial de Toulouse
18, avenue Edouard-Belin
F-31401 Toulouse Cedex 4
Telephone: (0)561273131
Telefax: (0)561283315
WWW: http://www.cnes.fr/
Founded: 1962 (CNES)
Activities: see main entry
Periodicals: see main entry
City Reference Coordinates: 001°26'00"E 43°36'00"N

Centre Régional de Promotion de la Culture Scientifique, Technique et Industrielle
Rue Vercord
F-59650 Villeneuve d'Ascq
Telephone: (0)320193600
Telefax: (0)320193601
Founded: 1989
Activities: research * creation of cultural products * exhibitions * lectures * training * information centre * planetarium project
• Alternate name: "ALIAS"
City Reference Coordinates: 003°10'00"E 50°37'00"N

Centre Spatial Guyanais (CSG)
• See "Centre National d'Études Spatiales (CNES), Centre Spatial Guyanais (CSG)"

Cercle Scientifique Briochin
c/o Alain Le Gué
8, rue Xavier Grall
F-22000 Saint-Brieuc
WWW: http://perso.club-internet.fr/printant/astro/c22000.html
City Reference Coordinates: 002°45'00"W 48°31'00"N

CERGA
• See "Observatoire de la Côte d'Azur (OCA), Centre d'Études et de Recherches en Géodynamique et Astrométrie (CERGA)"

Cité de l'Espace
Avenue Jean-Gonord
Boîte Postale 5855
F-31506 Toulouse Cedex 5
Telephone: (0)562716480
(0)562714871 (Online infos)
Telefax: (0)561807470
WWW: http://www.cite-espace.com/
Founded: 1997
Staff: 60
Activities: hands-on exhibits * space exploration * planetarium
City Reference Coordinates: 001°26'00"E 43°36'00"N

Cité des Sciences et de l'Industrie de la Villette, Planétarium
F-75930 Paris Cedex 19
or :
30, avenue Corentin-Cariou
F-75019 Paris
Telephone: (0)140057022
Telefax: (0)140057118
Electronic Mail: planetar@world-net.sct.fr
WWW: http://www.cite-sciences.fr/francais/ala_cite/spectacl/planetar/global_fs.htm
Founded: 1986
Activities: shows * exhibitions
City Reference Coordinates: 002°20'00"E 48°52'00"N

Club Ajaccien des Amateurs d'Astronomie (C3A)
c/o Centre Scientifique de Vignola
Route des Sanguinaires
F-20000 Ajaccio
Telephone: (0)495218448
Founded: 1986
Membership: 74
Activities: observing * photography * popularization * education * Sun * planetarium
Coordinates: 008°39'03"E 41°54'44"N H50m (Observatoire de Vignola)
City Reference Coordinates: 008°44'00"E 41°55'00"N

Club Astro Alpha Centauri
c/o MJC de Carcassonne
91, rue Aimé-Ramond
F-11000 Carcassonne
Telephone: (0)468111700
Founded: 1981
Membership: 12
Activities: monthly lectures * visual observing * meteors * meetings
Coordinates: 002°08'40"E 43°04'10"N H255m (Pauligne, Limoux)
City Reference Coordinates: 002°21'00"E 43°13'00"N

Club Astro de Wittelsheim (CAW)
c/o Jean-Luc Garambois (President)
32, rue du Rempart
F-68190 Ensisheim
Telephone: (0)389810718
Founded: 1977
Membership: 66
Activities: observing * popularization * meetings
Periodicals: (6) "Procyon" (ISSN 0297-1038)
City Reference Coordinates: 007°20'00"E 47°51'00"N (Ensisheim)
007°20'00"E 47°49'00"N (Wittelsheim)

Club Astro Guynemer
c/o Comité d'Entreprise
Thomson CSF Optronique
Rue Guynemer
Boîte Postale 55
F-78283 Guyancourt Cedex
Founded: 1999
Membership: 56
Activities: instrument loan * education * circulating periodicals
City Reference Coordinates: 002°08'00"E 48°48'00"N (Versailles)

Club Astro Junior M67
10, rue des Vosges
F-67800 Hoenheim
Telephone: (0)621942944
Electronic Mail: caj_m67@yahoo.fr
WWW: http://www.multimania.com/caj/
Founded: 1986
Membership: 15
Activities: observing * meetings * photography * education
City Reference Coordinates: 007°45'00"E 48°35'00"N (Strasbourg)

Club Astronomie de Saint-Claude (CASC)
16, route de Chaumont
F-39200 Saint-Claude
Telephone: (0)384453156
Founded: 1973
Membership: 18
Activities: observing * photography
Coordinates: 005°48'09"E 46°15'38"N H1,180m
City Reference Coordinates: 005°52'00"E 46°23'00"N

Club Astronomie Nature du Valromey (CANV)
Observatoire du Col de la Lèbe
F-01260 Sutrieu
Telephone: (0)479876731
Telefax: (0)479876731
Electronic Mail: astroval@free.fr
WWW: http://www.astroval.free.fr/
Founded: 1993
Activities: education * astronomy * ornithology * botany
Coordinates: 005°41'00"E 45°54'00"N H914m (Col de la Lèbe)
City Reference Coordinates: 005°41'00"E 45°54'00"N (Champagne-en-Valromey)

Club Astronomique de la Région Lilloise (CARL)
c/o Maison de la Nature et de l'Environnement
23, rue Gosselet
F-59000 Lille
Telephone: (0)320859919
Telefax: (0)320861556

Electronic Mail: carl@nordnet.fr
WWW: http://asso.nordnet.fr/carl/
Founded: 1976
Membership: 105
Activities: promoting astronomy and related fields * popularization * observing * exhibitions * workshops * mobile planetarium
Periodicals: (6) "Astronomie pour Tous"
Observatories: 1 (Ferme du Héron, Villeneuve d'Ascq)
City Reference Coordinates: 003°04'00"E 50°38'00"N

Club Astronomique Véga de la Lyre
15, avenue Juncarret
F-33870 Vayres
Telephone: (0)557748100 (Secretariat)
          (0)557849947 (Observatory)
Electronic Mail: vega-lyre@astrosurf.com
WWW: http://astroclub.net/saturne/vega-lyre/
      http://www.astrosurf.com/vega-lyre/
Founded: 1985
Membership: 90
Activities: lectures * observing * introduction to astronomy
Coordinates: 000°21'00"W 44°52'00"N H33m
City Reference Coordinates: 000°21'00"W 44°52'00"N

Club Copernic
Maison des Associations
642, rue des Batteries
F-83600 Fréjus
Telephone: (0)494828361
Telefax: (0)494828361
WWW: http://www.chez.com/pstj/Astro/copernic.htm
Founded: 1974
Membership: 60
Activities: amateur astronomy and astrophysics * talks * lectures * visits * observing * education * popularization
City Reference Coordinates: 006°44'00"W 43°26'00"N

Club d'Astronomie Alpha Centaure (CAAC)
c/o Elisabeth Margaron-Invernizzi
Résidence "Le Parc"
1, avenue Dr. Bonnet
F-26100 Romans-sur-Isère
Telephone: (0)475430290
Telefax: (0)475431905
Electronic Mail: alpha-centaure@wanadoo.fr
WWW: http://assoc.wanadoo.fr/alpha-centaure/
Founded: 1979
Membership: 50
Activities: theoretical study of astronomy * observing natural, terrestrial and celestial phenomena
Coordinates: 005°06'00"E 45°28'00"N H630m (Rochefort-Samson)
City Reference Coordinates: 005°06'00"E 45°28'00"N

Club d'Astronomie de Chamonix
• See now "Nadir - Astronomie et Sciences"

Club d'Astronomie de Lyon-Ampère (CALA)
37, rue Paul-Cazeneuve
F-69008 Lyon
Telephone: (0)478012905
Telefax: (0)478749843
Electronic Mail: cala@cala.asso.fr
WWW: http://www.cala.asso.fr/
Founded: 1968
Membership: 150
Activities: education * observing * photography * instrument making * popularization * lectures * planetarium shows
Periodicals: (4) "Nouvelle Gazette du Club 69 (NGC 69)"
City Reference Coordinates: 004°51'00"E 45°45'00"N

Club d'Astronomie de Villemur (CAV)
c/o Jean-Louis Prieur
Bondigoux
F-31340 Villemur
Electronic Mail: prieur@obs-mip.fr (Jean-Louis Prieur, President)
WWW: http://webast.ast.obs-mip.fr/users/prieur/cav/cav.html
Founded: 1975
Membership: 12
Activities: observing * photography * instrumentation * popularization

Coordinates: 001°32'00"E 43°53'00"N
City Reference Coordinates: 001°32'00"E 43°53'00"N

Club d'Astronomie du Trégor
c/o Planétarium de Bretagne
Cosmopolis
F-22560 Pleumeur-Bodou
WWW: http://perso.club-internet.fr/printant/astro/c22560.html
City Reference Coordinates: 003°32'00"W 48°46'00"N

Club d'Astronomie Janus
1, place Saint-Just
F-92230 Gennevilliers
WWW: http://www.multimania.com/astrojan/
Founded: 1979
Membership: 22
Activities: lectures in schools * observing * computing * photography * CCD camera
Coordinates: 002°17'15"E 48°55'31"N
City Reference Coordinates: 002°18'00"E 48°06'00"N

Club d'Astronomie Miranda
36, rue de Rennes
F-35170 Bruz
WWW: http://perso.club-internet.fr/printant/astro/c35170.html
Founded: 1987
City Reference Coordinates: 001°44'00"W 48°01'00"N

Club d'Astronomie "Randonnée Céleste"
c/o ACS de Boissy-Fresnoy
Mairie
F-60440 Boissy-Fresnoy
or :
18, rue Jean-Charron
F-60440 Boissy-Fresnoy
Telephone: (0)344888300
Telefax: (0)344888300
Electronic Mail: bdussart@bigfoot.com (Bernard Dussart)
WWW: http://www.astrosurf.org/rceleste/
Founded: 1993
Membership: 23
Activities: observing * lectures * exhibitions * popularization * sundials
City Reference Coordinates: 002°49'00"E 49°08'00"N (Nanteuil le Haudouin)

Club d'Astronomie Spica
Maison des Associations
4, avenue de Verdun
F-06800 Cagnes-sur-Mer
Telephone: (0)493244969
WWW: http://www.chez.com/pstj/Astro/spica.htm
City Reference Coordinates: 007°09'00"E 43°04'00"N

Club d'Astronomie Uranie
Maison des Jeunes et de la Culture
2A, avenue de la Libération
F-42400 Saint-Chamond
Telephone: (0)477317115
Telefax: (0)477220422
Electronic Mail: cbreisse@free.fr
WWW: http://cbreisse.free.fr/uranie/
City Reference Coordinates: 004°30'00"E 45°28'00"N

Club d'Information Scientifique (CIS) de la Poste et de France Telecom
57, rue de la Colonie
F-75013 Paris
Telephone: (0)143132151
Electronic Mail: cis-colonie@netcourrier.com
WWW: http://www.astrosurf.org/cis/
Founded: 1976
Membership: 230
Activities: lectures * observing * photography * CCDs * education
Periodicals: "Regard de l'Astronome"
City Reference Coordinates: 002°20'00"E 48°52'00"N

Club Éclipse
c/o Thierry Midavaine

102, rue Vaugirard
F-75006 Paris
or :
c/o André Bradel
26, boulevard Jean-Jaurès
F-92100 Boulogne-Billancourt
Telephone: (0)146040791 (A. Bradel)
Electronic Mail: club-eclipse@egroups.com
WWW: http://www.astroclub.net/club_eclipse
Founded: 1979
Membership: 14
Activities: QSOs * comets * CCDs * image intensifiers * photography * Messier objects * minor planets
Periodicals: (4) "La Lettre du Club Éclipse"
City Reference Coordinates: 002°20'00"E 48°52'00"N (Paris)
002°15'00"E 48°50'00"N (Boulogne-Billancourt)

Club Jean Perrin (CJP)
Palais de la Découverte
Avenue F.D. Roosevelt
F-75008 Paris
Telephone: (0)140748108
Electronic Mail: hdevichi@enst.fr (Hadrien Devichi)
WWW: http://www.stud.enst.fr/~hdevichi/cjp/cjp.html
Founded: 1970
Membership: 25
Activities: introduction to theoretical astronomy and astrophysics * observing * photography * workshops
City Reference Coordinates: 002°20'00"E 48°52'00"N

CNRS Éditions
(Siège Social)
15, rue Malebranche
F-75005 Paris
or :
La Librairie de CNRS Éditions
151 bis, rue Saint-Jacques
F-75005 Paris
Telephone: (0)153102700
(0)153100505 (Librairie)
Telefax: (0)153102727
(0)153100507 (Librairie)
Electronic Mail: cnrseditions@cnrseditions.fr
librairie@cnrseditions.fr
WWW: http://www.cnrseditions.fr/
• Publisher
City Reference Coordinates: 002°20'00"E 48°52'00"N

CODATA
• See "Committee on Data for Science and Technology (CODATA)"

Collège J. Valéri, Association du Planétarium
128, avenue Saint-Lambert
F-06100 Nice
Telephone: (0)492090924
Telefax: (0)492090924
Electronic Mail: planet.valeri@wanadoo.fr
Founded: 1967
Membership: 90
Activities: shows * workshops * observing * mirror polishing
Periodicals: "Le Gluon"
City Reference Coordinates: 007°15'00"E 43°42'00"N

Comité de Liaison Enseignants Astronomes (CLEA)
c/o Annie Mercier
Laboratoire d'Astronomie
Bâtiment 470
Université de Paris-Sud
F-91405 Orsay Cedex
Telephone: (0)169157766
Telefax: (0)169156381
Electronic Mail: annie.mercier@df.cso.u-psud.fr
WWW: http://www.ac-nice.fr/clea/
Founded: 1979
Membership: 1000
Activities: training of teachers in astronomy * Starlab planetarium
Periodicals: (4) "Les Cahiers Clairaut" (ISSN 0758-234x, circ.: 1,800)

(0)153697236
Electronic Mail: <userid>@hq.esa.fr
WWW: http://www.esrin.esa.it/esa/descrip/estabs.htm
http://www.esrin.esa.it/
http://www.esa.int/
Founded: 1964
Staff: 2000
Activities: promoting cooperation among European states in space research and technology and their application for peaceful purposes
Periodicals: see "European Space Agency (ESA), ESTEC, ESA Publications Division (EPD)" (Netherlands)
City Reference Coordinates: 002°20'00"E 48°52'00"N

European Synchrotron Radiation Facility (ESRF)
Boîte Postale 220
F-38043 Grenoble Cedex
Telephone: (0)476882000
Telefax: (0)476882020
Electronic Mail: <userid>@esrf.fr
information@esrf.fr
WWW: http://www.esrf.fr/
Founded: 1988 (Opening: 1994)
Staff: 500
City Reference Coordinates: 005°43'00"E 45°10'00"N

European Telecommunications Standards Institute (ETSI)
650, route des Lucioles
F-06921 Sophia-Antipolis Cedex
Telephone: (0)492944200
Telefax: (0)493654716
Electronic Mail: secretariat@etsi.org
WWW: http://www.etsi.org/
Membership: 800
Activities: producing European telecommunications standards
City Reference Coordinates: 007°00'00"E 43°33'00"N (Cannes)

Euroscience
c/o Raymond Seltz (Secretary General)
8, rue des Écrivains
F-67000 Strasbourg
Telephone: (0)388241150
Telefax: (0)388247556
Electronic Mail: office@euroscience.ws
WWW: http://www.euroscience.org/
Founded: 1997
Activities: promotion of European science and technology
Periodicals: (4) "Euroscience News"
Awards: (1) "Rammal Medal"
City Reference Coordinates: 007°45'00"E 48°35'00"N

EUROSPACE
(European Industrial Space Study Group)
15-17 avenue de Ségur
F-75007 Paris
Telephone: (0)144420070
Telefax: (0)144420079
Electronic Mail: letterbox@eurospace.org
WWW: http://www.eurospace.org/
Founded: 1961
Activities: promoting European space activity
Periodicals: (1) "European Space Directory"
City Reference Coordinates: 002°20'00"E 48°52'00"N

Fédération Laïque d'Éducation Populaire (FLEP), Section Astronomie
Foyer Laïque d'Éducation Populaire
Château des Izards
F-24660 Coulounieix-Chamiers
or :
c/o Cyrille Debray (President)
2, impasse Suzanne-Lacorre
F-24660 Coulounieix-Chamiers
Telephone: (0)553352405
(0)553467520
Electronic Mail: astronomie24@perigord.tm.fr
WWW: http://www.astrosurf.org/astroflep/
Founded: 1991

Membership: 60
Activities: solar observing * education * photography
Periodicals: (12) "Astro Passion"
Coordinates: 000°42'24"E 45°09'17"N (La Rampisole)
City Reference Coordinates: 000°43'00"E 45°11'00"N (Périgueux)

Ferme des Étoiles (La_)
• See "La Ferme des Étoiles"

Fondation Européenne de la Science
• See "European Science Foundation (ESF)"

Fondation Louis de Broglie
23, rue Marsoulan
F-75012 Paris
Telephone: (0)140020008
Telefax: (0)140049707
Electronic Mail: fond_broglie@compuserve.com
WWW: http://www.fondationlouisdebroglie.org/
Founded: 1973
Staff: 12
Activities: seminars
Periodicals: (4) "Annales de la Fondation Louis de Broglie" (ISSN 0182-4295)
City Reference Coordinates: 002°20'00"E 48°52'00"N

Galaxy Contact
7, rue Gustave-Cuvelier
Boîte Postale 26
F-62101 Calais Cedex
Telephone: (0)321352515
Telefax: (0)321351784
Electronic Mail: contact@spacephotos.com
WWW: http://www.spacephotos.com/
Founded: 1983
Staff: 8
Activities: documentation, photographs, video tapes, etc. on space
City Reference Coordinates: 001°50'00"E 50°57'00"N

Galerie Alain Carion
6, rue Jean-du-Bellay
F-75004 Paris
or :
92, rue Saint-Louis-en-l'Île
F-75004 Paris
Telephone: (0)143260116
Telefax: (0)143259233
Founded: 1972
Activities: sales of minerals, fossils, meteorites, tectites, ...
City Reference Coordinates: 002°20'00"E 48°52'00"N

Gap Astronomie - Association Copernic
c/o Léone Pellenq
Mairie de Gap
Rue du Colonel Roux
F-05000 Gap
or :
c/o École Copernic
Haute Corréo
F-05400 La Roche des Arnauds
Telephone: (0)492578317 (Observatory)
Telefax: (0)492578317 (Observatory)
Electronic Mail: asso.copernic@netcourrier.com
WWW: http://www.astrosurf.com/copernic
Founded: 1985
Membership: 100
Activities: lectures * observing * photography * CCDs
Coordinates: 006°00'00"E 44°33'46"N H1,200m (Haute Corréo)
City Reference Coordinates: 006°05'00"E 44°34'00"N (Gap)
005°57'00"E 44°34'00"N (La Roche des Arnauds)

Géospace Observatoire d'Aniane
c/o Institut de Botanique
163, rue Auguste-Broussonnet
F-34090 Montpellier
Telephone: (0)467040222

Telefax: (0)467542675
Electronic Mail: geospace@cnusc.fr
WWW: http://astro.u-strasbg.fr/Obs/PLANETARIUM/aniane.html
Founded: 1982
Membership: 500
Coordinates: 003°36'11"E 43°41'01"N H176m
City Reference Coordinates: 003°53'00"E 43°36'00"N

Groupe Astronomique Hague Querqueville (GAHQ)
Ferme de la Rocambole
61, rue Roger-Glinel
F-50460 Querqueville
Telephone: (0)233033709
Telefax: (0)333033709
Electronic Mail: gahqhag@aol.com
Founded: 1980
Membership: 100
Activities: introduction to astronomy * instrument making * photography * CCD * image processing * history of astronomy * spectroscopy
Periodicals: "Mira"
Coordinates: 001°45'00"W 49°38'00"N
City Reference Coordinates: 001°45'00"W 49°38'00"N

Groupe d'Entraînement et de Recherche pour les Méthodes d'Éducation Active (GERMEA)
25, rue Montaigne
F-64000 Pau
or :
Domaine du Pignada
1, allée de l'Empereur
F-64600 Anglet
Telephone: (0)559522254
Telefax: (0)559401422
Electronic Mail: germea@wanadoo.fr
WWW: http://perso.wanadoo.fr/germea
Founded: 1983
Membership: 200
Activities: education * teacher training * activities in schools * popularization
City Reference Coordinates: 000°22'00"W 43°18'00"N (Pau)
001°30'00"W 43°29'00"N (Anglet)

Groupe Électronique Astronomie Informatique (GEAI)
c/o Claude Rivas
113, rue Paradis
F-13006 Marseille
Telephone: (0)491815585
(0)684541584
Electronic Mail: claude.rivas@club.francetelecom.fr
WWW: http://www.multimania.com/geai13
Founded: 1985
Membership: 15
Activities: radioastronomy * spectrography * photography * observing * Sun * Moon * planets * double stars * computing * electronics
Coordinates: 005°37'47"E 43°17'57"N H150m (Gemenos)
City Reference Coordinates: 005°24'00"E 43°18'00"N

Groupe Européen d'Observation Stellaire (GEOS)
c/o Michel Dumont (Secretary)
3, promenade Vénézia
F-78000 Versailles
Telephone: (0)130219023
Founded: 1973
Membership: 96
Activities: observing variable stars * analysis of collected data
Periodicals: (30) "Notes Circulaires"; (5) "GEOS Circulars"
City Reference Coordinates: 002°08'00"E 48°48'00"N

Groupement Astronomique Populaire de la Région d'Antibes (GAPRA)
18, boulevard Chancel
F-06600 Antibes
Electronic Mail: brunetto@pacwan.fr (Laurent Brunetto)
WWW: http://perso.pacwan.fr/brunetto/GAPRA/
Founded: 1975
Membership: 50
City Reference Coordinates: 007°07'00"E 43°35'00"N

Groupement d'Astronomie Populaire Sottevillais (GAPS)
c/o Maison pour Tous
2, rue Thiremberg
F-76200 Sotteville-lès-Rouen
Telephone: (0)235723105
(0)235898711
Telefax: (0)235722567
Electronic Mail: gaps@astrosurf.com
WWW: http://www.astrosurf.com/gaps/
Founded: 1990
Membership: 16
Activities: observing (Sun, planets, comets, Moon, variable and double stars) * radioastronomy * photography * instrument making * education * meetings * lectures * planetarium
Periodicals: (12) "Le Ciel du Mois"; (1) "Le Ciel de vos Vacances"
Coordinates: 001°05'00"E 49°25'00"N H10m
City Reference Coordinates: 001°05'00"E 49°25'00"N

Groupement des Astronomes Amateurs de la Gâtine (GAAG)
18, avenue de la Maladrerie
F-79200 Parthenay
Telephone: (0)549642301 (President)
Founded: 1983
Membership: 29
Activities: popularization * weekly meetings * observing * instrumentation * exhibitions * mobile planetarium (Constellarium E4)
Observatories: 2 (Pompaire, Le Pin)
City Reference Coordinates: 000°15'00"W 46°39'00"N

Groupement des Industries Françaises Aéronautiques et Spatiales (GIFAS)
4, rue Galilée
F-75782 Paris Cedex 16
Telephone: (0)144431700
Telefax: (0)140709141
Electronic Mail: infogifas@gifas.asso.fr
WWW: http://www.gifas.asso.fr/
Founded: 1908
Membership: 200 (companies)
Activities: union of French aerospace manufacturers
Periodicals: (11) "GIFAS Letter" (ISSN 0399-4864); (1) "Rapport Annuel"
City Reference Coordinates: 002°20'00"E 48°52'00"N

Groupement des Industries Françaises de l'Optique (GIFO)
39-41, rue Louis-Blanc
F-92400 Courbevoie
Telephone: (0)147176400
Telefax: (0)147176398
Electronic Mail: gifo@gifo.org
WWW: http://www.gifo.org/
Founded: 1896
Staff: 6
Activities: representing French optical industry
Periodicals: (4) "Infos GIFO" (ISSN 1149-168x)
City Reference Coordinates: 002°15'00"E 48°54'00"N

Groupement Français pour l'Observation et l'Étude du Soleil (GFOES)
Le Vauroux
F-37500 Chinon
Telephone: (0)247932744
Electronic Mail: gfoes@ifrance.com
WWW: http://www.astrosurf.com/gfoes/
Founded: 1989
Membership: 50
Activities: Sun observing * data processing
Periodicals: (2) "Helios" (circ.: 70)
City Reference Coordinates: 000°30'00"E 47°10'00"N

Images, Reflets, Initiation Scientifique (IRIS)
20, rue Diffonty
F-13600 La Ciotat
Telephone: (0)442714019
Founded: 1991
Membership: 70
Activities: lectures * observing
City Reference Coordinates: 005°36'00"E 43°10'00"N

IMASTRO Rhône-Alpes
c/o Eric Dubiel
"Béchanne"
F-01370 Saint-Étienne du Bois
Telephone: (0)474305896
Electronic Mail: eric.dubiel@ac-lyon.fr
WWW: http://www.bressenet.com/imastro/index.html
Founded: 1990
Membership: 25
Activities: CCDs * popularization * portable planetarium
City Reference Coordinates: 005°13'00"E 46°12'00"N (Bourg-en-Bresse)

Institut d'Astrophysique de Paris (IAP)
• See "Centre National de la Recherche Scientifique (CNRS), Institut d'Astrophysique de Paris (IAP)"

Institut d'Astrophysique Spatiale (IAS)
• See "Centre National de la Recherche Scientifique (CNRS), Institut d'Astrophysique Spatiale (IAS)"

Institut de l'Information Scientifique et Technique (INIST)
• See "Centre National de la Recherche Scientifique (CNRS), Institut de l'Information Scientifique et Technique (INIST)"

Institut de Mécanique Céleste et de Calcul des Éphémérides (IMCCE)
• See now "Observatoire de Paris, Institut de Mécanique Céleste et de Calcul des Éphémérides (IMCCE)"

Institut de Physique du Globe de Paris (IPGP)
Tour 24 B89
4, place Jussieu
F-75282 Paris Cedex 05
Telephone: (0)144272404
Telefax: (0)144273373
WWW: http://www.ipgp.jussieu.fr/
Founded: 1990
Activities: space geodynamics * geophysics * geodesy * seismology * planetology * magnetism * remote sensing * oceanic altimetry
City Reference Coordinates: 002°20'00"E 48°52'00"N

Institut de Radioastronomie Millimétrique (IRAM), Grenoble
300, rue de la Piscine
Domaine Universitaire de Grenoble
F-38406 Saint-Martin d'Hères Cedex
Telephone: (0)476824900
Telefax: (0)476515938
Electronic Mail: <userid>@iram.grenet.fr
WWW: http://iram.fr/
Founded: 1979
Staff: 96
Periodicals: (1) "Annual Report"; (6) "IRAM Newsletter"
Coordinates: 005°54'26"E 44°38'01"N H2,552m (Observatoire du Plateau de Bure)
003°23'58"W 37°04'06"N H2,870m (Pico Veleta, Spain)
• Joint facility of the "Centre National de la Recherche Scientifique (CNRS)", France, the "Max-Planck-Gesellschaft zur Förderung der Wissenschaften eV" (Max-Planck-Gesellschaft), Germany, and the "Instituto Geográfico Nacional (IGN)", Spain, to carry out mm-wavelength astronomy
City Reference Coordinates: 005°43'00"E 45°10'00"N (Grenoble)

Institut de Radioastronomie Millimétrique (IRAM), Observatoire du Plateau de Bure
L'Enclus
F-05250 Saint-Étienne-en-Devoluy
Telephone: (0)476538520
Telefax: (0)476538523
WWW: http://iram.fr/
Founded: 1979
Activities: mm-wavelength astronomy
Coordinates: 005°54'29"E 44°38'02"N H2,560m (interferometer)
• Mail should be sent to IRAM headquarters in Grenoble (see separate entry)
City Reference Coordinates: 005°43'00"E 45°10'00"N (Grenoble)

Institut des Sciences de la Terre, de l'Eau et de l'Espace de Montpellier (ISTEEM)
Place Eugène Bataillon
F-34095 Montpellier Cedex 5
Telephone: (0)467144593
Telefax: (0)467144785
Electronic Mail: isteem@dstu.univ-montp2.fr
WWW: http://www.dstu.univ-montp2.fr/
Founded: 1994

Staff: 220
Activities: research * education
Periodicals: "Mémoires de l'ISTEEM" (ISSN 1279-7782)
• Jointly run by the "Centre National de la Recherche Scientifique (CNRS)" and the "Université de Montpellier II Sciences et des Techniques du Languedoc"
City Reference Coordinates: 003°53'00"E 43°36'00"N

Institut des Sciences de la Terre, de l'Eau et de l'Espace de Montpellier (ISTEEM), Groupe de Recherche en Astronomie et Astrophysique du Languedoc (GRAAL)
Place Eugène Bataillon
F-34095 Montpellier Cedex 5
Telephone: (0)467143415
Telefax: (0)467144535
Electronic Mail: <userid>@graal.univ-montp2.fr
WWW: http://www.isteem.univ-montp2.fr/GRAAL/
Founded: 1989 (ISTEEM: 1994)
Staff: 14
Activities: observational cosmology * stellar physics * stellar atmospheres * late-type stars * analysis of large surveys
City Reference Coordinates: 003°53'00"E 43°36'00"N

Institut Gassendi pour la Recherche Astronomique en Provence (IGRAP)
• See now "Observatoire Astronomique Marseille-Provence"

Institut National de Physique Nucléaire et de Physique des Particules (IN2P3), Centre de Physique des Particules de Marseille (CPPM)
• See "Centre de Physique des Particules de Marseille (CPPM)"

Institut National de Physique Nucléaire et de Physique des Particules (IN2P3), Laboratoire de l'Accélérateur Linéaire (LAL)
• See "Laboratoire de l'Accélérateur Linéaire (LAL)"

Institut National de Physique Nucléaire et de Physique des Particules (IN2P3), Laboratoire de Physique Nucléaire des Hautes Énergies (LPNHE)
• See "Laboratoire de Physique Nucléaire des Hautes Énergies (LPNHE)"

Institut National des Sciences Appliquées (INSA), Club Astronomie
Bureau des Élèves
Maison des Étudiants
20, avenue Albert-Einstein
F-69621 Villeurbanne Cedex
Telephone: (0)472438229
Electronic Mail: cail@free.fr
WWW: http://www.insa-lyon.fr/Associations/ASTRO/
Founded: 1967
Membership: 25
Activities: photography * CCDs * observing (planets, deep sky) * lectures
City Reference Coordinates: 004°53'00"E 45°46'00"N

Institut National des Sciences de l'Univers (INSU)
3, rue Michel-Ange
Boîte Postale 287
F-75766 Paris Cedex 16
Telephone: (0)144964000 (switchboard)
          (0)144964338 (switchboard)
          (0)144964377
Telefax: (0)144965005
Electronic Mail: <userid>@cnrs-dir.fr
WWW: http://www.insu.cnrs-dir.fr/
Founded: 1985
City Reference Coordinates: 002°20'00"E 48°52'00"N

Institut National des Sciences de l'Univers (INSU), Centre d'Analyse des Images (CAI)
c/o Observatoire de Paris
Bâtiment Perrault
77, avenue Denfert-Rochereau
F-75014 Paris
Telephone: (0)140512098
Telefax: (0)140512090
Electronic Mail: jean.guibert@obspm.fr
WWW: http://dsmama.obspm.fr/
Founded: 1990
Staff: 8
Activities: digitalized processing of astronomical and remote-sensing images * data analysis and processing * automatic classification * astrometry * photometry * managing the "Machine Automatique à Mesurer pour l'Astronomie (MAMA)"

City Reference Coordinates: 002°20'00"E 48°52'00"N

Institut National des Sciences de l'Univers (INSU), Centre de Données astronomiques de Strasbourg (CDS)
• See "Centre de Données astronomiques de Strasbourg (CDS)"

International Academy of Astronautics (IAA)
Boîte Postale 1268-16
F-75766 Paris Cedex 16
or :
6, rue Galilée
F-75016 Paris
Telephone: (0)147238215
Telefax: (0)147238216
Electronic Mail: sgeneral@iaanet.org
WWW: http://www.iafastro.com/academy/academy.htm
http://www.iaanet.org/
Founded: 1960
Membership: 1160
Periodicals: (4) "Acta Astronautica"
Awards: "Theodore von Kármán Award"
City Reference Coordinates: 002°20'00"E 48°52'00"N

International Association of Universities (IAU)
c/o F. Eberhard
1, rue Miollis
F-75732 Paris Cedex 15
Telephone: (0)145682545
(0)147832339
Telefax: (0)147347605
Electronic Mail: iau@unesco.org
WWW: http://www.unesco.org/iau
Founded: 1950
Staff: 16
Activities: information * studies * research * meetings * IAU/UNESCO Information Centre on Higher Education * coordinating centre for WHED (World Higher Education Database)
Periodicals: (6) "IAU Newsletter" (ISSN 0020-6032); (4) "Higher Education Policy" (ISSN 0279-4631); (1/2) "International Handbook of Universities"; "World List of Universities and Other Institutions of Higher Education", "Issues in Higher Education"
City Reference Coordinates: 002°20'00"E 48°52'00"N

International Astronautical Federation (IAF)
3-5, rue Mario Nikis
F-75015 Paris
Telephone: (0)145674260
Telefax: (0)142732120
(0)142737537
Electronic Mail: iaf@wanadoo.fr
WWW: http://www.iafastro.com/
Founded: 1950
Membership: 143
Staff: 3
Activities: fostering development of astronautics for peaceful purposes * encouraging widespread dissemination of technical and other information concerning astronautics * stimulating public interest in and support for the development of all aspects of astronautics through the various media of mass communication * encouraging participation in astronautical research or other relevant projects by international and national research institutions, universities, commercial firms and individual experts * creating and fostering as activities of the Federation academies, institutes and commissions dedicated to continuing research in, and the fostering on, all aspects of the natural and social sciences relating to astronautics and the peaceful use of outer space * convoking and organizing with support of its respective academies, institutes and commissions, international astronautical congresses, symposia, colloquia and other scientific meetings * cooperating and advising with appropriate international and national governmental and non-governmental organizations and institutions on all aspects of the natural, engineering and social sciences related to astronautics and the peaceful uses of outer space
Sections: (committees) Allan D. Emil Memorial Award * Finance * Publications * Liaison with International Organizations and Developing Nations * Astrodynamics * Communications * Earth Observations * Education * Student Activities * Materials and Structures * Power * Propulsion * Space Exploration * Space Processing and Microgravity Applications * Space Station * Space Systems * Space Transportation * Life Sciences * Space Physiology and Medicine * Space and Planetary Biology and Biophysics * Human Factors * Biotechnology and Life Support * Space and Natural Disaster Reduction
Awards: "Allan D. Emil Memorial Award", "Frank J. Malina Award", "Student Award", "L.G. Napolitano Award"
City Reference Coordinates: 002°20'00"E 48°52'00"N

International Astronomical Union (IAU)
(Union Astronomique Internationale - UAI)
Secrétariat
98bis, boulevard Arago
F-75014 Paris
Telephone: (0)143258358

F-53000 Laval
Telephone: (0)243535958
Electronic Mail: m53mayenneastronomie@wanadoo.fr
WWW: http://www.astrosurf.com/m53astro/
Founded: 1994
City Reference Coordinates: 000°46'00"W 48°04'00"N

Maison de l'Astronomie
33-35, rue de Rivoli
F-75004 Paris
Telephone: (0)142779955
Telefax: (0)148874087
Electronic Mail: info@maison-astronomie.com
WWW: http://www.maison-astronomie.com/
Founded: 1928
Staff: 17
Activities: selling astronomical instruments and accessories, books, maps, atlases, pictures and posters * lectures * workshops * travels * Club Privilège
Periodicals: (12) "Astro News" (ISSN 0990-2862, circ.: 4,000)
City Reference Coordinates: 002°20'00"E 48°52'00"N

Météo France
1, quai Branly
F-75340 Paris Cedex 07
Telephone: (0)145567171
          (0)892680808 (on-line weather forecast)
Telefax: (0)145567005
Electronic Mail: <userid>@meteo.fr
WWW: http://www.meteo.fr/
Founded: 1945
Staff: 3500
Activities: data acquisition and processing * weather forecast * climatology * documentation * research * training * network of climatic stations all over the country
Periodicals: (52) "Météo-Hebdo"; (12) "Bulletin Climatique"; (4) "Met Mar" (ISSN 0222-5123, circ.: 2,000); (1) "Rapport d'Activité"; "meteo.fr", "Atmosphère et Climat"
City Reference Coordinates: 002°20'00"E 48°52'00"N

Musée de l'Air et de l'Espace
Aéroport du Bourget
Boîte Postale 173
F-93352 Le Bourget Cedex
Telephone: (0)149927199
          (0)149927171
Telefax: (0)149927095
WWW: http://www.mae.org/
Founded: 1919
Staff: 100
Activities: exhibitions * visits * planetarium shows * documentation center
Periodicals: (24) "Pégase" (ISSN 0399-9939)
City Reference Coordinates: 002°26'00"E 48°56'00"N

Musée de l'Instrumentation Optique
Le Capitole
Place de la Mairie
F-68600 Biesheim
Telephone: (0)389720159
Telefax: (0)389721449
Activities: museum of optical instrumentation * exhibitions * astronomy club
City Reference Coordinates: 007°33'00"E 48°01'00"N (Neuf-Brisach)

Musée des Arts et Métiers (MAM)
292, rue Saint-Martin
F-75003 Paris
Telephone: (0)153018220
          (0)153018200
          (0)140272331 (answering machine)
Electronic Mail: <userid>@cnam.fr
WWW: http://www.arts-et-metiers.net/
Founded: 1794
Staff: 65
City Reference Coordinates: 002°20'00"E 48°52'00"N

Muséum National d'Histoire Naturelle (MNHN)
57, rue Cuvier
F-75231 Paris Cedex 05

Telephone: (0)140793226
Telefax: (0)140793524
Electronic Mail: sgmnhn@mnhn.fr
WWW: http://www.mnhn.fr/
Founded: 1974
Staff: 1800
Activities: exhibitions * public outreach * education * research * sciences of the universe * physico-chemistry and biology * biodiversity * ecology
City Reference Coordinates: 002°20'00"E 48°52'00"N

Musicosmic
c/o Ch. Bruneau/UGC0311
3, rue de Wimphelinge
F-67000 Strasbourg
Telephone: (0)390242475
Founded: 1990
Membership: 8
Activities: astronomy-related music * musical historiography * contemporeanous creations
City Reference Coordinates: 007°45'00"E 48°35'00"N

Nadir - Astronomie et Sciences
Observatoire du Mont Blanc
67, lacets du Belvédère
F-74400 Chamonix
Telephone: (0)450534516
Electronic Mail: nadir@obs-montblanc.org
WWW: http://www.obs-montblanc.org/
Founded: 2001
Membership: 21
Activities: public presentations * lectures * exhibitions * photography * observing at high-altitude sites
Coordinates: 006°52'00"E 45°55'00"N H1,036m
• Formerly known as "Club d'Astronomie de Chamonix"
City Reference Coordinates: 006°52'00"E 45°55'00"N

Novespace SA
15, rue des Halles
F-75001 Paris
Telephone: (0)142334141
Telefax: (0)140260860
Electronic Mail: espace@novespace.fr
WWW: http://www.novespace.fr/
Founded: 1986
Staff: 12
Activities: technology transfer
City Reference Coordinates: 002°20'00"E 48°52'00"N

Observatoire Antarès
193, chemin des Eaux
F-83500 La-Seyne-sur-Mer
Telephone: (0)494877447
Telefax: (0)494877447
Founded: 1964
Membership: 22
Activities: observing * information handling
Coordinates: 005°53'00"E 43°06'00"N
City Reference Coordinates: 005°53'00"E 43°06'00"N

Observatoire Astronomique d'Aniane
• See now "Géospace Observatoire d'Aniane"

Observatoire Astronomique de Saint-Jean-le-Blanc
142, rue Demay
F-45650 Saint-Jean-le-Blanc
Telephone: (0)238510757
Electronic Mail: obsstjean@free.fr
WWW: http://obsstjean.free.fr/
Founded: 1985
Membership: 25
Activities: observing * education * popularization * instrumentation
Periodicals: (6) "L'Obs'session"
Coordinates: 001°56'04"E 47°52'04"N
City Reference Coordinates: 001°53'00"E 47°53'00"N

Observatoire Astronomique des Pises (OAP)
F-30750 Dourbies

Telephone: (0)467735558
Electronic Mail: jmlopez@pratique.fr (Jean-Marie Lopez, President)
WWW: http://www.bdl.fr/pises/pises.html
http://www.bdl.fr/s2p/pises/
Founded: 1988
Membership: 50
Activities: deep-sky observing * photography * SN and comet search
Coordinates: 003°27'00"E 44°02'22"N H1,300m (IAU Code 122)
City Reference Coordinates: 003°28'00"E 44°05'00"N

Observatoire Astronomique de Strasbourg
● See "Université de Strasbourg I, Observatoire Astronomique"

Observatoire Astronomique Marseille-Provence, Laboratoire d'Astrophysique de Marseille (LAM)
● See "Laboratoire d'Astrophysique de Marseille (LAM)"

Observatoire Astronomique Marseille-Provence, Observatoire de Haute Provence (OHP)
● See "Observatoire de Haute Provence (OHP)"

Observatoire de Besançon
Boîte Postale 1615
F-25010 Besançon Cedex
or :
41bis, avenue de l'Observatoire
F-25000 Besançon
Telephone: (0)381666900
Telefax: (0)381666944
Electronic Mail: <userid>@obs-besancon.fr
WWW: http://www.obs-besancon.fr/
Founded: 1882
Staff: 35
Activities: galactic structure and dynamics * time metrology * planetology
Coordinates: 005°59'18"E 47°14'58"N H315m (IAU Code 016)
City Reference Coordinates: 006°02'00"E 47°15'00"N

Observatoire de Bordeaux
● See "Université de Bordeaux 1, Observatoire"

Observatoire de Grenoble
● See "Université de Grenoble I, Observatoire de Grenoble, Groupe d'Astrophysique"

Observatoire de Haute Provence (OHP)
F-04870 Saint-Michel-l'Observatoire
Telephone: (0)492706400
Telefax: (0)492766295
Electronic Mail: <userid>@obs-hp.fr
WWW: http://www.obs-hp.fr/
Founded: 1936
Staff: 60
Activities: maintaining observing facilities for visiting astronomers * extra-solar planets * peculiar stars (cool, variable, X sources) * galaxies * QSOs * ISM
Periodicals: (1) "Rapport Annuel d'Activité"; "La Lettre de l'OHP", "Pré-publications"
Coordinates: 005°42'47"E 43°55'53"N H665m (IAU Code 511)
● Component of the "Observatoire Astronomique Marseille-Provence"
City Reference Coordinates: 005°43'00"E 43°56'00"N

Observatoire de la Côte d'Azur (OCA)
Boîte Postale 4229
F-06304 Nice Cedex 4
Telephone: (0)492003011
Telefax: (0)492003033
Electronic Mail: <userid>@obs-nice.fr
WWW: http://www.obs-nice.fr/
Founded: 1988 (1887 as Observatoire de Nice)
Staff: 220
Activities: astrometry * geodynamics * dynamics * fluid mechanics * stellar, galactic, and extragalactic astronomy * solar physics
Coordinates: 007°18'06"E 43°43'24"N H372m (IAU Code 020)
City Reference Coordinates: 007°15'00"E 43°42'00"N

Observatoire de la Côte d'Azur (OCA), Centre d'Études et de Recherches en Géodynamique et Astrométrie (CERGA)
Avenue Copernic
F-06130 Grasse

or :
c/o Observatoire de Nice
Boîte Postale 229
F-06004 Nice Cedex 4
Telephone: (0)493405353
Telefax: (0)493405333
Electronic Mail: <userid>@obs-azur.fr
WWW: http://www.obs-nice.fr/
Founded: 1974 (1988 as part of OCA)
Staff: 60
Activities: astrometry * celestial mechanics * reference frames * Earth-Moon system * space geodesy * geodynamics * laser techniques * history of sciences * time standards
Coordinates: 006°55'30"E 43°44'54"N H1,270m (Calern - see separate entry)
City Reference Coordinates: 006°55'00"E 43°40'00"N (Grasse)
007°15'00"E 43°42'00"N (Nice)

Observatoire de la Côte d'Azur (OCA), Station de Calern
2130, route de l'Observatoire
Caussols
F-06460 Saint-Vallier-de-Thiey
Electronic Mail: <userid>@obs-azur.fr
WWW: http://www.obs-nice.fr/
Founded: 1974 (1988 as part of OCA)
Staff: 40
Coordinates: 006°55'30"E 43°44'54"N H1,270m (IAU Code 010)
City Reference Coordinates: 006°15'00"E 43°42'00"N

Observatoire de l'Association Culturelle de Dax (OACD)
7, rue des Chênes
F-40100 Dax
Telephone: (0)558561447
WWW: http://www.iap.fr/saf/cldax.htm
http://www.astrosurf.com/obsdax/
http://ourworld.compuserve.com/homepages/DUPOUY_Philippe/
Founded: 1968
Membership: 50
Activities: introduction to astronomy * observing * photography * radioastronomy * meteorological imaging * image processing * planetarium
Coordinates: 001°01'46"W 43°41'35"N H35m (IAU Code 958)
City Reference Coordinates: 001°03'00"W 43°43'00"N

Observatoire de Lyon
● See now "Centre de Recherche Astronomique de Lyon (CRAL)"

Observatoire de Marseille
● See now "Laboratoire d'Astrophysique de Marseille (LAM)"

Observatoire de Meudon
● See "Observatoire de Paris"

Observatoire de Nice
● See now "Observatoire de la Côte d'Azur (OCA)"

Observatoire de Paris
61, avenue de l'Observatoire
F-75014 Paris
or :
5, place Jules-Janssen
F-92195 Meudon Cedex
Telephone: (0)140512221
Telefax: (0)143541804
Electronic Mail: <userid>@obspm.fr
WWW: http://www.obspm.fr/
Founded: 1667
Staff: 700
Activities: astrophysical and physical research * cosmology * Sun * planets * stars * galaxies * ISM * radioastronomy * space-time reference frames * space research * instrumentation * education
Periodicals: "Cartes Synoptiques de la Chromosphère Solaire", "Bulletin de l'IERS"
Coordinates: 002°20'14"E 48°50'10"N H67m (IAU Code 007)
002°13'53"E 48°48'18"N H162m (Meudon)
002°11'51"E 47°22'24"N H183m (Nançay - see separate entry)
City Reference Coordinates: 002°20'00"E 48°52'00"N (Paris)
002°14'00"E 48°48'00"N (Meudon)

Observatoire de Paris, Institut de Mécanique Céleste et de Calcul des Éphémérides (IMCCE)
77, avenue Denfert Rochereau
F-75014 Paris
Telephone: (0)140512128
    (0)140512270
Telefax: (0)146332834
Electronic Mail: <userid>@bdl.fr
    bdl@bdl.fr
WWW: http://www.bdl.fr/
Founded: 1795
Staff: 43
Activities: celestial mechanics * solar system * ephemeris * fundamental astronomy * astrometry
Periodicals: (1) "Éphémérides Astronomiques - Annuaire du Bureau des Longitudes", "La Connaissance des Temps" (ISSN 0181-3048), "Phénomènes et Configurations des Satellites Galiléens de Jupiter" (ISSN 0769-1033), "Configurations des Huit Premiers Satellites de Saturne" (ISSN 0769-1025), "Éphémérides des Satellites de Jupiter, Saturne et Uranus", "Éphémérides des Satellites Faibles de Jupiter et de Saturne" (ISSN 0769-1041), "Éphémérides Nautiques"; "Notes Scientifiques et Techniques du BDL"
• Formerly "Bureau des Longitudes (BDL)"
City Reference Coordinates: 002°20'00"E 48°52'00"N

Observatoire de Paris, Station de Radioastronomie de Nançay
F-18330 Nançay
Telephone: (0)248518241
    (0)248518606 (Director)
    (0)248518605 (Secretary)
Telefax: (0)248518318
Electronic Mail: nicolas.dubouloz@obs-nancay.fr (Nicolas Dubouloz, Director)
WWW: http://www.obs-nancay.fr/
Founded: 1953
Staff: 50
Activities: radioastronomical research
Coordinates: 002°11'48"E 47°22'48"N H150m
City Reference Coordinates: 002°15'00"E 47°19'00"N

Observatoire de Puimichel
c/o Danny Cardoen
Le Haut de la Chapelle
F-04700 Puimichel
Telephone: (0)492787922
    (0)492787969
WWW: http://www.insa-lyon.fr/Associations/ASTRO/fr/puimichel.html
    http://ping4.ping.be/eurospace/pui.htm
Coordinates: 006°01'00"E 43°58'30"N H750m
City Reference Coordinates: 006°01'00"E 43°59'00"N

Observatoire de Rouen (OR)
3, impasse Adrien-Auzout
F-76000 Rouen
Telephone: (0)235880196
Electronic Mail: observatoirederouen@wanadoo.fr
WWW: http://www.chez.com/observatoirederouen/
Founded: 1884
Membership: 60
Activities: observing * popularization (public, schools) * photography * workshops for teachers * public sessions * video recording * planetarium shows
Periodicals: (12) "Bulletin de l'Observatoire de Rouen"
Coordinates: 001°14'33"E 49°26'29"N H46m
City Reference Coordinates: 001°05'00"E 49°26'00"N

Observatoire des Alpes Maritimes (OAM)
• See now "Observatoire de la Côte d'Azur (OCA)"

Observatoire de Toulouse
• See "Observatoire Midi-Pyrénées (OMP)"

Observatoire du Pic du Midi
• See "Observatoire Midi-Pyrénées (OMP), Observatoire du Pic du Midi"

Observatoire Midi-Pyrénées (OMP)
14, avenue Edouard-Belin
F-31400 Toulouse
Telephone: (0)561332929

Planétarium de Bourbon-Lancy
Place Sénateur Turlier
F-71140 Bourbon-Lancy
Telephone: (0)385890978
Electronic Mail: verdenetmichel@minitel.net (Michel Verdenet)
WWW: http://astro.u-strasbg.fr/Obs/PLANETARIUM/bourbon/bourbon.htm
Founded: 1993
City Reference Coordinates: 003°46'00"E 46°37'00"N

Planétarium de Bretagne
Cosmopolis
F-22560 Pleumeur-Bodou
Telephone: (0)296158030
Telefax: (0)296158031
Electronic Mail: contact@planetarium-bretagne.fr
WWW: http://www.planetarium-bretagne.fr/
Founded: 1988
Staff: 6
Activities: popularization * shows
• Formerly know as "Planétarium du Trégor"
City Reference Coordinates: 003°32'00"W 48°46'00"N

Planétarium de Cholet
c/o Maison des Sciences
1, rue Lamarque
F-49300 Cholet
Telephone: (0)241624036
Telefax: (0)241719461
Electronic Mail: sla@mygale.org (Société des Sciences, Lettres et Arts de Cholet)
WWW: http://www.mygale.org/04/sla
Founded: 1973
Membership: 400
Staff: 20
Activities: shows * observing * instrument making * lectures
City Reference Coordinates: 000°53'00"W 47°04'00"N

Planétarium de Montpellier
Jardin des Plantes
Boîte Postale 1088
F-34007 Montpellier
Telephone: (0)467634322
Telefax: (0)467611008
Electronic Mail: planetarium@planetarium-montpellier.com
WWW: http://www.planetarium-montpellier.com/accueil.htm
Founded: 1989
Staff: 50
Activities: shows * education
Periodicals: "Le Planétarium" (ISSN 1243-4833)
City Reference Coordinates: 003°53'00"E 43°36'00"N

Planétarium de Nantes
8, rue des Acadiens
F-44100 Nantes
Telephone: (0)240739923
Telefax: (0)240419239
Electronic Mail: Teletel: 3614 Code: TELEM
WWW: http://tourisme.voila.fr/villes/nantes/fra/sit/ville/planetar/acc.htm
Founded: 1981
Staff: 5
Activities: shows related to astronomy and to space sciences and techniques
City Reference Coordinates: 001°33'00"W 47°13'00"N

Planétarium de Nîmes
• See "Association des Amis du Planétarium, Nîmes"

Planétarium de Parthenay
20, rue de la Citadelle
F-79200 Parthenay
Telephone: (0)549642301
Founded: 1996
Staff: 1
Activities: shows * workshops * observing
City Reference Coordinates: 000°15'00"W 46°39'00"N

Planétarium de Poitiers
• See "Planétarium - Lasérium"

Planétarium de Reims
Ancien Collège des Jésuites
1, place Museux
F-51100 Reims
Telephone: (0)326855150
Telefax: (0)326827863
Electronic Mail: planetica@wanadoo.fr
WWW: http://www.ville-reims.com/
http://assoc.wanadoo.fr/planetica/
Founded: 1980
Activities: education * public shows
City Reference Coordinates: 004°02'00"E 49°15'00"N

Planétarium de Saint-Étienne
Espace Fauriel
28, rue Pierre-et-Dominique Ponchardier
F-42100 Saint-Étienne
Telephone: (0)477255492
Telefax: (0)477333570
Electronic Mail: planetarium@sideral.com
WWW: http://www.sideral.com/
Founded: 1990
Staff: 8
City Reference Coordinates: 004°24'00"E 45°26'00"N

Planétarium de Vaulx-en-Velin
Place de la Nation
Boîte Postale 166
F-69512 Vaulx-en-Velin Cedex
Telephone: (0)478795010
(0)478795013
Electronic Mail: stars@planetariumvv.com
WWW: http://www.planetariumvv.com/
City Reference Coordinates: 004°56'00"E 45°47'00"N

Planétarium du Trégor
• See now "Planétarium de Bretagne"

Planétarium Observatoire de Montredon-Labessonnié
Mairie
F-81360 Montredon-Labessonnié
Telephone: (0)563756312
Telefax: (0)563751811
Electronic Mail: planetarn@wanadoo.fr
WWW: http://perso.wanadoo.fr/planetarn/
Founded: 1993
Membership: 11
Activities: popularization * shows * observing * education
Coordinates: 002°17'50"E 43°44'25"N
City Reference Coordinates: 002°18'00"E 44°45'00"N

Planétarium Peiresc
Château St-Mitre
7, rue des Robiniers
F-13090 Aix-en-Provence
or :
Parc Saint-Mitre
Avenue Jean- Monnet
F-13090 Aix-en-Provence
Telephone: (0)442204366
Telefax: (0)442204366
Electronic Mail: aix.planet@free.fr
WWW: http://aix.planet.free.fr/
Founded: 2002
• See also "Amis du Planétarium d'Aix-en-Provence (APAP)"
City Reference Coordinates: 005°26'00"E 43°32'00"N

Planète Sciences
c/o Secrétariat
16, place Jacques-Brel
F-91130 Ris-Orangis
Telephone: (0)169027610

Telefax: (0)169432143
Electronic Mail: astro@anstj.org
WWW: http://www.anstj.org/
Founded: 1962
Membership: 1200 (and 600 clubs)
Activities: education * science for young people * workshops * training of teachers * national coordination * observatory * holiday camps for young people
Periodicals: (5) "ANSTJ Bonjour"
Coordinates: 002°26'16"E 48°17'30"N (IAU Code 199)
• Formerly known as "Association Nationale Sciences Techniques Jeunesse (ANSTJ)"
City Reference Coordinates: 002°25'00"E 48°39'00"N

Prospace
34, rue des Bourdonnais
F-75001 Paris
Telephone: (0)144889930
Telefax: (0)144889939
Electronic Mail: prospace@prospace-fr.com
WWW: http://www.prospace-fr.com/
Founded: 1974
Membership: 52 (societies)
Staff: 6
Activities: promoting activities, products, means, and services of French space industry for export * prospecting space markets
Periodicals: "News from Prospace"
City Reference Coordinates: 002°20'00"E 48°52'00"N

Provence Sciences Techniques Jeunesse (PSTJ)
1257, route de Grasse
F-06580 Pégomas
Telephone: (0)493609177
Telefax: (0)493609277
Electronic Mail: pstj@chez.com
WWW: http://www.chez.com/pstj
  http://wwwrc.obs-azur.fr/pstj/pstj.htm
Founded: 1992
Membership: 200
Activities: education * astronomy * rocketry * robotics * computer science * environment
Periodicals: (6) "Scoop" (Circ.: 50)
City Reference Coordinates: 007°00'00"E 43°33'00"N (Cannes)

REOSC, Recherche Étude Optique
(Groupe SFIM)
Avenue de la Tour Maury
F-91280 Saint Pierre du Perray
Telephone: (0)169897200
Telefax: (0)169897220
Electronic Mail: reosc@sfim.fr
WWW: http://www.reosc.com/
Founded: 1937
Staff: 95
Activities: studying and manufacturing optical instrumentation for space, astronomy and laser systems * figuring, lightweighting and polishing large mirrors
City Reference Coordinates: 002°18'00"E 48°42'00"N (Longjumeau)

R.S. Automation Industrie (RSAI)
Boîte Postale 40
Z.I. de la Vaure
Rue des Mineurs
F-42290 Sorbiers
Telephone: (0)477533048
Telefax: (0)477533048
Electronic Mail: sales@rsacosmos.com
WWW: http://www.rsacosmos.com/
Founded: 1979
Staff: 10
Activities: manufacturing planetariums * producing software for simulations
Coordinates: 004°24'00"E 45°25'00"N H550m
City Reference Coordinates: 004°30'00"E 45°29'00"N (Izieux)

Salons Internationaux de l'Aéronautique et de l'Espace
4, rue Galilée
F-75116 Paris
Telephone: (0)153233333
Telefax: (0)147200086
Electronic Mail: info@salon-du-bourget.fr

WWW: http://www.salon-du-bourget.fr/
Founded: 1909
Staff: 15
Activities: organizing the biennial Paris air show
City Reference Coordinates: 002°20'00"E 48°52'00"N

SAT
* See "Société Anonyme de Télécommunications (SAT)"

Scot
8-10 rue Hermès
Parc Technologique du Canal
F-31526 Ramonville-Saint-Agne Cedex
Telephone: (0)561394600
Telefax: (0)561394610
Electronic Mail: contact@scot.cnes.fr
WWW: http://www.scot-sa.com/
Founded: 1987
Activities: design and consultancy in Earth resources
City Reference Coordinates: 001°26'00"E 43°36'00"N (Toulouse)

Service Hydrographique et Océanographique de la Marine (SHOM)
Boîte Postale 5
F-00307 Armées
or :
3, avenue Octave-Gréard
F-75007 Paris
Telephone: (0)144384116
Telefax: (0)140659998
Electronic Mail: <userid>@shom.fr
WWW: http://www.shom.fr/
Founded: 1720
Staff: 862
Activities: hydrography * oceanography * marine mapping
Periodicals: "Épidécides Lunaires" (ISSN 0240-8376), "Épiménides Nautiques" (ISSN 0240-8368), "Annales Hydrographiques" (ISSN 0373-3629), "Rapport Annuel" (ISSN 0989-5876)
City Reference Coordinates: 002°20'00"E 48°52'00"N (Paris)

Société Astronomique de Bordeaux (SAB)
Hotel des Sociétés Savantes
1, place Bardineau
F-33000 Bordeaux
Telephone: (0)556516932
Electronic Mail: sab33@wanadoo.fr
  guy-libante@wanadoo.fr (Guy Libante, Editor)
WWW: http://www.astrosurf.org/sab33
Founded: 1909
Membership: 70
Activities: general astronomy * meteorology * observing * lectures * slide shows * films
Periodicals: "Astronomie Passion" (ISSN 1283-3339)
City Reference Coordinates: 000°34'00"W 44°30'00"N

Société Astronomique de Bourgogne (SAB)
4, rue Chancelier de l'Hospital
F-21000 Dijon
Telephone: (0)380364413
Electronic Mail: sab@astrosurf.com
WWW: http://www.astrosurf.com/sab
Founded: 1975
Membership: 120
Activities: popularization * observing * education * lectures * Sun * Moon * deep sky * CCDs * light pollution
Periodicals: (4) "L'Observation du Ciel"
Coordinates: 004°57'44"E 47°18'11"N H429m (Hautes-Plates)
City Reference Coordinates: 005°02'00"E 47°20'00"N

Société Astronomique de France (SAF)
3, rue Beethoven
F-75016 Paris
Telephone: (0)142241374
Electronic Mail: saf@calva.net
WWW: http://www.iap.fr/saf/
Founded: 1887
Membership: 2500
Activities: amateur astronomy
Periodicals: (12) "L'Astronomie" (ISSN 0004-6302, circ.: 6,000); (4) "Observations et Travaux" (ISSN 0769-0878)

Awards: (1) "Prix Janssen"
Coordinates: 002°20'41"E 48°50'55"N (Sorbonne, Paris)
City Reference Coordinates: 002°20'00"E 48°52'00"N

Société Astronomique de France (SAF), Groupe d'Alsace (SAFGA)
c/o Observatoire Astronomique de Strasbourg
11, rue de l'Université
F-67000 Strasbourg
WWW: http://www.iap.fr/saf/
Founded: 1931
Membership: 120
Activities: lectures * education * colloquia * observing * CCDs
Periodicals: (12) "Alsace Astronomie"
Coordinates: 007°46'04"E 48°35'02"N H142m
City Reference Coordinates: 007°45'00"E 48°35'00"N

Société Astronomique de Haute-Marne (SAHM)
c/o Pietro Bergamini
Rue du Stade
F-52100 Valcourt
Telephone: (0)325062218 (Observatory)
Founded: 1985
Membership: 40
Activities: observing * lectures * exhibitions * education * meetings
Coordinates: 004°54'00"E 48°38'00"N H139m
City Reference Coordinates: 004°54'00"E 48°38'00"N

Société Astronomique de la Montagne de Lure
Route de Lure
F-04230 Saint-Étienne-des-Orgues
Telephone: (0)492731798
Founded: 1992
Activities: observing * popularization
City Reference Coordinates: 005°47'00"E 44°02'00"N

Société Astronomique de Lyon (SAL)
c/o Observatoire de Lyon
Avenue Charles André
F-69230 Saint-Genis-Laval
Telephone: (0)478595839
Electronic Mail: soas.lyon@wanadoo.fr
WWW: http://sal.ifrance.com/sal/
Founded: 1931
Membership: 140
Activities: observing * workshops * lectures * visits of observatories * mirror polishing
Periodicals: (4) "Société Astronomique de Lyon"
City Reference Coordinates: 004°51'00"E 45°45'00"N

Société Astronomique de Montpellier (SAM) "Pierre Vauriot"
66, boulevard de l'Observatoire
F-34000 Montpellier
Telephone: (0)467661214
Electronic Mail: nicolas.montviloff@wanadoo.fr (Nicolas Montviloff, President)
Founded: 1979
Membership: 30
Activities: popularization * observing * photography * CCDs * software * education
Observatories: 1 (Observatoire Astronomique des Pises - see separate entry)
City Reference Coordinates: 003°53'00"E 43°36'00"N

Société Astronomique du Haut-Rhin (SAHR)
21, rue du Bois
F-68570 Osenbach
Telephone: (0)389476248
WWW: http://astroclub.net/saturne/sahr/
Coordinates: 007°12'50"E 47°59'22"N H470m (IAU Code 630)
City Reference Coordinates: 007°13'00"E 47°59'00"N

Société Astronomique du Havre
5, passage Lenormand
F-76600 Le Havre
Telephone: (0)235515611
Electronic Mail: philippe.baudouin@wanadoo.fr
WWW: http://perso.wanadoo.fr/philippe.baudouin/
Founded: 1971
Membership: 30

Activities: telescope making * introduction to observing * video observing * double stars * variable stars * photography * computing
City Reference Coordinates: 000°08'00"E 49°30'00"N

Société Astronomique Populaire du Centre (SAPC)
Observatoire d'Arçay
40, Grande Rue
F-18340 Arcay
or :
25, place de la Pyrotechnie
F-18000 Bourges
Telephone: (0)248251001
Electronic Mail: pierre.durand@wanadoo.fr (Pierre Durand, President)
WWW: http://www.astrosurf.com/SAPC/
Founded: 1986
Membership: 80
Activities: education * observing
City Reference Coordinates: 002°24'00"E 47°05'00"N (Bourges)

Société d'Astronomie de Cannes (SACA)
Boîte Postale 125
F-06405 Cannes Cedex
Telephone: (0)493649884
         (0)493990345
Founded: 1991
Membership: 50
Activities: observing * lectures
City Reference Coordinates: 007°01'00"E 43°33'00"N

Société d'Astronomie de Nantes (SAN)
35, boulevard Louis-Millet
F-44300 Nantes
Telephone: (0)240689120
Telefax: (0)240938123
Electronic Mail: san@san-fr.com
WWW: http://www.san-fr.com/
Founded: 1971
Membership: 130
Activities: meetings * library * observing * computing * lectures * popularization
City Reference Coordinates: 001°33'00"W 47°13'00"N

Société d'Astronomie de Saône et Loire (SASL)
25, rue Pierre-Bridet
F-71100 Chalon-sur-Saône
Telephone: (0)385920706
Electronic Mail: sasl@astrosurf.com
WWW: http://www.astrosurf.com/sasl/
Founded: 1984
Membership: 46
Activities: education * observing * instrumentation * CCDs
Coordinates: 004°40'32"E 46°43'05"N H417m (Buxy)
City Reference Coordinates: 004°51'00"E 46°47'00"N

Société d'Astronomie Populaire (SAP)
c/o Observatoire de Jolimont
1, avenue Camille Flammarion
F-31500 Toulouse
Telephone: (0)561584201
Electronic Mail: sap@saptoulouse.net
WWW: http://saptoulouse.net/
Founded: 1910
Membership: 1000
Activities: observing * popularization
Periodicals: (6) "Pulsar" (ISSN 0154-4101); (1) "Éphémérides"
Observatories: 1 (Observatoire de Jolimont)
City Reference Coordinates: 001°26'00"E 43°36'00"N

Société Française de Physique (SFP)
33, rue Croulebarbe
F-75013 Paris
Telephone: (0)144086710
Telefax: (0)144086719
Electronic Mail: secretariat@sfpnet.org
WWW: http://sfp.in2p3.fr/
Founded: 1873

Membership: 3200
Periodicals: (5) "Bulletin" (ISSN 0037-9360); (1) "Annuaire" (ISSN 0081-1076)
Awards: "Gentner-Kastler" (jointly with the Deutsche Physikalische Gesellschaft), "Holweck" (jointly with the Institute of Physics), "Rammal Rammal", "Jean Ricard", "Félix Robin", "Jean Perrin", "Louis Ancel", "Aimé Cotton", "Paul Langevin", "Joliot-Curie", "Esclangon", "Foucault", "Daniel Guinier"
City Reference Coordinates: 002°20'00"E 48°52'00"N

Société Française des Spécialistes d'Astronomie (SFSA)
• See now "Société Française d'Astronomie et d'Astrophysique (SF2A)"

Société Française d'Astronomie et d'Astrophysique (SF2A)
c/o Institut d'Astrophysique de Paris
98bis, boulevard Arago
F-75014 Paris
Electronic Mail: sf2a@iap.fr
WWW: http://www.iap.fr/sf2a/
Founded: 1978
Membership: 520
Activities: organizing colloquia, courses, meetings * yearly "École de Goutelas"
Awards: (1) "Prix Compaq"
• Formerly known as "Société Française des Spécialistes d'Astronomie (SFSA)"
City Reference Coordinates: 002°20'00"E 48°52'00"N

Société Française d'Optique (SFO)
Bâtiment 503
Centre Universitaire d'Orsay
F-91403 Orsay Cedex
Telephone: (0)169358816
Telefax: (0)169853565
Electronic Mail: sfo@france-optique.org
WWW: http://www.france-optique.org/
Founded: 1983
Membership: 1650
Periodicals: (4) "Photoniques" (ISSN 1629-4475); (1) "Annuaire" (ISSN 1154-4317)
City Reference Coordinates: 002°11'00"E 48°48'00"N

Société Lorraine d'Astronomie (SLA)
c/o Lycée Saint Joseph
413, avenue de Boufflers
F-54520 Laxou
Telephone: (0)383933568
WWW: http://perso.wanadoo.fr/isabelle.berquand/
City Reference Coordinates: 006°08'00"E 48°41'00"N

SPOT Image
Boîte Postale 4359
F-31030 Toulouse Cedex
or :
5, rue des Satellites
F-31030 Toulouse Cedex
Telephone: (0)562194040
Telefax: (0)562194011
WWW: http://www.spotimage.com/
Founded: 1982
Staff: 190
Activities: distributing worldwide geographical data collected by SPOT satellites
Periodicals: (4) "SPOT Flash" (ISSN 1161-3289 & 1161-3297); (2) "SPOT Magazine" (ISSN 0764-048x)
City Reference Coordinates: 001°26'00"E 43°36'00"N

Station Astronomique Jansky (SAJ)
Prairie de Mérignan
F-45240 La Ferté Saint-Aubin
Telephone: (0)254988797
Telefax: (0)254988797
Electronic Mail: saj@astrosurf.com
WWW: http://www.astrosurf.com/saj/
Founded: 1987
Membership: 15
Activities: observing
Coordinates: 001°55'00"E 47°49'00"N H110m
City Reference Coordinates: 001°55'00"E 47°49'00"N

Station de Radioastronomie de Nançay
• See "Observatoire de Paris, Station de Radioastronomie de Nançay"

UNESCO
● See "United Nations Educational, Scientific and Cultural Organization"

Union Astronomique Internationale (UAI)
● See "International Astronomical Union (IAU)"

Union Internationale des Associations et Organismes Techniques (UATI)
c/o Maison de l'UNESCO
1, rue Miollis
F-75732 Paris Cedex 15
Telephone: (0)145669410
Telefax: (0)143062927
Electronic Mail: uati@unesco.org
WWW: http://www.unesco.org/uati/
Founded: 1951
Staff: 10
Activities: promoting and coordinating scientific and technical activities of concern to members
Periodicals: (2) "Bulletin"; (1) "Annuaire"
City Reference Coordinates: 002°20'00"E 48°52'00"N

Union Rhône-Alpes des Clubs d'Astronomie (URACA)
c/o Club d'Astronomie de Lyon-Ampère
37, rue Paul-Cazeneuve
F-69008 Lyon
Telephone: (0)478012905
Founded: 1985
Membership: 25 (clubs)
Activities: promoting astronomy in the area
City Reference Coordinates: 004°51'00"E 45°45'00"N

Université d'Aix-Marseille I, Laboratoire d'Astrophysique de Marseille (LAM)
● See "Laboratoire d'Astrophysique de Marseille (LAM)"

Université de Bordeaux 1, Observatoire
2, rue de l'Observatoire
Boîte Postale 89
F-33270 Floirac
Telephone: (0)557776100
Telefax: (0)557776110
Electronic Mail: <userid>@observ.u-bordeaux.fr
WWW: http://www.observ.u-bordeaux.fr/
Founded: 1879
Staff: 69
Activities: astrometry (automatic meridian circle, stars, planetary satellites, minor planets) * stellar kinematics and dynamics * radioastronomy (molecular clouds, star formation, VLBI) * solar convection * interplanetary matter * planetary atmospheres * Earth * exobiology
Coordinates: 000°31'42"W 44°50'06"N H73m (IAU Code 999)
City Reference Coordinates: 000°32'00"W 44°49'00"N

Université de Grenoble I, Laboratoire de Glaciologie et de Géophysique de l'Environnement (LGGE)
54, rue Molière
Boîte Postale 96
F-38402 Saint-Martin d'Hères Cedex
Telephone: (0)476824200
Telefax: (0)476824201
Electronic Mail: raynaud@glaciog.ujf-grenoble.fr (Dominique Raynaud, Director)
WWW: http://www.ujf-grenoble.fr/ujf/fr/recherche/labujf/lggefran.phtml
    http://glaciog.ujf-grenoble.fr/
Staff: 45
Activities: glaciology * climatology * atmospheric chemistry * planetology * experimental astrophysics * IR spectroscopy
● Alternate university name: "Université Joseph Fourier"
City Reference Coordinates: 005°43'00"E 45°10'00"N

Université de Grenoble I, Observatoire de Grenoble, Laboratoire d'Astrophysique (LAOG)
Boîte Postale 53X
F-38041 Grenoble Cedex 9
or :
414, rue de la Piscine
F-38042 Saint-Martin d'Hères
Telephone: (0)476514788
    (0)476514981
Telefax: (0)476448821
Electronic Mail: <userid>@obs.ujf-grenoble.fr
WWW: http://www-laog.obs.ujf-grenoble.fr/
Founded: 1985 (1979 as "Groupe d'Astrophysique")

Staff: 49
Activities: mm-wavelength astronomy * molecular clouds * star formation * astrochemistry * molecular excitation * ISM * QSOs * circumstellar envelopes * IR astronomy * high-angular resolution * very-low-mass stars * planetary disks * high-energy astrophysics
Coordinates: 005°54'26"E 44°38'01"N H2,552m (Observatoire du Plateau de Bure)
• Alternate university name: "Université Joseph Fourier"
City Reference Coordinates: 005°43'00"E 45°10'00"N

Université de Lille, Laboratoire d'Astronomie
1, impasse de l'Observatoire
F-59000 Lille
Telephone: (0)320524424
Telefax: (0)320580328
Electronic Mail: lille@bdl.fr
WWW: http://www.bdl.fr/Equipes/LILLE/eqLILLE-pres.html
Founded: 1930
Staff: 4
Activities: celestial mechanics * astrometry * Saturn's satellites
Coordinates: 003°04'14"E 50°36'57"N H32m
• Full university name: "Université des Sciences et Technologies de Lille"
City Reference Coordinates: 003°04'00"E 50°38'00"N

Université de Montpellier II Sciences et des Techniques du Languedoc, Institut des Sciences de la Terre, de l'Eau et de l'Espace de Montpellier (ISTEEM)
• See "Institut des Sciences de la Terre, de l'Eau et de l'Espace de Montpellier (ISTEEM)"

Université de Nice-Sophia-Antipolis, Département d'Astrophysique
Parc Valrose
F-06108 Nice Cedex 2
Telephone: (0)493529806
Telefax: (0)493529004
Electronic Mail: <userid>@obs-nice.fr
WWW: http://www-astro.unice.fr/
Founded: 1968
Staff: 32
Activities: Sun * solar and stellar seismology * atmospheric optics * imagery * speckle interferometry
City Reference Coordinates: 007°15'00"E 43°42'00"N

Université de Paris-Sud
• See "Université de Paris XI"

Université de Paris XI, Centre de Spectrométrie Nucléaire et de Spectrométrie de Masse (CSNSM)
Bâtiments 104-108
F-91405 Orsay Cedex
Telephone: (0)169415213
Telefax: (0)169415268
Electronic Mail: <userid>@csnsm.in2p3.fr
WWW: http://www-csnsm.in2p3.fr/
Founded: 1962
Staff: 100
Activities: nuclear physics * nucleosynthesis * meteorites * accelerator mass spectrometry * solid-state physics
Periodicals: (1) "Annual Report"
• Alternate university name: "Université de Paris-Sud"
City Reference Coordinates: 002°11'00"E 48°48'00"N

Université de Paris XI, Laboratoire d'Astronomie
Bâtiment 470
F-91405 Orsay Cedex
Telephone: (0)169417766
Telefax: (0)169416380
Electronic Mail: michele.presse@df.cso.u-psud.fr
WWW: http://www.u-psud.fr/
• Alternate university name: "Université de Paris-Sud"
City Reference Coordinates: 002°11'00"E 48°48'00"N

Université de Paris XI, Laboratoire de Physique des Gaz et des Plasmas (LPGP)
Bâtiment 212
F-91045 Orsay Cedex
Telephone: (0)169417251
Telefax: (0)169417844
Electronic Mail: deut@psisun.u-psud.fr
WWW: http://www.u-psud.fr/
Founded: 1959
Staff: 117

Activities: basic research * hot plasmas * strongly coupled plasmas * weakly ionized discharges * molecular plasmas * GAPHYOR database
• Alternate university name: "Université de Paris-Sud"
City Reference Coordinates: 002°11'00"E 48°48'00"N

Université de Picardie Jules Verne (UPJV), Laboratoire de Physique Théorique et d'Astrophysique
33, rue Saint-Leu
F-80039 Amiens Cedex
Telephone: (0)322827633
Electronic Mail: <userid>@u-picardie.fr
WWW: http://www.u-picardie.fr/
Founded: 1982
Staff: 2
Activities: education * PN * HII regions
City Reference Coordinates: 002°18'00"E 49°54'00"N

Université des Sciences et Technologies de Lille
• See Université de Lille

Université de Strasbourg I, Centre de Données astronomiques de Strasbourg (CDS)
• See "Centre de Données astronomiques de Strasbourg (CDS)"

Université de Strasbourg I, Observatoire Astronomique
11, rue de l'Université
F-67000 Strasbourg
Telephone: (0)390242416 (Director)
(0)390242410 (Secretary)
Telefax: (0)390242432
Electronic Mail: <userid>@astro.u-strasbg.fr
WWW: http://astro.u-strasbg.fr/Obs-e.html (English Homepage)
http://astro.u-strasbg.fr/Obs.html (French Homepage)
Founded: 1881
Staff: 60
Activities: education * research * statistical astronomy * statistical methodology * galaxies * stellar physics * high-energy astrophysics * databases * hosting the "Centre de Données astronomiques de Strasbourg (CDS)" (see separate entry) * planetarium
Coordinates: 007°46'04"E 48°35'02"N H142m (IAU Code 522)
• Alternate university name: "Université Louis Pasteur (ULP)"
City Reference Coordinates: 007°45'00"E 48°35'00"N

Université Denis Diderot
• See "Université de Paris VII"

Université Joseph Fourier
• See "Université de Grenoble I"

Université Louis Pasteur (ULP)
• See "Université de Strasbourg I"

Université Pierre et Marie Curie
• See "Université de Paris VI"

Uranoscope de France
94, rue de la Glacière
F-75013 Paris
Telephone: (0)164420002
Telefax: (0)164078604
Electronic Mail: uranos@club-internet.fr
WWW: http://perso.club-internet.fr/uranos/
Founded: 1995
Staff: 15
Activities: developing international relationships between amateur astronomers
City Reference Coordinates: 002°20'00"E 48°52'00"N

Uranoscope de l'Île de France
7, avenue Carnot
F-77220 Gretz-Armainvilliers
Telephone: (0)164420002
Telefax: (0)164078604
Electronic Mail: uranos@club-internet.fr
WWW: http://perso.club-internet.fr/uranos/
Founded: 1983
Membership: 125
Activities: observing * photography * introduction to astronomy * planetarium
Periodicals: (4) "Cosmos Express"
Observatories: 1 (Observatoire Astronomique de la Brie)
City Reference Coordinates: 002°44'00"E 48°44'00"N

Georgia

Abastumani Astrophysical Observatory (AbAO)
• See "Georgian Academy of Sciences, Abastumani Astrophysical Observatory (AbAO)"

Astronomical Society of Georgia (ASG)
Kazbegi Avenue 2a
380060 Tbilisi
Electronic Mail: gsal@dtapha.kheta.ge
WWW: http://www.gcci.org.ge/asgeo.htm
Founded: 1996
Membership: 60
Periodicals: "Letters in Astronomy"
City Reference Coordinates: 042°50'00"E 41°46'00"N

Georgian Academy of Sciences, Abastumani Astrophysical Observatory (AbAO)
Mount Kanobili
383762 Abastumani
Telephone: (0)32-955367
Electronic Mail: tenat@dtapha.kheta.ge
WWW: http://www.acnet.ge/abastumani.htm
Founded: 1932
Staff: 185
Activities: Sun * planetary systems * variable stars * flare stars * spectral classification * Galaxy * galaxies * plasma astrophysics * upper atmosphere * interplanetary dust * history of astronomy
Periodicals: "Abastumanskaya Astrofizicheskaya Observatorija Byulleten" (ISSN 0375-6644)
Coordinates: 042°49'16"E 41°45'15"N H1,650m (IAU Code 119)
City Reference Coordinates: 042°51'00"E 41°44'00"N

Georgian Academy of Sciences, Abastumani Astrophysical Observatory (AbAO), Town Department
Kazbegi Avenue 2a
380060 Tbilisi
Telephone: (0)32-375226
          (0)32-375228
Electronic Mail: tenat@dtapha.kheta.ge
WWW: http://www.acnet.ge/abastumani.htm
Founded: 1967
Activities: see main entry
Periodicals: see main entry
City Reference Coordinates: 042°50'00"E 41°46'00"N

State Department for Standardization, Metrology and Certification of Georgia
67 Chargali Street
380092 Tbilisi
Telephone: (0)32-612530
Telefax: (0)32-612530
Electronic Mail: gestand@caucasus.net
City Reference Coordinates: 042°50'00"E 41°46'00"N

Germany

Abbe-Stiftung (Ernst-)
• See "Zeiss-Planetarium der Ernst-Abbe-Stiftung"

Akademie der Wissenschaften und der Literatur Mainz
Geschwister-Scholl-Straße 2
D-55131 Mainz
Telephone: (0)6131-5770
Telefax: (0)6131-577206
Electronic Mail: generalsekretariat@adwmainz.de
WWW: http://www.adwmainz.de/
Founded: 1949
Membership: 213
Activities: interdisciplinary research * basic research * publishing * symposiums
City Reference Coordinates: 008°16'00"E 50°01'00"N

Albert-Einstein-Institut (AEI)
• See "Max-Planck-Institut für Gravitationsphysik"

Allgäuer Volkssternwarte Ottobeuren (AVSO) eV
Bürgermeister-Hasel-Straße 17
D-87724 Ottobeuren
Telephone: (0)8332-9366058
Telefax: (0)8332-936890
Electronic Mail: info@avso.de
WWW: http://www.avso.de/
Founded: 1966
Membership: 100
Activities: education * photography * computing * observing
Periodicals: (4) "Astro-Amateur"
Coordinates: 010°17'18"E 47°55'48"N H746m
City Reference Coordinates: 010°19'00"E 47°56'00"N

Altec
Ostallee 41
D-54290 Trier
Telephone: (0)651-9940820
Telefax: (0)651-9940821
Electronic Mail: info@altec.de
WWW: http://www.altec.de/
Founded: 1994 (in telescope distribution)
City Reference Coordinates: 006°28'00"E 49°45'00"N

Amateurastronomische Vereinigung Göttingen (AVG) eV
c/o Matthias Elsen
Bramwaldstraße 6a
D-37081 Göttingen
Telephone: (0)551-9899051
Electronic Mail: maelavg@aol.com
WWW: http://www.avgoe.de/
Founded: 1996
Membership: 60
Activities: observing * monthly public presentations * weekly meetings * education
Periodicals: (4) "Nachtschicht"
City Reference Coordinates: 009°55'00"E 51°32'00"N

Amateur- und Präzisionsoptik-Mechanik Markus Ludes
• See now "APM-Telescopes Markus Ludes"

Andromeda
Markenkamp 16
D-45721 Haltern
Telephone: (0)2364-8003
Telefax: (0)2364-169211
Electronic Mail: jelitte.andromeda@cityweb.de
Founded: 1988
Staff: 4
Activities: mail order dealer (NASA publications, English and German astronomical books, planetary maps, slides, videos, NASA decals, patches, ...)
City Reference Coordinates: 007°10'00"E 51°46'00"N

APM-Telescopes Markus Ludes
Goebenstraße 35
D-66117 Saarbrücken
Telephone: (0)681-9641161
Telefax: (0)681-9541169
Electronic Mail: apm-telescopes@web.de
WWW: http://www.apm-telescopes.de/
Founded: 1990
Staff: 3
Activities: designing and manufacturing telescopes and computer-controlled mounts * importing/exporting and distributing components and accessories
• Formerly "Amateur- und Präzisionsoptik-Mechanik Markus Ludes"
City Reference Coordinates: 006°58'00"E 49°15'00"N

Arbeitsgemeinschaft Walter-Hohmann-Sternwarte Essen eV
Wallneyer Straße 159
D-45133 Essen
Telephone: (0)201-493941 (answering machine)
Telefax: (0)201-493941
Electronic Mail: info@walter-hohmann-sternwarte.de
WWW: http://www.walter-hohmann-sternwarte.de/
Founded: 1969
Membership: 85
Activities: public lectures * photography * optical and radio observing * CCDs * minor planets
Periodicals: (4) "Nova"
Coordinates: 006°58'46"E 51°23'41"N (IAU Code 636)
City Reference Coordinates: 007°00'00"E 51°45'00"N

Arbeitskreis Astronomie Freiburg eV (AKAF)
Basler Landstraße 74b
D-79111 Freiburg-im-Breisgau
Telephone: (0)761-46700
Founded: 1981
Membership: 48
Activities: Sun * Moon * planets * galaxies * popularization
Periodicals: (4) "Astronomisches Informationsblatt der Volkssternwarte Freiburg" (circ.: 1,000)
Coordinates: 007°50'00"E 48°00'00"N (Volkssternwarte Freiburg)
City Reference Coordinates: 007°50'00"E 48°00'00"N

Arbeitskreis Meteore eV (AKM)
Postfach 60 01 18
D-14401 Potsdam
or :
Mehlbeerenweg 5
D-14469 Potsdam
Telephone: (0)331-520707
Electronic Mail: jrendtel@aip.de
WWW: http://www.tu-chemnitz.de/~smo/meteore/akm.html
Founded: 1978
Membership: 70
Activities: observing and analyzing data on meteors, haloes, noctilucent clouds, aurorae * photographic meteor patrol
Periodicals: (12) "Meteoros" (ISSN 1435-0424)
City Reference Coordinates: 013°04'00"E 52°24'00"N

Arbeitskreis Sternfreunde Lübeck (ASL) eV
Postfach 22 09
D-23510 Lübeck
or :
(Observatory)
Am Ährenfeld 2
D-23564 Lübeck
Telephone: (0)451-898547 (Uwe Freitag, President)
Electronic Mail: asl@astronomie-luebeck.de
sternwarte@astronomie-luebeck.de
polaris@astronomie-luebeck.de
WWW: http://www.astronomie-luebeck.de/
Founded: 1977 (as former "Arbeitskreis Sternwarte Lübeck")
Membership: 75
Activities: education * observing (Sun, planets, occultations, comets, variable stars, aurorae) * photography
Periodicals: (3) "Polaris" (ISSN 0930-4916)
Observatories: 1 (Fliegenfelde)
City Reference Coordinates: 010°40'00"E 53°52'00"N

Archenhold-Sternwarte
Alt-Treptow 1

D-12435 Berlin
Telephone: (0)30-5348080
Telefax: (0)30-5348083
Electronic Mail: info@astw.de
WWW: http://www.astw.de/
Founded: 1896
Staff: 13
Activities: popularization * education * lectures * conferences * hosting the Arbeitskreis Geschichte der Astronomie
Coordinates: 013°28'42"E 52°29'12"N H32m (IAU Code 604)
City Reference Coordinates: 013°24'00"E 52°31'00"N

Astroclub Radebeul eV (ACR)
Auf den Ebenbergen 10a
D-01445 Radebeul
Telephone: (0)351-8381907
Telefax: (0)351-8381907
Electronic Mail: mail@astroclub-radebeul.de
WWW: http://www.astroclub-radebeul.de/
   http://www.astronomie-sachsen.de/sternfreund/ (Der Sternfreund)
Founded: 1959
Membership: 60
Activities: Sun * meteors * planets * photography * variable stars * CCDs * radio astronomy
Periodicals: (6) "Der Sternfreund" (co-publisher)
Coordinates: 013°37'20"E 51°06'59"N H185m
City Reference Coordinates: 013°43'00"E 51°13'00"N

Astrocom GmbH
Lochhamer Schlag 5
D-82116 Gräfelfing
Telephone: (0)89-89889600
Telefax: (0)89-89889601
WWW: http://www.astrocom.de/
Founded: 1984
Activities: manufacturing and distributing telescopes and accessories
City Reference Coordinates: 011°35'00"E 48°08'00"N (München)

Astronomie Arbeitsgruppe der Universität Oldenburg
Carl von Ossietzky Universität Oldenburg
Fachschule Physik
FB 8 Physik
Postfach
D-26111 Oldenburg
or :
Ammerländer Heerstraße 114-118
D-26129 Oldenburg (for parcel delivery)
Telephone: 441-7980
Telefax: 441-7983000
Electronic Mail: michael.uhlemann@informatik.uni-oldenburg.de
WWW: http://www.infodrom.north.de/~muh/Astronomie/Astro-AG/
Founded: 1992
City Reference Coordinates: 010°52'00"E 54°17'00"N

Astronomie Software Service
c/o Daniel Roth
Strundener Straße 79
D-51069 Köln
Telephone: (0)221-687112
Telefax: (0)221-687112
Electronic Mail: roth@ph-cip.uni-koeln.de
Founded: 1986
Staff: 1
● Software distributor
City Reference Coordinates: 006°57'00"E 50°56'00"N

Astronomiestation Demmin
An den Tannen
Postfach 11 46
D-19101 Demmin
Telephone: (0)3998-222410
Electronic Mail: c.fischer.demmin@t-online.de
WWW: http://home.t-online.de/home/c.fischer.demmin/Astrostation_Demmin.html
Founded: 1978
Staff: 1
Activities: education * observing
Coordinates: 013°03'18"E 53°54'18"N H34m
City Reference Coordinates: 013°03'00"E 53°55'00"N

Astronomische Arbeitsgemeinschaft Aalen (A4) eV
c/o Ulrich Görze
Besselstraße 5
D-73447 Oberkochen
Telephone: (0)7364-7406
Electronic Mail: ulrich.goerze@cyberfun.de
WWW: http://www.sternwarte-aalen.de/
Founded: 2001
Membership: 30
Activities: popularization * education * amateur astronomy
Coordinates: 010°04'47"E 48°50'05"N H467m (Sternwarte Aalen)
City Reference Coordinates: 010°07'00"E 48°50'00"N (Aalen)

Astronomische Arbeitsgemeinschaft der Liebigschule
c/o Werner Ziegs
Kollwitzstraße 3
D-60488 Frankfurt/Main
Telephone: (0)69-21239480
Telefax: (0)69-21239479
Electronic Mail: werner.ziegs@coherentinc.com
WWW: http://ffm.junetz.de/astro
Founded: 1974
Membership: 25
Activities: education
City Reference Coordinates: 008°40'00"E 50°07'00"N

Astronomische Arbeitsgemeinschaft der Volkssternwarte Singen eV
c/o Wolfgang Bodenmüller (President)
Ostlandstraße 18
D-78315 Radolfzell
Telephone: (0)7732-2418
Telefax: (0)7732-942654
Electronic Mail: wo.bo@t-online.de
WWW: http://www.sternwarte-singen.de/
Founded: 1983
Membership: 50
Activities: meetings * education * slide and video shows * public observing * computing * photography
Coordinates: 008°50'40"E 47°45'15"N H437m
City Reference Coordinates: 008°50'00"E 47°45'00"N (Singen)

Astronomische Arbeitsgemeinschaft Heuchelheim eV
Bachstraße 61
D-35452 Heuchelheim
Electronic Mail: frage@aag-heuchelheim
WWW: http://www.aag-heuchelheim.de/
Founded: 1975
City Reference Coordinates: 008°24'00"E 50°35'00"N (Giessen)

Astronomische Arbeitsgemeinschaft Mainz eV
Postfach 11 64
D-55001 Mainz
Telephone: (0)6764-3660
     (0)6131-236940 (Observatory)
Telefax: (0)6764-3660
Electronic Mail: aagmainz@gmx.de
WWW: http://iphcip1.physik.uni-mainz.de/~astro/aag/Welcome.html
     http://iphcip1.physik.uni-mainz.de/~astro/pop/VSW.html (Observatory)
Founded: 1970
Membership: 110
Activities: observing (deep sky, occultations, comets, eclipses, haloes, ...) * CCDs
Periodicals: (6) "Mitteilungen astronomischer Vereinigungen Rhein-Main-Nahe" (circ.: 500)
City Reference Coordinates: 008°16'00"E 50°01'00"N

Astronomische Arbeitsgemeinschaft Pfaffenwinkel (AAP)
c/o Magnus Zwick
Döbendorferstraße 3
D-86956 Schongau
WWW: http://ourworld.compuserve.com/homepages/DBecker_AAP/
Coordinates: 010°47'10"E 47°50'11"N
City Reference Coordinates: 010°54'00"E 47°49'00"N

Astronomische Arbeitsgemeinschaft Waldhügel
c/o Georg Neumann
Birkhahnweg 8
D-48429 Rheine

Telephone: (0)5971-7406
Founded: 1988
Membership: 10
Activities: observing * photography * computing * lectures
Coordinates: 007°25'12"E 52°15'39"N (Waldhügel)
007°28'00"E 52°18'23"N H38m (Altenrheine)
City Reference Coordinates: 007°26'00"E 52°17'00"N

Astronomische Arbeitsgemeinschaft Wanne-Eickel/Herne eV
c/o Bernd Brinkmann
Bochumer Straße 226
D-44625 Herne
or :
c/o Dieter Rösener
Tupenweg 48
D-44651 Herne
or :
Am Böckenbusch 2a
D-44652 Herne
Telephone: (0)2325-77202
(0)2323-43628
Telefax: (0)2325-51864
Electronic Mail: info@sternwarte-herne.de
WWW: http://www.sternwarte-herne.de/
Founded: 1981
Membership: 40
Activities: observing * lectures * seminars
Coordinates: 007°10'34"E 51°31'40"N H55m
City Reference Coordinates: 007°12'00"E 51°32'00"N

Astronomische Arbeitsgruppe Laufen (AAL) eV
c/o Rudolf Reiser
Goethestraße 8
D-83410 Laufen
WWW: http://astronomy.meta.org/
Founded: 1986
City Reference Coordinates: 012°55'00"E 47°56'00"N

Astronomische Gesellschaft (AG)
c/o Reinhard E. Schielicke (Secretary)
Universitäts-Sternwarte Jena
Schillergäßchen 2
D-07745 Jena
Telephone: (0)3641-947526
Telefax: (0)3641-947502
Electronic Mail: schie@astro.uni-jena.de (Reinhard E. Schielicke, Secretary)
WWW: http://www.astro.uni-jena.de/Astron_Ges/ag0home.html
Founded: 1863
Membership: 800
Activities: scientific meetings * publication of scientific papers * support of young astronomers
Periodicals: (1) "Mitteilungen der Astronomischen Gesellschaft" (ISSN 0374-1958, circ.: 1,200); "Astronomische Gesellschaft Abstract Series" (ISSN 0934-4438), "Reviews in Modern Astronomy" (ISSN 0941-1445)
Awards: "Karl-Schwarzschild-Medaille", "Ludwig-Biermann-Förderpreis", "Bruno-H.-Bürgel-Preis", "Hans-Ludwig-Neumann-Preis"
City Reference Coordinates: 011°35'00"E 50°56'00"N

Astronomische Gesellschaft (AG), Arbeitskreis Astronomiegeschichte
c/o Wolfgang R. Dick (Secretary)
Vogelsang 35A
D-14478 Potsdam
Telephone: (0)331-863199
Telefax: (0)6592-985140
Electronic Mail: wdick@astrohist.org
WWW: http://www.astro.uni-bonn.de/~pbrosche/aa/aa.html
http://www.astro.uni-bonn.de/~pbrosche/astoria.html
Founded: 1992
Membership: 182
Activities: history of astronomy
Periodicals: (2) "Mitteilungen zur Astronomiegeschichte" (ISSN 0944-1999, circ.: 300); "Elektronische Mitteilungen zur Astronomiegeschichte / Electronic Newsletter for the History of Astronomy" (circ.: 400), "Acta Historica Astronomiae/Beiträge zur Astronomiegeschichte" (circ.: 300)
City Reference Coordinates: 013°04'00"E 52°24'00"N

Astronomische Gesellschaft Bochum-Melle
Oststraße 17
D-49324 Melle

Telephone: (0)5422-3986
Telefax: (0)5422-45413
City Reference Coordinates: 007°13'00"E 51°28'00"N (Bochum)
008°20'00"E 52°13'00"N (Melle)

Astronomische Gesellschaft Buchloe eV
Alois-Reiner-Straße 15b
D-86807 Buchloe
Telephone: (0)8241-7924
Electronic Mail: b_koch@t-online.de (Bernd Koch, President)
WWW: http://www.astronomie-buchloe.de/
Founded: 1997 (Observatory: 1987)
Membership: 90
Activities: popularization * observing (comets, minor planets, meteors, atmospheric optics, variable stars, deep sky) * photography * fighting light pollution
Periodicals: (1) "Faszinierendes Universum"; "Buchloer Astronomisches Zirkular"
Coordinates: 010°43'58"E 48°01'02"N H636m (IAU Code 215)
City Reference Coordinates: 010°45'00"E 48°04'00"N

Astronomische Gesellschaft Urania eV, Wiesbaden
c/o Sternwarte Wiesbaden
Bierstadter Straße 47
D-65189 Wiesbaden
or :
c/o Horst-Rainer Schneider
Feldbergstraße 22
D-65527 Niedernhausen
Telephone: (0)6127-7410 (H.R. Schneider)
(0)611-317438 (Observatory - Mondays, after 20:00)
Electronic Mail: horst.schneider@arcor.net
WWW: http://iphcip1.physik.uni-mainz.de/~astro/urania/
Founded: 1925
Membership: 100
Activities: observing (Sun, comets, minor planets, occultations) * photography * photometry * CCD * astrometry
Periodicals: (6) "Mitteilungen astronomischer Vereinigungen Rhein-Main-Nahe" (circ.: 500)
Coordinates: 008°15'47"E 50°04'58"N H198m (IAU Code 023)
City Reference Coordinates: 008°14'00"E 50°05'00"N (Wiesbaden)
008°20'00"E 50°09'00"N (Niedernhausen)

Astronomische Instrumente Stefan Thiele (AIT)
Walkmühlstraße 4
D-65195 Wiesbaden
Telephone: (0)611-407226
Telefax: (0)611-407226
Electronic Mail: office@ait-trading.com
WWW: http://www.ait-trading.com
Founded: 1980
Staff: 2
Activities: distributing telescopes, accessories, software and filters for H-alpha solar observing and photography * GPS hand-held navigators
City Reference Coordinates: 008°14'00"E 50°05'00"N

Astronomische Nachrichten (AN)
Astrophysikalisches Institut Potsdam
An der Sternwarte 16
D-14482 Potsdam
Telephone: (0)331-7499232
Telefax: (0)331-7499429
Electronic Mail: an@aip.de
WWW: http://www.wiley-vch.de/berlin/journals/an/
http://www.aip.de/AN/
Founded: 1821
Staff: 3
• Journal (6) (ISSN 0004-6337)
City Reference Coordinates: 013°04'00"E 52°24'00"N

Astronomischer Arbeitskreis der Heimvolksschule Schloß Dhaun
c/o Heimvolksschule Schloß Dhaun
D-55606 Hochstetten-Dhaun
Telephone: (0)6752-5374
Electronic Mail: wichb000@mzdmza.zdv.uni-mainz.de (Burkhard Wiche)
WWW: http://iphcip1.physik.uni-mainz.de/~astro/pop/Dhaun.html
Membership: 20
Periodicals: "Mitteilungen astronomischer Vereinigungen Rhein-Main-Nahe" (circ.: 550)
Coordinates: 007°29'44"E 49°48'53"N H365m
City Reference Coordinates: 007°31'00"E 49°48'00"N

Astronomischer Arbeitskreis Ingolstadt (AAI) eV
c/o Dieter Leistritz
Lilienthalstraße 137
D-85077 Manching
Telephone: (0)8459-6587
Electronic Mail: aai@bingo-ev.de
WWW: http://www.aai-ingolstadt.de/
Coordinates: 011°25'26"E 48°44'47"N
City Reference Coordinates: 011°31'00"E 48°43'00"N

Astronomischer Arbeitskreis Kassel eV (AAK)
c/o K.P. Haupt
Wilhelmshöher Allee 300a
D-34131 Kassel
Telephone: (0)561-311116
Telefax: (0)561-311116
Electronic Mail: kphaupt@aol.com
WWW: http://www.astronomie-kassel.de/
Founded: 1972
Membership: 185
Activities: observing (planets, Sun) * photography * popularization
Periodicals: (4) "Sternzeit" (co-publisher) (ISSN 0721-8168, circ.: 2,000); (3) "Korona" (circ.: 350)
Observatories: 1 (Sternwarte Calden)
City Reference Coordinates: 009°29'00"E 51°19'00"N

Astronomischer Arbeitskreis Merkur/Venus Göttingen (AAMVG)
c/o Detlev Niechoy
Bertheaustraße 26
D-37075 Göttingen
Telephone: (0)551-33830
Telefax: (0)551-33871
Electronic Mail: dniechoy@t-online.de
WWW: http://www.geocities.com/~dniechoy/
Founded: 1994
Membership: 4
Activities: inner planets * Sun * photometry * telescope drawings
Coordinates: 009°56'36"E 51°31'00"N
City Reference Coordinates: 009°55'00"E 51°32'00"N

Astronomischer Arbeitskreis Pforzheim (AAP) 1982 eV
c/o Jürgen Falkenberg
Daimlerstraße 1
D-75334 Straubenhardt-Feldrennach
Telephone: (0)7082-793730
Telefax: (0)7082-793731
Electronic Mail: aap@stuhlinger.de
WWW: http://www.mitglied.lycos.de/aap
Founded: 1982
Membership: 84
Activities: education * meetings * running a public observatory
Periodicals: (4) "Astro-News"
Coordinates: 008°42'13"E 48°47'55"N (Bieselsberg)
City Reference Coordinates: 008°42'00"E 48°54'00"N (Pforzheim)

Astronomischer Arbeitskreis Wetzlar eV
Lindenstraße 11
D-35606 Solms/Lahn
Telephone: (0)6442-1039
  (0)6442-927640 (astronomical information)
Telefax: (0)6442-927641
Electronic Mail: sternwarte@sternwarteburgsolms.de
WWW: http://www.sternwarteburgsolms.de/ (Sternw. Burgsolms – IAU Code 554)
City Reference Coordinates: 008°29'00"E 50°33'00"N (Wetzlar)

Astronomischer Freundeskreis Ostsachsen (AFO)
c/o Volks- und Schulsternwarte Bruno H. Bürgel
Zöllnerweg 12
D-02689 Sohland
Telephone: (0)35936-37270
WWW: http://ctch06.chm.tu-dresden.de/afo/afo_home.htm
  http://www.astronomie-sachsen.de/afo/
  http://www.astronomie-sachsen.de/sternfreund/ (Der Sternfreund)
Founded: 1992
Membership: 25
Periodicals: (6) "Der Sternfreund"
City Reference Coordinates: 014°25'00"E 51°02'00"N

Astronomischer Verein der Grafschaft Bentheim (AVGB) eV
Jahnstraße 3
D-49828 Neuenhaus
Telephone: (0)5941-990904
Telefax: (0)5941-920764
Electronic Mail: info@avgb.de
WWW: http://www.avgb.de/
Founded: 1990
City Reference Coordinates: 007°10'00"E 52°19'00"N (Bad Bentheim)

Astronomischer Verein der Volkssternwarte Papenburg eV
Wilhelm-Leuschnerstraße 48
D-26871 Papenburg
Telephone: (0)4961-1694
Founded: 1984
Membership: 50
Activities: observing * photography * instrument making * activities in schools * seminars * consulting * journeys
Periodicals: (4) "Sternzeit" (co-publisher) (ISSN 0721-8168, circ.: 2,000); "Astro-Nachrichten"
Coordinates: 007°23'30"E 53°03'20"N
City Reference Coordinates: 007°25'00"E 53°05'00"N

Astronomischer Verein Dortmund (AVD) eV
Postfach 30 01 30
D-44231 Dortmund
Telephone: (0)231-104076
Electronic Mail: vorstand@sternwarte-dortmund.de
WWW: http://www.avd-online.de/
    http://www.sternwarte-dortmund.de (Observatory)
Founded: 1913
Membership: 60
Activities: observing (variable stars, Sun, planets)
Periodicals: (4) "Sternzeit" (co-publisher) (ISSN 0721-8168, circ.: 2,000), "Raum und Zeit"
Observatories: 1 (Volkssternwarte Dortmund)
City Reference Coordinates: 007°28'00"E 51°31'00"N

Astronomischer Verein Hoyerswerda (AVH) eV
c/o Peter Schubert
Jan-Arnost-Smoler-Straße 3
D-02977 Hoyerswerda
Telephone: (0)3571-417020
Electronic Mail: arlev@t-online.de
WWW: http://www.germany.net/teilnehmer/100/142601/astro.htm
    http://home.germany.net/100-142601/astro.htm
City Reference Coordinates: 014°14'00"E 51°26'00"N

Astronomischer Verein Remscheid (AVRS) eV
Postfach 10 01 03
Palmstraße 29
D-42801 Remscheid
Telephone: (0)2191-75607
Electronic Mail: astro.rs@t-online.de
    dick.huetzluff@t-online.de (Dick Hützluff)
WWW: http://avrs.rscheid.de/
Founded: 1982
Membership: 75
Activities: observing * photography * popularization
Periodicals: (4) "Infoheft"
Coordinates: 007°10'06"E 51°11'00"N
City Reference Coordinates: 007°10'00"E 51°11'00"N

Astronomische Schulstation Adolph Diesterweg
G.-Petri-Straße 3
D-38855 Wernigerode
Telephone: (0)3943-632270
Telefax: (0)3943-632421
Founded: 1967
Staff: 10
Activities: observing * education * actualities
Coordinates: 010°47'00"E 51°50'00"N H240m
City Reference Coordinates: 010°47'00"E 51°50'00"N

Astronomisches Rechen-Institut (ARI)
Mönchhofstraße 12-14
D-69120 Heidelberg
Telephone: (0)6221-4050

Telefax: (0)6221-405297
Electronic Mail: <userid>@urz.uni-heidelberg.de
WWW: http://www.ari.uni-heidelberg.de/
http://www.ari.uni-heidelberg.de/index.eng.htm
Founded: 1700
Staff: 70
Activities: astrometry * stellar dynamics * ephemerides * bibliography
Periodicals: (1) "Apparent Places of Fundamental Stars", "Astronomische Grundlagen für den Kalender" (ISSN 0067-0014); "Veröffentlichungen des Astronomisches Rechen-Institut",
City Reference Coordinates: 008°43'00"E 49°25'00"N

Astronomische Station Heinrich S. Schwabe
c/o Walter-Gropius-Gymnasium
Peterholzstraße 58
D-06849 Dessau
Telephone: (0)340-8581858
WWW: http://www.wggdessau.de/index/sub_link.html
Founded: 1967
Staff: 1
Activities: observing * planetarium shows * education
City Reference Coordinates: 012°14'00"E 51°50'00"N

Astronomische Station Tycho Brahe
Nelkenweg 6
D-18057 Rostock
Telephone: (0)381-4934068
Electronic Mail: sternwarte.rostock@t-online.de
WWW: http://sternwarte.rostock.bei.t-online.de/
Founded: 1965
Staff: 3
Activities: education
Coordinates: 012°04'40"E 54°05'09"N
City Reference Coordinates: 012°07'00"E 54°05'00"N

Astronomische Sternwarte Nessa
Dorfstraße 11
D-06682 Nessa
Telephone: (0)34443-20960
Telefax: (0)34443-20961
Electronic Mail: nessa@t-online.de
WWW: http://www.nessa.via.t-online.de/
Founded: 1969
Staff: 1
Activities: CCD observing (variable stars, faint planet satellites)
Coordinates: 012°01'15"E 51°09'05"N H169m
City Reference Coordinates: 012°01'00"E 51°09'00"N

Astronomisches Zentrum Bruno-H.-Bürgel
Im Neuen Garten 6
D-14469 Potsdam
Telephone: (0)331-2702721
(0)331-2702724
Telefax: (0)331-292447
Electronic Mail: planetarium.potsdam@t-online.de
Founded: 1971
Staff: 2
Activities: popularization * teacher training * sundials * photography * computing * planetarium
Coordinates: 013°04'06"E 52°25'12"N
City Reference Coordinates: 013°04'00"E 52°24'00"N

Astronomisches Zentrum Burg
c/o Elke Sommer
Kirchofstraße 3
D-39288 Burg
Telephone: (0)3921-3014
WWW: http://lbs.st.schule.de/lbs/pl-burg.htm
City Reference Coordinates: 011°51'00"E 52°17'00"N

Astronomisches Zentrum Halberstadt (AZH)
Wilhelm-Trautewein-Straße 19
D-38820 Halberstadt
Telephone: (0)3941-600039
WWW: http://lbs.st.schule.de/lbs/pl-halb.htm
Founded: 1990
Activities: education * observing * planetarium shows

Coordinates: 011°06'00"E 51°09'00"N
City Reference Coordinates: 011°06'00"E 51°09'00"N

Astronomisches Zentrum Magdeburg (AZM)
Pablo-Picasso-Straße 21
D-39128 Magdeburg
Telephone: (0)391-445862
Telefax: (0)391-2445862
Electronic Mail: astromd@gmx.de
WWW: http://www.astronomie-magdeburg.de/
Founded: 1977
Activities: observing (Sun) * planetarium
Periodicals: (4) "Sternzeit" (co-publisher) (ISSN 0721-8168, circ.: 2,000)
Coordinates: 011°37'30"E 52°10'12"N H60m (Johannes Kepler Observatory)
City Reference Coordinates: 011°38'00"E 52°07'00"N

Astronomisches Zentrum Schkeuditz
Postfach 11 29
D-04431 Schkeuditz
or :
Bergbreite 1
D-04435 Schkeuditz
Telephone: (0)34204-62616
Telefax: (0)34204-62616
Electronic Mail: stern@www.uni-leipzig.de
WWW: http://www.uni-leipzig.de/~stern/
Founded: 1978
Staff: 1
Activities: education * planetarium shows * public observatory
Coordinates: 012°13'30"E 51°23'45"N H119m
City Reference Coordinates: 012°13'00"E 51°24'00"N

Astronomische Vereinigung Albstadt (AVA) eV
Hartmannstraße 140
D-72458 Albstadt-Ebingen
Telephone: (0)7431-72881
Telefax: (0)7431-72881
Electronic Mail: 0743172881-0001@t-online.de
WWW: http://home.t-online.de/home/0743172881-0001@t-online.de/
Founded: 1973
Membership: 15
Activities: photography * Moon * planets * Sun * comets * observing * planetarium shows (see separate entry) * CCD * instrumentation
Coordinates: 008°59'59"E 48°12'37"N H737m
City Reference Coordinates: 009°02'00"E 48°13'00"N

Astronomische Vereinigung Augsburg eV (AVA)
c/o Volkssternwarte
Pestalozzistraße 17a
D-86420 Diedorf
Telephone: (0)8238-7344
Telefax: (0)1212-513010417
Electronic Mail: info@sternwarte-diedorf.de
WWW: http://www.sternwarte-diedorf.de/
Founded: 1965
Membership: 170
Activities: public observing * education * photography * planetarium shows * star parties
Periodicals: "Uranus" (circ.: 1000), "AVA Aktuell" (circ.: 200)
Coordinates: 010°47'11"E 48°21'15"N H504m
City Reference Coordinates: 010°53'00"E 48°23'00"N (Augsburg)

Astronomische Vereinigung Karlsruhe (AVKa) eV
Krokusweg 49
D-76199 Karlsruhe
Telephone: (0)721-9430896
WWW: http://hikwww2.fzk.de/avka/
Founded: 1974
Membership: 75
Coordinates: 008°24'55"E 48°58'39"N H125m (Sternwarte Karlsruhe - IAU Code 021)
City Reference Coordinates: 008°24'00"E 49°03'00"N

Astronomische Vereinigung Nürtingen (AVN) eV
c/o Hans-Dieter Haas
Birkenweg 7
D-72622 Nürtingen

Telephone: (0)7022-33678
Telefax: (0)7022-31408
Electronic Mail: hdhaas@nuertingen.netsurf.net
WWW: http://www.avnev.nepustil.net
City Reference Coordinates: 009°20'00"E 48°37'00"N

Astronomische Vereinigung Tübingen (AVT) eV
Waldhäuser Straße 64
D-72076 Tübingen
Telephone: (0)7071-2978607
(0)7071-65349
Telefax: (0)7071-293458
Electronic Mail: aaptue@ait.physik.uni-tuebingen.de
WWW: http://magellan.tat.physik.uni-tuebingen.de/~avt/
Founded: 1972
Membership: 35
Activities: public lectures * star parties * popularization * education
Periodicals: "Sternwarte Tübingen"
City Reference Coordinates: 009°02'00"E 48°31'00"N

Astronomische Vereinigung Weikersheim eV
c/o Albert Hammer
Marienstraße 38
D-97980 Bad Mergentheim
Telephone: (0)7931-2120
Electronic Mail: info@sternwarte-weikersheim.de
WWW: http://www.sternwarte-weikersheim.de/
Founded: 1977
Membership: 72
Activities: observing * photography * lectures
Periodicals: (4) "Sternzeit" (co-publisher) (ISSN 0721-8168, circ.: 2,000)
City Reference Coordinates: 009°46'00"E 49°30'00"N

Astronomische Vereinigung West-München (AVWM)
c/o Bruno Wagner
Egelseestraße 21
D-86949 Windach
or :
Grasslfinger Straße
D-82194 Gröbenzell
Telephone: (0)8193-366
Electronic Mail: info@avwm.org
WWW: http://www.avwm.org/
Founded: 1978
Membership: 55
Activities: observing * meteors
Periodicals: (3) "Ganymed"
Coordinates: 011°02'18"E 48°03'33"N H595m
City Reference Coordinates: 011°03'00"E 48°04'00"N (Windach)
011°22'00"E 48°11'00"N (Gröbenzell)
011°35'00"E 48°08'00"N (München)

Astronomy Study Unit (ASU)
c/o Eckehard Schmidt
Postfach 46 16
D-90025 Nürnberg
or :
Brunhildstraße 1a
D-90461 Nürnberg
Telephone: (0)911-4720978
Telefax: (0)911-5865549
Founded: 1972
Membership: 61
Activities: gathering stamps with astronomical themes
Periodicals: (4) "Astrofax"
City Reference Coordinates: 011°04'00"E 49°27'00"N

Astrophysikalisches Institut Potsdam (AIP)
An der Sternwarte 16
D-14482 Potsdam
Telephone: (0)331-74990
Telefax: (0)331-7499200
Electronic Mail: <userid>@aip.de
WWW: http://www.aip.de/
Founded: 1700 (refounded: 1992)
Staff: 76

Activities: extragalactic astrophysics * cosmology * stellar activity * magnetohydrodynamics
Coordinates: 013°00'23"E 52°24'24"N H74m (Babelsberg – IAU Code 042)
City Reference Coordinates: 013°04'00"E 52°24'00"N

Astrophysikalisches Institut Potsdam (AIP), Observatorium für Solare Radioastronomie (OSRA)
D-14552 Tremsdorf
Telephone: (0)33205-2261
Telefax: (0)33205-2393
Electronic Mail: <userid>@aip.de
WWW: http://aipsoe.aip.de/~det/online_spectra.html
Founded: 1700 (AIP)
City Reference Coordinates: 013°04'00"E 52°24'00"N (Potsdam)

Astrophysikalisches Institut Potsdam (AIP), Sonnenobservatorium Einsteinturm (SOE)
Telegrafenberg
D-14473 Potsdam
Telephone: (0)331-2880
Telefax: (0)331-2882310
Electronic Mail: <userid>@aip.de
WWW: http://aipsoe.aip.de/
Founded: 1920
Staff: 10
Activities: solar physics * solar magnetic fields * sunspots * solar activity
Coordinates: 013°04'00"E 52°22'54"N H100m
City Reference Coordinates: 013°04'00"E 52°24'00"N

astro-shop
Eiffestraße 426
D-20537 Hamburg
Telephone: (0)40-5114348
Telefax: (0)40-5114594
Electronic Mail: astro@astro-shop.com
WWW: http://www.astro-shop.com/
Founded: 1978
City Reference Coordinates: 009°59'00"E 53°33'00"N

Astro-Versand
Birkenstraße 14
D-72145 Hirrlingen
Telephone: (0)7478-261613
Telefax: (0)7478-261614
Electronic Mail: versand1@aol.com
WWW: http://www.astro-versand.com/
Founded: 1979
Staff: 3
Activities: astronomical mail order company
City Reference Coordinates: 009°02'00"E 48°31'00"N (Tübingen)

Baader Planetarium GmbH
Zur Sternwarte
D-82291 Mammendorf
Telephone: (0)8145-8802
Telefax: (0)8145-8805
Founded: 1966
Staff: 12
Activities: manufacturing and distributing observatory domes, planetariums, spectroscopes, coronagraphs, H-alpha filters, telescopes, optical pieces and accessories
City Reference Coordinates: 011°09'00"E 48°12'00"N

Bayerische Akademie der Wissenschaften (BAW)
Marstallplatz 8
D-80539 München
Telephone: (0)89-230310
Telefax: (0)89-23031100
WWW: http://www.badw.de/
Founded: 1759
Membership: 90
Periodicals: (1) "Jahrbuch (ISSN 0084-6090); "Nova Kepleriana - Abhandlungen - Neue Folge" (ISSN 0078-2246), "Sitzungsberichte der Philosophische-historische Klasse" (ISSN 0342-5991), "Abhandlungen der Philosophische-historische Klasse" (ISSN 0005-710x), "Sitzungsberichte der Mathematische-naturwissenschaftliche Klasse" (ISSN 0340-7586), "Abhandlungen der Mathematische-naturwissenschaftliche Klasse" (ISSN 0005-6995)
City Reference Coordinates: 011°35'00"E 48°08'00"N

Bayerische Julius-Maximilians-Universität Würzburg
• See "Universität Würzburg"

Bayerische Volkssternwarte München eV
Rosenheimer Straße 145h
D-81671 München
Telephone: (0)89-406239
Telefax: (0)89-494987
Electronic Mail: volkssternwarte@lrz.tum.de
WWW: http://www.sternwarte-muenchen.de/
Founded: 1947
Membership: 600
Staff: 3
Activities: popularization * observing * planetarium shows
Periodicals: (4) "Blick ins All"
Coordinates: 011°36'31"E 48°07'21"N H572m (München)
 011°10'00"E 48°00'00"N H600m (Herrsching – IAU Code 564)
City Reference Coordinates: 011°35'00"E 48°08'00"N

Bayerische Volkssternwarte Neumarkt/Opf. eV
c/o Hans-Werner Neumann
Moorstraße 5
D-92318 Neumarkt
or :
Am Höhenberg 31
D-92318 Neumarkt
Telephone: (0)9181-44592
WWW: http://www.sternwarte-neumarkt.de/
Founded: 1969
Membership: 250
Activities: weekly observing evenings * lectures * open-door days * solar protuberances
Sections: Astrophotography * Space * Solar Protuberances
Periodicals: (1) "Jahresbericht"
Coordinates: 011°27'40"E 49°16'50"N
City Reference Coordinates: 011°28'00"E 49°17'00"N

Berliner Sternfreunde eV
c/o Thomas Grau
Puschkinstraße 20
D-16123 Bernau
Telephone: (0)175-4994300
 (0)3338-764881
Electronic Mail: info@berlinersternfreunde.de
WWW: http://www.berlinersternfreunde.de/
Membership: 35
City Reference Coordinates: 013°24'00"E 52°31'00"N (Berlin)

Bismarckschule Hannover, Planetarium
An der Bismarckschule 5
D-30173 Hannover
Telephone: (0)511-16843456
WWW: http://www.h.shuttle.de/h/bimsch/
City Reference Coordinates: 009°44'00"E 52°24'00"N

Bremerhavener Sternfreunde eV
c/o Günter Neumann
Buschkämpen 29
D-27576 Bremerhaven
WWW: http://www.bremerhavener-sternfreunde.de/
Founded: 1982
Membership: 20
Activities: observing * public lectures
City Reference Coordinates: 008°34'00"E 53°33'00"N

Bruno-H.-Bürgel-Sternwarte, Berlin
Heerstraße 531
D-13593 Berlin
Telephone: (0)30-3636242
Telefax: (0)30-48096923
Electronic Mail: bhb.sternwarte-berlin@gmx.de
WWW: http://mitglied.tripod.de/BHB_Sternwarte/
Founded: 1982
Membership: 120
Activities: education * lectures * observing * excursions * exhibitions * shows
Coordinates: 013°09'11"E 52°31'09"N (Hahneberg)
City Reference Coordinates: 013°24'00"E 52°31'00"N

Bruno-H.-Bürgel-Sternwarte, Hartha
Töpelstraße 49
D-04746 Hartha
Telephone: (0)34328-3158
Telefax: (0)34328-3158
Electronic Mail: sternwartehartha@lycos.de
WWW: http://www.sternwartehartha.de/
Founded: 1956
Activities: variable stars
Periodicals: "Mitteilungen der Bruno-H. Bürgel-Sternwarte Hartha", "Harthaer Beobachtungszirkulare", "Sonderdrucke"
Coordinates: 012°57'37"E 51°06'11"N H324m
City Reference Coordinates: 013°00'00"E 51°06'00"N

Bundesamt für Kartographie und Geodäsie (BKG), Außenstelle Leipzig
Karl-Rothe-Straße 10-14
D-04105 Leipzig
Telephone: (0)341-56340
Telefax: (0)341-5634415
WWW: http://www.leipzig.ifag.de/
Activities: international VLBI service
City Reference Coordinates: 012°20'00"E 51°19'00"N

Bundesamt für Kartographie und Geodäsie (BKG), Fundamentalstation Wettzell
Sackenrieder Straße 25
D-93444 Kötzting
Telephone: (0)9941-6030
Telefax: (0)9941-603222
WWW: http://www.wettzell.ifag.de/
Activities: international VLBI service
City Reference Coordinates: 012°52'00"E 49°11'00"N

Bundesamt für Seeschiffahrt und Hydrographie (BSH), Hamburg
Postfach 30 12 20
D-20305 Hamburg
or :
Bernhard-Nocht-Straße 78
D-20359 Hamburg
Telephone: (0)40-31900
Telefax: (0)40-31905000
Electronic Mail: posteingang@bsh.d400.de
WWW: http://www.bsh.de/
Periodicals: (1) "Nautisches Jahrbuch" (ISSN 0077-6211)
City Reference Coordinates: 009°59'00"E 53°33'00"N

Bundesamt für Seeschiffahrt und Hydrographie (BSH), Rostock
Dierkower Damm 45
D-18146 Rostock
Telephone: (0)381-45635
Telefax: (0)381-4563948
Electronic Mail: posteingang.rostock@bsh.d400.de
WWW: http://www.bsh.de/
City Reference Coordinates: 012°07'00"E 54°05'00"N

Bundesdeutsche Arbeitsgemeinschaft für Veränderliche Sterne eV (BAV)
Munsterdamm 90
D-12169 Berlin
Telephone: (0)30-7848453
Electronic Mail: braune.bav@t-online.de
WWW: http://thola.de/bav.html
Founded: 1950
Membership: 220
Activities: observing (variable stars)
Periodicals: (4) "BAV-Rundbrief" (ISSN 0405-5497, circ.: 320); (2) "BAV-Mitteilungen" (circ.: 320); (1) "BAV-Circular" (circ.: 280)
City Reference Coordinates: 013°24'00"E 52°31'00"N

Carl Zeiss Jena GmbH, Astronomische Geräte
Carl-Zeiss-Promenade 10
Postfach 125
D-07740 Jena
Telephone: (0)3641-640
         (0)3641-642382
Telefax: (0)3641-642023

Electronic Mail: astro@zeiss.de
                <userid>@zeiss.de
WWW: http://www.zeiss.de/astro/
Founded: 1897
Staff: 1650
Activities: designing and manufacturing astronomical instrumentation * professional telescopes * solar telescopes * observatories * enclosures * optical systems * focal instrumentation * planetariums
Periodicals: "Innovation" (ISSN 1431-8059)
City Reference Coordinates: 011°35'00"E 50°56'00"N

Carl-Zeiss-Planetarium Stuttgart
Mittlerer Schloßgarten
D-70173 Stuttgart
Telephone: (0)711-1629215
           (0)711-162920
Telefax: (0)711-2163912
Electronic Mail: planetarium@stuttgart.de
WWW: http://planetarium.stuttgart.de/
Founded: 1977
Staff: 33
Activities: education * shows
Periodicals: (8-12) "Planetarium Stuttgart, Program"
Coordinates: 009°11'51"E 48°47'01"N H354m (Schwäbische Sternw. - see separate entry)
             009°35'49"E 48°52'30"N H547m (Welzheim)
City Reference Coordinates: 009°11'00"E 48°46'00"N

Christian-Albrechts-Universität zu Kiel
• See "Universität Kiel"

Christian Jutz Volkssternwarte Berg eV
Sonnenweg 18
D-82335 Berg-Aufkirchen
or :
Lindenallee
D-82335 Berg-Aufkirchen
Telephone: (0)8151-979851
           (0)8151-51112
Electronic Mail: christian.jutz@t-online.de
WWW: http://people.freenet.de/ovsb/
Founded: 1992
Membership: 50
Activities: public observing
Periodicals: (4) "Sterngucker"
Coordinates: 011°22'00"E 47°58'00"N H 683m
City Reference Coordinates: 011°21'00"E 47°58'00"N

Commission on International Coordination of Space Techniques for Geodesy and Geodynamics (CSTG)
c/o Hermann Drewes (Secretary)
Center for Space Research
Deutsches Geodätisches Forschungsinstitut (DGFI)
Marstallplatz 8
D-80539 München
Telephone: (0)89-23031106
           (0)89-23031107
Telefax: (0)89-23031240
Electronic Mail: drewes@dgfi.badw-muenchen.de
WWW: http://dgfi2.dgfi.badw-muenchen.de/~cstg/
Founded: 1971
Activities: developing links between international groups engaged in space geodesy and geodynamics * coordinating their works * elaborating and proposing projects * reporting on their advancement and results
Periodicals: (2) "CSTG Bulletin"
City Reference Coordinates: 011°35'00"E 48°08'00"N

Copernicus Planetarium (Nicolaus_)
• See "Nicolaus Copernicus Planetarium"

Daimler Benz Aerospace
Postfach 80 11 09
D-81663 München
Telephone: (0)89-6070
           (0)89-60734361
Telefax: (0)89-60726481
        (0)89-60734373
WWW: http://www.dasa.de/

Activities: manufacturing i.a. space modules and instrumentation
- Formerly "Deutsche Aerospace AG (DASA)"
- Now part of the "European Aeronautic, Defense and Space Co. (EADS)"

City Reference Coordinates: 011°35'00"E 48°08'00"N

Deutsche Aerospace AG (DASA)
- See now "Daimler-Benz Aerospace"

Deutsche Agentur für Raumfahrtangelegenheiten (DARA) GmbH
- Now part of the "Deutsches Zentrum für Luft- und Raumfahrt (DLR) eV"

Deutsche Forschungsanstalt für Luft- und Raumfahrt (DLR) eV
- See now "Deutsches Zentrum für Luft- und Raumfahrt (DLR) eV"

Deutsche Forschungsgemeinschaft (DFG)
Kennedyallee 40
D-53175 Bonn
Telephone: (0)228-8851
Telefax: (0)228-8852777
(0)228-8852180 (PR)
WWW: http://www.dfg.de/
Founded: 1920 (refounded: 1949)
Membership: 87 (organizations)
Activities: promoting research in all fields * counseling parliaments * fostering international academic relationships
Periodicals: (4) "forschung - Mitteilungen der DFG" (ISSN 0172-1518); (3) "german research - Reports of the DFG" (ISSN 0172-1526); (1) "Jahresbericht"; "Statistik",
City Reference Coordinates: 007°05'00"E 50°44'00"N

Deutsche Geophysikalische Gesellschaft (DGG) eV
c/o W. Webers
GeoForschungsZentrum Potsdam
Telegrafenberg
D-14473 Potsdam
Telephone: (0)331-2881232
Telefax: (0)331-2881235
Electronic Mail: wigor@gfz-potsdam.de
WWW: http://www.dgg-online.de/
Founded: 1922
Membership: 900
Activities: professional society of geophysicists
Periodicals: "Geophysical Journal International", "DGG Mitteilungen"
Awards: "Emil-Wiecherte-Medaille", "Walter-Kertz-Medaille"
City Reference Coordinates: 013°04'00"E 52°24'00"N

Deutsche Gesellschaft für Chronometrie (DGC) eV
Ziehrerweg 8
D-71254 Ditzingen
Telephone: (0)7156-951640
Telefax: (0)7156-951640
WWW: http://www.ph-cip.uni-koeln.de/~roth/dgc.html
http://192.41.20.246/dgc/
Founded: 1949
City Reference Coordinates: 009°03'00"E 48°49'00"N

Deutsche Gesellschaft für Luft- und Raumfahrt (DGLR) eV
Godesberger Allee 70
D-53175 Bonn
Telephone: (0)228-308050
Telefax: (0)228-3080524
Electronic Mail: geschaeftsstelle@dglr.de
WWW: http://www.dglr.de/
Founded: 1967
Membership: 3300
Activities: meetings * fostering aeronautical and space research
Sections: (commissions) FG 1. Systemtechnik und Systemplanung * FA 1.1 Basismaterialien * FA 1.2 Projektführung * FA 1.3 Unterstützungsfunktionen * FA 2A. Luftfahrtzeuge * FA 2A.1 Transport- und Hochgeschwindigkeitsflugzeuge * FA 2A.2 Drehflüger * FA 2A.3 Segelflug und Sportflugzeuge * FA 2A.4 Flugbetrieb für Luftfahrtzeuge * FA 2A.5 Flugsysteme leichter als Luft * FG 2B. Flugkörper und Raumfahrt * FA 2B.1 Flugkörper, RPV's und Drohnen * FA 2B.2 Bemannte Raumfahrtsysteme * FA 2B.3 Satelliten und Sonden * FA 2B.4 Transportsysteme (einsch. Wiedereintrittssysteme) * FA 2B.5 Bodeneinrichtungen für Raumfahrtzeuge * FG 3. Fluid- und Thermodynamik * FA 3.1 Hydro-, Aero- und Gasdynamik * FA 3.2 Strömungsakustik/Fluglärm * FA 3.3 Versuchswesen der Fluid- und Thermodynamik * FG 4. Flugmechanik und Flugführung * FA 4.1 Flugleistungen und Bahnen * FA 4.2 Flugeigenschaften * FA 4.3 Flugregelung * FA 4.4 Ortung und Navigation * F14.5 Anthropotechnik * FA 4.6 Flugversuchstechnik * FA 4.7 Begriffsbestimmungen * FG 5. Antriebe * FA 5.1 Luftatmende Antriebe * FA 5.2 Raketenantriebe * FA 5.3 Elektrische und unkonventionelle Flugantriebe * FG 6. Energieversorgung, Schutzsysteme, Automation * FA 6.1 Energieversorgung * FA 6.2 Lebenserhaltungssysteme, Wärmeschutzsysteme * FA 6.3

Robotik, Automation, RVD, EVA * FA 6.4 Technologie der rechnergestützten Simulation * FG 7. Festigkeit, Bauweisen, Werkstoffe * FA 7.1 Festigkeit und Bauweisen * FA 7.2 Aeroelastik und Strukturdynamik * FA 7.3 Werkstoffe * FG 8. Elektronik-Nachrichtentechnik * FA 8.1 Telemetrie * FA 8.2 Avionik * FA 8.3 EMV/EMP * FA 8.4 Nachrichtenübertragung einsch. Signal- und Informationsverarbeitung * FA 8.5 Sensoren, Informationserfassung, Datenauswertung * FG 9. Flugmedizin und Biologie * FG 10. Luftraum- und Weltraumkunde * FA 10.1 Extraterrestrische Systeme * FA 10.2 Geowissenschaftliche Systeme * FA 10.3 Werkstoffforschung unter Schwerelosigkeit * FA 10.4 Medizin-/Biologieforschung unter Schwerelosigkeit * FG 11. Luftrecht und Weltraumrecht * FG 12. Geschichte der Luft- und Raumfahrt * FG 13. Dokumentation
Periodicals: (6) "Luft- und Raumfahrt" (ISSN 0173-6264), "Mitteilungen aus der DGLR"
Awards: (1) "Ludwig-Prandtl-Ring"
City Reference Coordinates: 007°05'00"E 50°44'00"N

Deutsche Meteorologische Gesellschaft (DMG) eV
c/o Sekretariat DMG
Institut für Meteorologie
Freie Universität Berlin
Carl-Heinrich-Becker-Weg 6-10
D-12165 Berlin
Telephone: (0)30-83871197
Telefax: (0)30-7919002
Electronic Mail: dmg@bibo.met.fu-berlin.de
WWW: http://www.met.fu-berlin.de/dmg
Founded: 1883
Activities: furthering meteorology in Germany
Periodicals: "Zeitschrift für Meteorologie"
Coordinates: 013°18'00"E 52°28'00"N H37m
City Reference Coordinates: 013°24'00"E 52°31'00"N

Deutsche Physikalische Gesellschaft (DPG) eV
Hauptstraße 5
D-53604 Bad Honnef
Telephone: (0)2224-92320
Telefax: (0)2224-923250
WWW: http://www.dpg-physik.de/
Founded: 1845
Membership: 30000
Periodicals: (12) "Physikalische Blätter" (ISSN 0031-9279); "Verhandlungen der Deutsche Physikalische Gesellschaft"
Awards: "Max-Planck-Medaille", "Stern-Gerlach-Preis", "Max-Born-Preis" (together with the "Institute of Physics"), "Gentner-Kastler-Preis" (together with the "Société Française de Physique"), "Robert-Wichard-Pohl-Preis", "Gustav-Hertz-Preis", "Walter-Schottky-Preis"
City Reference Coordinates: 007°13'00"E 50°39'00"N

Deutscher Wetterdienst (DWD)
Frankfurter Straße 135
D-63067 Offenbach/Main
Telephone: (0)69-80620
Telefax: (0)69-80622484
Electronic Mail: info@.dwd.de
WWW: http://www.dwd.de/
Founded: 1952
Staff: 3200
Periodicals: (daily) "Europäischer Wetterbericht" (ISSN 0341-2970), "Wetterkarte" (ISSN 0936-5818); (52) "Agrarmeteorologischer Wochenhinweis" (ISSN 0172-0570); (12) "Monatlicher Witterungsbericht" (ISSN 0435-7965), "Großwetterlagen Europas" (ISSN 0017-4645); (4) "Promet - meteorologische Fortbildung" (ISSN 0340-4552); (1) "Agrarmeteorologische Bibliographie" (ISSN 0515-6831), "Deutsches Meteorologisches Jahrbuch" (ISSN 0724-7125), "DWD Jahresbericht" (ISSN 0433-8251)
• Meteorological office
City Reference Coordinates: 008°47'00"E 50°08'00"N

Deutscher Wetterdienst (DWD), Referat Hydrometeorologische Entwicklungen und Anwendungen

Lindenberger Weg 24
D-13125 Berlin
Telephone: (0)30-9400940
Telefax: (0)30-9497324
Electronic Mail: hm2@dwd.de
WWW: http://www.dwd.de/
Founded: 1952 (DWD)
Activities: hydrometeorology
City Reference Coordinates: 013°24'00"E 52°31'00"N

Deutsches Elektronen-Synchrotron (DESY), Hamburg
Notkestraße 85
D-22607 Hamburg
Telephone: (0)40-89980
Telefax: (0)40-89983282
Electronic Mail: desyinfo@desy.de

<userid>@desy.de
WWW: http://www.desy.de/
Founded: 1959
Staff: 1550
Activities: basic research in particle physics * investigations with synchrotron radiation
City Reference Coordinates: 009°59'00"E 53°33'00"N

Deutsches Elektronen-Synchrotron (DESY), Zeuthen
Platanenallee 6
D-15738 Zeuthen
Telephone: (0)33762-770
Telefax: (0)33762-77330
Electronic Mail: desyinfo@ifh.de
        <userid>@ifh.de
WWW: http://www.ifh.de/
Founded: 1992 (DESY Hamburg)
Staff: 120
Activities: particle physics * detector development * neutrino astrophysics * parallel computing
• Formerly "Institut für Hochenergiephysik"
City Reference Coordinates: 013°37'00"E 52°20'00"N

Deutsches Fernerkundungsdatenzentrum (DFD)
• See "Deutsches Zentrum für Luft- und Raumfahrt (DLR) eV, Deutsches Fernerkundungsdatenzentrum (DFD)"

Deutsches Geodätisches Forschungsinstitut (DGFI)
Marstallplatz 8
D-80539 München
Telephone: (0)89-23031106
        (0)89-23031107
Telefax: (0)89-23031240
Electronic Mail: mailer@dgfi.badw-muenchen.de
WWW: http://dgfi2.dgfi.badw-muenchen.de/
Founded: 1952
Staff: 26
Activities: geodesy * theoretical geodesy * reference systems * space geodesy
Periodicals: "Veröffentlichungen der Deutschen Geodätischen Kommission"
City Reference Coordinates: 011°35'00"E 48°08'00"N

Deutsches Hydrographisches Institut (DHI)
• See now "Bundesamt für Seeschiffahrt und Hydrographie (BSH)"

Deutsches Institut für Normung (DIN)
D-10772 Berlin
or :
Burggrafenstraße 6
D-10787 Berlin
Telephone: (0)30-26010
Telefax: (0)30-26011231
Electronic Mail: <userid>@din.de
WWW: http://www.din.de/
Founded: 1917
Membership: 6200
Staff: 800
Activities: standardization
Periodicals: (12) "DIN Mitteilungen", "Elektronorm" (ISSN 0722-2912)
City Reference Coordinates: 013°24'00"E 52°31'00"N

Deutsches Klimarechenzentrum (DKRZ) GmbH
Bundesstraße 55
D-20146 Hamburg
Telephone: (0)40-411730 (Switchboard)
        (0)40-41173334 (Secretariat)
        (0)40-41173275 (User Information)
Telefax: (0)40-41173270
Electronic Mail: beratung@dkrz.de
        <userid>@dkrz.de
WWW: http://www.dkrz.de/
Founded: 1987
City Reference Coordinates: 009°59'00"E 53°33'00"N

Deutsches Museum München, Abteilung Astronomie
Museumsinsel 1
D-80538 München
Telephone: (0)89-2179350
        (0)89-2179456

Telefax: (0)89-2179513
Electronic Mail: as.astronomie@extern.lrz-muenchen.de
WWW: http://www.deutsches-museum.de/
Founded: 1903
Staff: 6
Activities: exhibitions * planetarium * public observatory * history of scientific publishing * popularization
Periodicals: "Kultur und Technik" (Museum publication)
Coordinates: 011°34'50"E 48°07'51"N H545m
City Reference Coordinates: 011°35'00"E 48°08'00"N

Deutsches Zentrum für Luft- und Raumfahrt (DLR) eV
Linder Höhe
D-51147 Köln
Telephone: (0)2203-6010
Telefax: (0)2203-67310
Electronic Mail: <userid>@dlr.de
WWW: http://www.dlr.de/
http://www.dlr.de/institute_d.html (DLR Institutes)
Founded: 1969
Staff: 4500
Activities: national research establishment performing scientific and technical research for the development and utilization of future aircraft and spacecraft * research in energy technology * constructing and operating large-scale test and simulation equipment, as well as space operation facilities * operating research centres with local branches and liaison offices in Paris and washington (see separate entries) * acting as German space agency
Periodicals: (8) "Aerospace, Science and Technology (AST)" (ISSN 1270-9638, circ.: 1,000); (4) "DLR-Nachrichten" (ISSN 0937-0420, circ.: 8,000)
Awards: "Wissenschaftliche Preise der DLR"
City Reference Coordinates: 006°57'00"E 50°56'00"N

Deutsches Zentrum für Luft- und Raumfahrt (DLR) eV, Institut für Planetenerkundung
Rudower Chaussee 5
D-12489 Berlin
Telephone: (0)30-69545300
Telefax: (0)30-69545303
Electronic Mail: <userid>@terra.pe.ba.dlr.de
WWW: http://www.ba.dlr.de/NE-PE/
http://www.ba.dlr.de/
http://www.dlr.de/
Founded: 1992
Staff: 98
Activities: planetary surfaces and atmospheres * physics of planets, comets and minor bodies * comparative planetology * modelling of planetary processes * image processing * designing and developing sensor electronics for cameras and spectrometers
City Reference Coordinates: 013°24'00"E 52°31'00"N

Deutsches Zentrum für Luft- und Raumfahrt (DLR) eV, Standort Oberpfaffenhofen
Postfach 11 16
D-82230 Weßling
Telephone: (0)8153-280
Telefax: (0)8153-281243
Electronic Mail: <userid>@dlr.de
WWW: http://www.dlr.de/
http://www.op.dlr.de/dlr_research_center_op_d.html
http://www.op.dlr.de/ne-hf/Welcome.html
Founded: 1969
Staff: 1070
Activities: space missions * telecommunications * radio frequency technology * atmospheric physics * central data processing * optoelectronics * robotics for space applications * dynamic systems research * research flight facilities * satellite data acquisition, processing and distribution * see also main entry
Periodicals: see main entry
City Reference Coordinates: 011°18'00"E 48°05'00"N

Die Sterne
• Now merged with "Sterne und Weltraum"

Dresdner Verein für Himmelskunde eV
c/o Achim Grünberg
Auf den Ebenbergen 10a
D-01445 Radebeul
WWW: http://home.t-online.de/home/Badicke/html3.htm
Founded: 1991
Membership: 20
Activities: meetings * excursions * observing
Periodicals: (6) "Der Sternfreund" (co-publisher)
City Reference Coordinates: 013°44'00"E 51°03'00"N (Dresden)

Dr. Remeis-Sternwarte
• See "Universität Erlangen-Nürnberg, Astronomisches Institut, Dr. Remeis-Sternwarte"

Dr. Vehrenberg KG
• See "Vehrenberg KG"

Eberhard-Karls-Universität Tübingen
• See "Universität Tübingen"

Erhard Friedrich Verlag GmbH & Co. KG
Postfach 10 01 50
D-30926 Seelze
or :
Im Brande 17
D-30926 Seelze/Velber
Telephone: (0)511-400040
Telefax: (0)511-40004110
Electronic Mail: info@friedrich-verlag.de
WWW: http://www.friedrich-verlag.de/
Periodicals: (6) "Astronomie + Raumfahrt im Unterricht"
• Publisher
City Reference Coordinates: 009°25'00"E 52°24'00"N

Ernst-Abbe-Stiftung
• See "Zeiss-Planetarium der Ernst-Abbe-Stiftung"

Ernst-Mach-Institut (EMI)
• See "Fraunhofer-Institut für Kurzzeitdynamik, Ernst-Mach-Institut (EMI)"

ESOC
• See "European Space Agency (ESA), European Space Operations Centre (ESOC)"

EUMETSAT
• See "European Organisation for the Exploitation of Meteorological Satellites"

European Association for Astronomy Education (EAAE)
c/o European Southern Observatory
Karl-Schwarzschild-Straße 2
D-85748 Garching
Electronic Mail: dps@eugenides_found.edu.gr (Dionysios P. Simopoulos, Chairman)
　　　　　　　　reichen@obs.unize.ch (Michael Reichen, Editor)
　　　　　　　　anders@astro.su.se (Anders Västerberg)
WWW: http://www.algonet.se/~sirius/eaae.htm
　　　　http://www.rz.uni-frankfurt.de/EAAE/
Founded: 1995
City Reference Coordinates: 011°40'00"E 48°14'00"N

European Asteroidal Occultation Network (EAON)
c/o Francis Delahaye
Font Darlan
Paillet
F-33550 Langoiran
Electronic Mail: francis.delahaye@wanadoo.fr
WWW: http://www.xcom.it/cana/EAON/welcome.htm
Founded: 1988
Membership: 200
Activities: observing (occultations by minor planets)
Periodicals: (3) "EAON Informations and Asteroidal Results"
City Reference Coordinates: 000°21'00"W 44°42'00"N

European Astronauts Centre (EAC)
• See "European Space Agency (ESA), European Astronauts Centre (EAC)"

European Council of Skeptical Organizations (ECSO)
c/o Amardeo Sarma
Kirchgasse 4
D-64380 Rossdorf
Telephone: (0)6154-695028
Telefax: (0)6154-695029
Electronic Mail: info@ecso.org
WWW: http://www.ecso.org/
Founded: 1995

Membership: 10
Activities: promoting science and critical thinking * organizing European conferences * exchange and dissemination of ideas and research results
City Reference Coordinates: 008°45'00"E 49°51'00"N

European Geophysical Society (EGS)
Max-Planck-Straße 13
D-37191 Katlenburg-Lindau
Telephone: (0)5556-1440
Telefax: (0)5556-4709
Electronic Mail: egs@copernicus.org
WWW: http://www.copernicus.org/EGS/EGS.html
Founded: 1971
Membership: 6000
Activities: pursuit of excellence in the geosciences and space sciences * organizing annual general assemblies, topical conferences and courses * publishing journals and books * supporting young scientists and colleagues from Eastern Europe
Periodicals: (25) "Physics and Chemistry of the Earth"; (12) "Annales Geophysicae" (see separate entry in France), "Geophysical Journal International", "Planetary and Space Science" (9) "Journal of Atmospheric Chemistry" (ISSN 0167-7764); (8) "Climate Dynamics" (ISSN 0930-7575), "Journal of Geodynamics"; (6) "Surveys in Geophysics" (ISSN 0169-3298), "Tectonics"; (4) "Newsletter"; "Hydrology and Earth System Sciences", "Nonlinear Processes in Geophysics" (ISSN 1023-5809)
City Reference Coordinates: 010°06'00"E 51°41'00"N

European Optical Society (EOS)
c/o Klaus-Dieter Nowitzki
Hollerithallee 8
D-30419 Hannover
Telephone: (0)511-2788115
Telefax: (0)511-2788100
Electronic Mail: no@lzh.de
WWW: http://www.europeanopticalsociety.org/
Founded: 1991
Membership: 800
Activities: conventions
Periodicals: (6) "Journal of the European Optical Society: Part A - Pure and Applied Optics", "Journal of the European Optical Society: Part B: Quantum Optics"; "EOS Newsletter" (in "Opto & Laser Europe")
City Reference Coordinates: 009°44'00"E 52°24'00"N

European Organisation for the Exploitation of Meteorological Satellites
(EUMETSAT)
Am Kavalleriesand 31
D-64295 Darmstadt
Telephone: (0)6151-807366
Telefax: (0)6151-807304
Electronic Mail: <userid>@eumetsat.de
WWW: http://www.eumetsat.de/
Founded: 1986
Membership: 17 (States)
Staff: 150
Activities: European intergovernmental organisation * exploitation of meteorological satellites
Periodicals: (1) "Annual Report" (ISSN 1013-3410); (2) "Image"; "Special Publications", "Proceedings of Conferences and Workshops",
City Reference Coordinates: 008°40'00"E 49°53'00"N

European Patent Office (EPO)
(Office Européen des Brevets - OEB)
Erhardtstraße 27
D-80331 München
Telephone: (0)89-23990
Telefax: (0)89-23994560
WWW: http://www.european-patent-office.org/
Founded: 1977
Staff: 3800
City Reference Coordinates: 011°35'00"E 48°08'00"N

European Planetarium Network (EuroPlaNet)
c/o Klaus Wörle
Hofweg 32
D-93053 Regensburg
Telephone: (0)941-944254
Telefax: (0)941-9449233
Electronic Mail: kw@fan.net
WWW: http://www-nw.uni-regensburg.de/~.wok13927.augen.klinik.uni-regensburg.de/epn.htm
http://www.artofsky.com/epn/
Founded: 1995
Staff: 3
Activities: planetarium resources * education

City Reference Coordinates: 012°06'00"E 49°01'00"N

European Radio Astronomy Club (ERAC)
Ziethen Straße 97
D-68259 Mannheim
Telephone: (0)621-794597
    (0)621-512470 (working hours)
Electronic Mail: erac@wegalink.com
WWW: http://www.erac.wegalink.com/
Founded: 1995
Activities: organizing international radioastronomy congress every third year
Periodicals: (4) "The ERAC Newsletter"
City Reference Coordinates: 008°29'00"E 48°29'00"N

European Southern Observatory (ESO), Headquarters
Karl-Schwarzschild-Straße 2
D-85748 Garching
Telephone: (0)89-320060 (Switchboard)
    (0)89-32006226 (Director General)
    (0)89-32006223 (Visiting Astronomers Office)
    (0)89-32006221 (Administration)
    (0)89-32006276 (Education and Public Relations)
    (0)89-32006252 (VLT Division)
    (0)89-32006356 (Instrumentation Division)
    (0)89-32006509 (Data Management Division)
Telefax: (0)89-3202362
Electronic Mail: <userid>@eso.org
WWW: http://www.eso.org/
Founded: 1962
Activities: intergovernmental organization (Member States: Belgium, Denmark, France, Germany, Italy, Netherlands, Sweden & Switzerland) * theoretical and observational astrophysics * instrumentation * observatories at La Silla and Paranal, Chile (see separate entries)
Periodicals: (4) "The Messenger" (ISSN 0722-6691); (1) "Annual Report" (ISSN 0531-4496); "ESO-MIDAS Courier" (ISSN 1018-3051), "ESO Conference and Workshop Proceedings"
City Reference Coordinates: 011°40'00"E 48°14'00"N

European Space Agency (ESA), European Astronauts Centre (EAC)
Linder Höhe
D-51147 Köln
Telephone: (0)2203-60010
Telefax: (0)2203-600166
WWW: http://www.estec.esa.nl/spaceflight/astronaut/
    http://www.esa.int/
Founded: 1990
Staff: 25
Activities: astronaut selection, recruitment, training and coordination
City Reference Coordinates: 006°57'00"E 50°56'00"N

European Space Agency (ESA), European Space Operations Centre (ESOC)
Robert-Bosch-Straße 5
D-64293 Darmstadt
Telephone: (0)6151-900
    (0)6151-902300
Telefax: (0)6151-90495
Electronic Mail: <userid>@esoc.esa.de
WWW: http://www.esoc.esa.de/
    http://www.esa.int/
Founded: 1967
Staff: 320
Activities: ground control of space operations * tracking stations in a worldwide network
City Reference Coordinates: 008°39'00"E 49°52'00"N

European Space Operations Centre (ESOC)
• See "European Space Agency (ESA), European Space Operations Centre (ESOC)"

Eurospace Technische Entwicklungen GmbH
Heinrich-Heine-Straße 5
D-09557 Flöha
Telephone: (0)3726-783300
Telefax: (0)3726-712378
Electronic Mail: eurospace@eurospace.de
WWW: http://www.eurospace.de/
Activities: high-technology consulting
City Reference Coordinates: 013°04'00"E 50°51'00"N

Fachhochschule Mannheim, Astronomische Arbeitsgemeinschaft
c/o W. Heinecke
Windeckstraße 110
D-68163 Mannheim
Telephone: (0)621-2926355
Telefax: (0)621-2926454
Electronic Mail: w.heinecke@fh-mannheim.de
Founded: 1976
Membership: 25
City Reference Coordinates: 008°29'00"E 48°29'00"N

Fachinformationszentrum (FIZ) Karlsruhe
(Gesellschaft für Wissenschaftlich-Technische Information mbH)
Postfach 24 65
D-76012 Karlsruhe
Telephone: (0)7247-808555
Telefax: (0)7247-808259
Electronic Mail: helpdesk@fiz-karlsruhe.de
WWW: http://www.fiz-karlsruhe.de/
Founded: 1977
Staff: 330
Activities: providing information and documentation in fields of science and technology including astronomy and astrophysics (in print and electronically)
Periodicals: (24) "Mathematics Abstracts"; (6) "STNews", "International Reviews on Mathematical Education", "Bibliographie Informatik für Schule, Hochschule und Weiterbildung"; (4) "Reports in the Fields of Science and Technology", "High-Energy Physics Index"; (1) "Jahresberichte des Bundesministers für Forschung und Entwicklung in der Biologie, Ökologie, Energie"; "Computer Theoretikum und Praktikum für Physiker", "Energy Data", "FIZ-KA Referenzserie", "FIZ-KA Berichte", "Physics Data", "STN International: Databases in Science and Technology" and others
• For parcels use: D-76344 Eggenstein-Leopoldshafen
City Reference Coordinates: 008°24'00"E 49°03'00"N

FAST COMTEC GmbH
Grünwalder Weg 28
D-82041 Oberhaching
Telephone: (0)89-66518050
Telefax: (0)89-66518040
Electronic Mail: support@fastcomtec.com
WWW: http://www.fastcomtec.com/
Founded: 1984
Staff: 12
Activities: manufacturing scanning photon counting systems and transient recorders
City Reference Coordinates: 011°37'00"E 48°02'00"N

Fehrenbach-Planetarium (Richard-_)
• See "Richard-Fehrenbach-Planetarium"

Fireball Data Center (FIDAC)
• See "International Meteor Organization (IMO), Fireball Data Center (FIDAC)"

Förderkreis Planetarium Göttingen (FPG) eV
c/o Karsten Bischoff
Universitäts-Sternwarte Göttingen
Geismarlandstraße 11
D-37083 Göttingen
Telephone: (0)551-395068
Telefax: (0)551-395043
Electronic Mail: fpg@uni-sw.gwdg.de
WWW: http://www.uni-sw.gwdg.de/FPG
Founded: 1994
Membership: 118
Activities: construction of a planetarium
City Reference Coordinates: 009°55'00"E 51°32'00"N

Forum der Technik Betriebsgesellschaft mbH, Planetarium
Museumsinsel 1
D-80538 München
Telephone: (0)89-211250
Telefax: (0)89-21125120
Electronic Mail: schwab@fdt.de
　　　　　　　 steblei@fdt.de
WWW: http://www.fdt.de/planetarium/index.phtml
Activities: producing i.a. educational programs, planetarium shows, music and laser shows
City Reference Coordinates: 011°35'00"E 48°08'00"N

Forum Weltraumforschung Aachen
• See "Technische Hochschule Aachen, Forum Weltraumforschung"

Fraunhofer Gesellschaft zur Förderung der angewandten Forschung eV
Leonrodstraße 54
D-80636 München
Telephone: (0)89-1205577
(0)89-1205544
Telefax: (0)89-1205317
Electronic Mail: info@fraunhofer.de
WWW: http://www.fraunhofer.de/
Founded: 1949
Staff: 11000
Activities: applied research (56 institutes)
Periodicals: "Fraunhofer Magazine", "Annual Report"
City Reference Coordinates: 011°35'00"E 48°08'00"N

Fraunhofer-Institut für Atmosphärische Umweltforschung
Kreuzeckbahnstraße 19
D-82467 Garmisch-Partenkirchen
Telephone: (0)8821-1830
Telefax: (0)8821-73573
Electronic Mail: info@ifu.fhg.de
<userid>@ifu.fhg.de
WWW: http://www.ifu.fhg.de/
Founded: 1974
Staff: 80
Activities: biogeochemical cycles of trace constituents in the atmosphere * climatic changes * pollution
Observatories: 4 (Garmisch-Partenkirchen - H740m; Wank-Gipfel - H1,780m; Zugspitze - H2,964m; Cape Point - South Africa)
City Reference Coordinates: 011°05'00"E 47°29'00"N

Fraunhofer-Institut für Kurzzeitdynamik, Ernst-Mach-Institut (EMI)
Eckerstraße 4
D-79104 Freiburg-im-Breisgau
Telephone: (0)761-27140
Telefax: (0)761-2714316
Electronic Mail: info@emi.fhg.de
<userid>@emi.fhg.de
WWW: http://www.emi.fhg.de/
Founded: 1949 (FhG)
Staff: 190
Activities: propulsion * impact physics * safety technology * numerical simulation * fluid dynamics * detonics * high-speed measurement techniques * system studies and analysis
City Reference Coordinates: 007°51'00"E 47°59'00"N

Fraunhofer-Institut Physikalische Meßtechnik (IPM)
Heidenhofstraße 8
D-79110 Freiburg-im-Breisgau
Telephone: (0)761-88570
Telefax: (0)761-8857224
Electronic Mail: info@ipm.fhg.de
<userid>@ipm.fhg.de
WWW: http://www.ipm.fhg.de/
http://www.fhg.de/
Founded: 1973
Membership: 100
Activities: optical measurement systems * optical spectroscopy * measurement systems for space science * optical intersatellite communication * contract research and development
• Formerly "Institut für Physikalische Weltraumforschung"
City Reference Coordinates: 007°51'00"E 47°59'00"N

Freie Universität Berlin
• See "Universität Berlin"

Freundeskreis der Himmelskunde Bad Salzschlirf eV
c/o Michael Passarge
Dr.-Martiny-Straße 1
D-36364 Bad Salzschlirf
Telephone: (0)6648-3272
Electronic Mail: 066483236-0001@t-online.de
WWW: http://observatoriumbs.rhoen.de/
Founded: 1991 (Observatory: 1983) (Solar Observatory: 1995)

Membership: 40
Activities: observing (Sun, planets, comets, deep sky) * photography * lectures
Periodicals: "Himmelskundliches aus Bad Salzschlirf"
Coordinates: 009°29'57"E 50°37'37"N H280m
009°29'57"E 50°38'06"N H367m (Solar Observatory)
City Reference Coordinates: 009°30'00"E 50°38'00"N

Friedrich-Alexander-Universität Erlangen-Nürnberg
• See "Universität Erlangen-Nürnberg"

Friedrich Ebert Visual And Radiotelescope Observatory (FEVARO)
c/o Friedrich Ebert Gymnasium
Alter Postweg 30-38
D-21075 Hamburg
Telephone: (0)40-771702048
Telefax: (0)40-7659275
Electronic Mail: feg@feg.hh.schule.de
WWW: http://www.christopher-siebert.de/feg/fevaro/
Founded: 1996
Membership: 18
Activities: comets * nebulae * planets * Sun * Moon
City Reference Coordinates: 009°59'00"E 53°33'00"N

Friedrich-Schiller-Universität Jena
• See "Universität Jena"

Friedrichs-Gymnasium der Stadt Herford, Sternwarte
Werrestraße 9
D-32049 Herford
Telephone: (0)5221-72924
(0)5221-189358
Telefax: (0)5221-189763
Founded: 1963
Staff: 3
Activities: popularization
Coordinates: 008°40'27"E 52°07'18"N H78m
City Reference Coordinates: 008°40'00"E 52°07'00"N

Georg-August-Universität Göttingen
• See "Universität Göttingen"

Gesellschaft für astronomische Bildung (GAB) eV
Peißnitzinsel 4a
D-06108 Halle
Telephone: (0)345-2028776
Electronic Mail: rfp@physik.uni-halle.de
WWW: http://www.planetarium.halle-aktuell.de/Html/GABeV/start_gabev.htm
Founded: 1990
Membership: 15
Activities: popularization * observing
Observatories: 1 (Weißenschirmbach)
City Reference Coordinates: 011°58'00"E 51°28'00"N

Gesellschaft für astronomische Bildung in Mecklenburg-Vorpommern
c/o Claus Fischer
Postfach 1146
D-17101 Demmin
or :
Asternweg 13
D-17109 Demmin
Telephone: (0)3998-222410
Telefax: (0)3998-222410
Electronic Mail: c.fischer.demmin@t-online.de
Founded: 1990
Membership: 30
Activities: popularization * education * teacher training * "Tag der Astronomie"
Periodicals: "Astroblick"
City Reference Coordinates: 013°02'00"E 53°54'00"N

Gesellschaft für volkstümliche Astronomie eV (GvA)
c/o Planetarium Hamburg
Hindenburgstraße Ö1
D-22303 Hamburg
Telephone: (0)40-516560
Electronic Mail: verwaltung@gva-hamburg.de

WWW: http://www.gva-hamburg.de/
Founded: 1964
Membership: 600
Activities: deep-sky observing * photography * CCDs * popularization
Periodicals: (4) "Sternkieker"
Observatories: 3 (Repsold-Sternwarte Hamburg, Max-Koch-Sternwarte Cuxhaven)
City Reference Coordinates: 009°59'00"E 53°33'00"N

Gesellschaft für Weltallkunde (GfW) eV
Postfach 20 51
D-40677 Erkrath
or :
c/o Michael Köchling
Sandheider Straße 70
D-40699 Erkrath
Telephone: (0)2104-47508
Telefax: (0)211-7982650
Electronic Mail: mkoechling@weltallkunde.de
WWW: http://www.weltallkunde.de/
Founded: 1986
Membership: 135
Activities: education * research
Periodicals: (4) "Weltallkunde"
City Reference Coordinates: 006°55'00"E 51°13'00"N

Gesellschaft zur wissenschaftlichen Untersuchung von Parawissenschaften eV (GWUP)
Arheilger Weg 11
D-64380 Rossdorf
Telephone: (0)6154-695021
Telefax: (0)6154-695022
Electronic Mail: info@gwup.org
WWW: http://www.gwup.org/
Founded: 1987
Membership: 330
Activities: conferences * investigations * press releases * publishing
Periodicals: (4) "Skeptiker" (ISSN 0936-9244)
City Reference Coordinates: 008°45'00"E 49°51'00"N

Gotenburgsternwarte
c/o Martin Dietrich
Augustusweg 101
D-01445 Radebeul
Telephone: (0)351-8301196
Telefax: (0)351-8301196
Electronic Mail: gotenburg@aol.com
WWW: http://members.aol.com/Gotensternwarte/
City Reference Coordinates: 013°43'00"E 51°13'00"N

Grasweg Sternwarte
c/o Rudolf A. Hillebrecht
Heinrichstraße 4
D-37581 Bad Gandersheim
Telephone: (0)5382-3020
Telefax: (0)5382-3020
Electronic Mail: 053822297-0001@t-online.de
Founded: 1980
Activities: observing (Sun, Moon, planets, comets, stars) * photography * CCDs
City Reference Coordinates: 010°01'00"E 51°22'00"N

Gymnasium Philippinum Sternwarte
Leopold-Lucas-Straße 18
D-35037 Marburg
Telephone: (0)6421-931805
Telefax: (0)6421-931804
Founded: 1989
Membership: 30
Activities: popularization * observing (Sun, satellites, variable stars, planets) * photography * planetarium
Coordinates: 008°08'00"E 50°08'00"N H146m
City Reference Coordinates: 008°08'00"E 50°08'00"N

Hamburger Sternwarte
• See "Universität Hamburg, Hamburger Sternwarte"

Hans-Nüchter Sternwarte
Domänenweg 2

D-36037 Fulda
Telephone: (0)661-65037
         (0)661-969400
Founded: 1977
Membership: 30
Activities: education * photography * observing * planetarium
Periodicals: (2) "Ganymed" (circ.: 100)
Coordinates: 009°41'57"E 50°33'21"N H300m
City Reference Coordinates: 009°41'00"E 50°33'00"N

Harzplanetarium
Walter-Rathenau-Straße 11
D-38855 Wernigerode
Telephone: (0)3943-603939
         (0)3943-603959
WWW: http://www.harzplanetarium.de/
Founded: 1972
Staff: 2
Activities: education * lectures
Coordinates: 010°45'00"E 51°49'00"N H240m
City Reference Coordinates: 010°45'00"E 51°49'00"N

Hermann-von-Helmholtz-Gemeinschaft Deutscher Forschungszentren (HGF)
Ahrstraße 45
D-53175 Bonn
Telephone: (0)228-308180
Telefax: (0)228-3081830
WWW: http://www.helmholtz.de/
Founded: 1970
Membership: 16 (research centres)
Periodicals: "Forschungsthemen", "HGF-Mitteilungen", "Handbuch der Helmholtz-Zentren"
• Formerly known as "Arbeitsgemeinschaft der Großforschungseinrichtungen"
City Reference Coordinates: 007°05'00"E 50°44'00"N

Institut für Physikalische Weltraumforschung, Freiburg-im-Breisgau
• See now "Fraunhofer-Institut für Physikalische Meßtechnik"

Intercon Spacetec GmbH
Gablinger Weg 9
D-86154 Augsburg
Telephone: (0)821-414081
Telefax: (0)821-414085
Electronic Mail: info@intercon-spacetec.com
WWW: http://www.intercon-spacetec.com/
Founded: 1981
Staff: 7
Activities: distributing telescopes, accessories, and astronomy software * manufacturing ICS Newtonian telescopes from 8" to 25"
City Reference Coordinates: 010°53'00"E 48°23'00"N

Interessengemeinschaft Astrofotografie Bochum (IAB)
Grillostraße 70
D-44799 Bochum 1
Telephone: (0)234-382771
WWW: http://members.tripod.de/Volker_Mette/
Founded: 1984
Activities: photography (deep sky, comets, zodiacal light, counterglow phenomenon) * image processing and analysis * darkroom techniques * hypersensitizing * excursions * education * amateur research projects
City Reference Coordinates: 007°13'00"E 51°28'00"N

Interessengemeinschaft Astronomie Crimmitschau (IGAC) eV
c/o Sternwarte "Johannes Kepler"
Lindenstraße 8
D-08451 Crimmitschau
Telephone: (0)3762-3730
Electronic Mail: info@sternwarte-crimmitschau.de
WWW: http://www.sternwarte-crimmitschau.de/
Founded: 1960 (Observatory: 1929)
Membership: 30
Activities: Sun * lunar occultations * CCD photography * deep sky * computing
Periodicals: (6) "Infosheet: Crimmitschauer Astronomische Nachrichten"
Coordinates: 012°23'07"E 50°48'52"N H273m
City Reference Coordinates: 012°23'00"E 50°49'00"N

Interessengemeinschaft Astronomie Diez und Umgebung eV
c/o Jola Grün
Jahnstraße 10
D-56412 Nentershausen
Telephone: (0)6485-911696
Telefax: (0)6485-911532
Electronic Mail: jola.gruen@sternwarte-diez.de
WWW: http://www.sternwarte-diez.de/index.htm
Activities: meetings * observing * Westerwälder Tag der Astronomie (WTA)
Coordinates: 008°01'28"E 50°21'47"N H170m
City Reference Coordinates: 009°56'00"E 51°02'00"N (Nentershausen)
008°01'00"E 50°22'00"N (Diez)

Internationale Amateur-Sternwarte (IAS) eV
c/o K.-L. Bath
Geranienstraße 2
D-79312 Emmendingen
Telephone: (0)7641-3492
Telefax: (0)7641-3492
Electronic Mail: ias@epost.de
k.-l.bath@t-online.de (Karl-Ludwig Bath, Secretary)
WWW: http://www.ias.webwide.de/
Founded: 1999
Membership: 50
Activities: amateur astronomy * observing * Southern sky (Namibia)
City Reference Coordinates: 007°51'00"E 48°07'00"N

International Earth Rotation Service (IERS)
c/o Bundesamt für Kartographie und Geodäsie
Richard-Strauß-Allee 11
D-60598 Frankfurt/Main
Telephone: (0)69-6333273
Telefax: (0)69-6333425
Electronic Mail: central_bureau@iers.org
WWW: http://www.iers.org/
Founded: 1988
Activities: Earth rotation * terrestrial and celestial reference systems
Periodicals: (52) "IERS Bulletin A" (circ.: 630); (12) "IERS Bulletin B" (circ.: 700); (2) "IERS Bulletin C" (circ.: 1000); (1) "IERS Annual Report" (ISSN 0068-4236, circ.: 850); "IERS Bulletin D" (circ.: 560), "IERS Technical Notes" (ISSN 1019-4568, circ.: 550), "IERS Messages" (circ.: 1600);
Coordinates: 008°39'51"E 50°05'22"N H163m
City Reference Coordinates: 008°40'00"E 50°07'00"N

International Meteor Organization (IMO)
c/o Jürgen Rendtel (President)
Seestraße 6
D-14476 Marquardt
WWW: http://www.imo.net/
http://www.amsmeteors.org/imo-mirror
Founded: 1988
Membership: 260
Activities: observing (visual, photographic, telescopic, radio and video) * databases * bibliography
Sections: (commissions) Radio * Telescopic * Visual * Camera Network * Video
Periodicals: (6) "WGN" (ISSN 1016-3115; circ.: 320); (1) "Annual Report", "IMO Proceedings"
City Reference Coordinates: 012°57'00"E 52°27'00"N (Marquardt)

Jenoptik AG
Carl-Zeiss-Straße 1
D-07743 Jena
Telephone: (0)3641-650
Telefax: (0)3641-424514
Founded: 1991 (1846 as "Carl Zeiss")
Staff: 7000
Activities: development, production and distribution of laser systems, optics and equipment * optical medical instruments * engineering metrology equipment * engineering of semiconductors * equipment for the telecommunication industry * facility management * clean-room automation * electromechanical systems
Awards: "Ernst-Abbe-Preis", "Otto-Schott-Preis"
City Reference Coordinates: 011°35'00"E 50°56'00"N

Johann-Wolfgang-Goethe-Universität Frankfurt/Main
• See "Universität Frankfurt/Main"

Joint Organization for Solar Observations (JOSO)
c/o P.N. Brandt
Kiepenheuer-Institut für Sonnenphysik
Schöneckstraße 6
D-79104 Freiburg-im-Breisgau
Telephone: (0)761-3198250
Telefax: (0)761-3198111
Electronic Mail: pnb@kis.uni-freiburg.de
pbrandt@solar.stanford.edu
WWW: http://joso.oat.ts.astro.it/
Founded: 1969
Membership: 26 (countries)
Activities: annual meetings * exchange of information on solar research in Europe and abroad
Periodicals: (1) "Annual Report" (circ.: 300)
City Reference Coordinates: 007°51'00"E 47°59'00"N

Jutz Volkssternwarte Berg eV (Christian_)
• See "Christian Jutz Volkssternwarte Berg eV"

Karl-Schwarzschild-Observatorium (KSO)
• See "Thüringer Landessternwarte (TLS) - Tautenburg, Karl-Schwarzschild-Observatorium (KSO)"

Kayser-Threde GmbH
Wolfratshauser Straße 48
D-81379 München
Telephone: (0)89-724950
Telefax: (0)89-72495291
Electronic Mail: info@kayser-threde.de
<userid>@kayser-threde.de
WWW: http://www.kayser-threde.de/
Founded: 1967
Staff: 170
Activities: manufacturing microgravity research instruments, small satellites, Earth observation and optical instrumentation, small reentry capsules, sounding rockets payloads, navigation, space telescopes and space science instrumentation, on-board data handling systems
City Reference Coordinates: 011°35'00"E 48°08'00"N

Kiepenheuer-Institut für Sonnenphysik (KIS)
Schöneckstraße 6
D-79104 Freiburg-im-Breisgau
Telephone: (0)761-31980
(0)7602-226 (Schauinsland Observatory)
Telefax: (0)761-3198111
(0)761-382280 (Schauinsland Observatory)
Electronic Mail: secr@kis.uni-freiburg.de
<userid>@kis.uni-freiburg.de
WWW: http://www.kis.uni-freiburg.de/
http://www.kis.uni-freiburg.de/kiswwwe.html (English page)
Founded: 1943
Staff: 13
Activities: solar research * observing * theoretical studies * instrumentation
Coordinates: 007°54'22"E 47°54'52"N H1,240m (Schauinsland)
016°30'30"W 28°17'50"N H2,395m (Izaña, Teide – IAU Code 954)
City Reference Coordinates: 007°51'00"E 47°59'00"N

Landessternwarte Heidelberg
Königstuhl
D-69117 Heidelberg
Telephone: (0)6221-5090
Telefax: (0)6221-509202
Electronic Mail: <userid>@lsw.uni-heidelberg.de
WWW: http://www.lsw.uni-heidelberg.de/
Founded: 1898
Staff: 45
Activities: observational and theoretical astrophysics
Coordinates: 008°43'18"E 49°23'54"N H564m (IAU Code 024)
City Reference Coordinates: 008°43'00"E 49°25'00"N

LOBO electronic GmbH
Robert-Bosch-Straße 100
D-73437 Aalen
Telephone: (0)7361-968710
Telefax: (0)7361-968730

Electronic Mail: mail@lobo.de
WWW: http://www.lobo.de/
Founded: 1982
Staff: 25
Activities: manufacturing transputer-controlled laser animation systems for i.a. planetariums
City Reference Coordinates: 010°05'00"E 48°50'00"N

Lohrmann-Observatorium
• See "Technische Universität Dresden, Institut für Planetare Geodäsie, Lohrmann-Observatorium"

Max-Planck-Gesellschaft (MPG)
Residenzstraße 1a
D-80333 München
Telephone: (0)89-21081
Telefax: (0)89-229850
WWW: http://www.mpg.de/
Founded: 1948
Periodicals: "MPG Spiegel" (ISSN 0341-7727), "Berichte und Mitteilungen" (ISSN 0341-7778)
City Reference Coordinates: 011°35'00"E 48°08'00"N

Max-Planck-Institut für Aeronomie (MPAe), Katlenburg-Lindau
Postfach 20
D-37189 Katlenburg-Lindau
or :
Max-Planck-Straße 2
D-37191 Katlenburg-Lindau
Telephone: (0)5556-9790
Telefax: (0)5556-979240
Electronic Mail: <userid>@linmpi.mpae.gwdg.de
WWW: http://www.mpae.gwdg.de/
    http://www.mpae.gwdg.de/mpae_projects/SOUSY/sousy_home.html (SOUSY project)
    http://www.mpae.gwdg.de/mpae_projects/STARE/STARE.html (STARE project)
Founded: 1936
Staff: 300
Activities: atmospheric physics * magnetospheric physics * planets * Sun * interplanetary medium
City Reference Coordinates: 010°06'00"E 51°41'00"N

Max-Planck-Institut für Astronomie (MPIA), Heidelberg
Königstuhl 17
D-69117 Heidelberg
Telephone: (0)6221-5280
Telefax: (0)6221-528246
Electronic Mail: <userid>@mpia-hd.mpg.de
WWW: http://www.mpia-hd.mpg.de/
    http://www.mpia-hd.mpg.de/MPIA/
Founded: 1969
Staff: 150
Activities: star forming regions * circumstellar matter * ISM * AGNs * evolution of galaxies * clusters of galaxies * IR observing from space
Periodicals: (1) "Annual Report"
Coordinates: 008°43'07"E 49°23'49"N H560m
City Reference Coordinates: 008°43'00"E 49°25'00"N

Max-Planck-Institut für Astrophysik (MPA)
Postfach 1523
D-85740 Garching
or :
Karl-Schwarzschild-Straße 1
D-85748 Garching
Telephone: (0)89-329900
Telefax: (0)89-32993235
Electronic Mail: <userid>@mpa-garching.mpg.de
WWW: http://www.mpa-garching.mpg.de/
Founded: 1958
Staff: 68
Activities: astrophysical research
City Reference Coordinates: 011°40'00"E 48°14'00"N

Max-Planck-Institut für extraterrestrische Physik (MPE)
Postfach 16 03
D-85740 Garching
or :
Giessenbachstraße
D-85748 Garching
Telephone: (0)89-329900

Telefax: (0)89-32993569
Electronic Mail: mpe@mpe.mpg.de
                  <userid>@mpe.mpg.de
WWW: http://www.mpe.mpg.de/
Founded: 1962
Staff: 355
Activities: in situ space data collection * astrophysical data collection * data analysis * theory
Periodicals: (1) "Jahresbericht" (ISSN 0947-8787)
Coordinates: 011°40'23"E 48°15'46"N H476m
City Reference Coordinates: 011°40'00"E 48°14'00"N

Max-Planck-Institut für Gravitationsphysik
(Albert-Einstein-Institut - AEI)
Schlaatzweg 1
D-14473 Potsdam
Telephone: (0)331-275370
Telefax: (0)331-2753798
Electronic Mail: office@aei-potsdam.mpg.de
                  <userid>@aei-potsdam.mpg.de
WWW: http://www.aei-potsdam.mpg.de/
      http://www.livingreviews.org/ (Living Reviews in Relativity)
Founded: 1995
Staff: 40
Activities: gravitation * relativity * astrophysics * quantum gravity * numerical relativity * black holes * mathematical relativity * gravitational waves * superconducting
Periodicals: "Living Reviews in Relativity" (ISSN 1433-8351)
City Reference Coordinates: 013°04'00"E 52°24'00"N

Max-Planck-Institut für Kernphysik (MPI-K), Bereich Astrophysik
Postfach 10 39 80
D-69029 Heidelberg
or :
Saupfercheckweg 1
D-69117 Heidelberg
Telephone: (0)6221-516295
Telefax: (0)6221-516549
       (0)6221-516324
Electronic Mail: <userid>@mpi-hd.mpg.de
                  mpik@mpi-hd.mpg.de
WWW: http://www.mpi-hd.mpg.de/
Founded: 1958
Staff: 22
Activities: gamma-ray astronomy * neutrino astronomy * ISM * IR astronomy * star formation * cosmic-ray physics
Periodicals: (1) "Jahresbericht"
City Reference Coordinates: 008°43'00"E 49°25'00"N

Max-Planck-Institut für Physik (MPP)
(Werner-Heisenberg-Institut - WHI)
Föhringer Ring 6
D-80805 München
Telephone: (0)89-323540
Telefax: (0)89-3226704
Electronic Mail: <userid>@mppmu.mpg.de
WWW: http://www.mppmu.mpg.de/
Founded: 1917 (as "Kaiser Wilhelm-Institut für Physik")
Activities: fundamental constituents of matter, their interactions, and the role they play in astrophysics
City Reference Coordinates: 011°35'00"E 48°08'00"N

Max-Planck-Institut für Radioastronomie (MPIfR)
Auf dem Hügel 69
D-53121 Bonn
Telephone: (0)228-5250
          (0)2257-3010 (Effelsberg)
Telefax: (0)228-525229
      (0)2257-30169 (Effelsberg)
Electronic Mail: <userid>@mpifr-bonn.mpg.de
WWW: http://www.mpifr-bonn.mpg.de/
Founded: 1967
Staff: 180
Activities: radioastronomy * spectroscopy * continuum radiation * VLBI * ISM structure and physics * extended galaxies * extragalactic sources * optical interferometry * developing new observing techniques
Coordinates: 006°53'06"E 50°31'36"N H369m (Bad Münstereifel/Effelsberg)
City Reference Coordinates: 007°05'00"E 50°44'00"N

Menke Planetarium und Sternwarte
Fachhochschule Flensburg

Founded: 1979
Coordinates: 009°33'52"E 49°25'52"N H360m
City Reference Coordinates: 009°34'00"E 49°26'00"N

Observatorium Wendelstein
• See "Universität München, Institut für Astronomie und Astrophysik, Observatorium Wendelstein"

Office Européen des Brevets (OEB)
• See "European Patent Office (EPO)"

Olbers-Planetarium
c/o Hochschule Bremen
Fachbereich Nautik
Werderstraße 73
D-28199 Bremen
Telephone: (0)421-706882
Telefax: (0)421-7942842
Electronic Mail: planetarium@hs-bremen.de
WWW: http://www.hs-bremen.de/planetarium/
Founded: 1952
Staff: 2
Activities: shows for students and public
City Reference Coordinates: 008°49'00"E 53°04'00"N

Optische und Elektronische Systeme (OES) GmbH
Dr. Neumeyer Straße 240
D-91349 Egloffstein
Telephone: (0)9197-698980
Telefax: (0)9197-698982
Electronic Mail: ff@fonline.de (Frank Fleischmann)
WWW: http://www.fonline.de/home/ff/
Founded: 1988
Staff: 4
Activities: manufacturing and distributing CCD cameras, image intensifiers, photon-counting devices, speckle-interferometric devices, telescope control systems and astronomical image processing software
Coordinates: 011°16'00"E 49°42'00"N H380m
City Reference Coordinates: 011°16'00"E 49°42'00"N

Palitzsch-Gesellschaft eV
Am Anger 20
D-01237 Dresden
Telephone: (0)351-2847765
Telefax: (0)351-2847765
Electronic Mail: pag@prohlis-online.de
WWW: http://www.palitzsch-gesellschaft.de/
City Reference Coordinates: 013°44'00"E 51°03'00"N

Physikalischer Verein Frankfurt, Volkssternwarte
Robert-Mayer-Straße 2-4
D-60054 Frankfurt/Main
Telephone: (0)69-704630
Telefax: (0)69-97981342
Electronic Mail: info@physikalischer-verein.de
WWW: http://www.physikalischer-verein.de/
Founded: 1824 (Verein)
Membership: 700
Activities: popularization * observing (PN, deep sky, Sun, planets, Moon, double stars)
Periodicals: (4) "Mitteilungen astronomischer Vereinigungen Rhein-Main-Nahe" (circ.: 550); (1) "Jahresbericht"
Awards: (1) "Philipp-Siedler-Preis", "Christian-Ernst-Neeff-Preis", "Samuel-Thomas-von Soemmering-Preis", "Eugen-Hartmann-Didaktik-Preis"
Coordinates: 008°39'12"E 50°07'05"N H115m
           008°26'43"E 50°13'22"N H825m (Feldberg)
City Reference Coordinates: 008°40'00"E 50°07'00"N

Physikalisch-Technische Bundesanstalt (PTB)
Bundesallee 100
D-38116 Braunschweig
Telephone: (0)531-5920
Telefax: (0)531-5929292
Electronic Mail: <userid>@ptb.de
WWW: http://www.ptb.de/
Founded: 1887
Staff: 1600
Activities: legal metrology * realization and dissemination of SI-units * determination of physical constants
Periodicals: (1) "Jahresbericht" (ISSN 0340-4366)

Awards: (1/3) "Helmholtz-Preis"
City Reference Coordinates: 010°31'00"E 52°16'00"N

Physik Instrumente (PI) GmbH & Co.
Polytec-Platz 5-7
D-76337 Waldbronn
Telephone: (0)7243-604100
Telefax: (0)7243-604145
WWW: http://www.physikinstrumente.com/
Founded: 1969
Staff: 40
Activities: manufacturing piezoelectric translators, tilting mirrors, positioning systems, step- and DC-motor controllers
Periodicals: "Movement & Positioning"
City Reference Coordinates: 008°25'00"E 48°56'00"N (Ettlingen)

Planetarium Aschersleben
Im Tierpark
Auf der Alten Burg 40
D-06449 Aschersleben
Telephone: (0)3473-2592
Telefax: (0)3473-3841
WWW: http://lbs.st.schule.de/lbs/pl-asch.htm
City Reference Coordinates: 011°28'00"E 51°46'00"N

Planetarium der Fachhochschule Kiel
Postfach 28 48
D-24027 Kiel
or :
Alte Chaussee 32
D-24107 Kiel
Telephone: (0)431-5198211
Electronic Mail: postfach@planetarium.fh-kiel.de
WWW: http://www.planetarium.fh-kiel.de/planetarium/
Founded: 1969
Staff: 10
Activities: shows for schools and public
City Reference Coordinates: 010°08'00"E 54°20'00"N

Planetarium der Fachhochschule Stralsund
Hainholzstraße 59
D-18435 Stralsund
Telephone: (0)3831-456528
          (0)3831-456529
Electronic Mail: rudi.wendorf@fh-stralsund.de
WWW: http://www-httpd.fh-stralsund.de/Allgemein/planet.html
City Reference Coordinates: 013°05'00"E 54°19'00"N

Planetarium der Stadt Wolfsburg GmbH
Uhlandweg 2
D-38440 Wolfsburg
Telephone: (0)5361-21939
Telefax: (0)5361-21272
Electronic Mail: info@planetarium-wolfsburg.de
WWW: http://www.planetarium-wolfsburg.de/
Founded: 1983
Activities: planetarium shows * concerts * narrations * education * lectures * seminars
City Reference Coordinates: 010°47'00"E 52°25'00"N

Planetarium Hamburg
Hindenburgstraße Ö1
D-22303 Hamburg
Telephone: (0)40-5149850
Telefax: (0)40-51498510
WWW: http://www.hamburg-cityguide.de/planeta.html
       http://www.hamburg.de/Behoerden/Kulturbehoerde/Planetarium/
Founded: 1930
City Reference Coordinates: 009°59'00"E 53°33'00"N

Planetarium Hoyerswerda
3. Mittelschule "Mittelschule am Planetarium" (WK VI)
Collins-Straße 29
D-02977 Hoyerswerda
Telephone: (0)3571-417020
WWW: http://www.germany.net/teilnehmer/101/57799/planetar.htm
Founded: 1969

Coordinates: 014°15'00"E 51°26'30"N
City Reference Coordinates: 014°14'00"E 51°26'00"N

Planetarium im Vonderau Museum
Jesuitenplatz 2
D-36037 Fulda
Telephone: (0)661-928350
Telefax: (0)661-9283513
Electronic Mail: museum@fulda.de
WWW: http://www.fulda.de/
Founded: 1990
Staff: 5
Activities: popularization
City Reference Coordinates: 009°41'00"E 50°33'00"N

Planetarium Lübz
Am Markt 22
D-19386 Lübz
or :
Neuer Teich 6
D-19386 Lübz
Telephone: (0)38731-23581
Telefax: (0)38731-22234
WWW: http://www.luebz.de/eindruecke.html#planetarium
http://ourworld.compuserve.com/homepages/as171/luebz.htm
Founded: 1980
Coordinates: 012°02'56"E 53°27'38"N H61m
City Reference Coordinates: 012°03'00"E 53°28'00"N

Planetarium Mannheim GmbH
Wilhelm-Varnholt-Allee 1
D-68165 Mannheim
Telephone: (0)621-419420
(0)621-415692
Telefax: (0)621-412411
WWW: http://www.mannheim.de/planetarium/
Founded: 1984
Staff: 9
Activities: shows for public and schools * cultural events
Coordinates: 008°29'38"E 49°28'42"N
City Reference Coordinates: 008°29'00"E 48°29'00"N

Planetarium Merseburg
c/o Helmut Conrad
Schulstraße 1a
D-06242 Braunsbedra
Telephone: (0)34633-20909
WWW: http://lbs.st.schule.de/lbs/pl-mersb.htm
Founded: 1966
Staff: 2
Activities: education
City Reference Coordinates: 011°53'00"E 51°18'00"N (Braunsbedra)
012°00'00"E 51°22'00"N (Merseburg)

Planetarium Senftenberg
An der Ingenieurschule
D-01968 Senftenberg
Telephone: (0)3573-2112
WWW: http://www.planetenjahn.de/planetarium.htm
Founded: 1966
Membership: 18
Activities: popularization * training * lectures * shows
Coordinates: 013°59'00"E 51°22'00"N H104m (IAU Code 640)
City Reference Coordinates: 013°59'00"E 51°22'00"N

Rat Deutscher Sternwarten (RDS)
(Council of German Observatories)
c/o Detlev Koester (Chairman)
Institut für Theoretische Physik und Astrophysik
Universität Kiel
Olshausenstraße 40
D-24098 Kiel
Telephone: (0)431-8804110
Telefax: (0)431-8804100
Electronic Mail: koester@astrophysik.uni-kiel.de

Founded: 1959
Membership: 35 (institutes, observatories and research groups)
Activities: representing nationally and internationally research at astronomical observatories in Germany * coordinating activities between these institutions
Periodicals: (1) "Annual Report"
City Reference Coordinates: 010°08'00"E 54°20'00"N

Raumflugplanetarium Cottbus
Lindenplatz 21
D-03042 Cottbus
Telephone: (0)355-713109
Telefax: (0)355-7295822
WWW: http://www-user.tu-cottbus.de/~embergb/planetarium/
Founded: 1996
City Reference Coordinates: 014°21'00"E 51°43'00"N

Raumflugplanetarium Halle (RFP)
Peißnitzinsel 4a
D-06108 Halle
Telephone: (0)345-8060317
Telefax: (0)345-8060317
Electronic Mail: rfp@physik.uni-halle.de
WWW: http://www.planetarium.halle-aktuell.de/
  http://www.halle.de/DEUTSCH/4/2/01625/01625.HTM
Founded: 1978
Staff: 2
Activities: education * shows * popularization
Coordinates: 011°56'59"E 51°29'47"N
City Reference Coordinates: 011°58'00"E 51°28'00"N

Regionale Astronomische Arbeitsgemeinschaft Euskirchen (RAAGE)
c/o Heinz-Jürgen Schäfer
Danziger Straße 29
D-53879 Euskirchen
WWW: http://home.t-online.de/home/willi-graf-realschule/raage.htm
Founded: 1982
Membership: 15
Activities: lectures * photography * computing
City Reference Coordinates: 006°47'00"E 50°39'00"N

Remeis-Sternwarte (Dr._)
• See "Universität Erlangen-Nürnberg, Astronomisches Institut, Dr. Remeis-Sternwarte"

Rheinische Friedrich-Wilhelms-Universität Bonn
• See "Universität Bonn"

Rheinische-Westfälische Technische Hochschule Aachen
• See "Technische Hochschule Aachen"

Richard-Fehrenbach-Planetarium
Friedrichstraße 51
D-79098 Freiburg-im-Breisgau
Telephone: (0)761-276099
Telefax: (0)761-276099
WWW: http://www.astrosurf.com/euroastronomie/D-79098.htm
Founded: 1975
Activities: shows for schools and public
City Reference Coordinates: 007°51'00"E 47°59'00"N

Robert-Mayer-Volks-und-Schulsternwarte Heilbronn eV
Bismarckstraße 10
D-74072 Heilbronn
Telephone: (0)7131-81299
Telefax: (0)7131-677777
Electronic Mail: info@sternwarte.org
WWW: http://www.sternwarte.org/
Founded: 1914 (Association: 1987)
Membership: 171
Activities: education * public observing * lectures * youth activities * working groups * trips
Coordinates: 009°13'40"E 49°08'25"N H188m
City Reference Coordinates: 009°14'00"E 49°08'00"N

Römer-Sternwarte Rheinhausen (RRS) eV (Rudolf-_)
• See "Rudolf-Römer-Sternwarte Rheinhausen (RRS) eV"

Rudolf-Römer-Sternwarte Rheinhausen (RRS) eV
Postfach 14 18 07
D-47208 Duisburg
or :
Schwarzenberger Straße 147
D-47226 Duisburg-Rheinhausen
Telephone: (0)2065-75012
Founded: 1971
Membership: 80
Activities: observing * photography * education
Periodicals: (4) "Duisburger Komet" (ISSN 0179-0730)
City Reference Coordinates: 006°46'00"E 51°25'00"N

Ruhr-Universität Bochum
• See "Universität Bochum"

Ruprecht-Karls-Universität Heidelberg
• See "Universität Heidelberg"

Schott Glas, Optics Division
Hattenbergstraße 10
D-55120 Mainz
Telephone: (0)6131-660
(0)6131-663844
Telefax: (0)6131-662000
(0)6131-662077
Electronic Mail: zerodur@schott.de
WWW: http://www.schott.de/
Activities: optical glas manufacturer
City Reference Coordinates: 008°16'00"E 50°01'00"N

Schulplanetarium Chemnitz
Nikolaus-Kopernikus-Mittelschule
Albert-Köhler-Straße 48
D-09122 Chemnitz
Telephone: (0)371-229111
Telefax: (0)371-229111
WWW: http://members.aol.com/ugbsoft/point/11_1.htm
http://www.tu-chemnitz.de/~wth/11_1.htm
Founded: 1981
Staff: 2
Activities: education * shows * lectures
City Reference Coordinates: 012°55'00"E 50°50'00"N

Schulsternwarte und Planetarium Rodewisch
Rützengrüner Straße 41A
D-08228 Rodewisch
Telephone: (0)3744-32313
Telefax: (0)3744-32815
Electronic Mail: info@sternwarte-rodewisch.de
WWW: http://www.sternwarte-rodewisch.de/
Founded: 1950
Activities: education * observing * planetarium shows
Coordinates: 012°24'56"E 50°31'42"N H498m
City Reference Coordinates: 012°25'00"E 50°32'00"N

Schul- und Volkssternwarte K.E. Ziolkowski
Postfach 30 05 05
D-98504 Suhl
or :
Auf dem Hoheloh 1
D-98527 Suhl
Telephone: (0)3681-723556
Telefax: (0)3681-303799
Electronic Mail: kretzer.sternwarte-suhl@t-online.de (Olaf Kretzer, Manager)
Founded: 1965
Staff: 1
Activities: education * popularization * observing * astronautics * planetarium
Coordinates: 010°42'08"E 50°36'14"N H527m
City Reference Coordinates: 010°43'00"E 50°37'00"N

Schul- und Volkssternwarte Leinfelden-Echterdingen eV
Immanuel-Kant-Gymnasium
Anemonenstraße 15
D-70771 Leinfelden-Echterdingen

Telephone: (0)711-1600500
Telefax: (0)711-1600503
WWW: http://www.sternwarte-le.de/
Founded: 1987
Membership: 50
City Reference Coordinates: 009°08'00"E 48°41'00"N

Schwäbische Sternwarte (SSW) eV
(Geschäftsstelle)
Seestraße 59A
D-70174 Stuttgart
or :
Sternwarte Stuttgart
Zur Uhlandshöhe 41
D-70188 Stuttgart
Telephone: (0)711-2260893 (Geschäftsstelle)
          (0)711-281871 (Observatory)
Telefax: (0)711-2260895 (Geschäftsstelle)
          (0)711-2624546 (Observatory)
WWW: http://www.sternwarte.de/
Founded: 1922 (Association: 1920)
Membership: 30
Activities: occultations * variable stars * photography * CCDs
Periodicals: "Sternwarte Stuttgart"
Coordinates: 009°11'51"E 48°47'01"N H354m (IAU Code 025)
City Reference Coordinates: 009°11'00"E 48°46'00"N

Schwarzschild-Observatorium (KSO) (Karl-_)
● See "Thüringer Landessternwarte (TLS) - Tautenburg, Karl-Schwarzschild-Observatorium (KSO)"

Scientific and Technical Information Network (STN) International
c/o Fachinformationszentrum Karlsruhe
Postfach 24 65
D-76012 Karlsruhe
Telephone: (0)7247-808555
Telefax: (0)7247-808131
Electronic Mail: hlpdeskk@fiz-karlsruhe.de
WWW: http://www.fiz-karlsruhe.de/
Founded: 1977
Staff: 330
Activities: operating the STN Karlsruhe centre * databanks including information in the fields of astronomy and astrophysics
Periodicals: (6) "STNews"
● For parcels use: D-76344 Eggenstein-Leopoldshafen
City Reference Coordinates: 008°24'00"E 49°03'00"N (Karlsruhe)
                            008°23'00"E 49°04'00"N (Eggenstein)

Sonnenobservatorium Einsteinturm (SOE)
● See "Astrophysikalisches Institut Potsdam (AIP), Sonnenobservatorium Einsteinturm (SOE)"

Spacetec GmbH (Intercon_)
● See "Intercon Spacetec GmbH"

Space Telescope - European Coordinating Facility (ST-ECF)
c/o E.S.O.
Karl-Schwarzschild-Straße 2
D-85748 Garching
Telephone: (0)89-32006291
Telefax: (0)89-32006480
Electronic Mail: <userid>@eso.org
WWW: http://www.stecf.org/
Founded: 1984
Staff: 18
Activities: European focal point of ST-related activities * coordinating the development of data analysis software in Europe and with the "Space Telescope Science Institute (STScI)" (USA-MD) * maintaining a copy of the ST archive * supporting European astronomers in the preparation of ST proposals * organizing HST-related meetings
Periodicals: "ST-ECF Newsletter"
● Joint facility of the "European Space Agency (ESA)" and the "European Southern Observatory (ESO)"
City Reference Coordinates: 011°40'00"E 48°14'00"N

Sparkassen-Planetarium Augsburg
Im Thäle 3
D-86152 Augsburg
Telephone: (0)821-3246740 (Information)
          (0)821-3246762 (Management)
          (0)821-314936

Telefax: (0)821-314946
Electronic Mail: s-planetarium@a-city.de
WWW: http://www.augsburg.baynet.de/s-planetarium/
http://www.a-city.de/s-planetarium/
City Reference Coordinates: 010°53'00"E 48°23'00"N

Springer-Verlag
Postfach 10 52 80
D-69042 Heidelberg
or :
Tiergartenstraße 17
D-69121 Heidelberg
Telephone: (0)6221-487360
Telefax: (0)6221-487150
Electronic Mail: springer@vax.ntp.springer.de
WWW: http://www.springer.de/
http://science.springer.de/
Founded: 1842
Staff: 1230
Periodicals: among many other journals: "Astronomy and Astrophysics Reviews" (ISSN 0935-4956 & 1432-0754)
• Publisher
City Reference Coordinates: 008°43'00"E 49°25'00"N

Starkenburg-Sternwarte eV Heppenheim
Niemöllerstraße 9
D-64646 Heppenheim
Telephone: (0)6252-798844
Electronic Mail: e.schwab@gsi.de (Erwin Schwab)
WWW: http://www.starkenburg-sternwarte.de/
Founded: 1970
Membership: 180
Activities: minor planets * radioastronomy * Sun * photography
Periodicals: (4) "Sirius"
Coordinates: 008°39'11"E 49°38'53"N H256m (IAU Code 611)
City Reference Coordinates: 008°39'00"E 49°38'00"N

Stefan Thiele (Astronomische Instrumente_)
• See "Astronomische Instrumente Stefan Thiele (AIT)"

Sterne (Die_)
• Now merged with "Sterne und Weltraum"

Sterne und Weltraum (SuW)
c/o Jakob Staude (Editor)
Redaktion SuW
MPI für Astronomie
Königstuhl 17
D-69117 Heidelberg
or :
Verlag Sterne und Weltraum
Hüthig GmbH
Im Weiher 10
D-69121 Heidelberg
Telephone: (0)6221-528229 (Editor)
(0)6221-4890 (Publisher)
Telefax: (0)6221-528377 (Editor)
(0)6221-489279 (Publisher)
Electronic Mail: quetz@mpia.de
WWW: http://www.mpia.de/suw/
http://www.sterne-und-weltraum.de/
Founded: 1962
Staff: 6
• Journal (11) (ISSN 0039-1263, circ.: 40,000)
City Reference Coordinates: 008°43'00"E 49°25'00"N

Sternfreunde Braunschweig-Hondelage eV
Ackerweg 1B
D-38108 Braunschweig
Telephone: (0)162-7551815
Telefax: (0)5309-1758
Electronic Mail: hans-w.zimmermann@t-online.de (Hans Zimmermann, President)
WWW: http://www.sternfreunde-braunschweig.de/
Founded: 1982
Membership: 90
Activities: observing * education

City Reference Coordinates: 010°31'00"E 52°16'00"N

Sternfreunde Breisgau (SFB) eV
c/o Karl-Ludwig Bath
Geranienstraße 2
D-79312 Emmendingen
Telephone: (0)7641-3492
Telefax: (0)7641-3492
WWW: http://www.kis.uni-freiburg.de/~ps/SFB/
Founded: 1973
Membership: 60
Activities: occultations * general amateur astronomy * popularization
Periodicals: (3) "Mitteilungen"
Coordinates: 007°54'20"E 47°54'53"N H1,240m (Schauinslandsternwarte)
City Reference Coordinates: 007°51'00"E 48°07'00"N

Sternfreunde Donzdorf eV
Gmünder Straße 12
D-73072 Donzdorf
or :
Beim Schulzentrum
D-73072 Donzdorf
Telephone: (0)7162-24713
Electronic Mail: sternwarte-donzdorf@gmx.de
WWW: http://www.sternwarte-donzdorf.de/ (Messelberg-Sternwarte)
Founded: 1985
Membership: 140
Activities: observing * photography * education * lectures
Periodicals: (4) "Galaxis"
Coordinates: 009°49'12"E 48°41'10"N H456m (Messelberg-Sternwarte)
City Reference Coordinates: 009°49'00"E 48°41'00"N

Sternfreunde Durmersheim und Umgebung eV
c/o Jürgen Linder
Im Eck 1/19
D-76448 Durmersheim
Telephone: (0)7245937594
Telefax: (0)7245937596
Electronic Mail: Juergen.Linder@t-online.de
WWW: http://www.sternfreunde-durmersheim.de/
Founded: 1988
Membership: 50
Activities: popularization * education * deep sky * comets * meteors * general astronomy
Periodicals: "Starlight"
City Reference Coordinates: 008°17'00"E 48°56'00"N

Sternfreunde Franken (SFF) eV
Dr. Neumeyer Straße 240
D-91349 Egloffstein
Telephone: (0)9197-698980
Telefax: (0)9197-698982
Electronic Mail: fo0107@fonline.de (Frank Fleischmann, President)
WWW: http://sff.coolworld.de/
Founded: 1995
Membership: 28
Activities: popularization * observing * photography * CCDs * instrumentation
Coordinates: 011°15'30"E 49°44'00"N H450m
City Reference Coordinates: 011°16'00"E 49°42'00"N

Sternfreunde im FEZ (SiFEZ)
c/o Steffen Janke
Eichgestell
D-12459 Berlin
Telephone: (0)30-53071445
Telefax: (0)30-53071445
Electronic Mail: astro@sjanke.in-berlin.de
WWW: http://www.sifez.de/
Founded: 1979
City Reference Coordinates: 013°24'00"E 52°31'00"N

Sternwarte Bautzen
• See "Sternwarte Johannes Franz Bautzen"

Sternwarte der Volkshochschule Aachen
Peterstraße 21-25

D-52062 Aachen
Telephone: (0)241-47920
Telefax: (0)241-406023
Electronic Mail: vhsac1@aol.com
WWW: http://www.sternwarte-aachen.de/
Founded: 1935
Activities: education
Coordinates: 006°04'15"E 50°45'36"N H211m
City Reference Coordinates: 006°05'00"E 50°47'00"N

Sternwarte des Bruder-Klaus-Heim
Sankt-Michael-Straße 15
D-86450 Altenmünster-Violau
Telephone: (0)8295-1097
Telefax: (0)8295-499
Electronic Mail: kcmayer@dillingen.baynet.de (Christoph Mayer, Manager)
WWW: http://home.t-online.de/home/082951097/stern.htm
Founded: 1962
Staff: 4
Activities: public observatory * popularization * photography * meteors * planetarium
Periodicals: (1) "Boundless Universe" (calendar)
Coordinates: 010°34'29"E 48°27'14"N H480m
City Reference Coordinates: 010°40'00"E 48°27'00"N (Welden)

Sternwarte des Max-Born-Gymnasiums
Johann-Sebastian-Bach-Straße 8
D-82110 Germering
Telephone: (0)89-843111
Telefax: (0)89-845790
WWW: http://www.mitr.de/astro/home.shtml
Founded: 1967
Membership: 35
Activities: education * popularization * observing
Coordinates: 015°23'09"E 48°06'18"N H535m
City Reference Coordinates: 015°21'00"E 48°07'00"N

Sternwarte des Rottmayr-Gymnasiums
Barbarossastraße 16
D-83410 Laufen
Telephone: (0)8682-89320
Telefax: (0)8682-893222
WWW: http://astronomy.meta.org/RGL/Welcome.html
Founded: 1970
Coordinates: 012°56'05"E 47°56'04"N
City Reference Coordinates: 012°55'00"E 47°56'00"N

Sternwarte Greifswald
Domstraße 10a
D-17489 Greifswald
Telephone: (0)3834-864708
Telefax: (0)3834-864701
Electronic Mail: sternwarte@physik.uni-greifswald.de
WWW: http://www.physik.uni-greifswald.de/~sterne/Observatory/
Founded: 1924
Staff: 1
Activities: popularization * education
Periodicals: "Mitteilungen der Greifswalder Sternwarte", "MV-Astroblick"
Coordinates: 013°22'27"E 54°05'40"N H35m
City Reference Coordinates: 013°24'00"E 54°06'00"N

Sternwarte Hof
Egerländerweg 25
D-95032 Hof
Telephone: (0)9281-95278
Telefax: (0)9281-79217
Electronic Mail: astro@sternwarte-hof.de
WWW: http://www.sternwarte-hof.de/
Founded: 1971
Staff: 15
Activities: education * observing * digital image processing * radioastronomy * photography (H-alpha, deep sky) * weather satellite station * automatic telescope positioning * software * meteors * CCD * tours * meetings * lectures * Zeiss-Starlab planetarium
Periodicals: (4) "Sternzeit" (co-publisher) (ISSN 0721-8168, circ.: 2,000); (2) "Program"; (1) "Annual Report"
Coordinates: 011°54'57"E 50°18'07"N H500m
    011°52'36"E 50°14'56"N H623m (remote station)
City Reference Coordinates: 011°56'00"E 50°19'00"N

Sternwarte Höfingen
Oberes Ende der Uhlandstraße
D-71229 Leonberg Höfingen
Electronic Mail: gdietze@uni-hohenheim.de (Gerald Dietze)
WWW: http://www.uni-hohenheim.de/~gdietze/astro.html
Founded: 1970
Membership: 15
City Reference Coordinates: 009°01'00"E 48°48'00"N

Sternwarte Johannes Franz Bautzen
Postfach 11 09
D-02607 Bautzen
or :
Czornebohstraße 82
D-02625 Bautzen
Telephone: (0)3591-47126
Telefax: (0)3591-44071
WWW: http://ctch06.chm.tu-dresden.de/afo/bzn_home.htm
http://www.astronomie-sachsen.de/bautzen/
http://www.astronomie-sachsen.de/sternfreund/ (Der Sternfreund)
Founded: 1872
Staff: 2
Activities: training for teachers * education * popularization * lectures * planetarium * meteorology
Periodicals: (6) "Der Sternfreund"
Coordinates: 014°27'30"E 51°09'48"N H207m
City Reference Coordinates: 014°29'00"E 51°11'00"N

Sternwarte Kronshagen
Hofbrook 64
D-24119 Kronshagen
Telephone: (0)431-581632
Founded: 1980
Staff: 1
Activities: education * planetarium
Coordinates: 010°04'44"E 54°20'33"N
City Reference Coordinates: 010°05'00"E 54°20'00"N

Sternwarte Lauenstein
c/o Alfred Gaitzsch
Altenberger Weg 4
D-01778 Lauenstein
City Reference Coordinates: 009°33'00"E 52°04'00"N

Sternwarte Neanderhöhe Hochdahl eV
Postfach 22 45
D-40679 Erkrath
Telephone: (0)2104-947666
Telefax: (0)2104-947667
Electronic Mail: info@snh.rp-online.de
WWW: http://snh.rp-online.de/
Founded: 1967
Staff: 3
Activities: public observing * planetarium shows * education
Periodicals: (4) "Sternzeit" (co-publisher) (ISSN 0721-8168, circ.: 2,000)
Coordinates: 006°58'59"E 51°12'37"N H135m
City Reference Coordinates: 006°55'00"E 51°13'00"N

Sternwarte-Planetarium Bochum
Castroper Straße 67
D-44777 Bochum
Telephone: (0)234-516060
(0)234-5160611
Telefax: (0)234-5160651
Electronic Mail: planetarium@bochum.de
WWW: http://bochum-info.ruhr.de/sci/planetar.htm
http://www.bochum.de/
Founded: 1947
Staff: 12
Activities: education * public observing * planetarium shows
Periodicals: (4) "Programm"
Coordinates: 007°13'24"E 51°27'54"N H132m
City Reference Coordinates: 007°13'00"E 51°28'00"N

Sternwarte Solingen
• See "Walter-Horn-Gesellschaft eV (WHG)"

Technische Universität München, Institut für Astronomische und Physikalische Geodäsie (IAPG)
D-80290 München
or :
Arcisstraße 21
D-80333 München
Telephone: (0)89-28923190 (Secretariat)
Telefax: (0)89-28923178
Electronic Mail: <userid>@bv.tum.de
WWW: http://step.iapg.verm.tu-muenchen.de/iapg/index.html
Founded: 1961
Staff: 11
Activities: theoretical geodesy * satellite geodesy * gravimetry
Periodicals: "Mitteilungen", "Jahresbericht" (ISSN 0938-846x)
Coordinates: 011°34'00"E 48°09'00"N H530m
City Reference Coordinates: 011°35'00"E 48°08'00"N

Teleskopschmiede
Kaiserin-Augusta-Straße 68
D-12103 Berlin
Telephone: (0)30-75652947
Telefax: (0)30-75652954
Electronic Mail: info@teleskopschmiede.de
WWW: http://www.teleskopschmiede.de/
Activities: distributing telescopes and accessories
City Reference Coordinates: 013°24'00"E 52°31'00"N

Teleskop Service Wolfgang Ransburg GmbH
Keferloher Marktstraße 19c
D-85640 Putzbrunn/Solalinden
Telephone: (0)89-1892870
Telefax: (0)89-18928710
Electronic Mail: info@teleskop-service.de
WWW: http://www.teleskop-service.de/
Founded: 1999
Staff: 5
Activities: distributing telescopes and accessories
City Reference Coordinates: 011°35'00"E 48°08'00"N (München)

Teleskoptechnik Halfmann
Gessertshausener Straße 8
D-86356 Neusäß-Vogelsang
Telephone: (0)821-483070
Telefax: (0)821-485999
WWW: http://alpha.uni-sw.gwdg.de/~hessman/MONET/misc/Halfmann.html
Founded: 1990
Staff: 25
Activities: manufacturing professional telescopes and instrumentation
City Reference Coordinates: 006°27'00"E 50°35'00"N

Thiele (Astronomische Instrumente Stefan_)
● See "Astronomische Instrumente Stefan Thiele"

Thüringer Landessternwarte (TLS) - Tautenburg, Karl-Schwarzschild-Observatorium (KSO)
Sternwarte 5
D-07778 Tautenburg
Telephone: (0)36427-8630
Telefax: (0)36427-86329
Electronic Mail: <userid>@tls-tautenburg.de
WWW: http://www.tls-tautenburg.de/
Founded: 1960
Staff: 24
Activities: star formation * YSOs * extrasolar planets * galaxy evolution * AGNs * gamma-ray bursts * magnetic stars
Coordinates: 011°42'48"E 50°58'54"N H331m (IAU Code 033)
City Reference Coordinates: 011°43'00"E 50°59'00"N

Turtle Star Observatory (TSO)
c/o Axel Martin
Friedhofstraße 15
D-45478 Mülheim/Ruhr
Telephone: (0)208-55151
Electronic Mail: axelm@bph.ruhr-uni-bochum.de
WWW: http://www.bph.ruhr-uni-bochum.de/~axelm/tso/tso.htm
Founded: 1995

Staff: 4
Activities: astrometry * minor planets * astrophotography * CCD
Periodicals: (4) "Spica"
Coordinates: 006°50'39"E 51°25'43"N H42m (IAU Code 628)
City Reference Coordinates: 006°54'00"E 51°24'00"N

Universität Bochum, Astronomisches Institut
Universitätsstraße 150
D-44780 Bochum
Telephone: (0)234-3225802
(0)234-3223454
Telefax: (0)234-3214412
(0)234-3214169
Electronic Mail: <userid>@astro.ruhr-uni-bochum.dbp.de
WWW: http://www.astro.ruhr-uni-bochum.de/
Founded: 1966
Staff: 21
Activities: stellar, galactic and extragalactic astronomy
Coordinates: 007°15'51"E 51°26'41"N H179m
070°44'16"W 29°15'17"S H2,340m (La Silla)
• Full university name: "Ruhr-Universität Bochum"
City Reference Coordinates: 007°13'00"E 51°28'00"N

Universität Bonn, Astronomische Institute, Institut für Astrophysik und Extraterrestrische Forschung
Auf dem Hügel 71
D-53121 Bonn
Telephone: (0)228-733676
Telefax: (0)228-733672
Electronic Mail: sek-iaef@astro.uni-bonn.de
<userid>@astro.uni-bonn.de
WWW: http://www.astro.uni-bonn.de/~webiaef
Founded: 1964
Staff: 11
Activities: ionosphere * exosphere * thermosphere * solar wind * interplanetary matter * ISM * SN physics * pulsars * QSOs * black holes * jets * cosmology * rocket and satellite experiments
• Full university name: "Rheinische Friedrich-Wilhelms-Universität Bonn"
City Reference Coordinates: 007°05'00"E 50°44'00"N

Universität Bonn, Astronomische Institute, Observatorium Hoher List (OHL)
D-54550 Daun
Telephone: (0)6592-2150
Telefax: (0)6592-985140
Electronic Mail: <userid>@astro.uni-bonn.de
WWW: http://www.astro.uni-bonn.de/~webstw/hl.html
http://www.astro.uni-bonn.de/
Founded: 1952
Staff: 15
Activities: optical observing * minor planets * stars and stellar systems
Coordinates: 006°51'02"E 50°09'49"N H533m (IAU Code 017)
• Full university name: "Rheinische Friedrich-Wilhelms-Universität Bonn"
City Reference Coordinates: 006°50'00"E 50°11'00"N

Universität Bonn, Astronomische Institute, Radioastronomisches Institut (RAIUB)
Auf dem Hügel 71
D-53121 Bonn
Telephone: (0)228-733658
Telefax: (0)228-731775
Electronic Mail: sek-rai@astro.uni-bonn.de
<userid>@astro.uni-bonn.de
WWW: http://www.astro.uni-bonn.de/~webrai
Founded: 1962
Staff: 10
Activities: solar, planetary, galactic and extragalactic radioastronomy * ISM * dark matter * radio technology * image and data processing in radioastronomy
Coordinates: 006°43'24"E 50°34'12"N H435m (Radioobservatorium Stockert)
• Full university name: "Rheinische Friedrich-Wilhelms-Universität Bonn"
City Reference Coordinates: 007°05'00"E 50°44'00"N

Universität Bonn, Astronomische Institute, Sternwarte
Auf dem Hügel 71
D-53121 Bonn
Telephone: (0)228-733655
Telefax: (0)228-733672
Electronic Mail: <userid>@astro.uni-bonn.de
WWW: http://www.astro.uni-bonn.de/~webstw

http://www.astro.uni-bonn.de/
Founded: 1846
Staff: 15
Activities: astrophysics * photometry * spectroscopy * stars * Magellanic clouds * ISM * planetoids * astrometry * galactic structure
Periodicals: "Veröffentlichungen der Astronomischen Institute der Universität Bonn"
Coordinates: 007°04'05"E 50°43'52"N H91m (IAU Code 520)
• Full university name: "Rheinische Friedrich-Wilhelms-Universität Bonn"
City Reference Coordinates: 007°05'00"E 50°44'00"N

Universität Erlangen-Nürnberg, Astronomisches Institut, Dr. Remeis Sternwarte
Sternwartestraße 7
D-96049 Bamberg
Telephone: (0)951-952220
Telefax: (0)951-9522222
Electronic Mail: <userid>@sternwarte.uni-erlangen.de
WWW: http://www.sternwarte.uni-erlangen.de/
Founded: 1889
Staff: 8
Activities: variable stars * stellar atmospheres
Coordinates: 010°53'24"E 49°53'06"N H288m (IAU Code 521)
• Full university name: "Friedrich-Alexander-Universität Erlangen-Nürnberg"
City Reference Coordinates: 010°53'00"E 49°53'00"N

Universität Frankfurt/Main, Institut für Theoretische Physik/Astrophysik
Robert-Mayer-Straße 10
D-60054 Frankfurt/Main
Telephone: (0)69-79822357
Telefax: (0)69-79828350
Electronic Mail: <userid>@astro.uni-frankfurt.de
WWW: http://www.astro.uni-frankfurt.de/
http://earth.astro.uni-frankfurt.de/
Founded: 1908
Staff: 8
Activities: ISM
• Full university name: "Johann-Wolfgang-Goethe-Universität Frankfurt/Main"
City Reference Coordinates: 008°40'00"E 50°07'00"N

Universität Hamburg, Hamburger Sternwarte
Gojenbergsweg 112
D-21029 Hamburg
Telephone: (0)40-72524112
Telefax: (0)40-72524198
Electronic Mail: <userid>@hs.uni-hamburg.de
WWW: http://www.physnet.uni-hamburg.de/fb12/stw.html
http://www.hs.uni-hamburg.de/
Founded: 1833
Staff: 18
Activities: theoretical astrophysics * stellar activity * stellar evolution * stellar atmospheres * astrometry * ISM * QSO survey * gravitational lenses * QSO absorption lines
Periodicals: "Abhandlungen aus der Hamburger Sternwarte" (ISSN 0374-1583)
Coordinates: 010°14'30"E 53°28'54"N H45m (Bergedorf – IAU Code 029)
City Reference Coordinates: 009°59'00"E 53°33'00"N

Universität Hannover, Institut für Erdmessung (IfE), Astronomische Station
Schneiderberg 50
D-30167 Hannover
Telephone: (0)511-7622475
Telefax: (0)511-7624006
Electronic Mail: <userid>@mbox.ife.uni-hannover.de
WWW: http://www.ife.uni-hannover.de/
Founded: 1963
Staff: 16
Activities: determining astronomical latitude, longitude and vertical deflections by portable zenith cameras * satellite geodesy * GPS * geoid and gravity field determination
Coordinates: 009°42'45"E 52°23'15"N H123m
City Reference Coordinates: 009°44'00"E 52°24'00"N

Universität Heidelberg, Institut für Theoretische Astrophysik (ITA)
Tiergartenstraße 15
D-69121 Heidelberg
Telephone: (0)6221-544837
Telefax: (0)6221-544221
Electronic Mail: ita@ita.uni-heidelberg.de
<userid>@ita.uni-heidelberg.de
WWW: http://www.ita.uni-heidelberg.de/

Founded: 1976
Activities: theoretical astrophysics * stellar spectra * chromospheres and coronae * star formation * accretion disks * evolution of galaxies * astrochemistry * galactic centre
• Full university name: "Ruprecht-Karls-Universität Heidelberg"
City Reference Coordinates: 008°43'00"E 49°25'00"N

Universität Jena, Astrophysikalisches Institut und Universitäts-Sternwarte
Schillergäßchen 2
D-07745 Jena
Telephone: (0)3641-947501
Telefax: (0)3641-947502
Electronic Mail: obs@astro.uni-jena.de
WWW: http://www.astro.uni-jena.de/Sternwarte/observ.html
Founded: 1813
Staff: 10
Activities: ISM * star formation * early stellar phases * photometry * laboratory astrophysics
Coordinates: 011°35'03"E 50°55'36"N H164 (IAU Code 032)
011°29'00"E 50°55'48"N H356m (Großschwabhausen – IAU Code 134)
• Full university name: "Friedrich-Schiller-Universität Jena"
City Reference Coordinates: 011°35'00"E 50°56'00"N

Universität Kiel, Institut für Theoretische Physik und Astrophysik
Olshausenstraße 40
D-24098 Kiel
Telephone: (0)431-8804110
Telefax: (0)431-8804100
Electronic Mail: <userid>@astrophysik.uni-kiel.de
WWW: http://www.astrophysik.uni-kiel.de/
http://www.astrophysik.uni-kiel.de/e-home.html
Founded: 1872
Staff: 30
Activities: stellar atmospheres * stellar spectroscopy * stellar evolution * Sun and solar-type stars * white dwarfs * galactic evolution * ISM evolution * chemo-dynamical evolution of galaxies * hydrodynamics * stellar dynamics
• Use postal code D-24118 for parcels.
• Formerly "Institut für Astronomie und Astrophysik"
• Full university name: "Christian-Albrechts-Universität zu Kiel"
City Reference Coordinates: 010°08'00"E 54°20'00"N

Universität Köln, Erstes Physikalisches Institut
Zülpicher Straße 77
D-50937 Köln
Telephone: (0)221-4703562
Telefax: (0)221-4705162
Electronic Mail: <userid>@ph1.uni-koeln.de
WWW: http://www.ph1.uni-koeln.de/
http://www.ph1.uni-koeln.de/kosma.html (KOSMA - see separate entry in Switzerland)
http://www.ph1.uni-koeln.de/kosma_observatory.html (KOSMA - see separate entry in Switzerland)
Founded: 1980
Staff: 50
Activities: molecular spectroscopy * microwave-IR, sub-mm interstellar molecular spectroscopy
Coordinates: 007°47'04"E 45°59'04"N H3,130m (KOSMA, Gornergrat)
City Reference Coordinates: 006°57'00"E 50°56'00"N

Universität München, Institut für Astronomie und Astrophysik, Observatorium Wendelstein
Wendelsteingipfel
D-83735 Bayrischzell
Telephone: (0)8023-406
Telefax: (0)8023-9141
Electronic Mail: <userid>@usm.uni-muenchen.de
WWW: http://bigbang.usm.uni-muenchen.de:8002/USM/WDST/
Founded: 1940
Staff: 4
Activities: photometry * spectrophotometry * direct imaging * spectroscopy
Coordinates: 012°00'49"E 47°42'16"N H1,841m
City Reference Coordinates: 012°01'00"E 47°42'00"N

Universität München, Institut für Astronomie und Astrophysik, Universitäts-Sternwarte München (USM)
Scheinerstraße 1
D-81679 München
Telephone: (0)89-21806001
Telefax: (0)89-21806003
Electronic Mail: <userid>@usm.uni-muenchen.de
adis@usm.uni-muenchen.de
WWW: http://www.usm.uni-muenchen.de/
Founded: 1816

Staff: 48
Activities: stellar spectroscopy * model atmospheres * stellar winds * stellar evolution * UV spectroscopy * cataclysmic variable stars * high-speed photometry * hydrodynamics * atomic physics * instrumentation * galaxies * large-scale structure * plasma astrophysics
Coordinates: 011°36'30"E 48°08'48"N H529m (München IAU Code 532)
012°00'50"E 47°42'16"N H1,841m (Wendelstein)
City Reference Coordinates: 011°35'00"E 48°08'00"N

Universität Münster, Institut für Planetologie (IfP)
Wilhelm-Klemm-Straße 10
D-48149 Münster
Telephone: (0)251-8333496
Telefax: (0)251-8339083
(0)251-8336301
Electronic Mail: <userid>@uni-muenster.de
WWW: http://ifp.uni-muenster.de/
Founded: 1986
Staff: 50
Activities: analytical planetology * planetary physics
• Full university name: "Westfälische Wilhelms-Universität Münster"
City Reference Coordinates: 007°37'00"E 51°58'00"N

Universität Potsdam, Lehrstuhl Astrophysik
Postfach 60 15 53
D-14415 Potsdam
or :
Am Neuen Palais 10
D-14469 Potsdam
Telephone: (0)331-9771054
Telefax: (0)331-9771107
Electronic Mail: office@astro.physik.uni-potsdam.de
WWW: http://www.astro.physik.uni-potsdam.de/
Founded: 1995
Staff: 6
City Reference Coordinates: 013°04'00"E 52°24'00"N

Universitäts-Sternwarte Bonn (USB)
• See "Universität Bonn, Astronomische Institute, Sternwarte"

Universitäts-Sternwarte Göttingen (USG)
Geismarlandstraße 11
D-37083 Göttingen
Telephone: (0)551-395042
(0)551-395053
Telefax: (0)551-395043
Electronic Mail: <userid>@uni-sw.gwdg.de
WWW: http://www.uni-sw.gwdg.de/
http://www.uni-sw.gwdg.de/Welcome_de.html (German version)
Founded: 1749
Staff: 40
Activities: solar physics * extragalactic research * theoretical astrophysics * stellar atmospheres * stellar evolution * high-energy astrophysics
Coordinates: 009°56'33"E 51°31'48"N H159m (IAU Code 528)
009°58'34"E 51°31'32"N H347m (Hainberg)
016°30'00"W 28°18'00"N H2,409m (Izaña, Teide – IAU Code 954)
• Full university name: "Georg-August-Universität Göttingen"
City Reference Coordinates: 009°55'00"E 51°32'00"N

Universitäts-Sternwarte Jena (USJ)
• See "Universität Jena, Astrophysikalisches Institut und Universitäts-Sternwarte"

Universitäts-Sternwarte München (USM)
• See "Universität München, Institut für Astronomie und Astrophysik, Universitäts-Sternwarte München (USM)"

Universität Tübingen, Institut für Astronomie und Astrophysik, Abteilung Astronomie
Sand 1
D-72076 Tübingen
Telephone: (0)7071-2972486
Telefax: (0)7071-293458
Electronic Mail: <userid>@astro.uni-tuebingen.de
WWW: http://astro.uni-tuebingen.de/
Founded: 1949
Staff: 20
Activities: optical, UV, EUV, X-ray, and gamma-ray astronomy * instrumentation * binaries * PN * AGNs * ISM * stellar atmospheres

Coordinates: 009°03'30"E 48°32'18"N H470m
• Full university name: "Eberhard-Karls-Universität Tübingen"
City Reference Coordinates: 009°02'00"E 48°31'00"N

Universität Tübingen, Lehr- und Forschungsbereich Theoretische Astrophysik
Auf der Morgenstelle 10c
D-72076 Tübingen
Telephone: (0)7071-292487
Telefax: (0)7071-295889
Electronic Mail: <userid>@tat.physik.uni-tuebingen.de
WWW: http://www.tat.physik.uni-tuebingen.de/
Founded: 1968
Staff: 60
Activities: theoretical astrophysics * plasma physics * atomic physics in strong fields * cosmic X-sources * computational physics * solar physics * scientific visualization * biomechanics * neural networks
• Full university name: "Eberhard-Karls-Universität Tübingen"
City Reference Coordinates: 009°02'00"E 48°31'00"N

Universität Würzburg, Astronomisches Institut
Am Hubland
D-97074 Würzburg
Telephone: (0)931-8885031
Telefax: (0)931-8884603
Electronic Mail: <userid>@astro.uni-wuerzburg.de
WWW: http://www.astro.uni-wuerzburg.de/
Founded: 1967
Staff: 16
Activities: solar physics * galactic and extragalactic astronomy * theoretical astrophysics
Observatories: 1 (Izaña, Teide – IAU Code 954)
• Full university name: "Bayerische Julius-Maximilians-Universität Würzburg"
City Reference Coordinates: 009°56'00"E 49°48'00"N

Vehrenberg KG
Meerbuscher Straße 64-78
D-40670 Meerbusch-Osterath
Telephone: (0)2159-520321
Telefax: (0)2159-520333
Electronic Mail: service@vehrenberg.de
WWW: http://www.vehrenberg.de/
Founded: 1912
Staff: 9
Activities: publishing * distributing Celestron, Vixen and Tele Vue products in Germany, Austria and Luxembourg
City Reference Coordinates: 006°47'00"E 51°12'00"N (Düsseldorf)

Verein der Amateurastronomen des Saarlandes (VAS) eV
Postfach 18 06
D-66409 Homburg
Electronic Mail: info@sternwarte-peterberg.de
WWW: http://www.sternwarte-Peterberg.de/ (Sternwarte Peterberg)
Founded: 1977
Membership: 155
Coordinates: 007°00'01"E 49°34'22"N H570m (Sternwarte Peterberg)
City Reference Coordinates: 007°20'00"E 49°20'00"N

Verein für Volkstümliche Astronomie Essen (VVA) eV
Weberplatz 1
D-45127 Essen
Telephone: (0)201-510401
         (0)201-554018
Founded: 1981
Membership: 25
Activities: observing * photography * annual "Astrobörse"
Periodicals: (1) "Aus Astronomie und Raumfahrt" (circ.: 2,000); (4) "Sternzeit" (co-publisher) (ISSN 0721-8168, circ.: 2,000)
City Reference Coordinates: 007°01'00"E 51°28'00"N

Verein Herzberger Sternfreunde eV
Zeiss-Kleinplanetarium
Lugstraße 3
D-04916 Herzberg/Elster
Telephone: (0)3535-22671 (Holger Knobloch)
Electronic Mail: planetarium.herzberg@lausitz.net
WWW: http://herzberger-sternfreunde-ev.de/
City Reference Coordinates: 013°14'00"E 51°42'00"N

Verein Historische Sternwarten Gotha eV
c/o O. Schwarz
Uthmannstraße 8
D-99867 Gotha
Electronic Mail: sternwgth@aol.com
WWW: http://members.aol.com/sternwGTH/
City Reference Coordinates: 010°43'00"E 50°57'00"N

Vereinigte Amateur-Astronomen Eschwege 1975 eV (VAAE)
c/o Tibor Kaulitzki
Hauptstraße 36
D-36205 Sontra-Ulfen
Telephone: (0)6221-509256
Electronic Mail: mdietric@mail.lsw.uni-heidelberg.de
WWW: http://www.eschwege.de/astro/vaae.htm
Founded: 1975
Membership: 15
Activities: meetings * observing (planets, Sun, deep sky) * popularization * trips * lectures
Periodicals: (4) "Sirius" (ISSN 0941-2352, circ.: 200), "Sternzeit" (co-publisher) (ISSN 0721-8168, circ.: 2,000)
Coordinates: 010°02'00"E 51°15'00"N (Meinhard-Hitzelrode)
010°00'06"E 51°10'00"N (Witzenhausen)
City Reference Coordinates: 010°03'00"E 51°11'00"N (Eschwege)
009°56'00"E 51°05'00"N (Sontra)

Vereinigung der Nordenhamer Sternfreunde eV
c/o Siegfried Lührs
Lutherplatz 2
D-26954 Nordenham
Telephone: (0)4731-23251
Electronic Mail: info@sternfreunde.nordenham.de
WWW: http://www.sternfreunde.nordenham.de/
City Reference Coordinates: 008°29'00"E 53°30'00"N

Vereinigung der Sternfreunde eV (VdS)
Am Tonwerk 6
D-64646 Heppenheim
Telephone: (0)6252-787154
Telefax: (0)6252-787220
Electronic Mail: service@vds-astro.de
WWW: http://www.vds-astro.de/
http://neptun.uni-sw.gwdg.de/sonne.html (Solar Section)
Founded: 1953
Membership: 2500
Activities: meetings * observing * popularization
Sections: (Fachgruppen) Amateurteleskope * astrophotographie * Atmosphärische Phänomene * CCD-Technik * Geschichte * Jugendarbeit * Kleine Planeten * Kometen * Dark Sky - Bewegung gegen Lichtverschmutzung * Meteore * Planeten * Pseudowissenschaften, Esoterik und Astrologie * UFOs * Radioastronomie * Rechnende Astronomie * Sonne * Spektroskopie * Sternbedeckungen * Veränderliche * Visuelle Deep-Sky Beobachtung * Volkssternwarten und Planetarien
Periodicals: (12) "VdS-Nachrichten" (in "Sterne und Weltraum" - see separate entry) "Provisional Sunspot Numbers"; (4) "Sonne" (ISSN 0721-0094, circ.: 500); (1) "Sonne Datenblatt" (circ.: 200); "New Sunspot Indices Bulletin" (ISSN 0934-8220), "Sonne Tageskarten", "Sonne Zirkular", "Astro Fax Zirkular"
City Reference Coordinates: 008°39'00"E 49°38'00"N

Vereinigung Krefelder Sternfreunde eV
Postfach 29 64
D-47729 Krefeld
or :
Waldhofstraße 132
D-47800 Krefeld
Telephone: (0)2151-503116
WWW: http://www.krefeld-city.de/vks/sternfreunde.html
Founded: 1966
Membership: 90
Activities: popularization
Periodicals: (4) "Sternenbote"
Coordinates: 006°34'00"E 51°19'00"N H80m
City Reference Coordinates: 006°34'00"E 51°19'00"N

Verkehr Raumfarht und Systemtechnik (VRS) GmbH
Walter-Köhn-Straße 1b
D-04356 Leipzig
Telephone: (0)341-526230
Telefax: (0)341-5262356

Electronic Mail: info@vrs.de
WWW: http://www.vrs.de/
Founded: 1993
Staff: 16
Activities: studies, analysis and development of subsystems for space research in the field of Earth observation, microgravity and small satellites
City Reference Coordinates: 012°20'00"E 51°19'00"N

Verlag Harri Deutsch
Gräfstraße 47
D-60486 Frankfurt/Main
Telephone: (0)69-775021
Telefax: (0)69-7073739
Electronic Mail: buchhandlung@harri-deutsch.de
            verlag@harri-deutsch.de
WWW: http://www.harri-deutsch.de/
Founded: 1961
Staff: 20
Activities: publishing * bookshop * software
City Reference Coordinates: 008°40'00"E 50°07'00"N

Volkshochschule der Stadt Fürstenfeldbruck eV, Arbeitsgemeinschaft Astronomie
c/o Michael A. Rappenglück
Heinrich-Feller-Straße 15
D-82275 Emmering
or :
Theodor-Heuss-Straße 2
D-82256 Fürstenfeldbruck
Telephone: (0)8141-4827
Telefax: (0)8141-4827
Electronic Mail: 100639.637@compuserve.com (Michael A. Rappenglück)
Founded: 1981
Membership: 30
Activities: popularization * photography * planets * history of astronomy * computing * observing (Sun * deep sky)
Coordinates: 011°14'10"E 48°09'52"N H535m
City Reference Coordinates: 011°15'00"E 48°10'00"N (Fürstenfeldbruck)

Volkshochschule der Stadt Soest, Astronomische Arbeitsgemeinschaft
c/o Ernst Fleischer
Steinkuhlenweg 6
D-59494 Soest
Telephone: (0)2921-77490
Founded: 1966
Membership: 35
Activities: education * observing
Coordinates: 008°05'37"E 51°34'23"N H96m (Volkssternwarte Soest)
City Reference Coordinates: 008°06'00"E 51°34'00"N

Volkssternwarte Adolph Diesterweg Radebeul
Auf den Ebenbergen 10a
D-01445 Radebeul
Telephone: (0)351-8305905
Telefax: (0)351-8381906
Electronic Mail: rattei@chemie.rmhs1.tu-dresden.d400.de
WWW: http://ctch06.chm.tu-dresden.de/stw-rdbl/
      http://www.astronomie-sachsen.de/radebeul/
Founded: 1959
Staff: 1
Activities: observing * popularization * planetarium
Coordinates: 013°37'20"E 51°06'59"N H185m
City Reference Coordinates: 013°43'00"E 51°13'00"N

Volkssternwarte Bonn (VSB), Astronomische Vereinigung eV
Poppelsdorfer Allee 47
D-53115 Bonn
Telephone: (0)228-222270
Electronic Mail: vorstand@volkssternwarte-bonn.de
WWW: http://www.volkssternwarte-bonn.de/
Founded: 1972
Membership: 130
Activities: education * public readings * observing (Sun, planets, deep sky)
Periodicals: (4) "Telescopium" (ISSN 0723-1121, circ.: 450), "Sternzeit" (co-publisher) (ISSN 0721-8168, circ.: 2,000)
Coordinates: 007°05'56"E 50°43'44"N H59m
City Reference Coordinates: 007°05'00"E 50°44'00"N

Volkssternwarte Darmstadt (VSD) eV
Am Blauen Stein 4
D-64295 Darmstadt
Telephone: (0)6151-130900
Telefax: (0)6151-51482
Electronic Mail: vorstand@vsda.de
WWW: http://www.vsda.de/
Founded: 1969
Activities: observing * popularization
Periodicals: (12) "Mitteilungen der Volkssternwarte Darmstadt"
Coordinates: 008°39'47"E 49°50'38"N
City Reference Coordinates: 008°39'00"E 49°52'00"N

Volkssternwarte Ennepetal eV
Am Hinnenberg 80
D-58247 Ennepetal
Telephone: (0)2333-62646
Electronic Mail: info@volkssternwarte-ennepetal.de
WWW: http://www.volkssternwarte-ennepetal.de/
City Reference Coordinates: 007°23'00"E 51°18'00"N

Volkssternwarte Erich Bär (VEB)
Stolpener Straße 74
D-01454 Radeberg
Telephone: (0)180-505255390836
Telefax: (0)180-505255390836
Electronic Mail: sternwarte-radeberg@web.de
WWW: http://www.canaletto.net/sites/f.schaefer/stwhome.htm
         http://www.astronomie-sachsen.de/sternfreund/ (Der Sternfreund)
Founded: 1964
Membership: 30
Activities: lectures * observing (occultations, Sun, Jupiter * meteors * deep sky) * photography * education
Periodicals: (6) "Der Sternfreund" (co-publisher); (1) "Sonne - Data Report"
Coordinates: 013°55'59"E 51°06'51"N H218m
City Reference Coordinates: 013°56'00"E 51°08'00"N

Volkssternwarte Erich Scholz
Hochwaldstraße 21c
D-02763 Zittau
WWW: http://ctch06.chm.tu-dresden.de/afo/zit_home.htm
         http://www.astronomie-sachsen.de/zittau/
Founded: 1968
City Reference Coordinates: 014°47'00"E 50°54'00"N

Volkssternwarte Geschwister Herschel Hannover eV
Am Lindener Berge 27
D-30449 Hannover
Telephone: (0)511-456290
Telefax: (0)511-456290
Electronic Mail: info@sternwarte-hannover.de
WWW: http://www.sternwarte-hannover.de/
Founded: 1968
Membership: 100
Activities: popularization * lectures * observing
Coordinates: 009°42'25"E 52°21'50"N
City Reference Coordinates: 009°44'00"E 52°24'00"N

Volkssternwarte Hagen eV
Postfach 146
D-58001 Hagen
or :
Eugen-Richter-Turm
D-58135 Hagen
Telephone: (0)2331-590790
Telefax: (0)2331-590791
Electronic Mail: 100276.1140@compuserve.com
WWW: http://www.sternwarte.bnet.de/
Founded: 1955
Membership: 45
City Reference Coordinates: 007°28'00"E 51°22'00"N

Volkssternwarte Hattingen
c/o Ingo B. Schmidt

Schonnefeldstraße 23
D-45236 Essen
Telephone: (0)201-8336082
WWW: http://www.sternwarte-hattingen.de/
Founded: 1988
City Reference Coordinates: 007°01'00"E 51°28'00"N (Essen)

Volkssternwarte im Volksbildungswerk Hofheim-Marxheim
c/o Hermann Minor
Lessingstraße 56
D-65719 Hofheim/Taunus
Telephone: (0)6192-3599
Electronic Mail: lingnau@tm.informatik.uni-frankfurt.de (Anselm Lingnau)
WWW: http://www.sternwarte-hofheim.de/
Founded: 1977
Membership: 60
Periodicals: "Mitteilungen astronomischer Vereinigungen Rhein-Main-Nahe" (circ.: 550)
City Reference Coordinates: 008°26'00"E 50°07'00"N

Volkssternwarte Jonsdorf
An der Sternwarte 3
D-02796 Jonsdorf
WWW: http://ctch06.chm.tu-dresden.de/afo/jon_home.htm
    http://www.astronomie-sachsen.de/jonsdorf/
Founded: 1962
City Reference Coordinates: 014°43'00"E 50°51'00"N

Volkssternwarte Köln
Postfach 86 01 84
D-51023 Köln
or :
c/o Schiller-Gymnasium
Nikolausstraße 55
D-50937 Köln
Telephone: (0)221-415467
WWW: http://www.volkssternwarte-koeln.de/
Founded: 1922
City Reference Coordinates: 006°57'00"E 50°56'00"N

Volkssternwarte Laupheim eV und Planetarium
Leibnizstraße 35
D-88471 Laupheim
Telephone: (0)7392-18055
Telefax: (0)7392-17464
Electronic Mail: contact@planetarium-laupheim.de
WWW: http://www.planetarium-laupheim.de/
Founded: 1975
Membership: 100
Activities: popularization * amateur astronomical fair * exhibitions * planetarium
Coordinates: 009°52'56"E 48°13'37"N H518m
    009°52'56"E 48°43'35"N H506m
City Reference Coordinates: 009°55'00"E 48°13'00"N

Volkssternwarte Manfred von Ardenne
Plattleite 27
D-01324 Dresden
or :
c/o Lars Stephan
Klenzestraße 3
D-17424 Seebad Heringsdorf
Telephone: (0)351-2637299
WWW: http://www.physik.uni-greifswald.de/~sterne/Volkssternwarte_heringsdorf/
    http://www.astronomie-sachsen.de/dresdenmva/
    http://ctch06.chm.tu-dresden.de/MvAObserv/
Founded: 1960
Staff: 1
Activities: observing (occultations, planets) * photography
Coordinates: 014°10'27"E 53°57'22"N H5m
City Reference Coordinates: 013°44'00"E 51°03'00"N (Dresden)
    014°10'00"E 53°58'00"N (Seebad Heringsdorf)

Volkssternwarte Marburg eV
c/o Udo Wittekindt
Pestalozzistraße 5a
D-35274 Kirchhain

or :
Dresdener Straße 18
D-35274 Kirchhain
Telephone: (0)6422-3947
Electronic Mail: mail@volkssternwarte-marburg.de
WWW: http://www.volkssternwarte-marburg.de/
Founded: 1975
Membership: 45
Activities: education * lectures * observing * popularization * photography
Coordinates: 008°55'27"E 50°49'51"N H225m
City Reference Coordinates: 008°36'00"E 50°49'00"N (Marburg – IAU Code 525)
008°55'00"E 50°49'00"N (Kirchhain)

Volkssternwarte Norderstedt (VSN) eV
c/o Harald Prahl
Breslauer Straße 7
D-22850 Norderstedt
Telephone: (0)40-5285383
Electronic Mail: fornasiero@physnet.uni-hamburg.de (Lirio Fornasiero, Chairman)
WWW: http://home.t-online.de/home/0406317305-0001/vsnhome1.htm
Founded: 1973
Membership: 26
Activities: photography * observing (occultations, comets) * lectures
Periodicals: (4) "Sternzeit" (co-publisher) (ISSN 0721-8168, circ.: 2,000)
• IAU Code 606
City Reference Coordinates: 008°59'00"E 53°43'00"N

Volkssternwarte Paderborn eV
Postfach 11 42
D-33041 Paderborn
or :
Schloß Neuhaus
Im Schloßpark
Marstallstraße 13
D-33104 Paderborn
Telephone: (0)5254-932042
Telefax: (0)5254-932043
Electronic Mail: mail@astroobspb.de
WWW: http://www.astroobspb.de/
Founded: 1971
Membership: 140
Activities: INTER-SOL programme * public observing * lectures * popularization (press, radio, TV) * photography
Periodicals: (3) "INTER-SOL Reports"
Coordinates: 008°42'39"E 51°44'50"N H119m
City Reference Coordinates: 008°44'00"E 51°43'00"N

Volkssternwarte Passau
c/o Rainer Klemm
Anton-Poetzl-Straße 6
D-94034 Passau
Telephone: (0)851-45453
Electronic Mail: kalz@passau.baynet.de
WWW: http://www.passau.baynet.de/~pa000127/vsw/
City Reference Coordinates: 013°28'00"E 48°35'00"N

Volkssternwarte Prenzlau
c/o Robert Ehrlich
Kirchstraße 6
D-17291 Roepersdorf
Telephone: (0)3984-802039
Electronic Mail: sternwarte-prenzlau@t-online.de
WWW: http://home.t-online.de/home/sternwarte-prenzlau/
Founded: 1962
Staff: 2
Activities: occultations
Coordinates: 013°51'51"E 53°18'38"N H57m
City Reference Coordinates: 013°52'00"E 53°19'00"N (Prenzlau)

Volkssternwarte Rothwesten
c/o Friedel Spitzer
Brüder-Grimm-Straße 24
D-34233 Fuldatal
Telephone: (0)5607-459
Electronic Mail: info@volkssternwarte-rothwesten.de
WWW: http://www.volkssternwarte-rothwesten.de/
Founded: 1963

Activities: observing * photography
Coordinates: 009°30'53"E 51°23'26"N H319m
City Reference Coordinates: 009°31'00"E 51°23'00"N

Volkssternwarte Tirschenreuth
c/o Peter Postler
Marienstraße 49
D-95643 Tirschenreuth
or :
Großenseeser Straße
D-95643 Tirschenreuth
Telephone: (0)9631-4460 (after 18:00)
WWW: http://www.sternwarte-tirschenreuth.de/
Founded: 1964
Activities: general observing * slide shows * education
Coordinates: 012°21'00"E 49°54'00"N
City Reference Coordinates: 012°21'00"E 49°54'00"N

Volkssternwarte und Planetarium Drebach
Straße der Jugend 14
D-09430 Drebach
Telephone: (0)37341-7435
Electronic Mail: kontakt@sternwarte-drebach.de
WWW: http://www.sternwarte-drebach.de/
http://www.tu-chemnitz.de/home/ods/stw_drebach/
Founded: 1969
Activities: photography * occultations * popularization
City Reference Coordinates: 013°01'00"E 50°40'00"N

Volkssternwarte und Planetarium Reutlingen
Karlstraße 40
D-72764 Reutlingen
Telephone: (0)7121-3360
WWW: http://home.t-online.de/home/B.Augustin/stwrtpln.htm
Founded: 1956
Membership: 18
Activities: education * photography * observing (Sun, lunar occultations)
Coordinates: 009°12'58"E 48°29'55"N H395m
City Reference Coordinates: 009°11'00"E 48°29'00"N

Volkssternwarte Urania Jena eV
Schillergäßchen 2a
D-07745 Jena
Telephone: (0)3641-363406 (Thomas Marold - after 18:00)
WWW: http://www.urania-sternwarte.de/
Founded: 1909
Membership: 50
Activities: observing (Sun, planets, deep sky, comets, variable stars) * photography * lectures * popularization
Coordinates: 011°35'00"E 50°55'35"N H174m (Urania-Volkssternwarte)
011°35'30"E 50°55'30"N H346m (Forst-Sternwarte)
City Reference Coordinates: 011°35'00"E 50°56'00"N

Volkssternwarte Wetterau eV
c/o Walter Groening
Gartenfeldstrasse 16
D-61231 Bad Nauheim
WWW: http://www.sternwarte-wetterau.de/
Coordinates: 008°43'41"E 50°21'59"N H265m
City Reference Coordinates: 008°44'00"E 50°21'00"N

Volkssternwarte Würzburg eV
c/o Josef Laufer
Peter-Wagner-Straße 4
D-97230 Estenfeld
or :
Cronthalstraße 25
D-97074 Würzburg
Telephone: (0)9305-993216
(0)931-73020
Electronic Mail: vstw@gmx.de
WWW: http://www.wuerzburg.de/vstw/
Founded: 1984
Membership: 85
Activities: observing * lectures
Coordinates: 009°57'34"E 49°46'33"N H271m

City Reference Coordinates: 009°56'00"E 49°48'00"N (Würzburg)

Volks- und Schulsternwarte Bartholomäus Scultetus
An der Sternwarte 1
D-02827 Görlitz
Telephone: (0)3581-78222
WWW: http://www.astronomie-sachsen.de/goerlitz
http://www.astronomie-sachsen.de/sternfreund/ (Der Sternfreund)
Founded: 1856
Activities: education * popularization * Sun * variable stars * occultations * local astronomy history * planetarium
Periodicals: (6) "Der Sternfreund"
Coordinates: 014°57'05"E 51°08'08"N H235m
City Reference Coordinates: 014°59'00"E 51°09'00"N

Volks- und Schulsternwarte Bruno H. Bürgel eV
Zöllnerweg 12
D-02689 Sohland
Telephone: (0)35936-37270
(0)35936-34012 (Observatory)
Electronic Mail: starklabk@aol.com (Matthias Stark)
WWW: http://ctch06.chm.tu-dresden.de/afo/shl_home.htm
http://members.aol.com/stwsohland/
Founded: 1963
Membership: 38
Activities: photography * lectures * education * variable stars
Coordinates: 014°25'00"E 51°12'00"N H335m
City Reference Coordinates: 014°25'00"E 51°12'00"N

Volks- und Schulsternwarte Geretsried eV
Adalbert-Stifter-Straße 14
D-82538 Geretsried
Telephone: (0)8171-932532
Telefax: (0)8171-90412
Electronic Mail: motl@ilo.de (Kurt Motl, Chairman)
WWW: http://www.sternwarte-geretsried.de/
Founded: 1976
Membership: 55
Activities: education * observing * photography * computing * weather satellite receiving system
Coordinates: 011°28'34"E 47°52'18"N H600m
City Reference Coordinates: 011°28'00"E 47°51'00"N

Volks- und Schulsternwarte Juri Gagarin
Mansberg 18
D-04838 Eilenburg
Telephone: (0)3423-603153
Telefax: (0)3423-603153
Founded: 1964
Staff: 2
Activities: popularization * education * observing (sunspots, lunar occultations) * planetarium
Coordinates: 012°37'39"E 51°27'05"N H120m
City Reference Coordinates: 012°37'00"E 51°27'00"N

VRS GmbH
• See "Verkehr Raumfarht und Systemtechnik (VRS) GmbH"

Walter-Horn-Gesellschaft eV (WHG)
Sternwarte Solingen
Postfach 19 05 50
D-42705 Solingen
or :
Sternstraße 5
D-42719 Solingen
Telephone: (0)212-232425
Telefax: (0)212-2324260
Electronic Mail: info@sternwarte-solingen.de
WWW: http://www.sternwarte-solingen.de/
Founded: 1921
Membership: 125
Activities: education * multimedia theater * observing
Periodicals: (4) "Sternzeit" (co-publisher) (ISSN 0721-8168, circ.: 2,000)
Coordinates: 007°01'19"E 51°10'32"N H145m (IAU Code 592)
City Reference Coordinates: 007°05'00"E 51°10'00"N

Weltastronomie Studienreisen (WAS)
c/o Eckehard Schmidt

Postfach 46 16
D-90025 Nürnberg
or :
Maxfeldstraße 50
D-90409 Nürnberg
Telephone: (0)911-586550
Telefax: (0)911-5865549
Electronic Mail: eckehard@orion.franken.de (Eckehard Schmidt)
Founded: 1990
Staff: 2
Activities: organizing travels to astronomy and space sites * study trips
Periodicals: "Astrotrotter"
City Reference Coordinates: 011°04'00"E 49°27'00"N

Werner-Heisenberg-Institut (WHI)
• See "Max-Planck-Institut für Physik (MPP)"

Westfälisches Museum für Naturkunde, Planetarium
Sentruper Straße 285
D-48161 Münster
Telephone: (0)251-59105
Telefax: (0)251-5916098
WWW: http://www.lwl.org/naturkundemuseum/
City Reference Coordinates: 007°37'00"E 51°58'00"N

Westfälische Volkssternwarte und Planetarium
Stadtgarten 6
D-45657 Recklinghausen
Telephone: (0)2361-23134
Telefax: (0)2361-23134
Electronic Mail: info@sternwarte-recklinghausen.de
WWW: http://www.sternwarte-recklinghausen.de/
Founded: 1953
Staff: 3
Activities: public lectures * planetarium shows * public observing * history of astronomy * education
Periodicals: (2) "Program of Activities" (circ.: 15,000)
Coordinates: 007°10'54"E 51°37'30"N H118m
City Reference Coordinates: 007°11'00"E 51°37'00"N

Westfälische Wilhelms-Universität Münster
• See "Universität Münster"

Wilhelm-Foerster-Sternwarte (WFS) eV
Munsterdamm 90
D-12169 Berlin
Telephone: (0)30-7900930
Telefax: (0)30-79009312
Electronic Mail: wfs@wfs.be.schule.de
wfs@bics.be.schule.de
WWW: http://www.be.schule.de/schulen/wfs/homepage.html
http://www.wfs.be.schule.de/
Founded: 1953
Staff: 7
Periodicals: "Veranstaltungsprogramm"
• IAU Code 544
City Reference Coordinates: 013°24'00"E 52°31'00"N

Wittenberger Planetarium
Falkstraße 83
D-06886 Lutherstadt Wittenberg
Telephone: (0)3491-403455
Telefax: (0)3491-403465
WWW: http://lbs.st.schule.de/lbs/pl-witt.htm
Founded: 1987
Staff: 1
Activities: lectures
City Reference Coordinates: 012°39'00"E 51°53'00"N

Zeiss Jena GmbH (Carl_)
• See "Carl Zeiss Jena GmbH

Zeiss-Grossplanetarium Berlin (ZGP)
Prenzlauer Allee 80
D-10405 Berlin
Telephone: (0)30-42184512

Telefax: (0)30-4251252
Electronic Mail: agp@astw.de
WWW: http://www.astw.de/zgp/ueber_zf.htm
Founded: 1987
Staff: 12
Activities: shows * education * public outreach * lectures * exhibitions * life concerts
Periodicals: (12) "Program"
City Reference Coordinates: 013°24'00"E 52°31'00"N

Zeiss-Planetarium der Ernst-Abbe-Stiftung
Am Planetarium 5
D-07743 Jena
Telephone: (0)3641-885488
Telefax: (0)3641-885420
WWW: http://www.jena.de/tourism/planet.htm
Founded: 1926
Staff: 8
Activities: public shows
Periodicals: "Programmheft"
City Reference Coordinates: 011°35'00"E 50°56'00"N

Zeiss-Kleinplanetarium Herzberg
Lugstraße 3
D-04916 Herzberg/Elster
Telephone: (0)3535-70057
Telefax: (0)3535-70057
Electronic Mail: planetarium.herzberg@lausitz.net
WWW: http://herzberger-sternfreunde-ev.de/planetarium.htm
City Reference Coordinates: 013°14'00"E 51°42'00"N

Gibraltar

Gibraltar Astronomical Society (GAS)
c/o J. M. Reyes
13 Silver Birch Lodge
Montagu Gardens
Telephone: 75763
Electronic Mail: gasjjgon@gibnet.gi
WWW: http://www.gibnet.gi/~gasjjgon/
City Reference Coordinates: 005°21'00"E 36°09'00"N

Greece

Academy of Athens, Research Center for Astronomy and Applied Mathematics
14 Anagnostopoulou Street
GR-106 73 Athens
Telephone: (0)1-3613589
Telefax: (0)1-3631606
WWW: http://www.academyofathens.gr/
Founded: 1959
Staff: 8
Activities: solar activity * variable stars * astrophysics * stars * planets
Periodicals: "Contributions", "Annual Report"
City Reference Coordinates: 023°43'00"E 37°58'00"N

Aristotle University of Thessaloniki
● See "University of Thessaloniki"

Eugenides Foundation, Planetarium
87 Leof. Syggrou
GR-175 64 Athens
Telephone: (0)9411181
Telefax: (0)9417372
Electronic Mail: library@eugenides_found.edu.gr
WWW: http://www.eugenides_found.edu.gr/
http://www.eugenfound.edu.gr/
City Reference Coordinates: 023°43'00"E 37°58'00"N

Hellenic Astronomical Society (HELAS)
c/o Section of Astrophysics, Astronomy and Mechanics
Department of Physics
National University of Athens
Panepistimiopolis
GR-157 84 Zografos
Telephone: (0)31-998173
Telefax: (0)31-995384
Electronic Mail: elaset@astro.auth.gr
WWW: http://www.hri.org/elaset/
http://www.astro.auth.gr/elaset/
Founded: 1993
Membership: 140
Activities: promoting astronomy * distributing astronomical information
Periodicals: (4) "Hellenic Astronomical Society Newsletter"
City Reference Coordinates: 023°43'00"E 37°58'00"N

Hellenic National Meteorological Service (HNMS)
14 E. Venizelou Street
GR-167 77 Hellinikon
Telephone: (0)1-9629415
Telefax: (0)1-9628952
Electronic Mail: director@hnms.gr
WWW: http://www.hnms.gr/english/
Founded: 1931
Staff: 700
Activities: meteorological support of national defence, national economy, safety of lives and properties
City Reference Coordinates: 023°43'00"E 37°58'00"N

Hellenic Organization for Standardization (ELOT)
313 Acharnon Street
GR-111 45 Athens
Telephone: (0)1-2280001
Telefax: (0)1-2020776
Electronic Mail: elotinfo@elot.gr
WWW: http://www.elot.gr/
Founded: 1976
Staff: 71
Activities: standardization * certification * quality control * testing * information on standards and technical regulations
Periodicals: (1) "Catalog of Hellenic Standards"; "Information Bulletin" (in Greek)
City Reference Coordinates: 023°43'00"E 37°58'00"N

Hellenic Physical Society
c/o C. Helmis (President)
6 Grivaion Street
GR-106 80 Athens

Telephone: (0)1-3635701
Telefax: (0)1-3610690
Electronic Mail: chelmis@atlas.uoa.gr
Membership: 65
City Reference Coordinates: 023°43'00"E 37°58'00"N

National Observatory of Athens (NOA), Institute of Astronomy and Astrophysics
P.O. Box 20048
GR-118 10 Athens
or :
Lofos Nymfon
Thission
Telephone: (0)1-3490106
         (0)1-8040619
Telefax: (0)1-3490140
        (0)1-8040453
Electronic Mail: <userid>@leon.ariadne-t.gr
                 ourania@astro.noa.gr
WWW: http://www.astro.noa.gr/
Founded: 1853
Staff: 25
Activities: extragalactic astronomy * variable stars * solar system * solar physics
Periodicals: "Memoirs of the National Observatory of Athens, Series I, Astronomy"
Coordinates: 022°37'18"E 37°58'24"N H1,005m (Kryonerion Station)
• IAU Code 066
City Reference Coordinates: 023°43'00"E 37°58'00"N (Athinai)

National University of Athens, Department of Physics, Section of Astrophysics, Astronomy and Mechanics
Panepistimiopolis
GR-157 83 Zografos
Telephone: (0)1-7243211
         (0)1-7243414
         (0)1-7235122
Telefax: (0)1-7238413
Electronic Mail: <userid>@cc.uoa.gr
WWW: http://www.uoa.gr/departs/physics/sectc_en.htm
     http://www.cc.uoa.gr/physics/sections/astrophysics/
Founded: 1896
Staff: 24
Activities: astronomy & astrophysics * nonlinear mechanics * cosmology * relativity * stellar structure and evolution * solar physics * stellar dynamics * observing
Sections: Laboratory of Astronomy * Laboratory of Astrophysics
Observatories: 1 (Athens University Observatory)
City Reference Coordinates: 023°43'00"E 37°58'00"N

University of Crete, Department of Physics, Astrophysics and Space Physics
P.O. Box 2208
GR-710 03 Heraklion
Telephone: (0)810-235014
Telefax: (0)810-239735
Electronic Mail: <userid>@physics.uch.gr
WWW: http://astrophysics.physics.uoc.gr/
Founded: 1986
Staff: 6
Activities: compact X-ray sources * gamma-ray sources * ISM * local-group galaxies * star formation * jets * solar physics * Earth magnetosphere * ionosphere
City Reference Coordinates: 025°09'00"E 35°20'00"N

University of Ionnina, Department of Physics, Section of Astro-Geophysics
GR-451 10 Ionnina
Telephone: (0)651-98471
         (0)651-98480
Telefax: (0)651-45697
        (0)651-98682
Electronic Mail: calissar@cc.uoi.gr (C.E. Alissandrak)
WWW: http://www.uoi.gr/
Founded: 1968
Activities: solar physics * chromospherically active dwarf stars * flare stars * galaxies * cosmology * education
Coordinates: 020°51'00"E 39°40'00"N H700m (Dourouti Observatory)
City Reference Coordinates: 020°51'00"E 39°40'00"N

University of Patras, Department of Physics
GR-261 10 Rion
Telephone: (0)61-997571
         (0)61-997636

Telefax: (0)61-997571
WWW: http://www.physics.upatras.gr/
Founded: 1970
Staff: 6
Activities: education * stellar dynamics * SETI * galactic nebulae * Seyfert galaxies
City Reference Coordinates: 021°44'00"E 38°15'00"N

University of Thessaloniki, Department of Physics, Section of Astrophysics, Astronomy and Mechanics
GR-540 06 Thessaloniki
Telephone: (0)31-998407
Telefax: (0)31-995384
Electronic Mail: <userid>@astro.auth.gr
WWW: http://www.astro.auth.gr/
Founded: 1943
Staff: 9
Activities: education * observing * theoretical astronomy
Periodicals: "Annual Report"
Coordinates: 022°57'30"E 40°37'00"N H28m
• Full university name: "Aristotle University of Thessaloniki"
City Reference Coordinates: 022°58'00"E 40°38'00"N

Guam

University of Guam (UoG), Planetarium
UoG Station
Mangilao, GU 96923
Telephone: 735-2783
Telefax: 734-1299
Electronic Mail: stars@kuentos.guam.net
WWW: http://www.guam.net/planet/
Founded: 1992
City Reference Coordinates: 144°47'00"E 13°28'00"N

Honduras

Servicio Meteorológico Nacional (SMN)
Apartado 30145
Tegucigalpa, MDC
Telephone: (0)338075
(0)331112
(0)331113
(0)331114
Telefax: (0)333683
City Reference Coordinates: 088°13'00"W 14°06'00"N

Universidad Nacional Autónoma de Honduras (UNAH), Sección de Astrofísica
Apartado Postal 3023
Tegucigalpa MDC
or :
Ciudad Universitaria
Tegucigalpa
Telephone: (0)322110 x230
Telefax: (0)327196
(0)314686
Electronic Mail: <userid>@ns.hondunet.net
WWW: http://www.fisica.unah.hondunet.net/Fisica.htm
Founded: 1990
Staff: 2
Activities: education * theoretical research * observatory under construction
Coordinates: 087°09'00"W 14°05'00"N H1,063m
City Reference Coordinates: 087°09'00"W 14°05'00"N

Hungary

Albireo Amateur Astronomy Society
Nemzetör út. 8
H-8900 Zalaegerszeg
Telephone: (0)92-313490
Electronic Mail: albireo@alpha.dfmk.hu
WWW: http://alpha.dfmk.hu/~albireo
Founded: 1971
Membership: 100
Activities: solar system * deep sky * double stars * eclipsing binaries * amateur meteorology
Periodicals: (4) "Albireo"
Coordinates: 016°50'18"E 46°28'12"N H120m
City Reference Coordinates: 016°51'00"E 46°53'00"N

Astronomical Foundation of Nógrád County
Móricz Zs. út. 9
H-3100 Salgótarján
Telephone: (0)32-310464
(0)32-310250
(0)30-9108868
Founded: 1992
Staff: 7
Activities: assisting local and national amateur astronomy activities * public observatory
Periodicals: (6) "A Csillagvizsgáló"
Coordinates: 019°48'00"E 48°40'09"N H255m
019°48'35"E 48°41'28"N H356m (Urania Observatory)
019°53'20"E 48°41'46"N H555m (ALFA Station)
City Reference Coordinates: 019°48'00"E 48°07'00"N

Astrotech Instruments & Computers KKT
P.O. Box 116
H-6501 Baja
or :
Szegedi út. III/70
H-6500 Baja
Telephone: (0)20-9370042
Telefax: (0)79-427001
Electronic Mail: hege@electra.bajaobs.hu
virtual@emitel.hu
WWW: http://www.bajaobs.hu/baja/astrotch.htm
Founded: 1993
Staff: 3
Activities: import-export of astronomical telescopes, accessories, CCD cameras, filters, educational material, computers and software
City Reference Coordinates: 018°57'00"E 46°11'00"N

Attila University (József_)
• See "József Attila University"

Baja Astronomical Observatory
Szegedi út. III/70
P.O. Box 766
H-6500 Baja
Telephone: (0)79-424027
Telefax: (0)79-427001
Electronic Mail: hege@electra.bajaobs.hu
baja@electra.bajaobs.hu
WWW: http://www.bajaobs.hu/
Founded: 1947
Staff: 6
Activities: photoelectric photometry * variable stars * eclipsing binaries * cataclysmic variables * CCD imaging * lightcurve analysis * period variations
Coordinates: 019°00'43"E 46°10'50"N H110m
City Reference Coordinates: 018°58'00"E 46°11'00"N

Budapest Planetarium
Ker. Népliget
P.O. Box 46
H-1476 Budapest
Telephone: (0)1-2650725
Telefax: (0)1-2633104
WWW: http://www.planetarium.hu/

Founded: 1977
Staff: 11
Activities: lectures * shows
City Reference Coordinates: 019°05'00"E 47°30'00"N

Debrecen Heliophysical Observatory
• See "Hungarian Academy of Sciences, Debrecen Heliophysical Observatory"

Eötvös Loránd Physical Society
P.O. Box 433
H-1371 Budapest
Telephone: (0)1-2018682
Telefax: (0)1-2018682
Electronic Mail: elft@rmk530.rmki.kfki.hu
    mail.elft@mtesz.hu
WWW: http://www.kfki.hu/~elfthp/
Membership: 700
City Reference Coordinates: 019°05'00"E 47°30'00"N

Eötvös Loránd University, Department of Astronomy
P.O. Box 32
H-1518 Budapest
or :
XI. Pázmány
Péter sétány 1/A
H-1000 Budapest
Telephone: (0)1-3722946
Telefax: (0)1-3722940
Electronic Mail: astro@astro.elte.hu
    <userid>@astro.elte.hu
WWW: http://astro.elte.hu/
Founded: 1753
Staff: 10
Activities: stellar astronomy * celestial mechanics * solar MHD * dynamo theory * star formation * planetology
Periodicals: (3) "Publications of the Astronomy Department of the Eötvös Loránd University" (ISSN 0238-2423)
City Reference Coordinates: 019°05'00"E 47°30'00"N

Eötvös Loránd University, Gothard Astrophysical Observatory (GAO)
Szent Imre Herceg út. 112
H-9707 Szombathely
Telephone: (0)94-313871
Telefax: (0)94-328324
Electronic Mail: obs@gothard.hu
WWW: http://www.gothard.hu/
Founded: 1881
Staff: 4
Activities: high-dispersion spectroscopy of emission-line stars * education * history of astronomy
Coordinates: 016°36'30"E 47°15'40"N H223m
City Reference Coordinates: 016°38'00"E 47°14'00"N

Gothard Astrophysical Observatory (GAO)
• See "Eötvös Loránd University, Gothard Astrophysical Observatory (GAO)"

Haynald Observatory
Hunyadi János út. 23-25
H-6300 Kalocsa
WWW: http://www.bajaobs.hu/kalocsa/obszhay.htm
Founded: 1878
Activities: education * public demonstrations
Coordinates: 018°58'30"E 46°31'42"N H117m
City Reference Coordinates: 019°00'00"E 46°31'00"N

Hungarian Academy of Sciences
Roosevelt tér 9
P.O. Box 6
H-1361 Budapest
Telephone: (0)1-3319353
    (0)1-3327176
Telefax: (0)1-3328943
Electronic Mail: <userid>@ella.hu
WWW: http://www.mta.hu/
Founded: 1825
Membership: 300
City Reference Coordinates: 019°05'00"E 47°30'00"N

Hungarian Academy of Sciences, Debrecen Heliophysical Observatory
P.O. Box 30
H-4010 Debrecen
Telephone: (0)52-311015
  (0)66-361553 (Gyula)
WWW: http://fenyi.sci.klte.hu/
Founded: 1958 (Gyula: 1972)
Staff: 13 (Gyula: 3)
Activities: sunspot positions and proper motions * solar flares * solar magnetic fields * photography (solar photosphere)
Periodicals: "Publications of the Debrecen Observatory" (ISSN 0209-7567)
Coordinates: 021°37'24"E 47°33'36"N H132m
  021°16'12"E 46°39'12"N H135m (Gyula Observing Station)
City Reference Coordinates: 021°38'00"E 47°32'00"N

Hungarian Academy of Sciences, Institute for Particle and Nuclear Physics
P.O. Box 49
H-1525 Budapest
or :
Konkoly-Thege út. 29-33
H-1121 Budapest
Telephone: (0)1-3922222
Telefax: (0)1-3959151
  (0)1-3922598
Electronic Mail: <userid>@rmk520.rmki.kfki.hu
WWW: http://www.rmki.kfki.hu/
Founded: 1991
Staff: 253
Activities: particle and nuclear physics * space physics * plasma physics * materials science * biophysics
City Reference Coordinates: 019°05'00"E 47°30'00"N

Hungarian Academy of Sciences, Konkoly Observatory
P.O. Box 67
H-1525 Budapest
or :
Konkoly-Tueje út. 15-17
H-1121 Budapest
Telephone: (0)1-3754122
Telefax: (0)1-2754668
Electronic Mail: <userid>@konkoly.hu
WWW: http://www.konkoly.hu/
Founded: 1899
Staff: 58
Periodicals: "Communications" (ISSN 0238-2091), "Information Bulletin on Variable Stars" (see separate entry)
Coordinates: 018°57'54"E 47°30'00"N H474m (IAU Code 053)
  019°54'00"E 47°55'00"N H946m (Piszkéstetö Mountain Stn – IAU Code 561)
City Reference Coordinates: 019°05'00"E 47°30'00"N

Hungarian Amateur Astronomical Society
(MACSIT)
P.O. Box 36
H-1387 Budapest 62
or :
Ujhegyi út. 4
H-8624 Kötcse
Electronic Mail: sta@ax.hu
  zsberend@hotmail.com (Zsolt Berend, Secretary)
Founded: 1987
Membership: 40
Activities: observing * popularization
Periodicals: (12) "Amatörcsillagászati Courier"
Coordinates: 017°52'45"E 46°45'10"N H290m (Kötcse)
City Reference Coordinates: 019°05'00"E 47°30'00"N (Budapest)

Hungarian Astronautical Society
Fö út. 68
H-1027 Budapest
or :
P.O. Box 433
H-1371 Budapest
Telephone: (0)1-2018443
Telefax: (0)1-3561215
Founded: 1956
Membership: 450
Activities: conferences * lectures * tours * contest * popularization * international relationships
Periodicals: "Ürkaleidoszkóp", "Asztronautikai Tájékoztató"
Awards: (1) "Fonó Albert Award", "Nagy Ernö Award"

City Reference Coordinates: 019°05'00"E 47°30'00"N

Hungarian Astronomical Association
P.O. Box 219
H-1461 Budapest
or :
Bartók Béla út. 11-13
H-1114 Budapest
Telephone: (0)1-2790429
          (0)30-8515364
Electronic Mail: mcse@mcse.hu
WWW: http://www.mcse.hu/
Founded: 1946 (refounded: 1989)
Membership: 2200
Activities: organizing amateur activities * fostering collaborations between amateur and professional astronomers
Sections: History of Astronomy * Moon * Astronomical Computing * Meteors * Variable Stars
Periodicals: (12) "Meteor" (ISSN 0133-249x): (1) "Yearbook" (ISSN 0866-2851)
Coordinates: 017°47'00"E 47°12'27"N H502m (Ráktanya)
City Reference Coordinates: 019°05'00"E 47°30'00"N

Hungarian Meteorological Service (HMS)
P.O. Box 38
H-1525 Budapest
or :
Kitaibel Pál út. 1
H-1024 Budapest
Telephone: (0)1-3464600
Telefax: (0)1-3464669
Electronic Mail: intrel@met.hu
            mets@.met.hu
WWW: http://www.met.hu/
      http://www.met.hu/index-e.html
Founded: 1870
Staff: 307
Activities: observing * telecommunications * weather forecast * aeronautical meteorology * climatology * agroclimatology * satellite meteorology * environmental protection
Periodicals: (4) "Idöjárás" (ISSN 0324-6329)
City Reference Coordinates: 019°05'00"E 47°30'00"N

Hungarian Space Office (HSO)
Szervita tér 8
H-1052 Budapest
Telephone: (0)1-3178717
Telefax: (0)1-2666728
Electronic Mail: elod.both@omfb.x400gw.itb.hu (Elöd Both, Director)
Periodicals: "Space Activities in Hungary" (ISSN 1217-7725)
City Reference Coordinates: 019°05'00"E 47°30'00"N

Hungarian Standards Institution
(Magyar Szabványügyi Testület - MSZT)
Üllöi út. 25
P.O. Box 24
H-1450 Budapest
Telephone: (0)1-2183011
Telefax: (0)1-2185125
WWW: http://www.mszt.hu/
Founded: 1921
Staff: 120
Activities: standards of development * training * certification * information centre
Periodicals: (12) "MSZT Bulletin"
City Reference Coordinates: 019°05'00"E 47°30'00"N

Information Bulletin on Variable Stars (IBVS)
c/o Konkoly Observatory
P.O. Box 67
H-1525 Budapest
Telephone: (0)1-3754122
Telefax: (0)1-2754668
Electronic Mail: ibvs@ogyalla.konkoly.hu
WWW: http://www.konkoly.hu/IBVS/IBVS.html
      http://www.konkoly.hu:80/IAUC27/
Founded: 1961
Staff: 5
• Journal (ISSN 0374-0676) of IAU Commissions 27 and 42
City Reference Coordinates: 019°05'00"E 47°30'00"N

Institute for Geodesy, Cartography and Remote Sensing, Satellite Geodetic Observatory (SGO)
P.O. Box 546
H-1373 Budapest
or :
Bosnyak tér 5
H-1149 Budapest
Telephone: (0)27-310980
Telefax: (0)27-310982
Electronic Mail: <userid>@sgo.fomi.hu
WWW: http://www.sgo.fomi.hu/
Founded: 1972
Staff: 15
Activities: satellite geodesy
Coordinates: 019°16'55"E 47°47'26"N H245m (Penc)
City Reference Coordinates: 019°05'00"E 47°30'00"N

József Attila University, Department of Physics, Astronomical Observatory
Dóm tér 9
H-6720 Szeged
Telephone: (0)62-420154
Telefax: (0)62-420154
Electronic Mail: <userid>@physx.u-szeged.hu
　　　　k.szatmary@physx.u-szeged.hu (Károly Szatmáry, Observatory Head)
WWW: http://www.jate.u-szeged.hu/obs/
Founded: 1992
Activities: variable stars * pulsating variable stars * binary stars * photoelectric photometry * automation of telescopes * education * astronomical software
Coordinates: 020°09'31"E 46°14'12"N H85m (IAU Code 629)
City Reference Coordinates: 020°09'00"E 46°15'00"N

Kecskemét Planetarium
Lánchíd utca 18/A
H-6000 Kecskemét
Telephone: (0)76-478994
Electronic Mail: kecsplan@externet.hu
WWW: http://planetarium.silicondreams.hu/
Founded: 1983
Staff: 7
Activities: shows * education
City Reference Coordinates: 019°43'00"E 46°46'00"N

Kiskunhalas Popular Observatory
Kossuth út. 43
H-6400 Kiskunhalas
Telephone: (0)77-423355
Founded: 1972
Membership: 31
Staff: 1
Activities: education * solar-activity impact
Coordinates: 019°00'00"E 46°06'00"N H136m
City Reference Coordinates: 019°00'00"E 46°06'00"N

Konkoly Observatory
• See "Hungarian Academy of Sciences, Konkoly Observatory"

Loránd University (Eötvös_)
• See "Eötvös Loránd University"

MACSIT
• See "Hungarian Amateur Astronomical Society"

Scientific Educational Society of Nógrádmegye
Mérleg út. 2
P.O. Box 156
H-3100 Salgótarján
Telephone: (0)32-314282
　　　　(0)32-311734
Founded: 1984
Membership: 2
Activities: observing (Sun, Moon, comets, deep sky) * photography
Coordinates: 019°44'04"E 48°07'58"N H356m
City Reference Coordinates: 019°48'00"E 48°07'00"N

Terkán Lajos Public Observatory (TELAPO)
Fürdö sor 3

H-8000 Székesfehérvár
Telephone: (0)22-314456
         (0)22-313028
         (0)22-311001
Electronic Mail: <userid>@mars.iif.hu
WWW: http://telapo.kodolanyi.hu/
Founded: 1967
Membership: 50
Activities: popularization * education
Periodicals: "Telapo"
Coordinates: 018°24'10"E 47°09'19"N H129m
City Reference Coordinates: 018°22'00"E 47°11'00"N

University of Budapest
• See "Eötvös Loránd University"

University of Szeged
• See "József Attila University"

Urania Observatory
Sánc út. 3/B
H-1016 Budapest
Telephone: (0)1-1869171
         (0)1-1869233
Telefax: (0)1-2671391
Electronic Mail: kondor@ludens.elte.hu
Founded: 1947
Membership: 21
Activities: popularization * lectures * telescope making * polishing mirrors * optical accessories
City Reference Coordinates: 019°05'00"E 47°30'00"N

Iceland

Amateur Astronomical Society of Seltjarnarnes
c/o Gudni G. Sigurdsson
Valhusaskola
170 Seltjarnarnes
Telephone: 561-2424
Electronic Mail: aquila@ismennt.is
WWW: http://rvik.ismennt.is/~aquila/
Membership: 70
City Reference Coordinates: 022°01'00"W 64°09'00"N

Icelandic Astronomical Society
Dunhaga 3
IS-107 Reykjavík
Telephone: (0)5254800
Telefax: (0)5528911
Electronic Mail: halo@raunvis.hi.is
Founded: 1988
Membership: 20
Activities: promoting astronomy and astrophysics in Iceland
City Reference Coordinates: 021°51'00"W 64°09'00"N

Icelandic Council for Standardization (STRÍ)
• See new "Icelandic Standards (IST)"

Icelandic Meteorological Office (IMO)
Bústadavegi 9
IS-150 Reykjavík
Telephone: (0)5226000
Telefax: (0)5226001
Electronic Mail: office@vedur.is
　　　　　　　　<userid>@vedur.is
WWW: http://www.vedur.is/
Founded: 1920
Staff: 100
Activities: meteorological service (about 300 weather stations in Iceland) * seismology * avalanche research
Periodicals: (12) "Vedráttan" (ISSN 0258-3836)
Coordinates: 021°51'00"W 64°08'00"N H52m
City Reference Coordinates: 021°51'00"W 64°08'00"N

Icelandic Physical Society
c/o Sveinbjorn Bjornsson (President)
Science Institute
University of Iceland
Dunhaga 3
IS-107 Reykjavík
Telephone: (0)694942
Telefax: (0)28911
Electronic Mail: svb@os.is
WWW: http://www.os.is/ei/
Founded: 1977
Membership: 80
City Reference Coordinates: 021°51'00"W 64°09'00"N

Icelandic Research Council
(Rannsóknarrád Íslands)
Laugavegi 13
IS-101 Reykjavík
Telephone: (0)5621320
Telefax: (0)5529814
Electronic Mail: rannis@rannis.is
WWW: http://www.rannis.is/
Founded: 1994
Membership: 11
City Reference Coordinates: 021°51'00"W 64°09'00"N

Icelandic Standards (IST)
Laugavegur 178
IS-105 Reykjavík
Telephone: (0)5207150
Telefax: (0)5207171

Electronic Mail: stadlar@stadlar.is
WWW: http://www.stri.is/
Founded: 1987
Staff: 10
Activities: Icelandic standards * representation of Iceland in international and regional standardization bodies
Periodicals: "Stadlatídindi"
City Reference Coordinates: 021°51'00"W 64°09'00"N

University of Iceland, Science Institute
Dunhaga 3
IS-107 Reykjavík
Telephone: (0)5254800
Telefax: (0)5528911
Electronic Mail: <userid>@raunvis.hi.is
WWW: http://www.raunvis.hi.is/RaunvisHomeE.html
Founded: 1966
Activities: almanac * STP * high-energy astrophysics * cosmology * cosmic rays
Coordinates: 021°57'24"W 64°08'24"N
City Reference Coordinates: 021°51'00"W 64°09'00"N

India

Aligarh Muslim University (AMU), Physics Department, Astrophysics Group
Aligarh 202 002
Telephone: (0)571-401001
Electronic Mail: agphys@amu.ernet.in
Founded: 1969
Staff: 5
Activities: theoretical astrophysics * ISM * history of astronomy in India and Islamic countries
City Reference Coordinates: 078°05'00"E 27°54'00"N

Amateur Astronomers' Association (Bombay)
c/o Physics Department
Saint Xavier's College
Mahapalika Marg
Mumbai 400 001
Telephone: (0)22-4306519 (c/o Aadil Desai)
Electronic Mail: aaa_bombay@hotmail.com
Founded: 1977
Membership: 400
Activities: monthly lectures * field trips * conferences * instrumentation * photography * film and slide shows * library * visits * observing * popularization * research * data collection and distribution
Periodicals: (6) "M51" (circ.: 500)
Awards: (1) "R. Kundaji Trophy"
Coordinates: 072°46'00"E 19°00'00"N H35m
City Reference Coordinates: 072°50'00"E 19°11'00"N

Amateur Astronomers Association of Delhi (AAAD)
c/o Nehru Planetarium
Teen Murti House
New Delhi 110 011
Electronic Mail: aaadel@nebulacorp.com
WWW: http://www.nebulacorp.com/org/aaadel/
http://www.nebulacorp.com/aaa/
City Reference Coordinates: 077°12'00"E 28°37'00"N

Amateur Astronomers' Association of Vadodara (AAAV)
52 Shri Satchidanand Nagar Society
b/h Swati Society
New Sama Road
Vadodara 390 008
Telephone: (0)265-782302
Electronic Mail: planet@ad1.vsnl.net.in (Mukesh Pathak, President)
Founded: 1996
Membership: 150
Activities: popularization * observing
City Reference Coordinates: 073°14'00"E 22°19'00"N

Astronomical Society of India (ASI)
c/o Ashok K. Pati (Secretary)
Indian Institute of Astrophysics
Koramangala
Bangalore 560 034
Electronic Mail: pati@iiap.ernet.in
WWW: http://www.rri.res.in/asi/
Founded: 1972
Membership: 500
Activities: holding scientific meetings * publishing * encouraging amateur astronomers
Periodicals: (4) "Bulletin of the Astronomical Society of India" (ISSN 0304-9523, circ.: 500)
Awards: "Vainu Bappu Memorial Award", "Young Astronomers Award"
City Reference Coordinates: 077°36'00"E 12°58'00"N

Astronomy and Space Science Association (ASSA)
c/o Astrophysics Group
Physics Department
Gauhati University
Guwahati 781 014
Telephone: (0)361-570412
Telefax: (0)361-570133
Electronic Mail: vcgu@gulib.iitg.ernet.in
Founded: 1993
Membership: 320
Activities: observing * popularization * research

Periodicals: (12) "The ASSA Newsletter"
Coordinates: 091°45'00"E 26°10'00"N H75m
City Reference Coordinates: 091°45'00"E 26°10'00"N

Birla Planetarium (B.M._)
- See "B.M. Birla Planetarium, Chennai"
- See also "B.M. Birla Planetarium, Hyderabad"
- See also "M.P. Birla Planetarium, Kolkata"

B.M. Birla Planetarium, Chennai
c/o Tamil Nadu Science and Technology Centre
Gandhi Mandapam Road
Engineering College P.O.
Chennai 600 025
Telephone: (0)44-416751
(0)44-4915250
Telefax: (0)44-4918787
Founded: 1988
Staff: 8
Activities: programmes for public and students * education * seminars * workshops * observing * outreach activities
City Reference Coordinates: 080°16'00"E 13°04'00"N

B.M. Birla Planetarium, Hyderabad
c/o Birla Archaeological, Cultural and Research Institute
Adarsh Nagar
Hyderabad 500 063
Telephone: (0)40-3235081
(0)40-3241067
Telefax: (0)40-3237266
Electronic Mail: birlasc@hd1.vnsl.net.in
WWW: http://www.birlavision.com/
Founded: 1985
Staff: 20
Activities: shows * camps * lectures * education * seminars * films
Awards: (1) "B.M. Birla Science Prize"
City Reference Coordinates: 078°29'00"E 17°23'00"N

Breakthrough Science Society (BSS)
9 Creek Row
Kolkata 700 014
Telephone: (0)33-2460563
Telefax: (0)33-2465114
Electronic Mail: breakthrough@ieee.org
WWW: http://tnp.saha.ernet.in/~basak/bss.html
Founded: 1995
Membership: 500
Activities: amateur astronomy * popularization * pollution * promoting scientific approach
Periodicals: (4) "Breakthrough: A Journal on Science and Society", "Prakriti: Vigyan O Samaj Vishayak Patrika"
City Reference Coordinates: 088°20'00"E 22°34'00"N

Bureau of Indian Standards (BIS)
Manak Bhavan
9 Bahadur Shah Zafar Marg
New Delhi 110 002
Telephone: (0)11-3311375
Telefax: (0)11-3314062
Electronic Mail: bisind@de12.vsn1.net.in
WWW: http://www.bis.org.in/
Founded: 1947
Staff: 2380
Activities: formulating standards * certifications
Periodicals: (12) "Standard India" (ISSN 0970-2628), "Standards Monthly Additions" (ISSN 0970-3985); (1) "Annual Report"; "Manakdoot", "Standards Worldover", "Current Published Information on Standardization", "EEC Norm Scan"
City Reference Coordinates: 077°15'00"E 28°36'00"N

Committee on Science and Technology in Developing Countries (COSTED)
24 Gandhi Mandap Road
Chennai 600 025
Telephone: (0)44-4901367
Telefax: (0)44-4914543
Electronic Mail: costed@vsnl.com
WWW: http://www.costed-icsu.org/
Founded: 1966
Membership: 30
Activities: promoting science and technology in developing countries

City Reference Coordinates: 080°16'00"E 13°04'00"N

Current Science Association
C.V. Raman Avenue
Sadashivanagar
P.O. Box 8001
Bangalore 560 080
Telephone: (0)80-3342310
Telefax: (0)80-3346094
Electronic Mail: currsci@ias.ernet.in
WWW: http://tejas.serc.iisc.ernet.in/~currsci/
Founded: 1932
Staff: 7
Periodicals: (24) "Current Science" (ISSN 0011-3891, circ.: 3,200) (in collaboration with the Indian Academy of Sciences)
City Reference Coordinates: 077°36'00"E 12°58'00"N

C.Z. Instruments India Pvt. Ltd.
• See now "Gordhandas Desai Pvt. Ltd."

Federation of Asian Scientific Academies and Societies (FASAS)
c/o FASAS Secretariat
Bahadur Shah Zafar Marg
New Delhi 110002
or :
c/o Indian Academy of Sciences
C.V. Raman Avenue
P.O. Box 8005
Sadashivanagar
Bangalore 560 080
Telephone: (0)11-3313153
         (0)80-3342546
         (0)80-3344592
Telefax: (0)11-3235648
         (0)80-3346094
Electronic Mail: insa@csird.ernet.in
              office@ias.ernet.in
Founded: 1984
Membership: 14 (countries)
City Reference Coordinates: 077°15'00"E 28°36'00"N (New Delhi)
                    077°36'00"E 12°58'00"N (Bangalore)

Gauhati University, Physics Department, Astrophysics Group
Guwahati 781 014
Telephone: (0)361-570531
Telefax: (0)361-570133
Electronic Mail: <userid>@iucaa.ernet.in
Founded: 1962
Activities: theoretical nuclear astrophysics * neutron stars * gravitational collapse * black holes * early universe * variable stars
City Reference Coordinates: 091°45'00"E 26°10'00"N

Goodwill Cryogenics Enterprises
213 Nirman Vyapar Kendra
Sector 17
Vashi
New Mumbai 400 703
Telephone: (0)22-7890642
Telefax: (0)22-7822769
Electronic Mail: vinochop@bomb2.vsnl.net.in
Founded: 1979
Activities: cryogenics vacuum and refrigeration * temperature and magnetic-field measurement instrumentation
City Reference Coordinates: 072°50'00"E 19°11'00"N

Gordhandas Desai Pvt. Ltd. (GDPL)
(C.Z. Instruments India Pvt. Ltd.)
P.O. Box 11108
Mumbai 400 020
or :
Court Chambers
35 Sir V. Thackersey Road
Mumbai
Telephone: (0)22-2004208
Telefax: (0)22-2000581
Electronic Mail: bomsales@gdpl.com
WWW: http://www.gdpl.com/

Founded: 1924
Staff: 51
Activities: supplying, erecting and maintaining planetariums, telescopes and related equipment * Carl Zeiss representative
City Reference Coordinates: 072°50'00"E 19°11'00"N

India Meteorological Department (IMD), New Delhi
Mausam Bhavan
Lodi Road
New Delhi 110 003
Telephone: (0)11-618242
(0)11-616602
Telefax: (0)11-699216
WWW: http://www.nic.in/snt/c9imd.htm
Founded: 1875
Staff: 8400
Activities: weather forecasting * aviation weather service * hydrology * seismology
Periodicals: "MAUSAM"
City Reference Coordinates: 077°15'00"E 28°36'00"N

India Meteorological Department (IMD), Weather Forecasting
Observatory
Shivajinagar
Pune 411 005
Telephone: (0)212-325211
(0)212-323403 x510
Telefax: (0)212-323201
WWW: http://www.nic.in/snt/c9imd.htm
Founded: 1875
Staff: 250
Activities: weather forecasting * training * R&D
Periodicals: "All India Weather Summary", "India Daily Weather Report", "India Weekly Weather Report", "Atlas of Tracks of Storms and Depressions in the Bay of Bengal and Arabian Sea"
City Reference Coordinates: 073°54'00"E 18°31'00"N

Indian Academy of Sciences (IASc)
C.V. Raman Avenue
P.O. Box 8005
Sadashivanagar
Bangalore 560 080
Telephone: (0)80-3612546
(0)80-3614592
Telefax: (0)80-3616094
Electronic Mail: office@ias.ernet.in
WWW: http://www.ias.ac.in/
Founded: 1934
Staff: 20
Activities: publishing scientific journals * organising scientific meetings
Periodicals: (24) "Current Science" (ISSN 0011-3891, circ.: 5,000) (in collaboration with the Current Science Association); (12) "Pramana - Journal of Physics" (ISSN 0304-4289, circ.: 1,250), "Resonance" (ISSN 0971-8044, circ.: 5,000); (6) "Proceedings (Chemical Sciences)" (ISSN 0253-4134, circ.: 1,000), "Sadhana (Engineering Science)" (ISSN 0256-2499, circ.: 800), "Bulletin of Materials Science" (ISSN 0250-4707, circ.: 2,400); (4) "Journal of Astrophysics and Astronomy" (ISSN 0250-6335, circ.: 900), "Proceedings (Earth and Planetary Sciences)" (ISSN 0253-4126, circ.: 800), "Proceedings (Mathematical Sciences)" (ISSN 0253-4142, circ.: 950); other journals in the field of life sciences
City Reference Coordinates: 077°36'00"E 12°58'00"N

Indian Institute of Astrophysics (IIA)
Koramangala
Bangalore 560 034
Telephone: (0)80-5530672
(0)80-5530673
(0)80-5530674
(0)80-5530675
(0)80-5530676
Telefax: (0)80-5534043
Electronic Mail: <userid>@iiap.ernet.in
WWW: http://www.iiap.ernet.in/~website/home_page.html
http://www.iiap.ernet.in/~vbo/ (Vainu Bappu Observatory)
http://www.iiap.ernet.in/~kodai/ (Kodaikanal Observatory)
http://www.iiap.ernet.in/~gauri/ (Gauribidanur Observatory)
http://www.iiap.ernet.in/~iao/ (Indian Astronomical Observatory)
Founded: 1786 (Kodaikanal Observatory)
Staff: 337
Activities: solar chromospheres * convection and magnetic fields * plasma processes in the corona * solar cycles and related topics * solar system, comets, minor planets * stellar atmospheres * abundances * Ap stars * binary stars * photometry of globular and galactic clusters * photometry of galaxies * galactic dynamics * pulsars * neutron stars * cosmology * ISM * radioastronomy (decametric wavelengths) * flux distributions

Coordinates: 077°36'00"E 12°58'00"N H921m
  078°49'54"E 12°24'32"N H725m (Vainu Bappu Obs., Kavalur – IAU Code 220)
  077°28'00"E 10°03'08"N H2,343m (Kodaikanal Obs.)
  077°26'07"E 13°36'12"N H687m (Radiotelescope Stn., Gauribidanur)
  078°57'05"E 32°46'46"N H4,517m (Indian Astron. Obs., Hanle)
City Reference Coordinates: 077°36'00"E 12°58'00"N

Indian Institute of Science (IISc), Joint Astronomy Programme (JAP)
Karnataka
Bangalore 560 012
Telephone: (0)80-3600411
Telefax: (0)80-3600683
Electronic Mail: regr@admin.iisc.ernet.in
WWW: http://physics.iisc.ernet.in/~jap/
Founded: 1909 (IISc)
Activities: graduate programme in astronomy in collaboration with the other astronomical institutions in India * theoretical astrophysics * galactic dynamics * formation and evolution of galaxies * solar physics
Periodicals: "Journal"
City Reference Coordinates: 077°36'00"E 12°58'00"N

Indian Skeptics
c/o B. Premanand (Convener)
11/7 Chettipalayam Road
Podanur 641 023
Telephone: (0)422-872423
WWW: http://www.indian-skeptic.org/html/
Founded: 1976
Activities: parascience debunking * education
Periodicals: (12) "Indian Skeptics"
City Reference Coordinates: 077°00'00"E 10°57'00"N

Indian Space Research Organization (ISRO)
Antariksh Bhavan
New BEL Road
Bangalore 560 094
Telephone: (0)80-3415474
  (0)80-3425275
Telefax: (0)80-3412253
Electronic Mail: <userid>@isro.ernet.in
WWW: http://www.isro.org/
Founded: 1972
Staff: 16700
Activities: providing operational space services in telecommunications, TV broadcast, radio networking, meteorology and remote sensing for natural resources survey and management * developing satellites, launch vehicles and associated ground systems
Periodicals: (4) "Space India"; (1) "Annual Report"
Awards: "Vikram Sarabhai Award"
City Reference Coordinates: 077°36'00"E 12°58'00"N

Indian Space Research Organization (ISRO), Satellite Centre (ISAC)
Airport Road
Vimanapura Post
Bangalore 560 017
Telephone: (0)80-5266251
Telefax: (0)80-5265407
Electronic Mail: <userid>@isac.ernet.in
WWW: http://www.isro.org/cen_isac.htm
Founded: 1972
Staff: 800
Activities: spacecraft design and development * space astronomy
Periodicals: (2) "Journal of Spacecraft Technology" (ISSN 0971-1600)
City Reference Coordinates: 077°36'00"E 12°58'00"N

Inter-University Centre for Astronomy and Astrophysics (IUCAA)
Post Bag 4
Ganeshkhind
Pune 411 007
or :
Meghnad Saha Road
Pune University Campus
Ganeshkhind
Pune 411 007
Telephone: (0)20-5691414
Telefax: (0)20-5690760
Electronic Mail: root@iucaa.ernet.in
  <userid>@iucaa.ernet.in

WWW: http://www.iucaa.ernet.in/
Founded: 1988
Staff: 100
Activities: fundamental research * Sun * planets * galactic and extragalactic astronomy * high-energy astrophysics * theoretical cosmology * gravitation and relativity * quantum gravity * quantum cosmology * education * observing * popularization
Periodicals: (4) "Khagol: The IUCAA Bulletin"
Coordinates: 073°58'00"E 18°34'00"N
City Reference Coordinates: 073°54'00"E 18°31'00"N

Mount Abu Infrared Observatory
• See "Physical Research Laboratory (PRL), Mount Abu Infrared Observatory"

M.P. Birla Planetarium, Kolkata
96 Jawaharlal Nehru Road
Kolkata 700 071
Telephone: (0)33-2231516
(0)33-2236610
Telefax: (0)33-2487988
Electronic Mail: planetarium@vsnl.com
WWW: http://education.vsnl.com/planetarium/
Founded: 1962
Periodicals: "The Journal of Birla Planetarium"
City Reference Coordinates: 088°20'00"E 22°34'00"N

National Centre for Radio Astrophysics (NCRA), Junnar Taluk
P.O. Box 6
Junnar Taluk 410 504
or :
Khodad Village
Narayangaon
Junnar Taluk 410 504
Telephone: (0)2132-52111
Telefax: (0)2132-52118
WWW: http://www.ncra.tifr.res.in/ncra_hpage/gmrt/
Activities: radio astronomy with the Giant Metrewave Radio Telescope (GMRT)
Coordinates: 074°03'00"E 19°05'30"N H650m (Junnar Taluk)
City Reference Coordinates: 073°58'00"E 19°15'00"N (Junnar)

National Centre for Radio Astrophysics (NCRA), Ooty
P.O. Box 8
Ooty 643 001
Telephone: (0)423-442032
(0)423-550334
(0)423-550335
Telefax: (0)423-442588
WWW: http://www.ncra.tifr.res.in/ncra_hpage/ort/ort.html
Founded: 1970
Staff: 15
Activities: observing at 327 MHZ * aperture synthesis * radio source structure * observational cosmology * pulsars * IPS * ISM * VLBI * SNR * galaxy clusters * galactic plane survey * protocluster search * image processing
Coordinates: 076°40'02"E 11°22'56"N H2,150m (Ooty Radio Telescope)
City Reference Coordinates: 076°43'00"E 11°25'00"N

National Centre for Radio Astrophysics (NCRA), Pune
c/o Pune University Campus
Post Bag 3
Ganeshkhind
Pune 411 007
Telephone: (0)20-5697107
(0)20-5691384
(0)20-5691385
Telefax: (0)20-5695149
Electronic Mail: root@ncra.tifr.res.in
WWW: http://www.ncra.tifr.res.in/
Founded: 1990
Staff: 200
Activities: radio astronomy * cosmology * pulsars * interplanetary scintillations * aperture synthesis * meterwavelength observing
Coordinates: 074°03'00"E 19°06'00"N H650m (Giant Meterwave Radio Telescope)
City Reference Coordinates: 073°54'00"E 18°31'00"N

National Physical Laboratory (NPL), Radio and Atmospheric Sciences Division (RASD)
Dr. K.S. Krishnan Road
New Delhi 110 012
Telephone: (0)11-5787657

Telefax: (0)11-5852678
(0)11-5764189
Electronic Mail: scgarg@csnpl.ren.nic.in (S.C. Garg, Head)
Founded: 1947
Staff: 80
Activities: planetary atmospheres * aeronomy * tropospheric and ionospheric communications * middle atmosphere * solar-terrestrial physics * upper atmosphere * ionosphere * global change
City Reference Coordinates: 077°15'00"E 28°36'00"N

National Remote Sensing Agency (NRSA)
Balanagar
Hyderabad 500 037
Telephone: (0)40-279572
(0)40-279573
(0)40-279574
(0)40-279575
(0)40-279576
(0)40-278360
Telefax: (0)40-278648
Electronic Mail: info@nrsa.gov.in
WWW: http://www.nrsa.gov.in/
Founded: 1975
Staff: 1025
Activities: satellite and aerial data acquisition * data processing and dissemination * remote sensing * R&D
Periodicals: (4) "Interface"
City Reference Coordinates: 078°29'00"E 17°23'00"N

Nehru Planetarium
Teen Murti House
New Delhi 110 011
City Reference Coordinates: 077°12'00"E 28°37'00"N

Optical Society of India (OSI)
c/o Department of Applied Physics
University of Kolkata
92 Acharya Prafulla Chandra Road
Kolkata 700 009
Electronic Mail: osi@cucc.ernet.in
WWW: http://www.ositvm.org/osi.htm
Founded: 1965
Membership: 350
Staff: 2
Activities: promotion of optics in India * publishing * symposia
Periodicals: "Journal of Optics"
City Reference Coordinates: 088°20'00"E 22°34'00"N

Osmania University, Centre of Advanced Study in Astronomy (CASA)
Hyderabad 500 007
Telephone: (0)7097306
(0)7682247
Founded: 1908
Staff: 10
Activities: education * galactic dynamics * photometry & spectroscopy of variable and binary stars * radio observing (Sun)
Periodicals: "Contributions from the Nizamiah and Japal-Rangapur Observatories"
Coordinates: 078°27'12"E 17°25'59"N H554m (Nizamiah Observatory)
078°43'42"E 17°05'54"N H695m (Japal-Rangapur Obs. – IAU Code 219)
City Reference Coordinates: 078°29'00"E 17°23'00"N

Physical Research Laboratory (PRL), Mount Abu Infrared Observatory, Headquarters
Navrangpura
Ahmedabad 380 009
Telephone: (0)79-462129
(0)79-6425037
Telefax: (0)79-6560502
Electronic Mail: <userid>@prl.ernet.in
WWW: http://www.prl.ernet.in/astronomy/
http://www.prl.ernet.in/astronomy/irtel.shtml
Founded: 1947
Staff: 85
Activities: AGNs * starburst galaxies * PNs * stellar physics * novae * star formation * HII regions * lunar occultations * comets * solar eclipses
Coordinates: 072°46'47"E 24°39'09"N H1,680m
City Reference Coordinates: 072°35'00"E 23°02'00"N

Physical Research Laboratory (PRL), Udaipur Solar Observatory (USO)
P.O. Box 198
Bari Road
Dewali
Udaipur 313 001
Telephone: (0)294-560626
Telefax: (0)294-526325
Electronic Mail: root@uso.ernet.in
WWW: http://www.prl.ernet.in/~shibu/uso.html
http://www.prl.ernet.in/
Founded: 1975
Staff: 18
Activities: solar astronomy
Coordinates: 073°42'45"E 24°35'08"N H301m
City Reference Coordinates: 073°47'00"E 24°36'00"N

Punjabi University, Department of Physics
Patiala 147 002
Telephone: (0)175-822161 x96
Electronic Mail: astro@pbi.ernet.in
WWW: http://www.universitypunjabi.org/pages/teaching/teaching28.html#PHYSICS
Founded: 1978
Staff: 16
Activities: stellar populations * space physics * Be stars * HII regions * ISM reddening
Coordinates: 076°45'00"E 30°45'00"N H253m
City Reference Coordinates: 076°45'00"E 30°45'00"N

Raman Research Institute (RRI)
C.V. Raman Avenue
Sadashivanagar
Bangalore 560 080
Telephone: (0)80-3610122
Telefax: (0)80-3610492
Electronic Mail: root@rri.res.in
<userid>@rri.res.in
WWW: http://www.rri.res.in/
Founded: 1948
Staff: 12 (astronomers and astrophysicists)
Activities: extragalactic astronomy * Galaxy * ISM * neutron stars * pulsars
Coordinates: 077°35'00"E 13°00'45"N (mm, Bangalore)
077°26'07"E 13°36'12"N H500m (dkm, Gauribidanur)
City Reference Coordinates: 077°36'00"E 12°58'00"N

Tata Institute of Fundamental Research (TIFR)
Homi Bhabha Road
Colaba
Mumbai 400 005
Telephone: (0)22-2152971
(0)22-2152311
Telefax: (0)22-2152110
Electronic Mail: <userid>@tifr.res.in
WWW: http://www.tifr.res.in/
Founded: 1945
Staff: 700
Activities: fundamental research in mathematics * theoretical and experimental physics and biology, including astronomy, nuclear and atomic physics, high-energy physics, cosmic-ray physics, condensed-matter physics, chemical physics, solid-state electronics and computer science
Periodicals: (1) "Annual Report"
Coordinates: 076°40'02"E 11°22'50"N H2,150m (RAC, Ooty - see separate entry)
074°03'00"E 19°06'00"N H655m (GMRT, Pune - see separate entry)
City Reference Coordinates: 072°50'00"E 19°11'00"N

Udaipur Solar Observatory (USO)
• See "Physical Research Laboratory (PRL), Udaipur Solar Observatory (USO)"

Uttar Pradesh State Observatory (UPSO)
Manora Peak
Naini Tal 263 169
Telephone: (0)5942-35136
(0)5942-35583
(0)5942-35053
Electronic Mail: <userid>@upso.ernet.in
Founded: 1954

Staff: 41
Activities: photometry of variable stars * spectrophotometry * open clusters * eclipsing binaries * occultations * comets * Sun * stellar atmospheres * Be stars * site survey
Coordinates: 079°27'24"E 29°21'36"N H1,950m
City Reference Coordinates: 079°26'00"E 29°22'00"N

Vikram Sarabhai Space Centre (VSSC)
Trivandrum 695 022
Telephone: (0)471-562444
　　　　　　(0)471-562555
Telefax: (0)471-79795
WWW: http://www.isro.org/centers/cen_vssc.htm
Founded: 1963
Staff: 5600
Activities: launch vehicle technology development
City Reference Coordinates: 076°57'00"E 08°28'00"N

Indonesia

Bandung Institute of Technology, Bosscha Observatory
Lembang 40391
Telephone: (0)22-2786001
Telefax: (0)22-2786001
Electronic Mail: bosscha@ibm.net
WWW: http://www.bosscha.itb.ac.id/
Founded: 1923
Staff: 11
Activities: binary stars * galactic structure * photometry * variable stars
Periodicals: "Contributions", "Publications", "Annals"
Coordinates: 107°37'00"E 06°49'30"S H1,400m (IAU Code 299)
City Reference Coordinates: 107°37'00"E 06°50'00"S

Bandung Institute of Technology, Department of Astronomy
Jalan Ganesha 10
Bandung 40132
Telephone: (0)22-2511576
(0)22-2509170
Telefax: (0)22-2509170
Electronic Mail: <userid>@as.itb.ac.id
WWW: http://www.as.itb.ac.id/
Founded: 1951
Staff: 24
Activities: galactic structure * cosmology * stellar physics * solar systems
Observatories: 1 (Bosscha Observatory - see separate entry)
City Reference Coordinates: 107°36'00"E 06°54'00"S

Bosscha Observatory
• See "Bandung Institute of Technology, Bosscha Observatory"

Indonesian Institute of Sciences (LIPI)
Gedung Widya Graha
Jalan Jend. Gatot Subroto 10
Selatan 12710
Jakarta
Telephone: (0)22-5225711
Telefax: (0)22-5207226
WWW: http://www.lipi.go.id/
Founded: 1967
Staff: 4750
Activities: research * development
City Reference Coordinates: 106°48'00"E 06°10'00"S

Meteorological and Geophysical Agency
(Badan Meteorologi dan Geofisika - BMG)
Jalan Angkasa I/2 Kemayoran
Jakarta Pusat 10720
Telephone: (0)21-4246321
(0)21-4246311
Telefax: (0)21-4246703
Electronic Mail: bmg@bmg.go.id
WWW: http://www.bmg.go.id/
Founded: 1980
City Reference Coordinates: 106°48'00"E 06°10'00"S

National Institute of Aeronautics and Space (LAPAN)
Jalan Pemuda Persil 1
Jakarta 13220
Telephone: (0)21-4892802
(0)21-4894941
Telefax: (0)21-4894815
WWW: http://www.lapan.go.id/
Founded: 1963
Staff: 1350
Activities: remote sensing * space technology * atmosphere * ionosphere * Sun
Periodicals: (4) "Majalah Lapan" (ISSN 0126-0480), "Warta Lapan" (ISSN 0216-9754)
Coordinates: 107°10'00"E 07°40'00"S (Pameungpeuk)
107°47'00"E 06°54'00"S H760m (Tanjung Sari)
113°30'00"E 06°55'00"S (Watukosek)
135°47'00"E 01°15'00"S (Biak)
City Reference Coordinates: 106°48'00"E 06°10'00"S

Standardization Council of Indonesia
(Dewan Standardisasi Nasional - DSN)
Jalan Jenderal Gatot Subroto 10
P.O. Box 3123
Jakarta 12710
Telephone: (0)21-5221686
Telefax: (0)21-5206574
Electronic Mail: pustan@rad.net.id
Founded: 1984
Staff: 55
Activities: services in standardization
Periodicals: "Warta Standardisasi"
City Reference Coordinates: 106°48'00"E 06°10'00"S

Iran

Biruni Observatory
• See "Shiraz University, Physics Department, Biruni Observatory"

Institute of Standards and Industrial Research of Iran (ISIRI)
(Headquarters)
P.O. Box 31585-163
Karaj
or :
(Central Office)
14 Shahamati Alley
Veli-e-Asr Avenue
Tehran 15946
Telephone: (0)261-227045 (Headquarters)
(0)261-226031 (Public and International Relations)
(0)21-899308 (Central Office)
Telefax: (0)261-225015 (Headquarters)
(0)21-8802276 (Central Office)
Founded: 1960
Staff: 1125
City Reference Coordinates: 050°59'00"E 35°48'00"N (Karaj)
051°26'00"E 35°40'00"N (Tehran)

Islamic Republic of Iran Meteorological Organization (IRIMO)
P.O. Box 13185-461
Mehrabad Airport
Tehran
Telephone: (0)21-6004026/8
Telefax: (0)21-6000417
(0)21-6469044
WWW: http://www.irimet.net/
Founded: 1955
Staff: 2200
Activities: meteorology * agrometeorology * climatology * statistics
Periodicals: (52) "Bulletin of Meteorological Data"; (4) "Journal of NIVAR"
Observatories: 154 synoptic stations
City Reference Coordinates: 051°26'00"E 35°40'00"N

Shiraz University, Physics Department, Biruni Observatory
Shiraz 71454
Telephone: (0)71-24609
Telefax: (0)71-20027
WWW: http://www.physics.susc.ac.ir/Biruni.html
Founded: 1975
Activities: photoelectric photometry of variable stars * theoretical study of stellar systems * cosmology
Coordinates: 052°30'00"E 29°35'00"N H1,600m
City Reference Coordinates: 052°34'00"E 29°38'00"N

Sirius Astronomical Society of Kermanshah
c/o Mohammad Ali Khodayari
P.O. Box 67145-1416
Kermanshah
Telephone: (0)831-722444
Telefax: (0)831-760863
Electronic Mail: mkhodayari@yahoo.com
WWW: http://hometown.aol.com/siriusask/
Founded: 1995
Membership: 250
Activities: education * observing
Periodicals: (6) "I & Heavens" (in Farsi)
Coordinates: 047°05'00"E 34°19'00"N H1200m
City Reference Coordinates: 047°05'00"E 34°19'00"N

Tabriz University, Center for Applied Physics and Astronomical Research
Tabriz 51664
Telephone: (0)41-342564
Telefax: (0)41-347050
Electronic Mail: rca@ark.tabrizu.ac.ir
WWW: http://www.tabrizu.ac.ir/
Founded: 1973
Activities: education * observing * solar physics * photometry of double stars * planetarium * Khadjeh Nassir Addin Toussi Observatory

Coordinates: 046°19'57"E 37°52'10"N (Gazan Mountains)
City Reference Coordinates: 046°18'00"E 38°05'00"N

Zarvan Co. Ltd.
P.O. Box 15875-1487
Tehran
or :
4 Seventh Alley
Balbochestan Street
Nasr Avenue
14469 Tehran
Telephone: (0)21-8271363
Telefax: (0)21-8272507
Electronic Mail: nojum@apadana.com
Founded: 1991
Staff: 18
Activities: publishing * popularization * manufacturing domes * telescope dealer
Periodicals: "Nojum Magazine" (ISSN 1019-584x)
City Reference Coordinates: 051°26'00"E 35°40'00"N

Iraq

Central Organization for Standardization and Quality Control (COSQC)
c/o Ministry of Planning
P.O. Box 13032
Aljadiria
Baghdad
Telephone: (0)1-7765180
Telefax: (0)1-7765781
City Reference Coordinates: 044°25'00"E 33°21'00"N

Iraqi Meteorological Organization
Almansoor
P.O. Box 6078
Baghdad
Telephone: (0)1-5560070
City Reference Coordinates: 044°25'00"E 33°21'00"N

Scientific Research Council, Space and Astronomy Research Center (SARC)
P.O. Box 2441
Jadiriyah
Baghdad
Telephone: (0)1-7756127
Founded: 1980
Activities: galaxies * ISM * variable stars * solar system
Periodicals: "Journal of Space and Astronomy Research"
Coordinates: 044°24'00"E 33°48'00"N H100m
City Reference Coordinates: 044°24'00"E 33°48'00"N

Ireland

Astronomy Ireland
P.O. Box 2888
Dublin 1
Telephone: (0)1-4598883
Telefax: (0)1-4599933
Electronic Mail: ai@iol.ie
WWW: http://www.astronomy.ie/
Founded: 1990
Membership: 4
Activities: meetings
Periodicals: (6) "Astronomy & Space Magazine" (ISSN 0781-8062, circ.: 5,000)
City Reference Coordinates: 006°15'00"W 53°20'00"N

Birr Castle
Demesne
Birr, Co. Offaly
Telephone: (0)509-20056
  (0)509-20336
Telefax: (0)509-21583
Electronic Mail: info@birrcastle.com
WWW: http://www.birrcastle.com/
Founded: 1984 (Birr Scientific & Heritage Foundation)
Staff: 1
Activities: Ireland's Historic Science Centre * museum * exhibitions * displays * lectures
Periodicals: "Focus"
City Reference Coordinates: 007°54'00"W 53°05'00"N

Cork Astronomy Club
c/o Charles Coughlan (Secretary)
12 Forest Ridge Crescent
Wilton, Cork
Telephone: (0)21-543669
Electronic Mail: chas@indigo.ie
WWW: http://indigo.ie/~chas/astronomy.html
City Reference Coordinates: 008°28'00"W 51°54'00"N

Cork Institute of Technology, Department of Applied Physics and Instrumentation
Rossa Avenue
Bishoptown, Co. Cork
Telephone: (0)21-4326214
Telefax: (0)21-4345191
Electronic Mail: phelanm@cit.ie
WWW: http://www.physics.cit.ie/PHYMAI.html
City Reference Coordinates: 008°28'00"W 51°54'00"N (Cork)

Dublin Institute for Advanced Studies (DIAS), School of Cosmic Physics, Astronomy Section
(Dunsink Observatory)
Castleknock
Dublin 15
Telephone: (0)1-8387911
  (0)1-8387959
Telefax: (0)1-8387090
Electronic Mail: astro@dunsink.dias.ie
WWW: http://www.dunsink.dias.ie/
Founded: 1947
Activities: extragalactic astronomy * AGNs * massive stars * X-ray observing * large-scale structure * archival research
Coordinates: 006°20'16"W 53°23'14"N H86m (IAU Code 982)
City Reference Coordinates: 006°15'00"W 53°20'00"N

Dublin Institute for Advanced Studies (DIAS), School of Cosmic Physics, Astrophysics Section
5 Merrion Square
Dublin 2
Telephone: (0)1-6621333
Telefax: (0)1-6621477
Electronic Mail: <userid>@cp.dias.ie
WWW: http://www.dias.ie/dias/cosmic/astrophysics/
Founded: 1947
Staff: 8
Activities: cosmic-ray abundances * cosmic-ray acceleration * IR * star formation
Periodicals: "Reprints"
Coordinates: 006°25'00"W 53°24'00"N

City Reference Coordinates: 006°25'00"W 53°24'00"N

Dunsink Observatory
• See "Dublin Institute for Advanced Studies (DIAS), School of Cosmic Physics, Astronomy Section"

Farran Technology Ltd. (FTL)
Ballincollig
Cork
Telephone: (0)21-4872814
Telefax: (0)21-4873892
Electronic Mail: sales@farran.com
WWW: http://www.farran.com/
Founded: 1984
Staff: 19
Activities: manufacturing mm waveguide components, mm-wavelength sources, Schottky barrier mixer and varactor diodes for mm/sub-mm wavelengths, mm/sub-mm radiometer systems, and quasi-optical 300-3000 GHz components
City Reference Coordinates: 008°28'00"W 51°54'00"N

FORBAIRT
(Irish Science and Technology Agency)
Glasnevin
Dublin 9
Telephone: (0)1-8370101
    (0)1-8082000
Telefax: (0)1-8370172
    (0)1-8082020
Electronic Mail: <userid>@forbairt.ie
WWW: http://www.forbairt.ie/
Founded: 1994
Staff: 480
Activities: state agency responsible for the development, application, and promotion of science and technology in Ireland * coordinating Ireland's industrial and scientific involvment in the European Space Agency (ESA) space programmes
Periodicals: "Technology Ireland" (ISSN 0040-1676), "S+T News" (ISSN 0079-2735)
Awards: (1) research grants
City Reference Coordinates: 006°15'00"W 53°20'00"N

Irish Astronomical Society (IAS)
P.O. Box 2547
Dublin 14
Telephone: (0)1-2981268 (John O'Neill, Secretary)
    (0)1-2980181 (James O'Connor, Secretary)
Electronic Mail: jgoneill@indigo.ie
WWW: http://indigo.ie/~stepryan/ias.htm
Founded: 1937
Membership: 120
Activities: lectures * observing * exhibitions * astronomy weekends * stars parties * beginners' classes * telescope making classes * visits * sidewalk astronomy * relativity classes
Periodicals: (6) "Orbit" (circ.: 580); (1) "Sky-High"
Coordinates: 006°13'00"W 53°14'00"N H300m
City Reference Coordinates: 006°15'00"W 53°20'00"N

Irish Meteorological Service
• See now "Met Éireann"

Irish National Committee for Astronomy and Space Research
c/o Royal Irish Academy
Academy House
19 Dawson Street
Dublin 2
Telephone: (0)1-6762570
Telefax: (0)1-6762346
Electronic Mail: admin@ria.ie
WWW: http://www.ria.ie/
Founded: 1969
Membership: 17
Activities: international collaboration * promotion of astronomy and space science within Ireland
City Reference Coordinates: 006°15'00"W 53°20'00"N

Irish National Committee for Physics
c/o Royal Irish Academy
Academy House
19 Dawson Street
Dublin 2
Telephone: (0)1-6762570
Telefax: (0)1-6762346

Electronic Mail: admin@ria.ie
WWW: http://www.ria.ie/
Founded: 1963
Membership: 17
Activities: international collaboration * promotion of physics within Ireland
City Reference Coordinates: 006°15'00"W 53°20'00"N

Irish Science and Technology Agency
● See now "FORBAIRT"

Met Éireann
(Irish Meteorological Service)
Glasnevin Hill
Dublin 9
Telephone: (0)1-8064234
Telefax: (0)1-8064247
Electronic Mail: meteireann@met.ie
Founded: 1936
Membership: 100
Activities: weather forecast * climatic data * geophysical data
Periodicals: "Met Éireann Technical Note Series"
Observatories: anout 14 stations throughout the country
Coordinates: 006°18'00"W 53°23'00"N
City Reference Coordinates: 006°15'00"W 53°20'00"N

National Standards Authority of Ireland (NSAI)
Glasnevin
Dublin 9
Telephone: (0)1-8073800
Telefax: (0)1-8073838
Electronic Mail: nsai@nsai.ie
WWW: http://www.nsai.ie/
Founded: 1996
Staff: 130
Periodicals: "Standards Bulletin"
City Reference Coordinates: 006°15'00"W 53°20'00"N

National University of Ireland, Galway, Department of Physics, Astrophysics and Applied Imaging Research Group
Galway
Telephone: (0)91-524411
Telefax: (0)91-525700
Electronic Mail: <userid>@epona.physics.ucg.ie
WWW: http://www.physics.ucg.ie/airg/
Founded: 1845 (University, as "University College Galway - UCG")
City Reference Coordinates: 009°03'00"W 53°16'00"N

National University of Ireland, Maynooth, Experimental Physics Department
Maynooth, Co. Kildare
Telephone: (0)1-7083641
Telefax: (0)1-6289277
Electronic Mail: physicsec@may.ie
WWW: http://www.may.ie/academic/physics/
Founded: 1795
Staff: 12
Activities: space science * mm and radio astronomy * ground-based gamma-ray astronomy
● Formerly "Saint Patrick's College"
City Reference Coordinates: 006°35'00"W 53°23'00"N

Royal Irish Academy
Academy House
19 Dawson Street
Dublin 2
Telephone: (0)1-6762570
Telefax: (0)1-6762346
Electronic Mail: admin@ria.ie
WWW: http://www.ria.ie/
Founded: 1785
Membership: 250
Activities: learned society of Ireland including IAU contact through the "Irish National Committee for Astronomy and Space Research" and IUPAP contact through the "Irish National Committee for Physics" (see separate entries)
Periodicals: "Proceedings. Sect. A (Mathematical, Astronomical and Physical Sciences)" and others
City Reference Coordinates: 006°15'00"W 53°20'00"N

Schull Planetarium
de Leyns
Colla Road
Schull, Co. Cork
Telephone: (0)28-28552
City Reference Coordinates: 009°33'00"W 51°32'00"N

Shannonside Astronomy Club (SAC)
26 Ballycannon Heights
Meelick, Co. Clare
Telephone: (0)61-453322
Telefax: (0)61-453322
WWW: http://members.xoom.com/shnastroclub/home.htm
Founded: 1986
Membership: 50
Activities: public educational programmes * annual Whirlpool Star Party * observing
Periodicals: (6) "Constellation"
Coordinates: 008°11'00"W 52°42'00"N H315m (Wood-Cock Hill)
City Reference Coordinates: 008°11'00"W 52°42'00"N

Trinity College Dublin (TCD), Department of Physics
College Green
Dublin 2
Telephone: (0)1-6081675
Telefax: (0)1-6711759
Electronic Mail: physics@tcd.ie
WWW: http://www2.tcd.ie/Physics/
Founded: 1592
Staff: 31
Activities: optoelectronics * condensed-matter physics * magnetism * computational physics (including astrophysics)
City Reference Coordinates: 006°15'00"W 53°20'00"N

Tullamore Astronomical Society (TAS)
145 Arden Vale
Tullamore, Co. Offaly
Telephone: (0)506-41983
Electronic Mail: seanmck@iol.ie
WWW: http://www.iol.ie/~seanmck/tas.htm
Founded: 1986
Membership: 45
Activities: lectures * education * observing * telescope and observatory building * star parties * Irish Astrofest * visits * photography * radiotelescope
Periodicals: (4) "Réalta"
Coordinates: 007°31'00"E 53°17'00"N H100m
City Reference Coordinates: 007°31'00"E 53°17'00"N

University College Dublin (UCD), Experimental Physics Department
Stillorgan Road
Belfield
Dublin 4
Telephone: (0)1-693244
Telefax: (0)1-837275
        (0)1-694409
Electronic Mail: <userid>@irlearn.bitnet
WWW: http://www.ucd.ie/~physics/main.html
Staff: 4
Activities: gamma-ray and optical astronomy
City Reference Coordinates: 006°15'00"W 53°20'00"N

Israel

Asher Space Research Institute
• See "Israel Institute of Technology, Asher Space Research Institute"

Florence and George Wise Observatory
• See "Tel Aviv University, Florence and George Wise Observatory"

Ben Gurion University (BGU), Department of Physics
P.O. Box 653
Beer-Sheva 84105
Telephone: (0)7-6461111
WWW: http://www.bgu.ac.il/phys/physics.html
City Reference Coordinates: 034°47'00"E 31°15'00"N

Hebrew University, Racah Institute of Physics
Jerusalem 91904
Telephone: (0)2-6584550
Telefax: (0)2-6584437
Electronic Mail: <userid>@vms.huji.ac.il
       <userid>@astro.huji.ac.il
WWW: http://www.fiz.huji.ac.il/
Founded: 1945
Staff: 40
Activities: cosmology * galaxies and large-scale structure * high-energy and relativistic astrophysics * relativity and gravitation * stellar structure and evolution
City Reference Coordinates: 035°13'00"E 31°46'00"N

Israel Academy of Sciences and Humanities
Albert Einstein Square
43 Jabotinsky Street
P.O. Box 4040
Jerusalem 91040
Telephone: (0)2-5676222
Telefax: (0)2-5666059
Electronic Mail: <userid>@academy.ac.il
WWW: http://www.academy.ac.il/
Founded: 1959
Membership: 70
Activities: cultivating and promoting scholarly and scientific endeavour * advising government * maintaining contact with similar bodies abroad * publishing
Periodicals: "Proceedings of the Israel Academy of Sciences and Humanities"
City Reference Coordinates: 035°13'00"E 31°46'00"N

Israel Institute of Technology, Asher Space Research Institute (ASRI)
Technion City
Haifa 32000
Telephone: (0)4-8293020
Telefax: (0)4-8230956
Electronic Mail: <userid>@tx.technion.ac.il
WWW: http://www.technion.ac.il/ASRI/
       http://www.technion.ac.il/shell/Research/Space-Institute.html
Founded: 1983
Staff: 24
Activities: space technology * space research
City Reference Coordinates: 035°00'00"E 32°49'00"N

Israel Meteorological Service
P.O. Box 25
Bet Dagan 50250
Telephone: (0)3-9682121
       (0)3-9682116
Electronic Mail: meteo@serv.gov.il
City Reference Coordinates: 034°40'00"E 32°00'00"N

Israel Physical Society (IPS)
P.O. BOX 4040
Jerusalem
or :
c/o Department of Physics
Ben-Gurion University
Beer-Sheva 84105

Telephone: (0)8-6461567
(0)8-6461764
Telefax: (0)8-6472904
(0)8-6472864
WWW: http://ory.ph.biu.ac.il/IPS/
Founded: 1954
Membership: 250
City Reference Coordinates: 035°13'00"E 31°46'00"N (Jerusalem)

Lasky Planetarium
c/o Eretz Israel Museum
P.O. Box 17068
Tel-Aviv 61170
or :
2 Haim Lebanon
Ramat-Aviv
Tel-Aviv
Telephone: (0)3-6415244
Telefax: (0)3-6412408
Founded: 1967
Staff: 5
Activities: astronomy and space educational multimedia presentations
City Reference Coordinates: 034°45'00"E 32°07'00"N

Racah Institute of Physics
• See "Hebrew University, Racah Institute of Physics"

Raymond and Beverly Sackler Institute of Astronomy
• See "Tel Aviv University, Department of Physics and Astronomy"

Standards Institution of Israel (SII)
42 Chaim Levanon Street
Tel-Aviv 69977
Telephone: (0)3-6465114
Telefax: (0)3-6465205
WWW: http://www.iso.co.il/sii/
Founded: 1945 (as "Israeli Institute of Standards")
City Reference Coordinates: 034°45'00"E 32°07'00"N

Tel Aviv University, Department of Geophysics and Planetary Science
Ramat-Aviv
Tel-Aviv 69978
Telephone: (0)3-6408633
Telefax: (0)3-6409282
WWW: http://www.tau.ac.il/geophysics/
Founded: 1968
Staff: 20
Activities: research
City Reference Coordinates: 034°45'00"E 32°07'00"N

Tel Aviv University, School of Physics and Astronomy
Ramat-Aviv
Tel-Aviv 69978
Telephone: (0)3-6408208
Telefax: (0)3-6408179
Electronic Mail: <userid>@wise.tau.ac.il
WWW: http://wise-obs.tau.ac.il/institute.html (Sackler Institute)
Founded: 1965
City Reference Coordinates: 034°45'00"E 32°07'00"N

Tel Aviv University, Florence and George Wise Observatory
Ramat-Aviv
Tel-Aviv 69978
or :
P.O. Box 90
Mitzpe Ramon 80650
Telephone: (0)3-6409279
(0)3-6408729
(0)7-6588133 (Mitzpe Ramon)
Telefax: (0)3-6408179
Electronic Mail: <userid>@wise.tau.ac.il
WWW: http://wise-obs.tau.ac.il/
Founded: 1971
Staff: 12
Activities: stars * galaxies * solar system * direct imaging * spectroscopy * photometry

Coordinates: 034°45'48"E 30°35'45"N H900m (IAU Code 097)
City Reference Coordinates: 034°45'00"E 32°07'00"N (Tel-Aviv)

Wise Observatory (Florence and George_)
• See "Tel Aviv University, Florence and George Wise Observatory"

(0)6-5662688
Electronic Mail: astro.altair@usa.net
WWW: http://altair.freeweb.org
City Reference Coordinates: 012°17'00"E 41°43'00"N

Associazione Astrofili Aurunca
c/o Scuola Media F. De Sanctis
Via G. Bruno 1
I-81037 Sessa Aurunca
Telephone: (0)823-938627
Telefax: (0)823-937117
Electronic Mail: pago@sessa.peoples.it (Pascuale Ago, President)
WWW: http://utenti.tripod.it/astrofili/
Founded: 1997
Membership: 30
Activities: research * popularization
Periodicals: "Altair"
Awards: (1) "Giovanni Bruno"
City Reference Coordinates: 013°56'00"E 41°14'00"N

Associazione Astrofili Bolognesi (AAB)
Casella Postale 313
I-40100 Bologna
or :
Via Polese 13
I-40122 Bologna
Telephone: (0)51-306583
Telefax: (0)51-750360
Electronic Mail: astrofil@iperbole.bologna.it
WWW: http://www.bo.astro.it/aab/aabhome.html
Founded: 1967
Membership: 100
Activities: education * observing * popularization
Periodicals: (4) "Giornale dell'AAB" (ISSN 0392-3932)
Coordinates: 011°09'13"E 44°21'28"N H651m (Felsina)
City Reference Coordinates: 011°20'00"E 44°29'00"N

Associazione Astrofili del Basso Vicentino (AABV) "Edmund Halley"
c/o Municipio di Sossano
Piazza Mazzini 1
I-36040 Sossano
WWW: http://digilander.libero.it/aabv/
Founded: 1992
City Reference Coordinates: 011°53'00"E 45°25'00"N (Asiago)

Associazione Astrofili del Gruppo Telecomitalia
Via Sebino 11
I-00199 Roma
Telephone: (0)6-8414338
Telefax: (0)6-8414338
WWW: http://utenti.lycos.it/astris/Astris.html
Founded: 1990
Membership: 80
Activities: popularization
Coordinates: 012°30'19"E 41°55'21"N H60m
City Reference Coordinates: 012°29'00"E 41°54'00"N

Associazione Astrofili di Piombino
c/o Meucci Stefano (President)
Via Cellini 18
I-57025 Piombino
Telephone: (0)565-220142
WWW: http://www.arcetri.astro.it/CDAs/homepages/home21.html
City Reference Coordinates: 010°32'00"E 42°56'00"N

Associazione Astrofili Fiorentini (AAF)
c/o Walter Benedetti
Piazzetta Valdambra 9
I-50127 Firenze
Telephone: (0)55-410264
Telefax: (0)55-604854
Electronic Mail: aaf@geocities.com
aaf1@dada.it
sanpolo@dada.it
WWW: http://www.geocities.com/~aaf

http://www.geocities.com/CapeCanaveral/1229/
Founded: 1958
Membership: 35
Activities: photometry * astrometry
Coordinates: 011°10'26"E 43°43'41"N H140m (San Polo a Mosciano – IAU Code 632)
City Reference Coordinates: 011°15'00"E 43°46'00"N

Associazione Astrofili "Galileo Galilei" (AAGG)
Via Vecchia Fiorentina 424
I-56015 Riglione
Telephone: (0)50-542514
WWW: http://www.arcetri.astro.it/CDAs/homepages/G_galilei/
Founded: 1982
Membership: 100
City Reference Coordinates: 010°20'00"E 43°43'00"N (Pisa)

Associazione Astrofili Garfagnana (AAG)
Via della Centrale 4
I-55032 Castelnuovo di Garfagnana
WWW: http://www.arcetri.astro.it/CDAs/homepages/Aag/home27.html
City Reference Coordinates: 010°24'00"E 44°06'00"N

Associazione Astrofili "Geminiano Montanari"
Via Concordia 200
I-41032 Cavezzo
Telephone: (0)535-58755
Telefax: (0)535-58755
Electronic Mail: gmengoli@arcanet.it (Giorgio Mengoli)
oss_astr_cav@arcanet.it
WWW: http://www.arcanet.it/oss_astronomico/
Founded: 1972
Activities: observing
Coordinates: 011°00'20"E 44°51'50"N H18m (IAU Code 107)
City Reference Coordinates: 011°02'00"E 44°50'00"N

Associazione Astrofili Imolesi (AAI)
Via Emilia 147
I-40026 Imola
or :
Via Comezzano 21
I-40026 Imola
Telephone: (0)542-684335
Electronic Mail: imoastro@freemail.it
WWW: http://www.angelfire.com/ct/imoastro
Founded: 1983
Membership: 100
Activities: popularization * photography (chemical and digital)
Periodicals: (1) "Almanacco Astronomico"
Coordinates: 011°38'15"E 44°20'16"N H250m (Osservatorio "A. Betti")
City Reference Coordinates: 011°43'00"E 44°22'00"N

Associazione Astrofili "Ionico Etnei"
c/o G. Catanzaro
Corso V. Bellini 84
I-95013 Fiumefreddo di Sicilia
Electronic Mail: gca@sunct.ct.astro.it
WWW: http://www.mclink.it/mclink/astro/ass/aaie/
Founded: 1989
Membership: 30
City Reference Coordinates: 015°13'00"E 37°47'00"N

Associazione Astrofili Mantovani (AAM)
c/o Luciano Luppi
Via Dugoni 24
I-46027 San Benedetto Po
Telephone: (0)376-615156
Electronic Mail: reald@tin.it
assoc.astrofili@polirone.mn.it
WWW: http://www.freeweb.org/associazioni/aam/home.htm
Founded: 1992
Activities: popularization
City Reference Coordinates: 010°55'00"E 45°03'00"N

Associazione Astrofili Monti della Tolfa (AAMT)
Via degli Orti 30

I-00053 Civitavecchia
Telephone: (0)766-542936
Electronic Mail: bocci@roma2.infn.it (Valerio Bocci)
   giuseppe@etruria.net (Giuseppe Fusco)
WWW: http://www.mclink.it/mclink/astro/ass/aamt/
   http://astrolink.mclink.it/mclink/ass/aamt/
   http://www.mclink.it/mclink/astro/ass/ass0004.htm
Founded: 1985
Membership: 30
Activities: observing * photography * education
Coordinates: 011°47'00"E 42°06'00"N H10m (Osservatorio San Pio X)
City Reference Coordinates: 011°48'00"E 42°06'00"N

Associazione Astrofili Pegaso
c/o Antonio Citati
Via San Antonio 2
I-28021 Borgomanero
Telephone: (0)322-836224
Telefax: (0)322-860717
Electronic Mail: info@astropegaso.org
WWW: http://www.astropegaso.org/
City Reference Coordinates: 008°27'00"E 45°42'00"N

Associazione Astrofili Sardi (AAS)
c/o Marco Massa
Vico IV San Giacomo 1
I-09033 Decimomannu
or :
Stazione Astronomica
Loc. Poggio dei Pini
Strada 54
I-09012 Capoterra
Telephone: (0)70-655454
Electronic Mail: utzeri@mbox.vol.it (Andrea Utzeri)
WWW: http://www.ca.astro.it/astrofili/
Founded: 1977
Membership: 50
Activities: meetings * lectures * education * star parties * photography * CCD observing (deep sky, planets, minor planets)
City Reference Coordinates: 008°58'00"E 39°19'00"N (Decimomannu)
   008°58'00"E 39°11'00"N (Capoterra)

Associazione Astrofili Segusini (AAS)
c/o Andrea Ainardi (President)
Corso Couvert 5
I-10059 Susa
or :
Corso Trieste 15
I-10059 Susa
Telephone: (0)122-622766
Telefax: (0)122-32060
Electronic Mail: ainardi@tin.it
WWW: http://www.mclink.it/mclink/astro/ass/grange/
   http://astrolink.mclink.it/ass/grange/
Founded: 1973
Membership: 45
Activities: lectures * observing * astrometry * photography
Periodicals: "Circolare Interna"
Observatories: 1 (Osservatorio Astronomico Grange - see separate entry)
City Reference Coordinates: 007°03'00"E 45°08'00"N

Associazione Astrofili Spezzini (AAS)
Casella Postale 11
I-19100 La Spezia
or :
c/o Istituto Tecnico Statale Nautico "N. Sauro"
Viale Italia 88
I-19100 La Spezia
Telephone: (0)187-502240
Electronic Mail: mc7316@mclink.it
   giscarfi@tin.it (Giulio Scarfi)
WWW: http://www.astrofilispezzini.org/
Founded: 1978
Membership: 35
Activities: popularization * minor-planet astrometry * SN CCD survey * digital image processing * archaeoastronomy * sundials
Periodicals: (3) "Astronomica"; "Comunicando"

Coordinates: 009°47'00"E 44°08'26"N H350m (Monte Vissegi – IAU Code 126)
City Reference Coordinates: 009°50'00"E 44°07'00"N

Associazione Astrofili Teatini (AAT)
Via Crociferi 33
I-66100 Chieti
Telephone: (0)871-348921
Founded: 1974
Membership: 20
Activities: observing (planets) * education * public lectures * exhibitions
Coordinates: 014°10'00"E 42°21'02"N H330m
City Reference Coordinates: 014°10'00"E 42°21'00"N

Associazione Astrofili Tethys (AAT)
Via Indipendenza 14
I-27055 Rivanazzano
Telephone: (0)383-944160
Telefax: (0)131-887268
Electronic Mail: giorgia.c@aznet.it
WWW: http://www.aat.idp.it/
Founded: 1987
Membership: 90
Activities: telescope making * computational astronomy * education
Periodicals: (4) "Albireo"; (1) "Almanacco Astronomico"
City Reference Coordinates: 009°01'00"E 44°56'00"N

Associazione Astrofili Trentini (AAT)
c/o Museo Tridentino di Scienze Naturali
Via Calepina 14
I-38100 Trento
Telephone: (0)461-270311
Telefax: (0)461-233820
Electronic Mail: aat@mtsn.tn.it
WWW: http://www.mtsn.tn.it/astrofili/
Founded: 1976
Membership: 500
Activities: popularization
Periodicals: (4) "Notiziario AAT"
City Reference Coordinates: 011°08'00"E 46°04'00"N

Associazione Astrofili Trevigiani
Borgo Covour 40
I-31100 Treviso
Telephone: (0)422-411725
Founded: 1974
Membership: 70
Activities: popularization * education * lectures * variable stars * meteorology * solar activity * planetarium (Collegio Pio X) * observing
Periodicals: (1) "Lezioni di Astronomia"
Coordinates: 012°13'00"E 45°40'00"N H25m
City Reference Coordinates: 012°13'00"E 45°40'00"N

Associazione Astrofili Valdinievole (AAV)
Casella Postale 156
I-51015 Monsummano Terme
or :
Via di Gragnano 349
I-51015 Monsummano Terme
Telephone: (0)572-51741
Telefax: (0)572-81192
Electronic Mail: aav@aavpieri.org
WWW: http://www.aavpieri.org/
Founded: 1979
Membership: 40
Activities: photography * meetings * observing * lectures * software
Periodicals: (1) "Appunti di Astronomia"
Coordinates: 010°48'35"E 43°52'21"N H20m
City Reference Coordinates: 010°48'00"E 43°52'00"N

Associazione Astrofili Valtellinesi (AAV)
Casella Postale 52
I-23100 Sondrio
or :
Biblioteca Civica
I-23026 Ponte in Valtellina

Telephone: (0)342-219111
(0)347-2461569
WWW: http://astrovalt.freeweb.supereva.it/
Founded: 1996
Membership: 47
Activities: observing * lectures * meetings
Coordinates: 009°57'00"E 46°10'00"N H485m
City Reference Coordinates: 009°52'00"E 46°11'00"N (Sondrio)

Associazione Astrofili Veneziani (AAV)
Casella Postale 433
I-30100 Venezia
or :
Convento San Nicolò
I-30126 Venezia Lido
Telephone: (0)41-770745 (Secretary)
WWW: http://www.mclink.it/mclink/astro/ass/venezia/
http://astrolink.mclink.it/ass/venezia/
Founded: 1976
Activities: observing * popularization * planetarium (Venezia Lido)
Periodicals: (12) "Circolare Informativa"
Coordinates: 012°38'00"E 45°42'00"N H5m (Lido)
City Reference Coordinates: 012°38'00"E 45°42'00"N

Associazione Astronomica Cassino (AAC)
c/o Gianni Fardelli
Via Cavatelle 6
I-03043 Cassino
Telephone: (0)776-337761
Electronic Mail: giafar@officine.it
WWW: http://www.officine.it/citylife/astro/astronom.htm
http://bertario.officine.it/citylife/aac/
Founded: 1995
City Reference Coordinates: 013°49'00"E 41°30'00"N

Associazione Astronomica Cortina (AAC)
Via Pecol 95
I-32043 Cortina d'Ampezzo
or :
Casella Postale 193
I-32043 Cortina d'Ampezzo
Electronic Mail: aac@sunrise.it
WWW: http://www.sunrise.it/associazioni/aac/
Founded: 1972
Membership: 100
Periodicals: "Cortina Astronomica"
Observatories: 1 (Col Drusciè, H1780m – IAU Code 154)
City Reference Coordinates: 012°08'00"E 46°32'00"N

Associazione Astronomica Feltrina "G.J. Rheticus"
Casella Postale 2
I-32032 Feltre
Telephone: (0)439-304366 (President)
WWW: http://www.mclink.it/mclink/astro/ass/ass0009.htm
http://astrolink.mclink.it/ass/ass0009.htm
Founded: 1989 (1973 as "Gruppo Astrofili Feltrini")
Membership: 90
Activities: popularization * comets * sundials
Periodicals: (4) "Rheticus"
Coordinates: 011°54'43"E 46°03'17"N H462m (Vignui)
City Reference Coordinates: 011°54'00"E 46°01'00"N

Associazione Astronomica Frusinate (AAF)
Via Fosse Ardeatine 234
I-03100 Frosinone
Telephone: (0)775-833737
Telefax: (0)775-211238
Founded: 1981
Observatories: 1 (Campo Catino - see separate entry)
City Reference Coordinates: 013°22'00"E 41°28'00"N

Associazione Astronomica Madonna di Campiglio (AMDIC)
c/o Matteo Maturi
Via Cima Tosa 24
I-38084 Madonna di Campiglio

Telephone: (0)465-441010
Electronic Mail: hotzeled@well.it
WWW: http://www.geocities.com/amdic2/
City Reference Coordinates: 010°49'00"E 46°13'00"N

Associazione Astronomica Milanese (AAM)
c/o Roberto Boccadoro
Viale Zara 118
I-20125 Milano
Telephone: (0)2-6686263
Electronic Mail: roberto_boccadoro@lotus.com
WWW: http://www.mclink.it/mclink/astro/ass/ass0003.htm
http://astrolink.mclink.it/ass/ass0003.htm
Founded: 1991
Membership: 50
City Reference Coordinates: 009°12'00"E 45°28'00"N

Associazione Astronomica Quasar
c/o Centro di Scienze Naturali (CSN)
Via di Galceti 74
I-59100 Prato
Telephone: (0)574-38960
(0)574-460503 (CSN)
Electronic Mail: quasar@comune.prato.it
WWW: http://www.comune.prato.it/csn/gen/htm/planet.htm (Planetarium)
City Reference Coordinates: 011°06'00"E 43°53'00"N

Associazione Cernuschese Astrofili (ACA)
Centro Cardinal Colombo
Piazza Matteotti 20
I-20063 Cernusco sul Naviglio
Telephone: (0)2-9231747
Electronic Mail: erusso@pointest.com (Emanuele Russo)
astral@freenet.hut.fi (Gabriele Barletta)
WWW: http://lasvegas.pointest.com/astrofili/
Founded: 1988
Membership: 40
Activities: observing * education
City Reference Coordinates: 009°19'00"E 45°31'00"N

Associazione Friulana di Astronomia e Meteorologia (AFAM)
Casella Postale 179
I-33100 Udine
or :
Via San Stefano
I-33047 Remanzacco
Telephone: (0)432-668176
Electronic Mail: sostero@elettra.trieste.it
afam@conecta.it
WWW: http://www.conecta.it/afam/dex.htm
Founded: 1969
Membership: 160
Activities: education * observing (visual, photographic, photometric) * variable stars * radioastronomy
Periodicals: (4) "L'Osservatorio" (circ.: 250)
Coordinates: 013°18'59"E 46°05'11"N H113m (Remanzacco – IAU Code 473)
City Reference Coordinates: 013°14'00"E 46°03'00"N (Udine)

Associazione IDRA
Piazzetta Arnella 4
I-80100 Napoli
or :
c/o Giuseppe Borrelli
Via Jannelli 18
I-80100 Napoli
Telephone: (0)81-5791228
Electronic Mail: mvivaldi@mbx.idn.it
WWW: http://astrolink.mclink.it/ass/idra/
City Reference Coordinates: 014°17'00"E 40°51'00"N

Associazione Ligure Astrofili Polaris
Via Galata 33/5
I-16121 Genova
Telephone: (0)10-5533045
Telefax: (0)10-5531775
Electronic Mail: bigatti@dima.unige.it (A.Bigatti)

WWW: http://www.mclink.it/mclink/astro/ass/polaris/
http://astrolink.mclink.it/ass/polaris/
http://members.xoom.com/astropolaris
Founded: 1994
Membership: 75
Activities: public lectures * observing * photography
Periodicals: "Notiziario"
City Reference Coordinates: 008°57'00"E 44°25'00"N

Associazione Ligure per lo Studio e la Divulgazione dell'Astronomia e dell'Astronautica
• See "Urania"

Associazione Marchigiana Astrofili (AMA)
c/o Massimo Morroni
Istituto Nautico A. Elia
Lungomare Vanvitelli 76
I-60100 Ancona
Telephone: (0)71-203444
Founded: 1969 (Planetarium: 1985)
Membership: 80
Activities: popularization * running a planetarium
Coordinates: 013°30'00"E 43°30'00"N H130m (Senigalliesi)
City Reference Coordinates: 013°30'00"E 43°38'00"N

Associazione Maremmana Studi Astronomici (AMSA)
Casella Postale 112
I-58100 Grosseto
Telephone: (0)564-36409
Electronic Mail: amsa@gol.grosseto.it
WWW: http://www.gol.grosseto.it/asso/amsa/
Founded: 1983
Membership: 40
Activities: public observing * photography * lectures
Periodicals: "Bollettino AMSA"
Coordinates: 011°10'15"E 42°48'44"N H100m
City Reference Coordinates: 011°08'00"E 42°46'00"N

Associazione Ogliastrina di Astronomia (AOA)
Casella Postale 87bis
I-08045 Lanusei
Founded: 1989
Membership: 60
Activities: popularization towards public and schools * Ferdinando Caliumi Observatory
Coordinates: 009°30'11"E 39°52'35"N H1,150m (Monte Armidda)
City Reference Coordinates: 009°34'00"E 39°52'00"N

Associazione per lo Studio e la Ricerca Astronomica (ASTRA)
c/o Osservatorio Astronomico "Galileo Galilei"
Via G. Carducci 172
I-73050 Salve
Telephone: (0)833-520426
(0)832-302712
WWW: http://astrolink.mclink.it/ass/astra/
http://astra.educations.net/
Founded: 1997
City Reference Coordinates: 018°11'00"E 40°23'00"N (Lecce)

Associazione Ravennate Astrofili Rheyta (ARAR)
c/o Planetario di Ravenna
Viale Santi Baldini 4A
I-48100 Ravenna
Telephone: (0)544-62534
Electronic Mail: arar@linknet.it
WWW: http://racine.ra.it/planet/
Founded: 1973
Membership: 33
Activities: popularization * planetarium * observing
Periodicals: (10) "Notiziario Rheyta"
Coordinates: 012°12'00"E 44°24'00"N
City Reference Coordinates: 012°12'00"E 44°24'00"N

Associazione Reggiana di Astronomia (ARA)
c/o Alessandro Geom. Guatteri
Via Cavagnola 1/1
I-42024 Castelnuovo di Sotto

or :
c/o Osservatorio Astronomico "Padre Angelo Secchi"
Via Prati Landi
I-42024 Castelnuovo di Sotto
Telephone: (0)522-682266
(0)522-683641
Telefax: (0)522-682235
Electronic Mail: andzmb@tin.it
ara@coopsette.it
WWW: http://www.bo.astro.it/~pigi/Astrofili_Reggiani/
Founded: 1976
Membership: 12
Activities: Sun (photospheric activity, Wolf number) * planets (visual and photographic observing) * astrometry of comets and minor planets * photography * popularization
Periodicals: (4) "Cielo" (circ.: 400)
Coordinates: 010°33'46"E 44°48'12"N H28m
City Reference Coordinates: 010°34'00"E 44°48'00"N

Associazione Romana Astrofili (ARA)
Casella Postale 4011
I-00100 Roma
Telephone: (0)6-79960075
WWW: http://www.ara-frasso-sabino.org/HOME_ARA.htm
Founded: 1982
City Reference Coordinates: 012°29'00"E 41°54'00"N

Associazione Sabina Astrofili (ASA)
Via del Vivaio 5A
I-02045 Limiti Di Greccio
Telephone: (0)746-204960
(0)746-274539
Electronic Mail: vascleri@geocities.com
WWW: http://www.geocities.com/CapeCanaveral/Hangar/9330/
Founded: 1995
Membership: 30
City Reference Coordinates: 012°51'00"E 42°24'00"N (Rieti)

Associazione Salentina Astrofili (ASA) "Edwin Hubble"
c/o Gianluca Ingrosso (Secretary)
Via Cialdini 10
I-73012 Campi Salentina
Telephone: (0)832-794666
Electronic Mail: astro.asa@cybergal.com
WWW: http://asa.astrofili.org/
Founded: 1996
Membership: 30
Activities: popularization * education * observing * photography
Coordinates: 018°01'01"E 40°23'52"N H37m
City Reference Coordinates: 018°01'00"E 40°23'00"N

Associazione Tuscolana di Astronomia (ATA)
c/o Istituto di Astrofisica Spaziale del CNR
Casella Postale 67
I-00044 Frascati
or :
Viale della Galassia 43
I-00040 Rocca Priora
Telephone: (0)6-9406339
WWW: http://www.ata.panservice.it/
Founded: 1995
Membership: 40
City Reference Coordinates: 012°29'00"E 41°54'00"N (Frascati)
012°45'00"E 41°48'00"N (Rocca Priora)

Associazione Valdostana Scienze Astronomiche (AVSA)
Regione Crou 17
I-11100 Aosta
Telephone: (0)165-33853
(0)165-555192
Electronic Mail: lravello@aostanet.com (Luciano Ravello, President)
WWW: http://www.aostanet.com/astro-radio/
Founded: 1986
Membership: 3
Activities: observing * education
City Reference Coordinates: 007°19'00"E 45°43'00"N

Associazione Vigevanese Divulgazione Astronomica (AVDA)
c/o Biblioteca Civica
Corso Cavour 82
I-27019 Vigevano
Telephone: (0)381-70149
Telefax: (0)381-70149
Electronic Mail: marcmoli@tin.it
WWW: http://www.vigevano.vol.it/cultura/avda/
http://www.geocities.com/CapeCanaveral/Campus/6640/avda5.htm
Founded: 1996
Membership: 60
Activities: education * observing
City Reference Coordinates: 008°51'00"E 45°19'00"N

Astro Club Voyager (ACV)
Via Grande 2/A
I-31030 Castello di Godego
WWW: http://digilander.iol.it/acv
City Reference Coordinates: 012°15'00"E 45°40'00"N (Treviso)

astronomia (L'_)
● See "L'astronomia"

Astronomical Observatory of Campo Catino (AOCC)
Colle Pannunzio
I-03016 Guarcino
Telephone: (0)775-833737
(0)775-435945
Telefax: (0)775-211238
Electronic Mail: oss.astronomico.campocatino@rtmol.stt.it
mario.di.sora@rtmol.stt.it (Mario Di Sora, Director)
WWW: http://www.rtmol.it/aocc/
Founded: 1987
Membership: 7
Coordinates: 013°19'47"E 41°49'16"N (IAU Code 468)
City Reference Coordinates: 013°19'00"E 41°48'00"N

Auriga Srl
Via Mario Fabio Quintiliano 30
I-20138 Milano
Telephone: (0)2-5097780
Telefax: (0)2-5097324
Electronic Mail: auriga@auriga.it
WWW: http://www.auriga.it/
Founded: 1983
Staff: 13
Activities: distributing planetariums, telescopes, accessories and optical products
City Reference Coordinates: 009°12'00"E 45°28'00"N

BeppoSAX Scientific Data Center (SDC)
c/o Nuova Telespazio
Via Corcolle 19
I-00131 Roma
Telephone: (0)6-40796307
Telefax: (0)6-40796291
Electronic Mail: helpdesk@sax.sdc.asi.it
WWW: http://www.sdc.asi.it/
Founded: 1995
Staff: 11
City Reference Coordinates: 012°29'00"E 41°54'00"N

Centro Astrofili Bolzano (CAB)
Via M.Longon 3
I-39100 Bolzano
Telephone: (0)471-273165
Electronic Mail: mfrasc@em.parsec.it
WWW: http://www.parsec.it/cab/
Founded: 1990
City Reference Coordinates: 011°22'00"E 46°30'00"N

Centro Astronomico "Neil Armstrong" (CANA)
Via Canali 17
I-84121 Salerno
Telephone: (0)81-5750414
(0)89-790803

Electronic Mail: astro@salerno.infn.it
astro@vaxsa.csied.unisa.it
WWW: http://www.gisa.it/nucleo/cana/cana.htm
http://www.xcom.it/cana/
Founded: 1982
Activities: popularization * star parties * occultations * SN search * variable stars
City Reference Coordinates: 014°47'00"E 40°41'00"N

Centro Astronomico Orione (CAO)
Via Due Giugno 31
I-00043 Ciampino
Telephone: (0)6-7912381
Electronic Mail: gubaca@tini.it (Guerrino [Rino] Bacaloni)
Founded: 1994
Membership: 11
Activities: annual meetings * observing * public meetings
Periodicals: (4) "Bollettino Astronomico"
Coordinates: 012°35'42"E 41°48'00"N H185m (Stazione Astronomica Rigel)
City Reference Coordinates: 012°36'00"E 41°48'00"N

Cimini Astronomical Observatory (CiAO)
(Osservatorio Astronomico dei Monti Cimini)
c/o Paolo Candy (Director)
Via Dei Monti Cimini 36
I-01100 Viterbo
Telephone: (0)338-1439673
(0)761-752275
Electronic Mail: simeis@hesnet.net
WWW: http://www.hesnet.net/candy/CiAO/CiAO.htm
Founded: 2000
Activities: observing * photography * lectures * education * publishing
Coordinates: 012°11'20"E 42°23'45"N H800m
City Reference Coordinates: 012°06'00"E 42°24'00"N

Circolo Astrofili Bergamaschi (CAB)
Via A. Maj 16/B
I-24121 Bergamo
Telephone: (0)35-223376
Telefax: (0)35-570641
Electronic Mail: cab@uninetcom.it
WWW: http://www.astrobg.com/home.html
Founded: 1974
Membership: 80
Activities: popularization * SN search * photography * variable stars * sundials * archaeoastronomy
Coordinates: 009°45'10"E 45°45'20"N H1300m (Selvino)
City Reference Coordinates: 009°43'00"E 45°41'00"N

Circolo Astrofili di Mestre "Guido Ruggieri"
Via Padre Egidio Gelain 7
I-30175 Marghera
Telephone: (0)41-936177
(0)41-951497
Electronic Mail: fasanta@tin.it
WWW: http://digilander.libero.it/astrofilimestre/
Founded: 1975
Membership: 30
Activities: variable stars * planets * popularization * observing
City Reference Coordinates: 012°15'00"E 45°29'00"N (Marghera)
012°14'00"E 45°30'00"N (Mestre)

Circolo Astrofili di Milano (CAM)
c/o Civico Planetario "Ulrico Hoepli"
Corso di Porta Venezia 57
I-20121 Milano
WWW: http://spazioinwind.libero.it/astrofilimilano/
Founded: 1932
City Reference Coordinates: 009°12'00"E 45°28'00"N

Circolo Astrofili Nord Sardegna
Via Mentana 34
I-07046 Porto Torres
Telephone: (0)79-510046
Founded: 1988
Membership: 10
Activities: popularization * education * photography

City Reference Coordinates: 008°24'00"E 40°51'00"N

Circolo Astrofili Talmassons
Via XXIV Maggio 10
I-33030 Talmassons
Telephone: (0)432-920670
Telefax: (0)432-920670
WWW: http://www.castfvg.it/
Founded: 1993
Membership: 100
Activities: SN search * meteors * Moon * planets * photography * CCDs * education * lectures
Periodicals: (4) "Notiziario"
Coordinates: 013°06'14"E 45°55'21"N H28m
City Reference Coordinates: 013°07'00"E 45°56'00"N

Circolo Astrofili Veronesi (CAV) "Antonio Cagnoli"
Casella Postale 2016
I-37100 Verona
or :
Centro d'Incontro Circoscrizione
2 Largo Stazione Vecchia
Parona
I-37100 Verona
Telephone: (0)45-574345 (S. Moltomoli)
Electronic Mail: grejbear@tin.it
                 peanuts@rcvr.vr.it
                 circolo.astrofili.veronesi@rcvr.org
WWW: http://www.rcvr.org/assoc/astro/main.htm
Founded: 1977
Membership: 130
Activities: astronomy * observing * photography * CCDs * popularization * astronautics
Periodicals: (3) "CAV Notiziario"
Coordinates: 010°56'53"E 45°28'47"N H70m
City Reference Coordinates: 011°00'00"E 45°27'00"N

Circolo Astronomico Dorico "Paolo Andrenelli"
Casella Postale 70
I-60100 Ancona
Telephone: (0)71-200913
           (0)71-2862087
Telefax: (0)71-2862087
Electronic Mail: stefano@ascu.unian.it (Stefano Rosoni, President)
Founded: 1990
Membership: 20
Coordinates: 013°30'00"E 43°37'00"N
City Reference Coordinates: 013°30'00"E 43°37'00"N

Circolo Casolese Astrofili "Betelgeuse"
c/o Maurizio Cabibbo
Via delle Querce 33
I-53031 Casole d'Elsa
Telephone: (0)577-948284
           (0)577-948373
WWW: http://www.arcetri.astro.it/CDAs/homepages/casole/chisiamo.html
Founded: 1995
Membership: 16
Activities: public observing * popularization * education
Coordinates: 011°02'00"E 43°20'00"N H420m
City Reference Coordinates: 011°02'00"E 43°20'00"N

Circolo Culturale Astronomico di Farra d'Isonzo (CCAF)
Strada della Colombara 11
I-34070 Farra d'Isonzo
Telephone: (0)481-888540
Electronic Mail: ccaf@ccaf.it
WWW: http://www.ccaf.it/
Founded: 1975
Activities: education * CCD astrometry of minor planets and comets
Periodicals: (1) "Lunario" (circ.: 2,000)
Coordinates: 013°31'33"E 45°54'57"N H53m (Colombara – IAU Code 595)
City Reference Coordinates: 013°31'00"E 45°55'00"N

Civico Planetario "Ulrico Hoepli"
Corso di Porta Venezia 57
I-20121 Milano

Telephone: (0)2-2895785
Telefax: (0)2-2047259
Electronic Mail: planet@imiucca.csi.unimi.it
WWW: http://www.brera.mi.astro.it/~planet/
Founded: 1929
Activities: education
City Reference Coordinates: 009°12'00"E 45°28'00"N

Comitato Italiano per il Controllo delle Affermazioni sul Paranormale (CICAP)
Casella Postale 1117
I-35100 Padova
Telephone: (0)426-22013
Telefax: (0)426-22013
Electronic Mail: info@cicap.org
WWW: http://www.cicap.org/
Founded: 1989
Membership: 19
Activities: lectures * investigations * media resource
Periodicals: (3) "Scienza & Paranormale" (circ.: 1,500)
City Reference Coordinates: 011°53'00"E 45°25'00"N

Consiglio Nazionale delle Ricerche (CNR)
Piazzale Aldo Moro 7
I-00185 Roma
Telephone: (0)6-49931
Telefax: (0)6-4461954
WWW: http://www.cnr.it/
Founded: 1923
City Reference Coordinates: 012°29'00"E 41°54'00"N

Consiglio Nazionale delle Ricerche (CNR), Istituto di Astrofisica Spaziale e Fisica Cosmica (IASFC)

Via del Fosso del Cavaliere
I-00133 Roma
Telephone: (0)6-49934472
          (0)6-49934473
Telefax: (0)6-20660188
Electronic Mail: <userid>@saturn.ias.rm.cnr.it
WWW: http://www.iasf.cnr.it/
Staff: 65
Activities: galactic and stellar evolution * high-energy astrophysics * diffuse matter in space * planetology * X-ray, gamma-ray, IR and UV astronomy * instrumentation data analysis
Periodicals: "Rapporto Interno IAS"
City Reference Coordinates: 012°41'00"E 41°41'00"N

Consiglio Nazionale delle Ricerche (CNR), Istituto di Cosmogeofisica
Corso Fiume 4
I-10133 Torino
Telephone: (0)11-6306811
Telefax: (0)11-6604056
WWW: http://www.area.to.cnr.it/IT/area/aicg.html
Founded: 1969
Staff: 12
Activities: cosmic rays * neutrinos * cosmogenesis * elementary particles * geophysics * climatology * oceanography * teledetection
City Reference Coordinates: 007°40'00"E 45°03'00"N

Consiglio Nazionale delle Ricerche (CNR), Istituto di Fisica Cosmica ed Applicazioni dell'Informatica (IFCAI)
Via Ugo La Malfa 153
I-90146 Palermo
Telephone: (0)91-6809690
          (0)91-6809577
Telefax: (0)91-6882258
Electronic Mail: <userid>@ifcai.pa.cnr.it
WWW: http://www.ifcai.pa.cnr.it/
Founded: 1981
Staff: 25
Activities: astrophysics * X-ray astronomy * gamma-ray astronomy * high-energy cosmic rays * space research * stratospheric balloons * data analysis methods
City Reference Coordinates: 013°21'00"E 38°07'00"N

Consiglio Nazionale delle Ricerche (CNR), Istituto di Fisica dello Spazio Interplanetario (IFSI)
Via del Fosso del Cavaliere
I-00133 Roma

Telephone: (0)6-49934488
(0)6-49934490
Telefax: (0)6-49934374
Electronic Mail: <userid>@hp.ifsi.fra.cnr.it
candidi@ifsi.rm.cnr.it (M. Candidi, Director)
WWW: http://www.ifsi.rm.cnr.it/
Staff: 48
City Reference Coordinates: 012°41'00"E 41°41'00"N (Frascati)
012°29'00"E 41°54'00"N (Roma)

Consiglio Nazionale delle Ricerche (CNR), Istituto di Metrologia "G. Colonnetti" (IMGC)
Strada delle Cacce 73
I-10135 Torino
Telephone: (0)11-39771
Telefax: (0)11-346761
Electronic Mail: <userid>@itoimgc.bitnet
WWW: http://www.imgc.to.cnr.it/
Founded: 1968
Staff: 105
Activities: metrology * measurements * instrument testing * thermodynamics * optics * optoelectronics * standards * fundamental constants * gravimetry
Periodicals: (1/2) "Report"; "Notizie di Metrologia"
City Reference Coordinates: 007°40'00"E 45°03'00"N

Consiglio Nazionale delle Ricerche (CNR), Istituto di Radioastronomia (IRA), Bologna
Via Gobetti 101
I-40129 Bologna
Telephone: (0)51-6399385
Telefax: (0)51-6399431
Electronic Mail: <userid>@ira.bo.cnr.it
WWW: http://www.ira.cnr.it/
Founded: 1970
Staff: 46
Activities: radioastronomy * cosmology * VLBI * instrumentation
City Reference Coordinates: 011°20'00"E 44°29'00"N

Consiglio Nazionale delle Ricerche (CNR), Istituto di Radioastronomia (IRA), Stazione di Medicina

Via Fiorentina Aia Cavicchio
I-40060 Villafontana
Telephone: (0)51-6965811
Telefax: (0)51-6965810
WWW: http://medvlbi.bo.cnr.it/
http://www.ira.bo.cnr.it/
Founded: 1983
Staff: 20
Activities: radioastronomical observing
Coordinates: 011°38'49"E 44°31'15"N H25m
City Reference Coordinates: 011°38'00"E 44°28'00"N

Consiglio Nazionale delle Ricerche (CNR), Istituto di Radioastronomia (IRA), Stazione di Noto
Casella Postale 141
I-96017 Noto
or :
Contrada Renna Bassa
Località Casa di Mezzo
I-96017 Noto
Telephone: (0)931-824111
Telefax: (0)931-824122
WWW: http://www.ira.noto.cnr.it/
http://www.ira.bo.cnr.it/
Founded: 1988
Activities: radioastronomical observing
Coordinates: 014°59'21"E 36°52'34"N H30m
City Reference Coordinates: 015°05'00"E 36°53'00"N

Consiglio Nazionale delle Ricerche (CNR), Istituto di Radioastronomia (IRA), Firenze
Largo Enrico Fermi 5
I-50125 Firenze
Telephone: (0)55-2752248
Telefax: (0)55-220039
Electronic Mail: <userid>@arcetri.astro.it
tirgo@arcetri.astro.it
WWW: http://www.arcetri.astro.it/irlab/tirgo/
Founded: 1981
Staff: 10

Activities: developing IR instrumentation * managing the TIRGO telescope located at the Gornergrat station in Switzerland * IR astronomical research
Coordinates: 007°47'04"E 45°59'04"N H3,130m (Gornergrat Station)
City Reference Coordinates: 011°15'00"E 43°46'00"N

Consiglio Nazionale delle Ricerche (CNR), Istituto di Tecnologie e Studio delle Radiazioni Extraterrestri (TESRE)
Via Gobetti 101
I-40129 Bologna
Telephone: (0)51-6398681
         (0)51-6398682
Telefax: (0)51-6398724
Electronic Mail: direzione@tesre.bo.cnr.it
                <userid>@tesre.bo.cnr.it
WWW: http://www.tesre.bo.cnr.it/
     http://www.cnr.it/
Founded: 1969
Staff: 37
Activities: X-ray, gamma-ray, IR astronomy * cosmic-ray physics * balloon and space technologies
City Reference Coordinates: 011°20'00"E 44°29'00"N

Consorzio Internazionale per l'Astrofisica Relativistica
• See "International Center for Relativistic Astrophysics (ICRA)"

D'Appolonia SpA
Via San Nazaro 19
I-16145 Genova
Telephone: (0)10-3628148
Telefax: (0)10-3621078
Electronic Mail: dappolonia@pn.itnet.it
WWW: http://www.dappolonia.it/
Founded: 1981
Activities: consultancy * technology transfer
City Reference Coordinates: 008°57'00"E 44°25'00"N

Denitron SnC
Via Roma 11
I-21027 Ispra
Telephone: (0)332-782398
Telefax: (0)332-782398
WWW: http://www.denitron.it/
Founded: 1988
Staff: 2
Activities: designing electronic instruments
City Reference Coordinates: 008°36'00"E 45°49'00"N

Edizioni Scientifiche Coelum
Via Appia 18
I-30173 Mestre
Telephone: (0)41-5321476
Telefax: (0)41-5327427
Electronic Mail: coelum@tin.it
                segretaria@coelum.com
WWW: http://www.coelum.com/
Activities: publishing monthly magazine, CD-ROMs, calendars, posters, books and manuals
Periodicals: (12) "Coelum Astronomia" (ISSN 1594-1299)
City Reference Coordinates: 012°14'00"E 45°30'00"N

Ente Nazionale Italiano di Unificazione (UNI)
Via Battistotti Sassi 11
I-20123 Milano
Telephone: (0)2-700241
Telefax: (0)2-70106106
Electronic Mail: uni@uni.com
WWW: http://www.uni.com/
City Reference Coordinates: 009°12'00"E 45°28'00"N

ESRIN
• See "European Space Agency (ESA), ESRIN"

Ettore Majorana Centre for Scientific Culture (EMCSC)
• See "Centro di Cultura Scientifica Ettore Majorana (CCSEM)"

European Association for Research Managers and Administrators (EARMA)
c/o INFM

Activities: popularization * education * research * Planetario "Galileo Galilei"
Periodicals: (1) "Annuario"
Coordinates: 011°53'48"E 45°24'00"N (Osservatorio G. Colombo)
City Reference Coordinates: 011°53'00"E 45°25'00"N

Gruppo Astrofili di Palermo
Casella Postale 1210
I-90100 Palermo
Telephone: (0)91-670231
WWW: http://astrolink.mclink.it/ass/gap/
Founded: 1993
City Reference Coordinates: 013°21'00"E 38°07'00"N

Gruppo Astrofili di Schio (GAS)
Via Tiziano Vecellio
Casella Postale 115
I-36015 Schio
WWW: http://dns.lead.it/AstroSchio/
http://www.lead.it/AstroSchio/
Founded: 1976
Membership: 50
Activities: observing * photography * popularization
Coordinates: 011°18'41"E 45°46'05"N H1,506m (Monte Novegno)
City Reference Coordinates: 011°21'00"E 45°43'00"N

Gruppo Astrofili di Rozzano (GAR)
c/o Biblioteca Civica
Via Togliatti
I-20089 Rozzano
Electronic Mail: info@astrofilirozzano.it
WWW: http://www.astrofilirozzano.it/
Founded: 1982
Membership: 40
Activities: deep-sky observing * lectures * education
City Reference Coordinates: 009°12'00"E 45°28'00"N (Milano)

Gruppo Astrofili Frentani
c/o studio Mancinone
Via Parma 2
I-66034 Lanciano
Electronic Mail: gruppoastrofilifrentani@yahoo.com
WWW: http://members.xoom.com/gaf97/
Founded: 1997
City Reference Coordinates: 014°23'00"E 42°13'00"N

Gruppo Astrofili Genovesi (GAG)
Casella Postale 836
I-16100 Genova
or :
Via A. Doria 9
I-16127 Genova
Telephone: (0)10-3460065
Electronic Mail: astro@gsi.it
WWW: http://www.gsi.it/astronomia/gag/
Founded: 1997
Membership: 36
Activities: meetings * astrophotography
Periodicals: "Superba"
City Reference Coordinates: 008°57'00"E 44°25'00"N

Gruppo Astrofili "Giovanni e Angelo Bernasconi"
Via San Giuseppe 34
I-21047 Saronno
Telephone: (0)331-830704
Telefax: (0)331-830704
WWW: http://www.hal.varese.it/~bernasconi/
Founded: 1965
Membership: 47
Activities: comets * planets * variable stars * instrumentation * lectures * scientific expeditions * photography * education
Periodicals: "Nihil Sub Astris Novum"
City Reference Coordinates: 009°02'00"E 45°38'00"N

Gruppo Astrofili Hipparcos
c/o CCCDS
Via Nomentana 175

I-00161 Roma
Telephone: (0)6-44250561
Electronic Mail: hipparcos.cds@mclink.it
WWW: http://diamante.uniroma3.it/hipparcos/index.htm
City Reference Coordinates: 012°29'00"E 41°54'00"N

Gruppo Astrofili "Isaac Newton"
Via Francesca Sud 329
I-56020 Santa Maria a Monte
Telephone: (0)587-706694 (Mauro Bachini)
(0)587-706813 (Marco Novi)
Electronic Mail: marco.novi@ipermedia.net
WWW: http://www.arcetri.astro.it/CDAs/homepages/home3.html
City Reference Coordinates: 010°20'00"E 43°43'00"N (Pisa)

Gruppo Astrofili "La Nuova Selene"
c/o Davide Salerno (President)
Via De Deo 21
I-72015 Fasano
Telephone: (0)330-701580
Telefax: (0)80-4392336
Electronic Mail: lanuovaselene@puglianet.it
WWW: http://www.puglianet.it/lanuovaselene/
Founded: 1997
Membership: 15
Activities: education
Coordinates: 017°19'00"E 40°45'00"N H450m
City Reference Coordinates: 017°21'00"E 40°50'00"N

Gruppo Astrofili Lariani (GAL)
Via Risorgimento 21
I-22038 Tavernerio
or :
c/o Circoscrizione 6
Via Grandi 21
I-22100 Como
Telephone: (0)31-4290487
(0)31-272196
Electronic Mail: sminardi@ing.unico.it
WWW: http://space.tin.it/associazioni/ytynesv/
Founded: 1974
Membership: 100
Activities: popularization * observing
Sections: Variable Stars * Planets
Periodicals: (4) "L'Astrofilo Lariano" (circ.: 300-500)
Observatories: 1 (Ramponio Verna H1,000m)
City Reference Coordinates: 009°05'00"E 45°47'00"N (como)

Gruppo Astrofili Manfredonia (GAM)
c/o Antonio Rubino
Lungomare del Sole 30
I-71043 Manfredonia
Telephone: (0)884-542883
Founded: 1975
Membership: 25
Activities: observing * education * introduction to astronomy
Periodicals: (12) "Bollettino Astronomico"
City Reference Coordinates: 015°55'00"E 41°38'00"N

Gruppo Astrofili Massesi (GAM)
Via Godola 42
I-54100 Massa
Telephone: (0)585-790594
Telefax: (0)585-790594
WWW: http://www.zia.ms.it/gam/
City Reference Coordinates: 010°09'00"E 44°02'00"N

Gruppo Astrofili Menkalinan (GAM)
Casa delle Culture
I-87100 Cosenza
Electronic Mail: polaris@diemme.it
orbiter@tin.it
WWW: http://www.geocities.com/Area51/Dimension/5189/gam.htm
Founded: 1997
City Reference Coordinates: 016°16'00"E 39°17'00"N

Gruppo Astrofili "N. Copernico"
Via Pulzona 1708
Santa Maria del Monte
I-47835 Saludecio
Telephone: (0)541-857026
Telefax: (0)541-21082
Electronic Mail: copernic@iper.net
WWW: http://www.iper.net/koppernick/
Founded: 1976
Membership: 12
Activities: education * observing
Periodicals: (2) "Notiziario di Astronomia"
Coordinates: 012°24'18"E 43°53'08"N H150m
City Reference Coordinates: 012°40'00"E 43°52'00"N

Gruppo Astrofili ORSA
• See "Organizzazione Ricerche e Studi di Astronomia (ORSA)"

Gruppo Astrofili Pavese (GAP)
c/o Davide Re
Via Torino 23
I-27100 Pavia
Telephone: (0)382-463030
Electronic Mail: calimero@venus.it (Michele Moroni)
WWW: http://www.asanet.it/ospiti/gap/
Founded: 1997
City Reference Coordinates: 009°10'00"E 45°10'00"N

Gruppo Astrofili Persicetani
c/o Romano Serra
Vicolo Baciadonne 1
I-40017 San Giovanni in Persiceto
Telephone: (0)51-827067
WWW: http://members.it.tripod.de/san_giovanni/San_giovanni/osservatorio.htm
    http://members.it.tripod.de/san_giovanni/San_giovanni/Planetario.htm
Membership: 25
City Reference Coordinates: 011°11'00"E 44°38'00"N

Gruppo Astrofili Pesarese (GAP)
c/o Maurizio Mucci (President)
Via Ferrari 7
I-61100 Pesaro
Telephone: (0)721-51935
Electronic Mail: maurmucc@tin.it
        acono@abanet.it (Fabio Arcidiacono)
WWW: http://www.comune.pesaro.ps.it/allegati/astrofili/cielo.htm
Founded: 1985
Membership: 20
City Reference Coordinates: 012°55'00"E 43°54'00"N

Gruppo Astrofili Piceni (GAP)
c/o Maurizio Morricone
Viale De Gasperi 101
I-63039 San Benedetto del Tronto
Telephone: (0)73583416
Electronic Mail: gap@insinet.it
WWW: http://www.insinet.it/gap/
City Reference Coordinates: 013°53'00"E 42°57'00"N

Gruppo Astrofili Reggini M31 del Dopolavoro Ferroviario
c/o Gaetano De Benedetto (President)
Via Ciccarello Traversa IV 14
Casella Postale 148
I-89100 Reggio di Calabria
Telephone: (0)965-622239
Telefax: (0)965-622239
Founded: 1976
Membership: 30
Activities: popularization * observing (occultations, minor planets, variable stars)
Coordinates: 015°19'05"E 38°06'25"N H93m (Osservatorio A. Righi)
City Reference Coordinates: 015°39'00"E 38°07'00"N

Gruppo Astrofili Rigel
c/o Zuffi Valerio
Via Adua 31

I-20010 Arluno
WWW: http://www.geocities.com/CapeCanaveral/Galaxy/3204/index.htm
City Reference Coordinates: 009°12'00"E 45°28'00"N (Milano)

Gruppo Astrofili Romani (GAR)
c/o Parrocchia San Filippo Neri
Via Martino V 28
I-00167 Roma
Telephone: (0)6-39730250 (President)
Electronic Mail: gianluca.rossi@agippetroli.emi.it (Gianluca Rossi, Presidente)
    memory.line@agora.stm.it (GAR News)
WWW: http://astrolink.mclink.it/ass/gar/
    http://www.freeweb.org/associazioni/GAR/
Founded: 1996
Membership: 26
Activities: education * observing * popularization * exhibitions * photography (deep sky, planets, Moon)
Periodicals: "GAR News"
City Reference Coordinates: 012°29'00"E 41°54'00"N

Gruppo Astrofili Savonesi (GAS)
Casella Postale 5
I-17100 Savona
Telephone: (0)19-822925 (Fabrizio Ciliberto, Secretary)
    (0)19-853110 (Roberto Bracco, President)
Electronic Mail: gas@ils.org
WWW: http://www.publinet.it/arte/gas/
Founded: 1969
Membership: 50
Activities: observing * popularization * education
Periodicals: (2) "Cielosservare" (circ.: 100)
Coordinates: 008°28'47"E 44°23'44"N H395m
City Reference Coordinates: 008°30'00"E 44°17'00"N

Gruppo Astrofili Soresinesi (GAS)
c/o Angelo Marchesini
Osservatorio Astronomico Pubblico
Via Matteotti 2
Casella Postale 21
I-26015 Soresina
Telephone: (0)374-43722 (Observatory)
    (0)374-70186 (President)
Telefax: (0)373-30901 (call for connection)
Founded: 1974
Membership: 100
Activities: popularization * image processing * software
City Reference Coordinates: 009°51'00"E 45°17'00"N

Gruppo Astrofili Vesuviano (GAV)
Via Mariotti 32
I-80047 San Giuseppe Vesuviano
Telephone: (0)81-5293941
Telefax: (0)81-5296195
Electronic Mail: astrofilivesuviani@yahoo.it
WWW: http://digilander.iol.it/gav/
Founded: 1998
Membership: 20
Activities: popularization * education * observing
City Reference Coordinates: 014°31'00"E 40°50'00"N

Gruppo Astrofili Vicentini (GAV) "Giorgio Abetti"
Centro Civico
Villa Lattes
Via Thaon di Revel
I-36100 Vicenza
Electronic Mail: gav@keycomm.it
WWW: http://www.keycomm.it/~gav/
Founded: 1986
Membership: 180
Activities: popularization
Coordinates: 011°32'09"E 45°29'50"N H190m
City Reference Coordinates: 011°33'00"E 45°33'00"N

Gruppo Astrofili "William Herschel" (GAWH)
Piazza del Monastero 6
I-10146 Torino

Telephone: (0)11-3835279
(0)11-354570
(0)11-336844
Electronic Mail: info@gawh.it
WWW: http://www.gawh.it/
Activities: meetings * observing
City Reference Coordinates: 007°40'00"E 45°03'00"N

Gruppo Astronomia Digitale (GAD)
c/o Claudio Lopresti
Via Castellazzo 8/D
I-19125 La Spezia
Telephone: (0)187-715391
Telefax: (0)187-715391
Electronic Mail: clop@libero.it
WWW: http://digilander.libero.it/clop33/gad/infogad.htm
Founded: 1992
Membership: 228
Activities: SM * image processing * digital astronomy * computing
Periodicals: "Notiziario GAD"
City Reference Coordinates: 009°50'00"E 44°07'00"N

Gruppo Astronomico Tradatese (GAT)
c/o Biblioteca Civica
Via Mameli 13
I-21049 Tradate
Telephone: (0)331-841820
Telefax: (0)331-820317
(0)331-810117
Electronic Mail: gatrad@gwtradate.tread.it (Lorenzo Comolli)
WWW: http://gwtradate.tread.it/tradate/gat/Welcome.html
Founded: 1975
Membership: 350
Activities: planets * Sun * CCD photography * exhibitions * education
Periodicals: (6) "Lettera ai Soci"
Awards: (1) "Premio Eros Benatti"
Observatories: 1 (Monte San Martino, Valcuvia)
City Reference Coordinates: 008°54'00"E 45°43'00"N

Gruppo Astronomico Viareggio (GAV)
c/o Martellini Davide
Casella Postale 406
I-55049 Viareggio
Telephone: (0)584-395895
Electronic Mail: gav.it@usa.net
WWW: http://members.tripod.it/gav/
Founded: 1973
Membership: 30
Activities: popularization * research
Periodicals: (6) "Astronews"
City Reference Coordinates: 010°14'00"E 43°52'00"N

Gruppo "G.E.D. Alcock"
c/o Marco Vincenzi
Via Accademia dei Virtuosi 4
I-00147 Roma
Telephone: (0)6-6281880
Electronic Mail: mc9648@mclink.it (Giovanni Guerrieri)
mc8255@mclink.it (Marco Vincenzi)
WWW: http://www.mclink.it/mclink/astro/ass/ass0002.htm
http://astrolink.mclink.it/ass/ass0002.htm
City Reference Coordinates: 012°29'00"E 41°54'00"N

Gruppo Marsicano Astrofili (GMA) "F. Angelitti"
c/o Paolo Maria Ruscitti
Via Opi 5
I-67051 Avezzano
Telephone: (0)863-414799
Electronic Mail: zauri@aquila.infn.it (Renato Zauri)
WWW: http://moloch.univaq.it/~zauri/astro/
Founded: 1985
City Reference Coordinates: 013°26'00"E 42°02'00"N

International Center for Relativistic Astrophysics (ICRA)
(Consorzio Internazionale per l'Astrofisica Relativistica)

c/o Dipartimento di Fisica
Università degli Studi di Roma
Piazzale Aldo Moro 5
I-00185 Roma
Telephone: (0)6-49914254 (Secretary)
 (0)6-49914304 (Director)
Telefax: (0)6-4454992
WWW: http://www.icra.it/
Founded: 1985
Activities: theoretical, experimental and observational astrophysics justifying international collaboration * managing instruments and equipment * appropriate methodologies and technologies
City Reference Coordinates: 012°29'00"E 41°54'00"N

International Centre for Theoretical Physics (ICTP) (Abdus Salam_)
• See now "Abdus Salam International Centre for Theoretical Physics"

International School for Advanced Studies (ISAS)
• See "Scuola Internazionale Superiore di Studi Avanzati (SISSA)"

International Supernovae Network (ISN)
c/o Stefano Pesci
Via Birolli 3
I-20125 Milano
Electronic Mail: peste@micronet.it
 villi@mbox.queen.it
WWW: http://www.queen.it/web4you/noprofit/isn/isn.htm
 http://www.supernovae.net/isn.htm
Founded: 1996
Membership: 110
Activities: SN search and observation
City Reference Coordinates: 009°12'00"E 45°28'00"N

Istituto di Astrofisica Spaziale e Fisica Cosmica (IASFC)
• See "Consiglio Nazionale delle Ricerche (CNR), Istituto di Astrofisica Spaziale e Fisica Cosmica (IASFC)"

Istituto di Fisica Cosmica ed Applicazioni dell'Informatica (IFCAI)
• See "Consiglio Nazionale delle Ricerche (CNR), Istituto di Fisica Cosmica ed Applicazioni dell'Informatica (IFCAI)"

Istituto di Fisica dello Spazio Interplanetario (IFSI)
• See "Consiglio Nazionale delle Ricerche (CNR), Istituto di Fisica dello Spazio Interplanetario"

Istituto di Metrologia G. Colonnetti (IMGC)
• See "Consiglio Nazionale delle Ricerche (CNR), Istituto di Metrologia G. Colonnetti (IMGC)"

Istituto di Radioastronomia (IRA)
• See "Consiglio Nazionale delle Ricerche (CNR), Istituto di Radioastronomia (IRA)"

Istituto di Tecnologie e Studio delle Radiazioni Extraterrestri (TESRE)
• See "Consiglio Nazionale delle Ricerche (CNR), Istituto di Tecnologie e Studio delle Radiazioni Extraterrestri (TESRE)"

Istituto e Museo di Storia della Scienza (IMSS), Planetario
Piazza dei Giudici 1
I-50122 Firenze
Telephone: (0)55-293493
 (0)55-2398876
Telefax: (0)55-288257
WWW: http://galileo.imss.firenze.it/indice.html (Italian)
 http://galileo.imss.firenze.it/general/ (English)
Founded: 1977 (Museum: 1930)
Staff: 12
Activities: education
Periodicals: (2) "Nuncius"
City Reference Coordinates: 011°15'00"E 43°46'00"N

Istituto Nazionale di Astrofisica (INAF)
Viale del Parco Mellini 84
I-00136 Roma
Telephone: (0)6-35340050
Telefax: (0)6-35343154
Electronic Mail: inaf@inaf.it
WWW: http://www.inaf.it/
Activities: managing astrophysical research in Italy
City Reference Coordinates: 012°29'00"E 41°54'00"N

Istituto Nazionale di Astrofisica (INAF), Osservatorio Astrofisico di Arcetri (OAA)
Largo Enrico Fermi 5
I-50125 Firenze
Telephone: (0)55-27521
Telefax: (0)55-220039
Electronic Mail: <userid>@arcetri.astro.it
WWW: http://www.arcetri.astro.it/
    http://lbtwww.arcetri.astro.it/ (Large Binocular Telescope - LBT)
Founded: 1872
Staff: 91
Activities: solar physics * ISM * extragalactic astrophysics * high-energy astrophysics * instrumentation * adaptive optics * LBT project * TIRGO telescope
Periodicals: "Arcetri Astrophysical Preprints", "Arcetri Technical Reports"; (1) "Arcetri Astrophysical Observatory Annual Report"
Coordinates: 011°15'19"E 43°45'14"N H180m (IAU Code 030)
    007°47'30"E 45°59'04"N H3,135m (TIRGO Telescope, Zermatt, Switzerland)
City Reference Coordinates: 011°15'00"E 43°46'00"N

Istituto Nazionale di Astrofisica (INAF), Osservatorio Astrofisico di Catania (OAC)
Via Santa Sofia 78
I-95123 Catania
Telephone: (0)95-7332111
Telefax: (0)95-330592
Electronic Mail: oacatania@ct.astro.it
WWW: http://woac.ct.astro.it/
Founded: 1880
Staff: 72
Activities: solar and stellar activity * interplanetary and interstellar matter * minor bodies of the solar system * relativity * cosmology * variable stars * detectors
Coordinates: 015°04'21"E 37°31'43"N H193m
    014°58'24"E 37°41'30"N H1,735m (Serra la Nave, Etna – IAU Code 559)
City Reference Coordinates: 015°06'00"E 37°30'00"N

Istituto Nazionale di Astrofisica (INAF), Osservatorio Astronomico di Bologna (OAB)
Via Ranzani 1
I-40127 Bologna
Telephone: (0)51-2095701
Telefax: (0)51-2095700
Electronic Mail: <userid>@astbo3.bo.astro.it
WWW: http://www.bo.astro.it/
    http://www.bo.astro.it/loiano/ (Loiano Observatory)
Founded: 1985
Staff: 63
Activities: galactic and stellar astrophysics * large-scale structures and cosmology * numerical methods for astronomy * instrumentation
Periodicals: "Bologna Astrophysical Preprints", "Technical Reports"
Coordinates: 011°20'12"E 44°15'30"N H785m (Loiano – IAU Code 598)
City Reference Coordinates: 011°20'00"E 44°29'00"N

Istituto Nazionale di Astrofisica (INAF) ,Osservatorio Astronomico di Brera (OAB), Merate
Via E. Bianchi 46
I-23807 Merate
Telephone: (0)39-999111
Telefax: (0)39-9991160
Electronic Mail: <userid>@merate.mi..astro.it
WWW: http://www.merate.mi.astro.it/
    http://albinoni.brera.unimi.it/
    http://www.mi.astro.it/
Founded: 1927
Staff: 35
Activities: stellar pulsations * observational cosmology * X-ray technology * celestial mechanics * meteorology * X-ray stellar sources * stellar evolution
Coordinates: 009°25'42"E 45°42'00"N H340m (IAU Code 096)
City Reference Coordinates: 009°25'00"E 45°42'00"N (Merate)
    009°12'00"E 45°28'00"N (Milano)

Istituto Nazionale di Astrofisica (INAF), Osservatorio Astronomico di Brera (OAB), Milano
Via Brera 28
I-20121 Milano
Telephone: (0)2-723201
Telefax: (0)2-72001600
Electronic Mail: <userid>@brera.mi.astro.it
WWW: http://www.brera.mi.astro.it/
    http://albinoni.brera.unimi.it/
Founded: 1760
Staff: 25

Activities: stellar pulsations * observational cosmology * X-ray technology * celestial mechanics * meteorology * X-ray stellar sources * stellar evolution
Coordinates: 009°11'30"E 45°28'00"N H146m (IAU Code 027)
City Reference Coordinates: 009°12'00"E 45°28'00"N (Milano)

Istituto Nazionale di Astrofisica (INAF), Osservatorio Astronomico di Cagliari (OAC)
Strada 54
Località Poggio dei Pini
I-09012 Capoterra
Telephone: (0)70-711801
Telefax: (0)70-71180222
Electronic Mail: oa-cagliari@ca.astro.it
　　　　　　　　<userid>@ca.astro.it
WWW: http://www.ca.astro.it/
Founded: 1899
Staff: 28
Activities: geodynamics * astrophysics
Coordinates: 008°58'24"E 39°08'12"N (Capoterra)
　　　　　　　008°18'44"E 39°08'09"N (Carloforte)
　　　　　　　009°19'47"E 39°59'42"N (Bruncu Spina)
City Reference Coordinates: 009°06'00"E 39°13'00"N (Cagliari)

Istituto Nazionale di Astrofisica (INAF), Osservatorio Astronomico di Capodimonte (OAC)
Via Moiariello 16
I-80131 Napoli
Telephone: (0)81-5575111 (Switchboard)
Telefax: (0)81-456710
Electronic Mail: <userid>@na.astro.it
WWW: http://oacosf.na.astro.it/
Founded: 1819
Staff: 50
Activities: stars * Sun * galaxies
Coordinates: 014°15'18"E 40°51'48"N H150m (IAU Code 044)
City Reference Coordinates: 014°17'00"E 40°51'00"N

Istituto Nazionale di Astrofisica (INAF), Osservatorio Astronomico di Padova
Vicolo dell'Osservatorio 5
I-35122 Padova
Telephone: (0)49-8293411
　　　　　　　(0)424-462032 (Monte Ekar)
Telefax: (0)49-8759840
　　　　　(0)424-462884 (Monte Ekar)
Electronic Mail: oa-padova@pd.astro.it
　　　　　　　　<userid>@pd.astro.it
WWW: http://www.pd.astro.it/
Founded: 1767
Staff: 70
Activities: solar system * stellar astrophysics * galactic and extragalactic astronomy * cosmology * high-energy astrophysics * space research * headquarters of the "Telescopio Nazionale Galileo (TNG - see separate entry in Spain)"
Coordinates: 011°52'18"E 45°24'00"N H38m (IAU Code 533)
　　　　　　　011°34'17"E 45°50'36"N H1,350m (Monte Ekar – IAU Codes 098 & 209)
City Reference Coordinates: 011°53'00"E 45°25'00"N

Istituto Nazionale di Astrofisica (INAF), Osservatorio Astronomico di Palermo (OAP) "Giuseppe S. Vaiana"
Piazza del Parlamento 1
I-90134 Palermo
Telephone: (0)91-233111
Telefax: (0)91-233444
Electronic Mail: astropa@unipa.it
WWW: http://www.astropa.unipa.it/
Founded: 1790 (re-established independent: 1989)
Staff: 21
Activities: high-energy astronomy * X-rays * solar and stellar coronae * history of astronomy
Periodicals: (1) "Elementi Astronomici"
Coordinates: 013°21'22"E 38°06'44"N H72m (IAU Code 535)
City Reference Coordinates: 013°21'00"E 38°07'00"N

Istituto Nazionale di Astrofisica (INAF), Osservatorio Astronomico di Roma (OAR), Monte Mario
Viale del Parco Mellini 84
I-00136 Roma
Telephone: (0)6-347056
Telefax: (0)6-3498236
WWW: http://oar.rm.astro.it/
　　　　http://www.rm.astro.it/
Founded: 1938

Staff: 21
Activities: cosmology * stellar evolution * AGNs
Coordinates: 013°33'37"E 42°26'35"N H2,200m (Campo Imperatore – IAU Code 599)
012°27'27"E 41°55'19"N H150m (Monte Mario – IAU Code 034)
City Reference Coordinates: 012°29'00"E 41°54'00"N

Istituto Nazionale di Astrofisica (INAF), Osservatorio Astronomico di Roma (OAR), Monteporzio Catone
Via di Frascati 33
I-00040 Monteporzio Catone
Telephone: (0)6-9428641
Telefax: (0)6-9447243
WWW: http://www.mporzio.astro.it/welc.html
Founded: 1938 (OAR)
Activities: stellar spectroscopy * stellar evolution * clusters of galaxies
Coordinates: 012°42'00"E 41°48'00"N H350m
City Reference Coordinates: 012°42'00"E 41°48'00"N

Istituto Nazionale di Astrofisica (INAF), Osservatorio Astronomico di Teramo "Vincenzo Cerulli"
Via Mentore Maggini
I-64100 Teramo
Telephone: (0)861-210490
Telefax: (0)861-210492
Electronic Mail: <userid>@astrte.te.astro.it
WWW: http://www.te.astro.it/
Founded: 1892
Staff: 18
Activities: stellar evolution * minor planets * clusters * nucleosynthesis * RR Lyrae stars * Cepheids * dwarf galaxies * population synthesis
Coordinates: 013°44'00"E 42°39'27"N H398m (Colluriana – IAU Code 037)
013°33'40"E 42°26'13"N H2,120m (Campo Imperatore – IAU Code 599)
City Reference Coordinates: 013°42'00"E 42°39'00"N (Teramo)

Istituto Nazionale di Astrofisica (INAF), Osservatorio Astronomico di Torino (OATo)
Strada Osservatorio 20
I-10025 Pino Torinese
Telephone: (0)11-8101900
Telefax: (0)11-8101030
Electronic Mail: <userid>@to.astro.it
WWW: http://www.to.astro.it/
Founded: 1789
Staff: 79
Activities: astronomy * astrophysics * planetology * astrometry
Coordinates: 007°46'29"E 45°02'16"N H622m (IAU Code 022)
City Reference Coordinates: 007°40'00"E 45°03'00"N

Istituto Nazionale di Astrofisica (INAF), Osservatorio Astronomico di Trieste (OAT)
Casella Postale Succursale Trieste 5
Via G.B. Tiepolo 11
I-34131 Trieste
Telephone: (0)40-3199241 (Secretary)
Telefax: (0)40-309418
Electronic Mail: <userid>@oat.ts.astro.it
WWW: http://www.oat.ts.astro.it/
http://www.oat.ts.astro.it/oat-home.html
Founded: 1753
Staff: 60
Activities: solar physics * solar radioastronomy * cometary physics * stellar physics * stellar evolution * stellar atmospheres * ISM * extragalactic astrophysics * cosmology * observational archives * image processing * supercomputing * astronomy and space technologies
Periodicals: (4) "Osservazioni Solari"
Coordinates: 013°01'22"E 45°38'35"N H77m (IAU Code 038)
City Reference Coordinates: 013°46'00"E 45°40'00"N

Istituto Nazionale di Fisica Nucleare (INFN), Laboratori Nazionali del Gran Sasso (LNGS)
S.S. 17/bis - Km 18.910
I-67010 Assergi
Telephone: (0)862-4371
Telefax: (0)862-410795
(0)862-437570
Electronic Mail: segretaria@lngs.infn.it
WWW: http://www.lngs.infn.it/
Founded: 1988
City Reference Coordinates: 013°42'00"E 42°27'00"N

Istituto Nazionale di Fisica Nucleare (INFN), Laboratori Nazionali di Frascati (LNF)
Via E. Fermi 4
I-00044 Frascati
Telephone: (0)6-94031
Telefax: (0)6-94032582
Electronic Mail: <userid>@lnf.infn.it
WWW: http://www.lnf.infn.it/
Founded: 1955
City Reference Coordinates: 012°41'00"E 41°41'00"N

Istituto Nazionale di Fisica Nucleare (INFN), Sezione di Napoli, Astrofisica Solare e Stellare
Mostra d'Oltremare
Pad. 20
I-80125 Napoli
Telephone: (0)81-7253111
Telefax: (0)81-2394508
Electronic Mail: <userid>@na.infn.it
WWW: http://www1.na.infn.it/wastro/
City Reference Coordinates: 014°17'00"E 40°51'00"N

Istituto Nazionale di Geofisica (ING)
Via di Vigna Murata 605
I-00143 Roma
Telephone: (0)6-518601
Telefax: (0)6-5041181
Electronic Mail: <userid>@ingrm.it
WWW: http://www.ingrm.it/
Founded: 1936
Staff: 220
Activities: physics of the Earth * seismology * geomagnetism * aeronomy
Periodicals: (12) "Annali di Geofisica"
City Reference Coordinates: 012°29'00"E 41°54'00"N

Istituto Spezzino Ricerche Astronomiche (IRAS)
c/o Claudio Lopresti
Via Castellazzo 8/D
I-19125 La Spezia
Telephone: (0)187-715391
Telefax: (0)187-715391
Electronic Mail: clop@libero.it
WWW: http://digilander.iol.it/clop33/iras/
Founded: 1991
Membership: 78
Activities: popularization * photometry of variable stars * SN * image processing * digital astronomy * computing
Periodicals: "Notiziario IRAS"
Coordinates: 010°06'22"E 46°11'21"N H178m (Gragnola)
City Reference Coordinates: 009°50'00"E 44°07'00"N

Istituto Tecnico Nautico "Artiglio", Planetario
Via dei Pescatori
I-55049 Viareggio
Telephone: (0)584-390282
Telefax: (0)584-392090
Electronic Mail: info@nauticoartiglio.lu.it
WWW: http://www.nauticoartiglio.lu.it/
　　　　http://www.nauticoartiglio.lu.it/planetario/planetar00.htm
City Reference Coordinates: 010°14'00"E 43°52'00"N

Istituto Tecnico Nautico Statale, Planetario
Via Mazzini 26
I-66026 Ortona
Telephone: (0)85-9063441
Telefax: (0)85-9063441
Electronic Mail: nautico.ortona@tin.it
WWW: http://www.nauticortona.it/
Founded: 1921
Staff: 6
Activities: education
Coordinates: 014°24'18"E 42°21'00"N H80m
City Reference Coordinates: 014°24'00"E 42°21'00"N

Istituto Universitario Navale, Cattedra di Astronomia Nautica
Via Acton 38

I-10053 Bussoleno
or :
Via M. d'Azeglio 34
I-10053 Bussoleno
Telephone: (0)122-640797
Telefax: (0)122-640797
Electronic Mail: grange@mclink.it
WWW: http://www.mclink.it/mclink/astro/ass/grange/grange1.htm
http://www.mclink.it/mclink/astro/ass/grange/english.htm
Founded: 1993
Staff: 2
Activities: astrometry of minor planets
Coordinates: 007°08'29"E 45°08'31"N H470m (IAU Code 476)
City Reference Coordinates: 007°09'00"E 45°08'00"N

Osservatorio Astronomico Pubblico, Soresina
c/o Erinio Pini
Via Matteotti 4
I-26015 Soresina
Telephone: (0)374-43722
Founded: 1974
Staff: 15
City Reference Coordinates: 009°51'00"E 45°17'00"N

Osservatorio Astronomico Santa Lucia di Stroncone
Via Santa Lucia 68
I-05039 Santa Lucia di Stroncone
Electronic Mail: a.vagnozzi@giaweb.it
• Private observatory (IAU Code 589)
City Reference Coordinates: 012°39'00"E 42°34'00"N (Terni)

Osservatorio Astronomico "Serafino Zani"
c/o Centro Studi e Ricerche Serafino Zani
Via Bosca 24
Casella Postale 104
I-25066 Lumezzane
Telephone: (0)30-872164
Telefax: (0)30-872545
Electronic Mail: info@serafinozani.it
Founded: 1993
Staff: 15
Activities: observing * photography * CCDs * education * projects * planetariums * star parties
Periodicals: (4) "Sagittario"
Awards: (1/2) "Shadows of Time"
Coordinates: 010°14'25"E 45°39'53"N H830m (IAU Code 130)
City Reference Coordinates: 010°13'00"E 45°33'00"N (Brescia)

Osservatorio Astronomico Sharru
Via Giovanni XXIII 13
I-24050 Covo
Telephone: (0)363-93102
Telefax: (0)363-93102
Founded: 1975
Staff: 2
Activities: visual and photoelectric observing (novae, SN, cataclysmic variable stars)
Coordinates: 009°46'15"E 45°29'50"N H115m
City Reference Coordinates: 009°43'00"E 45°41'00"N (Bergamo)

Osservatorio Bassano Bresciano
Via San Michele 4
I-25020 Bassano Bresciano
Electronic Mail: ulisse.quadri@libero.it (Ulisse Quadri)
WWW: http://web.tiscalinet.it/OsservatorioBassano
Founded: 1983
• Private observatory (IAU Code 565)
City Reference Coordinates: 010°13'00"E 45°33'00"N (Brescia)

Osservatorio San Giuseppe
Parr. San Donato a Livizzano
I-50056 Montelupo
Telephone: (0)571-671935
(0)571-671106
Telefax: (0)571-675835
Electronic Mail: mirediz@logo.it (Claudio Allegri, Observatory Manager)
Founded: 1990

Staff: 8
Activities: popularization * deep-sky observing
Coordinates: 011°10'00"E 43°41'40"N H196m (IAU Code 108)
City Reference Coordinates: 011°01'00"E 43°44'00"N

Osservatorio San Vittore
Via San Vittore 44
I-40136 Bologna
Electronic Mail: ermes.colombini@tin.it
WWW: http://www.giaweb.it/552/552.htm
Founded: 1969
Staff: 6
Activities: private observatory * astrometry of minor planets and comets
Coordinates: 011°20'24"E 44°28'05"N H280m (IAU Code 552)
City Reference Coordinates: 011°20'00"E 44°29'00"N

Osservatorio Astronomico di Sormano
Località Colma del Piano
I-22030 Sormano
or :
c/o Piero Sicoli (Director)
Via Valli 17
I-23846 Garbagnate Monastero
Electronic Mail: sormano@tin.it
WWW: http://www.brera.mi.astro.it/sormano/
Founded: 1986
Membership: 13
Activities: astrometry on minor planets and comets
Coordinates: 009°13'49"E 45°53'02"N H1128m (IAU Code 587)
City Reference Coordinates: 009°04'00"E 45°34'00"N

Pianoro Observatory
Viale della Resistenza 93
I-40065 Pianoro
Electronic Mail: bgorett@tin.it
Founded: 1996
Staff: 1
Activities: private observatory * photometry and astrometry of minor planets and comets * follow-up of NEOs
Coordinates: 011°20'35"E 44°23'30"N (IAU Code 610)
City Reference Coordinates: 011°21'00"E 44°23'00"N

Pieri Astronomy Observatory
c/o Alessandro Pieri
Via Fratelli Bandiera 19
I-04100 Latina
Telephone: (0)773-695322
Electronic Mail: alex@uni.net
WWW: http://www.mclink.it/mclink/astro/ass/pao/
      http://astrolink.mclink.it/ass/pao/
• Private observatory
City Reference Coordinates: 012°52'00"E 41°28'00"N

Planetario Comunale di Modena
c/o M.U. Lugli
Viale Jacopo Barozzi 31
Casella Postale 34
Succursale 2
I-41100 Modena
Telephone: (0)59-224726
Telefax: (0)59-224726
WWW: http://www.isesitalia.it/SCUOLA/PV1000/ASTRO.HTM
City Reference Coordinates: 010°55'00"E 44°39'00"N

Planetario Comunale di Pisa
Via Mario Lalli 4
I-56127 Pisa
Telephone: (0)50-580456
Telefax: (0)50-580456
Electronic Mail: planetario@comune.pisa.it
WWW: http://www.comune.pisa.it/doc/istruzione/mep.htm
Founded: 1989
Staff: 2
City Reference Coordinates: 010°20'00"E 43°43'00"N

Planetario del Museo Provinciale di Storia Naturale
Via Roma 234
I-57127 Livorno
Telephone: (0)586-802294
Founded: 1978
Staff: 30
Activities: education * popularization * education * lectures * observing
City Reference Coordinates: 010°19'00"E 43°33'00"N

Planetario di Ravenna
Viale Santi Baldini 4A
I-48100 Ravenna
Telephone: (0)544-62534
Telefax: (0)544-22928
WWW: http://racine.ra.it/planet/homeplan.html
Founded: 1985
Staff: 3
Activities: education * popularization
Periodicals: "Almanacco Astronomico"
City Reference Coordinates: 012°12'00"E 44°25'00"N

San Gersolè Planetary Group
Via San Gersolè 2
I-50020 Monteoriolo
Telephone: (0)55-678811
Electronic Mail: iei83_g_quarra@mclink.it (Giovanni Quarra)
      d.sarocchi@mclink.it (Damiano Sarocchi)
WWW: http://www.chim1.unifi.it/group/education/caat/sgpg/
    http://blu.chim1.unifi.it/group/education/caat/sgpg/
Founded: 1988
City Reference Coordinates: 011°15'00"E 43°46'00"N (Firenze)

San Marcello Pistoiese Astronomical Observatory
c/o Luciano Tesi
Viale Panoramico 949
I-51028 San Marcello Pistoiese
Telephone: (0)573-622220
Electronic Mail: luctesi@tin.it
WWW: http://space.tin.it/scienza/lutesi/
Founded: 1980
Membership: 8
Activities: popularization * minor planets (search, astrometry)
Periodicals: (3) "A Naso in Sù"
Coordinates: 010°48'15"E 44°03'17"N H980m (IAU Code 104)
City Reference Coordinates: 010°47'00"E 44°03'00"N

SAX Scientific Data Center (SDC)
• See "BeppoSAX Scientific Data Center (SDC)"

Scuola Internazionale Superiore di Studi Avanzati (SISSA)
(International School for Advanced Studies - ISAS)
Via Beirut 2-4
I-34014 Trieste
Telephone: (0)40-37871
Telefax: (0)40-3787528
Electronic Mail: <userid>@neumann.sissa.it
         <userid>@sissa.it
WWW: http://www.sissa.it/
    http://babbage.sissa.it/ (Preprint Server)
Founded: 1978
Staff: 46
Activities: physics * mathematics * biophysics * geometry * astrophysics * cognitive neurosciences
City Reference Coordinates: 013°46'00"E 45°40'00"N

Scuola Normale Superiore, Astrophysics Group
Piazza dei Cavalieri 7
I-56126 Pisa
Telephone: (0)50-509111
Telefax: (0)50-563513
Electronic Mail: <userid>@sns.it
WWW: http://www.sns.it/html/ClasseScienze/Gruppi/ASTRO/astro.html
Founded: 1810
Staff: 5

Activities: galactic dynamics * spiral and elliptical galaxies * cooling flows * dark matter * plasma astrophysics * n-body simulations * gravitational lensing * cluster of galaxies
City Reference Coordinates: 010°20'00"E 43°43'00"N

Società Astronomica "G.V. Schiaparelli"
c/o Salvatore Furia (President)
Via Andrea del Sarto 3
I-21100 Varese
or :
c/o Osservatorio Astronomico "G.V.Schiaparelli"
Campo dei Fiori
I-21100 Varese
Telephone: (0)332-235491
Telefax: (0)332-237143
Electronic Mail: astrogeo@astrogeo.va.it
WWW: http://www.astrogeo.va.it/astronom/astronom.htm
Founded: 1956
Membership: 400
Activities: popularization * astronomy * meteorology * botanics * seismology
Coordinates: 008°46'15"E 45°52'04"N (IAU Code 204)
City Reference Coordinates: 008°49'00"E 45°49'00"N

Società Astronomica Italiana (SAIt)
c/o The Secretary
Osservatorio Astrofisico di Arcetri
Largo Enrico Fermi 5
I-50125 Firenze
Telephone: (0)55-435939
Telefax: (0)55-220039
Electronic Mail: info@sait.it
WWW: http://www.sait.it/
Founded: 1920
Membership: 900
Periodicals: (4) "Memorie della Società Astronomica Italiana" (ISSN 0037-8720); "Il Giornale di Astronomia" (ISSN 0390-1106)
City Reference Coordinates: 011°15'00"E 43°46'00"N

Società Astronomica Urania (SAU)
c/o Centro Arte Pieve
Via Roma 95
I-15067 Novi Ligure
Telephone: (0)143-329565
Electronic Mail: fabry@mediacomm.it
Founded: 1994
Activities: popularization * observing * photography
Coordinates: 008°48'30"E 44°47'51"N H347m (IAU Code 579)
City Reference Coordinates: 008°47'00"E 44°46'00"N

Società Astronomica Versiliese (SAV)
Via San Agostino 1
I-55045 Pietrasanta
Telephone: (0)584-791122
Electronic Mail: bimbi@versilia.toscana.it (Marco Bimbi, President)
WWW: http://www.versilia.toscana.it/pietrasanta/sav/
Founded: 1976
Membership: 28
Activities: popularization * education
Coordinates: 010°13'14"E 43°57'31"N H22m (Osservatorio Spartaco Palla)
City Reference Coordinates: 010°14'00"E 43°57'00"N

Società Italiana di Fisica (SIF)
Via Castiglione 101
I-40136 Bologna
Telephone: (0)51-331554
Telefax: (0)51-581340
Electronic Mail: sif@bo.infn.it
WWW: http://www.sif.it/
Founded: 1897
Membership: 6300
Staff: 11
Activities: publishing * conventions * schools
Periodicals: "Il Nuovo Cimento" (ISSN 0392-6737), "La Rivista del Nuovo Cimento" (ISSN 0393-697x), "Giornale di Fisica", "Il Nuovo Saggiatore"
Awards: (1) "Prize for graduates in physics"
City Reference Coordinates: 011°20'00"E 44°29'00"N

Università degli Studi di Trieste
• See "Università di Trieste"

Università di Bologna, Dipartimento di Astronomia
Via Ranzani 1
I-40127 Bologna
Telephone: (0)516305727
Telefax: (0)516305700
Electronic Mail: <userid>@bo.astro.it
WWW: http://www.bo.astro.it/dip/
Founded: 1725
Staff: 26
Activities: stellar astrophysics * extragalactic astrophysics * instrumentation
Periodicals: "Bologna Astrophysical Preprints", "Bologna Technical Reports"
Coordinates: 011°20'12"E 44°15'32"N H785m
　　　　　　　011°20'04"E 44°15'23"N H795m
• Full university name: "Università degli Studi di Bologna"
City Reference Coordinates: 011°20'00"E 44°29'00"N

Università di Catania, Dipartimento di Astronomia
Via Santa Sofia 78
I-95123 Catania
Telephone: (0)95-7332111
Telefax: (0)95-330592
Electronic Mail: segretaria@alpha4.ct.astro.it
WWW: http://woac.ct.astro.it/
Founded: 1956
Staff: 15
Activities: solar and stellar activity * interplanetary and interstellar matter * solar-system minor bodies * relativity * cosmology * variable stars * nuclear astrophysics
Coordinates: 015°04'21"E 37°31'43"N H193m (IAU Code 156)
• Full university name: "Università degli Studi di Catania"
City Reference Coordinates: 015°06'00"E 37°30'00"N

Università di Firenze, Dipartimento di Astronomia e Scienza dello Spazio
Largo Enrico Fermi 5
I-50125 Firenze
Telephone: (0)55-27521
Telefax: (0)55-224193
Electronic Mail: <userid>@arcetri.astro.it
WWW: http://www.arcetri.astro.it/Dipartimento/
Founded: 1967
Activities: far-IR and mm astronomy and spectroscopy * atmospheric physics * molecular spectroscopy * interstellar clouds * cosmic dust * UV observing
Periodicals: "Internal Reports"
City Reference Coordinates: 011°15'00"E 43°46'00"N

Università di Lecce, Dipartimento di Fisica, Gruppo di Astrofisica
Via Per Arnesano
Casella Postale 193
I-73100 Lecce
Telephone: (0)832-320501
Telefax: (0)832-320505
Electronic Mail: <userid>@le.infn.it
WWW: http://www.unile.it/ateneo/
Founded: 1967
Staff: 8
Activities: IR astronomy * ISM * galactic dynamics
City Reference Coordinates: 018°11'00"E 40°23'00"N

Università di Milano, Dipartimento di Fisica
Via Celoria 16
I-20133 Milano
Telephone: (0)2-2392272
　　　　　　(0)2-2392206
Telefax: (0)2-70638413
　　　　　(0)2-2392205
Electronic Mail: <userid>@mi.infn.it
　　　　　　　　<userid>@uni.mi.astro.it
WWW: http://www.mi.infn.it/
Activities: plasma astrophysics * laboratory plasmas * stellar and galactic structure and evolution * cosmic microwave background and related technology * cosmic rays * astronomical databases * organizing workshops * education
Coordinates: 009°11'37"E 45°27'34"N

City Reference Coordinates: 009°12'00"E 45°28'00"N

Università di Padova, Dipartimento di Astronomia
Vicolo dell'Osservatorio 5
I-35122 Padova
Telephone: (0)49-8293411
Telefax: (0)49-8759840
Electronic Mail: <userid>@astrpd.pd.astro.it
            dipastro@www.pd.astro.it
WWW: http://www.pd.astro.it/~dipastro/
       http://www.pd.astro.it/asiago (Osservatorio Astrofisico di Asiago)
Founded: 1986
Staff: 43
Activities: solar system * stellar astrophysics * galactic and extragalactic astronomy * cosmology * high-energy astrophysics * space research
Coordinates: 011°52'18"E 45°24'00"N H38m (Osserv. Astrof. di Asiago – IAU Code 043)
• Full university name: "Università degli Studi di Padova"
City Reference Coordinates: 011°53'00"E 45°25'00"N

Università di Pavia, Dipartimento di Fisica Nucleare e Teorica (DFNT)
Via A. Bassi 6
I-27100 Pavia
Telephone: (0)382-507436
Telefax: (0)382-507752
Electronic Mail: dfnt@pv.infn.it
            <userid>@pv.infn.it
WWW: http://www.pv.infn.it/~dfntwww/dfnt.html
Founded: 1983
Staff: 100
Activities: theoretical and experimental nuclear and subnuclear physics * space physics * gravitation
City Reference Coordinates: 009°10'00"E 45°10'00"N

Università di Perugia, Osservatorio Astronomico
Via Bonfigli
I-06100 Perugia
Telephone: (0)75-20632
         (0)75-5853042
Electronic Mail: <userid>@vaxpg.pg.infn.it
WWW: http://wwwospg.pg.infn.it/
Founded: 1988
Activities: variable stars * stellar evolution * IR * history of astronomy
Periodicals: "Contributi Serie A", "Contributi Serie B"
Coordinates: 012°24'00"E 43°00'00"N H430m
City Reference Coordinates: 012°24'00"E 43°00'00"N

Università di Pisa, Dipartimento di Fisica, Sezione di Astronomia e Astrofisica
Piazza Torricelli 2
I-56100 Pisa
Telephone: (0)50-91111
Telefax: (0)50-48277
Electronic Mail: <userid>@df.unipi.it
WWW: http://astr17pi.difi.unipi.it/
Founded: 1730
Staff: 8
Activities: ISM * UV extinction * molecular clouds * OB stars * star formation * minor planets * galaxy evolution * stellar evolution * open and globular clusters
City Reference Coordinates: 010°20'00"E 43°43'00"N

Università di Pisa, Dipartimento di Matematica, Gruppo di Meccanica Spaziale
Via F. Buonarroti 2
I-56127 Pisa
Telephone: (0)50-599554
         (0)50-599552
Telefax: (0)50-599524
Electronic Mail: <userid>@icnucevm.bitnet
WWW: http://adams.dm.unipi.it/
Activities: celestial mechanics * planets * space geodesy * minor planets * small natural and artificial satellites
Periodicals: "Preprints"
City Reference Coordinates: 010°20'00"E 43°43'00"N

Università di Roma La Sapienza, Istituto Astronomico
Via G.M. Lancisi 29
I-00161 Roma
Telephone: (0)6-4403734
         (0)6-4403735

Electronic Mail: <userid>@astrm2.rm.astro.it
WWW: http://astrm2.rm.astro.it/home.html
http://astro1.astro.uniroma1.it/home.html
Founded: 1981
Staff: 15
Activities: stellar physics * high-energy astrophysics * AGNs * clusters of galaxies * structure and dynamics of the Galaxy
Coordinates: 012°12'20"E 41°48'49"N H380m (Monteporzio)
City Reference Coordinates: 012°29'00"E 41°54'00"N

Università di Roma Tor Vergata, Dipartimento di Fisica, Astrofisica
Via della Ricerca Scientifica 1
I-00133 Roma
Telephone: (0)6-72591
Telefax: (0)6-2023507
Electronic Mail: <userid>@roma2.infn.it
<userid>@itovf2.roma2.infn.it
WWW: http://itovf2.roma2.infn.it/
http://www.fisica.uniroma2.it/astro/
Founded: 1984
Staff: 12
Activities: cosmology * microwave background * AGNs * clusters of galaxies * solar physics * interplanetary plasma
City Reference Coordinates: 012°29'00"E 41°54'00"N

Università di Roma Tre, Dipartimento di Fisica
Via della Vasca Navale 84
I-00146 Roma
Telefax: (0)6-55179303
Electronic Mail: <userid>@fis.uniroma3.it
WWW: http://www.fis.uniroma3.it/new/
http://193.204.162.110/
http://193.204.162.110/~astro/
Activities: astrophysics * space physics * nuclear physics * particle physics * theoretical physics * environmental physics
City Reference Coordinates: 012°29'00"E 41°54'00"N

Università di Trieste, Dipartimento di Astronomia
Casella Postale Succursale Trieste 5
Via G.B. Tiepolo 11
I-34131 Trieste
Telephone: (0)40-3199255 (Secretary)
Telefax: (0)40-309418
Electronic Mail: <userid>@astrts.astro.it
WWW: http://www.oat.ts.astro.it/
Founded: 1985
Staff: 14
Activities: solar physics * stellar atmospheres * mass loss in stars * interacting binaries * globular clusters * stellar evolution * classification of stellar spectra * AGNs * clusters of galaxies * data analysis * image processing * astronomical technologies including space
• Full university name: "Università degli Studi di Trieste"
City Reference Coordinates: 013°46'00"E 45°40'00"N

Università Popolare Sestrese, Osservatorio Astronomico di Genova
Piazzetta dell'Università Popolare 4
I-16154 Genova
Telephone: (0)10-678368
WWW: http://astrolink.mclink.it/oss_gen.htm
Founded: 1984 (1907 as association)
Membership: 34
Activities: education * popularization * minor planets * comets * astrometry * variable stars * Sun
Periodicals: (4) "Circolare" (circ.: 300), "Contributo" (circ.: 100); (2) "Bollettino" (circ.: 150)
Coordinates: 008°50'13"E 44°26'03"N H124m (IAU Code 974)
City Reference Coordinates: 008°57'00"E 44°25'00"N

Urania
(Associazione Ligure per lo Studio e la Divulgazione dell'Astronomia e dell'Astronautica)
c/o Museo Civico di Storia Naturale "G.Doria"
Via Brigata Liguria 9
I-16121 Genova
or :
c/o Centro Astronomico
Osservatorio di Rovegno
I-16028 Rovegno
Telephone: (0)10-8352882
Electronic Mail: associazione.urania@tiscalinet.it
WWW: http://www.icom.it/freeweb/urania/
Founded: 1951
Membership: 125

Activities: public lectures * education * photography * planetology * Sun * comets * observing
Periodicals: (6) "Circolare"
Coordinates: 009°18'06"E 44°34'31"N H918 (Alta Val Trebbia)
City Reference Coordinates: 008°57'00"E 44°25'00"N (Genova)
009°17'00"E 44°35'00"N (Rovegno)

Japan

Aichi University of Education, Department of Physics and Astronomy
Hirosawa 1
Igaya-cho
Kariya 448-8542
Telephone: (0)566-36-3111
Telefax: (0)566-36-4337
Electronic Mail: <userid>@auephyas.aichi-edu.ac.jp
WWW: http://www.physics.aichi-edu.ac.jp/physics.htm
Founded: 1988
Staff: 5
Activities: education * research * spiral galaxies * magnetic fields * MHD * black holes * AGNs
City Reference Coordinates: 136°59'00"E 34°59'00"N

Akeno Observatory
• See "University of Tokyo, Institute for Cosmic Ray Research (ICCR), Akeno Observatory"

Astronomical Data Analysis Center (ADAC)
• See "National Astronomical Observatory of Japan (NAOJ), Astronomical Data Analysis Center (ADAC)"

Astronomical Society of Japan (ASJ)
c/o National Astronomical Observatory of Japan
2-21-1 Osawa
Mitaka
Tokyo 181-8588
Telephone: (0)422-31-1359
Telefax: (0)422-31-1359
Electronic Mail: <userid>@tenmon.or.jp
           pasj@tenmon.or.jp (PASJ)
WWW: http://www.tenmon.or.jp/
      http://www.tenmon.or.jp/pasj/ (PASJ)
Founded: 1908
Membership: 2800
Activities: developing and popularizing astronomy
Periodicals: (12) "The Astronomical Herald" (ISSN 0374-2466); (6) "Publications of the Astronomical Society of Japan (PASJ)" (ISSN 0004-6264)
City Reference Coordinates: 139°46'00"E 35°42'00"N

Astronomical Society of Wakabadai (ASW)
4-21-206 Wakabadai Asahiku
Yokohama 241
Electronic Mail: suchida@mvc.biglobe.ne.jp (Shigemi Uchida)
WWW: http://www2a.biglobe.ne.jp/~wakaba/
Founded: 1996
City Reference Coordinates: 139°39'00"E 35°27'00"N

Bisei Astronomical Observatory (BAO)
1723-70 Ohkura
Bisei
Oda-gun
Okayama 714-14
Telephone: (0)866-87-4222
Telefax: (0)866-87-4224
Electronic Mail: info@bao.go.jp
WWW: http://www.urban.ne.jp/home/bao/index-e.html
Founded: 1993
Coordinates: 133°34'27"E 34°40'36"N H516m
City Reference Coordinates: 133°55'00"E 34°49'00"N

Bisei Hydrographic Observatory
• See "Japan Hydrographic and Oceanographic Department (JHOD), Bisei Hydrographic Observatory"

Communications Research Laboratory (CRL), Hiraiso Solar-Terrestrial Research Center
3601 Isozaki-cho
Hitachinaka-shi
Ibaraki 311-12
Telephone: (0)292-65-7121
Telefax: (0)292-65-7209
Electronic Mail: <userid>@crl.go.jp
WWW: http://hiraiso.crl.go.jp/
Founded: 1915 (CRL: 1952)

Staff: 12
Activities: space weather forecasting * solar radio and optical observing * radio disturbance prediction * geomagnetic activities
Periodicals: (12) "Solar Radio Emission Hiraiso" (circ.: 40)
Coordinates: 140°37'30"E 36°21'54"N H26m
City Reference Coordinates: 140°26'00"E 36°17'00"N

Communications Research Laboratory (CRL), Inubo Radio Observatory
9961 Tennodai
Choshi
Chiba 288
Telephone: (0)479-22-0871
Telefax: (0)479-25-0675
Electronic Mail: <userid>@crl.go.jp
WWW: http://www.crl.go.jp/inb/overview.html
Founded: 1945 (CRL: 1952)
Coordinates: 140°51'30"E 35°42'12"N
City Reference Coordinates: 140°05'00"E 35°36'00"N

Communications Research Laboratory (CRL), Kansai Advanced Research Center (KARC)
588-2 Iwaoka
Nishi-ku
Kobe
Hyogo 651-24
Telephone: (0)78-969-2100
Telefax: (0)78-969-2200
Electronic Mail: <userid>@crl.go.jp
WWW: http://www-karc.crl.go.jp/
Founded: 1989 (CRL: 1952)
City Reference Coordinates: 135°00'00"E 35°00'00"N

Communications Research Laboratory (CRL), Kashima Space Research Center (KSRC)
893-1 Hirai
Kashima
Ibaraki 314-8501
Telephone: (0)299-82-1211
Telefax: (0)299-84-7159
Electronic Mail: <userid>@crl.go.jp
WWW: http://www.crl.go.jp/ka/
Founded: 1964 (CRL: 1952)
Staff: 12
Activities: VLBI * radio astronomy
Periodicals: (3) "Journal of the Communications Research Laboratory" (ISSN 0914-9260)
Coordinates: 140°39'46"E 35°57'15"N H79m (Kashima)
City Reference Coordinates: 140°26'00"E 36°17'00"N

Communications Research Laboratory (CRL), Okinawa Radio Observatory
829-3 Daigusukubaru
Aza-Kuba
Nakagusuku-son
Nakagami-gun
Okinawa 901-24
Telephone: (0)98-895-2045
Telefax: (0)98-895-4010
Electronic Mail: <userid>@crl.go.jp
WWW: http://www.crl.go.jp/okn/overview.html
Founded: 1972 (CRL: 1952)
Coordinates: 127°48'24"E 26°16'54"N
City Reference Coordinates: 127°59'00"E 26°31'00"N

Communications Research Laboratory (CRL), Tokyo
4-2-1 Nukuikitamachi
Koganei-shi
Tokyo 184-8795
Telephone: (0)423-21-1211
Telefax: (0)423-27-7596
Electronic Mail: <userid>@crl.go.jp
WWW: http://www.crl.go.jp/
Founded: 1952
Staff: 425
Activities: space communications * VLBI * SLR * space environment research * communications technology * remote sensing * time & frequency standards
Periodicals: (12) "Ionospheric Data in Japan" (ISSN 0021-0382), "Standard Frequency and Time Service Bulletin" (ISSN 0387-5857); (3) "Journal of the Communications Research Laboratory" (ISSN 0914-9260); (2) "Ionospheric Data at Syowa Station (Antarctica)" (ISSN 0389-8237)
Coordinates: 139°29'18"E 35°42'24"N (Space Optical Comm. Research Center)
140°40'00"E 35°57'12"N (Kashima - see separate entry)

140°37'30"E 36°22'54"N H26m (Hiraiso - see separate entry)
City Reference Coordinates: 139°46'00"E 35°42'00"N

Communications Research Laboratory (CRL), Wakkanai Radio Observatory
3-20 Midori 2-chome
Wakkanai
Hokkaido 097-004
Telephone: (0)162-23-3386
Telefax: (0)162-24-3227
Electronic Mail: <userid>@crl.go.jp
WWW: http://www2.crl.go.jp/dk/c233-235/wakkanai/index-e.html
Founded: 1946 (CRL: 1952)
Coordinates: 141°41'06"E 45°23'36"N
City Reference Coordinates: 141°40'00"E 45°25'00"N (Wakkanai)

Communications Research Laboratory (CRL), Yamagawa Radio Observatory
2719 Narikawa
Yamagawa-machi
Ibusuki-gun
Kagoshima 891-05
Telephone: (0)993-34-0077
Telefax: (0)993-35-2077
Electronic Mail: <userid>@crl.go.jp
WWW: http://www.crl.go.jp/yam/overview.html
Founded: 1946 (CRL: 1952)
Coordinates: 130°37'06"E 31°12'06"N
City Reference Coordinates: 130°33'00"E 31°36'00"N

Dodaira Observatory
• Closed in March 2000.

Fujitok Corp.
1-9-16 Kami-Jujo
Kita-ku
Tokyo 114-0034
Telephone: (0)3-3909-1791
Telefax: (0)3-3908-6450
WWW: http://www.barrassociates.com/
http://www.fujitok.co.jp/
Activities: designing and manufacturing custom astronomical, space-based and other precision optical filters
City Reference Coordinates: 139°46'00"E 35°42'00"N

Fukuoka University of Education, Department of Earth Sciences and Astronomy
1-729 Akama
Munakata
Fukuoka 811-41
Telephone: (0)940-35-1375
Telefax: (0)940-33-7730
Electronic Mail: <userid>@fueipc.fukuoka-edu.ac.jp
WWW: http://www.fukuoka-edu.ac.jp/index-e.html
Founded: 1949
Staff: 4
Activities: stellar spectroscopy * radio astronomy
Periodicals: (1) "Bulletin of the Fukuoka University of Education - Part III: Mathematics, Natural Sciences and Technology"

Coordinates: 130°35'48"E 33°48'42"N H70m
City Reference Coordinates: 130°24'00"E 33°35'00"N

Gotoh Planetarium and Astronomical Museum
9-25 Kaminoge 3-chome
Setagaya-ku
Tokyo 158-0093
Telephone: (0)3-3703-0062
WWW: http://www.f-space.co.jp/goto-planet/
Founded: 1957
City Reference Coordinates: 139°46'00"E 35°42'00"N

GOTO Optical Manufacturing Co.
4-16 Yazaki-cho
Fuchu-shi
Tokyo 183-8530
Telephone: (0)42-362-5311
Telefax: (0)42-361-9571
Electronic Mail: info@goto.co.jp
<userid>@goto.co.jp

WWW: http://www.goto.co.jp/
Founded: 1926
Staff: 185
Activities: manufacturing planetariums, telescopes, pano-hemispheric movie projectors
Coordinates: 138°18'27"E 35°53'25"N H1,035m (Yatsugakate)
City Reference Coordinates: 139°46'00"E 35°42'00"N

Hida Observatory
• See "Kyoto University, Kwasan and Hida Observatories, Hida Observatory"

Hiraiso Solar-Terrestrial Research Center
• See "Communications Research Laboratory (CRL), Hiraiso Solar-Terrestrial Research Center"

Hiroshima Planetarium
c/o Hiroshima Children's Museum
5-83 Motomachi
Nakaku
Hiroshima 730
Telephone: (0)82-222-5346
WWW: http://mothra.rerf.or.jp/ENG/Hiroshima-old/Art/Hiroshima-Children.html
http://www.tourism.city.hiroshima.jp/english/level7/h040100007.html
Founded: 1980
Activities: shows * lectures
Periodicals: (4) "Planetarium Show"
Coordinates: 132°27'23"E 34°23'43"N H4m
City Reference Coordinates: 132°27'00"E 34°24'00"N

Hitotsubashi University, Laboratory of Astronomy and Geophysics
Naka 2-1
Kunitachi
Tokyo 186-8601
Telephone: (0)425-72-1101
Telefax: (0)425-71-1893
Electronic Mail: nakajima@higashi.hit-u.ac.jp (Kaichi Nakajima)
WWW: http://iris.higashi.hit-u.ac.jp/
Founded: 1984
Staff: 2
Activities: education (astronomy, computer science, information science) * astronomical data service
City Reference Coordinates: 139°46'00"E 35°42'00"N

Hoshinoko Yakata Observatory
1470-24 Aoyama
Himeji
Hyogo 671-2222
Telephone: (0)792-67-3050
Telefax: (0)792-67-3055
Electronic Mail: hoshinoko@city.himeji.hyogo.jp
WWW: http://www.city.himeji.hyogo.jp/hoshinoko/index-e.html
Founded: 1992
Staff: 3
Activities: astronomy education for children
Coordinates: 134°37'55"E 34°50'59"N H71m
City Reference Coordinates: 135°00'00"E 35°00'00"N

Ibaraki University, Institute for Astrophysics and Planetary Science
2-1-1 Bunkyo
Mito 310-8512
Telephone: (0)29-228-8333
Telefax: (0)29-228-8407
Electronic Mail: <userid>@mito.ipc.ibaraki.ac.jp
WWW: http://orion.sci.ibaraki.ac.jp/sci2a-e.html
City Reference Coordinates: 140°28'00"E 36°22'00"N

Institute of Space and Astronautical Science (ISAS)
• See now "Japan Aerospace Exploration Agency (JAXA)"

Inubo Radio Observatory
• See "Communications Research Laboratory (CRL), Inubo Radio Observatory"

Itabashi Science and Education Center Planetarium
4-14-1 Tokiwadai
Itabashi-shi
Tokyo 174-0071
Telephone: (0)3-3559-6561
Telefax: (0)3-3559-6000

WWW: http://www.ecopolis.city.itabashi.tokyo.jp/edu/kagaku/pra/top.html
Founded: 1988
Staff: 30
Activities: planetarium shows * science exhibitions
City Reference Coordinates: 139°46'00"E 35°42'00"N

Japan Academy (The_)
• See "The Japan Academy"

Japan Aerospace Exploration Agency (JAXA)
World Trade Center Building
2-4-1 Hamamatsu-cho
Minato-ku
Tokyo 105-8060
Telephone: (0)3-3438-6111
WWW: http://www.jaxa.jp/
http://www.jaxa.jp/index_e.html
http://www.nasda.go.jp/ (previous NASDA page)
http://www.isas.ac.jp/ (previous ISAS page)
http://www.astro.isas.ac.jp/ (previous ISAS Astronomy Group)
Founded: 2003
Staff: 1500
Activities: implementing Japan's space activities * space science * space technology * scientific satellites * sounding rockets * balloons * launchers
Periodicals: (12) "ISAS News" (ISSN 0285-2861); (4) "NASDA Report"; "ISAS Report" (ISSN 0285-6808), "Uchuken Hokoku" (ISSN 0285-2853)
Coordinates: 131°04'45"E 31°15'00"N H210m (Kagoshima Space Center)
• Merging of the "Institute of Space and Astronautical Science (ISAS)", the "National Aerospace Laboratory (NAL)" and the "National Space Development Agency (NASDA)"
City Reference Coordinates: 139°38'00"E 35°28'00"N (Kanagawa)

Japan Association for Information Processing in Astronomy (JAIPA)
c/o S. Ichikawa
National Astronomical Observatory of Japan
2-21-1 Osawa
Mitaka
Tokyo 181-8588
Electronic Mail: jaipa@rl.mtk.nao.ac.jp
WWW: http://bandai.mtk.nao.ac.jp/jaipa/
Founded: 1990
Membership: 70
Activities: supporting ADAC/NAOJ (see separate entry) * developing astronomical software * exchanging information about computing, and so on
City Reference Coordinates: 139°46'00"E 35°42'00"N

Japanese Dark-Sky Association (JDA)
c/o Shigemi Uchida
4-21-206 Wakabadai Asahiku
Yokohama 241-0801
or :
c/o Jun-ichi Watanabe
National Astronomical Observatory of Japan
2-21-1 Osawa
Mitaka
Tokyo 181-8588
Telephone: (0)422-34-3644
Telefax: (0)422-34-3810
Electronic Mail: watanabe@sl9mtk.nao.ac.jp
Founded: 1993
Membership: 200
Activities: meetings * press releases * publishing
Periodicals: (2) "Kougai Bushin"
City Reference Coordinates: 139°39'00"E 35°27'00"N (Yokohama)
139°46'00"E 35°42'00"N (Tokyo)

Japanese Industrial Standards Committee (JISC)
c/o Standards Department
Agency of Industrial Science and Technology
Ministry of International Trade and Industry
1-3-1 Kasumigaseki
Chiyoda-ku
Tokyo 100-9001
Telephone: (0)3-3501-9295
(0)3-3501-9296
Telefax: (0)3-3580-1418
Electronic Mail: jisc@meti.go.jp

WWW: http://www.jisc.go.jp/
City Reference Coordinates: 139°46'00"E 35°42'00"N

Japan Hydrographic and Oceanographic Department (JHOD), Bisei Hydrographic Observatory
Bisei-cho
Oda-gun
Okayama 714-1045
Telephone: (0)8668-7-3355
Electronic Mail: bho@oka.urban.ne.jp
WWW: http://www.jhd.go.jp/
http://www.jhd.go.jp/jhd-E.html
Founded: 1949 (JHOD: 1871)
Activities: lunar occultations * satellite geodesy
Coordinates: 133°34'27"E 34°40'36"N H516m
City Reference Coordinates: 133°55'00"E 34°39'00"N

Japan Hydrographic and Oceanographic Department (JHOD), Geodesy and Geophysics Division
5-3-1 Tsukiji
Chuo-ku
Tokyo 104
Telephone: (0)3-3541-3816
Telefax: (0)3-3545-2885
Electronic Mail: tenmon@ws12.cue.jhd.go.jp
WWW: http://www.jhd.go.jp/
http://www.jhd.go.jp/jhd-E.html
Founded: 1871 (JHOD)
Staff: 31
Activities: computation of Japanese Ephemeris * services for International Lunar Occultation Center (ILOC) * satellite geodesy * satellite laser ranging * gravimetry * magnetometry * lunar occultations
Periodicals: (1) "Japanese Ephemeris" (ISSN 0373-3696), "Report of Hydrographic Observations - Series of Astronomy and Geodesy" (ISSN 0287-2633), "Report of Hydrographic Observations - Series of Satellite Geodesy" (ISSN 0914-5753), "Report of Hydrographic Researches" (ISSN 0373-3602), "Report of Lunar Occultation Observations" (ISSN 1341-7282), "Report of International Lunar Occultation Center"
Observatories: 3 (Bisei, Simosato and Sirahama Hydrographic Obs. - see separate entries)
Coordinates: 139°46'11"E 35°39'41"N H41m
City Reference Coordinates: 139°46'00"E 35°42'00"N

Japan Hydrographic and Oceanographic Department (JHOD), Shimosato Hydrographic Observatory (SHO)
Shimosato
Nachikatsuura-cho
Higashimuro-gun
Wakayama 649-5142
Telephone: (0)7355-8-0084
Telefax: (0)7355-8-1535
Electronic Mail: simosato@oak.ocn.ne.jp
WWW: http://www1.kaiho.mlit.go.jp/
http://www1.kaiho.mlit.go.jp/jhd-E.html
Founded: 1954 (JHOD: 1871)
Staff: 6
Activities: lunar occultations * satellite laser ranging * satellite geodesy
Coordinates: 135°56'23"E 33°34'27"N H63m
City Reference Coordinates: 135°11'00"E 34°13'00"N

Japan Hydrographic and Oceanographic Department (JHOD), Shirahama Hydrographic Observatory

Shirahama 3347
Simoda
Sizuoka 415-0012
Telephone: (0)5582-2-0865
Electronic Mail: sirasui@mail.wbs.ne.jp
WWW: http://www.jhd.go.jp/
http://www.jhd.go.jp/jhd-E.html
Founded: 1942 (JHOD: 1871)
Activities: lunar occultations * satellite geodesy
Coordinates: 138°59'20"E 34°42'47"N H172m
City Reference Coordinates: 138°23'00"E 34°58'00"N

Japan Information Center of Science and Technology (JICST)
• See now "Japan Science and Technology Corporation (JST)"

Japan Meteorological Agency (JMA)
1-3-4 Otemachi
Chiyoda-ku
Tokyo 100

Telephone: (0)3-3212-8341
(0)3-3211-4966
Telefax: (0)3-3211-3032
Electronic Mail: pro@hq.kishou.go.jp
WWW: http://www.kishou.go.jp/
http://www.kishou.go.jp/english/index.html
Activities: atmospheric and oceanographic observations * weather forecasts and oceanographic forecasts * numerical weather prediction * telecommunication and data processing * global environmental issues * aviation weather services * monitoring of earthquakes and volcanic activities and tsunami forecasting * research, education and training * international co-operation
Periodicals: "Geophysical Magazine" (ISSN 0016-8017), "Seismological Bulletin" (ISSN 0446-5059)
City Reference Coordinates: 139°46'00"E 35°42'00"N

Japan Physical Society (JPS)
• See "The Physical Society of Japan"

Japan Planetarium Society
c/o Suginami Science Education Center
3-3-13 Shimizu
Sugami-ku
Tokyo 167
Telephone: (0)3-3396-4391
Telefax: (0)3-3396-4393
Electronic Mail: khf11056@niftyserve.or.jp
City Reference Coordinates: 139°46'00"E 35°42'00"N

Japan Science and Technology Corporation (JST)
Kawaguchi Center Building
4-1-8 Honcho
Kawaguchi-shi
Saitama 332-0012
Telephone: (0)48-226-5630
Telefax: (0)48-226-5751
WWW: http://www.jst.go.jp/
http://www.jst.go.jp/EN/
Founded: 1996
Staff: 466
Activities: basic research * technological transfer * dissemination of scientific and technological information * public understanding of science and technology
Periodicals: (12) "Current Bibliography on Science and Technology (CBST)", "Journal of Information Processing and Management"; (1) "White Paper on Science and Technology"
• Formerly "Japan Information Center of Science and Technology (JICST)"
City Reference Coordinates: 139°30'00"E 36°00'00"N

Kagawa University, Department of Astronomy and Earth Sciences
Saiwaicho
Takamatsu-shi
Kagawa 760-8522
Telephone: (0)87-832-1466
Telefax: (0)87-832-1615
Electronic Mail: <userid>@ed.kagawa-u.ac.jp
WWW: http://www.ed.kagawa-u.ac.jp/
Founded: 1966
Staff: 2
Activities: astronomy * geophysics
Periodicals: "Memoirs of the Faculty of Education - Kagawa University (Part II)" (ISSN 0389-3057)
City Reference Coordinates: 134°02'00"E 34°15'00"N

Kamioka Underground Observatory
• See "University of Tokyo, Institute for Cosmic Ray Research (ICCR), Kamioka Underground Observatory"

Kanagawa University, Department of Physics
3-27-1 Rokkakubashi
Kanagawa-ku
Yokohama 221-8686
Telephone: (0)45-481-5661
(0)45-491-1701
Telefax: (0)45-791-4915
WWW: http://www.kanagawa-u.ac.jp/
Staff: 10
Activities: cosmic-ray astrophysics * solar-terrestrial physics * solid-state physics
City Reference Coordinates: 139°39'00"E 35°27'00"N

Kansai Advanced Research Center (KARC)
• See "Communications Research Laboratory (CRL), Kansai Advanced Research Center (KARC)"

Kashima Space Research Center (KSRC)
• See "Communications Research Laboratory (CRL), Kashima Space Research Center (KSRC)"

Katsushika City Museum, Planetarium
3-25-1 Shiratori
Katsushika-ku
Tokyo 125-0063
Telephone: (0)3-3838-1101
Telefax: (0)3-5680-0849
Electronic Mail: ldq05214@nifty.ne.jp (Tatsuyuki Arai, Astronomy Section)
WWW: http://www.obs.misato.wakayama.jp/~katusika/index-e.html
Founded: 1991 (Museum)
Activities: exhibitions * planetarium shows
City Reference Coordinates: 139°46'00"E 35°42'00"N

Kiso Observatory
• See "University of Tokyo, Institute of Astronomy, Kiso Observatory"

Kobe University, Department of Earth and Planetary Sciences
Rokkoudai-machi
Nada-ku
Kobe 657-8501
Telephone: (0)78-803-0529
(0)78-803-0540
(0)78-803-0541
Telefax: (0)78-803-5757
Electronic Mail: <userid>@kobe-u.ac.jp
WWW: http://www.planet.sci.kobe-u.ac.jp/index.html
Founded: 1973
Staff: 6
Activities: accretion * computational fluid dynamics * cosmic-gas and dust dynamics * observation of primitive bodies * origin and evolution of solar system * space exploration (Mars and asteroid)
City Reference Coordinates: 135°10'00"E 34°41'00"N

Kumamoto University, Department of Physics
2-39-1 Kurokami
Kumamoto 860-8555
Telephone: (0)96-344-2111
Telefax: (0)96-342-3320
WWW: http://www.eecs.kumamoto-u.ac.jp/
Founded: 1949
Staff: 1
Activities: cosmology * black holes * accretion disks * nucleosynthesis * stellar evolution
Periodicals: (1/2) "Physics Reports" (ISSN 0303-4070)
City Reference Coordinates: 130°43'00"E 32°48'00"N

Kwasan Observatory
• See "Kyoto University, Kwasan and Hida Observatories, Kwasan Observatory"

Kyoto Sangyo University, Department of Physics
Kita-ku
Kamigamo
Kyoto 603-8555
Telephone: (0)75-701-2151
Telefax: (0)75-705-1640
Electronic Mail: <userid>@cc.kyoto-su.ac.jp
WWW: http://www.kyoto-su.ac.jp/department/ph/index-j.html
Founded: 1965
Staff: 15
Activities: theoretical astrophysics * galaxy formation * distribution of matter in the universe * X-rays * celestial mechanics * mathematical physics
Periodicals: (4) "Acta Humanistica et Scientifica Universitatis Sangio Kyotiensis"
City Reference Coordinates: 135°45'00"E 35°00'00"N

Kyoto University, Department of Astronomy
Kitashirakawa
Sakyo-ku
Kyoto 606-8502
Telephone: (0)75-753-3890
Telefax: (0)75-753-3897
Electronic Mail: <userid>@kusastro.kyoto-u.ac.jp
WWW: http://www.kusastro.kyoto-u.ac.jp/
Founded: 1929
Staff: 9

Activities: accretion disks * circumstellar matter * early-type stars * eclipsing binaries * galaxies * ISM * stellar rotation * globular clusters * stellar spectroscopy * HII regions
Periodicals: "Contributions" (ISSN 0388-0230)
Coordinates: 135°47'00"E 35°01'48"N (Ouda Station)
City Reference Coordinates: 135°45'00"E 35°00'00"N

Kyoto University, Department of Astronomy, Ouda Station
Mochi
Ouda-cho
Uda
Nara 633-21
Telephone: (0)7458-3-3110
Electronic Mail: <userid>@kusastro.kyoto-u.ac.jp
WWW: http://www.kusastro.kyoto-u.ac.jp/
Founded: 1977
Activities: CCD imagery, photometry, and spectroscopy of galactic and extragalactic nebulae
Coordinates: 135°57'00"E 34°28'12"N H389m
City Reference Coordinates: 135°45'00"E 35°00'00"N (Kyoto)
135°50'00"E 34°41'00"N (Nara)

Kyoto University, Department of Physics
Kitashirakawa
Sakyo
Kyoto 606-8502
WWW: http://www.scphys.kyoto-u.ac.jp/index-e.html
http://www-tap.scphys.kyoto-u.ac.jp/index-e.html (Theoretical Astrophysics Group)
http://www-cr.scphys.kyoto-u.ac.jp/ (Cosmic-Ray Group)
Founded: 1898
City Reference Coordinates: 135°45'00"E 35°00'00"N

Kyoto University, Kwasan and Hida Observatories, Hida Observatory
Kamitakara
Gifu 506-1314
Telephone: (0)578-6-2311
Telefax: (0)578-6-2118
Electronic Mail: <userid>@kwasan.kyoto-u.ac.jp
WWW: http://www.kwasan.kyoto-u.ac.jp/Hida/Hida-j.html
http://www.kwasan.kyoto-u.ac.jp/Hida/Hida-e.html
Founded: 1968
Staff: 10
Activities: solar active phenomena * solar atmosphere * planets * comets * solar system
Periodicals: "Contributions of the Kwasan and Hida Observatories, Kyoto University" (ISSN 0388-2349)
Coordinates: 137°18'30"E 36°14'56"N H1,276m
City Reference Coordinates: 136°45'00"E 35°25'00"N

Kyoto University, Kwasan and Hida Observatories, Kwasan Observatory
Yamashina
Kyoto 607-8471
Telephone: (0)75-581-1235
Telefax: (0)75-593-9617
Electronic Mail: <userid>@kwasan.kyoto-u.ac.jp
WWW: http://www.kwasan.kyoto-u.ac.jp/Kwasan/Kwasan-j.html
Founded: 1929
Activities: solar and solar-system physics
Periodicals: "Contributions of the Kwasan and Hida Observatories, Kyoto University" (ISSN 0388-2349)
Coordinates: 135°47'46"E 34°59'26"N H221m (IAU Code 377)
City Reference Coordinates: 135°45'00"E 35°00'00"N (Kyoto)

Kyoto University, Yukawa Institute for Theoretical Physics (YITP)
Kitashirakawa-Owaike-cho
Sakyo-ku
Kyoto 606-8502
Telephone: (0)75-753-7008
Telefax: (0)75-753-7010
Electronic Mail: <userid>@yukawa.kyoto-u.ac.jp
WWW: http://www.yukawa.kyoto-u.ac.jp/
Founded: 1953
Staff: 24
Activities: fundamental theoretical physics
Periodicals: "Progress of Theoretical Physics" (ISSN 0033-068x)
City Reference Coordinates: 135°45'00"E 35°00'00"N

Minolta Planetarium Co. Ltd.
2-30 Toyotsu-cho
Suita

Osaka 564-0051
Telephone: (0)6386-2050
Telefax: (0)6386-2027
Electronic Mail: mp-osk@mom.minolta.co.jp
WWW: http://www.minolta.com/japan/mp/
Founded: 1984
Staff: 76
Activities: manufacturing i.a. planetariums
City Reference Coordinates: 135°30'00"E 34°40'00"N

Misato Observatory
Misato
Wakayama 640-1366
Telephone: (0)73-498-0305
Telefax: (0)73-498-0306
Electronic Mail: kenkyu@obs.misato.wakayama.jp
WWW: http://www.obs.misato.wakayama.jp/index-e.html
Founded: 1995
Staff: 5
Activities: education * popularization * observing
Coordinates: 135°24'34"E 34°08'29"N H430m (IAU Code 359)
City Reference Coordinates: 135°21'00"E 34°09'00"N

Mizusawa Astrogeodynamics Observatory
• See "National Astronomical Observatory of Japan (NAOJ), Mizusawa Astrogeodynamics Observatory"

Nagoya City Science Museum (NCSM), Astronomy Section
17-1 Sakae 2-chome
Naka-ku
Nagoya 460-0008
Telephone: (0)552-201-4486 x201
Telefax: (0)552-203-0788
Electronic Mail: astro@ncsm.city.nagoya.jp
WWW: http://www.ncsm.city.nagoya.jp/astro/astro.html
     http://www.ncsm.city.nagoya.jp/planet/index.html (Planetarium)
Activities: exhibitions * planetarium shows
City Reference Coordinates: 136°55'00"E 35°10'00"N

Nagoya University, Department of Physics and Astrophysics
Furo-cho
Chikusa-ku
Nagoya 464-8602
Telephone: (0)52-789-2840
Telefax: (0)52-789-2845
Electronic Mail: <userid>@a.phys.nagoya-u.ac.jp
WWW: http://www.a.phys.nagoya-u.ac.jp/
Coordinates: 138°19'12"E 36°31'18"N H1,280m (Sugadaira Station)
             138°36'42"E 35°26'36"N H1,000m (Fujigane Station)
             138°58'24"E 35°08'54"N H75m (Radio Astronomy Laboratory)
City Reference Coordinates: 136°55'00"E 35°10'00"N

Nagoya University, Solar-Terrestrial Environment Laboratory
3-13 Honohara
Toyokawa
Aichi 442-8507
Telephone: (0)5338-6-3154
Telefax: (0)5338-6-0811
Electronic Mail: <userid>@stelab.nagoya-u.ac.jp
WWW: http://www.stelab.nagoya-u.ac.jp/ste-www1/
Founded: 1990
Staff: 59
Activities: O3, NOx and aerosols observing * LIDAR * aurorae * ULF, VLF & LF-waves observing * interplanetary scintillation * cosmic rays * UHF telescopes * 3D MHD simulations
Coordinates: 130°43'00"E 31°29'00"N H20m (Kagoshima Observatory)
             137°03'00"E 34°44'00"N H3m (Sakushima Observatory)
             137°22'18"E 34°50'12"N H18m (Toyokawa Observatory)
             137°32'00"E 35°35'00"N H330m (Sakashita Observatory)
             137°38'00"E 35°48'00"N H1,100m (Kiso Station)
             138°37'00"E 35°26'00"N H1,000m (Fuji Observatory)
             138°19'00"E 36°31'00"N H1,300m (Sugadaira Station)
             142°16'00"E 44°22'00"N H290m (Moshiri Observatory)
City Reference Coordinates: 137°15'00"E 35°00'00"N

Nagoya University, Solar-Terrestrial Environment Laboratory, Cosmic-Ray Section
Furo-cho

Chikusa-ku
Nagoya 464-8601
Telephone: (0)52-789-4314
Telefax: (0)52-789-4313
Electronic Mail: <userid>@stelab.nagoya-u.ac.jp
WWW: http://www.stelab.nagoya-u.ac.jp/ste-www1/div3/CR/index.html
http://www.stelab.nagoya-u.ac.jp/omosaic/crle.html
Founded: 1947
Staff: 11
Activities: solar neutrons * cosmic rays * microlensing
Periodicals: "Annual Report"
City Reference Coordinates: 136°55'00"E 35°10'00"N

Nagoya University, X-Ray Astronomy Group
Furo-cho
Chikusa-ku
Nagoya 464-8602
Telephone: (0)52-789-2917
Telefax: (0)52-789-2919
Electronic Mail: ulab@u.phys.nagoya-u.ac.jp
WWW: http://www.u.phys.nagoya-u.ac.jp/uxg.html
City Reference Coordinates: 136°55'00"E 35°10'00"N

National Astronomical Observatory of Japan (NAOJ)
2-21-1 Osawa
Mitaka
Tokyo 181-8588
Telephone: (0)422-34-3600
Telefax: (0)422-34-3690
Electronic Mail: <userid>@mtk.nao.ac.jp
WWW: http://www.nao.ac.jp/
Founded: 1988
Staff: 255
Activities: almost all fields of ground-based, space, and theoretical astronomy
Periodicals: "Publications of the National Astronomical Observatory" (ISSN 0915-3640), "Report of the National Astronomical Observatory" (ISSN 0915-6321), "National Astronomical Observatory Reprint" (ISSN 0915-0021), "Annual Report of the National Astronomical Observatory of Japan" (ISSN 0915-66410)
Coordinates: 139°32'29"E 35°40'21"N H59m (Mitaka – IAU Code 388)
133°35'48"E 34°34'26"N H372m (Okayama - see separate entry)
137°33'19"E 36°06'49"N H2,876m (Norikura - see separate entry)
138°28'48"E 35°56'18"N H1,350m (Nobeyama - see separate entry)
141°07'52"E 39°08'03"N H61m (Mizusawa - see separate entry)
City Reference Coordinates: 139°46'00"E 35°42'00"N

National Astronomical Observatory of Japan (NAOJ), Astrometry and Celestial Mechanics Division
2-21-1 Osawa
Mitaka
Tokyo 181-8588
Telephone: (0)422-34-3633
Telefax: (0)422-34-3793
Electronic Mail: <userid>@nao.ac.jp
WWW: http://pluto.mtk.nao.ac.jp/
Founded: 1988
Staff: 24
Activities: celestial mechanics * non-linear dynamics * position astronomy * ephemerides * gravity-wave detection * relativistic astrometry * stellar dynamics * solar-system dynamics
City Reference Coordinates: 139°46'00"E 35°42'00"N

National Astronomical Observatory of Japan (NAOJ), Astronomical Data Analysis Center (ADAC)
2-21-1 Osawa
Mitaka
Tokyo 181-8588
Telephone: (0)422-34-3633
Telefax: (0)422-34-3793
Electronic Mail: <userid>@c1.mtk.nao.ac.jp
WWW: http://www.cc.nao.ac.jp/index-e.html
http://adac.mtk.nao.ac.jp/
Founded: 1988 (NAOJ)
City Reference Coordinates: 139°46'00"E 35°42'00"N

National Astronomical Observatory of Japan (NAOJ), Division of Earth Rotation
c/o Mizusawa Astrogeodynamics Observatory
2-12 Hoshigaoka-machi
Mizusawa-shi
Iwate 023-0861
Telephone: (0)197-22-7111

Telefax: (0)197-22-7120
Electronic Mail: <userid>@sinet.ad.jp
WWW: http://www.miz.nao.ac.jp/
Founded: 1988 (NAOJ)
Coordinates: 141°07'52"E 39°08'03"N H61m
City Reference Coordinates: 141°22'00"E 39°37'00"N

National Astronomical Observatory of Japan (NAOJ), Division of Theoretical Astrophysics
2-21-1 Osawa
Mitaka
Tokyo 181-8588
Telephone: (0)422-41-3740
Telefax: (0)422-41-3746
Electronic Mail: <userid>@th.nao.ac.jp
WWW: http://th.nao.ac.jp/
Founded: 1988
Staff: 11
Activities: cosmology * galaxy evolution * QSOs * star formation * planetary formation * ISM
City Reference Coordinates: 139°46'00"E 35°42'00"N

National Astronomical Observatory of Japan (NAOJ), Dodaira Observatory
• Closed in March 2000.

National Astronomical Observatory of Japan (NAOJ), Mizusawa Astrogeodynamics Observatory
2-12 Hoshigaoka-machi
Mizusawa-shi
Iwate 023-0861
Telephone: (0)197-22-7111
Telefax: (0)197-22-7120
Electronic Mail: <userid>@sinet.ad.jp
WWW: http://www.miz.nao.ac.jp/
Founded: 1988
Staff: 29
Activities: astronomy * Earth rotation * geodesy * geophysics
Periodicals: (1) "Technical Reports of the Mizusawa Kansoku Center National Astronomical Observatory" (ISSN 0915-3780), "Annual Report of the Mizusawa Astrogeodynamics Observatory - Time Service and Geophysical Observations" (ISSN 0916-6343)
Coordinates: 141°07'52"E 39°08'03"N H61m
           141°20'07"E 39°08'53"N H393m (Esashi Earth Tides Station)
City Reference Coordinates: 141°22'00"E 39°37'00"N

National Astronomical Observatory of Japan (NAOJ), Nobeyama Radio Observatory (NRO)
Nobeyama
Minamimaki-mura
Nagano 384-1305
Telephone: (0)267-63-4300
Telefax: (0)267-98-2884
        (0)267-98-2506
Electronic Mail: service@solar.nro.nao.ac.jp
                <userid>@nro.nao.ac.jp
WWW: http://solar.nro.nao.ac.jp/
     http://www.nro.nao.ac.jp/index-e.html
     http://www.nro.nao.ac.jp/cosmic-e.html
Founded: 1982
Staff: 62
Activities: solar and cosmic radio observing * molecular lines * continuum * polarization * solar and cosmic fine structures * dark clouds * SNR * galactic activities
Periodicals: "NRO Report" (ISSN 0911-5501), "NRO Newsletter" (ISSN 0911-5870)
Coordinates: 138°28'48"E 35°56'18"N H1,350m
City Reference Coordinates: 138°11'00"E 36°39'00"N

National Astronomical Observatory of Japan (NAOJ), Norikura Solar Observatory
Mount Norikura
Azumi-mura
Nagano 390-1500
Telephone: (0)90-2223-7010
WWW: http://solarwww.mtk.nao.ac.jp/norikura.html
Founded: 1949
Staff: 11
Activities: spectroscopic observing of solar corona, prominences, chromosphere and white-light flares * 5303Å emission corona
Coordinates: 137°33'19"E 36°06'49"N H2,876m (IAU Code 382)
City Reference Coordinates: 138°11'00"E 36°39'00"N

National Astronomical Observatory of Japan (NAOJ), Okayama Astrophysical Observatory (OAO)

Kamogata
Asakuchi
Okayama 719-0232
Telephone: (0)8654-4-2155
Telefax: (0)8654-4-2360
WWW: http://www.cc.nao.ac.jp/oao/
http://www.cc.nao.ac.jp/oao/e/index.html
Founded: 1960
Activities: optical and near-IR observing of stars and galaxies * solar magnetic field
Coordinates: 133°35'48"E 34°34'26"N H372m (Mount Chikurinji – IAU Code 371)
City Reference Coordinates: 133°55'00"E 34°39'00"N

National Astronomical Observatory of Japan (NAOJ), Optical and Infrared Astronomy Division
2-21-1 Osawa
Mitaka
Tokyo 181-8588
Telephone: (0)422-34-3520
Telefax: (0)422-34-3527
Electronic Mail: iye@optik.mtk.nao.ac.jp (Masanori Iye, Director)
<userid>@optik.mtk.nao.ac.jp
WWW: http://optik2.mtk.nao.ac.jp/index-e.html
Founded: 1888
Staff: 30
Activities: observing * theoretical studies * instrumentation * observatory operation
Periodicals: (1) "Annual Report" (ISSN 0915-6410)
Observatories: 2 (Hawaii, Okayama - see separate entry)
City Reference Coordinates: 139°46'00"E 35°42'00"N

National Astronomical Observatory of Japan (NAOJ), Radio Astronomy Division
c/o Nobeyama Radio Observatory
Nobeyama
Minamimaki-mura
Nagano 384-1305
Telephone: (0)267-98-2831
Telefax: (0)267-98-2884
WWW: http://www.nro.nao.ac.jp/index-e.html
Founded: 1988 (NAOJ)
City Reference Coordinates: 138°11'00"E 36°39'00"N

National Astronomical Observatory of Japan (NAOJ), Solar Physics Division
2-21-1 Osawa
Mitaka
Tokyo 181-8588
Telephone: (0)422-34-3716
Telefax: (0)422-41-3700
Electronic Mail: <userid>@solar.mtk.nao.ac.jp
WWW: http://solarwww.mtk.nao.ac.jp/
Founded: 1988
Staff: 24
Activities: solar atmosphere * solar activity (sunspots, flares, corona, magnetic fields)
City Reference Coordinates: 139°46'00"E 35°42'00"N

National Institute for Fusion Science (NIFS)
Oroshi-cho
Toki
Gifu 509-5292
Telephone: (0)572-58-2222
Telefax: (0)572-58-2601
WWW: http://www.nifs.ac.jp/
Founded: 1989
Staff: 300
Activities: nuclear fusion * plasma physics * atomic processes
Periodicals: (1) "Annual Report" (ISSN 0917-1185)
City Reference Coordinates: 137°10'00"E 35°20'00"N

National Institute of Polar Research (NIPR)
Kaga 1-9-10
Itabashi-ku
Tokyo 173-8515
Telephone: (0)3-3962-2214
Telefax: (0)3-3962-2224

WWW: http://www.nipr.ac.jp/
Founded: 1973
Staff: 140
Activities: organizing and carrying out Japanese antarctic research expeditions and arctic research programmes including atmospheric physics, meteorology, glaciology, Earth sciences and biology
Periodicals: "JARE Data Reports" (ISSN 0075-3343), "Advances in Polar Upper Atmospheric Research" (ISSN 1345-1065), "Polar Meteorology and Glaciology" (ISSN 1344-3437), "Polar Geoscience" (ISSN 1344-3194), "Antarctic Meteorite Research" (ISSN 1343-4284), "Polar Bioscience" (ISSN 1344-6231)
Coordinates: 039°35'00"E 69°00'00"S (Syowa Station)
  039°42'12"E 77°19'01"S H2,230m (Dome Fuji Station)
City Reference Coordinates: 139°46'00"E 35°42'00"N

National Institute for Research Advancement (NIRA)
Box 5004
Yebisu Garden Place Post Office
Shibuya-ku
Tokyo 150-6034
or :
34th Floor
Yebisu Garden Place Tower
4-20-3 Ebisu
Shibuya-ku
Tokyo 150-6034
Telephone: (0)3-5448-1700
Telefax: (0)3-5448-1743
WWW: http://www.nira.go.jp/
Founded: 1974
Activities: policy research organization
City Reference Coordinates: 139°46'00"E 35°42'00"N

National Science Museum, Department of Physical Sciences
Ueno Park
Taito-ku
Tokyo 110-8718
Telephone: (0)3-3822-0111
Telefax: (0)3-5814-9899
Activities: meteorites * variable stars * eclipsing binaries * observing (sunspots) * education
Periodicals: "Bulletin of the National Science Museum, Series E (Physical Sciences and Engineering)"
Coordinates: 139°46'48"E 35°42'47"N
City Reference Coordinates: 139°46'00"E 35°42'00"N

National Space Development Agency of Japan (NASDA)
• See now "Japan Aerospace Exploration Agency (JAXA)"

Nippon Meteor Society (NMS)
c/o Takatsugu Yoshida
110 Mukaiyama
Gyoyu-cho
Toyokawa-shi
Aichi 441-0211
Telephone: (0)533-88-6884
Telefax: (0)533-88-6884
Electronic Mail: lmj53851@biglobe.ne.jp
Founded: 1969
Membership: 350
Activities: observing (visual, telescopic, photographic, radio and video) and investigating meteors
Periodicals: "Astronomical Circular" (ISSN 0388-5852), "Ryusei-sokuho" (Meteor Advance Report), "Star Friends" (ISSN 0389-0341), "Journal of Meteor Observations"
Coordinates: 136°05'35"E 35°09'13"N H100m (Shiga)
City Reference Coordinates: 137°15'00"E 35°00'00"N

Nishi-Harima Astronomical Observatory (NHAO)
Sayo-cho
Hyogo 679-5313
Telephone: (0)790-82-3886
Telefax: (0)790-82-3514
Electronic Mail: harima@nhao.go.jp
WWW: http://www.nhao.go.jp/
Founded: 1990
Activities: popularization * education * observing * research
Periodicals: (1) "Annual Report of the Nishi-Harima Astronomical Observatory" (ISSN 0917-6926)
Coordinates: 134°20'19"E 35°01'21"N H446m
City Reference Coordinates: 135°00'00"E 35°00'00"N

Nobeyama Radio Observatory (NRO)
• See "National Astronomical Observatory of Japan (NAOJ), Nobeyama Radio Observatory (NRO)"

Norikura Solar Observatory
• See "National Astronomical Observatory of Japan (NAOJ), Norikura Solar Observatory"

Okayama Astrophysical Observatory (OAO)
• See "National Astronomical Observatory of Japan (NAOJ), Okayama Astrophysical Observatory (OAO)"

Okinawa Radio Observatory
• See "Communications Research Laboratory (CRL), Okinawa Radio Observatory"

Osaka Kyoiku University (OKU), Astronomical Institute
Asahigaoka
Kashiwara
Osaka 582-8582
Telephone: (0)729-76-3211 x3124
Telefax: (0)729-76-3269
Electronic Mail: <userid>@cc.osaka-kyoiku.ac.jp
WWW: http://quasar.cc.osaka-kyoiku.ac.jp/
      http://quasar.cc.osaka-kyoiku.ac.jp/astroe.html
Founded: 1949
Staff: 3
Activities: accretion disks * AGNs * astrodynamics * astrophysical jets * galaxy * stars
Periodicals: (2) "Memoirs of Osaka Kyoiku University"
Coordinates: 135°39'07"E 34°32'53"N
City Reference Coordinates: 135°30'00"E 34°40'00"N

Ouda Station
• See "Kyoto University, Department of Astronomy, Ouda Station"

Physical Society of Japan (The_)
• See "The Physical Society of Japan"

Preserving Starry-Sky Association (PSA)
c/o Shigemi Uchida (Section Organizer)
4-21-206 Wakabadai Asahiku
Yokohama 241
or :
c/o Satoru Ohotomo (Public Relations Officer)
3545 Kiyosato Takane-chu
Kitakoma-gun
Yamanashi 407-33
Telephone: (0)551-48-3822 (Shigemi Uchida)
         (0)45-921-2334 (Satoru Ohotomo)
Telefax: (0)551-48-3822 (Shigemi Uchida)
       (0)45-921-2334 (Satoru Ohotomo)
Electronic Mail: suchida@mxb.meshnet.or.jp
WWW: http://www2a.meshnet.or.jp/~wakaba/
Founded: 1996
City Reference Coordinates: 139°39'00"E 35°27'00"N (Yokohama)
                    138°40'00"E 35°40'00"N (Yamanashi)

Saji Astronomical Observatory
1071-1 Takayama
Saji-son
Yazu-gun
Tottiri 689-1312
Telephone: (0)858-89-1011
Telefax: (0)858-88-0103
Electronic Mail: sajinet@infosakyu.ne.jp
WWW: http://www.infosakyu.ne.jp/sajinet
Founded: 1994
• Private observatory (IAU Code 867)
City Reference Coordinates: 134°12'00"E 35°32'00"N

Science Council of Japan
22-34 Roppongi, 7-chome
Minato-ku
Tokyo 106-8555
Telephone: (0)3-3403-1949
Telefax: (0)3-3403-1942

Founded: 1949
Membership: 210
Activities: IAU contact through the "Japanese National Committee for Astronomy"
City Reference Coordinates: 139°46'00"E 35°42'00"N

Sendai Children's Space Center
1-8-6 Izumi-chuo
Izumi-ku
Sendai 980
Telephone: (0)22-373-0999
WWW: http://english.itp.ne.jp/jtd/jinfo/sendai.html
Activities: exhibitions * planetarium shows
City Reference Coordinates: 140°52'00"E 38°16'00"N

Shibayama Scientific Co. Ltd.
3 Chome 11-8
Minami-Otsuka
Toshima-ku
Tokyo 170-0005
Telephone: (0)3-3987-4151
Telefax: (0)3-3987-4155
WWW: http://www.shibayama.co.jp/
Activities: manufacturer * trading company * consultant for planetariums and museums
City Reference Coordinates: 139°46'00"E 35°42'00"N

Shiga University, Department of Earth Science
2-5-1 Hiratsu
Otsu 520-0862
Telephone: (0)775-37-7780
Telefax: (0)775-37-7839
WWW: http://www.sue.shiga-u.ac.jp/
Founded: 1949 (University)
Activities: solar activity (prominences, spicules, corona, oscillations)
Periodicals: (1) "Memoirs of the Faculty of Education, Shiga University"
Coordinates: 135°54'30"E 34°56'48"N
City Reference Coordinates: 135°52'00"E 35°00'00"N

Shimosato Hydrographic Observatory (SHO)
• See "Japan Hydrographic and Oceanographic Department (JHOD), Shimosato Hydrographic Observatory (SHO)"

Shirahama Hydrographic Observatory
• See "Japan Hydrographic and Oceanographic Department (JHOD), Shirahama Hydrographic Observatory"

Society of Geomagnetism and Earth, Planetary and Space Sciences (SGEPSS)
c/o Toshihiko Iyemori (Secretary General)
Data Analysis Center for Geomagnetism and Space Magnetism
Graduate School of Science
Kyoto University
Kitashirakawa
Oiwake-cho
Kyoto 606-8502
Telephone: (0)75-753-3949
Telefax: (0)75-722-7884
Electronic Mail: iyemori@kugi.kyoto-u.ac.jp
            sgepss@kurasc.kyoto-u.ac.jp
WWW: http://www.kurasc.kyoto-u.ac.jp/sgepss/index-e.html
Founded: 1946
Membership: 763
Periodicals: "Earth, Planets and Space" (ISSN 0022-1392)
City Reference Coordinates: 135°45'00"E 35°00'00"N

Sophia University, Department of Physics
7-1 Kioi-cho
Chiyoda-ku
Tokyo 102-8554
Telephone: (0)3-3238-3431
Telefax: (0)3-3238-3341
Electronic Mail: <userid>@hoffman.cc.sophia.ac.jp
WWW: http://www.sophia.ac.jp/
Founded: 1962
Staff: 24
Activities: education and research in physics (including astrophysics)
City Reference Coordinates: 139°46'00"E 35°42'00"N

Sunshine Planetarium
3-1-3 Higashi Ikebukuro
Toshima-ku
Tokyo 170-0013
Telephone: (0)3-3989-3475
WWW: http://www.princehotels.co.jp/english/info1/05000100/08001008.html
City Reference Coordinates: 139°46'00"E 35°42'00"N

The Japan Academy
7-32 Ueno Park
Taito-ku
Tokyo 110-0007
Telephone: (0)3-3822-2101
Telefax: (0)3-3822-2105
WWW: http://www.japan-acad.go.jp/english/e_index.htm
Founded: 1879
Membership: 150
Activities: prizes * publishing * international exchanges * public lectures
Periodicals: "Transactions of the Japan Academy" (ISSN 0388-0036), "Proceedings of the Japan Academy A - Mathematical Sciences" (ISSN 0386-2194), "Proceedings of the Japan Academy B - Physical and Biological Sciences" (ISSN 0386-2208)
Awards: "Imperial Prize", "The Japan Academy Prize", "The Duke of Edinburgh Prize"
City Reference Coordinates: 139°46'00"E 35°42'00"N

The Physical Society of Japan
Room 211
Kikai-Shinko Building
3-5-8 Shiba-Koen
Minato-ku
Tokyo 105-0011
Telephone: (0)3-3434-2671
Telefax: (0)3-3432-0997
Electronic Mail: <userid>@jps.or.jp
                    pubpub@jps.or.jp
WWW: http://wwwsoc.nacsis.ac.jp/jps/jps/
Founded: 1946
Membership: 19000
Staff: 13
Activities: holding scientific meetings * publishing journals and other documents * cooperation and exchange with other societies * documentation centre for members * sponsoring activities within the objectives of the society
Periodicals: (13) "Journal of the Physical Society of Japan (JPSJ)" (ISSN 0031-9015, circ.: 2,500); (12) "Butsuri" (ISSN 0029-0181, circ.: 19,000)
Awards: (1) "Article Prize"
City Reference Coordinates: 139°46'00"E 35°42'00"N

Tohoku University, Astronomical Institute
Aramaki Aza Aoba
Sendai
Aoba-ku
Sendai 980-8578
Telephone: (0)22-217-6512
Telefax: (0)22-217-6513
Electronic Mail: <userid>@jpntohok.bitnet
WWW: http://www.astr.tohoku.ac.jp/
Founded: 1934
Activities: stellar structure and evolution * pulsating stars * stellar winds and magnetospheres * accretion discs * emission-line stars * ISM * structure and dynamics of galaxies * protogalaxy * radio & optical observing
Periodicals: (3) "The Scientific Reports of the Tohoku University, Eighth Series: Physics and Astronomy" (ISSN 0388-5607); "Sendai Astronomiaj Raportoj"
Coordinates: 140°50'33"E 38°15'26"N H153m
City Reference Coordinates: 140°52'00"E 38°16'00"N (Sendai)

Tokai University, Research Institute of Civilization
1117 Kitakaname
Hiratsuka-shi
Kanagawa 259-1292
Telephone: (0)463-58-1211
Telefax: (0)463-35-2456
Electronic Mail: <userid>@keyaki.cc.u-tokai.ac.jp
                    <userid>@rh.u-tokai.ac.jp
WWW: http://www.u-tokai.ac.jp/English/
Founded: 1959
Staff: 36
Activities: stellar spectroscopy * SETI
Periodicals: "Bulletin of the Research Institute of Civilization" (ISSN 0285-0818)

City Reference Coordinates: 139°38'00"E 35°28'00"N

Tokushima Kainan Observatory
WWW: http://www-cc.ee.tokushima-u.ac.jp/kainan/
Founded: 1992
• Private observatory (IAU Code 872)
City Reference Coordinates: 134°34'00"E 34°03'00"N

Tokyo Astronomical Observatory
• See now "National Astronomical Observatory of Japan (NAOJ)"

Tokyo Gakugei University, Department of Astronomy and Earth Sciences
4-1-1 Nukuikitamachi
Koganei
Tokyo 184-8501
Telephone: (0)42-329-7111
Telefax: (0)42-329-7538
Electronic Mail: <userid>@u-gakugei.ac.jp
WWW: http://www.u-gakugei.ac.jp/~kouhou/div/e-index.html
Founded: 1949
Staff: 3 (astronomy group)
Activities: galaxies * clusters * variable stars * pre-main-sequence stars * ISM * education
Periodicals: (1) "Bulletin of the Tokyo Gakugei University - Section IV: Mathematics and Natural Sciences" (ISSN 0371-6813)

Coordinates: 139°30'00"E 35°42'00"N H95m
City Reference Coordinates: 139°30'00"E 35°42'00"N

Tokyo Metropolitan University (TMU), Department of Physics
1-1 Minami-Ohsawa
Hachioji
Tokyo 192-0397
Telephone: (0)426-77-2511
Telefax: (0)426-77-2483
Electronic Mail: <userid>@phys.metro-u.ac.jp
WWW: http://www.phys.metro-u.ac.jp/
http://www-astro.phys.metro-u.ac.jp/index-e.html
Founded: 1949
Staff: 50
Activities: research * education * high-energy physics * nuclear physics * theoretical astrophysics * experimental astrophysics * fundamental physics * statistical physics * nonlinear physics * condensed-matter physics * electronic theory of solids * atomic physics * solid-state spectroscopy * material science * magnetic resonance * neutron scattering
City Reference Coordinates: 139°46'00"E 35°42'00"N

Toyama Astronomical Observatory (TAO)
San-no-kuma 49-4
Toyama 930-0155
Telephone: (0)76-434-9098
Telefax: (0)76-434-9228
Electronic Mail: watanabe@tsm.toyama.toyama.jp (Watanabe Makoto, Staff Member)
WWW: http://www.tsm.toyama.toyama.jp/tao/
Founded: 1956
Staff: 6
Activities: annex of Toyama Science Museum (TSM - see separate entry) * observing * education * exhibitions * research
Coordinates: 137°06'06"E 36°39'30"N (IAU Code 908)
City Reference Coordinates: 137°13'00"E 36°41'00"N

Toyama Science Museum (TSM)
Nishinakano 1-8-31
Toyama 939-8084
Telephone: (0)76-491-2123
Telefax: (0)76-421-5950
Electronic Mail: watanabe@tsm.toyama.toyama.jp (Watanabe Makoto, Staff Member)
WWW: http://www.tsm.toyama.toyama.jp/
Founded: 1979
Staff: 27
Activities: exhibitions (astronomy, physics, chemistry, biology, geology, meteorology) * planetarium * public observatory (see separate entry)
Periodicals: "Bulletin of Toyama Science Museum", "Special Publication of Toyama Science Museum"
City Reference Coordinates: 137°13'00"E 36°41'00"N

Toyama University, Department of Physics
3190 Gofu-ku
Toyama 930-8555
Telephone: (0)764-45-6590
Telefax: (0)764-45-6549

Electronic Mail: <userid>@sci.toyama-u.ac.jp
WWW: http://www.sci.toyama-u.ac.jp/sci/phys/
Founded: 1949
Staff: 4
Activities: physics of molecules * interstellar molecules * ionized regions * radio astronomy
City Reference Coordinates: 137°13'00"E 36°41'00"N

University of Tokyo, Department of Astronomy
Hongo 7-3-1
Bunkyo-ku
Tokyo 113-0033
Telephone: (0)3-812-2111 x4251
Telefax: (0)3-3813-9439
        (0)3-5841-7644
Electronic Mail: <userid>@tansei.cc.u-tokyo.ac.jp
                <userid>@astron.s.u-tokyo.ac.jp
WWW: http://www.astron.s.u-tokyo.ac.jp/
Founded: 1877
Periodicals: "Contributions" (ISSN 0563-8038)
City Reference Coordinates: 139°46'00"E 35°42'00"N

University of Tokyo, Department of Earth and Planetary Physics
7-3-1 Hongo
Tokyo 113-0033
Telephone: (0)3-3812-2111
Telefax: (0)3-3818-3247
WWW: http://www.geoph.s.u-tokyo.ac.jp/
City Reference Coordinates: 139°46'00"E 35°42'00"N (Tokyo)

University of Tokyo, Department of Earth Science and Astronomy
3-8-1 Komaba
Meguro-ku
Tokyo 153
Telephone: (0)3-3467-1171
Telefax: (0)3-3485-2904
Electronic Mail: <userid>@tansei.cc.u-tokyo.ac.jp
                <userid>@astron.s.u-tokyo.ac.jp
WWW: http://www.esa.c.u-tokyo.ac.jp/
     http://grape.c.u-tokyo.ac.jp/
Activities: N-body simulations * stellar dynamics * stellar structure * supernovae * rotating stars * relativistic astrophysics
City Reference Coordinates: 139°46'00"E 35°42'00"N

University of Tokyo, Department of Physics, Theoretical Astrophysics Group
Hongo 7-3-1
Bunkyo-ku
Tokyo 113-0033
Electronic Mail: <userid>@phys.s.u-tokyo.ac.jp
WWW: http://www-utap.phys.s.u-tokyo.ac.jp/index-e.html
     http://www.phys.s.u-tokyo.ac.jp/
City Reference Coordinates: 139°46'00"E 35°42'00"N

University of Tokyo, Institute for Cosmic Ray Research (ICCR)
5-1-5 Kashiwanoha
Kashiwa
Chiba 277-8582
Telephone: (0)4-7136-3102
Telefax: (0)4-7136-3115
Electronic Mail: <userid>@icrr.u-tokyo.ac.jp
WWW: http://www.icrr.u-tokyo.ac.jp/
Founded: 1953
Staff: 26
Activities: cosmic-ray research * high-energy astrophysics * interplanetary dust * particle astrophysics * neutrino astronomy
Periodicals: "ICRR Report" (ISSN 0919-8296), "Annual Report"
Coordinates: 137°33'00"E 36°06'00"N H2,770m (Norikura)
             138°20'00"E 35°04'42"N H900m (Akeno - see separate entry)
             137°19'11"E 36°25'26"N H-1,000m (Kamioka - see separate entry)
             139°49'48"E 36°34'48"N H-80m (Ohya)
             138°44'00"E 35°22'00"N H3,750m (Mount Fuji)
             139°51'00"E 35°09'00"N H-180m (Nokogiriyama)
City Reference Coordinates: 140°05'00"E 35°36'00"N

University of Tokyo, Institute for Cosmic Ray Research (ICCR), Akeno Observatory
5259 Asao
Akeno-mura
Kitakoma-gun

Yamanashi 407-0201
Telephone: (0)551-25-2301
Telefax: (0)551-35-2303
WWW: http://www.icrr.u-tokyo.ac.jp/akeno/akeno.html
Founded: 1953 (ICCR)
Activities: see main entry
Periodicals: see main entry
City Reference Coordinates: 138°40'00"E 35°40'00"N

University of Tokyo, Institute for Cosmic Ray Research (ICCR), Kamioka Underground Observatory
456 Higashi-mozumi
Kamioka-cho
Yoshiki-gun
Gifu 506-11111
Telephone: (0)578-5-2116
Telefax: (0)578-5-2121
Electronic Mail: <userid>@suketto.icrr.u-tokyo.ac.jp
WWW: http://www-sk.icrr.u-tokyo.ac.jp/
Founded: 1981
Staff: 29
Activities: solar neutrinos
Periodicals: see main entry
Coordinates: 137°19'11"E 36°25'26"N H-1000m
City Reference Coordinates: 136°45'00"E 35°25'00"N

University of Tokyo, Institute of Astronomy, Kiso Observatory
Tarusawa
Mitake-mura
Kiso-gun
Nagano 397-0101
Telephone: (0)264-52-3360
Telefax: (0)264-52-3361
WWW: http://www.ioa.s.u-tokyo.ac.jp/kisohp/
Founded: 1974
Staff: 5
Activities: light variabilities * objective prism survey * optical infrared imaging
Periodicals: (1) "Annual Report of the Kiso Observatory" (ISSN 0915-5392)
Coordinates: 137°37'42"E 35°47'39"N H1,130m (IAU Code 381)
City Reference Coordinates: 138°11'00"E 36°39'00"N

University of Tokyo, Research Center for the Early Universe (RESCEU)
School of Science
7-3-1 Hongo
Bunkyo-ku
Tokyo 113-0033
Telefax: (0)3-5802-8691
Electronic Mail: <userid>@resceu.s.u-tokyo.ac.jp
WWW: http://www.resceu.s.u-tokyo.ac.jp/Welcome.html
Founded: 1995
Staff: 10
Activities: structure and evolution of early universe and galaxies
City Reference Coordinates: 139°46'00"E 35°42'00"N

University of Tokyo, Tokyo Astronomical Observatory
• See now "National Astronomical Observatory of Japan (NAOJ)"

Variable Star Observers League in Japan (VSOLJ)
c/o National Science Museum
Ueno Park
Taito-ku
Tokyo 110-8718
Electronic Mail: nah01147@niftyserve.or.jp (Makoto Watanabe, Database Manager)
WWW: http://www.kusastro.kyoto-u.ac.jp/vsnet/VSOLJ/vsolj.html
Periodicals: "VSOLJ Variable Star Bulletin"
City Reference Coordinates: 139°46'00"E 35°42'00"N

Vixen Optical Industries Ltd.
247 Hongo
Tokorozawa
Saitama 359-0022
Telephone: (0)42-944-4141
Telefax: (0)42-944-9722
Electronic Mail: vixen@mb.infoweb.ne.jp
WWW: http://www.vixen.co.jp/english/index_e.htm
Activities: manufacturers, among others, telescopes, binoculars and scopes

City Reference Coordinates: 139°30'00"E 36°00'00"N

Wakkanai Radio Observatory
• See "Communications Research Laboratory (CRL), Wakkanai Radio Observatory"

World Data Center C2 for Ionosphere
c/o Communications Research Laboratory
4-2-1 Nukuikitamachi
Koganei-shi
Tokyo 184-8795
Telephone: (0)423-27-7478
Telefax: (0)423-27-7606
Electronic Mail: <userid>@crl.go.jp
WWW: http://wdc-c2.crl.go.jp/index_eng.html
Periodicals: (1) "Catalog of Data in World Data Center C2 for Ionosphere" (ISSN 0389-8229)
City Reference Coordinates: 139°46'00"E 35°42'00"N

World Data Center for Solar Activity
c/o National Astronomical Observatory of Japan
2-21-1 Osawa
Mitaka
Tokyo 181-8588
Telephone: (0)422-34-3716
Telefax: (0)422-34-3700
Electronic Mail: sunspot@solar.mtk.nao.ac.jp
Founded: 1978
Staff: 6
Periodicals: (12) "Monthly Bulletin on Solar Phenomena"; (4) "Quarterly Bulletin on Solar Activity (QBSA)"
City Reference Coordinates: 139°46'00"E 35°42'00"N

Yamagawa Radio Observatory
• See "Communications Research Laboratory (CRL), Yamagawa Radio Observatory"

Yamaguchi Prefectural Museum
8-2 Kasuga-cho
Yamaguchi 753-0073
Telephone: (0)839-22-0294
Telefax: (0)839-22-0353
Electronic Mail: yamahaku@ymg.urban.ne.jp
WWW: http://www.pref.yamaguchi.jp/e2top1f.htm
Founded: 1912
Staff: 19
Activities: research and public education (astronomy, earth sciences, botany, zoology, engineering, Japanese history)
Periodicals: (1) "Bulletin of the Yamaguchi Museum" (ISSN 0288-4232, circ.: 500), "Yamaguchi-ken no shizen - Nature Study of Yamaguchi-ken" (ISSN 0288-4240, circ.: 800), "Annual Report of the Yamaguchi Museum" (circ.: 1300)
Coordinates: 131°28'32"E 34°10'46"N H55m
City Reference Coordinates: 131°29'00"E 34°10'00"N

Yokohama Science Center (YSC), Astronomy Section
5-2-1 Yokodai
Isogo-ku
Yokohama 235-0045
Telephone: (0)45-832-1166
Telefax: (0)45-832-1161
Electronic Mail: <userid>@ysc.go.jp
WWW: http://www.city.yokohama.jp/yhspot/ysc/ysc/ysc.html
        http://www.city.yokohama.jp/yhspot/ysc/ysc/e-menu.html
Founded: 1984
Staff: 8
Activities: increasing understanding and appreciation of science including astronomy * WWW-based astronomy information service for the public and amateur astronomers * planetarium * education
Coordinates: 139°35'52"E 35°22'26"N H60m
City Reference Coordinates: 139°39'00"E 35°27'00"N

Yukawa Institute for Theoretical Physics (YITP)
• See "Kyoto University, Yukawa Institute for Theoretical Physics (YITP)"

Jordan

Al al-Bayt University, Institute of Astronomy and Space Sciences
P.O. Box 130040
Mafraq
Telephone: (0)6-4871101 x.2330
　　　　　　(0)6-5699270
Telefax: (0)6-5699270
　　　　　(0)6-4871232
Electronic Mail: alnaimiy@yahoo.com (Hamid M.K. Al-Naimiy, Director)
WWW: http://www.aabu.edu.jo/space/main1.htm
Periodicals: "Arab Journal of Astronomy and Space Sciences"
City Reference Coordinates: 036°12'00"E 32°20'00"N

Jordanian Astronomical Society (JAS)
P.O. Box 141568
Amman 11814
Telephone: (0)6-5534754
Telefax: (0)6-5534826
Electronic Mail: odehjas@jas.org.jo (Mohammad Shawkat Odeh)
WWW: http://www.jas.org.jo/
Founded: 1987 (as former "Jordanian Amateur Astronomers Society")
Membership: 50
City Reference Coordinates: 035°56'00"E 31°77'00"N

Jordanian Institution for Standards and Metrology (JISM)
P.O. Box 941287
Amman 11194
Telephone: (0)6-680139
Telefax: (0)6-681099
City Reference Coordinates: 035°56'00"E 31°77'00"N

Jordan Meteorological Department
P.O. Box 341011
Marka
Amman Civil Airport
Amman
Telephone: (0)6-4892408
　　　　　　(0)6-4892409
Telefax: (0)6-4894409
Electronic Mail: meteo@amra.nic.gov.jo
Founded: 1967
Staff: 300
Activities: meteorological services
City Reference Coordinates: 035°56'00"E 31°77'00"N

Kazakhstan

Fesenkov Astrophysical Institute (V.G._)
• See "National Academy of Sciences, V.G. Fesenkov Astrophysical Institute"

National Academy of Sciences, Space Research Institute (SRI)
Shevchenko Ul. 15
480021 Almaty
Telephone: (0)3272-615853
Telefax: (0)3272-494355
Electronic Mail: zak@kaziki.alma-ata.su
WWW: http://smis.iki.rssi.ru/inform/sri-kaz.htm
Founded: 1991
Staff: 125
City Reference Coordinates: 076°57'00"E 43°15'00"N

National Academy of Sciences, V.G. Fesenkov Astrophysical Institute
Kamenskoye Plato
480068 Almaty
Telephone: (0)3272-648311
  (0)3272-624040
  (0)3272-677018
  (0)3272-255425
Periodicals: "Trudy"
Coordinates: 076°57'24"E 43°11'18"N H1,450m
City Reference Coordinates: 076°57'00"E 43°15'00"N

National Academy of Sciences, V.G. Fesenkov Astrophysical Institute, Tien-Shan Observatory
Kamenskoye Plato
480068 Almaty
Telephone: (0)3272-658040
  (0)3272-252092
Electronic Mail: kurt@afi.academ.alma-ata.su (Kenes S. Kuratov, Chief)
Activities: education * research * observing (IAU Code 210)
City Reference Coordinates: 076°57'00"E 43°15'00"N

Republican Palace of School Children Observatory
Lenin Prospekt 114
480051 Almaty
Telephone: (0)3272-643866
Founded: 1983
Staff: 5
Activities: education * observing (Sun, comets, variable stars)
Coordinates: 075°40'00"E 43°16'00"N
City Reference Coordinates: 076°57'00"E 43°15'00"N

State Information Center for Standardization
Ul. Pushkina 166
473000 Astana
Telephone: (0)3172-752991
Telefax: (0)3172-320540
Electronic Mail: gic@asdc.kz
Founded: 1997
Staff: 22
Activities: national norms and standards
City Reference Coordinates: 076°57'00"E 43°15'00"N (Almaty)

Tien-Shan Observatory
• See "National Academy of Sciences, V.G. Fesenkov Astrophysical Institute, Tien-Shan Observatory"

V.G. Fesenkov Astrophysical Institute
• See "National Academy of Sciences, V.G. Fesenkov Astrophysical Institute"

Kenya

Kenya Bureau of Standards (KEBS)
Off Mombasa Road
Behind Belle Vue Cinema
P.O. Box 54974
Nairobi
Telephone: (0)2-502210
　　　　　(0)2-502219
Telefax: (0)2-503293
Electronic Mail: kebs@users.africaonline.co.ke
WWW: http://www.kebs.org/
City Reference Coordinates: 036°49'00"E 01°17'00"S

Kenya Meteorological Department (KMD)
Dagoretti Corner
Ngong Road
P.O. Box 30259
Nairobi
Telephone: (0)2-567880
Telefax: (0)2-576955
Electronic Mail: director@lion.meteo.go.ke
WWW: http://www.meteo.go.ke/
Founded: 1920
Staff: 800
Activities: providing meteorological services
City Reference Coordinates: 036°49'00"E 01°17'00"S

Kenya National Academy of Sciences (KNAS)
P.O. Box 39450
Nairobi
Telephone: (0)2-721345
Telefax: (0)2-721138
Electronic Mail: knas@ken.healthnet.org
　　　　　　　　　knas@iconnect.co.ke
Founded: 1986
Staff: 10
Activities: conference and scientific meetings * project development * scientific documentation and information * promotion, creation and guidance of scientific bodies
Periodicals: (1) "Kenya Journal of Sciences (KJS) - Series A: Physical Sciences", "Kenya Journal of Sciences (KJS) - Series B: Biological Sciences", "Kenya Journal of Sciences (KJS) - Series C: Humanities and Social Sciences"
Awards: "Scholastic Awards", "General Award", "Distinguished Professional Contribution Award"
City Reference Coordinates: 036°49'00"E 01°17'00"S

Founded: 1976
Staff: 7
Coordinates: 128°27'27"E 36°56'04"N H1,378m (IAU Code 345)
City Reference Coordinates: 128°27'00"E 36°56'00"N

Korea Astronomy Observatory (KAO), Taeduk Radio Astronomy Observatory (TRAO)
San 36-1
Whaam-dong
Yusong-gu
Taejon 305-348
Telephone: (0)42-8653282
Telefax: (0)42-8615610
Electronic Mail: <userid>@hanul.issa.re.kr
WWW: http://w3.trao.re.kr/trao/
http://www.issa.re.kr/trao/
Founded: 1988
Staff: 8
Activities: radio observing * Galaxy * clusters * eclipsing binaries * sunspots
Periodicals: "Publications of the Korea Astronomy Observatory"
Coordinates: 127°22'19"E 36°23'53"N H120m
• Previously known as "Daeduk Radio Astronomy Observatory (DRAO)"
City Reference Coordinates: 127°26'00"E 36°20'00"N

Korean Astronomical Society (KAS)
c/o Department of Astronomy
Seoul National University
San 56-1
Shinrim-dong
Kwanak-ku
Seoul 151-742
Telephone: (0)42-8653255
Telefax: (0)42-8653207
Electronic Mail: kas@kao.re.kr
WWW: http://www-kas.or.kr/
Founded: 1965
Membership: 565
Activities: organizing conferences
Periodicals: "Journal of the Korean Astronomical Society"
City Reference Coordinates: 127°26'00"E 36°20'00"N

Korean Meteorological Administration
460-18 Sindaebang-dong
Dongjak-gu
Seoul 156-720
Telephone: (0)2-836-2385
Telefax: (0)2-836-2386
WWW: http://www.kma.go.kr/
Founded: 1990 (1948 as "Central Meteorological Office")
City Reference Coordinates: 126°58'00"E 37°33'00"N

Korean National Institute of Technology and Quality (KNITQ)
2 Joongang-dong
Kwachon
Kyunggi-do 427-010
Telephone: (0)2-5074369
Telefax: (0)2-5037977
Electronic Mail: int_coop@mail.nitq.go.kr
WWW: http://www.nitq.go.kr
City Reference Coordinates: 127°15'00"E 37°30'00"N

Korean Physical Society (KPS)
Yuksam-Dong 635-4
Kangnam-Ku
Seoul 135-703
Telephone: (0)2-5564737
Telefax: (0)2-5541643
Electronic Mail: office@mulli.kps.or.kr
WWW: http://www.kps.or.kr/
City Reference Coordinates: 126°58'00"E 37°33'00"N

Korean Space Science Society (KSSS)
c/o Department of Astronomy
Yonsei University
134 Shinchon
Seodaemun-ku

Seoul 120-749
Telephone: (0)2-3612693
Telefax: (0)2-3625135
Electronic Mail: ksss@csa.yonsei.ac.kr
WWW: http://csaweb.yonsei.ac.kr/~ksss/
Founded: 1984
Membership: 440
Staff: 2
Activities: promoting advancement of astronomy and space sciences * meetings
Periodicals: (2) "Journal of Astronomy and Space Sciences (JASS)" (ISSN 1225-052x), "Bulletin of the Korean Space Science Society"
City Reference Coordinates: 126°58'00"E 37°33'00"N

Kyungpook National University (KNU), Department of Astronomy and Atmospheric Sciences (DAAS)
Taegu 702-701
Telephone: (0)53-950-6360
Telefax: (0)53-950-6359
Electronic Mail: <userid>@bh.kyungpook.ac.kr
WWW: http://sirius.kyungpook.ac.kr/
Founded: 1988
City Reference Coordinates: 128°35'00"E 35°52'00"N

Sejong University, Daeyang Observatory
98 Koonja-dong
Sungdon-ku
Seoul 133-747
Telephone: (0)2-4600345
Telefax: (0)2-4600299
WWW: http://www.sejong.ac.kr/institutions/observatory/obserFram.html
City Reference Coordinates: 126°58'00"E 37°33'00"N

Seoul National University (SNU), Department of Astronomy
San 56-1
Shinrim-dong
Kwanak-ku
Seoul 151-742
Telephone: (0)2-8806621
Telefax: (0)2-8871435
Electronic Mail: <userid>@astro.snu.ac.kr
WWW: http://astro.snu.ac.kr/
Founded: 1958
Membership: 50
Activities: education * research
City Reference Coordinates: 126°58'00"E 37°33'00"N

Sobaeksan Optical Astronomy Observatory (SOAO)
• See "Korea Astronomy Observatory (KAO), Sobaeksan Optical Astronomy Observatory (SOAO)"

Taeduk Radio Astronomy Observatory (TRAO)
• See "Korea Astronomy Observatory (KAO), Taeduk Radio Astronomy Observatory (TRAO)"

Yonsei University, Center for Space Astrophysics
134 Shinchon
Seodaemun-ku
Seoul 120-749
Telephone: (0)2-21234143
Telefax: (0)2-3625136
Electronic Mail: <userid>@csa.yonsei.ac.kr
WWW: http://csaweb.yonsei.ac.kr/
Founded: 1997
Staff: 12
Activities: GALEX mission
Coordinates: 126°59'00"E 37°34'00"N H67m
            126°48'30"E 37°41'01"N (Ilsan Station)
City Reference Coordinates: 126°59'00"E 37°34'00"N

Yonsei University, Department of Astronomy
134 Shinchon
Seodaemun-ku
Seoul 120-749
Telephone: (0)2-21232680
          (0)2-21233439 (Observatory)
Telefax: (0)2-3135033
         (0)2-3625136

Electronic Mail: <userid>@galaxy.yonsei.ac.kr
WWW: http://galaxy.yonsei.ac.kr/
Founded: 1967
Staff: 4
Activities: stellar evolution * globular clusters * elliptical galaxies * blue compact dwarf galaxies * eclipsing binaries * celestial mechanics * artificial satellites
Coordinates: 126°59'00"E 37°34'00"N H67m
    126°48'30"E 37°41'01"N (Ilsan Station)
City Reference Coordinates: 126°59'00"E 37°34'00"N

Latvia

Latvian Academy of Sciences, Radioastrophysical Observatory
● See now "University of Latvia, Institute of Astronomy"

Latvian Academy of Sciences, Ventspils International Radioastronomy Centre (VIRAC)
Akademijas Laukums 1
LV-1050 Riga
or :
Ances Pagasts
LV-3612 Ventspils
Telephone: (0)722-8321
Telefax: (0)722-1153
Electronic Mail: berv@latnet.lv (Edgars Bervalds, Director)
WWW: http://www.lanet.lv/members/LU/astro/ast_ven.html
Founded: 1996
Staff: 13
Activities: radioastronomy * interferometry
Coordinates: 021°51'18"E 57°33'10"N H15m
City Reference Coordinates: 024°08'00"E 56°53'00"N

Latvian Astronomical Society
(Latvijas Astronomijas Biedriba - LAB)
Raina Boulevard 19
LV-1586 Riga
Telephone: (0)7223637
          (0)7223149
Telefax: (0)7820180
Electronic Mail: astro@acad.latnet.lv
WWW: http://www.astr.lu.lv/lab/
Founded: 1990 (1947 as Latvian Branch of All-Union Astronomical and Geodetic Society)
Membership: 130
Activities: Summer camps * popularization
Periodicals: "Astronomical Calendar"
Coordinates: 024°51'00"E 57°09'00"N (Sigulda)
City Reference Coordinates: 024°08'00"E 56°53'00"N

Latvian Hydrometeorological Agency (HMP)
165 Maskavas Street
LV-1019 Riga
Telephone: (0)7-113274
Telefax: (0)7-145154
Electronic Mail: epoc@meteo.lv
           <userid>@meteo.lv
WWW: http://www.meteo.lv/
City Reference Coordinates: 024°08'00"E 56°53'00"N

Latvian National Center for Standardization and Metrology (LVS)
Kr. Valdemara Street 157
LV-1013 Riga
Telephone: (0)2-378165
Telefax: (0)2-362805
Electronic Mail: lvs@mail.eunet.lv
WWW: http://www.lvs.lv/
City Reference Coordinates: 024°08'00"E 56°53'00"N

Latvian Physical Society (LPS)
Zellu iela 8
LV-1002 Riga
Telephone: (0)7229747
Telefax: (0)7033751
Electronic Mail: odocenko@lanet.lv (Olga Docenko, Secretary)
WWW: http://www.lfb.lanet.lv/
Founded: 1993
Membership: 70
Activities: seminars * annual conference
Periodicals: (1) "Proceedings of the LPS Conference"
City Reference Coordinates: 024°08'00"E 56°53'00"N

University of Latvia, Institute of Astronomy
Raina Bulvaris 19
LV-1586 Riga
Telephone: (0)7034580

Telefax: (0)7034582
Electronic Mail: astra@latnet.lv
WWW: http://www.astr.lu.lv/
http://www.astr.lu.lv/zvd/ (Zvaigžņotá Debess)
Founded: 1946 (1997 under its current name)
Staff: 25
Activities: astrometry * space geodesy (SLR, GPS) * education * instrumentation * carbon and related stars (photometry, spectroscopy, synthetic spectra, evolution) * solar radio emission from active regions
Periodicals: (4) "Zvaigžņotá Debess" (The Starry Sky) (ISSN 0135-129x); "Acta Universitatis Latviensis (Astronomy)", "Investigations of the Sun and Red Stars" (ISSN 0135-1303)
Coordinates: 024°07'00"E 56°57'08"N
024°24'15"E 56°46'36"N H75m (Baldone – IAU Code 069)
City Reference Coordinates: 024°08'00"E 56°53'00"N

Ventspils International Radioastronomy Centre (VIRAC)
• See "Latvian Academy of Sciences, Ventspils International Radioastronomy Centre (VIRAC)"

Lebanon

Lebanese Astronomical Society (LAS)
Beirut
Electronic Mail: lebast@geocities.com
WWW: http://www.geocities.com/CapeCanaveral/Hall/6865/
Founded: 1999
Activities: monthly meetings * education * observing
• Formerly know as "Lebanese Amateur Astronomical Society"
City Reference Coordinates: 035°30'00"E 33°52'00"N

Lebanese Standards Institution (LIBNOR)
P.O. Box 55120
Sin El-Fil
Beirut
or :
Gedco Center 3
Bloc B - 9th floor
Mkalles Hayed Avenue
Sin El-Fil
Beirut
Telephone: (0)1-485927
Telefax: (0)1-485929
Electronic Mail: libnor@libnor/com
WWW: http://www.libnor.com/
City Reference Coordinates: 035°30'00"E 33°52'00"N

Lithuania

Institute of Physics, Stellar Systems Department
A. Goštauto 12
LT-2600 Vilnius
Telephone: (0)2-612610
Telefax: (0)2-617070
Electronic Mail: wladas@ktl.mii.lt (Vladas Vansevicius, Head)
              <userid>@ktl.mii.lt
WWW: http://www.fi.lt/
Founded: 1990
Coordinates: 025°16'48"E 54°41'23"N
City Reference Coordinates: 025°19'00"E 54°41'00"N

Institute of Theoretical Physics and Astronomy
A. Goštauto 12
LT-2600 Vilnius
Telephone: (0)2-620668
Telefax: (0)2-225361
Electronic Mail: astro@itpa.fi.lt
              astro@itpa.lt
WWW: http://www.itpa.lt/
      http://www.itpa.lt/~astro/ba/ (Baltic Astronomy)
Founded: 1989
Staff: 140
Activities: theoretical physics * stellar astronomy * galactic structure * interstellar extinction * atomic and molecular spectra * comets
Periodicals: (4) "Baltic Astronomy" (ISSN 1392-0049)
Coordinates: 025°33'45"E 55°19'00"N H200m (Moletai - see separate entry)
           066°52'30"E 38°41'03"N H2,540m (Maidanak - Uzbekistan)
City Reference Coordinates: 025°19'00"E 54°41'00"N

Institute of Theoretical Physics and Astronomy, Moletai Astronomical Observatory (MAO)
Astronomijos Observatorija
LT-4150 Moletai
Telephone: (0)3-045425
Telefax: (0)3-045425
Electronic Mail: astro@itpa.lt
WWW: http://www.astro.lt/mao/
Founded: 1969
Staff: 10
Activities: stellar photometry * interstellar extinction * astronomical instrumentation * variable stars
Coordinates: 025°33'48"E 55°19'00"N H200m (IAU Code 152)
City Reference Coordinates: 025°25'00"E 55°14'00"N

Institute of Theoretical Physics and Astronomy, Planetarium
Ukmerges 12A
LT-2005 Vilnius
Telephone: (0)2-724177
Telefax: (0)2-724177
Electronic Mail: gintaras@itpa.lt
              planet@astro.lt
WWW: http://www.astro.lt/planet/
Founded: 1962
Staff: 16
Activities: lectures * education * shows
City Reference Coordinates: 025°19'00"E 54°41'00"N

Lithuanian Academy of Sciences
Gedimino Pr. 3
LT-2600 Vilnius
Telephone: (0)2-613651
Telefax: (0)2-618464
Electronic Mail: prezidum@ktl.mii.lt
WWW: http://neris.mii.lt/Akademija/inf.html
City Reference Coordinates: 025°19'00"E 54°41'00"N

Lithuanian Hydrometeorological Service (LHMS)
6 Rudnios Street
LT-2600 Vilnius
Electronic Mail: lhmt@meteo.lt
WWW: http://www.meteo.lt/
Staff: 500

City Reference Coordinates: 025°19'00"E 54°41'00"N

Lithuanian Physical Society (LPS)
(Lietuvos Fiziku Draugija - LFD)
c/o A. Bernotas (Secretary)
Institute of Theoretical Physics and Astronomy
A. Goštauto 12
LT-2600 Vilnius
Telephone: (0)5-2620668
Telefax: (0)5-2224694
Electronic Mail: lfd@itpa.lt
WWW: http://www.itpa.lt/~lfd/
Founded: 1963
Membership: 165
Activities: to contribute to physical sciences, studies and education * to unite all interested parties in physics and its applications
Periodicals: (6) "Lithuanian Journal of Physics" (ISSN 1392-1932)
City Reference Coordinates: 025°19'00"E 54°41'00"N

Lithuanian Standards Board (LST)
T. Kosciuškos g. 30
LT-2600 Vilnius
Telephone: (0)2-226962
Telefax: (0)2-226252
Electronic Mail: lstboard@lsd.lt
WWW: http://www.lsd.lt/
Founded: 1990
Staff: 46
Activities: standardization * accreditation * metrology
Periodicals: (12) "LST Bulletin" (circ.: 600)
City Reference Coordinates: 025°19'00"E 54°41'00"N

Moletai Astronomical Observatory (MAO)
• See "Institute of Theoretical Physics and Astronomy, Moletai Astronomical Observatory (MAO)"

Palace of Pupils, Technical Activities, Astronomical Laboratory
Zirmunu 1b
LT-2012 Vilnius
Telephone: (0)122-773614
Founded: 1976
Staff: 3
Activities: education * expeditions * Summer camps
City Reference Coordinates: 025°19'00"E 54°41'00"N

Vilnius Planetarium
• See "Institute of Theoretical Physics and Astronomy, Planetarium"

Vilnius University, Astronomical Observatory
Čiurlionio 29
LT-2009 Vilnius
Telephone: (0)2-333343
Telefax: (0)2-335648
Electronic Mail: <userid>@ff.vu.lt
WWW: http://www.ff.vu.lt/astro/
Founded: 1753
Staff: 10
Activities: stellar photometry * variable stars * ISM * galactic structure and evolution * instrumentation * stellar radial velocities
Coordinates: 025°17'12"E 54°41'00"N H122m (IAU Code 570)
          025°34'00"E 55°19'00"N H200m (Moletai - see separate entry)
City Reference Coordinates: 025°19'00"E 54°41'00"N

Luxembourg

Amateurs Astronomes du Luxembourg (AAL)
Boîte Postale 1711
L-1017 Luxembourg
Telephone: (0)395343
Telefax: (0)50501420
Electronic Mail: aal@aal.lu
WWW: http://www.aal.lu/
Founded: 1971
Membership: 150
Activities: public observing * lectures * education
City Reference Coordinates: 006°09'00"E 49°36'00"N

Service de la Météorologie et de l'Hydrologie (GD Luxembourg)
16, route d'Esch
Boîte Postale 1904
L-1019 Luxembourg
Telephone: (0)457172341
Telefax: (0)457172222
City Reference Coordinates: 006°09'00"E 49°36'00"N

Service de l'Énergie de l'État (SEE), Département de Normalisation
34, avenue de la Porte-Neuve
Boîte Postale 10
L-2010 Luxembourg
Telephone: (0)4697461
Telefax: (0)46974639
Electronic Mail: see.normalisation@.ed.etat.lu
WWW: http://www.etat.lu/SEE
Founded: 1967
Activities: certification * normalization * testing
City Reference Coordinates: 006°09'00"E 49°36'00"N

Macedonia

Hydrometeorological Institute of Macedonia
P.O. Box 213
91001 Skopje
WWW: http://www.meteo.gov.mk/
Staff: 230
City Reference Coordinates: 021°26'00"E 41°59'00"N

Society of Physicists of the Republic of Macedonia
c/o Viktor Urumov (President)
Faculty of Natural Sciences and Mathematics
Cyril and Methodius University
P.O. Box 162
91000 Skopje
Telephone: (0)91-261330
Telefax: (0)91-228141
Electronic Mail: urumov@innona.pmf.ukim.edu.mk
Membership: 300
City Reference Coordinates: 021°26'00"E 41°59'00"N

Zavod za Standardizacija i Metrologija (ZSM)
(Bureau of Standardization and Metrology)
Samoilova 10
91000 Skopje
Telephone: (0)91-131102
    (0)91-131160
Telefax: (0)91-110263
Electronic Mail: sofijak@lotus.mpt.com.mk
Founded: 1995
Staff: 22
Activities: standardization * metrology * related activities
Periodicals: "Bulletin"
City Reference Coordinates: 021°26'00"E 41°59'00"N

México, DF 07360
Telephone: (0)5-7540200
Telefax: (0)5-7548707
WWW: http://www.fis.cinvestav.mx/
Founded: 1961
Activities: biology * physics * engineering * mathematics * education
Periodicals: (6) "Avance y Perspectiva"
City Reference Coordinates: 099°09'00"W 19°24'00"N

Observatorio Astrofísico Guillermo Haro (OAGH)
• See "Instituto Nacional de Astrofísica, Optica y Electronica (INAOE), Observatorio Astrofísico Guillermo Haro (OAGH)"

Observatorio Astronómico Nacional (OAN), Mexico
• See "Universidad Nacional Autónoma de México (UNAM), Observatorio Astronómico Nacional (OAN)"

Planetario de Cuidad Victoria
Centro Cultural Tamaulipas
16 Morelos Esq.
Apartado Postal 242
Ciudad Victoria, TAMPS 87000
Telephone: (0)131-51215
Telefax: (0)131-51215
Electronic Mail: joseba@spin.com.mx
WWW: http://www.uat.mx/Vinculos/planeta/
Founded: 1982
Staff: 15
Activities: scientific and technological popularization
City Reference Coordinates: 099°08'00"W 23°44'00"N

Servicio Meteorológico Nacional (SMN)
Avenida Observatorio 192
Colonia Observatorio
México, DF 11860
Telephone: (0)6-268650
Telefax: (0)6-268695
Electronic Mail: desparza@gsmn.cna.gob.mx
WWW: http://smn.cna.gob.mx/SMN.html
Founded: 1877
Staff: 1058
Activities: weather forecast * climatological data center
Periodicals: "Daily Forecast Bulletin"
Observatories: 72 observatories, 3,500 meteorological stations, 15 radiosonde stations and 12 radars
City Reference Coordinates: 099°09'00"W 19°24'00"N

Sociedad Astronómica Amateur de Sinaloa (SAASIN)
Angel Flores 119-1 Norte
Los Mochis, SIN 81200
Telephone: (0)68-182327
Telefax: (0)68-153097
Electronic Mail: info@saasin.org
WWW: http://www.saasin.org/
Founded: 1999
Membership: 10
Activities: observing * study * meetings
City Reference Coordinates: 109°00'00"W 25°48'00"N

Sociedad Astronómica de Aragón Ilhuícatl
Rosas Moreno 114-201
Colonia San Rafael
México, DF 06470
Telephone: (0)915-592-1553
         (0)915-623-0832
Electronic Mail: astro@hp-720.aragon.unam.mx
WWW: http://informatica.aragon.unam.mx/ilhuicatl/
Membership: 55
Activities: optics * telescope control systems * image processing * education * space technologies
Periodicals: (6) "Ilhuícatl"
City Reference Coordinates: 099°09'00"W 19°24'00"N

Sociedad Astronómica de México (SAM) AC
Parque Felipe S. Xicoténcatl
Colonia Álamos
Apartado Postal M-9647
México, DF 03480
Telephone: (0)5-5194730

WWW: http://www.avantel.net/~socastmx/
Founded: 1902
Activities: meetings * lectures * library * observing
Coordinates: 099°08'30"W 19°23'55"N H2,245m (Observatorio Luis G. León)
　　　　　　　099°31'23"W 19°47'24"N H3,070m (Observatorio Cerro Las Ánimas)
City Reference Coordinates: 099°09'00"W 19°24'00"N

Universidad Autónoma Metropolitana-Iztapalapa (UAM-I), Departamento de Física, Area Gravitación y Astrofísica
Avenida Michoacan y Purisima s/n
Colonia Vicentina
Apartado Postal 55-534
México, DF 09340
Telephone: (0)5-7244623
　　　　　　(0)5-6860322
Telefax: (0)5-6861717
WWW: http://www.iztapalapa.uam.mx/iztapala.www/division.cbi/fisica/progravi.htm
Founded: 1974
Staff: 8
Activities: astrophysics * classical and quantum cosmology * physics education
City Reference Coordinates: 099°09'00"W 19°24'00"N

Universidad de Guadalajara, Instituto de Astronomía y Meteorología
Avenida Vallarta 2602
Guadalajara, JAL 44140
Telephone: (0)3-6164937
Telefax: (0)3-6159829
WWW: http://www.iam.udg.mx/
Founded: 1903
Coordinates: 103°23'09"W 20°40'32"N H1,583m
City Reference Coordinates: 103°20'00"W 20°40'00"N

Universidad de Guanajuato, Departamento de Astronomía
Apartado 144
Guanajuato, Gto 36000
Telephone: (0)473-7329548
　　　　　　(0)473-7329607
Telefax: (0)473-7320253
Electronic Mail: <userid>@astro.ugto.mx
WWW: http://www.astro.ugto.mx/
Founded: 1995
Staff: 12
Activities: variable stars * hot stars * binary stars * HI in galaxies * ISM * interactive galaxies * clusters of galaxies * radio sources * large-scale structure
City Reference Coordinates: 101°15'00"W 21°01'00"N

Universidad de Sonora, Centro de Investigación en Física, Área de Astronomía
Apartado Postal 5-088
Hermosillo, SON 83190
Telephone: (0)62-592156
Telefax: (0)62-126649
Electronic Mail: <userid>@cajeme.cifus.uson.mx
WWW: http://spin.cifus.uson.mx/unison/EOS/eos.htm
　　　　http://cosmos.cifus.uson.mx/eosdata.htm (Estación de Observación Solar - EOS)
Founded: 1990
Activities: solar physics * observing * extragalactic astronomy
City Reference Coordinates: 110°58'00"W 29°04'00"N

Universidad Nacional Autónoma de México (UNAM), Instituto de Astronomía
Apartado Postal 70-264
México, DF 04510
or :
Circuito de la Investigación Científica
Ciudad Universitaria
México, DF 04510
Telephone: (0)5-6223906/11
Telefax: (0)5-6160653
Electronic Mail: <userid>@astroscu.unam.mx
WWW: http://www.astroscu.unam.mx/
Founded: 1863 (originally called "Observatorio Astronómico Nacional")
Staff: 90
Activities: ISM * HII regions * kinematics and dynamics * galaxies * cosmology * instrumentation * stars
Periodicals: (2) "Revista Mexicana de Astronomía y Astrofísica" (ISSN 0185-1101); "Revista Mexicana de Astronomía y Astrofísica - Serie de Conferencias" (ISSN 1485-2059)
Awards: (1/2) "Premio Harold Johnson"; "Premio Guillermo Haro Barraza"

City Reference Coordinates: 099°09'00"W 19°24'00"N

Universidad Nacional Autónoma de México (UNAM), Instituto de Astronomía, Unidad Morelia
J.J. Tablada 1006
Lomas de Santa María
Morelia, MICH 58090
Telephone: (0)43-236162
Telefax: (0)43-236165
Electronic Mail: <userid>@astrosmo.unam.mx
WWW: http://www.astrosmo.unam.mx/
Founded: 1995
Staff: 16
Activities: star formation * radio astronomy * ISM * HII regions * gas dynamics * MHD * radiative transfer * atmospheric turbulence * infrared astronomy
City Reference Coordinates: 101°11'00"W 19°40'00"N

Universidad Nacional Autónoma de México (UNAM), Instituto de Geofísica
Circuito Interior s/n
Ciudad Universitaria
Coyoacán
México, DF 04510
Telephone: (0)5-6224122
(0)5-6162344
Telefax: (0)5-5502486
Electronic Mail: juf@tonatiuh.igeofcu.unam.mx (Jaime Urrutia Fucugauchi, Director)
WWW: http://www.igeofcu.unam.mx/
Founded: 1949
Staff: 110
Activities: impact craters * paleoclimatology * volcanology * paleomagnetism * geodynamics * hydrogeology * mathematical geophysics * remote sensing * solar-terrestrial relationships * cosmic rays * geomagnetism
Periodicals: (4) "Geofísica Internacional" (ISSN 0016-7169);
Awards: (1) "Julio Monges"
City Reference Coordinates: 099°09'00"W 19°24'00"N

Universidad Nacional Autónoma de México (UNAM), Observatorio Astronómico Nacional (OAN)
Apartado Postal 877
Ensenada, BC 22800
or :
Carretera de Tijuana Km. 106
Ensenada, BC 22860
Telephone: (0)617-44548
(0)617-44580
(0)617-44593
Telefax: (0)617-44607
Electronic Mail: oan@bufadora.astrosen.unam.mx
<userid>@bufadora.astrosen.unam.mx
WWW: http://bufadora.astrosen.unam.mx/
Founded: 1868
Staff: 39
Activities: ISM * HII regions * dust * kinematics and dynamics * galaxies * PNs * instrumentation * stars
Coordinates: 115°27'49"W 31°02'40"N H2,890m (San Pedro Mártir – IAU Code 679)
098°19'00"W 19°02,00"N H2,130m (Tonantzintla)
• See also addresses in USA-CA
City Reference Coordinates: 116°37'00"W 31°52'00"N

Universum - Museo de las Ciencias
c/o Jorge Flores Valdés
Cto. Mario de la Cueva s/n
Zona Cultural de la Ciudad Universitaria
Apartado Postal 70-487
México, DF 04510
Telephone: (0)5-6653761
Telefax: (0)5-6653769
Electronic Mail: universum@servidor.unam.mx
WWW: http://www.universum.unam.mx/
City Reference Coordinates: 099°09'00"W 19°24'00"N

Moldova

Department of Standards, Metrology and Technical Supervision
Coca 28
2064 Kishinev
Telephone: (0)2-748588
Telefax: (0)2-750581
Electronic Mail: moldovastandard@standart.mldnet.com
WWW: http://www.moldova.md/
City Reference Coordinates: 028°50'00"E 47°00'00"N

Integral Applied Studies Institute (IASI), Astrophysical Observatory
60 Mateevici Street
Building 4a
2009 Kishinev
Telephone: (0)2-577726
(0)2-577459
Telefax: (0)2-762365
Electronic Mail: iasi@usm.md
Founded: 1973
Staff: 10
Activities: variable stars * stellar photometry * atmospheric sciences * remote sensing * environment * optics * education
Coordinates: 028°30'00"E 47°00'00"N H400m
City Reference Coordinates: 028°50'00"E 47°00'00"N

Mongolia

Mongolian Academy of Sciences, Centre of Astronomy and Geophysics
P.O. Box 788
Ulaanbataar 210613
Telephone: (0)1-50218
　　　　　(0)1-358849
Telefax: (0)1-358849
Electronic Mail: gnoon@magicnet.mn
　　　　　　　　bechtur@csj.mn
Founded: 1997
Staff: 70
Activities: latitude * longitude * Earth rotation * microwave satellite observing * solar physics * radiation transfer * spectral-line formation * seismology * dynamic study of the Earth
Coordinates: 107°03'08"E 47°51'54"N (Khurel Togót Observatory)
City Reference Coordinates: 106°53'00"E 47°55'00"N

Mongolian National Institute for Standardisation and Metrology (MNISM)
Peace Street
Ulaanbataar 51
Telephone: (0)1-358032
　　　　　(0)1-53529
Telefax: (0)1-358032
Electronic Mail: mncsm@magicnet.mn
Founded: 1953
Staff: 120
Activities: standardization * metrology * quality assurance
Periodicals: (12) "Standards and Metrology"
City Reference Coordinates: 106°53'00"E 47°55'00"N

Netherlands

A.F. Philips Sterrenwacht (Dr._)
● See "Dr. A.F. Philips Sterrenwacht"

Artis Planetarium
Postbus 20164
NL-1000 HD Amsterdam
or :
Plantage Kerklaan 38-40
NL-1018 CZ Amsterdam
Telephone: (0)20-5233452
Telefax: (0)20-5233481
(0)20-5233518
Electronic Mail: planetarium@artis.nl
WWW: http://www.artis.nl/
Founded: 1988
Staff: 20
Activities: shows * exhibitions * lectures * shop
City Reference Coordinates: 004°54'00"E 52°22'00"N

Astronomische Vereniging Wega
Gen. de Wetstraat 31
NL-5021 TK Tilburg
Telephone: (0)13-5422534
Electronic Mail: spaninks@freemail.nl
Founded: 1972
Membership: 60
Activities: observing * lectures
Periodicals: (3) "de Lier"
City Reference Coordinates: 005°05'00"E 51°34'00"N

Centaurus-A
c/o B.A. Gerritsen (Secretary)
Wardstraat 33
NL-6681 CG Bemmel
Telephone: (0)481-461368
Telefax: (0)481-465289
Founded: 1962
Membership: 130
Activities: popularization * lectures * telescope making
City Reference Coordinates: 005°54'00"E 51°54'00"N

Centrum voor Constructie en Mechatronica (CCM)
Postbus 12
NL-5670 AA Nuenen
or :
De Pinehart 24
NL-5674 CC Nuenen
Telephone: (0)40-2834405
Telefax: (0)40-2837135
Founded: 1969
Staff: 85
Activities: technical research * development projects * product development * automation * signal processing * automatic instruments for biological experiments in space
City Reference Coordinates: 005°33'00"E 51°29'00"N

Centrum voor Wiskunde en Informatica (CWI)
Postbus 94079
NL-1090 GB Amsterdam
or :
Kruislaan 413
NL-1098 SJ Amsterdam
Telephone: (0)20-5929333
Telefax: (0)20-5924199
Electronic Mail: info@cwi.nl
WWW: http://www.cwi.nl/
Founded: 1946
Staff: 202
Activities: research in mathematics and computer sciences
Periodicals: (4) "CWI Quarterly"
City Reference Coordinates: 004°54'00"E 52°22'00"N

Corona Borealis
c/o H.J.M. v. Rongen (Secretary)
Schuttersweg 143
NL-7314 LG Apeldoorn
or :
Eiberstraat 14
NL-6883 EJ Velp
Telephone: (0)55-3551044
Electronic Mail: bvrongen@tip.nl
WWW: http://www.coronaborealis.nl/
Founded: 1977
Staff: 3
Activities: public observatory * lectures * working groups * radioastronomy * education
Periodicals: (4) "Corona Borealis"
Coordinates: 005°59'11"E 51°59'11"N H16m (Velp)
  003°11'44"E 43°42'29"N H270m (Lunas, France)
City Reference Coordinates: 005°58'00"E 52°13'00"N (Apeldoorn)
  005°59'00"E 52°00'00"N (Velp)

Delft Institute for Earth-Oriented Space Research (DEOS)
• See "Technische Universiteit Delft, Delft Institute for Earth-Oriented Space Research (DEOS)"

Delft Instruments NV
Postbus 103
NL-2600 AC Delft
or :
Röntgenweg 1
NL-2624 BD Delft
Telephone: (0)15-2601200
Telefax: (0)15-2601222
Electronic Mail: info@delftinstruments.com
WWW: http://www.delftinstruments.com/
Activities: handling technologies in the medical, space, industrial & scientific industry
City Reference Coordinates: 004°21'00"E 52°00'00"N

Delft University of Technology
• See "Technische Universiteit Delft"

De Zonnewijzerkring
Van Gorkumlaan 39
NL-5641 WN Eindhoven
Telephone: (0)40-2817818
Electronic Mail: ferdv@iae.nl
WWW: http://www.iae.nl/users/ferdv
Founded: 1978
Membership: 160
Activities: sundials (registration, construction, restoration, research, design, advice)
Periodicals: (3) "Bulletin"
City Reference Coordinates: 005°28'00"E 51°26'00"N

Dr. A.F. Philips Sterrenwacht
c/o P. Haagen
Burg. Mollaan 8
NL-5582 CK Waalre
or :
Alb.Thijmlaan 1
NL-5615 EB Eindhoven
Telephone: (0)40-2214355
  (0)40-2118409
Founded: 1936
Staff: 5
Periodicals: "De Sterrenkijker"
City Reference Coordinates: 005°26'00"E 51°24'00"N (Waalre)
  005°28'00"E 51°26'00"N (Eindhoven)

Dutch Meteor Society (DMS)
c/o Hans Betlem
Lederkarper 4
NL-2318 NB Leiden
Telephone: (0)71-5223817
Telefax: (0)71-5223817
Electronic Mail: betlem@strw.leidenuniv.nl
WWW: http://www.dmsweb.org/
Founded: 1979
Membership: 80

Activities: meteor research * observing * spectrography * photography * all-sky network * statistics * photoelectric meteor registration
Periodicals: (6) "Radiant" (ISSN 0925-8566)
City Reference Coordinates: 004°30'00"E 52°09'00"N

Dutch Occultation Association (DOA)
Frescobaldistraat 21
NL-1323 BB Almere
WWW: http://web.inter.NL.net/hcc/elimburg.doa/
Founded: 1946
Membership: 65
City Reference Coordinates: 005°12'00"E 52°22'00"N

Dutch Youth Association for Space and Astronomy
• See "Nederlandse Jeugdvereniging voor Ruimtevaart en Sterrenkunde (NJRS)"

Eise Eisinga's Planetarium
Eise Eisingastraat 3
NL-8801 KE Franeker
Telephone: (0)517-393070
Founded: 1781
Staff: 1
Activities: museum * shows * exhibitions
City Reference Coordinates: 005°32'00"E 53°11'00"N

Elsevier Science BV, Physics and Astronomy Department
Postbus 103
NL-1000 AC Amsterdam
or :
Sara Burgerhartstraat 25
NL-1005 KV Amsterdam
Telephone: (0)20-4853757
Telefax: (0)20-4853432
Electronic Mail: nlinfo@elsevier.com
    c.schwarz@elsevier.com (Carl Schwarz, Senior Publishing Editor)
WWW: http://www.elsevier.nl/
Founded: 1950
Staff: 1600
Activities: publishing, among many titles, "Advances in Space Research" (ISSN 0273-1177), "Astroparticle Physics" (ISSN 0927-6505), "Chinese Astronomy and Astrophysics" (ISSN 0275-1062), "Computer Physics Communications (CPC)" (ISSN 0010-4655), "COSPAR Information Bulletin", "Earth and Planetary Science Letters" (ISSN 0012-821x), "New Astronomy" (ISSN 1384-1076), "New Astronomy Reviews" (ISSN 1387-6473), "Physics Reports" (ISSN 0303-4070), "Planetary and Space Science" (ISSN 0032-0633),
City Reference Coordinates: 004°54'00"E 52°22'00"N

ESTEC
• See "European Space Agency (ESA), ESTEC"

Euregio Publiekscentrum voor Sterrenkunde (EPS)
Postbus 93
NL-7590 AB Denekamp
or :
Frensdorferweg 22
NL-2635 NK Lattrop
Telephone: (0)541-229700
Electronic Mail: sterrenwacht@cnt.antenna.nl
    eps@introweb.nl
WWW: http://www.introweb.nl/~eps/
Founded: 1960
Membership: 50
Activities: public observatory
City Reference Coordinates: 007°00'00"E 52°23'00"N (Denekamp)
    006°59'00"E 52°26'00"N (Lattrop)

European Association of Geoscientists and Engineers (EAGE)
Postbus 59
NL-3990 DB Houten
or :
Standerdmolen 10
NL-3995 AA Houten
Telephone: (0)30-6354055
Telefax: (0)30-6343524
Electronic Mail: eage@eage.nl
WWW: http://www.eage.nl/
Founded: 1995

Staff: 10
Activities: promoting exploration geophysics * fostering fellowship and cooperation among those working, studying or being otherwise interested in this field
Periodicals: "Geophysical Prospecting" (ISSN 0016-8025), "First Break", "Proceedings", "Petroleum Geoscience", "Extended Abstracts", "Yearbook" (ISSN 0531-2728)
Awards: (1) "Conrad Schlumberger Award", "Best Poster Award", "Best Paper Award", "Distinguished Lecturer Award"
City Reference Coordinates: 005°10'00"E 52°02'00"N

European Space Agency (ESA), ESTEC
(European Space Research and Technology Centre)
Postbus 2200
NL-2200 AG Noordwijk
or :
Keplerlaan 1
NL-2201 AZ Noordwijk
Telephone: (0)71-5656555
           (0)71-5656565
Telefax: (0)71-5657400
Electronic Mail: <userid>@estec.esa.nl
WWW: http://www.estec.esa.nl/
     http://www.esa.int/
Founded: 1963
Staff: 1500
Activities: European space scientific and technical centre
City Reference Coordinates: 004°26'00"E 52°14'00"N

European Space Agency (ESA), ESTEC, Astrophysics Division
Keplerlaan 1
Postbus 299
NL-2200 AG Noordwijk
Telephone: (0)71-5653557
Telefax: (0)71-5654690
Electronic Mail: astro@astro.estec.esa.nl
                <userid>@astro.estec.esa.nl
WWW: http://astro.estec.esa.nl/
     http://www.esa.int/
Founded: 1964
Staff: 30
Activities: space astrophysics * developing instruments * observational astronomy (space and ground based) in gamma rays, X-ray, UV, optical, IR and radio P * managing the ESA astrophysics archives
Periodicals: (3) "Astronews"; (1/2) "ISO Info"
City Reference Coordinates: 004°26'00"E 52°14'00"N

European Space Agency (ESA), ESTEC, ESA Publications Division (EPD)
Keplerlaan 1
Postbus 299
NL-2200 AG Noordwijk
Telephone: (0)71-5653408 (Secretary)
Telefax: (0)71-5655433
Electronic Mail: <userid>@estec.esa.nl
WWW: http://esapub.esrin.esa.it/publicat/epdi.htm
     http://www.esa.int/
Founded: 1964 (ESA)
Periodicals: (4) "Reaching for the Skies" (ISSN 1013-9044), "Earth Observation Quarterly" (ISSN 0256-596x), "Preparing for the Future"; (3) "Microgravity News from ESA"; (1) "ESA Annual Report"; "ESA Bulletin" (ISSN 0376-4265), "ESA Special Publications" (ISSN 0379-6566), "ESA Brochures" (ISSN 0250-1589), "Space Components Steering Board Newsletter", "ESA History Study Reports", "On Station"
City Reference Coordinates: 004°26'00"E 52°14'00"N

European Space Research and Technology Centre (ESTEC)
● See "European Space Agency (ESA), ESTEC"

Fédération Internationale d'Information et de Documentation (FID)
(International Federation for Information and Documentation)
Postbus 90402
NL-2509 LK Den Haag
or :
Prins-Willem-Alexanderhof 5
NL-2595 BE Den Haag
Telephone: (0)70-3140671
Telefax: (0)70-3140667
Electronic Mail: secretariat@fid.nl
                fid@fid.nl
WWW: http://www.fid.nl/
Founded: 1895 (as "Institut International de Bibliographie - IIB")
Staff: 5

Activities: information management * classification
Periodicals: (6) "FID Review"
City Reference Coordinates: 004°18'00"E 52°06'00"N

FOM, Institute for Atomic and Molecular Physics (AMOLF)
Kruislaan 407
NL-1098 SJ Amsterdam
Telephone: (0)20-6081234
Telefax: (0)20-6684106
Electronic Mail: secr@amolf.nl
                 <userid>@amolf.nl
WWW: http://www.amolf.nl/
Founded: 1949
Staff: 180
City Reference Coordinates: 004°54'00"E 52°22'00"N

Galaxis, Vereniging voor Amateur Astronomie
Piet Slagerstraat 50
NL-5213 XT 's-Hertogenbosch
Telephone: (0)73-6410257 (C.V. Huijkelom)
Founded: 1977
Membership: 58
Activities: monthly meetings * lectures * experience sharing * education
Periodicals: (4) "Galaxis"
City Reference Coordinates: 005°19'00"E 51°41'00"N

Gebiedsbestuur Aard- en Levenswetenschappen (ALW)
Postbus 93120
NL-2509 AC Den Haag
or :
Laan van Nieuw Oost Indië 131
NL-2593 BM Den Haag
Telephone: (0)70-3440780
          (0)70-3440619
Telefax: (0)70-3832173
       (0)70-3819033
Electronic Mail: alw@nwo.nl
WWW: http://www.nwo.nl/alw/
Founded: 1999
Staff: 21
Activities: funding agency for fundamental and strategic research regarding geosciences and life sciences
Periodicals: (12) "KNMG/ALW Nieuwsbrief"
City Reference Coordinates: 004°18'00"E 52°06'00"N

International Association for Statistical Computing (IASC)
428 Prinses Beatrixlaan
Postbus 950
NL-2270 AZ Voorburg
Telephone: (0)70-3375737
Telefax: (0)70-3860025
Electronic Mail: isi@cbs.nl
WWW: http://www.stat.unipg.it/iasc/
Founded: 1977
Membership: 900
Activities: fostering world-wide interest in effective statistical computing and exchange of technical knowledge * organising COMPSTAT meetings
Periodicals: (10) "Computational Statistics & Data Analysis" (ISSN 0167-9473); "Statistical Software Newsletter"
City Reference Coordinates: 004°23'00"E 52°05'00"N

International Federation for Information and Documentation
● See "Fédération Internationale d'Information et de Documentation (FID)"

International Federation of Library Associations and Institutions (IFLA)
Postbus 95312
NL-2509 CH Den Haag
or :
Prins Willem-Alexanderhof 5
NL-2595 BE Den Haag
Telephone: (0)70-3140884
Telefax: (0)70-3834827
Electronic Mail: ifla@ifla.org
WWW: http://www.ifla.org/
Founded: 1927
Membership: 1650

Membership: 19 (observatories)
Staff: 5
Activities: stimulating member observatories * setting up programs * raising funds * popularization
City Reference Coordinates: 006°33'00"E 53°13'00"N (Groningen)

Landelijk Samenwerkende Volkssterrenwachten (LSV)
• See now "Landelijk Samenwerkende Publiekssterrenwachten (LSPS)"

Leidse Sterrewacht (WLS) (Werkgroep_)
• See "Werkgroep Leidse Sterrewacht (WLS)"

Meteo Consult BV
Postbus 617
NL-6700 AP Wageningen
Telephone: (0)317-423300
Telefax: (0)317-423164
Electronic Mail: <userid>@meteocon.nl
WWW: http://www.weer.nl/index_ns.html
City Reference Coordinates: 005°40'00"E 51°58'00"N

Museum Boerhaave
• See "Rijksmuseum voor de Geschiedenis van de Natuurwetenschappen en van de Geneeskunde"

Nationaal Instituut voor Kernfysica en Hoge-Energie Fysica (NIKHEF)
Kruislaan 409
NL-1098 SJ Amsterdam
Telephone: (0)20-5922000
Telefax: (0)20-5922165
 (0)20-5925155
Electronic Mail: <userid>@nikhef.nl
WWW: http://www.nikhef.nl/
Founded: 1946
Staff: 270
Activities: sub-atomic physics
Coordinates: 004°57'05"E 52°22'24"N H-5m
City Reference Coordinates: 004°54'00"E 52°22'00"N

Nationaal Lucht- en Ruimtevaartlaboratorium (NLR)
(National Aerospace Laboratory - Netherlands)
Postbus 90502
NL-1006 BM Amsterdam
or :
Anthony Fokkerweg 2
NL-1059 CM Amsterdam
Telephone: (0)20-5113113
Telefax: (0)20-5113210
Electronic Mail: info@nlr.nl
WWW: http://www.nlr.nl/
Founded: 1937 (1919 as "Government Service for Aeronautical Studies")
Staff: 900
Activities: research and development * fluid dynamics * flight testing * simulation * structures and materials * remote sensing * space user support * information technology * electronics
Periodicals: "NLR News"
City Reference Coordinates: 004°54'00"E 52°22'00"N

Nederlands Comité Astronomie (NCA)
c/o J. Lub
Sterrewacht Leiden
Postbus 9513
NL-2300 RA Leiden
or :
Niels Bohrweg 2
NL-2333 CA Leiden
Telephone: (0)71-5275833
Telefax: (0)71-5275819
Electronic Mail: lub@strw.leidenuniv.nl
City Reference Coordinates: 004°30'00"E 52°09'00"N

Nederlandse Astronomenclub (NAC)
c/o M.H.M. Heemskerk (Secretary)
Kruislaan 403
NL-1098 SJ Amsterdam
Telephone: (0)20-5257491
Telefax: (0)20-5257484
Founded: 1918

Membership: 600
Activities: meetings with lectures on astronomical topics * annual Dutch astronomers conference
City Reference Coordinates: 004°54'00"E 52°22'00"N

Nederlandse Jeugdvereniging voor Ruimtevaart en Sterrenkunde (NJRS)
(Dutch Youth Association for Space and Astronomy)
Postbus 38
NL-5340 AA Oss
Telephone: (0)24-6419929
Electronic Mail: njrs@dse.nl
WWW: http://www.dse.nl/njrs/
Founded: 1974
Membership: 15
Activities: education * exhibitions * space camps * research * observing * meeting
Periodicals: "Astrum"
City Reference Coordinates: 005°31'00"E 51°46'00"N

Nederlandse Natuurkundige Vereniging (NNV)
(Dutch Physical Society)
Postbus 41882
NL-1009 DB Amsterdam
Telephone: (0)20-5922117
Electronic Mail: ed@nikhef.nl
WWW: http://www.nnv.nl/
Founded: 1921
Activities: meetings * conferences
Periodicals: "Nederlandse Tijdschrift voor Natuurkunde"
City Reference Coordinates: 004°54'00"E 52°22'00"N

Nederlandse Organisatie voor Toegepast-Natuurwetenschappelijk Onderzoek (TNO)
Schoemakerstraat 97
Postbus 6070
NL-2600 JA Delft
Telephone: (0)15-2696900
Telefax: (0)15-2612403
WWW: http://www.tno.nl/
Founded: 1932
Staff: 5000
Activities: research in industrial technology, space, energy, environment, nutrition and food, health, defence, building and infrastructure
City Reference Coordinates: 004°21'00"E 52°00'00"N

Nederlandse Organisatie voor Wetenschappelijk Onderzoek (NWO)
Postbus 93138
NL-2509 AC Den Haag
or :
Laan van Nieuw Oost Indië 131
NL-2593 BM Den Haag
Telephone: (0)70-3440640
Telefax: (0)70-3850971
Electronic Mail: nwo@nwo.nl
            <userid>@nwo.nl
WWW: http://www.nwo.nl/
Founded: 1950
Staff: 4500
Activities: initiation, stimulation and coordination of scientific research in the Netherlands by funding and advisory work
Periodicals: (10) "Onderzoekberichten" (ISSN 0928-6640); (6) "Research Reports from the Netherlands" (ISSN 0927-880x), "Forschungsnachrichten aus den Niederlanden" (ISSN 0928-6403), "Bulletin de Recherche des Pays-Bas" (ISSN 0928-6411); (4) "Hypothese" (ISSN 1381-5652); (1) "Jaarboek" (ISSN 0167-6792); "NWO FActs"
City Reference Coordinates: 004°18'00"E 52°06'00"N

Nederlandse Vereniging voor Ruimtevaart
c/o Stichting De Koepel
Zonnenburg 2
NL-3512 NL Utrecht
Telephone: (0)30-2311360
Electronic Mail: info@ruimtevaart-nvr.nl
WWW: http://www.ruimtevaart-nvr.nl/
Founded: 1951
Membership: 800
Activities: meetings on space sciences * collaboration with Stichting De Koepel
Periodicals: (6) "Ruimtevaart"
City Reference Coordinates: 005°08'00"E 52°05'00"N

Nederlandse Vereniging voor Weer- en Sterrenkunde (NVWS)
c/o J.A. de Boer (Secretary)
Prinses Irenelaan 1
NL-9765 AL Paterswolde
Telephone: (0)50-3094290 (after 19:00)
Electronic Mail: j.a.de.boer@fwn.rug.nl
WWW: http://www.astro.rug.nl/~nvws/
http://stkwww.fys.ruu.nl:8000/~mathlenr/Zenit.html (Zenit)
http://www.fys.ruu.nl/~wwwzenit/Zenit.html
Founded: 1901
Membership: 3000
Activities: meetings * excursions
Sections: Meteors * Star Occultations * Sun * Variable Stars * Comets * Satellites * Meteorology * Instrumentation * Special Group for Teenagers * Photography * Moon and Planets
Periodicals: (11) "Zenit" (ISSN 0165-0211); (1) "Sterrengids" (in collaboration with Stichting de Koepel)
City Reference Coordinates: 006°33'00"E 53°08'00"N

Nederlands Instituut voor Vliegtuigontwikkeling en Ruimtevaart (NIVR)
(Netherlands Agency for Aerospace Programmes)
Postbus 35
NL-2600 AA Delft
or :
Kluyverweg 4
NL-2629 HS Delft
Telephone: (0)15-2788025
Telefax: (0)15-2623096
Electronic Mail: info@nivr.nl
WWW: http://www.nivr.nl/
Founded: 1946
Staff: 32
Activities: promoting industrial aerospace activities primarily through R&D funding * advising Ministry of Economic Affairs on aerospace developments
City Reference Coordinates: 004°21'00"E 52°00'00"N

Nederlands Normalisatie-Instituut (NNI)
Kalfjeslaan 2
Postbus 5059
NL-2600 GB Delft
Telephone: (0)15-2690390
Telefax: (0)15-2690190
Electronic Mail: info@nni.nl
WWW: http://www.nni.nl/
Founded: 1916
Staff: 180
Activities: standardization * publishing * documentation services
Periodicals: (1) "Jaarverslag"; "Newsletter", "Catalogs", "CD-ROMs"
City Reference Coordinates: 004°21'00"E 52°00'00"N

Nederlands Onderzoekschool voor Astronomie (NOVA)
(Netherlands Research School for Astronomy)
P.O. Box 9513
NL-2300 RA Leiden
or :
J.H. Oort Gebouw
Niels Bohrweg 2
NL-2333 CA Leiden
Telephone: (0)71-5275852
Telefax: (0)71-5275743
Electronic Mail: nova@strw.leidenuniv.nl
WWW: http://www.strw.leidenuniv.nl/nova/
Activities: carrying out front-line astronomical research in the Netherlands * training young astronomers at the highest international level
City Reference Coordinates: 004°30'00"E 52°09'00"N

Netherlands Agency for Aerospace Programmes
• See "Nederlands Instituut voor Vliegtuigontwikkeling en Ruimtevaart (NIVR)"

Netherlands Foundation for Research in Astronomy (NFRA)
• See now "Stichting Astronomisch Onderzoek in Nederland (ASTRON)"

Netherlands Foundation for Radio Astronomy (NFRA)
• See now "Stichting Astronomisch Onderzoek in Nederland (ASTRON)"

Netherlands Industrial Space Organisation (NISO)
Postbus 32070
NL-2303 DB Leiden
or :
Newtonweg 1
NL-2333 CP Leiden
Telephone: (0)71-5245124
Telefax: (0)71-5245125
Electronic Mail: info@niso.nl
WWW: http://www.niso.nl/
Founded: 1989
Membership: 23 (Dutch companies)
Activities: association of Dutch industries and research institutes * defining common interests * developing and advocating common strategies * harmonizing Dutch governmental policies and industrial strategies * stimulating use and application of space facilities
City Reference Coordinates: 004°30'00"E 52°09'00"N

Netherlands Remote Sensing Board
• See "Beleidscommissie Remote Sensing (BCRS)"

Netherlands Research School for Astronomy
• See "Nederlands Onderzoekschool voor Astronomie (NOVA)"

Nonius BV
Postbus 811
NL-2600 AV Delft
or :
Röntgenweg 1
NL-2624 BD Delft
Telephone: (0)15-2698300
Telefax: (0)15-2627401
Electronic Mail: info@nonius.nl
WWW: http://www.nonius.com/
Founded: 1925
Staff: 90
Activities: developing, manufacturing, marketing and supporting X-ray diffraction instruments
City Reference Coordinates: 004°21'00"E 52°00'00"N

Noordwijk Space Expo (NSE)
Postbus 277
NL-2200 AG Noordwijk
or :
Keplerlaan 3
NL-2201 AZ Noordwijk
Telephone: (0)71-3646446
Telefax: (0)71-3646453
Electronic Mail: mailnse@estec.esa.nl
WWW: http://www.spaceexpo.nl/
Founded: 1990
Staff: 9
Activities: permanent space exhibition * education * information * group visits * films * guided tours * press conferences * meetings * visitors' centre for ESA/ESTEC
Periodicals: (1) "Annual Report"; "SpaceTalk"
City Reference Coordinates: 004°26'00"E 52°14'00"N

Nutssterrenwacht Ommen
Koperwiekstraat 4
NL-7731 ZH Ommen
Telephone: (0)529-451849
Founded: 1992
Staff: 4
Activities: public education on astronomy, meteorology and astronautics
Periodicals: "Stand van Zaken"
Coordinates: 006°25'00"E 50°32'00"N H8m
City Reference Coordinates: 006°25'00"E 50°32'00"N

Omniversum Space Theater and Digital Planetarium
President Kennedylaan 5
NL-2517 JK Den Haag
Telephone: (0)70-3547479
          (0)70-3545454
Telefax: (0)70-3524280
WWW: http://www.omniversum.nl/
Founded: 1984
Staff: 17

Activities: popularization of science in general and astronomy more particularly
City Reference Coordinates: 004°18'00"E 52°06'00"N

Paagman Sterrenwacht (JPS) (Jan_)
• See "Jan Paagman Sterrenwacht (JPS)"

Philips Sterrenwacht (Dr. A.F._)
• See "Dr. A.F. Philips Sterrenwacht"

Phoenix Public Observatory
Vordenseweg 6
NL-7241 SB Lochem
Telephone: (0)573-254310
WWW: http://www.phoenix.vuurwerk.nl/
City Reference Coordinates: 006°25'00"E 52°10'00"N

Planetron
Drift 11b
NL-7911 AA Dwingeloo
Telephone: (0)521-593535
Telefax: (0)521-593541
Electronic Mail: info@elders.nl
WWW: http://www.planetron.nl/
City Reference Coordinates: 006°22'00"E 52°49'00"N

Publiekssterrenwacht Schothorst
Kootwijkerzand 9
NL-3823 ZB Amersfoort
or :
Schothorsterlaan 27
NL-3822 NA Amersfoort
Telephone: (0)33-4559616 (after 13:30)
Electronic Mail: slooten@tiscali.nl
WWW: http://home-1.tiscali.nl/~slooten/schoth.htm
Founded: 1987
City Reference Coordinates: 005°23'00"E 52°09'00"N

Quasar
Bovenstraat 89
NL-4741 SK Hoeven
Telephone: (0)165-502439
          (0)165-502493
Telefax: (0)165-504959
WWW: http://www.quasarheelal.nl/
Founded: 1961
Staff: 3
Coordinates: 004°33'48"E 51°34'00"N H9m
City Reference Coordinates: 004°35'00"E 51°35'00"N

Radiosterrenwacht Dwingeloo
Postbus 2
NL-7990 AA Dwingeloo
or :
Oude Hoogeveensedijk 4
NL-7991 PD Dwingeloo
Telephone: (0)521-595100
Telefax: (0)521-597332
Electronic Mail: secretary@nfra.nl
              <userid>@nfra.nl
WWW: http://www.nfra.nl/
Founded: 1956
Staff: 90
Activities: observing * instrument building * data processing
Coordinates: 006°23'49"E 52°48'47"N H25m
City Reference Coordinates: 006°22'00"E 52°49'00"N

Radiosterrenwacht Westerbork
Schattenberg 1
NL-9433 TA Zwiggelte
Telephone: (0)593-592421
Telefax: (0)593-592486
Electronic Mail: <userid>@nfra.nl
WWW: http://www.nfra.nl/wsrt/wsrtpage.htm
Founded: 1968
Staff: 4

Activities: radio astronomical data acquisition * operating the Westerbork Synthesis Radio Telescope (WSRT)
Coordinates: 006°36'15"E 52°55'00"N H16m
City Reference Coordinates: 006°35'00"E 52°52'00"N

Raymond and Beverly Sackler Laboratorium voor Astrofysica
• See "Universiteit Leiden, Raymond and Beverly Sackler Laboratorium voor Astrofysica"

Rijksmuseum voor de Geschiedenis van de Natuurwetenschappen en van de Geneeskunde
(Museum Boerhaave)
Postbus 11280
NL-2301 EG Leiden
or :
Lange Sint Agnietenstraat 10
NL-2312 WC Leiden
Telephone: (0)71-5214224
Telefax: (0)71-5120344
WWW: http://www.museumboerhaave.nl/
Founded: 1928
Staff: 40
Activities: collection, exposition and scientific research of historical scientific instruments
City Reference Coordinates: 004°30'00"E 52°09'00"N

Rijksuniversiteit Groningen
• See "Universiteit Groningen"

Rijksuniversiteit Leiden
• See "Universiteit Leiden"

Rijksuniversiteit Utrecht
• See "Universiteit Utrecht"

Simon Stevin Public Astronomical Observatory
• See now "Quasar"

Space Research Organization Netherlands (SRON), Groningen
Postbus 800
NL-9700 AV Groningen
or :
Landleven 12
NL-9747 AD Groningen
Telephone: (0)50-3634074
Telefax: (0)50-3634033
Electronic Mail: <userid>@sron.nl
            secr@sron.nl
WWW: http://www.sron.nl/
Founded: 1984 (1966 as KNAW branch)
Staff: 40
Activities: designing, manufacturing and operating space instrumentation for atmospheric and astronomical photometry and spectroscopy in (far) IR and sub-mm wavelengths
Periodicals: (1) "Annual Report"
City Reference Coordinates: 006°33'00"E 53°13'00"N

Space Research Organization Netherlands (SRON), Utrecht
Sorbonnelaan 2
NL-3584 CA Utrecht
Telephone: (0)30-2535600
Telefax: (0)30-2540860
Electronic Mail: <userid>@sron.nl
            secr@sron.nl
WWW: http://www.sron.nl/
Founded: 1983
Activities: designing, developing and building space instruments * processing and interpreting satellite data in cooperation with university groups * X rays * IR * Earth observation
City Reference Coordinates: 005°08'00"E 52°05'00"N

StarTel BV
• See "Kapteyn Sterrenwacht - StarTel BV"

Sterrenkundig Instituut Anton Pannekoek
• See "Universiteit Amsterdam, Sterrenkundig Instituut Anton Pannekoek"

Sterrenwacht De Tiendesprong
Gen. de Wetstraat 31
NL-5021 TK Tilburg

Telephone: (0)13-5422534
Electronic Mail: spaninks@freemail.nl
Founded: 1988
Staff: 1
Activities: observing (Sun, Moon, planets) * education * popularization * lectures * shows
Coordinates: 005°04'48"E 51°32'49"N
City Reference Coordinates: 005°05'00"E 51°34'00"N

Sterrenwacht Halley
Postbus 110
NL-5384 ZJ Heesch
or :
Halleyweg 1
NL-5383 KT Heesch
Telephone: (0)412-452383
Founded: 1985
Membership: 250
Activities: popularization * observing
Periodicals: (4) "Halley Periodiek"
Coordinates: 005°29'16"E 51°42'15"N
City Reference Coordinates: 005°32'00"E 51°44'00"N

Sterrewacht Leiden
• See "Universiteit Leiden, Sterrewacht Leiden"

Sterrenwacht Schrieversheide
Schaapskooiweg 95
NL-6414 EL Heerlen
Telephone: (0)45-5225543
Telefax: (0)45-5630037
Electronic Mail: hercunet@cuci.nl
WWW: http://www.cuci.nl/~hercunet/
      http://www.sterrenwacht.nl/
Founded: 1976
Staff: 3
Activities: observing * exhibitions * holography * meteorology * multimedia productions
Periodicals: (11) "Hercules"
Observatories: 2
Coordinates: 005°59'00"E 50°54'00"N
• Formerly: Stichting Volkssterrewacht Hercules
City Reference Coordinates: 005°59'00"E 50°54'00"N

Stichting Astronomisch Onderzoek in Nederland (ASTRON)
Postbus 2
NL-7990 AA Dwingeloo
or :
Oude Hoogeveensedijk 4
NL-7991 PD Dwingeloo
Telephone: (0)521-595100
Telefax: (0)521-597332
Electronic Mail: secretary@nfra.nl
               <userid>@nfra.nl
WWW: http://www.nfra.nl/
Founded: 1949
Staff: 125
Activities: designing and commissioning instrumentation and software * operating Radiosterrenwacht Dwingeloo and Radiosterrenwacht Westerbork (see separate entries) * providing research grants for university research in astronomy
Periodicals: (2) "ASTRON/NFRA Newsletter"; (1) "Annual Report"
City Reference Coordinates: 006°22'00"E 52°49'00"N

Stichting De Koepel
c/o Sterrenwacht Sonnenborgh
Zonnenburg 2
NL-3512 NL Utrecht
Telephone: (0)30-2311360
Telefax: (0)30-2342852
Electronic Mail: info@sonnenborgh.nl (Museum Sterrenwacht Sonnenborgh)
WWW: http://stkwww.fys.ruu.nl:8000/~mathlenr/Zenit.html (Zenit)
      http://www.sonnenborgh.nl/ (Museum Sterrenwacht Sonnenborgh)
Founded: 1973
Membership: 6000
Staff: 2
Activities: coordinating amateur activities in Holland * publishing
Periodicals: (11) "Zenit" (ISSN 0165-0211); (10) "Informatieblad"; (1) "Sterrengids"
Coordinates: 005°07'48"E 52°05'12"N H14m (Sonnenborgh)
City Reference Coordinates: 005°08'00"E 52°05'00"N

Stichting Geologisch, Oceanografisch en Atmosferisch Onderzoek (GOA)
• See now "Gebiedsbestuur Aard- en Levenswetenschappen (ALW)"

Stichting Radiostraling van Zon en Melkweg (SRZM)
• Now merged with Stichting Astronomisch Onderzoek in Nederland (ASTRON) (see separate entry)

Stichting Skepsis
Postbus 2657
NL-3500 GR Utrecht
Telephone: (0)50-3129893
Electronic Mail: skepsis@wxs.n
WWW: http://www.skepsis.org/
Activities: promoting science and critical thinking * exchange and dissemination of ideas and research results * parascience debunking * education
City Reference Coordinates: 005°08'00"E 52°05'00"N

Stichting Volkssterrenwacht Saturnus
Frans Halsstraat 4
NL-1701 JL Heerhugowaard
or :
van Veenweg 98
NL-1701 HH Heerhugowaard
Telephone: (0)72-5745057
(0)72-5745323
Telefax: (0)72-5745323
Electronic Mail: info@st-saturnus.org
WWW: http://www.st-saturnus.nl/
Founded: 1976
Membership: 136
Activities: meetings * public observatory activities * telescope making * education * microscopy * weather satellite ground station * photography * electronics * computing
Periodicals: (12) "Titan"
City Reference Coordinates: 004°50'00"E 52°40'00"N

Stichting Volkssterrewacht Hercules
• See now "Sterrenwacht Schrieversheide"

Stichting Volkssterrewacht Nijmegen
Ruimtevaart-Sterrenkunde Centrum Saturnus
Gildekamp 22-26
NL-6545 KK Nijmegen
or :
Krekelstraat 10
NL- Nijmegen
Telephone: (0)24-3565283
(0)24-3785856
(0)24-3230293
WWW: http://www.telebyte.nl/~saturnus/
City Reference Coordinates: 005°52'00"E 51°50'00"N

Stork Product Engineering (SPE) BV
Postbus 379
NL-1000 AJ Amsterdam
or :
Czaar Peterstraat 229
NL-1018 PL Amsterdam
Telephone: (0)20-5563444
Telefax: (0)20-5563556
Electronic Mail: marketing@spe.storkgroup.com
WWW: http://www.spe.storkgroup.com/home.html
Founded: 1987
Staff: 70
Activities: carbon fibre technology * mechanical parts * ignition and gas generation systems * propulsion technology * electronic control units * electronic test and simulation equipment
City Reference Coordinates: 004°54'00"E 52°22'00"N

Technische Universiteit Delft, Delft Institute for Earth-Oriented Space Research (DEOS)
Postbus 5030
NL-2600 GA Delft
or :
Kluyverweg 1
NL-2629 HS Delft
Telephone: (0)15-2783289
Telefax: (0)15-2783711
Electronic Mail: <userid>@deos.tudelft.nl

WWW: http://www.deos.tudelft.nl/
Founded: 1945 (Faculty of Engineering)
Staff: 8
Activities: orbit mechanics * geophysics * oceanic physics * satellite systems * earth observation systems * remote sensing * tracking and satellite communications
City Reference Coordinates: 004°21'00"E 52°00'00"N

Universiteit Amsterdam, Instituut voor Hoge-Energie Astrofysica
Kruislaan 403
NL-1098 SJ Amsterdam
Telephone: (0)20-5257491
(0)20-5257492
Telefax: (0)20-5257484
Electronic Mail: <userid>@astro.uva.nl
WWW: http://www.uva.nl/onderzoek/instituten/hef.html
Founded: 1988
Activities: X-ray astronomy * compact stars * neutrino astrophysics * cosmology * plasma astrophysics
• This center gathers scientists from the University of Amsterdam, the University of Utrecht and the National Institute for Nuclear and High-Energy Physics
City Reference Coordinates: 004°54'00"E 52°22'00"N

Universiteit Amsterdam, Sterrenkundig Instituut Anton Pannekoek
Kruislaan 403
NL-1098 SJ Amsterdam
Telephone: (0)20-5257491
(0)20-5257492
Telefax: (0)20-5257484
Electronic Mail: secr-astro@astro.uva.nl
<userid>@astro.uva.nl
WWW: http://www.astro.uva.nl/
Founded: 1921
Staff: 23
• Full university name: "Universiteit van Amsterdam"
City Reference Coordinates: 004°54'00"E 52°22'00"N

Universiteit Amsterdam, Faculteit der Natuurkunde en Sterrenkunde
De Boelelaan 1081
NL-1081 HV Amsterdam
Telephone: (0)20-4447956
(0)20-4447957
Telefax: (0)20-4447899
Electronic Mail: <userid>@nat.vu.nl
secretariaat@nat.vu.nl
WWW: http://www.nat.vu.nl/
http://www.nat.vu.nl/vakgroepen/ster/english/
Staff: 11
Activities: light scattering in the solar system * laboratory astrophysics using lasers * light scattering in ISM
• Full university name: "Vrije Universiteit Amsterdam"
City Reference Coordinates: 004°54'00"E 52°22'00"N

Universiteit Groningen, Kapteyn Instituut
Landleven 12
NL-9747 AD Groningen
Telephone: (0)50-3634073
Telefax: (0)50-3636100
Electronic Mail: secr@astro.rug.nl
WWW: http://www.astro.rug.nl/
Founded: 1885
Staff: 70
Activities: galaxies * cosmic evolution * ISM and interaction with stars
Periodicals: (1) "Annual Report"
• Full university name: "Rijksuniversiteit Groningen"
City Reference Coordinates: 006°33'00"E 53°13'00"N

Universiteit Leiden, Raymond and Beverly Sackler Laboratorium voor Astrofysica
Sterrewacht Leiden
Postbus 9513
NL-2300 RA Leiden
or :
Niels Bohrweg 2
NL-2333 RA Leiden
Telephone: (0)71-5275801
Telefax: (0)71-5275819
Electronic Mail: <userid>@strw.leidenuniv.nl
WWW: http://www.strw.leidenuniv.nl/~Lab/
Founded: 1975

Staff: 4
Activities: spectroscopy and photochemistry of interstellar ice analogs * education * interstellar and circumstellar matter * comets * origin of life
Coordinates: 004°29'00"E 52°09'20"N H12m
• Full university name: "Rijksuniversiteit Leiden"
City Reference Coordinates: 004°30'00"E 52°09'00"N

Universiteit Leiden, Sterrewacht Leiden
Postbus 9513
NL-2300 RA Leiden
or :
Niels Bohrweg 2
NL-2333 CA Leiden
Telephone: (0)71-5275833
Telefax: (0)71-5275819
Electronic Mail: strwmail@strw.leidenuniv.nl
WWW: http://www.strw.leidenuniv.nl/
Founded: 1633
Staff: 50
Activities: ISM * galactic structure * active galaxies * cosmology
Coordinates: 004°29'02"E 52°09'20"N H6m (IAU Code 013)
• Full university name: "Rijksuniversiteit Leiden"
City Reference Coordinates: 004°30'00"E 52°09'00"N

Universiteit Nijmegen, Sterrenkunde
Postbus 9010
NL-6500 GL Nijmegen
or :
Toernooiveld
NL-6525 ED Nijmegen
Telephone: (0)24-3652080
         (0)24-3652121
Telefax: (0)24-3652191
Electronic Mail: kuijpers@hef.kun.nl (J. Kuijpers)
         gommers@sci.kun.nl (R. Gommers)
WWW: http://www-physics.sci.kun.nl/studiegids/
Coordinates: 005°52'06"E 51°49'31"N H62m
• Full university name: "Katholieke Universiteit Nijmegen"
City Reference Coordinates: 005°52'00"E 51°50'00"N

Universiteit Utrecht, Instituut voor Marien en Atmosferisch Onderzoek Utrecht (IMAU)
Postbus 80.005
NL-3508 TA Utrecht
or :
Princetonplein 5
NL-3584 CC Utrecht
Telephone: (0)30-2533275
Telefax: (0)30-2543163
Electronic Mail: imau@phys.uu.nl
WWW: http://www.phys.uu.nl/~wwwimau/
Founded: 1966
Staff: 20
Activities: meteorology * glaciology * oceanography * atmospheric chemistry * coastal dynamics
• Full university name: "Rijksuniversiteit Utrecht"
City Reference Coordinates: 005°08'00"E 52°05'00"N

Universiteit Utrecht, Sterrekundig Instituut
(Sterrekundig Instituut Utrecht - SIU)
Princetonplein 5
Postbus 80000
NL-3508 TA Utrecht
Telephone: (0)30-2535200
Telefax: (0)30-2535201
Electronic Mail: <userid>@phys.uu.nl
         astronomy@phys.uu.nl
WWW: http://stkwww.phys.uu.nl/
Founded: 1643
Staff: 14
Activities: research in astrophysics
Periodicals: (1) "Annual Report"
Coordinates: 005°07'48"E 52°05'12"N H14m (Sterrenwacht Sonnenborgh – IAU Code 015)
City Reference Coordinates: 005°08'00"E 52°05'00"N

Vereniging voor Weer- en Sterrenkunde Noord Drenthe
c/o G. van den Braak
Leenakkersweg 46

New Zealand

Astronautics Association of New Zealand (AANZ) Inc.
P.O. Box 11-734
Wellington
Telephone: (0)4-4773632 (Home)
            (0)4-5765158 (Office)
Telefax: (0)4-4773632 (Home)
Founded: 1979
Membership: 30
Activities: promoting awareness of space exploration
Periodicals: (6) "AANZ Journal" (ISSN 1170-7372)
City Reference Coordinates: 174°46'00"E 41°18'00"S

Astronomical Pocket Diary (APD)
c/o Norbert Haley
Poste Restante
P.O. Wellesley Street
Auckland
Electronic Mail: norb@kcbbs.gen.nz
                 norb@geocities.com
WWW: http://members.tripod.com/~apd2/apd.htm
Founded: 1989
Staff: 1
Activities: publishing pocket-size almanac
City Reference Coordinates: 174°46'00"E 36°55'00"S

Auckland Astronomical Society, Inc.
P.O. Box 24-187
Royal Oak
Auckland 6
Telephone: (0)9-622-2420
Electronic Mail: info@astronomy.org.nz
WWW: http://www.astronomy.org.nz/
Founded: 1922
Membership: 615
Activities: observing * research * lectures * education
Periodicals: (12) "Journal of the Auckland Astronomical Society"
Coordinates: 174°46'00"E 36°55'00"S (Auckland Obs. – IAU Code 467)
City Reference Coordinates: 174°46'00"E 36°55'00"S

Auckland Observatory and Planetarium Trust
P.O. Box 24-180
Auckland 6
or :
One Tree Hill Domain
Auckland 3
Telephone: (0)9-624-1246
            (0)9-625-6945 (Infoline)
Telefax: (0)9-625-2394
Electronic Mail: info@stardome.org.nz
WWW: http://www.stardome.org.nz/
Founded: 1967
Staff: 6
Activities: lectures * photoelectric photometry * eclipsing binaries * Miras * dwarf novae
Coordinates: 174°46'42"E 36°54'24"S H80m (IAU Code 467)
City Reference Coordinates: 174°46'00"E 36°55'00"S

Blaxall and Steven Ltd.
163 Saint Asaph Street
P.O. Box 25095
Christchurch
Telephone: (0)3-366-2828
Telefax: (0)3-365-2072
Electronic Mail: blaxall@voyager.co.nz (Michael T. Blaxall, General Manager)
Founded: 1954
Staff: 6
Activities: importing optical material
City Reference Coordinates: 172°38'00"E 43°42'00"S

Canterbury Astronomical Society (CAS)
P.O. Box 25-137
Victoria Street

Christchurch
Telephone: (0)3-374-9549
Telefax: (0)3-325-3865
Electronic Mail: l.hussey@lincoln.ac.nz
WWW: http://www.phys.canterbury.ac.nz/cas/
Founded: 1957
Membership: 150
Activities: observing (grazing occultations, minor-planet occultations, variable stars) * photometry * monthly meetings * public nights * public education
Periodicals: (11) "CASMAG"
Coordinates: 172°20'58"E 43°30'02"S H118m (Joyce Memorial Observatory)
City Reference Coordinates: 172°38'00"E 43°42'00"S

Canterbury Astronomical Society (CAS), Ashburton Branch
c/o Ken Lucas (President)
11 Queens Drive
Ashburton
Telephone: (0)3-308-9203
WWW: http://www.phys.canterbury.ac.nz/cas/ (CAS)
Founded: 1970
Membership: 12
Activities: amateur observing * monthly meetings with guest speakers * public nights * public education * seminars
Coordinates: 171°45'03"E 43°53'38"S
City Reference Coordinates: 171°46'00"E 43°54'00"S

Carter Observatory
P.O. Box 2909
Wellington 1
or :
40 Salamanca Road
Kelburn
Wellington
Telephone: (0)4-472-8167
Telefax: (0)4-472-8320
Electronic Mail: astronomy@carterobs.ac.nz
WWW: http://www.carterobs.ac.nz/
Founded: 1941
Staff: 13
Activities: National Observatory of New Zealand (IAU Code 485) * facilitation of astronomical research * planetarium shows * national astronomical heritage collection
Sections: public astronomy * education * astronomical heritage
Periodicals: (12) "Carter Observatory Newsletter" (ISSN 1173-8812); (1) "Annual Report" (ISSN 1173-5392),
City Reference Coordinates: 174°46'00"E 41°18'00"S

Gisborne Astronomical Society (GAS)
P.O. Box 678
Gisborne
Telephone: (0)6-867-7901
Electronic Mail: hac.gas@xtra.co.nz
City Reference Coordinates: 178°02'00"E 38°41'00"S

Hamilton Astronomical Society (HAS)
P.O. Box 9153
Hamilton 2001
or :
Next to 200 Brymer Road
Hamilton
Telephone: (0)7-849-8522
Electronic Mail: secretary@has.org.nz
WWW: http://www.has.org.nz/
Founded: 1933
Membership: 50
Activities: meetings * lectures * education * open public nights
Periodicals: (11) "Hamilton Astronomical Society Bulletin"
Coordinates: 175°13'40"E 37°46'32"S H100m
City Reference Coordinates: 175°15'00"E 37°45'00"S

Hawkes Bay Astronomical Society
7 Maadi Road
Napier
Telephone: 6-8439513
Telefax: 6-8439503
Founded: 1960
Membership: 40
Activities: popularization of astronomy and space exploration
City Reference Coordinates: 176°58'00"E 39°29'00"S

Hawkes Bay Holt Planetarium
Chambers Street
Napier
Telephone: 6-8344345
Telefax: 6-8344345
Electronic Mail: hb-holt-planetarium@xtra.co.nz
WWW: http://mysite.xtra.co.nz/~HoltPlanetarium/
Founded: 1998
Staff: 2
Activities: shows on astronomy and space exploration for schools and public
City Reference Coordinates: 176°58'00"E 39°29'00"S

Holt Planetarium (Hawkes Bay_)
• See "Hawkes Bay Holt Planetarium"

Institute of Geological and Nuclear Sciences (IGNS) Inc.
69B Gracefield Road
P.O. Box 30-368
Lower Hutt
Telephone: (0)4-5701444
Telefax: (0)4-5704600
WWW: http://www.gns.cri.nz/
Founded: 1865
Staff: 220
Activities: Earth sciences research (including seismology, volcanology, paleontology, isotopes, hydrocarbons)
Periodicals: "Monographs IGNS", "Geological Maps IGNS"
City Reference Coordinates: 174°46'00"E 41°18'00"S

Meteorological Service of New Zealand Ltd.
30 Salamanca Road
P.O. Box 722
Wellington 6015
Telephone: (0)4-4729-379
Telefax: (0)4-4735-231
Electronic Mail: <userid>@met.co.nz
            service@met.co.n
WWW: http://www.met.co.nz/
Founded: 1992
Staff: 177
Activities: weather forecasting * managing about 100 observing sites
City Reference Coordinates: 174°46'00"E 41°18'00"S

New Zealand Institute of Physics (NZIP)
c/o IRL
P.O. Box 31310
Lower Hutt
Electronic Mail: nzip@irl.cri.nz
WWW: http://nzip.rsnz.govt.nz/
City Reference Coordinates: 174°46'00"E 41°18'00"S

New Zealand Spaceflight Association (NZSA) Inc.
P.O. Box 5829
Wellesley Street
Auckland 1
Telephone: (0)9-480-7900
Electronic Mail: jeffg@kcbbs.gen.nz
WWW: http://homepages.ihug.co.nz/~mikehow/
Founded: 1977
Membership: 200
Activities: studying and promoting manned spaceflight and robotic craft
Periodicals: (6) "Liftoff"
City Reference Coordinates: 174°46'00"E 36°55'00"S

Pallasite Press
P.O. Box 33-1218
Takapuna
Auckland
Telephone: (0)9-486-2428
Telefax: (0)9-489-6750
Electronic Mail: j.schiff@auckland.ac.nz
WWW: http://meteor.co.nz/
Founded: 1994
Staff: 6
Periodicals: (4) "Meteorite!" (ISSN 1173-2245)
• Publisher

City Reference Coordinates: 174°46'00"E 36°55'00"S

Royal Astronomical Society of New Zealand (RASNZ) Inc.
P.O. Box 3181
Wellington
Electronic Mail: rasnz@rasnz.org.nz
WWW: http://www.rasnz.org.nz/
Founded: 1920
Membership: 220
Activities: publications * annual conference * observing
Sections: Aurorae and Sun * Comets and Minor Planets * Education * Occultations * Photometry * Variable Stars
Periodicals: (10) "RASNZ Newsletter" (ISSN 1173-8138); (4) "Southern Stars" (ISSN 0049-1640)
Awards: (1) "Murray Geddes Prize"
City Reference Coordinates: 174°46'00"E 41°18'00"S

Royal Society of New Zealand (RSNZ)
P.O. Box 598
Wellington
or :
4 Halswell Street
Thorndon
Wellington
Telephone: (0)4-472-7421
Telefax: (0)4-473-1841
Electronic Mail: ceo@rsnz.govt.nz
WWW: http://www.rsnz.govt.nz/
Founded: 1867
Membership: 20000
Staff: 31
Activities: statutory body incorporating the "New Zealand National Committee for Astronomy" and the "New Zealand Standing Committee on the Astronomical Sciences" (astronomy, space science, radio astronomy and astrophysics)
Periodicals: "Journal of The Royal Society of New Zealand", "New Zealand Journal of Agricultural Research", "New Zealand Journal of Botany", "New Zealand Journal of Crop and Horticultural Science", "New Zealand Journal of Geology and Geophysics", "New Zealand Journal of Marine and Freshwater Research", "New Zealand Journal of Zoology"
City Reference Coordinates: 174°46'00"E 41°18'00"S

Sky Laboratories International Ltd.
163 Saint Asaph Street
P.O. Box 25095
Christchurch
Telephone: (0)3-366-2827
Telefax: (0)3-365-2072
Electronic Mail: optics.orders@voyager.co.nz
WWW: http://www.skylab.co.nz/
Founded: 1979
Staff: 2
Activities: retailing optical material
City Reference Coordinates: 172°38'00"E 43°42'00"S

Southland Astronomical Society (SAS)
Southland Museum
P.O. Box 1168
Invercargill
Electronic Mail: trendyhendy@yahoo.com
WWW: http://ehp.iwarp.com/sas/
City Reference Coordinates: 168°21'00"E 46°26'00"S

Standards New Zealand (SNZ)
Private Bag 2439
Wellington 6020
or :
Standards House
155 The Terrace
Wellington 6001
Telephone: (0)4-498-5990
Telefax: (0)4-498-5994
Electronic Mail: snz@standards.co.nz
WWW: http://www.standards.co.nz/
Founded: 1932
Staff: 48
Activities: national standards authority
Periodicals: (11) "Standards New Zealand" (ISSN 0110-4667)
City Reference Coordinates: 174°46'00"E 41°18'00"S

The Phoenix Astronomical Society (TPAS)
P.O. Box 2217
Wellington
Electronic Mail: sacha.hall@dear.net.nz
WWW: http://www.phoenix.org.nz/
Founded: 1997
City Reference Coordinates: 174°46'00"E 41°18'00"S

University of Canterbury, Department of Physics and Astronomy
Private Bag 4800
Christchurch 8020
Telephone: (0)3-366-7001
Telefax: (0)3-364-2469
Electronic Mail: <userid>@canterbury.ac.nz
WWW: http://www.phys.canterbury.ac.nz/
http://www.canterbury.ac.nz/
Founded: 1872
Staff: 12
Activities: variable-star photometry * stellar spectroscopy * CCD photometry * minor-planet and comet astrometry
Observatories: 1 (Mount John University Observatory - see separate entry)
City Reference Coordinates: 172°38'00"E 43°42'00"S

University of Canterbury, Department of Physics and Astronomy, Mount John University Observatory (MJUO)
P.O. Box 56
Lake Tekapo 8770
Telephone: (0)3-680-6000
Electronic Mail: mjuo@phys.canterbury.ac.nz
WWW: http://www.phys.canterbury.ac.nz/
http://www.mjuo.canterbury.ac.nz/mjuo/
Founded: 1965
Staff: 8
Activities: variable-star photometry * stellar spectroscopy * solar-system astrometry * CCD photometry
Coordinates: 170°27'54"E 43°59'12"S H1,029m (IAU Code 474)
City Reference Coordinates: 170°29'00"E 44°02'00"S

University of Waikato, Mathematics Department
Private Bag 3105
Hamilton
Telephone: (0)7-856-2889
Telefax: (0)7-838-4666
Electronic Mail: <userid>@waikato.ac.nz
WWW: http://hoiho.math.waikato.ac.nz/
Founded: 1964
Staff: 8
Activities: theoretical astrophysics * MHD
City Reference Coordinates: 175°17'00"E 37°47'00"S

Wairarapa Astronomical Society
c/o Alison Adams (Secretary)
5 McKenzie Terrace
Carterton
or :
c/o B. Harvey
186d Colombo Road
Masterton
Telephone: (0)6-378-7865
Founded: 1985
Membership: 16
Activities: variable stars * occultations * comets * minor planets * aurorae * solar flares * photography * visits to schools * mobile observatory
City Reference Coordinates: 175°39'00"E 40°57'00"S (Masterton)

Wanganui Astronomical Society
17 Rata Street
Wanganui
b
c/o David Calder
12 Rees Street
Wanganui
Telephone: (0)6-345-6113
Founded: 1947
Membership: 100

Activities: public education * observing
City Reference Coordinates: 175°02'00"E 39°56'00"S

Wellington Astronomical Society (WAS)
P.O. Box 3126
Wellington
Telephone: (0)4-528-3468 (Mike Clear)
Electronic Mail: clearm@agriquality.co.nz
WWW: http://astronomy.wellington.net.nz/
Periodicals: (12) "Newsletter"
City Reference Coordinates: 174°46'00"E 41°18'00"S

Whangarei Astronomical Society Inc.
P.O. Box 436
Whangarei
Telephone: (0)4389630
Founded: 1963
Membership: 32
Activities: lectures * discussions * observing * grazes * talks
Periodicals: "Octantis"
City Reference Coordinates: 174°20'00"E 35°43'00"S

Nigeria

Meteorological Department
Oshodi
Lagos
or :
Private Mail Bag 12542
Lagos
Telephone: (0)1-631792
(0)1-631717
(0)1-2633371
Telefax: (0)1-2634489
(0)1-2636097
City Reference Coordinates: 003°24'00"E 06°27'00"N

Nigerian Academy of Sciences (NAS)
P.M.B. 1004
University of Lagos Post Office
Akoka
Yaba
Lagos
Telephone: (0)1-863874
Telefax: (0)1-863874
(0)2-8103118
Electronic Mail: library@ibadan.ac.ng
Founded: 1977
Membership: 85
Staff: 5
Activities: public lectures * conferences * symposia
Periodicals: "Proceedings" (ISSN 0794-7976)
Awards: (1) "MAN National Science Prize"
City Reference Coordinates: 003°24'00"E 06°27'00"N

Standards Organization of Nigeria (SON)
Federal Secretariat
Phase 1 - 9th Floor
Ikoyi
Lagos
Telephone: (0)1-2696178
Telefax: (0)1-2696176
City Reference Coordinates: 003°24'00"E 06°27'00"N

University of Nigeria, Department of Physics and Astronomy
Nsukka
Telephone: (0)42-770500
Telefax: (0)42-770644
Founded: 1960
Staff: 30
Activities: radio astronomy * high-energy astrophysics * solar energy * solid-state physics * optical astronomy * cosmology
Coordinates: 007°27'13"E 06°52'07"N
City Reference Coordinates: 007°29'00"E 06°51'00"N

Norway

Andøya Rocket Range (ARR)
• See "Norwegian Space Centre (NSC), Andøya Rocket Range (ARR)"

European Incoherent Scatter Facility (EISCAT), Ionospheric Heating Facility
Ramfjordmoen
N-9027 Ramfjordbotn
Telephone: 77692171
Telefax: 77692360
77692380
Electronic Mail: mike.rietveld@eiscat.uit.no (Mike Rietveld, Senior Scientist)
WWW: http://www.eiscat.uit.no/heating/
Founded: 1975 (EISCAT)
Staff: 2
Coordinates: 019°13'00"E 69°35'00"N
City Reference Coordinates: 019°00'00"E 69°40'00"N

European Incoherent Scatter Facility (EISCAT), Tromsø Station
Ramfjordmoen
N-9027 Ramfjordbotn
Telephone: 77692166
Telefax: 77692360
77692380
Electronic Mail: eiscat@eiscat.uit.no
<userid>@eiscat.uit.no
WWW: http://www.eiscat.uit.no/
Founded: 1975 (EISCAT)
Staff: 15
Activities: mesospheric, ionospheric and magnetospheric research using incoherent scatter radars (transmitter and receiver site)
Coordinates: 019°13'00"E 69°35'00"N H30m
City Reference Coordinates: 019°13'00"E 69°35'00"N

European Incoherent Scatter Facility (EISCAT), Svalbard Radar
P.O. Box 432
N-9170 Longyearbyen
Telephone: 79021236
Telefax: 79021751
Electronic Mail: esr@esr.eiscat.no
<userid>@esr.eiscat.no
WWW: http://www.esr.eiscat.no/
Founded: 1975 (EISCAT)
Staff: 4
City Reference Coordinates: 015°40'00"E 78°12'00"N

Maidanak Foundation (The_)
• See "The Maidanak Foundation"

Nordlysobservatoriet
• See "University of Tromsø, Department of Physics, Auroral Observatory"

Norsk Astronautisk Forening (NAF)
(Norwegian Astronautical Association)
Postboks 52
Blindern
N-0313 Oslo
Telephone: 88003130
Founded: 1951
Membership: 350
Periodicals: (8) "Smånytt om Romfart"; (4) "Nytt om Romfart" (ISSN 0332-5962)
City Reference Coordinates: 010°45'00"E 59°55'00"N

Norsk Astronomisk Selskap (NAS)
(Norwegian Astronomical Society - NAS)
Postboks 1029
Blindern
N-0315 Oslo
Telephone: 22456513
Telefax: 22856505
WWW: http://www.ifi.uio.no/~mikkels/nas/nas.html
http://www.ifi.uio.no/~mikkels/nas/InterNAS.html

http://www.astro.uio.no/nas/
Founded: 1938
Membership: 800
Activities: observing groups * publishing * lectures * excursions
Sections: Meteors * Sun * Variable Stars * Supernova Search * Occultations * Comets
Periodicals: (4) "Astronomi"
City Reference Coordinates: 010°45'00"E 59°55'00"N

Norske Videnskaps-Akademi
(Norwegian Academy of Science and Letters)
Drammensveien 78
N-0271 Oslo
Telephone: 2444296
Telefax: 2562656
WWW: http://www.dnva.no/
Founded: 1857
Membership: 610
Staff: 5
Activities: promoting science
Periodicals: (1) "Academy's Yearbook" (ISSN 0332-6909), "Kristian Birkeland Lecture"
Awards: "Fridtjof Nansen Medal"
City Reference Coordinates: 010°45'00"E 59°55'00"N

Norsk Fysisk Selskap (NFS)
(Norwegian Physical Society)
c/o T. Henriksen (President)
Department of Physics
University of Oslo
Postboks 1048
Blindern
N-0316 Oslo
Telephone: 22856426
Telefax: 22856422
WWW: http://www.norskfysikk.no/nfs//
Founded: 1953
Membership: 850
City Reference Coordinates: 010°45'00"E 59°55'00"N

Norsk Teknisk Museum (NTM)
(Norwegian Museum of Science and Technology)
Kjelsåsvn 143
N-0491 Oslo
Telephone: 22796000
Telefax: 22796100
Electronic Mail: jaa@tekniskmuseum.no
WWW: http://www.tekniskmuseum.no/
http://www.teknoteket.no/no/planetar.html (St.-Exupéry Planetarium)
Founded: 1914
Staff: 40
City Reference Coordinates: 010°45'00"E 59°55'00"N

Norwegian Academy of Science and Letters
• See "Norske Videnskaps-Akademi"

Norwegian Astronautical Association
• See "Norsk Astronautisk Forening (NAF)"

Norwegian Astronomical Society (NAS)
• See "Norsk Astronomisk Selskap (NAS)"

Norwegian Geotechnical Institute (NGI)
Sognsveien 72
Postboks 3930
Ullevaal Stadion
N-0806 Oslo
Telephone: 22023000
Telefax: 22230448
Electronic Mail: ngi@ngi.no
WWW: http://www.ngi.no/
Founded: 1953
Staff: 136
Activities: geotechnical engineering * arctic engineering * construction control * cross hole seismic * drainage design * earthquake engineering * engineering geology * environmental geotechnology * erosion control * feasibility studies * field instrumentation * foundation engineering * geochemistry * geodynamics * geological mapping * grouting design * in situ testing * model testing * numerical analysis * performance observing * permeability testing * permafrost investigations *

petroleum reservoir technology * pollution * risk analysis * rock blast design * rock mechanics * sampling * seepage control * site investigation * snow mechanics * soil dynamics * soil mechanics * stability analysis * vibration control
Periodicals: "NGI Publications Series" (ISSN 0078-1193)
City Reference Coordinates: 010°45'00"E 59°55'00"N

Norwegian Meteorological Institute (DNMI)
Niels Henrik Aabelsvei 40
Postboks 43
Blindern
N-0313 Oslo
Telephone: 22963000
Telefax: 22963050
Electronic Mail: met.inst@met.no
WWW: http://www.dnmi.no/
Founded: 1866
Staff: 450
Activities: national weather forecasting centre * meteorological research * NOAA and Meteosat receiving and processing
City Reference Coordinates: 010°45'00"E 59°55'00"N

Norwegian Physical Society
• See "Norsk Fysisk Selskap (NFS)"

Norwegian Space Centre (NSC)
Drammensveien 105
Postboks 113
Skoyen
N-0212 Oslo
Telephone: 22511800
Telefax: 22511801
Electronic Mail: spacecentre@spacecentre.no
               <userid>@spacecentre.no
WWW: http://www.spacecentre.no/
Founded: 1987
Staff: 75
Activities: national agency coordinating Norwegian space activities * operating the "Andøya Rocket Range (ARR)" and the "Tromsø Satellite Station (TSS)" (see separate entries)
Periodicals: (1) "Space Research in Norway", "Space Technology and Industries in Norway", "Annual Report", "National Long-Term Plan for Space Activities"
City Reference Coordinates: 010°45'00"E 59°55'00"N

Norwegian Space Centre (NSC), Andøya Rocket Range (ARR)
Postboks 54
N-8480 Andenes
Telephone: 76141644
Telefax: 76141857
Electronic Mail: info@rocketrange.no
WWW: http://www.rocketrange.no/
       http://www.spacecentre.no/
Founded: 1962
Staff: 31
Activities: launching sounding rockets and scientific balloons
Coordinates: 016°01'00"E 69°17'00"N
City Reference Coordinates: 016°08'00"E 69°16'00"N

Norwegian Space Centre (NSC), Tromsø Satellite Station (TSS)
• See "Tromsø Satellite Station (TSS)"

Norwegian Standards Association
(Norges Standardiseringsforbund - NSF)
Drammensveien 145
Postboks 353
Skøyen
N-0212 Oslo
or :
Hegdehaugsveien 31
Postboks 7020
Homansbyen
N-0306 Olso
Telephone: 22466094
          22049200
Telefax: 22464457
         22049211
Electronic Mail: firmapost@nsf.telemax.no
WWW: http://www.standard.no/nsf
Founded: 1923

City Reference Coordinates: 010°45'00"E 59°55'00"N

Norwegian University of Science and Technology, Institute of Physics
Haakon Magnussons gate 3
N-7491 Trondheim
or :
Høgskoleringen 5
N-7034 Trondheim
Telephone: 73591850
Telefax: 73591852
Electronic Mail: <userid>@phys.ntnu.no
WWW: http://www.phys.ntnu.no/
Founded: 1922
Staff: 15
Activities: theoretical physics * astrophysics (AGNs, compact stars, cosmology, gravitation) * elementary particles * field theory
City Reference Coordinates: 010°25'00"E 63°25'00"N

Research Council of Norway
Postboks 2700
St. Hanshaugen
N-0131 Oslo
or :
Stensberggata 26
N- Oslo
Telephone: 22037000
Telefax: 22037001
Electronic Mail: info@forskningsradet.no
WWW: http://www.forskningsradet.no/
Founded: 1993
Activities: bioproduction and processing * industry and energy * culture and society * medicine and health * environment and development * natural sciences and technology
City Reference Coordinates: 010°45'00"E 59°55'00"N

Skibotn Astronomical Observatory
N-9048 Skibotn
Telephone: 77715820
Coordinates: 020°22'00"E 69°21'00"N H155m (IAU Code 093)
City Reference Coordinates: 020°22'00"E 69°21'00"N

Stavanger Astronomiske Forening (SAF)
(Stavanger Astronomical Association)
Postboks 453
N-4002 Stavanger
Telephone: 51566813 (Wednesdays 19:00-23:00)
Electronic Mail: saf-styret@bsda.his.no
WWW: http://www.ux.his.no/saf/
Founded: 1969
Membership: 80
Activities: deep sky * planets * meteorites * minor planets * comets
Coordinates: 005°42'08"E 58°58'26"N H77m
City Reference Coordinates: 005°45'00"E 58°58'00"N

Svalsat
• See "Tromsø Satellite Station (TSS), SvalSat"

Teknoteket
• See now "Norsk Teknisk Museum (NTM)"

The Maidanak Foundation
Welhavens Gate 61
N-5006 Bergen
Telefax: 55962663
Electronic Mail: henrik.nilsen@fi.uib.no (Henrik E. Nilsen, Chairman)
WWW: http://www.maidanak.org/
Founded: 1999
Staff: 6
Activities: supporting scientific teams running Mount Maidanak Observatory (Uzbekistan) * providing funding for scientific equipment
City Reference Coordinates: 005°20'00"E 60°23'00"N

Tromsø Satellite Station (TSS)
Prestvannsveien 38
N-9291 Tromsø
Telephone: 4777600250

Telefax: 4777600299
Electronic Mail: tss@tss.no
WWW: http://www.tss.no/
Founded: 1967
Staff: 40
Activities: reception and distribution of data from polar orbiting satellites * search and rescue via satellites * monitoring services based on Earth observation data
Coordinates: 018°56'00"E 69°39'00"N
• Jointly and equally owned by the "Norwegian Space Centre (NSC)" and the "Swedish Space Corp. (SSC), Earth Observation Division"
City Reference Coordinates: 018°56'00"E 69°39'00"N

Tromsø Satellite Station (TSS), SvalSat
P.O. Box 458
N-9171 Longyearbyen
Telephone: 4779022550
Telefax: 4779023784
Electronic Mail: svalsat@tss.no
WWW: http://www.svalsat.com/
Founded: 1997
Activities: tracking and command of polar orbiting satellites
City Reference Coordinates: 015°40'00"E 78°12'00"N

University of Bergen, Department of Physics
Allégaten 55
N-5007 Bergen
Telephone: 55582806
Telefax: 55589440
Electronic Mail: <userid>@fi.uib.no
WWW: http://www.fi.uib.no/
Founded: 1948
Staff: 78
Activities: space physics * magnetospheric research
City Reference Coordinates: 005°20'00"E 60°23'00"N

University of Oslo, Department of Physics
Postboks 1048
Blindern
N-0316 Oslo
Telephone: 22856428
Telefax: 22856422
Electronic Mail: admin@fys.uio.no
    <userid>@fys.uio.no
WWW: http://www.fys.uio.no/english.html
Founded: 1964 (University: 1811)
Activities: general physics * plasma and space physics * high energy * solid state * electronics * nuclear physics * biophysics * materials * theoretical physics
City Reference Coordinates: 010°45'00"E 59°55'00"N

University of Oslo, Institute of Theoretical Astrophysics
Postboks 1029
Blindern
N-0315 Oslo
Telephone: 22856501
    22856502
Telefax: 22856505
Electronic Mail: mats.carlsson@astro.uio.no (Mats Carlsson, Director)
WWW: http://www.astro.uio.no/
Founded: 1934 (University: 1811)
Staff: 13
Activities: solar transition zone * prominences * sunspots * cool stars * solar wind * plasma physics * planetary systems * cosmology * gravitational lenses
Periodicals: "Institute of Theoretical Astrophysics, Reports" (ISSN 0078-6780)
City Reference Coordinates: 010°45'00"E 59°55'00"N

University of Tromsø, Department of Physics, Auroral Observatory
Prestvannsveien 40
N-9037 Tromsø
Telephone: 77644000
    77644001
    77645150
Telefax: 77648850
    77645580
Electronic Mail: <userid>@phys.uit.no
WWW: http://www.phys.uit.no/
    http://geo.phys.uit.no/

Founded: 1930
Staff: 25
Activities: cataclysmic variables * stellar pulsations * degenerate objects * interplanetary dust * cosmic geophysics * plasma physics * microelectronics and communications * molecular quantum mechanics * Skibotn Astronomical Observatory/Nordlysobservatoriet (see separate entry)
Periodicals: "Journal of Magnetic Observations" (ISSN 0373-4854)
Coordinates: 020°21'54"E 69°20'54"N (Skibotn Astron. Obs. – IAU Code 093)
City Reference Coordinates: 018°58'00"E 69°40'00"N

University of Tromsø, Department of Physics, Astrophysics Group
Nordlysobservatoriet
Prestvannsveien 40
N-9037 Tromsø
Telephone: 77645191
   77715820 (Skibotn Astronomical Observatory)
Telefax: 77645595
Electronic Mail: <userid>@phys.uit.no
WWW: http://www.phys.uit.no/fysikk/astro/index-e.html
Founded: 1972
Staff: 1
Activities: degenerate objects
Coordinates: 020°21'54"E 69°20'54"N H157m (Skibotn Astron. Obs. – IAU Code 093)
City Reference Coordinates: 018°58'00"E 69°40'00"N

Pakistan

Lahore Astronomical Society Pakistan (LAST)
36 Tariq Block
New Garden Town
Lahore
Telephone: (0)5831727
Electronic Mail: pulsar@brain.net.pk
WWW: http://www.geocities.com/Broadway/8701/
Founded: 1976
City Reference Coordinates: 074°18'00"E 31°35'00"N

Pakistan International Airlines (PIA), Planetarium, Lahore
University Ground
Lake Road
Lahore 54000
Telephone: (0)42-210810
Founded: 1987
Staff: 19
Activities: shows for public and groups of students and teachers
City Reference Coordinates: 074°18'00"E 31°35'00"N

Pakistan Meteorological Department (PMD)
P.O. Box 8454
University Road
Karachi 75270
Telephone: (0)21-8112223
Telefax: (0)21-8112885
Electronic Mail: pmd@paknet3.ptc.pk
WWW: http://met.gov.pk/
Founded: 1947
Staff: 2000
Activities: meteorology * climatology * hydrology * seismology * geophysics * atmospheric physics * environmental pollution
Periodicals: (12) "Monthly Climatic Summary of Pakistan"; "Agromet Bulletin of Pakistan"
City Reference Coordinates: 067°03'00"E 24°52'00"N

Pakistan Space & Upper Atmosphere Research Commission (SUPARCO)
Sector 28, Gulzar-e-Hijri
Off University Road
P.O. Box 8402
Karachi 75270
Telephone: (0)21-472630
(0)21-470158
Telefax: (0)21-4960553
Founded: 1961
Staff: 220
Activities: national space agency for the promotion of peaceful uses and applications of space sciences and technology * remote sensing * satellite communications * ionospheric research
Periodicals: (4) "Space Horizons" (ISSN 0259-9163)
City Reference Coordinates: 067°03'00"E 24°52'00"N

Pakistan Standards Institution (PSI)
39 Garden Road
Saddar
Karachi 74400
Telephone: (0)21-7729527
Telefax: (0)21-7728124
Electronic Mail: pakqltyk@super.net.pk
Founded: 1951
Staff: 152
Activities: drawing up standards and promoting their adoption
City Reference Coordinates: 067°03'00"E 24°52'00"N

Founded: 1982
Staff: 58
Activities: research * education * popularization
Periodicals: (3) "Boletín Informativo SAA"; (1) "Revista Peruana de Astronomía y Astrofísica"
Coordinates: 075°05'00"W 12°04'00"S H40m
City Reference Coordinates: 077°03'00"W 12°03'00"S

Philippines

Bureau of Product Standards (BPS)
Trade and Industry Building
361 Sen. Gil J. Puyat Avenue
P.O. Box 3328 MCPO
Makati 1200
Telephone: (0)2-8904965
Telefax: (0)2-8904926
Electronic Mail: dtibpsrp@mnl.sequel.net
WWW: http://www.dti.gov.ph/bps
Founded: 1964 (as former "Division of Standards")
Staff: 84
Activities: standards development * product certification and testing * information dissemination and training * laboratory accreditation * accreditation of conformity assessment bodies * registration of quality assessors
Periodicals: (4) "BPS Directions" (ISSN 0118-3648)
City Reference Coordinates: 121°01'00"E 14°34'00"N

Manila Observatory (MO)
P.O. Box 122
UP Post Office
1101 Quezon City
or :
Ateneo de Manila Campus
Loyola Heights
Quezon City
Telephone: (0)2-4265921
    (0)2-4265922
    (0)2-4265923
Telefax: (0)2-4266141
Electronic Mail: root@observatory.ph
WWW: http://www.observatory.ph/
Founded: 1865
Staff: 36
Activities: crustal deformation * inventory of greenhouse gases * climate change * air quality monitoring * upper atmosphere * focal mechanism of earthquakes
Periodicals: "Solar Patrol", "Ionospheric Data", "Solar Maps and Activity"
Coordinates: 121°04'35"E 14°38'10"N H58m
City Reference Coordinates: 121°02'00"E 14°39'00"N

National Museum Planetarium
P. Burgos Street
Rizal Park
P.O. Box 2659
Manila 2801
Telephone: (0)5271830
Electronic Mail: planet@fastmail.i-next.net
WWW: http://members.tripod.com/philmuseum/planetarium.htm
Founded: 1975
Staff: 11
Activities: planetarium shows * lectures * demonstrations
City Reference Coordinates: 120°59'00"E 14°35'00"N

Philippine Astronomical Society (PAS)
P.O. Box 122
UP Post Office 1101
Quezon City
Telephone: (0)2-9241751
Telefax: (0)2-9241751
WWW: http://www.pworld.net.ph/user/jkty/pas.html
Founded: 1971
Membership: 120
Activities: meetings * observing * instrumentation * lectures * consulting * astronomical clearinghouse
Periodicals: (12) "The Appulse"
Awards: (1) "Padre Faura Astronomy Award"
Coordinates: 121°04'35"E 14°38'10"N H58m
City Reference Coordinates: 121°02'00"E 14°39'00"N

Philippine Atmospheric, Geophysical and Astronomical Services Administration (PAGASA), Astronomy Research and Development Section
P.O. Box 2277
Manila
or :

Asia Trust Building
1424 Quezon Avenue
Quezon City
Metro Manila
Telephone: (0)9228416
Telefax: (0)9221872
Electronic Mail: pagasa@mail.ph.net
WWW: http://www.pagasa.dost.ph/
Founded: 1954
Staff: 33
Activities: astronomical observing * disseminating data/space science information * recording and collecting solar data * keeping and disseminating time * research in astronomy and related fields * planetarium
Periodicals: (1) "Philippines Astronomical Handbook", "Almanac for Geodetic Engineers", "Moonrise, Moonset, Sunrise, Sunset and Twilight Tables", "Bulletin of Astronomical Observations", "Calendar Data", "Annual Report"
Coordinates: 121°04'18"E 14°39'12"N H70m
City Reference Coordinates: 120°59'00"E 14°35'00"N (Manila)
121°00'00"E 14°38'00"N (Quezon City)

Poland

Acta Astronomica
Al. Ujazdowskie 4
PL-00-478 Warszawa
Telephone: (0)22-6295346
Telefax: (0)22-6294967
Electronic Mail: acta@sirius.astrouw.edu.pl
Founded: 1925
Staff: 3
• Journal (4) (ISSN 0001-5237)
City Reference Coordinates: 021°00'00"E 52°15'00"N

Copernicus Astronomical Center (NCAC) (N._)
• See "Polish Academy of Sciences, N. Copernicus Astronomical Center (NCAC)"

Copernicus Museum (Nicolas_)
• See "Nicolas Copernicus Museum"

Copernicus Planetarium and Observatory in Chorzów (Nicolas_)
• See "Nicolas Copernicus Planetarium and Observatory in Chorzów"

Frombork Planetarium and Observatory
Ul. Elblaska 2
P.O. Box 6
PL-14-530 Frombork
Telephone: (0)506-7392
WWW: http://www.frombork.art.pl/Pol08.htm
Founded: 1983
Membership: 67
Activities: Summer astronomical camps * assistance in operating the planetarium and observatory in Nicolas Copernicus Museum (see separate entry) * observing (meteors, comets)
Periodicals: (4) "Wiadomości Pulsara"
City Reference Coordinates: 019°41'00"E 54°22'00"N

Gdansk University, Institute of Theoretical Physics and Astrophysics
Ul. Wita Stwosza 57
PL-80-952 Gdansk
Telephone: (0)415241 x188
         (0)415241 x210
         (0)415241 x243
Electronic Mail: <userid>@halina.univ.gda.pl
WWW: http://www.univ.gda.pl/
Founded: 1981
Staff: 22
Activities: stellar spectroscopy * stellar atmospheres (observing and modelling) * late-type stars * RS CVn binaries * IUE spectra
City Reference Coordinates: 018°40'00"E 54°23'00"N

Institute of Geodesy and Cartography
Ul. Jasna 2/4
PL-00-950 Warszawa
Telephone: (0)22-270328
Telefax: (0)22-270328
Electronic Mail: jnstgeo@plearn.bitnet
WWW: http://www.gik.pw.edu.pl/
Founded: 1930
Staff: 11
Activities: geodetic astronomy * time service * satellite geodesy
Periodicals: (1) "Astronomical Yearbook"
Coordinates: 021°02'06"E 52°28'36"N H110m (Borowa Gòra)
City Reference Coordinates: 021°00'00"E 52°15'00"N

Institute of Meteorology and Water Management
Ul. Podleśna 61
PL-01-673 Warszawa
Telephone: (0)22-341651
Telefax: (0)22-345466
Electronic Mail: <userid>@imgw.pl
WWW: http://www.imgw.pl/
Founded: 1919
Staff: 1500

Activities: meteorology * hydrology * oceanology * water management * hydraulics * water engineering * water resources * water quality * network of stations around the country
Periodicals: (6) "Journal of the Institute of Meteorology and Water Management Observer" (ISSN 0208-4325, circ.: 4,000); (4) "Reports of the Institute of Meteorology and Water Management" (ISSN 0208-6263, circ.: 480); "Research Papers": "Hydrology and Oceanology" (ISSN 0239-6297, circ.: 250), "Meteorology" (ISSN 0239-6262, circ.: 300), "Water Engineering" (ISSN 0239-6254, circ.: 150), "Water Management and Water Protection" (ISSN 0239-6238, circ.: 250); "Bibliography of Hydrology and Oceanology" (ISSN 0239-6246, circ.: 150), "Bibliography of Meteorology" (ISSN 0239-6270, circ.: 150), "Bibliography of Water Management and Engineering" (ISSN 0239-622x, circ.: 150)
Coordinates: 020°58'00"E 52°17'00"N H99m
City Reference Coordinates: 021°00'00"E 52°15'00"N

Jagiellonian University
• See "Kraków Jagiellonian University"

Józefoslaw University of Technology
• See "Warsaw Józefoslaw University of Technology"

Kraków Jagiellonian University, Astronomical Observatory
Ul. Orla 171
PL-30-244 Kraków
Telephone: (0)12-4251294
Telefax: (0)12-4251318
Electronic Mail: <userid>@oa.uj.edu.pl
WWW: http://www.oa.uj.edu.pl/
Founded: 1792
Staff: 24
Activities: comets * variable stars * solar physics * AGNs * extragalactic radiosources * ISM * intergalactic matter * magnetic fields in galaxies * cosmology * cosmic rays
Periodicals: (12) "Monthly Report on Solar Radio Emission"; (1) "Rocznik Astronomiczny Obserwatorium Krakowskiego" (Ephemeris of eclipsing variable stars)
Coordinates: 019°57'34"E 50°03'52"N H221m (IAU Code 055)
　　　　　　019°49'38"E 50°03'16"N H348m (Fort Skala Stn – IAU Code 555)
　　　　　　022°15'00"E 49°10'00"N H840m (Bieszczady Mountains Station)
City Reference Coordinates: 019°58'00"E 50°03'00"N

Kraków Pedagogical University, Department of Physics, Mount Suhora Observatory
Ul. Podchorazych 2
PL-30-084 Kraków
Telephone: (0)12-6379747
　　　　　　(0)90-330126 (Observatory Dome)
Telefax: (0)12-6372243
Electronic Mail: <userid>@cyf-kr.edu.pl
WWW: http://www.as.wsp.krakow.pl/
Founded: 1987
Staff: 7
Activities: photometry of eclipsing binaries * cataclysmic variable stars * spot stars
Coordinates: 020°05'24"E 49°34'48"N H1,000m
City Reference Coordinates: 019°58'00"E 50°03'00"N

Maria Curie-Sklodowska University, Institute of Physics, Astrophysics and Didactics, Astronomy Laboratory
Maria Curie-Sklodowska Square 1
PL-20-031 Lublin
Telephone: (0)81-5376186
Telefax: (0)81-5376191
Electronic Mail: lgladysz@tytan.umcs.lublin.pl (Longin Gładyszewski, Head)
Founded: 1944
Staff: 4
Activities: education * meteorites * solar physics * solar radioastronomy
Periodicals: "Annales UMCS", "Annual Report"
Coordinates: 022°32'37"E 51°14'47"N H210m
City Reference Coordinates: 022°35'00"E 51°15'00"N

Mlodziezowe Obserwatorium Astronomiczne
c/o Aleksander Trebacz
Ul. M. Kopernika 2
PL-32-005 Niepolomice
Telephone: (0)12-811561
Telefax: (0)12-812131
Electronic Mail: moa@pkpf.if.uj.edu.pl
WWW: http://www.cyf-kr.edu.pl/~ufjochym/MOA/
City Reference Coordinates: 020°13'00"E 50°02'00"N

Mount Suhora Observatory
• See "Kraków Pedagogical University, Department of Physics, Mount Suhora Observatory"

N. Copernicus Astronomical Center (NCAC)
• See "Polish Academy of Sciences, N. Copernicus Astronomical Center (NCAC)"

Nicolas Copernicus Museum
Ul. Katedralna 8
PL-14-530 Frombork
Telephone: (0)55-2437218
Telefax: (0)55-2437218
Electronic Mail: frombork@softel.elblag.pl
Founded: 1948
Staff: 35
Activities: exhibitions * planetarium shows * observing
Periodicals: "Komentarze Fromborskie"
Coordinates: 019°40'59"E 54°20'31"N H47m
City Reference Coordinates: 019°40'00"E 54°21'00"N

Nicolas Copernicus Planetarium and Observatory in Chorzów
P.O. Box 10
PL-41-501 Chorzów
Telephone: (0)546330
Telefax: (0)413296
WWW: http://www.man.katowice.pl/katowice/informator/tekst/english/s08.shtml
Founded: 1955
Staff: 11
Activities: popularization * astrometry (comets, minor planets) * meteorological and seismological station * software production
Coordinates: 018°59'30"E 50°17'33"N H325m (IAU Code 553)
City Reference Coordinates: 018°56'00"E 50°19'00"N

Opole Pedagogical University, Department of Astrophysics
Ul. Oleska 48
PL-45-951 Opole
Telephone: (0)77-35841
Telefax: (0)77-38387
Electronic Mail: <userid>@sparc-1.wsp.opole.pl
WWW: http://www.uni.opole.pl/
Founded: 1950
Staff: 4
Activities: atomic spectroscopy * line profiles * stellar spectra * atomic partition functions * white dwarfs
City Reference Coordinates: 017°55'00"E 50°41'00"N

Planetarium and Astronomical Observatory, Grudziądz
Ul. Hoffmanna 1/7
PL-86-300 Grudziądz
Telephone: (0)56-4658384
WWW: http://www.grudziadz.pl/planetarium-astronomical-obserwatory.html
Founded: 1972
Staff: 1
Activities: popularization * seminars
Coordinates: 018°45'16"E 53°28'45"N H72m
City Reference Coordinates: 018°45'00"E 53°29'00"N

Planetarium and Astronomical Observatory, Olsztyn
Al. Marszałka Piłsudskiego 38
PL-10-450 Olsztyn
Telephone: (0)334951
         (0)335178
WWW: http://www.planetarium.olsztyn.pl/
Founded: 1973
Staff: 6
Activities: programs for public and schools * lectures * exhibitions * education for young people * public observing
Coordinates: 020°29'25"E 53°46'25"N H150m
City Reference Coordinates: 020°29'00"E 53°48'00"N

Polish Academy of Sciences, N. Copernicus Astronomical Center (NCAC)
Ul. Bartycka 18
PL-00-716 Warszawa
Telephone: (0)22-8410041
Telefax: (0)22-8410046
Electronic Mail: <userid>@camk.edu.pl
WWW: http://www.camk.edu.pl/
Founded: 1974
Staff: 48
Activities: binary stars * stellar oscillations * stellar atmospheres * circumstellar matter * neutron stars * high-energy astrophysics * cosmology * n-body dynamics * ionosphere plasma * elementary particles

City Reference Coordinates: 021°00'00"E 52°15'00"N

Polish Academy of Sciences, N. Copernicus Astronomical Center (NCAC), Department of Astrophysics I
Ul. Rabianska 8
PL-87-100 Torun
Telephone: (0)56-6219249
(0)56-6219319
(0)56-6219341
Telefax: (0)56-6219381
Electronic Mail: <userid>@ncac.torun.pl
WWW: http://www.ncac.torun.pl/
Founded: 1974 (NCAC)
Staff: 9
City Reference Coordinates: 018°35'00"E 53°02'00"N

Polish Academy of Sciences, Space Research Center
Ordona 21
PL-01-237 Warszawa
Telephone: (0)22-362885
Telefax: (0)22-368961
Electronic Mail: cbk@cbk.waw.pl
WWW: http://www.cbk.waw.pl/
Founded: 1977
Staff: 55
Activities: space physics * physics and dynamics of planets and comets * heliophysics * astrometry * satellite geodesy
Periodicals: "Artificial Satellites" (two series "Space Physics" and "Planetary Geodesy") (ISSN 0571-205x)
Coordinates: 017°04'30"E 52°16'36"N H85m (Borowiec – IAU Code 187)
City Reference Coordinates: 021°00'00"E 52°15'00"N

Polish Academy of Sciences, Space Research Center, Astrogeodynamic Observatory
Borowiec
Ul. Drapalka 4
PL-62-035 Kórnik
Telephone: (0)61-8170187
Telefax: (0)61-8170219
Electronic Mail: sch@cbk.poznan.pl (Stanislaw Schillak, Head of Observatory)
WWW: http://www.cbk.poznan.pl/
Founded: 1953
Staff: 15
Activities: satellite geodesy * space research * geodynamics * Earth rotation * GPS * SLR * time
Periodicals: (1) "Borowiec Laser Station Operational Report"
Coordinates: 017°04'30"E 52°16'36"N H85m (Borowiec – IAU Code 187)
• Formerly known as "Astronomical Latitude Observatory"
City Reference Coordinates: 017°04'00"E 52°17'00"N

Polish Academy of Sciences, Space Research Center, Solar Physics Department
Ul. Kopernika 11
PL-51-622 Wrocław
Telephone: (0)71-483238
Telefax: (0)71-729372
Electronic Mail: js@cbk.pan.wroc.pl
WWW: http://cbk.pan.wroc.pl/
Founded: 1969
Staff: 16
Activities: solar physics * X-rays * space instrumentation * data analysis
City Reference Coordinates: 017°00'00"E 51°06'00"N

Polish Amateur Astronomical Society
Ul. Sw. Tomasza 30/8
PL-31-027 Kraków
Telephone: (0)12-223892
Electronic Mail: ptma@oa.uj.edu.pl
WWW: http://www.oa.uj.edu.pl/~ptma/
Founded: 1919
Membership: 2040
Activities: popularization * observing (occultations, variable stars, sunspots, comets, planets, meteors, eclipses) * telescope making
Periodicals: (12) "Urania" (ISSN 0042-0794); "Biblioteka Uranii"
Awards: (1/3) "Gold Award", "Silver Award"; "Honorable Membership"
City Reference Coordinates: 019°58'00"E 50°03'00"N

Polish Astronomical Society
(Polskie Towarzystwo Astronomiczne – PTA)
Ul. Bartycka 18

PL-00-716 Warszawa
Telephone: (0)22-8410041 x76
Telefax: (0)22-8410828
Electronic Mail: library@camk.edu.pl
WWW: http://www.camk.edu.pl/pta/
Founded: 1923
Membership: 220
Activities: developing astronomical sciences * education and popularization of astronomy * organizing symposia, meetings and Summer schools
Periodicals: (4) "Postepy Astronomii"
Awards: (1/2) "W. Zonn Medal"; (1) "Prize for Young Astronomers"
City Reference Coordinates: 021°00'00"E 52°15'00"N

Polish Committee for Standardization
(Polski Komitet Normalizacyjny - PKN)
P.O. Box 411
PL-00-950 Warszawa
or :
Ul. Swietokrzyska 14
PL-00-050 Warszawa
Telephone: (0)22-5567755
Telefax: (0)22-5567416
WWW: http://www.pkn.pl/
Founded: 1994
Staff: 306
Periodicals: (12) "Normalizacja" (ISSN 0029-179x); "Biuletyn Informacyjny" (ISSN 1230-8242)
City Reference Coordinates: 021°00'00"E 52°15'00"N

Polish Geophysical Society
Ul. Podleśna 61
PL-01-673 Warszawa
Telephone: (0)341651 x528
           (0)341651 x321
Telefax: (0)345466
Founded: 1947
Membership: 420
Activities: meteorology * hydrology * geophysics
Periodicals: "Review of Geophysics" (ISSN 0033-2135)
Awards: (1/2) "Marian Molga Scientific Award", "Award of the Polish Geophysical Society"
City Reference Coordinates: 021°00'00"E 52°15'00"N

Polish Physical Society
Ul. Hoza 69
PL-00-681 Warszawa
Telephone: (0)2-6212668
Telefax: (0)2-6212668
Electronic Mail: ptf@fuw.edu.pl
WWW: http://www.fuw.edu.pl/~ptf/
Founded: 1920
Membership: 1800
City Reference Coordinates: 021°00'00"E 52°15'00"N

Poznań University A. Mickiewicza, Astronomical Observatory
Ul. Słoneczna 36
PL-60-286 Poznań
Telephone: (0)61-8292770
Telefax: (0)61-8292772
Electronic Mail: secretary@mail.astro.amu.edu.pl
WWW: http://www.astro.amu.edu.pl/welcome.html
Founded: 1919
Staff: 26
Activities: celestial mechanics * fundamental astrometry * theory of refraction * dynamics of artificial satellites * small bodies of the solar system * minor planets * stellar astrophysics
Coordinates: 016°52'50"E 52°23'53"N H85m (IAU Code 047)
City Reference Coordinates: 016°53'00"E 52°25'00"N

Société Européenne pour l'Astronomie dans la Culture (SEAC)
c/o S. Iwaniszewski (Secretary)
State Archaeological Museum
P.O. Box 69
PL-00-950 Warszawa
Telephone: (0)22-313221
Telefax: (0)22-315195
WWW: http://mikrob.com/seac2001/aboutseac.html
Founded: 1992
Membership: 80

Activities: promotion of research on astronomical practice in its cultural context
Periodicals: "Newsletter"
City Reference Coordinates: 021°00'00"E 52°15'00"N

Torun Planetarium
Ul. Franciszka 19
PL-87-100 Torun
Telephone: (0)56-25066
WWW: http://www.man.torun.pl/Planetarium/
City Reference Coordinates: 018°35'00"E 53°02'00"N

Torun University Nicolaus Copernicus, Centre for Astronomy
Ul. Gagarina 11
PL-87-100 Torun
Telephone: (0)56-6113002
Telefax: (0)56-6113008
Electronic Mail: <userid>@astro.uni.torun.pl
WWW: http://www.astri.uni.torun.pl/
Founded: 1997
Staff: 36
Activities: VLBI * pulsars * spectral lines * Sun Radio Patrol * cosmology * Wolf-Rayet stars * PNe * ISM * celestial mechanics * physics of peculiar stars * local galactic structure
Coordinates: 018°33'42"E 53°05'44"N H80m (Piwnice Obs. – IAU Code 092)
City Reference Coordinates: 018°35'00"E 53°02'00"N

Warsaw Józefoslaw University of Technology, Institute of Geodesy and Geodetic Astronomy
Pl. Politechniki 1
PL-00-661 Warszawa
Telephone: (0)22-6228515
(0)22-6607754
Telefax: (0)22-6210052
Electronic Mail: <userid>@gik.pw.edu.pl
WWW: http://www.gik.pw.edu.pl/JOZE/jozehistorye.html
Founded: 1949
Activities: GPS observing for IGS and EUREF * GPS evaluation and analysis centre * gravity measurements (absolute and tidal) * meteorology * VZT latitude observing
Periodicals: "Latitude Circulars", "Scientific Publications of Józefoslaw Warsaw University of Technology - Series: Geodesy", "Reports on Geodesy"
Coordinates: 021°02'00"E 52°05'50"N
City Reference Coordinates: 021°00'00"E 52°15'00"N

Warsaw University, Astronomical Observatory
Al. Ujazdowskie 4
PL-00-478 Warszawa
Telephone: (0)22-294011
Telefax: (0)22-294967
Electronic Mail: <userid>@sirius.astrouw.edu.pl
WWW: http://www.astrouw.edu.pl/
Founded: 1825
Staff: 15
Activities: variable stars (observing and theory) * stellar atmospheres * ISM dynamics * statistics of extragalactic objects
Coordinates: 021°25'12"E 52°05'23"N H138m (Ostrowik – IAU Code 060)
City Reference Coordinates: 021°00'00"E 52°15'00"N

Warsaw University of Technology
● See "Warsaw Józefoslaw University of Technology"

Wrocław University, Astronomical Institute
Ul. Kopernika 11
PL-51-622 Wrocław
Telephone: (0)71-3729373
Telefax: (0)71-3729378
Electronic Mail: <userid>@astro.uni.wroc.pl
WWW: http://www.astro.uni.wroc.pl/
Founded: 1791
Periodicals: "Contributions", "Reprints"
Coordinates: 017°05'18"E 51°06'42"N H117m (IAU Code 547)
City Reference Coordinates: 017°00'00"E 51°06'00"N

Portugal

Associação Portuguesa de Astronomos Amadores (APAA)
Rua Alexandre Herculano 57 - 4° Dto
P-1250 Lisboa
Telefax: (0)21-7784153
WWW: http://www.terravista.pt/FerNoronha/1475/
City Reference Coordinates: 009°08'00"W 38°43'00"N

Associação Portuguesa para o Ensino da Astronomia
Apartado 52503
Amial
P-4202-301 Porto
Electronic Mail: astroportugal@ip.pt
City Reference Coordinates: 008°36'00"W 41°11'00"N (Porto)

Centro de Astrofísica da Universidade do Porto (CAUP)
• See "Universidade do Porto, Centro de Astrofísica"

Centro de Observação Astronômica do Algarve (COAA)
Poio
P-8500-149 Portimão
Telephone: 282471180
Telefax: 282471516
Electronic Mail: coaa@mail.telepac.pt
WWW: http://www.ip.pt/coaa/
Founded: 1987
Staff: 3
Activities: providing facilities for visiting astronomers
Periodicals: "COAA News"; (1) "Annual Report"
Coordinates: 008°35'37"W 37°11'29"N H65m (IAU Code 965)
City Reference Coordinates: 008°32'00"W 37°08'00"N

Fundação para a Ciência e a Tecnologia (FCT)
Avenida D. Carlos I 126
P-1200 Lisboa
Telephone: 213924300
Telefax: 213907481
Electronic Mail: presidencia@fct.mct.pt
WWW: http://www.fct.mct.pt/
Founded: 1997
Staff: 150
Activities: promoting, following up and funding science and technology activities in Portugal
City Reference Coordinates: 009°08'00"W 38°43'00"N

Instituto de Meteorologia do Portugal
Rua C
Aeroporto de Lisboa
P-1700 Lisboa
Telephone: 218472880
218472890
Telefax: 218402370
Electronic Mail: im@meteo.pt
<userid>@meteo.pt
WWW: http://www.meteo.pt/
Founded: 1916
City Reference Coordinates: 009°08'00"W 38°43'00"N

Instituto para a Cooperação Científica e Tecnológica International (ICCTI)
Rua Castillo 5
4th Floor
P-1250-066 Lisboa
or :
Avenida Don Carlos I 126
P-1200 Lisboa Codex
Telephone: 213924300
Telefax: 213975144
Electronic Mail: iccti@mail.telepac.pt
WWW: http://www.iccti.mct.pt/
Founded: 1997
Staff: 40
Activities: supporting, guiding and coordinating all international collaborations in the fields of science and technology

City Reference Coordinates: 009°08'00"W 38°43'00"N

Instituto Português da Qualidade (IPQ)
Rua C à Avenida dos Três Vales
P-2825 Monte da Caparica
Telephone: 212948100
Telefax: 212948101
Electronic Mail: lcmds@correio.ipq.gtw-ms.mailpac.pt
　　　　　　　　ipqmail@mail.ipq.pt
WWW: http://www.ipq.pt/
Founded: 1986
Staff: 210
Activities: standardization * metrology * certification * accreditation
Periodicals: "OPÇÃO" (ISSN 0872-2884), "Qualirama","Metrologia", "Catálogo IPQ"
Awards: "Prémio Excelência"
City Reference Coordinates: 009°08'00"W 38°43'00"N (Lisboa)

Observatório Astronômico de Lisboa (OAL)
Tapada da Ajuda
P-1349-018 Lisboa
Telephone: 213616730
Telefax: 213621722
WWW: http://www.oal.ul.pt/OAL.html
Founded: 1861
Activities: astrometry * meridian astronomy * time and latitude * lunar occultations * double and multiple stars * astronomical information to the national community
Periodicals: (1) "Dados Astronômicos para os Almanaques"; "Bulletin of the Astronomical Observatory of Lisbon"
Coordinates: 009°11'10"W 38°42'31"N H111m (IAU Code 971)
City Reference Coordinates: 009°08'00"W 38°43'00"N

Observatório Astronômico, Coimbra
● See "Universidade de Coimbra, Observatório Astronômico"

Observatório Astronômico Prof. Manuel de Barros
● See "Universidade do Porto, Observatório Astronômico Prof. Manuel de Barros"

Observatório das Ciências e das Tecnologias (OCT)
Rua das Praças 13-B
P-1200-765 Lisboa
Telephone: 213926000
Telefax: 213950979
Electronic Mail: geral@oct.mct.pt
WWW: http://www.oct.mct.pt/
Founded: 1996
Staff: 40
Activities: collection, processing, analysis and publication of information on R&D and on the national S&T system
City Reference Coordinates: 009°08'00"W 38°43'00"N

Planetário Calouste Gulbenkian (PCG)
Praça do Império
P-1400-206 Lisboa
Telephone: 213610508
　　　　　　213610123
Telefax: 213636005
Electronic Mail: planetario@mail.telepac.pt
WWW: http://www.marinha.pt/planetario
Founded: 1965
Staff: 14
Activities: shows for children, students and public
City Reference Coordinates: 009°08'00"W 38°43'00"N

Rede Nacional de Observação Astronómica (RNOA)
Trv. 1° de Maio 16
P-2430 Marinha Grande
Telephone: 244566986 (Carlos Reis, Presidente)
　　　　　　244503820 (João Clérigo, Coordinator)
Electronic Mail: rnoa@mail.telepac.pt
　　　　　　　　info@rnoa.rcts.pt
WWW: http://www.anoa.pt/
　　　http://www.rnoa.rcts.pt/
Founded: 1995
Periodicals: "Boletim RNOA"
City Reference Coordinates: 008°45'00"W 39°45'00"N

Secção Portuguesa das Uniões Internacionais Astronômica e Geodesica e Geofisica (SPUIAGG)
Rua Artlharia Um. 107
P-1099-052 Lisboa
Telephone: 213819600
Telefax: 213819699
Founded: 1923
Staff: 9
Activities: coordinating astronomical, geodetical and geophysical activities * Portuguese representation at IAU and IGGU * seminars * conferences
City Reference Coordinates: 009°08'00"W 38°43'00"N

Sociedade Portuguesa de Física (SPF)
c/o J. Bessa Sousa (President)
Laboratório da Física
Universidade do Porto
Avenida da República 37-4°
P-1050-187 Lisboa
Telephone: 217993665
Telefax: 217952349
Electronic Mail: spf@nautilus.fis.uc.pt
jbsousa@fcl.fc.up.pt
WWW: http://spf.pt/
http://nautilus.fis.uc.pt/~spf/
Founded: 1974
Membership: 1000
Periodicals: "Gazeta de Física"
City Reference Coordinates: 009°08'00"W 38°43'00"N

Universidade da Madeira, Grupo de Astronomia
c/o Departamento de Matemática
Caminho da Penteada
P-9000-390 Funchal
Telephone: 291705150
Telefax: 291705199
WWW: http://www.uma.pt/Investigacao/Astro/Grupo/
Founded: 2000
Staff: 5
Activities: research * education * public outreach
Coordinates: 016°55'00"W 32°46'00"N
City Reference Coordinates: 016°55'00"W 32°39'00"N

Universidade de Coimbra, Observatório Astronômico
Caixa Postal 147
P-3002 Coimbra Codex
Telephone: 239814947
WWW: http://www.fis.uc.pt/
http://www.fis.uc.pt/old/df_home.html
Founded: 1772
Activities: spectroscopy of K line * astrolabe astrometry
Periodicals: (1) "Astronomical Ephemeris", "Anais" (ISSN 0870-2856); "Communications"
Coordinates: 008°26'37"W 40°11'53"N H99m
City Reference Coordinates: 008°25'00"W 40°12'00"N

Universidade de Évora, Departamento de Física, Grupo de Astronomia
Rua Romão Ramalho 59
P-7000 Évora
Telephone: 266744616/8
Telefax: 266744546
Electronic Mail: <userid>@astro.ce.uevora.pt
WWW: http://www.lca.uevora.pt/
Founded: 1994
Activities: galactic physics * pulsars * planet formation and evolution * interstellar jets * gas dynamics * ISM * computational astrophysics
City Reference Coordinates: 007°54'00"W 38°34'00"N

Universidade de Lisboa, Departamento de Física, Centro de Astronomia e Astrofísica
Observatório Astronômico de Lisboa (OAL)
Tapada da Ajuda
P-1349-018 Lisboa
Telephone: 213616739
Telefax: 213616752
Electronic Mail: <userid>@oal.ul.pt
WWW: http://www.oal.ul.pt/

Founded: 1992
Staff: 10
Activities: education * research * star formation * galactic structure and evolution * active galaxies * planetary astronomy
Coordinates: 009°09'20"W 38°45'28"N
City Reference Coordinates: 009°08'00"W 38°43'00"N

Universidade do Porto, Centro de Astrofísica
Rua das Estrelas
P-4150-762 Porto
Telephone: 226089830
Telefax: 226089839
Electronic Mail: www@astro.up.pt
WWW: http://www.astro.up.pt/
   http://www.astro.up.pt/planetario/ (Planetarium)
Founded: 1989
Staff: 22
Activities: research * education * popularization
City Reference Coordinates: 008°36'00"W 41°11'00"N

Universidade do Porto, Observatório Astronômico Prof. Manuel de Barros
Monte da Virgem
P-4430-146 Vila Nova de Gaia
Telephone: 227820404
Telefax: 227827253
Electronic Mail: observatorio@oa.fc.up.pt
WWW: http://www.fc.up.pt/oa
Founded: 1948
Staff: 14
Activities: astrophysics * solar radioastronomy * geodesy * remote sensing
Periodicals: "Publicações do Observatório Astronômico Prof. Manuel de Barros" (ISSN 0871-1542), "Informações do Observatório Astronômico Prof. Manuel de Barros" (ISSN 0871-1550)
Coordinates: 008°35'15"W 41°06'28"N H232m
City Reference Coordinates: 008°34'00"W 41°45'00"N

Puerto Rico

Arecibo Observatory
• See "National Astronomy and Ionosphere Center (NAIC), Arecibo Observatory"

National Astronomy and Ionosphere Center (NAIC), Arecibo Observatory
HC3 Box 53995
Arecibo, PR 00612
or :
Rt 625 Bo Esperanza
Arecibo, PR 00612
Telephone: 878-2612
Telefax: 878-1861
878-0662
Electronic Mail: <userid>@naic.edu
WWW: http://aosun.naic.edu/
Founded: 1963
Staff: 14
Activities: radioastronomy * planetary radar * atmospheric physics
Periodicals: "NAIC/Arecibo Observatory Newsletter"
Coordinates: 066°45'11"W 18°20'37"N H497m (IAU Code 251)
City Reference Coordinates: 066°44'00"W 18°29'00"N

Sociedad de Astronomia de Puerto Rico
c/o Ernesto E. Santiago-Jordan
34th Street P-2
Urbanización Bairoa
Caguas, PR 00725
WWW: http://fast.to/Astronomia
City Reference Coordinates: 066°04'00"W 18°14'00"N

University of Puerto Rico (UPR), Humacao, Observatory
Department of Physics
Humacao, PR 00791
Telephone: 850-9381
850-9336
Telefax: 852-4638
850-9471
Electronic Mail: observ@cuhwww.upr.clu.edu.
rj_muller@cuhac.upr.clu.edu (Rafael J. Muller, Observatory Director)
WWW: http://cuhwww.upr.clu.edu/~observ/new/observ.htm
Founded: 1985
Staff: 2
Activities: observing * CCDs * photometry
Coordinates: 065°50'00"W 18°08'00"N H0m
City Reference Coordinates: 065°50'00"W 18°08'00"N

Romania

Asociatia de Standardizare din Romania (ASRO)
Str. Mendeleev 21-25
RO-70168 Bucureşti
Telephone: (0)1-2113296
Telefax: (0)1-2100833
Electronic Mail: irs@kappa.ro
City Reference Coordinates: 026°06'00"E 44°26'00"N

Cluj-Napoca Astronomical Observatory
• See "Romanian Academy, Astronomical Institute, Cluj-Napoca Astronomical Observatory"

National Institute of Meteorology and Hydrology
Soseaua Bucureşti-Ploieşti 97
RO-71552 Bucureşti 18
Telephone: (0)1-2303116
Telefax: (0)1-2303143
Electronic Mail: info@meteo.inmh.ro
WWW: http://www.inmh.ro/
Founded: 1884
Activities: synoptic meteorology * dynamic meteorology * numerical forecasting * aeronautical meteorology * climatology * atmospheric physics * aerology * agrometeorology * air pollution * operational hydrology * underground hydrology
Periodicals: "Romanian Journal of Meteorology" (ISSN 1223-1118), "Romanian Journal of Hydrology and Water Resources" (ISSN 1223-1126), "Studii si Cercetari de Meteorologie", "Studii si Cercetari de Hidrologie"
City Reference Coordinates: 026°06'00"E 44°26'00"N

Romanian Academy, Astronomical Institute
Cuţitul de Argint 5
RO-75212 Bucureşti 28
Telephone: (0)1-3356892
         (0)1-3358010
Telefax: (0)1-3373389
Electronic Mail: <userid>@aira.astro.ro
WWW: http://www.astro.ro/
Founded: 1908
Staff: 48
Activities: variable stars * solar physics * photographic astrometry * reference systems * Earth rotation * cosmology * celestial mechanics * upper terrestrial atmosphere * history of astronomy
Periodicals: (2) "Romanian Astronomical Journal" (ISSN 1210-5168, circ.: 300); "Observations Solaires" (circ.: 230), "Anuarul Astronomic" (circ.: 300)
Coordinates: 026°05'47"E 44°24'50"N H80m (Bucureşti – IAU Code 073)
            023°35'53"E 46°42'48"N H730m (Cluj-Napoca)
            021°13'45"E 45°44'15"N H300m (Timişoara - see separate entry)
City Reference Coordinates: 026°06'00"E 44°26'00"N

Romanian Academy, Astronomical Institute, Cluj-Napoca Astronomical Observatory
Calea Cireşilor 19
RO-3400 Cluj-Napoca
Telephone: (0)64-194592
Telefax: (0)64-192820
Electronic Mail: academy1@mail.soroscj.ro
Founded: 1920
Staff: 14
Activities: celestial mechanics * variable stars
Coordinates: 023°35'40"E 46°45'36"N H400m
            023°35'52"E 46°42'48"N H750m (Faget Station)
City Reference Coordinates: 023°36'00"E 46°47'00"N

Romanian Academy, Astronomical Institute, Timişoara Astronomical Observatory
Piata Axente Sever 1
RO-1900 Timişoara
Telephone: (0)56-162838
Telefax: (0)56-162838
Electronic Mail: astro@astrotm.sorostm.ro
WWW: http://www.sorostm.ro/astro/
Founded: 1959
Coordinates: 021°24'55"E 45°44'15"N H88m
            021°13'45"E 45°44'15"N H300m
City Reference Coordinates: 021°13'00"E 45°45'00"N

Romanian Space Agency (ROSA)
21-25 Mendeleev Street

RO-70168 Bucureşti 1
Telephone: (0)1-6504222
Telefax: (0)1-3128804
Electronic Mail: <userid>@rosa.ro
WWW: http://www.rosa.ro/
Founded: 1995
City Reference Coordinates: 026°06'00"E 44°26'00"N

Societatea Română de Fizică
(Romanian Physical Society)
c/o Alexandru Calboreanu (General Secretary)
Tandem Laboratory
Institute of Atomic Physics
P.O. Box MG-6
RO-70000 Bucureşti
Telephone: (0)1-7807040
Telefax: (0)1-4231650
Electronic Mail: calbo@roifa.ifa.ro
Founded: 1890 (re-founded in 1990)
Membership: 1200
Activities: organizing national conferences * consulting * international representation
Periodicals: "Curierul de Fizică"
City Reference Coordinates: 026°06'00"E 44°26'00"N

Russia

A.F. Ioffe Physical Technical Institute
● See "Russian Academy of Sciences, A.F. Ioffe Physical Technical Institute"

Astrocosmical Association Sirius-86
c/o Yu.V. Golendukhin
Ul. Lenina 76/5
623750 Rezh
Telephone: (0)34364-24219
Electronic Mail: sirius@online.ural.ru
WWW: http://home.ural.ru/~sirius86/
Founded: 1986
Membership: 100
Activities: education * observing * research
Coordinates: 061°23'00"E 57°22'00"N H200m
City Reference Coordinates: 061°20'00"E 57°26'00"N

Astronomical Club Centaur
c/o Andrey V. Demidov
Public Institution of Additional Education
Pr. Leningradskiy 341, korp. 1
163001 Arkhangelsk
Telephone: (0)8182-414189
          (0)8182-475701
Electronic Mail: smis@agtu.ru
WWW: http://friends.pomorsu.ru/centaurus/
City Reference Coordinates: 040°32'00"E 64°34'00"N

Astronomical Club IKAR
c/o Alevtina M. Kireeva
Ul. Vershinina 17, kab 601
643041 Tomsk
Telephone: (0)3822-557770
Electronic Mail: kira@dtu.tsu.ru
WWW: http://www.dtu.tsu.ru:8101/ikar/
City Reference Coordinates: 085°05'00"E 56°30'00"N

Astronomical Club Parsec
Dvorec of Children and Youth
Ul. Irtyashskaya 1
456790 Ozersk 10
Telephone: (0)351-7179018
Founded: 1979
Membership: 15
Activities: photography * comets * meteors * Moon * Sun * stars
Coordinates: 060°29'51"E 55°40'48"N H281m
City Reference Coordinates: 060°30'00"E 55°41'00"N

Astronomical-Geodetical Society
Sadovaja-Kudrinskaya Ul. 24
103001 Moscow
Telephone: (0)095-2915896
Founded: 1932
Membership: 8000 (of which 1500 juniors)
Activities: astronomical and geodetical education and popularization * national development projects for astronomy, geodesy and cartography * publishing * meteors * meteorites * NLC * comets * variable stars * telescope designing
Periodicals: (6) "Zemlja y Vsselennaja" (ISSN 0044-3948), "Astronomicheskij Vestnik" (ISSN 0320-930x); (3-5) "Circulars" (ISSN 0201-7342), "Soobshcheniia" (ISSN 0235-3431)
City Reference Coordinates: 038°50'00"E 55°05'00"N

Astro Space Center (ASC)
● See "Russian Academy of Sciences, P.N. Lebedev Physics Institute, Astro Space Center (ASC)"

Baksan Neutrino Observatory (BNO)
● See "Russian Academy of Sciences, Institute for Nuclear Research, Baksan Neutrino Observatory (BNO)"

Center for Astronomical Data (CAD)
● See "Russian Academy of Sciences, Institute for Astronomy, Center for Astronomical Data (CAD)"

Engelhardt Astronomical Observatory
● See "Kazan State University, Engelhardt Astronomical Observatory"

Euro-Asian Astronomical Society (EAAS)
c/o Sternberg State Astronomical Institute
University Avenue 13
119899 Moscow
Telephone: (0)095-9328844
Telefax: (0)095-9328844
(0)095-9328841
Electronic Mail: boch@astronomy.msk.su
eaas@sai.msu.ru
WWW: http://www.issp.ac.ru/univer/astro/eaas_e.html
Founded: 1990
Membership: 750
Activities: maintaining the development of astronomy * reinforcing communication between national and foreign astronomers * organizing scientific conferences * supporting astronomical education
Periodicals: "Astronomical and Astrophysical Transactions" (ISSN 1055-6796), "Solar Data", "EAAS Bulletin", "Astrocourier", "Bulletin of Association of Planetaria of Russia", "Universe and Ourselves", "Zvezdociot"
City Reference Coordinates: 038°50'00"E 55°05'00"N

Fomalhaut Astronomical Society
c/o Andrey T. Fomin
Ul. Pogranichnaya 68, Kab 116
693000 Yuzhno-Sakhalinsk
Telephone: (0)4242-728619
Electronic Mail: andrey@sakhgu.sakhalin.ru
WWW: http://fomalhaut.metastock.ru/
City Reference Coordinates: 142°45'00"E 46°58'00"N

Institute for Physical-Technical and Radio-Technical Measurements (VNIIFTRI)
Mendeleevo
141570 Moscow
Telephone: (0)095-5358490
(0)095-5352401
Telefax: (0)095-5357386
Electronic Mail: admin@vniiftri.ru
WWW: http://www.vniiftri.ru/
Founded: 1956
Activities: Earth rotation parameters * data acquisition, processing and publishing
Periodicals: "Universal Time and Pole Coordinates": "Circular A" (ISSN 0135-2415), "Circular E" (ISSN 0234-1069)
City Reference Coordinates: 038°50'00"E 55°05'00"N

INTERSPUTNIK
2 Smolensky Per. 1/4
121099 Moscow
Telephone: (0)095-2440333
Telefax: (0)095-2539906
Electronic Mail: dir@intersputnik.com
WWW: http://intersputnik.com/
Founded: 1971
Staff: 15
Activities: international organization of space communications
City Reference Coordinates: 038°50'00"E 55°05'00"N

Ioffe Physical Technical Institute (A.F._)
• See "Russian Academy of Sciences, A.F. Ioffe Physical Technical Institute"

Irkutsk State University, Astronomical Observatory
Sovetskaya 119A
664009 Irkutsk
Telephone: (0)3952-270294
Electronic Mail: uuastra@astra.isu.runnet.ru
WWW: http://www.isu.ru/english/other/obs.htm
Founded: 1931
Staff: 12
Activities: Earth rotation parameters * NLC * upper atmosphere dynamics * education
Coordinates: 104°20'42"E 52°36'44"N H468m
City Reference Coordinates: 104°20'00"E 52°16'00"N

IZMIRAN
• See "Russian Academy of Sciences, Institute of Terrestrial Magnetism, Ionosphere and Radio Wave Propagation (IZMIRAN)"

Joint Scientific and Educational Center (JSEC)
Politechnicheskaya Street 21
194021 Saint Petersburg
Telephone: (0)812-2472223

Pushchino Radioastronomy Observatory (PRAO)
• See "Russian Academy of Sciences, P.N. Lebedev Physics Institute, Pushchino Radioastronomy Observatory (PRAO)"

Rostov State University (RSU), Department of Space Research
Zorge Ul. 5
344090 Rostov-na-Donu
Telephone: (0)8632-220858
Telefax: (0)8632-285044
Electronic Mail: physdep@rsu.rnd.runnet.ru
WWW: http://www.unird.ac.ru/
Founded: 1970
Staff: 10
Activities: stars * stellar atmospheres * structure, evolution, dynamics and chemical composition of galaxies * ISM * nucleosynthesis * large-scale structure * background radiation
City Reference Coordinates: 039°25'00"E 57°11'00"N

Rostov State University (RSU), Institute of Physics
Zorge Ul. 5
344090 Rostov-na-Donu
Telephone: (0)8632-220857
Telefax: (0)8632-220884
Electronic Mail: <userid>@phys.rsu.ru
WWW: http://www.phys.rsu.ru/
Founded: 1967
Staff: 14
Activities: galactic structure and evolution * ISM * abundances * cosmic background radiation
City Reference Coordinates: 039°25'00"E 57°11'00"N

Russian Academy of Sciences, A.F. Ioffe Physical Technical Institute, Division of Plasma Physics, Atomic Physics, and Astrophysics
Politechnicheskaya 26
194021 Saint Petersburg
Telephone: (0)812-2471017
Telefax: (0)812-2472245
Electronic Mail: <userid>@astro.ioffe.rssi.ru
WWW: http://www.ioffe.rssi.ru/
Founded: 1963
Staff: 22
Activities: QSOs * cosmology * ISM * gamma rays * star formation regions * neutron stars * education
Periodicals: (1) "Annual Report"
City Reference Coordinates: 030°15'00"E 59°55'00"N

Russian Academy of Sciences, Institute for Nuclear Research, Baksan Neutrino Observatory (BNO)
P/O Neutrino
Tyrnyauz
361609 Kabardino-Balkaria
Telephone: (0)866-2-477475
Telefax: (0)866-2-477475
Founded: 1967
Staff: 60
Activities: anisotropy and chemical composition of primary cosmic rays * collapse * dark matter * double beta decay * gamma astronomy * solar neutrino flux registration * gamma bursts
Coordinates: 043°32'00"E 43°40'40"N
City Reference Coordinates: 043°32'00"E 43°41'00"N

Russian Academy of Sciences, Committee on Meteorites
c/o Yuri A. Shukolyukov (Chairman)
Vernadsky Institute of Geochemistry and Analytical Chemistry
Kosygin Street 19
117975 Moscow
Telephone: (0)095-1374270
Telefax: (0)095-9382054
WWW: http://www.ras.ru/
Founded: 1939
Membership: 30
Activities: meteorites * meteors * meteoritic craters
Periodicals: "Meteoritika"
City Reference Coordinates: 038°50'00"E 55°05'00"N

Russian Academy of Sciences, Department of General Physics and Astronomy (DGPA)
Lenin Avenue 32a
117993 Moscow
Telephone: (0)095-9381695
 (0)095-9385500
Telefax: (0)095-9381714

Electronic Mail: <userid>@oofa.msk.su
　　　　　　　<userid>@gpad.ac.ru
WWW: http://www.ras.ru/RAS/oofa.html
　　　http://www.gpad.ac.ru/
Staff: 120
City Reference Coordinates: 038°50'00"E 55°05'00"N

Russian Academy of Sciences, Institute for Astronomy (INASAN)
Ul. Pyatnitskaya 48
119017 Moscow
Telephone: (0)095-9515461
Telefax: (0)095-2302081
Electronic Mail: <userid>@inasan.rssi.ru
WWW: http://www.inasan.rssi.ru/
Founded: 1936
Staff: 100
Activities: physics and evolution of stars * stellar spectroscopy * variable stars * astronomical catalogs * dynamics of stellar and planetary systems * satellite geodesy * geophysics * geodynamics * space research
Periodicals: "Peremennye Zvezdy" (ISSN 0373-7683) (circ.: about 700)
Coordinates: 036°46'26"E 55°41'40"N H173m (Zvenigorod – IAU Code 102)
City Reference Coordinates: 038°50'00"E 55°05'00"N

Russian Academy of Sciences, Institute for Astronomy, Center for Astronomical Data (CAD)
Ul. Pyatnitskaya 48
109017 Moscow
Telephone: (0)095-2315461
Telefax: (0)095-2302081
Electronic Mail: olgad@inasan.rssi.ru
Staff: 20
City Reference Coordinates: 038°50'00"E 55°05'00"N

Russian Academy of Sciences, Institute for Theoretical Astronomy (ITA)
Ul. Naberezhnaya Kutuzova 10
191187 Saint Petersburg
Telephone: (0)812-2788809
　　　　　　(0)812-2788810
　　　　　　(0)812-2790667
Telefax: (0)812-2797968
Electronic Mail: <userid>@iipah.spb.ru
WWW: http://www.ras.ru/
Founded: 1919
Staff: 60
Activities: celestial mechanics * ephemeris * minor planets * natural satellites * comet dynamics * satellite geodesy
Periodicals: "Trudy ITA" (ISSN 0568-6016); (2) "Bulletin ITA" (ISSN 0002-3302); (1) "Astronomical Yearbook", "Ephemeris of Minor Planets"
City Reference Coordinates: 030°15'00"E 59°55'00"N

Russian Academy of Sciences, Institute of Applied Astronomy
Zdanovskaya Ul. 8
197110 Saint Petersburg
Telephone: (0)812-2307414
Telefax: (0)812-2307413
Electronic Mail: iparan@ipa.rssi.ru
WWW: http://www.ipa.rssi.ru/
Founded: 1988
Staff: 232
Activities: radioastronomy * VLBI * astrometry * geodesy * geophysics
Periodicals: "Communications of the Institute of Applied Astronomy"
Coordinates: 029°46'54"E 60°32'00"N H80m (Svetloe)
City Reference Coordinates: 030°15'00"E 59°55'00"N

Russian Academy of Sciences, Institute of Applied Astronomy, Irkutsk Department
Lermontova Ul. 297-B-45
664033 Irkutsk
Telephone: (0)3952-460822
WWW: http://www.ipa.rssi.ru/
Founded: 1988
Staff: 8
Activities: radioastronomy * VLBI * astrometry * geodesy * geophysics
City Reference Coordinates: 104°20'00"E 52°16'00"N

Russian Academy of Sciences, Institute of Applied Astronomy, Zelenchuk Department
Kalinin Ul. 1a
357140 Zelenchukskaya
Telephone: (0)87878-43402

Electronic Mail: aps@ipa.rssi.ru
WWW: http://www.ipa.rssi.ru/
Founded: 1988
Staff: 28
City Reference Coordinates: 041°36'00"E 43°53'00"N

Russian Academy of Sciences, Institute of Applied Physics, Millimeter Astronomy Group
Uljanov Ul. 46
603600 Nizhny Novgorod
Telephone: (0)8312-367253
Telefax: (0)8312-362061
Electronic Mail: zin@appl.sci-nnov.ru
WWW: http://zin.appl.sci-nnov.ru/mm-astro/
Founded: 1977 (Institute)
Staff: 10
Activities: ISM * star formation * dense cores * mm-wave radio astronomy
City Reference Coordinates: 044°00'00"E 56°20'00"N

Russian Academy of Sciences, Institute of Cosmophysical Research and Aeronomy (IKFIA)
Lenin Avenue 31
677891 Yakutsk
Telephone: (0)411-2445551
Telefax: (0)441-2445551
Electronic Mail: ikfia@ysn.ru
WWW: http://www.ysn.ru/
Founded: 1962
Staff: 220
Activities: cosmic rays * solar wind * ionosphere * magnetosphere * atmosphere * astrophysics * STP
Coordinates: 129°40'00"E 62°02'00"N (Yakutsk)
           128°45'00"E 71°40'00"N (Tixie Bay)
City Reference Coordinates: 129°40'00"E 62°02'00"N

Russian Academy of Sciences, Institute of Spectroscopy
142092 Troitsk
Telephone: (0)095-3340579
Telefax: (0)095-3340886
Electronic Mail: <userid>@isan.msk.su
WWW: http://www.isan.troitsk.ru/
Founded: 1968
Staff: 270
Activities: optical spectroscopy
City Reference Coordinates: 037°19'00"E 55°28'00"N

Russian Academy of Sciences, Institute of Terrestrial Magnetism, Ionosphere and Radio Wave Propagation (IZMIRAN)
142190 Troitsk
Telephone: (0)095-3340120
Telefax: (0)095-3340124
Electronic Mail: kvd@charley.izmiran.rssi.ru
             <userid>@izmiran.rssi.ru
WWW: http://www.izmiran.rssi.ru/
Founded: 1940
Staff: 730
Activities: solar physics * STP * radiophysics * ionosphere * geophysics * geomagnetism * space
Coordinates: 037°19'00"E 55°28'00"N H190m
City Reference Coordinates: 037°19'00"E 55°28'00"N

Russian Academy of Sciences, Institute of Theoretical and Experimental Physics
Ul. B. Cheremushkinskaya 25
117259 Moscow
Telephone: (0)095-1230292
         (0)095-1236584
WWW: http://www.itep.ru/
Activities: SN * stellar structure * low-energy neutrinos
City Reference Coordinates: 038°50'00"E 55°05'00"N

Russian Academy of Sciences, Keldysh Institute of Applied Mathematics
Miusskaja 4
125047 Moscow
Telephone: (0)095-2581314
WWW: http://www.keldysh.ru/
Staff: 30
Activities: planets * celestial mechanics * cosmology * relativistic astrophysics
City Reference Coordinates: 038°50'00"E 55°05'00"N

Russian Academy of Sciences, Nuclear Physics Institute
Gatchina
188350 Saint Petersburg
Telephone: (0)812-2949132
 (0)812-2949146
 (0)812-2983538
Telefax: (0)812-2980257
 (0)812-7137196
Electronic Mail: <userid>@pnpi.ru
WWW: http://www.pnpi.spb.ru/
City Reference Coordinates: 030°15'00"E 59°55'00"N

Russian Academy of Sciences, P.N. Lebedev Physics Institute
Lenin Avenue 53
117924 Moscow
Telephone: (0)095-1357980
 (0)095-1357980
 (0)095-1326573
Telefax: (0)095-1357880
 (0)095-9382251
Electronic Mail: <userid>@sci.fian.msk.su
WWW: http://www.lpi.msk.su/
 http://www.lpi.msk.su/LPI/NPAD/NPAD.html (Nuclear Physics and Astrophysics Division)
City Reference Coordinates: 038°50'00"E 55°05'00"N

Russian Academy of Sciences, P.N. Lebedev Physics Institute, Astro Space Center (ASC)
Ul. Profsojuznaya 84/32
117810 Moscow
Telephone: (0)095-3332378
Telefax: (0)095-3332378
Electronic Mail: <userid>@asc.rssi.ru
WWW: http://www.asc.rssi.ru/
Activities: space physics * interferometry * solar atmosphere * UV * cosmic rays * upper atmosphere
City Reference Coordinates: 038°50'00"E 55°05'00"N

Russian Academy of Sciences, P.N. Lebedev Physics Institute, Pushchino Radioastronomy Observatory (PRAO)
142292 Pushchino
Telephone: (0)967-732780
Telefax: (0)967-732482
Electronic Mail: <userid>@prao.psn.ru
WWW: http://www.prao.psn.ru/
 http://psun32.prao.psn.ru/ (Pulsar Astrometry)
City Reference Coordinates: 037°52'00"E 56°01'00"N

Russian Academy of Sciences, P.N. Lebedev Physics Institute, Nuclear Physics and Astrophysics
Lenin Avenue 53
117924 Moscow
Telephone: (0)095-1354264
Telefax: (0)095-1357880
 (0)095-9382251
Electronic Mail: <userid>@lpi.msk.su
WWW: http://www.lpi.msk.su/
Founded: 1956
Staff: 33
City Reference Coordinates: 038°50'00"E 55°05'00"N

Russian Academy of Sciences, Pulkovo Observatory
65 Pulkovo
196140 Saint Petersburg
Telephone: (0)812-2982242
Electronic Mail: <userid>@gao.spb.su
WWW: http://www.gao.spb.ru/
 http://www.gao.spb.ru/english/index.html
Founded: 1839
Periodicals: "Izvestiia" (ISSN 0367-8466), "Solar Data Bulletin"
Coordinates: 030°19'36"E 59°46'24"N H75m (IAU Code 084)
City Reference Coordinates: 030°15'00"E 59°55'00"N

Russian Academy of Sciences, Pulkovo Observatory, Kislovodsk Astronomical Station
357700 Kislovodsk
Telephone: (0)86537-33088
 (0)812-1234096
Telefax: (0)812-1231922
 (0)812-3143360

Electronic Mail: <userid>@gaosun.spb.su
WWW: http://www.gao.spb.ru/english/staff/dsp/mas.html
Founded: 1948
Staff: 34
Activities: solar physics * astrometry
Periodicals: "Solnechnye Dannye"
Coordinates: 042°40'12"E 43°44'47"N H2,130m
City Reference Coordinates: 042°44'00"E 42°55'00"N

Russian Academy of Sciences, Siberian Division, Institute of Solar-Terrestrial Physics
Ul. Lermontov 126
P.O. Box 4026
664033 Irkutsk
Telephone: (0)3952-461865
Telefax: (0)3952-460265
Electronic Mail: post@iszf.irk.ru
WWW: http://www.iszf.irk.ru/
Founded: 1961
Staff: 547
Activities: Sun * STP * instrumentation * Siberian Solar Radio Telescope (SSRT)
Periodicals: (4) "Issledovania po Geomagnetismu, Aeronomii i Fisike Solntsa"
Coordinates: 100°55'30"E 51°37'19"N H2,012m (Sayan Solar Observatory)
104°55'00"E 51°50'45"N H600m (Baikal Astrophysical Observatory)
102°12'30"E 51°45'27"N H832m (Sayan Mountains Radiophysical Observatory)
City Reference Coordinates: 104°20'00"E 52°16'00"N

Russian Academy of Sciences, Space Research Institute (IKI)
Ul. Profsojuznaya 84/32
117810 Moscow
Telephone: (0)095-3333122
(0)095-3335212
Telefax: (0)095-3107023
(0)095-3335178
(0)095-9133040
Electronic Mail: <userid>@sovamsu.sovusa.com
WWW: http://www.iki.rssi.ru/
Founded: 1967
Staff: 1500
Activities: space physics * space plasma * astrophysics * solar system
Awards: "Zel'dovich Award"
City Reference Coordinates: 038°50'00"E 55°05'00"N

Russian Academy of Sciences, Special Astrophysical Observatory (SAO)
Nizhny Arkhyz
369167 Karachai-Cherkesia
Telephone: (0)87878-46436
Telefax: (0)901-4982931
Electronic Mail: adm@sao.ru
<userid>@sao.ru
WWW: http://www.sao.ru/
Founded: 1966
Staff: 120
Activities: observing (optical and radio) * cosmology * extragalactic astronomy * instrument building
Periodicals: "Bulletin of the Special Astrophysical Observatory" (ISSN 0320-9318)
Coordinates: 041°26'30"E 43°39'12"N H2,100m (6m altazimutal reflector)
041°35'25"E 43°49'32"N H973m (600m radiotelescope)
City Reference Coordinates: 041°57'00"E 43°44'00"N

Russian Academy of Sciences, Ussurijsk Astrophysical Observatory (UAO)
P/O Gornotayozhnoye
692533 Ussurijsk
Telephone: 42341-91121
Telefax: 42341-91121
Electronic Mail: baranov@ml.ussurijsk.ru
WWW: http://www.ras.ru/
Founded: 1954
Staff: 34
Activities: Sun * solar activity * comets
Periodicals: (1) "Solar Activity and its Influence on the Earth" (in Russian)
Coordinates: 132°10'00"E 43°40'00"N H250m
City Reference Coordinates: 131°99'00"E 43°88'00"N

Russian Federal Service for Hydrometeorology and Environmental Monitoring (ROSHYDROMET)
2 Novovagan'kovsky Ul.
123242 Moscow

Telephone: (0)095-2520808
(0)095-2552493
(0)095-2523067
Telefax: (0)095-2521158
(0)095-2552092
WWW: http://www.mecom.ru/roshydro/pub/index.htm
Founded: 1834
City Reference Coordinates: 038°50'00"E 55°05'00"N

Russian Space Agency
Shukin Street
Building 42
129857 Moscow
Electronic Mail: <userid>@rka.ru
WWW: http://www.rka.ru/
http://www.rka.ru/english/eindex.htm
City Reference Coordinates: 038°50'00"E 55°05'00"N

Saint Petersburg University, Astronomical Institute
Bibliotechnaia Pl. 2
Petrodvorets
198904 Saint Petersburg
Telephone: (0)812-4287129
(0)812-4284265
Telefax: (0)812-4284259
(0)812-4287129
Electronic Mail: ai@astro.spbu.ru
<userid>@astro.spbu.ru
WWW: http://www.astro.spbu.ru/
Founded: 1881
Staff: 90
Activities: stellar structure * radiation transfer * active galaxies * cosmic-gas dynamics * celestial dynamics * stellar dynamics * astrometry * Earth rotation parameters * fundamental stellar catalogs * solar radio astronomy
Periodicals: (1/2) "Transactions of the Saint Petersburg Astronomical Institute" (ISSN 0136-8109 & 0136-8141)
Coordinates: 030°17'45"E 59°56'32"N H0m (Saint Petersburg – IAU Code 584)
044°17'30"E 40°20'07"N H1,500m (Byurakan – see separate entry)
City Reference Coordinates: 030°15'00"E 59°55'00"N

Saint Petersburg University, Department of Applied Mathematics and Control Processes
Fakultet PM-PU
Bibliotechnaia Pl. 2
198904 Saint Petersburg
Telephone: (0)812-4287159
Telefax: (0)812-4287159
Electronic Mail: lapetr@robot.apmath.spb.su
WWW: http://www.apmath.spbu.ru/
Founded: 1969
Staff: 24
Activities: control in problems of celestial mechanics * Moon motion * N-body problem * stellar dynamics * star clusters * data processing
Periodicals: (4) "Vestnik Sankt-Peterburskogo Universiteta - Series of Mathematics, Mechanics, and Astronomy" (ISSN 0132-4624)
City Reference Coordinates: 030°15'00"E 59°55'00"N

Scientific and Research Centre on Space Hydrometeorology "Planeta"
7 Bolshoy Predtechensky Per.
123242 Moscow
Telephone: (0)95-2551263
(0)95-2523717
Telefax: (0)95-2004210
Electronic Mail: asmus@ns.planeta.rssi.ru (Vassily Asmus, Director)
WWW: http://sputnik.infospace.ru/
Founded: 1974
Staff: 300
Activities: research * remote sensing * environment * hydrometeorology * satellite operations * receiving, processing, archiving and disseminating data
City Reference Coordinates: 038°50'00"E 55°05'00"N

Special Astrophysical Observatory (SAO)
• See "Russian Academy of Sciences, Special Astrophysical Observatory (SAO)"

State Committee for Standardization and Metrology (GOST)
Leninsky Prospekt 9
Lenin Avenue 9
119991 Moscow

Telephone: (0)095-2360300
Telefax: (0)095-2366231
Electronic Mail: info@gost.ru
WWW: http://www.gost.ru/
City Reference Coordinates: 038°50'00"E 55°05'00"N

Sternberg State Astronomical Institute (SAI)
University Avenue 13
119899 Moscow
Telephone: (0)095-9392858
Telefax: (0)095-9398841
Electronic Mail: <userid>@sai.msu.su
WWW: http://www.sai.msu.su/
   http://comet.sai.msu.su/ (Radio Astronomy Department)
   http://xray.sai.msu.su/ (X-Ray Group)
   http://crydee.sai.msu.ru/ (Heliophysics)
Founded: 1931
Staff: 400
Activities: Earth * stars * Sun * galaxies * X-ray sources * planets * ISM * radio sources
Periodicals: "Trudy" (ISSN 0371-6791)
Coordinates: 037°32'40"E 55°42'00"N H195m (Moscow)
   034°01'30"E 44°33'42"N (Crimea, Ukraine)
City Reference Coordinates: 038°50'00"E 55°05'00"N

Tomsk Planetarium
Batenkova Street 3
634069 Tomsk
Telephone: (0)3822-909721
   (0)3822-232780
Electronic Mail: niipmm@urania.tomsk.su
Founded: 1948
Staff: 12
Activities: lectures * observing * meetings * expeditions
City Reference Coordinates: 085°05'00"E 56°30'00"N

Tomsk State University, Astronomical Observatory
P.O. Box 1106
634010 Tomsk
Telephone: (0)3822-410576
Electronic Mail: niipmm@urania.tomsk.su
WWW: http://phys.tsu.ru/
Founded: 1922
Staff: 15
Activities: small bodies (satellites, minor planets, comets, meteorites, meteoroids) * celestial mechanics * astrometry * geodesy
Periodicals: (1) "Astronomy and Geodesy" (ISSN 0201-5099)
Coordinates: 084°56'47"E 56°28'07"N H130m (IAU Code 236)
City Reference Coordinates: 085°05'00"E 56°30'00"N

Tsiolkovskiy Astronomical Club
c/o Sergey Savinov
Palace of Youth Activity
Ul. Kirova 17
426001 Izhevsk
Electronic Mail: ssa2000@izh.com
WWW: http://astroclub.vov.ru/
City Reference Coordinates: 053°11'00"E 56°49'00"N

Udmurt State University, Department of Astronomy and Mechanics
Universitetskaya Street 1
426034 Izhevsk
Telephone: (0)3412-761794
Electronic Mail: <userid>@uni.udm.ru
WWW: http://www.uni.udm.ru/
Founded: 1991
Staff: 7
Activities: stellat dynamics * celestial mechanics * cataclysmic variables * potential theory * planets
Periodicals: "Vestnik Udmurtskogo Universiteta"
Coordinates: 053°10'48"E 56°51'12"N
City Reference Coordinates: 053°11'00"E 56°49'00"N

Ufa Planetarium
Pl. Lenina 3
450075 Ufa
Telephone: (0)3472-357023
Telefax: (0)3472-528653 (c/o Planetarium)

Electronic Mail: postmaster@uctlgf.bashkiria.su (c/o Planetarium)
Founded: 1964
Staff: 15
Activities: education * shows
City Reference Coordinates: 055°56'00"E 54°44'00"N

Ural State University, Department of Astronomy and Geodesy
Pr. Lenina 51
620083 Ekaterinburg
Telephone: (0)223386
(0)220729
WWW: http://www.usu.ru/eng/usu/faculty/astron.htm
http://www.usu.ru/eng/usu/subdivisions/observ.htm (Kourov)
Founded: 1965
Staff: 50
Activities: stellar astronomy * solar physics * satellites
Coordinates: 059°32'00"E 57°02'00"N H500m
City Reference Coordinates: 059°32'00"E 57°02'00"N

Ussurijsk Astrophysical Observatory (UAO)
● See "Russian Academy of Sciences, Ussurijsk Astrophysical Observatory (UAO)"

VNIIFTRI
● See "Institute for Physical Technical and Radio-Technical Measurements (VNIIFTRI)"

World Data Center B for Solar-Terrestrial Physics
c/o Geophysical Institute
Russian Academy of Sciences
Ul. Molodezhnaya 3
117296 Moscow
Telephone: (0)095-9305619
Telefax: (0)095-9305509
Electronic Mail: <userid>@wdcb.ru
WWW: http://www.wdcb.rssi.ru/WDCB/wdcb_stp.shtml
Founded: 1957
Staff: 83
Activities: collecting, storing and exchanging solar activity data * distributing data upon request
Periodicals: (10) "Materials of the World Data Center B"
City Reference Coordinates: 038°50'00"E 55°05'00"N

Yaroslavl Astronomical and Geodetical Society "Meridian"
Astronomical Observatory
State Pedagogical University
Ul. Respublicanskaya 108
150000 Yaroslavl
Telephone: (0)852-728570
(0)852-354123
Telefax: (0)852-354777
Electronic Mail: perov@yspu.yar.ru (Nikolay Perov, Chairman)
Founded: 1948
Membership: 22
Activities: celestial mechanics * astrometry * astrophysics * solar system * galaxy * geodesy * education * public outreach * SETI * expeditions
Coordinates: 039°53'00"E 57°35'00"N
City Reference Coordinates: 039°52'00"E 57°34'00"N

Youth Astronomical School
c/o Aleksander G. Sergeev
Ul. Warsawskaya 122-125
196240 Saint Petersburg
Telephone: (0)812-2333112
Electronic Mail: yaseu@mail.ru
WWW: http://yaseu.da.ru/
http://yaseu.webprovider.com/
City Reference Coordinates: 030°15'00"E 59°55'00"N

Zabaikalsky State Pedagogical University
Babushkin Ul. 129
672045 Chita
Telephone: (0)302-267317
(0)302-234874
Telefax: (0)302-267935
WWW: http://zgpu.chita.ru/english/info.htm
Founded: 1938
Staff: 443 (one astronomer)

Activities: teacher training * physics * astronomy * mathematics * computing * chemistry
Coordinates: 113°29'00"E 52°03'00"N
City Reference Coordinates: 113°29'00"E 52°03'00"N

Saudi Arabia

King Abdulaziz City for Science and Technology (KACST)
P.O. Box 6086
Riyadh 11442
Telephone: (0)1-4813547
Telefax: (0)1-4813523
Electronic Mail: <userid>@kacst.edu.sa
WWW: http://www.kacst.edu.sa/
Founded: 1977
Staff: 25
Activities: formulating national science and technology policies * setting priorities for funding and coordinating research activities * establishing research institutes relevant to national development
Periodicals: "Science and Technology" (in Arabic)
Coordinates: 039°37'48"E 21°22'12"N H333m (Makka)
 036°04'12"E 29°10'12"N H330m (H. Ammar)
 036°22'48"E 26°27'00"N H100m (Wajh)
 040°54'36"E 27°15'00"N H1,000m (Hail)
 046°24'00"E 23°33'36"N H853m (Hariq)
 042°36'00"E 19°13'48"N H2,377m (Namass)
City Reference Coordinates: 046°43'00"E 24°38'00"N

King Abdulaziz University, Astronomy and Geophysics Research Institute (AGRI)
P.O. Box 9028
Jeddah 21413
Telephone: (0)2-6952284
WWW: http://www.kacst.edu.sa/
Founded: 1979
Activities: PN * radioastronomy * artificial satellites * sunspots * celestial mechanics * lunar occultations * stellar photometry * solar spectroscopy
Coordinates: 039°49'00"E 21°28'00"N
City Reference Coordinates: 039°49'00"E 21°28'00"N

King Saud University (KSU), College of Science, Astronomy Department
P.O. Box 2455
Riyadh 11451
Telephone: (0)1-4676324
Telefax: (0)1-4674253
WWW: http://www.mohe.gov.sa/ksu/
Founded: 1986
Staff: 5
Activities: education * research * astrophysics * external galaxies * history of astronomy
Coordinates: 046°43'02"E 24°37'58"N
City Reference Coordinates: 046°43'00"E 24°38'00"N

Meteorology and Environmental Protection Administration (MEPA)
P.O. Box 1358
Jeddah 21431
Telephone: (0)2-6512312
 (0)2-6710448
 (0)2-6710439
 (0)2-6713238
Telefax: (0)2-6511424
Electronic Mail: sidc@mepa.org.sa
WWW: http://www.mepa.org.sa/
City Reference Coordinates: 039°49'00"E 21°28'00"N

Saudi Arabian Standards Organization (SASO)
P.O. Box 3437
Riyadh 11471
Telephone: (0)1-4520000
 (0)1-4520132
Telefax: (0)1-4520086
 (0)1-4520133
Electronic Mail: sasoinfo@saso.org
WWW: http://www.sao.org/
Founded: 1972
Staff: 351
Activities: developing national standards * promoting and distributing them * conformity certification * quality marking * metrology * testing * technical information
City Reference Coordinates: 046°43'00"E 24°38'00"N

Serbia-Montenegro

Astronomical Observatory, Belgrade
Volgina 7
11050 Beograd
Serbia
Telephone: (0)11-419357
    (0)11-401320
Telefax: (0)11-419553
Electronic Mail: contact@aob.bg.ac.yu
WWW: http://www.aob.bg.ac.yu/
Founded: 1887
Staff: 50
Activities: astrometry * celestial mechanics * Earth rotation * solar photosphere * close and visual binaries * radiative transfer * stellar spectra
Periodicals: (1-2) "Bulletin" (ISSN 0373-3734), "Publications" (ISSN 0373-3742)
Coordinates: 020°30'45"E 44°48'10"N H253m (IAU Code 057)
City Reference Coordinates: 020°30'00"E 44°50'00"N

Belgrade University, Department of Astronomy
Studentski Trg. 16
11000 Beograd
Serbia
Telephone: (0)1-638715
    (0)11-187133
Telefax: (0)11-187133
Electronic Mail: ffh@ffh.bg.ac.yu
WWW: http://www.ffh.bg.ac.yu/
Periodicals: "Publications" (ISSN 0350-3283)
City Reference Coordinates: 020°30'00"E 44°50'00"N

Boskovitch Astronomical Society (Rudjer_)
• See "Rudjer Boskovitch Astronomical Society"

Federal Hydrometeorological Institute (FMHI)
Birčaninova 6
P.O. Box 604
11001 Beograd
Serbia
Telephone: (0)11-646555
Telefax: (0)11-646369
Electronic Mail: <userid>@meteo.yu
WWW: http://www.meteo.yu/
Founded: 1947
Staff: 153
Activities: meteorology * hydrology * air and water quality monitoring * national climatological centre
Periodicals: (1) "Meteorological and Hydrological Yearbook"
Observatories: network of meteorological and hydrological observing stations
City Reference Coordinates: 020°30'00"E 44°50'00"N

Institute of Physics
P.O. Box 57
11001 Beograd
Serbia
or :
Pregrevica 118
11080 Zemun
Serbia
Telephone: (0)11-3160260 (Switchboard)
    (0)11-3162067 (Director)
Telefax: (0)11-3162190
Electronic Mail: info@phy.bg.ac.yu
WWW: http://www.phy.bg.ac.yu/
Founded: 1961
Staff: 250
Activities: theoretical physics * solid-state physics * plasma physics * elementary-particle physics * theory of gravitation * atomic and molecular physics * laser physics * technical and applied physics
Periodicals: (4) "Sveske Fizičkih Nauka" (ISSN 0352-7859); "Sveske Fizičkih Nauka Series A: Conferences" (ISSN 0354-9291)
City Reference Coordinates: 020°30'00"E 44°50'00"N (Beograd)
    020°23'00"E 44°51'00"N (Zemun)

Institute of Standardization (SZS)
Kneza Milosa 20
P.O. Box 933
11000 Beograd
Serbia
Telephone: (0)11-3613150
Telefax: (0)11-3617341
Electronic Mail: jus@szs.sv.gov.yu
City Reference Coordinates: 020°30'00"E 44°50'00"N

Rudjer Boskovitch Astronomical Society
c/o Public Observatory
Kalemegdan
Gornji Grad 16
11000 Beograd
Serbia
Telephone: (0)11-624605
Electronic Mail: vladanc@mrsys1.mr-net.co.yu
astrorbo@eunet.yu
WWW: http://solair.eunet.yu/~astrorbo/asrb.htm
Founded: 1934
Membership: 1300
Activities: Sun * occultations * eclipses * double stars * photography * history of astronomy * dense-matter astrophysics * planetarium
Periodicals: (5) "Vasiona" (ISSN 0506-4295); "Publication of the Astronomical Society Rudjer Boskovitch"
City Reference Coordinates: 020°30'00"E 44°50'00"N

Union of Societies of Mathematicians, Physicists and Astronomers, Physics Section
c/o D. Bek-Uzarov
Institute Boris Kidric
P.O. Box 522
11001 Beograd
Serbia
Telephone: (0)11-4440871
Telefax: (0)11-4440195
City Reference Coordinates: 020°30'00"E 44°50'00"N

Singapore

Astronomical Society of Singapore (TASOS) (The_)
● See "The Astronomical Society of Singapore (TASOS)"

Astro Scientific Centre Pte. Ltd. (ASCPL)
P.O. Box 322
PSA Building
Singapore 911142
or :
c/o Singapore Science Centre
Omni Theatre
15 Science Centre Road
Singapore 609081
Telephone: (0)567-4163
Telefax: (0)567-4826
Electronic Mail: astro@astro.com.sg
sales@astro.com.sg
WWW: http://astro.com.sg/
Founded: 1987
Staff: 7
Activities: distributing and servicing telescopes, binoculars, ...
City Reference Coordinates: 103°48'00"E 01°22'00"N

Meteorological Service Singapore (MSS)
P.O. Box 8
Changi Airport
Singapore 918141
Telephone: (0)545-7193
(0)545-7190
(0)545-7191
Telefax: (0)545-7192
Electronic Mail: mss_operations@mss.gov.sg
WWW: http://www.gov.sg/metsin/
Staff: 125
City Reference Coordinates: 103°48'00"E 01°22'00"N

National University of Singapore (NUS), Department of Mathematics
5 Lower Kent Ridge Road
Singapore 119260
Telephone: (0)874-2738
Telefax: (0)779-5452
WWW: http://www.math.nus.edu.sg/
Founded: 1929
Staff: 2
Activities: education * research * radiative transfer in atmospheres
City Reference Coordinates: 103°48'00"E 01°22'00"N

Singapore Institute of Standards and Industrial Research (SISIR)
Kent Ridge
1 Science Park Drive
Singapore 118221
Telephone: (0)7787777
(0)2786666
Telefax: (0)7780086
(0)7781280
WWW: http://www.psb.gov.sg/
Founded: 1973
Staff: 500
Activities: materials technology advising * electronic and computer applications * consultancy * product and process development and design * standards and certification authority
Periodicals: (4) "SISIR News" (ISSN 0129-0908); (1) "Annual Report"
City Reference Coordinates: 103°48'00"E 01°22'00"N

Singapore Productivity and Standards Board
PSB Building
2 Bukit Merah Central
Singapore 159835
Telephone: (0)2786666
Telefax: (0)2786665
(0)2786667
Electronic Mail: queries@psb.gov.sg
WWW: http://www.psb.gov.sg/

City Reference Coordinates: 103°48'00"E 01°22'00"N

Singapore Science Centre, Omnitheatre and Observatory
Science Centre Road
Off Jurong Town Hall Road
Singapore 609081
Telephone: (0)560-3316
(0)425-2500
Telefax: (0)565-9533
Electronic Mail: <userid>@sci-ctr.edu.sg
WWW: http://www.sci-ctr.edu.sg/
Founded: 1977
Activities: omnimax movies * planetarium programmes * public lectures * observing sessions
Periodicals: (12) "Skytrack"
Coordinates: 103°44'15"E 01°20'03"N
City Reference Coordinates: 103°48'00"E 01°22'00"N

The Astronomical Society of Singapore (TASOS)
c/o Astro Scientific Centre Pte. Ltd.
Omni Theatre
15 Science Centre Road
Singapore 609081
Telephone: (0)567-4163
Telefax: (0)567-4826
Electronic Mail: tasos@tasos.org.sg
WWW: http://tasos.org.sg/
Founded: 1992
Membership: 200
Activities: observing * photography * lectures
Periodicals: (4) "Moonstarer"
City Reference Coordinates: 103°48'00"E 01°22'00"N

Periodicals: "Contributions of the Geophysical Institute of the Slovak Academy of Sciences", "Bulletin of the Slovak Seismological Stations", "Results of Geomagnetic Observations at the Hurbanovo Geomagnetic Observatory"
Observatories: 5 for geophysics (Hurbanovo, Sroborova, Vyhne, Bratislava & Skalnaté Pleso) and 3 for meteorology (Mlynany, Stará Lesná & Skalnaté Pleso)
Coordinates: 020°14'00"E 49°11'00"N H1,778m (Skalnaté Pleso)
          020°17'00"E 49°09'00"N H810m (Stará Lesná)
          018°20'00"E 48°19'00"N H195m (Mlynany)
City Reference Coordinates: 017°10'00"E 48°10'00"N

Slovak Astronomical Society (SAS)
059 60 Tatranská Lomnica
Telephone: (0)969-467866
Telefax: (0)967-467656
Electronic Mail: sas@ta3.sk
WWW: http://www.ta3.sk/sas/sas.html
Founded: 1959
Membership: 253
Activities: observing (Sun, comets, meteors, variable stars) * education * lectures * amateur-professional collaborations
City Reference Coordinates: 020°15'00"E 49°11'00"N

Slovak Central Observatory (SCO)
Komárnanská 134
947 01 Hurbanovo
Telephone: (0)818-7602484
Telefax: (0)818-7602487
Electronic Mail: suh@kemar.sk
          kozmos@netlab.sk (Kozmos)
WWW: http://www.geomag.sk/
          http://www.ta3.sk/kozmos/kozmos.html (Kozmos)
Founded: 1871
Membership: 29
Activities: observing (Sun, variable stars, meteors, occultations) * education * planetarium shows * lectures * public observing
Periodicals: (6) "Kozmos"; (1) "Astronomical Yearbook", "Astronomical Calendar"
Coordinates: 018°11'22"E 47°52'27"N H112m (IAU Code 551)
• Formerly "Slovak Centre of Amateur Astronomy"
City Reference Coordinates: 018°10'00"E 47°53'00"N

Slovak Hydrometeorological Institute
Jeséniova 17
P.O. Box 37
833 15 Bratislava
Telephone: (0)7-371247
          (0)7-3785226
Telefax: (0)7-374593
          (0)7-375670
          (0)7-376197
Electronic Mail: <userid>@shmuvax.shmu.sk
WWW: http://www.shmu.sk/
Founded: 1969
Staff: 700
Activities: hydrology * meteorology * climatology * environment * water quality and quantity * air pollution
Periodicals: "Bulletin of Air Pollution", "Collection of Papers", "Hydrological Yearbook"
Observatories: meteorological, hydrological, climatological and air pollution monitoring network of stations
City Reference Coordinates: 017°10'00"E 48°10'00"N

Slovak Office of Standards, Metrology and Testing (UNMS)
P.O. Box 75
810 05 Bratislava
or :
Štefanovičova 3
814 39 Bratislava
Telephone: (0)7-391085
Telefax: (0)7-391050
Electronic Mail: <userid>@normoff.gov.sk
WWW: http://www.normoff.gov.sk/unms_sr/
Founded: 1993
Staff: 56
Activities: national body for standardization, testing, certification and metrology
Periodicals: "Bulletin of the Slovak Office of Standards, Metrology and Testing", "Standardization", "Metrology and Testing"
City Reference Coordinates: 017°10'00"E 48°10'00"N

Slovak Physical Society
c/o Dalibor Krupa (President)
Institute of Physics
Dúbravská cesta 9
842 28 Bratislava

Telephone: (0)7-395676
Telefax: (0)7-395676
Electronic Mail: krupa@savba.sk
sfs@savba.sk
WWW: http://www.savba.sk/~fyzisfs
http://www.sfs.sk/
Founded: 1993
Membership: 280
Activities: promoting advancement of physics research, teaching and training
Periodicals: "Gradient"
Awards: (1) "Young Physicists Competition"
City Reference Coordinates: 017°10'00"E 48°10'00"N

Slovak Union of Amateur Astronomers
c/o Rimavská Sobota Observatory
Tomasovska 63
979 01 Rimavská Sobota
Telephone: (0)866-5624709
Telefax: (0)866-5624709
Electronic Mail: astrors@bb.telecom.sk
Founded: 1970
Membership: 350
Activities: photography * popularization * history of astronomy * observing (Sun, meteors, variable stars, eclipsing binaries, comets)
Periodicals: (6) "Kozmos" (see separate entry)
Coordinates: 020°00'25"E 48°22'29"N H228m (Rimavská Sobota - see separate entry)
City Reference Coordinates: 020°00'00"E 48°22'00"N

Slovak University of Technology, Observatory
Radlinského 11
813 68 Bratislava
Telephone: (0)7-52498047
Telefax: (0)7-52925476
Electronic Mail: <userid>@cvt.stuba.sk
hefty@cvt.stuba.sk (Ján Hefty, Director)
WWW: http://www.stuba.sk/
Founded: 1956
Staff: 4
Activities: Earth rotation * astrometry * geodynamics * GPS
Coordinates: 017°07'11"E 48°09'18"N H170m
City Reference Coordinates: 017°10'00"E 48°10'00"N

Žilina Observatory
Horný Val 20/41
P.O. Box B 153
012 42 Žilina
Telephone: (0)89-43780
Electronic Mail: rksza@za.sanet.sk
Founded: 1961
Staff: 5
Activities: popularization * observing (Sun, comets, variable stars, occultations, eclipses, meteors)
Coordinates: 018°45'13"E 49°12'22"N H404m (Malý Diel)
City Reference Coordinates: 018°40'00"E 49°14'00"N

Slovenia

Hydrometeorological Institute of Slovenia
Vojkova 1b
P.O. Box 2549
1001 Ljubljana
Telephone: (0)61-1784255
Telefax: (0)61-1331396
WWW: http://www.rzs-hm.si/
Founded: 1947
Staff: 210
Activities: weather forecast * climatology * observing * air pollution * hydrology * hydrometry * water quality * ground water * agrometeorology
Coordinates: 013°51'00"E 46°23'00"N H2,514m (Kredarica)
City Reference Coordinates: 014°31'00"E 46°03'00"N

Javornik Astronomical Society
Kolodvorska 6
1000 Ljubljana
Telephone: (0)61-315198
(0)61-446311
Electronic Mail: stane@nettaxi.com
WWW: http://www.javornik-drustvo.si/
Founded: 1979
Membership: 120
Activities: popularization * education * Summer camp * workshops * observing
Periodicals: (2) "Astronom"; "Circular of ADJ"
Coordinates: 014°03'52"E 45°39'24"N H1,140m
City Reference Coordinates: 014°31'00"E 46°03'00"N

Slovenian Academy of Sciences and Arts (SAZU)
Novi trg 3
61000 Ljubljana
Telephone: (0)61-1256-068
Telefax: (0)61-1253423
Electronic Mail: sazu@sazu.si
WWW: http://www.sazu.si/
Founded: 1938
Membership: 171
City Reference Coordinates: 014°31'00"E 46°03'00"N

Society of Mathematicians, Physicists and Astronomers of Slovenia
c/o Department of Mathematics
University of Ljubljana
Jadranska Ulica 19
1000 Ljubljana
Telephone: (0)61-1766500
Telefax: (0)61-2517281
Electronic Mail: tajnik@dmfa.si
WWW: http://www.dmfa.si/
Founded: 1949
Membership: 20
Activities: popularization of mathematics, physics and astronomy
Periodicals: (6) "Obzornik za Matematiko in Fiziko" (ISSN 0473-7466), "Presek" (ISSN 0351-6652)
City Reference Coordinates: 014°31'00"E 46°03'00"N

Standards and Metrology Institute of Slovenia (SMIS)
Kotnikova 6
61000 Ljubljana
Telephone: (0)61-1312322
(0)61-1783000
Telefax: (0)61-314882
(0)61-1783196
Electronic Mail: ic@usm.mzt.si
smis@usm.mzt.si
WWW: http://www.usm.mzt.si/
Founded: 1991
Staff: 70
Activities: national standards organization * legal metrology * national accreditation service
Periodicals: (12) "Sporočila/Messages"
City Reference Coordinates: 014°31'00"E 46°03'00"N

University of Ljubljana, Department of Physics, Astronomical and Geophysical Observatory
Pot na Golovec 25
1000 Ljubljana
Telephone: (0)1-5401353
Telefax: (0)1-5405370
Electronic Mail: <userid>@uni-lj.si
WWW: http://www.fiz.uni-lj.si/astro/
Founded: 1948
Staff: 4
Activities: education * photometry * spectroscopy * general relativity * pulsars
Periodicals: (1) "Astronomical Ephemeris" (ISSN 1318-0614)
Coordinates: 014°31'38"E 46°02'37"N H401m (IAU Code 103)
City Reference Coordinates: 014°31'00"E 46°03'00"N

South Africa

Astronomical Society of Southern Africa (ASSA)
P.O. Box 9
7935 Observatory
Telephone: (0)21-4470025
Electronic Mail: basbs@bremner.uct.ac.za (Brian Skinner, Honorary Secretary)
mgs@maties.sun.ac.za
mnassa@saao.ac.za (MNASSA)
WWW: http://www.saao.ac.za/sky/freq.html#assa
Founded: 1912
Membership: 1000
Activities: meetings * discussions * observations and searches
Sections: variable stars * comets and meteors * minor planets occultations * grazing occultations * nova search * computing * ordinary occultations * solar observing * double stars * imaging
Periodicals: (6) "Monthly Notes of the Astronomical Society of Southern Africa (MNASSA)" (ISSN 0024-8266); (1) "ASSA Handbook" (ISSN 0571-7191)
Awards: (1) "Gill Medal"
City Reference Coordinates: 018°22'00"E 33°55'00"S

Astronomical Society of Southern Africa (ASSA), Cape Centre
P.O. Box 13018
7705 Mowbray
Electronic Mail: micjohn@netactive.co.za (Secretary)
Founded: 1912
Activities: lectures * observing * instrumentation * education
Periodicals: (12) "Newsletter"; (4) "The Cape Observer"
Coordinates: 018°32'00"E 33°55'00"S
City Reference Coordinates: 018°32'00"E 33°55'00"S

Astronomical Society of Southern Africa (ASSA), Garden Route Centre
c/o Hans Daehne (Secretary)
9A Ironside Street
6529 George
Telephone: (0)44-8745902
Telefax: (0)44-3431736
Electronic Mail: janhers@pixie.co.za (Jan Hers, Chairman)
Founded: 1998
Membership: 35
Activities: monthly meetings * observing
City Reference Coordinates: 022°28'00"E 33°57'00"S

Astronomical Society of Southern Africa (ASSA), Johannesburg Centre
P.O. Box 93145
2143 Yeoville
or :
18A Gill Street
Observatory
Johannesburg
Telephone: (0)11-7163199
Telefax: (0)11-3392926
Electronic Mail: assa_jhb@aqua.co.za
WWW: http://www.aqua.co.za/assa_jhb.htm
Founded: 1922
Membership: 144
Activities: amateur astronomy * popularization * public observing
City Reference Coordinates: 028°00'00"E 26°15'00"S

Astronomical Society of Southern Africa (ASSA), Natal Centre
P.O. Box 5330
4001 Durban
Electronic Mail: sthomson@mweb.co.za
WWW: http://www.astronomical.lia.net/
City Reference Coordinates: 031°00'00"E 29°53'00"S

Astronomical Society of Southern Africa (ASSA), Natal Midlands Centre (Pietermaritzburg)
c/o The Secretary
P.O. Box 2106
3200 Pietermaritzburg
Telephone: (0)33-3433646
WWW: http://www.botany.unp.ac.za/nmc/nmc.htm
Founded: 1974
Membership: 50

Activities: variable stars * observing * occultations * education
Periodicals: (12) "Stardust"
Observatories: 1 (World's View Observatory)
City Reference Coordinates: 030°16'00"E 29°37'00"S

Astronomical Society of Southern Africa (ASSA), Pretoria Centre
c/o N.P. Young (Secretary)
201 Kritzinger street
Meyers Park
0184 Pretoria
or :
c/o Membership Secretary
P O Box 11151
Queerswood 0121
Telephone: (0)12-833765
Telefax: (0)12-3257534
WWW: http://mafadi.aero.csir.co.za/assa/
Founded: 1968
Membership: 51
Activities: amateur astronomy * observing * data gathering
Periodicals: (12) "Newsletter"
Coordinates: 028°16'03"E 25°44'11"S
City Reference Coordinates: 028°12'00"E 25°45'00"S (Pretoria)

Cederberg Observatory
P.O. Box 5203
8000 Cape Town
Electronic Mail: zasah52@iafrica.com (Bill Hollenbach)
WWW: http://www.cederbergobs.org.za/
Founded: 1986
Staff: 6
Coordinates: 019°15'06"E 33°29'46"S
City Reference Coordinates: 018°22'00"E 33°55'00"S

Center for Astronomical Observing Quality (CAOQ)
c/o Andre Erasmus
South African Astronomical Observatory (SAAO)
P.O. Box 9
7935 Observatory
Telephone: (0)21-4470025
Telefax: (0)21-4473639
Electronic Mail: erasmus@saao.ac.za
WWW: http://da.saao.ac.za/~erasmus/caoq/
Founded: 1997
Activities: monitoring and forecasting atmospheric effects at existing and potential telescope sites
City Reference Coordinates: 018°22'00"E 33°55'00"S

Council for Scientific and Industrial Research (CSIR)
P.O. Box 395
0001 Pretoria
Telephone: (0)12-8412911
Telefax: (0)12-3491153
WWW: http://www.csir.co.za/
Founded: 1945
Staff: 3000
City Reference Coordinates: 028°10'00"E 25°45'00"S

Durban Natural Science Museum, Astronomy Interest Group
P.O. Box 4085
4000 Durban
or :
City Hall - 1st Floor
Smith Street
Durban
Telephone: (0)31-3006212
Telefax: (0)31-3006308
Electronic Mail: mario@dimaggio.org (Mario Di Maggio, Education Officer)
WWW: http://www.kwazuzulwazi.co.za/AIG-default.htm
Founded: 1998
Membership: 106
Activities: multimedia presentations * demonstrations * observing
Periodicals: (12) "Newsletter" (circ.: 450)
City Reference Coordinates: 030°56'00"E 29°55'00"S

Foundation for Research Development (FRD)
• See now "National Research Foundation (NRF)"

Hartebeesthoek Radio Astronomy Observatory (HartRAO)
• See "National Research Foundation (NRF), Hartebeesthoek Radio Astronomy Observatory (HartRAO)"

National Research Foundation (NRF)
P.O. Box 2600
0001 Pretoria
Telephone: (0)12-4814000
Telefax: (0)12-3491179
Electronic Mail: info@nrf.ac.za
                             <userid>@frd.ac.za
WWW: http://www.nrf.ac.za/
Activities: supporting and promoting research
• Formerly "Foundation for Research Development (FRD)"
City Reference Coordinates: 028°10'00"E 25°45'00"S

National Research Foundation (NRF), Hartebeesthoek Radio Astronomy Observatory (HartRAO)

P.O. Box 443
1740 Krugersdorp
Telephone: (0)12-3260742
Telefax: (0)12-3260756
Electronic Mail: <userid>@hartrao.ac.za
WWW: http://www.hartrao.ac.za/
Founded: 1975
Staff: 33
Activities: national facility for radioastronomy * VLBI * spectroscopy * pulsars * masers * variable radio sources * surveys
Periodicals: (1) "Annual Report"
Coordinates: 027°41'07"E 25°53'23"S H1,416m
City Reference Coordinates: 027°35'00"E 26°05'00"S

Potchefstroom University for Christian Higher Education, Department of Physics, Space Research Unit (SRU)
2520 Potchefstroom
Telephone: (0)18-2992423
              (0)18-2992403
Telefax: (0)18-2992421
Electronic Mail: <userid>@puknet.puk.ac.za
WWW: http://www.puk.ac.za/fskdocs/ern_topE.html
       http://www.puk.ac.za/fskdocs/phys_topE.html
Founded: 1961
Staff: 11
Activities: cosmic rays * magnetospheric and heliospheric physics * observing (solar flares, galactic cosmic-rays) * gamma-ray astronomy * pulsars * X-ray binaries * cataclysmic variable stars
Observatories: 4 (Potchefstroom, Hermanus, Tsumeb - Namibia and Sanae - Antarctica)
Coordinates: 026°54'09"E 27°10'32"S H1,408m (Potchefstroom)
City Reference Coordinates: 027°06'00"E 27°42'00"S

Rhodes University, Astronomy Society
Grahamstown
Electronic Mail: queries@astrosoc.soc.ru.ac.za
WWW: http://astrosoc.soc.ru.ac.za/
Founded: 1991
Membership: 120
City Reference Coordinates: 026°31'00"E 33°19'00"S

Rhodes University, Department of Physics and Electronics
P.O. Box 94
6140 Grahamstown
Telephone: (0)461-318111
Telefax: (0)461-25049
Electronic Mail: <userid>@hippo.ru.ac.za
WWW: http://phlinux.ru.ac.za/physics/
Founded: 1957
Staff: 9
Activities: 2.3 GHz radio-continuum survey of the Southern sky * radio continuum mapping * ionospheric model for Southern African region
Coordinates: 026°30'00"E 33°32'00"S
City Reference Coordinates: 026°31'00"E 33°19'00"S

Sedgefield Observatory
P.O. Box 48
6573 Sedgefield
Telephone: (0)44-3431736
Telefax: (0)44-3431736
Electronic Mail: janhers@pixie.co.za (Jan Hers)
Founded: 1979
Staff: 1
Activities: variable stars (visual and photoelectric measures)
Coordinates: 022°47'46"E 34°00'44"S (IAU Code 581)
City Reference Coordinates: 022°48'00"E 34°01'00"S

South African Association of Science and Technology Centres (SAASTEC)
c/o Rudi Horak
Gold Fields Exploratorium
Faculty of Science
University of Pretoria
0002 Pretoria
Telephone: (0)12-4202865
Telefax: (0)12-4203874
Electronic Mail: rudi@gold.up.ac.za
WWW: http://www.saastec.co.za/
Founded: 1996
Membership: 20 (institutions) and 19 (individuals)
Activities: stimulating interest in and establishing more science and technology centres * education * popularization * workshops
City Reference Coordinates: 028°10'00"E 25°45'00"S

South African Astronomical Observatory (SAAO)
P.O. Box 9
7935 Observatory
Telephone: (0)21-4470025
Telefax: (0)21-4473639
Electronic Mail: <userid>@saao.ac.za
WWW: http://www.saao.ac.za/
        http://www.salt.ac.za/ (Southern African Large Telescope - SALT)
Founded: 1820
Staff: 42
Activities: optical region and IR photometry * spectroscopy * CCD photometry
Periodicals: (1) "SAAO Annual Report" (ISSN 0250-0671); "SAAO Circulars", "SAAO Newsletter"
Coordinates: 018°28'07"E 33°56'03"S H18m (Cape Town – IAU Code 051)
             020°48'07"E 32°22'07"S H1,771m (Sutherland - see separate entry)
City Reference Coordinates: 018°22'00"E 33°55'00"S (Cape Town)

South African Astronomical Observatory (SAAO), Sutherland Observing Station
c/o SAAO
P.O. Box 9
7935 Observatory
Telephone: (0)23-5711205
Telefax: (0)23-5711413
Electronic Mail: sutherland@saao.ac.za
WWW: http://www.saao.ac.za/
Founded: 1972
Staff: 18
Coordinates: 020°48'07"E 32°22'07"S H1,771m
City Reference Coordinates: 018°22'00"E 33°55'00"S

South African Bureau of Standards (SABS)
Private Bag X 191
0001 Pretoria
or :
1 Dr. Lategan Road
Groenkloof
Telephone: (0)12-4287911
Telefax: (0)12-3441568
Electronic Mail: info@sabs.co.za
WWW: http://www.sabs.co.za/
City Reference Coordinates: 028°10'00"E 25°45'00"S (Pretoria)

South African Institute of Physics (SAIP)
c/o Physics Department
University of Zululand
3886 Kwa Dlangezwa

Telephone: (0)24-93911
Telefax: (0)24-93571
Electronic Mail: <userid>@unizul1.uzulu.ac.za
            bspoelst@pan.uzulu.ac.za (B. Spoelstra, Honorary Secretary)
WWW: http://www.sun.ac.za/physics/saip/
Founded: 1955
Membership: 550
Awards: (1/2) "Gold Medal for Outstanding Achievement", "Silver Medal for Outstanding Achievement of Physicist under 35yrs"
City Reference Coordinates: 030°56'00"E 29°55'00"S (Durban)

South African Museum, Planetarium
P.O. Box 61
8000 Cape Town
or :
25 Queen Victoria Street
8001 Cape Town
Telephone: (0)21-4243330
Telefax: (0)21-4246716
Electronic Mail: t.ferreira@samuseum.ac.za (Theo Ferreira)
WWW: http://www.museums.org.za/sam/planet/planetar.htm
Founded: 1987
Staff: 10
Activities: public and school shows * education * concerts * special events * show kits
Periodicals: (1) "Annual Report"
City Reference Coordinates: 018°22'00"E 33°55'00"S

South African Weather Bureau (SAWB)
Private Bag X-097
0001 Pretoria
or :
Forum Building
Struben Street
0001 Pretoria
Telephone: (0)12-3093120
Telefax: (0)12-3093127
Electronic Mail: <userid>@cirrus.sawb.gov.za
WWW: http://cirrus.sawb.gov.za/
Founded: 1860
Staff: 310
Activities: meteorology * surface and upper air observations * climatology * forecasting * research * instrumentation
Periodicals: (12) "SAWB Daily Weather Bulletin" (ISSN 0011-5517), "Climate Summary of Southern Africa"; "SAWB Technical Papers", "SAWB Technical Reports"; (1) "SAWB Yearly Weather Report"
City Reference Coordinates: 028°10'00"E 25°45'00"S

Spreeufontein Observatory
c/o Albert G. Jansen
Markstraat 3
6930 Prince Albert
Telephone: (0)23-5411871
Telefax: (0)23-5411871
Electronic Mail: agjansen@ilink.nis.za
WWW: http://www.nis.za/~agjansen/spreeu.htm
Founded: 1995
Staff: 1
Activities: visual, photographic and CCD observing
Coordinates: 022°11'45"E 32°56'44"S H720m
City Reference Coordinates: 022°02'00"E 32°13'00"S

Sutherland Observing Station
• See "South African Astronomical Observatory (SAAO), Sutherland Observing Station"

University of Cape Town (UCT), Department of Applied Mathematics
Private Bag
7700 Rondebosch
Telephone: (0)21-6502340
Telefax: (0)21-6502334
Electronic Mail: <userid>@maths.uct.ac.uk
WWW: http://vishnu.mth.uct.ac.za/cosmos/ (Relativity and Cosmology Group)
Founded: 1974
Staff: 4
Activities: theoretical research * cosmology * relativity
City Reference Coordinates: 018°22'00"E 33°55'00"S (Cape Town)

University of Cape Town (UCT), Department of Astronomy
Private Bag
7700 Rondebosch
Telephone: (0)21-6502391
Telefax: (0)21-6503342
Electronic Mail: astro@artemisia.ast.uct.ac.za
WWW: http://artemisia.ast.uct.ac.za/~astro/
Founded: 1970
Staff: 6
Activities: high-speed photometry * variable stars (cataclysmic, rapid) * large-scale structure
Periodicals: "Publications of the Department of Astronomy of the University of Cape Town"
Coordinates: 018°28'36"E 33°56'03"S
City Reference Coordinates: 018°22'00"E 33°55'00"S (Cape Town)

University of Cape Town (UCT), Department of Physics
Private Bag
7700 Rondebosch
Telephone: (0)21-6503331
(0)21-6503326
Telefax: (0)21-6503342
Electronic Mail: <userid>@physci.uct.ac.za
WWW: http://www.uct.ac.za/depts/physics/
Founded: 1893
City Reference Coordinates: 018°22'00"E 33°55'00"S (Cape Town)

University of South Africa (UNISA), Department of Mathematics, Applied Mathematics and Astronomy
P.O. Box 392
0003 UNISA
Telephone: (0)12-4296345
(0)12-4296202 (Secretary)
(0)12-4293381 (Observatory)
Telefax: (0)12-4296064
Electronic Mail: dps@astro.unisa.ac.za
WWW: http://www.unisa.ac.za/dept/wis/
http://astro.unisa.ac.za/~uniobs (Observatory)
Founded: 1962 (Department) (Observatory: 1992)
Staff: 2 (Astronomy)
Activities: education * mathematics and applied mathematics * dust in galaxies * OH masers
Coordinates: 028°11'58"E 25°46'06"S H1,448m
City Reference Coordinates: 028°10'00"E 25°45'00"S (Pretoria)

University of the Witwatersrand, Department of Computational and Applied Mathematics
P.O. Wits
2050 Johannesburg
Johannesburg
Telephone: (0)11-7163808
Telefax: (0)11-3397620
Electronic Mail: <userid>@gauss.cam.wits.ac.za
block@gauss.cam.wits.ac.za (David L. Block, Astronomy)
WWW: http://www.cam.wits.ac.za/
Founded: 1922 (University)
Staff: 24
City Reference Coordinates: 028°00'00"E 26°15'00"S

University of the Witwatersrand, Department of Physics
P.O. Wits
2050 Johannesburg
Johannesburg
Telephone: (0)11-7162439
Telefax: (0)11-3397965
(0)11-3398262
Electronic Mail: <userid>@physnet.phys.wits.ac.za
WWW: http://www.wits.ac.za/wits/fac/science/physics/physics.html
Founded: 1922 (University)
Staff: 50
City Reference Coordinates: 028°00'00"E 26°15'00"S

Spain

Agrupación Astrofotografica "Quarks"
Cuesta Santa Lucia 8
E-23400 Úbeda
Telephone: 953770409
Founded: 1990
Membership: 31
Activities: instrument making * observing * education * meetings
Periodicals: (4) "Fenix"
Coordinates: 002°04'00"W 38°02'00"N H668m
City Reference Coordinates: 003°22'00"W 38°01'00"N

Agrupación Astronómica Aragonesa (AAA)
Avenida Navarra 54
Aula 2
E-50010 Zaragoza
Electronic Mail: komet@arrakis.es
WWW: http://aaa.kipelhouse.com/
Founded: 1989
Membership: 150
Activities: education * popularization * observing
Periodicals: (6) "Boletín"
Coordinates: 000°52'48"W 41°39'24"N
City Reference Coordinates: 000°53'00"W 41°38'00"N

Agrupación Astronómica Cántabra (AAC)
Apartado 573
E-39080 Santander
Electronic Mail: agrupacion@astrocantabria.org
WWW: http://www.astrocantabria.org/
Founded: 1982
Membership: 50
Activities: variable stars * solar physics * occultations * planets
Periodicals: (6) "Estela"
City Reference Coordinates: 003°48'00"W 43°28'00"N

Agrupación Astronómica Complutense (AAC)
Apartado 199
E-28880 Alcalá de Henares
or :
Casa de la Juventud
Paseo del Val 2
E-28800 Alcalá de Henares
Telephone: 918896612
Electronic Mail: felixm@wanadoo.es (Félix L. Moreno, Secretary)
Membership: 150
Activities: popularization * talks * meetings
Periodicals: (4) "AAC Actividades"
Coordinates: 003°22'05"W 40°28'53"N
City Reference Coordinates: 003°22'00"W 40°28'00"N

Agrupación Astronómica de Barcelona (ASTER)
Passeig de Gràcia 71 (àtic)
E-08008 Barcelona
Telephone: 932151531
Electronic Mail: aster@encomix.es
WWW: http://www.encomix.es/~aster/
Founded: 1948
Membership: 200
Activities: observing (deep sky, Sun, Moon, meteorites, planets, variable stars) * photography
Sections: Astrophotography * Computing * Sun * Planets * Meteorites * Meteorology * Variable Stars * Library
Periodicals: (2) "ASTER" (ISSN 0212-6036); (4) "Ephemerides"
Coordinates: 000°02'10"E 41°23'35"N H69m
City Reference Coordinates: 002°11'00"E 41°23'00"N

Agrupación Astronómica de Cáceres
c/o Gabino Muriel Brillo
Apartado 153
E-10080 Cáceres
WWW: http://www.geocities.com/CapeCanaveral/Hall/5175/
http://www.arrakis.es/~aacc
City Reference Coordinates: 006°23'00"W 39°29'00"N

Agrupación Astronómica de Córdoba (AAC)
C/ Huerto de San Pedro El Real 1
Apartado 701
E-14080 Córdoba
Telephone: 957456077
Electronic Mail: aacordoba@astrored.i-p.com
WWW: http://www.astrored.net/AAC/
Founded: 1982
Membership: 80
Activities: observing * lectures * collaborative projects * library
Periodicals: (2) "Boletín Informativo de la AAC"
Observatories: 1 (Santa María de Trassierra)
City Reference Coordinates: 004°46'00"W 37°53'00"N

Agrupación Astronómica de Fuerteventura (AAF)
C/ Los Coroneles 16
E-35640 La Oliva
Fuerteventura
Islas Canarias
Electronic Mail: cvera@arrakis.es (Carlos Vera)
WWW: http://www.arrakis.es/~cvera/aaf
http://www.astrored.net/aaf
Founded: 1995
Membership: 12
Activities: observing * photography * popularization * education
Coordinates: 013°56'00"W 28°37'00"N H200m
City Reference Coordinates: 013°53'00"W 28°36'00"N

Agrupación Astronómica de Gran Canaria (AAGC)
C/ Juan E. Doreste 14
E-35001 Las Palmas de Gran Canaria
Islas Canarias
Telephone: 928310718
Telefax: 928310718
Electronic Mail: aagc@terra.es
WWW: http://aagc.dis.ulpgc.es/
Founded: 1990
Membership: 80
Activities: meetings * observing * public outreach
Periodicals: "SPACEX 90 - Nuevos Horizontes"
City Reference Coordinates: 015°24'00"W 28°06'00"N

Agrupación Astronómica de Huesca (AAH)
Apartado 61
E-22080 Huesca
or :
Costanilla de Oteiza 1
E-22000 Huesca
Telephone: 974223255
WWW: http://dftuz.unizar.es/aah/aah.html
http://dftuz.unizar.es/externo/aah/aah.html
Founded: 1994
Membership: 100
City Reference Coordinates: 000°25'00"W 42°08'00"N

Agrupación Astronómica de la Autónoma
Universidad Autónoma de Barcelona
Aula del Cosmos
Paseo Mas Roig 57 Bajos
E-08190 Sant Cugat del Vallès
WWW: http://tau.uab.es/~a3/
City Reference Coordinates: 002°05'00"E 41°28'00"N

Agrupación Astronómica de la Palma (AAP)
Apartado 449
E-38700 Santa Cruz de la Palma
Islas Canarias
Telephone: 922411974
607592175
Telefax: 922412203
Electronic Mail: jafami@santandersupernet.com (José Antonio Fernandez Arozena, Secretary)
WWW: http://www.ing.iac.es/AAP/AAP.html
Founded: 1986
Membership: 7
Activities: meteors * variable stars * photoelectric photometry
Periodicals: "Circulars"

000°10'12"E 37°13'27"N H2,200m (Calar Alto – IAU Code 493)
City Reference Coordinates: 003°41'00"W 37°13'00"N

Consejo Superior de Investigaciones Científicas (CSIC), Instituto de Física de Cantabria
• See "Instituto de Física de Cantabria"

Consejo Superior de Investigaciones Científicas (CSIC), Observatorio del Ebro
• See "Universidad Ramón Llull (URL), Observatorio del Ebro"

Equipo Sirius
C/ Rafael Finat 34
E-28044 Madrid
Telephone: 917104379
Telefax: 917054304
Electronic Mail: astronomia@wanadoo.es
WWW: http://www.laeff.esa.es/~tribuna/sirius/sirius.html
Founded: 1985
Staff: 10
Activities: publishing astronomy books and journals
Periodicals: (12) "Tribuna de Astronomía" (ISSN 0213-5892, circ.: 8,500)
City Reference Coordinates: 003°41'00"W 40°24'00"N

European Space Agency (ESA), Villafranca Satellite Tracking Station (VILSPA), Infrared Space Observatory (ISO)
Apartado 50727
E-28080 Madrid
Telephone: 918131100
Telefax: 918131172
Electronic Mail: <userid>@vilspa.esa.es
WWW: http://isowww.estec.esa.nl/
     http://www.iso.vilspa.esa.es/
     http://www.ipac.caltech.edu/iso (ISO US Support Center)
     http://www.esa.int/
Founded: 1977
Activities: operations of the "Infrared Space Observatory" * astrophysics
Periodicals: "ISO Info"
City Reference Coordinates: 003°41'00"W 40°24'00"N

Grup d'Estudis Astronòmics (GEA)
(Grupo de Estudios Astronómicos - GEA)
Apartado 9481
E-08080 Barcelona
Telephone: 935930225
Telefax: 935930331
     935702148
Electronic Mail: gea@astro.gea.cesca.es
          info@astro.gea.cesca.es
          gea@montsant.cesca.es
WWW: http://astro.gea.cesca.es/texte.html
Founded: 1984
Membership: 40
Activities: photoelectric photometry (solar system, variable stars) * photography
Periodicals: (2) "Circular GEA"
Coordinates: 001°46'34"E 41°32'24"N H375m (Hostalets)
     002°12'29"E 41°32'24"N H70m (Mollet)
     000°22'44"W 41°38'31"N H465m (Monegrillo)
     001°45'18"E 41°31'23"N H324m (Piera – IAU Code 165)
     001°49'19"E 42°22'12"N H1,070m (Sampsor)
     002°53'53"E 39°38'49"N H200m (Sencelles)
City Reference Coordinates: 002°11'00"E 41°23'00"N

Grupo Astronómico Silos (GAS)
Apartado 200
E-50080 Zaragoza
Electronic Mail: jltrisan@lander.es (José Luis Trisan, Public Relations & Treasurer)
WWW: http://es.geocities.com/gas_astronomia/gas.html
Founded: 1983
Membership: 40
Activities: Sun * Moon * Jupiter * occultations * Messier objects * planets * variable stars * photography * CCDs * instrumentation * meteorology
Periodicals: "Boletín Informativo del GAS" (ISSN 1135-223x), "Planetary Circular"
Coordinates: 000°52'42"W 41°39'28"N
City Reference Coordinates: 000°53'00"W 41°38'00"N

Grupo Canario de Estrellas Variables (GCEV)
Las Palmas de Gran Canaria
WWW: http://aagc.dis.ulpgc.es/gcev.html
Founded: 1994
City Reference Coordinates: 015°24'00"W 28°06'00"N

Grupo de Estudios Astronómicos (GEA)
• See "Grup d'Estudis Astronomics (GEA)"

Grupo de Estudios Astronómicos de Puertollano (GEAP)
Travesma Alta 13
E-13500 Puertollano
Telephone: 926411806
Electronic Mail: geap@teleline.es
WWW: http://teleline.terra.es/personal/jaduque/
Founded: 1998
City Reference Coordinates: 004°07'00"W 38°41'00"N

Grupo Europeo de Observaciones Estelares (GEOS)
c/o Luis Rivas
Calle Colón 9 - 12°
E-46016 Tabernes Blanques
Telephone: 961850948
WWW: http://www.upv.es/geos/
Founded: 1978
Membership: 12
Activities: photoelectric and visual observing (variable stars) * participating in European observational campaigns * computing periods and orbits * searching new variable stars * holding a yearly international congress, a photoelectric photometry school and Summer camp * international meetings
Periodicals: "GEOS Circulares", "Notas Circulares", "Fichas Técnicas"
Coordinates: 000°49'54"W 38°15'37"N H275m (Observatorio CEMA, Crevillente)
City Reference Coordinates: 000°16'00"W 39°04'00"N

Grupo Joven de Astronomía Scorpius Elche
C/ Capitan Gaspar Ortiz 106, 2°-2
E-03201 Elche
Telephone: 970340219
WWW: http://www.geocities.com/CapeCanaveral/Launchpad/1627/scorpius.htm
City Reference Coordinates: 000°41'00"W 38°16'00"N

Infrared Space Observatory (ISO)
• See "European Space Agency (ESA), Villafranca Satellite Tracking Station (VILSPA), Infrared Space Observatory (ISO)"

Instituto de Astrofísica de Andalucia (IAA)
• See "Consejo Superior de Investigaciones Científicas (CSIC), Instituto de Astrofísica de Andalucia (IAA)"

Instituto de Astrofísica de Canarias (IAC)
C/ Vía Láctea s/n
E-38200 La Laguna
Islas Canarias
Telephone: 922605200
    922329100 (Teide Observatory)
    922405500 (Roque de los Muchachos Observatory - see this entry)
Telefax: 922605201
    922329117 (Teide Observatory)
Electronic Mail: <userid>@iac.es
WWW: http://www.iac.es/home.html
Founded: 1975
Staff: 164
Activities: Spanish research organization managing two international observatories * structure of the universe * cosmology * structure and evolution of galaxies * star formation and evolution * planetary systems * Sun * ISM * instrumentation
Periodicals: "Preprint Series", "Publicaciones del IAC", "Noticias del IAC" (ISSN 0213-893x), "Memoria del IAC"
Coordinates: 016°30'25"W 28°17'56"N H2,398m (Teide – IAU Code 954)
    017°52'34"W 28°45'34"N H2,400m (Roque de los Much. - see this entry)
City Reference Coordinates: 016°19'00"W 28°29'00"N

Instituto de Astrofísica de Canarias (IAC), Observatorio del Roque de los Muchachos (ORM)
Apartado 303
E-38700 Santa Cruz de la Palma
Islas Canarias
Telephone: 922405500
Telefax: 922405501
WWW: http://www.iac.es/gabinete/orm/orm1.htm

Founded: 1975 (IAC)
Coordinates: 017°52'34"W 28°45'34"N H2,400m
City Reference Coordinates: 017°46'00"W 28°41'00"N

Instituto de Astronomía y Geodesia (IAG)
c/o Facultad de Ciencias Matemáticas
Cuidad Universitaria
E-28040 Madrid
Telephone: 913944585
Telefax: 913944607
Electronic Mail: vieira@iagmat1.mat.ucm.es (R.Vieira Díaz, Director)
WWW: http://www.mat.ucm.es/deptos/iag/
Founded: 1947
Staff: 20
Activities: Earth tides * geodynamics * gravimetry * Earth rotation * spatial geodesy * geoid
Periodicals: "Publicaciones" (ISSN 0213-6198)
• Jointly operated by the "Consejo Superior de Investigaciones Científicas (CSIC)" and the "Universidad Complutense"
City Reference Coordinates: 003°41'00"W 40°24'00"N

Instituto de Física de Cantabria
c/o Facultad de Ciencias
Avenida de los Castros s/n
E-39005 Santander
Telephone: 942201461
Telefax: 942201402
Electronic Mail: <userid>@astro.unican.es
WWW: http://www.ifca.unican.es/
Founded: 1995
Staff: 6
Activities: extragalactic astronomy * observational cosmology
• Jointly operated by the "Consejo Superior de Investigaciones Científicas (CSIC)" and the "Universidad de Cantabria"
City Reference Coordinates: 003°48'00"W 43°28'00"N

Instituto de Radioastronomía Milimétrica (IRAM)
Núcleo Central
Avenida Divina Pastora 7
E-18012 Granada
Telephone: 958228899
          958226696
          958482002 (30m telescope)
Telefax: 958222363
          958481148 (30m telescope)
Electronic Mail: <userid>@iram.es
WWW: http://www.iram.es/irame_home.html
        http://iram.fr/
Founded: 1979
Staff: 31
Activities: radio observing at mm-wavelengths
Coordinates: 003°23'35"W 37°03'59"N H2,916m (Observatorio Pico de Veleta)
City Reference Coordinates: 003°41'00"W 37°13'00"N

Instituto Nacional de Meteorología (INM)
Apartado 285
E-28040 Madrid
or :
Camino de las Morenas s/n
E-28040 Madrid
Telephone: 915819630
          915819810 (Meteorological information)
Telefax: 915819811
Electronic Mail: <userid>@inm.es
WWW: http://www.inm.es/
City Reference Coordinates: 003°41'00"W 40°24'00"N

Instituto Nacional de Técnica Aerospacial (INTA), División de Investigaciones Espaciales
Carretera de Ajalvir Km. 4
E-28850 Torrejón de Ardóz
Telephone: 916270999
Telefax: 916270945
Electronic Mail: <userid>@inta.es
WWW: http://www.inta.es/homepage.html
Founded: 1991
Staff: 55
Activities: space science research * astrophysics * fundamental physics * remote sensing * Earth atmosphere
City Reference Coordinates: 003°29'00"W 40°27'00"N

Instituto Nacional de Técnica Aeroespacial (INTA), Laboratorio de Astrofísica Espacial y Física Fundamental (LAEFF)
Apartado 50727
E-28080 Madrid
or :
c/o VILSPA
E-28691 - Villanueva de la Cañada
Telephone: 918131161
Telefax: 918131160
Electronic Mail: <userid>@laeff.esa.es
WWW: http://www.laeff.esa.es/
Founded: 1990
Staff: 12
Activities: space astrophysics * stellar evolution * binary stars * galactic clusters * high-energy astrophysics * UV astronomy * fundamental physics * cosmology * astroparticle physics * gravity * elementary particles * radio astronomy
City Reference Coordinates: 003°41'00"W 40°24'00"N (Madrid)

Instituto Politécnico Marítimo Pesquero de Pasajes, Planetario
Marinos 2
Apartado 5
E-20110 Pasajes
Telephone: 943399340
Telefax: 943404177
Founded: 1970
Staff: 1
Activities: maritime astronomy * education
City Reference Coordinates: 001°55'00"W 43°20'00"N

Isaac Newton Group (ING), Observatorio del Roque de los Muchachos (ORM)
Apartado 321
E-38700 Santa Cruz de la Palma
Islas Canarias
Telephone: 922425400
922405459
Telefax: 922425401
922405646
Electronic Mail: <userid>@ing.iac.es
WWW: http://www.ing.iac.es/
http://www.ast.cam.ac.uk/ING/
Founded: 1984
Staff: 60
Activities: operating the Isaac Newton Group of telescopes: 4.2m William Herschel Telescope (WHT), 2.5m Isaac Newton Telescope (INT) and 1.0m Jacobus Kapteyn Telescope (JKT)
Periodicals: "ING Newsletter"
Coordinates: 017°52'53"W 28°45'38"N H2,332m
City Reference Coordinates: 017°46'00"W 28°41'00"N

Istituto Nazionale di Astrofisica (INAF), Telescopio Nazionale Galileo (TNG)
Apartado 565
E-38700 Santa Cruz de la Palma
Islas Canarias
Telephone: 922425043
Telefax: 922420508
Electronic Mail: wwwstaff@tng.iac.es
WWW: http://www.tng.iac.es/
Founded: 1990
Coordinates: 017°52'35"W 28°45'34"N H2,373m
City Reference Coordinates: 017°46'00"W 28°41'00"N

Museo de la Ciencia y el Cosmos (MCC)
C/ Vía Láctea s/n
E-38200 La Laguna
Islas Canarias
Telephone: 922263454
922315080
922315265
Telefax: 922263295
Electronic Mail: museo@cosmos.mcc.rcanaria.es
WWW: http://www.mcc.rcanaria.es/
http://www.mcc.rcanaria.es/inform/inform.htm#Planetario
Founded: 1993
Staff: 15
Activities: popularization * education * shows * workshops
Coordinates: 016°18'00"W 28°30'00"N H700m
City Reference Coordinates: 016°19'00"W 28°29'00"N

Museu de la Ciència de la Fundació "La Caixa"
Teodor Roviralta 55
E-08022 Barcelona
Telephone: 932126050
Telefax: 932537457
Electronic Mail: musciencia.fundacio@lacaixa.es
　　　　　　　info.fundacio@lacaixa.es
WWW: http://www.fundacio.lacaixa.es/
Founded: 1981
Activities: science museum * planetarium
Coordinates: 002°11'00"E 41°23'00"N H170m
City Reference Coordinates: 002°11'00"E 41°23'00"N

Nordic Optical Telescope Scientific Association (NOTSA)
c/o Observatorio del Roque de los Muchachos
Apartado 474
E-38700 Santa Cruz de la Palma
Islas Canarias
Telephone: 922425473
Telefax: 922425475
Electronic Mail: tau@not.iac.es
　　　　　　　<userid>@not.iac.es
WWW: http://www.not.iac.es/
Founded: 1984
Staff: 15
Activities: operating the Nordic Optical Telescope (NOT)
Coordinates: 017°52'48"W 28°45'30"N H2,372m
City Reference Coordinates: 017°46'00"W 28°41'00"N

Nuevas Tecnologías Observacionales (NTO)
Candido Soto 12
E-28223 Pozuelo de Alarcón
WWW: http://www.nto.org/
City Reference Coordinates: 003°49'00"W 40°26'00"N

Observatorio Astronómico del Garraf (OAG)
(Observatori Astronòmic del Garraf - OAG)
C/ Tetuan 6
E-08800 Vilanova i la Geltrú
Telephone: 938143189
Electronic Mail: rho@arrakis.es
WWW: http://www.arrakis.es/~jaumeplan/
Founded: 1990
Staff: 5
Activities: double stars * education
Coordinates: 001°43'59"E 41°13'29"N
City Reference Coordinates: 001°43'00"E 41°13'00"N

Observatorio Astronómico de Mallorca (OAM)
Camino de Observatorio s/n
E-07144 Costitx
Telephone: 971876019
Telefax: 971892380
Electronic Mail: astroam@bitel.es
WWW: http://www.oam.es/
Founded: 1989
Membership: 20
Activities: planetary atmospheres * comets * minor planets * astrometry * photometry
Periodicals: (3) "Astrosplai i Natura"
Coordinates: 002°57'05"E 39°38'38"N (IAU Code 620)
City Reference Coordinates: 002°57'00"E 39°39'00"N

Observatorio Astronómico de Valencia
• See "Universidad de Valencia, Observatorio Astronómico"

Observatorio Astronómico Nacional (OAN), Spain
Campus Universitario de Alcalá de Henares
Apartado 1143
E-28800 Alcalá de Henares
Telephone: 918855060
　　　　　　918855061
Telefax: 918855062
Electronic Mail: <userid>@oan.es
WWW: http://www.oan.es/
Founded: 1790

Staff: 60
Activities: astronomy (observations, theory, technical developments) * time signal distribution * museum * popularization
Periodicals: (1) "Anuario del Observatorio Astronómico Nacional"
Coordinates: 003°41'17"W 40°24'30"N H670m
          003°05'21"W 40°31'24"N H930m (Yebes/CAY – IAU Code 491)
          002°32'54"W 37°13'27"N H2,165m (Obs. de Calar Alto – IAU Code 493)
City Reference Coordinates: 003°22'00"W 40°29'00"N

Observatorio Astronómico Nacional (OAN), Centro Astronómico de Yebes (CAY)
Apartado 148
E-19080 Guadalajara
or :
Cerro de la Palera s/n
E-19041 Yebes
Telephone: 949290311
Telefax: 949290063
Electronic Mail: <userid>@cay.oan.es
WWW: http://www.oan.es/cay/
City Reference Coordinates: 003°10'00"W 40°37'00"N (Guadalajara)
                                     003°07'00"W 40°31'00"N (Yebes)

Observatorio Astronómico Nacional (OAN), Observatorio de Calar Alto
Apartado 793
E-04080 Almería
or :
c/o Delegación del Instituto Geográfico Nacional
Edificio de Servicios Múltiples
C/ Hermanos Machado 4, 6° planta
E-04004 Almería
Telephone: 951225576 x256
               951234507
               950521034 (Dome)
               950632596 (Dome)
Telefax: 950632585 (Dome)
Electronic Mail: eoca@oan.es
                     <userid>@caserv.caha.es
WWW: http://www.caha.es/
Founded: 1790 (OAN)
Activities: spectroscopy * photometry
Coordinates: 002°32'54"W 37°13'27"N H2,160m (IAU Code 493)
City Reference Coordinates: 002°27'00"W 36°50'00"N (Almería)

Observatorio Astronómico Ramón María Aller
● See "Universidad de Santiago de Compostela, Observatorio Astronómico Ramón María Aller"

Observatorio de Calar Alto
● See "Centro Astronómico Hispano-Alemán, Observatorio de Calar Alto"
● See also "Observatorio Astronómico Nacional (OAN), Observatorio de Calar Alto"

Observatorio del Ebro
● See Universidad Ramón Llull (URL), Observatorio del Ebro"

Observatorio del Roque de los Muchachos (ORM)
● See "Instituto de Astrofísica de Canarias (IAC), Observatorio del Roque de los Muchachos (ORM)"
● See also "Isaac Newton Group (ING), Observatorio del Roque de los Muchachos (ORM)"

Parque de las Ciencias
Avenida del Mediterraneo s/n
E-18006 Granada
Telephone: 958131900
Telefax: 958 133582
Electronic Mail: cpciencias@parqueciencias.com
WWW: http://www.parqueciencias.com/
       http://www.parqueciencias.com/planet.html (Planetarium)
City Reference Coordinates: 003°41'00"W 37°13'00"N

Planetario de Madrid
Avenida Planetario 16
E-28045 Madrid
Telephone: 914673461
               914673578
               914673898 (answering machine)
Telefax: 914681154
Electronic Mail: buzon@planetmad.es
WWW: http://www.planetmad.es/

Founded: 1986
Staff: 19
Activities: shows * scientific exhibitions
City Reference Coordinates: 003°41'00"W 40°24'00"N

Planetario de Pamplona
Sancho Ramirez s/n
E-31008 Pamplona
Telephone: 948260004
           948260056
Telefax: 948261919
Electronic Mail: armentia_javier@euskom.spritel.es
                 planetario@cin.es
WWW: http://www.ucm.es/OTROS/Astrof/pamplona/pp-casa.html
     http://www.ucm.es/info/Astrof/pamplona/pp-casa.html
Founded: 1993
Staff: 13
Activities: education * popularization * science centre * museum * exhibitions * lectures * conferences
City Reference Coordinates: 001°38'00"W 42°49'00"N

Real Instituto y Observatorio de la Armada (ROA)
Cecilio Pujazon s/n
E-11110 San Fernando
Telephone: 956599365
Telefax: 956599366
Electronic Mail: astro@roa.es
Founded: 1753
Staff: 17
Activities: ephemerides * astrometry * time and frequency * magnetism * sismology * artificial satellites * lasers * GPS
Periodicals: "Almanaque Náutico" (ISSN 0210-735x), "Efemérides Astronómicas" (ISSN 0080-5971), "Fenómenos Astronómicos" (ISSN 0210-8127), "Anales ROA"
Coordinates: 006°12'19"W 36°27'42"N H29m (IAU Code 983)
City Reference Coordinates: 006°12'00"W 36°28'00"N

Real Sociedad Española de Física (RSEF)
c/o Facultad de Ciencias Físicas
Universidad Complutense
Ciudad Universitaria s/n
E-28040 Madrid
Telephone: 913944359
Telefax: 915433879
Electronic Mail: rsef@fis.ucm.es
WWW: http://www.ucm.es/info/rsef
Founded: 1903
Membership: 1400
Periodicals: (5) "Revista Española de Física" (ISSN 0213-862x); "Anales de Física"
Awards: (1) "Medalla de la Real Sociedad Española de Física"; (3) "Premio a Investigadores Noveles de la Real Sociedad Española de Física"
City Reference Coordinates: 003°41'00"W 40°24'00"N

Royal Swedish Academy of Sciences, Research Station for Astrophysics
Grupo Sueco
Apartado 66
E-38700 Santa Cruz de la Palma
Islas Canarias
Telephone: 922405590
Telefax: 922405592
Electronic Mail: paco@not.iac.es (Paco Armas, Administrator)
WWW: http://www.astro.su.se/groups/solar/solar.html
Founded: 1979
Activities: high spatial resolution studies of the dynamic evolution of magnetic fields and velocity fields on the Sun
Coordinates: 017°52'48"W 28°45'30"N H2,372m
City Reference Coordinates: 017°46'00"W 28°41'00"N

Sociedad Astronómica de España y América (SADEYA)
Avenida Diagonal 377, 2°
E-08008 Barcelona
Telephone: 934160478
Electronic Mail: sadeya@sadeya.cesca.es
WWW: http://www.sadeya.cesca.es/
Founded: 1911
Membership: 750
Activities: popularizing astronomy and related sciences * lectures * education
Sections: comets * computing * observatory * optics * radioastronomy

Periodicals: (6) "Astronomía, Astrofotografía y Astronautica" (ISSN 0213-621x), "SADEYA"; (1) "Almanaque Astronómico"; "Circulars"
Coordinates: 000°56'54"E 42°22'00"N H1,400m (Antist)
City Reference Coordinates: 002°11'00"E 41°23'00"N

Sociedad Astronómica Granadina (SAG)
Apartado 195
E-18080 Granada
or :
Avenida Barcelona 20-10 C
E-18006 Granada
Telephone: 958136363
Electronic Mail: sag@goliat.ugr.es
WWW: http://www.ugr.es/~sag/frames.htm
Founded: 1981
Membership: 340
Activities: popularization * observing * education
Periodicals: (2) "Halley"
Coordinates: 003°35'00"W 37°10'00"N H680m (Cubillas)
City Reference Coordinates: 003°41'00"W 37°13'00"N

Sociedad Astronómica SYRMA
Apartado 5380
E-47080 Valladolid
Telephone: 983339210
Electronic Mail: gua@uva.es
WWW: http://www.gvi.uva.es/~gua
Founded: 1981
Membership: 83
Activities: popularization * education * observing * library * conferences
Periodicals: "Astrea"
Coordinates: 004°21'00"W 41°39'00"N H885m (Cogeces del Monte)
City Reference Coordinates: 004°43'00"W 41°39'00"N

Sociedad de Astronomía Balear (SAB)
Apartado 1206
E-07080 Palma de Mallorca
Telephone: 971284390
Founded: 1979
Membership: 100
Activities: popularization * observing
Periodicals: "Circular"
City Reference Coordinates: 002°39'00"E 39°34'00"N

Sociedad de Ciencias Aranzadi, Sección Astronomía
Alto Zorroaga s/n
E-20014 San Sebastian
Telephone: 943466142
Telefax: 943455811
Founded: 1975
Membership: 120
Activities: observing * photography * education
Periodicals: (3) "Boletín de Astronomía" (ISSN 1132-2306)
Coordinates: 001°59'03"W 43°17'46"N H105m (Torres Arbide)
City Reference Coordinates: 001°59'00"W 43°19'00"N

Sociedad de Observadores de Meteoros y Cometas de España (SOMYCE)
c/o Manuel Solano Ruiz (Secretary)
C/ Callao de Lima 1-1°
E-38003 Santa Cruz de Tenerife
Islas Canarias
Electronic Mail: orbesa@teleline.es
     msolano@teleline.es
WWW: http://www.astrored.net/somyce/
Founded: 1988
Membership: 47
Activities: observing (comets, meteors, minor planets) * analyzing meteor, comet and minor-planet data * orbit computations * visual, radio and photographic work
Periodicals: (6) "Meteors" (ISSN 1137-9111, circ.: 75)
City Reference Coordinates: 016°14'00"W 28°27'00"N

Sociedad Einstein de Astronomía (SEDA)
Oteros 8
E-23680 Alcalá la Real
Telephone: 953581535

Electronic Mail: barsoom@arrakis.es
Founded: 1983
Membership: 10
Activities: popularization * instrument making * observing
City Reference Coordinates: 003°55'00"W 37°28'00"N

Sociedad Española de Astronomía (SEA)
c/o Secretaría
Departamento de Astronomía y Meteorología
Facultad de Física
Universidad de Barcelona
C/ Martí i Franquès 1
E-08028 Barcelona
Telephone: 934021125
Telefax: 934021133
Electronic Mail: secretaria@sea.am.ub.es
WWW: http://sea.am.ub.es/
Founded: 1993
Membership: 385
Activities: contributing to the development of astronomy in Spain * disseminating information of interest to astronomers
Periodicals: (2) "Boletín SEA" (ISSN 1575-3476)
City Reference Coordinates: 002°11'00"E 41°23'00"N

Sociedad Española de Optica (SEDO)
C/ Serrano 121
E-28006 Madrid
Telephone: 915616070
Telefax: 915645557
Electronic Mail: sedo@fresno.csic.es
WWW: http://sedo.optica.csic.es/
Founded: 1969
Membership: 320
Activities: promoting optical research and development in Spain
Periodicals: "Optica Pura y Aplicada" (ISSN 0030-3917)
City Reference Coordinates: 003°41'00"W 40°24'00"N

Sociedad Malagueña de Astronomía (SMA)
Apartado de Correos 6072
E-29080 Málaga
or :
Centro Cultural José María Gutiérrez Romero
C/ República Argentina 9
Urbanización El Limonar
E-29016 Málaga
Electronic Mail: sma@laeff.esa.es
WWW: http://www.laeff.esa.es/~sma/
City Reference Coordinates: 004°25'00"W 36°43'00"N

Telescopio Nazionale Galileo (TNG)
● See "Istituto Nazionale di Astrofisica (INAF), Istituto Nazionale di Astrofisica (INAF)"

Universidad Autónoma de Madrid (UAM), Grupo de Astrofísica
Departamento de Física Teórica
Modulo C-XI
Campus de Cantoblanco
E-28049 Madrid
Telephone: 913974880 (Office)
           913978605 (Observatory)
Telefax: 913973936
Electronic Mail: <userid>@uam.es
WWW: http://pollux.ft.uam.es/astro/
Founded: 1987
Staff: 15
Activities: active galaxies * chemical evolution * extragalactic astrophysics * HII regions * Galaxy formation * numerical cosmology * star formation
City Reference Coordinates: 003°41'00"W 40°24'00"N

Universidad Complutense, Departamento de Astrofísica
c/o Facultad de Físicas
Ciudad Universitaria
E-28040 Madrid
Telephone: 913944577
           913944592
Telefax: 913945195
Electronic Mail: <userid>@astrax.fis.ucm.es

WWW: http://www.ucm.es/info/Astrof/
Staff: 9
Activities: stellar activity (RS CVn stars) * stellar outer atmosphere * eclipsing binaries * stellar photometry * Ba stars * compact galaxies * abundance and stellar content of galaxies
City Reference Coordinates: 003°41'00"W 40°24'00"N

Universidad Complutense, Instituto de Astronomía y Geodesia (IAG)
• See "Instituto de Astronomía y Geodesia (IAG)"

Universidad de Barcelona, Departamento de Astronomía y Meteorología
Avenida Diagonal 647
E-08028 Barcelona
Telephone: 934021125
Telefax: 934021133
Electronic Mail: <userid>@mizar.am.ub.es
<userid>@alcor.am.ub.es
secre@mizar.am.ub.es
WWW: http://www.am.ub.es/
http://mizar.am.ub.es/
Founded: 1970
Staff: 20
Activities: stellar kinematics and dynamics * SN * gravitational lenses * stellar formation * molecular clouds * HII regions * interplanetary particle events * galaxy clusters * Earth rotation * large-scale structures * nucleosynthesis * structure and evolution of galaxies * stellar clusters
City Reference Coordinates: 002°11'00"E 41°23'00"N

Universidad de Cantabria, Instituto de Física de Cantabria
• See "Instituto de Física de Cantabria"

Universidad de Granada, Departamento de Física Teórica y del Cosmos
c/o Facultad de Ciencias
Avenida Fuentenueva s/n
E-18002 Granada
Telephone: 958243305
Telefax: 958248529
Electronic Mail: <userid>@ugr.es
WWW: http://www.ugr.es/~fteorica/astro.html
Founded: 1982
Staff: 5
Activities: MHD in spiral galaxies * magnetic fields in intergalactic space * education * carbon stars * SN
City Reference Coordinates: 003°41'00"W 37°13'00"N

Universidad de las Islas Baleares (UIB), Departamento de Física
Edificio Mateu Orfila
E-07071 Palma de Mallorca
Telephone: 971173228
Telefax: 971173426
Electronic Mail: <userid>@eps.uib.es
WWW: http://www.uib.es/depart/dfs/
Founded: 1973
Staff: 25
Activities: education * research * solar physics * relativity * material sciences * condensed matter * nuclear physics * geophysical fluid dynamics
City Reference Coordinates: 002°39'00"E 39°34'00"N

Universidad de Santiago de Compostela, Observatorio Astronómico Ramón María Aller
Avenida de las Ciencias s/n
Apartado 197
E-15706 Santiago de Compostela
Telephone: 981592747
Telefax: 981597054
Electronic Mail: <userid>@usc.es
WWW: http://www.usc.es/astro/
Founded: 1943
Staff: 3
Activities: celestial mechanics * double stars
Periodicals: (4) "Publicaciones del Observatorio Astronómico Ramón María Aller" (ISSN 1130-0892)
Coordinates: 008°33'33"W 42°52'32"N H240m
City Reference Coordinates: 008°33'00"W 42°53'00"N

Universidad de Valencia, Departamento de Astronomía y Astrofísica
Dr. Moliner 50
E-46100 Burjassot
Electronic Mail: <userid>@vlbi.matapl.uv.es
WWW: http://vlbi.uv.es/

Founded: 1995
City Reference Coordinates: 000°22'00"W 39°28'00"N

Universidad de Valencia, Faculdad de Matemáticas, Grupo de Astronomía y Ciencias del Espacio (GACE)
E-46100 Burjassot
Telephone: 963864573
Telefax: 963864364
Electronic Mail: <userid>@pollux.uv.es
WWW: http://pollux.uv.es/
Founded: 1990
City Reference Coordinates: 000°22'00"W 39°28'00"N

Universidad de Valencia, Observatorio Astronómico
Avenida Blasco Ibañez 13
E-46010 Valencia
Telephone: 963864773
Telefax: 963864773
Electronic Mail: alvaro.lopez@uv.es
WWW: http://www.uv.es/obsast/
Founded: 1909
Staff: 4
Activities: astrometry of minor planets * PZT * celestial mechanics
Coordinates: 000°22'00"W 39°28'41"N H30m
City Reference Coordinates: 000°22'00"W 39°28'00"N

Universidad de Zaragoza, Grupo de Mecanica Espacial
Edificio Matemáticas
E-50009 Zaragoza
Telephone: 976761000
Telefax: 976761140
Electronic Mail: <userid>@posta.unizar.es
WWW: http://gme.unizar.es/
Founded: 1986
Staff: 8
Activities: celestial mechanics * binary stars * artificial-satellite theory * algebra computation
City Reference Coordinates: 000°53'00"W 41°38'00"N

Universidad Politécnica de Cataluña, Departamento de Física Aplicada
C/ Jordi Girona Salgado 1-3
Módulo B5
Campus Nord
E-08034 Barcelona
Telephone: 934016802
Telefax: 934016090
Electronic Mail: <userid>@etseccpb.upc.es
WWW: http://www-fa.upc.es/
Activities: white dwarfs * cooling * stars * evolution * SN * convection
City Reference Coordinates: 002°11'00"E 41°23'00"N

Universidad Ramón Llull (URL), Observatorio del Ebro
E-43520 Roquetes
Telephone: 977500511
Telefax: 977504660
WWW: http://www.readysoft.es/observebre/
Founded: 1904
Staff: 18
Activities: STP * solar physics * geomagnetism * ionosphere * meteorology * seismology
Periodicals: "Boletín del Observatorio del Ebro - Ionosfera" (ISSN 0211-5166), "Publicaciones del Observatorio del Ebro - Miscelánea" (ISSN 0211-4534), "Publicaciones del Observatorio del Ebro - Memoria"
Coordinates: 000°29'30"E 40°49'12"N H50m
• Associated with the "Consejo Superior de Investigaciones Científicas (CSIC)"
City Reference Coordinates: 000°30'00"E 40°50'00"N

Villafranca Satellite Tracking Station (VILSPA)
• See "European Space Agency (ESA), Villafranca Satellite Tracking Station (VILSPA)"

VILSPA
• See "European Space Agency (ESA), Villafranca Satellite Tracking Station (VILSPA)"

Sri Lanka

Department of Meteorology, Sri Lanka
383 Bauddhaloka Mawatha
Colombo 7
Telephone: (0)1-694846
Telefax: (0)1-691443
WWW: http://www.meteo.slt.lk/
Founded: 1907
Activities: weather forecast * limited astronomical service * seismological service * supplying meteorological and climatological data
Periodicals: (1) "Annual Report"
Coordinates: 079°52'00"E 06°54'00"N H7m
City Reference Coordinates: 079°52'00"E 06°54'00"N

Institute of Fundamental Studies (IFS)
Hantana Road
Kandy
Telephone: (0)8-232002
Telefax: (0)8-232131
Electronic Mail: ifs@ifs.ac.lk
Founded: 1981
Staff: 150
Activities: fundamental research * science dissemination
Periodicals: (12) "Pragna"
City Reference Coordinates: 080°38'00"E 07°18'00"N

National Science Foundation (NSF)
47/5 Maitland Place
Colombo 7
Telephone: (0)1-696771
(0)1-696772
(0)1-696773
Telefax: (0)1-691691
Electronic Mail: info@nsf.ac.lk
WWW: http://www.nsf.ac.lk/
Founded: 1968
Staff: 98
Activities: research funding * S&T information * S&T policies
Periodicals: (2) "Journal of the National Science Foundation of Sri Lanka" (ISSN 0300-9254), "Sri Lanka Journal of Social Science" (ISSN 0258-9710); (1) "Annual Report"
Awards: (1/2) "Merit Awards for Scientific Research"
City Reference Coordinates: 079°51'00"E 06°56'00"N

Sri Lanka Standards Institution (SLSI)
17 Victoria Place
Elvitigala Mawatha
Colombo 08
Telephone: (0)1-671567
(0)1-671572
Telefax: (0)1-671579
Electronic Mail: slsi@slsi.slt.lk
WWW: http://www.naresa.ac.lk/slsi/
City Reference Coordinates: 079°51'00"E 06°56'00"N

Sweden

Astromedia AB
Rodingsgatan 2A
SE-426 58 Vastra Frolunda
Telephone: (0)31-694500
Telefax: (0)31-694508
Electronic Mail: info@astromedia.se
WWW: http://www.astromedia.se/
Founded: 1988
Activities: manufacturing and distributing telescopes and accessories, books, and star catalogs * exclusive dealer for Celestron, Vixen, and Tele Vue in Sweden, and Vixen in Finland
Periodicals: (12) "TeleScoop"
City Reference Coordinates: 012°00'00"E 57°45'00"N (Göteborg)

Astronomiska Sällskapet Tycho Brahe (ASTB)
• "Tycho Brahe Astronomical Society"

Broman Planetarium AB
Kryddpepparg 63
SE-424 53 Angered
Telephone: (0)313302837
Telefax: (0)313302835
Electronic Mail: pbr@planetarium.euromail.se (Per Broman)
          lbr@planetarium.euromail.se (Lars Broman)
WWW: http://www.planetarium.se/
Founded: 1985 (Incorporated: 1990)
City Reference Coordinates: 012°00'00"E 57°45'00"N (Göteborg)

Chalmers Technical University
• See "Göteborg University, Chalmers Technical University"

CosmoNorr
c/o Christer Strand
Poppelvägen 11
SE-832 54 Frösön
Telephone: (0)63-43785
Telefax: (0)63-105181
Electronic Mail: christer.strandh@micro.se
WWW: http://www.jamtnet.se/cosmonorr/
Coordinates: 014°30'00"E 63°06'00"N (Valla Observatory)
City Reference Coordinates: 014°32'00"E 63°11'00"N

Cosmonova
c/o Naturhistoriska Riksmuseet
Frescativägen 40
Box 50007
SE-104 05 Stockholm
Telephone: (0)8-51955101
Telefax: (0)8-51955100
Electronic Mail: tom.callen@nrm.se (Tom Callen)
WWW: http://www.nrm.se/cosmonova/
Founded: 1992
Staff: 11
City Reference Coordinates: 018°03'00"E 59°20'00"N

Erna and Victor Hasselblad Foundation
Ekmansgatan 8
SE-412 56 Göteborg
Telephone: (0)31-7781990
Telefax: (0)31-7784640
Electronic Mail: info@hasselbladfoundation.o.se
WWW: http://www2.hasselbladfoundation.o.se/
Founded: 1979
Staff: 2
Activities: promoting research and academic teaching in the natural sciences and photography
Awards: (1) "Hasselblad International Award in Photography"
City Reference Coordinates: 012°00'00"E 57°45'00"N

Esrange
• See "Swedish Space Corp. (SSC), Esrange"

European Incoherent Scatter Facility (EISCAT), Headquarters
Box 812
SE-981 28 Kiruna
Telephone: (0)980-79153
Telefax: (0)980-79161
Electronic Mail: eiscat@eiscathq.irf.se
             <userid>@eiscathq.irf.se
WWW: http://snowflake.irf.se/
     http://www.eiscat.uit.no/eiscat.html (EISCAT Scientific Association)
Founded: 1975
Staff: 7 (all sites: 35)
Activities: incoherent scatter radar system site for atmospheric, ionospheric and magnetospheric research
Periodicals: "Annual Report" (ISSN 0349-2710), "Technical Reports"
Coordinates: 020°26'00"E 67°52'00"N H412m (Kiruna)
         019°14'00"E 69°35'00"N H30m (Tromsø, Norway - see separate entry)
         026°38'00"E 67°22'00"N H198m (Sodankylä, Finland - see separate entry)
         016°03'00"E 78°09'00"N (Longyearbyen, Norway - see separate entry)
City Reference Coordinates: 020°27'00"E 67°52'00"N

European Space Agency (ESA), Satellite Station, Kiruna
Box 815
SE-981 28 Kiruna
Telephone: (0)980-76000
Telefax: (0)980-17121
WWW: http://www.esrin.esa.it/
     http://www.esa.int/
Founded: 1962 (ESA)
City Reference Coordinates: 020°27'00"E 67°52'00"N

Föreningen för Astronomi och Astronautik (FAA)
• See "Society of Astronomy and Astronautics"

Framtidsmuseet
• See "Future's Museum"

Future's Museum
(Framtidsmuseet)
Jussi Björlingsväg 25
SE-784 32 Borlänge
Telephone: (0)243-793000
       (0)243-793900
Telefax: (0)243-226977
Electronic Mail: tnc@framtidsmuseet.se
WWW: http://www.framtidsmuseet.se/
Founded: 1986
Staff: 4
Activities: lectures * Kosmorama Rymdteater * multimedia shows
City Reference Coordinates: 015°25'00"E 60°29'00"N

Gislaved Astronomiska Sällskap (GAS) Orion
c/o Bo Ekström
Högåsstigen 9
SE-332 33 Gislaved
Telephone: (0)371-14213
Electronic Mail: bo.ekstrom@ebox.tninet.se
Founded: 1983
Membership: 40
Activities: lectures * observing
City Reference Coordinates: 013°30'00"E 57°19'00"N

Göteborg Astronomical Club
(Göteborgs Astronomiska Klubb - GAK)
c/o Naturhistoriska Museet
Box 7283
SE-402 35 Göteborg
Telephone: (0)31-144250
WWW: http://www.tripnet.se/gak/
Founded: 1955
Membership: 125
Activities: observing * lectures
Periodicals: (3) "Aurora" (ISSN 1101-1718, circ.: 200)
Coordinates: 012°06'34"E 57°38'22"N H110m (Lahall Observatory)
City Reference Coordinates: 012°00'00"E 57°45'00"N

Göteborgs Astronomiska Klubb (GAK)
- See "Göteborg Astronomical Club"

Göteborg University, Chalmers Technical University, Department of Astronomy and Astrophysics
SE-412 96 Göteborg
Telephone: (0)31-7723135
Telefax: (0)31-7723204
Electronic Mail: <userid>@fy.chalmers.se
WWW: http://fy.chalmers.se/~marek/Department/astro.html
http://www.chalmers.se/
Founded: 1985
Activities: interstellar abundances * red giant stars * molecular clouds * star formation processes * galaxy dynamics * accretion disks
City Reference Coordinates: 012°00'00"E 57°45'00"N

Göteborg University, Chalmers Technical University, Onsala Space Observatory (OSO)
SE-439 92 Onsala
Telephone: (0)31-7725500
Telefax: (0)31-7725590
Electronic Mail: <userid>@oso.chalmers.se
WWW: http://www.oso.chalmers.se/
Founded: 1949
Staff: 50
Activities: radioastronomy * mm-wave spectroscopy * ISM * circumstellar shells * galaxies * VLBI * QSOs * plate tectonics * GPS * aeronomy * geodesy
Coordinates: 011°53'35"E 57°23'45"N H59m
070°43'56"W 29°15'48"S H2,410m (La Silla, Chile)
City Reference Coordinates: 012°00'00"E 57°25'00"N

Halmstad Astronomical Association
(Halmstads Astronomiska Sällskap - HAS)
c/o T. Nilsson
Norra Vägen 9B
SE-302 31 Halmstad
Founded: 1958
Membership: 60
Activities: observing * education
City Reference Coordinates: 012°55'00"E 56°41'00"N

Halmstads Astronomiska Sällskap (HAS)
- See "Halmstad Astronomical Association"

Hasselblad Foundation (Erna and Victor _)
- See "Erna and Victor Hasselblad Foundation"

House of Technology
- See "Teknikens Hus"

Institutet för Rymdfysik (IRF)
- See "Swedish Institute of Space Physics"

Karlskrona Astronomi Förening (KAF)
- See "Karlskrona Astronomy Association"

Karlskrona Astronomy Association
(Karlskrona Astronomi Förening - KAF)
c/o Bernth Svensson
N. Möllebacksgränd 2
SE-371 34 Karlskrona
Telephone: (0)455-17340
Telefax: (0)455-54444
Founded: 1975
Membership: 20
Activities: observing
Periodicals: (6) "KAF-NYTT"
City Reference Coordinates: 015°35'00"E 56°10'00"N

Kvistaberg Station
- See "Uppsala University, Astronomical Observatory, Kvistaberg Station"

Lund Observatory
- See "Lund University, Lund Observatory"

Lund Planetarium
• See "Lund University, Lund Observatory, Planetarium"

Lund University, Department of Physics
Box 118
SE-221 00 Lund
or :
Professorsgatan 1
SE-223 62 Lund
Telephone: (0)46-126097
Telefax: (0)46-2224709
Electronic Mail: <userid>@fysik.lu.se
WWW: http://ferrum.fysik.lu.se/
Founded: 1950
Activities: basic and applied research in atomic spectroscopy * laboratory astrophysics * fusion research * stellar spectroscopy
City Reference Coordinates: 013°11'00"E 55°42'00"N

Lund University, Lund Observatory
Svanegatan 9
Box 43
SE-221 00 Lund
Telephone: (0)46-2227300
Telefax: (0)46-2224614
Electronic Mail: <userid>@astro.lu.se
WWW: http://www.astro.lu.se/
Founded: 1672
Staff: 30
Activities: solar physics * meteors * stellar physics * stellar evolution * magnetic phenomena * rapid and high-energy processes * star formation * double and multiple stars * stellar clusters * stellar dynamics * galactic structure * ISM * galactic evolution * astrometry * telescopes * ancillary instrumentation * planetary nebulae * radial velocities
Periodicals: "Lund Observatory Reports" (ISSN 0349-4217)
Coordinates: 013°11'12"E 55°41'54"N H34m (IAU Code 039)
　　　　　　　013°26'00"E 55°37'24"N H145m (Jävan Station)
City Reference Coordinates: 013°11'00"E 55°42'00"N

Lund University, Lund Observatory, Planetarium
Svanegatan 9
Box 43
SE-221 00 Lund
Telephone: (0)46-2227302
　　　　　　(0)46-2220153 (Eva Mezey, Planetarium Director)
Telefax: (0)46-2224614
Electronic Mail: planetariet@astro.lu.se
WWW: http://www.astro.lu.se/~planetariet/
Founded: 1978
Staff: 1
Activities: public shows * education * exhibitions * information service
City Reference Coordinates: 013°11'00"E 55°42'00"N

Malmö Astronomi & Rymfarts Sällskap (MARS)
• "Malmö Astronomy and Space Society"

Malmö Astronomy and Space Society
(Malmö Astronomi & Rymfarts Sällskap - MARS)
Box 5017
SE-200 71 Malmö
Telephone: (0)40-547012
WWW: http://www.ludat.lth.se/~dat93jso/mars.html
　　　　　http://www.mars.m.se/default.htm
　　　　　http://www.mars.m.se/mars.htm
Founded: 1962
Membership: 69
Activities: observing * photography * lectures * travels * deep sky
Periodicals: (4) "MARS-Bulletinen" (ISSN 0284-6667)
Observatories: 1 (Tycho Brahe Observatory, Malmö)
City Reference Coordinates: 013°00'00"E 55°36'00"N

Mariestads Astronomiska Klubb (MAK)
c/o Rune Fogelquist (President)
Borgmästaregatan 7
SE-542 33 Mariestad
Telephone: (0)501-18167
Electronic Mail: makastro@algonet.se
WWW: http://www.algonet.se/~makastro
Founded: 1978

Drottninggade 120
SE-113 60 Stockholm
WWW: http://www.astro.su.se/STAR/
Founded: 1988
Membership: 250
City Reference Coordinates: 018°03'00"E 59°20'00"N

Stockholm Observatory
- See "Stockholm University, Stockholm Observatory"

Stockholms Amatörastronomer (STAR)
- See "Stockholm Amateur Astronomers"

Stockholm University, Stockholm Observatory
SCFAB
Institutionen för astronomi
SE-106 91 Stockholm
or :
Roslagstullsbacken 21
SE-106 91 Stockholm
Telephone: (0)8-55378500
Telefax: (0)8-55378510
Electronic Mail: <userid>@astro.su.se
WWW: http://www.astro.su.se/home.html
Founded: 1753
Staff: 40
Activities: stellar structure and evolution * ISM and star formation * radiative transfers * galactic structure and dynamics * AGNs * IR astronomy
Coordinates: 018°18'30"E 59°16'18"N H60m (IAU Code 052)
- Use Roslagsvägen 30B for deliveries
City Reference Coordinates: 018°03'00"E 59°20'00"N

Sundsvalls Astronomiska Förening
c/o Johannes Nordvall
Juniskärsvägen 570
SE-862 91 Kvissleby
Telephone: (0)60-562351
Electronic Mail: johannes.nordvall@mailbox.swipnet.se
Founded: 1986
Coordinates: 017°25'24"E 62°17'47"N H35m
City Reference Coordinates: 017°22'00"E 62°22'00"N (Sundsvall)

Svensk Amatör Astronomisk Förening (SAAF)
- See "Swedish Amateur Astronomical Society"

Svenska Astronomiska Sällskapet (SAS)
- See "Swedish Astronomical Society"

Swedish Amateur Astronomical Society
(Svensk Amatör Astronomisk Förening - SAAF)
c/o Jan Persson
Eklanda Hage 31
SE-431 49 Mölndal
Telephone: (0)31-277820
Electronic Mail: jan.persson@space.se
WWW: http://www.saaf.se/
Founded: 1972
Membership: 760
Activities: photography * comets * Sun * meteors * deep sky * optics * telescope making * variable stars
Sections: Deep Sky * Optics and Telescope Manufacturing * Planets * Solar System * Variable Stars
Periodicals: (4) "Astro" (ISSN 0280-7173)
City Reference Coordinates: 012°01'00"E 57°39'00"N

Swedish Astronomical Society
(Svenska Astronomiska Sällskapet - SAS)
c/o Stockholm Observatory
SE-106 91 Saltsjöbaden
Telephone: (0)8-164458
         (0)8-164477
Telefax: (0)8-7174719
Electronic Mail: sas@astro.su.se
                dan@astro.su.se
WWW: http://www.astro.su.se/sas/sas.html
Founded: 1919
Membership: 1300

Activities: lectures * colloquia
Periodicals: (4) "Astronomisk Tidskrift" (ISSN 0004-6345, circ.: 1,700)
City Reference Coordinates: 018°18'00"E 59°17'00"N

Swedish Institute of Space Physics
(Institutet för Rymdfysik - IRF)
Box 812
SE-981 28 Kiruna
Telephone: (0)980-79000
Telefax: (0)980-79091
Electronic Mail: irf@irf.se
WWW: http://www.irf.se/
Founded: 1957
Staff: 100
Activities: research and education in space plasma physics * space technology * atmospheric physics
Periodicals: (1) "Annual Report" (ISSN 0284-169x); "Kiruna Geophysical Data" (ISSN 0453-9478), "Ionospheric Data Sweden", "Scientific Reports"
Coordinates: 020°24'00"E 67°50'00"N H420m (Kiruna)
 018°48'00"E 64°42'00"N (Lycksele)
 017°36'00"E 59°48'00"N (Uppsala)
City Reference Coordinates: 020°24'00"E 67°50'00"N

Swedish Institute of Space Physics, Kiruna Division
(Institutet för Rymdfysik - IRF, Kiruna)
Box 812
SE-981 28 Kiruna
Telephone: (0)980-79000
Telefax: (0)980-79050
Electronic Mail: <userid>@irf.se
WWW: http://www.irf.se/irfK.html
 http://www.irf.se/
Founded: 1957 (IRF)
City Reference Coordinates: 020°27'00"E 67°52'00"N

Swedish Institute of Space Physics, Solar-Terrestrial Physics Division
(Institutet för Rymdfysik - IRF, Lund)
Solar-Terrestrial Physics Division
Scheelevägen 17
SE-223 70 Lund
Electronic Mail: <userid>@irf.se
WWW: http://www.irfl.lu.se/HeliosHome/irflund.html
 http://www.irf.se/
Founded: 1957 (IRF)
City Reference Coordinates: 013°11'00"E 55°42'00"N

Swedish Institute of Space Physics, Umeå Division
(Institutet för Rymdfysik - IRF, Umeå)
c/o University of Umeå
SE-901 87 Umeå
Telephone: (0)90-130505
Telefax: (0)90-166673
Electronic Mail: <userid>@tp.umu.se
 <userid>@physics.umu.se
WWW: http://www.irf.se/LMV/
 http://www.irf.se/
Founded: 1957 (IRF)
Staff: 6
Activities: space plasma theory
City Reference Coordinates: 020°15'00"E 63°50'00"N

Swedish Institute of Space Physics, Uppsala Division
(Institutet för Rymdfysik - IRF, Uppsala)
Box 537
SE-751 21 Uppsala
or :
Ångström Laboratory
Lägerhyddsvägen 1
SE- Uppsala
or :
(for parcel delivery)
Regementsvägen 1
SE-752 37 Uppsala
Telephone: (0)18-4715901
 (0)18-4715903
Telefax: (0)18-4715905
Electronic Mail: irf@irfu.se

office@irfu.se
<userid>@irfu.se
WWW: http://www.irfu.se/
http://www.irf.se/
Founded: 1952 (IRF: 1957)
Staff: 31
Activities: magnetospheric and ionospheric physics * space plasma physics
Periodicals: "Annual Report"
City Reference Coordinates: 017°38'00"E 59°52'00"N

Swedish Meteorological and Hydrological Institute (SMHI)
Folkborgsvägen 1
SE-601 76 Norrköping
Telephone: (0)11-4958000
Telefax: (0)11-4958001
Electronic Mail: smhi@smhi.se
WWW: http://www.smhi.se/
Founded: 1873
Staff: 550
Activities: meteorological, hydrological and oceanographical service
Periodicals: "Meteorology Report", "Hydrology Report", "Oceanography Report"
City Reference Coordinates: 016°11'00"E 58°36'00"N

Swedish National Space Board (SNSB)
Box 4006
SE-171 04 Solna
or :
Albygatan 107
Solna
Telephone: (0)8-6276480
Telefax: (0)8-6275014
Electronic Mail: rymdstyrelsen@snsb.se
<userid>@snsb.se
WWW: http://nos.snsb.se/
Founded: 1972
Membership: 7
Staff: 10
Activities: initiating R&D * coordinating Swedish activities within the fields of space technology, research and remote sensing
City Reference Coordinates: 018°01'00"E 59°22'00"N

Swedish Natural Science Research Council
(Naturvetenskapliga Forskningsrådet - NFR)
Box 7142
SE-113 87 Stockholm
or :
Regeringsgatan 56
3rd floor
SE-113 87 Stockholm
Telephone: (0)8-4544200
Telefax: (0)8-4544250
Electronic Mail: nfr@nfr.se
WWW: http://www.nfr.se/
Founded: 1977
Staff: 38
Activities: allocating research grants * creating research positions * supporting international collaborations * research information
Periodicals: (4) "carpe scientiam" (ISSN 1403-3542); (1) "NFR Yearbook"
City Reference Coordinates: 018°03'00"E 59°20'00"N

Swedish Physical Society
c/o Håkan Danared (Secretary)
Manne Siegbahn Laboratory
Stockholm University
Frescativägen 24
SE-104 05 Stockholm
Telephone: (0)8-161038
Telefax: (0)8-158674
Electronic Mail: sfs@msi.se
WWW: http://sfs.msi.se/
Founded: 1920
Membership: 950
City Reference Coordinates: 018°03'00"E 59°20'00"N

Swedish Space Corp. (SSC), Esrange
Box 802
SE-981 28 Kiruna

Telephone: (0)980-72000
Telefax: (0)980-12890
Electronic Mail: <userid>@ssc.se
WWW: http://www.ssc.se/esrange/
Founded: 1966
Activities: launching sounding rockets * releasing scientific balloons * satellite receiving and control stations
Coordinates: 021°04'00"E 67°56'00"N
City Reference Coordinates: 021°04'00"E 67°56'00"N

Swedish Space Corp. (SSC), Satellitbild
Box 816
SE-981 28 Kiruna
or :
Rymdhuset Österleden 15
SE-981 28 Kiruna
Telephone: (0)980-67100
Telefax: (0)980-16044
Electronic Mail: <userid>@ssc.se
custsupp@ssc.se
WWW: http://www.ssc.se/sb
Founded: 1982
Staff: 55
City Reference Coordinates: 020°27'00"E 67°52'00"N

Swedish Space Corp. (SSC), Solna
Albygatan 107
Box 4207
SE-171 04 Solna
Telephone: (0)8-6276200
Telefax: (0)8-987069
Electronic Mail: <userid>@ssc.se
WWW: http://www.ssc.se/
Founded: 1972
Staff: 320
Activities: design and development (satellites, equipment for sounding rockets, balloons, satellite navigation, ...) * launches (sounding rockets, balloons) * operation and control (satellites) * telecommunication services * data collection, processing, archiving and dissemination * value-added products * mapping services * remote sensing * methodologies * airborne maritime surveillance systems
City Reference Coordinates: 018°01'00"E 59°22'00"N

Swedish Standards Institution (SIS)
(Standardiseringen i Sverige - SIS)
Box 6455
SE-113 82 Stockholm
or :
St Eriksgatan 115
SE-103 66 Stockholm
Telephone: (0)8-6103000
Telefax: (0)8-307757
Electronic Mail: info@sis.se
WWW: http://www.sis.se/
Founded: 1922
Staff: 75
Activities: standardization
Periodicals: "Månadens Standard", "Teknik & Standard"
City Reference Coordinates: 018°03'00"E 59°20'00"N

Teknikens Hus
(House of Technology)
Högskoleområdet
SE-971 87 Luleå
Telephone: (0)920-72205
Telefax: (0)920-72202
Electronic Mail: age@teknikens-hus.se (Ann-Gerd Eriksson, Science Educator)
WWW: http://www.teknikens-hus.se/
Founded: 1988
Staff: 20
Activities: science education * planetariums
City Reference Coordinates: 022°10'00"E 65°34'00"N

Tierps Astronomiska Klubb (TAK)
c/o Daniel Söderström
Sjukarby 2509
SE-815 92 Tierp
Telephone: (0)293-12254
Electronic Mail: astro.tak@swipnet.se

WWW: http://home1.swipnet.se/~w-12155
Founded: 1994
Membership: 20
City Reference Coordinates: 017°30'00"E 60°20'00"N

Tom Tits Experiment
Storgatan 33
SE-151 36 Södertälje
Telephone: (0)8-52252500
Telefax: (0)8-52252510
Electronic Mail: info@tomtit.se
WWW: http://www.tomtit.se/Planetary/Planetary_uk.html
Founded: 1987
City Reference Coordinates: 017°39'00"E 59°11'00"N

Tycho Brahe Astronomical Society
(Astronomiska Sällskapet Tycho Brahe - ASTB)
c/o Peter Linde
Box 43
SE-221 00 Lund
Telephone: (0)40-21 23 40
(0)418-19946 (Bengt Rosengren, Secretary)
Electronic Mail: peter@astro.lu.se
WWW: http://www.astro.lu.se/~tycho/
Founded: 1937
City Reference Coordinates: 013°11'00"E 55°42'00"N

Tycho Brahe Observatory (TBO)
Box 2
SE-238 21 Oxie
Telephone: (0)40-547012
Electronic Mail: jan@astro.lu.se (Jan Sonnvik)
WWW: http://www.astro.lu.se/~jan/tbobs.html
http://www.tbobs.lu.se/
Coordinates: 013°05'12"E 55°32'34"N
City Reference Coordinates: 013°04'00"E 55°33'00"N

Uppsala Amatörastronomer (UAA)
c/o Johan Warell (President)
Wennerbergsgatan 11
SE-754 21 Uppsala
Telephone: (0)18241342
Electronic Mail: johwar@astro.uu.se
WWW: http://www.astro.uu.se/uaa/
Founded: 1980
Membership: 60
Activities: observing * meetings * instrumentation * photography
Observatories: 1 (Sandvreten)
City Reference Coordinates: 017°38'00"E 59°52'00"N

Uppsala University, Astronomical Observatory
Box 515
SE-751 20 Uppsala
Telephone: (0)18-530265
Telefax: (0)18-527583
Electronic Mail: astro@astro.uu.se
<userid>@astro.uu.se
WWW: http://www.astro.uu.se/
Founded: 1650
Staff: 37
Activities: extragalactic research * galactic structure * stellar atmospheres * solar system * observational astrophysics
Periodicals: "Uppsala Astronomical Observatory Reports", "Uppsala Preprints in Astronomy"
Coordinates: 017°37'30"E 59°51'30"N H21m (IAU Code 549)
017°36'24"E 59°30'06"N H35m (Kvistaberg Station - see separate entry)
149°04'00"E 31°16'36"S H1,164m (Uppsala Southern Station)
City Reference Coordinates: 017°38'00"E 59°52'00"N

Uppsala University, Astronomical Observatory, Kvistaberg Station
SE-197 91 Bro
Telephone: (0)8-58240157
Telefax: (0)8-58240157
WWW: http://www.astro.uu.se/
Founded: 1944
Staff: 1
Activities: Schmidt observing * photoelectric photometry * CCDs * minor planets

Coordinates: 017°36'24"E 59°30'06"N H35m (IAU Code 049)
City Reference Coordinates: 017°38'00"E 59°31'00"N

Västerås Astronomi och Rymdforsknings Förening
c/o Sven-Erik Persson
Dragverksgatan 29
SE-724 74 Västerås
Telephone: (0)21-356783
Electronic Mail: vart@vasteras.mail.teia.com
WWW: http://w1.213.telia.com/~u21304183
Founded: 1989
Membership: 33
Activities: popularization
Periodicals: "A&R-bladet"
City Reference Coordinates: 016°32'00"E 59°36'00"N

Victor Hasselblad Foundation (Erna and_)
• See "Erna and Victor Hasselblad Foundation"

Switzerland

Académie Suisse des Sciences Naturelles (ASSN)
• See "Schweizerische Akademie der Naturwissenschaften (SANW)"

Accademia Svizzera di Scienze Naturali (ASSN)
• See "Schweizerische Akademie der Naturwissenschaften (SANW)"

Association des Amis de l'Observatoire des Creusets à Arbaz (AOCA)
c/o Lycée-Collège des Creusets
Rue St-Guérin 34
CH-1950 Sion
Telephone: (0)27-3222930
Telefax: (0)27-3237920
WWW: http://www.cobweb.ch/obs-creusets/
Founded: 1995
City Reference Coordinates: 007°22'00"E 46°14'00"N (Sion)

Astroclub Solaris Aarau (ASA)
c/o Alte Kantonsschule
Bahnhofstrasse 91
CH-5001 Aarau
Telephone: (0)62-8271177
Electronic Mail: christian.roth@ibpv.unil.ch (Christian Roth, Treasurer)
WWW: http://homer.span.ch/~spaw2173/ASA/welcome.html
    http://homer.span.ch/~spaw2173/ASA/nuetziweid/Nuetziweid.html (Volkssternwarte Nütziweid)
Founded: 1979
Membership: 113
Activities: education * trips * observing
Coordinates: 008°03'00"E 47°24'00"N H707m (Volkssternwarte Nütziweid)
City Reference Coordinates: 008°03'00"E 47°24'00"N

Astronomie-Verein Olten (AVO)
c/o Marcel Lips
Allmendstrasse 40
CH-4658 Däniken
Telephone: (0)62-2913259
Founded: 1977
Membership: 35
Activities: observing * meetings
Periodicals: "Newsletter"
Coordinates: 007°53'00"E 47°23'00"N H785m (Froburg)
City Reference Coordinates: 007°53'00"E 47°23'00"N

Astronomische Gesellschaft Baden (AGB)
c/o Jean-Marc Schweizer
Sooremattstrasse 6
CH-5212 Hausen bei Brugg
Telephone: (0)56-416703
WWW: http://www.astroinfo.ch/clubs/agb/
Founded: 1951
Membership: 70
Activities: lectures * displays * trips * discussions
Coordinates: 008°06'45"E 47°31'40"N H665m (Cheisacker)
City Reference Coordinates: 008°19'00"E 47°28'00"N (Baden)
    008°13'00"E 47°29'00"N (Brugg)

Astronomische Gesellschaft Bern (AGB)
Hangweg
CH-3148 Lanzenhäusern
Telephone: (0)31-7311604
Founded: 1923
Membership: 242
Activities: lectures * discussions * observing
Coordinates: 007°25'42"E 46°57'12"N H550m (Muesmatt)
City Reference Coordinates: 007°26'00"E 46°57'00"N (Bern)

Astronomische Gesellschaft Graubünden (AGG)
c/o Rolf Stauber
Carmennaweg 83
CH-7000 Chur
Electronic Mail: stauber@swissonline.ch

WWW: http://organisationen.freepage.de/agg/
City Reference Coordinates: 009°32'00"E 46°52'00"N

Astronomische Gesellschaft Oberwallis (AGO)
c/o Rudolf Arnold
Wierystrasse 101
CH-3902 Brig-Glis
Telephone: (0)27-9241387
WWW: http://oberwallis.astronomie.ch/
http://www.astroinfo.org/clubs/ago/
Founded: 1982
Membership: 62
Activities: observing * lectures
Periodicals: "AGO-Mitteilungen"
City Reference Coordinates: 008°00'00"E 46°19'00"N

Astronomische Gesellschaft Solothurn (AGS)
c/o Fred Nicolet
Jupiterstrasse 6
CH-4500 Solothurn
Telephone: (0)32-6223020
Founded: 1954
Membership: 30
City Reference Coordinates: 007°32'00"E 47°13'00"N

Astronomische Gesellschaft Zürcher Oberland (AGZO)
c/o Walter Brändli
Oberer Hömel 32
CH-8636 Wald
Telephone: (0)55-954212
(0)55-2461763
Telefax: (0)55-955162
Electronic Mail: astro@pax.eunet.ch
christoph.bosshard@astroinfo.ch (Christoph Bosshard)
WWW: http://www.astroinfo.ch/clubs/agzo/
Founded: 1967
Membership: 55
Activities: observing * education
Coordinates: 008°54'40"E 47°15'44"N H753m
City Reference Coordinates: 008°33'00"E 47°23'00"N

Astronomische Gesellschaft Zürcher Unterland (AGZU)
c/o Urs Stich
Gerstmattstrasse 41
CH-8172 Niederglatt
Telephone: (0)1-8506319
Telefax: (0)1-8506319
Electronic Mail: agzu@astronomie.ch
WWW: http://agzu.astronomie.ch/
Founded: 1970
Membership: 200
Activities: observing * lectures * excursions * special activities for young members
Periodicals: (6) "Himmelsbeobachter"
Observatories: 1 (Schul- und Volkssternwarte Bülach - see separate entry)
City Reference Coordinates: 008°31'00"E 47°29'00"N (Niederglatt)
008°33'00"E 47°23'00"N (Zürich)

Astronomische Gruppe des Kantons Glarus
Postfach 157
CH-8750 Glarus
Telephone: (0)55-6501620
Telefax: (0)55-6501623
Electronic Mail: villavanilla@freesurf.ch (Marc Meyer-Beck)
Founded: 1961
Membership: 34
Activities: observing * education
City Reference Coordinates: 009°04'00"E 47°03'00"N

Astronomische Vereinigung Frauenfeld (AVF)
Thundorferstraße 139
CH-8500 Frauenfeld
Electronic Mail: p_guhl@eudoramail.com
WWW: http://www.astroinfo.ch/clubs/avf/
Founded: 1995
Observatories: 1 (Oberherten)

City Reference Coordinates: 008°54'00"E 47°34'00"N

Astronomische Vereinigung Kreuzlingen (AVK)
c/o Robert Testa
Waldheimstrasse 1
CH-8280 Kreuzlingen
Telephone: (0)71-6725301
(0)71-6725855 (Observatory)
Telefax: (0)71-6725301
Founded: 1972
Membership: 250
Periodicals: (1) "Jahresbericht"
Coordinates: 009°09'41"E 47°38'33"N H481m
City Reference Coordinates: 009°11'00"E 47°39'00"N

Astronomische Vereinigung Sankt Gallen
c/o Rolf Burgstaller
Blattenstrasse 25
CH-9052 Niederteufen
Telephone: (0)71-3331374
Founded: 1955
Membership: 92
Activities: popularization * observing
City Reference Coordinates: 009°23'00"E 47°23'00"N (Teufen)

Astronomische Vereinigung Zürich (AVZ)
c/o Andreas Inderbitzin (President)
Winterthurerstrasse 420
CH-8051 Zürich
or :
c/o Dieter Späni
Bachmattstrasse 9
CH-8618 Oetwill-am-See
Telephone: (0)1-3067523
(0)1-9291127
Electronic Mail: inderbitzin.a@bluewin.ch
WWW: http://www.astroinfo.ch/clubs/avz/
Founded: 1949
Membership: 190
Activities: observing
Periodicals: (4) "AVZ-Mitteilung"
City Reference Coordinates: 008°33'00"E 47°23'00"N (Zürich)

Astrooptik Kohler (AOK)
Emmenweidstrasse 607/M4
CH-6020 Emmenbrücke
Telephone: (0)41-2601677
Telefax: (0)41-2601677
Electronic Mail: sales@aokswiss.ch
WWW: http://www.aokswiss.ch/
Founded: 1988
Staff: 1
Activities: developing and producing astronomical telescope systems * distributing telescopes and optical accessories
City Reference Coordinates: 008°17'00"E 47°04'00"N

Bielser Observatorien
Hauptstrasse 22
CH-4132 Muttenz
Telephone: (0)79-6590414
Telefax: (0)61-4618177
Electronic Mail: bielser.gerold@datacomm.ch
WWW: http://www.astroinfo.org/bielser/
Activities: manufacturing domes
City Reference Coordinates: 007°39'00"E 47°31'00"N

Bundesanstalt für Meteorologie und Klimatologie (MeteoSchweiz)
• See "MeteoSwiss"

Centre Européen pour la Recherche Nucléaire (CERN)
(European Organization for Nuclear Research)
Route de Meyrin
CH-1211 Genève 23
Telephone: (0)22-7676111 (central switchboard)
(0)22-7672210 (reception desk)
Telefax: (0)22-7677555 (central fax)

Electronic Mail: libdesk@cernvm.cern.ch (Scientific Information Service)
WWW: http://www.cern.ch/
http://info.cern.ch/
Founded: 1954
Staff: 10000
Activities: theoretical and experimental research in elementary-particle physics * accelerator and detector design and development
Periodicals: (6) "CERN Courier" (ISSN 0304-288x), "Courrier CERN" (ISSN 0374-2288), "Liste des Publications Scientifiques" (ISSN 0304-2871), "Annual Report" (ISSN 0304-2901), "Rapport Annuel" (ISSN 0304-291x), "CERN-HERA Reports" (ISSN 0366-5690), "CERN Reports" (ISSN 0007-8328), "CERN School of Physics Proceedings" (ISSN 0531-4283)
City Reference Coordinates: 006°09'00"E 46°12'00"N

Centre Européen pour la Recherche Nucléaire (CERN), Astronomy Club
• See "CERN Astronomy Club"

CERN Astronomy Club
c/o Claudia Parente (President)
Route de Meyrin
CH-1211 Genève 23
Electronic Mail: astronomy.club@cern.ch
WWW: http://z.home.cern.ch/z/zenith/www/
Founded: 2000
Membership: 20
City Reference Coordinates: 006°09'00"E 46°12'00"N

Club d'Astronomes Amateurs de Bienne (CAAB)
c/o Claudio Cerini
Route d'Aegerten 31
CH-2503 Bienne
Telephone: (0)32-3658074
Founded: 1981
Membership: 20
Activities: popularization * observing * photography
City Reference Coordinates: 007°12'00"E 47°10'00"N

Cryophysics SA
Rue Rotschild 66
CH-1202 Genève
Telephone: (0)22-7329520
Telefax: (0)22-7315588
Electronic Mail: cryophysicsch@compuserve.com
WWW: http://ourworld.compuserve.com/homepages/cryophysicsch
Founded: 1967
Staff: 80
Activities: manufacturing and distributing cryogenic equipment
Periodicals: "Cryophysics Newsletter"
City Reference Coordinates: 006°09'00"E 46°12'00"N

École Polytechnique Fédérale
• See "Eidgenössische Technische Hochschule (ETH)"

Eidgenössische Technische Hochschule (ETH), Institut für Astronomie
ETH-Zentrum
CH-8092 Zürich
Telephone: (0)1-6323813
Telefax: (0)1-6321205
Electronic Mail: <userid>@astro.phys.ethz.ch
WWW: http://www.astro.phys.ethz.ch/
Staff: 26
Activities: solar and stellar physics * radio astronomy * instrumentation
Coordinates: 009°40'06"E 46°47'00"N H2,050m
City Reference Coordinates: 008°33'00"E 47°23'00"N

Etel SA
Zône Industrielle
CH-2112 Môtiers
Telephone: (0)38-611858
Telefax: (0)38-612419
Electronic Mail: etel@etel.ch
WWW: http://www.etel.ch/
Founded: 1974
Staff: 33
Activities: designing and manufacturing large torque motors for i.a. telescopes
City Reference Coordinates: 006°37'00"E 46°55'00"N

European Astronomical Society (EAS)
Case Postale 82
CH-1213 Petit-Lancy 2
WWW: http://www.iap.fr/eas/
Founded: 1991
• See main entry in Czech Republic
City Reference Coordinates: 006°09'00"E 46°12'00"N (Genève)

Federal Office of Meteorology and Climatology (MeteoSwiss)
• See "MeteoSwiss"

Ferrovia Monte Generoso (FMG) SA
CH-6825 Capolago
Telephone: (0)91-6481105
Telefax: (0)91-6481107
Electronic Mail: info@montegeneroso.ch
WWW: http://www.montegeneroso.ch/
Founded: 1890
Staff: 50
Activities: touristic rack railway * sporting and recreational activities * excursions * restaurant * accommodation * public observatory
Coordinates: 008°59'00"E 45°55'00"N H1870m (IAU Code 179)
City Reference Coordinates: 008°59'00"E 45°55'00"N

Fondation de l'Observatoire de Vérossaz
c/o Bernard Deletroz
CH-1891 Vérossaz
Telephone: (0)79-2063157
          (0)79-3707054
Electronic Mail: verobservatoire@freesurf.ch
City Reference Coordinates: 007°01'00"E 46°14'00"N (Saint-Maurice)

Fondation de l'Observatoire François-Xavier Bagnoud
CH-3961 Saint-Luc
Telephone: (0)27-4755808
Telefax: (0)27-4755811
Electronic Mail: ofxb@bluewin.ch
WWW: http://www.ofxb.ch/
Founded: 1993
Staff: 1
Activities: observing (Sun, night sky) for tourists and schools * instrumentation available for good amateurs
Coordinates: 007°36'46"E 46°13'42"N H2,200m (IAU Code 175)
City Reference Coordinates: 007°33'00"E 46°18'00"N (Sierre)

Fondation Jungfraujoch-Gornergrat
• See "Internationale Stiftung Hochalpine Forschungsstationen Jungfraujoch und Gornergrat (HFSJG)"

Fondation Robert-A. Naef
Route du Petit-Épendes 45
CH-1731 Épendes
Telephone: (0)26-4131099
Electronic Mail: info@observatoire-naef.ch
WWW: http://www.observatoire-naef.ch/
Coordinates: 007°08'00"E 46°45'00"N H700m
City Reference Coordinates: 006°37'00"E 46°46'00"N

Fondo Nazionale Svizzero per la Ricerca Scientifica
• See "Schweizerischer Nationalfonds zur Förderung der Wissenschaftlichen Forschung"

Fonds National Suisse de la Recherche Scientifique
• See "Schweizerischer Nationalfonds zur Förderung der Wissenschaftlichen Forschung"

Forschungsstationen Jungfraujoch und Gornergrat (HFSJG) (Internationale Stiftung Hochalpine_)
• See "Internationale Stiftung Hochalpine Forschungsstationen Jungfraujoch und Gornergrat (HFSJG)"

Gesellschaft der Freunde der Urania-Sternwarte (GFUS)
c/o Volkshochschule des Kantons Zürich
Splügenstraße 10
CH-8002 Zürich
Telephone: (0)1-2058484
Telefax: (0)1-2059495
Electronic Mail: vhszh@access.ch
Founded: 1936
Activities: supporting and managing Urania Observatory

Coordinates: 008°32'26"E 47°22'32"N H408m
City Reference Coordinates: 008°33'00"E 47°23'00"N

International Electrotechnical Commission (IEC)
Rue de Varembe 1
Case Postale 131
CH-1211 Genève 20
Telephone: (0)22-9190211
Telefax: (0)22-9190300
Electronic Mail: info@iec.ch
WWW: http://www.iec.ch/
Founded: 1906
Membership: 50 (countries)
Activities: preparing and publishing international standards for all electrical, electronic and related technologies
City Reference Coordinates: 006°09'00"E 46°12'00"N

INTEGRAL Science Data Centre (ISDC)
Chemin d'Écogia 16
CH-1290 Versoix
Telephone: (0)22-9509100
Telefax: (0)22-9509133
Electronic Mail: isdc@obs.unige.ch
WWW: http://isdc.unige.ch/
Activities: acting as interface between the data of the INTEGRAL satellite and the scientific community
City Reference Coordinates: 006°09'00"E 46°12'00"N

Internationale Stiftung Hochalpine Forschungsstationen Jungfraujoch und Gornergrat (HFSJG)
Sidlerstrasse 5
CH-3012 Bern
Telephone: (0)31-6314052
Telefax: (0)31-6314405
Electronic Mail: louise.wilson@phim.unibe.ch
WWW: http://www.ifjungo.ch/
Founded: 1931
Activities: running the Jungfraujoch and Gornergrat scientific stations (see also these entries)
Coordinates: 007°59'02"E 46°32'53"N H3,450m (Jungfraujoch Sc. Stn) H3,580m (Obs.)
007°47'04"E 45°59'04"N H3,130m (Gornergrat)
City Reference Coordinates: 007°26'00"E 46°57'00"N

International Organization for Standardization (ISO)
Rue de Varembe 1
Case Postale 56
CH-1211 Genève 20
Telephone: (0)22-7490111
Telefax: (0)22-7333430
Electronic Mail: central@iso.org
WWW: http://www.iso.org/
Founded: 1947
Membership: 145 (countries)
Staff: 165
Activities: development of international standards and related activities in all fields (except electrical and electronic engineering)
Periodicals: (12) "ISO Bulletin" (ISSN 0303-805x); (1) "ISO Memento", "ISO Catalog"; "ISO Management Systems" (ISSN 1680-8096)
City Reference Coordinates: 006°09'00"E 46°12'00"N

International Space Science Institute (ISSI)
Hallerstrasse 6
CH-3012 Bern
Telephone: (0)31-6314896
Telefax: (0)31-6314897
Electronic Mail: secretary@issi.unibe.ch
WWW: http://www.issi.unibe.ch/
Founded: 1995
Staff: 12
Activities: contributing to the achievement of a deeper understanding of the results from space-research missions, adding value to those results through multi-disciplinary research in an atmosphere of international cooperation
Periodicals: "Space Sciences Series of ISSI (SSSI)", "ISSI Scientific Reports Series", "Pro-ISSI Spatium Series"
City Reference Coordinates: 007°26'00"E 46°57'00"N

International Telecommunication Union (ITU)
(Union Internationale des Télécommunications - UIT)
Place des Nations
CH-1211 Genève 20
Telephone: (0)22-7305111

Telefax: (0)22-7337256
(0)22-7305939
Electronic Mail: itumail@itu.int
WWW: http://www.itu.int/
Founded: 1934 (1865 as former "International Telegraph Union")
Membership: 189 (countries)
Staff: 800
Activities: maintaining and extending international cooperation between all members * promoting and offering technical assistance to developing countries in the field of telecommunications * promoting the development of technical facilities and their most efficient operation * harmonizing national actions to these ends * coordinating efforts to eliminate harmful interferences * improving the use made of the radio frequency spectrum * coordinating efforts to harmonizing the development of telecommunication space techniques
Periodicals: (10) "ITU News" (ISSN 0497-137x)
City Reference Coordinates: 006°09'00"E 46°12'00"N

Istituto Ricerche Solari Locarno (IRSOL)
Via Patocchi
CH-6605 Locarno-Monti
Telephone: (0)91-7434226
Telefax: (0)91-7434226
Electronic Mail: mbianda@ccsc.ch (Michele Bianda)
Founded: 1960
Staff: 2
Activities: solar research (visible spectrum)
Coordinates: 008°47'22"E 46°10'39"N H500m
City Reference Coordinates: 008°48'00"E 46°10'00"N

Les Pléiades, Société d'Astronomie de Saint-Imier
Passage de la Reine-Berthe 5
CH-2610 Saint-Imier
Telephone: (0)32-9401463
Telefax: (0)32-4813605
Electronic Mail: info@pleiades.ch
WWW: http://www.pleiades.ch/
Founded: 1998
Membership: 60
Activities: education * photography * popularization * exhibitions * visits * observing
Coordinates: 006°59'23"E 47°09'55"N (Observatoire Astronomique Mont-Soleil)
City Reference Coordinates: 007°00'00"E 47°09'00"N

MeteoSchweiz
• See "MeteoSwiss"

MétéoSuisse
• See "MeteoSwiss"

MeteoSvizzera
• See "MeteoSwiss"

MeteoSwiss
(Federal Office of Meteorology and Climatology)
(MeteoSchweiz - MétéoSuisse - MeteoSvizzera)
Krähbühlstrasse 58
Postfach 514
CH-8044 Zürich
Telephone: (0)1-2569111
Telefax: (0)1-2569278
WWW: http://www.sma.ch/
Founded: 1881
Staff: 201
Activities: meteorology * climatology
City Reference Coordinates: 008°33'00"E 47°23'00"N

Murmel Spielwerkstatt und Verlag
Hermann Greulich-Strasse 60
CH-8004 Zürich
Telephone: (0)1-4015156
(0)1-2421718
Telefax: (0)1-4015158
Founded: 1983
Activities: designing, producing and distributing games, including astronomical ones
City Reference Coordinates: 008°33'00"E 47°23'00"N

Musée Suisse des Transports et des Communications
• See "Verkehrshaus der Schweiz"

Observatoire Cantonal de Neuchâtel
Rue de l'Observatoire 58
CH-2000 Neuchâtel
Telephone: (0)32-8896870
Telefax: (0)32-8896281
Electronic Mail: observatoire.cantonal@ne.ch
WWW: http://www.ne.ch/adm/dep/ocne/
Founded: 1858
Staff: 30
Activities: time * frequencies * atomic clocks * meteorology * geophysics
Periodicals: (1) "Rapport Annuel"
Coordinates: 006°57'30"E 46°59'54"N H488m (IAU Code 019)
City Reference Coordinates: 006°55'00"E 46°59'00"N

Observatoire de Genève
Chemin des Maillettes 51
CH-1290 Sauverny
Telephone: (0)22-7552611
Telefax: (0)22-7553983
Electronic Mail: <userid>@obs.unige.ch
WWW: http://obswww.unige.ch/
Founded: 1772
Staff: 100
Activities: stellar evolution * exoplanets * stellar kinematics * stellar dynamics * astroseismology * high-energy astrophysics * AGNs * galaxies
Observatories: Jungfraujoch (Switzerland), Saint-Michel-l'Observatoire (France) and La Silla (Chile)
Coordinates: 006°04'00"E 46°18'24"N H455m
City Reference Coordinates: 006°09'00"E 46°12'00"N

Observatoire des Creusets
● See "Association des Amis de l'Observatoire des Creusets à Arbaz (AOCA)"

Observatoire François-Xavier Bagnoud (OFXB)
● See "Fondation de l'Observatoire François-Xavier Bagnoud"

Office Fédéral de Météorologie et de Climatologie (MétéoSuisse)
● See "MeteoSwiss"

Osservatorio Ticinese di Locarno-Monti
● See "Istituto Svizzero di Meteorologia, Osservatorio Ticinese di Locarno-Monti"

Organisation Météorologique Mondiale (OMM)
● See "World Meteorological Organization (WMO)"

Organisation Mondiale de la Propriété Intellectuelle (OMPI)
● See "World Intellectual Property Organization (WIPO)"

Physikalisch-Meteorologisches Observatorium Davos (PMOD), Weltstrahlungszentrum
(World Radiation Center - WRC)
Dorfstrasse 33
CH-7260 Davos-Dorf
Telephone: (0)81-4175111
Telefax: (0)81-4175100
Electronic Mail: <userid>@pmodwrc.ch
WWW: http://www.pmodwrc.ch/
Founded: 1907
Staff: 18
Activities: solar irradiance * solar variability * helioseismology * atmospheric physics
City Reference Coordinates: 009°50'00"E 46°48'00"N

Planetarium Zürich
Haldenstrasse 138
CH-8055 Zürich
Telephone: (0)1-4625500 (Infoline)
          (0)1-4623410 (Secretariat)
Telefax: (0)1-4625501
Electronic Mail: info@plani.ch
WWW: http://www.plani.ch/
Founded: 1990
Staff: 3
Activities: shows (movable planetarium) * development of instrumentation
City Reference Coordinates: 008°33'00"E 47°23'00"N

P. Wyss Photo-Video
Postfach
CH-8034 Zürich
or :
Dufourstrasse 124
CH-8008 Zürich
Telephone: (0)1-3830108
Telefax: (0)1-3830094
Electronic Mail: wyssproastro@access.ch
Founded: 1982
Staff: 6
• Dealer
City Reference Coordinates: 008°33'00"E 47°23'00"N

Rudolf Wolf Gesellschaft
c/o H.U. Keller (Secretary)
Kolbenhofstrasse 33
CH-8045 Zürich
Telephone: (0)1-4616814
Founded: 1992
Membership: 50
Activities: securing the continuation of Wolf's sunspot observations with Wolf's original telescope
Periodicals: (2) "Mitteilungen der Rudolf Wolf Gesellschaft" (ISSN 1021-8823)
City Reference Coordinates: 008°33'00"E 47°23'00"N

Schul- und Volkssternwarte Bülach
Postfach 282
CH-8180 Bülach
Telephone: (0)1-8608448 (Observatory)
          (0)1-8601221 (Bookings)
Telefax: (0)1-8604954
Electronic Mail: buelach@astronomie.ch
WWW: http://buelach.astronomie.ch/
Founded: 1983
Activities: observing * popularization
Coordinates: 008°34'22"E 47°31'13"N H550m (IAU Code 167)
City Reference Coordinates: 008°32'00"E 47°32'00"N

Schweizerische Akademie der Naturwissenschaften (SANW)
(Académie Suisse des Sciences Naturelles - ASSN)
(Accademia Svizzera di Scienze Naturali - ASSN)
(Swiss Academy of Sciences - SAS)
Bärenplatz 2
CH-3011 Bern
Telephone: (0)31-3123375
Telefax: (0)31-3123291
Electronic Mail: sanw@sanw.unibe.ch
WWW: http://www.sanw.ch/
       http://www.assn.ch/
Founded: 1815
Activities: research fostering
Periodicals: (1) "Jahrbuch/Annuaire"; "Mitteilungsblatt/Bulletin d'Information", "Interne Mitteilungen/Communications Internes"
City Reference Coordinates: 007°26'00"E 46°57'00"N

Schweizerische Astronomische Gesellschaft (SAG)
(Société Astronomique de Suisse - SAS)
c/o Sue Kernen (Secretary)
Gristenbühl
CH-9315 Neukirch
Telephone: (0)71-4771743
Electronic Mail: astro_mod_4@ezinfo.vmsmail.ethz.ch
          noel.cramer@obs.unige.ch (Noël Cramer, Orion Editor-in-Chief)
WWW: http://ezinfo.ethz.ch/ezinfo/astro/kontakt/SAS.html
       http://ezinfo.ethz.ch/ezinfo/astro/kontakt/kon2_0.html
Founded: 1938
Membership: 3000
Activities: education * popularization * promoting cooperation between amateur and professional astronomers, as well as between astronomical groups
Periodicals: (6) "Orion" (ISSN 0030-557x, circ.: 2,800)
City Reference Coordinates: 009°23'00"E 47°32'00"N

Schweizerische Gesellschaft für Astrophysik und Astronomie (SGAA)
● See "Société Suisse d'Astrophysique et d'Astronomie (SSAA)"

Schweizerische Normen-Vereinigung (SNV)
(Swiss Association for Standardization)
Bürglistrasse 29
CH-8400 Winterthur
Telephone: (0)52-2245454
Telefax: (0)52-224
Electronic Mail: info@snv.ch
WWW: http://www.snv.ch/
Founded: 1919
Membership: 600
Staff: 30
Activities: standardization (national, European, international)
Periodicals: (12) "Switec Information", (11) "SNV Bulletin"
City Reference Coordinates: 008°45'00"E 47°30'00"N

Schweizerische Physikalische Gesellschaft (SPG)
● See "Société Suisse de Physique (SSP)"

Schweizerischer Nationalfonds zur Förderung der Wissenschaftlichen Forschung
(Fonds National Suisse de la Recherche Scientifique)
(Fondo Nazionale Svizzero per la Ricerca Scientifica)
(Swiss National Science Foundation)
Wildhainweg 20
Postfach 8232
CH-3001 Bern
Telephone: (0)31-3082222
Telefax: (0)31-3013009
Electronic Mail: pri@snf.ch
WWW: http://www.snf.ch/
Founded: 1952
Staff: 93
Periodicals: (4) "Horizonte/Horizons"
City Reference Coordinates: 007°26'00"E 46°57'00"N

Società Astronomica Ticinese
c/o Specola Solare Ticinese
CH-6605 Locarno-Monti
Telephone: (0)91-7526376
          (0)91-7562376
Telefax: (0)91-7526310
WWW: http://www.karawari.com/sat/
Founded: 1961
Membership: 150
Activities: observing (Sun, meteors, variable stars, planets)
Periodicals: (6) "Meridiana" (circ.: 1000)
City Reference Coordinates: 008°48'00"E 46°10'00"N

Société Astronomique de Genève
Terreaux-du-Temple 6
CH-1201 Genève
Telephone: (0)22-7383322
Telefax: (0)22-7383322
Electronic Mail: sag@infomaniak.ch
WWW: http://www.astrosurf.com/sag/
Founded: 1923
Membership: 210
Activities: meetings * observing * popularization
Periodicals: "L'Observateur"
City Reference Coordinates: 006°09'00"E 46°12'00"N

Société Astronomique de Suisse (SAS)
● See "Schweizerische Astronomische Gesellschaft (SAG)"

Société d'Astronomie de Saint-Imier
● See "Les Pléiades - Société d'Astronomie de Saint-Imier"

Société d'Astronomie du Haut-Léman (SAHL)
c/o René Durussel
Rue des Communaux 19
CH-1800 Vevey

or :
c/o Observatoire
Avenue E. Bieler
CH-1800 Vevey
Telephone: (0)21-9228308 (President)
(0)21-9431538 (Secretary)
Founded: 1970
Membership: 60
Activities: observing * photography * CCDs
Periodicals: (4) "Le Courrier de l'Observatoire"
Coordinates: 006°51'04"E 46°28'00"N H461m
City Reference Coordinates: 006°51'00"E 46°28'00"N

Société d'Astronomie du Nord Vaudois (SANV)
Case Postale 683
CH-1401 Yverdon-les-Bains
or :
c/o Nicole Beuchat (Secretary)
Avenue Pierre-de-Savoie 49A
CH-1400 Yverdon-les-Bains
Telephone: (0)24-4253061
Electronic Mail: nicole.beuchat@freesurf.ch
WWW: http://www.eivd.ch/astronomie
Founded: 1995
Observatories: 1 (Chamblon)
City Reference Coordinates: 006°38'00"E 46°47'00"N

Société d'Astronomie du Valais Romand (SAVAR)
c/o Jacques Zufferey
Eaux-Vives 5
CH-3965 Chippis
Telephone: (0)27-4556085
(0)79-5198017
Electronic Mail: jacques.zufferey@tvs2net.ch
WWW: http://savar.astronomie.ch/
Founded: 1994
Membership: 70
Activities: observing * photography * CCDs
Periodicals: (6) "Bulletin de la Société d'Astronomie du Valais Romand"
City Reference Coordinates: 007°21'00"E 46°14'00"N (Sion)

Société de Physique et d'Histoire Naturelle (SPHN)
Case Postale 6434
CH-1211 Genève 6
or :
c/o Muséum d'Histoire Naturelle
Route de Malagnou 1
CH-1211 Genève 6
Telephone: (0)22-4186300
Telefax: (0)22-4186301
Electronic Mail: jean.wuest@nhn.ville-ge.ch (Jean Wuest, Secretary)
WWW: http://www.unige.ch/sphn/
Founded: 1790
Membership: 240
Staff: 12
Activities: publishing * excursions * interdisciplinary forums * popularization
Periodicals: (3) "Archives des Sciences"; "Mémoires de la Société de Physique et d'Histoire Naturelle"
Awards: "Prix Augustin-Pyramus de Caudolle", "Prix Marc-Auguste Pictet"
City Reference Coordinates: 006°09'00"E 46°12'00"N

Société Jurassienne d'Astronomie (SJA)
c/o Michel Ory
Rue du Béridier 30
CH-2800 Délémont
Telephone: (0)32-4233286
Electronic Mail: private@bluewin.ch (Michel Ory, President)
Founded: 1980
Membership: 120
Activities: observing * minor planet astrometry * photography * CCD * Observatoire Astronomique Jurassien (IAU Code 185)
Coordinates: 007°25'20"E 47°21'20"N H505m (Obs. Astron. Jurassien, Vicques)
City Reference Coordinates: 007°21'00"E 47°22'00"N

Société Neuchâteloise d'Astronomie (SNA)
c/o Bernard Nicolet (President)
Observatoire de Genève

CH-1290 Sauverny
Telephone: (0)22-7552611
Electronic Mail: bernard.nicolet@obs.unige.ch
WWW: http://obswww.unige.ch/~nicolet/sna/Welcome.html
Founded: 1979
Membership: 100
City Reference Coordinates: 006°55'00"E 46°59'00"N (Neuchâtel)

Société Suisse d'Astrophysique et d'Astronomie (SSAA)
(Schweizerische Gesellschaft für Astrophysik und Astronomie - SGAA)
c/o Arnold Benz (President)
Institut für Astronomie
Eidgenössische Technische Hochschule
ETH-Zentrum
CH-8092 Zürich
Electronic Mail: benz@astro.phys.ethz.ch
WWW: http://obswww.unige.ch/ssaa/
Founded: 1969
Membership: 180
Activities: national society of professional astronomers
City Reference Coordinates: 008°33'00"E 47°23'00"N

Société Suisse de Physique (SSP)
(Schweizerische Physikalische Gesellschaft - SPG)
(Swiss Physical Society - SPS)
c/o Institut für Physik
Universität Basel
Klingelbergstrasse 82
CH-4056 Basel
Telephone: (0)61-2673715
Telefax: (0)61-2673716
Electronic Mail: sps@ubaclu.unibas.ch
WWW: http://www.sps.ch/sps/
Founded: 1908
Membership: 1400
City Reference Coordinates: 007°35'00"E 47°33'00"N

Société Valaisanne des Sciences Naturelles "La Murithienne"
Case Postale 2251
CH-1950 Sion 2 Nord
Telephone: (0)27-6064731
Telefax: (0)27-6064734
Founded: 1861
Membership: 650
Activities: excursions * lectures
Periodicals: (1) "Bulletin de La Murithienne" (ISSN 0374-6402)
City Reference Coordinates: 007°21'00"E 46°14'00"N

Société Vaudoise d'Astronomie (SVA)
Case Postale 190
CH-1018 Lausanne
Founded: 1942
Membership: 285
Activities: photography * computing * education * CCD * popularization
Periodicals: (6) "Galaxie", "Bulletin SVA"
Coordinates: 006°37'28"E 46°32'07"N H595m
City Reference Coordinates: 006°08'00"E 46°18'00"N

Specola Solare Ticinese
CH-6605 Locarno-Monti
Telephone: (0)91-7562376
Telefax: (0)91-7562310
Electronic Mail: cortesi@webshuttle.ch (Sergio Cortesi, Director)
Founded: 1957
Staff: 3
Activities: observing (Sun, variable stars)
Coordinates: 008°47'21"E 46°10'23"N H367m
City Reference Coordinates: 008°48'00"E 46°10'00"N (Locarno)

Sternwarte Eschenberg
c/o Markus Griesser
Breitenstrasse 2
CH-8542 Wiesendangen
Telephone: (0)52-3372848
Telefax: (0)52-3372848

Electronic Mail: griesser@spectraweb.ch
WWW: http://www.spectraweb.ch/~griess/Sternwarte/
Founded: 1979
Staff: 1
Coordinates: 008°44'38"E 47°28'33"N H542m (IAU Code 151)
City Reference Coordinates: 008°48'00"E 47°32'00"N

Sternwarte Rümlang
• See "Verein Sternwarte Rotgrueb Rümlang (VSRR)"

Sternwarte Sursee
Berufsschulhaus Kotten
CH-6210 Sursee/Lu
Electronic Mail: ens@ens.ch (Peter Ens)
WWW: http://www.ens.ch/sternwarte/
City Reference Coordinates: 008°07'00"E 47°11'00"N

Sternwarte Urania Burgdorf
Frommgutweg 6
CH-3400 Burgdorf
Telephone: (0)34-4227612
Electronic Mail: urania@urania.ch
markus.buetikofer@spectraweb.ch
WWW: http://www.urania.ch/
Founded: 1920
City Reference Coordinates: 007°37'00"E 47°04'00"N

Stiftung Sternwarte Uitikon
c/o Andreas Inderbitzin
Winterthurerstrasse 420
CH-8051 Zürich
Telephone: (0)1-3228736 (Private)
(0)1-3067523 (Office)
Electronic Mail: inderbitzin.a@bluewin.ch
WWW: http://www.uitikon.ch/sternw.htm
City Reference Coordinates: 008°33'00"E 47°23'00"N (Zürich)

Swiss Association for Standardization
• See "Schweizerische Normen-Vereinigung (SNV)"

Swiss Museum of Transport and Communications
• See "Verkehrshaus der Schweiz"

Swiss National Science Foundation
• See "Schweizerischer Nationalfonds zur Förderung der Wissenschaftlichen Forschung"

Ufficio Federale di Meteorologia e Climatologia (MeteoSvizzera)
• See "MeteoSwiss"

Union Internationale des Télécommunications (UIT)
• See "International Telecommunication Union (ITU)"

Universität Basel, Astronomisches Institut
Venusstrasse 7
CH-4102 Binningen
Telephone: (0)61-2055454
Telefax: (0)61-2055455
Electronic Mail: <userid>@unibas.ch
WWW: http://www.astro.unibas.ch/
Founded: 1890
Staff: 20
Activities: stellar astronomy * stellar photometry * galactic structure * SN * extragalactic astronomy * cosmology
Periodicals: "Preprints"
Coordinates: 007°35'00"E 47°32'30"N H318m (Basel)
007°28'48"E 47°28'30"N H557m (Metzerlen – IAU Codes 577 & 590)
City Reference Coordinates: 007°35'00"E 47°33'00"N

Universität Basel, Theoretische Kern/Teilchen- und Astrophysik
Institut für Physik
Klingelbergstrasse 82
CH-4056 Basel
Telephone: (0)61-2673111
(0)61-2673691
Telefax: (0)61-2673784

Electronic Mail: <userid>@quasar.physik.unibas.ch
WWW: http://quasar.physik.unibas.ch/
Staff: 24
Activities: nuclear astrophysics * nucleosynthesis * stellar evolution * novae * SN * hydrodynamics * neutron stars * gamma and X-ray bursts
City Reference Coordinates: 007°35'00"E 47°33'00"N

Universität Bern, Astronomisches Institut
Sidlerstrasse 5
CH-3012 Bern
Telephone: (0)31-6318591
Telefax: (0)31-6313869
Electronic Mail: <userid>@aiub.unibe.ch
WWW: http://www.cx.unibe.ch/aiub/
Founded: 1921
Staff: 25
Activities: fundamental astronomy * satellite geodesy
Periodicals: (1) "Mitteilungen der Satellitenbeobachtungsstation Zimmerwald" (circ.: 100)
Coordinates: 007°25'43"E 46°57'13"N H563m (IAU Code 009)
             007°27'54"E 46°52'36"N H929m (Zimmerwald Stn – IAU Code 026)
City Reference Coordinates: 007°26'00"E 46°57'00"N

Universität Bern, Institut für Angewandte Physik
Sidlerstrasse 5
CH-3012 Bern
Telephone: (0)31-6318911
Telefax: (0)31-6313765
Electronic Mail: <userid>@sun.iap.unibe.ch
                iapemail@iap.unibe.ch
WWW: http://www.cx.unibe.ch/iap/
Founded: 1967
Activities: solar flares
City Reference Coordinates: 007°26'00"E 46°57'00"N

Université de Lausanne, Institut d'Astronomie
Bâtiment des Sciences Physiques
Dorigny
CH-1015 Lausanne
Telephone: (0)21-6922373
Electronic Mail: <userid>@obs.unige.ch
WWW: http://obswww.unige.ch/
Founded: 1818
Staff: 10
City Reference Coordinates: 006°38'00"E 46°31'00"N

Université de Lausanne, Institut d'Astronomie, Observatoire
CH-1290 Chavannes-des-Bois
Telephone: (0)22-7552611
Telefax: (0)22-7553983
Electronic Mail: <userid>@obs.unige.ch
WWW: http://obswww.unige.ch/
     http://obswww.unige.ch/~mermio/ia/
Activities: stellar photometry (calibration, data base, CP stars, stellar clusters)
Coordinates: 006°08'12"E 46°18'24"N H455m
City Reference Coordinates: 006°08'00"E 46°18'00"N

Urania-Sternwarte
Uraniastrasse 9
CH-8001 Zürich
Electronic Mail: urania@astroinfo.ch
WWW: http://urania.astronomie.ch/
Founded: 1907
Staff: 10
Activities: education * public observing
City Reference Coordinates: 008°33'00"E 47°23'00"N

Verein Sternwarte Rotgrueb Rümlang (VSRR)
c/o Walter Bersinger (President)
Obermattenstrasse 9
CH-8153 Rümlang
Telephone: (0)1-8172813
Electronic Mail: bersingerw@bluewin.ch
WWW: http://ruemlang.astronomie.ch/
Periodicals: (6) "VSRR-Infoblatt" (circ.: 200)
Coordinates: 008°31'28"E 47°26'22"N H495m

City Reference Coordinates: 008°32'00"E 47°27'00"N

Verkehrshaus der Schweiz
(Musée Suisse des Transports et des Communications)
(Swiss Museum of Transport and Communications)
Lidostrasse 5
CH-6006 Luzern
Telephone: (0)41-3704444
Telefax: (0)41-3706168
Electronic Mail: mail@verkehrshaus.org
  planetarium@verkehrshaus.org (Planetarium)
WWW: http://www.verkehrshaus.org/
Founded: 1959 (Planetarium: 1969)
Activities: museum of transport and communication * planetarium * IMAX theatre
Coordinates: 008°20'14"E 47°03'13"N H435m
City Reference Coordinates: 008°18'00"E 47°03'00"N

Volkssternwarte Schanfigg Arosa (VSA)
Postfach 13
CH-7029 Peist
Telephone: (0)61-6927146
Telefax: (0)61-2673012
Electronic Mail: volkssternwarte@mail.com
WWW: http://www.astro.arosa.ch/
City Reference Coordinates: 009°41'00"E 46°47'00"N

Wolf Gesellschaft (Rudolf_)
• See "Rudolf Wolf Gesellschaft"

World Intellectual Property Organization (WIPO)
(Organisation Mondiale de la Propriété Intellectuelle - OMPI)
Chemin des Colombettes 34
CH-1211 Genève 20
Telephone: (0)22-7309111
Telefax: (0)22-7335428
Electronic Mail: wipo.mail@wipo.int
WWW: http://www.wipo.int/
Founded: 1970
Membership: 173 (states)
Staff: 750
Activities: United Nations specialized agency responsible for the promotion of the protection of intellectual property throughout the world through cooperation among States and for the administration of various multilateral treaties dealing with the legal and administrative aspects of intellectual property
City Reference Coordinates: 006°09'00"E 46°12'00"N

World Meteorological Organization (WMO)
(Organisation Météorologique Mondiale - OMM)
Avenue de la Paix 7bis
Case Postale 2300
CH-1211 Genève 2
Telephone: (0)22-7308111
Telefax: (0)22-7308181
Electronic Mail: wmo@gateway.wmo.ch
WWW: http://www.wmo.ch/
Founded: 1950
Membership: 185 (states & territories)
Activities: meteorology * climatology * operational hydrology
Periodicals: (4) "World Meteorological Organization Bulletin" (ISSN 0042-9767); (1) "World Meteorological Organization Annual Report"
Awards: (1) "IMO Prize", "Vilho Väisälä Award", "Norbert Gerbier-Mumm International Award", "WMO Research Award for Young Scientists"
City Reference Coordinates: 006°09'00"E 46°12'00"N

World Radiation Center
• See "Physikalisch-Meteorologisches Observatorium Davos (PMOD), Weltstrahlungszentrum"

Wyss Photo-Video (P._)
• See "P. Wyss Photo-Video"

Syria

Syrian Arab Organization for Standardization and Metrology (SASMO)
P.O. Box 11836
Damascus
Telephone: (0)11-5128225
Telefax: (0)11-5128214
WWW: http://www.syr-industry.org/standard.htm
City Reference Coordinates: 036°18'00"E 33°30'00"N

Syrian Cosmological Society (SCS)
P.O. Box 13187
Damascus
Telephone: (0)11-2776729
 (0)11-5113004
Electronic Mail: sulafa-s@scs-net.org
Founded: 1980
Membership: 150
Activities: lectures * observing * training * theoretical research * several observatories (Damascus, Alleppo, ...)
City Reference Coordinates: 036°18'00"E 33°30'00"N

Syrian Meteorological Department
P.O. Box 4211
Damascus
Telephone: (0)11-6620554
Telefax: (0)11-6620553
Founded: 1952
Staff: 600
Activities: meteorology * climatology * air pollution
Periodicals: (12) & (1) "Climatological Data"
Observatories: 21 synoptic, 98 climatological and 290 precipitation stations
City Reference Coordinates: 036°18'00"E 33°30'00"N

Taiwan (ROC)

Academia Sinica Institute of Astronomy and Astrophysics (ASIAA), Preparatory Office
P.O. Box 1-87
Taipei 11529
or :
128 Yen Chiou Yuan Road
Section 2
Nankang
Taipei 11529
Telephone: (0)2-26522020
Telefax: (0)2-27881106
Electronic Mail: asiaa@asiaa.sinica.edu.tw
    <userid>@asiaa.sinica.edu.tw
WWW: http://www.asiaa.sinica.edu.tw/
Founded: 1993
Staff: 40
Activities: radio-astronomy * optical observing * star and planetary system formation * nuclear astrophysics * cosmology * interacting galaxies * galactic nuclei * galactic dynamics * instrumentation
Coordinates: 121°37'00"E 25°03'00"N
City Reference Coordinates: 121°32'00"E 25°05'00"N

Central Weather Bureau (CWB), Astronomical Observatory
64, Kung Yuan Road
Taipei 10039
Telephone: (0)2-3491096
Electronic Mail: <userid>@cwb.gov.tw
WWW: http://www.cwb.gov.tw/
Founded: 1945
Staff: 5
Activities: solar observing
Periodicals: (2) "Report on Sunspot Observations"
Coordinates: 121°30'24"E 25°02'23"N
City Reference Coordinates: 121°32'00"E 25°05'00"N

Directorate General of Telecommunications (DGT), Lunping Observatory
180 Lunping
Kuanyin
Taoyuan 32814
Telephone: (0)3-4263467
Telefax: (0)3-4987444
Electronic Mail: lunping@dgt.gov.tw
Founded: 1965
Activities: sunspot * geomagnetism * ionosphere
Periodicals: (1) "Report of the Lunping Observatory - Sunspots" (ISSN 0254-3796), "Report of the Lunping Observatory - Geomagnetism" (ISSN 1019-8695)
Coordinates: 120°10'00"E 25°00'00"N H100m
City Reference Coordinates: 120°10'00"E 25°00'00"N

Lunping Observatory
• See "Directorate General of Telecommunications (DGT), Lunping Observatory"

National Central University (NCU), Institute of Astronomy
Chung-li 32054
Telephone: (0)3-4262302
Telefax: (0)3-4262304
Electronic Mail: astronomy@phyast.phy.ncu.edu.tw
WWW: http://www.phy.ncu.edu.tw/
Founded: 1992
Staff: 8
Activities: research in astronomy and astrophysics
Coordinates: 121°11'12"E 24°58'12"N H152m
City Reference Coordinates: 121°08'00"E 24°55'00"N

National Tsing Hua University, Department of Physics
101 Section 2 Kuang Fu Road
Hsinchu 300
Telephone: (0)3-574-2511
Telefax: (0)3-572-3052
Electronic Mail: <userid>@phys.nthu.edu.tw
WWW: http://www.phys.nthu.edu.tw/
Founded: 1965

Staff: 70
City Reference Coordinates: 120°58'00"E 24°48'00"N

Nick Enterprise Co. Ltd.
198, 12F-3, Sec. 2
Roosevelt Road
Taipei
Telephone: (0)2-3655790
Telefax: (0)2-3687854
WWW: http://www.nick.com.tw/
Founded: 1955
Staff: 25
Activities: importing and distributing scientific equipment and astronomical instrumentation * agent of GOTO Optical Manufacturing Co.
City Reference Coordinates: 121°32'00"E 25°05'00"N

Space Theater
c/o National Museum of Natural Science
1 Kuan Chien Road
P.O. Box 1412
Taichung
Telephone: (0)4-23226940
Telefax: (0)4-23222290
WWW: http://www.nmns.edu.tw/
Founded: 1986
Staff: 13
Activities: planetarium and OmniMax film shows
City Reference Coordinates: 120°41'00"E 24°09'00"N

Taipei Astronomical Museum (TAM)
363 Kee-Ho Road
Taipei (111)
Telephone: (0)2-2831-4551
Telefax: (0)2-2831-4578
Electronic Mail: webmaster@tam.gov.tw
WWW: http://www.tam.gov.tw/
Founded: 1997
Staff: 88
Activities: sky shows * education * exhibitions * laboratory activities * conferences * field observing
Periodicals: "Astronomy Newsletter", "Taipei Skylight", "Sunspot Observations", "Almanac" (in Chinese)
Coordinates: 121°31'04"E 25°05'45"N
• Formerly "Taipei Observatory"
City Reference Coordinates: 121°32'00"E 25°05'00"N

Taipei Observatory
• See now "Taipei Astronomical Museum"

Tajikistan

Tajik Academy of Sciences, Institute of Astrophysics
Bukhoro Sir. 22
734042 Dushanbe
Telephone: (0)3772-274614
(0)3770-231432
(0)3770-275483 (Mount Sanglok Observatory)
Electronic Mail: pulat@astro.td.silk.org (P.B. Babadzhanov, Director)
WWW: http://www.glasnet.ru/~ermdtaj/projects/iastro.htm
Founded: 1932
Staff: 11
Activities: physics and dynamics of small solar-system bodies * astrometry * variable stars * structure and dynamics of galaxies * Earth ionosphere
Periodicals: "Komety y Meteory" (ISSN 0568-6199), "Bulletin" (ISSN 0568-6865)
Coordinates: 068°38'00"E 38°29'00"N H730m (Hisar Astronomical Observatory)
069°00'00"E 38°16'00"N H2300m (Mt. Sanglok Obs. – IAU Code 193)
074°00'00"E 38°00'00"N H4350m (Pamir Branch)
City Reference Coordinates: 068°48'00"E 38°35'00"N

Thailand

Chiang Mai University (CMU), Department of Physics
Chiang Mai 50002
Telephone: (0)53-222699 x3367
Telefax: (0)53-222268
WWW: http://physics.science.cmu.ac.th/
Founded: 1977
Staff: 6
Activities: photoelectric and CCD photometry of eclipsing binaries
Coordinates: 098°38'00"E 18°36'00"N H784m
City Reference Coordinates: 098°58'00"E 18°46'00"N

Chulalongkorn University, Department of Physics
Bangkok 10330
Telephone: (0)2-2527985
         (0)2-2185280
Telefax: (0)2-2531150
         (0)2-2531130
WWW: http://www.sc.chula.ac.th/department/Physics/eindex.htm
     http://www.phys.sci.chula.ac.th/
Founded: 1934
Staff: 7
Activities: education * research * astronomy unit
Coordinates: 100°32'00"E 13°45'00"N H15m
City Reference Coordinates: 100°32'00"E 13°45'00"N

Thai Astronomical Society (TAS)
c/o Bangkok Planetarium
928 Sukhumvit Road
Bangkok 10110
Telephone: (0)3-3902301
         (0)3-3902305
Electronic Mail: thaiastro@inet.co.th
WWW: http://thaiastro.nectec.or.th/
     http://thaiastro.nectec.or.th/eng/index.html
Founded: 1981
Membership: 3000
Activities: star parties * quiz contests * astronomical projects contests * lectures
Periodicals: (6) "Tang Chang Pueg" (ISSN 0858-2637)
Awards: (1) contest awards
Coordinates: 100°30'00"E 13°45'00"N
City Reference Coordinates: 100°30'00"E 13°45'00"N

Thai Industrial Standards Institute (TISI)
Rama 6 Street
Ratchathewi
Bangkok 10400
Telephone: (0)2-20233004
Telefax: (0)2-2023415
Electronic Mail: thaistan@tisi.go.th
WWW: http://www.tisi.go.th/
Founded: 1968
Staff: 530
Activities: standardization * product and quality system certification * testing * laboratory accreditation
Periodicals: (4) "TISI Newsletter" (ISSN 0858-4648)
City Reference Coordinates: 100°32'00"E 13°45'00"N

Thai Meteorological Department
4353 Sukhumvit Road
Bang-Na
Phaakanong
Bangkok 10260
Telephone: (0)2-3989886
         (0)2-2580437
         (0)2-2587057
Telefax: (0)2-3989886
         (0)2-3994014
         (0)2-3984972
         (0)2-3989816
WWW: http://www.ncrt.go.th/htmlpages/met/wattana.html
     http://www.thaimet.tmd.go.th/
Founded: 1942

Staff: 1030
Activities: weather forecast * warning centre for severe weather conditions and floods
City Reference Coordinates: 100°32'00"E 13°45'00"N

Trinidad & Tobago

Caribbean Meteorological Organization (CMO)
P.O. Box 461
Port of Spain
Trinidad
Telephone: 624-4481
Telefax: 623-3634
Founded: 1951
Activities: meteorology * operational hydrology
City Reference Coordinates: 061°31'00"W 10°39'00"N

Meteorological Office, Crown Point
International Airport
Crown Point
Tobago
Telephone: 639-8780
Telefax: 639-9987
Founded: 1967
Staff: 22
Activities: weather observing * lectures * national emergency management centre
Coordinates: 060°50'00"W 11°09'00"N H3m
City Reference Coordinates: 060°50'00"W 11°09'00"N

Meteorological Office, Piarco
P.O. Box 2141
National Mail Centre
Piarco
or :
Piarco International Airport
Piarco
Trinidad
Telephone: 669-5465
Telefax: 669-4009
Electronic Mail: dirmet@tstt.net.tt
Founded: 1946
Staff: 69
Activities: weather forecast * meteorological activities for aviation, public and agriculture * climatology * tropical cyclone monitoring * upper air * equipment maintenance
Periodicals: (4) "Climate Summary"; (1) "Climate Summary"
Coordinates: 061°21'00"W 10°35'00"N
City Reference Coordinates: 061°21'00"W 10°35'00"N

Trinidad & Tobago Bureau of Standards (TTBS)
P.O. Box 467
Port of Spain
or :
Lot 1 Century Drive
Trincity Industrial Estate
Macoya
Tunapuna
Telephone: 662-8827
Telefax: 663-4335
Electronic Mail: ttbs@ttbs.org.tt
WWW: http://www.ttbs.org.tt/
Founded: 1974
Staff: 60
Activities: developing and monitoring national standards for non-food items * certification of products * quality assurance training
Periodicals: (4) "The Standard"
City Reference Coordinates: 061°31'00"W 10°39'00"N (Port of Spain)
　　　　　　　　　　　　　　　061°23'00"W 10°38'00"N (Tunapuna)

University of West Indies (UWI), Trinidad, Department of Physics
Saint Augustine
Trinidad
Telephone: 663-1369
Telefax: 663-9686
Electronic Mail: <userid>@uwimona.edu.jm
WWW: http://www.uwi.tt/
Founded: 1964
Staff: 10
Activities: education * research * solar energy * liquid crystals * computational physics * digital communications
City Reference Coordinates: 061°31'00"W 10°39'00"N (Port of Spain)

Turkey

Ankara University, Department of Astronomy and Space Science
Degol Cad.
Tandogan
06100 Ankara
Telephone: (0)312-2126720
(0)312-2121364
Telefax: (0)312-2232395
WWW: http://www.ankara.edu.tr/~astro
Founded: 1943
Staff: 17
Activities: close binary stars * stellar spectroscopy * photoelectric photometry * long-period variables
Coordinates: 032°46'48"E 39°50'37"N H1,257m
City Reference Coordinates: 032°52'00"E 39°56'00"N

Bilkent University, Department of Physics
Bilkent
06533 Ankara
Telephone: (0)312-2664397
Telefax: (0)312-2664579
Electronic Mail: astro@fen.bilkent.edu.tr
WWW: http://www.fen.bilkent.edu.tr/~physics/
http://www.fen.bilkent.edu.tr/~astro/
City Reference Coordinates: 032°52'00"E 39°56'00"N

Bogaziçi University, Kandilli Observatory and Earthquake Research Institute
Cengelköy
81220 Istanbul
Telephone: (0)216-3080514
Telefax: (0)216-3321711
Electronic Mail: kandil@boun.edu.tr
WWW: http://www.boun.edu.tr/
http://www.koeri.boun.edu.tr/astronomy/astronomy.html
Founded: 1911
Staff: 70
Activities: solar activity * flares * astrometry * history of astronomy * earthquakes and related topics * geodesy * geophysics
Coordinates: 029°03'39"E 41°03'49"N H120m
City Reference Coordinates: 028°58'00"E 41°01'00"N

Cukurova University, Department of Physics, Center for Space Sciences
Balcali
01330 Adana
Telephone: (0)71-326941
(0)71-326942 x2480
(0)71-326942 x2691
Telefax: (0)71-326945
(0)71-326070
WWW: http://www.cc.cu.edu.tr/
Founded: 1990
Staff: 6
Activities: radioastronomy * X- and gamma-ray astronomy * remote sensing applications * data analysis and image processing * telescope site selection * photometry
City Reference Coordinates: 035°18'00"E 37°01'00"N

Ege University, Department of Astronomy and Space Sciences
35100 Bornova
Telephone: (0)232-3880110 x2332
Electronic Mail: <userid>@ege.edu.tr
WWW: http://bornova.ege.edu.tr/
http://bornova.ege.edu.tr/~fenfak/astronomy/bolum.html
http://www.sci.ege.edu.tr/~euobs/
http://astronomy.sci.ege.edu.tr/
Founded: 1965
Staff: 20
Activities: photoelectric photometry * instrinsic and eclipsing variable stars * education * cosmology
Periodicals: "Publications of the Ege University Observatory"
Coordinates: 027°16'30"E 38°23'54"N H795m
027°10'03"E 38°23'36"N H20m
City Reference Coordinates: 027°15'00"E 38°28'00"N

Erciyes University, Department of Astronomy and Space Sciences
Kayseri

Telephone: (0)352-4374901
(0)352-4333110
Telefax: (0)352-4374933
Electronic Mail: <userid>@erciyes.edu.tr
WWW: http://fenedebiyat.erciyes.edu.tr/akademik/bolumler/Astronomi/astro.htm
City Reference Coordinates: 035°28'00"E 38°42'00"N

Inönü University, Physics Department
44069 Malatya
Telephone: (0)422-3410010
Telefax: (0)422-3410037
WWW: http://fef.inonu.edu.tr/fizik.html/
Founded: 1986
Activities: astrometry * stellar photometry * binary stars * variable stars
City Reference Coordinates: 038°19'00"E 38°21'00"N

Istanbul University, Astronomy and Space Sciences Department
34452 University
Telephone: (0)212-5283847
Electronic Mail: astronomy@istanbul.edu.tr
WWW: http://www.istanbul.edu.tr/fen/astronomy/
Founded: 1933
Staff: 23
Activities: solar astrophysics * stellar atmospheres * cataclysmic variables * galactic and extragalactic astronomy
Periodicals: "Publications of the Istanbul University Observatory", "Journal of Astronomy and Physics" (ISSN 1015-5295)
Coordinates: 028°57'52"E 41°00'40"N H65m (IAU 080)
City Reference Coordinates: 028°58'00"E 41°01'00"N (Istanbul)

Kandilli Observatory and Earthquake Research Institute
• See "Bogaziçi University, Kandilli Observatory and Earthquake Research Institute"

Marmara Research Center
• See "Scientific and Technical Research Council of Turkey (TÜBITAK), Marmara Research Center"

Marmara University, Department of Physics
81040 Fikirtepe
Telephone: (0)1-3459090 x145
Telefax: (0)1-3306022
Electronic Mail: physics@marun.edu.tr
WWW: http://www.marun.edu.tr/faculties/arts_sciences/physics/
City Reference Coordinates: 027°34'00"E 40°36'00"N (Marmara)

Middle East Technical University, Physics Department, Astrophysics Group
Inönü Bulvari
06531 Ankara
Telephone: (0)312-2103252
Telefax: (0)312-2101281
Electronic Mail: <userid>@astroa.physics.metu.edu.tr
WWW: http://astroa.physics.metu.edu.tr/
Founded: 1960
Staff: 20
Activities: stellar astronomy * stellar structure and evolution * X-ray astronomy * neutron stars * pulsars * photometry * spectroscopy * data analysis * pulsating stars * rotating stars
Coordinates: 032°46'28"E 39°54'10"N H954m
City Reference Coordinates: 032°52'00"E 39°56'00"N

Scientific and Technical Research Council of Turkey (TÜBITAK)
Atatürk Bulvari 221
Kavaklidere
06100 Ankara
Telephone: (0)312-4685300
Telefax: (0)312-4277489
WWW: http://www.tubitak.gov.tr/eng/index_eng.html
Founded: 1963
Activities: promoting, organising and coordinating research and development
Periodicals: (12) "Bilimve Teknik", "Içindekiler"
City Reference Coordinates: 032°52'00"E 39°56'00"N

Scientific and Technical Research Council of Turkey (TÜBITAK), Marmara Research Center, Space Sciences Department
P.O. Box 21
Gebze
41470 Kocaeli
Telephone: (0)262-6412300 x3300, x2165

Telefax: (0)262-6412309
WWW: http://www.mam.gov.tr/
Founded: 1991
Staff: 26
Activities: radio astronomy * remote sensing * astrophysics * radio physics * instrumentation
Periodicals: (1) "Annual Report"
Observatories: 1 (Marmara Radio Telescope Observatory, Gebze)
City Reference Coordinates: 029°55'00"E 40°46'00"N

TÜBITAK
• See "Scientific and Technical Research Council of Turkey (TÜBITAK)"

TÜBITAK Ulusal Gozlemevi (TUG)
• See "Turkish National Observatory"

Turkish National Observatory
(TÜBITAK Ulusal Gozlemevi - TUG)
Akdeniz Üniversitesi Kampüsü
Dumlupinar Bulvari
07058 Antalya
Telephone: (0)242-2278401
Telefax: (0)242-2278400
Electronic Mail: tug@tug.tubitak.gov.tr
WWW: http://www.tug.tubitak.gov.tr/
Founded: 1997
Staff: 29
Activities: stellar photometry * spectroscopy * astrometry
Coordinates: 030°20'08"E 36°49'30"N H2,550m
City Reference Coordinates: 030°42'00"E 36°53'00"N (Antalya)

Turkish Physical Society
c/o N. Enduran (General Secretary)
Physics Department
Istanbul University
Vezneciler
34459 Istanbul
Electronic Mail: erduran@istanbul.edu.tr
Founded: 1950
Membership: 50
City Reference Coordinates: 028°58'00"E 41°01'00"N

Turkish Standards Institution (TSE)
Necatibey Cad. 112
Bakanliklar
Ankara
Telephone: (0)312-4178330
Telefax: (0)312-4254399
Electronic Mail: didb@tse.org.tr
WWW: http://www.tse.org.tr/English/default.asp
Founded: 1954
Staff: 1000
Activities: preparing standards * certification * quality control * metrology * education * promotion * testing * applied industrial research * publishing
Periodicals: (12) "Standard"; (1) "Catalog of Turkish Standards"
City Reference Coordinates: 032°52'00"E 39°56'00"N

Turkish State Meteorological Service
P.O. Box 401
Ankara
Telephone: (0)312-3141616
Telefax: (0)312-3593430
WWW: http://www.meteor.gov.tr/
Founded: 1937
Staff: 3500
Activities: meteorological data collecting and archiving * forecasting * 270 climatic, 91 synoptic, 30 marine, 34 aeronautic and 7 Rawinsonde stations
Periodicals: (12) "Meteorological Bulletin"
City Reference Coordinates: 032°52'00"E 39°56'00"N

Ukraine

Chernigovka Amateur Astronomical Society
c/o School n° 1
Babenko Ul. 9
71200 Chernigovka
Telephone: (0)61-4092911
Founded: 1972
Membership: 15
Activities: comets * instrumentation * eclipses
Coordinates: 036°14'58"E 47°11'14"N H120m
City Reference Coordinates: 036°12'00"E 47°11'00"N

Crimean Amateur Astronomical Society (CAAS)
Gogol'a 26
P.O. Box 52
333000 Simferopol
Telephone: (0)652-252558
Telefax: (0)652-270829
Electronic Mail: <userid>@ssaa.cris.crimea.ua
WWW: http://www.cris.net/~astro
Founded: 1946
Membership: 200
Activities: observing (meteors, Sun) * education
Coordinates: 034°07'00"E 44°58'00"N
             034°59'55"E 44°55'00"N (Sidak)
City Reference Coordinates: 034°07'00"E 44°58'00"N

Crimean Astrophysical Observatory (CrAO)
Nauchny
Bahchisarai
334413 Crimea
Telephone: (0)6554-71161
           (0)655-471166
Telefax: (0)6554-40704
         (0)655-471754
Electronic Mail: <userid>@crao.crimea.ua
WWW: http://www.sai.crimea.ua/
Founded: 1945
Staff: 380
Activities: galaxies * stars * gas and dust nebulae * Sun * solar system * solar and stellar activity * gamma-ray astronomy
Periodicals: "Izvestiia" (ISSN 0367-8466); "Bulletin" (ISSN 0190-2717)
Coordinates: 034°01'00"E 44°43'42"N H550m (Nauchny - IAU Code 095)
City Reference Coordinates: 033°59'00"E 44°24'00"N

Crimean Astrophysical Observatory (CrAO), Radioastronomical Department, Katsiveli
334247 Crimea
Telephone: (0)654-727952
Telefax: (0)654-727961
Electronic Mail: <userid>@nad.crimea.ua
WWW: http://www.sai.crimea.ua/
Founded: 1945 (CrAO)
Coordinates: 033°58'47"E 44°23'53"N H25m
City Reference Coordinates: 033°59'00"E 44°24'00"N

Crimean Astrophysical Observatory (CrAO), Radioastronomical Department, Simeiz
Mount Koshka
334242 Crimea
Telephone: (0)654-771079
Telefax: (0)654-727961
Electronic Mail: <userid>@cat.crimea.ua
WWW: http://www.sai.crimea.ua/
Founded: 1945 (CrAO)
Coordinates: 033°59'48"E 44°24'12"N H346m (IAU Code 094)
City Reference Coordinates: 033°59'00"E 44°24'00"N

Donetsk Planetarium
Artema Ul. 165
340048 Donetsk
Telephone: (0)622-552508
Telefax: (0)622-557462
Electronic Mail: filippov@kinetic.ac.donetsk.ua (Irina Filippova, Director)
Founded: 1962

Staff: 8
Activities: shows * education * lectures
City Reference Coordinates: 037°48'00"E 48°00'00"N

International Centre of Astronomical, Medical and Ecological Research (ICAMER)
27 Zabolotnoho Street
03680 Kyiv
Telephone: (0)44-2662286
Telefax: (0)44-2662147
Electronic Mail: <userid>@mao.kiev.ua
WWW: http://www.mao.kiev.ua/icamer/
Founded: 1992
Staff: 80
Activities: kinematical and physical characteristics of celestial objects * developing and improving astronomical instrumentation * atmospheric and solar physics * high-altitude medicine
Coordinates: 042°29'59"E 43°16'34"N H3,100m (Terskol)
City Reference Coordinates: 030°30'00"E 50°25'00"N

Kalinenkov Astronomical Observatory
● See "Mykolaiv State Pedagogical Institute, Kalinenkov Astronomical Observatory"

Kharkov Amateur Astronomers' Society (KAAS)
P.O. Box 8857
61058 Kharkiv
Telephone: (0)572-437062
Electronic Mail: root@astro.org.ua
WWW: http://www.astro.org.ua/
Founded: 1999
Membership: 30
Activities: photography * comets * eclipses * meteor * history of astronomy * development of information resources * variable stars
Coordinates: 036°15'00"E 50°00'00"N H138m
City Reference Coordinates: 036°15'00"E 50°00'00"N

Kharkiv State University, Astronomical Observatory
Sumskaja Ul. 35
310022 Kharkiv
Telephone: (0)572-432428
Electronic Mail: <userid>@astron.kharkov.ua
WWW: http://khassm.virtualave.net/ (Multiwave Station of Solar Monitoring)
Founded: 1883
Staff: 50
Activities: minor planets * planets * Moon * Sun * astrometry * Earth rotation * speckle interferometry * atmospheric seeing * statistics of stellar systems * QSO statistics
Coordinates: 036°13'54"E 50°00'12"N H138m (IAU Code 101)
036°56'12"E 49°38'34"N (Chuguev – IAU Code 121)
City Reference Coordinates: 036°15'00"E 50°00'00"N

Kharkiv Y.A. Gagarin Planetarium
Per. Kravtsova 15
61003 Kharkiv
Telephone: (0)572-439190
Telefax: (0)572-433368
Electronic Mail: planeta@igrocom.ua
WWW: http://www.planetarium.com.ua/
Founded: 1957
Staff: 23
Activities: education * observing
Coordinates: 036°10'00"E 50°00'00"N H130m
City Reference Coordinates: 036°15'00"E 50°00'00"N

Kyiv University, Astronomical Observatory
Observatorna 3
04053 Kyiv
Telephone: (0)44-2162691
Telefax: (0)44-2162630
Electronic Mail: <userid>@aoku.freenet.kiev.ua
WWW: http://observ.univ.kiev.ua/
Founded: 1845
Staff: 60
Activities: astrometry * Sun (activity, MHD) * relativistic astrophysics * comets * meteors
Periodicals: "Visnyk Kyivskoho Universitetu. Astronomia", "Comet Circular"
Coordinates: 030°30'08"E 50°27'12"N H184m (IAU Code 085)
030°31'35"E 50°17'53"N H131m (Lisnyky)
029°55'00"E 50°35'00"N H140m (Pylypovychi)

• Full university name: "Kyiv National Taras Shevchenko University"
City Reference Coordinates: 030°30'00"E 50°25'00"N

Kyiv University, Astronomy and Space Physics Department
pr. Glushkova 6
03022 Kyiv
Telephone: (0)44-2664457
Telefax: (0)44-2664507
Electronic Mail: astdept@astrophys.ups.kiev.ua
WWW: http://space.univ.kiev.ua/
• Full university name: "Kyiv National Taras Shevchenko University"
City Reference Coordinates: 030°30'00"E 50°25'00"N

L'viv Astronomical Society
c/o Chair of Astrophysics
L'viv University
Kyrylo & Mephody 8
290005 L'viv
Telephone: (0)322-729088
Telefax: (0)322-729088
Electronic Mail: vita@astro.lviv.ua
Founded: 1997
Membership: 28
Activities: popularization * assistance to professional and amateur astronomers
City Reference Coordinates: 024°00'00"E 49°50'00"N

L'viv Polytechnical State University, Department of Geodesy and Astronomy
12 Bandera Street
290646 L'viv
Telephone: (0)322-398804
Telefax: (0)322-744300
Electronic Mail: <userid>@polynet.lviv.ua
WWW: http://www.polynet.lviv.ua/
Coordinates: 024°00'51"E 49°50'11"N H340m (IAU Code 068)
City Reference Coordinates: 024°00'00"E 49°50'00"N

L'viv University, Astronomical Observatory
8 Kyryla & Mefodiya Street
79005 L'viv
Telephone: (0)322-729088
Telefax: (0)322-729088
Electronic Mail: director@astro.lviv.ua
  star@astro.lviv.ua
WWW: http://www.franko.lviv.ua/ao/index_e.html
Founded: 1769
Staff: 30
Activities: solar physics and activity * variable stars * PNs * star clusters * cosmology * satellites
Coordinates: 023°57'11"E 49°55'04"N H325m (IAU Code 067)
City Reference Coordinates: 024°00'00"E 49°50'00"N

Main Astronomical Observatory (MAO)
• See "Ukrainian National Academy of Sciences, Main Astronomical Observatory (MAO)"

Mykolaiv Astronomical Observatory
1 Observatorna Street
54030 Mykolaiv
Telephone: (0)512-375714
Telefax: (0)512-565129
Electronic Mail: root@mao.nikolaev.ua
WWW: http://www.mao.nikolaev.ua/
Founded: 1821
Staff: 75
Activities: astrometry * instrumentation * time and frequency service * artificial and natural space objects * CCDs
Coordinates: 031°58'26"E 46°58'21"N H52m (IAU Code 089)
City Reference Coordinates: 032°00'00"E 46°57'00"N

Mykolaiv State Pedagogical Institute, Kalinenkov Astronomical Observatory
24 Nikolska Street
327000 Mykolaiv
Telephone: (0)512-352213
Electronic Mail: <userid>@comcent.nikolaev.ua
City Reference Coordinates: 032°00'00"E 46°57'00"N

National Centre of the Ukrainian Youth Airspace Education
26 Gagarina Street

P.O. Box 5595
3200005 Dniepropetrovsk
Telephone: (0)562-466032
Telefax: (0)562-466171
Electronic Mail: sirius@unaec.dts.iskra.dp.ua
Founded: 1996
Staff: 40
Activities: developing and conducting airspace science and technology education
City Reference Coordinates: 035°00'00"E 48°29'00"N

Nikolaev Astronomical Observatory (NAO)
• See now "Mykolaiv Astronomical Observatory"

Odessa State University, Astronomical Observatory
T.G. Shevchenko Park
270014 Odessa
Telephone: (0)482-228442
Telefax: (0)482-228442
Electronic Mail: astro@paco.odessa.ua
Founded: 1871
Staff: 150
Activities: variable stars * close binaries * cool stars * models of atmospheres and envelopes * comets * meteors * dust components * spectrophotometry * polarimetry * astrometry * instrumentation * artificial satellites * photometry * time series analysis * sky patrol
Periodicals: (1) "Odessa Astronomical Publications"
Coordinates: 030°16'15"E 46°23'50"N H19m (Mayaki – IAU Code 583)
                030°48'00"E 46°33'42"N H40m (Kryzhanovska)
                030°45'30"E 46°28'36"N H60m (Odessa – IAU Code 086)
City Reference Coordinates: 030°44'00"E 46°28'00"N

Poltava Gravimetrical Observatory
• See "Ukrainian National Academy of Sciences, Poltava Gravimetrical Observatory"

State Committee of Ukraine for Standardization, Metrology and Certification (DSTU)
174 Gorkiy Ul.
252650 Kyiv
Telephone: (0)44-2262971
Telefax: (0)44-2262970
Electronic Mail: iso@dstu1.kiev.ua
City Reference Coordinates: 030°30'00"E 50°25'00"N

Sudak Astronomical Club
c/o Nick P. Rogov
Sudak Town College
Ul. Majakovskogo 2a
334 882 Sudak
Telephone: (0)6566-23605
Telefax: (0)6522-76236
Electronic Mail: yudjin@cris.crimea.ua
Founded: 1965
Membership: 38
Activities: observing meteors
Observatories: 4 (Sudak, Simferopol, Alushta, Karadag)
City Reference Coordinates: 034°59'00"E 44°52'00"N

Ukrainian Astronomical Association (UAA)
3 Observatorna Street
04053 Kyiv
Telephone: (0)44-2162691
Telefax: (0)44-2246387
Electronic Mail: uaa@aoku.freenet.kiev.ua
WWW: http://www.mao.kiev.ua/uaa
Founded: 1991
Membership: 54
Staff: 30
Activities: supporting astronomy and space research in Ukraine * representing in Ukraine the International Astronomical Union (IAU) and other organizations * organizing conferences * coordinating activities from Ukrainian astronomical institutions
Periodicals: (3) "Ukrainian Astronomical Association Information Bulletin"
Awards: (1/2) "Outstanding Contribution to Development of Astronomy in Ukraine"
City Reference Coordinates: 030°30'00"E 50°25'00"N

Ukrainian National Academy of Sciences, Institute of Radio Astronomy
4 Chervonopraporna Street
61002 Kharkiv
Telephone: (0)572-451009

Telefax: (0)572-476506
Electronic Mail: rai@ira.kharkov.ua
WWW: http://www.ira.kharkov.ua/
Founded: 1985
Staff: 300
Activities: decameter radioastronomy * space radiophysics * mm radioastronomical equipment * ionospheric and space plasma research * radio wave propagation * antennae
Periodicals: (4) "Radio Physics and Radio Astronomy"
Coordinates: 036°54'00"E 49°42'00"N H150m (Grakovo)
City Reference Coordinates: 036°15'00"E 50°00'00"N

Ukrainian National Academy of Sciences, Main Astronomical Observatory (MAO)
27 Zabolotnoho Street
03680 Kyiv
or :
Golosiiv
252650 Kyiv 22
Telephone: (0)44-2663110
Telefax: (0)44-2662147
Electronic Mail: maouas@mao.kiev.ua
                     director@mao.kiev.ua
                     <userid>@mao.kiev.ua
WWW: http://www.mao.kiev.ua/
Founded: 1944
Staff: 200
Activities: positional astronomy * Earth rotation * planetary atmospheric physics * physics and evolution of stars and galaxies * solar physics * space plasma physics
Periodicals: "Kinematics and Physics of Celestial Bodies", "Space Science and Technology", "Astronomical Almanac"
Coordinates: 030°30'30"E 50°21'54"N H188m (IAU Code 083)
                 042°31'00"E 43°16'00"N H3,100m (High Altitude Station Terskol)
City Reference Coordinates: 030°30'00"E 50°25'00"N

Ukrainian National Academy of Sciences, Poltava Gravimetrical Observatory
27/29 Myasoedova Street
36029 Poltava
Telephone: (0)5322-72039
Electronic Mail: geo@geo.kot.poltava.ua
City Reference Coordinates: 034°35'00"E 49°35'00"N

Ukrainian Physical Society (UPS)
c/o V.O. Andreev (Executive Secretary)
Kyiv University
Academician Glushkov Avenue 6
252122 Kyiv
Telephone: (0)44-2664477
Telefax: (0)44-2208258
Electronic Mail: andreev@ipu.univ.kiev.ua
                     andreev@office.ups.kiev.ua
WWW: http://www.ups.kiev.ua/
Founded: 1993
City Reference Coordinates: 030°30'00"E 50°25'00"N

Ukrainian Youth Aerospace Association (UYAA) "Suzirya"
Head Office
P.O. Box 4406
49000 Dniepropetrovsk
or :
Institutska Street 14a
Kyiv
Telephone: (0)44-2540016
Telefax: (0)44-2540532
Electronic Mail: suzir@users.ukrsat.com
WWW: http://www.suzir.org.ua/
Founded: 1991
Membership: 14000
Staff: 15
Activities: popularization of space sciences and technologies
Periodicals: "Suzirya"
City Reference Coordinates: 035°00'00"E 48°29'00"N (Dniepropetrovsk)
                             030°30'00"E 50°25'00"N (Kyiv)

Uzhgorod State University, Laboratory for Space Research
2A Daleka Street
294000 Uzhgorod
Telephone: (0)3122-36065
Electronic Mail: space@univ.uzhgorod.ua

City Reference Coordinates: 022°15'00"E 48°38'00"N

Zaporozhye Astronomical Club Altair
c/o Viktor N. Gladkij
Mayakovskogo Prospekt 14
330035 Zaporozhye
Telephone: (0)333126
Founded: 1975
Staff: 25
Activities: observing (meteors, Moon, variable stars) * instrumentation
Coordinates: 035°08'00"E 47°48'00"N
City Reference Coordinates: 035°08'00"E 47°48'00"N

United Kingdom

Aberdeen College, Planetarium
Gallowgate Centre
Aberdeen AB25 1BN
Scotland
Telephone: (0)1224-612000
Telefax: (0)1224-612001
Electronic Mail: g.russell@abcol.ac.uk (G.A. Russell, Planetarium Development Officer)
WWW: http://www.abcol.ac.uk/planetarium/
Founded: 1978
Staff: 2
Activities: education * entertainment * theatre
City Reference Coordinates: 002°04'00"W 57°10'00"N

Abingdon Astronomical Society (AAS)
c/o Bob Dryden (Chairman)
21 Cross Road
Cholsey OX10 9PE
Telephone: (0)1491-201620
Electronic Mail: bob@drydenr.freeserve.co.uk
WWW: http://www2.prestel.co.uk/holtnet/aas/
City Reference Coordinates: 001°17'00"W 51°41'00"N (Abingdon)

Academia Europaea
31 Old Burlington Street
London W1X 1LB
Telephone: (0)20-77345402
Telefax: (0)20-72875115
Electronic Mail: acadeuro@compuserve.com
WWW: http://academia.darmstadt.gmd.de/
Founded: 1988
Membership: 1900
Staff: 4
Activities: annual meeting * study groups * conferences
Periodicals: (4) "European Review" (ISSN 1062-7987); (2) "Academia Europaea Newsletter" (ISSN 0968-3356); (1) "Academia Europaea Directory" (ISSN 1462-3854)
Awards: (1) "Erasmus Medal", "Gold Medal"
City Reference Coordinates: 000°10'00"W 51°30'00"N

Altrincham and District Astronomical Society (ADAS)
c/o Don Utton (Secretary)
34 Ollerbarrow Road
Hale
Altrincham WA15 9PQ
Telephone: (0)161-9414185
Electronic Mail: don.utton@btinternet.com
WWW: http://www.adas.u-net.com/
Founded: 1963
Membership: 31
Periodicals: "Ad Astra"
Coordinates: 002°21'00"W 53°24'00"N (Altrincham – IAU Code 495)
City Reference Coordinates: 002°21'00"W 53°24'00"N

Amateur Astronomy Centre (AAC)
• See now "Planet:Earth Centre"

Andover Astronomical Society (AAS)
c/o Chas Roach (Secretary)
Rose Cottage
1 The Rank
Fyfield
Telephone: (0)1264-771813
Electronic Mail: chas@andoverastronomy.org.uk
WWW: http://www.andoverastronomy.org.uk/
City Reference Coordinates: 001°28'00"W 51°13'00"N (Andover)

Armagh Observatory
College Hill
Armagh BT61 9DG
Northern Ireland
Telephone: (0)28-37522928
Telefax: (0)28-37527174

Weymouth DT4 9TE
Telephone: (0)1305-782871
Telefax: (0)1305-760670
Electronic Mail: bhcsales@bhc.co.uk
WWW: http://www.bhc.co.uk/
Founded: 1968
Activities: aluminium electrolytic capacitors for AC and DC applications * aluminium electrolytic energy discharge capacitors * microwave oven capacitor
City Reference Coordinates: 002°28'00"W 50°36'00"N

Birmingham Astronomical Society (BAS)
c/o Peter Bolas
4 Moat Bank
Burton-upon-Trent DE15 0QJ
Telephone: (0)1283-530786
Electronic Mail: pbolas@aol.com
WWW: http://www.birmingham-astronomy.co.uk/
Founded: 1950
Membership: 100
Activities: lectures * library * instrumentation
Periodicals: "Journal", "Communiqué"
Coordinates: 001°50'00"W 52°30'00"N
City Reference Coordinates: 001°50'00"W 52°30'00"N (Birmingham)
001°36'00"W 52°48'00"N (Burton-upon-Trent)

Blackpool and District Astronomical Society (BADAS)
c/o Terry Devon
30 Victory Road
Blackpool FY1 3JT
Telephone: (0)1253-625975
WWW: http://www.u-net.com/ph/nwgas/blackpl/blackpl.htm
Membership: 16
City Reference Coordinates: 003°03'00"W 53°50'00"N

Bolton Astronomical Society (BAS)
c/o Peter Miskiw
9 Hedley Street
Bolton BL1 3LF
Telephone: (0)1204-491568
Electronic Mail: petermiskiw@hotmail.com (Peter Miskiw, Secretary)
WWW: http://www.bolton-astronomical-society.co.uk/
Founded: 1979
Membership: 24
Activities: lectures * discussions * observing * trips
Periodicals: (4) "Newsletter"
City Reference Coordinates: 002°26'00"W 53°35'00"N

Border Astronomical Society
14 Shap Grove
Carlisle CA2 5QR
or :
Trinity School Observatory
Strand Road
Carlisle, Cumbria
Telephone: (0)1228-532724
Electronic Mail: david.pettitt@supanet.net (David Pettitt, Honorary Secretary)
Founded: 1971
Membership: 20
Activities: observing (planets, meteors, aurorae, NLCs, Sun, Moon, etc.) * instrumentation * astrophotography * magnetometry
Coordinates: 002°56'30"W 54°54'42"N H24m
City Reference Coordinates: 002°55'00"W 54°54'00"N

Bradford Astronomical Society (BAS)
c/o David Cooper
36 Pollard Lane
Undercliffe BD2 4RN
Electronic Mail: bas@akili.demon.co.uk
WWW: http://www.andybat.demon.co.uk/bas/
Founded: 1973
Membership: 40
City Reference Coordinates: 001°45'00"W 53°48'00"N (Bradford)

Braintree Astronomical Society (BAS)
c/o Iain Manning

11 Walnut Drive
Witham CM8 2ST
Telephone: (0)1376-521897
Electronic Mail: iain@manning.netlineuk.net
WWW: http://www.btinternet.com/~braintree.astro/
Founded: 1980
Membership: 45
Activities: monthly meetings
Periodicals: (12) "Braintree Astronomical Society Newsletter"
City Reference Coordinates: 000°32'00"E 51°53'00"N (Braintree)
000°38'00"E 51°48'00"N (Witham)

Breckland Astronomical Society (BAS)
18 Baxter Road
Hingham
Norwich NR9 4HY
Telephone: (0)1953-850571
Electronic Mail: sastrohing@aol.com
Founded: 1993
Membership: 154
Activities: monthly meetings * lectures
Periodicals: (12) "Breckland Astronomical Society Newsletter"
Coordinates: 000°59'28"E 52°34'50"N H58m
City Reference Coordinates: 001°14'00"E 52°37'00"N

Bridgend Astronomical Society
c/o Clive Down (Secretary)
10 Glan-y-Llyn
North Cornelly
Mid Glamorgan CF33 4EF
Wales
Telephone: (0)1656-740754
Electronic Mail: clivedown@btinternet.com
WWW: http://www.freenetpages.co.uk/hp/BridgendAS/
Founded: 1981
Membership: 55
Activities: observing * lectures
Periodicals: (4) "Bridgend Astronomical Society Newsletter" (circ.: 200)
City Reference Coordinates: 003°35'00"W 51°31'00"N (Bridgend)

Bridgwater Astronomical Society (BAS)
c/o G. MacKenzie
Dennathorne
Spurwells
Ilton
Ilminster TA19 9HP
Telephone: (0)1278-423571
Founded: 1969
Membership: 15
Activities: photography * lectures * observing * history of astronomy
Observatories: 1 (Charterhouse Observatory)
City Reference Coordinates: 003°00'00"W 51°08'00"N (Bridgwater)
002°55'00"W 50°56'00"N (Ilminster)

Bridport Astronomy Society (BAS)
3 The Green
Walditch
Bridport DT6 4LB
Telephone: (0)1305-456250
Founded: 1985
Membership: 15
Activities: meetings * lectures * discussions * visits
City Reference Coordinates: 002°46'00"W 50°44'00"N

Brighton and Hove Astronomical Society (BHAS)
c/o Gary Burford (Secretary)
27 Tangmere Road
Brighton BN1 8TJ
Telephone: (0)1273-883548 (evenings)
Electronic Mail: gary@cepheus.freeserve.co.uk
secretary@bhas.fsnet.co.uk
WWW: http://www.bhas.fsnet.co.uk/
City Reference Coordinates: 000°08'00"W 50°50'00"N (Brighton)
000°11'00"W 50°49'00"N (Hove)

Bristol Astronomical Society (BAS)
c/o Andrew Grasemann (Membership Secretary)
23 Chiltern Close
Whitchurch BS14 9RH
Telephone: (0)1275-545212
WWW: http://www.bristolastrosoc.freeserve.co.uk/
Membership: 100
City Reference Coordinates: 002°35'00"W 51°27'00"N (Bristol)

British Association of Planetaria (BAP)
c/o London Planetarium
Marylebone Road
London NW1 5LR
Telephone: (0)20-74870200 (Administration)
Telefax: (0)20-74650862
Founded: 1976
Membership: 44
City Reference Coordinates: 000°10'00"W 51°30'00"N

British Astronomical Association (BAA)
Burlington House
Piccadilly
London W1J 0DU
Telephone: (0)20-77344145
Telefax: (0)20-74394629
Electronic Mail: office@britastro.com
WWW: http://www.britastro.org/
Founded: 1890
Membership: 3000
Activities: observing (planets, Sun, Moon, meteors, comets, aurorae) * photography * telescope making * electronic imagining * artificial satellites * meetings * library
Periodicals: (6) "Journal of the British Astronomical Association" (ISSN 0007-0297); (1) "Handbook of the British Astronomical Association" (ISSN 0068-130x); "Circulars of the British Astronomical Association" (ISSN 0264-4185), "BAA Variable Star Section Circulars" (ISSN 0267-9272)
City Reference Coordinates: 000°10'00"W 51°30'00"N

British Geological Survey (BGS)
Kingsley Dunham Centre
Keyworth
Nottingham NG12 5GG
Telephone: (0)115-9363100
(0)115-9363578 (Public Relations and Media)
(0)115-9363241 (Sales Desk)
Telefax: (0)115-9363200
Electronic Mail: <userid>@nkw.ac.uk
WWW: http://www.bgs.ac.uk/
Founded: 1835
Staff: 1000
Activities: providing relevant and up-to-date geoscience information and advice for the UK, onshore, offshore and internationally
City Reference Coordinates: 001°01'00"W 52°58'00"N

British Interplanetary Society (BIS)
27/29 South Lambeth Road
London SW8 1SZ
Telephone: (0)20-77353160
Telefax: (0)20-78201504
Electronic Mail: bis.bis@virgin.net
WWW: http://www.bis-spaceflight.com/
Founded: 1933
Membership: 4500
Staff: 6
Activities: promoting and advancing knowledge relative to space research, technology and applications * reviewing national and international space activities and formulating forward-looking proposals for the advancement of space exploration and utilization
Periodicals: (12) "Spaceflight" (ISSN 0038-6340), "Journal of the British Interplanetary Society" (ISSN 0007-084x)
Awards: "Space Achievement Medals", "Patrick Moore Medals"
City Reference Coordinates: 000°10'00"W 51°30'00"N

British National Space Centre (BNSC)
151 Buckingham Palace Road
London SW1W 9SS
Telephone: (0)20-72150000
(0)20-72150807 (Information Line)
(0)20-72150806 (Press Office)
Telefax: (0)20-72150936

Electronic Mail: information@bnsc-hq.ccmail.compuserve.com
                            <userid>@bnsc-hq.ccmail.compuserve.com
                            bnscinfo@dti.gsi.gov.uk
WWW: http://www.bnsc.gov.uk/
Founded: 1985
Staff: 50
Activities: coordinating UK interests and activities in space
Periodicals: (2) "Space News"; (1) "UK Space Index", "UK Space Activities"
City Reference Coordinates: 000°10'00"W 51°30'00"N

British Standards Institution (BSI)
389 Chiswick High Road
London W4 4AL
Telephone: (0)20-89969000
Telefax: (0)20-89967400
Electronic Mail: info@bsi.org.uk
WWW: http://www.bsi.org.uk/
City Reference Coordinates: 000°10'00"W 51°30'00"N

British Sundial Society (BSS)
c/o Douglas A. Bateman (General Secretary)
4 New Wokingham Road
Crowthorne RG45 7NR
or :
c/o Robert. B. Sylvester (Membership Secretary)
Windycroft
Alexander Place
Askam-in-Furness LA16 7BT
Telephone: (0)1344-772303
              (0)1229-465536
WWW: http://www.sundialsoc.org.uk/
Founded: 1989
Membership: 630
Periodicals: (3) "Bulletin of the British Sundial Society"
City Reference Coordinates: 000°48'00"W 51°22'00"N (Crowthorne)
                                    002°45'00"W 54°37'00"N (Askam-in-Furness)

British UFO Research Association (BUFORA) Ltd.
BM Bufora
London WC1N 3XX
or :
70 High Street
Wingham CT3 1BJ
Telephone: (0)1227-722916
Telefax: (0)1227-722916
Electronic Mail: enquiries@bufora.org.uk
WWW: http://www.bufora.org.uk/
Founded: 1964
Membership: 800
Staff: 1
Activities: scientific investigation of the UFO phenomenon * monthly lectures * conferences
Periodicals: (8) "BUFORA Bulletin" (ISSN 1466-8017)
City Reference Coordinates: 000°07'00"W 50°57'00"N (London)
                                    001°13'00"E 51°17'00"N (Wingham)

Broadhurst Clarkson & Fuller Ltd.
Telescope House
63 Farringdon Road
London EC1M 3JB
Telephone: (0)20-74052156
              (0)20-74050448
Telefax: (0)20-74303471
Founded: 1750
Staff: 6
Activities: importing, renovating and distributing telescopes and accessories
City Reference Coordinates: 000°10'00"W 51°30'00"N

Cambridge Astronomical Association (CAA)
c/o Peter Howell (Membership Secretary)
18 High Street
Foxton
Cambridge CB2 6SP
Telephone: (0)1223-870665
WWW: http://www.caa-cambridge.co.uk/caahome.html
Founded: 1959
Membership: 60

Activities: monthly lectures * education * observing * instrument making
Periodicals: (6) "Capella"
Coordinates: 000°02'00"W 52°09'30"N H20m
City Reference Coordinates: 000°08'00"E 52°13'00"N

Cambridge University Astronomical Society (CUAS)
c/o Institute of Astronomy
Madingley Road
Cambridge CB3 0HA
Telephone: (0)1223-337548
Telefax: (0)1223-337523
Electronic Mail: ncd21@cam.ac.uk (James Keen)
WWW: http://www.cam.ac.uk/societies/cuas/
Founded: 1942
Membership: 450
Activities: weekly lectures in full term * observing facilities available for members * weekly social gatherings * visits to places of astronomical interest * quizes against other astronomical societies
Periodicals: (6) "Neptune"
Coordinates: 000°05'46"E 52°12'47"N H30m
City Reference Coordinates: 000°08'00"E 52°13'00"N

Cambridge University Press (CUP)
Edinburgh Building
Shaftesbury Road
Cambridge CB2 2RU
Telephone: (0)1223-325760
Telefax: (0)1223-315052
Electronic Mail: smitton@cup.cam.ac.uk
WWW: http://www.cup.cam.ac.uk/
Founded: 1534
● Publisher
City Reference Coordinates: 000°08'00"E 52°13'00"N

Cardiff Astronomical Society (CAS)
c/o Jackie Rudd (Membership Secretary)
19 Norris Close
Penarth
Vale of Glamorgan CF6 1QW
Wales
or :
c/o David Powell (Secretary)
1 Tal-y-Bont Road
Ely
Cardiff
Wales
Telephone: (0)29-20551704
Electronic Mail: cas@ilddat.demon.co.uk
WWW: http://www.astro.cf.ac.uk/cas/cas_home.html
        http://users.aol.com/astrocas/
        http://users.aol.com/astrocas/private/cas_home.html
        http://www.astro.cf.ac.uk/cas/cas_hom.html
Founded: 1975
Membership: 200
Activities: weekly lectures * observing * promoting public awareness
Periodicals: (5) "News Letter"
Awards: "Bill Sutherland Award"
City Reference Coordinates: 003°13'00"W 51°29'00"N (Cardiff)

Castle Point Astronomy Club (CPAC)
c/o Andrew P. Turner (Secretary)
3 Canewdon Hall close
Canewdon
Rochford SS4 3PY
Telephone: (0)1702-258640
Electronic Mail: apt3chc@aol.com (Andy Turner)
            100436,2022@compuserve.com (Ted Rodway)
WWW: http://www.cpac.freeserve.co.uk/
Founded: 1969
Periodicals: "Star News"
City Reference Coordinates: 000°43'00"E 51°36'00"N

Centronic Ltd.
Centronic House
King Henry's Drive
New Addington
Croydon CR9 0BG

Telephone: (0)1689-842121
Telefax: (0)1689-843053
Electronic Mail: rdsales@centronic.co.u
WWW: http://www.centronic.co.uk/
Founded: 1947
Staff: 150
Activities: manufacturing silicon photodiodes, Sun sensors, star mappers, remote sensors and detectors (UV, visible, IR, nuclear)
City Reference Coordinates: 000°10'00"W 51°30'00"N (London)

Chester Astronomical Society (CAS)
c/o Frank White (Secretary)
23 Cedar Grove
Hoole
Chester CH2 3LQ
Telephone: (0)1244-314265
WWW: http://www.u-net.com/ph/nwgas/chester/chester.htm
Founded: 1970
Membership: 60
Activities: lectures * observing * trips
Periodicals: (10) "CAS Newsletter"
City Reference Coordinates: 002°54'00"W 53°12'00"N

Chilbolton Observatory
• See "Council for the Central Laboratory of the Research Councils (CCLRC), Rutherford Appleton Laboratory (RAL), Chilbolton Observatory"

Clacton and District Astronomical Society
c/o David Pugh
40 Hawthorn Road
Clacton-on-Sea CO15 4QZ
Telephone: (0)1255-429849
City Reference Coordinates: 001°09'00"E 51°48'00"N

Cleethorpes and District Astronomical Society (CDAS)
112 Mill Road
Cleethorpes DN35 8JD
Telephone: (0)1472-692959
Founded: 1969
Membership: 40
Activities: education * observing
Coordinates: 000°02'00"W 53°33'00"N (Beacon Hill Observatory)
City Reference Coordinates: 000°02'00"W 53°33'00"N

Cleveland and Darlington Astronomical Society (CaDAS)
c/o John McCue
40 Bradbury Road
Norton
Stockton-on-Tees TS20 1LE
Telephone: (0)1642-892446
Electronic Mail: john.mccue@ntlworld.com
WWW: http://www.stocktonsfc.ac.uk/mccue/caseden.htm
Founded: 1979
Membership: 45
Activities: monthly meetings * lectures * observing (Castle Eden Public Observatory)
City Reference Coordinates: 001°34'00"W 54°31'00"N (Darlington)
001°19'00"W 54°34'00"N (Stockton-on-Tees)

Cockermouth Astronomical Society (CAS)
c/o Stuart Atkinson (Secretary)
2 Horsman Street
Cockermouth CA13 0HE
Electronic Mail: stuartatk@aol.com
WWW: http://www.cockermouthas.ic24.net/newsletter.htm
Founded: 1992
Membership: 30
City Reference Coordinates: 003°21'00"W 54°40'00"N

Cornwall Astronomical Society
c/o Steve Harris
26 Nanscober Place
Helston TR13 0SP
Telephone: (0)1326-569254
Electronic Mail: sph@eidonet.co.uk
City Reference Coordinates: 005°16'00"W 50°05'00"N

Coronado Filters
Coronado Manufacturing
Unit KC 1
Balthane Industrial Park
Ballasalla
Isle of Man IM9 2AH
Telephone: (0)1624-825410
Telefax: (0)1624-825420
Electronic Mail: info@coronadofilters.com
WWW: http://www.coronadofilters.com/
City Reference Coordinates: 004°36'00"W 54°05'00"N

Cotswold Astronomical Society (CAS)
c/o J.A. Daniel
103 The Bassetts
Cashes Green
Stroud GL5 4SL
Telephone: (0)1453-757026
Electronic Mail: cotswoldas@xoommail.com
WWW: http://members.xoom.com/CotswoldAS
Founded: 1982
Membership: 35
Activities: photography * SN patrol
Periodicals: (6) "Mercury"
City Reference Coordinates: 002°12'00"W 51°45'00"N

Council for the Central Laboratory of the Research Councils (CCLRC)
c/o Tony Buckley (Head of Public Relations)
Daresbury Laboratory
Warrington WA4 4AD
Telephone: (0)1925-603000
Telefax: (0)1925-603100
Electronic Mail: a.g.buckley@daresbury.ac.uk
WWW: http://www.cclrc.ac.uk/
Periodicals: (1) "Annual Report" (ISSN 1359-5865)
City Reference Coordinates: 002°37'00"W 53°24'00"N

Council for the Central Laboratory of the Research Councils (CCLRC), Rutherford Appleton Laboratory (RAL)
Chilton
Didcot OX11 0QX
Telephone: (0)1235-821900
(0)1235-445789
Telefax: (0)1235-445808
(0)1235-446665
Electronic Mail: <userid>@rl.ac.uk
WWW: http://www.clrc.ac.uk/
http://star-www.rl.ac.uk/ssd.html
Founded: 1965
Staff: 1400
Activities: research in astrophysics, space science, remote sensing, particle physics, lasers, computing * providing facilities for universities' research
Periodicals: (1) "Annual Report" (ISSN 0263-8355); "Rutherford Appleton Laboratory Report Series"
Coordinates: 001°18'41"W 51°34'19"N
City Reference Coordinates: 001°15'00"W 51°37'00"N

Council for the Central Laboratory of the Research Councils (CCLRC), Rutherford Appleton Laboratory (RAL), Chilbolton Observatory
Stockbridge SO20 6BJ
Telephone: (0)1264-860391
Telefax: (0)1264-860142
WWW: http://www.rl.ac.uk/home.html
http://www.rl.ac.uk/rutherford.html
Founded: 1965 (RAL)
Coordinates: 001°26'13"W 51°08'40"N H85m
City Reference Coordinates: 001°29'00"W 51°07'00"N

Council for the Central Laboratory of the Research Councils (CCLRC), Rutherford Appleton Laboratory (RAL), Space Science Department (SSD)
Chilton
Didcot OX11 0QX
Telephone: (0)1235-821900
Telefax: (0)1235-445848
Electronic Mail: <userid>@rl.ac.uk
WWW: http://star-www.rl.ac.uk/ssd.html
http://www.ssd.rl.ac.uk/

http://ast.star.rl.ac.uk/astro_home_page.html (Astrophysics Division)
Founded: 1965 (RAL)
Staff: 200
Activities: collaborating with universities in astronomy and geophysics research programmes on satellite instrument design, project management and other support activities * providing national facilities in data processing, engineering, test and satellite operations * collaborating with international agencies including ESA, NASA, and Russian and Japanese organisations * liaising technically with British National Space Centre (BNSC) partners
City Reference Coordinates: 001°15'00"W 51°37'00"N

Council for the Central Laboratory of the Research Councils (CCLRC), Rutherford Appleton Laboratory (RAL), STARLINK Project
Chilton
Didcot OX11 0QX
Telephone: (0)1235-821900
Telefax: (0)1235-445848
Electronic Mail: ussc@star.rl.ac.uk
WWW: http://www.starlink.rl.ac.uk/
Founded: 1979 (RAL: 1965)
Staff: 13
Activities: data reduction and analysis
Periodicals: "Starlink Documentation Series", "Starlink Software Collection", "Starlink Bulletin"
City Reference Coordinates: 001°15'00"W 51°37'00"N

Coventry and Warwickshire Astronomical Society (CAWAS)
Telephone: (0)1296-402414
Electronic Mail: lib@cawas.freeserve.co.uk
WWW: http://www.cawas.freeserve.co.uk/
City Reference Coordinates: 001°30'00"W 52°25'00"N (Coventry)

CPC Program Library
c/o Department of Applied Mathematics and Theoretical Physics
Queen's University of Belfast
Belfast BT7 1NN
Northern Ireland
Telephone: (0)28-90273222
Telefax: (0)29-90239182
Electronic Mail: cpc@qub.ac.uk
WWW: http://cpc.cs.qub.ac.uk/
Founded: 1969
Staff: 3
Activities: storing and distributing all computer programs described and published in "Computer Physics Communications"
City Reference Coordinates: 005°55'00"W 54°35'00"N

Crawley Astronomical Society (CAS)
c/o Neil Morrison (Chairman)
73 Stafford Road
Crawley RH11 7LB
or :
c/o David Roberts
53 Stagelands
Langley Green
Crawley RH11 7PG
Telephone: (0)1293-511086 (Neil Morrison)
(0)1293-511086
Founded: 1962
Membership: 35
Activities: promoting astronomy
Periodicals: (6) "Crawley Astronomical Society Newsletter"
City Reference Coordinates: 000°12'00"W 51°07'00"N

Crayford Manor House Astronomical Society (CMHAS)
c/o Roger Derek Pickard
28 Appletons
Hadlow TN11 0DT
or :
The Manor House
Mayplace Road East
Crayford DA1 4HB
Telephone: (0)1732-850663
Electronic Mail: rdp@star.ukc.ac.uk
WWW: http://www.astronomy.freeserve.co.uk/
Founded: 1961
Membership: 70
Activities: variable stars * photoelectric photometry * occultations * instrumentation * Hewitt Camera archive * collaborating with professionals
City Reference Coordinates: 000°20'00"E 51°14'00"N (Hadlow)

000°11'00"E 51°27'00"N (Crayford)

Croydon Astronomical Society (CAS)
c/o John Murrell (Chairman)
17 Dalmeny Road
Carshalton SM5 4PW
Telephone: (0)20-86475490
Electronic Mail: johnmurrell@compuserve.com
WWW: http://www.users.dircon.co.uk/~tangel2/
Founded: 1956
Membership: 150
Activities: fortnightly meetings * observing * informal practical meetings * education * social activities
Periodicals: (6) "Altair" (circ.: 150)
Coordinates: 000°06'09"W 51°18'15"N H168m (Kenley Observatory)
City Reference Coordinates: 000°10'00"W 51°22'00"N

Culture and Cosmos
P.O. Box 1071
Bristol BS99 1HE
Telephone: (0)117-9291303
Electronic Mail: culture@caol.demon.co.uk
WWW: http://www.cultureandcosmos.com/
Founded: 1997
Staff: 4
• Periodical (2) (ISSN 1368-6534)
City Reference Coordinates: 002°35'00"W 51°27'00"N

Darlington Astronomical Society (CaDAS) (Cleveland and_)
• See "Cleveland and Darlington Astronomical Society (CaDAS)"

DERA Astronomy Society
c/o Paul Alner (Chairman)
Defence Evaluation and Research Agency
Ively Road
Farnborough GU14 0LX
Electronic Mail: pdalner@mail.dera.gov.uk (Phil Alner, Chairman)
WWW: http://www.astronomy.dera.gov.uk/
Founded: 1998
Activities: meetings * lectures * observing
City Reference Coordinates: 000°46'00"W 51°17'00"N

Devon Astronomical Association (DAA)
c/o Lawrence Harris
5 Burnham Park Road
Peverell PL3 5QB
Electronic Mail: lawrenceh@peverell.demon.co.uk
WWW: http://www.peverell.demon.co.uk/daa.html
Founded: 1974
City Reference Coordinates: 004°10'00"W 50°23'00"N (Plymouth)

Dumfries Museum and Camera Obscura
The Observatory
Dumfries DG2 7SW
Scotland
Telephone: (0)1387-253374
Telefax: (0)1387-265081
Electronic Mail: info@dumfriesmuseum.demon.co.uk
WWW: http://www.dumfriesmuseum.demon.co.uk/
Founded: 1836
Activities: museum exhibition on the history of the Observatory and related activities
City Reference Coordinates: 003°37'00"W 55°04'00"N

Dundee Astronomical Society (DAS)
37 Polepark Road
Dundee DD1 5QT
Scotland
Telephone: (0)1382-825576
Founded: 1956
Membership: 40
Activities: lectures * debates * observing * education
Periodicals: (1) "Newsletter"
City Reference Coordinates: 003°00'00"W 56°28'00"N

Eastbourne Astronomical Society (EAS)
c/o Peter B.J. Gill

18 Selwyn House
Selwyn Road
Eastbourne BN21 2LF
Telephone: (0)1323-646853
Telefax: (0)1323-646853
Electronic Mail: pbj.gill@btinternet.com
Founded: 1960
Membership: 90
Activities: monthly lectures * observing
Periodicals: (10) "Orbit"
Coordinates: 000°16'00"E 50°45'25"N
City Reference Coordinates: 000°17'00"E 50°46'00"N

East Riding Astronomical Society (HERAS) (Hull and_)
• See "Hull and East Riding Astronomical Society (HERAS)"

Edinburgh Mathematical Society (EMS)
c/o The Honorary Secretary
James Clerk Maxwell Building
Mayfield Road
Edinburgh EH9 3JZ
Scotland
Electronic Mail: edmathsoc@maths.ed.ac.uk
WWW: http://www.maths.ed.ac.uk/~edmathsoc/
Founded: 1883
Periodicals: "Newsletter", "Proceedings"
City Reference Coordinates: 003°13'00"W 55°57'00"N

EM Electronics
The Rise
Brockenhurst SO42 7SJ
Telephone: (0)1590-622934
Telefax: (0)1590-622192
WWW: http://www.emelectronics.co.uk/
Founded: 1979
Staff: 3
Activities: designing and manufacturing scientific instruments, particularly ultra-low level DC measuring equipment
City Reference Coordinates: 001°34'00"W 50°49'00"N

Engineering and Physical Sciences Research Council (EPSRC)
Polaris House
North Star Avenue
Swindon SN2 1ET
Telephone: (0)1793-444000
         (0)1793-444100 (Helpline)
Electronic Mail: infoline@epsrc.ac.uk
           <userid>@epsrc.ac.uk
WWW: http://www.epsrc.ac.uk/
Staff: 290
Activities: UK's main government agency for funding research and postgraduate training in basic science, engineering and technology
Periodicals: "Newsline"
City Reference Coordinates: 001°47'00"W 51°34'00"N

Esk Valley College Planetarium (Jewel and_)
• See "Jewel and Esk Valley College Planetarium"

European Centre for Medium-Range Weather Forecasts (ECMWF)
Shinfield Park
Reading RG2 9AX
Telephone: (0)118-9499000
Telefax: (0)118-9869450
Electronic Mail: ecmwf-director@ecmwf.int
WWW: http://www.ecmwf.int/
Founded: 1975
Staff: 140
City Reference Coordinates: 000°59'00"W 51°28'00"N

Ewell Astronomical Society (EAS)
46 Stanton Close
Epsom KT19 9NP
or :
Bourne Hall
Spring Street
Ewell KT17 1UD

Telephone: (0)20-83970385
(0)1737-353608
WWW: http://www.shine.demon.co.uk/EAS/group.htm
Founded: 1966
Membership: 107
Activities: monthly meetings * observing * instruments for loan * exhibitions * visits * excursions * books, magazines, videos for loan
Periodicals: (6) "Janus" (circ.: 110)
City Reference Coordinates: 000°15'00"W 51°21'00"N (Ewell)
000°16'00"W 51°20'00"N (Epsom)

Exeter Astronomical Society (EAS)
Penn Hill Cottage
Dunchideock
Exeter EX6 7YE
Telephone: (0)1392 832311
Electronic Mail: jruddy@cix.co.uk (John Ruddy, Chairman)
WWW: http://www.compulink.co.uk/~jruddy/eas.html
Membership: 30
Periodicals: "Helios"
City Reference Coordinates: 003°31'00"W 50°43'00"N

Explorers Tours
223 Coppermill Road
Wraysbury TW19 5NW
Telephone: (0)1753-680237
Telefax: (0)1753-682660
Electronic Mail: astro@explorers.co.uk
WWW: http://www.explorers.co.uk/
Founded: 1981
Staff: 10
Activities: organizing inclusive tours, including for observing total solar eclipses, meteor showers, aurorae and foreign observatories
City Reference Coordinates: 000°33'00"W 51°27'00"N

Federation of Astronomical Societies (FAS)
c/o Clive Down
10 Glan-y-Llyn
North Cornelly
Mid Glamorgan CF33 4EF
Wales
Telephone: (0)1656-740754
Electronic Mail: clivedown@btinternet.com
WWW: http://www.fedastro.org.uk/
Founded: 1974
Membership: 150 (societies)
Activities: linking societies together * information service
Periodicals: (5) "FAS Newsletter" (ISSN 1361-4126); (1) "Handbook", "Astrocalendar"
City Reference Coordinates: 003°10'00"W 51°30'00"N (Cardiff)

Furness Astronomical Society
c/o Richard Alldridge
56 Hartington Street
Barrow-in-Furness LA14 5SR
Telephone: (0)1229-826864
Electronic Mail: richard@ralldridge.freeserve.co.uk
WWW: http://www.furness-astro-society.org.uk/
Founded: 1973
Membership: 25
Activities: promoting astronomy * public education
Coordinates: 003°10'40"W 54°08'40"N H65m (Newton-in-Furness)
City Reference Coordinates: 003°11'00"W 54°09'00"N

Galvoptics Ltd., Telescope Division
Harvey Road
Burnt Mills Industrial Estate
Basildon SS13 1ES
Telephone: (0)1268-728077
Telefax: (0)1268-590445
Electronic Mail: sales@galvoptics.fsnet.co.uk
WWW: http://www.galvoptics.fsnet.co.uk/
Founded: 1968
Staff: 29
Activities: manufacturing lenses, prisms, filters, mirrors * optical coatings on glass, low-expansion material, quartz, saphire, fused silica, and other optical materials * removal and recoat of mirrors
City Reference Coordinates: 000°25'00"E 51°35'00"N

Glasgow College of Nautical Studies (GCNS), Planetarium
21 Thistle Street
Glasgow G5 9XB
Scotland
Telephone: (0)141-4293201
(0)141-5652700
Telefax: (0)141-4201690
(0)141-5652599
Electronic Mail: maritime@glasgow-nautical.ac.uk
WWW: http://www.glasgow-nautical.ac.uk/
Founded: 1969
Activities: demonstrations to clubs, societies, schools and public
City Reference Coordinates: 004°15'00"W 55°53'00"N

Guildford Astronomical Society (GAS)
c/o Alan Langmaid (Secretary)
22 West Mount
The Mount
Guildford GU2 5HL
Telephone: (0)1483-538677
WWW: http://www.laverhome.demon.co.uk/GuildfordAS/
Founded: 1955
City Reference Coordinates: 000°35'00"W 51°14'00"N

Gwynedd Astronomical Society
Ael-y-Bryn
Newborough
Llanfairpwll
Gwynedd, Wales
or :
18 Lon y Gamfa
Menai Bridge
Gwynedd LL59 5QJ
Telephone: (0)1248-440395
Electronic Mail: oss041@bangor.clss1
Founded: 1974
Membership: 15
Activities: lectures * observing
City Reference Coordinates: 004°08'00"W 53°13'00"N (Bangor)

Hampshire Astronomical Group (HAG)
10 Marie Court
348 London Road
Waterlooville PO7 7SR
Telephone: (0)2392-232491
Electronic Mail: info@hantsastro.org.uk
WWW: http://www.hantsastro.org.uk/
Founded: 1960
Membership: 122
Activities: observing * education * visits * weekly meetings * monthly lectures
Periodicals: (4) "The Hampshire Observer"
Awards: "Honorary Membership", "Life Membership"
Coordinates: 001°01'05"W 50°56'14"N H161m (Clanfield Obs. – IAU Code 940)
City Reference Coordinates: 001°02'00"W 50°53'00"N

Harrogate Astronomical Society (HAS)
1 York Lane
Knaresborough HG5 0AJ
Telephone: (0)1423-862002
WWW: http://www.harrogate-astro.freeserve.co.uk/
Founded: 1986
Membership: 35
Activities: monthly meetings * observing * talks to local groups
Periodicals: (4) "Astro News" (circ.: 30)
Coordinates: 001°41'55"W 53°58'40"N (Washburn Valley)
City Reference Coordinates: 001°33'00"W 54°00'00"N (Harrogate)

Hastings and Battle Astronomical Society
c/o Keith Woodcock
24 Emmanuel Road
Hastings TN34 3LB
Telephone: (0)1424-443883
Electronic Mail: keith@habas.freeserve.co.uk
City Reference Coordinates: 000°36'00"E 50°51'00"N

Havering Astronomical Society
c/o Frances Ridgley
133 Severn Drive
Upminster, Essex
Telephone: (0)1708-227397
Electronic Mail: 100304.2143@compuserve.com
WWW: http://homepages.tesco.net/~nik.szymanek/havering.htm
Founded: 1994
Membership: 50
City Reference Coordinates: 000°15'00"E 51°33'00"N

Heart of England Astronomical Society (HOEAS)
c/o John Williams (Secretary)
100 Stanway Road
Shirley
Solihull B90 3JG
Electronic Mail: hoeas@aol.com
WWW: http://members.aol.com/hoeas
Founded: 1988 (1972 as "Chelmsley Astronomical Society")
City Reference Coordinates: 001°45'00"W 52°25'00"N

Hebden Bridge Literary and Scientific Society, Astronomy Section
c/o P.G.H. Jackson
Royal Nook
44 Gilstead Lane
Gilstead
Bingley BD16 3NP
Telephone: (0)1282-616294
Electronic Mail: j.hbas@argonet.co.uk
Founded: 1985
Activities: monthly meetings * lectures
City Reference Coordinates: 002°00'00"W 53°45'00"N (Hebden Bridge)

Helius Designs
The White House
Aldington
Evesham WR11 5UB
Telephone: (0)1386-830083
Telefax: (0)1386-830083
Electronic Mail: 100674.431@compuserve.com
Founded: 1980
Staff: 4
Activities: designing and manufacturing computer-controlled telescopes and CCD imaging systems
City Reference Coordinates: 001°56'00"W 52°06'00"N

Hencoup Enterprises
Collins Cottage
Lower Road
Loosley Row HP27 0PF
Telephone: (0)1844-347493
Telefax: (0)1844-274782
Electronic Mail: hencoup@hencoup.demon.co.uk
WWW: http://www.hencoup.com/
Founded: 1985
Activities: promoting astronomy through TV, radio, newspapers, magazines, bulletin boards, videos, etc. * consultancy work for astronomical institutions
City Reference Coordinates: 000°50'00"W 51°50'00"N (Aylesbury)

Herschel Society (William_)
• See "William Herschel Society"

H.H. Wills Physics Laboratory
• See "University of Bristol, H.H. Wills Physics Laboratory"

Highlands Astronomical Society (HAS)
c/o James F. Dick
8 Holm Park
Inverness IV2 4XT
Scotland
WWW: http://www.stoob.demon.co.uk/has.htm
City Reference Coordinates: 004°38'00"W 57°13'00"N

Huddersfield Astronomical and Philosophical Society (HAPS)
c/o Paul D. Harper (Treasurer)
45 Lydgate

Lepton HD8 0LT
Telephone: (0)1484-606832
Founded: 1968
Membership: 40
Activities: lectures * visits to other astronomical societies and places of interest * telescope making * elementary astronomy classes * observing * slide shows * videos * quizes * planetarium
Periodicals: (4) "Omega"
Awards: (1) photographic cup
Coordinates: 001°50'00"W 53°38'00"N H300m (Crosland Hill)
City Reference Coordinates: 001°47'00"W 53°39'00"N (Huddersfield)
001°50'00"W 53°38'00"N (Lepton)

Hull and East Riding Astronomical Society (HERAS)
c/o A.G. Scaife
15 Beech Road
Elloughton
Brough HU15 1JX
Telephone: (0)1482-668665 (Helen Marshall, Secretary)
Founded: 1953
Membership: 45
Activities: monthly educational meetings
City Reference Coordinates: 002°19'00"W 54°32'00"N

IGG Component Technology Ltd.
Grove Road
Cosham
Portsmouth PO6 1LX
Telephone: (0)23-92210051
Telefax: (0)23-92210058
Electronic Mail: info@igg.co.uk
WWW: http://www.igg.co.uk/
Founded: 1978
Activities: providing procurement, parts engineering expertise, testing and qualification services for space-quality electronic, electrical and electro-mechanical components
City Reference Coordinates: 001°05'00"W 50°48'00"N

Imperial College of Science, Technology and Medicine (ICSTM), Physics Department, Astrophysics Group
Prince Consort Road
London SW7 2BZ
Telephone: (0)20-75895111
Telefax: (0)20-75899463
Electronic Mail: astro@ic.ac.uk
WWW: http://icstar5.ph.ic.ac.uk/
Founded: 1984
Staff: 24
Activities: astro-particle physics * satellite IR and X-ray astronomy * extragalactic astrophysics * cosmology * SNs * stellar winds
City Reference Coordinates: 000°10'00"W 51°30'00"N

Imperial College of Science, Technology and Medicine (ICSTM), Space and Atmospheric Physics Group
Blackett Laboratory
Prince Consort Road
London SW7 2BW
Telephone: (0)20-75947770
Telefax: (0)20-75947772
Electronic Mail: <userid>@ic.ac.uk
WWW: http://www.sp.ph.ic.ac.uk/
Founded: 1986
Staff: 35
Activities: space plasma physics * atmospheric physics * instrumentation for space and atmospheric missions
City Reference Coordinates: 000°10'00"W 51°30'00"N

Imperial College Press (ICP)
57 Shelton Street
Covent Garden
London WC2H 9HE
Telephone: (0)20-78363954
Telefax: (0)20-78362002
Electronic Mail: edit@icpress.co.uk
WWW: http://www.icpress.co.uk/
Founded: 1995
Staff: 8
• Publisher
City Reference Coordinates: 000°10'00"W 51°30'00"N

INSPEC
c/o Institution of Electrical Engineers
Michael Faraday House
Six Hills Way
Stevenage SG1 2AY
Telephone: (0)1438-313311
Telefax: (0)1438-742840
Electronic Mail: inspec@iee.org.uk
WWW: http://www.iee.org.uk/publish/inspec/
Founded: 1898 (as former "Science Abstracts")
Staff: 140
Activities: producing INSPEC database (over 6.5M records) covering literature in physics (including astronomy and astrophysics), electronics and computing * service available internationally online, as abstracts journals and on CD-ROMs
Periodicals: "Physics Abstracts" (ISSN 0036-8091), "Current Papers in Physics" (ISSN 0011-3786), "Computer and Control Abstracts", "Electrical and Electronics Abstracts", "Current Papers on Electrical and Electronics Engineering", "Current Papers on Computers and Control", plus other titles in Key Abstracts
City Reference Coordinates: 000°14'00"W 51°55'00"N

Institute of Physics (IOP)
76 Portland Place
London W1N 3DH
Telephone: (0)20-74704800
Telefax: (0)20-74704848
Electronic Mail: physics@iop.org
WWW: http://www.iop.org/
Founded: 1874
Membership: 20000
Activities: publishing books and journals * organising meetings and conferences * education
Periodicals: (50) "Journal of Physics C: Condensed Matter" (ISSN 0953-8984); (40) "Public Understanding of Science" (ISSN 0963-6625); (24) "Journal of Physics A: Mathematical and General" (ISSN 0305-4470), "Journal of Physics B: Atomic, Molecular and Optical Physics" (ISSN 0953-4075); (12) "Classical and Quantum Gravity" (ISSN 0264-9381), "Journal of Physics D: Applied Physics" (ISSN 0022-3727), "Journal of Physics G: Nuclear Physics" (ISSN 0954-3889), "Measurement Science and Technology" (ISSN 0957-0233), "Physics in Medicine and Biology" (ISSN 0031-9155), "Physics World" (ISSN 0953-8585), "Plasma Physics and Controlled Fusion" (ISSN 0741-3335), "Reports on Progress in Physics" (ISSN 0034-4885), "Semiconductor Science and Technology" (ISSN 0268-1242), "Superconductor Science and Technology" (ISSN 0953-2048); (6) "European Journal of Physics" (ISSN 0143-0807), "Inverse Problems" (ISSN 0266-5611), "Physics Education" (ISSN 0031-9120), "Nonlinearity" (ISSN 0951-7715), "OLE - Opto and Laser Europe" (ISSN 0966-9809), "Pure and Applied Optics" (ISSN 0963-9659), "Quantum Optics" (ISSN 0954-8998); (5) "Modelling and Simulation in Materials Science and Engineering" (ISSN 0965-0393); (4) "Distributed Systems Engineering" (ISSN 0967-1846), "High-Performance Polymers" (ISSN 0954-0083), "Journal of Hard Materials" (ISSN 0954-027x), "Journal of Micromechanics and Microengineering" (ISSN 0960-1317), "Journal of Radiological Protection" (ISSN 0952-4746), "Nanotechnology" (ISSN 0957-4484), "Network: Computation in Neural Systems" (ISSN 0954-898x), "Physiological Measurements" (ISSN 0967-3334), "Plasma Sources Science and Technology" (ISSN 0936-0252), "Smart Materials and Structures" (ISSN 0964-1726), "Waves in Random Media" (ISSN 0959-7171)
City Reference Coordinates: 000°10'00"W 51°30'00"N

Institute of Physics (IOP) Publishing Ltd.
Dirac House
Temple Back
Bristol BS1 6BE
Telephone: (0)117-9297481
Telefax: (0)117-9294318
Electronic Mail: custserv@ioppublishing.co.uk
              <userid>@ioppublishing.co.uk
              listproc@listserver.ioppublishing.com
WWW: http://www.ioppublishing.com/
Founded: 1874 (IOP)
Staff: 100
• Publisher (see list of journals under "Institute of Physics")
City Reference Coordinates: 002°35'00"W 51°27'00"N

Institution of Electrical Engineers (IEE)
Savoy Place
London WC2E 0BL
Telephone: (0)20-72401871
Telefax: (0)20-72407735
Electronic Mail: <userid>@iee.org.uk
WWW: http://www.iee.org.uk/Welcome.html
       http://www.iee.org.uk/Industry/Aerospace
Founded: 1871
Membership: 140000
Staff: 450
Activities: promoting advancement of electrical, manufacturing and information engineering * facilitating exchange of knowledge and ideas * publishing * bibliographical information service (see INSPEC)
Periodicals: "Computer and Control Abstracts" (ISSN 0036-8113), "Computing and Control Engineering Journal" (ISSN 0956-3385), "Electrical and Electronics Abstracts" (ISSN 0036-8105), "Electronics and Communication Engineering Journal"

(ISSN 0954-0695), "Electronics Education" (ISSN 0265-0096), "Electronics Letters" (ISSN 0013-5194), "Engineering Management Journal" (ISSN 0960-7919), "Engineering Science and Education Journal" (ISSN 0963-7346), "IEE News" (ISSN 0308-0684), "IEE Proceedings - Circuits, Devices and Systems" (ISSN 1350-2409), "IEE Proceedings - Communications" (ISSN 1350-2425), "IEE Proceedings - Computers and Digital Techniques" (ISSN 1350-2387), "IEE Proceedings - Control Theory and Applications" (ISSN 1350-2379), "IEE Proceedings - Electric Power Applications" (ISSN 1350-2352), "IEE Proceedings - Generation, Transmission and Distribution" (ISSN 1350-2360), "IEE Proceedings - Microwaves, Antennas and Propagation" (ISSN 1350-2417), "IEE Proceedings - Optoelectronics" (ISSN 1350-2433), "IEE Proceedings - Radar, Sonar and Navigation" (ISSN 1350-2395), "IEE Proceedings - Science, Measurement and Technology" (ISSN 1350-2344), "IEE Proceedings - Software Engineering" (ISSN 0268-6961), "IEE Proceedings - Vision, Image and Signal Processing" (ISSN 1350-245x), "IEE Review" (ISSN 0013-5127), "Manufacturing Engineer" (ISSN 0956-9944), "Medical and Biological Engineering and Computing" (ISSN 0140-0118), "Microwaves, Antennas and Propagation" (ISSN 1350-2417), "Optoelectronics" (ISSN 1350-2433), "Physics Abstracts" (ISSN 0036-8091), "Power Engineering Journal" (ISSN 0950-3366), "Radar, Sonar and Navigation" (ISSN 1350-2395)
City Reference Coordinates: 000°10'00"W 51°30'00"N

International Association of Astronomical Artists (IAAA), European Office
c/o Jackie Burns
1 Park Side
Fort William Road
Vange
Basildon SS16 5JX
Telephone: (0)1273)-323370
Electronic Mail: artist@jackieburns.co.uk
WWW: http://www.iaaa.org/
Founded: 1982 (USA)
Membership: 7
City Reference Coordinates: 000°25'00"E 51°35'00"N

International Information Services for the Physics and Engineering Communities
• See "INSPEC"

International Maritime Organization (IMO)
4 Albert Embankment
London SE1 7SR
Telephone: (0)20-77357611
Telefax: (0)20-75873210
Electronic Mail: info@imo.org
WWW: http://www.imo.org/
Founded: 1948
Membership: 153 (states)
Staff: 300
Sections: (committees) Maritime Safety * Maritime Environment Protection * Legal * Technical Cooperation * Facilitation
Periodicals: (4) "IMO News"
• Until 1982, known as Inter-Governmental Maritime Consultative Organization (IMCO)
City Reference Coordinates: 000°10'00"W 51°30'00"N

IOP Publishing Ltd.
• See "Institute of Physics (IOP) Publishing Ltd."

Ipswich School Astronomical Society (ISAS)
Henley Road
Ipswich IP1 3SG
Telephone: (0)1473-408300
Telefax: (0)1473-400058
Electronic Mail: isas@dial.pipex.com
WWW: http://dspace.dial.pipex.com/town/parade/lg53/
Founded: 1998
Membership: 16
Activities: weekly meetings * observing * instrumentation
City Reference Coordinates: 001°10'00"E 52°04'00"N

Irish Astronomical Association (IAA)
3 Vaddegan Avenue
Glengormley BT36 7SP
or :
4 Ailsa Park
Bangor BT19 1EA
Northern Ireland
WWW: http://www.btinternet.com/~jimmyaquarius/
Founded: 1946
Membership: 220
Activities: meetings * observing * education
Periodicals: (4) "Stardust"
City Reference Coordinates: 005°40'00"W 54°40'00"N (Bangor)

Irish Astronomical Journal (IAJ)
c/o Editorial Office
12 Heather Lea Avenue
Dore
Sheffield S17 3DJ
Electronic Mail: iaj@star.arm.ac.uk
WWW: http://star.arm.ac.uk/iaj/
Founded: 1950
Staff: 2
• Journal (2) (ISSN 0021-1052, circ.: 300)
City Reference Coordinates: 001°30'00"W 53°23'00"N

Isaac Newton Institute for Mathematical Sciences
20 Clarkson Road
Cambridge CB3 0EH
Telephone: (0)1223-335999
Telefax: (0)1223-330508
Electronic Mail: info@newton.cam.ac.uk
WWW: http://www.newton.cam.ac.uk/
Founded: 1992
Staff: 50
Activities: research linked to mathematical sciences
City Reference Coordinates: 000°08'00"E 52°13'00"N

Island Planetarium (The_)
• See "The Island Planetarium"

Isle of Man Astronomical Society
c/o James W. Martin
Ballaterson Farm
Peel
Isle of Man 1M5 3AB
Telephone: (0)1624-842954
Electronic Mail: tremott@enterprise.net (Mike Kelly, Chairman)
            gary.kewin@bigfoot.com (Gary Kewin, Vice Chairman)
WWW: http://www.iomastronomy.org/
Founded: 1989
Membership: 70
Activities: meetings * star parties
Coordinates: 004°38'00"W 54°10'00"N (Foxdale - IAU Code 987)
City Reference Coordinates: 004°30'00"W 54°15'00"N

James Lockyer Planetarium
• See "Norman Lockyer Observatory and James Lockyer Planetarium"

Jewel and Esk Valley College Planetarium
24 Milton Road East
Edinburgh EH15 2PP
Scotland
Telephone: (0)131-6577284
Telefax: (0)131-6572276
WWW: http://members.aol.com/DMaccon125/Leonardo/jewel.htm
Founded: 1978
Activities: shows for public and societies * school visits * elementary education
City Reference Coordinates: 003°13'00"W 55°57'00"N

Jodrell Bank Observatory (JBO)
• See "University of Manchester, Department of Physics and Astronomy, Jodrell Bank Observatory"

Jodrell Bank Science Centre, Planetarium and Arboretum
Macclesfield SK11 9DL
Telephone: (0)1477-571339
Telefax: (0)1477-571695
Electronic Mail: visitorcentre@jb.man.ac.uk
WWW: http://www.jb.man.ac.uk/scicen/
Founded: 1966
Staff: 15
Activities: visitor attraction for general public and groups * education
City Reference Coordinates: 002°07'00"W 53°16'00"N

JRA Aerospace & Technology Ltd.
CED House
Taylors Close

Marlow SL7 1PR
Telephone: (0)1628-891105
Telefax: (0)1628-890519
Electronic Mail: mail@jratech.co.uk
WWW: http://www.jratech.co.uk/
Founded: 1988
Activities: technology transfer consultancy (technology audit, assessment, bid support and marketing services)
City Reference Coordinates: 000°48'00"W 51°35'00"N

Kendall Hyde Ltd.
Kingsland Industrial Park
Stroudley Road
Basingstoke RG24 0UG
Telephone: (0)1256-840830
Telefax: (0)1256-840443
Founded: 1973
Staff: 20
Activities: optical coating engineering
City Reference Coordinates: 001°05'00"W 51°15'00"N

Kettering and District Amateur Astronomers Society (KDAAS)
• See now "Northants Amateur Astronomers (NAA)"

Lancaster and Morecambe Astronomical Society
4 Bedford Place
Scotforth
Lancaster LA1 4EB
Telephone: (0)1524-36211
Electronic Mail: ehelenerob@btinternet.com (E. Robinson, Secretary)
Founded: 1978
Membership: 30
Activities: monthly meetings * lectures
City Reference Coordinates: 002°48'00"W 54°03'00"N (Lancaster)
002°53'00"W 54°04'00"N (Morecambe)

Lancaster University, Environmental Science Department, Planetary Science Research Group (PS-RG)
Environmental Science Department
Lancaster LA1 4YQ
Telephone: (0)1524-593889
Telefax: (0)1524-593985
WWW: http://www.es.lancs.ac.uk/es/research/psrg/psrg.htm
Founded: 1970
Staff: 12
Activities: planetary geology, especially volcanology and tectonics
City Reference Coordinates: 002°48'00"W 54°03'00"N

Leeds Astronomical Society (LAS)
c/o Mark Simpson (PRO)
37 Roper Avenue
Gledhow
Leeds LS8 1LG
or :
4 Belvedere Grove
Leeds LS17 8BP
Electronic Mail: r.emery@westview.demon.co.uk (Ray Emery, President)
astro@westview.demon.co.uk
WWW: http://www.dsellers.demon.co.uk/las/
http://www.westview.demon.co.uk/las/
Founded: 1859
Membership: 45
Activities: meetings * visits * observing
Periodicals: (4) "Nebula"
City Reference Coordinates: 001°35'00"W 53°50'00"N

Leicester Astronomical Society (LAS)
c/o Ann Bonell (Vice-President)
53 Wardens Walk
Leicester Forest East LE3 3GG
WWW: http://www.leicastrosoc.freeserve.co.uk/
Founded: 1952
Membership: 60
Activities: meetings * observing
City Reference Coordinates: 001°05'00"W 52°38'00"N

Leicester University Astronomy Society
c/o Department of Physics and Astronomy
University of Leicester
University Road
Leicester LE1 7RH
Electronic Mail: klp5@le.ac.uk (President)
WWW: http://www.le.ac.uk/astrosoc/
Membership: 100
Activities: observing evenings * social events
City Reference Coordinates: 001°05'00"W 52°38'00"N

Letchworth and District Astronomical Society (LDAS)
c/o Alec Wilkinson (Secretary)
54 Broadmead
Hitchin SG4 9LX
Telephone: (0)1462-623654
Electronic Mail: secretary@ldas.org.uk
WWW: http://www.ldas.org.uk/
Founded: 1991 (1985 as "Stevenage & District Astronomical Society" - SADAS)
Membership: 70
Activities: meetings * observing
City Reference Coordinates: 000°14'00"W 51°58'00"N (Letchworth)
000°17'00"E 51°57'00"N (Hitchin)

Liverpool Astronomical Society (LAS)
31 Sandymount Drive
Wallasey L45 0LJ
Telephone: (0)151-7945356
Electronic Mail: ggastro@liverpool.ac.uk (Gerard Gilligan)
WWW: http://www.liv.ac.uk/~ggastro/home.html
Founded: 1881
Activities: public star parties * meetings * education
Periodicals: (1) "Liverpool's Monthly Sky Diary", "Astronomical Events for the Liverpool Area"
Awards: "Silver Medal"
City Reference Coordinates: 002°55'00"W 53°25'00"N (Liverpool)
003°03'00"W 53°26'00"N (Wallasey)

Liverpool John Moores University, Astrophysics Research Institute
Twelve Quays House
Egerton Wharf
Birkenhead L41 1LD
Telephone: (0)151-2312337
Telefax: (0)151-2312475
Electronic Mail: <userid>@astro.livjm.ac.uk
WWW: http://www.livjm.ac.uk/astro/
Founded: 1992
Staff: 30
Activities: observational cosmology * galaxies * brown dwarfs * hot star environments * novae * star-forming regions * ISM * education
• See also "University of Liverpool"
City Reference Coordinates: 003°02'00"W 53°24'00"N

Liverpool Museum Planetarium
c/o Department of Earth and Physical Sciences
National Museums and Galleries on Merseyside
William Brown Street
Liverpool L3 8EN
Telephone: (0)151-2070001
Telefax: (0)151-4784390
WWW: http://www.nmgm.org.uk/livmus/planetariumframeset.html
Founded: 1969
Staff: 10
Activities: displays * planetarium shows * instrument collection * education * research
City Reference Coordinates: 002°55'00"W 53°25'00"N

Lockyer Observatory (Norman_)
• See "Norman Lockyer Observatory and James Lockyer Planetarium"

Lockyer Planetarium (James_)
• See "Norman Lockyer Observatory and James Lockyer Planetarium"

London Planetarium
Marylebone Road
London NW1 5LR
Telephone: (0)20-74870200 (Administration)

Telefax: (0)20-74650862
WWW: http://www.kidsnet.co.uk/places/planetar.shtml
Founded: 1958
Membership: 100
Activities: planetarium shows for public and schools * exhibitions
City Reference Coordinates: 000°10'00"W 51°30'00"N

LOT-Oriel Ltd.
1 Mole Business Park
Leatherhead KT22 7AU
Telephone: (0)1372-378822
Telefax: (0)1372-375353
Electronic Mail: info@lotoriel.co.uk
WWW: http://www.lotoriel.co.uk/
Founded: 1976
Staff: 12
Activities: distributing optical and electro-optical instruments
City Reference Coordinates: 000°20'00"W 51°18'00"N

Loughton Astronomical Society (LAS)
c/o Charles Munton (Secretary)
14a Manor Road
Wood Green
London N22 8YJ
Telephone: (0)20-8889-9253
Electronic Mail: charles@munton.u-net.com
 las@dial.pipex.com
WWW: http://www.ds.dial.pipex.com/allan.bell/las_home.htm
Founded: 1978
City Reference Coordinates: 000°10'00"W 51°30'00"N

Macclesfield Astronomical Society
c/o Cherry Moss
164A Chester Road
Macclesfield SK11 8PT
Electronic Mail: chris@motorlaw.co.uk (Christopher J. Rose, Vice-President)
WWW: http://www.employees.org/~macas/macc-as.htm
 http://www.g0-evp.demon.co.uk/
Founded: 1990
Membership: 150
City Reference Coordinates: 002°07'00"W 53°16'00"N

Maidenhead Astronomical Society
c/o T.V. Haymes
Hill Rise
Knowl Hill Common
Knowl Hill
Reading RG10 9YD
Telephone: (0)1628-822442
Electronic Mail: 101360.700@compuserve.com
Founded: 1957
Membership: 25
Activities: meetings * lectures * observing * exhibitions
City Reference Coordinates: 000°59'00"W 51°28'00"N (Reading)
 000°44'00"W 51°32'00"N (Maidenhead)

Manchester Astronomical Society (MAS)
c/o J.H.W. Davidson (Honorary Secretary)
The Godlee Observatory
Institute of Science and Technology
University of Manchester
Sackville Street
Manchester M60 1QD
Telephone: (0)161-2004977 (answering machine)
Telefax: (0)161-2287040
Electronic Mail: mike@ph.u-net.com
WWW: http://www.u-net.com/ph/mas/
Founded: 1903
Membership: 75
Activities: lectures * conventions * photography * instrumentation * Sun * Moon * meteors * occultations * popularization * advising
Periodicals: (2) "Current Notes"
Coordinates: 002°13'57"W 53°28'34"N H85m
City Reference Coordinates: 002°15'00"W 53°30'00"N

Mansfield and Sutton Astronomical Society (MSAS)
c/o Sherwood Observatory
Coxmoor Road
Sutton-in-Ashfield NG17 5LF
Telephone: (0)1623-552276
Electronic Mail: masas@innotts.co.uk
WWW: http://homepage.ntlworld.com/graham.shepherd/Sherwood_Observatory.html
Founded: 1970
Membership: 40
Activities: fostering interest in astronomy
Periodicals: (1) "Annual Reports and Accounts"
Coordinates: 001°18'21"W 53°06'50"N H188m
City Reference Coordinates: 001°11'00"W 53°09'00"N (Mansfield)
          001°15'00"W 54°14'00"N (Sutton)

Meteorological Office
London Road
Bracknell RG12 2SZ
Telephone: 0845 300 0300
     (0)1344-420242
     (0)1344-854260
Telefax: 0845 300 1300
   (0)1344-854412
   (0)1344-854948
Electronic Mail: enquiries@metoffice.com
       <userid>@meto.govt.uk
WWW: http://www.meto.govt.uk/
Founded: 1854
Staff: 2300
Activities: meteorology * climatology * atmospheric sciences * national and international weather services
Periodicals: (12) "Monthly Weather Report"; (1) "Annual Report"; "Marine Observer"
City Reference Coordinates: 000°45'00"W 51°26'00"N

Mexborough and Swinton Astronomical Society (MSAS)
c/o The Secretary
14 Sandalwood Drive
Swinton S64 8PN
WWW: http://www.msasuk.force9.co.uk/
Founded: 1978
City Reference Coordinates: 001°17'00"W 53°30'00"N (Mexborough)
          002°21'00"W 53°31'00"N (Swinton)

Mid-Kent Astronomical Society
92 North Street
Sittingbourne MG10 2NH
Telephone: (0)1795-420372
Telefax: (0)1795-420372
Electronic Mail: astroroadshow@lineone.net
WWW: http://ourworld.compuserve.com/homepages/DouglasHull/
Founded: 1976
City Reference Coordinates: 000°44'00"E 51°21'00"N

Midlands Spaceflight Society (MSS)
c/o Andy Salmon (Secretary)
Olympus Mons
13 Jacmar Crescent
Smethwick B67 7LF
or :
c/o Mike Bryce
16 Yellowhammer Court
Kidderminster DY10 4RR
Telephone: (0)121-5654845
Electronic Mail: andy_salmon@compuserve.com
WWW: http://wemfas.future.easyspace.com/mss/mss.html
Founded: 1990
Membership: 53
Periodicals: (6) "CapCom"
City Reference Coordinates: 001°58'00"W 52°30'00"N (Smethwick)
          002°14'00"W 52°23'00"N (Kidderminster)

Mills Observatory
Balgay Park
Glamis Road

Dundee DD2 2UB
Scotland
Telephone: (0)1382-435846
Telefax: (0)1382-435962
Electronic Mail: jeff.lashley@dundeecity.gov.uk (Jeff Lashley, Heritage Programme Officer)
WWW: http://www.mills-observatory.co.uk/
Founded: 1935
Staff: 3
Activities: public observing (planets, bright objects) * education * planetarium shows * visitor centre
Periodicals: (12) "The Mills Observatory Night Sky Guide"
Coordinates: 003°00'40"W 56°27'54"N H152m
City Reference Coordinates: 003°00'00"W 56°28'00"N

Milton Keynes Astronomical Society (MKAS)
19 Matilda Gardens
Shenley Church End
Milton Keynes MK5 6HT
Telephone: (0)1908-503692
Electronic Mail: mike-pat-leggett@shenley9.fsnet.co.uk (Michael John Leggett, Secretary)
WWW: http://www.mkas.org.uk/
Founded: 1972
City Reference Coordinates: 000°42'00"W 52°02'00"N

Mullard Radio Astronomy Observatory (MRAO)
Cavendish Laboratory
Madingley Road
Cambridge CB3 0HE
Telephone: (0)1223-337294
Telefax: (0)1223-354599
Electronic Mail: <userid>@mrao.cam.ac.uk
WWW: http://www.mrao.cam.ac.uk/
http://www.phy.cam.ac.uk/www/research/ra/mrao.home.html
Founded: 1946
Staff: 40
Activities: radio astronomy * radio galaxies * QSOs * cosmology * radio and optical interferometry * molecular clouds * ISM * interplanetary medium * SNR * mm-wavelength astronomy
Coordinates: 000°02'48"E 52°10'12"N H17m
City Reference Coordinates: 000°08'00"E 52°13'00"N

Mullard Space Science Laboratory (MSSL)
• See "University College London (UCL), Mullard Space Science Laboratory (MSSL)"

National Museum of Science and Industry (NMSI), Science Museum, Astronomy Section
Exhibition Road
London SW7 2DD
Telephone: (0)20-79424174
(0)20-79424173
Telefax: (0)20-79424102
Electronic Mail: j.wess@nmsi.ac.uk (Jane Wess, Curator)
k.johnson@nmsi.ac.uk (Kevin Johnson, Associate Curator)
WWW: http://www.nmsi.ac.uk/
Founded: 1857
Staff: 400 (Museum)
Activities: acquisition and display of astronomical instrumentation * history and education (ground-based and space astronomy)
Awards: "Science Book Prize" (set up by the Committee on the Public Understanding of Science - COPUS - and sponsored by Science Museum)
City Reference Coordinates: 000°10'00"W 51°30'00"N

National Remote Sensing Centre (NRSC) Ltd.
Delta House
Southwood Crescent
Southwood
Farnborough GU14 0NL
Telephone: (0)1252-541464
Telefax: (0)1252-375016
WWW: http://www.nrsc.co.uk/
Founded: 1989
Staff: 120
Activities: supplying data, products and services based on information derived from Earth observation satellites and aerial photography, including GIS and software engineering capabilities
Periodicals: "Albedo"
City Reference Coordinates: 000°46'00"W 51°17'00"N

National Space Science Centre (NSSC)
Mansion House
41 Guildhall Lane
Leicester LE1 5FQ
Telephone: (0)116-2530811
Telefax: (0)116-2616800
Electronic Mail: info@nssc.co.uk
WWW: http://www.spacecentre.co.uk/
http://www.nssc.co.uk/
Founded: 1997
Staff: 30
Activities: public understanding of space * developing visitor attraction
City Reference Coordinates: 001°05'00"W 52°38'00"N

Natural Environment Research Council (NERC)
Polaris House
North Star Avenue
Swindon SN2 1EU
Telephone: (0)1793-411500
Telefax: (0)1793-411501
Electronic Mail: <userid>@nerc.ac.uk
WWW: http://www.nerc.ac.uk/
Founded: 1965
City Reference Coordinates: 001°47'00"W 51°34'00"N

Natural History Museum, Meteoritic Research
Cromwell Road
London SW7 5BD
Telephone: (0)20-79389123
(0)20-79389445
Telefax: (0)20-79389268
Electronic Mail: mmg@nhm.ac.uk (Monica Grady)
WWW: http://www.nhm.ac.uk/mineralogy/intro/project4/
http://www.nhm.ac.uk/
City Reference Coordinates: 000°10'00"W 51°30'00"N

Nature
c/o Editorial Office
Porters South
4 Crinan Street
London N1 9XW
Telephone: (0)20-78334000
Telefax: (0)20-78434596
(0)20-78434597
WWW: http://www.nature.com/
Founded: 1869
• Journal (51) (ISSN 0028-0836)
City Reference Coordinates: 000°10'00"W 51°30'00"N

Newbury Amateur Astronomical Society (NAAS)
c/o Ann Davies
11 Sedgefield Road
Greenham
Newbury RG14 7TZ
Telephone: (0)1635-30598
WWW: http://www.naas.btinternet.co.uk/
City Reference Coordinates: 001°20'00"W 51°25'00"N

Newcastle-upon-Tyne Astronomical Society (NAS)
c/o A. Petty
7 Elmfield Park
Gosforth
Newcastle upon Tyne NE3 4UX
Telephone: (0)191-2851706
Founded: 1904
Staff: 50
Activities: lectures * observing * star parties
Periodicals: (3) "Newsletter"
Coordinates: 001°18'06"W 54°59'18"N H55m (Wylam)
City Reference Coordinates: 001°35'00"W 54°59'00"N

Newton Institute for Mathematical Sciences (Isaac_)
• See "Isaac Newton Institute for Mathematical Sciences"

Norman Lockyer Observatory and James Lockyer Planetarium
Salcombe Hill
Sidmouth EX10 0BS
Telephone: (0)1395-579941
Electronic Mail: g.e.white@exeter.ac.uk (Gerald White)
WWW: http://www.ex.ac.uk/nlo/welcome.htm
Founded: 1912
Membership: 268
Activities: education * astronomy * radio astronomy * meteorology * amateur radio
Coordinates: 003°13'07"W 50°41'16"N
City Reference Coordinates: 003°15'00"W 50°41'00"N

Northamptonshire Natural History Society (NNHS), Astronomy Section
The Humfrey Rooms
10 Castilian Terrace
Northampton NN1 1LD
Electronic Mail: ram@hamal.demon.co.uk (Bob Marriott, Secretary)
WWW: http://www.teg1.demon.co.uk/nnhs/
Founded: 1957 (NNHS: 1876)
Activities: meetings * observing
City Reference Coordinates: 000°54'00"W 52°14'00"N

Northants Amateur Astronomers (NAA)
6 Lucas Close
Irthlingsborough NN9 5AJ
or :
124 Senwick Drive
Wellingborough NN8 1RU
Telephone: (0)1933-272628
Electronic Mail: naastronomers@geocities.com
WWW: http://www.geocities.com/CapeCanaveral/Galaxy/6057/
http://www.geocities.com/naastronomy/
Founded: 1983
Membership: 35
Activities: meetings
Periodicals: (12) "In Focus"
● Formerly known as "Kettering and District Amateur Astronomers Society (KDAAS)"
City Reference Coordinates: 000°42'00"W 52°19'00"N

North Devon Astronomical Society (NDAS)
c/o P. G. Vickery
12 Brd. Park Crescent
Ilfracombe EX34 8DX
Telephone: (0)1271-863224
Electronic Mail: adavies915@aol.com
WWW: http://members.aol.com/NDASTRO/
Founded: 1973
City Reference Coordinates: 004°08'00"W 51°13'00"N

North East London Astronomical Society (NELAS)
c/o B. Beeston
38 Abbey Road
Bush Hill Park
Enfield EN1 2QN
or :
Wanstead House (Meetings)
21 The Green
Wanstead
London E11 2NT
Telephone: (0)20-83635696
Founded: 1956
Membership: 15
Activities: meetings * talks * slide shows * films * videos * observing
Periodicals: (11) "Newsletter"
City Reference Coordinates: 000°10'00"W 51°30'00"N (London)

North West Group of Astronomical Societies (NWGAS)
c/o Kevin Kilburn
66 Henshall Road
Bollington SK10 5DN
Telephone: (0)1625-572453
WWW: http://www.u-net.com/ph/nwgas
Founded: 1994
Membership: 12 (societies)
Activities: amateur astronomy
City Reference Coordinates: 002°06'00"W 53°18'00"N

Norwich Astronomical Society (NAS)
c/o Malcolm Jones (Secretary)
Tabor House
Norwich Road
Mulbarton NR14 8JT
Telephone: (0)1508-578392
Telefax: (0)1508-570986
Electronic Mail: 100257.1434@compuserve.com
WWW: htttp://nas.gurney.org.uk/
Founded: 1945
Membership: 120
Periodicals: (4) "Cygnus"
Coordinates: 001°14'00"E 52°37'00"N H40m (Seething Astronomical Observatory)
City Reference Coordinates: 001°14'00"E 52°37'00"N (Norwich)

Nottingham Astronomical Society (NAS)
40 Swindon Close
The Vale
Giltbrook
Nottingham NG16 2WD
Telephone: (0)115-8548687
Electronic Mail: carl.stella@virgin.net (Carl Brennan, Secretary)
Founded: 1946
Activities: amateur astronomy * lectures * observing
Periodicals: (12) "Journal of the Nottingham Astronomical Society"
City Reference Coordinates: 001°01'00"W 52°58'00"N

Nuffield Radio Astronomy Laboratories (NRAL)
• See now "University of Manchester, Department of Physics and Astronomy, Jodrell Bank Observatory (JBO)"

Observatory Magazine
c/o The Editors
Rutherford Appleton Laboratory
Building R25
Chilton
Didcot OX11 0QX
Telephone: (0)1235-446523
Telefax: (0)1235-445848
Electronic Mail: obs@ast.star.rl.ac.uk
WWW: http://www.ulo.ucl.ac.uk/obsmag/
Founded: 1877
Staff: 4 (Editors)
• Journal (3) (ISSN 0029-7704, circ.: 1,000)
City Reference Coordinates: 001°15'00"W 51°37'00"N

Old Royal Observatory (ORO)
• See now "Royal Observatory Greenwich"

Open University (OU), Department of Physics, Astronomy Research Group
Walton Hall
Milton Keynes MK7 6AA
Telephone: (0)1908-653229
Telefax: (0)1908-654192
Electronic Mail: <userid>@open.ac.uk
WWW: http://yan.open.ac.uk/ (Department of Physics)
    http://yan.open.ac.uk/research/astro/ (Astronomy Research Group)
Founded: 1997 (Department: 1969)
Staff: 30
Activities: interacting binaries * compact objects * accretion * Galactic chemical evolution * stellar abundances * plasma astrophysics * orbital dynamics * solar eclipses
City Reference Coordinates: 000°42'00"W 52°02'00"N

Open University (OU), Planetary Sciences Research Institute (PSRI)
Walton Hall
Milton Keynes MK7 6AA
Telephone: (0)1908-655169
Telefax: (0)1908-655910
Electronic Mail: psri@open.ac.uk
WWW: http://psri.open.ac.uk/
Founded: 1997
Staff: 30
Activities: instrumentation of space missions * extraterrestrial sample analysis * impact cratering research * meteorites
City Reference Coordinates: 000°42'00"W 52°02'00"N

Orpington Astronomical Society (OAS)
c/o Ian R. Carstairs (Membership Secretary)
28B Thicket Road
Pence
London SE20 8DD
Telephone: (0)20-86591096
Electronic Mail: oas@chocky.demon.uk
WWW: http://www.chocky.demon.co.uk/oas/
Founded: 1980
Membership: 100
Activities: lectures * observing * outings * visits
Periodicals: (5) "The Orpington Astronomical Society Times (TOAST)"
City Reference Coordinates: 000°10'00"W 51°30'00"N (London)
000°05'00"W 51°23'00"N (Orpington)

Orwell Astronomical Society Ipswich (OASI)
c/o Orwell Park School
Nacton
Ipswich IP10 0ER
or :
c/o R. Gooding (Secretary)
168 Ashcroft Road
Ipswich IP1 6AE
Electronic Mail: ipswich@ast.cam.ac.uk
WWW: http://www.ast.cam.ac.uk/~ipswich/
Founded: 1967
Membership: 70
Activities: observing
Periodicals: "Newsletter"
Coordinates: 001°13'57"E 52°00'33"N
City Reference Coordinates: 001°10'00"E 52°04'00"N

Oxford Instruments, Research Instruments
Tubney Woods
Abingdon OX13 5QX
Telephone: (0)1865-393200
Telefax: (0)1865-393333
Electronic Mail: info.ri@oxinst.co.uk
WWW: http://www.oxford-instruments.com/
Founded: 1959
Staff: 300
Activities: manufacturing scientific equipment
Periodicals: "Research Matters"
City Reference Coordinates: 001°17'00"W 51°41'00"N

Oxford University Space and Astronomical Society (OUSAS)
c/o Mark Roberts
Hertford College
Catte Street
Oxford OX1 3BW
Electronic Mail: space@sable.ox.ac.uk
WWW: http://users.ox.ac.uk/~space/
Founded: 1960
Membership: 100
Activities: promoting an interest in astronomy and space-related sciences among the students of Oxford University
City Reference Coordinates: 001°15'00"W 51°46'00"N

Oxford University Press (OUP)
Great Clarendon Street
Oxford OX2 6DP
Telephone: (0)1865-56767
Telefax: (0)1865-56646
Electronic Mail: enquiry@oup.co.uk
WWW: http://www.oup.co.uk/
http://www.comlab.ox.ac.uk/archive/publishers/oup.html
Founded: 1478
Staff: 800
• Publisher
City Reference Coordinates: 001°15'00"W 51°46'00"N

Particle Physics and Astronomy Research Council (PPARC)
Polaris House
North Star Avenue

Swindon SN2 1SZ
Telephone: (0)1793-442000
Telefax: (0)1793-442002
Electronic Mail: <userid>@pparc.ac.uk
WWW: http://www.pparc.ac.uk/
Staff: 100
Periodicals: "Frontiers"
City Reference Coordinates: 001°47'00"W 51°34'00"N

Particle Physics and Astronomy Research Council (PPARC), Royal Observatory Edinburgh (ROE)
Blackford Hill
Edinburgh EH9 3HJ
Scotland
Telephone: (0)131-6688100
Telefax: (0)131-6688264
Electronic Mail: <userid>@roe.ac.uk
WWW: http://www.roe.ac.uk/
          http://www.roe.ac.uk/ukstu/ukst.html (UK Schmidt Telescope)
Founded: 1822
Staff: 115
Periodicals: "Edinburgh Astronomy Preprints", "ROE Computing Newsletter", "Annual Report" (ISSN 0309-0108), "The JCMT-UKIRT Newsletter", "Spectrum" (ISSN 0963-2700)
Coordinates: 003°11'00"W 55°55'30"N H146m
                155°28'18"W 19°49'54"N H4,200m (Mauna Kea, USA-HI)
City Reference Coordinates: 003°13'00"W 55°57'00"N

Paton Hawksley Education Ltd. (PHEL)
Rockhill Laboratories
59 Wellsway
Keynsham BS18 1PG
Telephone: (0)1117-9862364
Telefax: (0)1117-9868285
Founded: 1953
Staff: 13
Activities: manufacturing diffraction gratings and spectroscopes
City Reference Coordinates: 002°30'00"W 51°26'00"N

Penny & Giles Controls Ltd.
36 Nine Mile Point Industrial Estate
Cwmfelinfach
Gwent NP11 7HZ
Wales
Telephone: (0)1495-202000
Telefax: (0)1495-202006
Electronic Mail: sales@pgcontrols.com
WWW: http://www.pgcontrols.com/index/index.asp
Founded: 1955
Activities: aerospace, industrial and audio/video products * rotary position sensors * joystick controllers * audio/video controllers
City Reference Coordinates: 003°00'00"W 51°30'00"N

Pick Travel Ltd.
11 Scraptoft Lane
Leicester LE5 2FD
Telephone: (0)116-2761818
Telefax: (0)116-2207018
Electronic Mail: aerotours@gptravel.co.uk
WWW: http://www.gptravel.co.uk/
Founded: 1984
Staff: 6
Activities: tour operators * educational visits to scientific institutions, museums, planetariums, and other astronomy and space sciences related fields
City Reference Coordinates: 001°05'00"W 52°38'00"N

Planet:Earth Centre
Swallows Barn
Clough Bank
Bacup Road
Todmorden OL14 7HW
Telephone: (0)1706-816964
Electronic Mail: planetearthcentre@btinternet.com
WWW: http://www.xenica.com/PlanetEarth
Founded: 1995
Staff: 2
Activities: education * observing * planetarium shows
Periodicals: "Calendar of the Night Sky"

Coordinates: 002°00'00"W 53°00'00"N H350m
• Formerly known as "Amateur Astronomy Centre (AAC)"
City Reference Coordinates: 002°00'00"W 53°00'00"N

Plymouth Astronomical Society (PAS)
c/o Lawrence D.J. Harris (Chairman)
5 Burnham Park Road
Peverell
Plymouth PL3 5QB
Telephone: (0)1752-775148
Telefax: (0)1752-775148
Electronic Mail: lawrence@itchycoo-park.freeserve.co.uk (Lawrence Harris, Chair)
            oakmount12@aol.com (Alan Penman, Secretary)
WWW: http://www.itchycoo-park.freeserve.co.uk/pas.html
Founded: 1965
Membership: 35
Activities: promoting and publicizing astronomy
City Reference Coordinates: 004°10'00"W 50°23'00"N

Preston and District Astronomical Society (PADAS)
c/o Jeremiah Horrocks Observatory
Preston PR1 2HE
Telephone: (0)1772-257181 (Keith Robinson)
Electronic Mail: padas@btinternet.com
WWW: http://www.btinternet.com/~roy.jackson/
Founded: 1979
Coordinates: 002°42'12"W 53°46'30"N H33m
City Reference Coordinates: 002°42'00"W 53°46'00"N

Queen Mary and Westfield College (QMWC)
• See "University of London, Queen Mary and Westfield College (QMWC)"

Queen's University of Belfast (QUB), Department of Applied Mathematics and Theoretical Physics
David Bates Building
Belfast BT7 1NN
Northern Ireland
Telephone: (0)28-90273177
Telefax: (0)28-90239182
Electronic Mail: <userid>@vi.am.qub.ac.uk
WWW: http://www.am.qub.ac.uk/
Staff: 46
Activities: calculation and use of atomic data in the interpretation of spectra from the Sun and stars
City Reference Coordinates: 005°55'00"W 54°35'00"N

Queen's University of Belfast (QUB), Department of Pure and Applied Physics, Astrophysics and Planetary Science Division
Belfast BT7 1NN
Northern Ireland
Telephone: (0)28-90245133 x3554
Telefax: (0)28-90438918
Electronic Mail: <userid>@qub.ac.uk
WWW: http://star.pst.qub.ac.uk/
Founded: 1967
Staff: 9
Activities: optical and UV spectroscopy of stars * ISM spectroscopy * diagnostics for solar and laboratory plasmas * spectroscopy and photometry of comets and minor planets
City Reference Coordinates: 005°55'00"W 54°35'00"N

Reading Astronomical Society (RAS)
c/o Ruth Sumner (Secretary)
22 Anson Crescent
Shinfield
Reading RG2 8JT
Telephone: (0)118-9873565
WWW: http://www.users.zetnet.co.uk/astro-reading/
Founded: 1972
Membership: 130
Activities: Meetings * visits
Periodicals: (4) "RASTAR"
City Reference Coordinates: 000°59'00"W 51°28'00"N

Remote Sensing Society (RSS)
c/o School of Geography
University of Nottingham
Nottingham NG7 2RD

Telephone: (0)115-9515435
Telefax: (0)115-9515249
Electronic Mail: rss@nottingham.ac.uk
WWW: http://www.the-rss.org/
Founded: 1974
Membership: 800
Activities: promoting remote-sensing activities and training
Periodicals: (4) "Remote Sensing Society Newsletter"; (1) "Annual Report"; "International Journal of Remote Sensing (IJRS)"

Awards: "Gold Medal", "Len Curtis European Award", "Student Award", "Taylor & Francis Best Letter Award"
City Reference Coordinates: 001°01'00"W 52°58'00"N

Richmond and Kew Astronomical Society (RKAS)
41A Bruce Road
Mitcham CR4 2BJ
Founded: 1986
Membership: 60
Activities: observing * lectures * visits
Periodicals: (4) "Newsletter"
City Reference Coordinates: 000°18'00"W 51°29'00"N (Kew)
000°09'00"W 51°24'00"N (Mitcham)
000°19'00"W 51°28'00"N (Richmond)

Riding Astronomical Society (Hull and East_)
• See "Hull and East Riding Astronomical Society"

Roditi International Corp. Ltd. (RICL)
Carrington House
130 Regent Street
London W1R 6BR
Telephone: (0)20-74394390
Telefax: (0)20-74340896
WWW: http://www.roditi.co.uk/
Founded: 1935
Staff: 12
Activities: distributing optical components, filters and laser crystals
City Reference Coordinates: 000°10'00"W 51°30'00"N

Roper Scientific
P.O. Box 1192
43 High Street
Marlow SL7 1GB UK
Telephone: (0)1628-890858
Telefax: (0)1628-898381
Electronic Mail: info@roperscientific.co.uk
WWW: http://www.roperscientific.co.uk/
Founded: 1970
Staff: 96
Activities: designing, developing and manufacturing CCD digital imaging systems
City Reference Coordinates: 000°48'00"W 51°35'00"N

Rossall-Assheton Astronomical Observatory
c/o Nick A. Lister
Rossall School
Broadway
Fleetwood FY7 8JW
Telephone: (0)1253-700654
Telefax: (0)1253-772052
Electronic Mail: nlister@btinternet.com
WWW: http://www.satsig.net/asshето1.htm
Founded: 1904
Activities: observing * photography * education
City Reference Coordinates: 003°01'00"W 53°56'00"N

Royal Astronomical Society (RAS)
Burlington House
Piccadilly
London W1J 0BQ
Telephone: (0)20-77343307
Telefax: (0)20-74940166
Electronic Mail: info@ras.org.uk
WWW: http://www.ras.org.uk/
Founded: 1820
Membership: 2970
Staff: 12

Activities: encouragement and promotion of astronomy and geophysics by publishing the results of research * maintaining a library * holding meetings
Periodicals: (24) "Monthly Notices of the Royal Astronomical Society (MNRAS)" (ISSN 0035-8711, circ.: 1,250) (12) "Geophysical Journal International" (ISSN 0955-419x); (4) "Astronomy & Geophysics" (ISSN 1366-8781, circ.: 3,500)
Awards: (1/3) "Eddington Medal", "Chapman Medal", "Herschel Medal", "Price Medal", "Hannah Jackson"; (1) "Gold Medal"; "Michael Penston Astronomy Prize"
City Reference Coordinates: 000°10'00"W 51°30'00"N

Royal Greenwich Observatory (RGO)
• Closed on 31 October 1998.

Royal Holloway and Bedford New College (RHBNC), Department of Physics
Egham Hill
Egham TW20 0EX
Telephone: (0)1784-443448
Telefax: (0)1784-472794
Electronic Mail: physics-department@rhbnc.ac.uk
                <userid>@rhbnc.ac.uk
WWW: http://www.ph.rhbnc.ac.uk/
Founded: 1884
Staff: 20
Activities: particle physics * 4D string theory * higher-dimensional cosmology
City Reference Coordinates: 000°34'00"W 51°26'00"N

Royal Meteorological Society (RMS)
104 Oxford Road
Reading RG1 7LL
Telephone: (0)1189-568500
Telefax: (0)1189-568571
Electronic Mail: execsec@royal-met-soc.org.uk
                <userid>@royal-met-soc.org.uk
WWW: http://itu.rdg.ac.uk/rms/rms.html
Founded: 1850
Membership: 3650
Staff: 9
Activities: meetings * education * awards * professional accreditation
Periodicals: (12) "International Journal of Climatology", "Weather" (ISSN 0043-1656); (4) "Quarterly Journal", "Meteorological Applications"
Awards: 3 annual, 5 biennial
City Reference Coordinates: 000°59'00"W 51°28'00"N

Royal Observatory Edinburgh (ROE)
• See "Particle Physics and Astronomy Research Council (PPARC), Royal Observatory Edinburgh (ROE)"

Royal Observatory Greenwich (ROG)
Greenwich
London SE10 9NF
Telephone: (0)20-88584422
Telefax: (0)20-83126734
Electronic Mail: astroline@nmm.ac.uk
WWW: http://www.rog.nmm.ac.uk/
Founded: 1675 (Museum: 1956)
Staff: 30
Activities: modern astronomy enquiry service * history of astronomy * school programmes * scientific instruments * Caird Planetarium
Coordinates: 000°00'00"E 51°28'38"N H47m
City Reference Coordinates: 000°10'00"W 51°30'00"N

Royal Society (RS)
6 Carlton House Terrace
London SW1Y 5AG
Telephone: (0)20-78395561
Telefax: (0)20-79302170
Electronic Mail: info@royalsoc.ac.uk
WWW: http://www.royalsoc.ac.uk/
Founded: 1660
Membership: 1200
Activities: promoting the exchange and development of scientific ideas and knowledge * recognizing excellence in scientific scholarship and research * encouraging scientific research and promoting its application * promoting and facilitating international scientific relations * providing a source of independent advice on scientific matters * initiating public debate on matters of public importance * helping focus and represent the views and interests of the scientific community * providing science education, awareness and understanding * providing information on and support for the history of scientific endeavour * providing research appointments and grants * organizing meetings and lectures * library and information services
Periodicals: (1) "Year Book", "Annual Report"; "Philosophical Transactions A - Mathematical and Physical Sciences" (ISSN 0080-4614), "Philosophical Transactions B - Biological Sciences", "Proceedings A - Mathematical and Physical Sciences",

"Proceedings B - Biological Sciences", "Biographical Memoirs of Fellows of the Royal Society", "Notes and Records", "Science and Public Affairs", "Reports"
Awards: (1) "Copley Medal", "Rumford Medal", "Royal Medals", "Davy Medal", "Darwin Medal", "Buchanan Medal", "Sylvester Medal", "Hughes Medal", "Leverhulme Medal", "Gabor Medal", "Mullard Award", "Esso Energy Award", "Wellcome Foundation Prize and Lecture", "Armourers & Brasiers' Award", "Michael Faraday Award"
City Reference Coordinates: 000°10'00"W 51°30'00"N

Royal Society of Chemistry (RSC)
Burlington House
Piccadilly
London W1V 0BN
Telephone: (0)20-74378656
Telefax: (0)20-74378883
Electronic Mail: info@rsc.org
WWW: http://www.rsc.org/
Founded: 1980
Membership: 46000
Staff: 300
Activities: result of the unification of the "Chemical Society" and the "Royal Institute of Chemistry" * advancement of chemistry and its applications * maintenance of high standards of competence and integrity among practising chemists * databases
Periodicals: (52) "Chemical Business Bulletins"; "Journal of the Chemical Society" subdivided into (24) "Chemical Communications" (ISSN 0022-4936), (24) "Dalton Transactions" (ISSN 0300-9246), (24) "Perkin Transactions I" (ISSN 0300-922x), (12) "Perkin Transactions II" (ISSN 0300-9580), and (24) "Faraday Transactions" (ISSN 0956-5000); (12) "Journal of Chemical Research" consisting of "Part S (Synopsis)" (ISSN 0308-2342), "Part M (Microfiche)" (ISSN 0308-2350), "Part M (Miniprint)" (ISSN 0000-037x), and "Vouchers"; (12) "Chemistry in Britain" (ISSN 0009-3106), "The Analyst" (ISSN 0003-2654), "Analytical Communications" (ISSN 1359-7337), "Chemical Hazards in Industry" (ISSN 0265-5721), "Laboratory Hazards Bulletin" (ISSN 0261-2917), "Methods in Organic Synthesis" (ISSN 0265-4245), "Natural Products Updates" (ISSN 0950-1711), "Current Biotechnology Abstracts" (ISSN 0264-3391), "Chemical Engineering Abstracts" (ISSN 0262-6438), "Mass Spectrometry Bulletin" (ISSN 0025-4738), "Chemical Business Update", "Journal of Materials Chemistry" (ISSN 0959-9428); (8) "Journal of Analytical Atomic Spectrometry"; (6) "Natural Products Report" (ISSN 0265-0568), "Education in Chemistry" (ISSN 0013-1350), "Theoretical Chemical Engineering Abstracts" (ISSN 0040-5787); (4) "Chemical Society Reviews" (ISSN 0306-0012); (2) "Faraday Discussions"; (1) "Annual Reports on the Progress of Chemistry" consisting of "Section A: Inorganic Chemistry" (ISSN 0260-1818), "Section B: Organic Chemistry" (ISSN 0060-3030), and "Section C: Physical Chemistry" (ISSN 0670-1826); "Analytical Abstracts" (ISSN 0003-2689)
City Reference Coordinates: 000°10'00"W 51°30'00"N

Royal Society of Edinburgh (RSE)
22-26 George Street
Edinburgh EH2 2PQ
Scotland
Telephone: (0)131-2405000
Telefax: (0)131-2405024
Electronic Mail: rse@rse.org.uk
WWW: http://www.ma.hw.ac.uk/RSE/
Founded: 1783
Periodicals: (4) "RSE News" (ISSN 1352-3325)
City Reference Coordinates: 003°13'00"W 55°57'00"N

Royal Statistical Society (RSS)
12 Errol Street
London EC1Y 8LX
Telephone: (0)20-76388998
Telefax: (0)20-72567598
Electronic Mail: rss@rss.org.uk
WWW: http://www.rss.org.uk/
Founded: 1834
Membership: 6000
Periodicals: (12) "RSS News"; "Journal of the Royal Statistical Society - Series A, B, C and D", "Annual Report"
City Reference Coordinates: 000°10'00"W 51°30'00"N

Rutherford Appleton Laboratory (RAL)
• See "Council for the Central Laboratory of the Research Councils (CCLRC), Rutherford Appleton Laboratory (RAL)"

Safelight
Unit 889
Cowley Road
London W3 7YD
Telephone: (0)20-87401516
Telefax: (0)20-87401011
Electronic Mail: service@safelight.com
WWW: http://www.safelight.com/
Activities: multimedia presentations
City Reference Coordinates: 000°10'00"W 51°30'00"N

Salford Astronomical Society (SAS)
59 Vancouver Quay
Salford Quays
Manchester M5 2TU
Telephone: (0)1204-531411 x3476
Telefax: (0)1204-528768
Electronic Mail: salfordac@ast.man.ac.uk
WWW: http://www.salfordastro.org.uk/
Founded: 1967
City Reference Coordinates: 002°15'00"W 53°30'00"N (Manchester)
002°16'00"W 53°30'00"N (Salford)

Salisbury Astronomical Society (SAS)
c/o Rita Collins
Mountains
3 Fairview Road
Salisbury SP1 1JX
Telephone: (0)1722-332892
Electronic Mail: ritacollins@compuserve.com
Founded: 1965
Membership: 25
Activities: observing (planets, Moon, occultations, deep sky) * computing
Periodicals: (12) "The Stargazer"
Coordinates: 001°55'00"W 51°02'00"N (Bishopstone)
001°50'00"W 51°10'00"N (Stonehenge)
001°39'00"W 51°05'00"N (Winterslow)
City Reference Coordinates: 001°55'00"W 51°02'00"N

Scarborough and District Astronomical Society
c/o Sheila Anderson
25 Highfield
Scarborough YO12 4AW
WWW: http://www.baryte.demon.co.uk/astro/welcome.htm
Founded: 1976
City Reference Coordinates: 000°24'00"W 54°17'00"N

Science History Publications Ltd.
c/o M.A. Hoskin (Director)
16 Rutherford Road
Cambridge CB2 2HH
Telephone: (0)1223-565532
Telefax: (0)1223-565532
Electronic Mail: shpltd@aol.com
Founded: 1969
Staff: 2
Periodicals: (5) "Journal for the History of Astronomy (JHA)" (ISSN 0021-8286, circ.: 750); (2) "Archaeoastronomy" (ISSN 0142-7253, circ.: 750)
• Publisher
City Reference Coordinates: 000°08'00"E 52°13'00"N

Science Museum
• See "National Museum of Science and Industry (NMSI), Science Museum"

Scottish Astronomers Group (SAG)
c/o Bill Ward (Information Officer)
The Rankine Building
University of Glasgow
Oakfield Avenue
Glasgow G12 8LT
Scotland
Telephone: (0)1382-667138
Electronic Mail: b.ward@elec.gla.ac.uk
WWW: http://www.elec.gla.ac.uk/~aurora/sag/saginfo.html
Founded: 1970
Membership: 50 + 11 affiliated societies
Activities: inter-society communication * lecture meetings * observing
Periodicals: (4) "SAG Newsletter"
City Reference Coordinates: 004°15'00"W 55°53'00"N

Sheffield Astronomical Society (SAS)
102 Sheffield Road
Woodhouse
Sheffield S13 7EM

or :
Mayfield Education Centre
David Lane
Fullwood
Sheffield
Telephone: (0)114-2692291
Electronic Mail: sheffieldastro@hotmail.com
WWW: http://www.sheffieldastro.org.uk/
Founded: 1934
Membership: 58
Activities: monthly meetings * public observing * public outreach
Periodicals: (12) "Skywatch"
Coordinates: 001°33'50"W 53°21'38"N
City Reference Coordinates: 001°30'00"W 53°23'00"N

Shropshire Astronomical Society (SAS)
c/o David Woodward
20 Station Road
Condover
Shrewsbury SY5 7BQ
Wales
Telephone: (0)1743-872991 (David Woodward, Chairman)
Electronic Mail: d.woodward@virgin.net
　　　　　　　　g.privett@astro.cf.ac.uk
WWW: http://www.astro.cf.ac.uk/sas/sasmain.html
Founded: 1996
Membership: 50
Activities: amateur astronomy
Periodicals: "Hermes"
City Reference Coordinates: 002°45'00"W 52°43'00"N

Sira Ltd., Electro-Optics Division
South Hill
Chislehurst BR7 5EH
Telephone: (0)20-84672636
Telefax: (0)20-84676515
Electronic Mail: info@sira.co.uk
　　　　　　　　marketing@siraeo.co.uk
WWW: http://www.sira.co.uk/
Founded: 1918
Staff: 220
Activities: R&D in instrumentation and control, optics, electronics, expert systems, engineering
Periodicals: (2) "Spotlight"
City Reference Coordinates: 000°04'00"E 51°25'00"N

Skeptic (The_)
● See "The Skeptic"

Skymap Software
9 Severn Road
Culcheth WA3 5ED
Electronic Mail: support@skymap.com
WWW: http://www.skymap.com/
Founded: 1990
● Software producer
City Reference Coordinates: 002°32'00"W 53°27'00"N

Society for Popular Astronomy (SPA)
36 Fairway
Keyworth NG12 5DU
Electronic Mail: spastronomy@aol.com
WWW: http://www.popastro.com/
Founded: 1953
Membership: 2600
Staff: 14
Activities: quarterly meetings * outings * organisation for beginners of all ages
Sections: Aurora * Comet * Deep Sky * Lunar * Meteor * Occultation * Planetary * Solar * Variable Star
Periodicals: (4) "Popular Astronomy" (ISSN 0261-0892); (6) "News Circulars"
Awards: "Fred Best Award"
● Formerly "Junior Astronomical Society (JAS)"
City Reference Coordinates: 001°06'00"W 52°52'00"N

Society for the History of Astronomy (SHA)
c/o Stuart Williams
Flamsteed Villa

26 Matlock Road
Bloxwich WS3 3QD
Electronic Mail: flamsteed@v21mail.co.uk
City Reference Coordinates: 002°00'00"W 52°37'00"N

Society of British Aerospace Companies (SBAC) Ltd.
Duxbury House
60 Petty France
Victoria
London SW1H 9EU
Telephone: (0)20-72271000
Telefax: (0)20-72271067
Electronic Mail: post@sbac.co.uk
WWW: http://www.sbac.co.uk/
Membership: 350 (companies)
Periodicals: (1) "Annual Report"
City Reference Coordinates: 000°10'00"W 51°30'00"N

Solent Amateur Astronomers Astronomical Society (SAAS)
c/o Ken Medway
443 Burgess Road
Swaythling SO16 3BL
Telephone: (0)23-80582204
WWW: http://www.delscope.demon.co.uk/
Founded: 1972
Membership: 75
Activities: practical astronomy (equipment, observatory construction) * occultations * monthly meetings
Periodicals: (12) "Newsletter"
Observatories: 2 (Toothill, Itchen)
City Reference Coordinates: 001°25'00"W 50°55'00"N (Southampton)

Southampton Astronomical Society (SAS)
124 Winchester Road
Shirley SO16 6US
Telephone: (0)23-80327952
Electronic Mail: hollies@tcp.co.uk (Michael Hobbs)
WWW: http://home.clara.net/lmhobbs/sas.html
Founded: 1924
Membership: 68
Activities: fostering interest in astronomy and related subjects * lectures * discussions * shows * visits * observing * telescope making
Periodicals: (6) "Newsletter"; (2) "Journal"
City Reference Coordinates: 001°48'00"W 52°24'00"N (Shirley)
001°25'00"W 50°55'00"N (Southampton)

South Coast Telescopes
3 Beacon Mount
Park Gate
Southampton SO31 7GN
Telephone: (0)1489-584192
(0)976690928
Telefax: (0)1489-603318
Founded: 1996
Staff: 2
Activities: manufacturing mirror making kits
City Reference Coordinates: 001°25'00"W 50°55'00"N

South Downs Astronomical Society (SDAS)
c/o J.K.W. Green
46 Central Avenue
Bognor Regis PO21 5HH
Telephone: (0)1243-774400
Telefax: (0)1243-829868
Electronic Mail: sdownsplanet@btclick.com.uk
WWW: http://www.southdowns.org.uk/sdas/
http://www.southdowns.org.uk/sdpt/ (South Downs Planetarium)
Founded: 1971 (Planetarium: 1994)
Activities: lectures *popularization * circulation of literature * observing
Periodicals: (12) "SDAS Mercury"; (4) "South Downs Planetarium News"; (1) "Supernova"
Awards: (1) "Gemini Awards"
Coordinates: 000°46'00"W 50°49'00"N
City Reference Coordinates: 000°41'00"W 50°47'00"N

South East Kent Astronomical Society (SEKAS)
c/o Alan Woodgett

Wales
Telephone: (0)1792-466814
Electronic Mail: sjw@astrabio.demon.co.uk (S.J. Wainwright, Webmaster)
WWW: http://www.swan.ac.uk/astra/starpage.htm
Founded: 1948
City Reference Coordinates: 003°57'00"W 51°38'00"N

Tavistock Astronomical Society
c/o Science Department
Kelly College
Tavistock PL19 0HZ
WWW: http://www.tavastro.org.uk/
City Reference Coordinates: 004°08'00"W 50°33'00"N

Taylor & Francis Ltd.
11 New Fetter Lane
London EC4P 4EE
Telephone: (0)20-75839855
Telefax: (0)20-78422298
WWW: http://www.tandf.co.uk/
Founded: 1798
Periodicals: among others, "Journal of Modern Optics" (formerly Optica Acta) (ISSN 0950-0340)
City Reference Coordinates: 000°10'00"W 51°30'00"N

Telescope Technologies Ltd. (TTL)
1 Morpeth Wharf
Birkenhead CH41 1NQ
Telephone: (0)151-6503100
Telefax: (0)151-6503113
Electronic Mail: info@gnat.com
WWW: http://www.ngat.com/
Founded: 1995
Staff: 47
Activities: manufacturing telescopes and associated equipment
City Reference Coordinates: 003°02'00"W 53°24'00"N

The Astronomer
16 Westminster Close
Basingstoke RG22 4PP
Telephone: (0)1256-471074
Telefax: (0)1256-471074
Electronic Mail: editor@theastronomer.org
WWW: http://www.theastronomer.org/
Founded: 1964
Membership: 400
Activities: observing projects * rapid publication of results * e-mail news service
Periodicals: (12) "The Astronomer"; "Discovery Circulars", "E-Mail Circulars"
Awards: (1) "Best Contributor to Magazine", "Best Contributor to Cover Photograph"
City Reference Coordinates: 001°05'00"W 51°15'00"N

The Astronomy Roadshow
92 North Street
Sittingbourne MG10 2HN
Telephone: (0)1795-420372
Telefax: (0)1795-420372
Electronic Mail: astroroadshow@lineone.net
Founded: 1993
Staff: 3
Activities: operating mobile planetariums
City Reference Coordinates: 000°44'00"E 51°21'00"N

The Island Planetarium
Fort Victoria Country Park
Westhill Lane
Norton
Yarmouth PO41 0RR
Isle of Wight
Telephone: (0)1983-761555
Electronic Mail: paule@astrofvp.demon.co.uk (Paul England)
WWW: http://www.astrofvp.demon.co.uk/
Founded: 1992
City Reference Coordinates: 001°29'00"W 50°42'00"N

The Skeptic
P.O. Box 475

Manchester M60 2TH
Telephone: (0)7020-935370
Telefax: (0)7020-935372
Electronic Mail: edit@skeptic.org.uk
WWW: http://www.skeptic.org.uk/
Founded: 1987
• Journal (6) (ISSN 0959-5228, circ.: 600)
City Reference Coordinates: 002°15'00"W 53°30'00"N

The Spaceguard Centre
Llanshay Lane
Knighton
Powys LD7 1LW
Telephone: (0)1547-520247
Telefax: (0)1547-520247
Electronic Mail: spaceguard@spaceguarduk.com
WWW: http://www.spaceguarduk.com/
Founded: 2001
Staff: 2
Activities: observing * minor planets * comets * meteors * NEO follow-up * public information and education * planetarium

Periodicals: (4) "Impact"
Coordinates: 003°01'05"W 52°19'34"N H417m
• Formerly "Powys County Observatory"
City Reference Coordinates: 003°03'00"W 52°21'00"N

Torbay Astronomical Society (TAS)
c/o John Stapleton
24 Prospect Terrace
Newton Abbot TQ12 2LL
Telephone: (0)1803-323600
Electronic Mail: tas@halien.net
WWW: http://www.halien.com/TAS/
Founded: 1956
City Reference Coordinates: 003°36'00"W 50°32'00"N (Newton Abbot)

True Technology Ltd.
Woodpecker Cottage
Red Lane
Aldermaston RG7 4PA
Telephone: (0)1189-700777
Telefax: (0)1189-701661
Electronic Mail: truetech@dircon.co.uk
WWW: http://www.users.dircon.co.uk/~truetech
Founded: 1993
City Reference Coordinates: 001°09'00"W 51°23'00"N

United Kingdom Industrial Space Committee (UKISC)
c/o A.G. Hicks (Secretary General)
P.O. Box 14
Wisbech PE13 1JZ
Telephone: (0)1945-64975
Telefax: (0)1945-61988
Electronic Mail: icks.ukisc@btinternet.com
WWW: http://www.ukspace.com/
Founded: 1975
Membership: 40
Activities: representing the collective interests of the UK space industry
Periodicals: (1) "Annual Report"
City Reference Coordinates: 000°10'00"E 52°40'00"N

United Kingdom Students for the Exploration and Development of Space (UKSEDS)
c/o Space Education Trust
Royal Aeronautical Society
4 Hamilton Place
London W1V 0BQ
Telephone: (0)1795-521784
Telefax: (0)1795-520880
Electronic Mail: info@uk.seds.org
WWW: http://www.uk.seds.org/
Founded: 1988
Periodicals: "Ecliptic"
City Reference Coordinates: 000°10'00"W 51°30'00"N

University College London (UCL), Department of Physics and Astronomy
Gower Street
London WC1E 6BT
Telephone: (0)20-73877050
Telefax: (0)20-73807145
Electronic Mail: <userid>@star.ucl.ac.uk
WWW: http://www.phys.ucl.ac.uk/
Founded: 1828
Staff: 30
Activities: education * research in UV, optical and IR astronomy * space sciences * upper atmosphere * planetary geology
Coordinates: 000°14'24"W 51°36'00"N H75m
City Reference Coordinates: 000°10'00"W 51°30'00"N

University College London (UCL), Department of Physics and Astronomy, University of London Observatory (ULO)
Mill Hill Park
London NW7 2QS
Telephone: (0)20-89590421
Telefax: (0)20-89064161
Electronic Mail: <userid>@ulo.ucl.ac.uk
WWW: http://www.ulo.ucl.ac.uk/
         http://www.phys.ucl.ac.uk/
Founded: 1928
Staff: 10
Activities: interstellar spectroscopy * CP stars * education
Periodicals: "Communications from University of London Observatory" (ISSN 0458-2128)
Coordinates: 000°14'27"W 51°36'46"N H81m (IAU Code 998)
City Reference Coordinates: 000°10'00"W 51°30'00"N

University College London (UCL), Mullard Space Science Laboratory (MSSL)
Holmbury
Saint Mary
Dorking RH5 6NT
Telephone: (0)1306-70292
           (0)1483-274111
Telefax: (0)1306-70201
         (0)1483-278312
Electronic Mail: <userid>@mssl.ucl.ac.uk
WWW: http://www.mssl.ucl.ac.uk/
City Reference Coordinates: 000°20'00"W 51°14'00"N

University of Birmingham, School of Physics and Space Research, Astrophysics and Space Research Group
Edgbaston Park Road
Birmingham B15 2TT
Telephone: (0)121-4146453
Telefax: (0)121-4143722
Electronic Mail: <userid>@star.sr.bham.ac.uk
               bison@bison.ph.bham.ac.uk (Birmingham Solar Oscillations Network - BiSON)
WWW: http://www.sr.bham.ac.uk/
         http://www.birmingham.ac.uk/physics/
         http://www.bham.ac.uk/physics/
         http://bison.ph.bham.ac.uk/ (BiSON)
Staff: 40
Activities: gamma-ray astronomy * X-ray astronomy * solar physics * space technology
Observatories: 1 (Wast Hills Observatory)
City Reference Coordinates: 001°50'00"W 52°30'00"N

University of Bradford, Faculty of Engineering, Engineering in Astronomy (EIA) Group
Bradford BD7 1DP
Telephone: (0)1274-234024
           (0)1274-234070
Telefax: (0)1274-236600
Electronic Mail: <userid>@bradford.ac.uk
WWW: http://www.telescope.org/
         http://www.eia.brad.ac.uk/eia.html
         http://www.eia.brad.ac.uk/rti/ (Bradford Robotic Telescope)
Founded: 1990
Staff: 12
City Reference Coordinates: 001°45'00"W 53°48'00"N

University of Bristol, H.H. Wills Physics Laboratory
Tyndall Avenue

Bristol BS8 1TL
Telefax: (0)117-9255624
Electronic Mail: <userid>@bristol.ac.uk
WWW: http://www.star.bris.ac.uk/
http://www.bris.ac.uk/
Founded: 1876 (University)
Staff: 45
Activities: astrophysics * particle physics * polymer physics * microstructural physics * theoretical physics * health physics * liquid physics * ...
City Reference Coordinates: 002°35'00"W 51°27'00"N

University of Cambridge, Department of Applied Mathematics and Theoretical Physics (DAMTP)
Silver Street
Cambridge CB3 9EW
Telephone: (0)1223-337900
Telefax: (0)1223-337918
Electronic Mail: enquiries@damtp.cam.ac.uk
WWW: http://www.damtp.cam.ac.uk/
Founded: 1959
Staff: 50
City Reference Coordinates: 000°08'00"E 52°13'00"N

University of Cambridge, Institute of Astronomy (IoA)
The Observatories
Madingley Road
Cambridge CB3 0HA
Telephone: (0)1223-337548
Telefax: (0)1223-337523
Electronic Mail: <userid>@ast.cam.ac.uk
WWW: http://www.ast.cam.ac.uk/
http://www.ast.cam.ac.uk/IOA/IOA.html
http://www-xray.ast.cam.ac.uk/ (X-Ray Astronomy Group)
Founded: 1820
Staff: 70
Activities: stellar radial velocities * stellar spectroscopy * cosmology * QSOs * AGNs * clusters * galaxy * stars * compact objects * helioseismology * instrumentation * survey astronomy * gravitational lensing * high-redshift objects * binary systems
Coordinates: 000°05'41"W 52°12'52"N H22m (IAU Code 503)
City Reference Coordinates: 000°08'00"E 52°13'00"N

University of Cambridge, Isaac Newton Institute for Mathematical Sciences
• See "Isaac Newton Institute for Mathematical Sciences"

University of Central Lancashire, Department of Physics, Astronomy and Mathematics
Preston PR1 2HE
Telephone: (0)1772-893540
(0)1772-892000 (Center for Astrophysics)
Telefax: (0)1772-892996
(0)1772-892903 (Center for Astrophysics)
Electronic Mail: info4pasm@uclan.ac.uk
<userid>@uclan.ac.uk
WWW: http://www.uclan.ac.uk/facs/science/physastr/index.htm
http://sa1.star.uclan.ac.uk/ (Center for Astrophysics)
Founded: 1926
Staff: 10
Activities: interstellar dust * star formation * flare stars * solar physics * STP * close binaries * white dwarfs * active galaxies * observational cosmology * multiwaveband astronomy
Coordinates: 002°42'12"W 53°46'30"N H34m (Jeremiah Horrocks Observatory)
002°35'30"W 53°48'06"N H63m (Alston Observatory)
City Reference Coordinates: 002°42'00"W 53°46'00"N

University of Durham, Department of Physics
Rochester Building
Science Laboratories
South Road
Durham DH1 3LE
Telephone: (0)191-3742167
Telefax: (0)191-3743749
Electronic Mail: physics.office@durham.ac.uk
WWW: http://www.dur.ac.uk/~dph0www/
Staff: 30
Activities: observing * theoretical cosmology * cosmic rays * gamma rays * extragalactic astronomy * polarimetry * ISM * galactic structure * instrumentation * image sharpening * historical astronomy
City Reference Coordinates: 001°34'00"W 54°47'00"N

University of Edinburgh, Institute for Astronomy
Royal Observatory
Blackford Hill
Edinburgh EH9 3HJ
Scotland
Telephone: (0)131-6688356 (E. Gibson, Administrative Assistant)
(0)131-6688100 (Switchboard)
Telefax: (0)131-6688416
Electronic Mail: <userid>@roe.ac.uk
WWW: http://www.roe.ac.uk/ifa/
Staff: 21
Activities: education * research * mobile planetarium
City Reference Coordinates: 003°13'00"W 55°57'00"N

University of Edinburgh, School of Mathematics
James Clerk Maxwell Building
Kings Buildings
Mayfield Road
Edinburgh EH9 3JZ
Scotland
Telephone: (0)131-6505035
Telefax: (0)131-6506553
Electronic Mail: <userid>@ed.ac.uk
maths@ed.ac.uk
WWW: http://www.maths.ed.ac.uk/
Staff: 4
Activities: stellar dynamics * globular star clusters * three-body problem * n-body simulations * dynamics of binary stars * astrophysical fluid dynamics * binary neutron stars * planetary dynamics
City Reference Coordinates: 003°13'00"W 55°57'00"N

University of Exeter, School of Physics, Astrophysics Group
Stocker Road
Exeter EX4 4QL
Telephone: (0)1392-264151
Telefax: (0)1392-264111
Electronic Mail: <userid>@astro.ex.ac.uk
WWW: http://www.astro.ex.ac.uk/
City Reference Coordinates: 003°31'00"W 50°43'00"N

University of Glasgow, Department of Aerospace Engineering
James Watt Building
Glasgow G12 8QQ
Scotland
Telephone: (0)141-3398855
Telefax: (0)141-3305560
Electronic Mail: <userid>@aero.gla.ac.uk
WWW: http://www.aero.gla.ac.uk/
Founded: 1952
Staff: 12
Activities: aeronautics and space technology * orbital dynamics * spacecraft guidance and control * mission analysis * solar sails * subcontractor for NASA, ESA and BNSC
City Reference Coordinates: 004°15'00"W 55°53'00"N

University of Glasgow, Department of Physics and Astronomy, Astronomy and Astrophysics Group
Kelvin Building
Glasgow G12 8QQ
Scotland
Telephone: (0)141-3305182
Telefax: (0)141-3305183
Electronic Mail: <userid>@astro.gla.ac.uk
WWW: http://www.astro.gla.ac.uk/
http://www.physics.gla.ac.uk/Home.html
Founded: 1760
Staff: 7
Activities: solar flares * accreting binaries * jets * plasma astrophysics * cosmology * gravitational waves * statistical astronomy * solar and stellar spectropolarimetry
City Reference Coordinates: 004°15'00"W 55°53'00"N

University of Glasgow, Observatory
Acre Road
Glasgow G20 0TL
Scotland
Telephone: (0)141-9465213

Telefax: (0)141-3305183
Electronic Mail: <userid>@astro.gla.ac.uk
　　　　　　　d.clarke@astro.gla.ac.uk (David Clarke, Director)
Founded: 1967
Staff: 3
Activities: education * stellar photometry and polarimetry * developing small optical instrumentation * planetarium
Coordinates: 004°18'20"W 55°54'08"N H53m
　　　　　　　004°24'13"W 55°56'21"N H150m
City Reference Coordinates: 004°15'00"W 55°53'00"N

University of Hertfordshire, Department of Physical Sciences, Division of Physics and Astronomy
College Lane
Hatfield AL10 9AB
Telephone: (0)1707-284602
Telefax: (0)1707-284644
Electronic Mail: phyqinfo@herts.ac.uk
　　　　　　　<userid>@star.herts.ac.uk
WWW: http://www.herts.ac.uk/natsci/Physics/
Founded: 1968
Staff: 50
Activities: undergraduate and postgraduate education * research
City Reference Coordinates: 000°13'00"W 51°46'00"N

University of Hertfordshire, Observatory
Bayfordbury
Near Hertford SG13 8LD
Telephone: (0)1707-285560
Telefax: (0)1992-503498
WWW: http://www.herts.ac.uk/
　　　http://www.herts.ac.uk/astro-ub
　　　http://www.herts.ac.uk/natsci/Physics/Observatory.html
Founded: 1968
Staff: 20
Activities: education * spectroscopy * image processing * polarimetry * active galaxies * protostars
Coordinates: 000°05'34"W 51°46'28"N H100m
City Reference Coordinates: 000°05'00"W 51°48'00"N (Hertford)

University of Keele, Department of Physics
Lennard-Jones Laboratory
Keele ST5 5BG
Telephone: (0)1782-583342
Telefax: (0)1782-711093
Electronic Mail: <userid>@astro.keele.ac.uk
WWW: http://www.keele.ac.uk/depts/ph/phhome.html
　　　http://www.astro.keele.ac.uk/ (Astrophysics Group)
Founded: 1952
Staff: 6
Activities: interacting binaries * cataclysmic variable stars * circumstellar dust
City Reference Coordinates: 002°17'00"W 53°00'00"N

University of Kent, Electronic Engineering Laboratories, Radio Astronomy Group
Canterbury CT2 7NT
Telephone: (0)1227-764000
Telefax: (0)1227-456084
Electronic Mail: <userid>@ukc.ac.uk
　　　　　　　star@star.ukc.ac.uk
WWW: http://www-star.ukc.ac.uk/
Founded: 1969
Staff: 7
Activities: molecular-line astronomy * SIS receiver development
City Reference Coordinates: 001°05'00"E 51°17'00"N

University of Kent, Unit for Space Sciences and Astrophysics
Physics Laboratory
Canterbury CT2 7NR
Telephone: (0)1227-823248
Telefax: (0)1227-827558
Electronic Mail: <userid>@ukc.ac.uk
　　　　　　　m.j.burchell@ukc.ac.uk (M.J. Burchell, Unit Director)
WWW: http://www.ukc.ac.uk/physical-sciences/space/
Founded: 1985
Staff: 14
Activities: solar-system ices * astrobiology * hypervelocity impacts * impact ionization * space debris * star formation * ISM
City Reference Coordinates: 001°05'00"E 51°17'00"N

University of Leeds, Department of Physics and Astronomy
E.C. Stoner Building
Woodhouse Lane
Leeds LS2 9JT
Telephone: (0)113-3433860
Telefax: (0)113-3433900
Electronic Mail: info@physics.leeds.ac.uk
                <userid>@ast.leeds.ac.uk
                <userid>@sun.leeds.ac.uk
WWW: http://www.leeds.ac.uk/physics
        http://ast.leeds.ac.uk/ (Astronomy Group)
        http://ast.leeds.ac.uk/haverah/hav-home.html (High Energy Cosmic Ray Group)
Founded: 1996 (1876 as "Department of Physics")
Staff: 30 (8 in astronomy)
Activities: star formation * AGNs * TeV gamma-ray astronomy * ultra-high energy cosmic rays
Coordinates: 001°35'00"W 53°50'00"N
City Reference Coordinates: 001°35'00"W 53°50'00"N

University of Leicester, Astronomy Society
• "Leicester University Astronomy Society"

University of Leicester, Department of Physics and Astronomy
University Road
Leicester LE1 7RH
Telephone: (0)116-2522073
Telefax: (0)116-2522070
Electronic Mail: <userid>@star.le.ac.uk
WWW: http://www.le.ac.uk/physics/
        http://www.star.le.ac.uk/theory/ (Theoretical Astrophysics)
        http://ion.le.ac.uk/ (Radio and Space Plasma Physics)
        http://www.star.le.ac.uk/xra.html (X-Ray Group)
        http://ledas-www.star.le.ac.uk (Leicester Database and Archive Service - LEDAS)
        http://www.ukaff.ac.uk/ (UK Astrophysical Fluids Facility - UKAFF)
Founded: 1965
Staff: 80
Activities: QSOs * AGNs * IR astronomy * accretion in close binary systems * brown dwarfs * space plasma, ionospheric and atmospheric research * X-ray astrophysics * space instrumentation * data centre for ROSAT and GINGA * XMM Survey Science Centre
Coordinates: 001°06'00"W 52°37'00"N
City Reference Coordinates: 001°06'00"W 52°37'00"N

University of Liverpool, Department of Physics
P.O. Box 147
Liverpool L69 3BX
Telephone: (0)151-7943370
Telefax: (0)151-7943348
Electronic Mail: <userid>@ns.ph.liv.ac.uk
WWW: http://www.ph.liv.ac.uk/
Founded: 1881
Staff: 30
Activities: nuclear astrophysics * nuclear structure * particle physics * solid-state physics * magnetism * surface science
• See also "Liverpool John Moores University"
City Reference Coordinates: 002°55'00"W 53°25'00"N

University of London, Goldsmith's College, Department of Mathematical and Computer Sciences
New Cross
London SE14 6NW
Telephone: (0)20-86927171
Electronic Mail: <userid>@gold.lon.ac.uk
WWW: http://www-maths.gold.ac.uk/
Founded: 1905
Staff: 4
Activities: solar-system dynamics * three-body problem * statistics of minor-planet and comet distributions * early star formation * galactic dark clouds
City Reference Coordinates: 000°10'00"W 51°30'00"N

University of London, Queen Mary and Westfield College (QMWC), Astronomy Unit
Mile End Road
London E1 4NS
Telephone: (0)20-79755454
Telefax: (0)20-89819587
Electronic Mail: astronomy@qmw.ac.uk
WWW: http://www.maths.qmw.ac.uk/Astronomy/
      http://www-star.qmw.ac.uk/
      http://www.maths.qmw.ac.uk/~www/astro/home.html
Founded: 1983

Staff: 30
Activities: theoretical astronomy * observing * data analysis * space experiments * solar physics * magnetospheric physics * solar system * stars * galaxies * cosmology * gravitation * celestial and galactic dynamics * star formation
City Reference Coordinates: 000°10'00"W 51°30'00"N

University of London, Queen Mary and Westfield College (QMWC), Astrophysics Group
Mile End Road
London E1 4NS
Telephone: (0)20-79755555
Telefax: (0)20-89755500
Electronic Mail: <userid>@qmw.ac.uk
WWW: http://www-star.qmw.ac.uk/AstroGroup.html
Founded: 1890
Staff: 23
Activities: molecular clouds * star formation * starburst * IRAS * ISO * sub-mm detector and receiver systems * active galaxies * starburst galaxies * planetary atmospheres * IR and sub-mm astronomy
City Reference Coordinates: 000°10'00"W 51°30'00"N

University of London Observatory (ULO)
• See "University College London (UCL), Department of Physics and Astronomy, University of London Observatory (ULO)"

University of Manchester, Department of Physics and Astronomy
Oxford Road
Manchester M13 9PL
Telephone: (0)161-2754100
Telefax: (0)161-2754297
Electronic Mail: <userid>@ast.man.ac.uk
WWW: http://www.ph.man.ac.uk/physics/
    http://www.ast.man.ac.uk/ (Astronomy Group)
    http://www.jb.man.ac.uk/ (Jodrell Bank Obs. - see separate entry)
Staff: 9
Activities: galactic structure * galactic encounters * QSOs * active galaxies * SNR * novae * HII regions * HH objects * bipolar nebulae * pulsars
Periodicals: "Astronomical Contributions from the University of Manchester, Series III"
City Reference Coordinates: 002°15'00"W 53°30'00"N

University of Manchester, Department of Physics and Astronomy, Jodrell Bank Observatory (JBO)

Macclesfield SK11 9DL
Telephone: (0)1477-571321
Telefax: (0)1477-571618
Electronic Mail: <userid>@jb.man.ac.uk
WWW: http://www.jb.man.ac.uk/
    http://www.jb.man.ac.uk/merlin/ (MERLIN)
Activities: radio-astronomy research * planetarium (see separate entry)
Coordinates: 002°18'26"W 53°14'10"N H78m (Jodrell Bank)
            002°32'03"W 53°09'22"N H47m (Darnhall)
            002°08'35"W 52°06'01"N (Defford)
            002°59'45"W 52°47'24"N (Knockin)
            002°59'46"W 53°06'45"N (Wardle)
            002°26'38"W 53°17'18"N (Pickmere)
            000°02'20"E 52°09'59"N (Cambridge)
• Konwn previously as "Nuffield Radio Astronomy Laboratories (NRAL)"
City Reference Coordinates: 002°07'00"W 53°16'00"N

University of Manchester Institute of Science and Technology (UMIST), Department of Physics, Astrophysics Group
P.O. Box 88
Manchester M60 1QD
Telephone: (0)161-2003677
Telefax: (0)161-2004303
Electronic Mail: <userid>@ast.phy.umist.ac.uk
WWW: http://saturn.phy.umist.ac.uk:8000/
Founded: 1983
Staff: 4
Activities: interstellar chemistry and physics * interstellar and interplanetary dust * circumstellar chemistry * evolution of interstellar clouds * star formation * chemical data * sub-mm observing * late stages of stellar evolution * maser theory and observation * radiative transfer
City Reference Coordinates: 002°15'00"W 53°30'00"N

University of Newcastle-upon-Tyne, Department of Physics
6 Kensington Terrace
Newcastle-upon-Tyne NE1 7RU
Telephone: (0)191-2227411
Telefax: (0)191-2227361

WWW: http://www.phys.ncl.ac.uk/
Founded: 1871
Staff: 16
Activities: quantum gravity * cosmology * laboratory astrophysics
Awards: (1/2) "Robinson Prize"
Coordinates: 001°48'01"W 54°59'19"N H50m (Close House Obs. – IAU Code 488)
City Reference Coordinates: 001°35'00"W 54°59'00"N

University of Nottingham, School of Physics and Astronomy
University Park
Nottingham NG7 2RD
Telephone: (0)115-9515164
Telefax: (0)115-9515180
Electronic Mail: <userid>@nottingham.ac.uk
WWW: http://www.nottingham.ac.uk/physics
City Reference Coordinates: 001°01'00"W 52°58'00"N

University of Oxford, Department of Physics, Astrophysics
Nuclear and Astrophysics Laboratory
Keble Road
Oxford OX1 3RH
Telephone: (0)1865-273303
Telefax: (0)1865-273390
Electronic Mail: sec@astro.ox.ac.uk
                <userid>@astro.ox.ac.uk
WWW: http://www-astro.physics.ox.ac.uk/
       http://www.physics.ox.ac.uk/
Founded: 1960
Staff: 12
Activities: observational and theoretical cosmology * galactic dynamics * high-energy astrophysics * stellar winds * chromospheres * coronae * stellar and laboratory spectroscopy * instrumentation
City Reference Coordinates: 001°15'00"W 51°46'00"N

University of Oxford, Department of Physics, Atmospheric, Oceanic and Planetary Physics
Clarendon Laboratory
Parks Road
Oxford OX1 3PU
Telephone: (0)1865-272933
Telefax: (0)1865-272923
Electronic Mail: <userid>@atm.ox.ac.uk
WWW: http://www.atm.ox.ac.uk/
Founded: 1920
Staff: 50
Activities: atmospheric, oceanic and planetary physics * planetary and Earth observation * atmospheric modelling * ocean modelling * education
City Reference Coordinates: 001°15'00"W 51°46'00"N

University of Oxford, Department of Physics, Theoretical Physics
1 Keble Road
Oxford OX1 3NP
Telephone: (0)1865-273999
Telefax: (0)1865-273947
Electronic Mail: <userid>@astro.ox.ac.uk
                <userid>@ermine.ox.ac.uk
WWW: http://www.physics.ox.ac.uk/Theory.html
Activities: structure and evolution of galaxies * stellar dynamics * UV and X-ray spectroscopy * stellar chromospheres, coronae and winds
City Reference Coordinates: 001°15'00"W 51°46'00"N

University of Portsmouth, School of Computer Science and Mathematics, Relativity and Cosmology Group
Mercantile House
Hampshire Terrace
Portsmouth PO1 2EG
Telephone: (0)23-92843108
Telefax: (0)23-92843106
Electronic Mail: <userid>@port.ac.uk
WWW: http://www.gravity.port.ac.uk/
Founded: 1994
Staff: 12
Activities: cosmology * relativistic astrophysics
City Reference Coordinates: 001°05'00"W 50°48'00"N

University of Saint Andrews, Mathematics and Statistics Department, Solar Theory Group
Saint Andrews KY16 9SS

Scotland
Telephone: (0)1334-463718
Telefax: (0)1334-463748
Electronic Mail: <userid>@mcs.st-and.ac.uk
WWW: http://www-solar.mcs.st-and.ac.uk/
Founded: 1972 (University: 1410)
Staff: 30
Activities: research on theory of solar magnetic fields
City Reference Coordinates: 002°48'00"W 56°20'00"N

University of Saint Andrews, School of Physics and Astronomy
North Haugh
Saint Andrews KY16 9SS
Scotland
Telephone: (0)1334-463100
Telefax: (0)1334-463104
Electronic Mail: <userid>@star.st-and.ac.uk
physics@st-and.ac.uk
WWW: http://www.st-and.ac.uk/~www_pa/
http://star-www.st-and.ac.uk/astronomy/Welcome.html
Founded: 1939 (University: 1410)
Staff: 7
Activities: star formation * accretion in AGN and binaries * extra-solar planets * galaxy formation * observational cosmology * stellar magnetic fields * undergraduate and graduate education
Coordinates: 002°48'54"W 56°20'12"N H30m (IAU Code 482)
City Reference Coordinates: 002°48'00"W 56°20'00"N

University of Sheffield, Department of Physics and Astronomy
Hicks Building
Hounsfield Road
Sheffield S3 7RH
Telephone: (0)114-2223519
Telefax: (0)114-2728079
Electronic Mail: star@sheffield.ac.uk
WWW: http://www.shef.ac.uk/~phys/
http://www.shef.ac.uk/~phys/research/astro/ (Astronomy Group)
http://www.shef.ac.uk/~phys/research/pa/ (Particle Astrophysics)
Founded: 1883 (as "Department of Physics")
City Reference Coordinates: 001°30'00"W 53°23'00"N

University of Southampton, Physics and Astronomy Department
Highfield
Southampton SO17 1BJ
Telephone: (0)23-80592093
Telefax: (0)23-80593910
Electronic Mail: <userid>@astro.soton.ac.uk
WWW: http://www.astro.soton.ac.uk/
Founded: 1952
Staff: 10
Activities: gamma and hard X-ray astronomy * telescope development * observational and theoretical high-energy astrophysics * STP
City Reference Coordinates: 001°25'00"W 50°55'00"N

University of Sussex, Astronomy Centre
c/o School of Chemistry, Physics and Environmental Science
Falmer
Brighton BN1 9QJ
Telephone: (0)1273-606755
Telefax: (0)1273-677196
Electronic Mail: <userid>@star.pact.cpes.susx.ac.uk
<userid>@sussex.ac.uk
WWW: http://astronomy.sussex.ac.uk/
Founded: 1966
Staff: 20
Activities: education * research * cosmology * astro-particle physics * galaxy formation * clustering and evolution * cataclysmic binaries * stellar evolution
City Reference Coordinates: 000°08'00"W 50°50'00"N

University of Sussex, SPRU - Science and Technology Policy Research
Mantell Building
Brighton BN1 9RF
Telephone: (0)1273-686758
Telefax: (0)1273-685865
WWW: http://www.susx.ac.uk/spru/
Founded: 1966
Staff: 70

Activities: research * education
City Reference Coordinates: 000°08'00"W 50°50'00"N

University of Teesside, School of Computing and Mathematics
Middlesbrough
Cleveland TS1 3BA
Telephone: (0)1642-342625
Telefax: (0)1642-230527
Electronic Mail: <userid>@tees.ac.uk
                 scm@tees.ac.uk
WWW: http://wheelie.tees.ac.uk/
Activities: origin of solar system * numerical methods for dynamical astronomy
City Reference Coordinates: 001°14'00"W 54°35'00"N (Teesside)

University of Wales, Aberystwyth, Department of Physics
Ceredigion SY23 3BZ
Telephone: (0)1970-622802
Telefax: (0)1970-622826
Electronic Mail: <userid>@aber.ac.uk
WWW: http://www.aber.ac.uk/~dphwww/
Founded: 1884
Activities: solar-terrestrial physics * atmospheric physics * radio and space physics
City Reference Coordinates: 004°05'00"W 52°25'00"N

University of Wales, College of Cardiff (UWCC), Physics and Astronomy Department
P.O. Box 913
Cardiff CF2 3YB
Wales
or :
North Building
5 The Parade
Cardiff
Wales
Telephone: (0)29-20874458
Telefax: (0)29-20874056
Electronic Mail: <userid>@astro.cf.ac.uk
WWW: http://www.astro.cf.ac.uk/
       http://www.cm.cf.ac.uk/
Founded: 1988
Staff: 18
Activities: galaxy observing * ST research * gravitational waves * general relativity * star formation theory * cosmic abundances * telescope design * galaxy simulations * cosmic magnetic fields
City Reference Coordinates: 003°13'00"W 51°13'00"N

University of Warwick, Physics Department, Space and Astrophysics Group
Coventry CV4 7AL
Telephone: (0)24-76524916
Telefax: (0)24-76692016
Electronic Mail: <userid>@warwick.ac.uk
WWW: http://www.astro.warwick.ac.uk/
City Reference Coordinates: 001°30'00"W 52°25'00"N

Vectis Astronomical Society (VAS)
c/o Rosemary M. Pears (secretary)
1 Rockmount Cottages
Undercliff Drive
Saint Laurence
Ventnor PO38 1XG
Isle of Wight
Telephone: (0)1983-853126
Telefax: (0)1983-853126
Electronic Mail: may@tatemma.freeserve.co.uk
WWW: http://www.wightskies.fsnet.co.uk/
Founded: 1976
Membership: 100
Periodicals: (12) "The New Zenith"
City Reference Coordinates: 001°11'00"W 50°36'00"N

Wadhurst Astronomical Society (WAS)
c/o Uplands Community College
Wadhurst TN5 6BA
WWW: http://users.aol.com/wadastro/
Founded: 1997
Membership: 40
Activities: meetings * observing

City Reference Coordinates: 000°21'00"E 51°04'00"N

Walsall Astronomical Society
c/o Alan Ledbury (Secretary)
71 Tame Street East
Walsall WS1 3LB
Electronic Mail: g.ledbury@cableinet.co.uk (Alan Ledbury, Secretary)
WWW: http://wkweb5.cableinet.co.uk/G.Ledbury/WASINDEX.HTM
    http://wemfas.future.easyspace.com/walsallas.html
Founded: 1990
Activities: meetings * observing
City Reference Coordinates: 001°58'00"W 52°35'00"N

Webb Society
c/o Don Miles
96 Marmion Road
Southsea PO5 2BB
Telephone: (0)23-92591146
Telefax: (0)23-92862466
Electronic Mail: rwa@ast.cam.ac.uk (R.W. Argyle, President)
WWW: http://www.webbsociety.org/
Founded: 1967
Membership: 450
Activities: education * publishing * annual meeting
Sections: Nebulae & Clusters * Southern Sky * Double Stars * Galaxies
Periodicals: (4) "Webb Society Quarterly Journal" (ISSN 0043-1680); (2) "Observing Section Reports" (ISSN 0953-8100), "The Deep-Sky Observer" (ISSN 0967-6139)
Awards: (1) "Webb Society Award"; "Webb Society Graphic Award", "Webb Society Technical Award"
City Reference Coordinates: 001°05'00"W 50°46'00"N

Welsh Border Astronomers (WBA)
c/o Pete Williamson
The Observatory
Top Street
Whittington SY11 4DR
Telephone: (0)1691-662076
Electronic Mail: pjw@wba.org.uk (Pete Williamson)
WWW: http://www.wba.org.uk/
Founded: 1996
Membership: 30
City Reference Coordinates: 003°00'00"W 52°52'00"N

Wessex Astronomical Society (WAS)
c/o Leslie A. Fry (Secretary)
Flat 7
816 Christchurch Road
Boscombe BH7 6DF
Telephone: (0)1202-416657
Electronic Mail: was@aegis1.demon.co.uk
WWW: http://ourworld.compuserve.com/homepages/dstrange/was.htm
    http://www.wessex-astro-society.freeserve.co.uk/
Founded: 1970 (1948 as "Poole Astronomical Society" - PAG)
Membership: 95
Periodicals: (6) "Wessex Astronomical Society News & Notes"
City Reference Coordinates: 001°50'00"W 50°44'00"N

West of London Astronomical Society (WOLAS)
c/o D. Radbourne (Secretary)
28 Tavistock Road
Edgware HA8 6DA
WWW: http://ourworld.compuserve.com/homepages/howard_bg/wolas.htm
Founded: 1967
Membership: 170
Activities: monthly meetings * observing
City Reference Coordinates: 000°16'00"W 51°36'00"N

West Yorkshire Astronomical Society (WYAS)
c/o Rosse Observatory
Carleton Grange
Carleton Road
Pontefract WF8 3RJ
or :
c/o Ken Willoughby
11 Hardistry Drive
Pontefract WF8 4BU

Electronic Mail: ken.willoughby@btinternet.com
WWW: http://www.welcome.to/wyas/
Founded: 1973
Membership: 70
Activities: education * observing (Sun, planets, deep sky) * photography * computing
Periodicals: (12) "Phobos"
Awards: (1) "Ralph Emerson Memorial Award", "Joynes Award", "Photographic Award"
Coordinates: 001°17'14"W 53°40'34"N (Rosse Observatory)
City Reference Coordinates: 001°18'00"W 53°42'00"N

William Herschel Society
19 New King Street
Bath BA1 2BL
Telephone: (0)1225-311342
Electronic Mail: comp@glam.ac.uk (Francis Ring, Chairman)
WWW: http://www.bath-preservation-trust.org.uk/museums/herschel/
Founded: 1976 (Museum: 1981)
Activities: managing William Herschel Museum * promoting work of William Herschel
Periodicals: (4) "Quarterly Bulletin"
City Reference Coordinates: 002°22'00"W 51°23'00"N

Wills Physics Laboratory (H.H._)
• See "University of Bristol, H.H. Wills Physics Laboratory"

Wolverhampton Astronomical Society
c/o Michael J. Bryce (Secretary)
16 Yellowhammer Court
Kidderminster DY10 4RR
Telephone: (0)1562-742850
Telefax: (0)1562-865684
Electronic Mail: secretary@wolvas.org.uk
WWW: http://www.wolvas.org.uk/
Founded: 1951
Membership: 50
Activities: lectures * discussions * meetings twice a month
Periodicals: (6) "Lyra"
City Reference Coordinates: 002°14'00"W 52°23'00"N (Kidderminster)
                            002°08'00"W 52°36'00"N (Wolverhampton)

Worth Hill Observatory (WHO)
Seacombe House
Worth Matravers
Swanage BH19 3LF
Telephone: (0)1929-439210
Electronic Mail: david@dstrange.freeserve.co.uk (David Granham Strange)
WWW: http://dstrange.freeserve.co.uk/
Founded: 1976
Staff: 1
Activities: private observatory * comets * SNs * dwarf novae * open to school groups and public observing with the Wessex Astronomical Society (see separate entry)
Coordinates: 002°02'00"W 50°32'00"N H120m
City Reference Coordinates: 001°58'00"W 50°37'00"N

Worthing Astronomical Society (WAS)
15 Newham Lane
Steyning BN44 3LR
Telephone: (0)1903-814090
Electronic Mail: was@nquinn.demon.co.uk
WWW: http://www.nquinn.demon.co.uk/was/
       http://www.worthingastro.freeserve.co.uk/
Founded: 1965
Membership: 90
Activities: amateur astronomy * observing
Periodicals: (12) "WAS News"
Coordinates: 000°25'44"W 50°48'59"N H8m
City Reference Coordinates: 000°20'00"W 50°53'00"N (Steyning)
                            000°23'00"W 50°48'00"N (Worthing)

Wycombe Astronomical Society (WAS)
3 Queens Court
Queens Road
High Wycombe HP13 6BA
Electronic Mail: paul@wolf359.demon.co.uk (Paul S. Scott, Chairman)
WWW: http://wycombeastro.org.uk/
Founded: 1981

Periodicals: (2) "Cygnus"
City Reference Coordinates: 000°46'00"W 51°38'00"N

York Astronomical Society (YAS)
c/o Martin Whipp (Secretary)
5 Whitham Drive
Huntington YO3 9YD
Telephone: (0)7744-751277
Electronic Mail: yas@talk21.com
WWW: http://www.yorkastro.freeserve.co.uk/
Founded: 1972
Membership: 25
Activities: observing * talks * education * popularization
Periodicals: (4) "Algol"; "Skynotes"
City Reference Coordinates: 001°04'00"W 54°01'00"N (Huntington)
                                    001°05'00"W 53°58'00"N (York)

York Observatory
Museum Gardens
York YO1 7FR
Telephone: (0)1904-629745
Telefax: (0)1904-651221
Founded: 1831
Staff: 1
Activities: education * exhibitions * public observing
Coordinates: 001°05'12"W 53°57'40"N
City Reference Coordinates: 001°05'00"W 53°58'00"N

Telephone: (0)2-5258618 x321
Telefax: (0)2-5250580
Electronic Mail: gallardo@fisica.edu.uy
WWW: http://heavy.fisica.edu.uy/oalm/sua/
City Reference Coordinates: 056°11'00"W 34°53'00"S

Universidad de la República, Departamento de Astronomía
Iguá 4225
11.400 Montevideo
Telephone: (0)2-5258624
 (0)2-5258625
 (0)2-5258626
Telefax: (0)2-5250580
Electronic Mail: <userid>@fisica.edu.uy
WWW: http://www.fisica.edu.uy/
Founded: 1955
Activities: education * research in cometary physics and dynamics * minor planets * gravitational lenses
Periodicals: "Publicaciones del Departamento de Astronomía"
Coordinates: 056°12'45"W 34°54'30"S H24m
City Reference Coordinates: 056°11'00"W 34°53'00"S

USA - Alabama

Auburn Astronomical Society (AAS)
• See USA-GA

Auburn University, Department of Aerospace Engineering
211 Aerospace Engineering Building
Auburn, AL 36849-5338
Telephone: 334-844-4874
Telefax: 334-844-6803
Electronic Mail: <userid>@eng.auburn.edu
WWW: http://www.eng.auburn.edu/department/ae/
Staff: 17
Activities: education * research
City Reference Coordinates: 085°29'00"W 32°36'00"N

Auburn University, Department of Physics
206 Allison Laboratory
Auburn, AL 36849
Telephone: 334-844-4264
Telefax: 334-844-4613
Electronic Mail: <userid>@physics.auburn.edu
WWW: http://www.physics.auburn.edu/
Staff: 25
Activities: condensed matter * plasma physics and spectroscopy * atomic/radiative physics * space physics * non-linear dynamics * magnetospheric physics * solar physics
City Reference Coordinates: 085°29'00"W 32°36'00"N

Birmingham Astronomical Society (BAS)
c/o Keith Tenney
5032 Wendover Drive
Birmingham, AL 35223-1632
or :
1341 Stonehedge Drive
Birmingham, AL 35235
Telephone: 205-956-5631
        205-853-0663
Electronic Mail: starport@wwisp.com
          br00k5@aol.com (Brooks Lide, President)
WWW: http://www.bas-astro.com/
Founded: 1977
Membership: 60
Activities: star parties * public presentations
Periodicals: "Newscope"
Observatories: 1 (Chandler Mountain)
City Reference Coordinates: 086°49'00"W 33°31'00"N

Center for Space Power and Advanced Electronics
c/o Space Power Institute
Auburn University
231 Leach Center
Auburn, AL 36849-5320
Telephone: 344-844-5894
Telefax: 344-844-5900
WWW: http://spi.auburn.edu/ccdspage.html
     http://spi.auburn.edu/ (Space Power Institute)
Founded: 1987
Staff: 15
Activities: power electronics * controls * high temperature hybrid electronics * packaging * electrical energy storage and transfer * silicon-carbide-based electronics * thermophysical properties of materials * hypervelocity space debris effects
City Reference Coordinates: 085°29'00"W 32°36'00"N

Gayle Planetarium (W.A._)
• See "W.A. Gayle Planetarium"

George C. Marshall Space Flight Center (MFSC)
• See "National Aeronautics and Space Administration (NASA), George C. Marshall Space Flight Center (MSFC)"

LeSueur Manufacturing Co. Inc.
3220 Lorna Road
Birmingham, AL 35216
Telephone: 205-822-0720

Telefax: 205-822-0749
Electronic Mail: lesmfg@bellsouth.net
WWW: http://www.astropier.com/
Founded: 1945
Staff: 8
Activities: manufacturing telescope mounting piers
City Reference Coordinates: 086°49'00"W 33°31'00"N

Marshall Space Flight Center (MFSC) (George C._)
• See "National Aeronautics and Space Administration (NASA), George C. Marshall Space Flight Center (MSFC)"

Mobile Astronomical Society (MAS)
c/o Rod Mollise
1207 Selma Street
Mobile, AL 36604
Telephone: 334-432-7071
Electronic Mail: rmollise@aol.com
WWW: http://members.aol.com/RMOLLISE/
Membership: 20
City Reference Coordinates: 088°05'00"W 30°42'00"N

National Aeronautics and Space Administration (NASA), George C. Marshall Space Flight Center (MSFC), Space Science Laboratory
ES-62
Huntsville, AL 35812
Telephone: 205-544-7690
Telefax: 205-544-7754
WWW: http://science.msfc.nasa.gov/
http://www.msfc.nasa.gov/
Founded: 1968
Staff: 20
Activities: gamma-ray astronomy * cosmic-ray research * space radiation environment and effects
City Reference Coordinates: 086°35'00"W 34°44'00"N

Teledyne Brown Engineering (TBE)
Cummings Research Park
300 Sparkman Drive
P.O. Box 070007
Huntsville, AL 35807-7007
Telephone: 1-800-933-2091
256-726-1000
Telefax: 256-726-5126
WWW: http://www.tbe.com/
Founded: 1953
Staff: 2500
City Reference Coordinates: 086°35'00"W 34°44'00"N

Universities Space Research Association (USRA), Huntsville Program Office (HPO)
4950 Corporate Drive
Suite 100
Huntsville, AL 35805-6227
Telephone: 256-895-0582
Telefax: 256-895-9222
WWW: http://www.usra.edu/
Founded: 1969
Membership: 80 (universities)
Activities: atmospheric sciences * microgravity * astronomy
City Reference Coordinates: 086°35'00"W 34°44'00"N

University of Alabama, Birmingham, Department of Physics
UAB Station
Birmingham, AL 35294-1170
Telephone: 205-934-8068
Telefax: 205-934-8042
Electronic Mail: physics@uab.edu
<userid>@phy.uab.edu
WWW: http://www.phy.uab.edu/
Staff: 28
City Reference Coordinates: 086°49'00"W 33°31'00"N

University of Alabama, Huntsville, Center for Space Plasma, Aeronomy, and Astrophysics Research (CSPAAR)
301 Sparkman Drive
Huntsville, AL 35899
Telephone: 256-895-6167

Telefax: 256-895-6790
　　　　　256-895-6873
Electronic Mail: <userid>@cspar.uah.edu
WWW: http://cspar.uah.edu/
　　　　http://www.uah.edu/physics/
Founded: 1986
Activities: solar-flare physics * cosmology * high-energy astrophysics
City Reference Coordinates: 086°35'00"W 34°44'00"N

University of Alabama, Tuscaloosa, Department of Physics and Astronomy, Astronomy Program
206 Gallalee Hall
Box 870324
Tuscaloosa, AL 35487-0324
Telephone: 205-348-5050
Telefax: 205-348-5051
Electronic Mail: <userid>@hera.astr.ua.edu
WWW: http://www.astr.ua.edu/
　　　　http://bama.ua.edu/~physics/
Founded: 1831
Staff: 6
Activities: research * education * ringed galaxies * interacting galaxies * N-body simulations * MHD * plasma astrophysics * AGNs * cosmological distance scale
Coordinates: 087°32'30"W 33°12'36"N H87m
City Reference Coordinates: 087°33'00"W 33°13'00"N

University of North Alabama (UNA), Planetarium and Observatory
P.O. Box 5150
Florence, AL 35632
Telephone: 256-765-4334
Electronic Mail: dcurott@unanov.una.edu (D.R. Curott, Planetarium Director)
WWW: http://www2.una.edu/physics/plm.htm
Founded: 1967
Staff: 1
Activities: shows * observing
Coordinates: 087°40'00"W 34°40'00"N H200m
City Reference Coordinates: 087°40'00"W 34°40'00"N

US Space Camp/US Space and Rocket Center
P.O. Box 070015
Huntsville, AL 35807-7015
or :
One Tranquility Base
Huntsville, AL 35805
Telephone: 1-800-63-SPACE
Telefax: 205-837-6137
Electronic Mail: info@spacecamp.com
WWW: http://www.spacecamp.com/
Founded: 1982
City Reference Coordinates: 086°35'00"W 34°44'00"N

Von Braun Astronomical Society (VBAS)
P.O. Box 1142
Huntsville, AL 35807
or :
Monte Sano State Park
Huntsville, AL 35801
Telephone: 256-539-0316
　　　　　256-464-0945
Electronic Mail: vbas@vbas.org
WWW: http://www.vbas.org/
Founded: 1954
Membership: 150
Activities: planetarium shows * star parties * research
Periodicals: (12) "Via Stellaris"
City Reference Coordinates: 086°35'00"W 34°44'00"N

W.A. Gayle Planetarium
c/o Troy State University Montgomery
1010 Forest Avenue
Montgomery, AL 36106
Telephone: 334-241-4799
Telefax: 334-240-4309
Electronic Mail: planet@tsum.edu
WWW: http://www.tsum.edu/planet/
City Reference Coordinates: 086°20'00"W 32°22'00"N

USA - Alaska

Drake Planetarium (Marie_)
● See "Marie Drake Planetarium"

Marie Drake Planetarium
c/o Scott Tierman
10014 Crazy Horse Drive
Juneau, AK 99801
Electronic Mail: druid@alaska.net.
WWW: http://www.juneau.com/jweb/planet.htm
City Reference Coordinates: 134°27'00"W 58°20'00"N

University of Alaska in Anchorage (UAA), Department of Physics and Astronomy
3221 Providence Drive
Anchorage, AK 99508
Telephone: 907-786-1094
Telefax: 907-786-4607
Electronic Mail: <userid>@uaa.alaska.edu
WWW: http://local.uaa.alaska.edu/~afjtp/physics.html
Founded: 1982
Staff: 4
Activities: education * astrophysics * chemical physics
City Reference Coordinates: 149°53'00"W 61°13'00"N

USA - Arizona

Arizona Amateur Astronomy Society (AAAZ)
P.O. Box 5311
Mohave Valley, AZ 86446
Electronic Mail: aaasociety@usa.net
WWW: http://members.xoom.com/AAASociety/
Coordinates: 114°36'06"W 34°51'02"N
City Reference Coordinates: 114°04'00"W 35°12'00"N (Kingman)

Arizona Science Center
600 East Washington Street
Phoenix, AZ 85004-2303
Telephone: 602-716-2000
Telefax: 602-716-2099
Electronic Mail: info@azscience.org
WWW: http://www.azscience.org/
      http://www.azscience.org/PLANETARIUM/planetarium.html (Dorrance Planetarium)
City Reference Coordinates: 112°05'00"W 33°27'00"N

Arizona State University (ASU), Department of Geology
Box 871404
Tempe, AZ 85287-1404
Telephone: 480-965-5081
Telefax: 480-965-8102
Electronic Mail: geology@asu.edu
WWW: http://www-glg.la.asu.edu/
      http://esther.la.asu.edu/ (Planetary Exploration Laboratory)
      http://europa.la.asu.edu/ (Planetary Geology Group)
      http://violetta.la.asu.edu/ (Planetary Geophysics Laboratory)
      http://violetta.la.asu.edu/cms/html/home/home.html (Center for Meteorite Studies)
Founded: 1948
Staff: 21
City Reference Coordinates: 111°56'00"W 33°25'00"N

Arizona State University (ASU), Department of Physics and Astronomy, Astronomy Group
Box 871504
Tempe, AZ 85287-1504
or :
Bateman Physical Sciences F-Wing
PSF 470
Telephone: 480-965-3561
Telefax: 480-965-7954
Electronic Mail: phyast.info@asu.edu
      astro.info@asu.edu
WWW: http://ast.asu.edu/ (Astronomy)
      http://phy.asu.edu/ (Physics)
      http://www.asu.edu/clas/dopa/Plan_dir/Planetarium.html (Planetarium)
Founded: 1960
Staff: 12
Activities: stellar structure * stellar evolution * cataclysmic variable stars * novae * SN * stars * comets * galaxies * QSOs * AGNs * X-ray binaries * galaxy dynamics * large-scale structure * ISM * planetarium
City Reference Coordinates: 111°56'00"W 33°25'00"N

Astronomical Adventures
2532 North Fourth Street
Suite 200
Flagstaff, AZ 86004
Telephone: 602-485-4367
Electronic Mail: bolly1@home.com (Linda Woolley, President)
WWW: http://Astronomy-Mall.com/regular/products/astronomical-adventures/
Founded: 1993
Staff: 2
Activities: astronomy observing * guided geologic tours
Coordinates: 111°14'00"W 34°56'00"N H1,920m (Flying M Ranch)
City Reference Coordinates: 111°38'00"W 35°12'00"N

Astronomical Consultants and Equipment (ACE) Inc.
P.O. Box 91946
Tucson, AZ 85752-1946
or :
4261 West Ina Road
Suite 111

Tucson, AZ 85741
Telephone: 520-579-0698
Telefax: 520-579-8570
Electronic Mail: info@astronomical.com
support@astronomical.com
WWW: http://www.astronomical.com/
Founded: 1994
Staff: 5
Activities: telescope control systems * designing and manufacturing instrumentation * whole telescope fabrication * consulting
City Reference Coordinates: 110°58'00"W 32°13'00"N

Astronomy Club of Sun City West
c/o Jim Crisman
17802 North 131st Avenue
Sun City West, AZ 85375
Telephone: 623-584-0896
Electronic Mail: jamescrisman@speedchoice.com
WWW: http://www.geocities.com/scwac
City Reference Coordinates: 112°17'00"W 33°36'00"N (Sun City)

Braeside Observatory (BO)
P.O. Box 906
Flagstaff, AZ 86002
Telephone: 520-774-4222
Telefax: 520-774-4222
Electronic Mail: captain@braeside.org
WWW: http://www.braeside.org/
Founded: 1976
Staff: 4
Activities: CCD photometry * cataclysmic and other variable stars
Coordinates: 111°44'44"W 35°11'25"N H2,160m
City Reference Coordinates: 111°38'00"W 35°12'00"N

Center for Astronomical Adaptive Optics (CAAO)
• See "University of Arizona, Center for Astronomical Adaptive Optics (CAAO)"

Center for Image Processing in Education (CIPE)
P.O. Box 13750
Tucson, AZ 85732-3750
Telephone: 1-800-322-9884 (USA only)
Electronic Mail: info@cipe.com
<userid>@cipe.com
WWW: http://www.cipe.com/
Activities: promoting computer-aided visualization as a tool for teaching and learning
City Reference Coordinates: 110°58'00"W 32°13'00"N

Coconino Astronomers
c/o Ralph Aeschliman
HC-30 Box 14
Flagstaff, AZ 86001
Telephone: 520-779-3431 (home)
520-556-7354 (work)
520-774-6227 (work)
Electronic Mail: raeschliman@iflag2.wr.usgs.gov
raeschliman@aol.com
WWW: http://www.infomagic.com/~bright/cocoast/cocoast.html
Membership: 15
City Reference Coordinates: 111°38'00"W 35°12'00"N

Coronado Instrument Group (CIG) Inc.
HCI Box 398
Pearce, AZ 85625
Telephone: 1-800-987-1516 (USA only)
520-795-3149
Telefax: 520-824-3216
Electronic Mail: lunt@theriver.com (David Lunt)
lunt@vtc.net (Technical Support)
WWW: http://www.coronadofilters.com/
Founded: 1997
Staff: 13
Activities: designing and manufacturing solar filters and solar telescopes
Periodicals: (4) "The Solar Observer"
Observatories: 1 (Coronado Solar Observatory - under construction)
City Reference Coordinates: 109°50'00"W 31°53'00"N

Custom Scientific
3852 North 15th Avenue
Phoenix, AZ 85015
Telephone: 602-200-9200
Telefax: 602-200-9206
Electronic Mail: customsci@delphi.com
optics@customscientific.com
WWW: http://customscientific.com/
Founded: 1989
Staff: 3
Activities: manufacturing optical and interference filters * optical, mechanical and electrical engineering of telescope accessories and relay optics including filter wheels and focal reducers
City Reference Coordinates: 112°05'00"W 33°27'00"N

Discovery Park, Gov Aker Observatory
1651 West Discovery Park Boulevard
Safford, AZ 85546
Telephone: 928-428-6260
Telefax: 928-428-8081
Electronic Mail: discover@discoverypark.com
WWW: http://www.discoverypark.com/
http://www.discoverypark.com/astron.html
Founded: 1995
Activities: observing * education * interactive exhibits
Periodicals: (6) "Discovery Park Explorer"
City Reference Coordinates: 109°43'00"W 32°50'00"N

East Valley Astronomy Club (EVAC)
P.O. Box 2202
Mesa, AZ 85214-2202
or :
c/o Aaron McNeely
3222 North 38th Street
Suite 1
Phoenix, AZ 85018
Telephone: 602-954-3971
Electronic Mail: amcneely@aol.com
WWW: http://www.eastvalleyastronomy.org/
Founded: 1987
Membership: 100
Activities: monthly meetings * star parties * lectures
Periodicals: (12) "East Valley Astronomy Club Newsletter"
City Reference Coordinates: 115°50'00"W 33°25'00"N (Mesa)
112°05'00"W 33°27'00"N (Phoenix)

EOS Technologies Inc. (EOST)
925 West Grant Road
Tucson, AZ 85705
Telephone: 520-624-6399
Telefax: 520-624-1906
Electronic Mail: inquiries@eostech.com
WWW: http://www.eostech.com/
Founded: 1995
Staff: 21
Activities: engineering design, assembly and test of telescopes
City Reference Coordinates: 110°58'00"W 32°13'00"N

Flagstaff Dark Skies Coalition (FDSC)
c/o John Grahame
P.O. Box 999
Flagstaff, AZ 86002
Telephone: 520-606-1055
Electronic Mail: darkskies@flagstaff.az.us
City Reference Coordinates: 111°38'00"W 35°12'00"N

Flandrau Science Center and Planetarium (Grace H._)
• See "Grace H. Flandrau Science Center and Planetarium"

Focus Software Inc. (FSI)
P.O. Box 18228
Tucson, AZ 85731
Telephone: 520-733-0130
Telefax: 520-733-0135
Electronic Mail: sales@focus-software.com
support@focus-software.com

WWW: http://www.focus-software.com/
Activities: producing optical design software
City Reference Coordinates: 110°58'00"W 32°13'00"N

Fred Lawrence Whipple Observatory (FLWO)
• See "Harvard-Smithsonian Center for Astrophysics (CfA), Fred Lawrence Whipple Observatory (FLWO)"

Gemini Observatory, Tucson Project Office
950 North Cherry Avenue
Tucson, AZ 85719
Telephone: 520-318-8548
Telefax: 520-318-8590
Electronic Mail: gemini@gemini.edu
               <userid>@gemini.edu
WWW: http://www.gemini.edu/
      http://www.noao.edu/usgp/usgp.html
Founded: 1998
Activities: international partnership (USA, United Kingdom, Canada, Australia, Chile, Argentina, Brazil) aiming at setting up two high-performance 8m telescopes on Mauna Kea (USA-HI) and Cerro Pachón (Chile)
Periodicals: "The Gemini Newsletter"
Coordinates: 155°28'18"W 19°49'36"N (Gemini North)
               070°43'24"W 30°13'42"S (Gemini South)
City Reference Coordinates: 110°58'00"W 32°13'00"N

Global Network of Automatic Telescopes (GNAT)
c/o D.L. Crawford
2127 East Speedway
Suite 209
Tucson, AZ 85719
Telephone: 520-325-4505
Telefax: 520-327-1388
Electronic Mail: crawford@gnat.org
WWW: http://www.gnat.org/~ida/gnat/
Founded: 1992
City Reference Coordinates: 110°58'00"W 32°13'00"N

Gov Aker Observatory
• See "Discovery Park, Gov Aker Observatory"

Grace H. Flandrau Science Center and Planetarium
1601 East University Boulevard
Tucson, AZ 85719
Telephone: 520-621-STAR
Telefax: 520-621-8451
WWW: http://www.flandrau.org/
Founded: 1975
Staff: 14
Activities: public and school programs * laser and star shows * workshops * public viewing
Periodicals: (4) "AMES (Accent on Mathematics Engineering and Science) Newsletter"
City Reference Coordinates: 110°58'00"W 32°13'00"N

Grasslands Observatory
5100 North Sabino Foothills Drive
Tucson, AZ 85750
Telephone: 520-760-2100
Electronic Mail: mcgaha@skepticus.com
WWW: http://www.3towers.com/
Founded: 1987
Staff: 2
Activities: private observatory (IAU Code 651) * SN search * minor-planet astrometry and occultations * high-resolution imaging * study of NGC 2261 * CCD imaging * photometry * NEO follow-up
Coordinates: 110°34'49"W 31°40'11"N H1,540m
City Reference Coordinates: 110°58'00"W 32°13'00"N

Haag - Meteorites (Robert A._)
• See "Robert A. Haag - Meteorites"

Harvard-Smithsonian Center for Astrophysics (CfA), Fred Lawrence Whipple Observatory (FLWO)
P.O. Box 97
Amado, AZ 85645-0097
or :
670 Mount Hopkins Road
Amado, AZ 85645
Telephone: 520-670-5701
             617-495-7461

520-670-5707
Telefax: 520-670-5714
617-495-7326
Electronic Mail: <userid>@base.sao.arizona.edu
WWW: http://linmax.sao.arizona.edu/help/FLWO/whipple.html
Founded: 1968
Activities: spectroscopy (extragalactic, stellar and planetary bodies) * gamma- and cosmic-ray astronomy * IR photometry * CCD visible and IR imaging * IR interferometry
Periodicals: "Harvard-Smithsonian Preprint Series"
Coordinates: 110°52'39"W 31°40'51"N H2,608m (Mount Hopkins – IAU Code 696)
City Reference Coordinates: 110°53'00"W 31°41'00"N

Harvard-Smithsonian Center for Astrophysics (CfA), Multiple Mirror Telescope Observatory (MMTO)
• See "Multiple Mirror Telescope Observatory (MMTO)"

Hextek Corp.
P.O. Box 42943
Tucson, AZ 85733
or :
1665 East 18th Street
Suite 208
Tucson, AZ 85719
Telephone: 520-623-7647
Telefax: 520-882-7384
Electronic Mail: mvoevodsky@hextek.com (Michael Voevodsky, VP Business Development)
WWW: http://www.hextek.com/
Founded: 1985
Staff: 8
Activities: manufacturing lightweight glass mirror blank substrates
City Reference Coordinates: 110°58'00"W 32°13'00"N

Huachuca Astronomy Club
5539 South Shawnee Drive
Sierra Vista, AZ 85650
Telephone: 520-458-5973
Electronic Mail: astro@tron.cochise.cc.az.us
WWW: http://c3po.cochise.cc.az.us/astro
City Reference Coordinates: 110°18'00"W 31°33'00"N

Image Reduction and Analysis Facility (IRAF)
• See "National Optical Astronomy Observatories (NOAO), Image Reduction and Analysis Facility (IRAF)"

Infrared Laboratories Inc.
1808 East 17th Street
Tucson, AZ 85719-6505
Telephone: 520-622-7074
Telefax: 520-623-0765
Electronic Mail: infrared@irlabs.com
WWW: http://www.irlabs.com/
Founded: 1967
Staff: 35
Activities: manufacturing IR detectors, cryogenic research Dewars, closed-cycle cryostats, FPA cameras, FPA drive electronics
City Reference Coordinates: 110°58'00"W 32°13'00"N

International Association of Astronomical Artists (IAAA)
c/o Kim Poor (Membership Secretary)
P.O. Box 37197
Tucson, AZ 85740
Telephone: 520-888-2424
Telefax: 520-292-9852
Electronic Mail: kim@novaspace.com
admin@iaaa.org
WWW: http://www.iaaa.org/
Founded: 1982
City Reference Coordinates: 110°58'00"W 32°13'00"N

International Dark-Sky Association (IDA)
c/o D.L. Crawford
3225 North First Avenue
Tucson, AZ 85719
Telephone: 520-293-3198
Telefax: 520-293-3192
Electronic Mail: ida@darksky.org
WWW: http://www.darksky.org/

Founded: 1988
Membership: 9000
Activities: promoting good outdoor lighting to conserve energy and preserve night skies
Periodicals: (4) "IDA Newsletter"
City Reference Coordinates: 110°58'00"W 32°13'00"N

Jensen Tools Inc.
7815 South 46th Street
Phoenix, AZ 85044-5399
Telephone: 602-453-3169
　　　　　1-800-426-1194 (USA only)
Telefax: 602-438-1690
　　　　1-800-366-9662 (USA only)
Electronic Mail: info@jensentools.com
WWW: http://www.jensentools.com/
Founded: 1958
Staff: 250
Activities: catalog supplier of tool kits and cases
City Reference Coordinates: 112°05'00"W 33°27'00"N

Kitt Peak National Observatory (KPNO)
• See "National Optical Astronomy Observatories (NOAO), Kitt Peak National Observatory (KPNO)"

Lawrence Whipple Observatory (FLWO) (Fred_)
• See "Harvard-Smithsonian Center for Astrophysics (CfA), Fred Lawrence Whipple Observatory (FLWO)"

Lowell Observatory
1400 West Mars Hill Road
Flagstaff, AZ 86001
Telephone: 520-774-3358
Telefax: 520-774-6296
Electronic Mail: <userid>@lowell.edu
WWW: http://www.lowell.edu/
Founded: 1894
Staff: 50
Activities: Ap stars * minor planets * astrometry * comets * galactic evolution * galactic structure * image processing * Mars * multiple stars * occultations * open clusters * photometry * planetary atmospheres * planetary satellites * solar-type stars * solar variations * speckle interferometry * star formation * stellar populations * variable stars
Periodicals: (4) "The Lowell Observer"
Coordinates: 111°39'48"W 35°12'08"N H2,210m (IAU Code 699)
　　　　　　　111°32'09"W 35°05'49"N H2,198m (Anderson Mesa Stn – IAU Code 688)
City Reference Coordinates: 111°38'00"W 35°12'00"N

McGraw Hill Observatory
• See now "MDM Observatory"

MDM Observatory
HC 02 Box 7520
Sells, AZ 85634
Telephone: 510-318-8661
Telefax: 520-318-8664
Electronic Mail: mdm@astro.lsa.umich.edu
WWW: http://space.mit.edu/mdm.html
Founded: 1975
Staff: 19
Activities: optical observing
Coordinates: 111°37'00"W 31°57'00"N H1,925m (IAU Code 697)
• Jointly operated by the "University of Michigan", the "Dartmouth College", the "Ohio State University" and the "Columbia University"
City Reference Coordinates: 111°37'00"W 31°57'00"N

Meteor Crater Enterprises Inc.
P.O. Box 70
Flagstaff, AZ 86002-0070
Telephone: 520-289-5898
Electronic Mail: info@meteorcrater.com
WWW: http://www.flagstaff.az.us/meteor/
　　　　http://www.meteorcrater.com/
City Reference Coordinates: 111°38'00"W 35°12'00"N

Michigan-Dartmouth-MIT Observatory
• See now "MDM Observatory"

Multiple Mirror Telescope Observatory (MMTO)
P.O. Box 210065

University of Arizona
Tucson, AZ 85721-0065
Telephone: 520-621-1558
Telefax: 520-670-5740
Electronic Mail: <userid>@as.arizona.edu
WWW: http://sculptor.as.arizona.edu/foltz/www/mmt.html
http://cfa-www.harvard.edu/cfa/oir/MMT/mmto.html
Founded: 1979
Staff: 19
Activities: maintaining a 6.5m telescope for research in astrophysics and astronomy * telescope development
Periodicals: (6) "MMTO Technical Reports"
Coordinates: 110°53'04"W 31°41'20"N H2,606m (Mount Hopkins)
• Joint research facility of the "Smithsonian Observatory" and the "University of Arizona"
City Reference Coordinates: 110°58'00"W 31°13'00"N

National Optical Astronomy Observatories (NOAO)
950 North Cherry Avenue
P.O. Box 26732
Tucson, AZ 85726-6732
Telephone: 520-318-8000
Telefax: 520-318-8360
Electronic Mail: <userid>@noao.edu
WWW: http://www.noao.edu/
Founded: 1984
Staff: 378
Activities: research * support of telescopes for the astronomical community
Periodicals: (1) "NOAO Annual Report"; (4) "NOAO Newsletter"
City Reference Coordinates: 110°58'00"W 31°13'00"N

National Optical Astronomy Observatories (NOAO), Image Reduction and Analysis Facility (IRAF)
P.O. Box 26732
Tucson, AZ 85726-6732
or :
950 North Cherry Avenue
Tucson, AZ 85719
Telephone: 520-318-8160 (hot line)
520-318-8000
Telefax: 520-318-8360
Electronic Mail: iraf@noao.edu
WWW: http://iraf.noao.edu/
Founded: 1981
City Reference Coordinates: 110°58'00"W 32°13'00"N

National Optical Astronomy Observatories (NOAO), Kitt Peak National Observatory (KPNO)
P.O. Box 26732
Tucson, AZ 85726-6732
or :
950 North Cherry Avenue,
Tucson, AZ 85719
Telephone: 520-318-8000
520-318-8600 (Kitt Peak Mountain switchboard)
Telefax: 520-318-8360
Electronic Mail: kpno@noao.edu
<userid>@noao.edu
WWW: http://www.noao.edu/kpno/kpno.html
Founded: 1958
Staff: 36
Activities: observing * astronomical research (Sun, planets, stars, galaxies) * instrumentation
Periodicals: (1) "NOAO Annual Report"; (4) "NOAO Newsletter"; "Contributions from KPNO"
Coordinates: 111°35'42"W 31°57'30"N H2,120m (IAU Code 695)
City Reference Coordinates: 110°58'00"W 32°13'00"N

National Optical Astronomy Observatories (NOAO), National Solar Observatory (NSO), Tucson
P.O. Box 26732
Tucson, AZ 85726-6732
Telephone: 520-325-9294
Telefax: 520-325-9278
Electronic Mail: <userid>@noao.edu
WWW: http://www.nso.noao.edu/welcome.html
Founded: 1984
Staff: 16
Activities: observing facilities for solar astronomers * developing new instrumentation * solar physics
Periodicals: (1) "NOAO Annual Report"; (4) "NOAO Newsletter"
Coordinates: 105°49'13"W 32°47'12"N
City Reference Coordinates: 110°58'00"W 32°13'00"N

National Radio Astronomy Observatory (NRAO), Arizona Operations
Campus Building 65
949 North Cherry Avenue
Tucson, AZ 85721-0655
Telephone: 520-882-8250 (Office)
520-318-8670 (Telescope)
Telefax: 520-882-7955
Electronic Mail: <userid>@nrao.edu
WWW: http://www.tuc.nrao.edu/Tucson.html
http://www.nrao.edu/
http://www.mma.nrao.edu/ (ALMA)
Founded: 1967
Staff: 29
Activities: mm-wavelength observing
Coordinates: 111°36'51"W 31°57'14"N H1,938m
City Reference Coordinates: 110°58'00"W 32°13'00"N

National Solar Observatory (NSO)
● See "National Optical Astronomy Observatories (NOAO), National Solar Observatory (NSO)"

Northern Arizona Astronomy Association (NAAA)
Telephone: 520-523-8121
Electronic Mail: orion@ieee.org
WWW: http://www.physics.nau.edu/~naaa/
Founded: 1952
Activities: meetings * observing
City Reference Coordinates: 111°38'00"W 35°12'00"N (Flagstaff)

Northern Arizona University (NAU), Department of Physics and Astronomy
NAU Box 6010
Flagstaff, AZ 86011-6010
Telephone: 928-523-7170
Telefax: 928-523-1371
Electronic Mail: astro.physics@nau.edu
<userid>@nau.edu
WWW: http://www.physics.nau.edu/
http://www.nuro.nau.edu/ (National Undergraduate Research Observatory - NURO)
Founded: 1987 (1963 as "Department of Physics")
Staff: 20
Activities: education * observing * public outreach
Periodicals: "inside"
Coordinates: 111°39'12"W 35°11'06"N H2,110m (IAU Code 687)
111°32'09"W 35°05'55"N H2,209m (NURO - IAU Code 688)
City Reference Coordinates: 111°38'00"W 35°12'00"N

Novaspace Galleries
P.O. Box 37197
Tucson, AZ 85740
Telephone: 1-800-727-NOVA (USA only)
1-800-727-6683 (USA only)
520-888-2424
Telefax: 520-292-9852
Electronic Mail: staff@novaspace.com
WWW: http://www.novaspace.com/
Founded: 1978
Staff: 6
Activities: publishing and distributing space art on a number of supports including screen savers
● Formerly "Novagraphics Space Art Gallery"
City Reference Coordinates: 110°58'00"W 32°13'00"N

Pachart Foundation
1130 San Lucas Circle
Tucson, AZ 85704
Telephone: 520-297-4797
Founded: 1984
Activities: fostering scholarly and educational work in common areas of concern regarding science (in particular astronomy and cosmology), philosophy, and theology * operating the Center for Inderdisciplinary Studies (CIS) and the Pachart Publishing House (see these entries)
City Reference Coordinates: 110°58'00"W 32°13'00"N

Phoenix Astronomical Society (PAS)
10828 North Biltmore Drive
Suite 141
Phoenix, AZ 85029
Telephone: 602-547-2420 (Terri Finch, Editor)

Electronic Mail: alienstarstuff@yahoo.com
WWW: http://pastimes.homestead.com/pastimes.html
Founded: 1948
Membership: 70
Activities: monthly star parties * monthly meetings
Periodicals: (10) "PAStimes"
City Reference Coordinates: 112°05'00"W 33°27'00"N

Planetary Science Institute (PSI), Tucson Division
620 North 6th Avenue
Tucson, AZ 85705-8331
Telephone: 520-622-6300
Telefax: 520-622-8060
Electronic Mail: psikey@psi.edu
WWW: http://www.psi.edu/
Founded: 1971
Staff: 10
Activities: planetary science research
City Reference Coordinates: 110°58'00"W 32°13'00"N

Prescott Astronomy Club (PAC)
P.O. Box 1178
Mayer, AZ 86333-1178
Telephone: 928-632-7355
Electronic Mail: gfreynpo@mindspring.com (Gary Frey)
WWW: http://www.pacorg.net/
Founded: 1974
Activities: meetings * lectures * star parties * public outreach
City Reference Coordinates: 112°14'00"W 34°24'00"N

Random Factory (The_)
• See "The Random Factory"

Robert A. Haag - Meteorites
2990 East Michigan Avenue
P.O. Box 27527
Tucson, AZ 85726
Telephone: 520-882-8804
Telefax: 520-743-7225
Electronic Mail: bobhaag@meteoriteman.com
WWW: http://www.meteoriteman.com/
Founded: 1981
• Dealer and distributor
City Reference Coordinates: 110°58'00"W 32°13'00"N

Roper Scientific
3440 East Britannia Drive
Tucson, AZ 85706
Telephone: 520-889-9933
Telefax: 520-573-1944
Electronic Mail: info@roperscientific.com
WWW: http://www.roperscientific.com/
Founded: 1970
Staff: 96
Activities: designing, developing and manufacturing CCD digital imaging systems
City Reference Coordinates: 110°58'00"W 32°13'00"N

Sabino Canyon Observatory (SCO)
5100 North Sabino Foothills Drive
Tucson, AZ 85750
Telephone: 520-760-2100
Electronic Mail: mcgaha@skepticus.com
WWW: http://www.3towers.com/
Founded: 1984
Staff: 2
Activities: private observatory (IAU Code 854) * SN search * minor-planet astrometry and occultations * high-resolution imaging * study of NGC 2261 * CCD image processing * NEO follow-up
Coordinates: 110°49'00"W 32°18'00"N H838m
City Reference Coordinates: 110°58'00"W 31°13'00"N

Saguaro Astronomy Club (SAC)
c/o Steve Coe
1011 E. Rowlands Lane
Phoenix, AZ 85022
or :

University of Arizona, Multiple Mirror Telescope Observatory (MMTO)
• See "Multiple Mirror Telescope Observatory (MMTO)"

University of Arizona, Optical Sciences Center (OSC)
Building 94
Tucson, AZ 85721
Telephone: 520-621-6995
Telefax: 520-621-9613
Electronic Mail: <userid>@optics.arizona.edu
WWW: http://www.optics.arizona.edu/
Founded: 1967
Staff: 300
Activities: material science * statistical optics * quantum optics physics * non-linear optics * medical optics * remote sensing * optical-sciences engineering * applied optics * astronomical optics * optical-science education
City Reference Coordinates: 110°58'00"W 32°13'00"N

University of Arizona Press
1230 North Park Avenue
Suite 102
Tucson, AZ 85719
Telephone: 520-626-4218
    1-800-426-3797 (USA only)
Telefax: 520-621-8899
WWW: http://www.uapress.arizona.edu/
Founded: 1959
Staff: 28
• Publisher
City Reference Coordinates: 110°58'00"W 32°13'00"N

US Naval Observatory (USNO), Flagstaff Station (NOFS)
P.O. Box 1149
Flagstaff, AZ 86002-1149
Telephone: 520-779-5132
Telefax: 520-774-3626
Electronic Mail: usno@nofs.navy.mil
WWW: http://www.nofs.navy.mil/home.html
Founded: 1955
Staff: 20
Activities: astrometry * photometry
Coordinates: 111°44'24"W 35°11'00"N H2,316m
City Reference Coordinates: 111°38'00"W 35°12'00"N

Vatican Observatory Research Group (VORG)
c/o Steward Observatory
University of Arizona
Tucson, AZ 85721
Telephone: 520-621-3225
Telefax: 520-621-1532
Electronic Mail: vatican@as.arizona.edu
WWW: http://clavius.as.arizona.edu/vo/
Founded: 1981
Staff: 6
Activities: station of the Vatican Observatory * research * observing
Coordinates: 110°56'54"W 32°14'00"N
    109°53'30"W 32°42'05"N H3,200m (Mount Graham)
City Reference Coordinates: 110°58'00"W 32°13'00"N

Vega-Bray Observatory
5655 North Via Umbrosa
Tucson, AZ 85750-1357
or :
Serenity Ranch
1311 South Astronomer's Road
Benson, AZ
Telephone: 520-615-3886
    520-721-3815 (day telephone)
Telefax: 520-615-3886
Electronic Mail: skywatcher@communiverse.com
    vegasky@azstarnet.com
WWW: http://www.communiverse.com/skywatcher/
Founded: 1990
Staff: 1
Activities: observing * CCD imaging * education * GOTO planetarium
Coordinates: 110°14'27"W 31°56'27"N H1,176m
City Reference Coordinates: 110°58'00"W 32°13'00"N (Tucson)

Whipple Observatory (FLWO) (Fred Lawrence_)
• See "Harvard-Smithsonian Center for Astrophysics (CfA), Fred Lawrence Whipple Observatory (FLWO)"

USA - Arkansas

Arkansas-Oklahoma Astronomical Society (AOAS)
P.O. Box 31
Fort Smith, AR 72902-0031
Telephone: 501-474-4740
Electronic Mail: daleh@interserv.com (Dale Hall)
WWW: http://www.aoas.org/
City Reference Coordinates: 094°27'00"W 35°22'00"N

Arkansas Tech University (ATU), Observatory
1701 North Boulder
Russellville, AR 72801-2222
Telephone: 479-964-0548
Telefax: 479-964-0837
Electronic Mail: jeff.robertson@mail.atu.edu (Jeff Robertson, Observatory Director)
WWW: http://cosmos.atu.edu/
Staff: 1
Coordinates: 093°08'15"W 35°17'39"N H107m
City Reference Coordinates: 093°08'00"W 35°17'00"N

Central Arkansas Astronomical Society (CAAS)
c/o Chris Lasley
730 Cherub Drive
Conway, AR 72032
Telephone: 501-396-5430 (Skyline)
          501-372-1367 (BBS)
WWW: http://www.propermotion.com/caas/
City Reference Coordinates: 092°27'00"W 35°05'00"N

Christian Association of Stellar Explorers (CASE)
c/o Patrick C. Carr (President)
1005 Hunter Ridge
Siloam Springs, AR 72761
Telephone: 501-524-0322
Electronic Mail: thecarrs@tcac.net
WWW: http://www.christian-astronomy.org/
Activities: education * meetings * star parties
City Reference Coordinates: 094°32'00"W 36°11'00"N

Pomeroy Planetarium
• See "University of Arkansas, Monticello (UAM), Pomeroy Planetarium"

River Valley Stargazers
76 Observatory Lane
P.O. Box 261
Dover, AR 72837
Telephone: 501-331-3773
Electronic Mail: rvstargazer@yahoo.com
WWW: http://www.geocities.com/CapeCanaveral/Launchpad/2148/
City Reference Coordinates: 093°07'00"W 35°24'00"N

University of Arkansas, Fayetteville, Department of Physics
Fayetteville, AR 72701
Telephone: 501-575-2506
Telefax: 501-575-4580
Electronic Mail: <userid>@comp.uark.edu
WWW: http://www.uark.edu/depts/physics/
          http://www.uark.edu/depts/physics/research/astr.html
Founded: 1871 (University)
Activities: observational studies of close binary stars
Coordinates: 094°05'36"W 36°00'36"N H667m (James Wesley Droke Observatory)
City Reference Coordinates: 094°10'00"W 36°03'00"N

University of Arkansas, Little Rock (UALR), Department of Physics and Astronomy
2801 South University Avenue
Little Rock, AR 72204
Telephone: 501-569-3275
          501-569-3259 (Planetarium)
Telefax: 501-569-3314
Electronic Mail: <userid>@ualr.edu
WWW: http://www.physics.ualr.edu/

http://www.planetarium.ualr.edu/ (Planetarium)
Activities: infrared imaging of galaxies * time series photometry of hot evolved stars * planetarium
City Reference Coordinates: 092°15'00"W 34°44'00"N

University of Arkansas, Monticello (UAM), Pomeroy Planetarium
c/o Joe M. Guenter (Director)
School of Mathematics and Sciences
Monticello, AR 71656
Telephone: 870-460-1416
Telefax: 870-460-1316
Electronic Mail: guenterj@uamont.edu
WWW: http://cotton.uamont.edu/~guenterj/planet.html
Founded: 1975
City Reference Coordinates: 091°47'00"W 33°38'00"N

University of Central Arkansas (UCA), Department of Physics and Astronomy
Conway, AR 72032
Telephone: 501-450-5902
Electronic Mail: <userid>@mail.uca.edu
WWW: http://www.uca.edu/physics/
Founded: 1967 (Observatory) (Planetarium: 1987)
Staff: 11
Activities: education * student research * public awareness * Lewis Science Center Planetarium
Coordinates: 092°27'40"W 35°04'09"N H100m (Lewis Science Center Observatory)
City Reference Coordinates: 092°27'00"W 35°05'00"N

USA - California

Advanced Light Source (ALS)
c/o Gary F. Krebs (User Services Group Leader)
Mail Stop 6-2100
One Cyclotron Road
Berkeley, CA 94720
Telephone: 510-486-7745
Telefax: 510-486-4773
Electronic Mail: <userid>@lbl.gov
gfkrebs@lbl.gov
WWW: http://www-als.lbl.gov/
Founded: 1993
Membership: 180
Activities: synchrotron science * protein crystallography * materials science * environmental and Earth science
City Reference Coordinates: 122°18'00"W 37°57'00"N

Aerospace Corp.
P.O. Box 92957
Los Angeles, CA 90009-2957
or :
2350 East El Segundo Boulevard
El Segundo, CA 90245-4691
Telephone: 310-336-5000
Telefax: 310-336-7055
Electronic Mail: corpcomm@aero.org
<userid>@aero.org
WWW: http://www.aero.org/
Founded: 1960
Staff: 3000
Activities: independent non-profit corporation supporting the US government in planning, research, development, acquisition and operation of space, launch, amd associated ground systems, and in science and technology important to national security
City Reference Coordinates: 118°15'00"W 34°03'00"N (Los Angeles)
118°24'00"W 33°55'00"N (El Segundo)

Alexander Morrison Planetarium
c/o California Academy of Science
Golden Gate Park
San Francisco, CA 94118
Telephone: 415-750-7127
415-750-7141 (shows/sky)
Telefax: 415-750-7346
Electronic Mail: planetarium@casmail.calacademy.org
WWW: http://www.calacademy.org/planetarium/
Founded: 1952
Staff: 34
Activities: astronomy education for pre-school * public programs on astronomy and space science * lectures
Observatories: 1 (Hume Observatory)
City Reference Coordinates: 122°24'00"W 37°48'00"N

Ames Research Center (ARC)
• See "National Aeronautics and Space Administration (NASA), Ames Research Center (ARC)"

Annual Reviews Inc. (ARI)
4139 El Camino Way
P.O. Box 10139
Palo Alto, CA 94303-0139
Telephone: 650-493-4400
1-800-523-8635 (Canada and USA only)
Telefax: 650-424-0910
Electronic Mail: annrevu@class.org
service@annurev.org
WWW: http://www.annualreviews.org/ari/
Founded: 1929
Staff: 32
Activities: non-profit scientific publisher
Periodicals: (1) "Annual Review of Astronomy and Astrophysics" (ISSN 0066-4146), "Annual Review of Earth and Planetary Sciences" (ISSN 0084-6597), "Annual Review of Fluid Mechanics" (ISSN 0066-4189), "Annual Review of Nuclear and Particle Science" (ISSN 0163-8998), and 23 other topics
Awards: (1) "J. Murray Luck Award for Scientific Reviewing"
City Reference Coordinates: 122°09'00"W 37°27'00"N

Antelope Valley Astronomy Club (AVAC)
P.O. Box 4595
Lancaster, CA 93539
Telephone: 661-266-2202 (Doug Drake, President)
805-358-3613
Electronic Mail: info@avac.av.org
WWW: http://www.avac.av.org/
Founded: 1982
Membership: 75
City Reference Coordinates: 118°09'00"W 34°42'00"N

Apogee Instruments Inc.
11760 Atwood Road
Suite 4
Auburn, CA 95603
Telephone: 530-888-0500
Telefax: 530-888-0540
Electronic Mail: info@ccd.com
sales@ccd.com
WWW: http://www.ccd.com/
Founded: 1992
Activities: manufacturing CCD cameras
City Reference Coordinates: 110°58'00"W 32°13'00"N

Applied Geomechanics
1336 Brommer Street
Santa Cruz, CA 95062
Telephone: 831-462-2801
Telefax: 831-462-4418
Electronic Mail: applied@geomechanics.com
WWW: http://www.geomechanics.com/
Founded: 1983
Staff: 12
Activities: manufacturing tiltmeters, tilt sensors, readout units, data loggers, and pointing control instruments * software producer
City Reference Coordinates: 122°01'00"W 36°58'00"N

Artemis Society International (ASI)
4572 Keever Avenue
Long Beach, CA 90807
Electronic Mail: <userid>@asi.org
WWW: http://www.asi.org/
Activities: aiming at establishing a permanent, self-supporting community on the Moon
City Reference Coordinates: 118°11'00"W 33°46'00"N

Association of Lunar and Planetary Observers (ALPO)
P.O. Box 16131
San Francisco, CA 94116
Telephone: 415-566-5786
Telefax: 415-731-8242
Electronic Mail: dparker@netside.net (Donald C. Parker, Executive Director)
WWW: http://www.lpl.arizona.edu/alpo/
Founded: 1947
Membership: 600
Activities: encouraging, coordinating and publishing amateur and professional solar-system observing * annual convention
Sections: Solar * Lunar * Mercury * Venus * Mars * Minor Planets * Jupiter * Saturn * Remote Planets * Comets * Meteors * Instruments * Computing (provisional) * Mercury/Venus Transits
Periodicals: (4) "The Journal of the ALPO" (ISSN 0039-2502, circ.: 600); (1) "ALPO Solar System Ephemeris" (ISSN 0890-216x, circ.: 130)
Awards: (1) "Walter H. Haas Observing Award"
City Reference Coordinates: 122°24'00"W 37°48'00"N

Astromart
712 Hilton Road
Walnut Creek, CA 94595
Telephone: 510-988-9694
Telefax: 510-988-9694
Electronic Mail: robertf@best.com
WWW: http://www.astromart.com/
Founded: 1994
City Reference Coordinates: 122°04'00"W 37°55'00"N

Astronomical Association of Northern California (AANC)
4917 Mountain Boulevard
Oakland, CA 94619-3014
or :
c/o Aland Gould (Secretary)
Lawrence Hall of Science
University of California
Berkeley, CA 94720
Telephone: 415-376-3007
          415-642-5863 (Secretary)
Electronic Mail: agould@uclink4.berkeley.edu (Alan Gould)
WWW: http://www.lhs.berkeley.edu/SII/AANC/aanc.html
       http://www.aanc-astronomy.org/
Founded: 1972
Activities: promoting astronomy * providing resources material * holding conferences
City Reference Coordinates: 122°13'00"W 37°47'00"N (Oakland)
                              122°18'00"W 37°57'00"N (Berkeley)

Astronomical Society of the Central Coast (ASCC)
c/o Monterey Institute for Research in Astronomy (MIRA)
200 Eighth Street
Marina, CA 93933
Telephone: 831-883-1000
Telefax: 831-883-1031
Electronic Mail: ascc@mira.org
WWW: http://www.mira.org/ascc/
City Reference Coordinates: 121°05'00"W 36°37'00"N

Astronomical Society of the Desert (ASOD)
P.O. Box 71
Rancho Mirage, CA 92270
Telephone: 619-328-2601
Electronic Mail: 75574.1514@compuserve.com (K.S. Engel, Secretary/Treasurer)
WWW: http://www.astrorx.org/
Founded: 1968
Membership: 75
Activities: popularization * observing
Periodicals: (12) "Newsletter"
Coordinates: 116°25'00"W 33°45'00"N H1500m
City Reference Coordinates: 116°25'00"W 33°45'00"N

Astronomical Society of the Pacific (ASP)
390 Ashton Avenue
San Francisco, CA 94112
Telephone: 415-337-1100
Telefax: 415-337-5205
Electronic Mail: director@astrosociety.org
WWW: http://www.astrosociety.org/
       http://pasp.phys.uvic.ca/ (Publications of the ASP)
Founded: 1889
Membership: 5000
Staff: 20
Activities: increasing public understanding and appreciation of astronomy * lectures * meetings * mail order catalog of educational materials in astronomy * awards * information service for media * workshops for teachers * information kits for public * publishing the International Astronomical Union Symposium volumes
Sections: (committees) history * awards * amateur advisory * publications * development
Periodicals: (12) "Publications of the Astronomical Society of the Pacific" (ISSN 0004-6280); (6) "Mercury" (ISSN 0047-6773, circ.: 5,000); (4) "Universe in the Classroom Newsletter"; "Astronomical Society of the Pacific Conference Series"
Awards: (1) "Bruce Medal", "Klumpke-Roberts Prize", "Trumpler Prize", "Erich and Maria Muhlmann Prize", "Amateur Achievement Award", "Brennan Award", "Las Cumbres Award"
City Reference Coordinates: 122°24'00"W 37°48'00"N

Astronomical Unit (AU)
c/o Santa Barbara Museum of Natural History
2559 Puesta del Sol Road
Santa Barbara, CA 93105-2998
Telephone: 805-569-9743 (Greg Brinser, President)
Electronic Mail: 88eight8@home.com (Greg Brinser, President)
              macpuzl@west.net (Chuck McPartlin, Newsletter Editor)
WWW: http://www.sbau.org/
Founded: 1985
Membership: 100
Activities: monthly meetings * public observing * education
Periodicals: (12) "AU AstroNews"

Awards: (1) "Crutchfield Adair Memorial Astronomy Award"
City Reference Coordinates: 119°42'00"W 34°25'00"N

Beattie Planetarium (George F._)
• See "George F. Beattie Planetarium"

Beverly Hills High School (BHHS), Planetarium
241 Moreno Drive
Beverly Hills, CA 90212
Telephone: 310-551-5100
WWW: http://bhhs.beverlyhills.k12.ca.us/
City Reference Coordinates: 118°26'00"W 34°03'00"N

Big Bear Solar Observatory (BBSO)
• See "New Jersey Institute of Technology (NJIT), Big Bear Solar Observatory (BBSO)"

Berkeley-Illinois-Maryland Association (BIMA), Hat Creek Radio Observatory (HCRO)
42231 Bidwell Road
Hat Creek, CA 96040
Telephone: 530-335-2364
Telefax: 530-335-4944
Electronic Mail: obs@hat.astro.uiuc.edu
WWW: http://bima.astro.umd.edu/
Founded: 1960 (BIMA: 1987)
Staff: 7
Activities: astronomical research * radio astronomy
Coordinates: 121°28'08"W 40°49'02"N H1021m
City Reference Coordinates: 121°40'00"W 40°55'00"N (Burney)

Byers Co. (Edward R._)
• See "Edward R. Byers Co."

California Academy of Science (CAS)
Golden Gate Park
San Francisco, CA 94118
Telephone: 415-750-7145
Telefax: 415-750-7346
WWW: http://www.calacademy.org/
        http://www.calacademy.org/planetarium/ (Alexander Morrison Planetarium)
Founded: 1853
Staff: 250
Activities: museum * research * science education * Alexander Morrison Planetarium (see separate entry) * aquarium
Observatories: 1 (Hume Observatory)
City Reference Coordinates: 122°24'00"W 37°48'00"N

California Institute for Physics and Astrophysics (CIPA)
366 Cambridge Avenue
Palo Alto, CA 94306
Telephone: 650-327-6284
Telefax: 650-327-6294
Electronic Mail: admin@calphysics.org
WWW: http://www.calphysics.org/
Founded: 1999
Staff: 10
Activities: fundamental physics * cosmology * gravitation * quantum vacuum * zero-point field * high-energy astrophysics
City Reference Coordinates: 122°09'00"W 37°27'00"N

California Institute of Technology, Department of Astronomy
c/o Robinson Astrophysics Laboratory
Mail Code 105-24
Pasadena, CA 91125
Telephone: 626-395-4671
           626-395-4169
           626-395-2097
Telefax: 626-568-9352
Electronic Mail: <userid>@astro.caltech.edu
WWW: http://astro.caltech.edu/
Founded: 1936
Staff: 50
Activities: cosmology * galaxy evolution * galactic and extragalactic radio sources * QSOs * SN * pulsars * stellar structure and evolution * physics of solar phenomena * ISM * radio galaxies * gamma-ray bursts * digital sky surveys * gravitational lensing
Awards: "Jesse Greenstein Prize"
City Reference Coordinates: 118°09'00"W 34°09'00"N

California Institute of Technology, Department of Theoretical Astrophysics
Mail Code 130-33
Bridge Annex
1201 East California Boulevard
Pasadena, CA 91125
Telephone: 626-395-4597
Telefax: 626-796-5675
Electronic Mail: <userid>@caltech.edu/
WWW: http://www.cco.caltech.edu/~esp/tapir/
Founded: 1920 (CalTech)
Periodicals: "CalTech Astrophysics Abstracts"
City Reference Coordinates: 118°09'00"W 34°09'00"N

California Institute of Technology, Division of Geological and Planetary Sciences
Mail Code 170-25
1200 East California Boulevard
Pasadena, CA 91125
Telephone: 818-395-6112
Telefax: 818-585-1917
WWW: http://www.gps.caltech.edu/
Founded: 1920 (CalTech)
Staff: 50
City Reference Coordinates: 118°09'00"W 34°09'00"N

California Institute of Technology, Division of Physics, Mathematics and Astronomy
111 East Bridge
Mail Code 103-33
Pasadena, CA 91125
Telephone: 818-395-4241
WWW: http://www.pma.caltech.edu/
  http://www.cco.caltech.edu/~btsoifer/ira.html (Infrared Astronomy)
  http://astro.caltech.edu/~lgg/ (Observational Cosmology)
  http://astro.caltech.edu/~srk/ (Pulsar)
  http://www.srl.caltech.edu/srl/ (Space Radiation Laboratory - SRL)
  http://www.cco.caltech.edu/~cso/submm.html (Submillimeter Astrophysics)
  http://www.ccsf.caltech.edu/astro/heastro.html (High-Energy Astrophysics)
  http://www.cco.caltech.edu/~esp/tapir/ (Theoretical Astrophysics and Relativity)
  http://www.cithep.caltech.edu/ (High-Energy Physics)
  http://citnp1.caltech.edu/ (Neutrino Physics)
  http://www.theory.caltech.edu/ (Theoretical Particle Physics)
  http://astro.caltech.edu/~george/dposs/dposs.html (Digital Sky Survey)
Founded: 1920 (CalTech)
Staff: 400
City Reference Coordinates: 118°09'00"W 34°09'00"N

California Institute of Technology, Infrared Processing and Analysis Center (IPAC)
770 South Wilson Avenue
Mail Code 100-22
Pasadena, CA 91125
Telephone: 626-397-9502
Telefax: 626-397-9600
Electronic Mail: info@ipac.caltech.edu
WWW: http://www.ipac.caltech.edu/
  http://nedwww.ipac.caltech.edu/ (NASA/IPAC Extragalactic Database - NED)
Founded: 1920 (CalTech)
City Reference Coordinates: 118°09'00"W 34°09'00"N

California Institute of Technology, Jet Propulsion Laboratory (JPL)
4800 Oak Grove Drive
Pasadena, CA 91109
Telephone: 818-354-4321
Electronic Mail: <userid>@jpl.nasa.gov
WWW: http://www.jpl.nasa.gov/
Founded: 1920 (CalTech)
Staff: 5000
City Reference Coordinates: 118°09'00"W 34°09'00"N

California Institute of Technology, Jet Propulsion Laboratory (JPL), Deep Space Tracking Systems Group
4800 Oak Grove Drive
Mail Code 238-600
Pasadena, CA 91109-8099
Telephone: 818-354-2789
Telefax: 818-393-4965
Electronic Mail: <userid>@jpl.nasa.gov
WWW: http://www.jpl.nasa.gov/

Founded: 1992
Staff: 13
Activities: spacecraft tracking * radio and optical interferometry * astrometry * astronomy * geodesy
Coordinates: 148°58'48"E 35°24'08"S H664m (Tidbindilla, Australia)
    116°48'17"W 35°18'00"N H1,001m (Goldstone – IAU Codes 252 & 253)
    004°14'02"W 40°25'48"N H796m (Robledo de Chavela, Spain)
City Reference Coordinates: 118°09'00"W 34°09'00"N

California Institute of Technology, Jet Propulsion Laboratory (JPL), Table Mountain Observatory (TMO)
P.O. Box 367
Wrightwood, CA 92397
or :
24490 Table Mountain Road
Wrightwood, CA 92397
Telephone: 760-249-3650
    760-249-3551 (24" Telescope)
    760-249-6610 (48" Telescope)
Telefax: 760-249-5392
Electronic Mail: <userid>@jpl.nasa.gov
WWW: http://tmf-web.jpl.nasa.gov/
Coordinates: 117°40'48"W 34°22'54"N H2,287m (IAU Codes 654 & 673)
City Reference Coordinates: 117°38'00"W 34°21'00"N

California Institute of Technology, Jet Propulsion Laboratory (JPL), US Space VLBI Project
Mail Code 264-802
4800 Oak Grove Drive
Pasadena, CA 91109-8099
Telephone: 818-393-1002
Telefax: 818-393-0042
Electronic Mail: spacevlbi@sgra.jpl.nasa.gov
WWW: http://sgra.jpl.nasa.gov/us-space-vlbi/
Activities: supporting the VLBI Space Observatory Program (VSOP) and the RadioAstron mission
City Reference Coordinates: 118°09'00"W 34°09'00"N

California Institute of Technology, Laser Interferometer Gravitational-wave Observatory (LIGO) Project
Mail Code 18-34
102 East Bridge
Pasadena, CA 91125
Telephone: 818-395-2129
Telefax: 818-304-9834
Electronic Mail: ligo@ligo.caltech.edu
    <userid>@ligo.caltech.edu
WWW: http://www.ligo.caltech.edu/
Founded: 1920 (CalTech)
City Reference Coordinates: 118°09'00"W 34°09'00"N

California Institute of Technology, Owens Valley Radio Observatory (OVRO)
P.O. Box 968
Big Pine, CA 93513
or :
100 Leighton Lane
Big Pine, CA 93513
Telephone: 760-938-2075
Telefax: 760-938-2075
Electronic Mail: <userid>@ovro.caltech.edu
WWW: http://www.ovro.caltech.edu/
    http://www.mmarray.org/ (CARMA)
Founded: 1958
Periodicals: "Owens Valley Radio Observatory Preprints"
Coordinates: 118°16'54"W 37°13'58"N H1,236m (40m telescope)
    118°17'37"W 37°13'58"N H1,236m (27.4m telescope)
    118°16'56"W 37°14'03"N H1,236m (10m array)
City Reference Coordinates: 118°17'00"W 37°10'00"N

California Institute of Technology, Owens Valley Radio Observatory (OVRO), Headquarters
c/o Robinson Astrophysics Laboratory
Mail Code 105-24
Pasadena, CA 91125
Telephone: 818-395-4912
Electronic Mail: <userid>@ovro.caltech.edu
WWW: http://www.ovro.caltech.edu/
Founded: 1958
Periodicals: "Owens Valley Radio Observatory Preprints"
City Reference Coordinates: 118°09'00"W 34°09'00"N

California Institute of Technology, Palomar Observatory
Mail Code 105-24
Pasadena, CA 91125
or :
35899 Canfield Road
Palomar Mountain, CA 92060
Telephone: 626-395-4033 (CalTech Office)
          760-742-2100 (Observatory)
Telefax: 626-568-1517 (CalTech Office)
        760-742-1728 (Observatory)
Electronic Mail: palomar@caltech.edu
WWW: http://astro.caltech.edu/observatories/palomar/
Founded: 1948
Periodicals: "Palomar Observatory Preprints"
Coordinates: 116°51'50"W 33°21'22"N H1,706m (200" telescope)
              116°51'31"W 33°20'56"N H1,706m (60" telescope)
              116°51'32"W 33°21'26"N H1,706m (48" telescope)
              116°51'42"W 33°21'14"N H1,706m (18" telescope)
City Reference Coordinates: 118°09'00"W 34°09'00"N (Pasadena)

California Space Institute
c/o Scripps Institution of Oceanography
University of California
Mail Code A-016
La Jolla, CA 92093-0216
Telephone: 619-534-4937
WWW: http://deimos.ucsd.edu/calspace.html
      http://calspace.ucsd.edu/
Founded: 1979
Activities: remote sensing * climate and global change * advanced technologies * space resources * small grant program
City Reference Coordinates: 117°52'00"W 33°51'00"N

California State University, Chico, Roth Planetarium
c/o Department of Geosciences
Chico, CA 95929
Telephone: 916-895-5262
Electronic Mail: jregas@cavax.csuchico.edu (James L. Regas)
WWW: http://rigel.csuchico.edu/roth/roth.html
City Reference Coordinates: 121°50'00"W 39°44'00"N

California State University, Fresno, Physics Department
2345 East San Ramon Avenue
Fresno, CA 93740-0037
Telephone: 559-278-2371
Telefax: 559-278-7741
Electronic Mail: <userid>@zimmer.csufresno.edu
WWW: http://physics.csufresno.edu/
      http://www.downing-planetarium.org/ (Downing Planetarium - see separate entry)
Founded: 1911
Activities: education * elementary astronomy
City Reference Coordinates: 119°45'00"W 36°45'00"N

California State University, Fullerton, Department of Physics
P.O. Box 34080
Fullerton, CA 92634-9480
Telephone: 714-773-3366
Telefax: 714-449-5810
Electronic Mail: physics@fullerton.edu
              <userid>@fullerton.edu
WWW: http://chaos.fullerton.edu/physics.html
Founded: 1961
Staff: 15
Activities: education in physics and astronomy
City Reference Coordinates: 117°55'00"W 33°52'00"N

California State University, Long Beach, Department of Physics and Astronomy
1250 Bellflower Boulevard
Long Beach, CA 90840-3901
Telephone: 562-985-4924
Telefax: 562-985-7924
Electronic Mail: <userid>@csulb.edu
WWW: http://www.physics.csulb.edu/
Periodicals: "Newsletter"
City Reference Coordinates: 118°11'00"W 33°46'00"N

California State University, Northridge, Department of Physics and Astronomy
18111 Nordhoff Street
Northridge, CA 91330-8268
Telephone: 818-677-2775
Telefax: 818-677-3234
Electronic Mail: physics@csun.edu
                 <userid>@csun.edu
WWW: http://www.csun.edu/Physics&Astronomy/
       http://www.csun.edu/phys/announcements_and_planetarium/planetarium.html (Donald E. Bianchi Planetarium)

Founded: 1970
Staff: 20
Activities: education * research
City Reference Coordinates: 118°33'00"W 33°14'00"N

California State University, Northridge, San Fernando Observatory (SFO)
c/o Department of Physics and Astronomy
18111 Nordhoff Street
Northridge, CA 91330-8268
or :
14031 San Fernando Road
Sylmar, CA 91342
Telephone: 818-367-9333
Telefax: 818-367-1903
Electronic Mail: <userid>@csun.edu
WWW: http://www.csun.edu/SanFernandoObservatory
       http://davinci.csun.edu/~astro/sfo.html
Founded: 1976
Staff: 4
Activities: solar astronomy
Coordinates: 118°29'28"W 34°18'29"N H371m
City Reference Coordinates: 118°33'00"W 33°14'00"N (Northridge)

California State University, Sacramento, Department of Physics and Astronomy
6000 J Street
Sacramento, CA 95819-6041
Telephone: 916-278-6518
Telefax: 916-278-7686
Electronic Mail: physics@csus.edu
WWW: http://www.csus.edu/Physics/
City Reference Coordinates: 121°30'00"W 38°35'00"N

CalTech
• See "California Institute of Technology"

Carina Software
602 Morninghome Road
Danville, CA 94526
Telephone: 925-838-0695
Telefax: 925-838-0535
Electronic Mail: support@carinasoft.com
                 sales@carinasoft.com
WWW: http://www.carinasoft.com/
Founded: 1988
• Software producer
City Reference Coordinates: 122°00'00"W 37°49'00"N

Carnegie Institution of Washington (CIW), Observatories
813 Santa Barbara Street
Pasadena, CA 91101-1292
Telephone: 626-577-1122
Telefax: 626-795-8136
Electronic Mail: <userid>@ociw.edu
WWW: http://www.ociw.edu/
       http://www.ciw.edu/
Founded: 1904
Staff: 60
Activities: astronomical instrumentation * black holes * extragalactic distance * galactic chemical evolution * galaxy structure and evolution * galactic clusters * globular clusters * high-redshift objects * intergalactic medium * radio sources * SNs * X-ray astronomy
Periodicals: (1) "Carnegie Institution of Washington Year Book"
Coordinates: 070°44'00"W 28°53'00"S H2,282m (Las Campanas - see Chile)
City Reference Coordinates: 118°09'00"W 34°09'00"N

Daniel Oberti - Architectural Ceramics
3796 Twig Avenue
Sebastopol, CA 95472
Telephone: 707-829-0584
Telefax: 707-829-2136
Electronic Mail: doberti@monitor.net
WWW: http://www.monitor.net/~doberti/
Activities: ceramic sculpting including sundials
City Reference Coordinates: 122°49'00"W 38°24'00"N

David Chandler Co.
P.O. Box 999
Springville, CA 93265
Telephone: 1-800-516-9756 (USA only)
            559-539-0900
Telefax: 559-539-7033
Electronic Mail: david@davidchandler.com
WWW: http://www.davidchandler.com/
Founded: 1976
Staff: 2
Activities: producing low-distortion planispheres, star atlases, astronomy materials
City Reference Coordinates: 118°50'00"W 36°05'00"N

Davis Astronomy Club
• See "Explorit Science Center, Davis Astronomy Club"

Davis Instruments
3645 Diablo Avenue
Hayward, CA 94545
Telephone: 510-732-9229
            510-732-7814 (Technical Support)
Telefax: 510-670-0589
            510-293-3548 (automated fax-back service)
Electronic Mail: info@davisnet.com
WWW: http://www.davisnet.com/
Founded: 1964
Staff: 100
Activities: manufacturing marine equipment, precision weather instruments and automotive monitors
City Reference Coordinates: 122°05'00"W 37°40'00"N

DayStar Filters
3857 Schaefer Avenue
Suite D
Chino, CA 91710
Telephone: 909-591-4673
Telefax: 909-591-6886
WWW: http://www.daystarfilters.com/
City Reference Coordinates: 117°42'00"W 34°01'00"N

Discovery Museum Science Center
3615 Auburn Boulevard
Sacramento, CA 95821
Telephone: 916-575-3941
Telefax: 916-575-3925
Founded: 1951
Staff: 27
Activities: exhibitions * school programs * portable planetarium * courses * teacher training
City Reference Coordinates: 121°30'00"W 38°35'00"N

Discovery Telescopes Inc.
615 South Tremont Street
Oceanside, CA 92054
Telephone: 760-967-6798
Telefax: 760-967-6798
Electronic Mail: scopinit@msn.com
WWW: http://www.discovery-telescopes.com/
Activities: manufacturing telescopes
City Reference Coordinates: 117°23'00"W 33°12'00"N

Downing Planetarium
California State University
5320 North Maple Avenue
M/S DP132
Fresno, CA 93740
Telephone: 559-278-4121 (Show Info)

Telefax: 559-278-4070
Electronic Mail: kharriso@csufresno.edu
WWW: http://www.downing-planetarium.org/
Activities: education * elementary astronomy
City Reference Coordinates: 119°45'00"W 36°45'00"N

Drescher Planetarium (John_)
• See "Santa Monica College (SMC), John Drescher Planetarium"

Dryden Flight Research Center (DFRC)
• See "National Aeronautics and Space Administration (NASA), Dryden Flight Research Center (DFRC)"

Dryden Flight Research Facility (DFRF)
• See now "National Aeronautics and Space Administration (NASA), Dryden Flight Research Center (DFRC)"

Eastbay Astronomical Society (EAS)
c/o Chabot Observatory and Science Center
19047 Robinson Road
Sonoma, CA 95476-5517
or :
4917 Mountain Boulevard
Oakland, CA 94619
Electronic Mail: eas@cosc.org
WWW: http://www.eastbayastro.org/
Founded: 1924
City Reference Coordinates: 122°28'00"W 38°17'00"N (Sonoma)
122°13'00"W 37°47'00"N (Oakland)

Eddy Co.
13590 Niabi Road
Apple Valley, CA 92308
Telephone: 760-961-8457
Telefax: 760-961-8458
Electronic Mail: tech_support@eddyco.com
WWW: http://www.eddyco.com/
Founded: 1971
Staff: 6
Activities: manufacturing thin-film optical coating equipment
Periodicals: "Eddy Currents"
City Reference Coordinates: 117°14'00"W 34°32'00"N

Edward R. Byers Co.
29001 West Highway 58
Barstow, CA 92311
Telephone: 619-256-2377
Telefax: 619-256-9599
Activities: designing and manufacturing specialized scientific instruments, heliostats, equatorial mountings, precision worm gear drive assemblies and camera equatorial mountings
City Reference Coordinates: 117°01'00"W 34°54'00"N

El Camino College, Planetarium and Observatory
16007 Crenshaw Boulevard
Torrance, CA 90506
Telephone: 310-660-3246 (Hotline)
310-660-3373 (Planetarium)
Electronic Mail: svlloyd@mail.idt.net (Stephen Vincent Lloyd)
WWW: http://www.elcamino.cc.ca.us/astronomy/ecc_astro.htm
http://www.elcamino.cc.ca.us/astronomy/plntrm.htm
City Reference Coordinates: 118°19'00"W 33°59'00"N

Electro Optical Industries (EOI) Inc.
859 Ward Drive
Santa Barbara, CA 93111
Telephone: 805-964-6701 x280
Telefax: 805-967-8590
Electronic Mail: eoi@electro-optical.com
WWW: http://www.electro-optical.com/
Founded: 1964
Staff: 40
Activities: manufacturing IR test and calibration instruments including black bodies, differential temperature sources, collimators, simulators, target projectors, radiometers, choppers, cryogenic standards
City Reference Coordinates: 119°42'00"W 34°25'00"N

E.O. Lawrence Berkeley National Laboratory (LBNL), Institute for Nuclear and Particle Astrophysics (INPA)
Mail Stop 50-208
One Cyclotron Road
Berkeley, CA 94720
Telephone: 510-486-5074
Telefax: 510-486-6738
Electronic Mail: inpa@lbl.gov
WWW: http://www-inpa.lbl.gov/
Founded: 1931
Staff: 70
Activities: supernovae * neutrinos * cosmic microwave background radiation * dark matter * cosmic rays * pulsars * neutron stars * weak interactions in atomic and nuclear processes
• Operated by the University of California under contract with the US Department of Energy
City Reference Coordinates: 122°18'00"W 37°57'00"N

Equatorial Platforms
c/o Tom Osypowski
15736 McQuiston Lane
Grass Valley, CA 95945
Telephone: 530-274-9113
Electronic Mail: tomosy@nccn.net
WWW: http://www201.pair.com/resource/astro.html/regular/products/eq_platforms/
Founded: 1986
Staff: 2
Activities: drive systems for Dobsonian telescopes
City Reference Coordinates: 121°04'00"W 39°13'00"N

Everything in the Universe
185 John Street
Oakland, CA 94611
Telephone: 510-547-6523
Telefax: 510-547-6523
Electronic Mail: creator@everythingintheuniv.com
            nsperling@california.com (Norman Sperling)
WWW: http://www.everythingintheuniv.com/
Founded: 1973
Staff: 1
Activities: astronomical goods and services
City Reference Coordinates: 122°13'00"W 37°47'00"N

Exploratorium
3601 Lyon Street
San Francisco, CA 94123
Telephone: 415-EXP-LORE
            415-563-7337
Telefax: 415-561-0307
Electronic Mail: <userid>@exploratorium.edu
WWW: http://www.exploratorium.edu/
Founded: 1969
Membership: 10000
Staff: 300
Activities: interactive science exhibits * education * webcasts
Periodicals: (4) "Exploratorium"
City Reference Coordinates: 122°24'00"W 37°48'00"N

Explorit Science Center, Davis Astronomy Club
3141 Fifth Street
P.O. Box 1288
Davis, CA 95617-1288
Telephone: 916-756-0191
Telefax: 916-756-1227
Electronic Mail: explorit@dcn.davis.ca.us
WWW: http://www.dcn.davis.ca.us/~explorit/astronomy.html
        http://www.dcn.davis.ca.us/~explorit/main_index.html
Founded: 1988
Membership: 5
Coordinates: 121°44'22"W 38°32'42"N
City Reference Coordinates: 121°44'00"W 38°33'00"N

Express Optics Inc.
P.O. Box 127
10564 Fern Avenue

Unit B
Stanton, CA 90680
Telephone: 714-484-3200
Telefax: 714-484-7600
Electronic Mail: nigi@newportglass.com
Founded: 1987
Staff: 20
Activities: manufacturing lenses, mirrors, filters and optical systems * thin film coatings
City Reference Coordinates: 117°60'00"W 33°48'00"N

Extrasolar Research Corp.
569 South Marengo
Suite 301
Pasadena, CA 91101
Telephone: 626-568-9702
Telefax: 626-792-2574
Electronic Mail: terebey@extrasolar.com (Susan Terebey)
WWW: http://www.extrasolar.com/
Founded: 1996
Staff: 2
City Reference Coordinates: 118°09'00"W 34°09'00"N

Finley Holiday Film Corp.
P.O. Box 619
Whittier, CA 90608
or :
12607 East Philadelphia Street
Whittier, CA 90601
Telephone: 1-800-345-6707 (USA only)
          562-945-3325
Telefax: 562-693-4756
Electronic Mail: finley-holiday@finley-holiday.com
WWW: http://www.finley-holiday.com/
Founded: 1965
City Reference Coordinates: 118°02'00"W 33°59'00"N

FirstLight Astronomy Club
c/o Temecula Valley High School
31555 Rancho Vista
Temecula, CA 92592
Electronic Mail: frstlght@pe.net
WWW: http://www.firstlightastro.com/
City Reference Coordinates: 117°09'00"W 33°30'00"N

Fleet Space Theater and Science Center (Reuben H._)
• See "Reuben H. Fleet Space Theater and Science Center"

Foothill College, Astronomy Department
12345 El Monte Road
Los Altos Hills, CA 94022
Telephone: 650-949-7288 (Department)
          650-949-7334 (Observatory)
Telefax: 650-949-7375
Electronic Mail: fraknoi@admin.fhda.edu (Andrew Fraknoi, Department Chair)
              brian_-_day@cup.portal.com (Observatory)
WWW: http://wwwfh.fhda.edu/foothill/astronomy/fhobs.htm
     http://www.foothill.fhda.edu/ast/
Founded: 1965
Staff: 4
Coordinates: 122°06'00"W 37°21'00"N H100m
• Observatory operated by the "Peninsula Astronomical Society"
City Reference Coordinates: 122°06'00"W 37°23'00"N

Fremont Peak Observatory Association (FPOA)
Fremont Peak State Park
P.O. Box 787
San Juan Bautista, CA 95045
Telephone: 831-623-2465
WWW: http://home.att.net/~fpoa/
Founded: 1985
Membership: 175
Activities: public viewing * talks
Periodicals: "The Observer"
Coordinates: 121°30'00"W 36°45'00"N H1,000m
City Reference Coordinates: 121°32'00"W 36°51'00"N

Gemmary Inc. (The_)
• See "The Gemmary Inc."

George F. Beattie Planetarium
c/o San Bernardino Valley College
701 South Mount Vernon Avenue
San Bernardino, CA 92410
Telephone: 909-888-6511 x1458
Electronic Mail: bwilson@sbccd.cc.ca.us
Founded: 1977
Staff: 3
Activities: education * public shows
City Reference Coordinates: 117°17'00"W 34°06'00"N

George H. Clever Planetarium
c/o San Joaquin Delta College
5151 Pacific Avenue
Stockton, CA 95207
Telephone: 209-954-5051
Electronic Mail: tcox@sjdccd.cc.ca.us (Timothy Cox, Director)
WWW: http://www.sjdccd.cc.ca.us/dept/planetarium/
Founded: 1981
Staff: 2
City Reference Coordinates: 121°17'00"W 37°57'00"N

Griffith Observatory
2800 East Observatory Road
Los Angeles, CA 90027
Telephone: 323-664-1181 (Offices)
    323-664-1191 (Program Information)
    323-663-8171 (Sky Report)
Telefax: 323-663-4323
Electronic Mail: jmosley@earthlink.net (John Mosley, Program Supervisor)
WWW: http://www.griffithobs.org/
    http://www.griffithobservatory.org/
Founded: 1935
Staff: 17
Activities: public education and information * planetarium shows * museum
Periodicals: (12) "Griffith Observer" (ISSN 0195-3982)
Coordinates: 118°18'06"W 34°06'47"N H357m
City Reference Coordinates: 118°15'00"W 34°03'00"N

Grossmont College, Department of Physics, Astronomy and Physical Sciences
8800 Grossmont College Drive
El Cajon, CA 92020
Telephone: 619-644-7351
    619-465-1700
Electronic Mail: <userid>@mail.gcccd.cc.ca.us
WWW: http://www.grossmont.net/physics/
Founded: 1961
Activities: undergraduate education * photography (Moon, Sun, planets)
Coordinates: 116°58'00"W 32°48'00"N
City Reference Coordinates: 116°58'00"W 32°48'00"N

Group 70
701 Aldo Avenue
Suite 49
Santa Clara, CA 95054-2211
Telephone: 510-727-5442
WWW: http://group70.org/
Activities: Large Amateur Telescope (LAT) project
Coordinates: 121°56'54"W 37°23'07"N
City Reference Coordinates: 121°57'00"W 37°21'00"N

Halls Valley Astronomical Group (HVAG)
c/o Paul Bricmont
1127 Sierra Avenue
San Jose, CA 95126-2641
Telephone: 408-287-9486
Electronic Mail: info@hvag.sbay.org
WWW: http://www.snap-design.com/HVAG/
Founded: 1982
Activities: monthly observing * star parties
Periodicals: (4) "62"Woodland Star"
Observatories: 1 (Halley Hill Observatory)

City Reference Coordinates: 121°53'00"W 37°20'00"N

Harvey Mudd College (HMC), Department of Physics
Claremont, CA 91711
Telephone: 909-621-8024
Telefax: 909-621-8887
Electronic Mail: <userid>@hmc.edu
WWW: http://www.physics.hmc.edu/
City Reference Coordinates: 117°43'00"W 34°06'00"N

Hat Creek Observatory
• See "University of California at Berkeley (UCB), Radio Astronomy Laboratory (RAL), Hat Creek Observatory"

Hat Creek Radio Observatory (HCRO)
• See "Berkeley-Illinois-Maryland Association (BIMA), Hat Creek Radio Observatory (HCRO)"

High Desert Astronomical Society (HiDAS)
c/o Lewis Center for Educational Research
20702 Thunderbird Road
Apple Valley, CA 92307
Telephone: 619-242-3514
Telefax: 619-242-3783
Electronic Mail: ronell@caron.ati.com
Membership: 70
Activities: observing * research
City Reference Coordinates: 117°14'00"W 34°32'00"N

Hollywood General Machining Inc.
1033 North Sycamore Avenue
Los Angeles, CA 90038
Telephone: 323-462-2855
Telefax: 323-462-2682
Electronic Mail: scott@losmandy.com
WWW: http://www.losmandy.com/
Activities: manufacturing astronomical accessories
City Reference Coordinates: 118°15'00"W 34°03'00"N

Holt Planetarium (William Knox_)
• See "William Knox Holt Planetarium"

Hummingbird Astronomical Observatory
c/o Jeffery G. Potter
3811 Hummingbird Drive
Antioch, CA 94509-6440
Telephone: 625-706-1200
Electronic Mail: dgtalmn@ecis.com
WWW: http://www.hummingbirdobservatory.com/
City Reference Coordinates: 121°49'00"W 38°01'00"N

Independence High School, Planetarium
1776 Educational Park Drive
San Jose, CA 95133-1795
Telephone: 408-729-3911
Electronic Mail: chaidg@esuhsd.org (Gail Chaid, Planetarium Director)
          chaid@aol.com
WWW: http://ihnet.esuhsd.org/planet/planet.html
Founded: 1976
City Reference Coordinates: 121°53'00"W 37°20'00"N

Infrared Processing and Analysis Center (IPAC)
• See "California Institute of Technology, Infrared Processing and Analysis Center (IPAC)"

Interactive NASA Space Physics Ionospheric Research Experiment (INSPIRE)
c/o Bill Pine
Chaffey High School
1245 North Euclid Avenue
Ontario, CA 91762
Telefax: 909-931-0392
Electronic Mail: pine@ndadsb.gsfc.nasa.gov
WWW: http://image.gsfc.nasa.gov/poetry/inspire/
Founded: 1989
Staff: 3
Activities: space plasma physics
Periodicals: "The INSPIRE Journal"

City Reference Coordinates: 117°39'00"W 34°04'00"N

International Thermal Instrument (ITI) Co.
P.O. Box 309
Del Mar, CA 92014
Telephone: 858-755-4436
Telefax: 858-755-6878
Electronic Mail: info@iticompany.com
WWW: http://www.iticompany.com/
Founded: 1969
Staff: 10
Activities: manufacturing calorimetry equipment, thermopiles, conductivity analyzers
City Reference Coordinates: 117°16'00"W 32°58'00"N

Isotope Products Laboratories (IPL)
24937 Avenue Tibbitts
Valencia, CA 91355-3408
Telephone: 661-309-1010
Telefax: 661-257-8303
Electronic Mail: sales@isotopeproducts.com
WWW: http://www.isotopeproducts.com/
Founded: 1967
Staff: 35
Activities: manufacturing radioactive sources and standards
City Reference Coordinates: 118°18'00"W 34°12'00"N

Jameco Electronics
1355 Shoreway Road
Belmont, CA 94002
Telephone: 650-592-8097
       1-800-831-4242
Telefax: 650-592-2503
       1-800-237-6948
Electronic Mail: info@jameco.com
WWW: http://www.jameco.com/
Founded: 1974
Staff: 150
Activities: mail order electronic components and computer products company
City Reference Coordinates: 122°17'00"W 37°31'00"N

Jet Propulsion Laboratory (JPL)
• See "California Institute of Technology, Jet Propulsion Laboratory (JPL)"

John Drescher Planetarium
• See "Santa Monica College (SMC), John Drescher Planetarium"

Kern Astronomical Society (KAS)
c/o Joe Fram
6416 Sally Avenue
Bakersfield, CA 93308
or :
423 Oleander Avenue
Bakersfield, CA 93304
Telephone: 805-326-5113
Electronic Mail: joefram@zeus.org
WWW: http://users.aol.com/frankrip/kas/kas.html
Activities: monthly meetings * observing
Periodicals: (12) "The Syzygy"
City Reference Coordinates: 118°30'00"W 35°05'00"N

Kinohi Institute
530 South Lake Avenue
Pasadena, CA 91101
Electronic Mail: mike@kinohi.org (M.C. Storrie-Lombardi, Executive Director)
WWW: http://www.kinohi.org/
Activities: astrobiology * origins of life * complexity * astronomy * geobiology * geochemistry
City Reference Coordinates: 118°09'00"W 34°09'00"N

Kowa Optimed Inc.
20001 South Vermont Avenue
Torrance, CA 90502
Telephone: 310 327-1913
Telefax: 310 327-4177
WWW: http://www.kowascope.com/
Activities: distributing scopes and binoculars

City Reference Coordinates: 118°19'00"W 33°59'00"N

Laser Images Inc.
6911 Hayvenhurst Avenue
Van Nuys, CA 91406
Telephone: 818-997-6611
Telefax: 818-787-7952
Electronic Mail: laserium@earthlink.net
WWW: http://www.laserium.com/
Founded: 1971
Staff: 30
Activities: laser entertainment, including for planetariums
City Reference Coordinates: 118°26'00"W 34°11'00"N

Laser Interferometer Gravitational-wave Observatory (LIGO) Project
• See "California Institute of Technology, Laser Interferometer Gravitational-wave Observatory (LIGO) Project"

Lawrence Berkeley National Laboratory (LBNL) (E.O._)
• See "E.O. Lawrence Berkeley National Laboratory (LBNL)"

Lawrence Livermore National Laboratory (LLNL), Institute of Geophysics and Planetary Physics (IGPP)
7000 East Avenue
L-413
P.O. Box 808
Livermore, CA 94550
Telephone: 925-423-0666
Telefax: 925-423-0238
Electronic Mail: <userid>@igpp.ucllnl.org
WWW: http://www.llnl.gov/urp/IGPP
Founded: 1984
Staff: 20
Activities: theoretical astrophysics * high-energy astrophysics * dark matter * adaptive optics * AGNs * cosmology
City Reference Coordinates: 121°46'00"W 37°41'00"N

Lebow Co.
5960 Mandarin Avenue
Goleta, CA 93117
Telephone: 805-964-7117
Telefax: 805-964-7117
Founded: 1973
Staff: 3
Activities: manufacturing soft X-ray filters, XUV filters * mirror coating plants * thin-film services
City Reference Coordinates: 119°50'00"W 34°27'00"N

LensPlus
11969 Livona Lane
Redding, CA 96003
Telephone: 1-800-659-7770
Electronic Mail: info@lensadapter.com
WWW: http://www.lensadapter.com/
Activities: distributing lens adapters, digital cameras and accessories
City Reference Coordinates: 122°24'00"W 40°35'00"N

Leuschner Observatory
• See "University of California at Berkeley (UCB), Department of Astronomy, Leuschner Observatory"

Lewis Center for Educational Research (LCER)
17500 Mana Road
Apple Valley, CA 92307
Telephone: 760-946-5414
Telefax: 760-946-9193
WWW: http://www.lewiscenter.org/
Founded: 1991
Activities: operating Goldstone-Apple Valley Radio Telescope (GAVRT)
• Formerly "Apple Valley Science and Technology Center (AVSTC)"
City Reference Coordinates: 117°14'00"W 34°32'00"N

Lick Observatory
• See "University of California at Santa Cruz (UCSC), Lick Observatory"

LIGO Project
• See "California Institute of Technology, Laser Interferometer Gravitational-wave Observatory (LIGO) Project"

Local Group of Santa Clarita Valley (The_)
• See "The Local Group of Santa Clarita Valley"

Lockheed Martin Corp., Advanced Technology Center, Solar and Astrophysics Laboratory
Department H1-12
Building 252
3251 Hanover Street
Palo Alto, CA 94304
Telephone: 650-424-3449
Telefax: 650-424-3994
Electronic Mail: <userid>@lmco.com
WWW: http://www.lmsal.com/
Founded: 1957
Staff: 70
Activities: space astronomical observing * solar physics * experiment and theory * instrumentation
City Reference Coordinates: 122°09'00"W 37°27'00"N

Loral Space Systems
3825 Fabian Way
Palo Alto, CA 94303-4604
Telephone: 650-852-4000
Electronic Mail: <userid>@ssd.loral.com
WWW: http://ssloral.com/
Founded: 1990
Activities: satellite systems, products and integration
City Reference Coordinates: 122°09'00"W 37°27'00"N

Los Angeles Astronomical Society (LAAS)
c/o Griffith Observatory
2800 East Observatory Road
Los Angeles, CA 90027-1299
Telephone: 213-673-7355
213-664-1181
Telefax: 213-663-4323
Electronic Mail: timothy.j.thompson@jpl.nasa.gov (Timothy J. Thompson, President)
patmann@ix.netcom.com (Pat Mann, Bulletin Editor)
WWW: http://www.laas.org/
Founded: 1926
City Reference Coordinates: 118°15'00"W 34°03'00"N

Los Angeles City College (LACC), Department of Physics and Astronomy
855 North Vermont Avenue
Los Angeles, CA 90029
Telephone: 213-663-9141 x318
213-669-4319
Electronic Mail: <userid>@email.lacc.cc.ca.us
WWW: http://www.lacc.cc.ca.us/academic/departments/physics/physics.htm
Founded: 1929 (College)
City Reference Coordinates: 118°15'00"W 34°03'00"N

Los Angeles Sidewalk Astronomers
1946 Vedanta Place
Hollywood, CA 90068
Telephone: 213-960-1737 (Bill Scott)
818-848-4533 (Donna Smith)
818-841-0548 (Bill Alborzian)
Electronic Mail: webmaster@sidewalkastronomers.com
WWW: http://www.sidewalkastronomers.com/
Periodicals: "The Sidewalk Astronomers"
• See also "San Francisco Sidewalk Astronomers" and "Sidewalk Astronomers"
City Reference Coordinates: 118°21'00"W 34°06'00"N (Hollywood)
118°18'00"W 34°12'00"N (Burbank)

Los Angeles Valley College (LAVC), Planetarium
5800 Fulton Avenue
Van Nuys, CA 91401
Telephone: 818-947-2864
818-947-2335 (Planetarium Info)
Telefax: 818-947-2610
Electronic Mail: falkdj@laccd.cc.ca.us (David Falk, Director)
WWW: http://www.lavc.cc.ca.us/
Founded: 1962
City Reference Coordinates: 118°26'00"W 34°11'00"N

Losmandy Astronomical Products
• See now "Hollywood General Machining Inc."

Lumicon
2111 Research Drive
Suites 4 & 5
Livermore, CA 94550
Telephone: 1-800-767-9576 (USA only)
              925-447-9570
Telefax: 925-447-9589
WWW: http://www.lumicon.com/
Founded: 1979
Staff: 7
Activities: manufacturing optics * visual determination of magnitude of planetary nebulae
Coordinates: 121°28'59"W 37°24'03"N H755m (Digger Pines Observatory)
City Reference Coordinates: 121°46'00"W 37°41'00"N

Maxtek Inc.
11980 Telegraph Road
Suite 104
Santa Fe Springs, CA 90670-6084
Telephone: 562-906-1515
Telefax: 562-906-1622
Electronic Mail: info@maxtekinc.com
WWW: http://www.maxtekinc.com/
Founded: 1975
Staff: 23
Activities: manufacturing thin-film deposition equipments and real-time on-line plating monitors
City Reference Coordinates: 118°04'00"W 33°56'00"N

Meade Instruments Corp.
6001 Oak Canyon
Irvine, CA 92618-5200
Telephone: 1-800-626-3233 (USA only)
              949-451-1450
Telefax: 949-451-1460
WWW: http://www.meade.com/
Founded: 1972
Staff: 170
Activities: manufacturing astronomical telescopes and accessories
City Reference Coordinates: 117°46'00"W 33°41'00"N

Merced Society for Telescope and Astronomy Recreation (M-STAR)
c/o Doug Kuhn (President)
1535 Harvest Avenue
Livingston, CA 95334
Telephone: 209-394-2014
Electronic Mail: asterion@elite.net
WWW: http://www.mstarastronomy.org/
Founded: 1982
Membership: 30
Activities: outreach programs
Periodicals: (12) "M-STAR Newsletter"
Coordinates: 120°19'53"W 37°18'07"N
City Reference Coordinates: 120°29'00"W 37°17'00"N (Merced)
                              120°43'00"W 37°23'00"N (Livingston)

Milliken Planetarium (Daniel B._)
• See "Daniel B. Milliken Planetarium"

Mineralogical Research Co.
15840 East Alta Vista Way
San Jose, CA 95127-1737
Telephone: 408-923-6800
Telefax: 408-926-6015
Electronic Mail: xtls@minresco.com
WWW: http://www.minresco.com/
Founded: 1960
Staff: 4
Activities: research * locating specimens * displays * collections * sales of minerals, meteorites and tektites * books * specimen boxes
Periodicals: (6) "New Acquisitions Lists"
Observatories: 1 (Alta Vista Observatory)

City Reference Coordinates: 121°53'00"W 37°20'00"N

Minolta Planetarium
c/o De Anza College
21250 Stevens Creek Boulevard
Cupertino, CA 95014
Telephone: 408-864-8814
Telefax: 408-864-5643
Electronic Mail: vonahnen@admin.fhda.edu
WWW: http://planetarium.deanza.fhda.edu/pltwww/ghome.html
Founded: 1970
Staff: 1
Activities: public and school shows * field trips * lectures * laser shows
City Reference Coordinates: 122°02'00"W 37°19'00"N

Mission College, Astronomy Department
3000 Mission College Boulevard
Santa Clara, CA 95054-1897
Telephone: 408-855-5262
Telefax: 408-855-5030
Electronic Mail: clint_poe@wvmccd.cc.ca.us (Clint Poe, Department Chairman)
        \<userid\>@wvmccd.cc.ca.us
WWW: http://www.wvmccd.cc.ca.us/mc/depts/astro/astro.html
City Reference Coordinates: 122°02'00"W 37°16'00"N

MMR Technologies Inc.
1400 North Shoreline Boulevard
Suite A-5
Mountain View, CA 94043-1346
Telephone: 650-962-9620
Telefax: 650-962-9647
Electronic Mail: bobp@mmr.com (Robert L. Paugh)
WWW: http://www.mmr.com/
Founded: 1980
Staff: 18
Activities: manufacturing cryogenic materials characterization and research systems and J-T refrigerators for cooling detectors
City Reference Coordinates: 122°04'00"W 37°23'00"N

Monoptec Corp.
1045 Mission Street
Suite 15
San Francisco, CA 94103-5818
Telephone: 415-264-6519
Telefax: 415-355-0374
Electronic Mail: monoptecfsd@monoptec.com
WWW: http://www.monoptec.com/
      http://monoptec.com/
Founded: 1989
Activities: licensing fixed shutter domes
City Reference Coordinates: 122°24'00"W 37°48'00"N

Monterey Institute for Research in Astronomy (MIRA)
200 Eighth Street
Marina, CA 93933
Telephone: 831-883-1000
Telefax: 831-883-1031
Electronic Mail: mira@mira.org
WWW: http://www.mira.org/
Founded: 1972
Staff: 6
Periodicals: (4) "The MIRA Newsletter"
City Reference Coordinates: 121°05'00"W 36°37'00"N

Morrison Planetarium (Alexander_)
• See "Alexander Morrison Planetarium"

Mountain Instruments
1213 South Auburn Street
Colfax, CA 95713
Telephone: 530-346-9113
Electronic Mail: larry@mountaininstruments.com
WWW: http://www.mountaininstruments.com/
Founded: 1991
Activities: building German and fork equatorial mounts
City Reference Coordinates: 120°56'00"W 39°07'00"N

Mountain Skies Astronomical Society (MSAS)
P.O. Box 1169
Lake Arrowhead, CA 92352-1169
or :
2001 Observatory Way
Lake Arrowhead, CA 92352
Telephone: 909-336-1699
Telefax: 909-336-4497
Electronic Mail: stargazersmail@mountain-skies.org
WWW: http://www.mountain-skies.org/
Founded: 1989
Membership: 2900
Activities: public education * MSAS Astronomy Village, Observatory and Science Center * Stargazers Giftshop
Periodicals: (12) "Sky Maps"; (4) "Stellar Matters", "Stargazer's Discovery Updates"
Coordinates: 117°12'37"W 34°13'54"N H901m (IAU Code 677)
City Reference Coordinates: 117°12'00"W 34°15'00"N

Mount Diablo Astronomical Society (MDAS)
1725 Sunnyvale Avenue
Walnut Creek, CA 94596
Telephone: 925-691-MDOA
      925-691-6362
Electronic Mail: stars4paul@aol.com
WWW: http://members.aol.com/mdas101b/private/
      http://mdia.org/mdas.html
Founded: 1957
City Reference Coordinates: 122°04'00"W 37°55'00"N

Mount Diablo Observatory Association (MDOA)
P.O. Box 346
Walnut Creek, CA 94597-0346
Telephone: 925-927-7222
Telefax: 877-349-5016
Electronic Mail: stars4paul@aol.com
WWW: http://www.mdia.org/astronomy.htm
City Reference Coordinates: 122°04'00"W 37°55'00"N

Mount Tamalpais Observers
c/o Tinka Ross
89 Dominican Drive
San Rafael, CA 94901-1337
Telephone: 415-454-4715
Electronic Mail: tinka@marin.cc.ca.us (Tinka Ross)
      tinka@ix.netcom.com (" " ")
Founded: 1989
Membership: 30
Activities: lectures * observing
City Reference Coordinates: 122°31'00"W 37°59'00"N

Mount Wilson Institute (MWI)
Hale Solar Laboratory
740 Holladay Road
Pasadena, CA 91106
Telephone: 818-793-3100
Telefax: 818-793-4570
Electronic Mail: mwi@mtwilson.edu
WWW: http://www.mtwilson.edu/
Founded: 1986
Staff: 20
Activities: astrophysical research
Coordinates: 118°03'36"W 34°13'00"N H1,742m (IAU Code 672)
City Reference Coordinates: 118°09'00"W 34°09'00"N

Mount Wilson Observatory (MWO)
• See "Mount Wilson Institute (MWI)"

Mount Wilson Observatory Association (MWOA)
P.O. Box 70076
Pasadena, CA 91117-7076
Electronic Mail: info@mwoa.org
WWW: http://www.mtwilson.edu/Services/Organizations/MWOA/
      http://www.mwoa.org/
Activities: supporting Mount Wilson Observatory activities
City Reference Coordinates: 118°09'00"W 34°09'00"N

N.A. Richardson Astronomical Observatory
• See "San Bernardino Valley College, N.A. Richardson Astronomical Observatory"

NASA Astrobiology Institute (NAI), Outreach Office
Mail Stop 240-1
Moffett Field, CA 94035-1000
Telephone: 650-604-0809
Telefax: 650-604-4251
WWW: http://nai.arc.nasa.gov/
Founded: 1998
Activities: scientific study of the origin, distribution, evolution, and future of life in the universe
City Reference Coordinates: 122°03'00"W 37°24'00"N

National Aeronautics and Space Administration (NASA), Ames Research Center (ARC), Astrophysics Branch
Mail Stop 245-6
Moffett Field, CA 94035-1000
Telephone: 650-604-5528
Telefax: 650-604-6779
Electronic Mail: <userid>@mail.arc.nasa.gov
WWW: http://www-space.arc.nasa.gov/
http://www.arc.nasa.gov/
Founded: 1971
Staff: 55
Activities: IR astrophysics * IR laboratory astrophysics * planetary astronomy * ISM * IR observing (ISM, stars, solar-system objects) * IR instrumentation
Observatories: 1 (Stratospheric Observatory for Infrared Astronomy - SOFIA)
City Reference Coordinates: 122°03'00"W 37°24'00"N

National Aeronautics and Space Administration (NASA), Ames Research Center (ARC), NASA Astrobiology Institute (NAI)
• See "NASA Astrobiology Institute (NAI)"

National Aeronautics and Space Administration (NASA), Ames Research Center (ARC), Stratospheric Observatory for Infrared Astronomy (SOFIA)
Mail Stop 244-30
Moffett Field, CA 94035
Telephone: 650-604-5343
Telefax: 650-604-1094
Electronic Mail: info@sofia-usra.arc.nasa.gov
<userid>@sofia.arc.nasa.gov
WWW: http://sofia.arc.nasa.gov/
Founded: 1985
Staff: 15
Activities: airborne IR astronomy
Periodicals: "Airborne Astronomy Flyer"
City Reference Coordinates: 122°03'00"W 37°24'00"N

National Aeronautics and Space Administration (NASA), Dryden Flight Research Center (DFRC)
P.O. Box 273
Edwards, CA 93523
Telephone: 661-276-3311
Electronic Mail: <userid>@dfrc.nasa.gov
WWW: http://www.dfrc.nasa.gov/
Founded: 1958 (NASA)
• Formerly "Dryden Flight Research Facility (DFRF)"
City Reference Coordinates: 117°53'00"W 34°44'00"N

National Center for Science Education (NCSE)
P.O. Box 9477
Berkeley, CA 94709-0477
Telephone: 510-526-1674
1-800-290-6006 (USA only)
Telefax: 510-526-1675
Electronic Mail: ncse@natcenscied.org
editor@natcenscied.org
WWW: http://www.natcenscied.org/
Founded: 1981
Activities: preserving integrity of science education * central information and resource clearinghouse
Periodicals: (6) "Reports of the National Center for Science Education" (ISSN 1064-2358)
City Reference Coordinates: 122°18'00"W 37°57'00"N

New Jersey Institute of Technology (NJIT), Big Bear Solar Observatory (BBSO)
40386 North Shore Lane
Big Bear City, CA 92314-9672

Telephone: 909-866-5791
Telefax: 909-866-4240 (Dome)
    909-866-5791 x24 (Office)
Electronic Mail: <userid>@bbso.njit.edu
WWW: http://www.bbso.njit.edu/
Founded: 1970
Staff: 20
Activities: high-resolution observations of the Sun * solar research
Periodicals: "Big Bear Solar Observatory Preprints"
Coordinates: 116°54'51"W 34°15'14"N H2,067m (Big Bear Lake)
City Reference Coordinates: 116°51'00"W 34°16'00"N

Newport Industrial Glass Inc.
1631 Monrovia Avenue
Costa Mesa, CA 92627
Telephone: 714-645-1500
Telefax: 714-645-6800
Electronic Mail: nig@netcom.com
Founded: 1983
Staff: 12
Activities: manufacturing and distributing optical glass and accessories
City Reference Coordinates: 117°55'00"W 33°39'00"N

Noctilume
P.O. Box 3042
Culver City, CA 90231
Telephone: 310-967-4719
Electronic Mail: d_sensiper@excite.com
WWW: http://www.imall.com/stores/noctilume/
Founded: 1986
Staff: 2
Activities: manufacturing illumination sources
City Reference Coordinates: 118°24'00"W 34°01'00"N

North Valley Astronomers (NOVA)
Electronic Mail: wizz@mail.csuchico.edu
WWW: http://rigel.csuchico.edu/nova/
Observatories: 1 (James Schwartz Observatory)
City Reference Coordinates: 121°50'00"W 39°44'00"N (Chico)

Oberti (Daniel_)
● See "Daniel Oberti - Architectural Ceramics"

Observatorio Astronómico Nacional (OAN), Mexico
● See "Universidad Nacional Autónoma de México (UNAM), Observatorio Astronómico Nacional (OAN)"

Oceanside Photo and Telescope
1024 Mission Avenue
Oceanside, CA 92054
Telephone: 760-722-3348
    760-722-3343
    1-800-48-FOCUS (USA only)
    1-800-483-6287 (USA only)
Telefax: 760-722-8133
Electronic Mail: opt@optcorp.com
WWW: http://www.optcorp.com/
Founded: 1947
City Reference Coordinates: 117°23'00"W 33°12'00"N

Optical Coating Laboratory Inc. (OCLI)
2789 Northpoint Parkway
Santa Rosa, CA 95407-7397
Telephone: 707-525-6957
Telefax: 707-525-7841
Electronic Mail: ocli_sales@ocli.com
    <userid>@ocli.com
WWW: http://www.ocli.com/
Founded: 1949
Staff: 1100
Activities: manufacturing thin film coatings and electro-optical components and assemblies
City Reference Coordinates: 112°43'00"W 38°26'00"N

Optical Research Associates (ORA)
3280 East Foothill Boulevard
Pasadena, CA 91107-3103

Telephone: 626-795-9101
Telefax: 626-795-9102
         626-795-0184
Electronic Mail: service@opticalres.com (Software)
                 engr@optical.com (Engineering)
WWW: http://www.opticalres.com/
Founded: 1963
Staff: 47
Activities: optical design * engineering services * software design * consultancy
City Reference Coordinates: 118°09'00"W 34°09'00"N

Orange County Astronomers (OCA)
c/o Charlie Oostdyk (Treasurer)
P.O. Box 1762
Costa Mesa, CA 92628
Telephone: 714-751-5381
           714-751-6867 (Starline)
Electronic Mail: charlie@cccd.edu
WWW: http://www.ocastronomers.org/
Founded: 1967
Membership: 620
Activities: monthly meetings * monthly observing * public outreach * planetarium shows * annual seminars
Periodicals: (12) "Sirius Astronomer"
Coordinates: 116°43'14"W 33°29'02"N H1,332m (Anza Obs. – IAU Code 643)
City Reference Coordinates: 117°55'00"W 33°39'00"N

Orion Telescope & Binocular Center
2450 17th Avenue
P.O. Box 1815
Santa Cruz, CA 95061
Telephone: 1-800-447-1001 (Orders - USA only)
           408-763-7030 (Information)
           408-464-0446
Telefax: 408-464-0466
         408-763-7017
Electronic Mail: sales@oriontel.com
WWW: http://www.oriontel.com/
     http://www.telescope.com/
Founded: 1975
Staff: 25
Activities: discount sales of amateur astronomy equipment
City Reference Coordinates: 122°01'00"W 36°58'00"N

Owens Valley Radio Observatory (OVRO)
• See "California Institute of Technology, Owens Valley Radio Observatory (OVRO)"

Pacific Union College (PUC), Physics Department
One Angwin Avenue
Angwin, California 94508-9707
Telephone: 707-965-7269
Telefax: 707-965-6390
Electronic Mail: <userid>@puc.edu
WWW: http://www.puc.edu/Departments/Physics/
Founded: 1969
Staff: 7
Activities: education * public observing * CCD imaging * Young Observatory
Coordinates: 122°25'43"W 38°34'41"N H580m
City Reference Coordinates: 122°26'00"W 38°34'00"N

Palomar College, Planetarium
1140 West Mission Road
San Marcos, CA 92069-1487
Telephone: 760-744-1150 x2516
WWW: http://www.palomar.edu/astronomy/
Founded: 1964
Activities: shows * observing
Coordinates: 117°10'32"W 33°09'02"N H183m (Palomar College Observatory)
City Reference Coordinates: 117°10'00"W 33°09'00"N

Palomar Observatory
• See "California Institute of Technology, Palomar Observatory"

Parks Optical Co.
750 Easy Street
Simi Valley, CA 93065

Telephone: 805-522-6722
Telefax: 805-522-0033
Electronic Mail: parksoptical@parksoptical.com
WWW: http://www.parksoptical.com/
Founded: 1954
Staff: 63
Activities: manufacturing telescopes, binoculars, microscopes, eyepieces, accessories and filters
Periodicals: "Your Universe Awaits"
City Reference Coordinates: 118°05'00"W 34°10'00"N (Thousand Oaks)

Pasco Scientific
P.O. Box 619011
10101 Foothills Boulevard
Roseville, CA 95747-9011
Telephone: 1-800-772-8700 (USA only)
           916-786-3800
Telefax: 916-786-8905
Electronic Mail: techsupp@pasco.com
                 sales@pasco.com
WWW: http://www.pasco.com/
Founded: 1964
Staff: 65
Activities: manufacturing educational optics apparatus
City Reference Coordinates: 121°17'00"W 38°45'00"N

Peninsula Astronomical Society (PAS)
P.O. Box 4542
Mountain View, CA 94040
Telephone: 650-493-4742 (William Phelps, Vice President)
Electronic Mail: wm@netcom.com (same)
                 pasinfo@usa.net
WWW: http://www.foothill.fhda.edu/ast/pas.htm
Membership: 250
Activities: meetings * lectures * observing * public astronomy programs * loaning instruments * operating Foothill College Observatory (see separate entry)
Periodicals: "Nite Skies"
City Reference Coordinates: 122°04'00"W 37°23'00"N

Photosolve
21272 Chiquita Way
Saratoga, CA 95070-4259
WWW: http://www.photosolve.com/
Activities: distributing accessories and adapters for cameras
City Reference Coordinates: 122°02'00"W 37°16'00"N

Planetarium Institute
c/o Department of Physics and Astronomy
San Francisco State University
San Francisco, CA 94132
Telephone: 415-338-1852
Telefax: 415-338-2178
Founded: 1973
Staff: 1
Activities: planetarium shows for elementary schools * university astronomy classes
Coordinates: 122°28'33"W 37°43'26"N
City Reference Coordinates: 122°24'00"W 37°48'00"N

Planetary Science Institute (PSI), San Juan Capistrano Division
31882 Camino Capistrano
Suite 102
San Juan Capistrano, CA 92675
Telephone: 949-240-2010
Telefax: 949-240-0482
Electronic Mail: educator@sji.org
WWW: http://www.psi.edu/
Founded: 1971
Staff: 10
Activities: planetary science research
City Reference Coordinates: 117°39'00"W 33°30'00"N

Planetary Society
65 North Catalina Avenue
Pasadena, CA 91106-2301
Telephone: 626-793-5100
Telefax: 626-793-5528

Electronic Mail: tps@planetary.org
WWW: http://planetary.org/
Founded: 1980
Membership: 100000
Activities: advocating exploration of the solar system and SETI * publishing * audiovisual resources * educational programs * display materials * scholarships * merchandise
Periodicals: (6) "Planetary Report" (ISSN 0736-3680); "Mars Underground News", "Bioastronomy News"
Awards: "Mars Institute Contest", "New Millennium Committee Scholarship", "College Followship"
City Reference Coordinates: 118°09'00"W 34°09'00"N

Polaris Observatory Association (POA)
P.O. Box 734
Somis, CA 93066
Electronic Mail: polaris@frazmtn.com
WWW: http://www.frazmtn.com/~polaris/
City Reference Coordinates: 118°05'00"W 34°10'00"N (Thousand Oaks)

Pomona College, Department of Physics and Astronomy
610 North College Avenue
Claremont, CA 91711-6348
Telephone: 714-621-8000 x2945
Telefax: 714-621-8463
Electronic Mail: <userid>@pomona.edu
WWW: http://www.physics.pomona.edu/
http://www.astronomy.pomona.edu/
Founded: 1888
Activities: particle physics * magnetic domains * fractal networks * ocean acoustics * molecular spectroscopy * SN * general relativity * stellar photometry and spectroscopy * Millikan Planetarium
Coordinates: 117°42'24"W 34°05'24"N H368m (Franck P. Brackett Observatory)
117°40'42"W 34°22'54"N H2,286m (Table Mountain Observatory - TMO)
City Reference Coordinates: 117°43'00"W 34°06'00"N

Pomona Valley Amateur Astronomers (PVAA)
P.O. Box 162
Upland, CA 91785
Telephone: 909-985-1684
Electronic Mail: patrick@cyberg8t.com
WWW: http://www.cyberg8t.com/patrick/PVAA.htm
Membership: 100
City Reference Coordinates: 117°38'00"W 34°06'00"N

QSP Optical Technology Inc.
1712-F Newport Circle
Santa Ana, CA 92705
Telephone: 714-557-2299
Telefax: 714-557-2170
Electronic Mail: info@qsptech.com
WWW: http://www.qsptech.com/
Activities: film coating * filters * mirrors
City Reference Coordinates: 117°54'00"W 33°44'00"N

Quadrant Engineering Inc.
10138 Commercial Avenue
Penn Valley, CA 95946
Telephone: 530-432-5285
Telefax: 530-432-5439
Electronic Mail: quadsales@qei-motion.com
info@qei-motion.com
WWW: http://www.qei-motion.com/
Founded: 1989
Activities: motion control * pointing and tracking mounts * rotary tables and stages * precision worm gears * custom engineering
City Reference Coordinates: 121°34'00"W 39°31'00"N (Oroville)

Reuben H. Fleet Space Theater and Science Center
Balboa Park
P.O. Box 33303
San Diego, CA 92163-3303
or :
1875 El Prado Balboa Park
San Diego, CA
Telephone: 619-238-1233
Telefax: 619-685-5771
Electronic Mail: rstroup@rhfleet.org
membership@rhfleet.org

WWW: http://www.rhfleet.org/
Founded: 1973
Membership: 3750
Activities: omnimax films * planetarium multimedia programs * science exhibitions * lectures * special program openings for members
Periodicals: (6) "Science Scope"; (1) "Annual Report"
City Reference Coordinates: 117°09'00"W 32°43'00"N

Reynard Corp.
1020 Calle Sombra
San Clemente, CA 92673-6227
Telephone: 949-366-8866
Telefax: 949-498-9528
Electronic Mail: sales@reynardcorp.com
WWW: http://www.reynardcorp.com/
Founded: 1984
Staff: 23
Activities: manufacturing optical components * thin film coatings * photolithography
City Reference Coordinates: 117°37'00"W 33°26'00"N

Richardson Astronomical Observatory (N.A._)
• See "San Bernardino Valley College, N.A. Richardson Astronomical Observatory"

Rigel Systems
26850 Basswood Avenue
Rancho Palos Verdes
Palos Verdes, CA 90274
Telephone: 310-375-4149
WWW: http://pw2.netcom.com/~rigelsys/Rigelsys.html
Activities: manufacturing LED flashlights
City Reference Coordinates: 118°24'00"W 33°48'00"N

Rio Hondo Astronomical Society
c/o Roger Wilcox
14517 East Broadway
Whittier, CA 90604
Telephone: 310-941-9755
          909-985-1684
Electronic Mail: rwilcox@loop.com
City Reference Coordinates: 118°02'00"W 33°59'00"N

Riverside Astronomical Society (RAS)
P.O. Box 51222
Riverside, CA 92517-2222
Telephone: 909-342-2389
Electronic Mail: 73024.3333@compuserve.com
WWW: http://www.pe.net/~wpl/ras.html
Founded: 1956
Membership: 170
City Reference Coordinates: 117°22'00"W 33°59'00"N

Riverside Community College Planetarium
c/o Astronomy Department
4800 Magnolia Avenue
Riverside, CA 92506-1299
Telephone: 909-222-8000
Telefax: 909-222-8036
Electronic Mail: <userid>@rccd.cc.ca.us
WWW: http://www.rccd.cc.ca.us/
Founded: 1968
City Reference Coordinates: 117°22'00"W 33°59'00"N

Riverside Telescope Makers' Conference (RTMC) Inc.
8300 Utica Avenue
Suite 105
Rancho Cucamonga, CA 91730
Telephone: 909-948-2205
Electronic Mail: robert_stephens@eee.org
WWW: http://www.rtmc-inc.org/
Founded: 1968
Staff: 10
Activities: conference related to the art of telescope making
Awards: (1) "Clifford W. Holmes Award", "Clyde Tombaugh Telescope Innovation Award", "Warren Estes Memorial Award", "Telescope Competition Merit Award"
Observatories: 1 (Big Bear City, H2,450m)

Coordinates: 116°45'20"W 34°13'43"N
City Reference Coordinates: 117°56'00"W 34°04'00"N (West Covina)

Rolyn Optics Co.
706 Arrowgrand Circle
Covina, CA 91722-2199
Telephone: 626-915-5707
Telefax: 626-915-1379
Founded: 1925
Staff: 20
Activities: distributing optical components
City Reference Coordinates: 117°53'00"W 34°05'00"N

Roth Planetarium
• See "California State University, Chico, Roth Planetarium"

Sacramento Valley Astronomical Society (SVAS)
P.O. Box 15274
Sacramento, CA 95851-0274
Telephone: 916-SVAS-111
WWW: http://www.skywatchers.org/
Founded: 1945
City Reference Coordinates: 121°30'00"W 38°35'00"N

San Bernardino Valley Amateur Astronomers (SBVAA)
c/o Chris Clarke
San Bernardino County Museum
2024 Orange Tree Lane
Redlands, CA 92374
Telephone: 909-888-6511 x1458
WWW: http://www.sbvaa.org/
Founded: 1958
City Reference Coordinates: 117°10'00"W 34°03'00"N

San Bernardino Valley College, N.A. Richardson Astronomical Observatory
701 South Mount Vernon Avenue
San Bernardino, CA 92410
Telephone: 909-888-6511 x1458
Electronic Mail: bwilson@sbccd.cc.ca.us
WWW: http://www.sbccd.cc.ca.us/sbvc/
Founded: 1928
Staff: 3
Activities: education * public viewing
Coordinates: 117°18'00"W 34°05'00"N H233m
City Reference Coordinates: 117°18'00"W 34°05'00"N

San Diego Aerospace Museum
2001 Pan American Plaza
San Diego, CA 92101
Telephone: 619-234-8291
Telefax: 619-233-4526
WWW: http://www.aerospacemuseum.org/
Founded: 1961
Staff: 44
Activities: history of aviation and spaceflight
Periodicals: (6) "Newsletter"
City Reference Coordinates: 117°09'00"W 32°43'00"N

San Diego Astronomy Association (SDAA)
P.O. Box 23215
San Diego, CA 92193-3215
Telephone: 619-645-8940 (Office)
          619-766-9911 (Observatory)
Electronic Mail: sdaa@digcir.cts.com
                sdaainfo@digcir.cts.com
WWW: http://www.sdaa.org/
Founded: 1963
Membership: 375
Activities: observing * education * library * field trips
Periodicals: (12) "SDAA News & Notes"
Awards: "SDAA Certificate of Merit"
Coordinates: 116°40'25"W 32°22'19"N H1,500m (James Lipp Observatory, Tierra del Sol)
City Reference Coordinates: 117°09'00"W 32°43'00"N

San Diego State University (SDSU), Astronomy Department
San Diego, CA 92182-1221
Telephone: 619-594-6182
Telefax: 619-594-1413
Electronic Mail: <userid>@mintaka.sdsu.edu
WWW: http://mintaka.sdsu.edu/
Founded: 1897
Staff: 7
Activities: eclipsing binaries * galaxy surface photometry * galaxy clusters * PN * CCD instrumentation * photometry * planetarium
Coordinates: 116°25'30"W 32°50'24"N H1,859m (Mount Laguna Observatory)
City Reference Coordinates: 117°09'00"W 32°43'00"N

San Fernando Observatory (SFO)
• See "California State University, Northridge, San Fernando Observatory (SFO)"

San Francisco Amateur Astronomers (SFAA)
c/o Morrison Planetarium
California Academy of Science
Golden Gate Park
San Francisco, CA 94118
Telephone: 415-566-2357
Electronic Mail: chelleb@aol.com
WWW: http://www.zennla.com/sfaa/
http://members.aol.com/chelleb/sfaa.htm
Founded: 1952
Membership: 250
Periodicals: "Above the Fog"
Awards: (1) "Service Award", "Herman fast Award"
City Reference Coordinates: 122°24'00"W 37°48'00"N

San Francisco Sidewalk Astronomers
• See "Sidewalk Astronomers"

San Francisco State University (SFSU), Department of Physics and Astronomy
1600 Holloway Avenue
San Francisco, CA 94132
Telephone: 415-338-1852
415-338-1852 (Planetarium Hotline)
Telefax: 415-338-2178
Electronic Mail: <userid>@stars.sfsu.edu
letsch@stars.sfsu.edu (Kenneth Letsch, Planetarium Curator)
WWW: http://www.physics.sfsu.edu/
http://www.physics.sfsu.edu/planetarium/planetarium.html (Planetarium)
http://www.physics.sfsu.edu/observatory/observatory.html (Observatory)
Founded: 1973
Staff: 15
Activities: research * education * Charles F. Hagar Planetarium * public observatory
Coordinates: 122°28'33"W 37°43'26"N H100m
City Reference Coordinates: 122°24'00"W 37°48'00"N

San Jose Astronomical Association (SJAA)
c/o James Van Nuland
3509 Calico Drive
San Jose, CA 95124
Telephone: 408-371-1307
Electronic Mail: secretary@sjaa.net
WWW: http://www.sjaa.net/
http://www.whiteoaks.com/eph/ (Newsletter)
Founded: 1954
Membership: 220
Activities: lectures * observing * discussion groups * grazing occultations * photography * telescope making
Periodicals: (12) "SJAA Ephemeris"
Awards: (1) "A.B. Gregory Award"
City Reference Coordinates: 121°53'00"W 37°20'00"N

San Jose State University (SJSU), Department of Physics
One Washington Square
San Jose, CA 95192-0106
Telephone: 408-924-5210
Telefax: 408-924-2917
Electronic Mail: <userid>@email.sjsu.edu
WWW: http://www.physics.sjsu.edu/

or :
1946 Vedanta Place
Hollywood, CA 90068
Telephone: 415-567-2063
415-289-2007 (SF hotline)
805-782-9640
818-841-0548 (second address)
818-842-6484 (second address)
Electronic Mail: vedantas@aol.com (second address)
WWW: http://members.aol.com/raycash/sidewalk.htm
http://www.geocities.com/CapeCanaveral/6389/
http://www.sidewalkastronomers.com/
Founded: 1968
Membership: 250
Activities: public star parties * school visits * telescope making * education
Periodicals: (4) "Newsletter"
• See also "Los Angeles Sidewalk Astronomers" and "San Francisco Sidewalk Astronomers"
City Reference Coordinates: 122°24'00"W 37°48'00"N (San Francisco)
118°21'00"W 34°06'00"N (Hollywood)

Software Systems Consulting (SSC)
616 South El Camino Real
San Clemente, CA 92672
Telephone: 949-498-5784
Telefax: 949-498-0568
Electronic Mail: jhoot@exo.com (John E. Hoot)
support@ssccorp.com
WWW: http://www.sscorp.com/
Founded: 1980
Staff: 6
Activities: custom software development services in fields including radio communications, process control, imaging, astronomy, and data acquisition and analysis
City Reference Coordinates: 117°37'00"W 33°26'00"N

Sonoma County Astronomical Society (SCAS)
P.O. Box 183
Santa Rosa, CA 95402
Telephone: 707-795-1448
Electronic Mail: mdcombs@ix.netcom.com (Merlin Combs)
Founded: 1975
Membership: 210
Activities: annual "Striking Sparks" project
Periodicals: (12) "Sonoma Skies"
Observatories: 1 (Palmieri Observatory - H1,000m)
City Reference Coordinates: 122°43'00"W 38°28'00"N

Sonoma Instrument Co.
P.O. Box 9011
Santa Rosa, CA 95405
Telephone: 707-542-8569
Electronic Mail: amps@sonoma-instrument.com
WWW: http://www.sonoma-instrument.com/
Activities: manufacturing amplifiers
City Reference Coordinates: 122°43'00"W 38°28'00"N

Sonoma State University (SSU), Department of Physics and Astronomy
1801 East Cotati Avenue
Rohnert Park, CA 94928
Telephone: 707-664-2119
Telefax: 707-664-2505
Electronic Mail: <userid>@sonoma.edu
WWW: http://www.phys-astro.sonoma.edu/
http://yorty.sonoma.edu/observatory/ (Observatory)
Founded: 1961
Staff: 7
Activities: education * research * optical, radio and X-ray astronomy
Coordinates: 122°40'34"W 38°20'34"N H36m
City Reference Coordinates: 122°42'00"W 38°21'00"N

South Bay Astronomical Society (SBAS)
2377 Crenshaw Boulevard
Suite 350
Torrance, CA 90501
Electronic Mail: sbas@meteorite.com
WWW: http://www.meteorite.com/sbas/
City Reference Coordinates: 118°19'00"W 33°59'00"N

Southern Stars Systems
P.O. Box 3792
Saratoga, CA 95070
or :
12525 Saratoga Creek Drive
Saratoga, CA 95070
Telephone: 866-887-7827
Telefax: 240-371-8010
Electronic Mail: info@southernstars.com
sthnstar@southernstars.com
WWW: http://www.southernstars.com/
Founded: 1993
* Software producer
City Reference Coordinates: 122°02'00"W 37°16'00"N

Sovietski Collection
P.O. Box 81347
San Diego, CA 92138-1347
or :
3450 Kurtz Street
Suite C
San Diego, CA 92110
Telephone: 619-237-8000
619-294-2000 (Mitch Siegler, President)
1-800-442-0002 (USA only)
Telefax: 619-294-2500
619-237-8010
Electronic Mail: fulcrum@sovietski.com
info@sovietski.com
WWW: http://www.sovietski.com/
Founded: 1991
Activities: distributing clocks, watches, optical instruments and accessories
City Reference Coordinates: 117°09'00"W 32°43'00"N

Space Craft International (SCI)
P.O. Box 61027
Catalina Station
Pasadena, CA 91116-7027
Telephone: 626-398-4800
1-800-472-4548 (USA only)
1-800-4-SCIKIT (USA only)
Telefax: 626-398-9600
1-800-307-0007 (USA only)
Electronic Mail: dd@scikits.com
commerce@scikits.com
WWW: http://www.scikits.com/
Founded: 1987
Staff: 1
Activities: designing and distributing educational models of spacecraft * science products * supplying schools and planetariums
City Reference Coordinates: 118°09'00"W 34°09'00"N

SpaceDev
7940 Silverton Avenue
Suite 202
San Diego, CA 92126
Telephone: 619-684-3570
Telefax: 619-693-6932
Electronic Mail: info@spacedev.com
WWW: http://www.spacedev.com/
City Reference Coordinates: 117°09'00"W 32°43'00"N

Space Frontier Foundation
11350 Ventura Boulevard
Suite 100
Studio City, CA 91604-3140
Telephone: 1-800-78-SPACE (USA only)
1-800-787-7223 (USA only)
Electronic Mail: information@space-frontier.org
WWW: http://www.space-frontier.org/
Founded: 1988
City Reference Coordinates: 118°24'00"W 34°09'00"N

Spectra Astro Systems
20620 Lassen Street
Chatsworth, CA 91311
or :

6631 Wilbur Avenue
Suite 30
Reseda, CA 91335
Telephone: 818-343-1352
1-800-735-1352 (Orders)
Telefax: 818-996-7698
Electronic Mail: spectraast@aol.com
spectraastro@spectraastro.com
WWW: http://www.rahul.net/resource/spectra
http://www.spectraastro.com/
Founded: 1987
Activities: distributing astronomical instrumentation and accessories * consultancy
City Reference Coordinates: 118°36'00"W 34°15'00"N (Chatsworth)
118°30'00"W 34°13'00"N (Reseda)

Spectrum Sciences Inc. (SSI)
P.O. Box 1950
Nipomo, CA 93444-1950
or :
44426 Airport Road
Suite 160
California, MD 20619
Telephone: 408-727-1567
1-800-862-1993 (USA only)
Telefax: 408-727-1322
301-862-2634
Electronic Mail: spicuzza@spectrumsciences.com
WWW: http://spectrumsciences.com/
Founded: 1968
Staff: 23
Activities: manufacturing scientific instrumentation * ion implantation systems * nuclear systems * radiation effects services * mass spectrometers * ion implanter systems for flat panel displays * solid-state nuclear particle detectors
City Reference Coordinates: 121°57'00"W 37°21'00"N (Santa Clara)

Stanford Research Systems (SRS) Inc.
1290 D Reamwood Avenue
Sunnyvale, CA 94089
Telephone: 408-744-9040
Telefax: 408-744-9049
Electronic Mail: info@srsys.com
WWW: http://www.srsys.com/
Founded: 1980
Activities: manufacturing electronic signal recovery and test and measurement instruments
City Reference Coordinates: 122°01'00"W 37°23'00"N

Stanford University, Astronomy Program
Varian Room 302b
Stanford, CA 94305-4060
Telephone: 650-723-1435
Telefax: 650-723-4840
Electronic Mail: vahe@astronomy.stanford.edu (Vahé Petrosian)
WWW: http://www.stanford.edu/dept/astro/
Founded: 1977
Staff: 8
Activities: theoretical high-energy astrophysics and cosmology * solar physics and seismology * XUV, soft X-ray, and gamma-ray spectroscopy * radar planetary exploration
City Reference Coordinates: 122°08'00"W 37°25'00"N

Stanford University, Solar Observatories Group, Wilcox Solar Observatory (WSO)
355 Via Palou
HEPL Annex B208
Stanford, CA 94305-4085
Telephone: 650-723-1505
Telefax: 650-725-2333
Electronic Mail: wso@solar.stanford.edu
<userid>@solar.stanford.edu
WWW: http://sun.stanford.edu/~wso/wso.html
Founded: 1974
Staff: 28
Activities: solar magnetograms and dopplergrams * measurement of solar mean magnetic field * low-degree solar velocity oscillations * solar oscillation investigations for SOHO
Coordinates: 122°10'02"W 37°24'35"N H115m
City Reference Coordinates: 122°08'00"W 37°25'00"N

Starsplitter Telescopes
3228 Rikkard Drive

Thousand Oaks, CA 91362
Telephone: 805-492-0489
Telefax: 805-492-0489
WWW: http://www.starsplitter.com/
Activities: manufacturing telescopes
City Reference Coordinates: 118°50'00"W 34°10'00"N

Stellarium
4560 Petaluma Hill Road
Santa Rosa, CA 95404
Telephone: 707-586-0660
WWW: http://www.stellarium.com/
Founded: 1983
City Reference Coordinates: 122°43'00"W 38°28'00"N

Stellar Products
7387 Celata Lane
San Diego, CA 92129
Electronic Mail: info@stellarproducts.com
WWW: http://www.stellarproducts.com/
Activities: manufacturing adaptive-optics systems
City Reference Coordinates: 117°09'00"W 32°43'00"N

Stellar Software
P.O. Box 10183
Berkeley, CA 94709
Telephone: 510-845-8405
Telefax: 510-845-2139
Electronic Mail: ssw@dnai.com
WWW: http://www.stellarsoftware.com/
Founded: 1986
Staff: 2
Activities: manufacturing and distributing optical ray tracing software
City Reference Coordinates: 122°18'00"W 37°57'00"N

Stockton Astronomical Society (SAS)
P.O. Box 243
Stockton, CA 95201
Telephone: 209-478-4380
Electronic Mail: tatkinsn@mediaone.net (Trevor H. Atkinson)
WWW: http://astro.sci.uop.edu/~sas/
Founded: 1950
Membership: 120
Activities: monthly meetings * star parties * school star parties
Periodicals: (12) "Valley Skies"
City Reference Coordinates: 121°20'00"W 37°58'00"N

Stony Ridge Observatory (SRO)
c/o Tim Cann
37345 Avenida Bravura
Temecula, CA 92592
Telephone: 619-946-2851 (Tony Heinzman, President)
  818-248-0246
Electronic Mail: palostar@earthlink.net (Steve Brewster)
WWW: http://stony-ridge.org/
Founded: 1963
Coordinates: 117°59'45"W 34°17'55"N H1730m (IAU Code 671)
City Reference Coordinates: 117°09'00"W 33°30'00"N

Stratospheric Observatory for Infrared Astronomy (SOFIA)
• See "National Aeronautics and Space Administration (NASA), Ames Research Center (ARC), Stratospheric Observatory for Infrared Astronomy (SOFIA)"

Sunsearch
c/o Wayne P. Johnson
21870 Mary Street
Mead Valley, CA 92570
Telephone: 909-653-8813 (until midnight PST)
  909-279-2820
Electronic Mail: wjohnson@citrus.ucr.edu
WWW: http://www.chapman.edu/oca/benet/mrgalaxy.htm
Activities: SN search and follow-up
City Reference Coordinates: 117°12'00"W 33°48'00"N (Perris)

Table Mountain Observatory (TMO)
● See "California Institute of Technology, Jet Propulsion Laboratory (JPL), Table Mountain Observatory (TMO)"

Tech Museum of Innovation (The_)
● See "The Tech Museum of Innovation"

Tekton Industries
1571 San Elijo Avenue
Cardiff by the Sea, CA 92007
Telephone: 760-431-7511
Founded: 1990
Staff: 3
Activities: manufacturing telescope accessories
City Reference Coordinates: 117°17'00"W 33°04'00"N (Encinitas)

Tessmann Planetarium
c/o Santa Ana College
1530 West Seventeenth Street
Santa Ana, CA 92706
Telephone: 714-564 6356
Electronic Mail: wbonney@chapman.edu
WWW: http://www.chapman.edu/oca/tessmann/
Founded: 1967
City Reference Coordinates: 117°54'00"W 33°44'00"N

The Astronomy Connection
P.O. Box 134
Los Gatos, CA 95031-0134
Telephone: 408-356-1125
WWW: http://www.seds.org/TAC
City Reference Coordinates: 121°59'00"W 37°13'00"N

The Gemmary Inc.
P.O. Box 2560
Fallbrook, CA 92088
Telephone: 760-728-3321
Telefax: 760-728-3322
Electronic Mail: rcb@gemmary.com
WWW: http://www.gemmary.com/
Founded: 1979
Staff: 3
Activities: rare books * antique scientific instruments * catalog mail order
City Reference Coordinates: 117°15'00"W 33°23'00"N

The Local Group of Santa Clarita Valley (LGSCV)
25032 Walnut Street
Newhall, CA 91321
Electronic Mail: clark@qnet.com (Doug Clark)
info@lgscv.org
WWW: http://www.lgscv.org/
http://home.earthlink.net/~plumber016/localgrp.htm
Founded: 1984
City Reference Coordinates: 118°31'00"W 34°23'00"N

Thermionics Laboratory Inc. (TLI)
P.O. Box 3711
Hayward, CA 94540
Telephone: 415-538-3304
Telefax: 415-538-2889
Electronic Mail: info@thermionics.com
sales@thermionics.com
WWW: http://www.thermionics.com/
Founded: 1958
Staff: 100
Activities: manufacturing high and ultra high vacuum systems, components, and hardware
City Reference Coordinates: 122°05'00"W 37°40'00"N

The Tech Museum of Innovation
201 South Market Street
San Jose, CA 95113
Telephone: 408-294-TECH
408-795-6100
Telefax: 408-279-7167

Electronic Mail: info@thetech.org
WWW: http://www.thetech.org/
Awards: (1) "The Tech Museum Awards" (five categories)
City Reference Coordinates: 121°53'00"W 37°20'00"N

Thin Film Technology (TFT)
153 Industrial Way
P.O. Box 1942
Buellton, CA 93427-1942
Telephone: 805-688-4949
Telefax: 805-688-8487
Electronic Mail: info@thinfilmtechnology.com
WWW: http://www.thinfilmtechnology.com/
Founded: 1972
Staff: 25
Activities: coating services
City Reference Coordinates: 120°13'00"W 34°38'00"N

TL Systems
2184 Primrose Avenue
Vista, CA 92083
Telephone: 760-599-4219
Telefax: 760-599-4219
Electronic Mail: tlsystem@ix.netcom.com
WWW: http://www.netcom.com/~tlsystem/
Founded: 1976
Staff: 4
Activities: manufacturing telescope equatorial platforms and altazimuth platforms * producing telescope tracking software
City Reference Coordinates: 117°15'00"W 33°12'00"N

Trango Systems Inc.
9939 Via Pasar
San Diego, CA 92126
Telephone: 858-653-3900
Telefax: 858-621-2725
Electronic Mail: trango@zcomm.com
WWW: http://www.trangosys.com/
Activities: manufacturing wireless video systems for telescopes
City Reference Coordinates: 117°09'00"W 32°43'00"N

Tri-Valley Stargazers (TVS)
P.O. Box 2476
Livermore, CA 94551-2476
Telephone: 510-661-4249 (Dave Anderson, President)
Electronic Mail: tvs@trivalleystargazers.org
WWW: http://www.trivalleystargazers.org/
Founded: 1979
Membership: 185
Activities: meetings * public, park and school star parties * education * observing * telescope making * rental telescopes * robotic solar observatory
Periodicals: (12) "Prime Focus"
Coordinates: 121°45'00"W 37°40'00"N H175m (Eyes on the Skies)
           121°00'00"W 37°00'00"N (Sky Shack)
           121°28'34"W 37°24'02"N (Digger Pines)
City Reference Coordinates: 121°46'00"W 37°41'00"N

Tulare Astronomical Association (TAA)
P.O. Box 515
Tulare, CA 93274
Telephone: 559-685-0585
Electronic Mail: sixlights@earthlink.net
WWW: http://home.earthlink.net/~sixlights/
Founded: 1967
City Reference Coordinates: 119°22'00"W 36°10'00"N

Tule River Amateur Astronomers (TRAA)
c/o David Chandler
P.O. Box 999
Springville, CA 93265
Telephone: 559-539-0900
Telefax: 559-539-7033
Electronic Mail: traa@davidchandler.com
WWW: http://www.pc.cc.ca.us/Students/Astronomy/astro.html
City Reference Coordinates: 119°01'00"W 36°04'00"N (Porterville)

University of California at Los Angeles (UCLA), Department of Earth and Space Sciences
P.O. Box 951567
Los Angeles, CA 90095-1567
or :
595 Circle Drive East
3806 Geology Building
Los Angeles, CA 90024
Telephone: 310-825-3880
Telefax: 310-825-2779
Electronic Mail: <userid>@ess.ucla.edu
WWW: http://www.ess.ucla.edu/
Founded: 1919
Staff: 50
Activities: earthquakes * geochemistry * planetary physics * geology * space physics * paleontology * geophysics * meteoritics * seismology * tectonics * solar system * dynamics * spacecraft instrumentation * solar wind
City Reference Coordinates: 118°15'00"W 34°03'00"N

University of California at Los Angeles (UCLA), Department of Physics and Astronomy, Division of Astronomy
8371 Mathematical Sciences Building
Los Angeles, CA 90025-1562
Telephone: 310-825-4434
Telefax: 310-206-2096
Electronic Mail: <userid>@bonnie.astro.ucla.edu
WWW: http://www.astro.ucla.edu/
    http://www.ee.ucla.edu/people/hkn/planetarium/ (UCLA Planetarium)
Founded: 1932 (as "Department of Astronomy")
Staff: 14
Activities: education * research
City Reference Coordinates: 118°15'00"W 34°03'00"N

University of California at Los Angeles (UCLA), Institute of Geophysics and Planetary Physics (IGPP)
Los Angeles, CA 90095-1567
Telephone: 310-825-1580
Telefax: 310-206-3051
WWW: http://www-ssc.igpp.ucla.edu/
    http://www.igpp.ucla.edu/
Founded: 1946
Periodicals: (1) "Annual Report"
City Reference Coordinates: 118°15'00"W 34°03'00"N

University of California at Riverside (UCR), Institute of Geophysics and Planetary Physics (IGPP)
Riverside, CA 92521
Telephone: 909-787-4503
Telefax: 909-787-4509
Electronic Mail: <userid>@ucrphys.ucr.edu
WWW: http://cnas.ucr.edu/~igpp/home.html
    http://tigre.ucr.edu/astro.html (High-Energy Astrophysics)
Founded: 1967
Activities: research * education
City Reference Coordinates: 117°22'00"W 33°59'00"N

University of California at San Diego (UCSD), Center for Astrophysics and Space Sciences (CASS)
9500 Gilman Drive
La Jolla, CA 92093-0424
Telephone: 858-534-3460 (Arthur M. Wolfe, Director)
Telefax: 858-534-2294
Electronic Mail: <userid>@ucsd.edu
WWW: http://casswww.ucsd.edu/
Founded: 1979
Staff: 100
Activities: observational work (X-ray, UV, optical, IR) * theoretical astrophysics * space physics * solar physics * solar-system studies
City Reference Coordinates: 117°52'00"W 33°51'00"N

University of California at San Diego (UCSD), Physics Department
9500 Gilman Drive
Mail Code 0354
La Jolla, CA 92093-0319
Telephone: 619-534-6438
Telefax: 619-534-7419
Electronic Mail: <userid>@ucsd.edu
WWW: http://www-physics.ucsd.edu/
Founded: 1959
City Reference Coordinates: 117°52'00"W 33°51'00"N

University of California at Santa Barbara (UCSB), Physics Department
Broida Hall
Building 572
Santa Barbara, CA 93106-9530
Telephone: 805-893-3888
Telefax: 805-893-3307
Electronic Mail: physics@physics.ucsb.edu
WWW: http://www.physics.ucsb.edu/
http://www.deepspace.ucsb.edu/
Founded: 1960
City Reference Coordinates: 119°42'00"W 34°25'00"N

University of California at Santa Barbara (UCSB), Institute for Theoretical Physics
Kohn Hall
Santa Barbara, CA 93106-4030
Telephone: 805-893-4111
Telefax: 805-893-2431
Electronic Mail: <userid>@itp.ucsb.edu
WWW: http://www.itp.ucsb.edu/
Founded: 1979
Staff: 4
Activities: theoretical physics
City Reference Coordinates: 119°42'00"W 34°25'00"N

University of California at Santa Cruz (UCSC), Department of Astronomy and Astrophysics
1156 High Street
Santa Cruz, CA 95064
Telephone: 408-429-2844
Telefax: 408-426-3115
Electronic Mail: board@helios.ucsc.edu
<userid>@ucolick.org
WWW: http://www.astro.ucsc.edu/
Founded: 1964
City Reference Coordinates: 122°01'00"W 36°58'00"N

University of California at Santa Cruz (UCSC), Lick Observatory
Mount Hamilton Road
P.O. Box 85
Mount Hamilton, CA 95140
Telephone: 408-274-5062
831-459-2513
Telefax: 408-270-9653
831-426-3115
Electronic Mail: <userid>@ucolick.org
WWW: http://www.ucolick.org/
Founded: 1888
Staff: 75
Activities: QSOs * AGNs * star forming regions * stellar populations * galaxy dynamics and formation * development of new instrumentation
Periodicals: "Bulletin", "Technical Reports"
Coordinates: 121°38'12"W 37°20'36"N H1,290m (Mount Hamilton – IAU Code 662)
City Reference Coordinates: 121°38'00"W 37°21'00"N

University of California at Santa Cruz (UCSC), Lick Observatory, Headquarters
c/o Natural Sciences II
1156 High Street
Santa Cruz, CA 95064
Telephone: 831-459-2513
Telefax: 831-459-5244
Electronic Mail: director@ucolick.org
<userid>@ucolick.org
WWW: http://www.ucolick.org/
Founded: 1888
Staff: 150
Activities: education * research
Periodicals: "Lick Observatory Bulletins" (ISSN 0075-9325)
Coordinates: 121°38'12"W 37°20'36"N H1,290m (Mount Hamilton)
City Reference Coordinates: 122°01'00"W 36°58'00"N

University of Southern California (USC), Department of Physics and Astronomy
Los Angeles, CA 90089-0484
Telephone: 213-740-0848
Telefax: 213-740-6653
Electronic Mail: physdept@usc.edu
<userid>@usc.edu
WWW: http://physics.usc.edu/

Activities: helioseismology * stellar ages * solar-stellar connection * galactic evolution * history of astronomy
Observatories: 1 (Mount Wilson Observatory)
City Reference Coordinates: 118°15'00"W 34°03'00"N

University of Southern California (USC), Space Sciences Center
SHS 274 MC-1341
Los Angeles, CA 90089
Telephone: 213-740-6334
Telefax: 213-740-6342
Electronic Mail: <userid>@usc.edu
WWW: http://www.usc.edu/dept/space_science/sscindex.htm
City Reference Coordinates: 118°15'00"W 34°03'00"N

Valley of the Moon Observatory Association (VMOA)
P.O. Box 898
Glen Ellen, CA 95442
Telephone: 707-833-6979
Electronic Mail: gloyer@wco.com (George Loyer, President)
WWW: http://www.phys-astro.sonoma.edu/VMOA/
http://yorty.sonoma.edu/VMOA/
Founded: 1995
Membership: 300
Activities: popularization * education
Periodicals: (4) "Focused"
City Reference Coordinates: 122°31'00"W 38°22'00"N

Vanguard Research Inc., Astronomy and Astrophysics
5321 Scotts Valley Drive
Suite 204
Scotts Valley, CA 95066
Telephone: 408-438-7233
Telefax: 408-438-8930
WWW: http://www.vrisv.com/
Founded: 1981 (as "Jamieson Science and Engineering (JSE) Inc."")
Staff: 10
City Reference Coordinates: 122°02'00"W 37°03'00"N

Ventura County Astronomical Society (VCAS)
P.O. Box 982
Simi Valley, CA 93062
Telephone: 805-529-3184
805-520-9666
Electronic Mail: vcas@aol.com
WWW: http://www.vcas.org/
Founded: 1967
Membership: 210
Activities: meetings * lectures * observing sessions
Periodicals: (12) "Celestial Horizons"
City Reference Coordinates: 118°05'00"W 34°10'00"N (Thousand Oaks)

Victor Valley Community College (VVCC), Planetarium
c/o Scott Bryan (Director)
18422 Bear Valley Road
Victorville, CA 92392-5849
Telephone: 760-245-4271 x324
Telefax: 760-245-9745
Electronic Mail: tsbryan@aol.com (Scott Bryan, Planetarium Director)
bryans@victor.cc.ca.us
WWW: http://www.vvo.com/comm/fop.htm
City Reference Coordinates: 117°18'00"W 34°32'00"N

Western Observatorium Inc.
4141 Ball Road
Suite 212
Cypress, CA 90630-3400
Telephone: 714-220-1310
Electronic Mail: pyrrho@cogent.net
WWW: http://www.cogent.net/~pyrrho/wohome.htm
Founded: 1976
Membership: 30
Activities: education * star parties * popularization
Periodicals: "The Western Observer", "Deep-Sky Objects of the Month"
Coordinates: 116°23'56"W 34°17'00"N H1,000m (Landers)
City Reference Coordinates: 118°01'00"W 33°50'00"N

Westmont College, Department of Physics
955 La Paz Road
Santa Barbara, CA 93108-1099
Telephone: 805-565-6000
Telefax: 805-565-6220
Electronic Mail: kihlstr@westmont.edu (Kenneth Kihlstrom)
WWW: http://physics.westmont.edu/
http://physics.westmont.edu/research/observatory (Observatory)
Founded: 1957
Staff: 3
Activities: observing (George Carroll Observatory)
Coordinates: 119°39'36"W 34°26'54"N H902m
City Reference Coordinates: 119°42'00"W 34°25'00"N

What in The World
P.O. Box 1767
Lake Arrowhead, CA 92352
or :
28200 State Highway 189
Suite E-210
Lake Arrowhead, CA 92352
Telephone: 909-337-5080
Telefax: 909-336-9708
Electronic Mail: bobbell@js-net.com (Robert L. Bell)
WWW: http://www.whatintheworld.com/
Founded: 1992
Staff: 4
Activities: manufacturing astronomical accessories * distributing telescopes and accessories
Periodicals: (4) "Newsletter"
Coordinates: 117°11'42"W 34°15'24"N H1575m (Chateau du Lac Observatory)
City Reference Coordinates: 117°12'00"W 34°15'00"N

Wilcox Solar Observatory (WSO)
• See "Stanford University, Center for Space Sciences and Astrophysics, Wilcox Solar Observatory (WSO)"

William Knox Holt Planetarium
c/o Lawrence Hall of Science
University of California
Berkeley, CA 94720-5200
Telephone: 510-643-5082
Telefax: 510-642-1055
Electronic Mail: agould@uclink4.berkeley.edu
WWW: http://www.lhs.berkeley.edu/
http://www.lhs.berkeley.edu/Planetarium.html
Founded: 1973
Staff: 10
Activities: school and public shows * classes * astronomy curriculum development * bi-monthly public stargazing
Periodicals: (3) "The LHS Quarterly"; (1) "LHS Programs for Schools"
Coordinates: 122°09'00"W 37°46'00"N H300m (Lawrence Hall of Science Plaza)
City Reference Coordinates: 122°09'00"W 37°46'00"N

Woodstock Products Inc.
P.O. Box 2519
Beverly Hills, CA 90213
Telephone: 323-650-6602
Electronic Mail: info@woodstockprod.com
WWW: http://www.woodstockprod.com/
Activities: manufacturing large full-color prints of NASA's outer-space photography
City Reference Coordinates: 118°26'00"W 34°03'00"N

X-Ray Instrumentation Associates (XIA)
8450 Central Avenue
Newark, CA 94560
Telephone: 510-494-9020
Telefax: 510-494-9040
WWW: http://www.xia.com/
Activities: producing advanced X-ray and gamma-ray detector electronics and related instruments with applications in research and industry
City Reference Coordinates: 122°03'00"W 37°31'00"N

USA - Colorado

Air Force Academy Planetarium
c/o Center for Educational Multimedia
34 ES/CEMM
Cadet Drive 2120
USAF Academy
Colorado Springs, CO 80840-5566
Telephone: 719-472-2779
    719-472-2777
    719-472-2778
Telefax: 719-472-3494
WWW: http://www.frii.com/~mmoss/usafa_cemm.htm
    http://www.usafa.af.mil/dfp/research/astro/
Founded: 1959
Activities: supporting cadet education * programs for visitors and local schools
Coordinates: 104°52'30"W 39°00'24"N H2,187m
    104°53'33"W 39°00'34"N H2,187m
City Reference Coordinates: 104°49'00"W 38°50'00"N

American Educational Products Inc.
401 West Hickory Street
P.O. Box 2121
Fort Collins, CO 80522
Telephone: 970-484-7445
Telefax: 970-484-1198
Electronic Mail: suef@amep.com (Susan R. Fresnel, International Sales Manager)
WWW: http://amep.com/
Staff: 75
Activities: manufacturing and distributing educational products
City Reference Coordinates: 105°05'00"W 40°35'00"N

Apollo-Foco International Inc.
2806 1/2 Bookcliff Avenue
Grand Junction, CO 81506
Telephone: 970-255-9233
Telefax: 970-256-7940
Electronic Mail: <userid>@apollo-foco.com
WWW: http://www.apollo-foco.com/
Activities: manufacturing optical accessories
City Reference Coordinates: 108°33'00"W 39°05'00"N

Aspen Center for Physics (ACP)
600 West Gillepsie Street
Aspen, CO 81611-1243
Telephone: 970-925-2585
Telefax: 970-920-1167
Electronic Mail: aspen@andy.bu.edu
WWW: http://andy.bu.edu/aspen/
    http://www.aspenphys.org/
Founded: 1961
City Reference Coordinates: 106°49'00"W 39°11'00"N

AstroSystems Inc.
124 North 2nd Street
La Salle, CO 80645
Telephone: 970-284-9471
Telefax: 970-284-9473
Electronic Mail: astrosys@frii.com
WWW: http://www.frii.com/~astrosys/
Founded: 1987
Staff: 4
Activities: manufacturing and distributing telescope components
Awards: (4-6) "Door Prizes"
City Reference Coordinates: 104°42'00"W 40°21'00"N

Aurora Astronomical Association (A3)
c/o Leroy Guatney
11390 East Louisiana Avenue
Aurora, CO 80012-4170
Telephone: 303-873-6552
Electronic Mail: lwlg@usa.net
WWW: http://a_cubed.tripod.com/

Founded: 2001
City Reference Coordinates: 104°50'00"W 39°45'00"N

BioServe Space Technologies (BST)
c/o University of Colorado
Campus Box 429
Boulder, CO 80309
Telephone: 303-492-1005
Telefax: 303-492-8883
WWW: http://www.colorado.edu/engineering/BioServe/
Founded: 1987
Staff: 10
Activities: life sciences R&D in space * related apparatus development tasks * NASA Center for the Commercial Development of Space (CCDS)
Periodicals: (1) "Annual Report"
City Reference Coordinates: 105°17'00"W 40°01'00"N

Black Forest Observatory (BFO)
12815 Porcupine Lane
Colorado Springs, CO 80908-3510
Telephone: 719-495-3828
Electronic Mail: bfo@observatory.org
WWW: http://www.observatory.org/
Founded: 1986
Coordinates: 104°48'09"W 39°01'10"N H2470m
City Reference Coordinates: 104°49'00"W 38°50'00"N

Center for Astrophysics and Space Astronomy (CASA)
• "University of Colorado, Center for Astrophysics and Space Astronomy (CASA)"

Charles C. Gates Planetarium
c/o Denver Museum of Natural History (DMNH)
2001 Colorado Boulevard
Denver, CO 80205
Telephone: 303-370-6317
Telefax: 303-331-6492
Electronic Mail: dasquin@mercury.cair.du.edu
WWW: http://www.dmnh.org/plntrm.htm
Founded: 1955
City Reference Coordinates: 105°01'00"W 39°43'00"N

Collins Electro Optics
9025 East Kenyon Avenue
Denver, CO 80237
Telephone: 303-889-5910
Telefax: 303-779-8496
Electronic Mail: billc@ceoptics.com
WWW: http://www.ceoptics.com/
Activities: manufacturing image intensifiers
City Reference Coordinates: 105°01'00"W 39°43'00"N

Colorado College, Department of Physics
14 East Cache La Poudre
Colorado Springs, CO 80903
Telephone: 719-389-6577
Telefax: 719-389-6429
Electronic Mail: physics@cc.colorado.edu
WWW: http://www2.coloradocollege.edu/dept/PC/
http://www2.coloradocollege.edu/dept/PC/astronomy/
Observatories: 1 (Gerald Hughes Phipps Observatory)
City Reference Coordinates: 104°49'00"W 38°50'00"N

Colorado Springs Astronomical Society (CSAS)
P.O. Box 62022
Colorado Springs, CO 80917
Telephone: 719-591-4926
719-593-0801
Electronic Mail: bygrens@codenet.net
WWW: http://www.rmss.org/
Founded: 1985
Membership: 50
Activities: monthly meetings * monthly star parties * public star parties * school education
Periodicals: (12) "Hypoxic Observer"
Coordinates: 105°00'00"W 38°50'00"N H2,750m
City Reference Coordinates: 104°49'00"W 38°50'00"N

Colorado State University, Department of Physics
Fort Collins, CO 80523
Telephone: 303-491-6206
Telefax: 303-491-7947
WWW: http://www.physics.colostate.edu/
http://www.colostate.edu/Depts/Physics/
Founded: 1965
Activities: physical characteristics of cool stars
City Reference Coordinates: 105°05'00"W 40°35'00"N

Colutron Research Corp.
2321 Yarmouth Avenue
Boulder, CO 80301
Telephone: 303-443-5211
Telefax: 303-449-5050
Electronic Mail: sales@colutron.com
technical@colutron.com
WWW: http://www.colutron.com/products/
Founded: 1964
Activities: research in cosmology, atomic physics and atmospheric physics
City Reference Coordinates: 105°17'00"W 40°01'00"N

Deep Space Exploration Society (DSES)
c/o Rex Craig
5921 Niwot Road
Niwot, CO 80503
WWW: http://www.dses.org/
City Reference Coordinates: 105°06'00"W 40°10'00"N (Longmont)

Denver Astronomical Society (DAS)
c/o Chamberlin Observatory
2930 East Warren Avenue
Denver, CO 80208
Telephone: 303-871-5172
Electronic Mail: bxkz36a@prodigy.com (Debra Davis, Observer Editor)
WWW: http://members.tripod.com/denverastro
Founded: 1952
Membership: 250
Activities: meetings * public nights * open houses * star parties
Periodicals: (12) "Denver Observer"
Coordinates: 104°57'12"W 39°40'36"N H1,644m (Chamberlin Obs. – IAU Code 707)
City Reference Coordinates: 105°01'00"W 39°43'00"N

DFM Engineering Inc.
1035 Delaware Avenue
Unit D
Longmont, CO 80501
Telephone: 303-443-2031
303-678-8143
Electronic Mail: dfm42@csn.net
WWW: http://www.dfmengineering.com/
Founded: 1976
Staff: 10
Activities: manufacturing astronomical telescopes
City Reference Coordinates: 105°06'00"W 40°10'00"N

Fiske Planetarium and Science Center (Wallace_)
• See "Wallace Fiske Planetarium and Science Center"

Galaxy Optics
P.O. Box 2045
Buena Vista, CO 81211
Telephone: 719-395-8242
Electronic Mail: galaxy@chaffee.net
WWW: http://castle.chaffee.net/~Egalaxy/
Activities: telescope optics
City Reference Coordinates: 106°08'00"W 38°50'00"N

Gates Planetarium (Charles C._)
• See "Charles C. Gates Planetarium"

Geological Society of America (GSA)
P.O. Box 9140
Boulder, CO 80301-9140
or :

3300 Penrose Place
Boulder, CO 80301-1806
Telephone: 303-447-2020
          1-888-443-4472 (USA only)
Telefax: 303-447-1070
Electronic Mail: <userid>@geosociety.org
WWW: http://www.geosociety.org/
Founded: 1888
Membership: 16000
City Reference Coordinates: 105°17'00"W 40°01'00"N

High Altitude Observatory (HAO)
• See "National Center for Atmospheric Research (NCAR), High Altitude Observatory (HAO)"

International Association for Astronomical Studies (IAAS)
P.O. Box 55
Golden, CO 80402-0055
Telephone: 303-349-IAAS (Chris Rand)
          303-349-4227
Electronic Mail: iaas@iaas.org
WWW: http://www.iaas.org/
Activities: astronomy * youth education * rocketry * amateur radio
City Reference Coordinates: 105°15'00"W 39°45'00"N

International Association of Geomagnetism and Aeronomy (IAGA)
c/o Herb Kroehl (Secretary-General)
Solar-Terrestrial Physics Division
325 Broadway
Boulder, CO 80303
Telephone: 303-497-6323
Telefax: 303-497-6513
Electronic Mail: hkroehl@ngdc.noaa.gov
WWW: http://www.ngdc.noaa.gov/IAGA/iagahome.html
Founded: 1954 (1919 as "IUGG Section on Terrestrial Magnetism and Electricity")
Membership: 76 (countries)
Activities: promoting studies of magnetism, aeronomy, and space plasma physics of the Earth, planets and the interplanetary medium of the solar system
Periodicals: (1) "IAGA News"; "IAGA Bulletin"
Awards: "Long-Service Award"
City Reference Coordinates: 105°17'00"W 40°01'00"N

International Council of Scientific Unions (ICSU), Panel on World Data Centres
World Data Center
325 Broadway
MC E/GC
Boulder, CO 80303
Telephone: 303-497-7284
Telefax: 303-497-6513
Electronic Mail: wdc@ngdc.noaa.gov
WWW: http://www.ngdc.noaa.gov/wdc/
Founded: 1957
Activities: coordinating and providing direction to the worldwide network of WDC
Periodicals: "Guide to the World Data Center System"
City Reference Coordinates: 105°17'00"W 40°01'00"N

Jim's Mobile Inc. (JMI)
810 Quail Street
Unit E
Lakewood, CO 80215
Telephone: 303-233-5353
Telefax: 303-233-5359
Electronic Mail: info@jimsmobile.com
WWW: http://www.jimsmobile.com/
Founded: 1983
Activities: manufacturing telescopes and accessories
City Reference Coordinates: 105°06'00"W 39°44'00"N

Joint Institute for Laboratory Astrophysics (JILA)
c/o University of Colorado
Campus Box 440
Boulder, CO 80309-0440
Telephone: 303-492-7789
Telefax: 303-492-5235
Electronic Mail: <userid>@jila.colorado.edu
WWW: http://jilawww.colorado.edu/

Founded: 1962
Staff: 225
Activities: atomic and molecular physics * laser physics * chemical physics * optical resonance * radiative transfer * galactic and extragalactic astronomy * gravitational physics
• Jointly operated by the "National Institute of Standards and Technology (NIST)" and the "University of Colorado"
City Reference Coordinates: 105°17'00"W 40°01'00"N

Las Brisas Observatory (LBO)
P.O. Box 6069
Colorado Springs, CO 80934
Telephone: 719-632-2829
Electronic Mail: lbosky@aol.com
WWW: http://lbo.teuton.org/
Founded: 1982
Staff: 1
• Private observatory
City Reference Coordinates: 104°49'00"W 38°50'00"N

Little Thompson Observatory (LTO)
P.O. Box 930
Berthoud, CO 80513
WWW: http://www.starkids.org/
City Reference Coordinates: 105°05'00"W 40°19'00"N

Longmont Astronomical Society (LAS)
P.O. Box 806
Longmont, CO 80502-0806
Telephone: 303-776-1338 (Secretary)
303-440-5469 (Newsletter Editor)
WWW: http://laps.fsl.noaa.gov/cgi/albers_las.homepage.cgi
Founded: 1987
Membership: 35
Activities: observing * monthly meetings
Periodicals: (12) "Longmont Astronomical Society Journal"
City Reference Coordinates: 105°06'00"W 40°10'00"N

Mars Society (The_)
• See "The Mars Society"

National Center for Atmospheric Research (NCAR)
P.O. Box 3000
Boulder, CO 80307-3000
Telephone: 303-497-1000
Telefax: 303-497-1194
Electronic Mail: <userid>@ncar.ucar.edu
WWW: http://www.ncar.ucar.edu/
http://www.ucar.edu/
Founded: 1960
Staff: 1150
Activities: weather and climate studies * atmospheric chemistry * mesoscale and microscale meteorology * STP * providing service and research facilities to the research community beyond its walls such as supercomputing power and state-of-the-art instruments
Periodicals: (12) "Staff Notes"; (1) "Annual Scientific Report"; "NCAR Technical Notes"
• Managed by the University Corp. for Atmospheric Research (UCAR) (see separate entry)
City Reference Coordinates: 105°17'00"W 40°01'00"N

National Center for Atmospheric Research (NCAR), High Altitude Observatory (HAO)
1850 Table Mesa Drive
P.O. Box 3000
Boulder, CO 80307-3000
Telephone: 303-497-1527
Telefax: 303-497-1589
303-497-1568
Electronic Mail: <userid>@hao.ucar.edu
WWW: http://www.hao.ucar.edu/
http://www.hao.ucar.edu/public/research/mlso/mlso_homepage.html (MLSO)
Founded: 1940
Staff: 65
Activities: coronal and interplanetary physics * solar activity and magnetic fields * solar interior * terrestrial interactions
Observatories: 1 (Mauna Loa Solar Observatory, USA-HI - MLSO)
City Reference Coordinates: 105°17'00"W 40°01'00"N

National Geophysical Data Center (NGDC)
c/o Solar-Terrestrial Physics Division
National Oceanic and Atmospheric Administration (NOAA)

NOAA/NGDC
325 Broadway
Boulder, CO 80303-3328
Telephone: 303-497-6826
Telefax: 303-497-6513
Electronic Mail: info@ngdc.noaa.gov
WWW: http://www.ngdc.noaa.gov/ngdc.html
Founded: 1968
Staff: 17
Activities: collecting, archiving and disseminating solar, interplanetary, cosmic-ray, geomagnetic, ionospheric, and aurora data from the international network of observatories
Periodicals: (12) "Solar-Geophysical Data" (ISSN 0038-0911), "Solar Indices Bulletin" (ISSN 1046-1914), "Geomagnetic Indices Bulletin"; (1) "International Geophysical Calendar"; "UAG Report Series"
City Reference Coordinates: 105°17'00"W 40°01'00"N

National Institute of Standards and Technology (NIST), Joint Institute for Laboratory Astrophysics (JILA)
• See "Joint Institute for Laboratory Astrophysics (JILA)"

National Oceanic and Atmospheric Administration (NOAA), Environmental Research Laboratories (ERL), Cooperative Institute for Research in Environmental Sciences (CIRES)
• See "Cooperative Institute for Research in Environmental Sciences (CIRES)"

National Oceanic and Atmospheric Administration (NOAA), Solar-Terrestrial Physics Division
NOAA/NGDC
325 Broadway
Boulder, CO 80303-3328
Telephone: 303-497-6324
Telefax: 303-497-6513
Electronic Mail: <userid>@ngdc.noaa.gov
<userid>@nesdis (telemail)
WWW: http://www.noaa.gov/
http://www.ngdc.noaa.gov/ngdc.html
Founded: 1970 (NOAA)
City Reference Coordinates: 105°17'00"W 40°01'00"N

National Oceanic and Atmospheric Administration (NOAA), Space Environment Center (SEC)
325 Broadway
Boulder, CO 80303
Telephone: 303-497-5127
Telefax: 303-497-7392
303-497-3645
Electronic Mail: swo@sec.noaa.gov
<userid>@sec.noaa.gov
WWW: http://www.sec.noaa.gov/
Founded: 1965
Staff: 65
Activities: solar-terrestrial research (sunspots, flares, corona, solar magnetic fields, solar cycle, interplanetary medium, real-time prediction of solar-terrestrial disturbances) * space weather * 24hrs forecasting operations
Periodicals: (52) "Preliminary Report and Forecast of Solar-Geophysical Activity"
City Reference Coordinates: 105°17'00"W 40°01'00"N

North American Skies
6874 East Harvard Avenue
Denver, CO 80224-2502
Telephone: 303-691-2172
Electronic Mail: starman@usa.net
WWW: http://www.webcom.com/safezone/NAS/
Founded: 1990
City Reference Coordinates: 105°01'00"W 39°43'00"N

Northern Colorado Astronomical Society (NCAS)
c/o David Chamness (President)
2920 Mount Royal Court
Fort Collins, CO 80526
or :
c/o Dan Laszlo (Newsletter Editor)
2001 South Shields Street
Fort Collins, CO 80526
Telephone: 970-498-9226
Electronic Mail: djlaszlo@aol.com
WWW: http://ncastro.org/
Membership: 60
Activities: monthly meetings * star parties
Periodicals: "The Objective View"

City Reference Coordinates: 105°05'00"W 40°35'00"N

Picosecond Pulse Labs (PSPL) Inc.
2500 55th Street
Boulder, CO 80301
Telephone: 303-443-1249
Telefax: 303-447-2236
Electronic Mail: info@picosecond.com
WWW: http://www.picosecond.com/
Founded: 1980
Staff: 7
Activities: manufacturing pulse generators and broadband coaxial components
City Reference Coordinates: 105°17'00"W 40°01'00"N

Popular Science Magazine
P.O. Box 51282
Boulder, CO 80322-1282
Telephone: 1-800-289-9399
Electronic Mail: reader@popsci.com
WWW: http://www.popsci.com/
Founded: 1872
- Journal

City Reference Coordinates: 105°17'00"W 40°01'00"N

Scientific Committee on Solar-Terrestrial Physics (SCOSTEP)
c/o Joe H. Allen (Scientific Secretary)
NOAA/NGDC
325 Broadway
Boulder, CO 80303-3328
Telephone: 303-497-7284
Telefax: 303-497-6513
Electronic Mail: jallen@ngdc.noaa.gov
WWW: http://www.ngdc.noaa.gov/stp/SCOSTEP/scostep.html
Founded: 1966
Staff: 2
Periodicals: "STP Newsletter"
City Reference Coordinates: 105°17'00"W 40°01'00"N

Software Bisque Inc.
912 Twelfth Street
Golden, CO 80401-1114
Telephone: 303-278-4478
          1-800-843-7599 (USA only)
Telefax: 303-278-0045
Electronic Mail: <userid>@bisque.com
WWW: http://www.bisque.com/
Founded: 1984
Staff: 3
- Software producer

City Reference Coordinates: 105°13'00"W 39°46'00"N

Southwest Research Institute (SwRI), Department of Space Studies
1050 Walnut Street
Suite 426
Boulder, CO 80302
Telephone: 303-546-9670
Telefax: 303-546-9687
Electronic Mail: <userid>@boulder.swri.org
WWW: http://www.boulder.swri.edu/
Founded: 1994
Staff: 13
Activities: Sun * stars * dynamics * UV * impacts * craters * minor planets * comets
City Reference Coordinates: 105°17'00"W 40°01'00"N

Space Science Institute (SSI)
3100 Marine Street
Suite A353
Boulder, CO 80303-1058
Telephone: 303-492-3774
Telefax: 303-492-3789
Electronic Mail: info@spacescience.org
WWW: http://www.spacescience.org/
Founded: 1994
Staff: 25
Activities: space science research and education * museum exhibit design

City Reference Coordinates: 105°17'00"W 40°01'00"N

Star Chaser Observatory
c/o Tom Arnold
14729 West Creek Road
Sedalia, CO 80135
Telephone: 303-647-0484
Electronic Mail: tarnold@starchaser-obs.com
WWW: http://www.starchaser-obs.com
City Reference Coordinates: 104°57'00"W 39°25'00"N

The Mars Society
P.O. Box 273
Indian Hills, CO 80454
Electronic Mail: rzubrin@marssociety.org (Robert Zubrin, President)
WWW: http://www.marssociety.org/
Founded: 1998
Activities: furthering exploration and settlement of Mars
City Reference Coordinates: 105°01'00"W 39°43'00"N

United States Space Foundation
2860 South Circle Drive
Suite 2301
Colorado Springs, CO 80906-4184
Telephone: 719-576-8000
Telefax: 719-576-8801
WWW: http://www.spacefoundation.org/
Founded: 1983
Staff: 25
Activities: National Space Symposium * education
City Reference Coordinates: 104°49'00"W 38°50'00"N

University Corp. for Atmospheric Research (UCAR)
P.O. Box 3000
Boulder, CO 80307-3000
or :
1850 Table Mesa Drive
Boulder, CO 80303
Telephone: 303-497-1650
Telefax: 303-497-1654
Electronic Mail: <userid>@ucar.edu
WWW: http://www.ucar.edu/
Founded: 1959
Membership: 90 (institutions)
Staff: 950
Activities: group of 59 institutions with doctoral-level programs in atmospheric, meteorological and related sciences
Periodicals: "UCAR Newsletter", "UCAR Corporate Report"
City Reference Coordinates: 105°17'00"W 40°01'00"N

University of Colorado, Center for Astrophysics and Space Astronomy (CASA)
U.C.B. 389
Boulder, CO 80309-0389
Telephone: 303-492-4050
Telefax: 303-492-7178
Electronic Mail: site-info@casa.colorado.edu
WWW: http://casa.colorado.edu/
    http://lyra.colorado.edu/sbo/ (Sommers-Bausch Observatory - SBO)
Founded: 1985
Staff: 75
Activities: multiwavelength spectroscopy * space instrumentation * theoretical astrophysics
Periodicals: (4) "Preprints"
Coordinates: 105°15'45"W 40°00'13"N H1,653m (Sommers-Bausch Obs. – IAU Code 463)
City Reference Coordinates: 105°17'00"W 40°01'00"N

University of Colorado, Center for Astrophysics and Space Astronomy (CASA), Astrophysics Research Laboratory (ARL)
Campus Box 593
Boulder, CO 80309-0593
Telephone: 303-492-2331
Telefax: 303-492-5941
Electronic Mail: site-info@casa.colorado.edu
WWW: http://casa.colorado.edu/
    http://casa.colorado.edu/sbo/ (Sommers-Bausch Observatory - SBO)
    http://www.colorado.edu/
Founded: 1985

City Reference Coordinates: 105°17'00"W 40°01'00"N

University of Colorado, Cooperative Institute for Research in Environmental Sciences (CIRES)
• See "Cooperative Institute for Research in Environmental Sciences (CIRES)"

University of Colorado, Department of Aerospace Engineering Sciences
Campus Box 429
Boulder, CO 80309-0429
Telephone: 303-492-6417
Telefax: 303-492-7881
Electronic Mail: <userid>@spot.colorado.edu
WWW: http://aerospace.colorado.edu/
Founded: 1947
Staff: 75
Activities: astrodynamics * atmospheric and oceanic sciences * bioengineering * controls and systems * fluids * global positioning * remote sensing * spacecraft design * structures
City Reference Coordinates: 105°17'00"W 40°01'00"N

University of Colorado, Department of Astrophysical and Planetary Sciences (APS)
Campus Box 391
Boulder, CO 80309-0391
Telephone: 303-492-8913
          303-492-8915
Telefax: 303-492-5105
         303-492-3822
Electronic Mail: <userid>@colorado.edu
                 admin@aps.colorado.edu
WWW: http://aps.colorado.edu/
     http://www.colorado.edu/fiske/ (Wallace Fiske Planetarium - see separate entry)
     http://stripe.colorado.edu/~planet/Home.html (Wallace Fiske Planetarium)
     http://casa.colorado.edu/sbo/ (Sommers-Bausch Observatory - SBO)
     http://lyra.colorado.edu/sbo/ (Sommers-Bausch Observatory - SBO)
Founded: 1997 (1980 as "Astrophysical, Planetary and Atmospheric Sciences" - APAS)
Staff: 24
Activities: Sun * hot stars * cool stars * UV and radio astronomy * planets * ISM * extragalactic astronomy * theoretical astrophysics * Wallace Fiske Planetarium (see separate entry)
Awards: "Billings Award"
Coordinates: 105°15'45"W 40°00'13"N H1,653m (Sommers-Bausch Observatory)
City Reference Coordinates: 105°17'00"W 40°01'00"N

University of Colorado, Joint Institute for Laboratory Astrophysics (JILA)
• See "Joint Institute for Laboratory Astrophysics (JILA)"

University of Colorado, Laboratory for Atmospheric and Space Physics (LASP)
1234 Innovation Drive
Boulder, CO 80309-7814
or :
Campus Box 392
Boulder, CO 80309-0392
Telephone: 303-492-6412
Telefax: 303-492-6444
Electronic Mail: ssl@sslab.colorado.edu
WWW: http://lasp.colorado.edu/
     http://laspwww.colorado.edu:7777/lasp_homepage.html
     http://www.colorado.edu/
Founded: 1948
Staff: 25
Activities: planets * atmosphere * solar physics
City Reference Coordinates: 105°17'00"W 40°01'00"N

University of Colorado at Colorado Springs (UCCS), Department of Physics
1420 Austin Bluffs Parkway
Colorado Springs, CO 80933-7150
Telephone: 1-800-990-UCCS
           719-262-3164
Electronic Mail: physics@serf.uccs.edu
                 physics@mail.uccs.edu
WWW: http://www.uccs.edu/~physics/
     http://www.uccs.edu/~physics/obs/
     http://www.uccs.edu/~ppo/ (Pikes Peak Observatory - PPO)
Founded: 1981
City Reference Coordinates: 104°49'00"W 38°50'00"N

University of Denver, Department of Physics and Astronomy
2112 East Wesley Avenue

Denver, CO 80208-0202
Telephone: 303-871-2238
Telefax: 303-871-4405
Electronic Mail: <userid>@du.edu
WWW: http://www.du.edu/physastron/
http://www.du.edu/~rstencel/MtEvans (Mt. Evans Meyer-Womble Observatory)
Activities: education * public outreach
Coordinates: 104°57'12"W 39°40'36"N H1,644m (Chamberlin Obs. – IAU Code 707)
105°38'24"W 39°35'12"N H4,313m (Mt. Evans Meyer-Womble Observatory)
City Reference Coordinates: 105°01'00"W 39°43'00"N

Van Slyke Engineering
12815 Porcupine Lane
Colorado Springs, CO 80908
Telephone: 719-495-3828
Telefax: 719-495-3828
Electronic Mail: bfo@observatory.org
WWW: http://www.observatory.org/vsengr.htm
Activities: manufacturing telescope accessories
City Reference Coordinates: 104°49'00"W 38°50'00"N

Wallace Fiske Planetarium and Science Center
c/o University of Colorado
Campus Box 408
Boulder, CO 80309
Telephone: 303-492-5002
Telefax: 303-492-1725
Electronic Mail: planet@colorado.edu
WWW: http://www.colorado.edu/fiske
Founded: 1974
Activities: astronomy educational entertainment
City Reference Coordinates: 105°17'00"W 40°01'00"N

Western Colorado Astronomy Club (WCAC)
P.O. Box 55032
Grand Junction, CO 81505
Telephone: 970-256-9907
Electronic Mail: areid@mesastate.edu (Aaron Reid)
WWW: http://www.wic.net/WCAC/
Founded: 1989
Membership: 45
Activities: meetings * observing sessions for public and schools
City Reference Coordinates: 108°33'00"W 39°05'00"N

World Data Center A for Solar-Terrestrial Physics
c/o National Oceanic and Atmospheric Administration
NOAA/NGDC
NOAA Code E/GC2
325 Broadway
Boulder, CO 80303
Telephone: 303-497-6324
Telefax: 303-497-6513
Electronic Mail: info@ngdc.noaa.gov
WWW: http://web.ngdc.noaa.gov/stp/WDC/wdcstp.html
Founded: 1957
Staff: 20
Activities: STP data collection, processing, archival and dissemination
City Reference Coordinates: 105°17'00"W 40°01'00"N

USA - Connecticut

Astronomical Society of Greater Hartford (ASGH)
P.O. Box 2271
Hartford, CT 06145-2271
Telephone: 860-521-0474
Electronic Mail: vega1@aol.com
WWW: http://www.asgh.org/
http://www.cshore.com/royce/asgh/
Periodicals: (12) "Star Trails"
City Reference Coordinates: 072°45'00"W 41°46'00"N

Astronomical Society of New Haven (ASNH)
c/o Linda Marks
P.O. Box 16234
New Haven, CT 06516
Telephone: 203-874-4816 (President)
Telefax: 203-481-5769
Electronic Mail: prez@asnh.org
WWW: http://www.asnh.org/
Founded: 1937
Membership: 70
Activities: observing * education * research
Periodicals: (12) "Newsletter"
Coordinates: 073°00'00"W 41°25'37"N H150m (Bethany)
City Reference Coordinates: 072°56'00"W 41°18'00"N

Astroptx Wholesale Optics
59 Mine Hill Road
New Milford, CT 06776
Telephone: 860-355-3132
Telefax: 860-355-3132
Electronic Mail: astroptx@astroptx.com
WWW: http://www.astroptx.com/
Founded: 1977
City Reference Coordinates: 073°25'00"W 41°35'00"N

Bethany Sciences LLC
• See "The Universe Collection"

Connecticut College, Department of Physics, Astronomy, and Geophysics
270 Mohegan Avenue
P.O. Box 5441
New London, CT 06320-4196
Telephone: 860-439-2345
Telefax: 860-439-5011
Electronic Mail: <userid>@conncoll.edu
WWW: http://www.conncoll.edu/ccacad/physics.web/
http://www.conncoll.edu/academics/departments/physics.web/
http://www.conncoll.edu/ccacad/physics.web/obs.html (Olin Observatory)
http://www.conncoll.edu/academics/departments/physics.web/obs.html (" ")
Staff: 8
Activities: education * observing * research
Coordinates: 072°06'19"W 41°22'44"N (Olin Observatory)
City Reference Coordinates: 072°07'00"W 41°21'00"N

Discovery Museum, Henry B. duPont III Planetarium
4450 Park Avenue
Bridgeport, CT 06604
Telephone: 203-372-3521
Telefax: 203-374-1929
WWW: http://www.discoverymuseum.org/
http://www.discoverymuseum.org/Planetarium/planetarium.htm
City Reference Coordinates: 073°13'00"W 41°12'00"N

Foran High School, Planetarium (Joseph A._)
• See "Joseph A. Foran High School, Planetarium"

Gengras Planetarium
c/o Science Center of Connecticut
950 Trout Brook Drive
West Hartford, CT 06119

Telephone: 860-231-2824
Telefax: 860-232-0705
WWW: http://www.sciencecenterct.org/
Founded: 1967
Membership: 93
Activities: education for youth * laser shows * public shows
City Reference Coordinates: 072°45'00"W 41°46'00"N

Henry B. duPont III Planetarium
• See "Discovery Museum, Henry B. duPont III Planetarium"

Hughes Danbury Optical Systems (HDOS) Inc.
• See now "Raytheon Optical Systems Inc."

Joseph A. Foran High School, Planetarium
80 Foran Drive
Milford, CT 06460
Telephone: 203-783-3400
Telefax: 203-783-3635
Electronic Mail: lmarks@milforded.org (Linda Marks)
WWW: http://www.milforded.org/schools/foran/fhs/home.htm
Founded: 1973
Staff: 2
Activities: planetarium shows * observing programs for students
City Reference Coordinates: 073°04'00"W 41°13'00"N

Mystic Seaport Planetarium
75 Greenmanville Avenue
P.O. Box 6000
Mystic, CT 06355
Telephone: 860-572-5339
Telefax: 860-572-5315
Electronic Mail: <userid>@mysticseaport.org
WWW: http://www.mysticseaport.org/public/education/planetarium/home.html
Founded: 1960
City Reference Coordinates: 071°58'00"W 41°21'00"N

North American Sundial Society (NASS)
c/o Fred Sawyer (Editor)
8 Sachem Drive
Glastonbury, CT 06033
Telephone: 203-633-8655
Electronic Mail: 71541.1662@compuserve.com
WWW: http://www.shadow.net/~bobt/nass/nass.htm
Founded: 1994
Membership: 175
Activities: registration and documentation of sundials * annual convention
Periodicals: (4) "Compendium" (ISSN 1074-3197), "Digital Compendium" (ISSN 1074-8059)
City Reference Coordinates: 072°37'00"W 41°43'00"N

Omega Engineering Inc.
One Omega Drive
P.O. Box 4047
Stamford, CT 06907-0047
Telephone: 1-800-848-4286 (USA and Canada only)
         1-800-82-TC-OMEGA (" " ")
         203-359-1660
Telefax: 203-359-7700
Electronic Mail: info@omega.com
WWW: http://omega.com/
Activities: manufacturing process measurement and control products
City Reference Coordinates: 073°32'00"W 41°03'00"N

PIC Design
86 Benson Road
Middlebury, CT 06762
Telephone: 203-758-8272
         1-800-243-6125 (USA only)
Telefax: 203-758-8271
Electronic Mail: info@pic-design.com
               sales@pic-design.com
WWW: http://www.pic-design.com/
Founded: 1954
Staff: 105
Activities: manufacturing precision mechanical components

City Reference Coordinates: 073°07'00"W 41°32'00"N

Precision Industrial Components (PIC) Corp.
● See "PIC Design"

Radio Research Instrument (RRI) Co. Inc.
c/o E.B. Doyle
584 North Main Street
Waterbury, CT 06704-3506
Telephone: 203-753-5840
Telefax: 203-754-2567
Electronic Mail: radiores@prodigy.com
WWW: http://www.techexpo.com/WWW/radiores/radiores.html
Founded: 1952
Staff: 20
Activities: manufacturing microwave equipments and various ancillary type items including radar
City Reference Coordinates: 073°02'00"W 41°33'00"N

Raytheon Optical Systems Inc.
100 Wooster Heights Road
Danbury, CT 06810-7589
Telephone: 203-797-5266
Telefax: 203-797-5114
WWW: http://www.raytheon.com/
Founded: 1937
Staff: 700
Activities: designing, manufacturing and testing optics, telescopes, adaptive optical systems and scientific instruments
● Formerly "Hughes Danbury Optical Systems (HDOS) Inc."
City Reference Coordinates: 073°27'00"W 41°23'00"N

Robert K. Wickware Planetarium
c/o Eastern Connecticut State University (ECSU)
83 Windham Street
Willimantic, CT 06226
Telephone: 860-465-5300
Electronic Mail: pazameta@ecsuc.ctstateu.edu (Zoran Pazameta, Director)
WWW: http://nutmeg.ctstateu.edu/depts/phs/planets/frame.htm
Founded: 1972
Staff: 1
Activities: education * presentations
City Reference Coordinates: 072°13'00"W 41°43'00"N

Southern Connecticut State University (SCSU), Planetarium
501 Crescent Street
New Haven, CT 06515
Telephone: 203-392-5841
Electronic Mail: fullmer@scsu.ctstateu.edu
WWW: http://www.southernct.edu/
Founded: 1960
Staff: 1
Activities: shows * exhibits * weather station
Coordinates: 072°56'55"W 41°20'02"N H25m
City Reference Coordinates: 072°56'00"W 41°18'00"N

Talcott Mountain Science Center (TMSC)
Montevideo Road
Avon, CT 06001
Telephone: 860-677-8571
Electronic Mail: talcott@tmsc.org
WWW: http://www.tmsc.org/
Activities: planetarium shows
City Reference Coordinates: 072°50'00"W 41°49'00"N

Thames Amateur Astronomical Society (TAAS)
c/o Bob Coggeshall (President)
8 Old Fitch Hill Road
Uncasville, CT 06382
or :
c/o William Falcone
21 High Street
Noank, CT 06346
Telephone: 860-536-1358
Electronic Mail: cogzal@aol.com
              wbf@idt.net
WWW: http://pages.cthome.net/taas/

City Reference Coordinates: 072°06'00"W 41°26'00"N
072°01'00"W 41°19'00"N (Noank)

The Universe Collection
(Bethany Sciences LLC)
P.O. Box 3726
Amity Station
New Haven, CT 06525-0726
Telephone: 203-393-3395
1-800-525-1052 (USA only)
Telefax: 203-393-2457
Electronic Mail: bethanysci@aol.com
WWW: http://www.universecollection.com/
Activities: retail of meteorites, tektites and related educational materials * space jewelry * posters * videos * books
City Reference Coordinates: 072°56'00"W 41°18'00"N

United Technologies Research Center (UTRC)
Mail Stop 43
411 Silver Lane
East Hartford, CT 06108
Telephone: 860-610-7000
Telefax: 203-727-7134
Electronic Mail: <userid>@utrc.utc.com
WWW: http://utrcwww.utc.com/
Founded: 1929
Staff: 850
Activities: fluid mechanics * chemical sciences (including combustion and environmental sciences) * embedded electronic systems * materials and structures * product development and manufacturing * information technology * dynamic systems and controls
City Reference Coordinates: 072°39'00"W 41°46'00"N

Universe Collection (The_)
● See "The Universe Collection"

University of Connecticut, Department of Physics
2152 Hillside Road
Storrs, CT 06269-3046
Telephone: 860-486-3346
Telefax: 860-486-4915
Electronic Mail: <userid>@uconnvm.uconn.edu
WWW: http://www.physics.uconn.edu/
http://www.physics.uconn.edu/~cynthia/planet.html (Planetarium)
http://mjolnir.phys.uconn.edu/~cynthia/obs.html (Observatory)
Founded: 1940
City Reference Coordinates: 072°15'00"W 41°48'00"N

Van Vleck Observatory (VVO)
● See "Wesleyan University, Department of Astronomy, Van Vleck Observatory (VVO)"

Wesleyan University, Department of Astronomy, Van Vleck Observatory (VVO)
Middletown, CT 06459-0123
Telephone: 860-685-2130
Telefax: 860-685-2131
Electronic Mail: <userid>@astro.wesleyan.edu
WWW: http://www.astro.wesleyan.edu/
Founded: 1831
Staff: 7
Coordinates: 072°39'36"W 41°33'18"N H65m
City Reference Coordinates: 072°39'00"W 41°34'00"N

Western Connecticut Chapter of the Society for Amateur Scientists (WCCSAS)
P.O. Box 1144
New Milford, CT 06776
Telephone: 860-354-1595
Telefax: 860-354-1595
Electronic Mail: mrobson@snet.net (Monty Robson, President)
WWW: http://www.wccsas.org/
Founded: 1997
Activities: operating the John J. McCarthy Observatory (JJMO – IAU Code 932)
City Reference Coordinates: 073°25'00"W 41°35'00"N

Western Connecticut State University (WCSU), Department of Physics, Astronomy and Meteorology
181 White Street
Danbury, CT 06810

Telephone: 203-837-8669
Telefax: 203-837-8320
Electronic Mail: <userid>@wcsu.ctstateu.edu
WWW: http://www.wcsu.ctstateu.edu/physics/homepage.html
http://www.wcsu.ctstateu.edu/physics/astronomy.html
Activities: education * research * Midtown Observatory * Westside Observatory * planetarium shows
Coordinates: 073°26'42"W 41°24'00"N H128m
City Reference Coordinates: 073°27'00"W 41°23'00"N

Westport Astronomical Society Inc. (WAS)
182 Bayberry Lane
Westport, CT 06880-2802
Telephone: 203-227-0925
WWW: http://www.was.visionnet.com/
Founded: 1975
Membership: 100
Periodicals: (12) "Field of View"
Observatories: 1 (Rolnick Observatory)
City Reference Coordinates: 073°22'00"W 41°09'00"N

Wickware Planetarium (Robert K._)
• See "Robert K. Wickware Planetarium"

Yale University, Department of Astronomy
P.O. Box 208101
New Haven, CT 06520-8101
or :
c/o J. Willard Gibbs Research Laboratory
260 Whitney Avenue
New Haven, CT 06511
Telephone: 203-432-3000
Telefax: 203-432-5048
Electronic Mail: <userid>@astro.yale.edu
WWW: http://www.astro.yale.edu/
Founded: 1932
Staff: 16
Activities: research * education
Periodicals: "Transactions"
Coordinates: 072°56'00"W 41°18'00"N (IAU Code 797)
073°00'00"W 41°25'37"N H150m (Bethany Obs. – IAU Code 798)
069°18'00"W 31°47'57"S H2,550m (Leoncito Southern Stn – IAU Code 808)
City Reference Coordinates: 072°56'00"W 41°18'00"N

Yale University, Department of Physics
Whitney Avenue
New Haven, CT 06511
Electronic Mail: <userid>@yale.edu
WWW: http://www.yale.edu/physics/
http://wnsl.physics.yale.edu/ (A.W. Wright Nuclear Structure Laboratory)
Founded: 1882
Staff: 94
City Reference Coordinates: 072°56'00"W 41°18'00"N

Yale University Press
P.O. Box 209040
New Haven, CT 06520-9040
Telephone: 203-432-0960
1-800-987-7323 (Orders, USA only)
Telefax: 203-432-0948
WWW: http://www.yale.edu/yup/
Founded: 1908
• Publisher
City Reference Coordinates: 072°56'00"W 41°18'00"N

USA - Delaware

Bartol Research Institute (BRI)
• See "University of Delaware, Bartol Research Institute (BRI)"

Delaware Astronomical Society (DAS)
P.O.Box 652
Wilmington, DE 19899
or :
c/o Mount Cuba Astronomical Observatory
Hillside Mill Road
P.O. Box 3915
Greenville, DE 19807
Telephone: 302-234-0740
           302-654-6407 (Mount Cuba Astronomical Observatory)
Telefax: 302-451-4596 (c/o Billie Westergard)
WWW: http://www.physics.udel.edu/MCAO/das.htm
Founded: 1956
Membership: 155
Activities: lectures * field trips * education * amateur astronomer exchanges * international collaborations * Mount Cuba Astronomical Planetarium
Periodicals: (12) "The Focus"
Awards: "Amateur Astronomer of the Year"
Observatories: 1 (Mount Cuba Astronomical Observatory - see separate entry)
City Reference Coordinates: 075°33'00"W 39°44'00"N (Wilmington)

Delmarva Star Gazers
514 Marilyn Road
Smyrna, DE 19977
Telephone: 302-653-9445 (Don Surles, President)
Electronic Mail: surlesd@engg.dnet.dupont.com
WWW: http://www.dol.net/~frank.sheldon/gazers.html
       http://www.delmarvastargazers.org/
Founded: 1993
City Reference Coordinates: 075°36'00"W 39°18'00"N

Mount Cuba Astronomical Observatory (MCAO) Inc.
1610 Hillside Mill Road
P.O. Box 3915
Greenville, DE 19807
Telephone: 302-654-6407
Electronic Mail: mtcuba@udel.edu
WWW: http://www.physics.udel.edu/MCAO/
Founded: 1957
Staff: 2
Activities: providing facilities for research and education * minor-planet and comet astrometry * photoelectric photometry of flare stars * double stars (visual colours, spectral classification) * lunar occultations * planetarium
Periodicals: (1) "Annual Report"
Coordinates: 075°38'00"W 39°47'06"N H91m (IAU Code 788)
City Reference Coordinates: 075°38'00"W 39°47'00"N

Starwalk Planetarium
Colonial School District
Chase Avenue
New Castle, DE 19720
Telephone: 302-429-4013
Telefax: 302-429-4005
Electronic Mail: hbouchel@udel.edu (Henry E.W. Bouchelle III, Director)
WWW: http://users.eclipsetel.com/~hbouchelle/starwalk.htm
Founded: 1969
Staff: 1
City Reference Coordinates: 075°34'00"W 39°40'00"N

University of Delaware, Bartol Research Institute (BRI)
217 Sharp Laboratory
Newark, DE 19716
Telephone: 302-831-8111
Telefax: 302-831-1843
Electronic Mail: pittel@bartol.udel.edu (Stuart Pittel, Director)
               <userid>@bartol.udel.edu
WWW: http://www.bartol.udel.edu/
Founded: 1987 (1927 at Swarthmore, USA-PA)
Staff: 50

Activities: cosmic-ray and neutrino monitoring * space plasma physics * elementary-particle astrophysics
Observatories: 4 (South Pole, McMundo, Thule, Newark)
City Reference Coordinates: 075°45'00"W 39°41'00"N

University of Delaware, Department of Physics and Astronomy
223 Sharp Laboratory
Newark, DE 19716
Telephone: 302-831-2661
Telefax: 302-831-1637
Electronic Mail: <userid>@physics.udel.edu
WWW: http://www.physics.udel.edu/
Founded: 1833 (University)
Staff: 30
Activities: stellar activity * stellar evolution * variable stars * white dwarfs * extragalactic radio astronomy * stellar spectroscopy * photometry
City Reference Coordinates: 075°45'00"W 39°41'00"N

USA - District of Columbia

Aerospace Industries Association (AIA)
1250 Eye Street NW
Suite 1200
Washington, DC 20005-3924
Telephone: 202-371-8400
Telefax: 202-371-8470
Electronic Mail: <userid>@aia-aerospace.org
WWW: http://www.aia-aerospace.org/index.cfm
Founded: 1959 (1919 as "Aeronautical Chamber of Commerce of America")
City Reference Coordinates: 077°01'00"W 38°54'00"N

Air Force Office of Scientific Research (AFOSR)
Bolling AFB
Washington, DC 20332-8050
Telephone: 202-767-5017
Telefax: 202-767-0466
Electronic Mail: afosr-m@fie.com
WWW: http://www.afosr.afrl.af.mil/
Founded: 1947
Staff: 150
Activities: aerospace sciences * chemistry * electronics * life sciences * mathematics * physics
Periodicals: (1) "Research Interests", "Research Highlights"
City Reference Coordinates: 077°01'00"W 38°54'00"N

Albert Einstein Planetarium
• See "Smithsonian Institution, National Air and Space Museum (NASM), Albert Einstein Planetarium"

American Association for the Advancement of Science (AAAS)
1200 New York Avenue NW
Washington, DC 20005
Telephone: 202-326-6400
Telefax: 202-842-1711
Electronic Mail: info@aaas.org
WWW: http://www.aaas.org/
  http://www.sciencemag.org/ (Science Magazine)
Founded: 1848
Membership: 138000
Activities: furthering the work of scientists * facilitating cooperation between them * fostering scientific freedom and responsibility * improving the effectiveness of science in the promotion of human welfare * increasing public understanding and appreciation of the importance and promise of the methods of science in human progress * publishing books * organizing meetings, seminars, colloquia and workshops
Sections: Agriculture (O) * Anthropology (H) * Astronomy (D) * Atmospheric and Hydrospheric Sciences (W) * Biological Sciences (G) * Chemistry (C) * Dentistry (R) * Education (Q) * Engineering (M) * General Interest in Science & Engineering (Y) * Geology & Geography (E) * History & Philosophy of Science (L) * Industrial Science (P) * Information, Computing & Communication (T) * Mathematics (A) * Medical Sciences (N) * Pharmaceutical Sciences (S) * Physics (B) * Psychology (J) * Social, Economic, & Political Sciences (K) * Societal Impacts of Science & Engineering (X) * Statistics (U)
Periodicals: (52) "Science Magazine" (ISSN 0036-8075)
Awards: "Hilliard Roderick Proze", "Newcomb Cleveland Prize", "Philip Hauge Abelson Prize", "Prize for Behavorial Science Research", "Scientific Freedom and Responsibility Award", "Westinghouse Award for Public Understanding of Science and Technology", "Westinghouse Science Journalism Awards"
City Reference Coordinates: 077°01'00"W 38°54'00"N

American Astronomical Society (AAS)
2000 Florida Avenue NW
Suite 400
Washington, DC 20009-1231
Telephone: 202-328-2010
Telefax: 202-234-2560
Electronic Mail: aas@aas.org
WWW: http://www.aas.org/
  http://www.journals.uchicago.edu/ApJ/ (Astrophysical Journal)
  http://www.noao.edu/apjl/apjl.html (Astrophysical Journal Letters)
  http://www.astro.washington.edu/astroj/ (Astronomical Journal)
Founded: 1899
Membership: 5600
Activities: promoting advancement of astronomy and closely related branches of science * meetings * grant programs * public lectures * job services
Sections: (divisions) Planetary Sciences * Solar Physics * Dynamical Astronomy * High-Energy Astrophysics * Historical Astronomy
Sections: (committees) Annie J. Cannon Award * Astronomy News * Audit * Chrétien International Research Grant * Astronomy and Public Policy * Employment * Light Pollution, Radio Interference and Space Debris * Status of Women in

Activities: museum fund raising
Periodicals: "Flyer"
City Reference Coordinates: 077°01'00"W 38°54'00"N

National Capital Astronomers (NCA)
• See USA-MD

National Earth Orientation Service (NEOS)
• See "US Naval Observatory (USNO), National Earth Orientation Service (NEOS)"

National Research Council (NRC), Board on Physics and Astronomy (BPA)
2101 Constitution Avenue
Washington, DC 20418
Telephone: 202-334-3520
Telefax: 202-334-3575
Electronic Mail: bpa@nas.edu
WWW: http://www.nas.edu/bpa/
http://www.nas.edu/bpa/caa.html (CAA)
http://www.nas.edu/bpa/corf.html (CORF)
Founded: 1983 (CORF: 1961) (CAA: 1992)
Staff: 6
Activities: astronomy * astrophysics * critical issues * science policy * radio frequency matters * physics
Sections: (committees) Committee on Radio Frequencies (CORF) * US National Committee for the International Astronomical Union (IAU) * Committee on Astronomy and Astrophysics (CAA) * US National Committee for the International Union of Radio Science (URSI) * US Liaison Committee for the International Union of Pure and Applied Physics (IUPAP) * Plasma Science Committee * Committee on Atomic, Molecular, and Optical Sciences
Periodicals: "BPA News"
City Reference Coordinates: 077°01'00"W 38°54'00"N

National Space Society (NSS)
600 Pennsylvania Avenue SE
Suite 201
Washington, DC 20003
Telephone: 202-543-1900
Telefax: 202-546-4189
Electronic Mail: nsshq@nss.org
WWW: http://www.nss.org/
Founded: 1974
Membership: 23000
Staff: 6
Activities: supporting space exploration and development through public education and political activism * launch tours
Periodicals: (6) "Ad Astra" (ISSN 1041-102x)
City Reference Coordinates: 077°01'00"W 38°54'00"N

Naval Research Laboratory (NRL), Space Sciences Division (SSD)
Code 7600
4555 Overlook Avenue SW
Washington, DC 20375-5000
Telephone: 202-767-6343
Telefax: 202-404-7296
Electronic Mail: <userid>@bradbury.nrl.navy.mil
<userid>@ssd0.nrl.navy.mil
WWW: http://spacescience.nrl.navy.mil/
Founded: 1958
Staff: 133
Activities: upper atmospheric, solar, and astronomical research * Backgrounds Data Center (BDC) * X-ray Astronomy Branch * Gamma and Cosmic Ray Astrophysics Branch * Solar Physics Branch
City Reference Coordinates: 077°01'00"W 38°54'00"N

Optical Society of America (OSA)
2010 Massachusetts Avenue NW
Washington, DC 20036-1023
Telephone: 202-223-8130
1-800-582-0416
Telefax: 202-223-1096
Electronic Mail: osamem@osa.org
WWW: http://www.osa.org/
http://www.osa.org/homepage.html
Founded: 1916
Membership: 12000
Activities: all branches of pure and applied optics
Periodicals: (36) "Applied Optics" (see separate entry); (24) "Optics Letters" (ISSN 0146-9592); (12) "Journal of the Optical Society of America A: Optics, Image Science and Vision" (ISSN 0740-3232), "Journal of the Optical Society of America B: Optical Physics" (ISSN 0740-3224), "Journal of Optical Technology" (ISSN 0038-5514), "Optics & Photonics News" (ISSN

0098-907x), "Optics and Spectroscopy" (ISSN 0030-400x), "Journal of Lightware Technology" (ISSN 0733-8724); (6) "Chinese Journal of Lasers B"
Awards: 13 different awards
City Reference Coordinates: 077°01'00"W 38°54'00"N

Organization of American States (OAS), Office of Science and Technology
1889 F Street NW
Washington, DC 20006-4499
Telephone: 202-458-3368
Telefax: 202-458-3167
Electronic Mail: pi@oas.org
WWW: http://www.oas.org/
Founded: 1890
Activities: strengthening peace and security in the region * promoting cooperation in science and technology
Periodicals: "Monografías" (ISSN 0078-6322)
Awards: (1) "Bernard Houssay", "Manuel Noriega Morales"
City Reference Coordinates: 077°01'00"W 38°54'00"N

Rock Creek Park Nature Center, Planetarium
5200 Glover Road NW
Washington, DC 20015
Telephone: 202-426-6829
WWW: http://sdcd.gsfc.nasa.gov/ISTO/DLT/GUIDE/rock_creek.html
http://www.nps.gov/rocr/planetarium/
City Reference Coordinates: 077°01'00"W 38°54'00"N

Science Service Inc.
1719 N Street NW
Washington, DC 20036-2888
Telephone: 202-785-2255
Telefax: 202-785-1243
WWW: http://www.sciserv.org/
http://www.sciencenews.org/ (Science News Online)
Founded: 1921
Staff: 45
Periodicals: (52) "Science News" (ISSN 0036-8423, circ.: 250,000)
• Publisher
City Reference Coordinates: 077°01'00"W 38°54'00"N

Smithsonian Institution, Division of Meteorites
c/o National Museum of Natural History
Mail Stop 119
Washington, DC 20560
Telephone: 202-357-1478
Telefax: 202-357-2476
Electronic Mail: <userid>@si.edu
WWW: http://www.nmnh.si.edu/minsci/meteor.htm
Founded: 1846 (Smithsonian Institution)
Staff: 4
Activities: managing meteorites collections for research and educational purposes * US national meteorite collection * research * education
Periodicals: "Smithsonian Contributions to the Earth Sciences"
City Reference Coordinates: 077°01'00"W 38°54'00"N

Smithsonian Institution, Global Volcanism Program (GVP)
c/o National Museum of Natural History
Room E-421
Washington, DC 20560-0129
Telephone: 202-357-1511
Telefax: 202-357-2476
Electronic Mail: <userid>@volcano.si.edu
WWW: http://www.volcano.si.edu/gvp/
Founded: 1975
Membership: 1100 (correspondents)
Activities: gathering and reporting information about worldwide volcanic activity
Periodicals: (12) "Bulletin of the Global Volcanism Network"
City Reference Coordinates: 077°01'00"W 38°54'00"N

Smithsonian Institution, National Air and Space Museum (NASM)
6th Street & Independence Avenue SW
Washington, DC 20560-0321
Telephone: 202-357-1529
Telefax: 202-786-2262
Electronic Mail: info@info.si.edu
WWW: http://www.nasm.si.edu/

Founded: 1976
Staff: 320
Activities: museum of aeronautical and space exploration * Earth and planetary research * Albert Einstein Planetarium (see separate entry) * curatorial research * public information * exhibits
Periodicals: "Air & Space" (ISSN 0888-2257, circ.: 400,000)
Awards: (1) "National Air and Space Museum Trophy"
City Reference Coordinates: 077°01'00"W 38°54'00"N

Smithsonian Institution, National Air and Space Museum (NASM), Albert Einstein Planetarium
6th Street & Independence Avenue SW
Room 3338
Washington, DC 20560
Telephone: 202-357-1529
Telefax: 202-357-4579
Electronic Mail: cheryl.bauer@nasm.si.edu (Cheryl Bauer, Planetarium Director)
   <userid>@nasm.si.edu
WWW: http://www.nasm.edu/NASMDOCS/PLANETARIUM/
   http://www.nasm.edu/
Founded: 1976
Staff: 8
Activities: public lectures * The Stars Tonight * Monthly Sky Lecture * pre-programmed multimedia astronomy-type shows * live demonstrations * school demonstrations * teachers workshops * weekly starwatchers report
City Reference Coordinates: 077°01'00"W 38°54'00"N

Smithsonian Institution, National Air and Space Museum (NASM), Center for Earth and Planetary Studies (CEPS)
6th Street & Independence Avenue SW
Room 3389
Washington, DC 20560
WWW: http://ceps.nasm.edu:2020/homepage.html
   http://www.nasm.edu/
   http://www.si.edu/
Founded: 1983 (NASM: 1976)
City Reference Coordinates: 077°01'00"W 38°54'00"N

Smithsonian Institution, National Air and Space Museum (NASM), Space History Division
6th Street & Independence Avenue SW
Washington, DC 20560-0321
Telephone: 202-357-2828
Telefax: 202-786-2262
WWW: http://www.nasm.edu/nasm/dsh/shdiv.htm
Founded: 1976 (NASM)
Staff: 12
Activities: writing, lecturing, collecting artifacts, and preparing exhibitions in the following fields: rocketry, computers and avionics, human spaceflight, satellites and commercial spacecraft, military space, ground- and space-based astronomy, and foreign space programs
City Reference Coordinates: 077°01'00"W 38°54'00"N

Space Data Resources & Information (SDR&I)
P.O. Box 23883
Washington, DC 20026-3883
Telephone: 202-546-0363
Telefax: 202-546-0132
WWW: http://www.reston.com/sdri/sdri.html
Founded: 1956
Staff: 3
Activities: consulting * research * free-lance writing on space-related activities * weekly updates to various magazines and newspapers
City Reference Coordinates: 077°01'00"W 38°54'00"N

US Naval Observatory (USNO)
3450 Massachusetts Avenue NW
Washington, DC 20392-5420
Telephone: 202-762-1467
   202-762-1438 (Public Affairs Information)
   202-762-1401 (Time Voice Announcer)
   202-762-1069 (Time Voice Announcer)
   202-762-1594 (Digital Time for Modems)
   202-762-1610 (Automated Data Service)
   202-762-1602 (Automated Data Service)
   202-762-1503 (Automated Data Service)
   202-762-1425 (Astrometry Department)
   202-762-1481 (Time Service)
Telefax: 202-762-1516 (Astrometry Department)
   202-762-1511 (Time Service)
Electronic Mail: usno@usno.navy.mil

grc@usno.navy.mil (Geoff Chester, Public Affairs Office)
WWW: http://www.usno.navy.mil/
http://aa.usno.navy.mil/AA/ (Astronomy Applications Department)
http://ad.usno.navy.mil/ (Astrometry Department)
Founded: 1830
Staff: 120
Activities: astrometry * precise time * Earth rotation * celestial mechanics * solar-system ephemerides * almanacs * instrumentation
Periodicals: (1) "The Astronomical Almanac" (ISSN 0737-6421), "Astronomical Phenomena" (ISSN 0083-2421); (2) "The Nautical Almanac" (ISSN 0077-6194), "The Air Almanac" (ISSN 0400-8456); "Publications of the US Naval Observatory" (ISSN 0083-2448)
Coordinates: 077°04'02"W 38°55'18"N H92m (Washington – IAU Code 786)
111°44'24"W 35°11'03"N H2,316m (Flagstaff – IAU Code 689)
City Reference Coordinates: 077°01'00"W 38°54'00"N

US Naval Observatory (USNO), National Earth Orientation Service (NEOS)
3450 Massachusetts Avenue NW
Washington, DC 20392-5420
Telephone: 202-762-1565
Telefax: 202-762-1563
Electronic Mail: <userid>@maia.usno.navy.mil
WWW: http://www.usno.navy.mil/
http://maia.usno.navy.mil/
Founded: 1985
Staff: 16
Activities: Earth orientation data and their prediction
Periodicals: (52) "IERS Bulletin A"; (1) "Annual Report"
City Reference Coordinates: 077°01'00"W 38°54'00"N

USA - Florida

Alachua Astronomy Club (AAC) Inc.
P.O. Box 13744
Gainesville, FL 32604-1744
or :
c/o Howard L. Cohen (Editor)
1501 NW 28th Street
Gainesville, FL 32605-5037
Telephone: 352-481-5238
Electronic Mail: cohenba@astro.ufl.edu
        fcl@nersp.nerdc.ufl.edu
WWW: http://www.astro.ufl.edu/aac/
Founded: 1987
Membership: 60
Activities: amateur astronomy * education
Periodicals: (12) "FirstLight"
Awards: "G.H. Russell Price"
Coordinates: 082°21'00"W 29°39'00"N H50m
City Reference Coordinates: 082°20'00"W 29°40'00"N

Ancient City Astronomy Club (ACAC)
P.O. Box 546
Saint Augustine, FL 32085-0546
Telephone: 904-446-4338
WWW: http://www.rightguide.com/space/orgs/ancient.htm
Founded: 1974
Membership: 30
Activities: observing * public sessions
Periodicals: (6) "Cosmic Echoes"
Observatories: 1 (St. Augustine High School Observatory)
City Reference Coordinates: 081°19'00"W 29°54'00"N

Ash Enterprises International Inc.
c/o John Hare (President)
3602 23rd Avenue West
Bradenton, FL 34205
Telephone: 941-746-3522
Telefax: 941-750-9497
Electronic Mail: jlhare@aol.com
WWW: http://www.ash-enterprises.com/
Founded: 1970
Staff: 2
Activities: providing technical and design services to planetariums * manufacturing planetarium special effects
• See also USA-VA
City Reference Coordinates: 082°34'00"W 27°29'00"N

Astronaut Hall of Fame
6225 Vectorspace Boulevard
Titusville, FL 32780
Telephone: 407-269-6100
Telefax: 407-267-3970
Electronic Mail: ahof@spacecamp.com
WWW: http://www.astronauts.org/
City Reference Coordinates: 080°49'00"W 28°37'00"N

Astronaut Memorial Planetarium and Observatory (AMPO)
c/o Brevard Community College
1519 Clearlake Road
Cocoa, FL 32922
Telephone: 321-634-3732
Telefax: 321-634-3744
WWW: http://www.brevard.cc.fl.us/~planet/
Founded: 1976
Staff: 10
Activities: planetarium shows * US Space Camp * observing * laser concerts and demonstrations * Starlab portable planetarium
Periodicals: (4) "Focal Point"
Coordinates: 080°45'42"W 28°23'04"N H10m (IAU Code 758)
City Reference Coordinates: 080°44'00"W 28°21'00"N

Astronomical Society of the Palm Beaches (ASPB)
c/o Charlie Fredrickson

P.O. Box 19652
West Palm Beach, FL 33416
Telephone: 561-471-0476
561-471-0932
Electronic Mail: aspb@geocities.com
cfredri921@aol.com
WWW: http://www.gopbi.com/community/groups/aspb/
Membership: 100
City Reference Coordinates: 080°06'00"W 26°49'00"N

Barbara Moore Observatory (James and_)
• See "James and Barbara Moore Observatory"

Bishop Planetarium
201 10th Street West
Bradenton, FL 34205
Telephone: 941-746-4132
Telefax: 941-747-2556
Electronic Mail: info@sfmbp.org
WWW: http://www.sfmbp.org/
Founded: 1966
Staff: 6
Activities: educational programs for schools and general public * public observing
Periodicals: (12) "Bishop Planetarium Sky Reporter"; (4) "Navigator"
Coordinates: 082°34'00"W 27°30'00"N H14m
City Reference Coordinates: 082°34'00"W 27°30'00"N

Brevard Astronomical Society (BAS) Inc.
P.O. Box 1084
Cocoa, FL 32922
or :
c/o Brad Lythe
802 Kara Circle
Rockledge, FL 32955
Telephone: 407-452-2645
Electronic Mail: roystarman@aol.com (Roy Uyematsu, Vice President)
starladyva@aol.com (Virginia Gallenberger, Secretary)
WWW: http://personal.lig.bellsouth.net/lig/g/p/gpippin/bas.htm
Founded: 1983
Membership: 30
Activities: meetings * star parties * public education
Periodicals: (6) "Sky Scanner"
City Reference Coordinates: 080°44'00"W 28°21'00"N (Cocoa)
080°43'00"W 28°20'00"N (Rockledge)

Buehler Planetarium
c/o Broward Community College
Central Campus
3501 SW Davie Road
Davie, FL 33314
Telephone: 954-475-6681
Telefax: 954-475-2858
WWW: http://fs.broward.cc.fl.us/central/buehler/
Founded: 1966
Staff: 12
Activities: public and school shows * public observing * mobile astronomy program * science outreach program * lectures
Periodicals: (12) "Florida Skies" (circ.: 135)
Coordinates: 080°14'00"W 26°05'00"N H4m (Buehler Observatory)
City Reference Coordinates: 080°14'00"W 26°05'00"N

Calusa Nature Center and Planetarium
3450 Ortiz Avenue
Fort Myers, FL 33905
Telephone: 941-275-3435
Telefax: 941-275-9016
Electronic Mail: stars@calusanature.com
WWW: http://calusanature.com/
Founded: 1970 (Planetarium: 1986)
Staff: 10
Activities: education (environment, earth, space science)
City Reference Coordinates: 081°54'00"W 26°37'00"N

Central Florida Astronomical Society (CFAS)
c/o Orlando Science Center
810 East Rollins Street

Orlando, FL 32803
Telephone: 407-896-7151
407-366-3888 (David Hearn, President)
Electronic Mail: dhearn@oo.com (David Hearn, President)
dhearn@gate.net
observer@procyon-sys.com
cfas-db@america.com (Daytona Beach Affiliate)
WWW: http://www.physics.ucf.edu/cfas/
http://www.america.com/~cfas-db/ (Daytona Beach Affiliate)
Founded: 1962
Membership: 120
Periodicals: "Astrolog", "Skylight"
Observatories: 1 (Robinson Observatory)
City Reference Coordinates: 081°23'00"W 28°32'00"N (Orlando)
080°60'00"W 29°12'00"N (Daytona Beach)

Chiefland Astronomy Village (CAV)
c/o Tom & Jeannie Clark
5450 NW 52nd Court
Chiefland, FL 32626
Telephone: 352-490-9101
Electronic Mail: tom@amateurastronomy.com
WWW: http://www.c-av.com/
Activities: providing convenient dark-sky observing facility
City Reference Coordinates: 082°53'00"W 29°30'00"N

Cosmo Astronomical Society (CAS)
c/o John Martello
3415 Silverwood Drive
Pine Hills, FL 32808
Telephone: 407-293-1739
WWW: http://clubs.yahoo.com/clubs/cosmoastronomicalsociety/
City Reference Coordinates: 081°23'00"W 28°33'00"N (Orlando)

Escambia Amateur Astronomers Association (EAAA)
c/o John Wayne Wooten
6235 Omie Circle
Pensacola, FL 32504
Telephone: 850-484-1152
Telefax: 850-484-1822
Electronic Mail: wwooten@pjc.cc.fl.us
WWW: http://www.meteor.dotstar.net/
Founded: 1977
Membership: 200
Activities: observing * meetings
Periodicals: (12) "The Meteor"
Observatories: 1 (Fort Pickens Amphitheater)
City Reference Coordinates: 087°13'00"W 30°25'00"N

Everglades Astronomical Society (EAS)
P.O. Box 10406
Naples, FL 34101-0406
Telephone: 941-394-9123
Electronic Mail: uc007@aol.com
WWW: http://www.naples.net/clubs/eas
Founded: 1980
Membership: 70 * public education
Periodicals: (12) "Newsletter"
Coordinates: 081°42'00"W 25°56'00"N
City Reference Coordinates: 081°48'00"W 26°08'00"N

Florida Institute of Technology (FIT), Department of Physics and Space Sciences
150 West University Boulevard
Melbourne, FL 32901-6988
Telephone: 305-674-8098
Telefax: 407-674-8461
407-674-7482
Electronic Mail: <userid>@pss.fit.edu
WWW: http://pss.fit.edu/
Founded: 1958 (Florida Tech)
Staff: 11
Activities: observational astrophysics * white dwarf stars * cataclysmic variable stars
City Reference Coordinates: 080°37'00"W 28°05'00"N

Florida International University (FIU), Department of Physics
C.P. 204
11200 SW Eighth Street
Miami, FL 33199
Telephone: 305-348-2605
Telefax: 305-348-6700
Electronic Mail: <userid>@fiu.edu
WWW: http://www.fiu.edu/physics
　　　　http://www.fiu.edu/~folcika/astronomy/astronomy_main.html (Astronomy)
Founded: 1972
Staff: 22
City Reference Coordinates: 080°12'00"W 25°46'00"N

Florida Tech
● See "Florida Institute of Technology (FIT)"

Gibson Observatory Astronomy Group (GOAG)
c/o South Florida Science Museum (SFSM)
4801 Dreher Trail North
West Palm Beach, FL 33405
Telephone: 561-832-1988 x224
Telefax: 561-833-0551
Founded: 1991
Activities: public observing * star parties * Sun
Periodicals: "GOAG News"
Coordinates: 080°04'00"W 26°40'00"N H6m
City Reference Coordinates: 080°04'00"W 26°43'00"N

Highlands Stargazers Astronomy Group
Electronic Mail: highlandsstargazers@highlandsstargazers.8m.net
WWW: http://highlandstargaze.cjb.net/
Founded: 1889
City Reference Coordinates: 081°23'00"W 28°33'00"N (Orlando)

Honeywell, Space Systems Headquarters
13350 US Highway 19 North
Clearwater, FL 34624-7290
Telephone: 813-539-4000
WWW: http://content.honeywell.com/Space/
　　　　http://www.honeywell.com/
Founded: 1885
City Reference Coordinates: 082°48'00"W 27°58'00"N

Indian River Astronomical Society
c/o Richard Woodward
6055 College Lane
Vero Beach, FL 32966
Electronic Mail: wrwoodward@compuserve.com
City Reference Coordinates: 080°24'00"W 27°39'00"N

James and Barbara Moore Observatory
c/o Linda Jacobson
Charlotte County Campus
26300 Airport Road
Punta Gorda, FL 33950
Electronic Mail: milkyway@peganet.com
WWW: http://www.MooreObservatory.com/
City Reference Coordinates: 082°01'00"W 26°56'00"N

John F. Kennedy Space Center (KSC)
● See "National Aeronautics and Space Administration (NASA), John F. Kennedy Space Center (KSC)"

Kennedy Space Center (KSC) (John F._)
● See "National Aeronautics and Space Administration (NASA), John F. Kennedy Space Center (KSC)"

Local Group of Deep Sky Observers
c/o Vic Menard
2311 23rd Avenue West
Bradenton, FL 34205
Telephone: 813-747-8334
WWW: http://www.thelocalgroup.org/
City Reference Coordinates: 082°34'00"W 27°29'00"N

Miami Museum of Science and Space Transit, Planetarium
3280 South Miami Avenue
Miami, FL 33129
Telephone: 305-854-4242
Electronic Mail: info@miamisci.org
WWW: http://www.miamisci.org/
   http://www.miamisci.org/www/eventsplan.html
Founded: 1966
Activities: public presentations * multimedia shows * education * TV series
City Reference Coordinates: 080°12'00"W 25°46'00"N

Moore Observatory (James and Barbara_)
• See "James and Barbara Moore Observatory"

Murnaghan Instruments Corp.
1781 Primrose Lane
West Palm Beach, FL 33414
Telephone: 561-795-2201
Telefax: 561-795-9889
Electronic Mail: murni@gate.net
WWW: http://www.murni.com/
   http://www.e-scopes.cc/Murnaghan_Instruments_Corp56469.html
Founded: 1992
Staff: 5
Activities: manufacturing CCD cameras, filters, filter wheels, flip mirrors, beam splitters and image processing equipment, as well as Coulter Optical's Odyssey telescopes
City Reference Coordinates: 080°04'00"W 26°43'00"N

Museum of Arts and Sciences, Planetarium
1040 Museum Boulevard
Daytona Beach, FL 32114
Telephone: 386-255-0285
Telefax: 386-255-0285
Electronic Mail: rhmoasplanet@mindspring.com
WWW: http://www.moas.org/planetrn.htm
   http://www.moas.org/
Staff: 3
Activities: public star shows
Periodicals: (4) "Arts & Sciences"
City Reference Coordinates: 080°60'00"W 29°12'00"N

National Aeronautics and Space Administration (NASA), John F. Kennedy Space Center (KSC), Visitor Complex
Mail Code DNPS
Kennedy Space Center, FL 32899
Telephone: 407-452-2121
WWW: http://www.kennedyspacecenter.com/
   http://www.kscvisitor.com/
   http://www.ksc.nasa.gov/
Founded: 1958 (NASA)
Staff: 14000 (KSC)
Periodicals: "Spaceport News", "KSC Countdown", "The Orbiter"
City Reference Coordinates: 080°32'00"W 28°27'00"N

Northeast Florida Astronomical Society (NEFAS)
P.O. Box 5432
Jacksonville, FL 32247-5432
Electronic Mail: president@nefas.org
WWW: http://www.nefas.org/
City Reference Coordinates: 081°40'00"W 30°20'00"N

Optronic Laboratories Inc. (OLI)
4632 36th Street
Orlando, FL 32811
Telephone: 407-422-3171
   1-800-899-3171 (USA only)
Telefax: 407-648-5412
Electronic Mail: was@olinet.com
WWW: http://www.olinet.com/
Founded: 1970
Staff: 38
Activities: manufacturing spectroradiometric instrumentation * standards * calibration services
City Reference Coordinates: 081°23'00"W 28°32'00"N

Orlando Science Center (OSC)
777 East Princeton Street
Orlando, FL 32803-1291
Telephone: 407-514-2000
          1-888-OSC-4FUN (USA only)
Telefax: 407-514-2277
Electronic Mail: info@osc.org
WWW: http://www.osc.org/
Founded: 1960
Staff: 7
Activities: shows for schools and public * laser shows * sky shows * media information * John Young Planetarium * Carolyn Wine Observatory
Periodicals: (6) "Focus"
Coordinates: 081°23'00"W 28°33'00"N H26m
City Reference Coordinates: 081°23'00"W 28°33'00"N

Palermiti Observatory
c/o Mike Palermiti
16222 133rd Drive North
Jupiter, FL 33478
Telephone: 561-745-7353
Electronic Mail: mpal@gate.net
WWW: http://www.gopbi.com/community/groups/skyview/
Founded: 1979
Staff: 1
Activities: observational astronomy * technical support * education
● IAU Code 837
City Reference Coordinates: 080°08'00"W 26°57'00"N

Pat Thomas Planetarium
c/o Florida State University
7545 Old Saint Augustine Road
Tallahassee, FL 32311
WWW: http://www.physics.fsu.edu/Planetarium/ptplanet.htm
Founded: 1980
City Reference Coordinates: 084°16'00"W 30°25'00"N

PCP Inc.
2155 Indian Road
West Palm Beach, FL 33409-3287
Telephone: 561-683-0507
Telefax: 561-683-9215
Founded: 1975
Staff: 7
Activities: manufacturing ion mobility spectrometry instrumentation * trace chemical detection
City Reference Coordinates: 080°04'00"W 26°43'00"N

Pensacola Junior College (PJC), Department of Physical Sciences
1000 College Boulevard
Pensacola, FL 32504-8998
Telephone: 850-484-1117
Telefax: 850-484-2050
Electronic Mail: sst@pjc.cc.fl.us
WWW: http://www.pjc.cc.fl.us/academics/departments/physci/index.htm
Founded: 1993
Staff: 3
Activities: school and public programs
City Reference Coordinates: 087°13'00"W 30°25'00"N

Poinciana Planetarium
c/o Ponciana Elementary School
Boynton Beach, FL 33460
Electronic Mail: swanson_k@popmail.firn.edu (Kris Swanson, Director)
WWW: http://www.palmbeach.k12.fl.us/poincianaes/labs/planetarium/
City Reference Coordinates: 080°04'00"W 26°32'00"N

Riverview High School (RHS), Planetarium
One Ram Way
Sarasota, FL 34231
Telephone: 941 923-1484
WWW: http://www.sarasota-online.com/rhs/planet.htm
City Reference Coordinates: 082°34'00"W 27°20'00"N

Saint Petersburg Astronomy Club (SPAC), Inc.
594 59th Street South
Saint Petersburg, FL 33707
Telephone: 813-343-1594
WWW: http://www.concentric.net/~Choff/
http://www.cris.com/~choff/
http://home1.gte.net/hoffmanc/
Founded: 1927
Membership: 250
Activities: observing (deep sky, planets) * education * telescope making * grazing-occultation expeditions * meteor shower data recording
Periodicals: (12) "Saint Petersburg Astronomy Club Examiner - SPACE"
Coordinates: 082°17'49"W 28°28'43"N H85m (Raden Memorial Observatory, Hickory Hill)
City Reference Coordinates: 082°38'00"W 27°46'00"N

Saint Petersburg Junior College (SPJC), Planetarium
6605 Fifth Avenue North
Saint Petersburg, FL 33710-3489
Telephone: 727-341-4320
Telefax: 727-341-4770
Electronic Mail: breslinm@email.spjc.cc.fl.us (Michael Breslin, Director)
WWW: http://www.spjc.edu/spg/science/dns/Planetarium.html
Founded: 1959
Staff: 1
Activities: astronomy programs * school and public shows
City Reference Coordinates: 082°38'00"W 27°46'00"N

Saunders Planetarium
c/o Museum of Science and Industry
4801 East Fowler Avenue
Tampa, FL 33617
Electronic Mail: peche@firnvx.firn.edu (Alan Peche)
WWW: http://www2.tia.net/mosi/
http://www.mosi.org/planet.htm
City Reference Coordinates: 082°27'00"W 22°57'00"N

Science Center of Pinellas County Inc.
7701 22nd Avenue North
Saint Petersburg, FL 33710
Telephone: 813-384-0027
WWW: http://home1.gte.net/scicen/
Activities: planetarium shows
City Reference Coordinates: 082°38'00"W 27°46'00"N

Scientific Expeditions Inc.
227 West Miami Avenue
Suite 3
Venice, FL 34285
Telephone: 941-484-3884
1-800-344-6867 (USA only)
Telefax: 941-485-0647
Electronic Mail: mwxw71a@prodigy.com
WWW: http://www.skypub.com/store/scinex/scinex.html
Activities: organizing i.a. solar eclipse cruises
City Reference Coordinates: 082°27'00"W 27°06'00"N

ScopeTronix
1423 SE 10th Street
Unit 1A
Cape Coral, FL 33990
Telephone: 941-945-6763
1-866-458-7658 (USA only)
Electronic Mail: info@scopetronix.com
WWW: http://www.scopetronix.com/
Activities: accessories, especially for Meade telescopes
City Reference Coordinates: 081°77'00"W 26°35'00"N

Seminole Community College (SCC), Planetarium
100 Weldon Boulevard
Sanford, FL 32773-6199
Telephone: 407-328-2360
407-328-2409
Telefax: 407-328-2407
Electronic Mail: pelleril@scc-fl.com
mcadaml@scc-fl.com

WWW: http://www.scc-fl.com/planet/
Founded: 1987
Staff: 3
Activities: public and school shows * astronomy events * outreach
Periodicals: (4) "Night Sights"
Coordinates: 081°11'25"W 28°44'41"N
City Reference Coordinates: 081°16'00"W 28°48'00"N

SkyComm Engineering
4630 North University Drive
Suite 329
Coral Springs, FL 33067
Telephone: 954-345-8726
Electronic Mail: info@skyeng.com
WWW: http://skyeng.com/
Activities: manufacturing telescope pointing devices
City Reference Coordinates: 080°08'00"W 26°14'00"N (Margate)

Southeastern Planetarium Association (SEPA)
c/o Bishop Planetarium
201 10th Street West
Bradenton, FL 34205
Telephone: 941-746-4132
Telefax: 941-747-2556
WWW: http://www.sepadomes.org/
Founded: 1971
Staff: 4
Activities: providing professional information to planetarium operators * annual conferences
Periodicals: (4) "Southern Skies"
City Reference Coordinates: 082°34'00"W 27°29'00"N

Southern Cross Astronomical Society (SCAS) Inc.
c/o Barbara Yager (Secretary)
10221 SW 116th Avenue
Miami, FL 33176
Telephone: 305-661-1375 (hot line)
Electronic Mail: barbyager@aol.com
WWW: http://www.scas.org/
Founded: 1922
Membership: 285
Activities: observing * lectures * organizing the Winter Star Party (WSP)
Periodicals: (12) "Coal Sack"
City Reference Coordinates: 080°12'00"W 25°46'00"N

South Florida Amateur Astronomers Association (SFAAA)
c/o Fox Observatory
16001 West State Road 84
Sunrise, FL 33313
Telephone: 954-384-0442
Electronic Mail: sfaaa@aol.com
          chuckfa@mediaone.net (Chuck Faranda, President)
WWW: http://members.aol.com/sfaaa/
     http://www.geocities.com/CapeCanaveral/Lab/1333/ (Fox Observatory)
Founded: 1969
Membership: 85
Activities: observing * lectures
Periodicals: "The Meridian"
City Reference Coordinates: 080°08'00"W 26°07'00"N (Fort Lauderdale)

South Florida Science Museum (SFSM)
4801 Dreher Trail North
West Palm Beach, FL 33405
Telephone: 561-832-1988
Telefax: 561-833-0551
WWW: http://www.sfsm.org/
Founded: 1964
Membership: 3000
Staff: 2
Activities: education * exhibits * shows * Aldrin Planetarium * Gibson Observatory
Periodicals: (4) "Imaginings"
Coordinates: 080°04'00"W 26°40'00"N H6m
City Reference Coordinates: 080°04'00"W 26°43'00"N

Spectrum Coatings
c/o Paul Zacharias

Telephone: 630-859-3434
Telefax: 630-859-8692
Electronic Mail: <userid>@scitech.com
WWW: http://scitech.mus.il.us/
Founded: 1988
Staff: 44
Activities: hands-on science center * public-access solar telescope * education * Summer camps * observing
City Reference Coordinates: 088°19'00"W 42°46'00"N

Sirius Instruments
141 North Charles Avenue
Villa Park, IL 60181
Telephone: 1-800-288-CWIP
        708-782-5819
Electronic Mail: siriusinst@aol.com
WWW: http://www.htennant.com/sirius/
Activities: distributing CWIP cameras
City Reference Coordinates: 087°59'00"W 41°53'00"N

Skokie Valley Astronomers (SVA)
c/o Gretchen Patti
1 South 751 Avon Drive
Warrenville, IL 60555
Telephone: 630-393-7929
Electronic Mail: gretchen@adichi.com
WWW: http://ourworld.compuserve.com/homepages/E_Neuzil/
Founded: 1982
Membership: 60
Activities: meetings * lectures * observing * education
Periodicals: (12) "Newsletter"
Coordinates: 087°54'46"W 42°10'43"N H187m (Ryerson Nature Center)
City Reference Coordinates: 087°46'00"W 42°02'00"N (Skokie)
                            088°11'00"W 41°49'00"N (Warrenville)

Southern Illinois University (SIU), Carbondale, Department of Physics
Carbondale, IL 62901-4401
Telephone: 618-453-2643
Telefax: 618-453-1056
Electronic Mail: physics@physics.siu.edu
WWW: http://www.physics.siu.edu/
Staff: 14
Activities: experimental and theoretical physics
City Reference Coordinates: 089°13'00"W 37°44'00"N

Staerkel Planetarium (William M._)
• See "William M. Staerkel Planetarium"

Strickler Planetarium
c/o Olivet Nazarene University
P.O. Box 592
Kankakee, IL 60901-0592
Telephone: 815-939-5267
Telefax: 815-939-0153
Electronic Mail: tstone@olivet.edu
WWW: http://www.olivet.edu/planetarium/
Founded: 1967
Staff: 1
Activities: planetarium shows
City Reference Coordinates: 087°52'00"W 41°07'00"N

Time Museum
P.O. Box 7327
Rockford, IL 61126
or :
7801 East State Street
Rockford, IL 61125
Telephone: 815-398-6000
Telefax: 815-398-4700
Electronic Mail: info@timemuseum.com
WWW: http://www.timemuseum.com/
Founded: 1970
Activities: extensive collection of time-measuring devices (sundials, chronometers, clocks, watches, and other scientific instruments)
City Reference Coordinates: 089°06'00"W 42°17'00"N

Trackman Planetarium (Herbert_)
• See "Herbert Trackman Planetarium"

Twin City Amateur Astronomers (TCAA) Inc.
c/o ISU Planetarium
Campus Box 4560
Normal, IL 61790-4560
Telephone: 309-438-2496
Electronic Mail: cjwennin@ilstu.edu
WWW: http://twincityamateurastronomers.org/
http://twincityamateurastronomers.org/sgo/sgo.html (Sugar Grove Obs.)
Founded: 1960
Membership: 60
City Reference Coordinates: 088°59'00"W 40°31'00"N

University of Chicago, Center for Astrophysical Thermonuclear Flashes
5640 South Ellis
RI 468
Chicago, IL 60637
Telephone: 773-834-2057
Telefax: 773-834-3230
WWW: http://flash.uchicago.edu/
Activities: physics of exploding stars
City Reference Coordinates: 087°38'00"W 41°53'00"N

University of Chicago, Department of Astronomy and Astrophysics
5640 South Ellis Avenue
Chicago, IL 60637
Telephone: 773-702-8203
Telefax: 773-702-8212
Electronic Mail: <userid>@uchicago.edu
WWW: http://astro.uchicago.edu/
Founded: 1892
Staff: 32
Activities: cosmology * galactic dynamics * stellar physics * adaptive optics * QSOs * theoretical astrophysics * relativistic astrophysics * astrophysical fluid dynamics * nuclear astrophysics * particle astrophysics * cosmic microwave background * Sloan Digital Sky Survey * astrophysics from Antarctica
Coordinates: 088°33'19"W 42°34'12"N
City Reference Coordinates: 087°38'00"W 41°53'00"N

University of Chicago, Department of Mathematics
5734 South University Avenue
Chicago, IL 60637
Telephone: 773-702-7329
Electronic Mail: <userid>@math.uchicago.edu
ucmath@math.uchicago.edu
WWW: http://www.math.uchicago.edu/
Founded: 1892
City Reference Coordinates: 087°38'00"W 41°53'00"N

University of Chicago, Department of Physics
5640 South Ellis Avenue
Chicago, IL 60637
Telephone: 773-702-7006
Electronic Mail: physics@uchicago.edu
WWW: http://physics.uchicago.edu/
Founded: 1893
Staff: 50
City Reference Coordinates: 087°38'00"W 41°53'00"N

University of Chicago, Enrico Fermi Institute (EFI)
5640 South Ellis Avenue
Chicago, IL 60637
Telephone: 773-702-7823
Telefax: 773-702-8038
Electronic Mail: <userid>@uchicago.edu
WWW: http://rainbow.uchicago.edu/efi/
http://ulysses.uchicago.edu/
Founded: 1945 (University: 1892)
Staff: 125
Activities: among others, cosmology * theoretical astrophysics * experimental astrophysics * space physics * infrared and optical astronomy * cosmic microwave background * general relativity * solar energy * high-energy physics * string theory * elementary particles
Awards: (1) "Fermi Fellowship", "McCormick Fellowship"
Coordinates: 088°33'19"W 42°34'12"N

City Reference Coordinates: 087°38'00"W 41°53'00"N

University of Chicago Press (UCP)
1427 East 60th Street
Chicago, IL 60637-2954
Telephone: 312-753-3347 (Subscriptions)
          312-702-7600
          312-753-3372
          312-753-3373 (Production)
          312-753-4225 (Editorial)
Telefax: 312-753-0811
Electronic Mail: marketing@press.uchicago.edu
                 <userid>@press.uchicago.edu
WWW: http://press-gopher.uchicago.edu/
     http://www.journals.uchicago.edu/
     http://www.journals.uchicago.edu/ApJ/ (Astrophysical Journal)
Founded: 1891
• Publisher (i.a. of the "Astrophysical Journal" and of the "Astrophysical Journal Supplement Series" - see separate entries)
City Reference Coordinates: 087°38'00"W 41°53'00"N

University of Illinois, Astronomical Society (UIAS)
Astronomy Building
1002 West Green Street
Urbana, IL 61801
Telephone: 217-333-8417
Electronic Mail: uias@uiuc.edu
WWW: http://www.astro.uiuc.edu/~uias/
Membership: 100
Activities: promoting astronomy on campus * open houses at Observatory * observing at darker sites
Periodicals: "Sidereal Messenger"
City Reference Coordinates: 088°12'00"W 40°07'00"N

University of Illinois at Springfield, Department of Astronomy-Physics
Shepherd Road
Springfield, IL 62793-9243
Telephone: 217-786-6720
Telefax: 217-786-7279
Electronic Mail: schweighauser.charles@uis.edu
WWW: http://www.uis.edu/Welcome.html
Founded: 1977
Staff: 2
Activities: education * research * observing
Coordinates: 089°53'30"W 39°51'30"N H190m (Henry R. Barber Research Observatory)
City Reference Coordinates: 089°40'00"W 39°47'00"N

University of Illinois at Urbana, Department of Astronomy
103 Astronomy Building
1002 West Green Street
Urbana, IL 61801
Telephone: 217-333-3090
Telefax: 217-244-7638
Electronic Mail: astronomy@uiuc.edu
WWW: http://www.astro.uiuc.edu/main.html
City Reference Coordinates: 088°12'00"W 40°07'00"N

University of Illinois at Urbana, Laboratory for Astronomical Imaging (LAI)
102 West Green Street
Urbana, IL 61801
Telephone: 217-333-3090
Telefax: 217-244-7638
Electronic Mail: <userid>@sirius.astro.uiuc.edu
WWW: http://www.astro.uiuc.edu/projects/lai/
City Reference Coordinates: 088°12'00"W 40°07'00"N

University of Illinois at Urbana, Laboratory for Computational Astrophysics (LCA)
5257 Beckman Institute
405 North Mathews Avenue
Urbana, IL 61801
Telephone: 217-244-6099
Telefax: 217-244-2909
Electronic Mail: lca@ncsa.uiuc.edu
WWW: http://zeus.ncsa.uiuc.edu/lca_home_page.html
Founded: 1993
Staff: 4
Activities: developing and disseminating theoretical modelling software * astrophysical fluid dynamics

City Reference Coordinates: 088°12'00"W 40°07'00"N

University of Illinois Press
1325 South Oak Street
Champaign, IL 61820-6975
Telephone: 217-244-0626
Telefax: 217-244-8082
Electronic Mail: uipress@uillinois.edu
WWW: http://www.press.uillinois.edu/
Founded: 1918
• Publisher
City Reference Coordinates: 088°14'00"W 40°07'00"N

Waubonsie Valley High School, Planetarium
2590 Ogden Avenue
Aurora, IL 60504
Telephone: 630-375-3247
Telefax: 630-375-3301
Electronic Mail: mary_schindewolf@ipsd.org
WWW: http://planetarium.ipsd.org/
Founded: 1976
City Reference Coordinates: 088°19'00"W 42°46'00"N

Wheaton College, Department of Physics and Astronomy
Wheaton, IL 60187
Telephone: 630-752-5007
Telefax: 630-752-5996
Electronic Mail: physics@wheaton.edu
           <userid>@wheaton.edu
WWW: http://www.wheaton.edu/Physics
Founded: 1860
Activities: education
Coordinates: 088°06'15"W 41°52'20"N H233m
City Reference Coordinates: 088°06'00"W 41°52'00"N

William M. Staerkel Planetarium
c/o Parkland College
2400 West Bradley Avenue
Champaign, IL 61821-1899
Telephone: 217-351-2581
Telefax: 217-351-2567
Electronic Mail: starman@prairienet.org (David C. Leake)
           chuckg@prairienet.org (Chuck Greenwood)
WWW: http://www.parkland.cc.il.us/coned/pla/
Founded: 1987
City Reference Coordinates: 088°14'00"W 40°07'00"N

William Rainey Harper College, Observatory
1200 West Algonquin Road
Palatine, IL 60067
Telephone: 847-925-6000
WWW: http://www.planets.org/hobs.htm
       http://www.harper.cc.il.us/
Founded: 1989
City Reference Coordinates: 088°03'00"W 42°07'00"N

USA - Indiana

Ball State University (BSU), Department of Physics and Astronomy
Muncie, IN 47306
Telephone: 765-285-8860
Telefax: 765-285-5674
Electronic Mail: <userid>@bsu.edu
WWW: http://www.bsu.edu/csh/physics/
http://www.bsu.edu/physics/astro/astronomy-frame.html
Activities: galactic structure * objective prism surveys * red dwarf stars * stellar kinematics * interactive binary stars * emission-lines objects * planetarium shows
Coordinates: 085°24'39"W 40°11'59"N
City Reference Coordinates: 085°23'00"W 40°11'00"N

Bowen Productions
748 East Bates Street
Suite 300W
Indianapolis, IN 46202
Telephone: 317-226-9650
Telefax: 317-226-9651
Electronic Mail: bowenprod@aol.com
WWW: http://www.bowenproductions.com/
Founded: 1985
Activities: designing visual and audio productions, i.a. for planetariums
City Reference Coordinates: 086°09'00"W 39°46'00"N

Butler University, Department of Physics and Astronomy
4600 Sunset Avenue
Indianapolis, IN 46208
Telephone: 317-940-9951
Telefax: 317-940-9950
Electronic Mail: physics@butler.edu
cprice@butler.edu (Chris Price, Secretary)
<userid>@butler.edu
WWW: http://www.butler.edu/physics/
City Reference Coordinates: 086°09'00"W 39°46'00"N

Butler University, J.I. Holcomb Observatory and Planetarium (JIHOP)
4600 Sunset Avenue
Indianapolis, IN 46208
Telephone: 317-940-8333
317-940-9333 (recorded info)
Telefax: 317-940-9950
Electronic Mail: cprice@butler.edu (Chris Price, Secretary)
WWW: http://www.butler.edu/holcomb/
Founded: 1954
Staff: 9
Activities: weekly tours * planetarium shows * two semiannual public programs
City Reference Coordinates: 086°09'00"W 39°46'00"N

Calumet Astronomical Society (CAS) Inc.
P.O. Box 851
Griffith, IN 46319-0851
Telephone: 219-924-7974
Electronic Mail: info@casonline.org
WWW: http://www.casonline.org/
Founded: 1976
Membership: 150
Activities: public outings * monthly meetings * school programs * field trips
Periodicals: (12) "The Mirror"
City Reference Coordinates: 087°25'00"W 41°32'00"N

Children's Museum of Indianapolis (The_)
● See "The Children's Museum of Indianapolis"

De Pauw University, Physics and Astronomy Department
Greencastle, IN 46135
Telephone: 317-658-4661
317-658-4654
Telefax: 317-658-4177
Electronic Mail: <userid>@depauw.edu
WWW: http://www.depauw.edu/phys/
http://www.depauw.edu/univ/mckim/ (McKim Observatory)

Founded: 1981 (McKim Observatory: 1884)
Activities: education * occultation timing * solar transit observing * photoelectric photometry
Coordinates: 086°51'06"W 39°38'47"N H267m
City Reference Coordinates: 086°52'00"W 39°38'00"N

Earlham College, Department of Physics and Astronomy
Richmond, IN 47374-4095
Telephone: 765-983-1239
Telefax: 765-983-1301
Electronic Mail: <userid>@earlham.edu
WWW: http://www.earlham.edu/~physics/physics.html
Founded: 1887
City Reference Coordinates: 084°54'00"W 39°50'00"N

E.C. Schouweiler Planetarium
c/o University of Saint Francis
2701 Spring Street
Fort Wayne, IN 46808
Telephone: 219-434-3278
Telefax: 219-434-7404
Electronic Mail: cexner@sf.edu
Founded: 1971
Staff: 2
Activities: school and public shows
Coordinates: 085°09'00"W 41°04'00"N H200m
City Reference Coordinates: 085°09'00"W 41°04'00"N

Evansville Astronomical Society (EAS)
P.O. Box 3474
Evansville, IN 47733
Telephone: 812-922-5681 (Observatory)
WWW: http://www.geocities.com/astronomy_eas/
Founded: 1959
Membership: 60
Activities: observing * meetings * public programs
Periodicals: (12) "The Observer"
Coordinates: 087°20'00"W 38°15'00"N (Wahnsiedler Observatory)
City Reference Coordinates: 089°56'00"W 38°05'00"N

Evansville Museum of Arts and Science
411 SE Riverside Drive
Evansville, IN 47713
Telephone: 812-425-2406
Telefax: 812-421-7507
WWW: http://www.emuseum.org/
     http://www.emuseum.org/shows.html (Planetarium)
Founded: 1904 (Planetarium: 1959)
Staff: 30 (Planetarium: 2)
Activities: public and school programs * exhibitions * classes * Koch Planetarium
Periodicals: (4) "Copia"
City Reference Coordinates: 089°56'00"W 38°05'00"N

Fort Wayne Astronomical Society (FWAS), Inc.
P.O. Box 11093
Fort Wayne, IN 46855
Telephone: 219-744-6575
Electronic Mail: thefwas@aol.com
WWW: http://members.aol.com/theFWAS/
Founded: 1959
Periodicals: (12) "The Eyepiece"
Observatories: 1 (Fox Island County Park)
City Reference Coordinates: 085°09'00"W 41°04'00"N

Holcomb Observatory and Planetarium (JIHOP) (J.I._)
● See "Butler University, J.I. Holcomb Observatory and Planetarium (JIHOP)"

Hook Memorial Observatory (John C._)
● See "Indiana State University, Department of Geography and Geology, John C. Hook Memorial Observatory"

Indiana Astronomical Society (IAS) Inc.
c/o J. Philip May
2 Wilson Drive
Mooresville, IN 46158
or :
P.O. Box 703

Mooresville, IN 46158
Telephone: 317-831-8387
Electronic Mail: dgoins@scican.net (Dan Goins, President)
WWW: http://www.a1.com/ias/
Founded: 1933
Membership: 150
City Reference Coordinates: 086°22'00"W 39°36'00"N

Indiana State University, Department of Geography and Geology, John C. Hook Memorial Observatory
Terre Haute, IN 47809
Telephone: 812-237-2271
          812-237-2444 (Robert C. Howe)
Telefax: 812-237-8029
Electronic Mail: <userid>@geosun.indstate.edu
WWW: http://thunder.indstate.edu/h9/bigdaddy/astro/
     http://mama.indstate.edu/users/jchook/astro/
     http://astro.indstate.edu/
Founded: 1959
Staff: 3
Activities: public viewing facility * photography
Coordinates: 087°24'30"W 39°28'10"N H175m
City Reference Coordinates: 087°24'00"W 39°28'00"N

Indiana University, Department of Astronomy
Swain West 319
727 East 3rd Street
Bloomington, IN 47405-7105
Telephone: 812-855-6911
Telefax: 812-855-8725
Electronic Mail: <userid>@astro.indiana.edu
                request@astro.indiana.edu
WWW: http://astrowww.astro.indiana.edu/
     http://www.astro.indiana.edu/
Founded: 1947
Staff: 15
Activities: education * research * stellar spectroscopy * stellar atmospheres * interacting binaries * star formation * cluster dynamics * high-energy astrophysics * active galaxies * galaxy photometry * planetary rings * planetarium
Coordinates: 086°23'42"W 39°33'00"N H300m (Goethe Link Obs. – IAU Code 760)
             086°31'18"W 39°09'56"N H238m (Kirkwood Observatory)
             086°26'23"W 39°18'49"N (Morgan-Monroe Station)
City Reference Coordinates: 086°32'00"W 39°10'00"N

Indiana University - Purdue University at Indianapolis (IUPUI), Department of Physics
402 North Blackford Street
Indianapolis, IN 46202-3273
Telephone: 317-274-6901
Telefax: 317-274-2393
Electronic Mail: <userid>@indyvax.iupui.edu
WWW: http://www.iupui.edu/
Founded: 1969
Staff: 14
Activities: physics * astronomy * education * biophysics * optics * magnetic resonance * material science
City Reference Coordinates: 086°09'00"W 39°46'00"N

Jefferson High School Planetarium
1801 South 18th Street
Lafayette, IN 47905
Telephone: 765-449-3400 x5036
Telefax: 765-449-3413
WWW: http://www.lsc.k12.in.us/huston/astro/planetarium.html
City Reference Coordinates: 086°53'00"W 40°25'00"N

J.I. Holcomb Observatory and Planetarium (JIHOP)
• See "Butler University, J.I. Holcomb Observatory and Planetarium (JIHOP)"

John C. Hook Memorial Observatory
• See "Indiana State University, Department of Geography and Geology, John C. Hook Memorial Observatory"

L.S. Noblitt Planetarium
East High School
230 South Marr Road
Columbus, IN 47201
Telephone: 812-376-4350
Founded: 1972

Staff: 1
Activities: school programs
City Reference Coordinates: 085°55'00"W 39°13'00"N

Merrillville Community Planetarium (MCP)
c/o Clifford Pierce Middle School
199 East 70th Avenue
Merrillville, IN 46410
Telephone: 219-736-4856
WWW: http://www.calunet.com/org/MCP/
http://www.mcp.home.ml.org/
Founded: 1973
Staff: 1
Activities: school, group, and public programs
City Reference Coordinates: 087°15'00"W 41°30'00"N

Michiana Astronomical Society (MAS)
P.O. Box 262
South Bend, IN 46624-0262
Electronic Mail: jperry@iusb.edu (Jeff Perry, President)
WWW: http://michiana.org/MFNetSci/MAS/
http://www.michianaastrosociety.homestead.com/
Founded: 1973
Membership: 26
Activities: promoting amateur astronomy
Periodicals: "The Sirius Observer"
City Reference Coordinates: 086°15'00"W 41°41'00"N

Noblitt Planetarium (L.S._)
• See "L.S. Noblitt Planetarium"

Northern Indiana Astronomical Group (NIAG)
1308 Sunset Drive
Winona Lake, IN 46590
WWW: http://clubs.kconline.com/was/niag.htm
Activities: gathering together clubs of Northern Indiana
City Reference Coordinates: 085°49'00"W 41°14'00"N

Pegasus Productions
713 Cushing
South Bend, IN 46616
Telephone: 219-282-1885
WWW: http://www.pegasusconcerts.com/
Founded: 1990
Staff: 1
Activities: planetarium shows * education
City Reference Coordinates: 086°15'00"W 41°41'00"N

Precision Cryogenic Systems Inc.
7804 Rockville Road
Indianapolis, IN 46214
Telephone: 317-273-2800
Telefax: 317-273-2802
Electronic Mail: prcry@iquest.net
WWW: http://www.precisioncryo.com/
Founded: 1982
Staff: 26
Activities: manufacturing cryogenic equipment
City Reference Coordinates: 086°09'00"W 39°46'00"N

Purdue University, Indianapolis
• See "Indiana University - Purdue University at Indianapolis (IUPUI)"

Purdue University, West Lafayette, Department of Physics
1396 Physics Building
West Lafayette, IN 47907-1306
Telephone: 317-494-3000
Telefax: 317-494-0706
Electronic Mail: <userid>@physics.purdue.edu
WWW: http://www.physics.purdue.edu/
http://www.physics.purdue.edu/department/astrophysics.html
http://earth.physics.purdue.edu/astro/astro.html (High-Energy Astrophysics)
Founded: 1874
Activities: variable stars * photometry * radial velocities * gamma-ray astronomy * general relativity * X-ray astronomy * pulsars

City Reference Coordinates: 086°53'00"W 40°25'00"N

Schouweiler Planetarium (E.C._)
• See "E.C. Schouweiler Planetarium"

The Children's Museum of Indianapolis
3000 North Meridian Street
Indianapolis, IN 46208
Telephone: 317-334-3322
Telefax: 317-921-4019
Electronic Mail: community@childrensmuseum.org
WWW: http://www.childrensmuseum.org/
Founded: 1989
Activities: exhibits * education * SpaceQuest Planetarium
City Reference Coordinates: 086°09'00"W 39°46'00"N

Tipton Planetarium
c/o Tipton Community School Corp.
817 South Main Street
Tipton, IN 46072-9775
Telephone: 317-675-2147
Telefax: 317-675-3857
Electronic Mail: parkerde@aol.com
Founded: 1972
Staff: 1
Activities: school and public programs * education
City Reference Coordinates: 086°02'00"W 40°17'00"N

Turkey Run State Park Nature Center Planetarium
Rural Route 1
Box 164
Marshall, IN 47859
Telephone: 765-597-2654
WWW: http://www.state.in.us/acin/dnr/statepar/turkeyru.html
Founded: 1986
Activities: shows
City Reference Coordinates: 087°24'00"W 39°52'00"N (Newport)

University of Notre Dame, Department of Physics, Astrophysics Group
225 Nieuwland Science Hall
Notre Dame, IN 46556
Telephone: 219-239-7588
Telefax: 219-631-5952
Electronic Mail: physics@nd.edu
          astro@iron.helios.nd.edu
          <userid>@nd.edu
WWW: http://www.science.nd.edu/
      http://www.nd.edu/~astro/
Founded: 1987
Activities: cosmic-ray extensive air showers * Project GRAND (Gamma-Ray Astrophysics at Notre Dame)
Coordinates: 086°13'09"W 41°42'21"N H222m
City Reference Coordinates: 086°14'00"W 41°42'00"N

Valparaiso University, Department of Physics and Astronomy
Valparaiso, IN 46483
Telephone: 219-464-5379
          219-464-5369
Telefax: 219-464-5489
Electronic Mail: physics@valpo.edu
          <userid>@valpo.edu
WWW: http://www.physics.valpo.edu/
      http://physics.valpo.edu/
      http://kepler.valpo.edu/
Founded: 1859 (University)
Activities: education * binary stars * evolved stars * observing
City Reference Coordinates: 087°03'00"W 41°28'00"N

Wabash Valley Astronomical Society (WVAS)
P.O. Box 2020
West Lafayette, IN 47906-0020
Electronic Mail: ed@mdbs.com (Edmund Hartmann, President)
WWW: http://www.stargazing.net/wvas/
Founded: 1971
Membership: 34
Periodicals: "The Nebula"

City Reference Coordinates: 086°53'00"W 40°25'00"N

Warsaw Astronomical Society (WAS)
1308 Sunset Drive
Winona Lake, IN 46590
Telephone: 219-269-6289 (Kurt Eberhardt, President)
Electronic Mail: kurt.eberhardt@kconline.com
WWW: http://clubs.kconline.com/was/
Founded: 1980
Periodicals: (12) "ViewFinder"
City Reference Coordinates: 085°49'00"W 41°14'00"N

USA - Iowa

Amateur Astronomers of Central Iowa (AACI)
c/o Jim Bonser
1009 Harding Street
Tama, IA 50173
Electronic Mail: jbonser@aol.com
WWW: http://www.freeyellow.com/members5/aaci/
City Reference Coordinates: 092°35'00"W 41°58'00"N

Ames Area Amateur Astronomers (AAAA)
3912 Brookdale Circle
Ames, IA 50010
Telephone: 515-233-0223
Electronic Mail: cosmosed@home.com
WWW: http://www.amesastronomers.org/
Founded: 1979
Membership: 85
Activities: monthly meetings * observing * stars parties * education
Sections: Dark-Sky Preservation * Educational Outreach
Periodicals: (12) "Newsletter of the AAAA"
Coordinates: 093°34'08"W 42°05'33"N H318m (McFarland Park Observatory)
City Reference Coordinates: 093°37'00"W 42°02'00"N

Cedar Amateur Astronomers (CAA) Inc.
c/o Doug Slauson (Newsletter Editor)
73 Summit Avenue NE
Swisher, IA 52338
Telephone: 319-857-4674
Electronic Mail: dslauson@cedar-rapids.net
WWW: http://www.cedar-astronomers.org/
Founded: 1979
Membership: 70
Activities: public astronomical outreach * observing
Periodicals: (12) "The Prime Focus"
Observatories: 1 (Palisades-Dows Observatory)
City Reference Coordinates: 091°40'00"W 41°59'00"N (Cedar Rapids)

Central Iowa Astronomers
117 West Hillcrest
Indianola, IA 50125
Telephone: 515-961-0452
Electronic Mail: jardaj1@msn.com (Roger D. Berry, Project Director)
Founded: 2000
Membership: 6
Activities: public education
Coordinates: 093°33'37"W 41°22'50"N H290m
093°34'49"W 41°17'55"N H300m
City Reference Coordinates: 093°32'00"W 41°22'00"N

Des Moines Astronomical Society (DMAS) Inc.
P.O. Box 6114
Des Moines, IA 50309
or :
2307 49th Street
Des Moines, IA 50310-2538
Telephone: 515-277-2739
Electronic Mail: being@dreamsenses.com
WWW: http://www.sciowa.org/~dmas
http://science.raccoon.com/~dmas
Founded: 1970
Membership: 65
Activities: lectures * observing
Periodicals: (12) "Starlight Journal"
Coordinates: 093°17'18"W 41°48'50"N (Ashton Observatory)
City Reference Coordinates: 093°37'00"W 41°35'00"N

Donald A. Schaefer Planetarium
c/o Bettendorf High School
3333 18th Street
Bettendorf, IA 52722
Telephone: 319-332-7001
Founded: 1973

Staff: 2
City Reference Coordinates: 090°30'00"W 41°32'00"N

Drake University, Department of Physics and Astronomy
Harvey Ingham Hall of Science
25th & University
Des Moines, IA 50311
Telephone: 515-271-3141
　　　　　　515-271-2011
Telefax: 515-271-3977
Electronic Mail: lawrence.staunton@drake.edu (Chair)
　　　　　　　　<userid>@drake.edu
WWW: http://www.drake.edu/artsci/physics/physics.html
Founded: 1908
Staff: 8
Activities: far-IR astronomy * active and starburst galaxies
Observatories: 1 (Drake University Municipal Obs. – IAU Code 742)
City Reference Coordinates: 093°37'00"W 41°35'00"N

Gale Observatory (Grant O._)
● See "Grant O. Gale Observatory"

Grant O. Gale Observatory
c/o Grinnell College
P.O. Box 805
Grinnell, IA 50112-0810
Telephone: 641-269-3016
Telefax: 641-269-4285
Electronic Mail: cadmus@ac.grin.edu (Robert R. Cadmus Jr.)
WWW: http://www.grin.edu/www/tour/galeobsv.html
Founded: 1983
Staff: 1
Activities: education * research * public programs
Periodicals: "Grinnell Magazine"
Coordinates: 092°43'12"W 41°45'21"N H318m
City Reference Coordinates: 092°43'00"W 41°45'00"N

Grout Museum of History and Science
503 South Street
Waterloo, IA 50701
Electronic Mail: grout@cedarnet.org
WWW: http://www.cedarnet.org/grout/groutmuseum.html
Founded: 1956
Activities: exhibitions * planetarium * library
City Reference Coordinates: 092°20'00"W 42°30'00"N

Heitkamp Memorial Planetarium
c/o Physics Department
Loras College
Dubuque, IA 52004-0178
Telephone: 319-588-7154
Electronic Mail: kmclaugh@loras.edu
WWW: http://www.loras.edu/~PLANET
Founded: 1966
Staff: 1
Activities: astronomy programs
City Reference Coordinates: 090°41'00"W 42°30'00"N

Iowa Academy of Science (IAS)
c/o Jane Vander Linden (Office Manager)
University of Northern Iowa
Cedar Falls, IA 50614-0508
Telephone: 319-273-2021
Telefax: 319-273-2807
Electronic Mail: jane.vanderlinden@uni.edu
WWW: http://www.iren.net/ias/
Founded: 1875
Periodicals: "Journal of the Iowa Academy of Science", "Iowa Science Teachers Journal"
City Reference Coordinates: 092°27'00"W 42°32'00"N

Iowa State University (ISU), Department of Aerospace Engineering and Engineering Mechanics
304 Town Engineering Building
Ames, IA 50011
Telephone: 515-294-3776
Telefax: 515-294-3262

Electronic Mail: uipress@uiowa.edu
WWW: http://www.uiowa.edu/~uipress
City Reference Coordinates: 091°32'00"W 41°40'00"N

University of Northern Iowa (UNI), Department of Earth Science
Cedar Falls, IA 50614-0335
Telephone: 319-273-6063
           319-273-2759
Telefax: 319-273-7124
Electronic Mail: <userid>@uni.edu
WWW: http://www.earth.uni.edu/
      http://www.earth.uni.edu/astro_fac.html
Founded: 1983
Activities: geology * astronomy * meteorology * earth science education * photometry of variable stars
Coordinates: 092°27'09"W 42°30'27"N (Hillside Observatory)
City Reference Coordinates: 092°27'00"W 42°32'00"N

USA - Kansas

Astronomy Associates of Lawrence (AAL)
c/o Department of Physics and Astronomy
Kansas University
1082 Malott Hall
Lawrence, KS 66045
Telephone: 785-864-3166 (Tony Crass)
Electronic Mail: aal@raven.cc.ukans.edu
WWW: http://www.ukans.edu/~aal/
City Reference Coordinates: 095°14'00"W 38°58'00"N

A Tech
629 SE Quincy
Suite 204
Topeka, KS 66601-3927
Telephone: 877-66A-TECH
877-662-8324
785-235-5252
WWW: http://www.astronomytech.com/
City Reference Coordinates: 095°41'00"W 39°03'00"N

Benedictine College, Department of Physics and Astronomy
c/o Douglas Brothers (Chair)
Second and Division Streets
Atchison, KS 66002-1499
Telephone: 913-367-5340 x527
Telefax: 913-367-6102
Electronic Mail: brothers@benedictine.edu
<userid>@benedictine.edu
WWW: http://www.benedictine.edu/departments/PcAs/PcAsMain.html
Founded: 1858
Staff: 3
Activities: education * stellar atmosphere * pulsating variable stars * stellar populations * local group of galaxies
Coordinates: 095°06'53"W 39°34'21"N
City Reference Coordinates: 095°07'00"W 39°34'00"N

Bushnell Corp., Sport Optics Division
9200 Cody
Overland Park, KS 66214
Telephone: 931-752-3400
Telefax: 913-752-3550
WWW: http://www.bushnell.com/
Founded: 1947
Activities: manufacturing binoculars and telescopes among other products
City Reference Coordinates: 094°41'00"W 38°57'00"N

Celestaire Inc.
416 South Pershing Street
Wichita, KS 67218
Telephone: 316-686-9785
1-800-727-9785 (USA only)
1-888-NAVIGATE (USA only)
Telefax: 316-686-8926
Electronic Mail: info@celestaire.com
WWW: http://www.celestaire.com/
Founded: 1972
Staff: 4
Activities: supplying celestial navigation equipment, books and software
Periodicals: (1) "Catalog"
City Reference Coordinates: 097°20'00"W 37°41'00"N

Emporia State University (ESU), Division of Physical Sciences
1200 Commercial
Emporia, KS 66801-5087
Telephone: 316-341-5330
Telefax: 316-341-6055
Electronic Mail: tabaresg@emporia.edu
keithron@emporia.edu (Ron Keith, Planetarium Director)
WWW: http://www.emporia.edu/physci/physci.htm
http://www.emporia.edu/physci/planet/planet.htm (Peterson Planetarium)
Founded: 1959 (Planetarium)
Staff: 1

Activities: education * public outreach
City Reference Coordinates: 096°11'00"W 38°24'00"N

Exploration Place
300 North McLean Boulevard
Wichita, KS 67203
Telephone: 316-263-3373
Electronic Mail: mratcliffe@exploration.org (Martin Ratcliffe, Director)
WWW: http://www.exploration.org/
Founded: 2000
Staff: 40
Activities: hands-on public science center * planetarium
City Reference Coordinates: 097°20'00"W 37°41'00"N

Heartland Astronomical Research Team (HART)
P.O. Box 3938
Topeka, KS 66604-6938
Telephone: 913-232-3693
Electronic Mail: hart@skygazer.org
WWW: http://www.skygazer.org/hart/
Founded: 1991
Activities: solar system * variable stars
Coordinates: 095°45'00"W 39°00'00"N H300m
City Reference Coordinates: 095°41'00"W 39°03'00"N

International Occultation Timing Association (IOTA)
c/o Craig A. McManus (Secretary)
2760 SW Jewell Avenue
Topeka, KS 66611-1614
Telephone: 785-232-3693
Electronic Mail: iota@inlandnet.net
WWW: http://www.occultations.org/
Founded: 1975
Membership: 300
Activities: occultation and eclipse prediction, observing and analysis
Periodicals: (4) "Occultation Newsletter" (ISSN 0737-6766, circ.: 300)
City Reference Coordinates: 095°41'00"W 39°03'00"N

Kansas Astronomical Observers (KAO)
P.O. Box 49013
Wichita, KS 67202
Electronic Mail: pgoedken@wichita.fn.net (Paul Goedken, President)
         jkfuz@msn.com
WWW: http://home.kscable.com/thekao/
Founded: 1982
City Reference Coordinates: 097°20'00"W 37°41'00"N

Kansas Astrophotographers and Observers Society (KAOS)
c/o Walt Robinson
515 West Kump
Bonner Springs, KS 66012
Telephone: 913-422-1262
Electronic Mail: robinson@sky.net (Rob Robinson, President)
Founded: 1988
Membership: 17
Activities: organizing the annual Great Plains Star Party (GPSP)
Periodicals: "The Scopeville Informer"
City Reference Coordinates: 094°37'00"W 39°05'00"N (Kansas City)

Kansas Cosmosphere and Space Center (KCSC)
1100 North Plum
Hutchinson, KS 67501
Telephone: 316-662-2305
Telefax: 316-662-3693
Electronic Mail: cosmo@cosmo.org
WWW: http://www.cosmo.org/
Founded: 1962
Staff: 50
Activities: OmniMax and Planetarium theatres * Hall of Space * Summer space camps * spacecraft restoration * spacecraft models * Justice Planetarium
City Reference Coordinates: 097°56'00"W 38°05'00"N

Lake Afton Public Observatory (LAPO)
1845 Fairmount
Wichita, KS 67260-0032

Telephone: 316-WSU-STAR
	316-978-7827
	316-978-3191
Telefax: 316-978-3350
Electronic Mail: observatory@wichita.edu
WWW: http://webs.wichita.edu/lapo/
Founded: 1979
Staff: 5
Activities: public education * observing
Periodicals: (4) "Skylights" (circ.: 200)
Coordinates: 097°37'37"W 37°37'20"N H305m
City Reference Coordinates: 097°20'00"W 37°41'00"N

Northeast Kansas Amateur Astronomers League (NEKAAL) Inc.
P.O. Box 951
Topeka, KS 66601
Telephone: 785-235-5467
Electronic Mail: nekaal@yahoo.com
WWW: http://www.nekaal.org/
Founded: 1978
Membership: 50
Activities: monthly observing * education * research * public information
Periodicals: (12) "NEKAAL Newsletter"
Awards: (1) "Astronomer of the Year", "Rising Star"
Observatories: 2 (Keene, Eskridge)
City Reference Coordinates: 095°41'00"W 39°03'00"N

Pittsburg State University, Physics Department
1701 South Broadway
Pittsburg, KS 66762
Telephone: 316-235-4292
Telefax: 316-235-4050
WWW: http://www.pittstate.edu/
	http://www.pittstate.edu/services/scied/kelce.htm (L. Russell Kelce Planetarium)
Founded: 1964
City Reference Coordinates: 094°42'00"W 37°25'00"N

Salina Astronomy Club
803 Mike Drive
Salina, KS 67401
Telephone: 785-827-6004
Electronic Mail: k2r2@midusa.net
Membership: 25
City Reference Coordinates: 097°37'00"W 38°50'00"N

StarMaster Portable Telescopes
c/o Rick Singmaster
2160 Birch Road
Arcadia, KS 66711
Telephone: 620-638-4743
Electronic Mail: starmaster@ckt.net
WWW: http://www.icstars.com/starmaster/
Founded: 1995
Activities: manufacturing telescopes
City Reference Coordinates: 094°37'00"W 37°38'00"N

University of Kansas, Department of Physics and Astronomy
Lawrence, KS 66045
Telephone: 785-864-4626
Telefax: 785-864-5262
Electronic Mail: <userid>@ukans.edu
WWW: http://www.phsx.ukans.edu/
	http://kusmos.phsx.ukans.edu/ (Cosmology Group)
Founded: 1885
Staff: 8
Activities: stars * clusters * galaxies * cosmology * space physics
Coordinates: 095°15'00"W 38°57'36"N H323m (Clyde W. Tombaugh Observatory)
City Reference Coordinates: 095°14'00"W 38°58'00"N

Washburn University of Topeka, Department of Physics and Astronomy
1700 College
Stoffer Science Hall
Topeka, KS 66621
Telephone: 913-231-1010 x1330
Telefax: 913-231-1089

Electronic Mail: <userid>@acc.wuacc.edu
WWW: http://www.washburn.edu/cas/physics/
    htttp://www.washburn.edu/cas/physics/crane/
Founded: 1865 (University)
Staff: 6
Activities: observing * photography * open houses for observing * planetarium
Coordinates: 095°41'48"W 39°02'12"N H306m (Zenas Crane Observatory)
City Reference Coordinates: 095°41'00"W 39°03'00"N

Wichita State University, Physics Department
1845 North Fairmount
Wichita, KS 67260-0032
Telephone: 316-978-3190
Telefax: 316-978-3350
Electronic Mail: <userid>@wichita.edu
WWW: http://www.wichita.edu/physics/
Founded: 1971
Staff: 5
Activities: low-temperature opacities * cool stars * active galaxies * education
City Reference Coordinates: 097°20'00"W 37°41'00"N

USA - Kentucky

Arnim D. Hummel Planetarium
c/o Eastern Kentucky University
Richmond, KY 40475
Telephone: 606-622-1547
Telefax: 606-622-6205
WWW: http://www.lexinfo.com/museum/hummel.html
http://www.planetarium.acs.eku.edu/
Founded: 1979
Staff: 7
Activities: shows for schools and public
Observatories: 1 (Smith Park Observatory)
City Reference Coordinates: 084°18'00"W 37°45'00"N

Big South Fork Star Gazers
c/o James Paulson
HC 82 Box 366-A
Pine Knot, KY 42635
Telephone: 606-354-1365
Electronic Mail: jpaulson@highland.net
WWW: http://sfsg.8k.com/
Membership: 10
City Reference Coordinates: 084°51'00"W 36°50'00"N (Monticello)

Blue Grass Amateur Astronomical Club (BGAAC)
c/o B.C. Canon
1016 Della Drive
Lexington, KY 40504
Telephone: 859-278-6155
Electronic Mail: pol140@ukcc.uky.edu
WWW: http://www.ms.uky.edu/~bgaac/
Founded: 1959
Membership: 31
Activities: observing * meeting programs
Periodicals: "The Practical Observer"
City Reference Coordinates: 084°30'00"W 38°03'00"N

Gheens Science Hall and Rauch Planetarium
c/o University of Louisville
Louisville, KY 40292
Telephone: 502-852-5855
Telefax: 502-852-0831
Electronic Mail: shawn.laatsch@louisville.edu (Shawn A. Laatsch, Planetarium Director)
WWW: http://www.louisville.edu/planetarium/
Founded: 1962
Staff: 4
Activities: education * planetarium programs * laser shows
Periodicals: (4) "Eye on the Sky"
Coordinates: 085°45'00"W 38°15'00"N H155m
City Reference Coordinates: 085°45'00"W 38°16'00"N

Golden Pond Planetarium and Observatory
c/o Land Between The Lakes
100 Van Morgan Drive
Golden Pond, KY 42211
Telephone: 502-924-2000
1-800-455-5897 (USA only)
Electronic Mail: lbl-planetarium@tva.gov
WWW: http://www2.lbl.org/lbl/PLGate.html
Founded: 1985 (LBL: 1963)
Staff: 2
City Reference Coordinates: 088°04'00"W 36°48'00"N

Hilltopper Astronomy Club (HAC)
c/o Department of Physics and Astronomy
Western Kentucky University
One Big Red Way
Bowling Green, KY 42101-3576
Telephone: 270-745-4537
Telefax: 270-745-2014
Electronic Mail: wku-astronomy@yahoogroups.com
WWW: http://www.geocities.com/CapeCanaveral/Cockpit/5715/

Founded: 1999
Membership: 4
Activities: public education
Coordinates: 086°26'57"W 36°59'12"N H198m
City Reference Coordinates: 086°37'00"W 37°00'00"N

Hummel Planetarium (Arnim D._)
• See "Arnim D. Hummel Planetarium"

Johnson County Middle School (JCMS), Rocket and Astronomy Club
251 North Mayo Trail
Paintsville, KY 41240
Telephone: 606-789-4133
Telefax: 606-789-4135
Electronic Mail: astronomyclub@yahoo.com
WWW: http://www.angelfire.com/ky/astronomyclub/
City Reference Coordinates: 082°48'00"W 37°48'00"N

Kentucky Space Grant Consortium (KSGC)
c/o Richard Hackney (Director)
TCCW 246
Department of Physics and Astronomy
Western Kentucky University
One Big Red Way
Bowling Green, KY 42101
Telephone: 270-745-4156/4044
Telefax: 270-745-4255
Electronic Mail: ksgc@wku.edu
richard.hackney@wku.edu
WWW: http://wkuweb1.wku.edu/Dept/Academic/Ogden/Phyast/k0_.htm
Founded: 1992
City Reference Coordinates: 086°37'00"W 37°00'00"N

Louisville Amateur Astronomers (LAA)
8302 Creek Trail Court
Louisville, KY 40291
Telephone: 502-239-9488
Electronic Mail: louamatastro@prodigy.net
WWW: http://www.geocities.com/lvl_astro
Founded: 1997
Membership: 21
Activities: sidewalk astronomy * public education * observing
Periodicals: (4) "Louisville Amateur Astronomers Newsletter"
City Reference Coordinates: 085°45'00"W 38°16'00"N

Louisville Astronomical Society (LAS) Inc.
P.O. Box 701043
Louisville, KY 40270-1043
Electronic Mail: president@louisville-astro.org
WWW: http://www.louisville-astro.org/
Founded: 1933
Membership: 75
Activities: observing
Periodicals: (12) "Starword"
City Reference Coordinates: 085°45'00"W 38°16'00"N

Louisville Science Center (LSC)
727 West Main Street
Louisville, KY 40202
Telephone: 502-561-6100
WWW: http://lsclouisnet.org/
City Reference Coordinates: 085°45'00"W 38°16'00"N

Midwestern Astronomers (MWA)
1720 Monticello Drive
Fort Wright, KY 41011
Telephone: 606-344-0264
Founded: 1977
Membership: 22
Activities: planets * deep sky * photometry * public education * classes * star watches * photography
Periodicals: "Newsletter"
Observatories: 2 (Milford, USA-OH & Bracken County, USA-KY)
City Reference Coordinates: 084°30'00"W 39°04'00"N (Covington)

Moore Observatory
● See "University of Louisville, Moore Observatory"

National Deep Sky Observers Society (NDSOS)
1607 Washington Boulevard
Louisville, KY 40242
Telephone: 502-426-4399
Electronic Mail: deepskyspy1@cs.com
WWW: http://www.cismall.com/deepsky/
Founded: 1976
Membership: 120
Activities: promoting deep-sky observing * offering observing challenges * assisting amateur research * information products
Periodicals: (4) "The Deep Sky", "The Practical Observer"
City Reference Coordinates: 085°45'00"W 38°16'00"N

Night Sky Observers (NSO)
c/o Chris S. Thomas
3621 Brownsboro Road
Suite 402D
Louisville KY 40207
or :
7004 Harvest Gold Way
Suite 7
Louisville, KY 40291
Telephone: 502-721-8794
WWW: http://www.angelfire.com/ky/astronomers/nso.html
Founded: 2000
Observatories: 1 (Otter Creek Observatory)
City Reference Coordinates: 085°45'00"W 38°16'00"N

Radio-Sky Publishing
P.O.Box 3552
Louisville, KY 40201-3552
Electronic Mail: radiosky@radiosky.com
WWW: http://www.radiosky.com/
Founded: 1991
Staff: 1
Activities: radio astronomy software and books * Jupiter storm predictions * amateur radiotelescope building
City Reference Coordinates: 085°45'00"W 38°16'00"N

Rauch Planetarium
● See "Gheens Science Hall and Rauch Planetarium"

University of Kentucky, Department of Physics and Astronomy
177 Chemistry/Physics Building
Lexington, KY 40506-0055
Telephone: 606-257-6722
Telefax: 606-323-2846
Electronic Mail: <userid>@pop.uky.edu
WWW: http://www.pa.uky.edu/
Staff: 51
Activities: research and education in physics and astronomy
City Reference Coordinates: 084°30'00"W 38°03'00"N

University of Louisville, Department of Physics
102 Natural Sciences Building
Louisville, KY 40292
Telephone: 502-852-6790
Telefax: 502-852-0742
Electronic Mail: physics@physics.louisville.edu
<userid>@physics.louisville.edu
WWW: http://www.physics.louisville.edu/
Founded: 1798 (University)
Activities: astrophysics and atomic physics * condensed matter theory * scanning tunneling microscopy * high-energy physics * experimental nuclear physics
City Reference Coordinates: 085°45'00"W 38°16'00"N

University of Louisville, Moore Observatory
8000 Old Zaring Road
Crestwood, KY 40014
Telephone: 502-241-7841
Electronic Mail: <userid>@moondog.astro.louisville.edu
WWW: http://www.astro.louisville.edu/

Founded: 1978
Coordinates: 085°31'48"W 38°20'06"N H216m
City Reference Coordinates: 085°38'00"W 38°19'00"N

Weatherford Planetarium
c/o Berea College
C.P.O. 2326
Berea, KY 40404
Telephone: 859-986-9341 x6240
Electronic Mail: smithtpowell@berea.edu (Smith T. Powell, Director)
WWW: http://physics.berea.edu/department/plantary.html
Founded: 1985
City Reference Coordinates: 084°17'00"W 37°34'00"N

Western Kentucky Amateur Astronomers (WKAA)
c/o Ross Workman
508 Park Avenue
Dawson Springs, KY 42408
Telephone: 270-797-8959
Electronic Mail: planetman@vci.net
WWW: http://www.wkaa.net/
Membership: 30
Activities: public outreach * observing * organizing Twin Lakes Star Party (TLSP)
City Reference Coordinates: 087°40'00"W 37°09'00"N

Western Kentucky University (WKU), Department of Physics and Astronomy
Ogden College
1526 Russellville Road
Bowling Green, KY 42101-3576
Telephone: 502-745-4357
Telefax: 502-745-6471
Electronic Mail: <userid>@wku.edu
WWW: http://www.wku.edu/Dept/Academic/Ogden/Phyast/p0_.htm
    http://www.wku.edu/Dept/Academic/Ogden/Phyast/p5_.htm (Observatory)
    http://www.wku.edu/Dept/Academic/Ogden/Phyast/p4_.htm (Hardin Planetarium)
Founded: 1877
Staff: 16
Activities: education * research
Coordinates: 086°36'40"W 36°55'11"N H238m
City Reference Coordinates: 086°37'00"W 37°00'00"N

USA - Louisiana

Amateur Astronomers of Acadiana (AAofA)
637 Girard Park Drive
Lafayette, LA 70503
Telephone: 318-291-5544
Telefax: 318-291-5464
Electronic Mail: dexterledoux@linknet.net (Dexter J. LeDoux, President)
Founded: 1984
Membership: 8
Activities: photography * photoelectric photometry * meteors * solar system * public and school telescope viewings
Coordinates: 091°49'42"W 30°08'49"N
City Reference Coordinates: 092°01'00"W 30°14'00"N

Baton Rouge Astronomical Society (BRAS)
c/o Craig Brenden
6348 Double Tree Court
Baton Rouge, LA 70817-1685
Telephone: 225-751-1685
Electronic Mail: walt@premier.net (Walt Cooney, Vice-President)
WWW: http://www.bro.lsu.edu/bras/
Founded: 1981
City Reference Coordinates: 091°11'00"W 30°23'00"N

Kenner Historical Museum, Planetarium
2100 3rd Street
Suite 11
Kenner, LA 70062
Telephone: 504-468-7228
WWW: http://pop3.dcc.edu/~mdoyle/kenner.htm
City Reference Coordinates: 090°15'00"W 29°59'00"N

Lafayette Natural History Museum (LNHM), Planetarium
637 Girard Park Drive
Lafayette, LA 70503
Telephone: 318-291-5544
Telefax: 318-291-5464
Electronic Mail: davehostetter@linknet.net (David E. Hostetter, Planetarium Curator)
WWW: http://www.lnhm.org/
Founded: 1969
Staff: 8 (Planetarium: 2)
Activities: public and school planetarium programs * classes * workshops * aerospace education
Periodicals: (6) "Plants & Planets"
City Reference Coordinates: 092°01'00"W 30°14'00"N

Louisiana Arts and Science Center
P.O. Box 70821
Baton Rouge, LA 70821
or :
100 South River Road
Baton Rouge, LA 70802
Telephone: 504-344-5272
Telefax: 504-344-9477
WWW: http://www.tfaoi.com/newsmu/nmus123.htm
City Reference Coordinates: 091°11'00"W 30°23'00"N

Louisiana Nature Center and Planetarium
P.O. Box 870610
New Orleans, LA 70187-0610
or :
10601 Dwyer Road
New Orleans, LA
Telephone: 504-246-STAR
Electronic Mail: mtrotter@auduboninstitute.org (Mark A. Trotter, Planetarium Director)
    dcowles@auduboninstitute.org (Dennis J. Cowles, Ass. Planetarium Director)
WWW: http://home.gs.verio.net/~strotter/
Staff: 5 (Planetarium)
Activities: public and group programs * laser shows * meteorite collection
City Reference Coordinates: 090°03'00"W 30°00'00"N

Louisiana State University (LSU), Department of Physics and Astronomy
Baton Rouge, LA 70803-4001
Telephone: 225-578-2261

Telefax: 225-578-5855
Electronic Mail: <userid>@rouge.phys.lsu.edu
WWW: http://www.physics.lsu.edu/
http://www.astronomy.lsu.edu/
http://spdsch.phys.lsu.edu/ (Space Science and Particle Astrophysics)
http://mleesun.phys.lsu.edu/observatory/ (Baton Rouge Observatory)
http://www.phys.lsu.edu/observatory/ (Baton Rouge Observatory)
Founded: 1934
Staff: 9
Periodicals: "Louisiana State University Observatory Contributions"
Coordinates: 090°58'06"W 30°48'06"N H70m (IAU Code 747)
City Reference Coordinates: 091°11'00"W 30°23'00"N

Louisiana State University Press
P.O. Box 25053
Baton Rouge, LA 70894-5053
Telephone: 1-800-861-3477 (Orders and Customer Service only)
Telefax: 504-388-6461
1-800-305-4416 (Orders and Customer Service only)
Electronic Mail: lsupress@lsu.edu
WWW: http://www.lsu.edu/lsupress/
Founded: 1935
Staff: 33
• Publisher
City Reference Coordinates: 091°11'00"W 30°23'00"N

LSU Press
• See "Louisiana State University Press"

Pontchartrain Astronomy Society (PAS)
409 Williams Boulevard
Kenner, LA 70062
or :
c/o Michael D. Sandras (President)
948 Avenue "E"
Westwego, LA 70094-4411
Telephone: 504-468-7229
504-340-0256
Electronic Mail: astrox@ix.netcom.com
WWW: http://www.nightskydesign.com/PAS/Index.html
Founded: 1959
Membership: 179
Activities: monthly meetings * star parties
Periodicals: (12) "PAStimes"
City Reference Coordinates: 090°15'00"W 29°59'00"N (Kenner)
090°09'00"W 29°55'00"N (Westwego)

Saint Charles Parish Library, Planetarium
105 Lakewood Drive
P.O. Box 949
Luling, LA 70070
Telephone: 504-785-8471
Telefax: 504-785-8499
WWW: http://www.stcharles.lib.la.us/planetarium/homepage.html
City Reference Coordinates: 097°39'00"W 29°51'00"N

Shreveport-Bossier Astronomical Society (SBAS) Inc.
353 Ockley Drive
Shreveport, LA 71105
Telephone: 318-797-1524
Electronic Mail: clucas@softdisk.com (Cran Lucas)
WWW: http://www.lsus.edu/nonprofit/sbas/sbas.htm
Founded: 1959
Observatories: 1 (Ralph A. Worley Observatory)
City Reference Coordinates: 093°45'00"W 32°30'00"N

Shreveport Parks and Recreation Planetarium
• See "SPAR Planetarium"

SPAR Planetarium
2820 Pershing Boulevard
Shreveport, LA 71109
Telephone: 318-673-STAR
Electronic Mail: daveastar1@abcstar.com
WWW: http://www.abcstar.com/planetarium.htm

City Reference Coordinates: 093°45'00"W 32°30'00"N

University of Louisiana at Lafayette, Department of Physics
Broussard Hall
Room 103
P.O. Box 44210
Lafayette, LA 70504-4210
Telephone: 337-482-6691
Telefax: 337-482-6699
Electronic Mail: <userid>@louisiana.edu
WWW: http://www.louisiana.edu/Academic/Sciences/PHYS/
City Reference Coordinates: 092°01'00"W 30°14'00"N

University of Southwestern Louisiana, Department of Physics
• See now "University of Louisiana at Lafayette"

USA - Maine

Astronomical Society of Northern New England (ASNNE)
P.O. Box 1338
Kennebunk, ME 04043-1338
Telephone: 207-967-5945
Electronic Mail: info@asnne.org
WWW: http://www.asnne.org/
Founded: 1982
Membership: 80
Activities: star parties * open house events * lectures and shows at local schools * working at reducing light pollution
Periodicals: (12) "Skylights"
Observatories: 2 (Starfield & East Sky Observatories)
City Reference Coordinates: 070°28'00"W 43°22'00"N

Bates College, Department of Physics and Astronomy
Carnegie Science Building
44 Campus Avenue
Lewiston, ME 04240
Telephone: 207-786-6490
Telefax: 207-786-8334
Electronic Mail: <userid>@abacus.bates.edu
WWW: http://www.bates.edu/Faculty/Physics/
Founded: 1855
Staff: 9
Activities: education * research * planetarium
City Reference Coordinates: 070°13'00"W 44°06'00"N

Bowdoin College, Department of Physics and Astronomy
Brunswick, ME 04011
Telephone: 207-725-3308
Telefax: 207-725-3638
Electronic Mail: <userid>@bowdoin.edu
WWW: http://www.bowdoin.edu/dept/physics/
Founded: 1802
Staff: 9
Coordinates: 069°57'48"W 43°54'36"N H25m
City Reference Coordinates: 069°58'00"W 43°55'00"N

Downeast Amateur Astronomers (DEAA)
c/o Charles Sawyer
P.O. Box K
Pembroke, ME 04666
Telephone: 207-726-4621
Electronic Mail: csawyer@nemaine.com
WWW: http://www.deaa.2ya.com/
Membership: 10
City Reference Coordinates: 067°10'00"W 44°57'00"N

Galileo Society
c/o David Gay
1381 River Road
Livermore, ME 04253
Telephone: 207-897-3794
Electronic Mail: galileosociety@geocities.com
davidgay@bigfoot.com
WWW: http://www.geocities.com/CapeCanaveral/6506/
Founded: 1996
City Reference Coordinates: 068°47'00"W 44°49'00"N (Bangor)
070°11'00"W 44°28'00"N (Livermore Falls)

Jordan Planetarium and Observatory (Maynard F._)
• See "University of Maine, Maynard F. Jordan Planetarium and Observatory"

Maine Astronomical Society
c/o Richard Kahn
Hatch Hill Road
Greene, ME 04236
Electronic Mail: drquattrocchi@earthlink.net
WWW: http://www.maineastronomicalsociety.com/
City Reference Coordinates: 070°08'00"W 44°11'00"N

Maynard F. Jordan Planetarium and Observatory
• See "University of Maine, Maynard F. Jordan Planetarium and Observatory"

Penobscot Valley Star Gazers (PVSG)
c/o Ralph Mallett (Secretary)
76 Eastern Avenue
Brewer, ME 04412
Telephone: 207-989-2074
Electronic Mail: pres@mainestargazers.org
WWW: http://mainestargazers.org/
Founded: 1990
Membership: 38
Activities: public and private star parties * education
City Reference Coordinates: 068°44'00"W 44°48'00"N

Project Pluto
168 Ridge Road
Bowdoinham, ME 04008
Telephone: 207-666-5750
        1-800-777-5886 (USA only)
        1-800-PR-PLUTO (USA only)
Telefax: 207-666-3149
Electronic Mail: pluto@projectpluto.com
WWW: http://www.projectpluto.com/
Founded: 1992
• Software producer
City Reference Coordinates: 070°27'00"W 43°29'00"N (Biddeford)

Southworth Planetarium
c/o University of Southern Maine
96 Falmouth Street
Portland, ME 04103
Telephone: 207-780-4249
Telefax: 207-780-4051
Electronic Mail: deines@portland.maine.edu (Planetarium Questions)
        rgallant@portland.maine.edu (Roy A. Gallant, Director)
WWW: http://www.usm.maine.edu/~planet/
    http://www.usm.maine.edu/planet/
Founded: 1970
Staff: 2
Activities: astronomy and laser light shows
City Reference Coordinates: 070°17'00"W 43°39'00"N

University of Maine, Department of Physics and Astronomy
5709 Bennett Hall
Orono, ME 04469-5709
Telephone: 207-581-1016
Electronic Mail: galaxy@maine.maine.edu
        <userid>@maine.maine.edu
WWW: http://inferno.asap.um.maine.edu/physics/
Staff: 3
Activities: computer models of galactic dynamics and evolution * observing large-scale structures among clusters of galaxies
City Reference Coordinates: 068°40'00"W 44°53'00"N

University of Maine, Maynard F. Jordan Planetarium and Observatory
5781 Wingate Hall
Orono, ME 04469-5781
Telephone: 207-581-1341
Telefax: 207-581-1314
Electronic Mail: aland@maine.maine.edu (Alan Davenport, Director)
WWW: http://www.umainesky.com/
Founded: 1954
Staff: 7
Activities: public observatory * public planetarium * education
Coordinates: 068°39'41"W 44°54'05"N H61m
City Reference Coordinates: 068°40'00"W 44°53'00"N

Youreyeball Observatory
c/o John Stetson
63 Ledgewood Drive
Falmouth, ME 04105
Electronic Mail: jstetson@maine.rr.com
City Reference Coordinates: 070°16'00"W 43°39'00"N

USA - Maryland

American Association of Physics Teachers (AAPT)
One Physics Ellipse
College Park, MD 20740-3845
Telephone: 301-209-3333
Telefax: 301-209-0845
Electronic Mail: aapt-exec@aapt.org
WWW: http://www.aapt.org/
    http://www.psrc-online.org/ (Physical Sciences Resource Center - PSRC)
    http://www.kzoo.edu/ajp/ (American Journal of Physics - AJP)
    http://www.aapt.org/tpt/ (The Physics Teacher - TPT)
Founded: 1930
Periodicals: (12) "American Journal of Physics (AJP)" (ISSN 0002-9505); (9) "The Physics Teacher (TPT)" (ISSN 0031-921x); (4) "Announcer" (ISSN 1042-0851)
City Reference Coordinates: 076°55'00"W 39°00'00"N

American Center for Physics (ACP)
One Physics Ellipse
College Park, MD 20740
Electronic Mail: <userid>@acp.org
WWW: http://www.acp.org/
Founded: 1994
City Reference Coordinates: 076°55'00"W 39°00'00"N

American Institute of Physics (AIP), Headquarters
One Physics Ellipse
College Park, MD 20740-3843
Telephone: 301-209-3100
Telefax: 301-209-0843
    301-209-3133
Electronic Mail: admin@aip.org
    aipinfo@aip.org
WWW: http://www.aip.org/
    http://www.aip.org/pt/ (Physics Today)
    http://www.oclc.org/oclc/menu/aplo.htm (Applied Physics Letters)
Founded: 1931
Periodicals: (52) "Applied Physics Letters (APL)" (ISSN 0003-6951); (24) "General Physics Advance Abstracts", "JETP Letters", "Journal of Applied Physics" (ISSN 0021-8979), "Journal of Chemical Physics" (ISSN 0021-9606), "Physics Briefs" (ISSN 0170-7434); (12) "Journal of Mathematical Physics" (ISSN 0022-2488), "Physica Scripta", "Physics of Fluids A", "Physics of Fluids B", "Physics Today" (ISSN 0031-9228), "Review of Scientific Instruments", "Soviet Journal of Low-Temperature Physics", "Soviet Journal of Nuclear Physics", "Soviet Journal of Plasma Physics", "Soviet Journal of Quantum Electronics", "Soviet Physics - Doklady", "Soviet Physics - JETP", "Soviet Physics - Semiconductors", "Soviet Physics - Solid State", "Soviet Physics - Technical Physics", "Soviet Physics - Uspekhi", "Soviet Technical Physics Letters", "Superconductivity"; (6) "Computers in Physics" (see separate entry), "Journal of Physical and Chemical Reference Data", "Soviet Astronomy", "Soviet Astronomy Letters", "Soviet Journal of Particles and Nuclei", "Soviet Physics - Acoustics", "Soviet Physics - Crystallography"; (4) "Chinese Physics", "Current Physics Index" (ISSN 0098-9819); "AIP Conference Proceedings" (ISSN 0094-243x), "American Journal of Physics" (see separate entry), "American Physical Society Bulletin" (ISSN 0003-0503), "Applied Optics" (ISSN 0003-6935), "Physical Review A: General Physics" (ISSN 0556-2791), "Physical Review B: Condensed Matter" (ISSN 0163-1824), "Physical Review C: Nuclear Physics" (ISSN 0556-2813), "Physical Review D: Particle and Fields" (ISSN 0556-2821), "Physical Review Abstracts" (ISSN 0048-4024), "Physical Review Index" (ISSN 0099-0003), "Physical Review Letters" (ISSN 0031-9007), "Physical Review Supplements" (ISSN 0066-5495), "Physics Teacher" (ISSN 0031-921x), and many other journals and translations
City Reference Coordinates: 076°55'00"W 39°00'00"N

American Physical Society (APS)
One Physics Ellipse
College Park, MD 20740-3844
Telephone: 301-209-3200
Telefax: 301-209-0865
Electronic Mail: <userid>@aps.org
WWW: http://www.aps.org/
Founded: 1899
Periodicals: "Physical Review A: General Physics" (ISSN 0556-2791), "Physical Review B: Condensed Matter" (ISSN 0163-1824), "Physical Review C: Nuclear Physics" (ISSN 0556-2813), "Physical Review D: Particle and Fields" (ISSN 0556-2821), "Physical Review Abstracts" (ISSN 0048-4024), "Physical Review Letters" (ISSN 0031-9007), "Reviews of Modern Physics", "Bulletin of the American Physical Society", "News of the American Physical Society"
City Reference Coordinates: 076°55'00"W 39°00'00"N

American Society for Information Science (ASIS)
8720 Georgia Avenue
Suite 501
Silver Spring, MD 20910

Telephone: 301-495-0900
Telefax: 301-495-0810
Electronic Mail: asis@asis.org
WWW: http://www.asis.org/
Founded: 1937
Staff: 7
Activities: conferences * publishing
Periodicals: (10) "Journal of the American Society for Information Science" (ISSN 0002-8231); (6) "Bulletin of the American Society for Information Science" (ISSN 0095-4403); (1) "Annual Review of Information Science and Technology" (ISSN 0066-4200)
Awards: (1) "Watson Davis Award", "Award of Merit", "Research Award"
City Reference Coordinates: 077°03'00"W 39°02'00"N

Arthur Storer Planetarium
600 Dares Beach Road
Prince Frederick, MD 20678
Telephone: 410-535-7425 (Nick Graziano, Planetarium Director)
Electronic Mail: planet@calvertnet.k12.md.us
WWW: http://www.calvertnet.k12.md.us/schools/planetarium/planetarium.html
Founded: 1977
Staff: 2
Activities: programs for school children * public programs * astronomy resources
Periodicals: (12) "Newsletter"
Coordinates: 076°53'00"W 38°32'10"N H43m
City Reference Coordinates: 076°35'00"W 38°33'00"N

Astronomical Data Center (ADC)
• Closed on 01 October 2002.

Astrophysics Data Facility (ADF)
• See "National Aeronautics and Space Administration (NASA), Goddard Space Flight Center (GSFC), Astrophysics Data Facility (ADF)"

Baltimore Astronomical Society (BAS)
c/o Maryland Science Center (MSC)
601 Light Street
Baltimore, MD 21230
Telephone: 410-685-2370 x427
Electronic Mail: stargzer@baltastro.org
WWW: http://www.baltastro.org/
Founded: 1881
Membership: 150
Activities: promoting and encouraging interest in astronomy * observing parties
Periodicals: (12) "The Newsletter of the Baltimore Astronomical Society"
City Reference Coordinates: 076°37'00"W 39°17'00"N

Banneker Planetarium (Benjamin_)
• See "Benjamin Banneker Planetarium"

Benjamin Banneker Planetarium
c/o Community College of Baltimore County
Catonsville Campus
800 South Rolling Road
Baltimore, MD 21228
Telephone: 410-455-4560
Founded: 1966
Staff: 5
Activities: shows for college classes, school groups and community organizations
City Reference Coordinates: 076°37'00"W 39°17'00"N

Brish Planetarium (William M._)
• See "William M. Brish Planetarium"

Bryan Planetarium (K. Price_)
• See "K. Price Bryan Planetarium"

Cambridge Scientific Abstracts (CSA)
7200 Wisconsin Avenue
Suite 601
Bethesda, MD 20814-4823
Telephone: 301-961-6700
          301-961-6750
Telefax: 301-961-6720
WWW: http://www.csa.com/
Staff: 110

Activities: publishing abstract journal and databases (on-line and CD-ROM) in aquatic, environmental, life and physical sciences
City Reference Coordinates: 077°06'00"W 38°59'00"N

Center for Archaeoastronomy
P.O. Box X
College Park, MD 20741-3022
Telephone: 301-864-6637
Telefax: 301-699-5337
WWW: http://www.wam.umd.edu/~tlaloc/archastro/
Founded: 1978
City Reference Coordinates: 076°55'00"W 39°00'00"N

Company Seven (C-VII), Astro-Optics Division
Box 2587
Montpelier, MD 20709-2587
or :
14300 Cherry Lane Court
Laurel, MD 20707
Telephone: 301-953-2000
Telefax: 301-470-3083
Electronic Mail: info@company7.com
WWW: http://www.company7.com/
Founded: 1974
Staff: 12
Activities: designing, manufacturing and training on telescopes (particularly apochromatic refractors) * mountings * CCD-guiding systems * accessories
Periodicals: "The C-VII Journal"
Awards: (1) "Quality/Design Awards to Manufacturers"
Observatories: 2 (Big Meadows, Shenandoah National Park, USA-VA & Bowie Park, USA-MD)
City Reference Coordinates: 076°51'00"W 39°06'00"N (Laurel)

Computers in Physics (CIP)
c/o American Institute of Physics
One Physics Ellipse
College Park, MD 20740-3843
Telephone: 301-209-3001
Telefax: 301-209-0842
Electronic Mail: rhonda@aip.org
WWW: http://www.aip.org/cip/
Founded: 1987
Staff: 20
Activities: publishing news, feature articles, columns and papers focusing on computers in physics and computational science
• Journal (6) (ISSN 0894-1866)
City Reference Coordinates: 076°55'00"W 39°00'00"N

Cumberland Astronomy Club (CAC)
c/o Stephen Luzader
59 Centennial Street
Frostburg, MD 21532
Telephone: 301-689-1976
Electronic Mail: sluzader@frostburg.edu
WWW: http://antoine.fsu.umd.edu/phys/luzader/cac/
City Reference Coordinates: 078°56'00"W 39°39'00"N

Eclipse Edge Expeditions
P.O. Box 15186
Chevy Chase, MD 20825
Telephone: 1-800-898-3343 (USA only)
         202-362-9176
WWW: http://eclipseedge.org/
City Reference Coordinates: 077°05'00"W 38°58'00"N

EG&G Inc.
900 Clopper Road
Suite 200
Gaithersburg, MD 20878
Telephone: 301-840-3000
Telefax: 301-840-3123
WWW: http://www.egginc.com/
Founded: 1947
Staff: 35000
Activities: manufacturing scientific instruments * management services
City Reference Coordinates: 077°12'00"W 39°09'00"N

Frostburg State University (FSU), Planetarium
101 Braddock Road
Frostburg, MD 21532-1099
Telephone: 301-687-4270
Telefax: 301-687-7966
Electronic Mail: r_doyle@fre.fsu.umd.edu
WWW: http://www.fsu.umd.edu/planet.html
Founded: 1969
Staff: 1
City Reference Coordinates: 078°56'00"W 39°39'00"N

Garrett County Board of Education Planetarium
40 South 4th Street
Oakland, MD 21550
Telephone: 301-334-8932
WWW: http://www.ga.k12.md.us/
Founded: 1967
Staff: 1
Activities: education
City Reference Coordinates: 079°24'00"W 39°25'00"N

Goddard Astronomy Club (GAC)
c/o NASA Goddard Space Flight Center
Code 912
Greenbelt, MD 20771
Telephone: 301-614-6282
Electronic Mail: evans@agnes.gsfc.nasa.gov (Keith Evans)
WWW: http://gewa.gsfc.nasa.gov/~astro/gac.html
Founded: 1961
Membership: 40
Activities: star parties * meetings * Astronomy Day
Periodicals: (12) "Newsletter"
Observatories: 1 (Goddard Optical Research Facility)
City Reference Coordinates: 076°53'00"W 39°01'00"N

Goddard Space Flight Center (GSFC)
• See "National Aeronautics and Space Administration (NASA), Goddard Space Flight Center (GSFC)"

Goucher College, Department of Physics and Astronomy
1021 Dulaney Valley Road
Baltimore, MD 21204
Telephone: 410-337-6320
Telefax: 410-337-6408
Electronic Mail: <userid>@goucher.edu
WWW: http://www.goucher.edu/physics/
City Reference Coordinates: 076°37'00"W 39°17'00"N

Greenbelt Astronomy Club (GAC)
c/o Owens Science Center
P.O. Box 727
Greenbelt, MD 20768
or :
9601 Greenbelt Road
Lanham, MD 20706
Telephone: 301-918-8750
Telefax: 301-918-8753
Electronic Mail: rwaugh@umd5.umd.edu (Russel R. Waugh)
WWW: http://lheawww.gsfc.nasa.gov/docs/outreach/gac/GAC.html
Founded: 1994
City Reference Coordinates: 076°52'00"W 38°58'00"N

Harford County Astronomical Society (HCAS)
P.O. Box 906
Bel Air, MD 21014
Telephone: 301-836-4155
Electronic Mail: virgoastronomics@erols.com
WWW: http://www.harfordastro.org/
Founded: 1970
Membership: 65
Activities: monthly meetings * weekly photography lab * weekly discussion night * star parties
Coordinates: 076°16'59"W 39°33'50"N H140m
City Reference Coordinates: 076°21'00"W 39°32'00"N

Hopkins University (JHU) (Johns_)
• See "Johns Hopkins University (JHU)"

Howard Astronomical League of Central Maryland
c/o Malcolm M. Wilette
2407 Birch Drive
Baltimore, MD 21207
WWW: http://www.howardastro.org/
City Reference Coordinates: 076°37'00"W 39°17'00"N

Howard B. Owens Science Center (HBOSC), Planetarium
9601 Greenbelt Road
Lanham, MD 20706
Telephone: 301-918-8750
Telefax: 301-918-8753
WWW: http://www.pgcps.pg.k12.md.us/~hbowens/
Founded: 1978
Staff: 15
Activities: planetarium education and programs * laboratory experiences * Challenger Learning Center
City Reference Coordinates: 076°52'00"W 38°58'00"N

International Occultation Timing Association (IOTA)
c/o David Dunham
7006 Megan Lane
Greenbelt, MD 20770-3012
Telephone: 301-474-4945
301-474-4722
Telefax: 301-953-6556
Electronic Mail: david_dunham@jhuapl.edu
WWW: http://www.occultations.org/
Founded: 1975
Membership: 250
Activities: occultation and eclipse prediction, observing and analysis
Periodicals: (4) "Occultation Newsletter" (ISSN 0737-6766, circ.: 300)
City Reference Coordinates: 076°53'00"W 39°01'00"N

International Planetarium Society (IPS)
c/o Shawn Laatsch (Treasurer)
Arthur Storer Planetarium
600 Dares Beach Road
Prince Frederick, MD 20678
Telephone: 410-535-7339
Telefax: 410-535-7200
Electronic Mail: 102424.1032@compuserve.com
WWW: http://www.ips-planetarium.org/
Founded: 1970
Membership: 700
Activities: international confederation of regional planetarium groups * conference every two years * circulating information among its members and upgrading the planetarium profession
Periodicals: (4) "The Planetarian" (ISSN 0090-3213)
City Reference Coordinates: 076°35'00"W 38°33'00"N

JHU Press
• See "Johns Hopkins University Press"

Johns Hopkins University (JHU), Department of Earth and Planetary Sciences
Olin Hall
Charles & 34th Streets
Baltimore, MD 21218
Telephone: 301-338-7034
Telefax: 301-338-7933
WWW: http://www.jhu.edu/~eps/
Founded: 1968
City Reference Coordinates: 076°37'00"W 39°17'00"N

Johns Hopkins University (JHU), Department of Physics and Astronomy
3400 North Charles Street
Baltimore, MD 21218-2686
or :
(for courier-delivered packages)
3701 San Martin Drive
Baltimore, MD 21218
Telephone: 410-516-7346
Telefax: 410-516-7239
WWW: http://www.pha.jhu.edu/
City Reference Coordinates: 076°37'00"W 39°17'00"N

Johns Hopkins University (JHU), Department of Physics and Astronomy, Center for Astrophysical Sciences (CAS)
Charles and 34th Streets
Baltimore, MD 21218
Telephone: 410-516-8497
Telefax: 410-516-8260
Electronic Mail: <userid>@pha.jhu.edu
WWW: http://eta.pha.jhu.edu/
http://tarkus.pha.jhu.edu/
http://muse.mse.jhu.edu/spot.jhu.html
Founded: 1985
Staff: 70
Activities: research * education
City Reference Coordinates: 076°37'00"W 39°17'00"N

Johns Hopkins University Press
(JHU Press)
2715 North Charles Street
Baltimore, MD 21218-4363
Telephone: 410-516-6900
Telefax: 410-516-6968
Electronic Mail: jlorder@chaos.press.jhu.edu
WWW: http://www.press.jhu.edu/
http://muse.jhu.edu/
Founded: 1878
Staff: 100
• Publisher
City Reference Coordinates: 076°37'00"W 39°17'00"N

K. Price Bryan Planetarium
c/o College of Notre Dame of Maryland
4701 North Charles Street
Baltimore, MD 21210
Telephone: 410-532-5702
Electronic Mail: jdirienzi@ndm.edu (Joseph Di Rienzi, Professor)
WWW: http://www.ndm.edu/faculty/PHYSIC.HTM
Founded: 1967
Staff: 1
Activities: class and group shows
City Reference Coordinates: 076°37'00"W 39°17'00"N

Maryland Science Center (MSC)
601 Light Street
Baltimore, MD 21230
Telephone: 410-685-2370
410-545-5918 (Starline)
410-685-5225 (Science Center/Planetarium information)
410-545-2999 (Observatory)
Telefax: 410-545-5974
WWW: http://www.mdsci.org/
http://www.mdsci.org/davis.html (Alan C. Davis Planetarium)
http://www.mdsci.org/obs.htm (Crosby Ramsey Memorial Observatory)
Founded: 1976
Staff: 11
Activities: exhibitions * lectures * science theater * demonstrations * trips * classes * planetarium shows * observing
City Reference Coordinates: 076°37'00"W 39°17'00"N

Maryland Space Grant Consortium (MSGC)
c/o Johns Hopkins University
Bloomberg Center for Physics and Astronomy
Room 203
3400 North Charles Street
Baltimore, MD 21218-2686
Telephone: 410-516-7351
Telefax: 410-516-4109
Electronic Mail: anne@pha.jhu.edu
WWW: http://msx4.pha.jhu.edu/msgc/
City Reference Coordinates: 076°37'00"W 39°17'00"N

MMI Corp.
2950 Wyman Parkway
P.O. Box 19907
Baltimore, MD 21211
Telephone: 410-366-1222
Telefax: 410-366-6311
Electronic Mail: mmicorp@aol.com

WWW: http://members.aol.com/mmicorp/
Founded: 1973
Staff: 6
Activities: educational material for astronomy * slides * videos * laser discs * globes * models * portable and permanent planetariums * observatory domes * US representative for GOTO planetariums
Periodicals: (1/2) "Astronomy Resource Manual", "Geology and Earth Resource Manual"
City Reference Coordinates: 076°37'00"W 39°17'00"N

Montgomery College Planetarium
Takoma Avenue & Fenton Street
Takoma Park, MD 20912-4197
Telephone: 301-650-1463
Telefax: 301-650-1550
Electronic Mail: hwilliam@mc.cc.md.us (Harold Williams)
WWW: http://www.mc.cc.md.us/Departments/planet/
Founded: 1959 (College: 1947)
Staff: 1
Activities: education * field trips * teacher training
City Reference Coordinates: 077°00'00"W 38°59'00"N

National Aeronautics and Space Administration (NASA), Center for Aerospace Information (CASI)
7121 Standard Drive
Hanover, MD 21076-1320
Telephone: 301-621-0390
Telefax: 301-621-0134
Electronic Mail: help@sti.nasa.gov
WWW: http://www.apollosaturn.com/frame-casi.htm
Founded: 1958 (NASA)
Periodicals: "Conference Publications", "Reference Publications", "Contractor Reports", "Special Publications" (ISSN 0091-0805), "Technical Memorandums" (ISSN 0499-9320), "Technical Notes" (ISSN 0499-9339), "Technical Papers"
City Reference Coordinates: 076°42'00"W 39°11'00"N

National Aeronautics and Space Administration (NASA), Goddard Space Flight Center (GSFC), Astronomical Data Center (ADC)
• Closed on 01 October 2002.

National Aeronautics and Space Administration (NASA), Goddard Space Flight Center (GSFC), Astrophysics Data Facility (ADF)
Code 631
Greenbelt, MD 20771
Telephone: 301-286-9392
Telefax: 301-286-1771
Electronic Mail: pisarski@gsfc.nasa.gov (Ryszard Pisarski, ADF Head)
　　　　　　　　　nancy.a.oliversen@gsfc.nasa.gov (Nancy A. Oliversen)
WWW: http://adf.gsfc.nasa.gov/adf/adf.html
　　　　http://www.gsfc.nasa.gov/
Founded: 1992
Staff: 30
Activities: processing, archiving, and distributing NASA astrophysics satellite data * providing remote access to NASA astrophysics data
City Reference Coordinates: 076°53'00"W 39°01'00"N

National Aeronautics and Space Administration (NASA), Goddard Space Flight Center (GSFC), Laboratory for High-Energy Astrophysics (LHEA)
Code 660
Greenbelt, MD 20771
Telephone: 301-286-8801
Electronic Mail: <userid>@lheamail.gsfc.nasa.gov
　　　　　　　　　<userid>@lheavx.gsfc.nasa.gov
WWW: http://lheawww.gsfc.nasa.gov/
　　　　http://heasarc.gsfc.nasa.gov/
　　　　http://www.gsfc.nasa.gov/
Founded: 1960
Staff: 250
Activities: space astronomical observing
City Reference Coordinates: 076°53'00"W 39°01'00"N

National Aeronautics and Space Administration (NASA), Goddard Space Flight Center (GSFC), National Space Science Data Center (NSSDC)
Code 633.4
Greenbelt, MD 20771
Telephone: 301-286-6695 (Request Coordinator)
　　　　　　 301-286-8310 (Astronomical Data Center)
　　　　　　 301-286-7355 (Director)
Telefax: 301-286-1771

Electronic Mail: request@nssdca.gsfc.nasa.gov
WWW: http://nssdc.gsfc.nasa.gov/
http://www.gsfc.nasa.gov/
Founded: 1966
Staff: 70
Periodicals: (4) "NSSDC News"
City Reference Coordinates: 076°53'00"W 39°01'00"N

National Aeronautics and Space Administration (NASA), Goddard Space Flight Center (GSFC), Solar Physics Branch
Code 682
Greenbelt, MD 20771
Telephone: 301-286-8811
Telefax: 301-286-8709
Electronic Mail: <userid>@uvsp.gsfc.nasa.gov
WWW: http://umbra.nascom.nasa.gov/ (Solar Data Analysis Center - SDAC)
http://www.gsfc.nasa.gov/GSFC_homepage.html (GSFC Homepage)
http://www.gsfc.nasa.gov/
Founded: 1958 (NASA)
Staff: 21
Activities: solar-physics research * high-energy solar physics * EUV imaging spectroscopy * solar-plasma theory
City Reference Coordinates: 076°53'00"W 39°01'00"N

National Aeronautics and Space Administration (NASA), Goddard Space Flight Center (GSFC), Space Science Data Operations Office (SSDOO)
Code 630
Greenbelt, MD 20771
Telephone: 301-286-7354
Telefax: 301-286-1771
Electronic Mail: green@bolero.gsfc.nasa.gov
green@nssdca.gsfc.nasa.gov
WWW: http://ssdoo.gsfc.nasa.gov/
http://www.gsfc.nasa.gov/c630/ssdoo_main.html
http://www.gsfc.nasa.gov/
Founded: 1958 (NASA)
Activities: data processing * archiving and dissemination of space sciences mission data
City Reference Coordinates: 076°53'00"W 39°01'00"N

National Aeronautics and Space Administration (NASA), Goddard Space Flight Center (GSFC), Visitor Center
Code 130
Greenbelt, MD 20771
Telephone: 301-286-8981
WWW: http://pao.gsfc.nasa.gov/vc/vc.htm
http://www.gsfc.nasa.gov/ (GSFC)
Founded: 1958 (NASA)
Staff: 3
Activities: public tours of NASA facilities * teacher resource laboratory * model rocket launches * star watches
Periodicals: (4) "Calendar of Events"
City Reference Coordinates: 076°53'00"W 39°01'00"N

National Capital Astronomers (NCA)
c/o Leith Holloway
10500 Rockville Pike
Appartment M10
Rockville, MD 20852
Electronic Mail: jleithh@aol.com
haroldw@umd5.umd.edu (Harold Williams, Vice-President)
WWW: http://capitalastronomers.org/
Founded: 1937
City Reference Coordinates: 077°10'00"W 39°05'00"N (Rockville)
077°01'00"W 38°54'00"N (Washington)

National Information Standards Organization (NISO)
4733 Bethesda Avenue
Suite 300
Bethesda, MD 20814
Telephone: 301-654-2512
Telefax: 301-654-1721
Electronic Mail: nisohq@niso.org
WWW: http://www.niso.org/
City Reference Coordinates: 077°06'00"W 38°59'00"N

National Institute of Standards and Technology (NIST), Administrative Offices
100 Bureau Drive

Stop 3460
Gaithersburg, MD 20899-3460
Telephone: 301-975-NIST
301-975-6478
Telefax: 301-926-1630
Electronic Mail: inquiries@nist.gov
WWW: http://www.nist.gov/
Founded: 1988 (1901 as "National Bureau of Standards")
City Reference Coordinates: 077°12'00"W 39°09'00"N

National Oceanic and Atmospheric Administration (NOAA), National Weather Service (NWS)
1325 East-West Highway
Silver Spring, MD 20910-3283
Telephone: 301-427-7689
301-713-0689
Telefax: 301-587-4524
Electronic Mail: <userid>@nws.noaa.gov
WWW: http://www.nnic.noaa.gov/weather.html
http://www.nws.noaa.gov/
Founded: 1870 (NOAA: 1970)
City Reference Coordinates: 077°03'00"W 39°02'00"N

National Space Science Data Center (NSSDC)
• See "National Aeronautics and Space Administration (NASA), Goddard Space Flight Center (GSFC), National Space Science Data Center (NSSDC)"

National Weather Service (NWS)
• See "National Oceanic and Atmospheric Administration (NOAA), National Weather Service (NWS)"

Owens Science Center (HBOSC) (Howard B._), Planetarium
• See "Howard B. Owens Science Center (HBOSC), Planetarium"

Rayleigh Optical Corp. (ROC)
3720 Commerce Drive
Suite 1112
Baltimore, MD 21227
Telephone: 410-247-6666
Telefax: 410-247-6666
Electronic Mail: info@rayleighoptical.com
WWW: http://www.rayleighoptical.com/
Activities: manufacturing reflective and refractive optical systems
City Reference Coordinates: 076°37'00"W 39°17'00"N

Raytheon Technical Services Co., Information Technology and Scientific Services (ITSS)
4400 Forbes Boulevard
Lanham, MD 20706-4392
Telephone: 301-794-5000
1-800-877-7072 (USA only)
WWW: http://itss.raytheon.com/
Activities: providing professional support services to federal government and international clients * providing scientific, engineering and software support in the areas of space and Earth science, scientific data management, remote sensing, satellite ground systems, aerospace engineering, transportation management technologies, computer networking and software maintenance
City Reference Coordinates: 076°52'00"W 38°58'00"N

Society for Applied Spectroscopy (SAS)
201B Broadway Street
Frederick, MD 21701-6501
Telephone: 301-694-8122
Telefax: 301-694-6860
Electronic Mail: sasoffice@aol.com
WWW: http://www.s-a-s.org/
Founded: 1958
Periodicals: "Applied Spectroscopy Journal"
City Reference Coordinates: 077°25'00"W 39°25'00"N

Southampton Planetarium
c/o Harford County Public Schools
Moores Mill Road
Bel Air, MD 21014
Telephone: 410-638-4150
City Reference Coordinates: 076°21'00"W 39°32'00"N

Southern Maryland Astronomical Society (SMAS)
P.O. Box 1727

White Plains, MD 20695-1727
Electronic Mail: mconte@crosslink.net (Mike Conte)
WWW: http://www.ameritel.net/users/hemlock/smas/
Founded: 1995
Membership: 30
City Reference Coordinates: 076°56'00"W 38°38'00"N (Waldorf)

Space Science Data Operations Office (SSDOO)
• See "National Aeronautics and Space Administration (NASA), Goddard Space Flight Center (GSFC), Space Science Data Operations Office (SSDOO)"

Space Telescope Science Institute (STScI)
Homewood Campus
3700 San Martin Drive
Baltimore, MD 21218
Telephone: 410-338-4700
Telefax: 410-338-4767
Electronic Mail: <userid>@stsci.edu
　　　　　　　　help@stsci.edu
WWW: http://www.stsci.edu/
Founded: 1981
Staff: 500
Activities: exploiting the Hubble Space Telescope (HST) * software development * seminars * workshops * symposia
Periodicals: "STScI Newsletter", "Preprint Series"; (4) "Observer"
City Reference Coordinates: 076°37'00"W 39°17'00"N

Storer Planetarium (Arthur_)
• See "Arthur Storer Planetarium"

Towson State University (TSU), Department of Physics, Astronomy, and Geosciences
Smith Hall - Room 445
8000 York Road
Towson, MD 21204-7097
Telephone: 410-830-3020
　　　　　　410-830-3906 (Watson-King Planetarium)
Telefax: 410-830-3511
Electronic Mail: <userid>@midget.towson.edu
WWW: http://www.towson.edu/~physics/
　　　　http://www.towson.edu/~physics/planet.html (Watson-King Planetarium)
Founded: 1970
City Reference Coordinates: 076°36'00"W 39°24'00"N

Technical Innovations Inc.
22500 Old Hundred Road
Barnesville, MD 20838-9725
Telephone: 301-972-8040
Telefax: 301-349-2441
Electronic Mail: domepage@erols.com
WWW: http://www.homedome.com/
Founded: 1991
Staff: 8
Activities: manufacturing domes, accessories and dome automation systems
Periodicals: "The Home-Dome Star"
City Reference Coordinates: 077°10'00"W 39°05'00"N (Rockville)

Tri-State Astronomers
c/o Frank Moon
7210 East Sundown Court
Frederick, MD 21702
or :
c/o Jim Stanicek
1028 Corbett Street
Hagerstown, MD21740
or :
c/o Washington County Planetarium and Space Science Center
823 Commonwealth Avenue
Hagerstown, MD 21740
Telephone: 301-695-5587
　　　　　　301-791-4172
　　　　　　301-790-3270
WWW: http://tristateastronomers.org/
Founded: 1985
Membership: 40
Activities: education * field trips * observing * planetarium shows
Periodicals: (12) "The Observer"

City Reference Coordinates: 077°43'00"W 39°39'00"N (Hagerstown)

Universities Space Research Association (USRA), Corporate Office
10227 Wincopin Circle
Suite 212
Columbia, MD 21044-3498
Telephone: 410-730-2656
Telefax: 410-730-3496
Electronic Mail: info@hq.usra.edu
WWW: http://www.usra.edu/
Founded: 1969
Membership: 90 (institutions)
Staff: 20
City Reference Coordinates: 076°52'00"W 39°13'00"N

University of Maryland, Department of Astronomy
Computer and Space Sciences Building
College Park, MD 20742-2421
Telephone: 301-405-1543
Telefax: 301-314-9067
Electronic Mail: <userid>@astro.umd.edu
WWW: http://www.astro.umd.edu/
Founded: 1807
Staff: 120
Activities: research * education
Coordinates: 076°57'24"W 39°00'06"N H53m
City Reference Coordinates: 076°55'00"W 39°00'00"N

University of Maryland, Department of Astronomy, Laboratory for Millimeter-Wave Astronomy
College Park, MD 20742
Telephone: 301-405-1502
Telefax: 301-314-9067
Electronic Mail: <userid>@astro.umd.edu
WWW: http://bima.astro.umd.edu/general/lma.html
http://bima.astro.umd.edu/bima/home.html (BIMA)
Founded: 1988
Activities: East coast observing site for BIMA mm-interferometer * software production * intercampus networking
City Reference Coordinates: 076°55'00"W 39°00'00"N

University of Maryland Baltimore County (UMBC), Physics Department
1000 Hilltop Circle
Baltimore, MD 21250
Telephone: 410-455-2513
Telefax: 410-455-1072
WWW: http://www.physics.umbc.edu/
http://www.jca.umbc.edu/ (Joint Center for Astrophysics)
City Reference Coordinates: 076°37'00"W 39°17'00"N

US Naval Academy (USNA), Physics Department
572 Holloway Road
Annapolis, MD 21402-5026
Telephone: 410-293-6658
410-293-6662
Telefax: 410-293-3729
Electronic Mail: <userid>@nadn.navy.mil
WWW: http://dlt.gsfc.nasa.gov/Stars/usna.html
http://physics.nadn.navy.mil/
Founded: 1970
Staff: 30 (astronomers: 2)
Activities: education * research * planetarium * observing
City Reference Coordinates: 076°30'00"W 38°59'00"N

Washington County Planetarium and Space Science Center
• See now "William M. Brish Planetarium"

Westminster Astronomical Society (WAS)
113 Walgrove Road
Reisterstown, MD 21136
Electronic Mail: roelle@erols.com
WWW: http://www.westminsterastro.org/
Founded: 1984
Membership: 80
Activities: monthly meetings * star parties * planetarium shows
Periodicals: (12) "Mason-Dixon Astronomer"
City Reference Coordinates: 077°00'00"W 39°35'00"N (Westminster)

076°50'00"W 39°28'00"N (Reisterstown)

Wilcoxon Research Inc.
21 Firstfield Road
Gaithersburg, MD 20878
Telephone: 301-330-8811
Telefax: 301-330-8873
Electronic Mail: sensors@wilcoxon.com
WWW: http://www.wilcoxon.com/
Founded: 1960
Staff: 110
Activities: manufacturing vibration instrumentation
City Reference Coordinates: 077°12'00"W 39°09'00"N

William M. Brish Planetarium
820 Commonwealth Avenue
Hagerstown, MD 21740
Telephone: 301-766-2898
WWW: http://www.wcboe.k12.md.us/mainfold/curric/planetarium/
Founded: 1969
Staff: 1
Activities: public and school programs
Periodicals: "Our Friendly Skies" in local newspapers
• Formerly "Washington County Planetarium and Space Science Center"
City Reference Coordinates: 077°43'00"W 39°39'00"N

World Data Center A for Rockets and Satellites
c/o National Aeronautics and Space Administration (NASA)
Goddard Space Flight Center (GSFC)
Code 633
Greenbelt, MD 20771
Telephone: 301-286-7355
Telefax: 301-286-1771
Electronic Mail: request@nssdca.gsfc.nasa.gov
WWW: http://nssdc.gsfc.nasa.gov/about/about_wdc-a.html
    http://nssdc.gsfc.nasa.gov/GSFC_homepage.html (GSFC Homepage)
Founded: 1957
Staff: 2
Periodicals: "SPACEWARN Bulletin"
City Reference Coordinates: 076°53'00"W 39°01'00"N

Coordinates: 071°10'10"W 42°01'48"N H57m
City Reference Coordinates: 071°13'00"W 42°02'00"N

Astronomical Society of Southern New England (ASSNE) Inc.
P.O. Box 477
Rehoboth, MA 02769
Electronic Mail: assne@aol.com
assne@ma.ultranet.com
WWW: http://www.ultranet.com/~assne/
Activities: education * popularization
City Reference Coordinates: 071°15'00"W 41°51'00"N

Atmospheric and Environmental Research (AER) Inc.
131 Hartwell Avenue
Lexington, MA 02421-3126
Telephone: 781-761-2288
Telefax: 781-761-2299
Electronic Mail: <userid>@aer.com
WWW: http://www.aer.com/
Founded: 1977
Activities: environmental research and consulting resource
City Reference Coordinates: 071°14'00"W 42°27'00"N

Automatic Systems Laboratories (ASL) Inc.
187 Ballardvale Street
Wilmington, MA 01877
Telephone: 508-658-0000
Telefax: 508-658-5444
Electronic Mail: aslinc@aol.com
WWW: http://www.aslco.com/
http://www.aslinc.com/
City Reference Coordinates: 077°55'00"W 34°14'00"N

Barr Associates Inc.
Two Liberty Way
Westford, MA 01886
Telephone: 973-692-7513
Telefax: 973-692-7443
Electronic Mail: barr@barrassociates.com
WWW: http://www.barrassociates.com/
Founded: 1972
Staff: 140
Activities: designing and manufacturing custom astronomical, space-based and other precision optical filters
City Reference Coordinates: 071°26'00"W 42°35'00"N

Bay Colony Astronomical Society (BCAS)
c/o Robert A. Johnson
7 Morton Park Road
P.O. Box 1418
Plymouth, MA 02362-1418
Electronic Mail: bcas@hotmail.com
astronomer77@galaxy5.com
WWW: http://clubs.yahoo.com/clubs/baycolonyastroonline
http://maxpages.com/starastronomer/home
Founded: 1990
Membership: 13
Activities: observing * meetings * instrumentation
City Reference Coordinates: 070°41'00"W 41°58'00"N

Bentley College, Observatory
175 Forest Street
Waltham, MA 02452-4705
Telephone: 781-891-3450
Telefax: 781-891-2838
Electronic Mail: baghassi@bentley.edu (Badri M. Aghassi, Director)
WWW: http://www.bentley.edu/
Founded: 1974
Activities: education
City Reference Coordinates: 071°14'00"W 42°23'00"N

Blake Planetarium (WRBP) (Dr. W. Russell_)
• See "Dr. W. Russell Blake Planetarium (WRBP)"

Boston University Astronomical Society (BUAS)
c/o Department of Astronomy

725 Commonwealth Avenue
Boston, MA 02215
Electronic Mail: esha@bu.edu (Alicia Eccles, President)
WWW: http://bu-ast.bu.edu/buas/buas.html
City Reference Coordinates: 071°04'00"W 42°21'00"N

Boston University, Astronomy Department
725 Commonwealth Avenue
Boston, MA 02215
Telephone: 617-353-2625 (Main Office)
          617-353-5700 (Curator/Observatory)
          617-353-3644 (Library)
          617-353-2630 (recorded Observatory Information)
Telefax: 617-353-5704
Electronic Mail: <userid>@bu.edu
WWW: http://www.bu.edu/astronomy/
     http://www.bu.edu/iar/ (Institute for Astrophysical Research - IAR)
     http://www.bu.edu/csp/ (Center for Space Physics - CSP)
Founded: 1839 (Observatory: 1890)
Staff: 23
Activities: education * galactic astronomy * high-energy astrophysics * stellar and galactic evolution * magnetospheric physics * ionospheric physics * radio astronomy * QSOs * VLBI * planetary atmospheres * instrumentation
Coordinates: 071°11'23"W 42°06'24"N (Judson Broadman Coit Memorial Obs.)
City Reference Coordinates: 071°04'00"W 42°21'00"N

Brandeis University, Department of Physics, Radio-Astronomy Group
Mail Stop 057
Waltham, MA 02254-9110
Telephone: 781-736-2848
Telefax: 781-736-2915
Electronic Mail: <userid>@vlbi.astro.brandeis.edu
WWW: http://pc.astro.brandeis.edu/
     http://www.physics.brandeis.edu/
City Reference Coordinates: 071°14'00"W 42°23'00"N

Cambridge Research and Instrumentation (CRI) Inc.
80 Ashford Street
Boston, MA 02134-1816
Telephone: 617-491-2627
Telefax: 617-864-3730
Electronic Mail: pfoukal@world.std.com
Founded: 1985
Staff: 14
Activities: solar research * manufacturing radiometers, liquid crystal tunable filters, laser intensity stabilizers
City Reference Coordinates: 071°04'00"W 42°21'00"N

Cape Cod Astronomical Society (CCAS)
P.O. Box 56
Harwich, MA 02645
Telephone: 508-432-4316
Electronic Mail: jcarlson@capecod.net
Founded: 1985
Membership: 60
City Reference Coordinates: 070°01'00"W 41°41'00"N

Center for Astrophysics (CfA) (Harvard-Smithsonian_)
• See "Harvard-Smithsonian Center for Astrophysics (CfA)"

Central Bureau for Astronomical Telegrams (CBAT)
60 Garden Street
Cambridge, MA 02138
Telephone: 617-495-7244
          617-495-7440
          617-495-7444
Electronic Mail: cbat@cfa.harvard.edu
WWW: http://cfa-www.harvard.edu/iau/cbat.html
Founded: 1920
Staff: 2
Activities: IAU office receiving, verifying and disseminating reports on transient astronomical phenomena * IAU Circular Computer Service
Periodicals: "IAU CBAT Circulars" (ISSN 0081-0304); (1) "Catalog of Cometary Orbits"
City Reference Coordinates: 071°06'00"W 42°22'00"N

Chandra X-ray Center (CXC)
c/o Harvard-Smithsonian Center for Astrophysics

60 Garden Street
Cambridge, MA 02138
Telephone: 617-495-7258
Telefax: 617-495-7356
Electronic Mail: <userid>@asc.harvard.edu
     axafnews@asc.harvard.edu (Newsletter)
WWW: http://asc.harvard.edu/
Founded: 1991
Activities: providing support required by the science community to realize fully the potential of the Chandra X-ray Observatory (AXAF)
Periodicals: "CXC Newsletter"
• Previously "AXAF Science Center (ASC)"
City Reference Coordinates: 071°06'00"W 42°22'00"N

Charles Hayden Planetarium
c/o Museum of Science
Science Park
Boston, MA 02114-1099
Telephone: 617-723-2500 (Information & Tickets)
     617-589-0270 (Planetarium Office)
Telefax: 617-589-0454
Electronic Mail: information@mos.org
WWW: http://www.mos.org/
Founded: 1958
Staff: 11
Activities: public and school programs * education
Observatories: 1 (Gilliland Observatory)
City Reference Coordinates: 071°04'00"W 42°21'00"N

Dr. W. Russell Blake Planetarium (WRBP)
P.O. Box 986
Plymouth, MA 02362
or :
PCIS Long Pond Road
Plymouth, MA 02360
Telephone: 508-830-4470
Telefax: 508-830-4462
WWW: http://www.eduzone.com/plycurriculum/science/planetarium/dailypla.htm
Founded: 1973
Staff: 3
Activities: education * public shows
City Reference Coordinates: 070°41'00"W 41°58'00"N

DynaVac
110 Industrial Park Road
Hingham, MA 02043
Telephone: 781-740-8600
Telefax: 781-740-9996
Electronic Mail: sales@dynavac.com
WWW: http://www.dynavac.com/
Founded: 1981
Staff: 40
Activities: manufacturing vacuum systems for space simulations, thermal tests, optical coating and vacuum metallizing
City Reference Coordinates: 070°53'00"W 42°14'00"N

Earthwatch
3 Clock Tower Place
Suite 100
Box 75
Maynard, MA 01754
Telephone: 1-800-776-0188 (USA/Canada only)
     978-461-0081
Telefax: 978-461-2332
Electronic Mail: info@earthwatch.org
WWW: http://www.earthwatch.org/
Founded: 1971
Membership: 70000
Activities: sponsoring expeditions worldwide * archaeoastronomy * variable-star observing
Periodicals: "Earthwatch Magazine" (ISSN 8750-0183)
City Reference Coordinates: 071°33'00"W 42°30'00"N

EcoTarium, Norton Observatory
222 Harrington Way
Worcester, MA 01604
Telephone: 508-929-2712
WWW: http://www.ecotarium.org/astro/astro101/

City Reference Coordinates: 071°48'00"W 42°16'00"N

F.C. Meichsner Co.
182 Lincoln Street
Boston, MA 02111
Telephone: 1-800-321-VIEW (USA only)
Telefax: 617-426-0837
Electronic Mail: meichs@aol.com
WWW: http://www.meichsner.com/
Activities: optical equipment and accessories
City Reference Coordinates: 071°04'00"W 42°21'00"N

Five College Radio Astronomy Observatory (FCRAO)
c/o University of Massachusetts
Graduate Research Tower 619I
Amherst, MA 01003
or :
New Salem, MA 01355
Telephone: 413-545-0789
Telefax: 413-545-4223
Electronic Mail: <userid>@fcrao1.phast.umass.edu
WWW: http://www-fcrao.phast.umass.edu/
       http://www.astro.umass.edu/~fcrao/
Founded: 1969
Staff: 30
Activities: astrochemistry * extragalactic studies * galactic molecular clouds * graduate student training * low-noise electronics * planetary science * QSOs * AGNs * star formation * stellar envelopes * sub-mm technology * VLBI
Periodicals: "Preprints"
Coordinates: 072°20'42"W 42°23'30"N H314m (Quabbin, New Salem)
City Reference Coordinates: 072°31'00"W 42°23'00"N (Amherst)

Harvard College Observatory (HCO)
• See "Harvard-Smithsonian Center for Astrophysics (CfA)"

Harvard-Smithsonian Center for Astrophysics (CfA)
60 Garden Street
Cambridge, MA 02138
Telephone: 617-495-7000
          617-495-7461 (SAO/Public Affairs Office)
          617-495-9059 (HCO/Administrative Office)
Telefax: 617-495-7105
        617-495-7468 (SAO)
        617-496-8018 (HCO)
Electronic Mail: <userid>@cfa.harvard.edu
WWW: http://cfa-www.harvard.edu/
       http://cfa-www.harvard.edu/sao-home.html
       http://cfa-www.harvard.edu/hco-home.html
       http://linmax.sao.arizona.edu/FLWO/FLWO/whipple.html (Whipple Obs. - see USA-AZ)
       http://cfa-www.harvard.edu/cfa/oir/OakRidge/oak.ridge.html (Oak Ridge Observatory)
       http://hea-www.harvard.edu/SIMBAD/simbad.home.html (SIMBAD USA)
       http://adsabs.harvard.edu/ (NASA Astrophysics Data System - ADS)
       http://www.si.edu/
Founded: 1839 (HCO) (SAO: 1890)
Staff: 600
Activities: ground-based, space, and laboratory astrophysics * atomic and molecular physics * high energy astrophysics * optical and IR astronomy * planetary sciences * radio and geoastronomy * solar and stellar physics * theoretical astrophysics * operating the Fred Lawrence Whipple Observatory (Amado, AZ), and the Oak Ridge Observatory (Harvard, MA)
Periodicals: "CfA Preprint Series"
Coordinates: 071°07'08"W 42°22'08"N H24m (CfA – IAU Code 802)
           071°33'30"W 42°30'18"N H185m (Oak Ridge Obs. – IAU Code 801)
           110°52'39"W 31°40'51"N H2,608m (Whipple Obs. – IAU Code 696)
• Joint facility of "Harvard College Observatory (HCO)" and the "Smithsonian Astrophysical Observatory (SAO)"
City Reference Coordinates: 071°06'00"W 42°22'00"N

Harvard University, Department of Astronomy
Mail Stop 46
60 Garden Street
Cambridge, MA 02138
Telephone: 617-495-3752
Telefax: 617-495-7105
Electronic Mail: department@cfa.harvard.edu
              <userid>@cfa.harvard.edu
WWW: http://cfa-www.harvard.edu/hco/astro/home.html
Founded: 1855
Activities: education * research
City Reference Coordinates: 071°06'00"W 42°22'00"N

Harvard University, Department of Physics
Jefferson Laboratory
17A Oxford Street
Cambridge, MA 02138
Telephone: 617-495-2872
Telefax: 617-495-0416
Electronic Mail: <userid>@phys2.harvard.edu
WWW: http://www.physics.harvard.edu/
City Reference Coordinates: 071°06'00"W 42°22'00"N

Harvard University Press (HUP)
79 Garden Street
Cambridge, MA 02138
Telephone: 617-495-2600
　　　　　　 1-800-448-2242 (USA and Canada only)
Telefax: 617-495-5898
Electronic Mail: contact_hup@harvard.edu
WWW: http://www.hup.harvard.edu/
Founded: 1913
• Publisher
City Reference Coordinates: 071°06'00"W 42°22'00"N

Hayden Planetarium (Charles_)
• See "Charles Hayden Planetarium"

Haystack Observatory
• See "Northeast Radio Observatory Corp. (NEROC), Haystack Observatory"

International Astronomical Union (IAU), Central Bureau for Astronomical Telegrams (CBAT)
• See "Central Bureau for Astronomical Telegrams (CBAT)"

International Astronomical Union (IAU), Minor Planet Center (MPC)
• See "Minor Planet Center (MPC)"

International Comet Quarterly (ICQ)
c/o Daniel W.E. Green (Editor)
Smithsonian Astrophysical Observatory
60 Garden Street
Cambridge, MA 02138
Telephone: 617-495-7440
Electronic Mail: green@cfa.harvard.edu
WWW: http://cfa-www.harvard.edu/icq/icq.html
Founded: 1973
• Journal (4) (ISSN 0736-6922)
City Reference Coordinates: 071°06'00"W 42°22'00"N

International Light (IL) Inc.
17 Graf Road
Newburyport, MA 01950
Telephone: 508-465-5923
Telefax: 508-462-0759
Electronic Mail: ilsales@intl-light.com
WWW: http://www.intl-light.com/
Founded: 1965
Staff: 20
Activities: manufacturing radiometers, photometers and spectroradiometers
City Reference Coordinates: 070°53'00"W 42°49'00"N

International Science Writers Association (ISWA)
c/o James Cornell (President)
Smithsonian Astrophysical Observatory
60 Garden Street
Cambridge, MA 02138
Telephone: 617-495-7461
Telefax: 617-495-7468
Electronic Mail: jcornell@cfa.harvard.edu
　　　　　　　　 sciwriters@aol.com (Howard Lewis, Secretary-Treasurer)
Founded: 1967
Membership: 210
Activities: promoting and protecting journalists' rights worldwide * improving science communication
Periodicals: "Newsletter", "Conference and Symposia Proceedings"
City Reference Coordinates: 071°06'00"W 42°22'00"N

Kluwer Academic Publishers (KAP), Boston Office
101 Philip Drive

Assinippi Park
Norwell, MA 02061
Telephone: 781-871-6600
Telefax: 781-871-6528
Electronic Mail: kluwer@wkap.com
WWW: http://www.wkap.com/
Founded: 1978
• Publisher
• See main entry in the Netherlands
City Reference Coordinates: 070°48'00"W 42°10'00"N

Laurin Publishing Co. Inc.
Berkshire Common
2 South Street
P.O. Box 4949
Pittsfield, MA 01202-4949
Telephone: 413-499-0514
Telefax: 413-442-3180
Electronic Mail: photonics@laurin.com
WWW: http://www.laurin.com/
          http://www.photonics.com/
Founded: 1959
Staff: 47
Periodicals: (12) "Photonics Spectra Magazine"; (3) "Photonics Product Portfolio", "Photonics Postcard Deck"; (2) "Photonics Literature Showcase"; (1) "Photonics Directory Set: Corporate Guide - Buyers Guide - Handbook & Dictionary", "European Directory", "Asian & Pacific Rim Directory"
• Publisher
City Reference Coordinates: 073°15'00"W 42°27'00"N

Learning Technologies Inc.
40 Cameron Avenue
Somerville, MA 02144
Telephone: 1-800-537-8703 (USA only)
          617-628-1459
Telefax: 617-628-8606
Electronic Mail: starlab@starlab.com
WWW: http://www.starlab.com/
Founded: 1977
Staff: 20
Activities: manufacturing Starlab portable planetariums * distributing Project STAR materials
City Reference Coordinates: 071°06'00"W 42°23'00"N

LightWedge
P.O. Box 396
Nantucket, MA 02554
Telephone: 1-877-777-9334 (USA only)
          508-228-9334
Telefax: 508-374-8512
Electronic Mail: service@lightwedge.com
WWW: http://www.lightwedge.com/
Activities: distributing dimm lighters
City Reference Coordinates: 070°06'00"W 41°17'00"N

Lincoln Laboratory
• See "Massachusetts Institute of Technology (MIT), Lincoln Laboratory"

Loch Ness Productions (LNP)
P.O. Box 1159
Groton, MA 01450-3159
Telephone: 978-448-3666
          1-888-4-NESSIE (USA only)
Telefax: 978-448-3799
Electronic Mail: info@lochness.com
WWW: http://www.lochness.com/
Founded: 1978
Activities: producing planetarium products
City Reference Coordinates: 071°34'00"W 42°34'00"N

L-3 Communications ESSCO
Old Powder Mill Road
Concord, MA 01742
Telephone: 978-369-7200
Telefax: 978-369-7641
Electronic Mail: info@esscoradomes.com
WWW: http://esscoradomes.com/

Founded: 1961
Staff: 60
Activities: designing, manufacturing and installing radomes and radiotelescopes, precision antenna systems for use in radio astronomy, mm-wave radar and satellite communications
City Reference Coordinates: 071°21'00"W 42°27'00"N

Maria Mitchell Observatory (MMO)
3 Vestal Street
Nantucket, MA 02554
Telephone: 508-228-9273
Electronic Mail: vladimir@mmo.org (Vladimir Strelnitski, Director)
<userid>@mmo.org
WWW: http://www.mmo.org/
Founded: 1908
Staff: 8
Activities: astronomical research * undergraduate research * Summer public programs * science education
Periodicals: (1) "Annual Report of the Nantucket Maria Mitchell Association"
Coordinates: 070°06'17"W 41°16'50"N H20m (IAU Code 811)
City Reference Coordinates: 070°06'00"W 41°17'00"N

Martha's Vineyard Amateur Astronomy Club (MVAAC)
P.O. Box 471
Edgartown, MA 02539
Telephone: 508-696-9000
508-693-8264
Electronic Mail: keltmon@tiac.net
mvaac@hotmail.com
WWW: http://www.angelfire.com/ma2/mvaac/mvaac.html
Membership: 15
Periodicals: "Event Horizon"
City Reference Coordinates: 070°30'00"W 41°23'00"N

Massachusetts Institute of Technology (MIT), Center for Space Research (CSR)
Building 37
70 Vassar Street
Cambridge, MA 02139-4307
Telephone: 617-253-1456
617-253-6104
Telefax: 617-253-0861
Electronic Mail: <userid>@mit.edu
WWW: http://space.mit.edu/
http://space.mit.edu/csr.html
http://heasarc.gsfc.nasa.gov/docs/asca/asca2.html (ASCA)
http://acis.mit.edu/ (ACIS)
http://hea-www.harvard.edu/asc/ascinfo/AXAF-description.html (AXAF)
http://sst.lanl.gov/projects/hete.html (HETE)
http://space.mit.edu/XTE/XTE.html (XTE)
http://heasarc.gsfc.nasa.gov/docs/xte/xte.html (XTE)
Founded: 1968
Staff: 160
Activities: X-ray astronomy * space sciences * interplanetary plasmas * space VLBI * gravity wave detection * planetary surfaces * geodesy * space life sciences * theoretical astrophysics * Advanced Satellite for Cosmology and Astrophysics (ASCA) * AXAF CCD Imaging Spectrometer (ACIS) * High-Energy Transient Experiment (HETE) * X-Ray Timing Explorer (XTE) * space plasma
City Reference Coordinates: 071°06'00"W 42°22'00"N

Massachusetts Institute of Technology (MIT), Department of Earth, Atmospheric and Planetary Science
77 Massachusetts Avenue
Cambridge, MA 02139
Telephone: 617-253-1000
617-253-2127
617-253-9317 (WAO)
508-692-4744 (WAO)
Telefax: 617-253-2886 (WAO)
Electronic Mail: <userid>@mit.edu
WWW: http://www-eaps.mit.edu/
http://space.mit.edu/wallace.html (Wallace Astrophysical Observatory - WAO)
Founded: 1971 (WAO)
Staff: 50
Activities: theoretical astrophysics * X-ray, optical, IR and radio astronomy * planets * VLBI * molecular spectroscopy * gravitation wave astronomy * gravitational lenses * X-ray binaries * microwave background * accretion processes * comets
Coordinates: 111°37'00"W 31°37'00"N (McGraw Hill Observatory on Kitt Peak)
City Reference Coordinates: 071°06'00"W 42°22'00"N

Massachusetts Institute of Technology (MIT), Department of Physics
77 Massachusetts Avenue
Room 6-113
Cambridge, MA 02139-4307
Telephone: 617-253-4801
Telefax: 617-253-8554
Electronic Mail: physics@mit.edu
<userid>@mit.edu
WWW: http://web.mit.edu/physics/
http://web.mit.edu/physics/research/astrophysics_research.htm (Astrophysics)
Founded: 1865
Activities: theoretical astrophysics * X-ray, optical, IR and radio astronomy * planets * interplanetary plasmas * VLBI * gravitation wave astronomy * gravitational lenses * microwave background * cosmology
City Reference Coordinates: 071°06'00"W 42°22'00"N

Massachusetts Institute of Technology (MIT), Lincoln Laboratory
244 Wood Street
Lexington, MA 02173-9108
Telephone: 781-981-5500
Telefax: 781-981-9057
Electronic Mail: <userid>@mit.edu
WWW: http://www.ll.mit.edu/
Founded: 1951
Activities: applying science and advanced technology to critical problems of national security
City Reference Coordinates: 071°14'00"W 42°27'00"N

Massachusetts Institute of Technology (MIT), Michigan Dartmouth MIT Observatory
• See "MDM Observatory (USA-AZ)"

Massachusetts Institute of Technology (MIT), Press
• See "MIT Press"

Meichsner Co. (F.C._)
• See "F.C. Meichsner Co."

Millstone Hill Observatory (MHO)
c/o MIT Haystack Observatory
Atmospheric Sciences Group
Route 40
Westford, MA 01886
Telephone: 617-981-5621 (Director)
Telefax: 617-981-5766
Electronic Mail: <userid>@hyperion.haystack.edu
WWW: http://hyperion.haystack.edu/homepage.html
Founded: 1960
City Reference Coordinates: 071°26'00"W 42°35'00"N

Milton Academy, Science Department
170 Centre Street
Milton, MA 02186
Telephone: 617-898-1798
WWW: http://www.milton.edu/academics/science.html
Observatories: 1 (Ayer Observatory)
City Reference Coordinates: 071°05'00"W 42°16'00"N

Minor Planet Center (MPC)
60 Garden Street
Cambridge, MA 02138
Telephone: 617-495-7244
617-495-7444
617-495-7440
Electronic Mail: <userid>@cfa.harvard.edu
WWW: http://cfa-www.harvard.edu/iau/mpc.html
Founded: 1947
Activities: IAU office receiving astrometric observations of minor planets and comets * calculating orbits * publishing and maintaining databases of these and related material
Periodicals: "MPC Circulars" (ISSN 0736-6884)
City Reference Coordinates: 071°06'00"W 42°22'00"N

Mitchell Observatory (MMO) (Maria_)
• See "Maria Mitchell Observatory (MMO)"

MIT Press
5 Cambridge Center
4th Floor
Cambridge, MA 02142-1493
Telephone: 617-253-5646
Telefax: 617-258-6779
Electronic Mail: <userid>@mitpress.mit.edu
WWW: http://www-mitpress.mit.edu/
Founded: 1962
Staff: 100
• Publisher
City Reference Coordinates: 071°06'00"W 42°22'00"N

New England Light Pollution Advisory Group (NELPAG)
c/o Daniel W.E. Green
Mail Stop 18
Harvard-Smithsonian Center for Astrophysics
60 Garden Street
Cambridge, MA 02138
Telephone: 617-495-7440
Electronic Mail: nelpag-request@yeehah.merk.com
green@cfa.harvard.edu
WWW: http://cfa-www.harvard.edu/cfa/ps/nelpag.html
http://cfa-www.harvard.edu/~graff/nelpag.html
Founded: 1993
Membership: 25
Activities: education regarding good outdoor night lighting
Periodicals: (5) "NELPAG Circulars"
City Reference Coordinates: 071°06'00"W 42°22'00"N

New England Meteoritical Services
P.O. Box 440
Mendon, PA 01756
Telephone: 508-478-4020
Telefax: 508-478-5104
Electronic Mail: nemsusa@delphi.com
staff@meteorlab.com
WWW: http://www.meteorlab.com/
Founded: 1989
Staff: 6
City Reference Coordinates: 071°33'00"W 42°06'00"N

Northeastern University, Department of Physics
111 Dana Research Center
110 Forsyth Street
Boston, MA 02115
Telephone: 617-373-2902
Telefax: 617-373-2943
Electronic Mail: <userid>@neu.edu
WWW: http://www.physics.neu.edu/
Founded: 1960 (University: 1898)
Staff: 50
Activities: experimental condensed matter physics * theoretical elementary particle physics * cosmology
City Reference Coordinates: 071°04'00"W 42°21'00"N

Northeast Radio Observatory Corp. (NEROC), Haystack Observatory
Off Route 40
Westford, MA 01886
Telephone: 978-692-4764
Electronic Mail: info@haystack.mit.edu
WWW: http://www.haystack.mit.edu/
Founded: 1970
Staff: 50
Activities: spectral line radio astronomy * VLBI
Coordinates: 071°29'18"W 42°37'24"N H146m (Tyngsboro)
City Reference Coordinates: 071°26'00"W 42°35'00"N

North Shore Amateur Astronomy Club (NSAAC)
c/o Richard Bickerton (Vice President)
58 Campmeeting Road
Topsfield, MA 01983
Telephone: 978-887-8533
Electronic Mail: president@nsaac.org
membership@nsaac.org
WWW: http://www.star.net/people/~nsaac
Founded: 1989

Membership: 100
Activities: supporting public observing at local colleges and star parties
Periodicals: (12) "The Celestial Observer"
City Reference Coordinates: 070°57'00"W 42°38'00"N

Norton Observatory
• See "EcoTarium, Norton Observatory"

Optical Corp. of America (OCA)
• See now "Corning Incorporated (USA-NY)"

Optometrics USA Inc.
Stony Brook Industrial Park
6 Nemco Way
Ayer, MA 01432
Telephone: 978-772-1700
Telefax: 978-772-0017
Electronic Mail: opto@optometrics.com
WWW: http://www.optometrics.com/
Activities: manufacturing filters
City Reference Coordinates: 071°35'00"W 42°34'00"N

PerkinElmer Optoelectronics
35 Congress Street
Salem, MA 01970
Telephone: 978-745-3200
Telefax: 978-745-0894
Electronic Mail: <userid>@perkinelmer.com
WWW: http://www.perkinelmer.com/
Founded: 1980
Staff: 25
Activities: manufacturing rubidium frequency standards
City Reference Coordinates: 070°55'00"W 42°31'00"N

Pittsfield Astronomy Club
c/o Richard Knower
28 Parker Street
Pittsfield, MA 01201
Electronic Mail: chm027@vgernet.net
WWW: http://www.vgernet.net/chm027/
   http://pulsar.netfirms.com/
Founded: 1995
City Reference Coordinates: 073°15'00"W 42°27'00"N

Russell Blake Planetarium (WRBP) (Dr. W._)
• See "Dr. W. Russell Blake Planetarium (WRBP)"

Seymour Planetarium
• See "Springfield Science Museum, Seymour Planetarium"

Skylight Astronomical Society (SAS) Inc.
P.O.Box 561
Acton, MA 01720
Electronic Mail: sas@sasobservatory.org
WWW: http://www.sasobservatory.org/
Observatories: 1 (Rockbottom Observatory)
City Reference Coordinates: 071°26'00"W 42°29'00"N

Sky Publishing Corp.
49 Bay State Road
Cambridge, MA 02138
Telephone: 617-864-7360
   1-800-253-0245 (USA & Canada only)
Telefax: 617-864-6117
Electronic Mail: skytel@skypub.com
WWW: http://www.skyandtelescope.com/
   http://www.skypub.com/
Founded: 1941
Staff: 50
Periodicals: (12) "Sky & Telescope" (ISSN 0037-6604, circ.: 125,000), "SkyWatch"
• Publisher
City Reference Coordinates: 071°10'00"W 42°24'00"N (Belmont)
   071°06'00"W 42°22'00"N (Cambridge)

University of Michigan, Department of Aerospace Engineering
François-Xavier Bagnoud Building
1320 Beal Avenue
Ann Arbor, MI 48109-2140
Telephone: 734-764-3310
Telefax: 734-763-0578
Electronic Mail: <userid>@engin.umich.edu
WWW: http://www.engin.umich.edu/dept/aero/
Founded: 1914
City Reference Coordinates: 083°45'00"W 42°18'00"N

University of Michigan, Department of Astronomy
830 Dennison Building
Ann Arbor, MI 48109-1090
Telephone: 734-764-3440
Telefax: 734-763-6317
Electronic Mail: astro@astro.lsa.umich.edu
WWW: http://www.astro.lsa.umich.edu/
Founded: 1854
Periodicals: "Publications" (ISSN 0076-8421)
Coordinates: 083°43'49"W 42°16'49"N H282m (IAU Code 767)
City Reference Coordinates: 083°45'00"W 42°18'00"N

University of Michigan, Department of Atmospheric, Oceanic and Space Sciences
Space Research Building
2455 Hayward Street
Ann Arbor, MI 48109-2143
Telephone: 734-936-0482
Telefax: 313-764-4585
Electronic Mail: aoss.um@umich.edu
              <userid>@umich.edu
WWW: http://aoss.engin.umich.edu/
       http://solar-heliospheric.engin.umich.edu/ (Solar and Heliospheric Research Group)
Founded: 1954 (as "Civil Engineering Department") (Solar Group: 1963)
Staff: 48
Activities: education * research
City Reference Coordinates: 083°45'00"W 42°18'00"N

University of Michigan, Department of Natural Sciences
4901 Evergreen Road
Dearborn, MI 48128-1491
Telephone: 313-593-5277
Telefax: 313-593-4937
Electronic Mail: dbord@umd.umich.edu
WWW: http://www.umd.umich.edu/
       http://www.umd.umich.edu/casl/natsci/physics/
Activities: education * stellar astronomy * spectroscopy of CP and long-period variable stars
City Reference Coordinates: 083°10'00"W 42°18'00"N

University of Michigan, Exhibit Museum of Natural History
1109 Geddes Avenue
Ann Arbor, MI 48109-1079
Telephone: 734-764-0478
Electronic Mail: mlinke@umich.edu
WWW: http://www.exhibits.lsa.umich.edu/
       http://www.exhibits.lsa.umich.edu/Planetarium/PlanetSchedule.html (Planetarium)
Founded: 1958
City Reference Coordinates: 083°45'00"W 42°18'00"N

University of Michigan, Michigan Dartmouth MIT Observatory
• See "MDM Observatory (USA-AZ)"

University of Michigan Press
P.O. Box 1104
Ann Arbor, MI 48106-1104
or :
839 Greene Street
Ann Arbor, MI 48104
Telephone: 734-764-4388
Telefax: 734-615-1540
Electronic Mail: umpress@umich.edu
WWW: http://www.press.umich.edu/
Founded: 1930
• Publisher
City Reference Coordinates: 083°45'00"W 42°18'00"N

University of Michigan, Student Astronomical Society
Electronic Mail: sas.astro@umich.edu
bmendez@astro.lsa.umich.edu (Bryan Mendez, President)
WWW: http://www.astro.lsa.umich.edu/sas/
City Reference Coordinates: 083°45'00"W 42°18'00"N (Ann Arbor)

University Optics
P.O. Box 1205
Ann Arbor, MI 48106
Telephone: 1-800-521-2828 (USA only)
734-665-3575
Telefax: 734-665-1815
Electronic Mail: uoptics@aol.com
WWW: http://www.universityoptics.com/
City Reference Coordinates: 083°45'00"W 42°18'00"N (Ann Arbor)

Warren Astronomical Society (WAS)
P.O. Box 1505
Warren, MI 48090-1505
Telephone: 810-447-2424
WWW: http://www.boonhill.net/was/
Founded: 1962
Activities: monthly meetings
Periodicals: "Warren Astronomical Society Paper" (WASP)
Coordinates: 082°55'04"W 42°45'29"N H330m (Stargate Observatory, Romeo)
City Reference Coordinates: 083°01'00"W 42°28'00"N

Wayne State University (WSU), Department of Physics and Astronomy
Detroit, MI 48202
Electronic Mail: <userid>@physics.wayne.edu
WWW: http://hal.physics.wayne.edu/
http://hal.physics.wayne.edu/wsu/planet.html (Planetarium)
City Reference Coordinates: 083°05'00"W 42°23'00"N

USA - Minnesota

Alworth Planetarium (Marshall W._)
- See "Marshall W. Alworth Planetarium"

Carleton College, Department of Physics and Astronomy
Northfield, MN 55057
Telephone: 507-646-4367
  507-646-4383
  507-646-4429
Telefax: 507-646-4384
Electronic Mail: <userid>@carleton.edu
WWW: http://physics.carleton.edu/
  http://physics.carleton.edu/Faculty/Joel/svpage.html (Sogn Valley Radio Obs.)
Activities: pulsars * AGNs * gravitation * ISM
Coordinates: 093°09'00"W 44°27'42"N (Goodsell Obs. – IAU Code 741)
City Reference Coordinates: 093°09'00"W 44°27'00"N

College of St. Catherine, Physics Department, Observatory
2004 Randolph Avenue
Saint Paul, MN 55105
Electronic Mail: askastro@stkate.edu
WWW: http://www.stkate.edu/physics/observatory.html
Founded: 1985
City Reference Coordinates: 093°07'00"W 44°58'00"N

Concordia College, Department of Physics, Observatory
901 8th Street South
Moorhead, MN 56562
Telephone: 218-299-3391
Telefax: 218-299-4308
Electronic Mail: <userid>@gloria.cord.edu
WWW: http://www.cord.edu/faculty/gealy/obs.html
Founded: 1992
City Reference Coordinates: 096°45'00"W 46°53'00"N

Feder Observatory (Paul P._)
- See "Paul P. Feder Observatory"

Futurestar Corp.
6529 Cecilia Circle
Bloomington, MN 55439
Telephone: 952-942-8388
Telefax: 952-942-8661
Electronic Mail: solutions@futurestarcorp.com
WWW: http://www.futurestarcorp.com/
Founded: 1984
Staff: 13
Activities: manufacturing teflon flow control devices
City Reference Coordinates: 093°17'00"W 44°50'00"N

Headwaters Science Center (HSC), Planetarium
P.O. Box 1176
Bemidji, MN 56601
or :
413 Beltrami Avenue
Bemidji, MN 56601
Telephone: 218-751-1110
WWW: http://www.spacestar.net/users/oishsc/
Founded: 1994
Activities: exhibitions * Starlab planetarium
City Reference Coordinates: 094°52'00"W 47°29'00"N

Hopkins Technology, LLC
421 Hazel Lane
Hopkins, MN 55343-7116
Telephone: 1-800-397-9211 (USA only)
  952-931-9376
Telefax: 952-931-9377
Electronic Mail: info@hoptechno.com
WWW: http://www.hoptechno.com/
Founded: 1986

• Software producer
City Reference Coordinates: 093°24'00"W 44°56'00"N

Mankato Area Astronomy Association (MAAA)
MSU 59
P.O. Box 8400
Mankato, MN 56002-8400
Electronic Mail: ptolemy@krypton.mankato.msus.edu
WWW: http://krypton.mnsu.edu/~maaa/html/welcome.html
Founded: 1994
City Reference Coordinates: 094°01'00"W 44°10'00"N

Mankato State University (MSU)
• See now "Minnesota State University at Mankato"

Marshall W. Alworth Planetarium
10 University Drive
Duluth, MN 55812-2496
Telephone: 218-726-7129
Electronic Mail: planet@d.umn.edu
WWW: http://www.d.umn.edu/~planet/
Founded: 1965
City Reference Coordinates: 092°06'00"W 46°47'00"N

Minneapolis Planetarium
300 Nicollet Mall
Minneapolis, MN 55401
Telephone: 612-630-6150
Telefax: 612-630-6180
Electronic Mail: rjbonadurer@mpls.lib.mn.us (Robert J. Bonadurer, Director)
WWW: http://www.mplanetarium.org/
Founded: 1950
Staff: 10
Activities: education * shows
City Reference Coordinates: 093°13'00"W 44°59'00"N

Minnesota Astronomical Society (MAS)
P.O. Box 583011
Minneapolis, MN 55458-3011
or :
30 East Tenth Street
Saint Paul, MN 55101
or :
1615 East River Parkway
Minneapolis, MN 55414-3627
Telephone: 651-649-4861 (Info Line)
Electronic Mail: info@mnastro.org
               onaninfo@mnastro.org (Onan Observatory)
WWW: http://www.mnastro.org/
      http://www.mnastro.org/onan/ (Onan Observatory)
Founded: 1967
Membership: 330
Activities: monthly lectures * monthly star parties * promoting astronomy * observing (occultations, variable stars) * assisting and conducting university-level education * telescope making classes
Periodicals: (6) "Gemini" (circ.: 200)
Coordinates: 092°49'19"W 44°56'16"N H300m (Metcalf Nature Center)
              093°56'23"W 44°48'34"N H320m (Onan Obs., Baylor Regional Park)
              092°51'43"W 44°11'50"N H320m (Cherry Grove Obs.)
City Reference Coordinates: 093°13'00"W 44°59'00"N (Minneapolis)
                                   093°07'00"W 44°58'00"N (St. Paul)
                                   093°24'00"W 44°51'00"N (Edina)

Minnesota State University at Mankato, Department of Physics and Astronomy
141 Trafton Science Center N
Mankato, MN 56001
Telephone: 507-389-5743
Telefax: 507-389-1095
Electronic Mail: <userid>@mankato.msus.edu
WWW: http://www.mankato.msus.edu/dept/astro/physast.html
Founded: 1867
Staff: 3
Activities: photometry * astrometry * CCD imaging * long-period variable stars * IR * gamma-ray bursts * galactic structure * Ba and Am stars
Coordinates: 093°59'49"W 44°08'18"N H302m (Andreas Observatory)
              093°59'49"W 44°08'28"N H302m (Standeford Observatory)

- Formerly known as "Mankato State University (MSU)"
City Reference Coordinates: 094°01'00"W 44°10'00"N

Moorhead State University (MSU), Department of Physics and Astronomy
Moorhead, MN 56563
Telephone: 218-236-2141
Telefax: 218-236-2290
Electronic Mail: <userid>@mhd1.moorhead.msus.edu
WWW: http://physweb.mnstate.edu/Department/
http://physweb.moorhead.msus.edu/
Staff: 8
Activities: education * planetarium shows * observing
City Reference Coordinates: 096°46'00"W 46°52'00"N

Paul P. Feder Observatory
Regional Science Center
Moorhead State University
Moorhead, MN 56563
Telephone: 218-236-2904
Telefax: 218-299-5864
WWW: http://www.moorhead.msus.edu/regsci/
City Reference Coordinates: 096°46'00"W 46°52'00"N

Paulucci Space Theatre (PST)
c/o Hibbing Community College
1502 East 23rd Street
Hibbing, MN 55746
Telephone: 218-262-6720
Telefax: 218-262-6719
Electronic Mail: p.davis@hcc.mnscu.edu (Peter Davis, Director)
WWW: http://www.hcc.mnscu.edu/SL2.html
Founded: 1980
Staff: 7
Activities: astronomy and science-related planetarium shows
City Reference Coordinates: 092°56'00"W 47°25'00"N

Saint Cloud State University, Department of Physics, Astronomy and Engineering Science
324 Mathematics and Science Center
720 Fourth Avenue South
Saint Cloud, MN 56301-4498
Telephone: 320-255-2011
Telefax: 320-255-4728
Electronic Mail: physics@condor.stcloudstate.edu
<userid>@stcloudstate.edu
WWW: http://condor.stcloudstate.edu/~physics/
Founded: 1974
Staff: 12
Activities: undergraduate and public education * planetarium * observatory
Coordinates: 094°08'59"W 45°33'05"N H330m
City Reference Coordinates: 094°10'00"W 45°33'00"N

Science Museum of Minnesota (SMM)
30 East 10th Street
Saint Paul, MN 55101
Telephone: 651-221-9444
WWW: http://www.smm.org/
Founded: 1907
City Reference Coordinates: 093°07'00"W 44°58'00"N

Southwest State University (SSU), Planetarium
SM 178
1501 State Street
Marshall, MN 56258
Telephone: 507-537-6175
507-537-6178
Telefax: 507-537-6151
Electronic Mail: kmurphy@southwest.msus.edu (Kenneth L. Murphy, Director)
WWW: http://www.southweststate.edu/kennethmurphy/planetarium/shows.htm
Founded: 1967 (University)
Staff: 1
City Reference Coordinates: 095°47'00"W 44°27'00"N

Steele County Astronomical Society (SCAS)
c/o Gene Kispert
112 - 15th Street NE

Owatonna, MN 55060
Telephone: 507-451-2365
Electronic Mail: gkispert@ii.net
WWW: http://scas.jumpoutdoors.com/
City Reference Coordinates: 093°14'00"W 44°05'00"N

University of Minnesota, Institute for Mathematics and its Applications (IMA)
400 Lind Hall
207 Church Street SE
Minneapolis, MN 55455-0436
Telephone: 612-624-6066
Telefax: 612-626-7370 (IMA East, 400 Lind Hall)
        612-624-4163 (IMA West, Vincent Hall)
Electronic Mail: staff@ima.umn.edu
WWW: http://www.ima.umn.edu/
City Reference Coordinates: 093°13'00"W 44°59'00"N

University of Minnesota, School of Physics and Astronomy
116 Church Street SE
Minneapolis, MN 55455
Telephone: 612-624-0211
Telefax: 612-626-2029
Electronic Mail: astdept@astro.spa.umn.edu
WWW: http://astro.umn.edu/
Founded: 1974
Staff: 16
Activities: computational astrophysics * infrared astronomy * radioastronomy * cosmology * space physics
Observatories: 2 (Marine-on-St.-Croix, USA-MN & Tucson, USA-AZ)
City Reference Coordinates: 093°13'00"W 44°59'00"N

USA - Mississippi

Davis Planetarium (Russell C._)
• See "Russell C. Davis Planetarium)

French Camp Academy, Rainwater Observatory and Planetarium
French Camp, MS 39745
Telephone: 662-547-6865
Telefax: 662-547-6790
Electronic Mail: jhill@astronomers.org (James G. Hill, Director)
rainwater@astronomers.org
WWW: http://rainwater.astronomers.org/
Founded: 1985
Staff: 2
Activities: education * observing * Astronomy Education Materials Center
Periodicals: (12) "The Rainwater Observer" (circ.: 450)
Coordinates: 089°23'00"W 33°17'00"N
City Reference Coordinates: 089°23'00"W 33°17'00"N

Jackson Astronomical Association (JAA)
P.O. Box 766
Clinton, MS 39060
or :
280 West Lorenz Boulevard
Jackson, MS 39213-7058
Telephone: 601-982-2388
WWW: http://jackson.astronomers.org/
Founded: 1979
Membership: 45
Activities: Messier & Herschel search * photography * occultations * astronomy software * instrumentation
Periodicals: (12) "The Observer II"
Awards: "Elliott-Yeates Award"
Coordinates: 090°11'00"W 32°17'00"N H250m (Butts-Clark Observatory)
City Reference Coordinates: 090°20'00"W 32°20'00"N (Clinton)
090°12'00"W 32°18'00"N (Jackson)

Mississippi State University (MSU), Department of Physics and Astronomy
Mississippi State, MS 39762-5167
Telephone: 601-325-2806
Telefax: 601-325-8898
Electronic Mail: physics@msstate.edu
<userid>@ra.msstate.edu
WWW: http://www.msstate.edu/Dept/Physics/
http://www.msstate.edu/Dept/Physics/html/observatory.html (Howell Observatory)
Founded: 1949
Staff: 21
Activities: high-energy astrophysics * gamma-ray observatory
Coordinates: 088°00'00"W 34°00'00"N H100m
City Reference Coordinates: 088°00'00"W 34°00'00"N

National Aeronautics and Space Administration (NASA), Stennis Space Center (SSC), Public Affairs Office
PA00
Stennis Space Center, MS 39529
Telephone: 228-688-3341
Electronic Mail: pao@ssc.nasa.gov
WWW: http://www.ssc.nasa.gov/
Activities: rocket propulsion testing * partnering with industry to develop and implement remote sensing technology
City Reference Coordinates: 089°47'00"W 30°17'00"N (Slidell)
089°20'00"W 30°19'00"N (Bay St Louis)

Observa-Dome Inc.
371 Commerce Park Drive
Jackson, MS 39213
Telephone: 601-982-3333
Electronic Mail: odl@misnet.com
WWW: http://observa-dome.com/
Founded: 1961
Activities: manufacturing observatory domes, missile tracking, and laser ranging shelters
City Reference Coordinates: 090°12'00"W 32°18'00"N

Rainwater Astronomical Association (RAA)
c/o Craig Hodges
114 Cole Street
Suite 11
Starkville, MS 39759
or :
c/o Rainwater Observatory and Planetarium
French Camp Academy
One Fine Place
French Camp, MS 39745
Telephone: 601-547-6865
Telefax: 601-547-6790
Electronic Mail: jhill@kopower.com (James G. Hill)
WWW: http://www.kopower.com/~jhill
Founded: 1985
Membership: 25
Activities: education * hosting Mid-South Star Gaze
Periodicals: "The Rainwater Observer"
Coordinates: 089°26'00"W 33°17'00"N (French Camp)
City Reference Coordinates: 089°50'00"W 33°27'00"N (Starkville)
  089°26'00"W 33°17'00"N (French Camp)

Rainwater Observatory and Planetarium
• See "French Camp Academy, Rainwater Observatory and Planetarium"

Russell C. Davis Planetarium
201 East Pascagoula Street
Jackson, MS 39201-4115
Telephone: 601-960-1550
Telefax: 601-960-1555
WWW: http://www.city.jackson.ms.us/Govt/planetarium.html
Founded: 1978
Staff: 11
Activities: planetarium programs and large-format films for school groups and public
City Reference Coordinates: 090°12'00"W 32°18'00"N

University of Mississippi, Department of Physics and Astronomy
108 Lewis Hall
P.O. Box 1848
University, MS 38677-1848
Telephone: 662-915-7046
Telefax: 662-915-5045
Electronic Mail: physics@phy.olemiss.edu
  <userid>@phy.olemiss.edu
WWW: http://www.olemiss.edu/depts/physics_and_astronomy/
Founded: 1859
Staff: 15
Activities: high-energy physics
Coordinates: 088°31'48"W 34°22'13"N H161m (Kennon Observatory)
City Reference Coordinates: 088°32'00"W 34°22'00"N

USA - Missouri

Astronomical League (AL)
c/o Jackie Beucher (Executive Secretary)
9201 Ward Parkway
Suite 100
Kansas City, MO 64114
Telephone: 816-333-7759
Electronic Mail: execsec@astroleague.org
　　　　　　　　m31@sky.net
WWW: http://www.astroleague.org/
Founded: 1947
Membership: 17000 individuals + 277 societies
Activities: promoting astronomy on the individual as well as the club level * national and regional conventions
Sections: activity certificates and Messier club * book service * observing programs information center * publications committee * education committee * Astronomy Day Headquarters (see separate entry)
Periodicals: (4) "The Reflector" (ISSN 0034-2963, circ.: 20,000)
Awards: (1) "Astronomical League Award"; "Herschel Award", "Messier Club Award", "Leslie Peltier Award", "Horkheimer Award" "National Outstanding Young Astronomer Award"
● See also the "Astronomy Day" entry in USA-MI
City Reference Coordinates: 094°35'00"W 39°05'00"N

Astronomical Society of Jefferson County
(Pythagorica)
c/o Joe Bohnert
11761 CC Highway
Festus, MO 63028
Telephone: 314-937-4764
WWW: http://home.att.net/~klossner/asojcm.html
City Reference Coordinates: 090°24'00"W 38°13'00"N

Astronomical Society of Kansas City (ASKC), Inc.
P.O. Box 400
Blue Springs, MO 64013
or :
1919 SW Fifth Street
Blue Springs, MO 64014
Telephone: 816-228-4238
Electronic Mail: askc@sound.net
WWW: http://www.askconline.org/
Founded: 1935
Membership: 320
Activities: education * observing (Powell Observatory) * meetings
Periodicals: (12) "Cosmic Messenger"
Awards: (1) "Helen A. Warkoczewski Award"
Coordinates: 094°41'45"W 38°38'46"N H356m (Powell Obs. – IAU Code 649)
City Reference Coordinates: 096°40'00"W 40°09'00"N

Central Methodist College (CMC)
411 CMC Square
Fayette, MO 65248
Telephone: 660-248-3391
　　　　　　888-CMC-1854
Telefax: 660-248-2287
　　　　　660-248-1872
WWW: http://www.cmc.edu/
Staff: 1
Activities: education * Morrison Observatory
Coordinates: 092°41'48"W 39°09'06"N H228m
City Reference Coordinates: 092°41'00"W 39°09'00"N

Central Missouri Astronomical Association (CMAA)
P.O. Box 1251
Jefferson City, MO 65102-1251
Telephone: 573-761-1108
Telefax: 573-634-7854
Electronic Mail: andy@www.system.missouri.edu (Andy Steere)
WWW: http://www.system.missouri.edu/cmaa/
Founded: 1958
City Reference Coordinates: 092°10'00"W 38°33'00"N

Earth and Space Science Center Planetarium
c/o Rock Bridge Senior High School

4303 South Providence Road
Columbia, MO 65203-7198
Telephone: 573-886-2560
573-886-2587
Telefax: 573-886-2031
Founded: 1974
Activities: programs for elementary and secondary students
City Reference Coordinates: 092°20'00"W 38°57'00"N

Eastern Missouri Dark Sky Observers (EMDSO)
c/o Richard Schwentker
75 Greenshire Road
Marthasville, MO 63357
or :
c/o Les Kemp
8247 Highway YY
New Haven, MO 63068
Telephone: 636-433-5479
Electronic Mail: emdso@mo-biz.com
WWW: http://www.mo-biz.com/emdso/
Periodicals: (12) "The Dark-Sky Observer"
City Reference Coordinates: 091°15'00"W 38°37'00"N (New Haven)

Great Plains Star Party (GPSP)
304 East Alger Drive
Blue Springs, MO 64014
Telephone: 816-229-2399
Electronic Mail: clearskyalways@hotmail.com
WWW: http://www.greatplainsstarparty.org/
City Reference Coordinates: 096°40'00"W 40°09'00"N

Kansas City Museum
3218 Gladstone Boulevard
Kansas City, MO 64123-1199
Telephone: 816-483-8300
Electronic Mail: universe@tyrell.net
WWW: http://www.kcmuseum.com/pltm.html
Activities: exhibits * hands-on experiments * planetarium
City Reference Coordinates: 094°35'00"W 39°05'00"N

Old Orchard Observatory (OOO)
310 Calvert Avenue
Saint Louis, MO 63119-4204
Telephone: 314-962-5443
Founded: 1987
Staff: 1
Activities: visual and photoelectric photometry of variable stars * occultations
Coordinates: 090°20'26"W 38°35'06"N H166m (Webster Groves)
City Reference Coordinates: 090°11'00"W 38°38'00"N

Pattonville Observatory and Planetarium
c/o Science Coordinator
Instructional Coordinating Center
Pattonville School District
11001 Saint Charles Rock Road
Saint Ann, MO 63074-1509
Telephone: 314-298-4487
Telefax: 314-298-4511
WWW: http://nightsky.psdr3.org/
Founded: 1962
Staff: 2
Activities: education
City Reference Coordinates: 090°23'00"W 38°43'00"N

Pythagorica
• See "Astronomical Society of Jefferson County"

Saint Louis Astronomical Society (SLAS)
c/o Cook Feldman (Treasurer)
1316 Tahiti Drive
Crestwood, MO 63126
or :
c/o Bruce Logan
8125 South Laclede Station Road
Saint Louis, MO 63123

Electronic Mail: scopes777@aol.com (Steve Sands)
    kronkg@medicine.wustl.edu (Gary W. Kronk)
WWW: http://www.slasonline.org
Founded: 1936
City Reference Coordinates: 090°11'00"W 38°38'00"N (Saint Louis)

Saint Louis Science Center (SLSC)
5050 Oakland Avenue
Saint Louis, MO 63110-1460
Telephone: 1-800-456-SLSC (USA only)
    314-289-4400
Telefax: 314-289-4440
WWW: http://www.slsc.org/
    http://www.slsc.org/docs/mod2/proginfo/pgmg005.htm (Planetarium)
Founded: 1972
Membership: 190
Activities: public education * exhibits * planetariums (one portable)
Periodicals: (4) "New Science"
Coordinates: 090°16'15"W 38°37'57"N H144m
City Reference Coordinates: 090°11'00"W 38°38'00"N

Seiler Instrument & Manufacturing Company Inc., Planetarium Division
P.O. Box 948
Narberth, PA 19072-0948
Telephone: 1-800-726-8805 (USA only)
    610-664-0308
Telefax: 610-664-0308
Electronic Mail: zeiss@seilerinst.com
WWW: http://www.seilerinst.com/
Founded: 1945
Activities: distributing Zeiss line of planetariums
Periodicals: (4) "Projections"
City Reference Coordinates: 075°07'00"W 39°57'00"N (Philadelphia)

Southwest Missouri State University (SMSU), Department of Physics, Astronomy, and Materials Science
901 South National Avenue
Springfield, MO 65804
Telephone: 417-836-5131
Telefax: 417-836-6934
Electronic Mail: <userid>@mail.smsu.edu
WWW: http://astronomy.smsu.edu/
Founded: 1971
Staff: 15 (Astronomy: 3)
Observatories: 1 (Baker Observatory)
City Reference Coordinates: 093°17'00"W 37°14'00"N

Springfield Astronomical Society (SAS)
c/o Ted Beresky
Cedar Hill Observatory
222 Four Cedars Lane
Rogersville, MO 65742
Telephone: 417-753-4260
Electronic Mail: tlberesky@hotmail.com
WWW: http://pages.prodigy.net/tberesky/
Membership: 25
Activities: meetings * observing * education
City Reference Coordinates: 093°05'00"W 37°07'00"N (Rogersville)
    093°17'00"W 37°14'00"N (Springfield)

University of Missouri, Columbia, Department of Physics and Astronomy
223 Physics Building
Columbia, MO 65211
Telephone: 314-882-3335
Telefax: 314-882-4195
Electronic Mail: <userid>@mizzou1.missouri.edu
WWW: http://www.missouri.edu/~physwww/physics.html
Founded: 1880
Staff: 30
Activities: education * public viewing
Coordinates: 092°20'00"W 38°27'00"N H240m (Laws Observatory)
City Reference Coordinates: 092°20'00"W 38°57'00"N

University of Missouri, Saint Louis (UMSL), Department of Physics and Astronomy
503 Benton Hall

8001 Natural Bridge Road
Saint Louis, MO 63121
Telephone: 314-516-5931
Telefax: 314-516-5931
Electronic Mail: <userid>@umsl.edu
WWW: http://www.umsl.edu/~physics/
Founded: 1966
Staff: 15
Activities: astrophysics * atomic physics * materials science * biophysics * particle physics
City Reference Coordinates: 090°11'00"W 38°38'00"N

Washington University in Saint Louis (WUStL), Department of Earth and Planetary Sciences
Campus Box 1169
One Brookings Drive
Saint Louis, MO 63130-4899
Telephone: 314-935-5610
Telefax: 314-935-7361
Electronic Mail: <userid>@wuphys.wustl.edu
WWW: http://epsc.wustl.edu/
City Reference Coordinates: 090°11'00"W 38°38'00"N

Washington University in Saint Louis (WUStL), Department of Physics
Campus Box 1105
One Brookings Drive
Saint Louis, MO 63130
Telephone: 314-935-6225
314-935-6257
Telefax: 314-935-4083
Electronic Mail: <userid>@wuphys.wustl.edu
WWW: http://www.physics.wustl.edu/
http://presolar.wustl.edu/ (Laboratory for Space Sciences)
http://www.physics.wustl.edu/mcdonnell/ (McDonnell Center for the Space Sciences)
Founded: 1974
Staff: 55
Activities: experimental and theoretical research in astrophysics, space sciences
City Reference Coordinates: 090°11'00"W 38°38'00"N

USA - Nebraska

Association of Astronomy Educators
c/o Katherine Becker
5103 Burt Street
Omaha, NE 68132
Telephone: 402-556-0082
Telefax: 402-280-2140
Electronic Mail: kbecke@bluejay.creighton.edu (Katherine E. Becker)
City Reference Coordinates: 095°57'00"W 41°16'00"N

Burke Planetarium
• See "Harry A. Burke High School, Planetarium"

Dale Planetarium (Fred G._)
• See "Fred G. Dale Planetarium"

Fred G. Dale Planetarium
c/o Wayne State College
Carhart Science Building
1111 Main Street
Wayne, NE 68787
Telephone: 402-375-7329
WWW: http://www.wsc.edu/academic/mathsci/planetar.htm
Founded: 1970
City Reference Coordinates: 097°03'00"W 42°43'00"N

Great Plains Planetarium Association (GPPA)
c/o Jack Dunn (President)
Ralph Mueller Planetarium
University of Nebraska
210 Morrill Hall
Lincoln, NE 68588-0375
Telephone: 402-472-2641
Telefax: 402-472-8899
Electronic Mail: jdunn@spacelaser.com
WWW: http://www.spacelaser.com/gppa/
Membership: 80
Activities: promoting planetaria and astronomy education
Periodicals: "GPPA Newsletter"
City Reference Coordinates: 096°42'00"W 40°48'00"N

Harry A. Burke High School, Planetarium
12200 Burke Boulevard
Omaha, NE 68154
Telephone: 402-557-3200
WWW: http://www.ops.org/burke/10planet/default-planet01.htm
Founded: 1968
City Reference Coordinates: 095°57'00"W 41°16'00"N

Hastings College, Sachtleben Observatory
900 East 9th Street
Hastings, NE 68901
Telephone: 402-4-OBSERV
402-462-7378
WWW: http://www.blackstarpress.com/arin/sach/
Founded: 2000
Coordinates: 098°21'36"W 40°35'24"N H362m
City Reference Coordinates: 098°22'00"W 40°37'00"N

Hastings Museum of Natural and Cultural History
P.O. Box 1286
Hastings, NR 68902-1286
or :
14th and Burlington Avenue
Hastings, NR 68901
Telephone: 402-461-IMAX
1-800-508-IMAX (USA only)
Electronic Mail: hm11706@alltel.net
WWW: http://www.hastingsnet.com/museum/welcome/welcome.htm
http://www.hastingsnet.com/museum/mcdonald.html (J.M. McDonald Planetarium)
Founded: 1958

City Reference Coordinates: 098°22'00"W 40°37'00"N

Hyde Memorial Observatory
c/o Lincoln Parks and Recreation Department
2740 A Street
Lincoln, NE 68502
or :
Holmes Lake (South Shore Road)
3701 South 70th Street
Lincoln, NE
Telephone: 402-441-7895
Telefax: 402-441-6468
Electronic Mail: hyde-wm@blackstarpress.com
nature@nrcdec.nrc.state.ne.us (attn.: Hyde Observatory)
WWW: http://www.blackstarpress.com/arin/hyde/
Founded: 1976
Staff: 59
Activities: public education * shows * observing * lectures * talks
Coordinates: 096°38'11"W 40°46'40"N H425m
City Reference Coordinates: 096°42'00"W 40°48'00"N

Jensen Planetarium
c/o Physics Department
Nebraska Wesleyan University
5000 Saint Paul Avenue
Lincoln, NE 68504-2796
Telephone: 402-465-2235
Electronic Mail: physics@nebrwesleyan.edu
WWW: http://physics.nebrwesleyan.edu/
Founded: 1969
Staff: 1
Activities: education * community service
City Reference Coordinates: 096°42'00"W 40°48'00"N

Lueninghoener Planetarium
c/o Midland Lutheran College
900 North Clarkson Street
Fremont, NE 68025
Telephone: 402-721-5480 (Gary A. Carlson)
Electronic Mail: carlson@campus.mlc.edu
WWW: http://www.mlc.edu/Events/planet.html
City Reference Coordinates: 096°30'00"W 41°26'00"N

Nebraska Star Party (NSP)
c/o Eric Balcom
P.O. Box 540307
Omaha, NE 68154-0307
Telephone: 402-489-8197
Electronic Mail: nsp@4w.com
WWW: http://www.nebraskastarparty.org/
Founded: 1994
City Reference Coordinates: 095°57'00"W 41°16'00"N

Northeast Nebraska Astronomy Club (NNAC)
c/o Dean Stange
204 South 17th
Norfolk, NE 68701
or :
c/o Gregg Adams (Secretary/Treasurer)
710 South 4th
Norfolk, NE 68701
Telephone: 402-371-0891
Electronic Mail: icstars@hotmail.com
gadams@kdsi.net
Founded: 1994
Activities: observing
City Reference Coordinates: 097°25'00"W 42°02'00"N

Omaha Astronomical Society (OAS)
c/o Rich Merten
2918 Bridgeford Road
Omaha, NE 68124
Electronic Mail: rmerten@concentric.net
cdcheney@aol.com (Deb Cheney, Secretary)
WWW: http://www.top.net/cdcheney/

http://www.omahaastro.com/
http://216.22.205.58/stella ("Stella Online!" newsletter)
Founded: 1962
City Reference Coordinates: 095°57'00"W 41°16'00"N

Platte Valley Astronomical Observers (PVAO)
c/o Richard Karohl
1319 North Cedar Street
Grand Island, NE 68801
Telephone: 308-382-3265
Electronic Mail: rlkstars@juno.com
WWW: http://204.234.8.16/~7hills/pvao/pvao.htm
City Reference Coordinates: 098°21'00"W 40°56'00"N

Prairie Astronomy Club (PAC)
P.O. Box 5585
Lincoln, NE 68505-5585
or :
5827 Lasalle Street
Lincoln, NE 68516
Telephone: 402-483-5639
Electronic Mail: pac@4w.com
WWW: http://www.prairieastronomyclub.org/
http://www.blackstarpress.com/arin/hyde/ (Hyde Memorial Observatory)
Founded: 1960
Membership: 62
Activities: monthly meetings * stars parties * observing
Periodicals: (12) "The Prairie Astronomer"
Coordinates: 096°38'13"W 40°46'40"N H356m (Hyde Memorial Observatory)
City Reference Coordinates: 096°42'00"W 40°48'00"N

Sachtleben Observatory
● See "Hastings College, Sachtleben Observatory"

Seven Hills Observatory
c/o Mark Urwiller
4711 Heather Lane
Kearney, NE 68847
Telephone: 308-234-6536
Electronic Mail: murwille@genie.esu10.k12.ne.us
WWW: http://genie.esu10.k12.ne.us/~murwille/7hills.htm
Founded: 1996
Staff: 8
Activities: private observatory * education * public visits
City Reference Coordinates: 099°05'00"W 40°42'00"N

University of Nebraska, Kearney (UNK), Department of Physics and Physical Science
905 West 25th Street
Kearney NE 68849
Telephone: 308-234-8277
Electronic Mail: <userid>@unk.edu
WWW: http://rip.physics.unk.edu/
http://rip.physics.unk.edu/Astronomy/
Staff: 10
Activities: research * education
Coordinates: 099°05'43"W 40°41'59"N H743m
City Reference Coordinates: 099°05'00"W 40°42'00"N

University of Nebraska, Lincoln (UNL), Department of Physics and Astronomy
Lincoln, NE 68588-0111
Telephone: 402-472-2770
Telefax: 402-472-2879
Electronic Mail: <userid>@unlinfo.unl.edu
WWW: http://www.physics.unl.edu/
http://www.unl.edu/uevents/calendar.html
http://www.unl.edu/physics/astronomy/PublicNight.html
http://physics.unl.edu/directory/lee/bo/bo-hp.html (Observatory)
http://www-museum.unl.edu/ (Planetarium)
http://www.spacelaser.com/ (Planetarium)
Founded: 1972
Staff: 4
Activities: variable and binary stars * star clusters * astronomical instrumentation * ISM * photometry * pulsation theory * Ralph Mueller Planetarium * Behlen Observatory
Coordinates: 096°26'48"W 41°10'18"N H362m
City Reference Coordinates: 096°42'00"W 40°48'00"N

University of Nebraska, Department of Physics
60th and Dodge Streets
Omaha, NE 68182-0266
Telephone: 402-554-2511
    888-UNO-NASA
Telefax: 402-554-2244
    888-554-3100
WWW: http://www.physics.unomaha.edu/
    http://www.physics.unomaha.edu/planet (Kountze Planetarium)
Founded: 1914
Staff: 14
Activities: research in physics and related fields * education * Kountze Planetarium
Coordinates: 095°57'00"W 41°18'00"N
City Reference Coordinates: 095°57'00"W 41°16'00"N

USA - Nevada

Astronomical Society of Nevada (ASN)
1403 Eastwood Drive
Reno, NV 89509-2208
or :
c/o Tony Berendsen
4520 Lynnfield Court
Reno, NV 89509
or :
c/o Fleischmann Planetarium
University of Nevada, Reno/272
Reno, NV 89557-0010
Telephone: 775-329-8809
Telefax: 775-329-8809
WWW: http://www.scs.unr.edu/planet/ASN.html
　　　　http://members.theglobe.com/astronomynv/
　　　　http://www.ramblin.com/ASN-LC_Menu.htm (Lahontan Chapter)
City Reference Coordinates: 119°48'00"W 39°31'00"N

Community College of Southern Nevada (CCSN), Planetarium
3200 East Cheyenne Avenue
North Las Vegas, NV 89030-4296
Telephone: 702-651-4SKY
　　　　　　702-651-4759
　　　　　　702-651-4138
Electronic Mail: drdale@nevada.edu (Dale Etheridge, Director)
WWW: http://www.ccsn.nevada.edu/planetarium/
Founded: 1977
Staff: 3
Periodicals: (12) "onOrbit Magazine"
City Reference Coordinates: 115°07'00"W 36°12'00"N

Fleischmann Planetarium and Science Center
c/o University of Nevada
Mail Stop 272
Reno, NV 89557-0012
Telephone: 775-784-4812
　　　　　　775-784-4811 (Show Information)
　　　　　　775-784-1SKY (Sky Information)
Telefax: 775-784-4822
Electronic Mail: keithj@unr.edu (Keith Johnson)
WWW: http://www.scs.unr.edu/planet/
Founded: 1963
Staff: 5
Activities: public and school shows * museum * observatory
Periodicals: (12) "The Fleischmann Flyer"
Coordinates: 119°49'05"W 39°32'46"N H1,400m
City Reference Coordinates: 119°48'00"W 39°31'00"N

Las Vegas Astronomical Society (LVAS)
c/o Community College of Southern Nevada
3200 East Cheyenne Avenue
North Las Vegas, NV 89030
Telephone: 702-361-1356
Electronic Mail: scopehed@skylink.net (William Vorce, President)
　　　　　　　　jbar2@vegas.infi.net (James R. Mellor, Newsletter Editor)
WWW: http://www.ccsn.nevada.edu/LVAS
　　　　http://www.ccsn.nevada.edu/other/LVAS/
Founded: 1980
Membership: 115
Activities: meetings * observing
Periodicals: (12) "LVAS Newsletter"
City Reference Coordinates: 115°07'00"W 36°12'00"N

Pacific Planetarium Association (PPA)
c/o Keith Johnson (Secretary/Treasurer)
Fleischmann Planetarium
University of Nevada
Reno, NV 89557
Electronic Mail: keithj@unr.edu
WWW: http://www.ccsn.nevada.edu/planetarium/PPA/
Founded: 1980

Membership: 130
Activities: holding regional planetarium conferences (twice per year)
Periodicals: (2) "Panorama" (after each conference)
City Reference Coordinates: 119°48'00"W 39°31'00"N

Sierra Nevada College, Department of Astronomy
Incline Village, NV 89450
Telephone: 702-831-1314
Telefax: 702-831-1347
WWW: http://www.sierranevada.edu/
Founded: 1969 (College)
Coordinates: 119°55'42"W 39°17'42"N H2,746m (McLean Observatory)
City Reference Coordinates: 119°46'00"W 39°10'00"N (Carson City)

University of Nevada, Department of Physics
4505 South Maryland Parkway
Las Vegas, NV 89154-4002
Telephone: 702-739-3563
Telefax: 702-739-0804
Electronic Mail: <userid>@physics.unlv.edu
            <userid>@nevada.edu
            <userid>@unlv.edu
WWW: http://www.physics.unlv.edu/
Founded: 1958
Staff: 10
Coordinates: 115°08'12"W 36°06'34"N (Las Vegas)
         110°52'39"W 31°41'03"N (Mount Hopkins, USA-AZ)
City Reference Coordinates: 115°08'00"W 36°11'00"N

USA - New Hampshire

Centorr - Vacuum Industries Inc.
55 Northeastern Boulevard
Nashua, NH 03063
Telephone: 603-595-7233
    1-800-962-8631 (USA only)
Telefax: 603-595-9220
Electronic Mail: sales@centorr.com
WWW: http://www.centorr.com/
Founded: 1954
Activities: manufacturing high-temperature and controlled-atmosphere furnaces
City Reference Coordinates: 071°27'00"W 42°46'00"N

Christa McAuliffe Planetarium (CMP)
3 Institute Drive
Concord, NH 03301
Telephone: 603-271-7827
Telefax: 603-271-7832
Electronic Mail: j-geruls@tec.nh.us (Jeanne T. Gerulskis, Executive Director)
WWW: http://www.cmp.state.nh.us/
    http://www.starhop.com/
Founded: 1990
Staff: 24
Activities: shows * exhibits on astronomy and space science * educational workshops * gift shop
City Reference Coordinates: 071°32'00"W 43°12'00"N

Cobblestone Publishing Inc.
30 Grove Street
Suite C
Peterborough, NH 03458
Telephone: 603-924-7209
    1-800-821-0115 (USA only)
Telefax: 603-924-7380
Electronic Mail: custsvc@cobblestone.mv.com
WWW: http://cobblestonepub.com/
Founded: 1980
Staff: 14
Periodicals: (10) "Odyssey" (ISSN 0163-0946), "Cobblestone" (ISSN 0199-5197); (9) "Faces" (ISSN 0749-1387); (5) "Calliope" (ISSN 1050-7086)
• Publisher
City Reference Coordinates: 071°57'00"W 42°53'00"N

Creare Inc.
Etna Road
P.O. Box 71
Hanover, NH 03755
Telephone: 603-643-3800
Telefax: 603-643-4657
Electronic Mail: info@creare.com
WWW: http://www.creare.com/
Founded: 1961
Staff: 85
Activities: applied research * analysis and experimentation * engineering design * new product development * expertise in heat transfer, cryogenics, turbo machinery, fluid mechanics, multiphase flow
City Reference Coordinates: 072°18'00"W 43°42'00"N

Dartmouth College, Department of Physics and Astronomy
6127 Wilder Laboratory
Hanover, NH 03755-3528
Telephone: 603-646-2854
Telefax: 603-646-1446
Electronic Mail: physics@dartmouth.edu
    <userid>@dartmouth.edu
WWW: http://www.dartmouth.edu/~physics/
Founded: 1854
Staff: 17
Activities: redshift surveys * cataclysmic variable stars * white dwarfs
Coordinates: 072°17'00"W 43°42'00"N H183m (Shattuck Observatory)
City Reference Coordinates: 072°17'00"W 43°42'00"N

Dartmouth College, Michigan Dartmouth MIT Observatory
• See "MDM Observatory (USA-AZ)"

Dartmouth College, Wilder Laboratory, Department of Physics and Astronomy
6127 Wilder Laboratory
Hanover, NH 03755-3528
Telephone: 603-646-2359
Telefax: 603-646-1446
Electronic Mail: physics.department@dartmouth.edu
WWW: http://www.dartmouth.edu/artsci/physics/
Founded: 1769
Staff: 6
Activities: research
City Reference Coordinates: 072°18'00"W 43°42'00"N

Grainger Observatory
• See "Phillips Exeter Academy (PEA), Grainger Observatory"

Keene Amateur Astronomers, Inc.
c/o Robert K. Esslinger
RR1 Box 256-A
Alstead, NH 03602
Telephone: 603-835-6182
Electronic Mail: juniemoon@cheshire.net
WWW: http://www.geocities.com/keeneastro/
Membership: 15
City Reference Coordinates: 072°24'00"W 43°10'00"N

Labsphere Inc.
Shaker Street
P.O. Box 70
North Sutton, NH 03260
Telephone: 603-927-4266
Telefax: 603-927-4694
Electronic Mail: labsphere@labsphere.com
WWW: http://www.labsphere.com/
Activities: diffuse reflectance technology
City Reference Coordinates: 071°32'00"W 43°12'00"N (Concord)

New Hampshire Astronomical Society (NHAS)
P.O. Box 1001
Manchester, NH 03105-1001
Telephone: 603-669-9841
Electronic Mail: info@nhastro.com
                nhas@compuserve.com (Mike Stebbins, President)
                lopez@mv.mv.com (Larry Lopez, Treasurer)
WWW: http://www.nhastro.com/
Founded: 1984
Membership: 130
Activities: education * observing * telescope making
Periodicals: (12) "The Observer"
Coordinates: 071°59'41"W 43°08'35"N H260m (Hillsboro)
City Reference Coordinates: 071°28'00"W 42°59'00"N

Phillips Exeter Academy (PEA), Grainger Observatory
Exeter, NH 03833
Telephone: 603-772-4311
Telefax: 603-778-9563
WWW: http://science.exeter.edu/
Founded: 1989
Staff: 3
Activities: education * variable stars * minor planets
Coordinates: 070°56'24"W 42°58'48"N H12m
City Reference Coordinates: 070°57'00"W 42°59'00"N

Plymouth State College, Planetarium
Plymouth, NH 03264
Telephone: 603-536-5000
Telefax: 603-536-1896
WWW: http://www.plymouth.edu/
      http://oz.plymouth.edu/~dennisma/ (Dennis E. Machnik, Planetarium Director)
Activities: public and school programs
City Reference Coordinates: 071°41'00"W 43°45'00"N

Rivers Camera Shop Inc.
454 Central Avenue
Dover, NH 03820
Telephone: 603-742-4888

Telefax: 603-742-5456
Electronic Mail: rivers@ttls.net
WWW: http://www.riverscamera.com/
Founded: 1970
Staff: 24
Activities: sales of telescopes and accessories
City Reference Coordinates: 070°56'00"W 43°12'00"N

Sky-Skan Inc.
51 Lake Street
Nashua, NH 03060-4513
Telephone: 603-880-8500
          1-800-880-8500 (USA only)
Telefax: 603-882-6522
Electronic Mail: office@skyskan.com
WWW: http://www.skyskan.com/
Founded: 1967
Staff: 25
Activities: manufacturing special effects and automation systems for domed theatres
City Reference Coordinates: 071°27'00"W 42°46'00"N

Stocker & Yale Inc. (S&Y)
32 Hampshire Road
Salem, NH 03079
Telephone: 603-893-8778
Telefax: 603-893-5604
Electronic Mail: stocker@postoffice.worldnet.att.net
WWW: http://www.stkr.com/
Founded: 1946
Activities: manufacturing compasses, watches, borescopes, self-luminous dials
City Reference Coordinates: 071°12'00"W 42°47'00"N

University of New Hampshire, Physics Department, Observatory
DeMeritt Hall
Durham, NH 03824
Telephone: 603-862-1950
          603-862-2669
Telefax: 603-862-2030
      603-862-2998
Electronic Mail: <userid>@einstein.unh.edu
WWW: http://www.physics.unh.edu/
      http://www.unh.edu/ur-phys.html
      http://wwwgro.unh.edu/ (High-Energy Astrophysics Group)
City Reference Coordinates: 070°56'00"W 43°08'00"N

University of New Hampshire, Space Science Center
Morse Hall
Durham, NH 03824
Telephone: 603-862-1950
          603-862-2669
Telefax: 603-862-2030
Electronic Mail: <userid>@unhedi1.unh.edu
WWW: http://leonardo.unh.edu:index.html/
      http://www-ssg.sr.unh.edu/
      http://www.physics.unh.edu/
City Reference Coordinates: 070°56'00"W 43°08'00"N

Wilder Laboratory
• See "Dartmouth College, Wilder Laboratory"

USA - New Jersey

Alice and Leonard Dreyfuss Planetarium
c/o Newark Museum
49 Washington Street
Newark, NJ 07101-0540
Telephone: 973-596-6529
Telefax: 973-642-0459
WWW: http://www.newarkmuseum.org/planetarium/
Founded: 1953
Staff: 3
Activities: education
City Reference Coordinates: 074°10'00"W 40°44'00"N

Amateur Astronomers Association of Princeton (AAAP)
P.O. Box 2017
Princeton, NJ 08543
Telephone: 609-737-2575 (Observatory)
Electronic Mail: wjm@sarnoff.com (Bill Murray, Director)
WWW: http://www.princetonastronomy.org/
http://www.princetonol.com/eye/aaap.html
Founded: 1963
Membership: 90
Activities: monthly meetings * lectures * star parties * field trips * public observing
Periodicals: (12) "Sidereal Times"
Observatories: 2 (John W.H. Simpson Observatory, Washington Crossing & Jenny Jump State Park)
City Reference Coordinates: 074°40'00"W 40°21'00"N

Amateur Astronomers Inc. (AAI)
c/o William Miller Sperry Observatory
1033 Springfield Avenue
Cranford, NJ 07016
Telephone: 908-549-0615
908-276-STAR (automated message)
908-709-7520 (observatory)
Electronic Mail: gchaplenko@aol.com (George Chaplenko, Corresp. Secretary)
WWW: http://www.asterism.org/
Founded: 1949
Membership: 300
Activities: popularization * telescope making * public observing * star parties * lectures * solar eclipse expeditions * photography * photometry
Periodicals: (12) "Asterism"; "Sperry Observations"
City Reference Coordinates: 074°19'00"W 40°39'00"N

Arcturus Observatory
1200 Coolidge Avenue
Whiting, NJ 08759
Telephone: 732-350-2102
Telefax: 732-350-2102
Electronic Mail: paulgitto@aol.com (Paul Gitto)
WWW: http://www.cometman.com/
Founded: 1996
Staff: 1
Activities: CCD observing (comets, asteroids)
City Reference Coordinates: 074°23'00"W 39°58'00"N

Astrolab Inc.
4 Powder Horn Drive
Warren, NJ 07059-5105
Telephone: 732-560-3800
Telefax: 732-560-9570
Electronic Mail: support@astrolab.com
sales@astrolab.com
WWW: http://www.astrolab.com/
Founded: 1961
Staff: 50
Activities: supplying high quality, high reliability RF and microwave transmission line connectors, adapters, precision phase shifters, coaxial cable and cable assemblies for military and commercial applications
City Reference Coordinates: 074°30'00"W 40°37'00"N

Astronomical Society of the Toms River Area (ASTRA)
c/o Robert J. Novins Planetarium
Ocean County College

P.O. Box 2001
Toms River, NJ 08754-2001
Telephone: 732-255-0343 (weekdays 09:30-16:00)
Electronic Mail: zimmermann@monmouth.com (Erik Zimmermann)
WWW: http://astra-nj.org/
Founded: 1977
Membership: 70
Activities: observing * monthly meetings with films, talks by members or guest speakers * field trips
Periodicals: (12) "Astral Projections"
City Reference Coordinates: 074°12'00"W 39°58'00"N

Bell Laboratories
• See "Lucent Technologies, Bell Laboratories"

Concordia Astronomy Club (CAC)
P.O. Box 22
Jamesburg, NJ 08831-0022
Telephone: 609-395-8251
Electronic Mail: panson@ix.netcom.com (Eli Drapkin)
WWW: http://www.erols.com/njastro/orgs/cac.htm
Founded: 1988
Membership: 100
Activities: monthly meetings * star parties * field trips * excursions
Periodicals: (12) "Concordia Constellation"
City Reference Coordinates: 074°26'00"W 40°21'00"N

County College of Morris (CCM), Planetarium
214 Center Grove Road
Randolph, NJ 07869
Telephone: 973-328-5076 (Reservations)
973-328-5755 (Jennifer Hedden, Astronomer)
Electronic Mail: jhedden@ccm.edu
WWW: http://www.ccm.edu/campuslife/planetarium.htm
City Reference Coordinates: 074°34'00"W 40°53'00"N (Dover)

Dreyfuss Planetarium (Alice and Leonard_)
• See "Alice and Leonard Dreyfuss Planetarium"

Electrim Corp.
356 Wall Street
Princeton, NJ 08540
Telephone: 609-683-5546
1-800-683-5546 (USA only)
Telefax: 609-683-5882
Electronic Mail: info@electrim.com
WWW: http://electrim.com/
Founded: 1988
Staff: 7
• Manufacturer
City Reference Coordinates: 074°40'00"W 40°21'00"N

Glenfield Planetarium
c/o Glenfield Middle School
25 Maple Avenue
Montclair, NJ 07042-4513
Telephone: 973-509-4174
Telefax: 973-509-4179
Electronic Mail: miller@montclair.k12.nj.us (Michael Miller)
Founded: 1970
City Reference Coordinates: 074°13'00"W 40°49'00"N

Henry Laboratories (Joseph_)
• See "Princeton University, Joseph Henry Laboratories"

Icon Isotopes
19 Ox Bow Lane
Summit, NJ 07901
Telephone: 201-273-0449
1-800-322-4266 (USA only)
Telefax: 908-273-0449
Electronic Mail: services@iconisotopes.com
WWW: http://www.iconisotopes.com/
Founded: 1984
Staff: 10
Activities: stable isotopes labeled compounds * multi-labeled compounds * noble-gas isotopes * reference standards

City Reference Coordinates: 074°22'00"W 40°43'00"N

Inrad Inc.
181 Legrand Avenue
Northvale, NJ 07647
Telephone: 201-767-1910
Telefax: 201-767-9644
Electronic Mail: sales@inrad.com
WWW: http://www.inrad.com/
Founded: 1973
Staff: 61
Activities: manufacturing electro-optic devices
City Reference Coordinates: 073°57'00"W 41°00'00"N

Institute for Advanced Study (IAS), School of Natural Sciences
Princeton, NJ 08540-0631
Telephone: 609-734-8000
Telefax: 609-924-8399
Electronic Mail: <userid>@sns.ias.edu
WWW: http://www.sns.ias.edu/
    http://www.sns.ias.edu/Main/astro.html
Founded: 1930
Staff: 100
Activities: research
City Reference Coordinates: 074°40'00"W 40°21'00"N

Joseph Henry Laboratories
• See "Princeton University, Joseph Henry Laboratories"

Liberty Science Center (LSC)
Liberty State Park
251 Phillip Street
Jersey City, NJ 07305
Telephone: 201-200-1000
Electronic Mail: mpeteet@lsc.org
WWW: http://www.lsc.org/
City Reference Coordinates: 074°03'00"W 40°43'00"N

Lucent Technologies, Bell Laboratories
600 Mountain Avenue
Murray Hill, NJ 07974
Telephone: 201-949-3803
Electronic Mail: <userid>@lucent.com
WWW: http://www.bell-labs.com/
    http://www.lucent.com/
Founded: 1996
Activities: mm observing of molecular clouds * galactic structure * outflow regions * microwave background isotropy
Coordinates: 074°11'12"W 40°23'31"N H100m (Crawford Hill)
City Reference Coordinates: 074°11'00"W 40°24'00"N

Minolta Planetarium Co. Ltd., US Representation
c/o Shigeki Ogawa
101 Williams Drive
Ramsey, NJ 07446
Telephone: 201-934-4732
Telefax: 201-818-0498
Electronic Mail: sogawa@minolta.com
WWW: http://www.minolta.com/japan/mp/
Activities: manufacturing i.a. planetariums
City Reference Coordinates: 074°08'00"W 41°03'00"N

Monmouth Mobile Observer's Group (MMOG)
Electronic Mail: mmog@monmouth.com
WWW: http://www.monmouth.com/~govega/mmogindex.htm
City Reference Coordinates: 074°16'00"W 40°16'00"N (Freehold)

Montclair State University (MSU), College of Science and Mathematics
Upper Montclair, NJ 07043
Telephone: 973-655-7266
Telefax: 973-655-7686
Electronic Mail: wwwteam@www.csam.montclair.edu
WWW: http://www.csam.montclair.edu/
City Reference Coordinates: 074°13'00"W 40°49'00"N (Montclair)

Morris Museum Astronomical Society
c/o Ronald Russo
6 Normand Heights Road
Morristown, NJ 07960
Telephone: 201-386-1848
Electronic Mail: apisano@ix.netcom.com
WWW: http://members.tripod.com/mmastrosociety
City Reference Coordinates: 074°29'00"W 40°48'00"N

New Jersey Astronomical Association (NJAA)
Voorhees State Park
P.O. Box 214
High Bridge, NJ 08829-0214
Telephone: 908-638-8500 (Observatory)
Electronic Mail: njaamail@aol.com
WWW: http://www.njaa.org/
Founded: 1965
Membership: 170
Activities: observing * meetings * classes * Edwin E. Aldrin Astronomical Center
Periodicals: (12) "Astronotes" (circ.: 200)
Coordinates: 074°53'59"W 40°40'54"N H290m (Paul Robinson Observatory)
City Reference Coordinates: 074°54'00"W 40°40'00"N

New Jersey Institute of Technology (NJIT), Center for Solar Research (CFSR)
University Heights
Newark, NJ 07102
Electronic Mail: <userid>@solar.njit.edu
WWW: http://solar.njit.edu/
Founded: 1997
Activities: operating "Big Bear Solar Observatory (BBSO)" and the solar radio array at "Owens Valley Radio Observatory (OVRO)" (see separate entries in USA-CA)
City Reference Coordinates: 074°10'00"W 40°44'00"N

New Jersey State Museum, Planetarium
205 West State Street
CN 530
Trenton, NJ 08625-0530
Telephone: 609-292-6333
          609-292-6303
Telefax: 609-599-4098
WWW: http://www.state.nj.us/state/museum/muspss.html
Founded: 1965
Staff: 10
Activities: planetarium and laser shows * observing
City Reference Coordinates: 074°45'00"W 40°13'00"N

Nippon Silica Glass USA Inc.
1952 US Highway 22
Bound Brook, NJ 08805-1520
Telephone: 908-725-0440
Telefax: 908-725-1485
WWW: http://nsg.mine.ne.jp/
Founded: 1984
Staff: 9
Activities: manufacturing fused quartz and fused silica
City Reference Coordinates: 074°32'00"W 40°34'00"N

North Jersey Astronomical Group (NJAG)
P.O. Box 1472
Clifton, NJ 07015-1472
Telephone: 973-680-8420 (StarLine)
Electronic Mail: west@astro.montclair.edu (Mary Lou West)
WWW: http://www.csam.montclair.edu/~west/njag.html
City Reference Coordinates: 074°09'00"W 40°53'00"N

Novins Planetarium (Robert J._)
● See "Robert J. Novins Planetarium"

Passaic County Astronomical Association (PCAA)
c/o Crowley Nature Center
Passaic County Parks Department
311 Pennsylvania Avenue
Paterson, NJ 07503

Telephone: 201-523-0024
WWW: http://www.users.nac.net/gburke/PCAA/
City Reference Coordinates: 074°09'00"W 40°56'00"N

Peddie School, Planetarium
c/o Nicholas R. Guilber
South Main Street
Box A
Highstown, NJ 08520
Telephone: 609-490-7536
Electronic Mail: nguilber@peddie.org
WWW: http://www.peddie.org/
City Reference Coordinates: 074°33'00"W 40°16'00"N

Princeton Plasma Physics Laboratory (PPPL)
● See "Princeton University, Princeton Plasma Physics Laboratory (PPPL)"

Princeton Research Instruments (PRI), Inc.
P.O. Box 1174
Princeton, NJ 08542-1174
Telephone: 609-924-0570
Telefax: 609-924-4970
Electronic Mail: info@prileeduhv.com
WWW: http://www.prileeduhv.com/
Founded: 1980
Staff: 9
Activities: manufacturing electron diffraction equipment, Rheed equipment, in-vacuum motion systems and in-vacuum stepping motors
City Reference Coordinates: 074°40'00"W 40°21'00"N

Princeton University, Department of Astrophysical Sciences
Peyton Hall - Ivy Lane
Princeton, NJ 08544-1001
Telephone: 609-258-3801
          609-258-3860 (Observatory)
Telefax: 609-258-1020
Electronic Mail: <userid>@astro.princeton.edu
WWW: http://astro.princeton.edu/
      http://www.astro.princeton.edu/
Founded: 1746 (University as "College of New Jersey")
Staff: 32
Activities: ISM * stellar structure and evolution * cosmology * galaxy formation
Periodicals: "Princeton Observatory Preprints"
Coordinates: 074°38'54"W 40°20'48"N (FitzRandolph Observatory)
City Reference Coordinates: 074°40'00"W 40°21'00"N

Princeton University, Joseph Henry Laboratories, Physics Department
Jadwin Hall
Princeton, NJ 08544-0708
Telephone: 609-452-4400
Telefax: 609-258-6853
Electronic Mail: <userid>@pupgg.princeton.edu
WWW: http://pupgg.princeton.edu/
      http://pulsar.princeton.edu/ (Pulsar Group)
Founded: 1904
Staff: 20 (astronomy group)
City Reference Coordinates: 074°40'00"W 40°21'00"N

Princeton University Press
41 William Street
Princeton, NJ 08540
Telephone: 609-258-4900
Telefax: 609-258-6305
WWW: http://pup.princeton.edu/
Founded: 1905
Staff: 100
● Publisher
City Reference Coordinates: 074°40'00"W 40°21'00"N

Princeton University, Princeton Plasma Physics Laboratory (PPPL)
James Forrestal Campus
P.O. Box 451
Princeton, NJ 08543-0451
Telephone: 609-243-2750
Telefax: 609-243-2751

Electronic Mail: pppl_info@pppl.gov
                 <userid>@pppl.gov
WWW: http://www.pppl.gov/
Founded: 1951
City Reference Coordinates: 074°40'00"W 40°21'00"N

Raritan Valley Community College, Planetarium
Route 28 and Lamington Road
North Branch, NJ 08876
Telephone: 908-231-8805
Telefax: 908-526-7938
Electronic Mail: planet@raritanval.edu
WWW: http://www.raritanval.edu/planetarium/
City Reference Coordinates: 074°40'00"W 40°36'00"N

Richard Stockton College of New Jersey (RSCNJ), Observatory
c/o Hal Taylor
P.O. Box 195
Pomona, NJ 08240-0195
Telephone: 609-652-4471 (office)
          609-829-7034 (home)
Telefax: 609-748-5515 (office)
       609-829-0870 (home)
Electronic Mail: hal@astro.stockton.edu
WWW: http://www2.stockton.edu/academics/undergraduate/natural_and_math_science/labs/physics/html/special_facilities.html
Founded: 1974 (College: 1971)
City Reference Coordinates: 074°35'00"W 39°29'00"N

Robert J. Novins Planetarium
c/o Ocean County College
P.O. Box 2001
Toms River, NJ 08754-2001
Telephone: 732-255-0343 (office - weekdays 09:30-16:00)
          732-255-0342 (show schedule)
WWW: http://www.ocean.cc.nj.us/planet
Founded: 1974
Staff: 8
Activities: school programs * public shows * education
City Reference Coordinates: 074°12'00"W 39°58'00"N

Rockland Astronomy Club (RAC) of New York
c/o Don Urban
73 Haring Street
Closter, NJ 07624-1709
Telephone: 201-768-6575
Electronic Mail: donaldurban@juno.com (Donald Urban)
WWW: http://www.erols.com/njastro/orgs/rac.htm
       http://www.rocklandastronomy.com/
Founded: 1958
Membership: 225
Activities: public star parties
Periodicals: (12) "Newsletter"
City Reference Coordinates: 073°58'00"W 40°59'00"N

Rutgers University, Department of Physics and Astronomy
136 Frelinghuysen Road
Piscataway, NJ 08855-0819
Telephone: 732-445-2501
Telefax: 732-445-4343
Electronic Mail: <userid>@physics.rutgers.edu
WWW: http://www.physics.rutgers.edu/ast/group-ast.html
Staff: 12
Activities: extragalactic observing * galactic dynamics * stellar dynamics * X-rays
Coordinates: 074°27'54"W 40°31'23"N H30m (Serin Observatory)
City Reference Coordinates: 074°27'00"W 40°34'00"N

SETI League Inc.
P.O. Box 555
Little Ferry, NJ 07643
or :
433 Liberty Street
Little Ferry, NJ 07643
Telephone: 201-641-1770
          1-800-TAU-SETI
Telefax: 201-641-1771

Electronic Mail: info@setileague.org
WWW: http://www.setileague.org/
Founded: 1994
Staff: 3
Activities: electromagnetic search for extra-terrestrial intelligence
Periodicals: (4) "Search Lites"
Awards: (1) "Giordano Bruno Memorial Award"
City Reference Coordinates: 074°03'00"W 40°51'00"N

South Jersey Astronomical Club (SJAC)
681 Port Elizabeth - Cumberland Road
Millville, NJ 08332
Telephone: 856-293-1584
Electronic Mail: fschaaf@aol.com (Fred Schaaf, President)
WWW: http://members.aol.com/sjastroc/
Founded: 1989
Activities: public outreach * observing * organizing Star Jersey Star Party (SJSP)
City Reference Coordinates: 075°02'00"W 39°24'00"N

Space Studies Institute (SSI)
P.O. Box 82
Princeton, NJ 08542
Telephone: 609-921-0377
Telefax: 609-921-0389
Electronic Mail: ssi@ssi.org
WWW: http://www.ssi.org/
Founded: 1977
Activities: international R&D organization providing practical and innovative solutions to (a) create new lands in the high frontier of space, (b) harness the energy and material resources of space, and (c) preserve Earth's biosphere and expand life's ecological rand beyond Earth's boundaries
Periodicals: (6) "SSI Update" (ISSN 0898-8242)
City Reference Coordinates: 074°40'00"W 40°21'00"N

S*T*A*R Astronomy Society Inc.
P.O. Box 863
Red Bank, NJ 07701
Electronic Mail: jlevantino@lucent.com (Joe Levantino, President)
WWW: http://www.monmouth.com/~ksears/
http://www.starastronomy.org/
Founded: 1950
City Reference Coordinates: 075°10'00"W 39°52'00"N

Stockton College of New Jersey (RSCNJ) (Richard_)
● See "Richard Stockton College of New Jersey (RSCNJ)"

Trailside Nature and Science Center (TNSC), Planetarium
452 New Providence Road
Mountainside, NJ 07092
Telephone: 908-789-3670
Electronic Mail: 73633.2300@compuserve.com
WWW: http://www.fieldtrip.com/nj/87893670.htm
Founded: 1969
Activities: shows * exhibitions * observing * sales * annual "Astronomy Sunday"
Periodicals: (12) "Our Parks"
Coordinates: 074°22'30"W 40°40'58"N
City Reference Coordinates: 074°21'00"W 40°40'00"N

United Astronomy Clubs of New Jersey Inc.
c/o Barry Malpas
20 Helen Street
Warren, NJ 07059
Telephone: 908-755-6932
Electronic Mail: njastro@erols.com
WWW: http://www.erols.com/njastro
City Reference Coordinates: 074°30'00"W 40°37'00"N

Willingboro Astronomical Society (WAS)
c/o Joy Crist
P.O. Box 371
Gloucester, NJ 08030-0371
Telephone: 609-461-9226
Electronic Mail: 104324.1232@compuserve.com
WWW: http://www.wasociety.org/
City Reference Coordinates: 075°10'00"W 39°50'00"N

Zeiss Historica Society of America (ZHSA)
c/o M.E. Zubatkin
300 Waxwing Drive
Cranbury, NJ 08512
or :
c/o Marc James Small
P.O. Box 2901
Roanoke, VA 24001-2901
Telephone: 540-981-1036
Telefax: 540-343-7315
Electronic Mail: msmall@roanoke.infi.net
Founded: 1980
Membership: 1
Activities: study, collection and publication of information on works manufactured by Carl Zeiss optical workshops and related companies
Periodicals: (4) "Journal of the Zeiss Historica Society"
City Reference Coordinates: 074°31'00"W 40°19'00"N (Cranbury)
                               079°57'00"W 37°16'00"N (Roanoke)

USA - New Mexico

Air Force Research Laboratory (AFRL), Office of Public Affairs
3550 Aberdeen Avenue SE
Kirtland AFB, NM 87117-5776
Telephone: 505-846-1911
Telefax: 505-846-0423
WWW: http://www.de.afrl.af.mil/pa
City Reference Coordinates: 106°40'00"W 35°05'00"N (Albuquerque)

Alamogordo Amateur Astronomers
c/o Paul Carnes
1502 Jefferson Avenue
Alamogordo, NM 88310
Telephone: 505-437-4505
WWW: http://www.zianet.com/mikemosier/astro/astro.htm
Founded: 1986
Membership: 20
Activities: observing * education
Coordinates: 105°00'00"W 32°00'00"N H1,310m
City Reference Coordinates: 105°57'00"W 32°54'00"N

Alamogordo Astronomy Club (AAC)
c/o Jackie Diehl
International Space Hall of Fame
Highway 2001
Alamogordo, NM 88310
Telephone: 505-437-4857 x2840
　　　　　505-437-5857
Electronic Mail: garfield@sunspot.noao.edu (Brian Armstrong)
WWW: http://www.zianet.com/mikemosier/astro/astro.htm
　　　　http://www.zianet.com/aacwp/
Activities: education * observing * flighting light pollution
City Reference Coordinates: 105°57'00"W 32°54'00"N

Albuquerque Astronomical Society (TAAS) (The_)
● See "The Albuquerque Astronomical Society (TAAS)"

Apache Point Observatory (APO)
2001 Apache Point Road
P.O. Box 59
Sunspot, NM 88349-0059
Telephone: 505-437-6822
Telefax: 505-434-5555
Electronic Mail: <userid>@galileo.apo.nmsu.edu
WWW: http://www.apo.nmsu.edu/
Founded: 1984
Staff: 7
Coordinates: 105°49'14"W 32°46'49"N H2788m (IAU Code 645 & 705)
City Reference Coordinates: 105°49'00"W 32°47'00"N

Astronomical Society of Las Cruces (ASLC)
P.O. Box 921
Las Cruces, NM 88004
Electronic Mail: adbailey@acca.nmsu.edu
WWW: http://www.zianet.com/aslc/
Founded: 1951
Membership: 56
City Reference Coordinates: 106°47'00"W 32°18'00"N

CapellaSoft
P.O. Box 1182
Cloudcroft, NM 88317
Telephone: 505-682-1183
　　　　　1-888-4-CAPELLA (USA only)
Electronic Mail: info@skyhound.com
WWW: http://www.skyhound.com/
Founded: 1992
Staff: 1
Coordinates: 105°44'02"W 32°57'11"N
● Software producer
City Reference Coordinates: 105°46'00"W 32°58'00"N

Clovis Astronomy Club
3517 Corlington Lane
Clovis, NM 88101-3012
Telephone: 505-763-7455
Founded: 1988
Membership: 11
Activities: star parties * education
Periodicals: (12) "Newsletter"
Coordinates: 103°11'45"W 34°28'66"N H1326m
City Reference Coordinates: 103°12'00"W 34°24'00"N

Goddard Planetarium (Robert H._)
• See "Roswell Museum and Art Center, Robert H. Goddard Planetarium"

Institute of Meteoritics (IOM)
• See "University of New Mexico (UNM), Department of Geology, Institute of Meteoritics (IOM)"

Langmuir Laboratory for Atmospheric Research
• See "New Mexico Institute of Mining and Technology (NMIMT), Langmuir Laboratory for Atmospheric Research"

LightPath Technologies Inc.
3819 Osuna Road NE
Albuquerque, NM 87109
Telephone: 505-342-1100
　　　　1-800-GRADIUM (USA only)
　　　　1-800-472-3486 (USA only)
Telefax: 505-342-1111
　　　　520-884-8611
Electronic Mail: <userid>@light.net
WWW: http://www.light.net/
　　　http://www.lightpath.com/
Founded: 1985
Activities: glass manufacturer
City Reference Coordinates: 106°40'00"W 35°05'00"N

LodeStar Project
801 University Boulevard NE
Albuquerque, NM 87106
Telephone: 505-272-7595
Electronic Mail: lodestar@lodestar.phys.unm.edu
WWW: http://lodestar.phys.unm.edu/
Founded: 1994
Staff: 20
Activities: education * observing * research
Periodicals: "The Star"
City Reference Coordinates: 106°40'00"W 35°05'00"N

Los Alamos National Laboratory (LANL), Space Science and Technology Division, Space Astronomy and Astrophysics Group
SST-9
Mail Stop D436
Los Alamos, NM 87545
Telephone: 505-667-5127
Telefax: 505-665-4414
Electronic Mail: <userid>@essdp2.xnet@lanl (arpanet)
WWW: http://sst.lanl.gov/
　　　http://sst.lanl.gov/nis2astro.html
　　　http://www.lanl.gov/
　　　http://mentor.lanl.gov/
　　　http://mioruilt.lanl.gov/ (Los Alamos Astrophysics Group - LAA)
　　　http://mioruilt.lanl.gov/fho/ (Fenton Hill Observatory - FHO)
City Reference Coordinates: 106°19'00"W 35°53'00"N

Los Alamos National Laboratory (LANL), T-6 Division (Theoretical Astrophysics)
P.O. Box 1663
Mail Stop B275
Los Alamos, NM 87545
Telephone: 505-667-2987
Telefax: 505-665-3003
Electronic Mail: <userid>@laln.gov
WWW: http://www.lanl.gov/
　　　http://mentor.lanl.gov/
Founded: 1980
Activities: theoretical astrophysics * modelling stellar collisions * stellar evolution * structure and evolution of the universe * SN * gamma-ray sources * general relativity

City Reference Coordinates: 106°19'00"W 35°53'00"N

Meteorite Museum
c/o Institute of Meteoritics
University of New Mexico
200 Yale NE
Albuquerque, NM 87131-1126
Telephone: 505-277-1644
Telefax: 505-277-3577
Electronic Mail: rjones@unm.edu (Rhian Jones, Curator)
slentz@unm.edu (Sarah Lentz)
WWW: http://www.unm.edu/
http://epswww.unm.edu/iom/iomcoll.htm
Founded: 1944
Staff: 16
Activities: collection and curation of meteorite samples
City Reference Coordinates: 106°40'00"W 35°05'00"N

National Optical Astronomy Observatories (NOAO), National Solar Observatory (NSO), Sacramento Peak
P.O. Box 62
Sunspot, NM 88349-0062
Telephone: 505-434-7000
Telefax: 505-434-7029
505-434-7009
Electronic Mail: sp@sunspot.noao.edu
<userid>@sunspot.noao.edu
WWW: http://www.sunspot.noao.edu/
Founded: 1983
Staff: 50
Activities: solar physics * instrumentation
Periodicals: (2) "NSO Update"; "Sacramento Peak Observatory Contributions", "Sacramento Peak Observatory Research Notes"
Coordinates: 105°49'02"W 32°47'12"N H2,800m
City Reference Coordinates: 105°49'00"W 32°47'00"N

National Radio Astronomy Observatory (NRAO), Array Operations Center (AOC)
P.O. Box 0
Socorro, NM 87801-0387
Telephone: 505-772-4011 (VLA site)
505-835-7000 (Array Operations Center)
Telefax: 505-772-4243 (VLA site)
505-835-7027 (Array Operations Center)
Electronic Mail: <userid>@nrao.edu
WWW: http://www.aoc.nrao.edu/vla/html/VLAhome.shtml
http://www.aoc.nrao.edu/vlba/html/VLBA.html
http://www.nrao.edu/
Founded: 1981
Staff: 195
Activities: aperture synthesis radio observing with the Very Large Array (VLA) and Very Long Baseline Array (VLBA) telescopes
Periodicals: "NRAO Newsletter"
Coordinates: 107°37'04"W 34°04'44"N H2,124m
064°35'03"W 17°45'31"N H16m (Saint Croix, Virgin Islands)
071°59'12"W 42°56'01"N H309m (Hancock, USA-NH)
091°34'26"W 41°46'17"N H241m (North Liberty, USA-IA)
103°56'00"W 30°19'00"N H1,615m (Fort Davis, USA-TX)
106°14'42"W 35°46'30"N H1,967m (Los Alamos, USA-NM)
108°07'07"W 34°18'04"N H2,371m (Pie Town, USA-NM)
111°36'42"W 31°57'22"N H1,916m (Kitt Peak, USA-AZ)
118°16'34"W 37°13'54"N H1,207m (Owens Valley, USA-CA)
119°40'55"W 48°07'53"N H255m (Brewster, USA-WA)
155°27'29"W 19°48'31"N H3,725m (Mauna Kea, USA-HI)
City Reference Coordinates: 106°54'00"W 34°04'00"N

National Solar Observatory (NSO)
• See "National Optical Astronomy Observatories (NOAO), National Solar Observatory (NSO)"

New Mexico Institute of Mining and Technology (NMIMT), Department of Physics, Astrophysics Research Center (ARC)
Socorro, NM 87801
Telephone: 505-835-5328
Telefax: 505-835-5707
Electronic Mail: <userid>@kestrel.nmt.edu
<userid>@nmt.edu
WWW: http://www.nmt.edu/~astro

http://www.physics.nmt.edu/
Founded: 1985
Staff: 8
Activities: education * research in optical and radio astronomy * plasma physics * instrumentation
Observatories: 2 (Etscorn Campus Obs. – IAU Code 719 – & Magdalena Ridge Obs.)
City Reference Coordinates: 106°54'00"W 34°04'00"N

New Mexico Institute of Mining and Technology (NMIMT), Langmuir Laboratory for Atmospheric Research
801 Leroy Place
Socorro, NM 87801
Telephone: 505-835-5423
Telefax: 505-835-5913
Electronic Mail: langmuir@ee.nmt.edu
WWW: http://www.ee.nmt.edu/~langmuir/
Founded: 1963
Staff: 10
Activities: lightning and thunderstorm research * future site of Magdalena Ridge Observatory Interferometer
Awards: "Langmuir Award"
Coordinates: 107°10'51"W 33°58'31"N H3,255m
City Reference Coordinates: 106°54'00"W 34°04'00"N

New Mexico State University (NMSU), Department of Astronomy
P.O. Box 30001
Dept. 4500
Las Cruces, NM 88003-8001
Telephone: 505-646-4438
Telefax: 505-646-1602
Electronic Mail: astro@nmsu.edu
WWW: http://charon.nmsu.edu/
http://ganymede.nmsu.edu/
http://astro.nmsu.edu/
Founded: 1970
Staff: 17
Activities: education * planetary atmospheres * variable stars * star clusters * active galaxies * stellar populations * spectroscopy * galaxy clusters * cosmology * ISM in galaxies
Coordinates: 106°41'48"W 32°16'35"N (Campus Observatory)
106°41'51"W 32°17'35"N H1,505m (Tortugas Mountain Station)
105°45'12"W 32°46'50"N (Apache Point Observatory - see separate entry)
City Reference Coordinates: 106°47'00"W 32°18'00"N

New Mexico Tech
• See "New Mexico Institute of Mining and Technology (NMIMT)"

New Mexico Tech, Astronomy Club
c/o Jennifer Rinard
Box 3032 C/S
Socorro, NM 87801
Electronic Mail: tobias@mailhost.nmt.edu
astro@nmt.edu
WWW: http://www.nmt.edu/~astro/
Observatories: 1 (Etscorn Campus Observatory, ECO – IAU Code 719)
City Reference Coordinates: 106°54'00"W 34°04'00"N

Pajarito Astronomers
P.O. Box 1092
Los Alamos, NM 87544
Telephone: 505-667-8964 (Stephen Becker, President)
505-665-4223 (David Hollowell, Vice-President)
505-667-0983 (Jerry Foropoulos, Secretary)
Electronic Mail: daveh@lanl.gov (David Hollowell)
WWW: http://www.la.unm.edu:8001/~beach/pajarito.html
Founded: 1970
Membership: 45
Activities: bimonthly meetings * monthly star parties during Summer
Periodicals: (12) "Newsletter"
City Reference Coordinates: 106°19'00"W 35°53'00"N

Robert H. Goddard Planetarium
• See "Roswell Museum and Art Center, Robert H. Goddard Planetarium"

Roswell Astronomy Club
Electronic Mail: roswellastroclub@starband.com
WWW: http://www.roswellastronomyclub.com
City Reference Coordinates: 104°33'00"W 33°24'00"N

Roswell Museum and Art Center, Robert H. Goddard Planetarium
100 West Eleventh Street
Roswell, NM 88201
Telephone: 505-624-6744
WWW: http://www.roswellmuseum.org/planet/frm_plnt.htm
City Reference Coordinates: 104°33'00"W 33°24'00"N

Santa Fe Community College (SFCC), Planetarium
c/o Suzanne Chippindale
6401 Richards Avenue
Santa Fe, NM 87505
Telephone: 505-428-1707
Telefax: 505-428-1302
Electronic Mail: schippi@santa-fe.cc.mn.us
WWW: http://www.santa-fe.cc.nm.us/
City Reference Coordinates: 105°57'00"W 35°41'00"N

Sollid Optics Inc.
365 Valle del Sol
Los Alamos, NM 87544-3563
Telephone: 505-672-3880
Telefax: 505-672-1771
Electronic Mail: sollid@earthlink.net
WWW: http://www.sollidoptics.com/
Founded: 1985
Staff: 3
Activities: optical metrology * process diagnostics * laser beam transport * holographic interferometry * laser diagnostics
City Reference Coordinates: 106°19'00"W 35°53'00"N

Space Center
Top of NM Highway 2001
P.O. Box 533
Alamogordo, NM 88311-0533
Telephone: 1-800-545-4021 (USA only)
  505-437-2840
Telefax: 505-434-2245
Electronic Mail: spacepr@zianet.com
WWW: http://www.spacefame.org/
Founded: 1973
Staff: 28
Activities: space Museum * planetarium * Imax theater * International Space Hall of Fame * public school programs * Clyde W. Tombaugh Space Theater * laser light shows * exhibits * educational outreach * lectures * guided tours
Periodicals: "Space Log"
City Reference Coordinates: 105°57'00"W 32°54'00"N

Star Hill Inn
Box 107
Sapello, NM 87745
Telephone: 505-425-5605
Electronic Mail: stay@starhillinn.com
WWW: http://www.starhillinn.com/
Founded: 1988
Staff: 2
Activities: "an astronomer's retreat in the rockies" accommodating observers * observing * photography * bird watching * workshops
Coordinates: 105°17'00"W 35°46'00"N H2,195m
City Reference Coordinates: 104°59'00"W 35°47'00"N

The Albuquerque Astronomical Society (TAAS)
P.O. Box 50581
Albuquerque, NM 87181-0581
Telephone: 505-296-0549 (Michael Pendley, President)
Electronic Mail: taas@phys.unm.edu
WWW: http://www.taas.org/
Founded: 1958
Membership: 320
Activities: public education * star parties * observing sessions * monthly meetings * field trips * telescope making
Periodicals: "Sidereal Times"
Awards: "John Dobson Award", "General Nathan Twining Award", "Colonel Leonard Browline Award"
Coordinates: 106°50'42"W 34°31'20"N H1,700m (General Nathan Twining Obs. - GNTO)
City Reference Coordinates: 106°40'00"W 35°05'00"N

University of New Mexico (UNM), Department of Physics and Astronomy
800 Yale NE
Albuquerque, NM 87131

Telephone: 505-277-2616
Telefax: 505-277-1520
Electronic Mail: <userid>@unm.edu
WWW: http://panda.unm.edu/
Founded: 1892
Observatories: 1 (Capilla Peak Observatory)
City Reference Coordinates: 106°40'00"W 35°05'00"N

University of New Mexico (UNM), Institute for Astrophysics
800 Yale Boulevard NE
Albuquerque, NM 87131
Telephone: 505-277-2726
Telefax: 505-277-9657
Electronic Mail: <userid>@astro.unm.edu
WWW: http://www.phys.unm.edu/ifa/
Staff: 30
Activities: education * LodeStar Project (see separate entry)
City Reference Coordinates: 106°40'00"W 35°05'00"N

University of New Mexico (UNM), Institute of Meteoritics (IOM)
200 Yale NE
Albuquerque, NM 87131-1126
Telephone: 505-277-2747
Telefax: 505-277-3577
Electronic Mail: meteor@unm.edu
          <userid>@unm.edu
WWW: http://eps.unm.edu/iom/home.htm
Founded: 1944
Staff: 12
Activities: study of lunar samples and meteorites * planetary volcanological processes
Periodicals: "Special Publications" (ISSN 0085-3968)
Coordinates: 106°40'00"W 35°05'00"N
City Reference Coordinates: 106°40'00"W 35°05'00"N

USA - New York

Acoustical Society of America (ASA)
c/o Elaine Moran (ASA Office Manager)
Suite 1N01
2 Huntington Quadrangle
Melville, NY 11747-4502
Telephone: 516-576-2360
Telefax: 516-576-2377
Electronic Mail: asa@aip.org
WWW: http://asa.aip.org/
Founded: 1929
Membership: 7000
City Reference Coordinates: 073°25'00"W 40°49'00"N

Adelphi University, Department of Physics
Blodgett Hall
1 South Avenue
Garden City, NY 11530
Telephone: 516-877-4880
Telefax: 516-877-4887
WWW: http://www.adelphi.edu/
City Reference Coordinates: 073°37'00"W 40°43'00"N

Adirondack Video Astronomy (AVA)
c/o John E. Cordiale
35 Stephanie Lane
Queensbury, NY 12804
Telephone: 1-888-799-0107
          518-761-0390
Telefax: 518-745-4114
Electronic Mail: avaastro@compuserve.com
WWW: http://ourworld.compuserve.com/homepages/AVAastro/
     http://www.astrovid.com/
Activities: manufacturing video imaging systems * distributing Starlight Xpress cameras * distributing flip mirrors, software and other astronomical imaging related products
City Reference Coordinates: 073°41'00"W 43°17'00"N (Glens Falls)

Adorama Camera Inc.
42 west 18th Street
New York, NY 10011
Telephone: 1-800-223-2500 (USA only)
          212-741-0052
Telefax: 212-463-7223
Electronic Mail: info@adoramacamera.com
WWW: http://www.adorama.com/
Activities: retail of cameras, telescopes, accessories, ...
City Reference Coordinates: 074°01'00"W 40°43'00"N

Albany Area Amateur Astronomers, Inc. (AAAA)
c/o Bob Mulford
1020 Mohegan Road
Niskayuna, NY 12309
Telephone: 518-374-8460
          518-374-8744
Electronic Mail: ramulford@aol.com
              bobmulford@juno.com
Founded: 1956
Membership: 140
Activities: public star parties * meetings * lectures * public information
Periodicals: (12) "Celestial Times"
Coordinates: 073°54'00"W 42°48'00"N (Johnson-Weil Observatory)
City Reference Coordinates: 073°33'00"W 44°54'00"N (Scotia)

Alfred University, Stull Observatory
c/o G. David Toot (Director)
Saxon Drive
Alfred, NY 14802
Telephone: 607-871-2207
          607-871-2270
Electronic Mail: ftoot@bigvax.alfred.edu
WWW: http://www.alfred.edu/map/stull.html
     http://merlin.alfred.edu/stull.html

13 Lincoln Avenue
Massapequa, NY 11758
Telephone: 516-781-6261 (Storch)
   516-795-3072 (Rizzi)
WWW: http://pw2.netcom.com/~alan-5/asli.html
   http://www.netcom.com/~alan-5/asli.html
   http://pweb.netcom.com/~alan-5/asli.html
Founded: 1965
Membership: 120
Activities: observing * public sessions * lectures
Periodicals: (12) "ASLI News" (circ.: 140)
Coordinates: 073°33'00"W 40°38'00"N
City Reference Coordinates: 073°34'00"W 40°41'00"N (North Merrick)
   073°29'00"W 40°40'00"N (Massapequa)

Astronomical Society of New York City
c/o Thomas Hamilton
153 Arlo Road
Staten Island, NY 10301
Telephone: 718-727-1967
Telefax: 718-727-1967
Membership: 20
City Reference Coordinates: 074°09'00"W 40°36'00"N

Atomic Data and Nuclear Data Tables (ADNDT)
c/o Angela Li-Scholz (Editor)
Department of Physics
State University of New York
Albany, NY 12222
Telephone: 518-442-4521
Telefax: 518-442-5260
WWW: http://www.academicpress.com/adndt
Founded: 1965
Staff: 3
• Journal (6) (ISSN 0092-640x)
City Reference Coordinates: 073°45'00"W 42°39'00"N

Aviation Week & Space Technology
1221 Avenue of the Americas
New York, NY 10020
Telephone: 212-512-4117
Telefax: 212-512-6068
Electronic Mail: aviation@mcgraw-hill.com
WWW: http://www.awgnet.com/aviation/
Founded: 1916
Staff: 100
• Journal (52) (ISSN 0005-2175)
City Reference Coordinates: 074°01'00"W 40°43'00"N

AVS The Science and Technology Society
120 Wall Street
32nd Floor
New York, NY 10005-3993
Telephone: 212-248-0200
Telefax: 212-248-0245
Electronic Mail: avsnyc@vacuum.org
WWW: http://www.avs.org/
Founded: 1953
Membership: 6000
Staff: 6
Activities: nonprofit organization which promotes communication, dissemination of knowledge, recommended practices, research, and education in the use of vacuum and other controlled environments to develop new materials, process technology, devices, and related understanding of material properties for the betterment of humanity
Periodicals: (12) "Journal of Vacuum Science and Technology A" (ISSN 0734-2101), (12) "Journal of Vacuum Science and Technology B" (ISSN 0734-211x); (4) "Surface Science Spectra" (ISSN 1055-5269)
Awards: (1) "Gaede-Langmuir", "Nerken", "Peter Mark", "Thornton", "Welch"
• Formerly "American Vacuum Society (AVS)"
City Reference Coordinates: 074°01'00"W 40°43'00"N

Baruch College (Bernard_)
• See "City University of New York, Bernard Baruch College"

Bernard Baruch College
• See "City University of New York, Bernard Baruch College"

Brailsford & Co. Inc.
670 Milton Road
P.O. Box 811
Rye, NY 10580
Telephone: 914-967-1820
Telefax: 914-967-1836
WWW: http://www.brailsfordco.com/
Founded: 1944
Staff: 25
Activities: manufacturing motors, blowers and pumps for aerospace applications
City Reference Coordinates: 073°41'00"W 40°59'00"N

Bronx Planetarium (Northeast_)
• See "Northeast Bronx Planetarium"

Brookhaven National Laboratory (BNL)
P.O. Box 5000
Upton, NY 11973-5000
Telephone: 631-344-8000 (Switchboard)
631-344-2345 (Public Inquiries)
Electronic Mail: pubaf@bnl.gov (Public Inquiries)
<userid>@bnl.gov
WWW: http://www.bnl.gov/
Founded: 1947
Staff: 3000
City Reference Coordinates: 072°40'00"W 40°55'00"N (Riverhead)

Brooklyn Thermometer Co. Inc.
90 Verdi Street
Farmingdale, NY 11735
Telephone: 516-694-7610
Telefax: 516-694-6329
Electronic Mail: paulrt@msn.com (Paul R. Teichert)
Founded: 1903
Staff: 9
Activities: manufacturing liquid-in-glass thermometers
City Reference Coordinates: 073°27'00"W 40°44'00"N

Buffalo Astronomical Association (BAA)
c/o Buffalo Museum of Science
1020 Humboldt Parkway
Buffalo, NY 14211
Telephone: 716-457-3104
Electronic Mail: alanfgag@aol.com
WWW: http://members.aol.com/BuffAstro
Membership: 105
Activities: education * CCD imaging * gamma-ray bursts
Periodicals: (6) "The Spectrum"
Coordinates: 078°24'00"W 42°40'00"N H500m (Beaver Meadow)
City Reference Coordinates: 078°48'00"W 42°58'00"N

Burleigh Instruments Inc.
7647 Main Street Fishers
Victor, NY 14564-8909
Telephone: 716-924-9355
Telefax: 716-924-9072
Electronic Mail: info@burleigh.com
WWW: http://www.burleigh.com/
Founded: 1972
Staff: 130
Activities: lasers wavelength measurement * laser spectral analysis * piezoelectric micropositioning
City Reference Coordinates: 077°26'00"W 42°59'00"N

Cambridge University Press (CUP)
40 West 20th Street
New York, NY 10011-4211
Telephone: 212-924-3900
Telefax: 212-691-3239
Electronic Mail: information@cup.org
<userid>@cup.org
WWW: http://www.cup.org/
http://www.cup.cam.ac.uk/
Founded: 1534 (UK Head Office)
• Publisher
City Reference Coordinates: 074°01'00"W 40°43'00"N

Casper - Meteorites Inc. (Michael I._)
• See "Michael I. Casper - Meteorites Inc."

Catskills Astronomy Club
P.O. Box 252
Lake Huntington, NY 12752
Electronic Mail: catskills_astronomy_club@yahoo.com
WWW: http://www.geocities.com/catskills_astronomy_club/
City Reference Coordinates: 074°60'00"W 41°41'00"N

Center for Backyard Astrophysics (CBA)
c/o Department of Astronomy
Columbia University
1316 Pupin Physics Laboratories
550 West 120th Street
New York, NY 10027
Telephone: 212-854-3276
Telefax: 212-854-8121
Electronic Mail: info@cba.phys.columbia.edu
cba@astro.columbia.edu
WWW: http://cba.phys.columbia.edu/
http://cba.astro.columbia.edu/~cba
City Reference Coordinates: 074°01'00"W 40°43'00"N

Ceramaseal
1033 State Route 20
New Lebanon, NY 12125
Telephone: 518-794-7800
1-800-752-SEAL (USA only)
Telefax: 518-794-8080
Electronic Mail: info@ceramaseal.com
sales@ceramaseal.com
WWW: http://www.ceramaseal.com/
Founded: 1951
Activities: manufacturing vacuum feedthrus, connectors, cables, viewports and hardware for high and ultra-high vacuum * instrumentation connectors and headers for low- and high-pressure devices * custom-engineered ceramic-metal sealed devices
City Reference Coordinates: 073°24'00"W 42°28'00"N

City College of New York (CCNY), Planetarium
c/o Department of Physics
138 Street and Convent Avenue
New York, NY 10031
Telephone: 212-650-6832
Electronic Mail: chung@scisun.sci.ccny.cuny.edu (Victor Chung, Director)
WWW: http://www.sci.ccny.cuny.edu/~jbaltz/planet.html
Founded: 1972
City Reference Coordinates: 074°01'00"W 40°43'00"N

City University of New York (CUNY), Bernard Baruch College, Department of Physics and Astronomy
17 Lexington Avenue
New York, NY 10010
Telephone: 212-802-3080
212-802-3082
Electronic Mail: <userid>@cuny.cuny.edu
WWW: http://www.baruch.cuny.edu/
Founded: 1898
Staff: 11
City Reference Coordinates: 074°01'00"W 40°43'00"N

City University of New York (CUNY), Herbert H. Lehman College, Department of Physics and Astronomy
Bronx, NY 10468
Telephone: 212-960-8542
Telefax: 212-960-8627
Electronic Mail: <userid>@lehman.cuny.edu
WWW: http://www.lehman.cuny.edu/departments/physics.html
City Reference Coordinates: 073°56'00"W 40°49'00"N

City University of New York (CUNY), Hunter College, Department of Physics and Astronomy
695 Park Avenue
Room 1200 HN
New York, NY 10021
Telephone: 212-772-5248
Telefax: 212-772-5393

Electronic Mail: <userid>@atlas.ph.hunter.cuny.edu
WWW: http://www.ph.hunter.cuny.edu/
City Reference Coordinates: 074°01'00"W 40°43'00"N

CKC Power
23 Graywood Drive
Orangeburg, NY 10962
Telephone: 845-627-1055
Telefax: 845-627-6642
Electronic Mail: sales@ckcpower.com
WWW: http://www.ckcpower.com/
Activities: distributing digital cameras and accessories
City Reference Coordinates: 073°57'00"W 43°03'00"N

Clarkson University, Department of Physics
Science Center 125
Department of Physics
Box 582
Potsdam, NY 13699-5820
Telephone: 315-268-2396
Telefax: 315-268-6610
Electronic Mail: physics@clarkson.edu
WWW: http://www.clarkson.edu/~physics/
Founded: 1934
Staff: 13
Activities: statistical physics * condensed matter
Periodicals: (1) "Annual Report"
City Reference Coordinates: 074°59'00"W 44°40'00"N

Colgate University, Department of Physics and Astronomy
13 Oak Drive
Hamilton, NY 13346
Telephone: 315-824-7723
          315-228-7215 (Foggy Bottom Obs./FBO)
Telefax: 315-824-7187
Electronic Mail: <userid>@colgate.edu
WWW: http://departments.colgate.edu/physics/
     http://astronomy.colgate.edu/
Founded: 1951 (Observatory) (University: 1819)
Activities: education * research
Observatories: 1 (Foggy Bottom Obs./FBO – IAU Code 776)
City Reference Coordinates: 075°33'00"W 42°50'00"N

College of Staten Island (CSI), Astrophysical Observatory
2800 Victory Boulevard
Staten Island, NY 10314-9881
Telephone: 718-982-3260
           718-982-2818
Electronic Mail: robbins@postbox.csi.cuny.edu (Irving K. Robbins, Director)
WWW: http://supernova7.apsc.csi.cuny.edu/
Founded: 1972
Staff: 4
Activities: education * public lectures * public observing * asteroid astrometry
Coordinates: 074°09'00"W 40°36'10"N
City Reference Coordinates: 074°09'00"W 40°36'00"N

Collins Observatory (Eileen_)
● See "Corning Community College, Eileen Collins Observatory"

Columbia Astrophysics Laboratory
● See "Columbia University, Columbia Astrophysics Laboratory"

Columbia University, Columbia Astrophysics Laboratory
Pupin Physics Laboratories
Mail Code 5247
550 West 120th Street
New York, NY 10027
Telephone: 212-854-3257
Telefax: 212-854-8121
Electronic Mail: <userid>@astro.columbia.edu
WWW: http://www.astro.columbia.edu/
Founded: 1754
Staff: 17
Activities: research * education
City Reference Coordinates: 074°01'00"W 40°43'00"N

Columbia University, Department of Astronomy
Pupin Physics Laboratories
Mail Code 5246
550 West 120th Street
New York, NY 10027
Telephone: 212-854-3278
Telefax: 212-854-8121
Electronic Mail: <userid>@astro.columbia.edu
WWW: http://www.astro.columbia.edu/
Founded: 1754
Staff: 17
Activities: research * education
City Reference Coordinates: 074°01'00"W 40°43'00"N

Columbia University, Department of Physics
538 West 120 Street
New York, NY 10027
Telephone: 212-280-3349
Electronic Mail: <userid>@phys.columbia.edu
WWW: http://phys.columbia.edu/
Founded: 1892
City Reference Coordinates: 074°01'00"W 40°43'00"N

Committee for the Scientific Investigations of Claims of the Paranormal (CSICOP)
P.O. Box 703
Amherst, NY 14226-0703
Telephone: 716-636-1425
Telefax: 716-636-1733
Electronic Mail: info@csicop.org
WWW: http://www.csicop.org/
Founded: 1976
Staff: 17
Activities: investigating claims of the paranormal with a scientific viewpoint * conferences * seminars * research library * media resource
Sections: (Subcommittees) Astrology * Electronic Communication * Health Claims * Parapsychology * UFO
Periodicals: (6) "The Skeptical Inquirer" (ISSN 0194-6730, circ.: 57,500), "The Skeptical Briefs" (circ.: 2,500)
City Reference Coordinates: 078°48'00"W 42°58'00"N

Cornell University, Center for Radiophysics and Space Research (CRSR)
318 Space Sciences Building
Ithaca, NY 14853
Telephone: 607-255-8544
Telefax: 607-255-9002
Electronic Mail: <userid>@cornell.edu
WWW: http://www.osp.cornell.edu/VPR/CenterDir/CRSR.html
Founded: 1959
Staff: 63
Activities: radio and IR astronomy * planets * theoretical astrophysics
City Reference Coordinates: 076°30'00"W 42°27'00"N

Cornell University, Department of Astronomy
512 Space Sciences Building
Ithaca, NY 14853-6801
Telephone: 607-255-4935
Telefax: 607-255-1767
Electronic Mail: <userid>@astrosun.tn.cornell.edu
WWW: http://www.astro.cornell.edu/
Founded: 1959
Staff: 59
Activities: theoretical astrophysics * radio and radar astronomy * optical and IR astronomy * planetary sciences
Coordinates: 076°23'00"W 42°27'30"N (Fuertes & Hartung Boothroyd Obs.)
City Reference Coordinates: 076°30'00"W 42°27'00"N

Cornell University Press
Sage House
512 East State Street
Ithaca, NY 14850
Telephone: 1-800-666-2211 (USA only)
　　　　　　607-277-2338 x251
Telefax: 607-277-2397
Electronic Mail: cupress-sales@cornell.edu
　　　　　　　orderbook@cupserv.org
WWW: http://www.cornellpress.cornell.edu/
Founded: 1869
● Publisher
City Reference Coordinates: 076°30'00"W 42°27'00"N

Corning Community College, Eileen Collins Observatory
1 Academic Drive
Corning, NY 14830
Telephone: 607-962-9494 (Deborah Dann, Observatory Director)
Electronic Mail: observatory@corning-cc.edu
WWW: http://www.corning-cc.edu/observatory/
Activities: education * observing (IAU Code 911)
City Reference Coordinates: 077°04'00"W 42°09'00"N

Corning Incorporated
One Riverfront Plaza
Corning, NY 14831
Telephone: 607-974-9000
          607-974-7709 (Sales Manager)
          607-4418 (Technical Reference)
Telefax: 607-974-7210
Electronic Mail: info@corning.com
                 <userid>@corning.com
WWW: http://www.corning.com/
Activities: manufacturing ultra-low-expansion and fused silica mirror blanks
City Reference Coordinates: 077°04'00"W 42°09'00"N

Cryomech Inc.
113 Falso Drive
Syracuse, NY 13211
Telephone: 315-455-2555
Telefax: 315-455-2544
Electronic Mail: specs@cryomech.com
                 service@cryomech.com
WWW: http://www.cryomech.com/lnp.html
Founded: 1963
Staff: 8
Activities: cryorefrigeration
City Reference Coordinates: 076°09'00"W 43°03'00"N

Custer Institute Inc., Observatory
Main Bayview Road
P.O. Box 1204
Southold, NY 11971
Telephone: 516-765-2626
Electronic Mail: starwatchr@aol.com
WWW: http://www.custerobservatory.org/
Founded: 1930
Staff: 150
Activities: observing * lectures * astronomy programs
Periodicals: (12) "Newsletter"
Coordinates: 072°26'36"W 41°35'00"N
City Reference Coordinates: 072°26'00"W 41°04'00"N

CVC Products Inc.
525 Lee Road
Rochester, NY 14606
Telephone: 716-458-2550
        1-800-448-5900 (USA only, outside NY State)
        1-800-962-5252 (in NY State)
Telefax: 716-458-0424
WWW: http://www.cvc.com/
Founded: 1973
Activities: high-vacuum diffusion pumps * mechanical pumps * fluids and lubricants * high-vacuum valves * baffles * gauges and sensors * deposition systems and accessories * thin-film coating
City Reference Coordinates: 077°36'00"W 43°10'00"N

Digitec Optical
2841 Jerusalem Avenue
Wantagh, NY 11793
Telephone: 1-888-327-5759 (USA only)
Telefax: 516-908-3958
Electronic Mail: digitec@optonline.net
WWW: http://www.digitecoptical.com/
Activities: distributing telescopes, accessories, software, books, etc.
City Reference Coordinates: 073°30'00"W 40°40'00"N

Discover Magazine
114 5th Avenue
New York, NY 10011

Founded: 1935
Activities: creation and performance of a wide variety of programs on astronomy and space science for general public as well as children from age 3 * education on astronomy, meteorology, and navigation * Richard S. Perkin Library in astronomy, astrophysics and space sciences
City Reference Coordinates: 074°01'00"W 40°43'00"N

Herbert H. Lehman College
• See "City University of New York (CUNY), Herbert H. Lehman College"

Hofstra University, Department of Physics and Astronomy
Hempstead, NY 11549-1330
Telephone: 516-463-5582
Telefax: 516-463-3059
Electronic Mail: <userid>@hofstra.edu
hofstra@hofstra.edu
WWW: http://www.hofstra.edu/Academics/HCLAS/Physics/index_Physics.cfm
Founded: 1935
Staff: 18
City Reference Coordinates: 073°37'00"W 40°42'00"N

Hubble Planetarium (Edwin P._)
• See "Edwin P. Hubble Planetarium"

Hunter College
• See "City University of New York (CUNY), Hunter College"

Icarus
c/o Editorial Office
Cornell University
Space Sciences Building
Ithaca, NY 14853-6801
Telephone: 607-255-4875
Telefax: 607-255-6354
Electronic Mail: office@icarus.cornell.edu
WWW: http://icarus.cornell.edu/
Founded: 1962
• Journal (12) (ISSN 0019-1035)
City Reference Coordinates: 076°30'00"W 42°27'00"N

ILC Data Device Corp.
105 Wilbur Place
Bohemia, NY 11716
Telephone: 631-567-5700
1-800-DDC-5757
631-567-5600
Telefax: 631-563-4331
Electronic Mail: literatu@ilcddc.com
WWW: http://www.ilcddc.com/
Founded: 1965
Staff: 700
Activities: manufacturing microelectronics components
Periodicals: "DDC News"
City Reference Coordinates: 073°07'00"W 40°46'00"N

Ithaca Amateur Astronomers (IAA)
c/o Rob West
274 Asbury Road
Ithaca, NY 14882
Telephone: 607-257-3788
Electronic Mail: wmti@aol.com
WWW: http://members.aol.com/rw76367/IAA.html
City Reference Coordinates: 076°30'00"W 42°27'00"N

Joerger Enterprises Inc.
166 Laurel Road
East Northport, NY 11731
Telephone: 631-757-6200
Telefax: 631-757-6201
Electronic Mail: joerger@joergerinc.com
WWW: http://www.joergerinc.com/
Founded: 1972
Staff: 5
Activities: designing, manufacturing and distributing data acquisition and process control modules
City Reference Coordinates: 073°19'00"W 40°52'00"N

Junior Museum Planetarium
282 Fifth Avenue
Troy, NY 12180
Telephone: 518-235-2120
WWW: http://www.rpi.edu/web/museum/
City Reference Coordinates: 073°40'00"W 42°43'00"N

Kohl Observatory
c/o Ronald W. Kohl
3923 Farview Lane
Lakewood, NY 14750-9653
Telephone: 716-763-8488
Telefax: 716-488-1803
Electronic Mail: ronccd@alltel.net
WWW: http://home.alltel.net/ronccd/
Founded: 1991
Staff: 1
Activities: private observatory * hosting school and group visits * astronomical imaging
Coordinates: 079°19'00"W 42°04'40"N
City Reference Coordinates: 079°21'00"W 42°06'00"N

Kopernik Astronomical Society (KAS)
c/o Kopernik Space Education Center
698 Underwood Road
Vestal, NY 13850
Telephone: 607-772-0660
Electronic Mail: blackhole@kopernik.org
WWW: http://www.kopernik.org/kas.htm
Founded: 1960
Membership: 75
City Reference Coordinates: 076°03'00"W 42°05'00"N

Kopernik Space Education Center (KSEC)
698 Underwood Road
Vestal, NY 13850
Telephone: 607-748-3685
           607-772-0660
Telefax: 607-748-3222
Electronic Mail: <userid>@kopernik.org
WWW: http://www.kopernik.org/
Founded: 1973
Staff: 4
City Reference Coordinates: 076°03'00"W 42°05'00"N

Lasermax Inc.
3495 Winton Place
Building B
Rochester, NY 14623-2807
Telephone: 1-800-LASER-03 (USA only)
           716-272-5420
Telefax: 716-272-5427
Electronic Mail: <userid>@lasermax-inc.com
WWW: http://www.lasermax-inc.com/
Activities: laser technology
City Reference Coordinates: 077°36'00"W 43°10'00"N

Lehman College (Herbert H._)
● See "City University of New York (CUNY), Herbert H. Lehman College"

Liebert Inc. (Mary Ann_)
● See "Mary Ann Liebert Inc."

Link Planetarium (Edwin A._)
● See "Roberson Museum and Science Center, Edwin A. Link Planetarium"

Longwood Regional Planetarium
c/o Longwood Middle School
41 Yaphank Middle Island Road
Middle Island, NY 11953-2369
Telephone: 516-345-2741
Telefax: 516-345-9296
Electronic Mail: caprijl@webtu.net
Founded: 1971
Staff: 1
Activities: education * shows

City Reference Coordinates: 072°56'00"W 40°53'00"N

Martz Astronomical Association (MAA) Inc.
120 East 3rd Street
Jamestown, NY 14701
or :
176 Robin Hill Road
Frewsburg, NY 14738
Telephone: 716-664-4506
Electronic Mail: bemusabord@aol.com
WWW: http://members.aol.com/bemusabord/
Founded: 1978
Membership: 60
Activities: education * observing
Periodicals: (12) "The Probe"
Coordinates: 079°04'00"W 42°00'30"N H634m (Martz Observatory)
City Reference Coordinates: 079°15'00"W 42°05'00"N (Jamestown)
079°11'00"W 42°03'00"N (Frewsburg)

Mary Ann Liebert Inc.
2 Madison Avenue
Larchmont, NY 10538-1961
Telephone: 914-834-3100
1-800-M-LIEBERT (USA only)
Telefax: 914-834-1388
Electronic Mail: mliebert@liebertpub.com
WWW: http://www.liebertpub.com/
Periodicals: (4) "Astrobiology" (ISSN 1531-1074)
City Reference Coordinates: 073°47'00"W 40°57'00"N

Michael I. Casper - Meteorites Inc.
Post Office Drawer J
Ithaca, NY 14851
Telephone: 607-257-5349
Telefax: 607-266-7904
Electronic Mail: info@meteorites.com
WWW: http://www.meteorites.com/
City Reference Coordinates: 076°30'00"W 42°27'00"N

Mid-Hudson Astronomy Association (MHAA)
c/o Arlene Schoonmaker
26 Smith Pond Road
Campbell Hall, NY 10916-3436
or :
c/o Physics Department
State University of New York
New Paltz, NY 12561
Telephone: 845-485-5669 (Hotline)
Electronic Mail: trankin@mhv.net
mhaa@geocities.com
WWW: http://www.geocities.com/CapeCanaveral/5679/
http://jump.to/mhaa
Founded: 1985
City Reference Coordinates: 074°05'00"W 41°45'00"N (New Paltz)

Mohawk Valley Astronomical Society (MVAS)
c/o Richard F. Somer
P.O. Box 52
Clinton, NY 13323
Telephone: 315-859-4122
Telefax: 315-859-4632
Electronic Mail: rfsomer@borg.com
WWW: http://www.dreamscape.com/corvus/MVAS/
Founded: 1989
Membership: 120
Activities: monthly meetings * star parties * instrumentation * education
Periodicals: (6) "Telescopic Topics"
City Reference Coordinates: 075°23'00"W 43°03'00"N

Mountain Observers
c/o Patrick Ferrick
P.O. Box 206
Old Forge, NY 13420
Electronic Mail: ferrick@telenet.net
City Reference Coordinates: 075°44'00"W 41°21'00"N

National Aeronautics and Space Administration (NASA), Goddard Institute for Space Studies (GISS)
2880 Broadway
New York, NY 10025
Telephone: 212-678-5500
Telefax: 212-678-5622
Electronic Mail: <userid>@giss.nasa.gov
WWW: http://www.giss.nasa.gov/
Founded: 1958 (NASA)
Staff: 150
Activities: climate research * atmospheric sciences * planetary research
City Reference Coordinates: 074°01'00"W 40°43'00"N

National Astronomy and Ionosphere Center (NAIC), Headquarters
c/o Cornell University
Space Sciences Building
Ithaca, NY 14853-6801
Telephone: 607-255-3735
Telefax: 607-255-8803
Electronic Mail: <userid>@astrosun.cornell.tn.edu
WWW: http://www.naic.edu/
Founded: 1963
Staff: 140
Activities: research * providing visitor facilities for scientists in the fields of astronomy and atmospheric sciences * operating the Arecibo Observatory (Puerto Rico - see separate entry) * atmospheric sciences * radar and radio astronomy
Periodicals: "NAIC Report Series"
City Reference Coordinates: 076°30'00"W 42°27'00"N

National Photocolor Corp. (NPC)
428 Waverly Avenue
P.O. Box 586
Mamaroneck, NY 10543-0586
Telephone: 914-698-8111
Telefax: 914-698-3629
Electronic Mail: npc@westnet.com
WWW: http://www.nationalphotocolor.com/
Founded: 1935
Staff: 4
Activities: manufacturing pellicle-type beamsplitters and windows
City Reference Coordinates: 073°44'00"W 40°57'00"N

New York Academy of Sciences (NYAS)
2 East 63rd Street
New York, NY 10021
Telephone: 1-800-843-6927 (USA only)
          212-838-0230
Telefax: 212-888-2894
Electronic Mail: nyas@nyas.org
WWW: http://www.nyas.org/
Founded: 1817
Membership: 50000
Activities: conferences * meetings
Periodicals: (6) "The Sciences"; (4) "Academy Update"
City Reference Coordinates: 074°01'00"W 40°43'00"N

New York Hall of Science (NYHoS)
47-01 111th Street
Flushing Meadows
Corona Park, NY 11368
Telephone: 718-699-0005
Telefax: 718-699-1341
Electronic Mail: mweiss@nyhallsci.org
WWW: http://www.nyhallsci.org/
Founded: 1964
Membership: 2500
Staff: 200
Activities: interactive science exhibitions * science workshops for schools and public * teacher training * portable planetarium
Periodicals: (3) "Newsletter"
City Reference Coordinates: 073°51'00"W 40°45'00"N

New York University, Physics Department
4 Washington Place
New York, NY 10003
Telephone: 212-998-7700
Telefax: 212-995-4016
Electronic Mail: <userid>@physics.nyu.edu
WWW: http://www.physics.nyu.edu/

Founded: 1893
Staff: 27
Activities: theory and observing of ISM * relativity and cosmology
City Reference Coordinates: 074°01'00"W 40°43'00"N

Nikon USA
1300 Walt Whitman Road
Melville, NY 11747
Telephone: 1-800-NIKON-US (USA only)
           1-800-645-4687 (USA only)
WWW: http://www.nikonusa.com/
Activities: distributing, among others, binoculars and scopes
City Reference Coordinates: 073°25'00"W 40°49'00"N

Northeast Bronx Planetarium
750 Baychester Avenue
Bronx, NY 10475
Telephone: 718-904-5520
Telefax: 718-904-5502
Electronic Mail: nebronx@netzero.net (Terry Buchalter, Director)
Founded: 1974
Staff: 3
Activities: providing planetarium programs to New York City School District 11 and to outside groups on request * 25 programs available targeted to a specific grade's curriculum
City Reference Coordinates: 073°56'00"W 40°49'00"N

Plainedge Planetarium
Sylvia Packard School
North Idaho Avenue
North Massapequa, NY 11758
Telephone: 516-755-5723
Electronic Mail: srusso@plainedge.ourschools.org (Steven Russo)
WWW: http://www.plainedge.ourschools.org/planetarium/
City Reference Coordinates: 073°30'00"W 40°40'00"N

Plenum Publishing Corp.
• See now "Kluwer Academic Publishers (KAP)"

Rensselaer Astrophysical Society
Box 24
Rensselaer Union
Troy, NY 12180-3599
Telephone: 518-276-6090
Electronic Mail: astro@rpi.edu
WWW: http://astro.union.rpi.edu/
Founded: 1948
City Reference Coordinates: 073°40'00"W 42°43'00"N

Rensselaer Polytechnic Institute (RPI), Department of Physics, Applied Physics and Astronomy
110 Eighth Street
Troy, NY 12180-3590
Telephone: 518-276-6310
Telefax: 518-276-6680
Electronic Mail: physics@rpi.edu
WWW: http://www.rpi.edu/dept/phys/Astro/astro.html
     http://www.rpi.edu/dept/phys/physics.html
Founded: 1940 (Institute: 1824)
City Reference Coordinates: 073°40'00"W 42°43'00"N

Richardson Grating Laboratory (RGL)
705 St Paul Street
Rochester, NY 14605
Telephone: 716-262-1331
           1-800-654-9955 (USA only)
Telefax: 716-454-1568
Electronic Mail: information@gratinglab.com
WWW: http://www.gratinglab.com/
Founded: 1947
Staff: 24
Activities: designing and manufacturing ruled and holographic diffraction gratings
City Reference Coordinates: 077°36'00"W 43°10'00"N

Roberson Museum and Science Center, Edwin A. Link Planetarium
30 Front Street
Binghamton, NY 13905-4704

Telephone: 607-772-0660
Telefax: 607-771-8905
Electronic Mail: linkpl@juno.com (Jim Rienhardt, Director)
WWW: http://www.roberson.org/
http://www.roberson.org/planetarium/planetarium.htm
Founded: 1965
Activities: space education * public shows
City Reference Coordinates: 075°54'00"W 42°08'00"N

Rochester Academy of Science, Astronomy Section (ASRAS)
P.O. Box 20292
Rochester, NY 14602
Telephone: 585-987-5330
Electronic Mail: info@rochesterastronomy.org
WWW: http://www.rochesterastronomy.org/
Founded: 1932
Membership: 150
Activities: observing * education
Periodicals: (12) "The Astronomer"
Coordinates: 077°47'37"W 43°17'17"N H85m
City Reference Coordinates: 077°36'00"W 43°10'00"N

Rochester Institute of Technology (RIT), Department of Physics
85 Lomb Memorial Drive
Rochester, NY 14623-5603
Telephone: 716-475-2538
Electronic Mail: jrksps@rit.edu (James Kern)
WWW: http://www.rit.edu/~674www/
http://www.cis.rit.edu/research/astro/
http://www.rit.edu/~ritobs/ (Observatory)
Founded: 1997
Coordinates: 077°77'32"W 43°00'05"N (IAU Code 920)
City Reference Coordinates: 077°36'00"W 43°10'00"N

Saint Lawrence University, Department of Physics
Bewkes Hall
Canton, NY 13617
Telephone: 315-229-5491
Telefax: 315-229-7421
Electronic Mail: physics@music.stlawu.edu
WWW: http://it.stlawu.edu:80/~physics/
Founded: 1856 (University)
Staff: 7
City Reference Coordinates: 075°11'00"W 44°35'00"N

Schenectady Museum and Planetarium
Nott Terrace Heights
Schenectady, NY 12308
Telephone: 518-382-7890
Telefax: 518-382-7893
WWW: http://tardis.union.edu/community/project95/SCHmuseum/permanent.html
City Reference Coordinates: 073°53'00"W 42°47'00"N

Sciencenter
601 First Street
Ithaca, NY 14850
Telephone: 607-272-0600
Telefax: 607-277-7469
Electronic Mail: info@sciencenter.org
WWW: http://www.sciencenter.org/
City Reference Coordinates: 076°30'00"W 42°27'00"N

Silverman Planetarium
c/o Milton J. Rubenstein Museum of Science & Technology
500 South Franklin Street
Syracuse, NY 13202
Telephone: 315-425-9068
Telefax: 315-425-9072
WWW: http://www.syracusecvb.org/Visitor/Fun/Museums/most.html
http://www.most.org/
Founded: 1979
Membership: 4300 (Museum)
Staff: 53 (Museum)
Activities: shows * observing * exhibitions
Periodicals: "Technologist Magazine", "MOST News"

Awards: (1) "Teacher of the Year"
City Reference Coordinates: 076°09'00"W 43°03'00"N

Society of Amateur Radio Astronomers (SARA) Inc.
c/o Vincent Caracci (Secretary)
247 North Linden Street
Massapequa, NY 11758
Telephone: 516-798-8459
Electronic Mail: vinhell@juno.com
WWW: http://irsociety.com/0c:/sara.html
http://www.bambi.net/sara.html
Founded: 1981
Membership: 250
Activities: providing information on radioastronomers workings * education
Periodicals: "Radio Astronomy"
City Reference Coordinates: 073°30'00"W 40°40'00"N

Springer-Verlag, New York
175 Fifth Avenue
New York, NY 10010
Telephone: 1-800-SPRINGER (USA only)
1-800-777-4643 (USA only)
212-460-1515
Telefax: 212-473-6272
Electronic Mail: service@springer-ny.com
tomv@springer-ny.com (Thomas von Foerster)
WWW: http://www.springer-ny.com/
Founded: 1964
Staff: 200
• Publisher
City Reference Coordinates: 074°01'00"W 40°43'00"N

St. Lawrence University, Department of Physics
• See "Saint Lawrence University, Department of Physics"

State University of New York (SUNY), Binghamton, Department of Physics and Astronomy
P.O. Box 6016
Binghamton, NY 13902-6016
Telephone: 607-777-2217
Telefax: 607-777-2546
Electronic Mail: physics@binghamton.edu
<userid>@binghamton.edu
WWW: http://physics.adm.binghamton.edu/
Staff: 20
City Reference Coordinates: 075°54'00"W 42°08'00"N

State University of New York (SUNY), Buffalo, Department of Physics
239 Fronzack Hall
Amherst, NY 14260-1500
Telephone: 716-645-2017
Telefax: 716-645-2507
Electronic Mail: ubphysics@buffalo.edu
<userid>@physics.buffalo.edu
WWW: http://electron.physics.buffalo.edu/
City Reference Coordinates: 078°48'00"W 42°58'00"N

State University of New York (SUNY), Geneseo, Department of Physics and Astronomy
Greene Science Building
1 College Circle
Geneseo, NY 14454-1484
Telephone: 716-245-5281
Telefax: 716-245-5288
Electronic Mail: <userid>@uno.cc.geneseo.edu
WWW: http://physics.sci.geneseo.edu/welcome.html
Founded: 1964
Activities: education * stellar astrophysics * meteor physics * numerical analysis
City Reference Coordinates: 077°49'00"W 42°48'00"N

State University of New York (SUNY), Potsdam, Planetarium
c/o Geology Department
Pierrepont Avenue
Potsdam, NY 13676
Telephone: 315-267-2289
Telefax: 315-267-2695
Electronic Mail: revettfa@potsdam.edu (Frank Revetta)

Founded: 1963
Activities: shows * education
City Reference Coordinates: 074°59'00"W 44°40'00"N

State University of New York (SUNY), Stony Brook, Department of Physics and Astronomy
Stony Brook, NY 11794-3800
Telephone: 631-632-8100
         631-632-8221
Telefax: 631-632-8176
        631-632-8240
Electronic Mail: <userid>@sbast1.ess.sunysb.edu
WWW: http://www.physics.sunysb.edu/physics/
     http://www.ess.sunysb.edu/astro/ (Astronomy Program)
Founded: 1965
Staff: 11
Activities: observational and theoretical astronomy * planetary sciences * stellar astronomy * star formation * galactic structure * ISM * nuclear astrophysics * cosmology
Coordinates: 073°07'35"W 40°54'53"N H60m
City Reference Coordinates: 073°09'00"W 40°56'00"N

State University of New York (SUNY), Stony Brook, Institute for Terrestrial and Planetary Atmospheres (ITPA)
Marine Sciences Research Center
Stony Brook University
Nicolls Road
Stony Brook, NY 11794-5000
Telephone: 631-632-8009
Telefax: 631-632-6251
Electronic Mail: <userid>@notes.cc.sunysb.edu
WWW: http://129.49.28.99/GEC/HTML/ITPA.html
Founded: 1988
City Reference Coordinates: 073°09'00"W 40°56'00"N

Strasenburgh Planetarium
c/o Rochester Museum and Science Center
657 East Avenue
Rochester, NY 14607
Telephone: 716-271-4320
Telefax: 716-271-5935
WWW: http://www.rmsc.org/
Founded: 1968
Staff: 6
Activities: star shows * laser shows * large-format motion pictures
City Reference Coordinates: 077°36'00"W 43°10'00"N

Stull Observatory
• See "Alfred University, Stull Observatory"

Sunstone Inc.
P.O. Box 788
Cooperstown, NY 13326
Telephone: 1-800-327-0306 (USA only)
         607-547-8207
Telefax: 607-547-8338
Founded: 1983
Staff: 3
Activities: publishing children's astronomy activity books
City Reference Coordinates: 074°56'00"W 42°42'00"N

Syracuse Astronomical Society (SAS) Inc.
500 South Salina Street
Suite 1010
Syracuse, NY 13202
or :
c/o Rosemary Kennett
1115 East Colvin Street
Syracuse, NY 13210
or :
c/o Arnis V. Sprancmanis
105 Kristin Road
North Syracuse, NY 13212
Telephone: 315-474-4334
         315-475-7032
Electronic Mail: avspranc@mailbox.syr.edu
WWW: http://www.pshrink.com/sas/

http://www.syracuse-astro.org/
Founded: 1970
Membership: 100
Activities: observing
Periodicals: "Newsletter"
Coordinates: 076°11'12"W 42°47'42"N (Darling Hill Obs., Vesper)
City Reference Coordinates: 076°09'00"W 43°03'00"N

Syracuse University, Department of Physics
201 Physics Building
Syracuse, NY 13244-1130
Telephone: 315-443-3901
Telefax: 315-443-9103
Electronic Mail: physics@physics.syr.edu
<userid>@physics.syr.edu
WWW: http://www.phy.syr.edu/
Founded: 1870
Staff: 33
Activities: general relativity * high-energy physics (experimental and theoretical) * condensed-matter physics * gravitational-wave detection
Coordinates: 076°08'24"W 43°02'12"N H160m
City Reference Coordinates: 076°09'00"W 43°03'00"N

Tele Vue Optics, Inc.
32 Elkay Drive
Chester, NY 10918
Telephone: 845-469-4551
WWW: http://www.televue.com/
Founded: 1977
Activities: designing and manufacturing apochromatic refracting telescopes, eyepieces, mountings and accessories
City Reference Coordinates: 075°21'00"W 39°51'00"N

The Amateur Sky Survey (TASS)
c/o Michael Richmond
Department of Physics
Rochester Institute of Technology
85 Lomb Memorial Drive
Rochester, NY 14623-5603
Telephone: 716-475-2538
Electronic Mail: tass@wwa.com
WWW: http://www.tass-survey.org/
Founded: 1995
Membership: 20
Activities: designing, building and using CCD cameras * stellar photometry * catalogs
Coordinates: 088°20'00"W 41°50'00"N H50m
072°40'00"W 44°03'00"N H460m
076°53'00"W 39°09'00"N H147m
083°52'00"W 39°48'00"N H50m
City Reference Coordinates: 077°36'00"W 43°10'00"N

Trek Inc.
P.O. Box 728
Medina, NY 14103-0728
or :
11601 Maple Ridge Road
Medina, NY 14103
Telephone: 716-798-3140
Telefax: 716-798-3106
Founded: 1968
Staff: 123
Activities: designing and manufacturing voltmeters, power supplies, amplifiers and EOS/ESD instrumentation
City Reference Coordinates: 078°23'00"W 43°13'00"N

Union College, Department of Physics
Schenectady, NY 12308
Telephone: 518-388-6254
Telefax: 518-388-6947
Electronic Mail: <userid>@union.edu
WWW: http://www.union.edu/PUBLIC/PHYDEPT/
http://www.union.edu/PUBLIC/PHYDEPT/observatory.htm (Observatory)
Activities: astrophysics * atomic spectroscopy * biophysics * condensed matter physics * medical physics * non-linear physics
City Reference Coordinates: 073°53'00"W 42°47'00"N

Unitron Inc.
170 Wilbur Place

P.O. Box 469
Bohemia, NY 11716-0469
Telephone: 631-589-6666
Telefax: 631-589-6975
Electronic Mail: info@unitronusa.com
WWW: http://www.unitronusa.com
Founded: 1952
Activities: manufacturing instrumentation
City Reference Coordinates: 073°07'00"W 40°46'00"N

University of Rochester, Department of Physics and Astronomy
Bausch & Lomb Hall
P.O. Box 270171
600 Wilson Boulevard
Rochester, NY 14627-0171
Telephone: 585-275-4351
Telefax: 585-273-3237
Electronic Mail: <userid>@pas.rochester.edu
WWW: http://www.pas.rochester.edu/
Founded: 1935
Activities: IR astronomy * instrumentation * theoretical astrophysics
Periodicals: "C.E. Kenneth Mees Observatory Reprints"
Coordinates: 077°24'30"W 42°42'00"N H701m (C.E. Kenneth Mees Observatory)
City Reference Coordinates: 077°36'00"W 43°10'00"N

Vanderbilt Museum, Planetarium
180 Little Neck Road
Centerport, NY 11787
Telephone: 516-854-5555
Telefax: 516-854-5527
Electronic Mail: juliek@juno.com (Julie M. Kostas, Office Manager)
  info@vanderbiltmuesum.org
WWW: http://www.webscope.com/vanderbilt/
  http://www.vanderbiltmuseum.org/planetarium.html
Founded: 1971
Staff: 7
Activities: planetarium shows * education * observing * workshops
City Reference Coordinates: 073°22'00"W 40°54'00"N

Vassar College, Department of Physics and Astronomy
124 Raymond Avenue
Poughkeepsie, NY 12604
Telephone: 914-437-7340
Electronic Mail: <userid>@vassar.edu
WWW: http://noether.vassar.edu/
Staff: 7
Activities: research * education * observing (IAU Code 794)
City Reference Coordinates: 073°56'00"W 41°42'00"N

Velmex Inc.
7550 State Route 5 & 20
Bloomfield, NY 14469
Telephone: 716-657-6151
  1-800-642-6446 (USA only)
Telefax: 716-657-6153
Electronic Mail: info@velmex.com
WWW: http://www.velmex.com/
Founded: 1967
Staff: 38
Activities: manufacturing and distributing mechanical slide assemblies and rotary tables for positioning
City Reference Coordinates: 077°26'00"W 42°54'00"N

Westchester Amateur Astronomers (WAA)
P.O. Box 208
Rye, NY 10580
Telephone: 877-456-5778
Electronic Mail: waa@dorsai.org
WWW: http://www.dorsai.org/~waa
Founded: 1984
Membership: 120
Activities: lectures * instrumentation * observing
Periodicals: (12) "Newsletter"
Coordinates: 073°36'00"W 41°15'00"N H106m (Ward Pound Ridge Reservation, Cross River)
  074°07'00"W 41°18'00"N H421m (Columbia University, Harriman Observatory)
City Reference Coordinates: 073°41'00"W 40°59'00"N

Whitworth Ferguson Planetarium
c/o Buffalo State College
1300 Elmwood Avenue
Buffalo, NY 14222-1095
Telephone: 716-878-4911
Telefax: 716-878-4009
WWW: http://www.buffalostate.edu/
Founded: 1966
Staff: 10
Activities: service to the college, local schools and public
City Reference Coordinates: 078°53'00"W 42°54'00"N

Wright Observatory (Ernest B._)
• See "Hartwick College, Ernest B. Wright Observatory"

USA - North Carolina

Appalachian State University, Department of Physics and Astronomy
Room 231 – CAP Building
525 Rivers Street
Boone, NC 28608
Telephone: 828-262-3090
828-262-2446
Telefax: 828-262-2049
Electronic Mail: <userid>@appstate.edu
WWW: http://www.phys.appstate.edu/
Founded: 1981
Activities: photographic and photoelectric photometry of binaries * photoelectric occultation timing
Coordinates: 081°24'53"W 36°15'09"N (Dark Sky Observatory)
City Reference Coordinates: 081°41'00"W 36°13'00"N

Astronomical Society of Rowan County (ASRC)
c/o Ellen H. Trexler
3021 Old Mockville Road
Salisbury, NC 28144-9076
Telephone: 704-636-1399
Electronic Mail: asrc@dialstar.com
ettympani@dialstar.com
WWW: http://members.dialstar.com/asrc/
Founded: 1981
Membership: 23
Activities: public star parties * education
Periodicals: (6) "Rowan Sky Line"
City Reference Coordinates: 080°29'00"W 35°40'00"N

Cape Fear Astronomical Society (CFAS)
5807 Dorothy Avenue
Wilmington, NC 28403
Telephone: 910-799-2255
Electronic Mail: capefearastro@yahoogroups.com
WWW: http://www.egroups.com/group/capefearastro
Founded: 1985
Staff: 5
Activities: public observing * education
Periodicals: "Cape Fear Skies"
Coordinates: 078°08'00"W 34°29'00"N H15m
City Reference Coordinates: 077°55'00"W 34°14'00"N

Catawba Valley Astronomy Club (CVAC)
c/o Greg Kirby
5026 Poplar View Lane
Granite Falls, NC 28630-8647
Telephone: 828-396-7656
Electronic Mail: tgkirby@msn.com (T. Gregory Kirby, Secretary)
WWW: http://users.vnet.net/heafnerj/cvac.html
http://communities.msn.com/CatawbaValleyAstronomyClub/_whatsnew.msnw
Founded: 1976
Membership: 70
Activities: public observing sessions * annual Blue Ridge Parkway Regional Stargaze (BRPRS) * school groups * stargaze * field trips * Astronomy Day exhibition * operating Lucile Miller Observatory (see separate entry)
Periodicals: (12) "After Dark"
Coordinates: 081°17'36"W 35°32'30"N H360m (Hickory, NC)
081°50'00"W 36°04'00"N H1,030 (Parkway Stargaze location)
City Reference Coordinates: 081°26'00"W 35°48'00"N

Center for Networked Information Discovery and Retrieval (CNIDR)
MCNC Information Technologies Division
3021 Cornwallis Road
Research Triangle Park, NC 27709-2889
Telephone: 919-248-1415
Telefax: 919-248-1101
Electronic Mail: information@cnidr.org
WWW: http://www.cnidr.org/
Founded: 1992
Staff: 15
City Reference Coordinates: 078°54'00"W 35°59'00"N (Durham)

Chapel Hill Astronomical and Observational Society (CHAOS)
P.O. Box 842
Chapel Hill, NC 27514-0842
WWW: http://rtpnet.org/chaos/
Founded: 1996
City Reference Coordinates: 079°04'00"W 35°55'00"N

Charlotte Amateur Astronomers Club (CAAC)
c/o Gayle Riggsbee
245 Timber Lane
Matthews, NC 28105
Telephone: 704-846-3136
Electronic Mail: inochuck@aol.com (Chuck Daffin, Secretary/Newsletter Editor)
WWW: http://www.335.com/caac/
Founded: 1954
City Reference Coordinates: 080°50'00"W 35°14'00"N (Charlotte)

Cleveland County Astronomical Society (CCAS)
c/o Craven E. Williams Observatory
Gardner-Webb University (GWU)
Boiling Springs, NC 28017
Telephone: 704-434-4433 (Tom English)
Electronic Mail: ngc@gardner-webb.edu
WWW: http://www.gardner-webb.edu/GWU/NaturalSci/physics/obs.htm
Founded: 1986
Membership: 25
Activities: observing * education
City Reference Coordinates: 081°40'00"W 35°16'00"N

Craven E. Williams Observatory
• See "Gardner-Webb University (GWU), Craven E. Williams Observatory"

Duke University, Department of Physics
Physics Building
Science Drive
Box 90305
Durham, NC 27708
Telephone: 919-660-2500
Telefax: 919-660-2525
Electronic Mail: <userid>@phy.duke.edu
WWW: http://www.phy.duke.edu/
City Reference Coordinates: 078°54'00"W 35°59'00"N

Foothills Astronomical Society (FAS)
c/o Tom Linder
P.O. Box 584
Columbus, NC 722
Electronic Mail: tlinder@aol.com
WWW: http://members.aol.com/TLINDER/FAS.html
City Reference Coordinates: 082°12'00"W 35°15'00"N

Forsyth Astronomical Society (FAS)
P.O. Box 30128
Winston-Salem, NC 27130-0128
Telephone: 910-945-0558 (ask for Wayne or Chris Ketner)
Founded: 1937
City Reference Coordinates: 080°15'00"W 36°06'00"N

Gardner-Webb University (GWU), Craven E. Williams Observatory
Boiling Springs, NC 28017
Telephone: 704-453-1017 (Tom English)
Electronic Mail: ngc@gardner-webb.edu
WWW: http://www.naturalsci.gardner-webb.edu/department/natl_sci/obs/obs.htm
Founded: 1990
Staff: 4
City Reference Coordinates: 081°40'00"W 35°16'00"N

Greensboro Astronomy Club (GAC)
c/o Natural Science Center
4301 Lawndale Drive
Greensboro, NC 27455
Telephone: 336-668-4502 (Roger Joyner)
Electronic Mail: jeffswanson@mindspring.com

WWW: http://www.greensboro.com/astronomy/index_1.html
Founded: 1947
Membership: 50
Activities: meetings * observing
Periodicals: "The Aperture"
City Reference Coordinates: 079°47'00"W 36°04'00"N

Kelly Planetarium
c/o Discovery Place
301 North Tryon Street
Charlotte, NC 28202
Telephone: 704-372-6261
   1-800-935-055 (USA only)
WWW: http://www.discoveryplace.org/
   http://www.discoveryplace.org/pframe.htm
Founded: 1964 (re-established: 1991)
City Reference Coordinates: 080°50'00"W 35°14'00"N

Lucile Miller Observatory (LMO)
Maiden High School
Maiden, NC 28650
Telephone: 704-328-4534 (Tom Curtis)
   704-428-9955 (Hot Line)
Electronic Mail: lucilem@twave.net
WWW: http://users.twave.net/jafowiz/lmo.htm
Founded: 1976
City Reference Coordinates: 081°12'00"W 35°34'00"N

Margaret C. Woodson Planetarium
1636 Parkview Circle
Salisbury, NC 28144
Telephone: 704-639-3004
Telefax: 704-639-3015
Electronic Mail: wilsonpk@rss.k12.nc.us (Patsy K. Wilson, Planetarium Educator)
Founded: 1967
Staff: 1
Activities: school programs
City Reference Coordinates: 080°29'00"W 35°40'00"N

Mid-Atlantic Star Party (MASP)
244 Deerfield Road
Apex, NC 27502
Telephone: 919-362-5194 (John Dilday)
Electronic Mail: masp@carolinaskies.com
WWW: http://www.carolinaskies.com/masp/home.htm
Founded: 1995
City Reference Coordinates: 078°52'00"W 35°57'00"N

Miller Observatory (LMO) (Lucile_)
• See "Lucile Miller Observatory (LMO)"

Morehead Planetarium
c/o University of North Carolina
Campus Box 3480
East Franklin Street
Chapel Hill, NC 27599-3480
Telephone: 919-962-1236
Telefax: 919-962-1238
Electronic Mail: mhplanet@unc.edu
WWW: http://www.morehead.unc.edu/
Founded: 1949
Membership: 800
Activities: star shows * exhibits * classes * science shop
Periodicals: (4) "Sundial"
Coordinates: 079°03'06"W 35°54'48"N
City Reference Coordinates: 079°04'00"W 35°55'00"N

National Climatic Data Center (NCDC)
NOAA/NESDIS E/CC3
151 Patton Avenue
Room 120
Asheville, NC 28801-5001
Telephone: 828-271-4800
Telefax: 828-271-4876
Electronic Mail: ncdc.info@noaa.gov

orders@ncdc.noaa.gov
WWW: http://lwf.ncdc.noaa.gov/oa/ncdc.html
Founded: 1951 (as "National Weather Records Center")
City Reference Coordinates: 082°33'00"W 35°34'00"N

North Carolina A&T State University, Department of Physics
101 Marteena Hall
Greensboro, NC 27411
Telephone: 336-334-7646
Telefax: 334-334-7423
Electronic Mail: physics@unicorn.ncat.edu
WWW: http://www.physics.ncat.edu/
Activities: research * education * planetarium
City Reference Coordinates: 079°47'00"W 36°04'00"N

North Carolina Museum of Life and Science (NCMLS)
433 Murray Avenue
P.O. Box 15190
Durham, NC 27704
Telephone: 919-220-5429
Telefax: 919-220-5575
Electronic Mail: exhibits@ncmls.org
WWW: http://www.herald-sun.com/ncmls
Founded: 1946
Staff: 65
Periodicals: (6) "Adventures"
City Reference Coordinates: 078°54'00"W 35°59'00"N

North Carolina State University (NCSU), Department of Physics, Astrophysics Group
P.O. Box 8202
Raleigh, NC 27695-8202
Telephone: 919-515-2521
Telefax: 919-515-6538
Electronic Mail: astrophysics@ncsu.edu
    john_blondin@ncsu.edu (John Blondin, Assistant Professor)
    <userid>@ncsu.edu
WWW: http://wonka.physics.ncsu.edu/www/Astro
Founded: 1985
Staff: 4
Activities: research * education
City Reference Coordinates: 078°39'00"W 35°47'00"N

Physics Academic Software (PAS)
c/o John S. Risley (Editor)
Department of Physics
North Carolina State University
Raleigh, NC 27695-8202
or :
Bureau of Mines Building
North Carolina State University
Raleigh, NC 27695
Telephone: 1-800-955-8275
    919-515-7229
Telefax: 919-515-2682
Electronic Mail: pas@ncsu.edu
WWW: http://www.aip.org/pas/
Founded: 1985
Staff: 4
Activities: publishing educational software for physics teaching and learning
City Reference Coordinates: 078°39'00"W 35°47'00"N

Pisgah Astronomical Research Institute (PARI)
1 PARI Drive
Rosman, NC 28772-9614
Telephone: 828-862-5554
Telefax: 828-862-5877
Electronic Mail: csosborne@citcom.net (Charles S. Osborne, Technical Director)
    info@pari.edu
WWW: http://www.pari.edu/
Activities: optical and radio astronomy
City Reference Coordinates: 082°49'00"W 35°03'00"N

Raleigh Astronomy Club (RAC)
P.O. Box 10643
Raleigh, NC 27605

Telephone: 919-460-7900
Electronic Mail: rac@rtpnet.org
WWW: http://rtpnet.org/~rac/
City Reference Coordinates: 078°39'00"W 35°47'00"N

Robeson Planetarium
c/o James A. Hooks (Director)
212 Tartan Road
Lumberton, NC 28358
Telephone: 910-671-6015 (Reservations)
Electronic Mail: jamhooks@interpath.com
WWW: http://home.interpath.net/jamhooks/
Founded: 1969
City Reference Coordinates: 079°03'00"W 34°37'00"N

Schiele Museum of Natural History
1500 East Garrison Boulevard
Gastonia, NC 28054-5199
Telephone: 704-866-6908
Telefax: 704-866-6041
Electronic Mail: info@schielemuseum.org
WWW: http://www.schielemuseum.org/
Founded: 1960
Staff: 32
Activities: guided tours * lectures * films * radio programs * regional educational TV program * Summer workshops * planetarium shows
Periodicals: (4) "Newsletter"; (1) "Annual Report"
City Reference Coordinates: 081°11'00"W 35°16'00"N

SciWorks Planetarium
400 West Hanes Mill Road
Winston-Salem, NC 27105
Telephone: 919-767-6730
Telefax: 919-661-1777
WWW: http://www.sciworks.org/planetarium.htm
Founded: 1982
Staff: 2
Activities: shows for public and school groups
City Reference Coordinates: 080°15'00"W 36°06'00"N

Siebert Optics
200 Short Johnson Road
Clayton, NC 27520
Telephone: 919-553-3980
Telefax: 919-553-4135
Electronic Mail: info@siebertoptics.com
WWW: http://www.siebertoptics.com/
Activities: distributing telescopes, binoculars and accessories
City Reference Coordinates: 078°28'00"W 35°39'00"N

Sigma Xi - The Scientific Research Society
99 Alexander Drive
P.O. Box 13975
Research Triangle Park, NC 27709
Telephone: 1-800-243-6534
         919-549-4691
Telefax: 919-549-0090
Electronic Mail: memberinfo@sigmaxi.org
WWW: http://www.sigmaxi.org/
     http://www.amsci.org/amsci/ (American Scientist)
Founded: 1886
Membership: 90000
Periodicals: "American Scientist"
City Reference Coordinates: 078°54'00"W 35°59'00"N (Durham)

Tar River Astronomy Club (TRAC)
c/o James Lamm
1107 Robin Hill Road
Wilson, NC 27896
Telephone: 252-237-3173
Electronic Mail: james.lamm@bbandt.com
                starscop@earthlink.net
WWW: http://home.earthlink.net/~starscope/
City Reference Coordinates: 077°55'00"W 35°44'00"N

Three College Observatory (TCO)
• See "University of North Carolina, Department of Physics and Astronomy, Three College Observatory (TCO)"

Triad Action Astronomy Club (TAAC)
305 Gregg Street
High Point, NC 27263-3303
Telephone: 910-431-5063
Electronic Mail: oatesla@hpe.infi.net (Arthur Oates)
WWW: http://www.geocities.com/CapeCanaveral/8366
Founded: 1997
Activities: meetings * observing
Periodicals: "The Meteor"
City Reference Coordinates: 080°01'00"W 35°58'00"N

University of North Carolina, Department of Physics and Astronomy
Campus Box 3255
Phillips Hall
Chapel Hill, NC 27599-3255
Telephone: 919-962-2078
Telefax: 919-962-0480
Electronic Mail: info@physics.unc.edu
　　　　　　　　<userid>@physics.unc.edu
WWW: http://www.physics.unc.edu/
　　　http://www.physics.unc.edu/research/astro/astro_morehead.html (Morehead Obs.)
　　　http://www.physics.unc.edu/soar/ (SOAR Project)
Founded: 1973
Staff: 7
Activities: research * education * Morehead Planetarium (see separate entry)
Coordinates: 079°03'02"W 35°54'50"N H161m (Morehead Observatory)
City Reference Coordinates: 079°04'00"W 35°55'00"N

University of North Carolina in Greensboro (UNCG), Department of Physics and Astronomy, Three College Observatory (TCO)
P.O. Box 26170
Greensboro, NC 27402-6170
Telephone: 336-334-5844
Electronic Mail: <userid>@uncg.edu
WWW: http://www.uncg.edu/phy/
　　　http://www.uncg.edu/phy/tco/
Founded: 1981
Staff: 3
Activities: astronomical imaging with low light level video * public education
City Reference Coordinates: 079°47'00"W 36°04'00"N

Wake Forest University (WFU), Department of Physics
308 Olin Physical Laboratory
Box 7507
Winston-Salem, NC 27109-7507
Telephone: 336-758-5337
Telefax: 336-758-6142
Electronic Mail: <userid>@wfu.edu
WWW: http://www.wfu.edu/Academic-departments/Physics/
Founded: 1899
Staff: 11
City Reference Coordinates: 080°15'00"W 36°06'00"N

Williams Observatory (Craven E._)
• See "Gardner-Webb University (GWU), Craven E. Williams Observatory"

Winston-Salem Astronomical League (WSAL)
c/o Robert B. Thompson
4231 Witherow Road
Winston-Salem, NC 27106
Electronic Mail: info@wsal.org
WWW: http://www.wsal.org/
City Reference Coordinates: 080°15'00"W 36°06'00"N

Woodson Planetarium (Margaret C._)
• See "Margaret C. Woodson Planetarium"

World Data Center A for Meteorology
c/o National Climatic Data Center
Federal Building
151 Patton Avenue
Asheville, NC 28801-5001

Telephone: 704-271-4474
             704-271-4994
Telefax: 704-271-4246
Electronic Mail: wdca@ncdc.noaa.gov
WWW: http://www.ngdc.noaa.gov/wdc/webbook/wdca/wdca_meteor.html
Founded: 1953
Activities: data archive * GARP data sets * foreign data files * world climate program data sets
City Reference Coordinates: 082°33'00"W 35°34'00"N

USA - North Dakota

Dakota Astronomical Society (DAS)
c/o John Leppert
1101 Westwood Street
Suite 209
Bismarck, ND 58504-6292
or :
c/o John Gramza (President)
757 Munich Drive
Bismarck, ND 58504
Telephone: 701-222-3283
        701-223-2117
Electronic Mail: denebobs@btigate.com
            rutschke@sendit.nodak.edu
Founded: 1987
Membership: 50
Activities: monthly meetings * lectures * sky tours * annual meeting and observing session
Periodicals: (6) "Stellar Messenger"
Awards: "Dakota Astronomer of the Year"
Observatories: 1 (Killdeer Mountains)
City Reference Coordinates: 100°47'00"W 46°48'00"N

Minot State University (MSU), Observatory
500 University Avenue West
Minot, ND 58707
Telephone: 1-800-777-0750 (USA only)
        701-857-3071
Telefax: 701-839-6933
Electronic Mail: martin@warp6.cs.misu.nodak.edu
WWW: http://warp6.cs.misu.nodak.edu/
Founded: 1967
Staff: 1
Activities: education
Coordinates: 101°18'07"W 48°15'04"N H524m
City Reference Coordinates: 101°19'00"W 48°16'00"N

Northern Skies Astronomical Society (NSAS)
Box 9008
University Station
Grand Forks, ND 58202-9008
Electronic Mail: deasmith@prairie.nodak.edu (Dean Smith, President)
WWW: http://aurora.physics.und.nodak.edu/~dean/nsas/nsas.html
City Reference Coordinates: 097°03'00"W 47°55'00"N

University of North Dakota (UND), Department of Space Studies
P.O. Box 9008
Grand Forks, ND 58202-9008
Telephone: 701-777-2480
Telefax: 701-777-3711
Electronic Mail: <userid>@space.edu
WWW: http://www.space.edu/
      http://volcano.und.edu/
Founded: 1987
Staff: 12
Activities: interdisciplinary education in space science and policy * research in remote sensing, planetary science, space policy, and global change
City Reference Coordinates: 097°03'00"W 47°55'00"N

Valley City State University (VCSU), Planetarium
Valley City, ND 58072
Telephone: 701-845-7522
        1-800-532-8641 x37522 (USA only)
Electronic Mail: eileen_starr@mail.vcsu.nodak.edu (Eileen Starr, Director)
WWW: http://www.vcsu.nodak.edu/offices/mst/faculty/Eileen_Starr/indexpla.htm
City Reference Coordinates: 097°58'00"W 46°57'00"N

USA - Ohio

Advanstar Communications Inc.
7500 Old Oak Boulevard
Cleveland, OH 44130-3369
Telephone: 440-243-8100
Telefax: 440-826-2833
Electronic Mail: information@advanstar.com
WWW: http://www.advanstar.com/
Founded: 1990
Activities: providing information and communications products
Periodicals: (12) "GPS World" (ISSN 1048-5104)
• Publisher
City Reference Coordinates: 081°41'00"W 41°30'00"N

Astronomy Club of Akron (ACA)
c/o Richard Ruggles
704 South Sheraton Circle
Akron, OH 44319-1955
Telephone: 330-644-5912
WWW: http://www.acorn.net/aca/
Founded: 1949
Membership: 100
Activities: monthly meetings * observing * lectures * public outreach
Periodicals: (12) "The Night Sky"
Observatories: 1 (Portage Lakes State Park)
City Reference Coordinates: 081°31'00"W 41°05'00"N

Aurora Astronomical Society (AAS)
c/o Herbert Gillespie
9848 Sunny Lane
Streetsboro, OH 44241
WWW: http://www.angelgire.com/space/aasmember/
City Reference Coordinates: 081°21'00"W 41°19'00"N (Aurora)

Beecher Planetarium (Ward_)
• See "Ward Beecher Planetarium"

Black River Astronomical Society (BRAS)
c/o Michael Harkey
1631 Maple Drive
Lorain, OH 44052
or :
19951 Mosher Road
Wellington, OH 44090
Telephone: 216-647-5509
          216-288-8556
Electronic Mail: dlengy@leeca8.leeca.ohio.gov (Dave Lengyel)
WWW: http://junior.apk.net/~arstar50/BlackRiver.index.html
Founded: 1949
Membership: 40
Activities: lectures * star parties * photography * eclipses
City Reference Coordinates: 082°13'00"W 41°10'00"N (Wellington)
                                082°11'00"W 41°28'00"N (Lorain)

Bowling Green State University (BGSU), Department of Physics and Astronomy
Bowling Green, OH 43403
Telephone: 419-372-2421
          419-372-8666 (Planetarium)
Telefax: 419-372-9938
Electronic Mail: <userid>@bgnet.bgsu.edu
WWW: http://physics.bgsu.edu/dept/
      http://physics.bgsu.edu/planetarium/ (Planetarium)
Founded: 1978
Activities: minor planets * stellar spectroscopy * stellar pulsations * galactic structure * education * planetarium shows
Coordinates: 083°38'29"W 41°22'47"N
City Reference Coordinates: 083°40'00"W 41°22'00"N

Case Western Reserve University (CWRU), Department of Astronomy, Warner and Swasey Observatory
Smith Building
10900 Euclid Avenue
Cleveland, OH 44106-7215

Activities: publishing and distributing astronomical and scientific books
City Reference Coordinates: 083°05'00"W 40°09'00"N

Denison University, Department of Physics and Astronomy
Granville, OH 43023
Telephone: 740-587-6223
Telefax: 740-587-6240
Electronic Mail: <userid>@denison.edu
WWW: http://www.denison.edu/physics/
Founded: 1909
Staff: 6
Activities: education * research
Coordinates: 082°31'32"W 40°04'18"N H360m (Swasey Observatory – IAU Code 825)
City Reference Coordinates: 085°46'00"W 42°54'00"N

Drake Planetarium
c/o Norwood High School
2020 Sherman Avenue
Cincinnati, OH 45212
Telephone: 513-396-5578
Telefax: 513-731-1093
Electronic Mail: drake@iglou.com
WWW: http://www.iglou.com/drake/info.html
City Reference Coordinates: 084°31'00"W 39°06'00"N

Glenn Research Center (GRC) (John H._)
• See "National Aeronautics and Space Administration (NASA), John H. Glenn Research Center (GRC)"

Great Lakes Planetarium Association (GLPA)
c/o Gene Zajac (Membership Chair)
Shaker Heights High School
15911 Aldersyde Drive
Shaker heights, OH 44120
Telephone: 216-295-4251
Telefax: 216-295-4277
Electronic Mail: starman@stratos.net
WWW: http://www.pa.msu.edu/abrams/glpa.html
Founded: 1965
Membership: 259 (176 planetariums represented)
Activities: annual conferences * state meetings * aiding planetariums in developing educational material and programs * organizing and distributing materials to members
Periodicals: (4) "GLPA Newsletter"; (1) "Proceedings of Annual Meetings"
Awards: "Honorary Life Membership", "Fellow of GLPA"
City Reference Coordinates: 086°36'00"W 41°29'00"N

Hoover-Price Planetarium
McKinley Museum of History, Science, and Industry
800 McKinley Monument Drive
Canton, OH 44708
Telephone: 330-455-7043
Electronic Mail: mm@neow.lrun.com
WWW: http://www.mckinleymuseum.org/
City Reference Coordinates: 075°11'00"W 44°35'00"N

Huron Valley Astronomers Group
69 South Ridge
Monroeville, OH 44847
Telephone: 419-465-2997
Electronic Mail: tddi@accnorwalk.com
WWW: http://www.accnorwalk.com/~tddi/tech2000/skytour/
City Reference Coordinates: 082°42'00"W 42°15'00"N

John H. Glenn Research Center (GRC)
• See "National Aeronautics and Space Administration (NASA), John H. Glenn Research Center (GRC)"

Kent State University (KSU), Physics Department, Planetarium
Smith Hall
Kent, OH 44242
Telephone: 216-672-2771
Telefax: 216-672-2959
Electronic Mail: anderson@ksuvxd.kent.edu (Bryon D. Anderson, Director)
WWW: http://physics17.kent.edu/observatory/planet.html
Founded: 1968
Staff: 2
Activities: shows

City Reference Coordinates: 081°22'00"W 41°09'00"N

Kenyon College, Department of Physics
Gambier, OH 43022-9623
Electronic Mail: <userid>@kenyon.edu
WWW: http://www2.kenyon.edu/depts/physics/
    http://www2.kenyon.edu/depts/physics/astro/reflnk.htm
Staff: 5
City Reference Coordinates: 082°24'00"W 40°23'00"N

Lewis Research Center (LeRC)
• See now "National Aeronautics and Space Administration (NASA), John H. Glenn Research Center (GRC)"

Mahoning Valley Astronomical Society (MVAS)
1076 State Route 534 NW
Newton Falls, OH 44444-9514
WWW: http://mvobservatory.com/
Founded: 1939
Membership: 60
Activities: annual convention * public star parties
Periodicals: (12) "The Meteorite"
Coordinates: 080°58'35"W 41°14'30"N (Mahoning Valley Cortese Observatory - MVCO)
City Reference Coordinates: 080°59'00"W 41°11'00"N

Medina County Astronomical Society (MCAS)
c/o Matt Oltersdorf
4532 Ledgewood Drive
Medina, OH 44256-9028
Telephone: 330-722-0887
Electronic Mail: matto@apk.net
WWW: http://junior.apk.net/~matto/mcas.htm
Founded: 1997
City Reference Coordinates: 081°52'00"W 41°08'00"N

Miami Valley Astronomical Society (MVAS)
3346 Fair Oaks Drive
Beavercreek, OH 45434
or :
2629 Ridge Avenue
Dayton, OH 45414
Telephone: 513-275-7541
Electronic Mail: ron.sherman@mvas.org
WWW: http://www.mvas.org/
Founded: 1935
Membership: 90
Activities: public stargazes * lectures * amateur observing
Periodicals: "The Amateur Astronomer"
Observatories: 2 (Apollo Observatory & John Bryan State Park Observatory)
City Reference Coordinates: 084°15'00"W 39°45'00"N (Dayton)

Millstream Astronomy Club (MAC)
c/o Steve Rice
232 East Bigelow Avenue
Findlay, OH 45840
Telephone: 419-424-0685
Electronic Mail: macprez2@aol.com
WWW: http://www.bright.net/~jshiley/
Founded: 1991
Membership: 35
Activities: monthly public observing * annual astronomy day
Periodicals: (12) "Stardust"
City Reference Coordinates: 083°40'00"W 41°02'00"N

Mount Union College (MUC), Department of Physics and Astronomy
1972 Clark Avenue
Alliance, OH 44601
Telephone: 216-823-3657
Telefax: 216-823-8531
Electronic Mail: physics@muc.edu
WWW: http://www.muc.edu/ph/
Founded: 1846
Staff: 2
Activities: education
City Reference Coordinates: 081°06'00"W 40°55'00"N

National Aeronautics and Space Administration (NASA), John H. Glenn Research Center (GRC)
Lewis Field
21000 Brookpark Road
Cleveland, OH 41135
Telephone: 216-433-2004
WWW: http://www.lerc.nasa.gov/
Founded: 1958 (NASA)
Activities: research development in space aeronautics
City Reference Coordinates: 081°41'00"W 41°30'00"N

National Aeronautics and Space Administration (NASA), John H. Glenn Research Center (GRC), Visitor Center
Lewis Field
Mail Stop 8-1
21000 Brookpark Road
Cleveland, OH 44135
Telephone: 216-433-2000
Telefax: 216-433-2001
WWW: http://www.lerc.nasa.gov/
Founded: 1941
Staff: 13
City Reference Coordinates: 081°41'00"W 41°30'00"N

National Association of Space Simulating Educators (NASSE)
8371 Locust Drive
Kirkland, OH 44094
Electronic Mail: nasse@us.edu
WWW: http://www.us.edu/nasse/
Founded: 1993
City Reference Coordinates: 081°22'00"W 41°42'00"N (Mentor)

Neil Armstrong Air & Space Museum
P.O. Box 1978
Wapakoneta, OH 45895-0978
or :
500 South Apollo Drive
Wapakoneta, OH 45895
Telephone: 1-800-860-0142 (USA only)
Telefax: 419-738-3361
Electronic Mail: namv@ohiohistory.org
WWW: http://www.ohiohistory.org/
Founded: 1972
Staff: 7
City Reference Coordinates: 084°12'00"W 40°34'00"N

Northwest Ohio Visual Astronomers (NOVA)
c/o Karen Kerr
6341 Loe Road
Cygnet, OH 43413
Telephone: 419-655-2391
Electronic Mail: kmkerr@wcnet.org
WWW: http://darwin.mco.edu/~jpatrick/nova
City Reference Coordinates: 083°39'00"W 41°14'00"N

Oberlin College, Department of Physics
Wright Laboratory of Physics
110 North Professor Street
Oberlin, OH 44074-1088
Telephone: 440-775-8330
Telefax: 440-775-8960
Electronic Mail: chair@physics.oberlin.edu
  <userid>@physics.oberlin.edu
WWW: http://www.oberlin.edu/physics/
Founded: 1890 (College: 1833)
City Reference Coordinates: 082°13'00"W 41°18'00"N

Oberwerk Corp.
3997 North Lakeshore Drive
Jamestown, OH 45335
Telephone: 937-675-6792
Telefax: 937-675-6793
Electronic Mail: info@oberwerk
WWW: http://www.oberwerk.com/
Activities: distributing big binoculars and scopes
City Reference Coordinates: 079°15'00"W 42°05'00"N

Ohio Aerospace Institute (OAI)
22800 Cedar Point Road
Brook Park, OH 44142
Telephone: 440-962-3000
Telefax: 440-962-3200
Electronic Mail: <userid>@oai.org
WWW: http://www.oai.org/
Founded: 1989
Staff: 60
City Reference Coordinates: 081°48'00"W 41°24'00"N

Ohio State University (OSU), Department of Astronomy
4055 McPherson Laboratory
140 West 18th Avenue
Columbus, OH 43210-1173
Telephone: 614-292-1773
Telefax: 614-292-2928
Electronic Mail: <userid>@payne.mps.ohio-state.edu
WWW: http://www-astronomy.mps.ohio-state.edu/
       http://www-astronomy.mps.ohio-state.edu/planetarium.html (Planetarium)
       http://www.astronomy.ohio-state.edu/planetarium.html (Planetarium)
Founded: 1962
Staff: 34
Activities: spectroscopy and photometry of stars, nebulae and galaxies * theoretical astrophysics * operating the Perkins Observatory (see separate entry)
City Reference Coordinates: 083°00'00"W 39°57'00"N

Ohio State University (OSU), Radio Observatory (OSURO)
2015 Neil Avenue
Columbus, OH 43210
Telephone: 614-292-6789
Electronic Mail: bob_dixon@osu.edu
WWW: http://www.bigear.org/ (Big Ear Observatory)
       http://www.naapo.org/ (North American Astrophysical Observatory/NAAPO)
Founded: 1952
Staff: 20
Activities: radio astronomy * SETI * astronomical data distribution
Coordinates: 083°02'54"W 40°15'06"N H282m (near Delaware, OH)
City Reference Coordinates: 083°00'00"W 39°57'00"N

Ohio Turnpike Astronomers Association (OTAA)
c/o Dawn Jenkins (Editor)
1494 Lakeland Avenue
Lakewood, OH 44107
Electronic Mail: cygnus@en.com
WWW: http://www.astras-stargate.com/whatis.htm
Founded: 1956
Periodicals: (6) "The Asteroid Belt"
City Reference Coordinates: 081°48'00"W 41°29'00"N

Ohio University, Department of Physics and Astronomy
Athens, OH 45701-2979
Telephone: 740-593-1718
Telefax: 740-593-0433
Electronic Mail: <userid>@helios.phy.ohiou.edu
WWW: http://www.phy.ohiou.edu/
Founded: 1804
Staff: 20
City Reference Coordinates: 082°06'00"W 39°20'00"N

Ohio Wesleyan University (OWU), Department of Physics and Astronomy
Delaware, OH 43015
Telephone: 740-368-3770
Telefax: 740-368-3999
WWW: http://physics.owu.edu/
Founded: 1842 (OWU) (Student Observatory: 1896)
Staff: 4
Activities: education * operating Perkins Observatory (see separate entry)
Coordinates: 083°04'36"W 40°17'56"N H280m (Student observatory)
City Reference Coordinates: 083°06'00"W 40°18'00"N

Otterbein College, Department of Physics and Astronomy
155 West Main Street
Westerville, OH 43081
Telephone: 614-823-1316

Telefax: 614-823-1968
Electronic Mail: dmoore@otterbein.edu (Diane Moore, Administrative Assistant)
WWW: http://www.otterbein.edu/dept/PHYS/
http://www.otterbein.edu/dept/PHYS/weitkamp.html (Observatory and Planetarium)
Founded: 1984
Staff: 20
Activities: education * research * maintaining and operating a 110m radiotelescope (Weitkamp Observatory) * planetarium

Periodicals: (12) "Signals"
City Reference Coordinates: 082°56'00"W 40°08'00"N

Perkins Observatory
61 South Sandusky Street
Box 449
Delaware, OH 43015
Telephone: 740-363-1257
Telefax: 740-363-1258
Electronic Mail: <userid>@osu.edu
WWW: http://www.perkins-observatory.org/
Founded: 1923
Staff: 4
Activities: public education in astronomy * facility depending from the Ohio State University (OSU) and the Ohio Wesleyan University (OWU)
Periodicals: "Night Times"
Coordinates: 083°03'15"W 40°15'04"N H280m
City Reference Coordinates: 083°06'00"W 40°18'00"N

ProtoStar
P.O. Box 258
Worthington, OH 43085
Telephone: 614-785-0245
Telefax: 614-785-0401
Electronic Mail: protostar@fpi-protostar.com
WWW: http://www.fpi-protostar.com/
Activities: diagonal mounts * spiders
City Reference Coordinates: 083°01'00"W 40°05'00"N

Ralph Mueller Planetarium
Cleveland Museum Natural History
1 Wade Oval Drive
University Circle
Cleveland, OH 44106
Telephone: 216-231-4600
1-800-317-9155 (USA only)
Telefax: 216-231-5919
Electronic Mail: info@cmnh.org
WWW: http://www.cmnh.org/
Founded: 1959
City Reference Coordinates: 081°41'00"W 41°30'00"N

Richland Astronomical Society (RAS)
P.O. Box 1118
Mansfield,OH 44901
or :
c/o Keith A. Moore
331 South Market Street
Galion, OH 44833
Telephone: 419-468-3542
Electronic Mail: kamoore@richnet.net
WWW: http://www.wro.org/ras/ras.htm
Founded: 1960
Periodicals: "M-111"
City Reference Coordinates: 082°31'00"W 40°46'00"N (Mansfield)
082°46'00"W 40°44'00"N (Galion)

Ritter Astrophysical Research Center (RARC)
• See "University of Toledo, Department of Physics and Astronomy, Ritter Astrophysical Research Center (RARC)"

Sandusky Valley Amateur Astronomy Club (SVAAC)
Electronic Mail: hotrod@friendlynet.com (Dave North, Vice President)
WWW: http://www.friendlynet.com/astronomy/
Founded: 1983
Observatories: 1 (Ballreich Observatory)
City Reference Coordinates: 083°11'00"W 41°07'00"N (Tiffin)

Schuele Planetarium
c/o Lake Erie Nature and Science Center
28728 Wolf Road
Bay Village, OH 44140
Telephone: 440-871-2900
Telefax: 440-871-2901
WWW: http://www.efohio.org/LENSC.htm
City Reference Coordinates: 084°53'00"W 35°10'00"N (Cleveland)

Scientific and Technical Information Network (STN) International
2540 Olentangy River Road
P.O. Box 3012
Columbus, OH 43210-0012
Telephone: 614-447-3600
         1-800-753-4227 (USA only)
Telefax: 614-447-3713
        614-447-3798
Electronic Mail: help@cas.org
                <userid>@cas.org
WWW: http://www.cas.org/
Founded: 1983
Staff: 1300
Activities: producing and supplying online databases of scientific information * cooperative network of more than 200 scientific databases
Periodicals: (6) "STNews"
City Reference Coordinates: 083°00'00"W 39°57'00"N

Spirus Co.
2314 Augustine Drive
Parma, OH 44134-4744
Telephone: 440-888-3639
Electronic Mail: spirus@stratos.net
WWW: http://www.stratos.net/spirus/
Founded: 1994
Staff: 2
Activities: manufacturing wheel accessory for amateur telescopes
City Reference Coordinates: 081°44'00"W 41°24'00"N

Stillwater Stargazers
c/o Gary Pike
Brukner Nature Center
5995 Horseshoe Bend Road
Troy, OH 45373
Telephone: 513-548-1731
Electronic Mail: garypike@bright.net
WWW: http://www.dma.org/~wagner
City Reference Coordinates: 073°40'00"W 42°43'00"N

STN International
• See "Scientific and Technical Information Network (STN) International"

Swasey Observatory
• See "Denison University, Department of Physics and Astronomy, Swasey Observatory"
• See also "Case Western Reserve University, Department of Astronomy, Warner and Swasey Observatory"

Tech 2000
3349 SR 99 South
Monroeville, OH 44847
Telephone: 419-465-2997
Telefax: 419-465-2484
Electronic Mail: tddi@accnorwalk.com
WWW: http://www.accnorwalk.com/~tddi/tech2000/
Activities: manufacturing telescope accessories
City Reference Coordinates: 082°42'00"W 42°15'00"N

Triplett Corp.

One Triplett Drive
Bluffton, OH 45817
Telephone: 1-800-874-7538 (USA only)
　　　　　419-358-5015
Telefax: 419-358-7956
WWW: http://www.triplett.com/
Founded: 1904
Staff: 7
Activities: manufacturing electrical and electronic test equipment and instrumentation
City Reference Coordinates: 083°54'00"W 40°54'00"N

University of Akron, Department of Physics
250 Buchtel Commons
Akron, OH 44325-4001
Telephone: 216-972-7078
Telefax: 216-972-5301
Electronic Mail: <userid>@nebula.physics.uakron.edu
WWW: http://www.physics.uakron.edu/
Founded: 1893
Staff: 10
Activities: education * research * electron tunneling * spectroscopy * nuclear magnetic resonance * statistical and polymer physics * field theory * surface physics
City Reference Coordinates: 081°31'00"W 41°05'00"N

University of Cincinnati, Department of Physics
P.O. Box 210011
Cincinnati, OH 45221-0011
or :
400 Geology/Physics Building
345 College Court
Cincinnati, OH 45221
Telephone: 513-556-0501
　　　　　513-321-5186
Telefax: 513-556-3425
Electronic Mail: astro@physunc.phy.ec.edu
WWW: http://www.physics.uc.edu/
　　　　http://www.physics.uc.edu/observatory/observ.html
Staff: 30
Coordinates: 084°25'24"W 39°08'24"N H247m
City Reference Coordinates: 084°31'00"W 39°06'00"N

University of Toledo, Department of Physics and Astronomy, Ritter Astrophysical Research Center (RARC)
2801 West Bancroft Street
Toledo, OH 43606
Telephone: 419-530-2650
Telefax: 419-530-5167
Electronic Mail: office@physics.utoledo.edu
　　　　　　<userid>@astro.utoledo.edu
　　　　　　rpbo@utphya.panet.utoledo.edu (Ritter Planetarium & Brooks Observatory)
WWW: http://www.physics.utoledo.edu/www/homepage.html
　　　　http://www.physics.utoledo.edu/www/ritter/ritter.html
　　　　http://www.physics.utoledo.edu/www/ritter/ritter_obs.html (Ritter Observatory)
　　　　http://www.physics.utoledo.edu/~rpbo/menu.html (Ritter Planetarium & Brooks Observatory)
Founded: 1967
Staff: 8
Activities: stellar spectroscopy * theoretical stellar atmospheres * ISM * Martian atmosphere * planetarium shows
Coordinates: 083°36'48"W 41°39'42"N H201m
City Reference Coordinates: 083°32'00"W 41°39'00"N

Ward Beecher Planetarium
c/o Youngstown State University
410 Wick Avenue
Youngstown, OH 44555
Telephone: 330-742-3616
　　　　　330-742-7278
Telefax: 330-742-3121
Electronic Mail: wardbeecherplanetarium@hotmail.com
WWW: http://cc.ysu.edu/physics-astro/planet.html
Founded: 1967
Staff: 3
Coordinates: 080°40'17"W 41°06'15"N
City Reference Coordinates: 080°39'00"W 41°06'00"N

Warner and Swasey Observatory
● See "Case Western Reserve University, Department of Astronomy, Warner and Swasey Observatory"

Warren Rupp Observatory (WRO)
c/o Barrie McConnell
P.O. Box 1118
Mansfield, OH 44901
Telephone: 419-524-7814
Electronic Mail: wro@wro.org
　　　　　　　　michael@wro.org
　　　　　　　　<userid>@wro.org
WWW: http://www.wro.org/home.htm
Coordinates: 082°26'00"W 40°37'30"N
• Operated by the "Richland Astronomical Society (RAS)" (see separate entry)
City Reference Coordinates: 082°31'00"W 40°46'00"N

Wilderness Center Astronomy Club (WCAC)
9877 Alabama Avenue
P.O. Box 202
Wilmot, OH 44689
Telephone: 330-359-5235
Electronic Mail: info@twcac.org
　　　　　　　　wcastro@neo.rr.com
WWW: http://www.twcac.org/
Founded: 1983
Activities: education * observing * planetarium shows
City Reference Coordinates: 081°41'00"W 40°40'00"N

Youngstown State University (YSU), Department of Physics and Astronomy
Youngstown, OH 44555
Telephone: 330-742-3616
Telefax: 330-742-1998
　　　　　330-742-3121
Electronic Mail: amphys01@ysub.ysu.edu
　　　　　　　　amphys02@ysub.ysu.edu
WWW: http://cc.ysu.edu/physics-astro/
City Reference Coordinates: 080°39'00"W 41°06'00"N

USA - Oklahoma

astronomics
680 SW 24th Avenue
Norman, OK 73069
Telephone: 405-364-0858
　　　　　　1-800-422-7876 (USA only)
Telefax: 405-447-3337
Electronic Mail: questions@astronomics.com
WWW: http://www.astronomics.com/
Activities: telescopes and accessories
City Reference Coordinates: 097°27'00"W 35°14'00"N

Astronomy Club of Tulsa
P.O. Box 470611
Tulsa, OK 74147-0611
Telephone: 918-688-6277
　　　　　　918-688-MARS
Electronic Mail: info@astrotulsa.com
WWW: http://astrotulsa.com/
Founded: 1937 (Incorporated: 1986)
Membership: 150
Periodicals: "Observer"
Observatories: 1 (Ronald McDonald's Children's Charities - RMCC - Observatory)
City Reference Coordinates: 095°58'00"W 36°09'00"N

Kirkpatrick Planetarium
• See "Kirkpatrick Science and Air Space Museum at Omniplex, Kirkpatrick Planetarium"

Kirkpatrick Science and Air Space Museum at Omniplex, Kirkpatrick Planetarium
2100 NE 52nd Street
Oklahoma City, OK 73111
Telephone: 405-602-3761
Telefax: 405-602-3768
Electronic Mail: wayne.wyrick@juno.com (Wayne Wyrick, Director)
WWW: http://www.omniplex.org/planetarium.html
Founded: 1958
Staff: 6
Activities: education * observing * student research * teacher training
Coordinates: 097°28'28"W 35°31'24"N
City Reference Coordinates: 097°32'00"W 35°28'00"N

Leonardo's Star Quest Astronomy Club
Leonardo's Discovery Warehouse
200 East Maple Street
Enid, OK 73701
Telephone: 580-233-2787
Telefax: 580-237-7574
Electronic Mail: bkillam@swbell.net (Robert J. Killam, President)
WWW: http://www.leonardos.org/
Founded: 1996
Membership: 18
Activities: education
City Reference Coordinates: 097°54'00"W 36°24'00"N

Oklahoma Astronomical Society (Arkansas-_)
• See USA-AR

Oklahoma Baptist University (OBU), Planetarium
500 West University
Box 61772
Shawnee, OK 74801
Telephone: 405-878-2090
　　　　　　405-878-2166
Telefax: 405-878-2050
Electronic Mail: sheila_strawn@mail.okbu.edu (Sheila Strawn, Director)
WWW: http://www.okbu.edu/obu_nat_sci/planet/
　　　　http://www.okbu.edu/academics/natsci/planet/
Founded: 1996
Staff: 6
City Reference Coordinates: 096°55'00"W 35°20'00"N

Oklahoma City Astronomy Club (OKCAC)
P.O. Box 21221
Oklahoma City, OK 73156
or :
2100 NE 52nd Street
Oklahoma City, OK 73111
Telephone: 405-424-5545
Telefax: 405-424-5106
Electronic Mail: wayne.wyrick@baremetl.com (Wayne Wyrick)
WWW: http://www.okcastroclub.com/
Founded: 1973
Membership: 100
Activities: public star parties
Periodicals: "Observer"
Awards: (1) "E. Ken Owen Award"
Coordinates: 098°18'06"W 35°25'00"N H527m (Canyon Camp)
City Reference Coordinates: 097°32'00"W 35°28'00"N

Oklahoma State University, Department of Physics, Astrophysics Group
Stillwater, OK 74078-3072
Telephone: 405-744-5796
          405-744-5785
Telefax: 405-744-6811
Electronic Mail: <userid>@okstate.edu
WWW: http://www.physics.okstate.edu/
Founded: 1890
Staff: 22
Activities: SNR * protoplanetary disks * nebulae
City Reference Coordinates: 097°04'00"W 36°07'00"N

Omniplex
● See now "Kirkpatrick Science and Air Space Museum at Omniplex"

University of Oklahoma, Department of Physics and Astronomy
400 West Brooks Street
Norman, OK 73019-0225
Telephone: 405-325-3961
Telefax: 405-325-7557
Electronic Mail: <userid>@mail.nhn.ou.edu
WWW: http://www.nhn.ou.edu/
Founded: 1908
Activities: SN * nuclear synthesis * galaxies * binary stars * outer solar system
Coordinates: 097°26'36"W 35°12'06"N H363m
City Reference Coordinates: 097°27'00"W 35°14'00"N

USA - Oregon

Eugene Astronomical Society (EAS)
c/o Barbara Shaw (President)
61 West 34th Avenue
Eugene, OR 97405
Telephone: 541-461-4665 (Fred Domineack, President)
541-683-1381 (Rick Kang, Vice President & Editor)
Electronic Mail: rkang@efn.org
WWW: http://www.eugeneastro.org/
Founded: 1970
Membership: 70
Activities: monthly meetings * instrumentation
Periodicals: "Io"
City Reference Coordinates: 123°05'00"W 44°02'00"N

Hardin Optical
1450 Oregon Avenue
Bandon, OR 97411
Telephone: 1-800-394-3307 (USA only)
Telefax: 541-347-8176
Electronic Mail: telescopeshop@hardin-optical.com
WWW: http://www.hardin-optical.com/
Activities: distributing telescopes and accessories
City Reference Coordinates: 124°25'00"W 43°07'00"N

Hinds Instruments Inc.
3175 NW Aloclek Drive
Hillsboro, OR 97124
Telephone: 503-690-2000
Telefax: 503-690-3000
Electronic Mail: info@hindspem.com
WWW: http://www.hindspem.com/
Founded: 1971
Staff: 30
Activities: manufacturing and marketing photoelastic modulators and birefringence measurement systems
Periodicals: (4) "Inside PEM"
City Reference Coordinates: 122°59'00"W 45°31'00"N

Lane Education Service District (ESD) Planetarium
2300 Leo Harris Parkway
Eugene, OR 97401
Telephone: 541-461-8227
Telefax: 541-687-6459
Electronic Mail: planetarium@lane.k12.or.us
WWW: http://www.laneplanetarium.org/
Founded: 1979
Staff: 3
Activities: school and public shows * observing
Coordinates: 123°13'00"W 44°07'00"N H129m
City Reference Coordinates: 123°13'00"W 44°07'00"N

Mount Hood Community College (MHCC), Planetarium
26000 SE Stark Street
Gresham, OR 97030
Telephone: 503-491-7297
WWW: http://www.starstuff.com/stars.htm
City Reference Coordinates: 122°26'00"W 45°30'00"N

Mount Hood Observatory Association (MHOA)
c/o Jack Breshears
2508 SE Clark Court
Troutdale, OR 97060
or :
10200 SE Orient Drive
Boring, OR 97009
Telephone: 503-663-9630
WWW: http://www.starstuff.com/stars.htm
http://www.starstuff.com/mhoa.htm
City Reference Coordinates: 122°23'00"W 45°26'00"N (Boring)

Murdock Sky Theater
• See "Oregon Museum of Science and Industry (OMSI), Murdock Sky Theater"

North West Astronomy Group (NWAG)
55371 McDonald Road
Vernonia, OR 97064
Telephone: 503-429-2430
Electronic Mail: marcc@easystreet.com
WWW: http://www.nwag.portland.or.us/
Observatories: 2 (Vernonia Peak, Hamill Observatory)
City Reference Coordinates: 123°11'00"W 45°52'00"N

Oregon Museum of Science and Industry (OMSI), Murdock Sky Theater
1945 SE Water Street
Portland, OR 97214-3354
Telephone: 503-797-4000
    503-797-4610 (Planetarium)
WWW: http://www.omsi.edu/visit/planetarium/
    http://www.omsi.edu/
Founded: 1968
City Reference Coordinates: 122°36'00"W 45°33'00"N

Oregon Star Party (OSP)
16016 SE Division
P.M.B. 307
Portland, OR 97236
Telephone: 503-306-2992
WWW: http://www.oregonstarparty.org/
Founded: 1988
City Reference Coordinates: 122°36'00"W 45°33'00"N

Rose City Astronomers (RCA)
c/o Oregon Museum of Science and Industry
1945 SE Water Street
Portland, OR 97214-3354
Telephone: 503-255-2016
Electronic Mail: rca@teleport.com
WWW: http://www.rca-omsi.org/rca
    http://www.omsi.edu/ (Oregon Museum of Science and Industry)
Founded: 1990
Membership: 300
Periodicals: (12) "Rosette Gazette"
City Reference Coordinates: 122°36'00"W 45°33'00"N

Southern Oregon Skywatchers
P.O. Box 4092
Medford, OR 97501
or :
880 Glendower Street
Ashland, OR 97520
Telephone: 541-482-0783
Electronic Mail: galaxygl@mind.net
WWW: http://www.starstuff.com/southor.htm
Founded: 1993
Activities: monthly meetings * lectures * observing
Periodicals: (12) "The Orbiter"
City Reference Coordinates: 071°07'00"W 42°25'00"N (Medford)
    122°42'00"W 42°12'00"N (Ashland)

Space: The Future Frontier (S:TFF)
P.O. Box 68158
Portland, OR 97268
Telephone: 503-723-8217
    503-267-8148
Electronic Mail: stff@teleport.com
    thanna@teleport.com (Thomas C. Hanna, Director)
WWW: http://www.teleport.com/~stff/
Founded: 1988
Membership: 40
Staff: 4
Activities: exhibits * lectures * consulting * school visits
City Reference Coordinates: 122°36'00"W 45°33'00"N

Association of Research Astronomers (ARA)
c/o Jason A. Cardelli
Department of Astronomy and Astrophysics
Villanova University
Villanova, PA 19085
WWW: http://www.phy.vill.edu/astro/faculty/ara/ara_home.htm
City Reference Coordinates: 075°21'00"W 40°02'00"N

Astro Cards
P.O. Box 35
Natrona Heights, PA 15065
Telephone: 724-295-4128
Electronic Mail: astrocards@aol.com
WWW: http://www.rahul.net/Astro-Mall/astro-cards.html
http://www.m31mall.com/astrocards/
Founded: 1975
Staff: 2
Activities: retailer of astronomy-related material
City Reference Coordinates: 079°44'00"W 40°38'00"N

Astronomical Society of Harrisburg (ASH)
c/o Donald Jones (Treasurer)
2320 Spring Road
Carlisle, PA 17013
or :
Edward Naylor Observatory
670 Observatory Drive
Lewisberry, PA 17339-9550
Telephone: 717-938-6041
Electronic Mail: astronomerted@aol.com
lthieble@epix.net
WWW: http://www.astrohbg.org/
Founded: 1955
Membership: 160
Periodicals: "Stardust"
Observatories: 1 (Edward Naylor Observatory)
City Reference Coordinates: 076°52'00"W 40°16'00"N (Harrisburg)
077°12'00"W 40°12'00"N (Carlisle)
076°52'00"W 40°08'00"N (Lewisberry)

Beaver Valley Astronomy Club (BVAC)
c/o Eric Vondra (President)
36 South Harrison Avenue
Suite 27
Pittsburgh, PA 15202
Telephone: 412-734-0653
Founded: 1975
Membership: 19
City Reference Coordinates: 080°00'00"W 40°26'00"N

Berks County Amateur Astronomical Society (BCAAS)
P.O. Box 6150
Wyomissing, PA 19610
Telephone: 610-921-0173
WWW: http://www.geocities.com/capecanaveral/hangar/3933/
http://www.geocities.com/bcaas_99
Founded: 1973
Membership: 100
Activities: meetings * observing * programs for school and scout groups
Periodicals: (6) "Pegasus"
City Reference Coordinates: 075°58'00"W 40°20'00"N

Brashear LP
615 Epsilon Drive
Pittsburgh, PA 15238
Telephone: 412-967-7700
Telefax: 412-967-7973
Electronic Mail: sales@brashearlp.com
WWW: http://www.brashearlp.com/
Founded: 1997
Activities: optics technology * electro-optical and electro-mechanical systems and instrumentation including precision telescopes
City Reference Coordinates: 080°00'00"W 40°26'00"N

Bucknell University, Department of Physics
Lewisburg, PA 17837
Telephone: 570-577-1207
Telefax: 570-577-3153
Electronic Mail: <userid>@bucknell.edu
WWW: http://www.eg.bucknell.edu/physics/
  http://www.eg.bucknell.edu/physics/astronomy/observatory/ (Observatory)
Staff: 11
Activities: comets * diffuse interstellar clouds * meteorites * numerical simulations * optical properties of small particles
Coordinates: 076°55'43"W 40°57'21"N
City Reference Coordinates: 076°55'00"W 40°58'00"N

Bucks-Mont Astronomical Association (BMAA)
c/o Ed Radomski
36 Far View Road
Chalfont, PA 18914
Telephone: 215-822-8312
Electronic Mail: bmaa@bmaa.freeyellow.com
WWW: http://bmaa.freeyellow.com/
Founded: 1985
Membership: 80
Activities: monthly meetings * star parties * Stella-Della-Valley weekends
Periodicals: (12) "Constellation"
City Reference Coordinates: 075°13'00"W 40°17'00"N

Buhl Jr. Planetarium and Observatory (Henry_)
• See "Henry Buhl Jr. Planetarium and Observatory"

Carnegie Mellon University (CMU), Department of Physics
5000 Forbes Avenue
Pittsburgh, PA 15213-3890
Telephone: 412-268-2740
Telefax: 412-681-0648
Electronic Mail: physics@andrew.cmu.edu
  <userid>@cmu.edu
WWW: http://info.phys.cmu.edu/
City Reference Coordinates: 080°00'00"W 40°26'00"N

Central Bucks Planetarium
c/o Central Bucks East High School
2804 Holicong Road
Doylestown, PA 18901-1400
Telephone: 215-794-7481
Telefax: 215-794-5446
Founded: 1969
Staff: 1
City Reference Coordinates: 075°08'00"W 40°18'00"N

Central Columbia School District Planetarium
4777 Old Berwick Road
Bloomsburg, PA 17815
Telephone: 717-784-6103 (Olin L. Shotwell)
Founded: 1972
Staff: 1
Activities: educational programs
City Reference Coordinates: 076°27'00"W 41°00'00"N

Central Dauphin School District Planetarium
4600 Locust Lane
Harrisburg, PA 17109
Telephone: 717-545-4703
Founded: 1970
Staff: 3
Activities: education
City Reference Coordinates: 076°52'00"W 40°16'00"N

Central Pennsylvania Observers (CPO)
P.O. Box 947
State College, PA 16801
Telephone: 814-339-7244
Electronic Mail: cpo@pop.psu.edu
WWW: http://www.cpo.homepad.com/
Founded: 1997

City Reference Coordinates: 077°52'00"W 40°48'00"N (Sate College)

Chesmont Astronomical Society (CAS)
c/o Karl Krasley
141 Azalea Circle
Limerick, PA 19468
Telephone: 610-495-0867
WWW: http://www.chesmontastro.org/
City Reference Coordinates: 075°32'00"W 41°14'00"N

Chester County Astronomical Society (CCAS)
c/o Peter LaFrance
413 Church Street
Avondale, PA 19311
Telephone: 610-268-2616
WWW: http://members.tripod.com/~ccas_2/ccas.html
Periodicals: "Observations"
City Reference Coordinates: 075°47'00"W 39°50'00"N

Contemporary Laboratory Experiences in Astronomy (CLEA)
c/o Gettysburg College
Gettysburg, PA 17325
Telephone: 717-337-6028
Telefax: 717-337-6666
Electronic Mail: clea@gettysburg.edu
WWW: http://www.gettysburg.edu/project/physics/clea/CLEAhome.html
Founded: 1992
Staff: 8
Activities: developing educational astronomy software
Periodicals: (4) "CLEA Newsletter"
City Reference Coordinates: 077°14'00"W 39°50'00"N

Delaware Valley Amateur Astronomers (DVAA)
P.O. Box 662
Kimberton, PA 19442-0662
Telephone: 610-933-0497
Electronic Mail: dvaaprez@voicenet.com (Marilyn Michalski, President)
WWW: http://www.dvaa.org/
Founded: 1976
Membership: 189
Activities: observing * star parties * instrumentation * education
Periodicals: (12) "The Delaware Valley Amateur Astronomer"
City Reference Coordinates: 075°34'00"W 40°08'00"N (Kimberton)
                           075°18'00"W 40°09'00"N (East Norriton)

Dickinson College, Department of Physics and Astronomy
Carlisle, PA 17013-2896
Telephone: 717-245-1413
Telefax: 717-245-1642
Electronic Mail: physics@dickinson.edu
                <userid>@dickinson.edu
WWW: http://physics.dickinson.edu/
Founded: 1773
Activities: undergraduate education in physics and astronomy * high-resolution laboratory and astronomical IR spectroscopy * archaeoastronomy * plasma physics * Paul Kanev Planetarium
Awards: (2) "Glover Award"; (1) "Priestley Award"
Coordinates: 077°11'44"W 40°12'11"N (Michael L. Britton Memorial Observatory)
City Reference Coordinates: 077°12'00"W 40°12'00"N

Drexel University, Department of Physics
3141 Chestnut Street
Philadelphia, PA 19104
or :
Disque Hall - Room 816
32nd and Chestnut Streets
Philadelphia, PA 19104
Telephone: 215-895-2708
Telefax: 215-895-5934
Electronic Mail: physics@einstein.drexel.edu
                <userid>@physics.drexel.edu
WWW: http://einstein.drexel.edu/
Founded: 1891 (University)
City Reference Coordinates: 075°07'00"W 39°57'00"N

Eastern College, Observatory and Planetarium
1300 Eagle Road
Saint Davids, PA 19087-3696
Telephone: 610-341-5800
610-341-5945 (David Bradstreet)
Electronic Mail: <userid>@eastern.edu
dbradstr@eastern.edu
WWW: http://www.eastern.edu/
Founded: 1972 (Planetarium) 1996 (Observatory)
Activities: education * observing (IAU Code 923)
City Reference Coordinates: 075°22'00"W 40°02'00"N

Edinboro University Planetarium (EUP)
Edinboro, PA 16444
Telephone: 814-732-2493
Telefax: 814-732-2422
Electronic Mail: dhurd@edinboro.edu (David Hurd, Director)
WWW: http://www.edinboro.edu/cwis/geosci/dhurd/planet.html
Founded: 1967
Staff: 4
Activities: education
Coordinates: 080°07'30"W 41°52'00"N H370m
City Reference Coordinates: 080°08'00"W 41°53'00"N

Erie County Mobile Observing Group (ECMOG)
c/o Beverly Whiting
5840 Georgetown Drive
Erie, PA 16509
Telephone: 814-866-6710
Electronic Mail: bwhiting@epix.net
WWW: http://www.ncinter.net/~alonmac/ecmog
City Reference Coordinates: 080°04'00"W 42°08'00"N

Europtik Ltd.
c/o W.D.M. Optical
P.O. Box 539
Dunmore, PA 18512
Telephone: 570-848-1794
Telefax: 570-842-0750
Electronic Mail: europtik@ptd.net
WWW: http://www.europtik.com/
Founded: 1986
Activities: distributing high-quality optical products
City Reference Coordinates: 075°38'00"W 41°25'00"N

Franklin and Marshall College, Department of Physics and Astronomy
P.O. Box 3003
Lancaster, PA 17604-3003
Telephone: 717-291-4136
717-291-4132
Telefax: 717-358-4474
Electronic Mail: <userid>@fandm.edu
WWW: http://www.fandm.edu/Departments/Astronomy/Astronomy.html
Founded: 1884
Staff: 3 (astronomers)
Activities: IR astronomy * star and planet formation * radio astronomy * pulsars * observational tests of general relativity
Coordinates: 076°20'00"W 40°03'09"N (J.R. Grundy Observatory)
City Reference Coordinates: 076°19'00"W 40°02'00"N

Franklin Institute Science Museum
222 North 20th Street
Philadelphia, PA 19103
Telephone: 215-448-1200
Telefax: 215-448-1117
WWW: http://www.fi.edu/tfi/
http://sln.fi.edu/tfi/info/fels.html (Fels Planetarium)
Founded: 1933
Staff: 4
Activities: science * astronomy * education * planetarium shows
Coordinates: 075°10'24"W 39°57'30"N H15m
City Reference Coordinates: 075°07'00"W 39°57'00"N

Gettysburg College, Observatory
300 North Washington Street
Gettysburg, PA 17325
Telephone: 717-337-6036 (Astro Hotline)
Telefax: 717-337-6666
WWW: http://www.gettysburg.edu/academics/physics/clea/obshome.html
http://www.gettysburg.edu/academics/physics/clea/ASTRO.html
http://www.gettysburg.edu/~s371624/hatter/ (Hatter Planetarium)
Founded: 1996
Coordinates: 077°14'26"W 39°50'25"N (IAU Code 842)
City Reference Coordinates: 077°14'00"W 39°50'00"N

Gordon and Breach Science Publishers
820 Town Center Drive
Langhorne, PA 19047
Telephone: 215-750-2642
Telefax: 215-750-6343
Electronic Mail: info@gbhap.com
WWW: http://www.gbhap.com/
Founded: 1961
Staff: 21
Periodicals: among many others: (4) "Earth Space Review (ESR)" (ISSN 1060-1848)
• Publisher
City Reference Coordinates: 074°55'00"W 40°10'00"N

Greater Hazleton Area Astronomical Society (GHAAS)
c/o Gary Honis
P.O. Box 1032
Conyngham, PA 18219
Telephone: 570-788-3446
Electronic Mail: ghonis@epix.net
WWW: http://users.nni.com/ghaas/
Founded: 1977
City Reference Coordinates: 076°04'00"W 40°59'00"N

Haverford College, Department of Physics and Astronomy
Haverford, PA 19041
Telephone: 610-896-1144
610-896-1145
610-896-1146
Telefax: 610-896-1224
Electronic Mail: <userid>@haverford.edu
WWW: http://www.haverford.edu/physics-astro/pahome.html
Founded: 1833
Activities: observational cosmology (especially microwave background) * radio astronomy * IR CCD observing * stellar pulsations
Coordinates: 075°18'12"W 40°00'42"N H116m (Strawbridge Observatory)
City Reference Coordinates: 075°18'00"W 40°01'00"N

Henry Buhl Jr. Planetarium and Observatory
c/o Carnegie Science Center
One Allegheny Avenue
Pittsburgh, PA 15212-5850
Telephone: 412-237-3397
Telefax: 412-237-3395
WWW: http://www.csc.clpgh.org/
http://www.csc.clpgh.org/exhibits/planet_skywatch.asp
Founded: 1939
Staff: 14
Activities: public education * planetarium shows * weather observations and forecast
Coordinates: 080°00'47"W 40°26'48"N H223m
City Reference Coordinates: 080°00'00"W 40°26'00"N

Institute for Scientific Information (ISI)
3501 Market Street
Philadelphia, PA 19104
Telephone: 215-386-0100
215-386-4394 (orders)
1-800-523-1850 (USA only)
1-800-336-4474 (USA only)
Telefax: 215-386-2911
215-386-6362
Electronic Mail: help@isinet.com

sales@isinet.com
WWW: http://www.isinet.com/
Founded: 1958
Activities: publishing citation and table-of-contents databases linking the research community to scientific, technical and medical peer-reviewed journals
Periodicals: (52) "Current Contents (Physical, Chemical and Earth Sciences)" (ISSN 0163-2574)
City Reference Coordinates: 075°07'00"W 39°57'00"N

Institute of Physics (IOP) Publishing Inc.
The Public Ledger Building
Suite 1035
150 South Independence Mall West
Philadelphia, PA 19106
Telephone: 215-627-0880
Telefax: 215-627-0879
Electronic Mail: info@ioppubusa.com
WWW: http://www.iop.org/
Founded: 1874 (IOP)
Staff: 8
• Publisher - see main entry in United Kingdom
City Reference Coordinates: 075°07'00"W 39°57'00"N

IOP Publishing Inc.
• See "Institute of Physics (IOP) Publishing Inc."

Keystone College, Astronomy Observatory
One College Green
Laplume, PA 18440-0200
Telephone: 570-945-5141 x3110 (Thomas G. Cupillari, Observatory Director)
Telefax: 570-945-7977
Electronic Mail: tcup@keystone.edu
Founded: 1973
Staff: 3
Activities: observing (students and public) * public lectures * astrometry * photometry * photography
Coordinates: 075°40'42"W 41°35'48"N H419m
City Reference Coordinates: 075°46'00"W 41°33'00"N

Keystone Oaks High School, Planetarium
c/o C.J. Rodkey
1000 Kelton Avenue
Pittsburgh, PA 15216
Telephone: 412-571-6042
Telefax: 412-571-6048
Electronic Mail: rodkey@kosd.org
WWW: http://www.kosd.org/
City Reference Coordinates: 080°00'00"W 40°26'00"N

Lackawanna Astronomical Society (LAS)
1112 Fairview Road
Clarks Summit, PA 18411
Electronic Mail: 74034.2640@compuserve.com
              sabiajohn@aol.com
WWW: http://members.aol.com/SabiaJohn/LAS.html
Founded: 1959
Membership: 100
Activities: observing * library * monthly meetings
Periodicals: (6) "The Ecliptic"
City Reference Coordinates: 075°42'00"W 41°30'00"N

Lafayette College, Department of Physics
124 Hugel Science Center
Easton, PA 18042-1782
Telephone: 610-330-5213
Telefax: 610-330-5714
Electronic Mail: physics@lafayette.edu
              <userid>@lafayette.edu
WWW: http://www.lafayette.edu/~physics/
Founded: 1915
Staff: 7
Activities: education * radioastronomy * X-ray astronomy
City Reference Coordinates: 075°12'00"W 41°42'00"N

Lehigh Valley Amateur Astronomical Society (LVAAS) Inc.
c/o George Maurer
Joe Grady Planetarium

WWW: http://www.astro.psu.edu/outreach/astroclb.html
Founded: 1973
Membership: 50
Activities: monthly meetings * observing
City Reference Coordinates: 077°52'00"W 40°48'00"N (Sate College)

Pennsylvania State University (PSU), Department of Astronomy and Astrophysics
525 Davey Laboratory
University Park, PA 16802
Telephone: 814-865-0418
Telefax: 814-863-3399
Electronic Mail: <userid>@astro.psu.edu
WWW: http://www.astro.psu.edu/
Founded: 1964
Staff: 22
Activities: education * research * ground-based telescope and satellite instrumentation development * planetarium
Periodicals: (1) "Annual Report"; "Reprints and Observatory Publications", "Views from Astronomy and Astrophysics"
Coordinates: 077°00'00"W 41°00'00"N H300m (Black Moshannon Observatory)
City Reference Coordinates: 077°52'00"W 40°48'00"N (Sate College)

Pennsylvania State University (PSU), Department of Statistics, Statistical Consulting Center for Astronomy (SCCA)
326 Classroom Building
University Park, PA 16802-2111
Telephone: 814-865-3631
Telefax: 814-863-7114
Electronic Mail: scca@stat.psu.edu
   <userid>@stat.psu.edu
WWW: http://www.stat.psu.edu/~mga/scca/
Founded: 1994
City Reference Coordinates: 077°52'00"W 40°48'00"N (Sate College)

Pocono Mountain Optics Inc.
104 NP 502 Plaza
Moscow, PA 18444
Telephone: 1-800-569-4323 (USA only)
   570-842-1500
Telefax: 570-842-8364
Electronic Mail: pocmtnop@ptdprolog.net
   pwo@poconoscopes.com
WWW: http://astronomy-mall.com/regular/products/pocono/pg1.htm
   http://www.pocmtnop.com/
City Reference Coordinates: 075°31'00"W 41°20'00"N

Questar Corp.
6204 Ingham Road
New Hope, PA 18938
Telephone: 1-800-247-9607 (USA only)
   215-862-5277
Telefax: 215-862-0512
Electronic Mail: questar@erols.com
WWW: http://www.questar-corp.com/
Founded: 1950
Staff: 20
Activities: manufacturing astronomy and birding telescopes, long-range surveillance optics, remote measurement systems and high-resolution stereo microscopes
City Reference Coordinates: 074°56'00"W 40°22'00"N

Reading Public Museum and Planetarium
500 Museum Road
Reading, PA 19611-1425
Telephone: 610-371-5850 x244
Telefax: 610-371-5632
Electronic Mail: readingplanetarium@usa.net
WWW: http://www.berks.net/museum/planet5.htm
   http://www.readingpublicmuseum.org/
Founded: 1969
Staff: 2
Activities: school and public shows * private parties * laser light shows
City Reference Coordinates: 075°56'00"W 40°20'00"N

Schuylkill Valley Astronomy Club
c/o Ryan Hannahoe
1056 Mahlon Drive
Leesport, PA 19533

Electronic Mail: hstinst@aol.com
WWW: http://svastronomyclub.homestead.com
City Reference Coordinates: 075°58'00"W 40°27'00"N

Sir Isaac Newton Astronomical Society (SINAS)
P.O. Box 591
Kane, PA 16735
Telephone: 814-837-8395
WWW: http://users.penn.com/~ksmith/sinas.html
Founded: 1979
Membership: 25
Activities: promoting astronomy * education
Periodicals: (4) "The Newtonian"
City Reference Coordinates: 078°49'00"W 41°40'00"N

Society for Industrial and Applied Mathematics (SIAM)
3600 University City Science Center
Philadelphia, PA 19104-2688
Telephone: 215-382-9800
Telefax: 215-386-7999
Electronic Mail: siam@siam.org
service@siam.org
pubs@siam.org
WWW: http://www.siam.org/
Founded: 1952
Membership: 9200
Staff: 50
Activities: publishing books and journals * conducting conferences
Periodicals: "SIAM Journal on Applied Mathematics", "SIAM Journal on Computing", "SIAM Journal on Control and Optimization", "SIAM Journal on Discrete Mathematics", "SIAM Journal on Mathematical Analysis", "SIAM Journal on Matrix Analysis and Applications", "SIAM Journal on Numerical Analysis", "SIAM Journal on Optimization", "SIAM Journal on Scientific Computing", "SIAM News", "SIAM Review", "Theory of Probability and its Applications"
Awards: "John von Neumann Lecture", "Richard C. DiPrima Prize", "George Polya Prize", "Theodore van Kármán Prize", "James H. Wilkinson Prize", "W.T. and Idalia Reid Prize", "SIAM Prize for Distinguished Service to the Profession", "SIAM Award in the Mathematical Contest in Modeling"; with the American Mathematical Society: "George David Birkhoff Prize" and "Norbert Wiener Prize"; with the Mathematical Programming Society: "George B. Dantzig Prize"
City Reference Coordinates: 075°07'00"W 39°57'00"N

Spitz Inc.
US Route One
P.O. Box 198
Chadds Ford, PA 19317-0198
Telephone: 610-459-5200
Telefax: 610-459-3830
Electronic Mail: spitz@spitzinc.com
WWW: http://www.spitzinc.com/
Founded: 1945
Staff: 70
Activities: manufacturing planetariums and domes
City Reference Coordinates: 075°35'00"W 39°52'00"N

Sproul Observatory
• See "Swarthmore College, Department of Physics and Astronomy, Sproul Observatory"

Starlight Astronomy Club
c/o Kristopher Kafka
Penn-Mont Academy
131 Holiday Hills Drive
Hollidaysburg, PA 16648
Electronic Mail: astroclub@alsenco.com
WWW: http://ehilbert.wao.net/Starlight/
City Reference Coordinates: 078°24'00"W 40°26'00"N

State Museum of Pennsylvania, Planetarium
P.O. Box 1026
Harrisburg, PA 17108-1026
Telephone: 717-783-9914
Telefax: 717-783-1073
Electronic Mail: jsmith@phmc.state.pa.us (Jeffrey L. Smith, Director)
Founded: 1965
Staff: 5
Activities: shows for school groups and public * exhibitions
City Reference Coordinates: 076°52'00"W 40°16'00"N

Susquehanna Valley Amateur Astronomers (SVAA)
c/o Bucknell University Observatory
Lewisburg, PA 17837
Telephone: 717-524-3767
Electronic Mail: aquila@csrlink.net (Al Francis)
WWW: http://www.geocities.com/CapeCanaveral/Hangar/2999/
Founded: 1990
City Reference Coordinates: 076°55'00"W 40°58'00"N

Swarthmore College, Department of Physics and Astronomy, Sproul Observatory
Swarthmore, PA 19081
Telephone: 610-328-8258 (Department)
610-328-8272 (Observatory)
Telefax: 610-328-7895 (Department)
Electronic Mail: <userid>@swarthmore.edu
WWW: http://laser.swarthmore.edu/
http://www.swarthmore.edu/
Founded: 1910 (College: 1864)
Staff: 2
Coordinates: 075°21'24"W 39°54'18"N H63m
City Reference Coordinates: 075°21'00"W 39°54'00"N

Taylor & Francis Inc.
325 Chestnut Street
Philadelphia, PA 19106
Telephone: 1-800-354-1420 (USA only)
Telefax: 215-625-8914
WWW: http://www.taylorandfrancis.com/
Founded: 1798
Periodicals: among others, "Journal of Modern Optics" (formerly Optica Acta) (ISSN 0950-0340)
• Publisher
City Reference Coordinates: 074°52'00"W 40°06'00"N

University of Pennsylvania, Department of Physics and Astronomy
David Rittenhouse Laboratory Building
209 South 33rd Street
Philadelphia, PA 19104-6394
Telephone: 215-898-8176
Telefax: 215-898-9336
Electronic Mail: <userid>@pennsas.upenn.edu
WWW: http://dept.physics.upenn.edu/
http://observatory.astro.upenn.edu/ (Flower and Cook Observatory)
Founded: 1892
Staff: 9
Activities: instrumental development * solar neutrinos * binary stars
Observatories: 1 (Flower and Cook Observatory, Malvern – IAU Code 791)
City Reference Coordinates: 075°07'00"W 39°57'00"N

University of Pittsburgh, Allegheny Observatory
P.O. Box 7658
Observatory Station
Pittsburgh, PA 15214
Telephone: 412-321-2400
Telefax: 412-321-0606
Electronic Mail: <userid>@vms.cis.pitt.edu
WWW: http://www.pitt.edu/~aobsvtry
Founded: 1860
Activities: astrometry * stellar positions * planet search
Periodicals: "Publications"
Coordinates: 080°01'21"W 40°28'58"N H370m (IAU Code 778)
City Reference Coordinates: 080°00'00"W 40°26'00"N

University of Pittsburgh, Department of Geology and Planetary Science
321 Engineering Hall
Pittsburgh, PA 15260
Telephone: 412-624-8780
Telefax: 412-624-3914
Electronic Mail: geology@pitt.edu
WWW: http://www.geology.pitt.edu/
Staff: 11
Activities: remote sensing of planetary surfaces * planetary geology * Antarctic meteorite recovery * thermoluminescence of meteorites
City Reference Coordinates: 080°00'00"W 40°26'00"N

University of Pittsburgh, Department of Physics and Astronomy
100 Allen Hall
3941 O'Hara Street
Pittsburgh, PA 15260
Telephone: 412-624-9000
Telefax: 412-624-9163
Electronic Mail: <userid>@pitt.edu
WWW: http://www.phyast.pitt.edu/
Founded: 1975
Staff: 37 (Faculty)
City Reference Coordinates: 080°00'00"W 40°26'00"N

Villanova Astronomical Society (VAS)
c/o Joseph Depasquale
Department of Astronomy and Astrophysics
Villanova University
Villanova, PA 19085
Telephone: 610-519-7485
Electronic Mail: vas@ast.vill.edu
WWW: http://www.phy.vill.edu/astro/vas/vas.htm
City Reference Coordinates: 075°21'00"W 40°02'00"N

Villanova University, Department of Astronomy and Astrophysics
Villanova, PA 19085
Telephone: 215-645-4820
Telefax: 215-645-7465
Electronic Mail: astronom@ucis.vill.edu
<userid>@ast.vill.edu
WWW: http://astro4.ast.vill.edu/
Founded: 1961
Activities: education * photometry of variable stars
Periodicals: "Observatory Contributions"
Coordinates: 075°20'30"W 40°02'24"N
City Reference Coordinates: 075°21'00"W 40°02'00"N

Widener University, Department of Physics
Kirkbride Hall
One University Place
Chester, PA 19013
Telephone: 215-499-4002
Telefax: 215-499-4059
Electronic Mail: <userid>@science.widener.edu
WWW: http://www.science.widener.edu/
Founded: 1821
Activities: experimental solid-state physics * meteorology * astronomy
City Reference Coordinates: 075°21'00"W 39°51'00"N

York County Astronomical Society (YCAS)
400 Mundis Race Road
York, PA 17402
Telephone: 717-840-7440
Electronic Mail: jlj276@aol (Jeri L. Jones, Asst. Secretary)
WWW: http://home1.gte.net/dmdewey/ycas.html
Founded: 1988
Membership: 52
Activities: Summer star party * public viewing sessions
Observatories: 1 (Rudy Park Observatory)
City Reference Coordinates: 076°44'00"W 39°58'00"N

USA - Rhode Island

Amateur Astronomical Society of Rhode Island
• See "Skyscrapers, Inc."

American Mathematical Society (AMS)
201 Charles Street
Providence, RI 02904-2294
Telephone: 1-800-321-4AMS (USA and Canada only)
　　　　　　401-455-4000
Telefax: 401-331-3842
Electronic Mail: cust-serv@ams.org
　　　　　　　　<userid>@ams.org
WWW: http://www.ams.org/
Founded: 1888
Membership: 28000
Staff: 250
Activities: publishing journals and book series in all areas of mathematics * sponsoring meetings, symposia, conferences and seminars * producing CD-ROMs * on-line databases
Periodicals: (12) "Notices of the American Mathematical Society" (ISSN 0002-9920)
Awards: with SIAM: "George David Birkhoff Prize" and "Norbert Wiener Prize"
City Reference Coordinates: 071°25'00"W 41°50'00"N

Brown University, Physics Department
Box 1843
Providence, RI 02912
or :
Barus-Holley Building
184 Hope Street
Providence, RI
Telephone: 401-863-2641
　　　　　　401-521-5680 (Ladd Observatory)
Telefax: 401-863-2024
Electronic Mail: <userid>@brownvm.brown.edu
WWW: http://www.physics.brown.edu/
　　　　http://www.physics.brown.edu/astro/
Founded: 1764 (Observatory: 1891)
Staff: 10
Activities: education * cosmic microwave background radiation * solar neutrinos * theory * cosmology * observing (Ladd Observatory)
City Reference Coordinates: 071°25'00"W 41°50'00"N

Cormack Planetarium
• See "Museum of Natural History and Cormack Planetarium"

Frosty Drew Observatory
Ninigret Park
P.O. Box 160
Charlestown, RI 02813
WWW: http://www.frostydrew.org/
City Reference Coordinates: 071°45'00"W 41°23'00"N

Museum of Natural History and Cormack Planetarium
Roger Williams Park
Providence, RI 02905
Telephone: 401-785-9450
　　　　　　401-785-9457
Telefax: 401-461-5146
Electronic Mail: museum@osfn.org
WWW: http://www.osfn.org/museum/
　　　　http://www.osfn.org/museum/planet.html (Cormack Planetarium)
Founded: 1952
Staff: 1
City Reference Coordinates: 071°25'00"W 41°50'00"N

Skyscrapers, Inc.
(Amateur Astronomical Society of Rhode Island)
47 Peep Toad Road
North Scituate, RI 02857-1328
Telephone: 401-934-0980
Electronic Mail: cormack_pl@ids.net
WWW: http://chandra.cis.brown.edu/astro/skyscrapers/
Founded: 1932

Membership: 120
Activities: meetings * observing
Observatories: 1 (Seagrave Memorial Obs. – IAU Code 814)
City Reference Coordinates: 071°35'00"W 41°50'00"N

Society for Amateur Scientists (SAS)
5600 Post Road
Suite 114-341
East Greenwich, RI 02818
Telephone: 401-823-7800
Telefax: 401-823-6800
Electronic Mail: info@sas.org
WWW: http://www.sas.org/
Founded: 1994
City Reference Coordinates: 071°27'00"W 41°40'00"N

Greenville, SC 29615-4229
Telephone: 864-281-1188
Telefax: 864-459-7034
Electronic Mail: <userid>@ropermountain.org
WWW: http://www.ropermountain.org/
Founded: 1985
Activities: T.C. Hooper Planetarium * Charles E. Daniel Observatory
City Reference Coordinates: 082°23'00"W 34°51'00"N

Settlemyre Planetarium
c/o Museum of York County
4621 Mount Gallant Road
Rock Hill, SC 29732-866
Telephone: 803-329-2121
803-981-9182
WWW: http://www.yorkcounty.org/planetarium/
City Reference Coordinates: 081°01'00"W 34°55'00"N

Stanback Planetarium
c/o South Carolina State University
P.O. Box 7636
Orangeburg, SC 29117
Electronic Mail: starman@scsu.edu
WWW: http://www.draco.scsu.edu/
Founded: 1979
City Reference Coordinates: 080°52'00"W 33°30'00"N

The Citadel, Department of Physics
Charleston, SC 29409-0270
Telephone: 843-953-5122
Telefax: 843-953-1382
Electronic Mail: <userid>@citadel.edu
WWW: http://www.citadel.edu/citadel/otherserv/phys/physics.html
http://www.citadel.edu/physics/
http://www.citadel.edu/physics/astra/index.html (ASTRA project)
Founded: 1842
Staff: 7
Activities: education * research
City Reference Coordinates: 079°57'00"W 32°48'00"N

University of South Carolina (USC), Department of Physics and Astronomy
Columbia, SC 29208
Telephone: 803-777-6466
Telefax: 803-777-3065
Electronic Mail: <userid>@psc.psc.scarolina.edu
WWW: http://astro.physics.sc.edu/
http://astro.physics.sc.edu/htmlpages/Melton/Melton.html (Melton Memorial Obs.)
Founded: 1928
Staff: 6
Activities: education * occultations * radio Sun
Coordinates: 081°01'36"W 33°59'51"N H97m (optical)
081°01'52"W 33°59'47"N H127m (radio)
City Reference Coordinates: 081°00'00"W 34°00'00"N

University of South Carolina Press
718 Devine Street
Columbia, SC 29208
Telephone: 1-800-768-2500
803-777-5243
Telefax: 1-800-868-0740
803-777-0160
Electronic Mail: <userid>@uscpress.scarolina.edu
WWW: http://www.sc.edu/uscpress/
Founded: 1944
• Publisher
City Reference Coordinates: 081°00'00"W 34°00'00"N

USA - South Dakota

Badlands Observatory
c/o Ron Dyvig
P.O. Box 37
Quinn, SD 57775
Telephone: 605-386-2105
Electronic Mail: rondyv@gwtc.net
WWW: http://www.sdsmt.edu/space/bo.htm.
Activities: private observatory (IAU Code 918)
City Reference Coordinates: 102°07'00"W 44°00'00"N

Black Hills Astronomical Society (BHAS)
c/o Steve Parker
824 6th Street
Rapid City, SD 57701-3612
Telephone: 605-348-4838
Electronic Mail: parkerrcsd@aol.com
WWW: http://www.sdsmt.edu/space/BHAS.htm
Observatories: 1 (Hidden Valley Observatory)
City Reference Coordinates: 103°14'00"W 44°05'00"N

Sioux Empire Astronomy Club
c/o Jim Morris
3705 North 9th Avenue
Sioux Falls, SD 57101
Electronic Mail: info@inskysd.org
WWW: http://www.inskysd.org/
Coordinates: 096°34'48"W 43°12'26"N
City Reference Coordinates: 096°44'00"W 43°32'00"N

USA - Tennessee

Akima Planetarium
c/o East Tennessee Discovery Center
P.O. Box 6204
Knoxville, TN 37914-0204
or :
516 North Beaman Street
Chilhowee Park
Knoxville, TN 37914
Telephone: 615-637-1121
WWW: http://funnelweb.utcc.utk.edu/~loganj/etdc/etdc.html
Founded: 1966
Staff: 7
Activities: school and public shows
Periodicals: "Newsletter"
City Reference Coordinates: 083°56'00"W 35°58'00"N

Anderson Planetarium (M.D._)
• See "M.D. Anderson Planetarium"

Arnold Engineering Development Center (AEDC)
Public Affairs Office
100 Kindel Drive
Suite B-213
Arnold AFB, TN 37389-2213
or :
676 2nd Street
Arnold AFB, TN 37389-4000
Telephone: 615-454-5599
WWW: http://www.arnold.af.mil/
Founded: 1951
City Reference Coordinates: 084°53'00"W 35°10'00"N (Cleveland)

Barnard Astronomical Society (BAS)
c/o F. Randolph Helms
838 Belvoir Hills Drive
Chattanooga, TN 37412-2016
Telephone: 423-629-6094
Electronic Mail: bas@chattanooga.net
WWW: http://www.chattanooga.net/bas/
Founded: 1923
Membership: 51
Activities: observing * photography * education
Periodicals: (12) "The Barnard Star"
City Reference Coordinates: 085°19'00"W 35°03'00"N

Barnard-Seyfert Astronomical Society (BSAS)
c/o A.M. Heiser
Arthur J. Dyer Observatory
1000 Oman Drive
Brentwood, TN 37027-4143
Telephone: 615-373-4897
WWW: http://www.bsasnashville.com/
Founded: 1928
Membership: 90
Periodicals: (12) "Newsletter"
Coordinates: 086°47'29"W 36°03'09"N H345m (Arthur J. Dyer Observatory)
City Reference Coordinates: 086°48'00"W 36°09'00"N (Nashville)

Bays Mountain Astronomy Club
c/o Bays Mountain Planetarium
853 Bays Mountain Park Road
Kingsport, TN 37660
Telephone: 615-229-9447
Electronic Mail: baysmtn@tricon.net
WWW: http://home.tricon.net/baysmtn/planetdept/astrclub.html
http://www.baysmountain.com/planetdept/astrclub.html
Founded: 1980
Membership: 60
Activities: meetings * observing * community awareness projects
Periodicals: (12) "The Celestial Observer"
Awards: (1) "Academic Achievement Award", "Service Award"

Coordinates: 082°36'57"W 36°30'03"N H558m (Bays Mountain Observatory)
City Reference Coordinates: 082°33'00"W 36°32'00"N

Bays Mountain Planetarium
853 Bays Mountain Park Road
Kingsport, TN 37660
Telephone: 423-229-9447
Electronic Mail: bmplanet@tricon.net
WWW: http://www.baysmountain.com/planetdept/planetarium.html
City Reference Coordinates: 082°33'00"W 36°32'00"N

Burke Observatory
• See "King College, Burke Observatory"

Cleveland State Community College (ClSCC), Department of Physics and Astronomy
3535 Adkisson Drive
P.O. Box 3570
Cleveland, TN 37320-3570
Telephone: 423-472-7141
Telefax: 423-478-6255
Electronic Mail: <userid>@clscc.cc.tn.us
WWW: http://www.clscc.cc.tn.us/
Founded: 1967
Staff: 200
Activities: education
Coordinates: 084°55'25"W 35°11'41"N H300m
City Reference Coordinates: 084°53'00"W 35°10'00"N

Comstock Inc.
1005 Alvin Weinberg Drive
Oak Ridge, TN 37830
Telephone: 865-483-7690
Telefax: 865-481-3884
Electronic Mail: salesinfo@comstockinc.com
WWW: http://www.comstockinc.com/
Founded: 1979
Staff: 15
Activities: manufacturing electrostatic energy analyzers, time-of-flight mass spectrometers, etc.
City Reference Coordinates: 084°16'00"W 36°01'00"N

Craigmont High School Planetarium
3333 Covington Pike
Memphis, TN 38128-3902
Telephone: 901-385-4319
Telefax: 901-385-4340
Electronic Mail: starman@netten.net
WWW: http://tqd.advanced.org/3461/
       http://www.people.memphis.edu/~dthomas/planet.html
       http://www.craigmont.org/skylights.html (Skylights Online Newsletter)
Founded: 1974
Staff: 2
Periodicals: (6) "Skylights", "Twinkles" (circ.: 1000)
City Reference Coordinates: 090°03'00"W 35°08'00"N

Cumberland Astronomical Society (CAS)
c/o Tom Golden
P.O. Box 1733
Gallatin, TN 37066
Telephone: 615-264-2504
Electronic Mail: tgolden1@juno.com
WWW: http://firefly.vscc.cc.tn.us/cas
       http://mistal2.tripod.com/CAS/
City Reference Coordinates: 086°27'00"W 36°24'00"N

CyberSphere Planetarium
c/o The Renaissance Center
719 East College Street
Dickson, TN 37055
Telephone: 615-446-4450
Telefax: 615-446-9168
WWW: http://www.rcenter.org/
City Reference Coordinates: 087°22'00"W 36°03'00"N

East Tennessee State University (ETSU), Department of Physics
Box 70652

Johnson City, TN 37614-0002
Telephone: 423-439-4231
Telefax: 423-439-6905
Electronic Mail: <userid>@etsu.edu
   <userid>@physics.etsu-tn.edu
WWW: http://www.etsu-tn.edu/physics/
   http://www.etsu-tn.edu/physics/etsuobs/obslocal.htm (Observatory)
   http://www.etsu-tn.edu/physics/plntrm/planetarium.html (Planetarium)
Founded: 1911
Staff: 6
Activities: education * stellar astronomy * planetarium
City Reference Coordinates: 082°21'00"W 36°19'00"N

International Amateur-Professional Photoelectric Photometry (IAPPP)
c/o Douglas S. Hall
Arthur J. Dyer Observatory
Vanderbilt University
Nashville, TN 37235
Telephone: 615-373-4897
Telefax: 615-343-7263
Electronic Mail: hall@astro.dyer.vanderbilt.edu
WWW: http://www.iappp.vanderbilt.edu/
Founded: 1980
Membership: 800
Activities: assisting small observatories in equipping themselves for photoelectric observing
Periodicals: (4) "IAPPP Communications" (ISSN 0886-6961)
Awards: (1) "IAPPP Richard D. Lines Special Award"
City Reference Coordinates: 086°48'00"W 36°09'00"N

King College, Burke Observatory
Bristol, TN 37620
Telephone: 615-652-4833
   615-652-6028
Telefax: 615-968-4456
WWW: http://www.king.edu/kingnews/120297c.htm
Founded: 1951
Staff: 2
Activities: observing * research * public education
City Reference Coordinates: 082°11'00"W 36°36'00"N

Knoxville Observers (KO)
c/o John Sparks
1708 Gillepsie Avenue
Knoxville, TN 37914
Electronic Mail: astrofrk@hotmail.com
WWW: http://www.knoxvilleobservers.org/
City Reference Coordinates: 083°56'00"W 35°58'00"N

Linda G. Sharpe Planetarium
• See "Memphis Pink Palace Museum and Linda G. Sharpe Planetarium"

M.D. Anderson Planetarium
Lambuth University
705 Lambuth Boulevard
Jackson, TN 38301-5296
Telephone: 901-425-3283
WWW: http://erc.jscc.cc.tn.us/jackson_other.html
City Reference Coordinates: 088°50'00"W 35°37'00"N

Memphis Astronomical Society (MAS)
P.O. Box 11301
Memphis, TN 38111
or :
1229 Pallwood Road
Memphis, TN 38122
Telephone: 901-682-2003
Electronic Mail: info@memphisastro.org
   rich@magibox.net (Ric Honey)
WWW: http://www.memphisastro.org/
Founded: 1953
Membership: 12
Activities: observing * education
Periodicals: (12) "Meteor Write"
City Reference Coordinates: 090°03'00"W 35°08'00"N

Memphis Pink Palace Museum and Linda G. Sharpe Planetarium
3050 Central Avenue
Memphis, TN 38111-3399
Telephone: 901-320-6320
Telefax: 901-320-6391
WWW: http://www.memphismuseums.org/
http://www.memphismuseums.org/planet.htm
http://www.memphisguide.com/PPalace.html
Founded: 1930 (Planetarium: 1949)
Staff: 100 (Museum) 10 (Planetarium)
City Reference Coordinates: 090°03'00"W 35°08'00"N

Middle Tennessee Astronomical Society (MTAS)
c/o Ron Landrum
614 Country Estates Drive
Winchester, TN 37398-4607
or :
c/o Donald W. Male
1305 Sycamore Street
Manchester, TN 37355
Telephone: 931-967-5774
615-728-7321
Electronic Mail: rlandrum@edge.net
Founded: 1966
Membership: 15
City Reference Coordinates: 086°09'00"W 35°10'00"N (Winchester)
086°05'00"W 35°29'00"N (Manchester)

Middle Tennessee State University (MTSU), Department of Physics and Astronomy
MTSU Box 71
Murfreesboro, TN 37132
or :
Wiser-Patten Building
Room 219
1301 East Main Street
Murfreesboro, TN 37132
Telephone: 615-898-2130
Telefax: 615-898-5303
Electronic Mail: dept@physics.mtsu.edu
<userid>@physics.mtsu.edu
WWW: http://physics.mtsu.edu/
Founded: 1995
Staff: 11
Activities: binary stars * cataclysmic variable stars * clusters * diffractive optics * optoelectronics * scattering and surface physics * solid-state physics
Coordinates: 086°23'00"W 35°51'00"N H187m (Wikswo Observatory)
City Reference Coordinates: 086°23'00"W 35°51'00"N

Oak Ridge Isochronous Observation Network (ORION)
P.O. Box 7114
Oak Ridge, TN 37831-7114
WWW: http://www.korrnet.org/wattec/orion/orion2.html
Founded: 1974
Membership: 35
Activities: research * radioastronomy * public lectures * balloon research * amateur radio project support
Coordinates: 083°57'00"W 36°00'00"N
City Reference Coordinates: 084°16'00"W 36°01'00"N

Rhodes College, Department of Physics
2000 North Parkway
Memphis, TN 38112
Telephone: 901-843-3915
Telefax: 901-843-3117
Electronic Mail: <userid>@rhodes.edu
WWW: http://www.physics.rhodes.edu/
Founded: 1848
Staff: 8
Activities: education * research
Coordinates: 089°59'00"W 35°09'00"N
City Reference Coordinates: 090°03'00"W 35°08'00"N

Roane State Community College (RSCC), Tamke-Allen Observatory
c/o David E. Fields
276 Patton Lane
Harriman, TN 37748-5011
Telephone: 865-882-4533

Electronic Mail: fieldsde@aol.com
WWW: http://www.rscc.cc.tn.us/obs/
Founded: 1988
Activities: observing * education
City Reference Coordinates: 084°32'00"W 35°57'00"N

Sharpe Planetarium (Linda G._)
• See "Memphis Pink Palace Museum and Linda G. Sharpe Planetarium"

Smoky Mountain Astronomical Society (SMAS)
P.O. Box 53265
Knoxville, TN 37950
or :
c/o East Tennessee Discovery Center
P.O. Box 6204
Knoxville, TN 37914-0204
or :
516 North Beaman Street
Chilhowee Park
Knoxville, TN 37914
Telephone: 615-637-1121
Electronic Mail: ngc457@aol.com (Shawn Grant)
WWW: http://www.smokymtnastro.org/
Founded: 1977
Membership: 40
Activities: observing * public education
Periodicals: (12) "Newsletter"
City Reference Coordinates: 083°56'00"W 35°58'00"N

Society of Low-energy Observers (SLO)
c/o William J. Cupo Jr.
3260 Spottswood Avenue
Memphis, TN 38111
or :
4277 Park Forest Drive
Memphis, TN 38141-7126
Telephone: 901-366-4201
Founded: 1989
Membership: 31
Activities: monthly observing * special event sessions
Periodicals: "The Refractor"
Coordinates: 090°00'00"W 35°00'00"N (Big Hill Pond State Park)
City Reference Coordinates: 090°03'00"W 35°08'00"N

Southern Adventist University, Department of Physics
College Drive
P.O. Box 370
Collegedale, TN 37315-0370
Telephone: 615-238-2869
Telefax: 615-238-3546
         615-396-3001
Electronic Mail: <userid>@southern.edu
WWW: http://www.physics.southern.edu/
Founded: 1951
Staff: 4
Activities: periodic systems of molecules * periodic law * prediction of data for small molecules of astrophysical interest using trends in data
City Reference Coordinates: 085°03'00"W 35°04'00"N

Sudekum Planetarium
c/o Cumberland Science Museum
800 Fort Negley Boulevard
Nashville, TN 37203
Telephone: 615-401-5077 (direct)
         615-862-5160 (general information)
         615-401-5092 (Astroline/astronomy information)
Telefax: 615-401-5086
Electronic Mail: planetarium@csmisfun.com
         mccallk@ten-nash.ten.k12.tn.us (Kris McCall, Director)
WWW: http://www.csmisfun.com/
         http://nashville.citysearch.com/E/V/NASTN/0002/59/32/6.html
         http://www.sudekumplanetarium.com/
Founded: 1952 (Museum: 1944)
Staff: 5
Activities: school and public programs * public observing * camps * astronomy information resource * portable planetarium * producing planetarium shows for sale

Periodicals: (12) "The Current", "Starcharts"
City Reference Coordinates: 086°48'00"W 36°09'00"N

Tamke-Allen Observatory
• See "Roane State Community College (RSCC), Tamke-Allen Observatory"

Tennessee State University (TSU), Center of Excellence in Information Systems
330 Tenth Avenue North
Box 139
Nashville, TN 37203-3401
Telephone: 615-963-7012
Telefax: 615-963-7027
Electronic Mail: <userid>@coe.tsuniv.edu
WWW: http://coe.tsuniv.edu/
Founded: 1986
Staff: 25
Activities: research * automatic telescopes * automated astronomy * advanced control systems * systems identification * applied mathematics * management information systems
City Reference Coordinates: 086°48'00"W 36°09'00"N

University of Memphis, Department of Physics
Campus Box 526670
Memphis, TN 38152-6670
Telephone: 901-678-2410
Telefax: 901-678-4733
Electronic Mail: <userid>@cc.memphis.edu
WWW: http://www.people.memphis.edu/~physics/
Founded: 1966
City Reference Coordinates: 090°03'00"W 35°08'00"N

University of Tennessee at Chattanooga (UTC), Department of Physics, Geology and Astronomy
615 McCallie Avenue
Chattanooga, TN 37403-2598
Telephone: 615-755-4546
Electronic Mail: <userid>@utc.edu
WWW: http://www.utc.edu/physics/
   http://www.utc.edu/~jonesobs/ (Clarence T. Jones Observatory)
Founded: 1940
Staff: 2
Activities: education * slide shows * videotapes * observing (planets, nearby galaxies) * Clarence T. Jones Planetarium
Awards: (1) "Clarence T. Jones Prize for Astronomy", "Karl and Harriet Hujer Award"
Coordinates: 085°18'00"W 35°02'44"N (Clarence T. Jones Observatory)
City Reference Coordinates: 085°19'00"W 35°03'00"N

University of Tennessee at Knoxville (UTK), Department of Physics and Astronomy
401 Nielsen Physics Building
Knoxville, TN 37996-1200
Telephone: 865-974-3342
Telefax: 865-974-7843
Electronic Mail: <userid>@utk.edu
   physics@utk.edu
WWW: http://www.phys.utk.edu/
City Reference Coordinates: 083°56'00"W 35°58'00"N

University of Tennessee at Tullahoma (UTT), Space Institute (UTSI)
B.H. Goethert Parkway
Tullahoma, TN 37388-9700
Telephone: 1-888-822-UTSI
   615-393-3864
Telefax: 615-393-7346
Electronic Mail: <userid>@utsi.edu
WWW: http://www.utsi.edu/
Founded: 1964
Observatories: 1 (South Point Observatory)
City Reference Coordinates: 086°11'00"W 35°22'00"N

University of Tennessee at Tullahoma (UTT), Space Institute (UTSI), Astronomy Club
B.H. Goethert Parkway
Tullahoma, TN 37388
Telephone: 615-393-3864
Telefax: 615-393-7346
Electronic Mail: cvandeko@utsi.edu (Christel van de Kolk)
Membership: 15
City Reference Coordinates: 086°11'00"W 35°22'00"N

University of Tennessee Press
293 Communications Building
Knoxville, TN 37996-0325
Electronic Mail: utpress2@utk.edu
WWW: http://gopher.lib.utk.edu:70/1/UTKgophers/UT-PRESS
• Publisher
City Reference Coordinates: 083°56'00"W 35°58'00"N

Vanderbilt University, Department of Physics and Astronomy
P.O. Box 1803
Station B
Nashville, TN 37235
or :
c/o Arthur J. Dyer Observatory
1000 Oman Drive
Brentwood, TN 37027
Telephone: 615-322-2828
         615-373-4897 (Observatory)
Telefax: 615-343-7263
Electronic Mail: hallxxds@ctrvax.vanderbilt.edu
WWW: http://comped1.cas.vanderbilt.edu/
     http://www.dyer.vanderbilt.edu/
Founded: 1953
Staff: 8
Activities: photoelectric photometry * objective prism and wide-angle photography * astronomical instrumentation * automatic telescopes * binary stars * variable stars * associations * globular clusters
Coordinates: 086°47'29"W 36°03'09"N H345m (Arthur J. Dyer Observatory)
City Reference Coordinates: 086°48'00"W 36°09'00"N (Nashville)

Vanderbilt University Press
VU Station B 351813
Nashville, TN 37235-1813
or :
112 21st Avenue South
Suite 201
Nashville, TN 37203
Telephone: 615-322-3585
Telefax: 615-343-8823
Electronic Mail: vupress@vanderbilt.edu
WWW: http://www.vanderbilt.edu/vupress
Founded: 1940
• Publisher
City Reference Coordinates: 086°48'00"W 36°09'00"N

USA - Texas

American Association of Amateur Astronomers (AAAA)
P.O. Box 7981
Dallas, TX 75209-0981
Electronic Mail: aaaa@corvus.com
WWW: http://www.corvus.com/
Founded: 1996
City Reference Coordinates: 096°48'00"W 32°47'00"N

Angelo State University (ASU), Planetarium
Vincent Science Building
San Angelo, TX 76909
Telephone: 915-942-2136
Telefax: 915-942-2188
Electronic Mail: mark.sonntag@angelo.edu (Mark S. Sonntag, Director)
              <userid>@angelo.edu
WWW: http://quark.angelo.edu/~msonntag/asuastronomy/planetarium.htm
      http://www.angelo.edu/asu-map/NPS.HTM
      http://www.angelo.edu/dept/phys/astro/planet.htm
Founded: 1985
Staff: 1
Activities: education * public and school programs
City Reference Coordinates: 100°26'00"W 31°28'00"N

Association of Amateur Astronomers (AAA)
c/o Department of Physics
Texas A & M University
College Station, TX 77843-4242
Telephone: 409-845-0536
Electronic Mail: astro@tamu.edu
WWW: http://www.isc.tamu.edu/~astro/aaa.html
Activities: monthly meetings
City Reference Coordinates: 096°21'00"W 30°37'00"N

Astronomical Society of East Texas (ASET)
c/o Hudnall Planetarium
Tyler Junior College
P.O. Box 9020
Tyler, TX 75711-9020
Telephone: 903-510-2200
Electronic Mail: kmears@tyler.net (Helen R. Mears, Secretary)
WWW: http://aset.tamu.edu/
Founded: 1982
Membership: 46
Activities: public star parties * monthly meetings * service projects * dark-sky observing
Periodicals: "Polaris"
City Reference Coordinates: 095°18'00"W 32°22'00"N

Austin Astronomical Society (AAS)
P.O. Box 12831
Austin, TX 78711-2831
Telephone: 512-252-2966
Electronic Mail: aas@kirchhof.com
WWW: http://www.austinastro.org/
Founded: 1969
Periodicals: (12) "Sidereal Times"
City Reference Coordinates: 097°45'00"W 30°16'00"N

Austin State University (Stephen F._)
• See "Stephen F. Austin State University"

Baker Planetarium (Burke_)
• See "Burke Baker Planetarium"

Big Bend Astronomical Society (BBAS)
c/o James T. Walker (Secretary)
HC 65 Box 14
Alpine, TX 79830
Electronic Mail: jwalker@brooksdata.net
WWW: http://www.brooksdata.net/personal/bbastro/
Membership: 51

Founded: 1970
Staff: 1
Activities: education
City Reference Coordinates: 096°48'00"W 32°47'00"N

Hill Country Astronomers (HCA)
c/o Russell Doescher
Physics Department
Southwest Texas State University
NSB #280
601 University Drive
San Marcos, TX 78666
Telephone: 512-245-8373
Electronic Mail: rdt@swt.edu
WWW: http://www.lbjsc.swt.edu/hca/
City Reference Coordinates: 097°57'00"W 29°53'00"N

Hobby-Eberly Telescope (HET)
c/o McDonald Observatory
HC75 Box 1337-MCD
Fort Davis, TX 79734-5020
Telephone: 915-426-3664
Telefax: 915-426-3636
WWW: http://www.as.utexas.edu/mcdonald/het/het.html
Founded: 1997
Staff: 12
Activities: research * education * public outreach
City Reference Coordinates: 103°54'00"W 30°35'00"N

Houston Astronomical Society (HAS)
P.O. Box 20332
Houston, TX 77225-0332
Telephone: 281-568-9340
Electronic Mail: goldberg@sccsi.com (Steve Goldberg)
WWW: http://www.astronomyhouston.org/
Founded: 1955
Membership: 250
Periodicals: (12) "Guidestar"
City Reference Coordinates: 095°22'00"W 29°46'00"N

Houston State University (SHSU) (Sam_)
• See "Sam Houston State University (SHSU)"

Hudnall Planetarium
c/o Tyler Junior College
P.O. Box 9020
Tyler, TX 75711
Telephone: 903-510-2312
          903-510-2662
Electronic Mail: info@tjc.tyler.cc.tx.us
WWW: http://www.tyler.cc.tx.us/planet/planet.htm
Founded: 1963
Staff: 1
Activities: shows * education * observing * exhibits * library
City Reference Coordinates: 095°18'00"W 32°21'00"N

Johnson Space Center (JSC) (Lyndon B._)
• See "National Aeronautics and Space Administration (NASA), Lyndon B. Johnson Space Center (JSC)"

Johnson Space Center Astronomical Society (JSCAS)
c/o Ken Lester
609 Wicklow Street
Deer Park, Texas 77536
Telephone: 281-479-1102
Electronic Mail: jscas@ghg.net
             starscan@swbell.net
WWW: http://www.ghg.net/cbr/jscas/
Founded: 1968
Membership: 250
Periodicals: (12) "STARSCAN"
City Reference Coordinates: 095°22'00"W 29°46'00"N (Houston)
                       095°08'00"W 29°42'00"N (Deer Park)

Lake Whitney Astronomy Association
7005 Southampton

North Richard Hills, TX 76180-4433
Electronic Mail: distglow@hotmail.com
WWW: http://www.whitney-astro.com/
City Reference Coordinates: 097°14'00"W 32°48'00"N

Land, Sea and Sky Inc.
3110 South Shepherd
Houston, TX 77098
Telephone: 713-529-3551
Telefax: 713-529-3108
WWW: http://www.lsstnr.com/
Founded: 1940
• Formerly "Texas Nautical Repair Co."
City Reference Coordinates: 095°22'00"W 29°46'00"N

Lunar and Planetary Institute (LPI)
• See "Universities Space Research Association (USRA), Lunar and Planetary Institute (LPI)"

Lyndon B. Johnson Space Center (JSC)
• See "National Aeronautics and Space Administration (NASA), Lyndon B. Johnson Space Center (JSC)"

Marian Blakemore Planetarium
c/o Museum of the Southwest
1705 West Missouri
Midland, TX 79701
Telephone: 915-683-2882
Telefax: 915-570-7770
WWW: http://www.museumsw.org/planet.html
Founded: 1972
City Reference Coordinates: 102°05'00"W 32°00'00"N

Marsh Rice University (William_)
• See "William Marsh Rice University"

McDonald Observatory
• See "University of Texas at Austin, McDonald Observatory"

Museum of Texas Tech University
4th and Indiana Avenues
P.O. Box 43191
Lubbock, TX 79409-3191
Telephone: 806-742-2432
WWW: http://www.ttu.edu/~museum/
http://www.ttu.edu/~museum/pltmSum01.html (Moody Planetarium)
Founded: 1976
Staff: 2
Activities: exhibits * public shows * school programs * observing
City Reference Coordinates: 101°51'00"W 33°35'00"N

National Aeronautics and Space Administration (NASA), Lyndon B. Johnson Space Center (JSC), Public Affairs Office
2101 NASA Road
Houston, TX 77058
Telephone: 713-483-8694
Telefax: 713-483-4876
Electronic Mail: <userid>@jsc.nasa.gov
WWW: http://www.jsc.nasa.gov/pao/
Founded: 1962
Staff: 42
Activities: media services * public information education * exhibits
Periodicals: "NASA Facts"
City Reference Coordinates: 095°22'00"W 29°46'00"N

Network Cybernetics Corp. (NCC)
5353 Alpha Road
Suite 205
Dallas, TX 75240
Telephone: 972-404-0248
Telefax: 972-404-0269
Electronic Mail: info@ncc.com
WWW: http://www.ncc.com/
Founded: 1992
Staff: 5
Activities: publishing CD-ROMs, including with astronomy-related ones
City Reference Coordinates: 096°48'00"W 32°47'00"N

Noble Planetarium
c/o Fort Worth Museum of Science and History
1501 Montgomery Street
Fort Worth, TX 76107
Telephone: 817-255-9300
    1-888-255-9300 (USA only)
WWW: http://www.fwmuseum.org/noble.html
City Reference Coordinates: 097°20'00"W 32°45'00"N

North Houston Astronomy Club (NHAC)
c/o Paul Downing
4522 Natural Bridge Drive
Kingwood, TX 77345
Telephone: 281-360-6562
Electronic Mail: paul.downing@bakerhughes.com
WWW: http://www.astronomyclub.org/
City Reference Coordinates: 095°18'00"W 30°00'00"N

North Texas Skywatch
c/o Michael Hibbs
113 Valley View
Azle, TX 76020
Telephone: 817-238-9883
Electronic Mail: mikehibbs@aol.com
WWW: http://members.aol.com/Mikehibbs/Skywatch/skywatch.html
City Reference Coordinates: 097°33'00"W 32°54'00"N

Rice University (William Marsh_)
• See "William Marsh Rice University"

Richard King High School, Planetarium
5225 Gollihar
Corpus Christi, TX 78412
Telephone: 512-994-6917
Telefax: 512-994-6918
Electronic Mail: wollbob@aol.com
WWW: http://members.aol.com/aggiebill/king/
Founded: 1967
City Reference Coordinates: 097°26'00"W 27°47'00"N

Richard R. Russell Planetarium
2501 Memorial
Mesquite, TX 75149
Telephone: 972-882-7750
Telefax: 972-882-7753
Electronic Mail: jrusk@tenet.edu (Director)
    jrusk@mesquiteisd.org (Director)
WWW: http://users.why.net/jrusk/
    http://www.why.net/jrusk/
    http://www.mesquite.isd.tenet.edu/planet/
Founded: 1977
City Reference Coordinates: 096°36'00"W 32°46'00"N

Richardson Independent School District (RISD), Planetarium
9465 Whitehurst Drive
Dallas, TX 75243-7567
Telephone: 214-503-2490
Telefax: 214-503-2493
Electronic Mail: jim.mcconnell@risd.org (Jim McConnell, Planetarium Director)
Founded: 1973
Staff: 3
Activities: public education
City Reference Coordinates: 096°48'00"W 32°47'00"N

Richland College, Planetarium
12800 Abrams Road
Dallas, TX 75243-2199
Telephone: 972-238-6013
    972-238-6001
WWW: http://www.rlc.dcccd.edu/ce/planet.htm
City Reference Coordinates: 096°48'00"W 32°47'00"N

Russell Planetarium (Richard R._)
• See "Richard R. Russell Planetarium"

Sam Houston State University (SHSU), Department of Physics
Huntsville, TX 77341
Telephone: 936-294-1601
Telefax: 936-294-1585
Electronic Mail: <userid>@shsu.edu
WWW: http://158.135.202.22/physics/
http://158.135.202.22/physics/astro.html
Activities: education * planetarium * observing
City Reference Coordinates: 095°33'00"W 30°43'00"N

San Angelo Amateur Astronomy Association (SAAAA)
P.O. Box 60391
San Angelo, TX 76906
WWW: http://members.aol.com/warhawk57/saaaa.html
http://www.saaaa.org/
Founded: 1962
Membership: 30
City Reference Coordinates: 100°26'00"W 31°28'00"N

San Antonio Astronomical Association (SAAA)
P.O. Box 701261
San Antonio, TX 78270-1261
Telephone: 210-497-5180 (Bob Gent, Membership Officer)
WWW: http://www.sanantonioastronomy.org/
Founded: 1974
Membership: 225
Observatories: 1 (South Texas Astro Retreat)
City Reference Coordinates: 098°31'00"W 29°28'00"N

San Antonio Space Society (SASS)
c/o Joe and Carol Redfield
609 Ridge View Drive
San Antonio, TX 78253
Telephone: 210-679-7625
Telefax: 210-679-6764
Electronic Mail: credfield@stmarytx.edu
Founded: 1987
Membership: 50
Activities: monthly meetings * education on space development
City Reference Coordinates: 098°31'00"W 29°28'00"N

Science Place
P.O. Box 151469
Dallas, TX 75315-1469
or :
1620 First Avenue
Dallas, TX 75210
Telephone: 214-428-5555 x392
Telefax: 214-428-3653
Electronic Mail: wilgus.burton@chrysalis.org
wilgusb@aol.com
WWW: http://www.scienceplace.org/
http://www.scienceplace.org/PL/PLindex.html (Planetarium)
Founded: 1956
Staff: 14
Activities: shows * star parties * special events
Periodicals: (4) "The Science Place Pages"
City Reference Coordinates: 096°48'00"W 32°47'00"N

Scobee Planetarium
c/o San Antonio College
1300 San Pedro
San Antonio, TX 78212
Telephone: 210-733-2910
Electronic Mail: bsnow@accd.edu (Bryan A. Snow, Planetarium Coordinator)
WWW: http://www.accd.edu/sac/ce/scobee/
Founded: 1961
City Reference Coordinates: 098°31'00"W 29°28'00"N

South Plains Astronomy Club (SPAC)
c/o Wayne Lewis
4803 76th Street
Lubbock, TX 79424-2148
Telephone: 806-742-2584 (office)
806-794-8766 (home)

Telefax: 806-742-1112
Electronic Mail: wlewis@math.ttu.edu
WWW: http://www.math.ttu.edu/~wlewis/astro.html
http://www.thespac.org/
Founded: 1960
Membership: 35
Activities: monthly meetings * observing
Periodicals: "Syzygy"
City Reference Coordinates: 101°51'00"W 33°35'00"N

Southwest Association of Planetariums (SWAP)
c/o Donna C. Pierce
Highland Park ISD Planetarium
4220 Emerson
Dallas, TX 75205
Telephone: 214-523-1836
Telefax: 214-522-4515
WWW: http://www.tyler.cc.tx.us/SWAP/
Founded: 1964
Membership: 50
Activities: conferences * workshops
Periodicals: (4) "Newsletter"
Awards: "H. Rich Calvird Award", "Bent Cage Award"
City Reference Coordinates: 096°48'00"W 32°47'00"N

Southwestern University, Department of Physics
University Avenue
Georgetown, TX 78626
Telephone: 512-863-1633
Telefax: 512-863-1696
Electronic Mail: <userid>@southwestern.edu
WWW: http://www.southwestern.edu/academic/physics.dept/physics.home.html
http://www.southwestern.edu/academic/physics.dept/astro1.html (Fountainwood Observatory)
Founded: 1882
Staff: 3
Coordinates: 097°40'00"W 30°38'03"N
City Reference Coordinates: 097°42'00"W 30°39'00"N

Southwest Research Institute (SwRI), Instrumentation and Space Research Division
P.O. Drawer 28510
6220 Culebra Road
San Antonio, TX 78288-0510
Telephone: 210-522-3236
Telefax: 210-520-9935
Electronic Mail: <userid>@swri.org
WWW: http://www.swri.org/4org/d15/d15home.htm
http://www.swri.edu/
Founded: 1947
Staff: 190
Activities: R&D
Periodicals: "Technology Today"
City Reference Coordinates: 098°31'00"W 29°28'00"N

Space Vacuum Epitaxy Center (SVEC)
• See "University of Houston, Space Vacuum Epitaxy Center (SVEC)"

Spaceweek International Association
1110 NASA Road One
Suite 100
Houston, TX 77058
Telephone: 281-333-3627
1-800-20-SPACE (USA only)
281-335-0229
Electronic Mail: admin@spaceweek.or
WWW: http://www.spaceweek.org/
Founded: 1980
Staff: 1
Activities: international headquarters for all Spaceweek events * promoting, developing and supporting event holder * coordinating event activities
Periodicals: (4) "Newsletter"
City Reference Coordinates: 095°22'00"W 29°46'00"N

Stephen F. Austin State University (SFASU), Department of Physics and Astronomy
P.O. Box 13044
SFA Station

Nacogdoches, TX 75962-3044
Telephone: 936-468-3001
Telefax: 936-468-4448
Electronic Mail: astro@sfasu.edu
                <userid>@sfasu.edu
WWW: http://www.physics.sfasu.edu/observatory/obs.htm (Observatory)
        http://www.physics.sfasu.edu/planetarium/ (Planetarium)
Founded: 1976
Activities: eclipsing binary dynamics * minor-planet search * telescope automation * observing * planetarium
Coordinates: 094°39'41"W 31°45'35"N (IAU Code 740)
City Reference Coordinates: 094°39'00"W 31°36'00"N

Tarrant County College, Astronomy Club
c/o Marles L. McCurdy
828 Harwood Road
Hurst, TX 76054
Telephone: 817-515-6559
Telefax: 817-515-6601
Electronic Mail: marles.mccurdy@tccd.net
WWW: http://www.tccd.net/campus_ne/nsc/astrclub/
Founded: 2000
Membership: 30
Activities: astronomy promotion among public and students * star parties
City Reference Coordinates: 097°11'00"W 32°47'00"N

Texas A & M University, Center for Space Power (CSP)
Room 223, WERC
College Station, TX 77843-3118
Telephone: 409-845-8768
Telefax: 409-847-8857
WWW: http://www.csp.tamu.edu/
Founded: 1987
Staff: 15
Activities: research * consulting * advising * thermal management * energy conversion and generation * power transmission * energy storage * power systems * materials * environment * space structures * communication systems & devices
City Reference Coordinates: 096°21'00"W 30°37'00"N

Texas A & M University, Department of Physics
College Station, TX 77843-4242
Telephone: 979-845-7717
Telefax: 979-845-2590
Electronic Mail: astro@tamu.edu
WWW: http://www.physics.tamu.edu/
Activities: experimental and theoretical research in atomic, nuclear and low-temperature/solid state physics * theory of elementary particle interactions, atmospheric physics, quantum optics and experimental high-energy physics
City Reference Coordinates: 096°21'00"W 30°37'00"N

Texas Astronomical Society (TAS) of Dallas
P.O. Box 25162
Preston Station
Dallas, TX 75225-1125
Telephone: 972-238-3788
Electronic Mail: tasclub@airmail.net
WWW: http://www.texasastro.org/
Founded: 1947
Membership: 700
Activities: observing
City Reference Coordinates: 096°48'00"W 32°47'00"N

Texas Christian University (TCU), Department of Physics and Astronomy
TCU Box 298840
Fort Worth, TX 76129
Telephone: 817-257-7375
Telefax: 817-257-7742
Electronic Mail: physics@tcu.edu
                <userid>@tcu.edu
WWW: http://www.phys.tcu.edu/
Founded: 1873 (TCU)
Staff: 7
Activities: astrochemistry * molecular and atomic spectroscopy * theoretical chemistry * statistical physics
City Reference Coordinates: 097°20'00"W 32°45'00"N

Texas Nautical Repair Co.
● See now "Land, Sea and Sky Inc."

Texas Star Party (TSP) Inc.
c/o TSP Registrar
4812 Twin Valley Drive
Austin, TX 78731-3539
Electronic Mail: tspreg@metronet.com
WWW: http://www.metronet.com/~tsp/index1.html
Founded: 1979
City Reference Coordinates: 097°45'00"W 30°16'00"N

Universities Space Research Association (USRA), Center for Advanced Space Studies (CASS)
3600 Bay Area Boulevard
Houston, TX 77058-1113
Telephone: 713-486-2139
Telefax: 713-486-2162
Electronic Mail: <userid>@cass.jsc.nasa.gov
WWW: http://cass.jsc.nasa.gov/CASS_home.html
         http://www.usra.edu/
Founded: 1991
Staff: 85
Activities: space exploration
City Reference Coordinates: 095°22'00"W 29°46'00"N

Universities Space Research Association (USRA), Division of Educational Programs (DEP)
3600 Bay Area Boulevard
Houston, TX 77058-1113
Telephone: 713-486-2139
Telefax: 713-486-2162
Electronic Mail: <userid>@cass.jsc.nasa.gov
WWW: http://www.sop.usra.edu/
         http://www.usra.edu/
Founded: 1991
City Reference Coordinates: 095°22'00"W 29°46'00"N

Universities Space Research Association (USRA), Division of Space Life Sciences (DSLS)
3600 Bay Area Boulevard
Houston, TX 77058-1113
Telephone: 713-486-2139
Telefax: 713-486-2162
Electronic Mail: <userid>@cass.jsc.nasa.gov
WWW: http://www.dsls.usra.edu/
         http://www.usra.edu/
Founded: 1991
Periodicals: (4) "Space Life Sciences"
City Reference Coordinates: 095°22'00"W 29°46'00"N

Universities Space Research Association (USRA), Lunar and Planetary Institute (LPI)
3600 Bay Area Boulevard
Houston, TX 77058-1113
Telephone: 281-486-2139
Telefax: 281-486-2162
Electronic Mail: <userid>@lpi.ursa.edu
WWW: http://www.lpi.usra.edu/
Founded: 1969
Staff: 65
Activities: origin of solar system * lunar and planetary research
Periodicals: (3) "Lunar and Planetary Information Bulletin (LPIB)"
City Reference Coordinates: 095°22'00"W 29°46'00"N

University of Houston, Department of Physics
4800 Calhoun Boulevard
Room 617 - S&R1 Building
Houston, TX 77204-5506
Telephone: 713-743-3550
Telefax: 713-743-8589
Electronic Mail: <userid>@uh.edu
WWW: http://www.uh.edu/~phys/
Founded: 1926
Staff: 30
City Reference Coordinates: 095°22'00"W 29°46'00"N

University of Houston, Space Physics Group
4800 Calhoun Boulevard
Room 617 - S&R1 Building
Houston, TX 77204-5506
Electronic Mail: <userid>@shasta.phys.uh.edu

WWW: http://www.uh.edu/research/spg/spgmain.html
Founded: 1968
City Reference Coordinates: 095°22'00"W 29°46'00"N

University of Houston, Space Vacuum Epitaxy Center (SVEC)
4800 Calhoun
Science and Research One
Room 724
Houston, TX 77204-5507
Telephone: 713-743-3621
Telefax: 713-747-7724
Electronic Mail: <userid>@uh.edu
WWW: http://www.svec.uh.edu/
Founded: 1986
Staff: 50
Activities: molecular and chemical beam epitaxy thin film growth * space hardware development
City Reference Coordinates: 095°22'00"W 29°46'00"N

University of North Texas (UNT), Department of Physics
P.O. Box 311427
Denton, TX 76203
Telephone: 940-565-2626
Electronic Mail: <userid>@unt.edu
WWW: http://www.phys.unt.edu/
Staff: 23
Activities: applied physics * astronomy * atomic and molecular physics * microwave spectroscopy * nuclear magnetic resonance * optics * solid-state physics * ultrasonic techniques
City Reference Coordinates: 097°10'00"W 33°14'00"N

University of North Texas (UNT), Sky Theater
P.O. Box 310559
Denton, TX 76203
Telephone: 940-369-7655
940-565-2994
Electronic Mail: skytheater@unt.edu (Sky Theater)
WWW: http://skytheater.unt.edu/ (Sky Theater)
City Reference Coordinates: 097°10'00"W 33°14'00"N

University of Texas at Arlington (UTA), Astronomical Society
c/o Department of Physics
Arlington, TX 76019
Telephone: 817-272-2467
817-795-9585
Telefax: 817-272-3637
Electronic Mail: sky@why.net (Kelly Mayfield, President)
WWW: http://www.uta.edu/student_orgs/astronomy/
http://www.olympusmons.org/
http://www.olympusmons.org/planetarium/ (UTA Planetarium)
Activities: public observing * education
Periodicals: (4) "The Aether"
City Reference Coordinates: 097°07'00"W 32°44'00"N

University of Texas at Arlington (UTA), Department of Physics
Arlington, TX 76019
Telephone: 817-272-2467
817-795-9585
Telefax: 817-272-3637
WWW: http://www.uta.edu/
http://www.olympusmons.org/planetarium/ (UTA Planetarium)
City Reference Coordinates: 097°07'00"W 32°44'00"N

University of Texas at Austin, Astronomy Department
2511 Speedway
RLM 15.308
Mail Code C1400
Austin, TX 78712-1083
Telephone: 512-471-3000
Telefax: 512-471-6016
Electronic Mail: <userid>@astro.as.utexas.edu
WWW: http://www.as.utexas.edu/
Staff: 28
Activities: extragalactic * QSOs * ISM * mm, IR, and optical spectroscopy * cosmology * photometry * stars * nebulae * novae * SN * solar system * instrumentation * operating McDonald Observatory
Periodicals: (1) "Annual Report"; "The University of Texas Publications in Astronomy" (ISSN 0276-1106)
City Reference Coordinates: 097°45'00"W 30°16'00"N

University of Texas at Austin, Center for Space Research (CSR)
3925 West Braker Lane
Suite 200
Austin, TX 78759-5321
Telephone: 512-471-5573
Telefax: 512-471-3570
Electronic Mail: info@csr.utexas.edu
WWW: http://www.csr.utexas.edu/
Founded: 1981
Staff: 65
Activities: space research * orbit determination * satellite remote sensing * satellite laser ranging * geodynamics * oceanography * astrometry * satellite altimetry and radiometry * geodesy * image processing
City Reference Coordinates: 097°45'00"W 30°16'00"N

University of Texas at Austin, Department of Physics
RLM 5.208
Austin, TX 78712
Telephone: 512-471-1153
Telefax: 512-471-9637
Electronic Mail: <userid>@physics.utexas.edu
WWW: http://www.ph.utexas.edu/
City Reference Coordinates: 097°45'00"W 30°16'00"N

University of Texas at Austin, Institute for Fusion Studies (IFS)
RLM 11.218
Dean Keaton and Speedway
Austin, TX 78712
Telephone: 512-471-1322
Telefax: 512-471-6715
Electronic Mail: <userid>@mail.utexas.edu
<userid>@physics.utexas.edu
WWW: http://hagar.ph.utexas.edu/ifs/
Founded: 1980
Staff: 25
Activities: plasma physics theory * magnetic confinement physics * fusion * plasma astrophysics * non-linear dynamics
Periodicals: "Preprints"
City Reference Coordinates: 097°45'00"W 30°16'00"N

University of Texas at Austin, McDonald Observatory
HC 75, Box 1337-MCD
Fort Davis, TX 79734-5020
Telephone: 915-426-3263
Telefax: 512-426-3641
Electronic Mail: <userid>@astro.as.utexas.edu
WWW: http://www.as.utexas.edu/mcdonald/
Founded: 1939
Staff: 55
Activities: optical and IR research base for the University of Texas at Austin
Coordinates: 104°01'19"W 30°40'19"N H2,075m (Mount Locke – IAU Code 711)
City Reference Coordinates: 103°54'00"W 30°35'00"N

University of Texas at Austin, McDonald Observatory, Administrative and Research Offices
c/o Astronomy Department
1 University Station
Mail Code C1400
Austin, TX 78712
Telephone: 512-471-3303
512-471-5285 (Public Information Office)
Telefax: 512-471-6016
512-471-5060 (Public Information Office)
Electronic Mail: <userid>@astro.as.utexas.edu
WWW: http://www.as.utexas.edu/mcdonald/mcdonald.html
http://stardate.org/ (StarDate Online)
Founded: 1939
Staff: 25
Activities: research * education
Periodicals: (6) "StarDate"; "McDonald Observatory Preprints"
City Reference Coordinates: 097°45'00"W 30°16'00"N

University of Texas at Dallas (UTD), William B. Hanson Center for Space Science (CSS)
2601 North Floyd Road
P.O. Box 830688 FO22
Richardson, TX 75083-0688
Telephone: 972-883-2851
Telefax: 972-883-2761
Electronic Mail: cssmail@utdssa.utdallas.edu

WWW: http://utd500.utdallas.edu/
Founded: 1969
Staff: 32
Activities: space instrumentation * data analysis and interpretation * education
City Reference Coordinates: 096°44'00"W 32°57'00"N

University of Texas Press
P.O. Box 7819
Austin, TX 78713-7819
or :
2100 Comal
Austin, TX 78722
Telephone: 1-800-252-3206 (USA and Canada only)
　　　　　 512-471-7233
Telefax: 512-232-7178
Electronic Mail: utpress@uts.cc.utexas.edu
WWW: http://www.utexas.edu/utpress/
Founded: 1950
• Publisher
City Reference Coordinates: 097°45'00"W 30°16'00"N

West Texas Astronomers (WTA)
c/o Marian Blakemore Planetarium
1705 West Missouri
Midland, TX 79701-6516
or :
c/o Robert Weil
2607 East 11th Street
Odessa, TX 79761
or :
c/o Bruce Barton (Secretary)
2801 Metz Drive
Midland, TX 79705
Telephone: 915-683-2882
Telefax: 915-570-7077
WWW: http://nonprofit.apex2000.net/wta/
　　　 http://www.museumsw.org/wta.html
Founded: 1972
Membership: 41
Activities: providing astronomical viewing for the public of West Texas and schools
Periodicals: "Astronews"
City Reference Coordinates: 102°22'00"W 31°51'00"N (Odessa)
　　　　　　　　　　　　　 102°05'00"W 32°00'00"N (Midland)

William B. Hanson Center for Space Science
• See "University of Texas at Dallas (UTD), William B. Hanson Center for Space Science (CSS)"

William Marsh Rice University, Physics and Astronomy Department
6100 Main Street
Mail Stop 61
Houston, TX 77005-1892
Telephone: 713-348-4938
Telefax: 713-348-4150
Electronic Mail: physics@rice.edu
WWW: http://dacnet.rice.edu/depts/ricephys/
Founded: 1963
City Reference Coordinates: 095°22'00"W 29°46'00"N

Williamson County Astronomy Club (WCAC)
c/o C.D. Sandberg
901 South Church Street
Georgetown, TX 78626
Electronic Mail: wcac@atm-workshop.com
WWW: http://www.atm-workshop.com/wcac/
Founded: 1980
Membership: 35
Activities: monthly meetings * star parties
City Reference Coordinates: 097°42'00"W 30°39'00"N

Lindon, UT 84042
Telephone: 801-785-4403
Electronic Mail: rtenney@uvaa.org
WWW: http://jove.prohosting.com/~uvaa
Membership: 80
Activities: monthly star parties
City Reference Coordinates: 111°24'00"W 41°44'00"N

Weber State University (WSU), Department of Physics
2508 University Circle
Ogden, UT 84408-2508
Telephone: 801-626-6163
Telefax: 801-626-7445
Electronic Mail: <userid>@weber.edu
WWW: http://astrophysics.weber.edu/physics.html
City Reference Coordinates: 111°58'00"W 41°14'00"N

Young University (Brigham_)
• See "Brigham Young University"

USA - Vermont

Antiquaries Manasek
P.O. Box 1204
Norwich, VT 05055
Telephone: 802-649-1722
Telefax: 802-649-2256
Electronic Mail: manasekinc@aol.com
WWW: http://www.antiquaries.com/
Activities: rare books, maps and atlases
City Reference Coordinates: 072°21'00"W 43°41'00"N (White River Junction)

Fairbanks Museum and Planetarium
1302 Main Street
St. Johnsbury, VT 05819-2248
Telephone: 802-748-2372
Telefax: 802-748-1893
WWW: http://www.fairbanksmuseum.org/
Founded: 1961
Activities: education facility for regional schools * public shows * local center for astronomical information
City Reference Coordinates: 072°01'00"W 44°25'00"N

G.B. Manasek Inc.
• See now "Antiquaries Manasek"

Janos Technology Inc.
1068 Grafton Road
Townshend, VT 05353-7702
Telephone: 802-365-7714
Telefax: 802-365-4596
Electronic Mail: <userid>@janostech.com
WWW: http://www.janostech.com/
Founded: 1970
Staff: 50
Activities: manufacturing crysaline and CVD optics for use in IR and UV * lenses * mirrors * beamsplitters * subassemblies
City Reference Coordinates: 072°35'00"W 42°58'00"N (Putney)

Manasek Inc. (G.B._)
• See now "Antiquaries Manasek"

Marlboro College, Department of Physics and Astronomy
Marlboro, VT 05344
Telephone: 802-257-4333 x255
Electronic Mail: <userid>@marlboro.edu
WWW: http://www.marlboro.edu/~mahoney/courses/Physics.html
Founded: 1975
Activities: education
Observatories: 1 (McArthur Observatory)
City Reference Coordinates: 072°35'00"W 42°58'00"N (Putney)

Middlebury College, Department of Physics
Middlebury, VT 05753-6151
Telephone: 802-443-5428
Telefax: 802-443-2072
Electronic Mail: physics@middlebury.edu
            <userid>@mail.middlebury.edu
WWW: http://www.middlebury.edu/~physics/
Founded: 1800 (College)
Activities: X-ray astrophysics * solar physics * SNR * stellar photometry
City Reference Coordinates: 073°10'00"W 44°01'00"N

Omega Optical Inc.
210 Main Street
Brattleboro, VT 05301
Telephone: 1-866-488-1064 (USA only)
            802-254-2690
Telefax: 802-254-3937
Electronic Mail: info@omegafilters.com
            astronomysales@omegafilters.com
WWW: http://www.omegafilters.com/
Founded: 1968
Staff: 119

Activities: manufacturing filters
City Reference Coordinates: 072°33'00"W 42°51'00"N

Parallax Instruments
c/o Joe Nastasi
P.O. Box 303
Montgomery Center, VT 05471
Telephone: 802-326-3140
Telefax: 802-326-3140
Electronic Mail: tscopes@sover.net
WWW: http://www.parallaxinstruments.com/
City Reference Coordinates: 072°41'00"W 44°59'00"N

Springfield Telescope Makers (STM) Inc.
P.O. Box 601
Reading, VT 05156
or :
Breezy Hill Road
Springfield, VT
Telephone: 802-484-5991
Electronic Mail: stellafane@vtnet.com
WWW: http://www.stellafane.com/
Founded: 1923
Membership: 80
Activities: monthly meetings * observing * telescope making * annual Stellafane convention of amateur telescope makers
Coordinates: 072°31'10"W 43°16'41"N H391m
City Reference Coordinates: 072°29'00"W 43°18'00"N (Springfield)

University of Vermont, Physics Department
A405 Cook Building
Burlington, VT 05405
Telephone: 802-656-2644
Telefax: 802-656-8429
Electronic Mail: office@physics.uvm.edu
          <userid>@uvm.edu
WWW: http://www.uvm.edu/~physics/
Founded: 1942
Staff: 2 (astronomy group)
Activities: pulsars * ISM
City Reference Coordinates: 073°14'00"W 44°28'00"N

Vermont Astronomical Society (VAS)
P.O. Box 782
Williston, VT 05495
Telephone: 802-985-3269
Electronic Mail: franky97@classic.msu.com (Frank Pakulski, President)
          paulwaav@together.net (Paul R. Walker, Secretary)
WWW: http://www.uvm.org/vas/
Founded: 1964
Membership: 57
Activities: research * public education
Periodicals: "Morning Star"
Coordinates: 073°12'00"W 44°28'00"N H34m (Green Mountain Observatory)
City Reference Coordinates: 073°14'00"W 44°28'00"N (Burlington)

USA - Virginia

AeroAstro
520 West Huntmar Park Drive
Herndon, VA 20170
Telephone: 703-709-2240
Telefax: 703-709-0790
Electronic Mail: info@aeroastro.com
WWW: http://www.aeroastro.com/
Founded: 1988
Staff: 20
Activities: micro and nano satellite development
City Reference Coordinates: 077°23'00"W 38°58'00"N

Amateur Satellite Observers of Southeast Virginia (ASOSV)
c/o James E. Byrd
44 Sandra Drive
Newport News, VA 23608
Telephone: 804-874-3104 (home)
            804-864-5961 (office)
Electronic Mail: jbyrd@seva.net
WWW: http://www.seva.net/reg/satellite
Founded: 1995
Membership: 105
Activities: satellite observing
City Reference Coordinates: 076°28'00"W 37°04'00"N

American Astronautical Society (AAS)
6352 Rolling Mill Place
Suite 102
Springfield, VA 22152-2354
Telephone: 703-866-0020
Telefax: 703-866-3526
Electronic Mail: info@astronautical.org
WWW: http://www.astronautical.org/
Founded: 1954
Membership: 1500
Staff: 2
Activities: advancement of astronautical sciences * exchange of ideas and information throughout the world space community * annual meetings and symposia * publications of technical journals, newsletter and books
Periodicals: (4) "Journal of the Astronautical Sciences" (ISSN 0021-9142); (6) "Space Times"
Awards: "Space Flight Award", "Flight Achievement Award", "Victor A. Prather Award", "Lloyd V. Berkner Award", "W. Randolph Lovelace II Award", "Melbourne W. Boynton Award", "Dirk Brouwer Award", "John F. Kennedy Astronautics Award", "Eugene M. Emme Astronautical Literature Award", "Military Astronautics Award"
City Reference Coordinates: 077°13'00"W 38°45'00"N

American Institute of Aeronautics and Astronautics (AIAA)
1801 Alexander Bell Drive
Suite 500
Reston, VA 20191-4344
Telephone: 703-264-7500
            1-800-NEW-AIAA (USA only)
Telefax: 703-264-7551
Electronic Mail: <userid>@aiaa.org
WWW: http://www.aiaa.org/
Founded: 1963
Membership: 30000
Staff: 89
Activities: aeronautics * astronautics * aerospace information resource
Periodicals: (12) "Aerospace America" (ISSN 0740-722x), "AIAA Journal"; (6) "Journal of Aircraft", "Journal of Guidance, Control and Dynamics", "Journal of Propulsion and Power", "Journal of Spacecraft and Rockets"; (4) "Journal of Thermophysics and Heat Transfer", "AIAA Student Journal"
City Reference Coordinates: 077°21'00"W 38°58'00"N

American Statistical Association (ASA)
1429 Duke Street
Alexandria, VA 22314-3415
Telephone: 703-684-1221
         1-888-231-3473 (USA only)
Telefax: 703-684-2037
Electronic Mail: asainfo@amstat.org
                 <userid>@amstat.org
WWW: http://www.amstat.org/

Founded: 1839
Membership: 18000
Periodicals: "Amstat News" (ISSN 0163-9617), "Chance", "Current Index to Statistics", "Journal of Agricultural, Biological, and Environmental Statistics" (ISSN 1085-7117), "Journal of Business and Economic Statistics" (ISSN 0735-0015), "Journal of Computational and Graphical Statistics" (ISSN 1061-8600), "Journal of the American Statistical Association" (ISSN 0162-1459), "Stats", "Technometrics" (ISSN 0040-1706), "The American Statistician" (ISSN 0003-1305)
Awards: various statistical and academic related awards
City Reference Coordinates: 077°03'00"W 38°48'00"N

Arlington Public Schools Planetarium
1426 North Quincy Street
Arlington, VA 22207
Telephone: 703-358-6070
Telefax: 703-358-6188
WWW: http://www.geocities.com/Athens/Acropolis/2148/planet.htm
Staff: 5
Activities: programs for public and schools * professional seminars
Periodicals: (2) "The Arlington Astroletter"; "Skye Stuff"
Coordinates: 077°50'00"W 38°53'00"N
City Reference Coordinates: 077°05'00"W 38°52'00"N

Ash Enterprises
c/o Eric Melenbrink
1221 Stanhope Avenue
Richmond, VA 23227
Telephone: 804-264-8888
Telefax: 804-266-7966
Electronic Mail: e.melenbrink@att.net
WWW: http://www.ash-enterprises.com/
Founded: 1970
Staff: 2
Activities: providing technical and design services to planetariums * manufacturing planetarium special effects
• See also USA-FL
City Reference Coordinates: 077°28'00"W 37°30'00"N

Associated Universities Inc. (AUI)
c/o National Radio Astronomy Observatory (NRAO)
520 Edgemont Road
Charlottesville, VA 22903
Telephone: 804-296-0211
Telefax: 804-296-0278
Electronic Mail: <userid>@nrao.edu
WWW: http://www.aui.edu/
Founded: 1957
Membership: 60
Activities: management group for the NRAO
City Reference Coordinates: 078°29'00"W 38°02'00"N

Association for the Advancement of Computing in Education (AACE)
P.O. Box 3728
Norfolk, VA 23514
Telephone: 757-623-7588
Telefax: 703-997-8760
Electronic Mail: aace@virginia.edu
WWW: http://www.aace.org/
Founded: 1981
Periodicals: "Educational Technology Review (ETR)", "Journal of Educational Multimedia and Hypermedia (JEMH)", "Journal of Computers in Mathematics and Science Teaching (JCMST)", "Journal of Interactive Learning Research (JILR)", "Journal of Technology and Teacher Education (JTATE)", "Journal of Computing in Childhood Education (JCCE)", "International Journal of Educational Telecommunications (IJET)"
City Reference Coordinates: 076°14'00"W 36°40'00"N

Astronomy Club of Virginia Tech
Blacksburg, VA 24061-0435
Telephone: 703-231-6544
Telefax: 703-231-7511
WWW: http://www.phys.vt.edu/~astroclb/
Founded: 1974
City Reference Coordinates: 080°25'00"W 37°14'00"N

Back Bay Amateur Astronomers (BBAA)
c/o Lelane Arneson (Treasurer)
245 Overholt Drive
Virginia Beach, VA 23462
or :

c/o Glendon L. Howell
2808 Flag Road
Chesapeake, VA 23323-2102
Telephone: 757-485-4242
Electronic Mail: twforte@aol.com
WWW: http://groups.hamptonroads.com/BBAA/
Founded: 1978
City Reference Coordinates: 075°58'00"W 36°51'00"N (Virginia Beach)
076°15'00"W 36°43'00"N (Chesapeake)

Blue Ridge Astronomy Club
c/o Jean Krason
123 Linden Avenue
Lynchburg, VA 24503
Electronic Mail: jlkrason@aol.com
WWW: http://www.blueridgeastronomy.org/
City Reference Coordinates: 079°10'00"W 37°24'00"N

Brackbill Planetarium (M.T._)
• See "M.T. Brackbill Planetarium"

Bristol Astronomy Club (BAC)
c/o R. Kenneth Childress
4559 Reedy Creek Road
Bristol, VA 24202-1501
Telephone: 540-466-8404
Electronic Mail: ballrl@tn.freei.net (Roger L. Ball, Secretary & Treasurer)
WWW: http://www.bristolastronomy.homepage.com/
Founded: 1957
Membership: 27
Activities: observing * public classes * telescope making * AAVSO variable stars * ALPO Mars section
Observatories: 1 (Burke Observatory, USA-TN - see separate entry)
City Reference Coordinates: 082°11'00"W 36°36'00"N

Carl Sandburg Middle School Planetarium
8428 Fort Hunt Road
Alexandria, VA 22308
Telephone: 703-799-6169
WWW: http://www.fcps.k12.va.us/DIS/OHSICS/planet/contacts/sandburg.htm
Founded: 1970
City Reference Coordinates: 077°03'00"W 38°48'00"N

Carl Zeiss Optical Inc.
13017 North Kingston Avenue
Chester, VA 23836
Telephone: 804-530-8300
1-800-338-2984 (USA only)
Telefax: 804-530-8311
1-800-561-2016 (USA only)
Electronic Mail: <userid>@zeiss.com
WWW: http://www.zeiss.com/optical/
Founded: 1889 (Carl Zeiss/Germany)
City Reference Coordinates: 077°27'00"W 37°22'00"N

Challenger Center for Space Science Education
1250 North Pitt Street
Alexandria, VA 22314
Telephone: 703-683-9740
Telefax: 703-683-7546
Electronic Mail: <userid>@challenger.org
WWW: http://www.challenger.org/
Founded: 1986
City Reference Coordinates: 077°03'00"W 38°48'00"N

Charlottesville Astronomy Society (CAS)
c/o Martin Wacksman
470 Willwood Drive
Earlysville, VA 22936
WWW: http://members.aol.com/kharker/cas/
Membership: 31
Periodicals: "Newsletter"
City Reference Coordinates: 078°29'00"W 38°02'00"N (Charlottesville)

Chesapeake Planetarium
310 Shea Drive

City Reference Coordinates: 078°58'00"W 36°34'00"N

National Aeronautics and Space Administration (NASA), Langley Research Center (LaRC)
Mail Stop 107
Hampton, VA 23665
Telephone: 804-965-2000
   804-864-6000 (Visitor Center)
Electronic Mail: <userid>@larc.nasa.gov
   opa@larc.nasa.gov (Office of External Affairs)
WWW: http://www.larc.nasa.gov/
Founded: 1917 (NASA: 1958)
Activities: aeronautics * earth science * space technology * structures and materials
City Reference Coordinates: 076°22'00"W 37°01'00"N

National Aeronautics and Space Administration (NASA), Wallops Flight Facility, Visitor Center
Wallops Island, VA 23337
Telephone: 757-824-2298
WWW: http://www.wff.nasa.gov/
Founded: 1982
Staff: 2
Activities: shows on past, present and future flights * displays on NASA research activities
City Reference Coordinates: 075°27'00"W 37°52'00"N

National Radio Astronomy Observatory (NRAO), Charlottesville
520 Edgemont Road
Charlottesville, VA 22903-2475
Telephone: 804-296-0211
Telefax: 804-296-0385
Electronic Mail: info@nrao.edu
   <userid>@nrao.edu
WWW: http://www.nrao.edu/
   http://www.cv.nrao.edu/
   http://info.aoc.nrao.edu/
   http://info.cv.nrao.edu/cv-home.html
   http://info.cv.nrao.edu/nrao-hq.html
Founded: 1958
Staff: 75
Activities: administrative headquarters of NRAO * electronics development * astronomical data analysis * astronomical research
Periodicals: (4) "NRAO Newsletter"; "Preprint Series"
City Reference Coordinates: 078°29'00"W 38°02'00"N

National Science Foundation (NSF)
4201 Wilson Boulevard
Arlington, VA 22230
Telephone: 703-306-1820
   703-306-1820 (Director, Division of Astronomical Sciences)
   703-306-1820 (Executive Officer, Division of Astronomical Sciences)
   703-306-1828 (Advanced Technology and Instrumentation)
   703-306-1819 (Education, Human Resources and Special Programs)
   703-306-1823 (Electromagnetic Spectrum Unit)
   703-306-1820 (Electromagnetic Spectrum Unit)
   703-306-1833 (Extragalactic Astronomy and Cosmology)
   703-306-1826 (Galactic Astronomy)
   702-306-1820 (Planetary Astronomy)
   703-306-1826 (Planetary Astronomy)
   703-306-1822 (NAIC Program Manager)
   703-306-1829 (NOAO Program Manager)
   703-306-1822 (NRAO Program Manager)
   703-306-1825 (Stellar Astronomy and Astrophysics)
Telefax: 703-306-0525 (Division of Astronomical Sciences)
   703-306-0910 (Division of Astronomical Sciences)
Electronic Mail: <userid>@nsf.gov
WWW: http://www.nsf.gov/
Founded: 1950
Staff: 20
Activities: providing awards for research in sciences and engineering
Periodicals: (10) "NSF Bulletin" (ISSN 0145-0670)
City Reference Coordinates: 077°05'00"W 38°52'00"N

National Science Teachers Association (NSTA)
1840 Wilson Boulevard
Arlington, VA 22201-3000
Telephone: 703-243-7100
   1-800-722-NSTA
Telefax: 703-243-7177

Electronic Mail: <userid>@nsta.org
WWW: http://www.nsta.org/
Founded: 1944
Membership: 53000
Periodicals: (6) "Quantum" (ISSN 1048-8820)
City Reference Coordinates: 077°05'00"W 38°52'00"N

National Technical Information Service (NTIS)
5285 Port Royal Road
Springfield, VA 22161
Telephone: 703-605-6000
Telefax: 703-321-8547
Electronic Mail: info@ntis.gov
WWW: http://www.ntis.gov/
Founded: 1946
• Publisher and software producer
City Reference Coordinates: 077°13'00"W 38°45'00"N

Norfolk Astronomical Society (NAS)
c/o Glendon L. Howell
2808 Flag Road
Chesapeake, VA 23323-2102
Telephone: 757-485-4242
Electronic Mail: nas2000@hamptonroads.com
WWW: http://groups.hamptonroads.com/nas/
City Reference Coordinates: 076°15'00"W 36°43'00"N

Northern Virginia Astronomy Club (NOVAC)
c/o Pete Johnson
5554 Sequoia Farms Drive
Centreville, VA 20120
or :
c/o Kevin Brown (Secretary)
5755 Walnut Wood Lane
Burke, VA 22015
Telephone: 703-758-4455
           703-494-0413
           703-978-0946
Electronic Mail: pjohnson1@mgfairfax.rr.com
WWW: http://www.novac.com/
Founded: 1980
Membership: 163
Activities: observing * trips to observatories * meetings
Periodicals: (4) "NOVAC Newsletter"
City Reference Coordinates: 077°26'00"W 38°50'00"N (Centreville)
                            077°16'00"W 38°48'00"N (Burke)

Oakton High School Planetarium
2900 Sutton Road
Oakton, VA 22180-2050
Telephone: 703-319-2735
WWW: http://www.fcps.k12.va.us/DIS/OHSICS/planet/contacts/oakton.htm
Founded: 1968
City Reference Coordinates: 077°18'00"W 38°53'00"N

Old Dominion University (ODU), Physics Department
Norfolk, VA 23529-0116
Telephone: 757-683-3468
           757-683-4108 (Planetarium)
Electronic Mail: <userid>@oduvm.cc.odu.edu
                 dbh100f@oduvm.cc.odu.edu (Bruce Hanna, Planetarium Director)
WWW: http://www.physics.odu.edu/
     http://www.physics.odu.edu/htmlstuf/docs/pretlow.htm (Pretlow Planetarium)
Founded: 1960
Staff: 15
Activities: education * research * Mary D. Pretlow Planetarium
City Reference Coordinates: 076°14'00"W 36°40'00"N

Orbital Sciences Corp. (OSC)
McLean, VA 22102
Telephone: 703-406-5000
Electronic Mail: <userid>@oscsystems.com
WWW: http://www.orbital.com/
Founded: 1982
Staff: 4500

Activities: space systems * launch vehicles * satellites * space sensors and electronics * ground stations * satellite access products * satellite-based communications * information services
City Reference Coordinates: 077°13'00"W 38°57'00"N (Herndon)

Pittsylvania County Schools' Planetarium
c/o Educational and Cultural Center
37 Pruden Avenue SE
Chatham, VA 24531
Telephone: 804-432-2761
804-793-1624
804-656-6248
Telefax: 804-432-9560
WWW: http://eclipse.dtl.pcs.k12.va.us/public/pcsplan.html
Founded: 1967
Activities: educational programs for public schools * amateur observing * computer modelling * video graphics
City Reference Coordinates: 079°24'00"W 36°50'00"N

Radford University, Planetarium
c/o Department of Chemistry and Physics
P.O. Box 6949
Radford, VA 24142-6949
Telephone: 540-831-6443
Electronic Mail: rherman@runet.edu (Rhett Herman)
WWW: http://www.runet.edu/~chem-web/physics/planetarium.html
Founded: 1971
City Reference Coordinates: 080°34'00"W 37°07'00"N

Randolph-Macon College (RMC), Keeble Observatory
P.O. Box 5005
Ashland, VA 23005-550
Telephone: 804-752-7344
Telefax: 804-752-4724
Electronic Mail: physics@rmc.edu
WWW: http://www.rmc.edu/academics/phys/keeble/
Founded: 1963
Staff: 2
Activities: education * photography * UBV photometry * low-dispersion spectroscopy * theory (radiation transport models of the ISM) * low-resolution 21cm observing
Coordinates: 077°29'00"W 37°46'00"N H67m
City Reference Coordinates: 077°29'00"W 37°46'00"N

Randolph-Macon Woman's College (RMWC), Physics Department
Lynchburg, VA 24503
Telephone: 804-947-8489
Electronic Mail: <userid>@rmwc.edu
tmichalik@rmwc.edu (Thomas R. Michalik, Observatory Director)
WWW: http://www.rmwc.edu/Academics/physics.html
http://faculty.rmwc.edu/tmichalik/abtwnfre.htm (Observatory)
Founded: 1891 (College) (Observatory: 1923)
Staff: 2
Activities: education * public viewing * variable stars
Coordinates: 079°10'23"W 37°26'22"N H220m (Winfree Observatory)
City Reference Coordinates: 079°10'00"W 37°24'00"N

Richmond Astronomical Society (RAS)
c/o Henry W. Stockmar, III
2218 Martin Street
Richmond, VA 23228
or :
c/o Terry Barker (Editor)
11536 Smoketree Drive
Richmond, VA 23236
Telephone: 804-379-8175 (T. barker)
Electronic Mail: tbarker@i2020.net
WWW: http://www.richastro.org/
Founded: 1949
Membership: 65
Periodicals: "RAS News"
City Reference Coordinates: 077°28'00"W 37°30'00"N

Roanoke Valley Astronomical Society (RVAS)
2607 Oregon Avenue
Roanoke, VA 24015
Telephone: 540-774-5651
Electronic Mail: gclose@pen.k12.va.us (Gary Close)

WWW: http://rvas.home.att.net/
Membership: 80
Activities: monthly meetings * observing * trips
City Reference Coordinates: 079°57'00"W 37°16'00"N

Sandburg Middle School Planetarium (Carl_)
• See "Carl Sandburg Middle School Planetarium"

Science Museum of Western Virginia (SMWV), Hopkins Planetarium
One Market Square
Roanoke, VA 24011
Telephone: 540-342-5710
Telefax: 540-224-1240
Electronic Mail: hopkins_planetarium@smwv.org
WWW: http://www.smwv.org/
http://www.smwv.org/skyinfo.html
Founded: 1983
City Reference Coordinates: 079°57'00"W 37°16'00"N

Shenandoah Astronomical Society (SAS)
c/o Jeffrey M. West (Treasurer)
276 Burnt Church Road
Winchester, VA 22604
Telephone: 540-662-9306
Electronic Mail: shenastro@worldnet.att.net
WWW: http://home.att.net/~shenastro/
Membership: 45
Periodicals: (12) "Newsletter"
City Reference Coordinates: 078°12'00"W 39°11'00"N

Signal Analytics Corp.
• See now "Scanalytics Inc."

Skywatchers Astronomy Club
c/o Marilyn E. Ogburn
124 Bowstring Drive
Williamsburg, VA 23185-4952
Telephone: 757-864-1175
Electronic Mail: meogburn@aol.com
m.e.ogburn@larc.nasa.gov
WWW: http://www.student.seas.gwu.edu/~aiaa/astronomy.html
Founded: 1958
Membership: 20
Activities: monthly meetings * observing sessions * popularization * education
Periodicals: "Skyline News"
Coordinates: 076°22'35"W 37°05'12"N (Hampton)
City Reference Coordinates: 079°43'00"W 37°16'00"N

Society for Scientific Exploration (SSE)
P.O. Box 3818
Charlottesville, VA 22903-0818
Telephone: 804-924-4905
Telefax: 804-924-3104
WWW: http://www.scientificexploration.org/
http://www.jse.com/ (Journal of Scientific Exploration)
Founded: 1981
Membership: 750
Activities: interdisciplinary studies of anomalous phenomena and events
Periodicals: (12) "The Explorer"; (2) "Journal of Scientific Exploration (JSE)"
Awards: "Dinsdale Prize"
City Reference Coordinates: 078°29'00"W 38°02'00"N

Society of Satellite Professionals International (SSPI)
225 Reinekers Lane
Suite 600
Alexandria, VA 22314
Telephone: 703-549-8696
Telefax: 703-549-9728
Electronic Mail: sspi@erols.com
WWW: http://www.sspi.org/
Founded: 1982
Membership: 1100
Staff: 2
Periodicals: (6) "Orbiter"
City Reference Coordinates: 077°03'00"W 38°48'00"N

Space Transportation Association (STA)
2800 Shirlington Road
Suite 405
Arlington, VA 22206
Telephone: 703-671-4111
Telefax: 703-931-6432
Electronic Mail: spacetra@erols.com
WWW: http://www.spacetransportation.org/
Activities: representing the interests of organizations and people who are engaged in developing, building, operating, and using space transportation vehicles, systems, and services to provide reliable, economical, safe, and routine access to space for private users and government, civil, and military users
City Reference Coordinates: 077°05'00"W 38°52'00"N

SPOT Image Corp.
1897 Preston White Drive
Reston, VA 22091-4326
Telephone: 703-715-3100
            1-800-ASK-SPOT (USA only)
Telefax: 703-645-1813
Electronic Mail: info@spot.com
              techsupport@spot.com
WWW: http://www.spot.com/spot-us.htm
Founded: 1982
Staff: 50
Activities: sales and marketing of SPOT satellite products and services
Periodicals: (4) "SPOTLight" (ISSN 0888-5850)
City Reference Coordinates: 077°21'00"W 38°58'00"N

Thomas Jefferson High School for Science and Technology Planetarium
6550 Braddock Road
Alexandria, VA 22310-1719
Telephone: 703-750-8380
Electronic Mail: lahennig@pen.k12.va.us (Lee Ann Hennig, Planetarium Teacher)
WWW: http://www.fcps.k12.va.us/DIS/OHSICS/planet/contacts/tjhsst.htm
Founded: 1967
City Reference Coordinates: 077°03'00"W 38°48'00"N

Triangulum Astronomical Society (TAS)
c/o Mark Slade
P.O. Box 868
Spotsylvania, VA 22553
or :
P.O. Box 7464
Fredericksburg, VA 22404
Telephone: 703-972-0023
            540-972-0023
Electronic Mail: marks@interserf.net (Mark Slade)
              mslade@erols.com
City Reference Coordinates: 077°35'00"W 38°12'00"N (Spotsylvania)
                          077°30'00"W 38°18'00"N (Fredericksburg)

University of Virginia, Department of Astronomy
P.O. Box 3818
University Station
Charlottesville, VA 22903-0818
or :
530 McCormick Road
Charlottesville, VA 22903
Telephone: 434-924-7494
Telefax: 434-924-3104
Electronic Mail: <userid>@virginia.edu
WWW: http://www.astro.virginia.edu/
Founded: 1877 (Observatory: 1885)
Staff: 21
Activities: theoretical astrophysics * astrometry * space astronomy * IR astronomy * galactic structure and evolution * extragalactic astronomy
Coordinates: 078°31'24"W 38°02'00"N H264m (Leander McCormick Obs. – IAU Code 780)
            078°41'36"W 37°52'42"N H566m (Fan Mountain Station)
City Reference Coordinates: 078°29'00"W 38°02'00"N

University Science Books (USB)
P.O. Box 605
Herndon, VA 20172
Telephone: 703-661-1572
Telefax: 703-661-1501
Electronic Mail: univscibks@igc.org

WWW: http://www.uscibooks.com/
Founded: 1978
• Publisher
City Reference Coordinates: 077°13'00"W 38°57'00"N

US Geological Survey (USGS)
National Center
Mail Stop 915
Reston, VA 22092
Telephone: 1-888-ASK-USGS (USA only)
1-888-275-8747 (USA only)
Telefax: 703-648-4888
WWW: http://www.usgs.gov/
Founded: 1879
Activities: providing reliable scientific information to describe and understand the Earth; to minimize loss of life and property from natural disasters; to manage water, biological, energy, and mineral resources; and to enhance and protect our quality of life
City Reference Coordinates: 077°21'00"W 38°58'00"N

Virginia Air and Space Center (VASC)
600 Settlers Landing Road
Hampton, VA 23669-4033
Telephone: 757-727-0900
Telefax: 757-727-0898
Electronic Mail: <userid>@vasc.org
WWW: http://www.vasc.org/
Founded: 1992
Staff: 55
Activities: official visitor center for NASA Langley Research Center * educational and interactive air and space exhibits * IMAX theater
Periodicals: (4) "Center Line"
City Reference Coordinates: 076°22'00"W 37°01'00"N

Virginia Beach City Public Schools Planetarium
3080 South Lynnhaven Road
Virginia Beach, VA 23452
Telephone: 757-431-4067
757-431-5331
Telefax: 757-473-5143
WWW: http://www.plazams.vbcps.k12.va.us/planetarium/
Founded: 1969
City Reference Coordinates: 075°58'00"W 36°51'00"N

Virginia Living Museum (VLM)
524 J. Clyde Morris Boulevard
Newport News, VA 23601
Telephone: 804-595-1900
Telefax: 804-599-4897
WWW: http://www.valivingmuseum.org/
http://www.valivingmuseum.org/site_pages/planetarium.htm (Peninsula Planetarium)
Founded: 1966
Activities: planetarium shows * daily H-alpha viewing of the Sun * night public viewing * major sky events monitoring
Periodicals: (12) "Paws and Reflect"
Coordinates: 076°28'00"W 37°05'00"N H0m (Abbitt Observatory)
City Reference Coordinates: 076°28'00"W 37°05'00"N

Virginia Military Institute (VMI), Department of Physics and Astronomy
Lexington, VA 24450
Telephone: 703-464-7225
Electronic Mail: <userid>@vmi.edu
WWW: http://academics.vmi.edu/physics/
Founded: 1990 (Observatory: 1960)
Staff: 1
Activities: education * variable stars * CCD photometry
City Reference Coordinates: 079°27'00"W 37°47'00"N

Virginia Polytechnic Institute and State University, Department of Physics
Robeson Hall
Blacksburg, VA 24061-0435
Telephone: 540-231-6544
Telefax: 540-231-7511
Electronic Mail: <userid>@astro.phys.vt.edu
WWW: http://www.phys.vt.edu/
Founded: 1904
Activities: radio astronomy research * wide-field H-alpha imaging * education

Activities: meetings * star parties * public outreach
Periodicals: "The Stargazer"
City Reference Coordinates: 122°14'00"W 47°59'00"N

Geer Planetarium (Willard_)
• See "Bellevue Community College (BCC), Willard Geer Planetarium"

Goldendale Observatory State Park (GOSP)
1602 Observatory Drive
Goldendale, WA 98620-3315
Telephone: 509-773-3141
Electronic Mail: goldobs@gorge.net
WWW: http://www.wolfenet.com/~per/gosp.html
Founded: 1973
Staff: 1
Activities: observing * tours * education
City Reference Coordinates: 120°50'00"W 45°49'00"N

Inglemoor High School Planetarium
15500 Simonds Road NE
Kenmore, WA 98028-4430
Telephone: 425-489-6500
Activities: shows for students
City Reference Coordinates: 122°14'00"W 47°46'00"N

International Society for Optical Engineering
(Society of Photo-Optical Instrumentation Engineers - SPIE)
P.O. Box 10
Bellingham, WA 98227-0010
or :
1000-20th Street
Bellingham, WA 98225
Telephone: 360-676-3290
Telefax: 360-647-1445
Electronic Mail: spie@spie.org
WWW: http://www.spie.org/
Founded: 1955
Membership: 11500
Activities: non-profit society dedicated to advancing engineering and scientific applications of optical, electro-optical, optoelectronic instrumentation, systems and technologies * providing the means for communicating new developments and applications to the scientific, engineering, user communities through its publications, symposia, and education
Sections: (working groups) Adaptive Optics * BACUS (Photomask Technologies) * Biomedical Optics * Electronic Imaging * Fiber Optics * High Speed Photography, Videography, and Photonics * Holography * Laser Communications * Lens Design * Noninvasive Inspection * Optical Computing * Optical Materials * Optical Signal and Image Processing * Optomechanical and Instrument Design * Penetrating Radiation * Photolithography * Robotics and Machine Perception * Smart Structures and Materials * Thermosense (Thermal Infrared Sensing for Diagnostics and Control)
Periodicals: (12) "Optical Engineering" (ISSN 0091-3286), "Optical Engineering Reports" (ISSN 1048-6879, circ.: 50,000); (4) "Journal of Electronic Imaging" (ISSN 1017-9909); (1) "Optics in Education"; "Proceedings of the SPIE" (ISSN 0277-788x)

City Reference Coordinates: 122°29'00"W 48°46'00"N

Laser Fantasy International (LFI)
8411 154th Avenue NE
Redmond, WA 98052
Telephone: 425-881-5356
Telefax: 425-883-5356
Electronic Mail: info@laserfantasy.com
WWW: http://www.laserfantasy.com/
Founded: 1977
Staff: 30
Activities: producing laser shows and manufacturing laser projection systems for museums, sciences centers and planetariums
City Reference Coordinates: 122°20'00"W 47°46'00"N (Seattle)

Olympic Astronomical Society (OAS)
P.O. Box 458
Keyport, WA 98345
Telephone: 360-698-0381
Electronic Mail: jbetteley@silverlink.net
WWW: http://www.silverlink.net/oas/
Periodicals: (12) "The Olympic Scope"
City Reference Coordinates: 122°38'00"W 47°42'00"N

Pacific Science Center, Willard W. Smith Planetarium
200 Second Avenue North
Seattle, WA 98109

Telephone: 206-443-2001
              206-223-2920 (Planetarium)
Telefax: 206-443-3631
WWW: http://www.pacsci.org/planetarium/
Staff: 390 (Planetarium: 9)
Activities: exhibits * interactive demonstrations * planetariums shows
City Reference Coordinates: 122°20'00"W 47°46'00"N

Quantum Technology Corp.
PMB 183
250 H Street
Blaine, WA 98230
Telephone: 604-938-0030
              1-888-271-9466 (USA only)
Telefax: 604-938-0061
              1-888-711-5222 (USA only)
Electronic Mail: quantum@quantum-technology.com
WWW: http://www.quantum-technology.com/
Founded: 1981
Staff: 10
Activities: manufacturing cryogenic instruments, helium liquefiers, liquid nitrogen cryostats and closed-cycle systems
City Reference Coordinates: 122°44'00"W 48°59'00"N

Remote Measurement Systems (RMS) Inc.
2633 Eastlake Avenue East
Suite 20
Seattle, WA 98102
Telephone: 206-328-2255
Telefax: 206-328-1787
Electronic Mail: rmstechinfo@measure.com
WWW: http://www.measure.com/
Founded: 1983
Activities: sensors * data acquisition * temperature monitoring * environmental measurements * analog to digital conversions * instrument interfaces * relative humidity * data loggers * control * weather measurements
City Reference Coordinates: 122°20'00"W 47°46'00"N

Seattle Astronomical Society (SAS)
P.O. Box 31746
Seattle, WA 98103-1746
Telephone: 206-523-ASTR
              206-523-2787
Electronic Mail: info@seattleastro.org
WWW: http://www.seattleastro.org/
Founded: 1948
Membership: 225
City Reference Coordinates: 122°20'00"W 47°46'00"N

Smith Planetarium (Willard W._)
• See "Pacific Science Center, Willard W. Smith Planetarium"

Society of Photo-Optical Instrumentation Engineers (SPIE)
• See "International Society for Optical Engineering"

Southwest Washington Astronomical Society (SWWAS) Inc.
c/o Dennis McHugo
6909 43rd Loop SE
Olympia, WA 98501
Electronic Mail: swwas@home.com
WWW: http://members.home.com/swwas/
City Reference Coordinates: 122°53'00"W 47°03'00"N

Spokane Astronomical Society (SAS)
P.O. Box 8114
Spokane, WA 99203-8114
Telephone: 509-448-1801 (Mary Singer, Media Contact)
Electronic Mail: mary@spokaneastronomical.org
WWW: http://www.runway.net/a/sas/
Founded: 1932
Membership: 150
City Reference Coordinates: 117°23'00"W 47°40'00"N

Table Mountain Star Party Association (TMSPA)
P.O. Box 785
Puyallup, WA 98371
WWW: http://www.tmspa.com/

Founded: 1980
Activities: organizing the Table Mountain Star Party (TMSP)
City Reference Coordinates: 122°18'00"W 47°11'00"N

Tacoma Astronomical Society (TAS)
c/o Pettinger-Guiley Observatory
6103 132nd Street East
Puyallup, WA 98373
Electronic Mail: taspgo@geocities.com
WWW: http://www.geocities.com/taspgo/
Founded: 1931
City Reference Coordinates: 122°18'00"W 47°11'00"N

Thermionics Vacuum Products
231-B Otto Street
Port Townsend, WA 98368
Telephone: 360-385-7707
          1-800-962-2310 (USA only)
Telefax: 360-385-6617
Electronic Mail: sales@thermionics.com
WWW: http://www.thermionics.com/
City Reference Coordinates: 122°46'00"W 48°07'00"N

Tri-City Astronomy Club (TCAC)
P.O. Box 651
Richland, WA 99352
WWW: http://www.tri-cityastronomyclub.org/
Activities: education * public star parties
City Reference Coordinates: 119°18'00"W 46°17'00"N

University of Washington, Department of Astronomy
Box 351580
Seattle, WA 98195-1580
Telephone: 206-543-2888
Telefax: 206-685-0403
Electronic Mail: <userid>@astro.washington.edu
WWW: http://www.astro.washington.edu/
Founded: 1965
Staff: 30
Activities: stellar and extragalactic astronomy * solar system dust
City Reference Coordinates: 122°20'00"W 47°46'00"N

University of Washington, Department of Physics
P.O. Box 351560
Seattle, WA 98195-1560
or :
C121 Physics Astronomy Building
FM-15
Seattle, WA 98195
Telephone: 206-543-2770
Telefax: 206-685-0635
Electronic Mail: <userid>@phys.washington.edu
WWW: http://www.phys.washington.edu/
City Reference Coordinates: 122°20'00"W 47°46'00"N

Washington State University (WSU), Astronomy Club
c/o Scott Austin
Program in Astronomy
Pullman, WA 99164-3113
Electronic Mail: astclub@delta.math.wsu.edu
WWW: http://www.sci.wsu.edu/math/students/astclub/astclub.html
City Reference Coordinates: 117°09'00"W 46°46'00"N

Washington State University (WSU), Program in Astronomy
Pullman, WA 99164-2814
Telephone: 509-335-1698
Telefax: 509-335-7816
Electronic Mail: astronomy@wsu.edu
WWW: http://astro.wsu.edu/
Founded: 1980
Staff: 3
Activities: stallar populations * gravitational radiation * observational cosmology * interferometry * planetarium
Coordinates: 117°10'00"W 46°44'00"N H760m (Jewett Observatory)
City Reference Coordinates: 117°09'00"W 46°46'00"N

Western Washington University (WWU), Physics/Astronomy Department
Bellingham, WA 98225
Telephone: 360-650-3818
Telefax: 360-650-7788
Electronic Mail: <userid>@physics.wwu.edu
skywise@cc.wwu.edu (Planetarium)
WWW: http://www.ac.wwu.edu/~physics/
http://www.wwu.edu/~skywise/ (Planetarium)
Founded: 1960
Staff: 13
Activities: education * research * planetarium shows
City Reference Coordinates: 122°29'00"W 48°46'00"N

Whatcom Association of Celestial Observers (WACO)
P.O Box 1721
Ferndale, WA 98248-1721
Telephone: 360-384-1281 (Gary D. Sell, Treasurer)
360-676-9530 (Dennis Hoofnagle)
Telefax: 360-676-4065 (Dennis Hoofnagle)
Electronic Mail: dhoof@az.com (Dennis Hoofnagle)
sherkhan@earthlink.net (Gary D. Sell)
WWW: http://www.ac.wwu.edu/~trekka/
Founded: 1988
Activities: meetings * observing * public outreach
Periodicals: (12) "WACO News"
City Reference Coordinates: 122°36'00"W 48°51'00"N (Ferndale)
122°38'00"W 48°55'00"N (Custer)

Whitman College, Department of Astronomy
Walla Walla, WA 99362
Telephone: 509-527-5225
Telefax: 509-527-5904
Electronic Mail: <userid>@whitman.edu
WWW: http://www.whitman.edu/astronomy/
Staff: 2
Activities: education * Charles F. Clise Planetarium
Coordinates: 118°19'00"W 46°04'00"N
City Reference Coordinates: 118°20'00"W 46°08'00"N

Willard Geer Planetarium
• See "Bellevue Community College (BCC), Willard Geer Planetarium"

Willard W. Smith Planetarium
• See "Pacific Science Center, Willard W. Smith Planetarium"

Yakima Valley Astronomy Club (YVAC)
143 Sole Road
Selah, WA 98942
Telephone: 509-457-6410 (Mac Knight, President)
Telefax: 509-457-0575 (prior notice needed)
Electronic Mail: mknight@wolfenet.com
WWW: http://www.perr.com/yvac.html
Founded: 1990
Membership: 25
Activities: education * observing * monthly meetings
Periodicals: (12) "What's up in Yakima?"
Coordinates: 120°30'39"W 46°36'06"N
City Reference Coordinates: 120°31'00"W 46°36'00"N (Yakima)
120°31'00"W 46°40'00"N (Selah)

USA - West Virginia

Astrolabe Astronomy Club of Wheeling
WLSC SMART - Center
1610 Warwood Avenue
Wheeling, WV 26003
Telephone: 304-277-2308
Telefax: 304-277-3505
Electronic Mail: strongro@wlsvax.wvnet.edu
WWW: http://www.neofoundation.org/astrolabe
City Reference Coordinates: 080°42'00"W 40°05'00"N

Astronomical Society of West Virginia
c/o Thomas R. Willmitch
The Science Center of West Virginia
500 Bland
Bluefield, WV 24701
Telephone: 304-325-8855
Telefax: 304-324-0513
Electronic Mail: twillmitch@citlink.net
WWW: http://www.homestead.com/wvastronomy
City Reference Coordinates: 081°17'00"W 37°14'00"N

Benedum Planetarium
Oglebay Park
Wheeling, WV 26003
Telephone: 304-243-4034
Telefax: 304-243-4110
Electronic Mail: stargate@hgo.net
Founded: 1978
Staff: 6
Activities: school and public planetarium programs * laser shows
City Reference Coordinates: 080°42'00"W 40°05'00"N

Berkeley County Planetarium
c/o Hedgesville High School
109 Ridge Road North
Hedgesville, WV 25427
Telephone: 304-754-3354 x3140
Telefax: 304-754-7445
Founded: 1977
Staff: 1
Activities: programs for school children and civic groups
Periodicals: (1) "Programs"
City Reference Coordinates: 078°01'00"W 39°33'00"N

Central Appalachian Astronomy Club
c/o Joe Gonzalez or James King
P.O. Box 1862
Clarksburg, WV 26301
Telephone: 304-745-4842
WWW: http://www.caacwv.org/
City Reference Coordinates: 080°22'00"W 29°16'00"N

Kanawha Valley Astronomical Society (KVAS)
c/o Bill Evans
226 Scenic Drive
Saint Albans, WV 25177-3444
or :
c/o Chuck Spann
916 Edgewood Drive
Charleston, WV 25302
or :
c/o Danny Blair (Treasurer)
2108 Woodhill Place
Saint Albans, WV 25177
Telephone: 304-346-3422
Electronic Mail: kvas@earth1.net
　　　　　　　　bill_evans@charter.net
　　　　　　　　cwspann@aol.com,
WWW: http://www.kvas.org/
City Reference Coordinates: 079°57'00"W 32°48'00"N (Charleston)
　　　　　　　　　　　　　　081°53'00"W 38°24'00"N (Saint Albans)

National Radio Astronomy Observatory (NRAO), Green Bank
P.O. Box 2
Green Bank, WV 24944-0002
Telephone: 304-456-2011
Telefax: 304-456-2271
Electronic Mail: <userid>@nrao.edu
WWW: http://info.gb.nrao.edu/
http://www.nrao.edu/
Founded: 1957
Staff: 80
Activities: radio observing * interferometry
Periodicals: "Reprints"
Coordinates: 079°50'30"W 38°25'48"N H836m
City Reference Coordinates: 079°51'00"W 38°26'00"N

Ohio Valley Astronomical Society (OVAS)
c/o Paul Brown
1827 Enslow Boulevard
Huntington, WV 25701
or :
c/o Jeffery Ball
32 Heritage Park
Huntington, WV 25704
Telephone: 304-525-3849
WWW: http://www.ovas.org/
Founded: 1974
Activities: observing * lectures * slide shows
City Reference Coordinates: 082°26'00"W 38°25'00"N

Shepherd College, Institute for Environmental Studies
P.O. Box 3210
Shepherdstown, WV 25443
Telephone: 304-876-5000
304-876-5227
Telefax: 304-876-5028
Electronic Mail: iesweb@shepherd.edu
WWW: http://www.shepherd.edu/iesweb/
Founded: 1997
Activities: environmental studies astronomy * physics
City Reference Coordinates: 077°48'00"W 39°26'00"N

Sunrise Museum
746 Myrtle Road
Charleston, WV 25314
Telephone: 304-344-8035
Telefax: 304-344-8038
Electronic Mail: info@sunrisemuseum.org
WWW: http://www.sunrisemuseum.org/
Activities: exhibits * planetarium shows
City Reference Coordinates: 081°40'00"W 38°23'00"N

West Virginia Space Grant Consortium (WVSGC)
c/o Majid Jaraiedi (Director)
108A Engineering Space Grant Consortium
P.O. Box 6107
Morgantown, WV 26506-6107
Telephone: 304-293-4099
Telefax: 304-293-4970
Electronic Mail: jaraiedi@cemr.wvu.edu
u31A0@wvnvm.wvnet.edu
WWW: http://www.cemr.wvu.edu./~wwwie/wvsgc/wvsgc.html
Founded: 1961
City Reference Coordinates: 079°57'00"W 39°38'00"N

West Virginia University (WVU), Astronomy Club
425 Hodges Hall
P.O. Box 6315
Morgantown, WV 26506-6315
Telephone: 304-293-3422 x1443
Telefax: 304-293-5732
WWW: http://www.as.wvu.edu/~jel/astroclub.html
City Reference Coordinates: 079°57'00"W 39°38'00"N

West Virginia University (WVU), Department of Physics
P.O. Box 6315

Morgantown, WV 26506-6315
Telephone: 304-293-3498
Electronic Mail: <userid>@wvnvm.wvnet.edu
WWW: http://www.as.wvu.edu/phys/astroph.html
  http://www.as.wvu.edu/~planet (Tomchin Planetarium)
Founded: 1881
Staff: 2
Activities: research * education * planetarium shows
City Reference Coordinates: 079°57'00"W 39°38'00"N

USA - Wisconsin

Barlow Planetarium
University of Wisconsin - Fox Valley
1478 Midway Road
Menasha, WI 54952
Telephone: 920-832-2600
Telefax: 920-832-2674
Electronic Mail: kklamczy@uwc.edu (Karen Klamczynski, Director)
WWW: http://www.fox.uwc.edu/barlow/
Founded: 1998
Staff: 6
Activities: public, private and school planetarium shows and special events * promoting astronomy and general science education
• Formerly "Fox Valley Planetarium"
City Reference Coordinates: 088°26'00"W 44°13'00"N

Beloit College, Thompson Observatory
700 College Street
Beloit, WI 53511
Telephone: 608-363-2258
Telefax: 608-363-2052
Electronic Mail: <userid>@beloit.edu
WWW: http://physics.beloit.edu/observatory/
Staff: 3
Coordinates: 089°01'54"W 42°30'18"N H255m
City Reference Coordinates: 089°04'00"W 42°31'00"N

Center for Astrophysical Research in Antarctica (CARA)
c/o Yerkes Observatory
373 West Geneva Street
P.O. Box 258
Williams Bay, WI 53191-0258
Telephone: 414-245-5555
Telefax: 414-245-9805
Electronic Mail: <userid>@yerkes.uchicago.edu
WWW: http://astro.uchicago.edu/cara/home.html
Founded: 1991
Staff: 24
Activities: Antarctic astronomy
Periodicals: "Antarctic Astrophysics"
City Reference Coordinates: 088°33'00"W 42°35'00"N

Chippewa Valley Astronomical Society (CVAS)
c/o Beaver Creek Reserve
S1 County Road K
Fall Creek, WI 54742
Electronic Mail: starman@discover-net.net
WWW: http://www.cvastro.org/
Membership: 50
Coordinates: 091°16'18"W 44°48'57"N H285m (Hobbs Obs. – IAU Code 750)
City Reference Coordinates: 091°17'00"W 44°46'00"N

Door Peninsula Astronomical Society (DPAS)
Crossroads at Big Creek
2041 Michigan Avenue
Sturgeon Bay, WI 54235
Telephone: 920-743-7812
Electronic Mail: president@doorastronomy.org
WWW: http://www.doorastronomy.org/
City Reference Coordinates: 087°21'00"W 44°51'00"N

Eagle Optics
2120 West Greenview Drive
Suite 4
Middleton, WI 53562
Telephone: 608-271-4751 (Technical Assistance)
          1-800-289-1132 (Orders)
          608-836-7172
Telefax: 608-836-4416
Electronic Mail: tomhall@eagleoptics (Thomas Hall)
          <userid>@eagleoptics.com
WWW: http://www.eagleoptics.com/

Founded: 1988
Staff: 12
Activities: distributing optical material
City Reference Coordinates: 089°31'00"W 43°06'00"N

East Central Wisconsin Astronomers (ECWA)
c/o Dave Schliepp
133 Jackson Street
Berlin, WI 54923
Telephone: 920-361-2847
Electronic Mail: deepspacedave@hotmail.com
WWW: http://www.vbe.com/~duwayne/ecwa.html
City Reference Coordinates: 088°57'00"W 43°58'00"N

Hubbard Scientific
1120 Halbleib Road
P.O. Box 760
Chippewa Falls, WI 54729
Telephone: 1-800-323-8368 (USA only)
715-723-4427
Telefax: 715-723-8021
Electronic Mail: pperkins@amep.com (Paula Perkins, International Sales Representative)
WWW: http://www.amep.com/ScottHubbard.htm
Activities: manufacturing scientific educational resources
City Reference Coordinates: 091°24'00"W 44°66'00"N

Kalmbach Publishing Co.
21027 Crossroads Circle
P.O. Box 1612
Waukesha, WI 53187-1612
Telephone: 262-796-8776
1-800-446-5489 (USA only)
Telefax: 262-796-1142
Electronic Mail: customerservice@kalmbach.com
WWW: http://www.kalmbach.com/
http://www.astronomy.com/ (Astronomy magazine)
Founded: 1934
Activities: publishing magazines, books and calendars
Periodicals: (12) "Astronomy" (ISSN 0091-6338, circ.: 172,000)
City Reference Coordinates: 088°14'00"W 43°01'00"N

Knollwood Books
P.O. Box 197
Oregon, WI 53575-0197
Telephone: 608-835-8861
Telefax: 608-835-8421
Electronic Mail: books@wi.rr.com
WWW: http://www.knollwoodbooks.com/
Founded: 1992
Staff: 2
Activities: selling used, out of print and antiquarian books on astronomy, meteorology, and space exploration
Periodicals: (4) "Catalog"
City Reference Coordinates: 089°23'00"W 42°56'00"N

La Crosse Area Astronomical Society (LCAAS)
c/o Robert Allen
P.O. Box 2041
La Crosse, WI 54602
or :
Cowley Hall
Department of Physics
University of Wisconsin
La Crosse, WI 54601
Telephone: 608-785-8669
Telefax: 608-785-8403
Electronic Mail: allen.robe@uwlax.edu
WWW: http://perth.uwlax.edu/planetarium/laas.html
Founded: 1978
Membership: 50
Activities: monthly meetings * observing sessions * lectures * videos
Periodicals: (12) "Newsletter"
City Reference Coordinates: 091°15'00"W 43°49'00"N

Madison Astronomical Society (MAS) Inc.
c/o Robert P. Manske

404 Prospect Road
Waunakee, WI 53597
Telephone: 608-849-5287
Electronic Mail: stargazer@njackn.com
WWW: http://t5L.adp.wisc.edu/~madastro/
http://www.madisonastro.org/
Founded: 1930
Membership: 90
Activities: observing (variable stars, Moon, occultations, planets)
Periodicals: "Capitol Skies"
Coordinates: 089°26'16"W 42°47'29"N H307m (Yanna Research Stn - IAU Code 927)
City Reference Coordinates: 089°27'00"W 43°11'00"N

Madison Metropolitan School District (MMSD), Planetarium and Observatory
201 South Gammon Road
Madison, WI 53717-1499
Telephone: 608-829-4053
Telefax: 608-829-1501
Electronic Mail: gholt@madison.k12.wi.us (Geoff Holt, Planetarium Director)
bsenson@madison.k12.wi.us (Ben Senson, Observatory Director)
WWW: http://www.madison.k12.wi.us/planetarium/
Founded: 1966 (Planetarium) 1999 (Observatory)
Staff: 2
Periodicals: (10) "Madison Skies"
City Reference Coordinates: 089°22'00"W 43°05'00"N

Manfred Olson Planetarium
• See "University of Wisconsin, Milwaukee (UWM), Department of Physics, Manfred Olson Planetarium"

Milwaukee Astronomical Society (MAS)
c/o Carlos Garces
16430 Melody Drive
New Berlin, WI 53151
or :
c/o Julie K. Frey (Membership Chairman)
11040 West Meinecke Avenue
Suite 4
Wauwatosa, WI 53226-1247
Telephone: 262-786-2623
Electronic Mail: masmemb@aol.com
WWW: http://www.milwaukeeastro.org/
Founded: 1932
City Reference Coordinates: 087°55'00"W 43°02'00"N (Milwaukee)

Milwaukee Public Museum (MPM)
800 West Wells Street
Milwaukee, WI 53233
Telephone: 414-278-2700
WWW: http://www.mpm.edu/
City Reference Coordinates: 087°55'00"W 43°02'00"N

Network and Systems Professionals Association (NaSPA) Inc.
7044 13th Street
Oak Creek, WI 53154
Telephone: 414-768-8000
Telefax: 414-768-8001
Electronic Mail: mbrship@nascom.com
<userid>@nascom.com
WWW: http://www.nascom.com/
Founded: 1986
Activities: association of corporate computing technical professionals
Periodicals: (12) "Technical Support" (ISSN 1079-3135)
City Reference Coordinates: 087°55'00"W 43°02'00"N (Milwaukee)

Neville Public Museum Astronomical Society (NPMAS)
c/o Ron Parmentier (Treasurer)
161 Rosemont Drive
Green Bay, WI 54301
or :
210 Museum Place
Green Bay, WI 54303
Telephone: 414-448-4460
Telefax: 414-448-4458
WWW: http://www.npmas.com/
Founded: 1966

Membership: 100
Activities: occultations * photography * variable stars * mirror making * general observing * deep sky * CCD imaging
City Reference Coordinates: 088°01'00"W 44°40'00"N

Northeast Wisconsin Stargazers (NEWSTAR)
P.O. Box 1611
Oshkosh, WI 54903-1611
or :
c/o Terry Becker
514 Union Avenue
Apartment E
Oshkosh, WI 54901
or :
c/o Andrea Gianopoulos
University of Wisconsin
1478 Midway Road
P.O. Box 8002
Menasha, WI 54952-8002
Telephone: 414-832-2848
            920-426-2286
Telefax: 414-832-2674
WWW: http://www.new-star.org/
Founded: 1986
Membership: 50
Activities: meetings * popularization
Periodicals: (12) "The Gazer's Gazette"
City Reference Coordinates: 088°33'00"W 44°01'00"N (Oshkosh)
                              088°26'00"W 44°13'00"N (Menasha)

Northern Cross Science Foundation (NCSF)
1418 Trillium Court
West Bend, WI 53095
or :
4163 W. River's Edge Circle
Suite 105
Brown Deer, WI 53209
or :
c/o Jeffrey S. Setzer
8142 North 66th Street
Brown Deer, WI 53223-3400
or :
1219 12th Avenue
Grafton, WI 53024-1923
Telephone: 414-355-3698
            262-338-8614
Electronic Mail: ncsf@gxsc.com
WWW: http://www.gxsc.com/ncsf/
Founded: 1974
Activities: promoting astronomy and related sciences * fostering public education in these fields
City Reference Coordinates: 087°59'00"W 43°10'00"N (Brown Deer)
                              087°56'00"W 43°19'00"N (Grafton)

Obsession Telescopes
P.O. Box 804
Lake Mills, WI 53551
Telephone: 414-648-2328
Telefax: 414-648-2328
Electronic Mail: obsessionscp@globaldialog.com
WWW: http://www.globaldialog.com/~obsessiontscp/OBHP.html
Founded: 1989
Staff: 1
Activities: manufacturing altazimuth telescopes
City Reference Coordinates: 088°55'00"W 43°05'00"N

Olson Planetarium (Manfred_)
• See "University of Wisconsin, Milwaukee (UWM), Department of Physics, Manfred Olson Planetarium"

Pine Bluff Observatory (PBO)
• See "University of Wisconsin, Madison, Pine Bluff Observatory (PBO)"

Racine Astronomical Society (RAS)
112 63rd Drive
Union Grove, WI 53408
c/o Bill Uminski
P.O.Box 085694

Racine, WI 53408
Electronic Mail: rasastro@wi.net
WWW: http://users.wi.net/~rasastro/
http://users.wi.net/~rasastro/Observatory.htm (Observatory)
Founded: 1956
Observatories: 1 (Modine-Benstead Observatory - MBO)
City Reference Coordinates: 087°48'00"W 42°43'00"N (Racine)
088°03'00"W 42°41'00"N (Union Grove)

Rock Valley Astronomical Society (RVAC)
c/o Steve Tuma
1425 Greenwich Lane
Janesville, WI 53545-1219
Founded: 1975
City Reference Coordinates: 089°01'00"W 42°41'00"N

Sheboygan Astronomical Society (SAS)
c/o Kurt Petersen
6405 Paradise Lane
Sheboygan Falls, WI 53085
Telephone: 414-467-2257
WWW: http://www.shebastro.org/
Founded: 1973
Membership: 25
Activities: observing
Periodicals: "Newsletter"
City Reference Coordinates: 087°36'00"W 43°46'00"N

Space Science and Engineering Center (SSEC)
• See "University of Wisconsin, Madison, Space Science and Engineering Center (SSEC)"

Thompson Observatory
• See "Beloit College, Thompson Observatory"

University of Chicago, Yerkes Observatory
373 West Geneva Street
P.O. Box 258
Williams Bay, WI 53191-0258
Telephone: 262-245-5555
Telefax: 262-245-9805
Electronic Mail: <userid>@yerkes.uchicago.edu
WWW: http://astro.uchicago.edu/yerkes/
Founded: 1897
Staff: 34
Activities: observational astronomy * ISM * dynamical astronomy * extragalactic studies * IR astronomy * education * Antarctic astronomy * instrumentation * adaptive optics * history of astronomy
Periodicals: "Publications of the Yerkes Observatory"
Coordinates: 088°33'24"W 42°34'13"N H334m (IAU Code 754)
City Reference Coordinates: 088°33'00"W 42°35'00"N

University of Wisconsin, Eau Claire, Department of Physics and Astronomy
Eau Claire, WI 54702-4004
Telephone: 715-836-3148
Telefax: 715-836-2380
Electronic Mail: <userid>@uwec.edu
gstecher@uwec.edu (George Stecher, Planetarium Director)
WWW: http://athena.phys.uwec.edu/thomas/uwecpa.html
http://www.phys.uwec.edu/
http://www.phys.uwec.edu/planetarium/ (L.E. Phillips Planetarium)
Staff: 15
Activities: research * education
Coordinates: 091°16'18"W 44°48'57"N H285m (Hobbs Obs. – IAU Code 750)
091°29'59"W 44°47'45"N H268m (Casey Observatory)
City Reference Coordinates: 091°31'00"W 44°49'00"N

University of Wisconsin, La Crosse, Planetarium
c/o Department of Physics
Cowley Hall
La Crosse, WI 54601
Telephone: 608-785-8669
Telefax: 608-785-8403
Electronic Mail: allen.robe@uwlax.edu
WWW: http://perth.uwlax.edu/planetarium/
Founded: 1966
Staff: 3

Activities: public, school and private programs
Coordinates: 091°15'00"W 43°48'00"N H200m
City Reference Coordinates: 091°15'00"W 43°49'00"N

University of Wisconsin, Madison, Department of Astronomy
475 North Charter Street
Madison, WI 53706-1582
Telephone: 608-262-3071
          608-262-9274 (hotline for viewing information)
Telefax: 608-263-0361
Electronic Mail: <userid>@astro.wisc.edu
WWW: http://www.astro.wisc.edu/
       http://www.astro.wisc.edu/Washburn/ (Washburn Observatory)
       http://www.astro.wisc.edu/wiyn/wiyn.html (WIYN Observatory)
       http://www.sal.wisc.edu/PBO/ (Pine Bluff Observatory - see separate entry)
Founded: 1878
Staff: 13
Activities: education * planetarium
Periodicals: "Wisconsin Astrophysics Preprint Series"
Coordinates: 089°24'30"W 43°04'36"N H292m (Washburn Obs. – IAU Code 753)
City Reference Coordinates: 089°22'00"W 43°05'00"N

University of Wisconsin, Madison, Department of Physics
2531 Sterling Hall
1150 University Avenue
Madison, WI 53706
Telephone: 608-262-7782
Telefax: 608-262-3077
Electronic Mail: physics@macc.wisc.edu
              office@wisc.physics.wisc.edu
WWW: http://www.physics.wisc.edu/
Founded: 1849
Activities: X-ray astronomy and instrumentation * Fabry-Perot spectroscopy and instrumentation
Periodicals: "Wisconsin Astrophysics Preprint Series"
City Reference Coordinates: 089°22'00"W 43°05'00"N

University of Wisconsin, Madison, Pine Bluff Observatory (PBO)
4065 Observatory Road
Cross Plains, WI 53528
Telephone: 608-262-3071
Telefax: 608-263-0361
Electronic Mail: <userid>@astro.wisc.edu
WWW: http://www.sal.wisc.edu/PBO/
Founded: 1958
Staff: 12
Activities: spectropolarimetry
Coordinates: 089°41'06"W 43°04'42"N H366m
City Reference Coordinates: 089°39'00"W 43°07'00"N

University of Wisconsin, Madison, Space Science and Engineering Center (SSEC)
1225 West Dayton Street
Madison, WI 53706
Telephone: 608-263-6730
Telefax: 608-262-5974
Electronic Mail: <userid>@ssec.wisc.edu
WWW: http://www.ssec.wisc.edu/
       http://cimss.ssec.wisc.edu/
Founded: 1965
Staff: 188
Activities: atmospheric studies of Earth and other planets * interactive computing, data access, image processing and scientific visualization * spaceflight hardware design and fabrication * scientific instrumentation
City Reference Coordinates: 089°22'00"W 43°05'00"N

University of Wisconsin, Milwaukee (UWM), Department of Physics, Manfred Olson Planetarium
c/o Department of Physics
1900 East Kenwood
P.O. Box 413
Milwaukee, WI 53201
Telephone: 414-229-4474
Telefax: 414-229-5589
WWW: http://www.uwm.edu/Dept/Physics/
       http://www.uwm.edu/UWM/Map/B_Planetarium.html
City Reference Coordinates: 087°55'00"W 43°02'00"N

University of Wisconsin, Oshkosh, Department of Physics and Astronomy
Oshkosh, WI 54901
Telephone: 414-424-4429
　　　　　　414-424-4433
Telefax: 414-424-7317
Electronic Mail: <userid>@maxwell.phys.uwosh.edu
WWW: http://maxwell.phys.uwosh.edu/
　　　　http://www.mio.uwosh.edu/tour/buckstaff.html (Buckstaff Planetarium)
City Reference Coordinates: 088°33'00"W 44°01'00"N

University of Wisconsin Press
2537 Daniels Street
Madison, WI 53718-6712
Telephone: 608-224-3900
Telefax: 608-224-3907
Electronic Mail: uwisc.press@macc.wisc.edu
WWW: http://www.wisc.edu/wisconsinpress/
Founded: 1936
• Publisher
City Reference Coordinates: 089°22'00"W 43°05'00"N

University of Wisconsin, River Falls (UWRF), Department of Physics
River Falls, WI 54022
Telephone: 715-425-3196
Telefax: 715-425-0652
Electronic Mail: <userid>@uwrf.edu
WWW: http://www.uwrf.edu/physics/
　　　　http://www.uwrf.edu/physics/astroview.html (Planetarium & Observatory)
Founded: 1874
Staff: 7
Activities: education * research * viewing sessions * planetarium shows
City Reference Coordinates: 092°38'00"W 44°52'00"N

University of Wisconsin, Stevens Point (UWSP), Department of Physics and Astronomy
Science Center
740 Reserve Street
Stevens Point, WI 54481
Telephone: 715-346-2139
　　　　　　715-346-2208
Telefax: 715-346-2944
Electronic Mail: <userid>@uwsp.edu
　　　　　　rolson@uwsp.edu (Randy W. Olson, Planetarium Director)
WWW: http://www.uwsp.edu/physastr/
　　　　http://www.uwsp.edu/physastr/plan_obs/ (Planetarium and Observatory)
Founded: 1945
Staff: 9
Coordinates: 089°34'16"W 44°32'20"N H340m
City Reference Coordinates: 089°34'00"W 44°31'00"N

Wehr Astronomical Society (WAS) Inc.
c/o Wehr Nature Center
9701 West College Avenue
Franklin, WI 53132
WWW: http://www.wehrastro.org/
Founded: 1981
City Reference Coordinates: 088°03'00"W 42°54'00"N

Yerkes Observatory
• See "University of Chicago, Yerkes Observatory"

USA - Wyoming

Campbell County School District Planetarium
c/o Sage Valley Junior High School
1000 Lakeway Drive
Gillette, WY 82716
Telephone: 307-682-4307
Telefax: 307-687-7614
Electronic Mail: nwilliams@ccsd.k12.wy.us (Nello Williams, Director)
Founded: 1980
Staff: 1
Activities: education * public shows
City Reference Coordinates: 105°30'00"W 44°18'00"N

Casper Planetarium
904 North Poplar Street
Casper, WY 82601
Telephone: 307-577-0310
Electronic Mail: stars@trib.com
WWW: http://www.trib.com/WYOMING/NCSD/PLANETARIUM/planetarium.html
Founded: 1966
Staff: 2
Activities: education * public programs * observing * gift shop
City Reference Coordinates: 106°19'00"W 42°51'00"N

Central Wyoming Astronomical Society (CWAS)
c/o Casper Planetarium
970 Glenn Road
Casper, WY 82601
or :
904 North Poplar Street
Casper, WY 82601
Telephone: 307-577-0310
Electronic Mail: cwas@coffey.com
Founded: 1989
Membership: 12
Activities: observing * public education * supporting professional activities
Coordinates: 106°19'00"W 42°51'30"N H1,519m to H2,520m
City Reference Coordinates: 106°19'00"W 42°51'00"N

Cheyenne Astronomical Society (CAS)
3539 Luther Place
Cheyenne, WY 82001
Telephone: 307-635-5944
Electronic Mail: mcurran@sisna.com (Martin D. Curran)
WWW: http://www.sisna.com/users/mcurran/
http://users.sisna.com/mcurran/
Founded: 1986
Membership: 100
Activities: monthly meetings * observing sessions * educational programs * star parties
Periodicals: (12) "Cosmic Babbler"
Observatories: 1 (Cheyenne Botanic Gardens)
City Reference Coordinates: 104°49'00"W 41°08'00"N

First Magnitude Corp. (FMC)
519 South Fifth Street
Laramie, WY 82070
Telephone: 307-745-3743
Telefax: 307-745-3743
Electronic Mail: firstmag@delphi.com
first@delphi.com
Activities: manufacturing digital cameras
City Reference Coordinates: 105°35'00"W 41°19'00"N

Jackson (Hole) Astronomy Club
c/o Mel Tucker
Box 123
Moran, WY 83013
505 Ponderosa Drive
Jackson, WY 83001
Telephone: 307-733-2173
Electronic Mail: wfarmer@wyoming.com
WWW: http://msnhomepages.talkcity.com/studiorow/wyomovieguy/astro.htm

Founded: 1993
Membership: 50
Activities: meetings * Astronomy Day
Periodicals: "Jackson (Hole) Astronomy Club Newsletter"
City Reference Coordinates: 110°35'00"W 43°52'00"N (Moran)
110°38'00"W 43°29'00"N (Jackson)

Laramie Astronomical Society and Space Observers (LASSO)
1860 North 23rd
Laramie, WY 82070
Electronic Mail: deepsky@lariat.org (Curtis MacDonald, President)
tjedgar@uwyo.edu (Thomas Edgar, Secretary)
WWW: http://www.lariat.org/LASSO/
http://www.geocities.com/capecanaveral/launchpad/1528/ (Mountain Skies Obs.)
Founded: 1996
Observatories: 1 (Mountain Skies Observatory)
City Reference Coordinates: 105°35'00"W 41°19'00"N

Rocky Mountain Planetarium Association (RMPA)
c/o Nello Williams
Campbell County School District Planetarium
Sage Valley Junior High School
1000 Lakeway Drive
Gillette, WY 82716
Telephone: 307-682-4307
Telefax: 307-687-7614
Electronic Mail: nwilliams@ccsd.k12.wy.us
WWW: http://www.azscience.org/planetarium/stars.html
City Reference Coordinates: 105°30'00"W 44°18'00"N

University of Wyoming, Department of Physics and Astronomy
P.O. Box 3905
University Station
Laramie, WY 82071
Telephone: 307-766-6150
Telefax: 307-766-2652
Electronic Mail: <userid>@uwyo.edu
edison@corral.uwyo.edu
WWW: http://faraday.uwyo.edu/
http://faraday.uwyo.edu/wiro/ (Wyoming Infrared Observatory - WIRO)
http://rbo.uwyo.edu/ (Red Buttes Observatory - RBO)
Founded: 1899 (Planetarium: 1972) (WIRO: 1977) (EISO: 1990)
Staff: 25
Activities: IR astrophysical research * education * planetarium shows * Wyoming Infrared Observatory (WIRO) * Red Buttes Observatory (RBO)
Coordinates: 105°58'35"W 41°05'53"N (WIRO)
City Reference Coordinates: 105°35'00"W 41°19'00"N

Vatican

Specola Vaticana
V-00120 Città del Vaticano
Telephone: (0)6-69885266
　　　　　　 (0)6-69885226
Telefax: (0)6-69884671
Electronic Mail: vatican@as.arizona.edu
WWW: http://clavius.as.arizona.edu/vo/
Founded: 1891
Staff: 10
Activities: astronomical observing * theoretical research
Periodicals: (1) "Annual Report"; "Proceedings of Meetings"
Coordinates: 012°39'06"E 41°44'48"N H450m (Castel Gandolfo, Italy – IAU Code 036)
　　　　　　　 109°53'30"W 32°42'05"N H3,130m (Mount Graham, USA-AZ – IAU Code 290)
• See also "Vatican Observatory Research Group (VORG), USA-AZ"
City Reference Coordinates: 012°27'00"E 41°54'00"N

Vatican Observatory
• See "Specola Vaticana"

Venezuela

Academia de Ciencias de América Latina (ACAL)
c/o Instituto Internacional de Estudios Avanzados
Apartado 17606
Caracas 1015-A
Telephone: (0)2-9763490
Telefax: (0)2-9763490
Electronic Mail: accal@reacciun.ve
WWW: http://www.acal-scientia.org/
Founded: 1982
Membership: 140
Periodicals: "Informes de Reuniones", "Boletín Ciencia en América Latina", "Boletín Electrónico Tecnología en América Latina", "Directorio de Instituciones Científicas de América Latina"
City Reference Coordinates: 066°56'00"W 10°30'00"N

Centro de Estudios de Astronomía (CEDA)
C/ Negrín
Edificio Davolca 9
Sabana Grande
Caracas 1050
Telephone: (0)2-718529
Telefax: (0)2-718529
Electronic Mail: <userid>@dino.conicit.ve
         jpineirov@etheron.net
Founded: 1986
Membership: 12
Activities: popularization through radio programmes * education * lectures * planetarium shows
City Reference Coordinates: 066°56'00"W 10°30'00"N

Centro de Investigaciones de Astronomía (CIDA)
Urb. Alto Chama
Calle 8 "Rio Mucujún"
Quinta CIDA
Mérida 5101
Telephone: (0)74-713883
         (0)74-712780
Telefax: (0)74-712459
Electronic Mail: info@cida.ve
WWW: http://www.cida.ve/
Founded: 1975
Activities: astrometry * stellar atmospheres * evolution of galaxies * comets * photometry of galaxies * star formation * instrumentation * education
Coordinates: 070°52'00"W 08°47'24"N H3,610m (Llano del Hato Observatory)
City Reference Coordinates: 071°08'00"W 08°36'00"N

Comisión Panamericana de Normas Técnicas (COPANT)
Avenida Andrés Bello
Edificio Torre Fondo Común - Piso 11
Caracas 1050
Telephone: (0)2-5754111
Telefax: (0)2-5741312
Electronic Mail: copant@cantv.net
WWW: http://www.copant.org/
Founded: 1949 (beginning of operations: 1961)
City Reference Coordinates: 066°56'00"W 10°30'00"N

Comisión Venezolana de Normas Industriales (COVENIN)
Avenida Andrés Bello
Edificio Torre Fondo Común - Piso 12
Caracas 1050
Telephone: (0)2-5752298
Telefax: (0)2-5741312
Electronic Mail: fonnor19@reacciun.ve
City Reference Coordinates: 066°56'00"W 10°30'00"N

Consejo Nacional de Investigaciones Científicas y Técnologicas (CONICIT)
Apartado de Correos 70.617
Caracas 1071-A
or :
Final de la Avenida Principal de los Cortijos de Lourdes
Edificio Maploca 1
Los Ruices

Telephone: (0)2-2397791
Electronic Mail: conicit@conicit.gov.ve
WWW: http://www.conicit.gov.ve/
Founded: 1967
Activities: planification, coordination and promotion of national R&D
City Reference Coordinates: 066°56'00"W 10°30'00"N (Caracas)

COPANT
• See "Comisión Panamericana de Normas Técnicas (COPANT)"

Dirección de Hidrología y Meteorología (DHM)
Apartado 2197
Maracay 2101-A
Telephone: (0)43-338043
WWW: http://marnr.gov.ve/dhym.htm
City Reference Coordinates: 067°36'00"W 10°15'00"N

Frente Unido Pro Astronomía Zuliana (FUPAZ)
Apartado Postal 10673
Maracaibo, Edo. Zúlia
Telephone: (0)61-524441
Telefax: (0)61-984174
Founded: 1997
Membership: 10
Activities: popularization * public observing * exhibitions * Sun * cosmology
Coordinates: 071°37'08"W 10°38'41"N
City Reference Coordinates: 071°37'00"W 10°40'00"N

Liga Venezolana de Astronomía (LIVA)
• See "Observatorio Astronómico EMVIC"

Observatorio Astronómico EMVIC
Apartado Postal 30109
Caracas 1030-A
Telephone: (0)2-4426636
Telefax: (0)2-4426636
Founded: 1969
Staff: 25
Activities: solar physics * occultations * planets * host of the "Liga Venezolana de Astronomía (LIVA)"
City Reference Coordinates: 066°56'00"W 10°30'00"N

Planetario Humboldt
Apartado Postal 6745
Parque del Este
Carmelitas
Caracas
Telefax: (0)2349188
Electronic Mail: 104721.1224@compuserve.com
WWW: http://intrade.impsat.com.ve/anmd/humboldt.htm
City Reference Coordinates: 066°56'00"W 10°30'00"N

Sociedad Universitaria de Astronomía (SUNA)
c/o Escuela de Geologia, Minas y Geofísica
Universidad Central de Venezuela
Caracas
Telephone: (0)6053190
Telefax: (0)6053120
Electronic Mail: hsalas@sagi.ucv.edu.ve (Henry Salas, Presidente)
WWW: http://www.ing.ucv.ve/suna/suna.htm
Founded: 1986
City Reference Coordinates: 066°56'00"W 10°30'00"N

Universidad de los Andes (ULA), Departamento de Física
Mérida 5101
Telephone: (0)74-442580 x235
(0)74-791202
(0)74-792660
Telefax: (0)74-441723
(0)74-791893
Electronic Mail: <userid>@ciens.ula.ve
WWW: http://www.ciens.ula.ve/departamentos/fisica.html
Founded: 1976
Activities: hydrodynamics * ISM * star formation * galactic structure * solar system * yearly school on relativity and theoretical astrophysics
City Reference Coordinates: 071°08'00"W 08°36'00"N

Zimbabwe

Astronomical Society of Southern Africa (ASSA), Harare Centre
c/o Mike Begbie
Prince Edward School
P.O. Box CY 418
Causeway
Harare
or :
P.O. Box UA 428
Union Avenue
Harare
Telephone: (0)14-29415
(0)14-46617
WWW: http://www.samara.co.zw/peschool/astronomy/assa.html
Founded: 1975
Membership: 40
Activities: popularization * observing
Periodicals: (12) "Notice"; "Cloudy Nights"
Coordinates: 031°00'23"E 17°45'30"S H1,450m (Belvedere)
City Reference Coordinates: 031°03'00"E 17°50'00"S

Department of Meteorological Services
P.O. Box BE150
Belvedere
Harare
Telephone: (0)4-778173
(0)4-778174
(0)4-778176
Telefax: (0)4-778161
Electronic Mail: zimmeteo@weather.utande.co.zw
WWW: http://weather.utande.co.zw/
Founded: 1925
Staff: 250
Activities: weather observing and forecasting
City Reference Coordinates: 031°03'00"E 17°50'00"S

Standards Association of Zimbabwe (SAZ)
P.O. Box 2259
Harare
Telephone: (0)4-8855112
(0)4-8820179
(0)4-8820212
Telefax: (0)4-882020
Electronic Mail: sazinfo@samara.co.zw
Founded: 1957
Staff: 100
Activities: standardization * quality improvement * technical services for testing goods and material * product and system certification
Periodicals: (4) "Fulcrum" (ISSN 1024-1787)
City Reference Coordinates: 031°03'00"E 17°50'00"S

Telephone and Telefax National Codes

Telephone and telefax codes for countries and territories round the world are provided in the following pages. They should be preceded by the international access number of the country or territory from which the call is placed (00, 011, and so on).

Telephone and telefax codes for countries and territories round the world are provided below. They should be preceded by the international access number of the country or territory from which the call is placed (00, 011, and so on).

Afghanistan	93
Albania	355
Algeria	213
American Samoa	684
Andorra	376
Angola	244
Anguilla	1809
Antarctica	672
Antigua and Barbuda	1268
Antilles (French_)	596
Antilles (Netherlands_)	599
Argentina	54
Armenia	374
Aruba	297
Ascension Island	247
Australia	61
Austria	43
Azerbaijan	994
Azores	351
Bahamas	1809
Bahrain	973
Bangladesh	880
Barbados	1246
Barbuda (Antigua and_)	1268
Belarus	375
Belgium	32
Belize	501
Benin	229
Bermuda	1441
Bhutan	975
Bolivia	591
Bophuthatsawa	27
Bosnia-Hercegovina	387
Botswana	267
Brazil	55
Brunei	673
Bulgaria	359
Burkina Faso	226
Burma (now Myanmar)	95
Burundi	257
Caicos Islands (Turks and_)	1809
Cambodia	855
Cameroon	237
Canada	1
Canary Islands	34
Cape Verde Islands	238
Cayman Islands	1809
Central African Republic	236
Chad	235
Channel Islands	44
Chile	56
China (PRC)	86
China (ROC)	886
Christmas Islands	6724
Ciskei	27
Cocos Islands	672
Cocos-Keeling Islands	61
Colombia	57
Commonwealth of Independent States	7
Comoro Islands	269
Congo	242
Cook Islands	682
Costa Rica	506
Côte d'Ivoire	225
Croatia	385
Cuba	53
Cyprus	357
Czech Republic	420
Denmark	45
Diego Garcia	246
Djibouti	253
Dominica	1809
Dominican Republic	1809
Dutch Antilles	599
Ecuador	593
Egypt	20
El Salvador	503
Equatorial Guinea	240
Eritrea	291
Estonia	372
Ethiopia	251
Falkland Islands	500
Faroe Islands	298
Fiji	679
Finland	358
France	33
French Antilles	596
French Guyana	594
French Polynesia	689
Futuna (Wallis and_)	681
Gabon	241
Gambia	220
Georgia	995
Germany	49
Ghana	233
Gibraltar	350
Greece	30
Greenland	299
Grenada	1809
Guadeloupe	590
Guam	671
Guatemala	502
Guinea	224
Guinea Bissau	245
Guinea Equatorial	240
Guyana	592
Guyana (French_)	594
Haiti	509
Hawaii	1808
Honduras	504
Hong Kong	852
Hungary	36
Iceland	354
India	91
Indonesia	62
Iran	98
Iraq	964
Ireland	353
Israel	972
Italy	39
Ivory Coast	225
Jamaica	1809
Japan	81
Jordan	962
Kazakhstan	7
Keeling Islands (Cocos-_)	61
Kenya	254
Kiribati	686
Kitts	1809
Korea (North_) (DPRK)	850
Korea (South_) (ROK)	82
Kuwait	965
Kyrgyzstan	371
Laos	856
Latvia	371
Lebanon	961
Lesotho	266
Liberia	231
Libya	218
Liechtenstein	41

Lithuania	370	St. Lucia	1758
Luxemburg	352	St. Pierre and Miquelon	508
Macao	853	St. Vincent	1809
Macedonia	389	Saipan	670
Madagascar	261	Samoa (American_)	684
Madeira	351	Samoa (Western_)	685
Malawi	265	San Marino	378
Malaysia	60	Sao Tome and Principe	239
Maldives	960	Saudi Arabia	966
Mali	223	Senegal	221
Malta	356	Seychelles	248
Mariana Islands	670	Sierra Leone	232
Marshall Islands	692	Singapore	65
Martinique	596	Slovakia	421
Mauritania	222	Slovenia	386
Mauritius	230	Solomon Islands	677
Mayotte	269	Somalia	252
Mexico	52	South Africa	27
Micronesia	691	South Korea (ROK)	82
Midway Islands	808	Spain	34
Miquelon (St. Pierre and_)	508	Sri Lanka	94
Moldova	373	Sudan	249
Monaco	33	Suriname	597
Mongolia	976	Swaziland	268
Montserrat	1664	Sweden	46
Morocco	212	Switzerland	41
Mozambique	258	Syria	963
Myanmar	95	Taiwan (ROC)	886
Namibia	264	Tajikistan	7
Nauru	674	Tanzania	255
Nepal	977	Thailand	66
Netherlands	31	Tobago (Trinidad and_)	1809
Netherlands Antilles	599	Togo	228
Nevis (St. Christopher-_)	1809	Tonga	676
New Caledonia	687	Transkei	27
New Zealand	64	Trinidad and Tobago	1809
Nicaragua	505	Tunisia	216
Niger	227	Turkey	90
Nigeria	234	Turkmenistan	7
Niue Island	683	Turks and Caicos Islands	1809
Norfolk Islands	6723	Tuvalu	688
North Korea (DPRK)	850	Uganda	256
Norway	47	Ukraine	380
Oman	968	United Arab Emirates	971
Pakistan	92	United Kingdom	44
Palau	680	United States of America	1
Palestinian Authority Territory	970	Uruguay	598
Panama	507	USA	1
Papua New Guinea	675	Uzbekistan	7
Paraguay	595	Vanuatu	678
Peru	51	Vatican	396
Philippines	63	Venda	27
Poland	48	Venezuela	58
Polynesia (French_)	689	Vietnam	84
Portugal	351	Virgin Islands	1809
Principe (Sao Tome and_)	239	Wake Island	1808
Puerto Rico	1787	Wallis and Futuna	681
Qatar	974	Western Samoa	685
Reunion Island	262	Yemen (former Arab Republic)	967
Romania	40	Yemen (former PDR)	969
Russia	7	Yugoslavia	381
Rwanda	250	Zaire	243
St. Christopher-Nevis	1809	Zambia	260
St. Helena	290	Zimbabwe	263

Abbreviations, Acronyms, Contractions, and Symbols

Besides the acronyms mentioned in the subsequent index, please refer also to the sister publication *StarBriefs Plus - A Dictionary of Abbreviations, Acronyms, and Symbols in Astronomy and Related Space Sciences* (ISBN 1-4020-1925-4).

It gathers together about 200,000 entries encountered in the literature relative to astronomy and space sciences as well as to related fields such as aeronautics, aeronomy, astronautics, atmospheric sciences, chemistry, communications, computer sciences, data processing, education, electronics, energetics, engineering, environment, geodesy, geophysics, information handling, management, mathematics, meteorology, optics, physics, remote sensing, and so on.

Abbreviations, acronyms, contractions and symbols in common use and/or of general interest have also been included where appropriate. The travelling scientist has not been forgotten either (codes of locations, currencies, and so on). Separate sections are devoted to Greek letters, mathematical symbols, special signs and characters, as well as to entries with a numerical beginning.

Index

This exhaustive index gives a breakdown not only by different designations and acronyms, but also by location and major terms in names. A search for information in the directory would normally begin with consultation of this index.

The information is also duplicated in thematic subindices of academies, awards, bibliographical services, consultants, data centres, dealers and distributors, funding organizations, IAU observatory codes, Internet service providers, ISS-Numbers, manufacturers, meteorological offices, norms and standards institutes, observatories, periodicals, planetariums, publishers, science museums, software producers, and so on.

5

5As
    Amherst Area Amateur Astronomers Association, 747

A

A3
    Aurora Astronomical Association, 652
A4
    Astronomische Arbeitsgemeinschaft Aalen, 210
AAA
    Agrupación Astronómica Aragonesa, 470
    Amateur Astronomers Association, 806
        Brooklyn Chapter, 806
        Staten Island Chapter, 806
    Asociación Argentina de Astronomía, 6
    Asociación de Aficionados a la Astronomía, 586
    Association Astronomique d'Anjou, 156
    Association Astronomique de l'Ain, 156
    Association of Amateur Astronomers, 877
    Associazione Acquavivese Astrofili, 303
AAAA
    Albany Area Amateur Astronomers, 805
    American Association of Amateur Astronomers, 877
    Ames Area Amateur Astronomers, 716
    Asociación Argentina Amigos de la Astronomía, 6
    Association des Astronomes Amateurs d'Auvergne, 158
    Astronomische Arbeitsgemeinschaft Aalen, 210
AAAAT
    Association des Astronomes Amateurs Abitibi-Témiscamingue, 78
AAAD
    Amateur Astronomers Association of Delhi, 281
AAAF
    Association Aéronautique et Astronautique de France, 155
AAAP
    Amateur Astronomers Association of Pittsburgh, 851
    Amateur Astronomers Association of Princeton, 791
AAAS
    American Association for the Advancement of Science, 669
    Arizona Amateur Astronomy Society, 593
AAAV
    Amateur Astronomers' Association of Vadodara, 281
AAB
    Association d'Astronomie El-Battani, 4
    Associazione Astrofili Bolognesi, 304
AABV
    Associazione Astrofili del Basso Vicentino, 304
AAC
    Agrupación Astronómica Cántabra, 470
    Agrupación Astronómica Complutense, 470
    Agrupación Astronómica de Córdoba, 471
    Alachua Astronomy Club, 678
    Alamogordo Astronomy Club, 799
    Amateur Astronomy Centre, 531, 563
    Asesores Astronómicos Cacereños, 474
    Associazione Astronomica Cortina, 308
    Atlanta Astronomy Club, 687
AACE
    Association for the Advancement of Computing in Education, 896
Aachen, 248, 251
AACI
    Amateur Astronomers of Central Iowa, 716
AACO
    Association Astronomique de la Côte d'Or, 156
AAD
    Agrupación Astronómica de los Dolores, 472
AAE
    Association for Astronomy Education, 532
AAF
    Agrupación Astronómica de Fuerteventura, 471
    Associazione Astrofili Fiorentini, 304
    Associazione Astronomica Frusinate, 308
AAFC
    Association Astronomique de Franche-Comté, 156
AAG
    Associazione Astrofili Garfagnana, 305
AAGC
    Agrupación Astronómica de Gran Canaria, 471
AAGG
    Associazione Astrofili "Galileo Galilei", 305
AAH
    Agrupación Astronómica de Huesca, 471
AAI
    Amateur Astronomers Inc., 791
    Association Astronomique de l'Indre, 157
    Associazione Astrofili Imolesi, 305
    Astronomischer Arbeitskreis Ingolstadt, 213
AAJ
    Amateur Astronomers of Jackson, 761
AAK
    Astronomischer Arbeitskreis Kassel eV, 213
AAL
    Amateurs Astronomes du Luxembourg, 376
    Astronomische Arbeitsgruppe Laufen, 211
    Astronomy Associates of Lawrence, 721
AALC
    Association Astronomique de Loir-et-Cher, 157
Aalen, 210, 235
Aalter, 53
AAM
    Agrupación Astronómica de Madrid, 472
    Agrupación Astronómica de Manresa, 472
    Associazione Astrofili Mantovani, 305
    Associazione Astronomica Milanese, 309
AAM31
    Asociación Astronómica M31, 474
AAMT
    Associazione Astrofili Monti della Tolfa, 305
AAMV
    Amateurastronomen Max Valier, 303
AAMVG
    Astronomischer Arbeitskreis Merkur/Venus Göttingen, 213
AANC
    Astronomical Society of Northern California, 610
AANZ
    Astronautics Association of New Zealand, 408
AAO
    Agrupación Astronómica d'Osona, 473
    Anglo-Australian Observatory
        Coonabarabran, 14
        Epping Laboratory, 14
AAofA
    Amateur Astronomers of Acadiana, 729
AAP
    Agrupación Astronómica de la Palma, 471
    Associazione Amici dei Planetari, 303
    Astronomische Arbeitsgemeinschaft Pfaffenwinkel, 210
    Astronomischer Arbeitskreis Pforzheim, 213
AAPT
    American Association of Physics Teachers, 734
AARAO
    Ålborg Amateur Radio Astronomy Observatory, 134
Aarau, 500
AARU
    Astronomy and Atmospheric Research Unit (USM), 379
        Sheikh Tahir Astronomical Centre, 378
AAS
    Abingdon Astronomical Society, 531
    Agrupació Astronòmica de Sabadell, 472
    American Astronautical Society, 646, 895
    American Astronomical Society, 669
    Andover Astronomical Society, 531
    Association Astronomique du Soissonnais, 157
    Associazione Astrofili Sardi, 306
    Associazione Astrofili Segusini, 306

Associazione Astrofili Spezzini, 306
Astronomical Association - Sofia, 74
Astronomischer Arbeitskreis Salzkammergut, 34
Auburn Astronomical Society, 687
Aurora Astronomical Society, 835
Austin Astronomical Society, 877
Australian Academy of Science, 17
Ayrshire Astronomical Society, 535
AASI
    Astronomical Association of Southern Illinois, 698
AASZ
    Astronomical Astronautical Society Zadar, 121
AAT
    Agrupación Astronómica de Tenerife, 473
    Anglo-Australian Telescope, 14
    Associazione Astrofili Teatini, 307
    Associazione Astrofili Tethys, 307
    Associazione Astrofili Trentini, 307
AAUA
    Agrupación Astronómica Universitaria de Alicante, 473
AAV
    Association Astronomique de la Vallée, 156
    Associazione Astrofili Valdinievole, 307
    Associazione Astrofili Valtellinesi, 307
    Associazione Astrofili Veneziani, 308
AAVSO
    American Association of Variable Stars Observers, 746
ABA
    Asociación Boliviana de Astronomía, 62
AbAO
    Abastumani Astrophysical Observatory, 206
        Town Department, 206
Abastumani, 206
Abastumani Astrophysical Observatory (AbAO), 206
    Town Department, 206
Abbe-Stiftung (Ernst_), 207
    Zeiss-Planetarium, 266
Abbotsford, 88
ABC
    Academia Brasileira de Ciencias, 65
Abdis Salam International Centre for Theoretical Physics (ICTP), 336
Abdus Salam International Centre for Theoretical Physics (ICTP), 302, 336
Aberdeen, 531
Aberdeen College
    Planetarium, 531
Abingdon, 561
Abingdon Astronomical Society (AAS), 531
Ablis, 162
ABNT
    Associação Brasileira de Normas Técnicas, 65
Abrams Planetarium, 761, 762
Abu (Mount_), 287
ACA
    Associazione Cernuschese Astrofili, 309
    Astro Club Aubagnais, 161
    Astronomie Centre Ardenne, 47
    Astronomy Club of Akron, 835
    Astronomy Club of Augusta, 687
ACAC
    Ancient City Astronomy Club, 678
Academia Brasileira de Ciencias (ABC), 65
Academia de Ciencias de América Latina (ACAL), 927
Academia de Ciencias de Cuba (ACC), 123
Academia de la Investigación Científica (AIC), 382
Academia Europaea, 531
Academia Mexicana de Ciencias (AMC), 382
Academia Nacional de Ciencias de Bolivia (ANCB), 62
Academia Sinica, 112
Academia Sinica Institute of Astronomy and Astrophysics (ASIAA)
    Preparatory Office, 516
Académie des Sciences, 154
Académie Royale des Sciences, des Lettres et des Beaux-Arts de Belgique, 47
Academies

AAS, 17
Abastumani, 206
ABC, 65
ACC, 123
Alma-Ata, 364
AMC, 382
América Latina, 927
Amsterdam, 395
ANCB, 62
Armenia, 12
ASRT, 142
ASSN, 508
Astronautics, 184
Athens, 268
Athinai, 268
Australia, 17
Austria, 37
Azerbaijan, 43
Baku, 43
Bangalore, 283, 284
Bavaria, 218
Bayern, 218
Beijing, 113, 114
Belarus, 46
Belgium, 47, 54
Bern, 508
Bolivia, 62
Bratislava, 459
Brazil, 65
Brussel, 54
Bruxelles, 47
București, 438
Budapest, 274, 275
Bulgaria, 75, 76
Byurakan, 12
Cairo, 142
Canberra, 17
Caracas, 927
China (PRC), 112–115
Cuba, 123
Czech Republic, 124
Debrecen, 275
Delingha, 114
Denmark, 139
Dublin, 296, 297
Dushanbe, 518
Egypt, 142
Estonia, 144
Europaea, 531
Finland, 146
France, 154
Georgia, 206
Helsinki, 146
Hungary, 274–276
India, 283, 284
Ireland, 296, 297
Irkutsk, 445, 448
Israel, 299
Italy, 302
Japan, 358
Jerusalem, 299
Jilin, 113
Kanobili (Mount_), 206
Karachaevo, 448
Kazakhstan, 364
Kenya, 365
Kharkiv, 528
Kharkov, 528
Kiev, 529
Kislovodsk, 447
Kitab, 925
KNAS, 365
KNAW, 395
København, 139
Korea (DPRK), 366
Kórnik, 430
Kunming, 115

Kyiv, 529
Lagos, 414
La Paz, 62
Latin America, 927
Latvia, 371
Lincei, 302
Ljubljana, 462
Mainz, 207
México, 382
Minsk, 46
Mongolia, 388
Moscow, 444–448, 451
Mount Kanobili, 206
Mt. Kanobili, 206
München, 218
Nairobi, 365
Nanjing, 114
NAS, 414
Netherlands, 395
Nigeria, 414
Nizhny Novgorod, 446
Norway, 416
Ondřejov, 124
Oslo, 416
Paris, 154, 184
Poland, 429, 430
Praha, 124
Pyongyang, 366
Riga, 371
Rio de Janeiro, 65
Roma, 302
Romania, 438
Russia, 444–448, 451
Saint Petersburg, 444, 445, 447
SANW, 508
SAS, 508
SAZU, 462
Shaanxi, 114
Shanghai, 114
Shemakha, 43
Slovak Republic, 459
Slovenia, 462
Smolyan, 75
Sodankylä, 146
Sofia, 75, 76
Stockholm, 493
St. Petersburg, 444, 445, 447
Sweden, 482, 493
Switzerland, 508
Taipei, 516
Taiwan, 516
Tajikistan, 518
Tallinn, 144
Tartu, 144
Tashkent, 925
Tatranská Lomnica, 459
Tbilisi, 206
Third World, 336
Timişoara, 438
Tokyo, 358
Torino, 302
Toruń, 430
Trieste, 336
Troitsk, 446
Turnov, 124
Ukraine, 528, 529
USA, 672, 673
Uzbekistan, 925
Warszawa, 429, 430
Washington, 672, 673
Wien, 37
Wrocław, 430
Yakutsk, 446
Zelenchuk, 445
Académie Suisse des Sciences Naturelles (ASSN), 500, 508
Academy of Athens
    Research Center for Astronomy and Applied Mathematics, 268
Academy of Sciences of the Czech Republic, 124
    Astronomical Institute
        Ondřejov, 124
        Praha, 124, 127
    Institute of Physics, 124
    Institute of Plasma Physics, 124
    Optical Development Workshop, 124
Academy of Sciences of the DPRK
    Pyongyang Astronomical Observatory (PAO), 366
Academy of Scientific Research and Technology (ASRT), 142

Acadiana (Amateur Astronomers of_), 729
ACAL
    Academia de Ciencias de América Latina, 927
ACC
    Academia de Ciencias de Cuba, 123
Accademia delle Scienze di Torino, 302
Accademia Nazionale dei Lincei, 302
Accademia Svizzera di Scienze Naturali (ASSN), 500, 508
ACDA
    Asociación Colombiana de Estudios Astronómicos, 119
ACE
    Astronomical Consultants and Equipment, 593
ACHAYA
    Asociación Chilena de Astronomía y Astronáutica, 107
ACIS
    AXAF CCD Imaging Spectrometer, 754
ACM
    Association for Computing Machinery, 807
Acoustical Society of America (ASA), 805
ACP
    American Center for Physics, 734
    Aspen Center for Physics, 652
    Association Canadienne des Physiciens, 80
Acquaviva, 303
ACR
    Astroclub Radebeul, 209
ACS
    American Chemical Society, 670
Acta Astronomica, 427
Acta Astronomica Sinica, 112
Acta Astrophysica Sinica, 112
Acton, 17, 22, 25, 757
ACV
    Astro Club Voyager, 312
ADA
    Association Drômoise d'Astronomie, 158
ADAC
    Astronomical Data Analysis Center, 352
Adalbert Jeszenkowitsch Gesellschaft (Dr._), 34, 35
Adana, 522
Adaptive Optics Associates (AOA) Inc., 746
ADAS
    Altrincham and District Astronomical Society, 531
ADC
    Astronomy Data Center, 740
Addin Toussi Observatory (Khadjeh Nassir_), 293
Adelaide, 14, 16, 29, 30
Adelaide Planetarium, 14
Adelphi University
    Department of Physics, 805
ADF
    Astrophysics Data Facility, 740
ADION
    Association pour le Développement International de l'Observatoire de Nice, 160
Adirondack Video Astronomy (AVA), 805
Adler Planetarium and Astronomy Museum, 698
ADNDT
    Atomic Data and Nuclear Data Tables, 808
Adorama Camera Inc., 805
ADR Spacelink & Commercialization Inc., 78
ADS
    Astrophysics Data System, 751
Advanced Light Source (ALS), 608
Advanced Mechanical and Optical Systems (AMOS), 47

Advanced Satellite for Cosmology and Astrophysics (ASCA), 754
Advanced Telescope Supplies, 14
Advanced X-ray Astrophysics Facility (AXAF), 750, 754
Advanstar Communications Inc., 835
AEAAC
    Association Éducative des Amateurs d'Astronomie du Centre, 158
AEDC
    Arnold Engineering Development Center, 870
AEI
    Albert-Einstein-Institut, 237
AENOR
    Asociación Española de Normalización y Certificación, 474
AER
    Atmospheric and Environmental Research, 748
AeroAstro, 895
Aerospace Corp., 608
Aerospace Industries Association (AIA), 669
Aerospatiale, 154
Aerovox
    Massachusetts, 746
    New Bedford, 746
    United Kingdom, 535
    USA-MA, 746
    Weymouth, 535
Aerovox Inc., 746
AFA
    Association Française d'Astronomie, 158
AFAM
    Associazione Friulana di Astronomia e Meteorologia, 309
AFAS
    Association Française pour l'Avancement des Sciences, 159
AFB
    Air Force Base
        Arnold, 870
        Bolling, 669
AFDA
    Australian Defence Force Academy
        School of Physics, 18
A.F. Ioffe Physical Technical Institute, 440
    Division of Plasma Physics, Atomic Physics, and Astrophysics, 444
AFNOR
    Association Française de Normalisation, 159
AFO
    Astronomischer Freundeskreis Ostsachsen, 213
AFOEV
    Association Française des Observateurs d'Étoiles Variables, 159
AFOSR
    Air Force Office of Scientific Research, 669
A.F. Philips Sterrenwacht (Dr._), 389, 390
Africa
    (South_), 464
    (Southern_), 464, 465, 929
AFRL
    Air Force Research Laboratory
        Maui Optical Station, 692
AG
    Astronomische Gesellschaft, 211
        Arbeitskreis Astronomiegeschichte, 211
AGAA
    Association Guadeloupéenne d'Astronomes Amateurs, 159
AGB
    Astronomische Gesellschaft Baden, 500
    Astronomische Gesellschaft Bern, 500
Agence Spatiale Canadienne (ASC), 78
    Programme des Sciences Spatiales, 82
Agence Spatiale Européenne (ASE), 154
    Allemagne, 228, 245
        Centre de Coordination pour le Télescope Spatial, 245
        Centre des Astronautes Européens, 228
        Centre d'Opérations Spatiales, 228
        Darmstadt, 228
        Division Astrophysique, 392
        Division des Publications, 392
        ECSL, 177
        EPD, 392
        Espagne, 476
        ESRIN, 318
        ESTEC, 392
            Division Astrophysique, 392
            Division des Publications, 392
            Noordwijk Space Expo, 399
        European Centre for Space Law, 177
        France, 177
        Frascati, 318
        Garching, 245
        Information Retrieval System, 318
        Italie, 318
        Kiruna, 489
        Köln, 228
        Madrid, 476
        Netherlands, 392
        Noordwijk, 392
        Noordwijk Space Expo, 399
        Paris, 177
        Pays-Bas, 392, 399
        Siège Principal, 177
        Station de Kiruna, 489
        Station de Villafranca, 476
        ST-ECF, 245
        Suède, 489
        VILSPA, 476
Agenzia Spaziale Italiana (ASI)
    Base di Lancio, 302
    Centro di Geodesia Spaziale "Giuseppe Colombo", 302
    Matera, 302
    Roma, 303
    Sede Legale, 303
    Trapani, 302
AGG
    Astronomische Gesellschaft Graubünden, 500
Agnes Scott College (ASC)
    Bradley Observatory, 687
AGO
    Astronomische Gesellschaft Oberwallis, 501
Agoura Hills, 646
AGRI
    Astronomy and Geophysics Research Institute (King Abdulaziz University), 453
Agrupació Astronòmica de Sabadell (AAS), 472
Agrupación Astrofotografica "Quarks", 470
Agrupación Astronómica Aragonesa (AAA), 470
Agrupación Astronómica Cántabra (AAC), 470
Agrupación Astronómica Complutense (AAC), 470
Agrupación Astronómica de Barcelona (ASTER), 470
Agrupación Astronómica de Cáceres, 470
Agrupación Astronómica de Córdoba (AAC), 470
Agrupación Astronómica de Fuerteventura (AAF), 471
Agrupación Astronómica de Gran Canaria (AAGC), 471
Agrupación Astronómica de Huesca (AAH), 471
Agrupación Astronómica de la Autónoma, 471
Agrupación Astronómica de la Palma (AAP), 471
Agrupación Astronómica de la Rioja, 472
Agrupación Astronómica de los Dolores (AAD), 472
Agrupación Astronómica de Madrid (AAM), 472
Agrupación Astronómica de Manresa (AAM), 472
Agrupación Astronómica de Sabadell (AAS), 472
Agrupación Astronómica de Santa Pola, 472
Agrupación Astronómica de Setenil "Stellae", 472
Agrupación Astronómica de Tenerife (AAT), 473
Agrupación Astronómica d'Osona (AAO), 473
Agrupación Astronómica "Stephen Hawking", 473
Agrupación Astronómica Tamix, 473
Agrupación Astronómica Universitaria de Alicante (AAUA), 473
Agrupación Vallisoletana de Astronomía (AVA), 483
AGS

Astronomische Gesellschaft Solothurn, 501
AGU
American Geophysical Union, 670
Aguilar (Observatorio Astronómico Felix_), 10
AGZO
Astronomische Gesellschaft Zürcher Oberland, 501
AGZU
Astronomische Gesellschaft Zürcher Unterland, 501
AHAS
Allegheny Highland Astronomical Society, 851
Ahmedabad, 287
AIA
Aerospace Industries Association (AIA), 669
AIAA
American Institute of Aeronautics and Astronautics, 895
AIC
Academia de la Investigación Científica, 382
Aichi, 351, 355
Aichi University of Education
Department of Physics and Astronomy, 342
AIG
Association Internationale de Géodésie, 138
Aiken, 866
Ain, 156
AIP
American Institute of Physics, 736
Headquarters, 734
Publishing Center, 806
Astrophysikalisches Institut Potsdam, 212, 217
Observatorium für Solare Radioastronomie, 218
Sonnenobservatorium Einsteinturm, 218
Australian Institute of Physics, 18
Air and Space Europe, 154
Airdrie Observatory, 532
Air Force Academy Planetarium, 652
Air Force Base (AFB)
Arnold, 870
Bolling, 669
Air Force Office of Scientific Research (AFOSR), 669
Air Force Research Laboratory (AFRL)
Maui Optical Station, 692
Office of Public Affairs, 799
AIT
Astronomische Instrumente Stefan Thiele, 212
Aix-en-Provence, 154, 197
Ajaccio, 169
AJP
American Journal of Physics, 734
Akademie der Wissenschaften und der Literatur Mainz, 207

AKAF
Arbeitskreis Astronomie Freiburg eV, 208
Akeno Observatory, 342, 360
Akima Planetarium, 870
AKM
Arbeitskreis Meteore, 208
Akron, 835, 844
AL
Astronomical League, 776
Alabama, 589–591
(North_), 591
Alachua Astronomy Club (AAC) Inc., 678
Al al-Bayt University
Institute of Astronomy and Space Sciences, 363
Alama-Ata, 364
Alamogordo, 799, 803
Alamogordo Amateur Astronomers, 799
Alamogordo Astronomy Club (AAC), 799
Alan C. Davis Planetarium, 739
Alaska, 592
Albacete, 474
Albania, 3
Albanian Physical Society, 3
Albany, 691, 808
Albany Area Amateur Astronomers, Inc. (AAAA), 805
Alberta, 101

Albert-Einstein-Institut (AEI), 207, 237
Albert Einstein Planetarium, 669, 676
Albi, 154
Albireo Amateur Astronomy Society, 273
Albiréo - Astronomes Amateurs Tarnais, 154
Ålborg, 134
Ålborg Amateur Radio Astronomy Observatory (AARAO), 134
Albstadt-Ebingen, 216, 250
Albuquerque, 799–801, 803, 804
Albuquerque Astronomical Society (TAAS) (The_), 799, 803

Alcalá de Henares, 470, 480
Alcalá la Real, 483
Alcatel
Études Techniques et Constructions Aérospatiales (ETCA), 47
Alcock (Gruppo G.E.D._), 324
Alden Planetarium, 746
Aldermaston, 573
Aldrin Astronomical Center (Edwin A._), 794
Aldrin Planetarium, 685
Além-Paraíba, 69
Alenia Spazio SpA, 303
Alexander Morrison Planetarium, 608, 611, 637
Alexandria, 895, 897, 898, 903, 904
Alfred, 805
Alfred University
Stull Observatory, 805
Algarve, 433
Alger, 4, 5
Algeria, 4
Algonquin, 704
Alicante, 473
Alice and Leonard Dreyfuss Planetarium, 791
Alice G. Wallace Planetarium, 746
Alice Springs, 15
Aligarh, 281
Aligarh Muslim University (AMU)
Physics Department
Astrophysics Group, 281
Allegheny College
Physics Department, 851
Allegheny Highland Astronomical Society (AHAS), 851
Allegheny Observatory, 851, 862
Allentown, 851, 858
Allentown School District Planetarium, 851
Aller (Observatorio Astronómico Ramón María_), 485
Allgäuer Volkssternwarte Ottobeuren (AVSO) eV, 207
Alliance, 839
Allison University (Mount_), 92
Alloway Observatory, 20
ALMA
Atacama Large Millimeter Array, 600
Alma-Ata, 364
Almaty, 364
Almería, 475, 481
Almere, 391
Alpena, 764
Alpes, 182, 203
Alpha Centaure (CAAC) (Club d'Astronomie_), 171
Alpine, 877
ALPO
Association of Lunar and Planetary Observers, 609
ALPSP
Association of Learned and Professional Society Publishers, 532
ALS
Advanced Light Source, 608
American Lunar Society, 851
Alsace, 200
Alstead, 789
Alston Observatory, 575
Alta Val Trebbia, 341
Altec, 207
Altenmünster, 248
Alternativa Racional a las Pseudociencias (ARP), 473

Altrincham, 531
Altrincham and District Astronomical Society (ADAS), 531

Alushta, 528
ALW
    Aard- en Levenswetenschappen, 393
Alworth Planetarium (Marshall W._), 770, 771
AMA
    Associazione Marchigiana Astrofili, 310
Amado, 596
AMANDA
    Antarctic Muon and Neutrino Detector Array, 646
Amarillo, 878
AMAS
    Association Marseillaise d'Astronomie, 159
Amatérská Prohlídka Oblohy (APO), 130
Amateurastronomen Max Valier (AAMV), 303
Amateur Astronomers Association (AAA), 806
    Brooklyn Chapter, 806
    Staten Island Chapter, 806
Amateur Astronomers' Association (Bombay), 281
Amateur Astronomers Association of Delhi (AAAD), 281
Amateur Astronomers Association of Pittsburgh (AAAP), 851
Amateur Astronomers Association of Princeton (AAAP), 791

Amateur Astronomers' Association of Vadodara (AAAV), 281
Amateur Astronomers Inc. (AAI), 791
Amateur Astronomers of Acadiana (AAofA), 729
Amateur Astronomers of Central Iowa (AACI), 716
Amateur Astronomers of Eastern Jutland, 134
Amateur Astronomers of Jackson (AAJ), 761
Amateur Astronomers Society Seulaset, 146, 150
Amateur Astronomical Club Canopus, 74
Amateur Astronomical Society of Rhode Island, 864
Amateur Astronomical Society of Seltjarnarnes, 279
Amateurastronomische Vereinigung Göttingen (AVG) eV, 207

Amateur Astronomy Centre (AAC), 531, 563
Amateur Observers' Society (AOS) of New York Inc., 806
Amateurs Astronomes du Luxembourg (AAL), 376
Amateur Satellite Observers of Southeast Virginia (ASOSV), 895
Amateur Telescope Maker's Association (ATMA), 907
Amateur Telescope Makers of Boston (ATMOB), 746
Amateur- und Präzisionsoptik-Mechanik Markus Ludes, 207, 208
AMC
    Academia Mexicana de Ciencias, 382
AMDIC
    Associazione Astronomica Madonna di Campiglio, 308
America, 675
    (North_), 657, 663, 867
Americana, 69
American Association for the Advancement of Science (AAAS), 669
American Association of Amateur Astronomers (AAAA), 877
American Association of Physics Teachers (AAPT), 734
American Association of Variable Stars Observers (AAVSO), 746
American Astronautical Society (AAS), 646, 895
American Astronomical Society (AAS), 669
American Center for Physics (ACP), 734
American Chemical Society (ACS), 670
American Educational Products Inc., 652
American Geophysical Union (AGU), 670
American Institute of Aeronautics and Astronautics (AIAA), 895
American Institute of Physics (AIP), 736
    Headquarters, 734
    Publishing Center, 806
American Journal of Physics (AJP), 734
American Lunar Society (ALS), 851
American Mathematical Society (AMS), 864
American Meteorological Society (AMS), 747

American Meteor Society (AMS), 806
American Museum of Natural History, 815
American National Standards Institute (ANSI), 670
American Nuclear Society (ANS), 698
American Physical Society (APS), 734
    Editorial Offices, 806
American Society for Engineering Education (ASEE), 671
American Society for Information Science (ASIS), 734
American Statistical Association (ASA), 895
American University (AU)
    Physics Department, 671
American Vacuum Society (AVS), 806, 808
AMES
    Accent on Mathematics Engineering and Science, 596
Ames, 716–718
Ames Area Amateur Astronomers (AAAA), 716
Ames Research Center (ARC), 608
    Astrophysics Branch, 630
    NASA Astrobiology Institute (NAI), 630
    Stratospheric Observatory for Infrared Astronomy (SOFIA), 630
Amherst, 747, 751, 759, 807, 812, 822
Amherst Area Amateur Astronomers Association (5As), 747

Amiens, 157, 205
Amis du Planétarium d'Aix-en-Provence (APAP), 154
Amman, 363
Ammar (H._), 453
AMOLF
    Institute for Atomic and Molecular Physics, 393
AMOS
    Advanced Mechanical and Optical Systems, 47
AMPO
    Astronaut Memorial Planetarium and Observatory, 678
AMS
    American Mathematical Society, 864
    American Meteorological Society, 747
    American Meteor Society, 806
AMSA
    Associazione Maremmana Studi Astronomici, 310
AMSEE
    Association Méditerranéenne des Sciences de l'Environnement et de l'Espace, 160
Amsterdam, 389, 391, 393, 395–397, 403, 404, 406
Amsterdamse Weer- en Sterrenkundige Kring (AWSK), 406

AMU
    Aligarh Muslim University, 281
AN
    Astronomische Nachrichten, 212
Anacortes, 907
Anacortes Telescopes, 907
Analogic Corp., 747
ANAP
    Association Narbonnaise d'Astronomie Populaire, 160
ANASA
    Azerbaijanian National Aerospace Agency, 43
ANCB
    Academia Nacional de Ciencias de Bolivia, 62
Anchorage, 592
Ancient City Astronomy Club (ACAC), 678
Ancón, 423
Ancona, 310, 314
Andenes, 417
Anderson Mesa Station, 598
Anderson Planetarium (M.D._), 870, 872
Andover Astronomical Society (AAS), 531
Andøya, 417
Andøya Rocket Range (ARR), 415, 417
Andromeda, 207
Andromeda Software Inc., 807
Andromède, 154
Andrus Planetarium, 807
Angelo State University (ASU)
    Planetarium, 877
Angénieux, 155
Angered, 488

Angers, 156, 157
Anglet, 180
Angleur, 47, 57, 58
Anglo-Australian Observatory (AAO)
    Coonabarabran, 14
    Epping Laboratory, 14
Angwin, 632
Animaciencia, 474
Anjou, 156
Ankara, 522–524
Ankara University
    Department of Astronomy and Space Science, 522
ANL
    Argonne National Laboratory, 698
Annales Geophysicae, 155
Annapolis, 744
Ann Arbor, 761, 765, 767–769
Annecy, 156
Annual Reviews Inc. (ARI), 608
ANS
    American Nuclear Society, 698
ANSI
    American National Standards Institute, 670
ANSTJ
    Association Nationale Sciences Techniques Jeunesse, 198
Antalya, 524
Antarctica, 466, 707, 915
Antarctic Muon and Neutrino Detector Array (AMANDA), 646
Antelope Valley Astronomy Club (AVAC), 609
Antibes, 180
Antioch, 623
Antiquaries Manasek, 893
Antique Telescope Society (ATS), 687
Antist, 483
Antofagasta, 108–110
Anton Pannekoek (Sterrenkundig Instituut_), 404
Antwerpen, 52, 54, 58, 60
ANU
    Australian National University
        Department of Physics and Theoretical Physics, 17
ANU, Australian National University
    CSIRO Office of Space Science and Applications, 22
    Earth Observation Centre, 22
    Mount Stromlo and Siding Spring Observatories, 18
    Research School of Astronomy and Astrophysics, 18
    Research School of Earth Sciences, 18
    Research School of Physical Sciences
        Department of Theoretical Physics, 18
    Research School of Physical Sciences and Engineering
        Plasma Research Laboratory, 19
Anvers, 52, 54, 58, 60
AOA
    Adaptive Optics Associates, 746
    Associazione Ogliastrina di Astronomia, 310
AOAS
    Arkansas-Oklahoma Astronomical Society, 606
AOCA
    Amis de l'Observatoire des Creusets à Arbaz, 500
AOCC
    Astronomical Observatory of Campo Catino, 312
AOK
    Astrooptik Kohler, 502
AOS
    Amateur Observers' Society, 806
Aosta, 311
APAA
    Associação Portuguesa de Astronomos Amadores, 433
Apache Point Observatory (APO), 799, 802
APAP
    Amis du Planétarium d'Aix-en-Provence, 154
APAS
    Astrophysical, Planetary and Atmospheric Sciences, 660
APAU
    Asociación de Profesores de Astronomía de Uruguay, 586
APD

Astronomical Pocket Diary, 408
Apeldoorn, 390
Apex, 829
APHELIE
    Astronomie, Physique, Élaboration, Instrumentation et Observation, 162
ApJ
    Astrophysical Journal, 708
        Supplement Series, 708
APL
    Applied Physics Letters, 734
APM-Telescopes Markus Ludes, 207
APM Telescopes USA, 747
APO
    Amatérská Prohlídka Oblohy, 130
    Apache Point Observatory, 799, 802
Apogee Instruments Inc., 609
Apollo-Foco International Inc., 652
Apollo Observatory, 839
Appalachian State University
    Department of Physics and Astronomy, 827
Apple Valley, 619, 623, 625
Apple Valley Science and Technology Center (AVSTC), 625

Applied Geomechanics, 609
APS
    American Physical Society, 734
        Editorial Offices, 806
    Australasian Planetarium Society, 17
    Department of Astrophysical and Planetary Sciences, 660
AQ
    AstroQueyras, 163
AR
    Annual Reviews Inc., 608
ARA
    Association of Research Astronomers, 852
    Associazione Reggiana Astrofili, 310
    Associazione Romana Astrofili, 311
Arabia (Saudi_), 453
ARAR
    Associazione Ravennate Astrofili Rheyta, 310
Arbeitsgemeinschaft Astronomie an der Volkshochschule Stadt Fürstenfeldbruck eV, 259
Arbeitsgemeinschaft Astronomie und Raumfahrt Urania-Sternwarte, 263
Arbeitsgemeinschaft der Großforschungseinrichtungen, 233
Arbeitsgemeinschaft Walter-Hohmann-Sternwarte Essen eV, 208
Arbeitsgruppe für Astronomie Haus der Natur, 34
Arbeitskreis Astronomie Freiburg eV (AKAF), 208
Arbeitskreis Geschichte der Astronomie, 209
Arbeitskreis Meteore eV (AKM), 208
Arbeitskreis Sternfreunde Lübeck (ASL) eV, 208
ARC
    Ames Research Center
        Astrophysics Branch, 630
        NASA Astrobiology Institute, 630
        Stratospheric Observatory for Infrared Astronomy, 630
    Astrophysics Research Center (NMIMT), 801
Arcadia, 723
Arcane, 155
Arcay, 201
Arcetri, 326, 338
Archenhold-Sternwarte, 208
Arcturus Observatory, 791
Ardmore, 859
Arecibo, 437
Arecibo Observatory, 437, 819
Arezzo, 318
Argentina, 6
Argonne, 698
Argonne National Laboratory (ANL), 698
Argus Planetarium, 761
Århus, 134, 139, 140
Århus University

History of Science Department, 134
Institute of Physics and Astronomy, 134
ARI
    Astronomisches Rechen-Institut, 214
Arianespace
    Évry, 155
    France, 155
    Head Office, 155
Arica, 111
Aristotle University of Thessaloniki, 268
    Laboratory of Astronomy, 270
Arizona, 593–604
    (Northern_), 600
Arizona Amateur Astronomy Society (AAAZ), 593
Arizona Science Center, 593
Arizona State University (ASU)
    Center For Meteorite Studies, 593
    Department of Geology, 593
    Department of Physics and Astronomy
        Astronomy Group, 593
    Planetary Exploration Laboratory, 593
    Planetary Geology Group, 593
    Planetary Geophysics Laboratory, 593
Arkansas, 606, 607
Arkansas-Oklahoma Astronomical Society (AOAS), 606
Arkansas Tech University (ATU)
    Observatory, 606
Arkhangelsk, 440
ARL
    Astrophysics Research Laboratory, 659
Arlington, 887, 896, 899, 900, 904
Arlington Public Schools Planetarium, 896
Arluno, 323
Armagh, 531–533
Armagh Observatory, 531, 533
Armagh Planetarium, 532
Armainvilliers (Gretz-_), 205
Armées, 199
Armenia, 12
Armenian Academy of Sciences
    Byurakan Astrophysical Observatory, 12
    Garny Space Astronomy Institute (GSAI), 12
Armenian Physical Society, 12
Armidda (Monte_), 310
Arnim D. Hummel Planetarium, 725
Arnold AFB, 870
Arnold Engineering Development Center (AEDC), 870
Arosa, 514
ARP
    Alternativa Racional a las Pseudociencias, 473
ARPA
    Association de Recherche de Phénomènes Astronomiques, 158
ARR
    Andøya Rocket Range, 417
Array Operations Center (AOC), NRAO, 801
ARSEC
    Association of Roman Catholics for the promotion of Space Exploration and Colonization, 748
Artemis Society International (ASI), 609
Arthur J. Dyer Observatory, 870, 872, 876
Arthur Storer Planetarium, 735, 738
Artigas, 587
Artis Planetarium, 389
Arunah Hill, 758
ASA
    Acoustical Society of America, 805
    American Statistical Association, 895
    Association Sammielloise d'Astronomie, 161
    Association Scientifique d'Astronomie, 4
    Associazione Sabina Astrofili, 311
    Associazione Salentina Astrofili, 311
    Astroclub Solaris Aarau, 500
    Astronomical Society of Australia, 15
    Austrian Space Agency, 37
ASAA
    Association Saranaise des Astronomes Amateurs, 161

ASAFI
    Asociación de Astrónomos Aficionados de Colombia, 119
ASASAC
    Asociación de Astrónomos Autodidactos de Colombia, 119
ASC
    Agence Spatiale Canadienne
        Laboratoire David Florida, 82
        Programme des Sciences Spatiales, 82
    Agnes Scott College
        Bradley Observatory, 687
    Astro Space Center (P.N. Lebedev Physics Institute), 447
    AXAF Science Center, 750
ASCA
    Advanced Satellite for Cosmology and Astrophysics, 754
ASCC
    Astronomical Society of the Central Coast, 610
Aschersleben, 241
ASCPL
    Astro Scientific Centre Pte. Ltd., 456
ASCT
    Association Sportive et Culturelle Toussaintaise Section Astronomie, 161
ASE
    Astronomical Society of Edinburgh, 533
ASE, Agence Spatiale Européenne
    Allemagne, 228, 245
    Centre de Coordination pour le Télescope Spatial, 245
    Centre des Astronautes Européens, 228
    Centre d'Opérations Spatiales, 228
    Darmstadt, 228
    Division Astrophysique, 392
    Division des Publications, 392
    Espagne, 476
    ESRIN, 318
    ESTEC, 392
    European Centre for Space Law, 177
    France, 177
    Frascati, 318
    Garching, 245
    Information Retrieval System, 318
    Italie, 318
    Kiruna, 489
    Köln, 228
    Madrid, 476
    Noordwijk, 392
    Noordwijk Space Expo, 399
    Paris, 177
    Pays-Bas, 392, 399
    Siège Principal, 177
    Station de Kiruna, 489
    Station de Villafranca, 476
    Suède, 489
    VILSPA, 476
ASEE
    American Society for Engineering Education, 671
Asesores Astronómicos Cacereños (AAC), 474
ASET
    Astronomical Society of East Texas, 877
ASF
    Astronomical Society of Frankston, 15
ASFM
    Astronomical Society of Fort McMurray, 78
ASG
    Astronomical Society of Georgia, 206
    Astronomical Society of Glasgow, 533
ASGH
    Astronomical Society of Greater Hartford, 662
ASGI
    Astronomical Science Group of Ireland, 533
ASGRG
    Australasian Society for General Relativity and Gravitation, 17
ASH

Astronomical Society of Haringey, 534
Astronomical Society of Harrisburg, 852
Astronomical Society of the Hunter, 16
Ashburton, 409
Ashburton Branch, Canterbury Astronomical Society, 409
Ash Enterprises, 896
    Bradenton, 678
    Florida, 678
    Richmond, 896
    USA-FL, 678
    USA-VA, 896
    Virginia, 896
Ash Enterprises International Inc., 678
Asher Space Research Institute, 299
Asher Space Research Institute (ASRI), 299
Asheville, 829, 832
Ashland, 849
Ash Manufacturing Co., 698
Ashton Observatory, 716
ASI
    Agenzia Spaziale Italiana
        Base di Lancio, 302
        Centro di Geodesia Spaziale "Giuseppe Colombo", 302
        Matera, 302
        Roma, 303
        Sede, 303
        Trapani, 302
    Artemis Society International, 609
    Astronomical Society of India, 281
ASIAA
    Academia Sinica's Institute of Astronomy and Astrophysics
        Preparatory Office, 516
Asiago, 304
ASICC
    Australian Space Industry Chamber of Commerce, 19
ASIS
    American Society for Information Science, 734
    Astronomy and Space Information Service, 162
ASJ
    Astronomical Society of Japan, 342
Askam-in-Furness, 539
ASKC
    Astronomical Society of Kansas City, 776
ASL
    Arbeitskreis Sternwarte Lübeck, 208
    Automatic Systems Laboratories
        Massachusetts, 748
        Milton Keynes, 534
        United Kingdom, 534
        USA-MA, 748
        Wilmington, 748
ASLC
    Astronomical Society of Las Cruces, 799
ASLI
    Astronomical Society of Long Island, 807
ASLIB
    Association for Information Management, 532
ASM
    Austrian Society for Aerospace Medicine, 35
ASN
    Astronomical Society of Nevada, 786
ASNAV
    Association Nationale pour l'Amélioration de la Vue, 160
ASNH
    Astronomical Society of New Haven, 662
ASNNE
    Astronomical Society of Northern New England, 732
ASNORA
    Association Normande d'Astronomie, 160
ASNSW
    Astronomical Society of New South Wales, 15
ASO
    Astronomía Sigma Octante, Centro de Investigación y Estudio en Astronomía, 62

Asociación Argentina Amigos de la Astronomía (AAAA), 6
Asociación Argentina de Astronomía (AAA), 6
Asociación Astronómica M31 (AAM31), 474
Asociación Boliviana de Astronomía (ABA), 62
Asociación Chilena de Astronomía y Astronáutica (ACHAYA), 107
Asociación Colombiana de Estudios Astronómicos (ACDA), 119
Asociación de Aficionados a la Astronomía, 422
Asociación de Aficionados a la Astronomía (AAA), 586
Asociación de Astrónomos Aficionados de Colombia (ASAFI), 119
Asociación de Astrónomos Autodidactos de Colombia (ASAS-AC), 119
Asociación de Profesores de Astronomía de Uruguay (APAU), 586
Asociación de Variabilistas de España (AVE), 474
Asociación Española de Normalización y Certificación (AENOR), 474
Asociación Salvadoreña de Astronomía, 143
Asociatia de Standardizare din Romania (ASRO), 438
ASOD
    Astronomical Society of the Desert, 610
ASOSV
    Amateur Satellite Observers of Southeast Virginia, 895
ASP
    Astronomical Society of the Pacific, 610
ASPB
    Astronomical Society of the Palm Beaches, 678
Aspen, 652
Aspen Center for Physics (ACP), 652
ASRAS
    Astronomy Section of the Rochester Academy of Sciences, 821
ASRC
    Astronomical Society of Rowan County, 827
ASRI
    Asher Space Research Institute, 299
    Australian Space Research Institute, 19
ASRO
    Asociatia de Standardizare din Romania, 438
ASRT
    Academy of Scientific Research and Technology, 142
ASSA
    Astronomical Society of Southern Africa, 464
        Cape Centre, 464
        Garden Route Centre, 464
        Johannesburg Centre, 464
        Natal Centre, 464
        Natal Midlands Centre, 464
        NMC, 464
        Pretoria Centre, 465
    Astronomy and Space Science Association, 281
ASSA Natal Midlands Centre (NMC), 464
Assergi, 328
ASSN
    Académie Suisse des Sciences Naturelles, 508
    Accademia Svizzera di Scienze Naturali, 508
ASSNE
    Astronomical Society of Southern New England, 748
Associação Astronômica da Poços de Caldas, 65
Associação Brasileira de Normas Técnicas (ABNT), 65
Associação Portuguesa de Astronomos Amadores (APAA), 433
Associação Portuguesa para o Ensino da Astronomia, 433
Associated Universities Inc. (AUI), 896
    Headquarters, 671
Association À Ciel Ouvert, 155
Association Aéronautique et Astronautique de France (AAAF), 155
Association Astronomie Tycho Brahe, 156
Association Astronomique d'Anjou (AAA), 156
Association Astronomique de Franche-Comté (AAFC), 156
Association Astronomique de la Côte d'Or (AACO), 156
Association Astronomique de l'Ain (AAA), 156
Association Astronomique de La Réunion, 156
Association Astronomique de la Vallée (AAV), 156

Association Astronomique de l'Indre (AAI), 157
Association Astronomique de Loir-et-Cher (AALC), 157
Association Astronomique du Soissonnais (AAS), 157
Association Astronomique Picarde M80, 157
Association Canadienne des Physiciens (ACP), 78, 80
Association Ciel d'Anjou (ACA), 157
Association Copernic (Gap Astronomie -_), 157, 179
Association d'Astronomie El-Battani (AAB), 4
Association d'Astronomie Véga, 157
Association de Recherche de Phénomènes Astronomiques (ARPA), 158
Association des Amis de l'Observatoire des Creusets à Arbaz (AOCA), 500
Association des Astronomes Amateurs Abitibi-Témiscamingue (AAAAT), 78
Association des Astronomes Amateurs d'Auvergne (AAAA), 158
Association des Utilisateurs de Détecteurs Électroniques (AUDE), 158
Association Drômoise d'Astronomie (ADA) "Les Pléiades", 158
Association du Planétarium du Collège Valéri, 173
Association Éducative des Amateurs d'Astronomie du Centre (AEAAC), 158
Association for Astronomy Education (AAE), 532
Association for Computing Machinery (ACM), 807
Association for Information Management (ASLIB), 532
Association for the Advancement of Computing in Education (AACE), 896
Association Française d'Astronomie (AFA), 158
Association Française de Normalisation (AFNOR), 159
Association Française des Observateurs d'Étoiles Variables (AFOEV), 159
Association Française pour l'Avancement des Sciences (AFAS), 159
Association Guadeloupéenne d'Astronomes Amateurs (AGAA), 159
Association in Scotland to Research into Astronautics (ASTRA) Ltd., 532
Association Internationale de Géodésie (AIG), 134, 138
Association Lunairienne d'Astronomie, 159
Association Marseillaise d'Astronomie (AMAS), 159
Association Méditerranéenne des Sciences de l'Environnement et de l'Espace (AMSEE), 160
Association Narbonnaise d'Astronomie Populaire (ANAP), 160
Association Nationale pour l'Amélioration de la Vue (ASNAV), 160
Association Nationale Sciences Techniques Jeunesse (ANSTJ), 160, 198
Association Normande d'Astronomie (ASNORA), 160
Association Novae, 160
Association of Amateur Astronomers (AAA), 877
Association of Astronomy Educators, 782
Association of Learned and Professional Society Publishers (ALPSP), 532
Association of Lunar and Planetary Observers (ALPO), 609

Association of Observatories and Planetariums, 125
Association of Research Astronomers (ARA), 851
Association of Roman Catholics for the promotion of Space Exploration and Colonization (ARSEC), 747
Association of Science-Technology Centers (ASTC) Inc., 671

Association of Universities for Research in Astronomy (AURA), Inc., 671
Association pour le Développement International de l'Observatoire de Nice (ADION), 160
Association Sammielloise d'Astronomie (ASA), 161
Association Saranaise des Astronomes Amateurs (ASAA), 161
Association Scientifique d'Astronomie (ASA) El-Bouzdjani, 4
Association Sirius d'Astronomie, 4
Association Sportive et Culturelle Toussaintaise (ASCT) Section Astronomie, 161
Association Stéphanoise d'Astronomie M42, 161

Association Sterenn
    Groupement d'Études Astronomiques de Queven (GEAQ), 161
Associazione Acquavivese Astrofili (AAA) "Hertzsprung-Russell", 303
Associazione Amici dei Planetari (AAP), 303
Associazione Astrofili Altair, 303
Associazione Astrofili Aurunca, 304
Associazione Astrofili Bolognesi (AAB), 304
Associazione Astrofili del Basso Vicentino (AABV) "Edmund Halley", 304
Associazione Astrofili del Gruppo Telecomitalia, 304
Associazione Astrofili di Piombino, 304
Associazione Astrofili Fiorentini (AAF), 304
Associazione Astrofili "Galileo Galilei" (AAGG), 305
Associazione Astrofili Garfagnana (AAG), 305
Associazione Astrofili "Geminiano Montanari", 305
Associazione Astrofili Imolesi (AAI), 305
Associazione Astrofili "Ionico Etnei", 305
Associazione Astrofili Mantovani (AAM), 305
Associazione Astrofili Monti della Tolfa (AAMT), 305
Associazione Astrofili Pegaso, 306
Associazione Astrofili Sardi (AAS), 306
Associazione Astrofili Segusini (AAS), 306
Associazione Astrofili Spezzini (AAS), 306
Associazione Astrofili Teatini (AAT), 307
Associazione Astrofili Tethys (AAT), 307
Associazione Astrofili Trentini (AAT), 307
Associazione Astrofili Trevigiani, 307
Associazione Astrofili Valdinievole (AAV), 307
Associazione Astrofili Valtellinesi (AAV), 307
Associazione Astrofili Veneziani (AAV), 308
Associazione Astronomica Cassino (AAC), 308
Associazione Astronomica Cortina (AAC), 308
Associazione Astronomica Feltrina "G.J. Rheticus", 308
Associazione Astronomica Frusinate (AAF), 308
Associazione Astronomica Madonna di Campiglio (AMDIC), 308
Associazione Astronomica Milanese (AAM), 309
Associazione Astronomica Quasar, 309
Associazione Cernuschese Astrofili (ACA), 309
Associazione Friulana di Astronomia e Meteorologia (AFAM), 309
Associazione IDRA, 309
Associazione Ligure Astrofili Polaris, 309
Associazione Ligure per lo Studio e la Divulgazione dell'Astronomia e dell'Astronautica, 310
Associazione Ligure per lo Studio e la Divulgazione dell'Astronomia e dell'Astronautica, 340
Associazione Marchigiana Astrofili (AMA), 310
Associazione Maremmana Studi Astronomici (AMSA), 310
Associazione Ogliastrina di Astronomia (AOA), 310
Associazione per lo Studio e la Ricerca Astronomica (ASTRA), 310
Associazione Ravennate Astrofili Rheyta (ARAR), 310
Associazione Reggiana di Astronomia (ARA), 310
Associazione Romana Astrofili (ARA), 311
Associazione Sabina Astrofili (ASA), 311
Associazione Salentina Astrofili (ASA) "Edwin Hubble", 311

Associazione Tuscolana di Astronomia (ATA), 311
Associazione Valdostana Scienze Astronomiche (AVSA), 311

Associazione Vigevanese Divulgazione Astronomica (AVDA), 311
ASSW
    Astronomical Society of the South West, 16
AST
    Aerospace, Science and Technology, 225
    Astronomical Society of Tasmania, 16
Astana, 364
ASTB
    Astronomiska Sällskapet Tycho Brahe, 498
ASTC
    Association of Science-Technology Centers, 671
Asten, 394
ASTER

Agrupación Astronómica de Barcelona, 470
Asti, 319
ASTRA
    Association in Scotland to Research into Astronautics, 532
    Associazione per lo Studio e la Ricerca Astronomica, 310
    Astronomical Society of the Toms River Area, 791
    Automated Spectrophotometric Telescope Research Associates, 868
Astral Press, 14
AstroArt, 533
Astro Cards, 852
Astro Club Aubagnais (ACA), 161
Astro-Club de Limagne-Sud, 162
Astro-Club du Collège de Lévis, 78
Astroclub Radebeul eV (ACR), 209
Astroclub Solaris Aarau (ASA), 500
Astro Club Voyager (ACV), 312
Astrocom GmbH, 209
Astro CORAM, 47
Astrocosmical Association Sirius-86, 440
Astroday Institute, 692
Astrodome, 533
AstroDomes, 15
Astrofest, 699
Astro I/S, 134
Astro Klub Kostkov, 125
Astrolabe Astronomy Club of Wheeling, 912
Astrolab Inc., 791
Astromart, 609
Astromedia AB, 488
ASTRON
    Stichting Astronomisch Onderzoek in Nederland, 402
Astronaut Hall of Fame, 678
Astronautics Association of New Zealand (AANZ) Inc., 408
Astronaut Memorial Planetarium and Observatory (AMPO), 678
Astronomía Sigma Octante (ASO), Centro de Investigación y Estudio en Astronomía, 62
Astronomers in Lelekovice, 125
Astronomer (The_), 533, 572
Astronomes Amateurs Tarnais, 154, 162
astronomia (L'_), 312, 330
Astronomical Adventures, 593
Astronomical Association of Northern California (AANC), 610
Astronomical Association of Southern Illinois (AASI), 698
Astronomical Association - Sofia (AAS), 74
Astronomical Association Tampereen Ursa, 146, 150
Astronomical Astronautical Society Zadar (AASZ), 121
Astronomical Club Centaur, 440
Astronomical Club IKAR, 440
Astronomical Club Parsec, 440
Astronomical Consultants and Equipment (ACE) Inc., 593
Astronomical Data Analysis Center (ADAC), 342, 352
Astronomical Data Center (ADC), 735
Astronomical Foundation of Nógrád County, 273
Astronomical-Geodetical Society, 440
Astronomical League (AL), 776
    Astronomy Day, 761
Astronomical Observatory
    Belgrade, 454
    Sliven, 74
    Uppsala, 498
        Kvistaberg Station, 498
    Vilnius, 375
Astronomical Observatory and Planetarium N. Copernicus, 74
Astronomical Observatory of Campo Catino (AOCC), 312
Astronomical Observatory "Slavey Zlatev", 74
Astronomical Pocket Diary (APD), 408
Astronomical Science Group of Ireland (ASGI), 533
Astronomical Society "Gea X", 121
Astronomical Society Ivan Štefek, 121
Astronomical Society Leo Brenner, 121
Astronomical Society of Alice Springs Inc., 15

Astronomical Society of Australia (ASA), 15
Astronomical Society of East Texas (ASET), 877
Astronomical Society of Edinburgh (ASE), 533
Astronomical Society of Fort McMurray (ASFM), 78
Astronomical Society of Frankston (ASF) Inc., 15
Astronomical Society of Georgia (ASG), 206
Astronomical Society of Glasgow (ASG), 533
Astronomical Society of Greater Hartford (ASGH), 662
Astronomical Society of Haringey (ASH), 534
Astronomical Society of Harrisburg (ASH), 852
Astronomical Society of India (ASI), 281
Astronomical Society of Japan (ASJ), 342
Astronomical Society of Jefferson County, 776
Astronomical Society of Kansas City (ASKC), Inc., 776
Astronomical Society of Las Cruces (ASLC), 799
Astronomical Society of Long Island, Inc. (ASLI), 807
Astronomical Society of Melbourne, 15
Astronomical Society of Nevada (ASN), 786
Astronomical Society of New Haven (ASNH), 662
Astronomical Society of New South Wales (ASNSW) Inc., 15
Astronomical Society of New York City, 808
Astronomical Society of Northern New England (ASNNE), 732
Astronomical Society of Rowan County (ASRC), 827
Astronomical Society of Singapore (TASOS) (The_), 456, 457
Astronomical Society of South Australia (ASSA) Inc., 16
Astronomical Society of Southern Africa (ASSA), 464
    Cape Centre, 464
    Garden Route Centre, 464
    Harare Centre, 929
    Johannesburg Centre, 464
    Natal Centre, 464
    Natal Midlands Centre (NMC), 464
    Natal Midlands Centre (Pietermaritzburg), 464
    Pretoria Centre, 465
Astronomical Society of Southern New England (ASSNE) Inc., 748
Astronomical Society of South Zealand, 134
Astronomical Society of Tasmania, Inc. (AST), 16
Astronomical Society of the Central Coast (ASCC), 610
Astronomical Society of the Desert (ASOD), 610
Astronomical Society of the Hunter (ASH), 16
Astronomical Society of the Pacific (ASP), 610
    PASP, 610
Astronomical Society of the Palm Beaches (ASPB), 678
Astronomical Society of the South West (ASSW) Inc., 16
Astronomical Society of the Toms River Area (ASTRA), 791

Astronomical Society of Victoria (ASV) Inc., 16
Astronomical Society of Wakabadai (ASW), 342
Astronomical Society of Western Australia (ASWA) Inc., 17

Astronomical Society of West Virginia, 912
Astronomical Society Pitomaca, 121
Astronomical Unit (AU), 610
astronomics, 846
Astronomie Arbeitsgruppe der Universität Oldenburg, 209
Astronomie Centre Ardenne (ACA), 47
Astronomie en Chinonais, 162
Astronomie en Touraine et Centre-Ouest (ATCO), 162
Astronomie Magazine, 162
Astronomie, Physique, Élaboration, Instrumentation et Observation (APHELIE), 162
Astronomie Software Service, 209
Astronomiestation Demmin, 209
Astronomie, Techniques et Communication (ATCO), 162
Astronomie-Verein Olten (AVO), 500
Astronomische Arbeitsgemeinschaft Aalen (A4) eV, 209
Astronomische Arbeitsgemeinschaft an der Fachhochschule Mannheim, 229
Astronomische Arbeitsgemeinschaft an der Volkshochschule der Stadt Soest, 259
Astronomische Arbeitsgemeinschaft der Liebigschule, 210
Astronomische Arbeitsgemeinschaft der Volkssternwarte Singen eV, 210

Astronomische Arbeitsgemeinschaft Heuchelheim eV, 210
Astronomische Arbeitsgemeinschaft Mainz eV, 210
Astronomische Arbeitsgemeinschaft Pfaffenwinkel (AAP), 210

Astronomische Arbeitsgemeinschaft Waldhügel, 210
Astronomische Arbeitsgemeinschaft Wanne-Eickel/Herne eV, 211
Astronomische Arbeitsgruppe Laufen (AAL) eV, 211
Astronomische Gesellschaft (AG), 211
 Arbeitskreis Astronomiegeschichte, 211
Astronomische Gesellschaft Baden (AGB), 500
Astronomische Gesellschaft Bern (AGB), 500
Astronomische Gesellschaft Bochum-Melle, 211
Astronomische Gesellschaft Buchloe eV, 212
Astronomische Gesellschaft Graubünden (AGG), 500
Astronomische Gesellschaft Oberwallis (AGO), 501
Astronomische Gesellschaft Solothurn (AGS), 501
Astronomische Gesellschaft Urania eV, Wiesbaden, 212
Astronomische Gesellschaft Zürcher Oberland (AGZO), 501

Astronomische Gesellschaft Zürcher Unterland (AGZU), 501

Astronomische Gruppe des Kantons Glarus, 501
Astronomische Instrumente Stefan Thiele (AIT), 212
Astronomische Nachrichten (AN), 212
Astronomischer Arbeitskreis der Heimvolksschule Schloß Dhaun, 212
Astronomischer Arbeitskreis Ingolstadt (AAI) eV, 212
Astronomischer Arbeitskreis Kassel eV (AAK), 213
Astronomischer Arbeitskreis Merkur/Venus Göttingen (AAM-VG), 213
Astronomischer Arbeitskreis Pforzheim (AAP) 1982 eV, 213

Astronomischer Arbeitskreis Salzkammergut (AAS), 34
Astronomischer Arbeitskreis Wetzlar eV, 213
Astronomischer Freundeskreis Ostsachsen (AFO), 213
Astronomischer Jugendclub Dingi-Vindemiatrix, 34
Astronomischer Verein der Grafschaft Bentheim (AVGB) eV, 213
Astronomischer Verein der Volkssternwarte Papenburg eV, 214
Astronomischer Verein Dortmund (AVD) eV, 214
Astronomischer Verein Hoyerswerda (AVH) eV, 214
Astronomischer Verein Remscheid (AVRS) eV, 214
Astronomisches Büro, 38
Astronomische Schulstation Adolph Diesterweg, 214
Astronomisches Rechen-Institut (ARI), 214
Astronomische Station Heinrich S. Schwabe, 215
Astronomische Station Tycho Brahe, 215
Astronomische Sternwarte Nessa, 215
Astronomisches Zentrum Bruno-H.-Bürgel, 215
Astronomisches Zentrum Burg, 215
Astronomisches Zentrum Halberstadt (AZH), 215
Astronomisches Zentrum Magdeburg (AZM), 216
Astronomisches Zentrum Schkeuditz, 216
Astronomische Vereinigung Albstadt (AVA) eV, 216
Astronomische Vereinigung Augsburg eV (AVA), 216
Astronomische Vereinigung Frauenfeld (AVF), 501
Astronomische Vereinigung Karlsruhe (AVKa) eV, 216
Astronomische Vereinigung Kärntens (AVK), 34
Astronomische Vereinigung Kreuzlingen (AVK), 502
Astronomische Vereinigung Nürtingen (AVN) eV, 216
Astronomische Vereinigung Sankt Gallen, 502
Astronomische Vereinigung Tübingen (AVT) eV, 217
Astronomische Vereinigung Weikersheim eV, 217
Astronomische Vereinigung West-München (AVWM), 217
Astronomische Vereinigung Zürich (AVZ), 502
Astronomische Vereniging Wega, 389
Astronomiska Sällskapet Tycho Brahe (ASTB), 488, 498
Astronomisk Forening for Sydsjaelland (AFS), 134, 135
Astronomisk Forlag, 135
Astronomisk Selskab i Danmark, 135, 136
Astronomsko Astronautičko Društvo Zadar, 121
Astronomy, 916
 Astronomy and Atmospheric Research Unit (AARU), 379
  Sheikh Tahir Astronomical Centre, 378
Astronomy and Space Information Service (ASIS), 162

Astronomy and Space Science Association (ASSA), 281
Astronomy Associates of Lawrence (AAL), 721
Astronomy Club of Akron (ACA), 835
Astronomy Club of Augusta (ACA), 687
Astronomy Club of Sun City West, 594
Astronomy Club of Tulsa, 846
Astronomy Club of Virginia Tech, 896
Astronomy Data Center (ADC), 740
Astronomy Day
 Headquarters, 761
Astronomy Ireland, 295
Astronomy Now, 534
Astronomy Roadshow (The_), 534, 572
Astronomy Study Unit (ASU), 217
Astronomy Technology Centre (ATC), 534
Astrooptik Kohler (AOK), 502
Astrophysical Journal (ApJ), 708
 Supplement Series, 708
Astrophysics Data Facility (ADF), 735, 740
Astrophysics Data System (ADS), 751
Astro-Physics Inc., 698
Astrophysics Research Laboratory (ARL), 659
Astrophysikalisches Institut Potsdam (AIP), 212, 217
 Observatorium für Solare Radioastronomie (OSRA), 218
 Sonnenobservatorium Einsteinturm (SOE), 218
Astroptx Wholesale Optics, 662
AstroQueyras (AQ), 163
Astrorama, 195
Astro Scientific Centre Pte. Ltd.
 The Astronomical Society of Singapore (TASOS), 457
Astro Scientific Centre Pte. Ltd. (ASCPL), 456
astro-shop, 218
Astro Space Center (ASC), 440, 447
AstroSystems Inc., 652
Astroteam Mariazellerland, 34
Astrotech Instruments & Computers KKT, 273
Astro-Terre, 163
Astro-Versand, 218
Astrovisuals, 17
ASU
 Angelo State University
  Planetarium, 877
 Arizona State University
  Astronomy Group, 593
  Center For Meteorite Studies, 593
  Department of Geology, 593
  Department of Physics and Astronomy, 593
  Planetary Exploration Laboratory, 593
  Planetary Geology Group, 593
  Planetary Geophysics Laboratory, 593
 Astronomy Study Unit, 217
Asunción, 422
ASW
 Astronomical Society of Wakabadai, 342
ASWA
 Astronomical Society of Western Australia Inc., 17
ATA
 Associazione Tuscolana di Astronomia, 311
Atacama Large Millimeter Array (ALMA), 600
ATC
 Astronomy Technology Centre (ROE), 534
Atchison, 721
ATCO
 Astronomie en Touraine et Centre-Ouest, 162
 Astronomie, Techniques et Communication, 162
A Tech, 721
Atelier d'Helios (L'_), 163, 188
Athens, 268, 269, 691, 841
Athens University Observatory, 269
Athinai, 268, 269
Atibaia, 70
Atlanta, 688–691
 (North_), 689
Atlanta Astronomy Club (AAC), 687
Atlantic
 (Middle_), 858
 (North_), 55

Atlantic Space Sciences Foundation (TASSF) Inc. (The_), 78, 100
ATMA
    Amateur Telescope Maker's Association, 907
ATMOB
    Amateur Telescope Makers of Boston, 746
Atmospheric and Environmental Research (AER) Inc., 748
ATNF
    Australia Telescope National Facility
        Compact Array, 22
        Epping, 21
        Headquarters, 21
        Mopra Telescope, 21
        Narrabri, 22
        Parkes Radio Observatory, 22
Atomic Data and Nuclear Data Tables (ADNDT), 808
ATS
    Antique Telescope Society, 687
    Suomen Avaruustutkimusseura, 149
Attard, 380
Attila University (József_), 273
    Astronomical Observatory, 277
    Department of Physics, 277
ATU
    Arkansas Tech University
        Observatory, 606
AU
    American University
        Physics Department, 671
        Astronomical Unit, 610
Aubagne, 161
Aubière, 158
Auburn, 589, 609
Auburn Astronomical Society (AAS), 589, 687
Auburn University
    Center for the Commercial Development of Space Power and Advanced Electronics, 589
    Department of Aerospace Engineering, 589
    Department of Physics, 589
    Space Power Institute, 589
Auckland, 408, 410
Auckland Astronomical Society, Inc., 408
Auckland Observatory, 408
Auckland Observatory and Planetarium Trust, 408
AUDE
    Association des Utilisateurs de Détecteurs Électroniques, 158
Augsburg, 216, 233, 245
Augusta, 687
Augustana College, 705
    John Deere Planetarium, 702
AUI
    Associated Universities Inc., 896
        Headquarters, 671
AURA
    Association of Universities for Research in Astronomy, 671
Auriga Srl, 312
Aurillac, 176
Aurora, 701, 705, 709, 835
Aurora Astronomical Association (A3), 652
Aurora Astronomical Society (AAS), 835
Auspace Ltd., 17
Austin, 877, 879, 886–889
Austin Astronomical Society (AAS), 877
Austin State University (Stephen F._), 877
Australasian Planetarium Society (APS), 17
Australasian Society for General Relativity and Gravitation (ASGRG), 17
Australia, 14
    (Western_), 17, 31
Australian Academy of Science (AAS), 17
Australian Defence Force Academy (AFDA)
    School of Physics, 18
Australian Institute of Physics (AIP), 18
Australian National University (ANU)
    CSIRO Office of Space Science and Applications (COSSA), 22
    Department of Physics and Theoretical Physics, 17
    Earth Observation Centre (EOC), 22
    Mount Stromlo and Siding Spring Observatories (MSSSO), 18
    Plasma Research Laboratory (PRL), 19
    Research School of Astronomy and Astrophysics (RSAA)
        Mount Stromlo and Siding Spring Observatories (MSSSO), 18
    Research School of Earth Sciences (RSES), 18
    Research School of Physical Sciences
        Department of Theoretical Physics, 18
    Research School of Physical Sciences and Engineering
        Plasma Research Laboratory (PRL), 18
Australian Planetarium Association, 19
Australian Space Industry Chamber of Commerce (ASICC), 19
Australian Space Research Institute (ASRI), 19
Australia Telescope National Facility (ATNF), 19
    Compact Array, 22
    Epping, 21
    Headquarters, 21
    Mopra Telescope, 21
    Narrabri, 22
    Parkes Radio Observatory, 22
Austria, 34
Austrian Academy of Sciences, 35, 37
Austrian Science Fund, 35
Austrian Society for Aerospace Medicine (ASM), 35
Austrian Space Agency (ASA), 37
Automated Spectrophotometric Telescope Research Associates (ASTRA), 868
Automatic Systems Laboratories (ASL)
    Massachusetts, 748
    Milton Keynes, 534
    United Kingdom, 534
    USA-MA, 748
    Wilmington, 748
Automatic Systems Laboratories (ASL) Inc., 748
Automatic Systems Laboratories (ASL) Ltd., 534
Auvergne, 158
AVA
    Adirondack Video Astronomy, 805
    Astronomische Vereinigung Albstadt, 216
    Astronomische Vereinigung Augsburg, 216
AVAC
    Antelope Valley Astronomy Club, 609
AVBG
    Astronomischer Verein der Grafschaft Bentheim, 214
AVD
    Astronomischer Verein Dortmund, 214
AVDA
    Associazione Vigevanese Divulgazione Astronomica, 312
AVE
    Asociación de Variabilistas de España, 474
Avezzano, 324
AVF
    Astronomische Vereinigung Frauenfeld, 501
AVG
    Amateurastronomischen Vereinigung Göttingen, 207
AVH
    Astronomischer Verein Hoyerswerda, 214
Aviation Week & Space Technology, 808
Avila, 472
AVK
    Astronomische Vereinigung Kärntens, 34
    Astronomische Vereinigung Kreuzlingen, 502
AVKa
    Astronomische Vereinigung Karlsruhe, 216
AVN
    Astronomische Vereinigung Nürtingen, 216
AVO
    Astronomie-Verein Olten, 500
Avon, 664
Avondale, 854
Avren, 74
AVRS

Astronomischer Verein Remscheid, 214
AVS
    American Vacuum Society, 808
AVSA
    Associazione Valdostana Scienze Astronomiche, 311
AVSO
    Allgäuer Volkssternwarte Ottobeuren, 207
AVSTC
    Apple Valley Science and Technology Center, 625
AVS The Science and Technology Society, 808
AVT
    Astronomische Vereinigung Tübingen, 217
AVWM
    Astronomische Vereinigung West-München, 217
AVZ
    Astronomische Vereinigung Zürich, 502
Awards
    Abbe (Cleveland_), 747
    Abbe (Ernst_), 234
    Abdus Salam, 336
    Abelson (Philip Hauge_), 669
    A.B. Gregory, 637
    Academic Achievement, 870
    Achievement (AURA), 671
    Adair (Crutchfield_), 611
    Adam Fleming (John_), 670
    Adams (Roger_), 670
    ADION, 161
    Adjudant Hubert Lefebvre, 47
    Adolphe Wetrems, 47
    Agathon De Potter, 47
    Agilent Technologies, 177
    Agostinelli (Gili-_), 302
    Aimé Cotton, 202
    Albert Brachet, 47
    Albert Einstein, 383
    Albert (Fonó_), 275
    Alfred Bader, 670
    Alfred Burger, 670
    Allan D. Emil, 184
    Amateur Achievement (ASP), 610
    Amateur Astronomer of the Year, 148, 667
    Amateur Astronomers, 806
    Aminoff (Gregori_), 493
    A.M. Thuring, 807
    Ancel (Louis_), 202
    Annie J. Cannon, 670
    Appleton, 53
    Armourers & Brasiers, 566
    Arthur C. Clarke, 758
    Arthur C. Cope, 670
    Article Prize, 358
    Astral, 15
    Astronomer of the Year, 380, 723
    Astronomical League, 776
    Astronomical Observation, 118
    Astronomical Poetry, 160
    Astronomy Day (Sky and Telescope _), 761
    Auguste Sacré, 47
    Augustin-Pyramus de Caudolle, 510
    Bader (Alfred_), 670
    Baes (Professeur Louis_), 47
    Baldwin (F.W._), 80
    Balthasar van der Pol, 53
    Bappu (Vainu_), 281
    Barnes (Earle B._), 670
    Baron Nicolet, 47
    Baron van Ertborn, 47
    Barraza (Guillermo Haro_), 385
    Barringer, 695
    Bart and Priscilla Bok, 603
    Battan (Louis J._), 747
    Beals (Carlyle S._), 80
    Beatrice M. Tinsley, 670
    Belgica, 47
    Benatti (Eros_), 324
    Bent Cage, 884

Bergan, 814
Berkner (Lloyd V._), 895
Bernard Houssay, 675
Best Contributor (The Astronomer), 572
Best (Fred_), 568
Biermann (Ludwig_), 211
Billings, 660
Bill Sutherland, 540
Birkhoff (George David_), 861, 864
Birla (B.M._), 282
B.M. Birla, 282
Bok, 15
Bok (Bart and Priscilla_), 603
Bomford, 138
Bonavera, 302
Booker, 53
Born (Max_), 223
Bourlart (Dr. A. De Leeuw-Damry_), 52
Bowie (William_), 670
Boynton (Melbourne W._), 895
Brachet (Albert_), 47
Brasiers (Armourers &_), 566
Brennan, 610
Bressa, 302
Brooks (Charles Franklin_), 747
Brooks (Kendall P._), 762
Brouwer (Dirk_), 670, 895
Browline (Colonel Leonard_), 803
Bruce, 610
Bruno (Giordano_), 797
Bruno (Giovanni_), 304
Bruno H. Bürgel, 211
Bruno Rossi, 670
Bryant Conant (James_), 670
Buchanan, 566
Bucher (Walter H._), 670
Bürgel (Bruno H._), 211
Burger (Alfred_), 670
Buttgenbach (Henri_), 47
Cage (Bent_), 884
Callebaut (O._), 54
Calvird (H. Rich_), 884
Camille Liégeois, 47
Cannon (Annie J._), 670
CAP Medals, 80
Carl-Gustaf Rossby, 747
Carl H. Gamble, 705
Carlyle S. Beals, 80
Carrière (Jean P._), 99
Catalan (Eugène_), 47
Caudolle (Augustin-Pyramus de_), 510
C. de Clercq, 54
C.D. Howe, 80
Chant, 97
Chapman, 565
Charles A. Whitten, 670
Charles Franklin Brooks, 747
Charles Lagrange, 47
Charles Lathrop Parsons, 670
Charles Lemaire, 47
Charney (Jule G._), 747
Chilton (Ken_), 97
Chrétien, 670
Christian-Ernst-Neeff, 240
Clarence T. Jones, 875
Clarke (Arthur C._), 758
Claude S. Hudson, 670
Cleveland Abbe, 747
Cleveland (Newton_), 669
Clifford W. Holmes, 635
Clyde Tombaugh Telescope Innovation, 635
Colonel Leonard Browline, 803
Conant (James Bryant_), 670
Conrad Schlumberger, 392
Cope (Arthur C._), 670
Copley, 566
Coppens (J._), 54

COSPAR, 174
  Distinguished Service, 174
Cotton (Aimé-), 202
Crafoord, 493
Craftsmans, 146
Crutchfield Adair, 611
Cumbres (Las-), 610
Curie (Joliot-_), 202
Curtis (Len-), 564
Dakota Astronomer, 834
Damry-Bourlart (Dr. A. De Leeuw-), 52
Daniel Guinier, 202
Dannie Heinemann, 670
Dantzig (George B.-), 861
Darwin, 566
David Birkhoff (George-), 861, 864
Davis (Watson-), 735
Davy, 566
Debauque (Dubois-), 47
de Boelpaepe, 47
Debye (Peter-), 670
de Caudolle (Augustin-Pyramus-), 510
de Clercq (C.-), 54
De Donder (Théophile-), 47
Dehalu, 58
De Leeuw-Damry-Bourlart (Dr. A.-), 52
Dellinger (John Howard-), 53
Delwart (Fondation Jean-Marie-), 47
De Meyer (Jean-), 47
De Potter (Agathon-), 47
Deruyts
  (François-), 47
  (Jacques-), 47
de Selys Longchamps (Edmond-), 47
Digital, 202
Dinsdale, 903
DiPrima (Richard C.-), 861
Dirac, 302
Dirk Brouwer, 670, 895
DLR, 225
Dobson (John-), 803
Door, 652
D'Or (Louis-), 57
Dr. A. De Leeuw-Damry-Bourlart, 52
Dubois-Debauque, 47
Dudley, 814
Duke of Edinburgh, 358
Dupont (Octave-), 47
Earle B. Barnes, 670
E.B. Hershberg, 670
Eckert-Mauchly, 807
Eddington, 565
Edmond de Selys Longchamps, 47
Édouard Mailly, 47
EFS Prize, 144
EFS Student Prize, 144
Einstein (Albert-), 383
E. Ken Owen, 847
Ellery, 15
Ellery Hale (George-), 670
Elliot (Lloyd G.-), 80
Elliott-Yeates, 774
Emerson (Ralph-), 584
Emil (Allan D.-), 184
Émile Laurent, 47
Emil-Wiecherte, 222
Emme (Eugene M.-), 895
Erasmus Medal, 531
Eric and Maria Muhlmann, 610
Ernest F. Fullam, 814
Ernest Guenther, 670
Ernest-John Solvay, 52
Ernö (Nagy-), 275
Ernst Abbe, 234
Ernst Julius Öpik, 144
Eros Benatti, 324
Errera (Léo-), 47

Esclangon, 202
Esso Energy, 566
Estes Memorial (Warren-), 635
Étoile d'Argent, 99
Eugène Catalan, 47
Eugene M. Emme, 895
Eugen-Hartmann-Didaktik, 240
E. van Beneden, 57
E.V. Murphree, 670
Ewing (Maurice-), 670
Excelência (IPQ), 434
Faraday (Michael-), 566
Fast (Herman-), 637
Faura (Padre-), 425
Félix Robin, 202
Fellow of GLPA, 838
Fermi, 707
Field (Frank H.-), 670
Fields, 68
Finnish Academy, 146
Flack Norris (James-), 670
Fleming (John Adam-), 670
Flight Achievement, 895
Fondation Jacques et Yvonne Ochs-Lefêbvre, 47
Fondation Jean-Marie Delwart, 47
Fonó Albert, 275
Foucault, 202
Fourmarier (Paul-), 47
François Deruyts, 47
Frank H. Field, 670
Frank J. Malin, 184
Franklin Brooks (Charles-), 747
Franklin (Joe L.-), 670
Fred Best, 568
Fredericq (Léon et Henri-), 47
Frederic Stanley Kipping, 670
Frédéric Swarts, 47
Fridtjof Nansen, 416
Fullam (Ernest F.-), 814
F. Van der Mueren, 54
F.W. Baldwin, 80
Gabor, 566
Gaede-Langmuir, 808
Galilei, 60
Gamble (Carl H.-), 705
Garvan, 670
Gautieri, 302
Geddes (Murray-), 411
Geh (Hans-Peter-), 394
Gemini, 569
General Nathan Twining, 803
Gentner-Kastler, 202, 223
Geoff Welch, 26
George B. Dantzig, 861
George C. Pimentel, 670
George David Birkhoff, 861, 864
George Ellery Hale, 670
George Polya, 861
Georges Van der Linden, 47
Georgette-Lemoyne, 99
George Van Biesbroeck, 670
Gerard P. Kuiper, 670
Gerbier-Mumm (Norbert-), 514
Gerlach (Stern--), 223
G.H. Russell, 678
Gili-Agostinelli, 302
Gillis (J.-), 54
Gill Medal, 464
Giordano Bruno, 797
Giovanni Bruno, 304
Glover, 854
Gluge (Théophile-), 47
Godeaux (Lucien-), 57
Golden Saturn, 34
Gold Medal
  (Academia Europaea), 531
  (Balthasar van der Pol), 53

(Booker), 53
(Canadian Aeronautics and Space Institute), 80
(John Howard Dellinger), 53
(Remote Sensing Society), 564
(Royal Astronomical Society), 565
(Royal Astronomical Society of Canada), 97
(South African Institute of Physics), 468
(Union of Czech Mathematicians and Physicists, Physics Section), 132
Grady (James T._), 670
Greenstein (Jesse_), 611
Gregori Aminoff, 493
Gregory (A.B._), 637
Guenther (Ernest_), 670
Guillermo Haro Barraza, 385
Guinier (Daniel_), 202
Gustav Hertz, 223
Gustav Hofmann, 394
Guust van Wesemael, 394
Haas (Walter H._), 609
Hale (George Ellery_), 670
Hampshire Astron. Group Honorary Membership, 547
Hampshire Astron. Group Life Membership, 547
Hannah Jackson, 565
Hans-Ludwig Neumann, 211
Hans-Peter Geh, 394
Haro Barraza (Guillermo_), 385
Harold C. Urey, 670
Harold Johnson, 385
Harriet Hujer (Karl and_), 875
Harry H. Hess, 670
Hartmann-Didaktik (Eugen-_), 240
Hasselblad, 488
Hauge Abelson (Philip_), 669
Heinemann (Dannie_), 670
Helen A. Warkoczewski, 776
Helen B. Warner, 670
Helen Sawyer Hogg, 80
Helmholtz, 241
Henri Buttgenbach, 47
Henry G. Houghton, 747
Henry Hildebrand (Joel_), 670
Henry H. Storch, 670
Henry Norris Russell, 670
Herbert C. Pollock, 814
Herlitzka, 302
Herman Fast, 637
Herschel, 565, 776
Hershberg (E.B._), 670
Hertz (Gustav_), 223
Herzberg Medal, 80
Hessberg Astronomy Campership, 814
Hess (Harry H._), 670
Heß (Viktor_), 38
Hildebrand (Joel Henry_), 670
Hilliard Roderick, 669
Hirschmann (Ralph F._), 670
H.L. Vanderlinden, 54
Hofmann (Gustav_), 394
Hogg (Helen Sawyer_), 80
Holmes (Clifford W._), 635
Holweck, 202
Horkheimer, 776
Horton (Robert E._), 670
Houghton (Henry G._), 747
Houssay (Bernard_), 675
Howard Dellinger (John_), 53
Howe (C.D._), 80
H. Rich Calvird, 884
H. Schouteden, 54
Hubert Lefebvre (Adjudant_), 47
Hudson (Claude S._), 670
Hughes, 566
Hujer (Karl and Harriet_), 875
ICTP, 302
IMO, 514
Imperial, 358

International Cooperation, 174
Ipatieff, 670
Irving Langmuir, 670
Issac Koga, 53
Italgas, 302
Jackson (Hannah_), 565
Jacques Deruyts, 47
Jacques et Yvonne Ochs-Lefêbvre (Fondation_), 47
James B. Macelwane, 670
James Bryant Conant, 670
James Flack Norris, 670
James H. Stack, 670
James H. Wilkinson, 861
James T. Grady, 670
Janssen, 200
Japan Academy, 358
J. Coppens, 54
Jean De Meyer, 47
Jean Lebrun, 47
Jean-Marie Delwart (Fondation_), 47
Jean P. Carrière, 99
Jean Perrin, 202
Jean Ricard, 202
Jean-Servais Stas, 47
Jesse Greenstein, 611
J. Gillis, 54
Jim Maybury, 16
J. Murray Luck, 608
Joe L. Franklin, 670
Joel Henry Hildebrand, 670
John Adam Fleming, 670
John Dobson, 803
John F. Kennedy, 895
John Howard Dellinger, 53
Johnson (Harold_), 385
John von Neumann, 861
Joliot-Curie, 202
Jones (Clarence T._), 875
Joseph Maisin, 52
Joseph Schepkens, 47
José Vasconcelos, 383
Joynes, 584
J.S. Plaskett, 80
Jule G. Charney, 747
Julio Monges, 386
Karl and Harriet Hujer, 875
Karl Schwarzschild, 211
Karlstrom (Karl V._), 807
Karl V. Karlstrom, 807
Kastler (Gentner-_), 202, 223
Ken Chilton, 97
Kendall P. Brooks, 762
Kennedy (John F._), 895
Ken Owen (E._), 847
Kertz (Walter_), 222
Kieviet, 702
Kipping (Frederic Stanley_), 670
Klumpke-Roberts, 610
KNAW, 395
Koga (Issac_), 53
Korkosz, 758
Kuiper (Gerard P._), 670
Kundaji (R._), 281
Lacy Pierce (Newton_), 670
Lagrange (Charles_), 47
Lamarck, 47
Langevin (Paul_), 202
Langmuir, 802
Langmuir (Gaede-_), 808
Langmuir (Irving_), 670
Las Cumbres, 610
Lathrop Parsons (Charles_), 670
Laurent (Émile_), 47
Lebrun (Jean_), 47
Leeuw-Damry-Bourlart (Dr. A. De_), 52
Lefebvre (Adjudant Hubert_), 47
Lemaire (Charles_), 47

Lemoyne (Georgette-_), 99
Lenaerts (R._), 54
Len Curtis, 564
Léo Errera, 47
Leonard, 695
Leonard Browline (Colonel), 803
Leonardo da Vinci, 383
Léon et Henri Fredericq, 47
Leslie Peltier, 776
Les Pléiades, 88
Levallois, 138
Leverhulme, 566
Leviton (Robert_), 747
L.G. Napolitano, 184
Liégeois (Camille_), 47
Lines (Richard D._), 872
Lloyd G. Elliot, 80
Lloyd V. Berkner, 895
Lorimer, 533
Louis Ancel, 202
Louis Baes (Professeur_), 47
Louis D'Or, 57
Louis J. Battan, 747
Louis Melsens, 47
Lovelace II (W. Randolph_), 895
Lucien Godeaux, 57
Luck (J. Murray_), 608
Ludwig Biermann, 211
Ludwig-Prandtl, 223
Macelwane (James B._), 670
MacLeod, 54
Mailly (Édouard_), 47
Maisin (Joseph_), 52
Malin (Frank J._), 184
Manuel Noriega Morales, 675
Marc-Auguste Pictet, 510
Maria Luisa Ferrari Soave e Dottore Luigi Soave, 302
Marian Molga, 431
Mark (Peter_), 808
Mars Institute, 634
Martinetto, 302
Massey, 174
Mauchly (Eckert-_), 807
Maurice Ewing, 670
Max Born, 223
Max Planck, 223
Max Poll, 47
Maybury (Jim_), 16
McCall, 851
McCormick, 707
McCurdy, 80
Medalla de la Real Sociedad Española de Física, 482
Meduza Certificates, 129
Meisinger, 747
Melbourne W. Boynton, 895
Melsens (Louis_), 47
Meritas, 88
Messier, 851
Messier Club, 776
Messier Club Member, 66
Michael Faraday, 566
Michael Penston, 565
Military Astronautics, 895
Modesto Panetti, 302
Molga (Marian_), 431
Monges (Julio_), 386
Moore (Patrick_), 538
Morales (Manuel Noriega_), 675
Muhlmann (Eric and Maria_), 610
Mullaney, 851
Mullard, 566
Murphree (E.V._), 670
Murray Geddes, 411
Murray Luck (J._), 608
Nagy Ernö, 275
Nansen (Fridtjof_), 416
Napolitano (L.G._), 184

Nathan Twining (General_), 803
National Air and Space Museum Trophy, 676
National Outstanding Young Astronomer, 776
Neeff (Christian-Ernst-_), 240
Nerken, 808
Neumann (Hans-Ludwig_), 211
Nevanlinna (Rolf_), 68
Newcomb (Simon_), 97
New Millennium, 634
Newton Cleveland, 669
Newton Lacy Pierce, 670
Nicolet (Baron_), 47
Nier, 695
Nobel
    Chemistry, 493
    Economic Sciences, 493
    Laureate Signature, 670
    Physics, 493
Norbert Gerbier-Mumm, 514
Norbert Wiener, 861, 864
Nordberg (William_), 174
Noriega Morales (Manuel_), 675
Norris (James Flack_), 670
Norris Russell (Henry_), 670
Norther, 16
Observer of the Year, 837
O. Callebaut, 54
Ochs-Lefêbvre (Fondation Jacques et Yvonne_), 47
Octave Dupont, 47
Öpik (Ernst Julius_), 144
Oscar Schwiglhofer, 532
Otto Schott, 234
Outstanding Contribution to Development of Astronomy in Ukraine, 528
Owen (E. Ken_), 847
Padre Faura, 425
Page, 15
Panetti (Modesto_), 302
Parsons (Charles Lathrop_), 670
Patrick Moore, 538
Paul et Marie Stroobant, 47
Paul Fourmarier, 47
Paul Langevin, 202
Pellati (Ravani-_), 302
Peltier (Leslie_), 776
Penston (Michael_), 565
Perrin (Jean_), 202
Peter Debye, 670
Peter Mark, 808
Petrie (R.M._), 80
Philip Hauge Abelson, 669
Philipp-Siedler, 240
Physik, 38
Pictet (Marc-Auguste_), 510
Pierce (Newton Lacy_), 670
Pimentel (George C._), 670
P.-J. et Édouard Van Beneden, 47
Planck (Max_), 223
Plaskett, 97
Plaskett (J.S._), 80
Pléiades (Les_), 88
Pohl (Robert-Wichard-_), 223
Pol et Christiane Swings, 47
Polish Geophysical Society, 431
Pollini, 302
Poll (Max_), 47
Pollock (Herbert C._), 814
Pol Swings, 57
Polya (George_), 861
Prandtl (Ludwig-_), 223
Prather (Victor A._), 895
Premio a Investigadores Noveles de la Real Sociedad Española de Física, 482
Price, 565
Priestley, 670, 854
Professeur Louis Baes, 47
P. van Oye, 54

Quality/Design, 736
Ralph Emerson, 584
Ralph F. Hirschmann, 670
Rammal, 178
Rammal Rammal, 202
Randolph Lovelace II (W._), 895
Ravani-Pellati, 302
Reid (W.T. and Idalia_), 861
Ricard (Jean_), 202
Richard C. DiPrima, 861
Richard D. Lines, 872
Rich Calvird (H._), 884
Rising Star, 723
R. Kundaji, 281
R. Lenaerts, 54
R.M. Petrie, 80
Robert E. Horton, 670
Robert Leviton, 747
Roberts (Klumpke-_), 610
Robert-Wichard-Pohl, 223
Robin (Félix_), 202
Robinson, 580
Roderick (Hilliard_), 669
Roger Adams, 670
Rolf Nevanlinna, 68
Rolf Schock, 493
Roman-Ulrich-Sexl, 38
Romeo Vachon, 80
Rossby (Carl-Gustaf_), 747
Rossi (Bruno_), 670
Royal, 566
Rumford, 566
Ruschi, 65
Russell (G.H._), 678
Russell (Henry Norris_), 670
Sacré (Auguste_), 47
Salam (Abdus_), 336
Samuel-Thomas-von Soemmering, 240
Sarabhai (Vikram_), 174, 285
SASI, 28
Sawyer Hogg (Helen_), 80
Scanlon, 851
Schepkens (Joseph_), 47
Schlumberger (Conrad_), 392
Schock (Rolf_), 493
Schottky (Walter_), 223
Schott (Otto_), 234
Schouteden (H._), 54
Schwarzschild (Karl_), 211
Schwiglhofer (Oscar_), 532
Science Book, 557
SDAA Certificate of Merit, 636
Service, 97, 870
Sexl (Roman-Ulrich_), 38
SFAA Service Award, 637
SFSA, 202
Shadows of Time, 332
Siedler (Philipp-_), 240
SIGRAV, 336
Silver Medal
   (Liverpool Astronomical Society), 554
   (Royal Danish Academy of Sciences and Letters), 139
   (South African Institute of Physics), 468
   (Union of Czech Mathematicians and Physicists, Physics Section), 132
Simon Newcomb, 97
Sky and Telescope Astronomy Day, 761
Smith (Waldo E._), 670
Soave (Maria Luisa Ferrari_ e Dottore Luigi_), 302
Solvay (Ernest-John_), 52
Space Achievement, 538
Space Camp Scholarship, 878
Space Flight, 895
Spenser Space Campership, 814
Stack (James H._), 670
Stanley Kipping (Frederic_), 670

Stas (Jean-Servais_), 47
Stella Arcti, 152
Stern-Gerlach, 223
Storch (Henry H._), 670
Stroobant (Paul et Marie_), 47
Sutherland (Bill_), 540
Swarts (Frédéric_), 47
Swings
   (Pol_), 57
   (Pol et Christiane_), 47
Sylvester, 566
Taylor & Francis Best Letter Award, 564
Teacher of the Year, 822
Telescope Competition Merit, 635
Theodore von Kármán, 184, 861
Théophile De Donder, 47
Théophile Gluge, 47
Third World Academy of Sciences, 336
Thornton, 808
Tinsley (Beatrice M._), 670
Tombaugh Telescope Innovation (Clyde_), 635
Trumpler, 610
Turing (A.M._), 807
Twining (General Nathan_), 803
Ulrich-Sexl (Roman_), 38
Urey (Harold C._), 670
Vachon (Romeo_), 80
Vainu Bappu, 281
Väisälä (Vilho_), 514
Vallauri, 302
van Beneden (E._), 57
Van Beneden (P.-J. et Édouard_), 47
Van Biesbroeck (George_), 670
Van der Linden (Georges_), 47
Vanderlinden (H.L._), 54
Van der Mueren (F._), 54
van der Pol (Balthasar_), 53
van Ertborn (Baron_), 47
van Oye (P._), 54
van Wesemael (Guust_), 394
Vasconcelos (Jose_), 383
Victor A. Prather, 895
Vikram Sarabhai, 174, 285
Viktor-Heß, 38
Vilho Väisälä, 514
Vinci (Leonardo da_), 383
von Kármán (Theodore_), 184, 861
von Neumann (John_), 861
von Soemmering (Samuel-Thomas-_), 240
Von Wippel, 858
WAITRO Honorary, 140
Waldo E. Smith, 670
Walter H. Bucher, 670
Walter H. Haas, 609
Walter Kertz, 222
Walter Schottky, 223
Warkoczewski (Helen A._), 776
Warner (Helen B._), 670
Warren Estes Memorial, 635
Watson Davis, 735
Webb Society, 583
Welch, 808
Welch (Geoff._), 26
Wellcome, 566
Westinghouse, 669
Wetrems (Adolphe_), 47
Whitten (Charles A._), 670
Wichard-Pohl (Robert-_), 223
Wiecherte (Emil_), 222
Wiener (Norbert_), 861, 864
Wilkinson (James H._), 861
William Bowie, 670
William Nordberg, 174
Wippel (Von_), 858
WMO, 514
W. Randolph Lovelace II, 895
W.T. and Idalia Reid, 861

W. Zonn, 431
Yeates (Elliott-), 774
Young Astronomers, 281
Zel'dovich, 448
Zonn (W..), 431
AWR Technology, 534
AWSK
    Amsterdamse Weer- en Sterrenkundige Kring, 406
A.W. Wright Nuclear Structure Laboratory (WNSL), 666
AXAF
    Advanced X-ray Astrophysics Facility, 750, 754
AXAF CCD Imaging Spectrometer (ACIS), 754
AXAF Science Center (ASC), 750
Aydat, 158
Ayer Observatory, 755
Aylesbury, 535, 548
Aylesbury Astronomical Society, 535
Ayr, 535
Ayrshire Astronomical Society (AAS), 535
Azerbaijan, 43
Azerbaijan Academy of Sciences, 43
    Shemakha Astrophysical Observatory, 43
Azerbaijanian National Aerospace Agency (ANASA), 43
Azerbaijan State Standardization and Metrology Centre, 43

AZH
    Astronomisches Zentrum Halberstadt, 215
Azle, 882
AZM
    Astronomisches Zentrum Magdeburg, 216

B

BAA
    British Astronomical Association, 538
        New South Wales Branch, 20
    Buffalo Astronomical Association, 809
    Burgenländische Amateurastronomen, 35
Baader Planetarium GmbH, 218
BAAS
    Bulletin of the American Astronomical Society, 670
Babelsberg, 218
BAC
    Beijing Astrophysics Center, 112
    Bristol Astronomy Club, 897
Bacchus (Col de), 158
Back Bay Amateur Astronomers (BBAA), 896
Backgrounds Data Center (BDC), 674
Badan Meteorologi dan Geofisika (BMG), 290
BADAS
    Blackpool and District Astronomical Society, 536
Bad Bentheim, 214
BADC
    Beijing Astronomical Data Center, 113
Baden, 500
Bad Gandersheim, 232
Bad Honnef, 223
Badlands Observatory, 869
Bad Mergentheim, 217
Bad Münstereifel, 237
Bad Nauheim, 263
Bad Salzschlirf, 230
Bad Sauerbrunn, 35
Baghdad, 294
Bahagian Kajian Sains Angkasa (BAKSA), 378
Baikal Astrophysical Observatory, 448
Bainbridge Island, 907
Baja, 273
Baja Astronomical Observatory, 273
Baker Observatory, 778
Baker Planetarium (Burke-), 877, 878
Bakersfield, 624
Bakker's Farm, 103
BAKSA
    Bahagian Kajian Sains Angkasa, 378
Baksan Neutrino Observatory (BNO), 440, 444

Baku, 43
Bâle, 512
Baleares (Islas-), 483
Ballaarat Astronomical Society (BAS), Inc., 19
Ballarat, 19
Ballarat Observatory, 19
Ballasalla, 542
Ballincollig, 296
Ball State University (BSU)
    Department of Physics and Astronomy, 710
Balmoral, 84
Baltimore, 735, 737–739, 742–744
Baltimore Astronomical Society (BAS), 735
Baltus Astronomical Observatory, 858
Bamberg, 254
Bandon, 848
Bandung, 290
Bandung Institute of Technology
    Bosscha Observatory, 290
    Department of Astronomy, 290
Bangalore, 281, 283–285, 288
Bangi, 378
Bangkok, 519
Bangkok Planetarium, 519
Bangladesh, 44
Bangladesh Meteorological Department (BMD), 44
Bangladesh Standards and Testing Institution (BSTI), 44
Bangor, 547, 551, 732
Banneker Planetarium (Benjamin-), 735
Banská Bystrica, 458
Banská Bystrica Astronomical Observatory, 458
BAO
    Beijing Astronomical Observatory, 113, 115
        Beijing astronomical Data Center, 113
    Bisei Astronomical Observatory, 342
BAP
    British Association of Planetaria, 538
Bappu Observatory (VBO) (Vainu-), 285
Barbados, 45
Barbados Astronomical Society (BAS), 45
Barbados National Standards Institution (BNSI), 45
Barbara Moore Observatory (James and-), 679, 681
Barber Observatory (Walter-), 687
Barber Research Observatory (Henry R..), 708
Barcelona, 470, 471, 476, 480, 482, 484–486
Barlow Planetarium, 915
Barnard Astronomical Society (BAS), 870
Barnard-Seyfert Astronomical Society (BSAS), 870
Barnesville, 743
Barr Associates
    Massachusetts, 748
    Newbury, 535
    United Kingdom, 535
    USA-MA, 748
    Westford, 748
Barr Associates Inc., 748
Barr Associates Ltd., 535
Barrie, 78, 97
Barrie Astronomy Club, 78
Barrow-in-Furness, 546
Barstow, 619
Bartol Research Institute (BRI), 667
Barton, 23
Baruch College (Bernard-), 808
    Department of Physics and Astronomy, 810
Barzago, 318
BAS
    Ballaarat Astronomical Society, 19
    Baltimore Astronomical Society, 735
    Barbados Astronomical Society, 45
    Barnard Astronomical Society, 870
    Bedford Astronomical Society, 535
    Birmingham Astronomical Society, 536, 589
    Boise Astronomical Society, 697
    Bolton Astronomical Society, 536
    Bradford Astronomical Society, 536
    Braintree Astronomical Society, 536

Breckland Astronomical Society, 537
Brevard Astronomical Society, 679
Bridgwater Astronomical Society, 537
Bridport Astronomy Society, 537
Brisbane Astronomical Society, 20
Bristol Astronomical Society, 538
Basel, 511, 512
Basildon, 546, 551
Basingstoke, 553, 572
Basle, 512
Bassano Bresciano, 332
Batavia, 700, 701
Bates College
    Department of Physics and Astronomy, 732
Bath, 584, 859
Baton Rouge, 729, 730
Baton Rouge Astronomical Society (BRAS), 729
Baton Rouge Observatory, 730
Battelle Planetarium, 836
Battle Creek, 763
Battle Point Astronomical Association (BPAA), 907
Bauduen, 163
Bausch Observatory (SBO) (Sommers-_), 659, 660
Bautzen, 249
BAV
    Bundesdeutsche Arbeitsgemeinschaft für Veränderliche
        Sterne eV, 220
Bavaria, 218, 219, 257
BAW
    Bayerische Akademie der Wissenschaften, 218
Bay City, 762
Bay Colony Astronomical Society (BCAS), 748
Bayerische Akademie der Wissenschaften (BAW), 218
Bayerische Julius-Maximilians-Universität Würzburg, 218
Bayerische Volkssternwarte München eV, 218
Bayerische Volkssternwarte Neumarkt/Opf. eV, 219
Bayern, 218, 219, 257
Bayfordbury, 577
Bayley Observatory (Harry_), 45
Baylor Regional Park, 771
Bay (North_), 95
Bayrischzell, 255
Bay Saint Louis, 774
Bays Mountain Astronomy Club, 870
Bays Mountain Observatory, 871
Bays Mountain Planetarium, 870, 871
Bay Village, 843
BBAA
    Back Bay Amateur Astronomers, 896
BBAS
    Big Bend Astronomical Society, 877
BBSO
    Big Bear Solar Observatory, 630, 794
BCAAS
    Berks County Amateur Astronomical Society, 852
BCAS
    Bay Colony Astronomical Society, 748
BCC
    Brevard Community College
        Planetarium and Observatory, 678
BDAS
    Bendigo District Astronomical Society, 19
BDC
    Backgrounds Data Center, 674
BDL
    Bureau des Longitudes, 163, 193
BDS
    Bulgarian Committee for Standardization and Metrology, 76
Beacon Hill, 541
Beacon Hill Observatory, 541
Beacon Hill Telescopes, 535
Beattie Planetarium (George F._), 611, 622
Beaumont-lès-Valence, 158
Beauvoir, 86
Beauzelle, 158
Beaver Meadow, 809

Beaver Valley Astronomy Club (BVAC), 852
Bedford Astronomical Society (BAS), 535
Beecher Planetarium (Ward_), 835, 844
Beenleigh, 27
Beernem, 54
Beer-Sheva, 299
Beg Astronomical Institute (Ulugh_), 925
    Space Geodynamics Department, 925
Behlen Observatory, 784
Beijing, 112–115, 118
Beijing Ancient Observatory, 112
Beijing Astronomical Data Center (BADC), 112, 113
Beijing Astronomical Observatory (BAO), 112, 113, 115
    Beijing astronomical Data Center (BDC), 113
Beijing Astrophysics Center (BAC), 112
Beijing Normal University
    Department of Astronomy, 112
Beijing Planetarium, 112
Beijing University
    Beijing Astrophysics Center (BAC), 112
    Department of Geophysics
        Astrophysics Division, 112
Beirut, 373
Bel Air, 737, 742
Belarus, 46
Belarusian Physical Society (BPS), 46
Belfast, 543, 563
Belgian Physical Society (BPS), 48
Belgian Space Industry Association (BELGOSPACE), 48
Belgisch Instituut voor Ruimte-Aëronomie (BIRA), 48
Belgium, 47
BELGOSPACE, 48
    Belgian Space Industry Association, 48
Belgrade, 454, 455
Belgrade University
    Department of Astronomy, 454
Bella Unión, 587
Bellevue, 907
Bellevue Community College (BCC)
    Willard Geer Planetarium, 907
Bellingham, 908, 911
Bell Laboratories, 792, 793
Belmont, 624
Beloeil, 83
Belogradtchik, 75
Belo Horizonte, 72
Beloit, 915
Beloit College
    Thompson Observatory, 915
BELST
    Belarusian Committee for Standardization, Metrology
        and Certification, 46
Belvedere, 929
Beman Observatory, 761, 766
Bemidji, 770
Bemmel, 389
Benátky nad Jizerou, 125
Benátky nad Jizerou Popular Observatory, 125
Bend
    (South_), 713
Bendigo, 19
Bendigo District Astronomical Society (BDAS) Inc., 19
Benedictine College
    Department of Physics and Astronomy, 721
Benedum Planetarium, 912
Ben-Gurion University
    Department of Physics, 299
Ben Gurion University (BGU)
    Department of Physics, 299
Benjamin Banneker Planetarium, 735
Benson, 602, 604
Bentheim (Bad_), 214
Bentleigh, 15
Bentley College
    Observatory, 748
Beograd, 454, 455
BeppoSAX Scientific Data Center (SDC), 312

Berazategui, 8
Berea, 728
Berea College
    Weatherford Planetarium, 728
Bergamo, 313, 332
Berg-Aufkirchen, 221
Bergen, 419
Bergerac, 188
Berkeley, 608, 610, 620, 630, 639, 643, 646, 647, 651
Berkeley County Planetarium, 912
Berkeley-Illinois-Maryland Association (BIMA), 647, 744
    Hat Creek Radio Observatory (HCRO), 611
Berkley, 762
Berks County Amateur Astronomical Society (BCAAS), 852

Berlin, 209, 219, 220, 223–225, 247, 251, 252, 265, 916
Berliner Sternfreunde eV, 219
Bern, 500, 505, 508, 509, 513
Bernard Baruch College, 808
    Department of Physics and Astronomy, 810
Bernau, 219
Berne, 500, 505, 508, 509, 513
Bernice P. Bishop Museum
    Kilohani Planetarium, 692
Beron Observatory (Dr. Petur_), 74
Berthoud, 656
Besançon, 156, 191
Besser Museum and Sky Theatre (Jesse_), 761, 764
Bet Dagan, 299
Bethany, 662
Bethany Observatory, 666
Bethany Sciences LLC, 662, 665
Bethesda, 735, 741
Bettendorf, 716
Beverly Hills, 611, 651
Beverly Hills High School (BHHS)
    Planetarium, 611
Beyrouth, 373
BFO
    Black Forest Observatory, 653
BGAAC
    Blue Grass Amateur Astronomical Club, 725
BGI
    Bureau Gravimétrique International, 163, 186
BGO
    Burke-Gaffney Observatory, 97
BGS
    British Geological Survey, 538
BGSU
    Bowling Green State University
        Department of Physics and Astronomy, 835
        Planetarium, 835
BGU
    Ben Gurion University
        Department of Physics, 299
BHAS
    Black Hills Astronomical Society, 869
    Brighton and Hove Astronomical Society, 537
BHC Aerovox Ltd., 535
BHHS
    Beverly Hills High School
        Planetarium, 611
BHS
    Burke High School
        Planetarium, 782
Biak, 290
Bianchi Planetarium (Donald E.), 615
Bibliographical services
    ADF, 740
    ARI, 214
    Astronomisches Rechen-Institut, 214
    Astrophysics Data Facility, 740
    BADC, 113
    Beijing Astronomical Data Center, 113
    Bibliothèque Nationale du Canada, 78
    BNC, 78
    Brussel, 55
    Bruxelles, 48
    Cambridge Scientific Abstracts, 735
    Canadian Institute for Scientific and Technical Information, 93
    CAS, 836
    CDS, 165
    Centre de Données astronomiques de Strasbourg, 165
    Centre National de Documentation Scientifique et Technique, 48
    Chemical Abstracts Service, 836
    CISTI, 93
    CNDST, 48
    CNRC, 93
    CNRS, 167
    CSA, 735
    Eggenstein, 229, 245
    ERS, 318
    ESIS, 318
    ESRIN, 318
    Fachinformationszentrum Karlsruhe, 229, 245
    FIZ Karlsruhe, 229, 245
    Frascati, 318
    Goddard Space Flight Center (GSFC)
        National Space Science Data Center, 740
        World Data Center A for Rockets and Satellites, 745
    GSFC
        World Data Center A for Rockets and Satellites, 745
    ICIST, 93
    IEE, 550
    INIST, 167
    INSPEC, 550
    Institut Canadien de l'Information Scientifique et Technique, 93
    Institut de l'Information Scientifique et Technique, 167
    Institute for Scientific Information, 856
    Institution of Electrical Engineers, 550
    IRS, 318
    ISI, 856
    Japan Science and Technology Corporation, 348
    JST, 348
    Karlsruhe, 229, 245
    LEDA, 318
    Leopoldshafen, 229, 245
    NASA, National Aeronautics and Space Administration
        National Space Science Data Center, 740
        World Data Center A for Rockets and Satellites, 745
    Nationaal Centrum voor Wetenschappelijke en Technische Dokumentatie, 55
    National Aeronautics and Space Administration (NASA)
        World Data Center A for Rockets and Satellites, 745
    National Library of Canada, 92
    National Space Science Data Center, 740
    NCWTD, 55
    NLC, 92
    NRCC, 93
    NSSDC, 740
    Ottawa, 93
    Scientific and Technical Information Network
        Columbus, 843
        Germany, 245
        Karlsruhe, 245
        Ohio, 843
        USA-OH, 843
    Stevenage, 550
    STN
        Columbus, 843
        Germany, 245
        Karlsruhe, 245
        Ohio, 843
        USA-OH, 843
    WDC
        A for Rockets and Satellites, 745
    World Data Center
        A for Rockets and Satellites, 745
Bibliothèque Nationale du Canada (BNC), 78
Bickley, 26

Biddeford, 733
Biel, 503
Bielefeld, 239
Bielser Observatorien, 502
Bienne, 503
Biesheim, 189
Bieszczady Mountains Station, 428
Big Bear City, 630, 635
Big Bear Solar Observatory (BBSO), 611, 630, 794
Big Bend Astronomical Society (BBAS), 877
Big Ear Observatory, 841
Big Hill Pond State Park, 874
Big Pine, 613
Big Sky Astronomical Society, 79
Big Sky Observatory (BSO), 79
Big South Fork Star Gazers, 725
Bilbao, 474
Bilkent University
    Department of Physics, 522
Billings, 781
BIMA
    Berkeley-Illinois-Maryland Association, 647, 744
        Hat Creek Radio Observatory, 611
        HCRO, 611
Binghamton, 820, 822
Bingley, 548
Binningen, 512
Binocular and Telescope Shop, 19
BioServe Space Technologies (BST), 653
BIPM
    Bureau International des Poids et Mesures, 163
BIRA
    Belgisch Instituut voor Ruimte-Aëronomie, 48
Birkenhead, 554, 572
Birla Planetarium (B.M._), 282
    Chennai, 282
    Hyderabad, 282
    Madras, 282
Birla Planetarium (M.P._), Kolkata, 286
Birmingham, 533, 536, 571, 574, 589, 590
Birmingham Astronomical Society (BAS), 536, 589
Birmingham Solar Oscillations Network (BiSON), 574
Birr, 295
Birr Castle, 295
Biruni Observatory, 292
BIS
    British Interplanetary Society, 538
    Bureau of Indian Standards, 282
Bisei Astronomical Observatory (BAO), 342
Bisei Hydrographic Observatory, 342
Bishop Museum
    Kilohani Planetarium (Bernice P._), 692
Bishop Planetarium, 679
Bishopstone, 567
Bishoptown, 295
Bismarck, 834
Bismarckschule Hannover
    Planetarium, 219
BiSON
    Birmingham Solar Oscillations Network, 574
BJU
    Bob Jones University
        Howell Memorial Planetarium, 866
BKG
    Bundesamt für Kartographie und Geodäsie
        Außenstelle Leipzig, 220
        Frankfurt/Main, 234
        Fundamentalstation Wettzell, 220
        IERS, 234
Black Forest Observatory (BFO), 653
Black Hills Astronomical Society (BHAS), 869
Black Moshannon Observatory, 860
Blackpool, 536
Blackpool and District Astronomical Society (BADAS), 536

Black River Astronomical Society (BRAS), 835
Blacksburg, 896, 905

Blagnac, 158
Blaine, 909
Blakemore Planetarium (Marian_), 878, 881, 889
Blake Planetarium (WRBP) (Dr. W. Russell_), 748, 750
Blanc (Mont._), 190
Blaxall and Steven Ltd., 408
Blénod Animation Loisirs
    Section Astronomie, 163
Blénod-lès-Pont-à-Mousson, 163
Blindern, 419
Bloomfield, 825
Bloomington, 702, 712, 770
Bloomsburg, 853
Bloxwich, 569
Bluefield, 912
Blue Grass Amateur Astronomical Club (BGAAC), 725
Blue Mountain Observatory, 781
Blue Ridge Astronomy Club, 897
Blue Ridge Parkway Regional Stargaze (BRPRS), 827
Blue Springs, 776, 777
Bluffton, 844
BMAA
    Bucks-Mont Astronomical Association, 853
B.M. Birla Planetarium
    Chennai, 282
    Hyderabad, 282
BMD
    Bangladesh Meteorological Department, 44
BMG
    Badan Meteorologi dan Geofisika, 290
BNL
    Brookhaven National Laboratory, 809
BNM
    Bureau National de Métrologie
        Laboratoire Primaire du Temps et des Fréquences, 164
BNO
    Baksan Neutrino Observatory, 444
BNQ
    Bureau de Normalisation du Québec, 79
BNSC
    British National Space Centre, 538, 543
BO
    Braeside Observatory, 594
BOAO
    Bohyunsan Optical Astronomy Observatory, 367
Bob Jones University (BJU)
    Howell Memorial Planetarium, 866
Bochum, 212, 233, 249, 253
Bogaziçi University
    Kandilli Observatory and Earthquake Research Institute, 522
Bognor Regis, 569
Bogotá, 119, 120
    (Santafé de_), 119
Bohemia, 816, 825
Bohr Institute (NBI) (Niels_), 135
Bohyunsan Optical Astronomy Observatory (BOAO), 367
Boiling Springs, 828
Boise, 697
Boise Astronomical Society (BAS), 697
Boissy-Fresnoy, 172
Bolivia, 62
Bolling Air Force Base, 669
Bollington, 559
Bologna, 304, 316, 317, 326, 333, 335, 338
Bolton, 536
Bolton Astronomical Society (BAS), 536
Bolzano, 303, 312
Bombay, 281, 283, 288
    (New_), 283
Bondar Planetarium (Roberta_), 79
Bondi Junction, 27
Bonn, 222, 233, 237, 253, 259
Bonner Springs, 722
Boone, 827
Boothroyd Observatory (Hartung), 812

Bordeaux, 199, 203
Border Astronomical Society, 536
Borgomanero, 306
Boring, 848
Borlänge, 489, 492
Bornova, 522
Borowa Gòra, 427
Borowiec, 430
Boscombe, 583
Boskovice, 125
Boskovice Observatory, 125
Boskovitch Astronomical Society (Rudjer_), 454, 455
Bosque Alegre Astrophysical Station, 10
Bosscha Observatory, 290
Boston, 746, 747, 749–751, 756, 759
Boston University
    Astronomical Society, 748
    Astronomy Department, 749
    Center for Space Physics (CSP), 749
    Department of Astronomy, 748
    Institute for Astrophysical Research (IAR), 749
Boston University Astronomical Society (BUAS), 748
Boulder, 653–661
Boulogne-Billancourt, 173
Bound Brook, 794
Bourbon-Lancy, 196
Bourg-en-Bresse, 156, 182
Bourges, 201
Bourgogne, 199
Bournemouth, 584
Bouzareah, 4
Bowdoin College
    Department of Physics and Astronomy, 732
Bowdoinham, 733
Bowen Productions, 710
Bowling Green, 725, 726, 728, 835
Bowling Green State University (BGSU)
    Department of Physics and Astronomy, 835
    Planetarium, 835
Boyd Observatories, 761
Boynton Beach, 683
Bozeman, 780, 781
Bozen, 303, 312
BPA
    Board on Physics and Astronomy, 674
BPAA
    Battle Point Astronomical Association, 907
BPS
    Belarusian Physical Society, 46
    Belgian Physical Society, 48
    Bureau of Product Standards of the Philippines, 425
Brackbill Planetarium (M.T._), 897, 899
Brackett Observatory (Frank P._), 634
Bracknell, 556
Bradenton, 678, 679, 681, 685
Bradford, 574
Bradford Astronomical Society (BAS), 536
Bradley Amateur Astronomy Club, 699
Bradley Observatory, 687
Bradley Planetarium, 687
Bradley University
    Bradley Amateur Astronomy Club, 699
Braeside Observatory (BO), 594
BRAG
    Brandeis Radio Astronomy Group, 749
Brahe Planetarium (Tycho_), 135, 140
Brailsford & Co. Inc., 808
Braine-l'Alleud, 50
Braintree, 759
Braintree Astronomical Society (BAS), 536
Brandeis Radio Astronomy Group (BRAG), 749
Brandeis University
    Department of Physics
        Radio-Astronomy Group, 749
Brandon, 79
Brandon University
    Department of Physics and Astronomy, 79

BRAS
    Baton Rouge Astronomical Society, 729
    Black River Astronomical Society, 835
Brashear LP, 852
Brasília, 66, 67
Brasschaat, 54
Bratislava, 458–461
Brattleboro, 893
Braunsbedra, 242
Braunschweig, 240, 246, 251
Brazil, 65
Breakthrough Science Society (BSS), 282
Breckland Astronomical Society (BAS), 537
Breezy Point, 806
Breisgau, 247
Bremen, 240
Bremerhaven, 219
Bremerhavener Sternfreunde eV, 219
Brentwood, 870, 876
Brera, 326
Brescia, 303, 332, 336
Breslau, 430, 432
Bretagne, 196
Brevard Astronomical Society (BAS) Inc., 679
Brevard Community College (BCC)
    Planetarium and Observatory, 678
Brewer, 733
Brewster, 801
BRI
    Bartol Research Institute, 667
Briars (The_), 15
Bridgend, 537
Bridgend Astronomical Society, 537
Bridgeport, 662
Bridgetown, 45
Bridgwater, 537
Bridgwater Astronomical Society (BAS), 537
Bridport, 537
Bridport Astronomy Society (BAS), 537
Brie, 205
Brig-Glis, 501
Brigham Young University
    Astronomical Society, 890
    Department of Physics and Astronomy, 890
    Sarah B. Summerhays Planetarium, 890
Brighton, 537, 581
Brighton and Hove Astronomical Society (BHAS), 537
Brisbane, 27, 30
Brisbane Astronomical Society (BAS), Inc., 20
Brisbane Planetarium (STBP) (Sir Thomas_), 20, 27
Brish Planetarium (William M._), 735, 745
Bristol, 544, 550, 575, 872, 897
Bristol Astronomical Society (BAS), 537
Bristol Astronomy Club (BAC), 897
British Association of Planetaria (BAP), 538
British Astronomical Association (BAA), 538
    New South Wales Branch, 20
British Geological Survey (BGS), 538
British Interplanetary Society (BIS), 538
British National Space Centre (BNSC), 538, 543
British Standards Institution (BSI), 539
British Sundial Society (BSS), 539
British UFO Research Association (BUFORA) Ltd., 539
Britton Memorial Observatory (Michael L._), 854
Brno, 126, 127, 129
Brno N. Copernicus Observatory and Planetarium, 126, 127, 129
Bro, 498
Broadhurst Clarkson & Fuller Ltd., 539
Broadman Coit Memorial Observatory (Judson_), 749
Brockenhurst, 545
Brockville, 79
Brockville Astronomical Society, 79
Broman Planetarium AB, 488
Brønshøj, 140
Bronx, 810, 820
    (Northeast_), 820

Bronx Planetarium (Northeast-), 809
Brookhaven National Laboratory (BNL), 809
Brookline, 746
Brooklyn, 814, 815
Brooklyn Observatory, 806
Brooklyn Thermometer Co. Inc., 809
Brook Park, 841
Brooks Astronomical Observatory, 762
Brooks Observatory, 844
Brough, 549
Brown Deer, 918
Brown University
    Ladd Observatory, 864
    Physics Department, 864
BRPRS
    Blue Ridge Parkway Regional Stargaze, 827
Bruce County Astronomical Society, 79
Brugg, 500
Brugge, 60
Bruncu Spina, 327
Bruno-H.-Bürgel-Sternwarte
    Berlin, 219
    Hartha, 220
Brunswick, 732
Brussel, 47–57, 59, 60
Brussels, 47–57, 59, 60
Bruxelles, 47–57, 59, 60
Bruz, 172
Bryan Planetarium (K. Price-), 735, 739
Bryan State Park Observatory (John-), 839
BSAS
    Barnard-Seyfert Astronomical Society, 870
BSH
    Bundesamt für Seeschiffahrt und Hydrographie
        Hamburg, 220
        Rostock, 220
BSI
    British Standards Institution, 539
BSNI
    Barbados National Standards Institution, 45
BSO
    Big Sky Observatory, 79
BSS
    Breakthrough Science Society, 282
    British Sundial Society, 539
BST
    BioServe Space Technologies, 653
BSTI
    Bangladesh Standards and Testing Institution, 44
BSU
    Ball State University
        Department of Physics and Astronomy, 710
BUAS
    Boston University Astronomical Society, 748
Bucarest, 438, 439
Bucharest, 438, 439
Buchloe, 212
Bucknell University
    Department of Physics, 853
    Observatory, 862
Bucks-Mont Astronomical Association (BMAA), 853
Buckstaff Planetarium, 921
București, 438, 439
Budapest, 273–278
Budapest Planetarium, 273
Buehler Planetarium, 679
Buellton, 645
Buena Vista, 654
Buenos Aires, 6–10
Buffalo, 809, 822, 826
Buffalo Astronomical Association (BAA), 809
Buffalo State College
    Whitworth Ferguson Planetarium, 826
BUFORA
    British UFO Research Association, 539
Buhl Jr. Planetarium and Observatory (Henry-), 853, 856, 858

Bülach, 508
Bulgaria, 74
Bulgarian Academy of Sciences
    Central Laboratory for Geodesy, 74
    Geophysical Institute, 75
    Institute of Astronomy and National Astronomical Observatory
        Smolyan, 75
        Sofia, 75
    National Institute of Meteorology and Hydrology (NIMH), 75
    Space Research Institute, 76
Bunbury, 16
Bundaberg, 20
Bundaberg Astronomical Society Inc., 20
Bundesamt für Kartographie und Geodäsie (BKG)
    Außenstelle Leipzig, 220
    Frankfurt/Main, 234
    Fundamentalstation Wettzell, 220
    IERS, 234
Bundesamt für Meteorologie und Klimatologie (MeteoSchweiz), 506
Bundesamt für Seeschiffahrt und Hydrographie (BSH)
    Hamburg, 220
    Rostock, 220
Bundesanstalt für Meteorologie und Klimatologie, 502
Bundesanstalt für Meteorologie und Klimatologie (MeteoSchweiz), 502
Bundesdeutsche Arbeitsgemeinschaft für Veränderliche Sterne eV (BAV), 220
Bundoora, 24
Bureau de Normalisation du Québec (BNQ), 79
Bureau des Longitudes (BDL), 163, 193
Bureau Gravimétrique International (BGI), 163, 186
Bureau International des Poids et Mesures (BIPM), 163
Bureau National de Métrologie (BNM), Laboratoire Primaire du Temps et des Fréquences (LPTF), 164
Bureau of Indian Standards (BIS), 282
Bureau of Product Standards (BPS), 425
Bure (Observatoire du Plateau de-), 182, 204
Burg, 215
Burgdorf, 512
Burgenländische Amateurastronomen (BAA), 35
Burgenländische Landessternwarte, 35
Burgum, 395, 406
Burjassot, 485, 486
Burke, 901
Burke Baker Planetarium, 878
Burke-Gaffney Observatory (BGO) (Rev. M.W.-), 79, 97
Burke High School (BHS)
    Planetarium, 782
Burke Observatory, 871, 872
Burke Planetarium, 782
Burleigh Instruments Inc., 809
Burlington, 98, 894
    (West-), 719
Burnett (Mount-), 25
Burney, 611
Burton-upon-Trent, 536
Bushnell Corp.
    Sport Optics Division, 721
Bussloo, 406
Bussoleno, 332
Butler University
    Department of Physics and Astronomy, 710
    J.I. Holcomb Observatory and Planetarium (JIHOP), 710
Butts-Clark Observatory, 774
Buxy, 201
BVAC
    Beaver Valley Astronomy Club, 852
Byers Co. (Edward R.-), 611, 619
Byron, 699
Byron Forest Preserve District
    Weiskopf Observatory, 699
Byurakan, 12, 449
Byurakan Astrophysical Observatory, 12

C

C3A
    Club Ajaccien des Amateurs d'Astronomie, 169
CAA
    Cambridge Astronomical Association, 539
    Cedar Amateur Astronomers, 716
    Committee on Astronomy and Astrophysics, 674
    Cuyahoga Astronomical Association, 837
CAAB
    Club d'Astronomes Amateurs de Bienne, 503
CAAC
    Capital Area Astronomy Club, 762
    Charlotte Amateur Astronomers Club, 828
    Club d'Astronomie Alpha Centaure, 171
CAACM
    Club des Astronomes Amateurs Centre Mauricie, 85
CAAL
    Club des Astronomes Amateurs de Laval, 85
    Club des Astronomes Amateurs de Longueuil, 85
CAAO
    Center for Astronomical Adaptive Optics, 603
CAAS
    Central Arkansas Astronomical Society, 606
    Club des Astronomes Amateurs de Sherbrooke, 86
    Crimean Amateur Astronomical Society, 525
CAAW
    Cercle des Astronomes Amateurs de Waterloo, 50
CAB
    Centro Astrofili Bolzano, 312
    Cercle Astronomique de Bruxelles, 49
    Circolo Astrofili Bergamaschi, 313
CAC
    Concordia Astronomy Club, 792
    Crossroads Astronomy Club, 878
    Cumberland Astronomy Club, 736
Cáceres, 470, 474
Cachoeira Paulista, 67
CAD
    Center for Astronomical Data, 445
    Club d'Astronomie de Drummondville, 83
CaDAS
    Cleveland and Darlington Astronomical Society, 541
CADC
    Canadian Astronomy Data Centre, 93
Caen, 160
Cagliari, 327
Cagnes-sur-Mer, 172
Caguas, 437
CAI
    Centre d'Analyse des Images, 183
Caird Planetarium, 565
Cairo, 142
Cairo University
    Department of Astronomy and Meteorology, 142
CALA
    Club d'Astronomie de Lyon-Ampère, 171, 203
Calais, 179
Calán (Cerro_), 111
Calar Alto, 475, 476, 481
Calar Alto Observatory, 474
Calcutta, 282, 286, 287
Calern, 192
Calgary, 80, 101, 102
Calgary Centennial Planetarium, 80
Calgary Science Center, 79
Cali, 119
California, 608–651
    (Northern_), 610
    (Southern_), 649, 650
California Academy of Science (CAS), 611
California Association for Research in Astronomy (CARA)
    W.M. Keck Observatory, 692
California Institute for Physics and Astrophysics (CIPA), 611
California Institute of Technology
    CalTech Submillimeter Observatory (CSO), 692
    Department of Astronomy, 611
    Department of Theoretical Astrophysics, 611
    Division of Geological and Planetary Sciences, 612
    Division of Physics, Mathematics and Astronomy, 612
    Infrared Processing and Analysis Center (IPAC), 612
    Jet Propulsion Laboratory (JPL), 612
        Deep Space Tracking Systems Group, 612
        Table Mountain Observatory (TMO), 613
        US Space VLBI Project, 613
    Laser Interferometer Gravitational-wave Observatory (LIGO) Project, 613
    Owens Valley Radio Observatory (OVRO), 613
        Headquarters, 613
    Palomar Observatory, 613
California Space Institute, 614
California State University
    Chico
        Roth Planetarium, 614
    Fresno
        Downing Planetarium, 618
        Physics Department, 614
    Fullerton
        Department of Physics, 614
    Long Beach
        Department of Physics and Astronomy, 614
    Northridge
        Department of Physics and Astronomy, 614
        Planetarium, 615
        San Fernando Observatory (SFO), 615
    Sacramento
        Department of Physics and Astronomy, 615
Callaghan, 26
Callaway Gardens, 689
Calouste Gulbenkian (PCG) (Planetário_), 434
CalTech, 615
CalTech, California Institute of Technology
    Department of Astronomy, 611
    Department of Theoretical Astrophysics, 612
    Division of Geology and Planetary Sciences, 612
    Division of Physics, Mathematics and Astronomy, 612
    Infrared Processing and Analysis Center (IPAC), 612
    Jet Propulsion Laboratory (JPL), 612
        Deep Space Tracking Systems Group, 612
        Table Mountain Observatory (TMO), 613
        US Space VLBI Project, 613
    Laser Interferometer Gravitational-wave Observatory Project, 613
    Owens Valley Radio Observatory, 613
        Headquarters, 613
    Palomar Observatory, 614
    Submillimeter Observatory, 692
CalTech Submillimeter Observatory (CSO), 692
Calumet Astronomical Society (CAS) Inc., 710
Calusa Nature Center and Planetarium, 679
Calvin College
    Department of Physics and Astronomy, 761
CAM
    Cercle Astronomique Mosan, 49
    Circolo Astrofili di Milano, 313
Cambridge, 539, 540, 552, 557, 567, 575, 579, 746, 749–752, 754–758, 760
Cambridge Astronomical Association (CAA), 539
Cambridge Research and Instrumentation (CRI) Inc., 749
Cambridge Scientific Abstracts (CSA), 735
Cambridge University Astronomical Society (CUAS), 540
Cambridge University Press (CUP), 540, 809
Campbell County School District Planetarium, 922, 923
Campbell Hall, 818
Campinas, 69, 72
Campi Salentina, 311
Campo Imperatore, 328
CANA
    Centro Astronomico Neil Armstrong, 312
Canada, 78
Canada-France-Hawaii Telescope Corp. (CFHT), 692
Canadian Aeronautics and Space Institute (CASI), 80
Canadian Association of Physicists (CAP), 80
Canadian Astronomical Society (CASCA), 80

Canadian Astronomy Data Centre (CADC), 81, 93, 94
Canadian Coast Guard College Planetarium, 81
Canadian Council of Science Centres (CCSC), 81
Canadian Institute for Advanced Research (CIAR), 81
Canadian Institute for Scientific and Technical Information (CISTI), 81, 93
Canadian Institute for Theoretical Astrophysics (CITA), 81

Canadian Meteorological Centre, 81
Canadian Space Agency (CSA)
    David Florida Laboratory (DFL), 81
    Space Science Program, 82
Canadian Space Resource Centre
    Atlantic, 82
    Ontario, 82
    Prairies, 105
    Québec, 87
Cananea, 383
Canberra, 17–23, 25, 30
Canberra Astronomical Society (CAS), Inc., 20
Canberra Deep Space Communication Complex (CDSCC), 20
Canberra Semiconductor nv, 48
Canberra Space Dome and Observatory (CSDO), 21
Cannes, 178, 198, 201
Cantabria, 470
Canterbury, 409, 412, 570, 577
Canterbury Astronomical Society (CAS), 408
    Ashburton Branch, 409
Canton, 821, 838
CANV
    Club Astronomie Nature du Valromey, 170
Canyelles, 475
Canyon Camp, 847
CAO
    Centro Astronomico Orione, 313
    Club d'Astronomie d'Ottignies, 50
CAOQ
    Center for Astronomical Observing Quality, 465
CAP
    Canadian Association of Physicists, 80
    Club de Astrofísica del Paraguay, 422
Cape Breton Astronomical Society (CBAS), 82
Cape Cod Astronomical Society (CCAS), 749
Cape Coral, 684
Cape Fear Astronomical Society (CFAS), 827
CapellaSoft, 799
Cape Town, 464, 465, 467–469
Capilla Peak Observatory, 804
Capital Area Astronomy Club (CAAC), 762
Capodimonte, 327
Capolago, 504
Capoterra, 306, 327
Cappelle-la-Grande, 195
Cap-Rouge, 95
CAR
    Club Astronomique Rochefortois, 50
CARA
    California Association for Research in Astronomy
        W.M. Keck Observatory, 692
    Center for Astrophysical Research in Antarctica, 915
Caracas, 927, 928
Carbondale, 698, 706
Carcassonne, 170
Cardiff, 546, 582
Cardiff Astronomical Society (CAS), 540
Cardiff by the Sea, 644
Carelia
    (South.), 146
Caribbean, 521
Caribbean Institute for Meteorology and Hydrology (CIMH), 45
Caribbean Meteorological Organization (CMO), 521
Carina Software, 615
CARJ
    Clube de Astronomia de Rio de Janeiro, 66
CARL

Club Astronomique de la Région Lilloise, 170
Carleton College
    Department of Physics and Astronomy, 770
Carleton University
    Centre for Research in Particle Physics (CRPP), 82
    Department of Physics
        High-Energy Physics Group, 83
Carlisle, 536, 852, 854
Carloforte, 327
Carl Sandburg Middle School Planetarium, 897
Carl Schurz Park, 806
Carl Zeiss, 234
    Chester, 897
    Germany, 220
    Jena, 220
    USA-VA, 897
    Virginia, 897
Carl Zeiss Jena GmbH
    Astronomische Geräte, 220
Carl Zeiss Optical Inc., 897
Carl-Zeiss-Planetarium Stuttgart, 221
CARMA
    Combined Array for Research in Millimeter-wave Astronomy, 613
Carmen de Carupa, 119
Carnegie Institution of Washington (CIW), 672
    Department of Terrestrial Magnetism (DTM), 672
    Las Campanas Observatory, 107
    Observatories, 615
        Las Campanas Observatory (LCO), 107
Carnegie Mellon University (CMU)
    Department of Physics, 853
Carnegie Science Center, 856, 858
Carolina
    (North.), 827–832
    (South.), 866–868
Carolyn Wine Observatory, 683
Carp, 96
Carrasco, 587
Carroll Observatory (George.), 651
Carshalton, 544
Carson City, 787
Carter Observatory, 409
Carterton, 412
CAS
    California Academy of Science, 611
    Calumet Astronomical Society, 710
    Canberra Astronomical Society, 20
    Canterbury Astronomical Society, 408
        Ashburton Branch, 409
    Cardiff Astronomical Society, 540
    Center for Astrophysical Sciences (JHU), 739
    Charlottesville Astronomy Society, 897
    Chemical Abstracts Service, 836
    Chesmont Astronomical Society, 854
    Chester Astronomical Society, 541
    Cheyenne Astronomical Society, 922
    Chicago Astronomical Society, 699
    Chinese Astronomical Society, 115
    Cincinnati Astronomical Society, 836
    Cockermouth Astronomical Society, 541
    Columbus Astronomical Society, 837
    Cotswold Astronomical Society, 542
    Crawley Astronomical Society, 543
    Croydon Astronomical Society, 544
    Cumberland Astronomical Society, 871
    Czech Astronomical Society, 126
CASA
    Center for Astrophysics and Space Astronomy, 659
        Astrophysics Research Laboratory, 659
    Centre of Advanced Study in Astronomy, 287
Casa de las Ciencias
    Planetario, 474
CASC
    Club Astronomie de Saint-Claude, 170
CASCA
    Canadian Astronomical Society - Société Canadienne

d'Astronomie, 80
Cascina, 318
CASE
    Christian Association of Stellar Explorers, 606
Case Western Reserve University (CWRU)
    Department of Astronomy
        Warner and Swasey Observatory, 835
    Department of Physics, 836
CASI
    Canadian Aeronautics and Space Institute, 80
    Center for Aerospace Information, 740
CASLEO
    Complejo Astronómico El Leoncito, 7, 10
Casole d'Elsa, 314
Casper, 922
Casper - Meteorites Inc. (Michael I._), 809, 818
Casper Planetarium, 922
CASS
    Center for Advanced Space Studies, 886
    Center for Astrophysics and Space Sciences, 648
    Center for Atmospheric and Space Sciences (USU), 891
Cassini Astronomie, 164
Cassino, 308
Castelldefels, 473
Castello di Godego, 312
Castelnuovo di Garfagnana, 305
Castelnuovo di Sotto, 311
Castle Eden Public Observatory, 541
Castle Point Astronomy Club (CPAC), 540
CAT
    Cercle Astronomique de Tournai, 49
Catalana de Telescopios, 475
Catania, 319, 326, 338
Catawba Valley Astronomy Club (CVAC), 827
Catholic University of America (CUA)
    Department of Physics
        Institute for Astrophysics and Computational Sciences (IACS), 672
        Institute for Astrophysics and Computational Sciences (IACS), 672
Catskills Astronomy Club, 810
CAUP
    Centro de Astrofísica da Universidade do Porto, 436
Caussols, 192
Cauthen Educational Media Center (John K._), 866
CAV
    Centre Astronomique Vendéen, 164
    Chiefland Astronomy Village, 680
    Circolo Astrofili Veronesi, 314
    Club d'Astronomie de Villemur, 171
Cavendish Laboratory
    Mullard Radio Astronomy Observatory, 557
Cavezzo, 305
CAW
    Club Astro de Wittelsheim, 170
CAWAS
    Coventry and Warwickshire Astronomical Society, 543
CAY
    Centro Astronómico de Yebes, 481
CBA
    Center for Backyard Astrophysics, 810
CBAA
    Comité Belge des Astronomes Amateurs, 50
CBAS
    Cape Breton Astronomical Society, 82
CBAT
    Central Bureau for Astronomical Telegrams, 749
CBST
    Current Bibliography on Science and Technology, 348
CCA
    Centre Culturel de l'Astronomie, 164
CCAF
    Circolo Culturale Astronomico di Farra d'Isonzo, 314
CCAS
    Cape Cod Astronomical Society, 749
    Central Coast Astronomical Society, 616
    Chester County Astronomical Society, 854

Cleveland County Astronomical Society, 828
CCCS
    Conseil Canadien des Centres des Sciences, 81, 86
CCDS
    Center for the Commercial Development of Space
        BioServe Space Technologies, 653
CCLRC, Council for the Central Laboratory of the Research Councils, 542
    Daresbury Laboratory, 542
    Rutherford Appleton Laboratory, 542, 560
        Chilbolton Observatory, 542
        Space Science Department, 542
        STARLINK Project, 543
CCM
    Centrum voor Constructie en Mechatronica, 389
    Consejo Cultural Mundial, 382
    County College of Morris
        Planetarium, 792
CCN
    Conseil Canadien des Normes, 86, 99
CCNY
    City College of New York
        Department of Physics, 810
        Planetarium, 810
CCSC
    Canadian Council of Science Centres, 81
CCSN
    Community College of Southern Nevada
        Planetarium, 786
CCSQ
    Commission des Cadrans Solaires du Québec, 86
CCSSC
    Coca-Cola Space Science Center, 688
CCST
    Centre de Culture Scientifique et Technique, 164
CDA
    Centro de Divulgação da Astronomia, 65
CDAS
    Cleethorpes and District Astronomical Society, 541
CDS
    Centre de Données astronomiques de Strasbourg, 165
CDSCC
    Canberra Deep Space Communication Complex, 20
CEA
    Center for Extreme Ultraviolet Astrophysics (UCB), 646
    Clube Estudantil de Astronomia, 66
    Commissariat à l'Énergie Atomique
        Département d'Astrophysique, de Physique des Particules, de Physique Nucléaire et de l'Instrumentation Associée, 174
CEAAL
    Centro de Estudios Astronômicos de Alagoas, 65
CECS
    Centro de Estudios Científicos de Santiago, 107
CEDA
    Centro de Estudios de Astronomía, 927
Cedar Amateur Astronomers (CAA) Inc., 716
Cedar Falls, 717, 720
Cedar Hill Observatory, 778
Cedar Rapids, 716, 718, 719
Cederberg Observatory, 465
Ceduna, 31
C.E. Kenneth Mees Observatory, 825
C.E. Kenneth Mees Solar Observatory, 693, 696
Celestaire Inc., 721
Celestial Observers, 616
Celestron International, 616
CEN
    Comité Européen de Normalisation, 51
    Comité Européen de Normalisation Électrotechnique, 51
Cengelköy, 522
Centaurus-A, 389
Center for Advanced Space Studies (CASS), 878, 886
Center for Aerospace Information (CASI), 740
Center for Archaeoastronomy, 736

Center for Astronomical Adaptive Optics (CAAO), 594, 603

Center for Astronomical Data (CAD), 440, 445
Center for Astronomical Observing Quality (CAOQ), 465
Center for Astrophysical Research in Antarctica (CARA), 915
Center for Astrophysics and Space Astronomy (CASA), 653, 659
    Astrophysics Research Laboratory (ARL), 659
Center for Astrophysics and Space Sciences (CASS), 648
Center for Astrophysics (CfA) (Harvard-Smithsonian_), 749, 751, 752
    Central Bureau for Astronomical Telegrams, 749
    Chandra X-ray Center (CXC), 749
    Fred Lawrence Whipple Observatory, 597
    Minor Planet Center, 755
    New England Light Pollution Advisory Group, 756
Center for Backyard Astrophysics (CBA), 810
Center for Earth and Planetary Studies (CEPS), 676
Center for Educational Multimedia, 652
Center for Extreme Ultraviolet Astrophysics (CEA), 646
Center for High Angular Resolution Astronomy (CHARA), 687, 688
Center for Image Processing in Education (CIPE), 594
Center for Interdisciplinary Studies (CIS), 600
Center for Networked Information Discovery and Retrieval (CNIDR), 827
Center for Planetary Science, 135
Center for Space Plasma, Aeronomy, and Astrophysics Research (CSPAAR), 590
Center for Space Power and Advanced Electronics, 589
Center for Space Research (CSR), 754, 888
Center for the Commercial Development of Space (CCDS)
    BioServe Space Technologies (BST), 653
Center of Science and Industry (COSI)
    Columbus, 836
    Toledo, 836
Centerport, 825
Centorr - Vacuum Industries Inc., 788
Central Appalachian Astronomy Club, 912
Central Arkansas Astronomical Society (CAAS), 606
Central Bucks Planetarium, 853
Central Bureau for Astronomical Telegrams (CBAT), 749
Central Coast Astronomical Society (CCAS), 616
Central Columbia School District Planetarium, 853
Central Dauphin School District Planetarium, 853
Central Florida Astronomical Society (CFAS), 679
    Daytona Beach Affiliate, 680
Central Iowa Astronomers, 716
Central Methodist College (CMC), 776
    Morrison Observatory, 776
Central Michigan University (CMU)
    Brooks Astronomical Observatory, 762
    Physics Department, 762
Central Missouri Astronomical Association (CMAA), 776
Central Montana Astronomy Society (CMAS), 780
Central Organization for Standardization and Quality Control (COSQC), 294
Central Pennsylvania Observers (CPO), 853
Central Texas Astronomical Society (CTAS) Inc., 878
Central Valley Astronomers, Inc. (CVA), 616
Central Weather Bureau (CWB)
    Astronomical Observatory, 516
Central Wyoming Astronomical Society (CWAS), 922
Centre Astronomique Vendéen (CAV), 164
Centre Culturel de l'Astronomie (CCA), 164
Centre d'Analyse des Images (CAI), 164, 183
Centre d'Astronomie, 164
Centre de Culture Scientifique et Technique (CCST), 164
Centre de Données astronomiques de Strasbourg (CDS), 165, 205
Centre de Physique des Particules de Marseille (CPPM), 165

Centre de Recherche Astronomique de Lyon (CRAL), 165
Centre de Recherches en Astronomie, Astrophysique et Géophysique (CRAAG), 4
Centre de Recherches en Physique de l'Environnement Terrestre et Planétaire (CRPE), 165, 166
Centre de Spectrométrie Nucléaire et de Spectrométrie de Masse (CSNSM), 165, 204
Centre d'Étude des Environnements Terrestre et Planétaires (CETP)
    Saint-Maur-des-Fossés, 165
    Vélizy, 166
Centre d'Études et de Recherches de Toulouse (CERT), 166, 194
Centre d'Études et de Recherches en Géodynamique et Astrométrie (CERGA), 166
    Grasse, 191
Centre d'Étude Spatiale des Rayonnements (CESR), 155, 166
Centre d'Observation et de Recherche Astronomique Mandeville (CORAM), 47
Centre Européen de Recherche et de Formation Avancée en Calcul Scientifique (CERFACS), 166
Centre Européen pour la Recherche Nucléaire (CERN), 502
    Astronomy Club, 503
Centre for Climate and Global Change Research (C2GCR), 83
Centre for Research in Earth and Space Technology (CRESTech), 83
Centre National de Documentation Scientifique et Technique (CNDST), 48
Centre National de la Recherche Scientifique (CNRS), 166
    Centre de Physique des Particules de Marseille (CPPM), 165, 167
    Centre de Physique Théorique (CPT), 167
    Délégation aux Systèmes d'Information (DSI), 167
    Éditions, 167
    Institut d'Astrophysique de Paris (IAP), 167
    Institut d'Astrophysique Spatiale (IAS), 167
    Institut de l'Information Scientifique et Technique (INIST), 167
    Institut de Radioastronomie Millimétrique (IRAM), 182
    Institut des Sciences de la Terre, de l'Eau et de l'Espace de Montpellier (ISTEEM), 168, 183
    Laboratoire de l'Accélérateur Linéaire (LAL), 168, 187
    Observatoire de Haute Provence (OHP), 168, 191
    Service d'Aéronomie, 168
Centre National d'Enseignement à Distance (CNED), 168
Centre National d'Études Spatiales (CNES), 168
    Centre Spatial de Toulouse (CST), 169
    Centre Spatial Guyanais (CSG), 168
    Délégation à la Communication et à l'Éducation, 168
Centre Permanent d'Étude la Nature (CPEN)
    Observatoire, 48
Centre Régional de Promotion de la Culture Scientifique, Technique et Industrielle, 169
Centre Spatial de Liège (CSL), 49, 58
Centre Spatial de Toulouse (CST), 169
Centre Spatial Guyanais (CSG), 168, 169
Centreville, 901
Centro Astrofili Bolzano (CAB), 312
Centro Astronómico de Yebes (CAY), 475, 481
Centro Astronómico Hispano-Alemán
    Observatorio de Calar Alto, 475
Centro Astronomico "Neil Armstrong" (CANA), 312
Centro Astronomico Orione (CAO), 313
Centro Cultural Alfa, 382
Centro de Astrofísica da Universidade do Porto (CAUP), 433, 436
Centro de Astronomia e Astrofísica de Lisboa, 435
Centro de Ciencias de Sinaloa, 382
Centro de Convenciones y Exposiciones de Morelia
    Planetario, 382
Centro de Divulgação da Astronomia (CDA), 65
Centro de Estudios Astronômicos de Alagoas (CEAAL), 65
Centro de Estudios Científicos de Santiago (CECS), 107
Centro de Estudios de Astronomía (CEDA), 927
Centro de Investigaciones de Astronomía (CIDA), 927
Centro de Investigaciones en Optica (CIO), 382
Centro de Investigación y Difusión Aeronáutico-Espacial (CIDA-E), 586

Centro de Observação Astronômica do Algarve (COAA), 433

Centro de Previsão de Tempo e Estudos Climáticos (CPTEC), 65, 67
Centro de Radio-Astronomia e Aplicações Espaciais (CRAAE), 65, 68
Centronic Ltd., 540
Centro Regional de Investigaciones Científicas y Tecnológicas (CRICYT), 6
Centrum voor Constructie en Mechatronica (CCM), 389
Centrum voor Wiskunde en Informatica (CWI), 389
CEPS
    Center for Earth and Planetary Studies, 676
Ceramaseal, 810
Cercle Astronomique de Bruxelles (CAB), 49
Cercle Astronomique de Tournai (CAT), 49
Cercle Astronomique Mosan (CAM), 49
Cercle d'Astronomie Olympus Mons, 49
Cercle d'Astronomie Tycho, 49
Cercle des Astronomes Amateurs de Waterloo (CAAW), 50
Cercle Scientifique Briochin, 169
Ceredigion, 582
CERFACS
    Centre Européen de Recherche et de Formation Avancée en Calcul Scientifique, 166
CERGA, 169
    Centre d'Études et de Recherches en Géodynamique et Astrométrie
        Grasse, 191
CERN
    Centre Européen pour la Recherche Nucléaire, 502
        Astronomy Club, 503
Cernan Earth and Space Center, 699
CERN Astronomy Club, 503
Cernusco sul Naviglio, 309
Cerro Calán, 111
Cerro de la Mariquita, 383
Cerro Las Ánimas, 385
Cerro Pachón, 109, 596, 693
Cerro Pochoco, 107
Cerro Tololo Interamerican Observatory (CTIO), 107, 109
CERT
    Centre d'Études et de Recherches de Toulouse, 194
České Budějovice, 126, 128
České Budějovice Observatory and Planetarium, 126
CESR
    Centre d'Étude Spatiale des Rayonnements, 155, 166
CESRA
    Community of European Solar Radio Astronomers, 174
CETP
    Centre d'Étude des Environnements Terrestre et Planétaires
        Saint-Maur-des-Foss és, 166
        V élizy, 166
Ceylon, 487
CfA
    Center for Astrophysics (Harvard-Smithsonian_), 751
        Central Bureau for Astronomical Telegrams, 749
        Chandra X-ray Center, 749
        Fred Lawrence Whipple Observatory, 597
        Minor Planet Center, 755
        New England Light Pollution Advisory Group, 756
CFAS
    Cape Fear Astronomical Society, 827
    Central Florida Astronomical Society, 679
        Daytona Beach Affiliate, 680
CFHT
    Canada-France-Hawaii Telescope Corp., 692
CFSR
    Center for Solar Research, 794
Chabot College
    Astronomy Department, 616
Chabot Observatory and Science Center (COSC), 616, 619
Chabot Space and Science Center (CSSC), 616
Chadds Ford, 861
Chaffee Planetarium (Roger B._), 762, 767
    Astronomy Day, 761

Chaffey College
    Daniel B. Milliken Planetarium, 617
Chagrin Falls, 836
Chagrin Valley Astronomical Society (CVAS), 836
Chalfont, 853
Challenger Center for Space Science Education, 897
Chalmers Technical University, 488
    Department of Astronomy and Astrophysics, 490
    Onsala Space Observatory, 490
Chalon-sur-Saône, 201
Chamberlin Observatory, 654, 660
Chamblon, 510
Chamiers, 178
Chamonix, 190
Champagne-en-Valromey, 170
Champaign, 699, 702, 703, 709
Champaign-Urbana Astronomical Society (CUAS), 699
Chandler Co. (David_), 616, 618
Chandler Mountain, 589
Chandra X-ray Center (CXC), 749
Chandra X-ray Observatory (CXO), 750
Changchun Observatory, 113
Changi, 456
Channel Islands, 106
Chantilly, 899
CHAOS
    Chapel Hill Astronomical and Observational Society, 828
Chapel Hill, 828, 829, 832
Chapel Hill Astronomical and Observational Society (CHAOS), 828
CHARA
    Center for High Angular Resolution Astronomy, 688
Charles C. Gates Planetarium, 653
Charles E. Daniel Observatory, 868
Charles Evans & Associates, 616
Charles F. Clise Planetarium, 911
Charles F. Hagar Planetarium, 637
Charles Hayden Planetarium, 750
Charleston, 866, 868, 912, 913
Charlestown, 864
Charles University
    Astronomical Institute, 126
    Centre for Theoretical Study (CTS), 126
Charlotte, 828, 829
Charlotte Amateur Astronomers Club (CAAC), 828
Charlottenlund, 137
Charlottesville, 896, 897, 900, 903, 904
Charlottesville Astronomy Society (CAS), 897
Charterhouse Observatory, 537
Chateau du Lac Observatory, 651
Chateauroux, 157
Châtenay-Malabry, 162, 175
Chatham, 902
Châtillon, 194
Chatsworth, 616, 641
Chattanooga, 870, 875
Chatterton Astronomy Department, 31
Chavannes-des-Bois, 513
Cheisacker, 500
Chelmsford, 758
Chelmsley Astronomical Society, 548
Chemical Abstracts Service (CAS), 836
Chemnitz, 244
Cheney, 907
Chennai, 282
Chernigovka, 525
Chernigovka Amateur Astronomical Society, 525
Cherokee, 718, 719
Cherry Grove Observatory, 771
Cherry Memorial Planetarium (Jim_), 688
Chesapeake, 897, 898, 901
Chesapeake Planetarium, 897
Chesmont Astronomical Society (CAS), 854
Chester, 541, 824, 863, 897
Chester Astronomical Society (CAS), 541
Chester County Astronomical Society (CCAS), 854

Chevy Chase, 736
Cheyenne, 922
Cheyenne Astronomical Society (CAS), 922
Chiang Mai, 519
Chiang Mai University (CMU)
    Department of Physics, 519
Chiba, 343, 360
Chicago, 698, 699, 702, 703, 705, 707, 708
Chicago Astronomical Society (CAS), 699
Chico, 614, 631
Chiefland, 680
Chiefland Astronomy Village (CAV), 680
Chieti, 307
Chikurinji (Mount.), 354
Chilbolton Observatory, 541, 542
Children's Museum of Indianapolis (The.), 710, 714
Chile, 107
Chilton, 542, 543, 560
China
    (PRC), 112
    (ROC), 516
China Lake Astronomical Society (CLAS), 617
China Meteorological Administration (CMA), 113
China State Bureau of Technical Supervision (CSBTS), 113

Chinese Academy of Sciences
    Beijing Astronomical Data Center (BADC), 113
    Beijing Astronomical Observatory (BAO), 113, 115
    Beijing Astrophysics Center (BAC), 112
    Changchun Observatory, 113
    Institute of High-Energy Physics
        Laboratory of Cosmic-Ray and High-Energy Astrophysics, 114
    National Astronomical Observatories, 114
    Purple Mountain Observatory, 112, 114
        Qinghai Radio Observatory, 114
    Shaanxi Astronomical Observatory (CSAO), 114
    Shanghai Astronomical Observatory (SHAO), 114
    Urumqi Astronomical Observatory (UAO), 115
    Yunnan Observatory, 115
Chinese Astronomical Society (CAS), 115
Chinese Journal of Astronomy and Astrophysics (CJAA), 115
Chino, 618
Chinon, 162, 181
Chippewa Falls, 916
Chippewa Valley Astronomical Society (CVAS), 915
Chippis, 510
Chislehurst, 568
Chita, 451
CHMI
    Czech Hydrometeorological Institute, 127
Cholet, 196
Cholsey, 531
Chorzów, 429
Christa McAuliffe Planetarium (CMP), 788
Christchurch, 408, 409, 411, 412
Christian-Albrechts-Universität zu Kiel, 221
Christian Association of Stellar Explorers (CASE), 606
Christian Jutz Volkssternwarte Berg eV, 221
Chuguev, 526
Chulalongkorn University
    Department of Physics, 519
Chungbuk, 367
Chung-li, 516
Chungnam National University (CNU)
    Department of Astronomy and Space Science, 367
Chur, 500
Ciampino, 313
CiAO
    Cimini Astronomical Observatory, 313
CIAR
    Canadian Institute for Advanced Research, 81
CICAP
    Comitato Italiano per il Controllo delle Affermazioni sul Paranormale, 315
CIDA
    Centro de Investigaciones de Astronomía, 927
CIDA-E
    Centro de Investigación y Difusión Aeronáutico-Espacial, 586
CIG
    Coronado Instrument Group, 594
CIMHaribbean Institute for Meteorology and Hydrology, 45

Cimini Astronomical Observatory (CiAO), 313
Cincinnati, 837, 838, 844
Cincinnati Astronomical Society (CAS), 836
Cincinnati Observatory Center, 837
Cinisello Balsamo, 319
CINVESTAV
    Centro de Investigación y de Estudios Avanzados, 383
CIO
    Centro de Investigaciones en Optica, 382
CIP
    Computers in Physics, 736
CIPA
    California Institute for Physics and Astrophysics, 611
CIPE
    Center for Image Processing in Education, 594
Circolo Astrofili Bergamaschi (CAB), 313
Circolo Astrofili di Mestre "Guido Ruggieri", 313
Circolo Astrofili di Milano (CAM), 313
Circolo Astrofili Nord Sardegna, 313
Circolo Astrofili Talmassons, 314
Circolo Astrofili Veronesi (CAV) "Antonio Cagnoli", 314
Circolo Astronomico Dorico "Paolo Andrenelli", 314
Circolo Casolese Astrofili "Betelgeuse", 314
Circolo Culturale Astronomico di Farra d'Isonzo (CCAF), 314
CIS
    Center for Interdisciplinary Studies, 600
    Club d'Information Scientifique (Poste et France Telecom), 172
CISTI
    Canadian Institute for Scientific and Technical Information, 93
CITA
    Canadian Institute for Theoretical Astrophysics, 81
Citadel (The.), 866
Cité de l'Espace, 169
Cité des Sciences et de l'Industrie de la Villette, 159
    Planétarium, 169
Città del Vaticano, 926
City College of New York (CCNY)
    Department of Physics, 810
    Planetarium, 810
City College of San Francisco
    Department of Astronomy, 617
City Observatory, Edinburgh, 533
City University of New York (CUNY)
    Bernard Baruch College
        Department of Physics and Astronomy, 810
    Herbert H. Lehman College
        Department of Physics and Astronomy, 810
    Hunter College
        Department of Physics and Astronomy, 810
Ciudad Victoria, 384
Ciutat de les Arts i les Ciències de València, 475
Civico Planetario "Ulrico Hoepli", 313, 314
Civitavecchia, 306
CIW
    Carnegie Institution of Washington, 672
        Department of Terrestrial Magnetism, 672
        Las Campanas Observatory, 107
        Observatories, 615
CJAA
    Chinese Journal of Astronomy and Astrophysics, 115
CJP
    Club Jean Perrin, 173
CKC Power, 811
Clacton and District Astronomical Society, 541
Clacton-on-Sea, 541
Clanfield Observatory, 547

Clare, 298
Claremont, 623, 634
Clarence T. Jones Observatory, 875
Clarence T. Jones Planetarium, 875
Clarksburg, 912
Clark Science Center, 758
Clarkson University
    Department of Physics, 811
Clarks Summit, 857
CLAS
    China Lake Astronomical Society, 617
Clayton, 831
CLEA
    Comité de Liaison Enseignants Astronomes, 173
    Contemporary Laboratory Experiences in Astronomy, 854
Clear Night Productions, 762
Clearwater, 681
Cleethorpes, 535, 541
Cleethorpes and District Astronomical Society (CDAS), 541

Clemson, 866
Clemson University
    Department of Physics and Astronomy, 866
Clermont, 156
Cleveland, 582, 835, 836, 840, 842, 843, 870, 871
Cleveland and Darlington Astronomical Society (CaDAS), 541
Cleveland County Astronomical Society (CCAS), 828
Cleveland Museum Natural History (CMNH), 842
Cleveland State Community College (ClSCC)
    Department of Physics and Astronomy, 871
Clever Planetarium (George H._), 617, 622
Cleves, 837
Clifton, 794
Climenhaga Observatory, 104
Clinton, 774, 818
Clise Planetarium (Charles F._), 911
Closter, 796
Cloudcroft, 799
Clovis, 800
Clovis Astronomy Club, 799
ClSCC
    Cleveland State Community College
        Department of Physics and Astronomy, 871
Club Ajaccien des Amateurs d'Astronomie (C3A), 169
Club Astro Alpha Centauri, 170
Club Astro de Wittelsheim (CAW), 170
Club Astro Guynemer, 170
Club Astro Junior M67, 170
Club Astronomie de l'Institut National des Sciences Appliquées, 183
Club Astronomie de Saint-Claude (CASC), 170
Club Astronomie Nature du Valromey (CANV), 170
Club Astronomique de la Région Lilloise (CARL), 170
Club Astronomique Rochefortois (CAR), 50
Club Astronomique Véga de la Lyre, 171
ClubAstro of Mission San Jose, 617
Club Copernic, 171
Club d'Astronomes Amateurs de Bienne (CAAB), 503
Club d'Astronomie Alpha Centaure (CAAC), 171
Club d'Astronomie d'Amateurs "Orion", 50
Club d'Astronomie de Beloeil, 83
Club d'Astronomie de Chamonix, 171, 190
Club d'Astronomie de Dorval, 83
Club d'Astronomie de Drummondville (CAD), 83
Club d'Astronomie de la Péninsule Acadienne, 83
Club d'Astronomie de l'Observatoire "L'Étoile d'Acadie", 84

Club d'Astronomie de l'Université du Québec à Rimouski, 84
Club d'Astronomie de Lyon-Ampère (CALA), 171
    Union Rhône-Alpes des Clubs d'Astronomie (URAC-A), 203
Club d'Astronomie de Rimouski, 84
Club d'Astronomie de Sept-Îles, 84
Club d'Astronomie de Villemur (CAV), 171

Club d'Astronomie d'Ottignies (CAO), 50
Club d'Astronomie du Trégor, 172
Club d'Astronomie Janus, 172
Club d'Astronomie Jupiter, 84
Club d'Astronomie Les Almucantars, 84
Club d'Astronomie Miranda, 172
Club d'Astronomie Orion, 85
Club d'Astronomie Pégase des Laurentides, 85
Club d'Astronomie "Randonnée Céleste", 172
Club d'Astronomie Sirius du Saguenay, 85
Club d'Astronomie Spica, 172
Club d'Astronomie Uranie, 172
Club de Astrofísica del Paraguay (CAP), 422
Club de Astronomía Felix Aguilar, 6
Club des Astronomes Amateurs Centre Mauricie (CAACM), 85
Club des Astronomes Amateurs de Laval (CAAL), 85
Club des Astronomes Amateurs de Longueuil (CAAL), 85
Club des Astronomes Amateurs de Sherbrooke (CAAS), 86
Club d'Information Scientifique (CIS) de la Poste et de France Telecom, 172
Club Éclipse, 172
Clube de Astronomia do Rio de Janeiro (CARJ), 66
Clube Estudantil de Astronomia (CEA), 66
Club Jean Perrin (CJP), 173
Cluj-Napoca, 438
Cluj-Napoca Astronomical Observatory, 438
Clyde W. Tombaugh Observatory, 723
Clyde W. Tombaugh Space Theater, 803
CMA
    China Meteorological Administration, 113
CMAA
    Central Missouri Astronomical Association, 776
CMAS
    Central Montana Astronomy Society, 780
CMC
    Central Methodist College
        Morrison Observatory, 776
CMHAS
    Crayford Manor House Astronomical Society, 543
CMNH
    Cleveland Museum Natural History, 842
CMO
    Caribbean Meteorological Organization, 521
CMP
    Christa McAuliffe Planetarium, 788
CMU
    Carnegie Mellon University
        Department of Physics, 853
    Central Michigan University
        Brooks Astronomical Observatory, 762
        Physics Department, 762
    Chiang Mai University
        Department of Physics, 519
CN
    Comité Estatal de Normalización de Cuba, 123
CNDST
    Centre National de Documentation Scientifique et Technique, 48
CNED
    Centre National d'Enseignement à Distance, 168
CNES
    Centre National d'Études Spatiales, 168
        Centre Spatial de Toulouse, 169
        Centre Spatial Guyanais, 168
        Délégation à la Communication et à l'Éducation, 169
CNIDR
    Center for Networked Information Discovery and Retrieval, 827
CNPq
    Conselho Nacional de Pesquisas, 66
CNRC, Conseil National de Recherches du Canada, 86
    Institut Canadien de l'Information Scientifique et Technique, 93
    Institut Herzberg d'Astrophysique
        Canadian Astronomy Data Centre, 93
        Observatoire Fédéral d'Astrophysique, 94

Observatoire Fédéral de Radioastrophysique, 94
CNR, Consiglio Nazionale delle Ricerche, 315
    Istituto di Astrofisica Spaziale e Fisica Cosmica, 315
    Istituto di Cosmogeofisica, 315
    Istituto di Fisica Cosmica ed Applicazioni dell'Informatica, 315
    Istituto di Fisica dello Spazio Interplanetario, 315
    Istituto di Metrologia "G. Colonnetti", 316
    Istituto di Radioastronomia
        Bologna, 316
        Firenze, 316
        Stazione di Medicina, 316
        Stazione di Noto, 316
    Istituto di Tecnologie e Studio delle Radiazioni Extraterrestri, 317
CNRS, Centre National de la Recherche Scientifique, 166
    Centre de Physique des Particules de Marseille, 165
    Centre de Physique Théorique, 167
    Délégation aux Systèmes d'Information, 167
    Éditions, 173
    Institut d'Astrophysique de Paris, 167
    Institut d'Astrophysique Spatiale, 167
    Institut de l'Information Scientifique et Technique, 167
    Institut de Radioastronomie Millimétrique, 182
    Institut des Sciences de la Terre, de l'Eau et de l'Espace de Montpellier (ISTEEM), 183
    Laboratoire de l'Accélérateur Linéaire, 187
    Observatoire de Haute Provence, 191
    Service d'Aéronomie, 168
CNRS Éditions, 173
CNU
    Chungnam National University
        Department of Astronomy and Space Science, 367
COAA
    Centro de Observação Astronômica do Algarve, 433
Cobblestone Publishing Inc., 788
Coca-Cola Space Science Center (CCSSC), 688
Cochabamba, 62
Cockermouth, 541
Cockermouth Astronomical Society (CAS), 541
Cocoa, 678, 679
Coconino Astronomers, 594
CODATA, 173
    Committee on Data for Science and Technology, 174
Coe College
    Physics Department, 719
Cogeces del Monte, 483
Coimbra, 435
Cointe, 58
Coit Memorial Observatory (Judson Broadman_), 749
Col de Bacchus, 158
Col de la Lèbe, 170
Col Drusciè, 308
Coldwater, 765
Colégio Magno
    Observatório, 66
Coleman Sr. Planetarium (George E._), 690
Colfax, 628
Colgate University
    Department of Physics and Astronomy, 811
Colin Hunt Observatory, 535
Collegedale, 874
Collège de Lévis (Astro-Club_), 78
Collège J. Valéri
    Association du Planétarium, 173
College of Cardiff
    Physics and Astronomy Department, 582
College of Charleston
    Department of Physics and Astronomy, 866
College of Southern Idaho
    Herrett Center for Arts and Science
        Earl and Hazel Faulkner Planetarium, 697
College of Staten Island (CSI)
    Astrophysical Observatory, 811
College of St. Catherine
    Physics Department
        Observatory, 770

College Park, 734, 736, 744
College Station, 877, 885
Collingwood, 23
Collins Electro Optics, 653
Collins Observatory (Eileen_), 811, 813
Colluriana, 328
Colmar, 195
Cologne, 209, 225, 228, 238, 250, 255, 261
Colombara, 314
Colombia, 119
Colombo, 487
Colonia, 587
Colorado, 652–661
    (Northern_), 657
    (Western_), 661
Colorado College
    Department of Physics, 653
Colorado Springs, 652, 653, 656, 659–661
Colorado Springs Astronomical Society (CSAS), 653
Colorado State University
    Department of Physics, 653
Columbia, 744, 777, 778, 866–868
    (District of_), 669
Columbia Astrophysics Laboratory, 811
Columbia University
    Center for Backyard Astrophysics (CBA), 810
    Columbia Astrophysics Laboratory, 811
    Department of Astronomy, 810, 811
    Department of Physics, 812
    Harriman Observatory, 825
    MDM Observatory, 598
Columbus, 688, 689, 712, 836, 837, 841, 843
Columbus Astronomical Society (CAS), 837
Colutron Research Corp., 654
Combined Array for Research in Millimeter-wave Astronomy (CARMA), 613
Comenius University
    Institute of Astronomy, 458
        Astronomical and Geophysical Observatory, 458
Comet Rapid Announcement Service (CRAS), 837
Comisión Nacional de Actividades Espaciales (CONAE), 6
Comisión Nacional de Astronomía
    Centro Astronómico Hispano-Alemán, 475
    Observatorio de Calar Alto, 475
Comisión Nacional de Investigaciones Científicas y Técnologicas (CONICYT), 107
Comisión Panamericana de Normas Técnicas (COPANT), 927
Comisión Venezolana de Normas Industriales (COVENIN), 927
Comitato Italiano per il Controllo delle Affermazioni sul Paranormale (CICAP), 315
Comité Belge des Astronomes Amateurs (CBAA), 50
Comité Belge pour l'Investigation Scientifique des Phénomènes Réputés Paranormaux (Comité PARA), 50
Comité de Liaison Enseignants Astronomes (CLEA), 173
Comité Estatal de Normalización (CN), 123
Comité PARA, Comité Belge pour l'Investigation Scientifique des Phénomènes Réputés Paranormaux, 50
Commission des Cadrans Solaires du Québec (CCSQ), 86
Commission on International Coordination of Space Techniques for Geodesy and Geodynamics (CSTG), 221

Committee for Standardization and Metrology (BDS), 76
Committee for Standardization, Metrology and Certification (BELST), 46
Committee for Standardization of the DPRK (CSK), 366
Committee for the Scientific Investigations of Claims of the Paranormal (CSICOP), 812
Committee on Astronomy and Astrophysics (CAA), 674
Committee on Atomic, Molecular, and Optical Sciences, 674

Committee on Data for Science and Technology (CODATA), 174
Committee on Radio Frequencies (CORF), 674
Committee on Science and Technology in Developing Countries (COSTED), 282

Committee on Space Research (COSPAR), 174
Committee on the Peaceful Uses of Outer Space (COPUOS), 40
Committee on the Public Understanding of Science (COPUS), 557
Commonwealth Bureau of Meteorology
    Head Office, 21
Commonwealth Scientific and Industrial Research Organisation (CSIRO)
    Australia Telescope National Facility (ATNF)
        Epping, 21
        Headquarters, 21
        Mopra Telescope, 21
        Narrabri, 21
        Parkes Radio Observatory, 22
    Earth Observation Centre (EOC), 22
    Editorial Services, 22
    Office of Space Science and Applications (COSSA), 22
    Paul Wild Observatory, 22
    Wild Observatory (Paul_), 22
Communications Research Laboratory (CRL)
    Hiraiso Solar-Terrestrial Research Center, 342
    Inubo Radio Observatory, 343
    Kansai Advanced Research Center (KARC), 343
    Kashima Space Research Center (KSRC), 343
    Okinawa Radio Observatory, 343
    Tokyo, 343, 362
    Wakkanai Radio Observatory, 344
    World Data Center C2 for Ionosphere, 362
    Yamagawa Radio Observatory, 344
Community College of Southern Nevada (CCSN)
    Planetarium, 786
Community of European Solar Radio Astronomers (CESRA), 174
Como, 321
Comox, 86
Comox Valley Astronomy Club (CVAC), 86
Company Seven (C-VII)
    Astro-Optics Division, 736
Compiègne, 175
Complejo Astronómico El Leoncito (CASLEO), 7, 10
Composants et Systèmes de Précision (CSP), 175
Comprehensive TeX Archive Network (CTAN)
    England, 850
    Germany, 850
    USA-TX, 850
Computer Sciences Corp. (CSC)
    Corporate Headquarters, 617
Computers in Physics (CIP), 736
Comstock Inc., 871
CONACYT
    Consejo Nacional de Ciencia y Tecnología (El Salvador), 143
    Consejo Nacional de Ciencia y Tecnología (Mexico), 383
CONAE
    Comisión Nacional de Actividades Espaciales (Argentina), 6
Concepción, 111
Concord, 753, 788, 789
Concordia Astronomy Club (CAC), 792
Concordia College
    Department of Physics
        Observatory, 770
CONCYTEC
    Consejo Nacional de Ciencia y Tecnología (Peru), 423
Conestoga Girl Scout Camp, 705
CONICET
    Consejo Nacional de Investigaciones Científicas y Técnicas (Argentina), 7
CONICIT
    Consejo Nacional de Investigaciones Científicas y Técnologicas (Venezuela), 927
CONICYT
    Comisión Nacional de Investigaciones Científicas y Técnologicas (Chile), 107
    Consejo Nacional de Investigaciones Científicas y Técnicas (Uruguay), 586
Connecticut, 662–666
    (Eastern_), 664
    (Southern_), 664
    (Western_), 665, 666
Connecticut College
    Department of Physics, Astronomy, and Geophysics, 662
Conseil Canadien des Centres des Sciences (CCCS), 81, 86
Conseil Canadien des Normes (CCN), 86, 99
Conseil National de Recherches du Canada (CNRC), 86
    Institut Canadien de l'Information Scientifique et Technique (ICIST), 86, 93
    Institut Herzberg d'Astrophysique
        Canadian Astronomy Data Centre, 93
        Observatoire Fédéral d'Astrophysique (OFA), 87, 94
        Observatoire Fédéral de Radioastrophysique (OFR), 86, 94
Consejo Cultural Mundial (CCM), 382
Consejo Nacional de Ciencia y Tecnología (CONACYT), 143, 383
Consejo Nacional de Ciencia y Tecnología (CONCYTEC), 423
Consejo Nacional de Investigaciones Científicas y Técnicas (CONICET), 7
Consejo Nacional de Investigaciones Científicas y Técnicas (CONICYT), 586
Consejo Nacional de Investigaciones Científicas y Técnologicas (CONICIT), 927
Consejo Superior de Investigaciones Científicas (CSIC), 475
    Instituto de Astrofísica de Andalucia (IAA), 475
    Instituto de Astronomía y Geodesia (IAG), 475, 478
    Instituto de Física de Cantabria, 476, 478
    Observatorio del Ebro, 476, 486
Conselho Nacional Desenvolvimento Scientifico e Tecnologico, 66
Consiglio Nazionale delle Ricerche (CNR), 315
    Istituto di Astrofisica Spaziale e Fisica Cosmica (IASFC), 315
    Istituto di Cosmogeofisica, 315
    Istituto di Fisica Cosmica ed Applicazioni dell'Informatica (IFCAI), 315
    Istituto di Fisica dello Spazio Interplanetario (IFSI), 315
    Istituto di Metrologia "G. Colonnetti" (IMGC), 316
    Istituto di Radioastronomia (IRA)
        Bologna, 316
        Firenze, 316
        Stazione di Medicina, 316
        Stazione di Noto, 316
    Istituto di Tecnologie e Studio delle Radiazioni Extraterrestri (TESRE), 317
Consortium for Undergraduate Research and Education in Astronomy (CUREA), 700
Consorzio Internazionale per l'Astrofisica Relativistica, 317, 324
Constantine, 4
Constellarium E4, 181
Consultants
    ACE, 593
    Astronomical Consultants and Equipment, 593
    Computer Sciences Corp., 617
    CSC, 617
    Euroconsult, 177
    Ferberts Associates, 602
    Hencoup Enterprise, 548
    Infinity Infocus, 378
    Intespace, 186
    Lancexport, 187
    Optical Research Associates, 631
    ORA, 631
    Scot, 199
    SDR&I, 676
    Seagoat Consulting, 602
    Shibayama Scientific, 357
    Software Systems Consulting, 640
    Sollid Optics, 803

Space Data Resources & Information, 676
Spectra Astro Systems, 641
SSC, 640
Verkehr Raumfarht und Systemtechnik, 258
VRS, 258
Contemporary Laboratory Experiences in Astronomy (CLEA), 854
Conway, 606, 607
Conyngham, 856
Cook Center for Arts, Science and Technology
Planetarium, 878
Cook Observatory (Flower and_), 862
Coonabarabran, 14, 29
Cooperstown, 823
Coot-tha (Mount_), 27
COPANT, 928
Comisión Panamericana de Normas Técnicas, 927
Copenhagen, 135-140
Copenhagen Astronomical Observatory, 135, 136
Copenhagen Astronomical Society, 135
Copenhagen University
Niels Bohr Institute for Astronomy, Physics and Geophysics (NBIfAFG), 135
Astronomical Observatory, 136
Danish Center for Earth System Science (DCESS), 137
Department of Geophysics, 138
Niels Bohr Institute (NBI), 136
Ørsted Laboratory, 136
Copernicus Astronomical Center (NCAC) (N._), 427, 429
Department of Astrophysics I, 430
Copernicus Museum (Nicolas_), 427, 429
Copernicus Planetarium, 837
Copernicus Planetarium and Observatory in Chorzów (Nicolas_), 427
Copernicus Planetarium (Nicolaus_), 221, 239
COPUOS
Committee on the Peaceful Uses of Outer Space, 40
COPUS
Committee on the Public Understanding of Science, 557
Coral Springs, 685
CORAM
(Astro_), 47
Centre d'Observation et de Recherche Astronomique Mandeville, 47
Córdoba, 6, 10, 471
CORF
Committee on Radio Frequencies, 674
Cork, 295, 296
Cork Astronomy Club, 295
Cork Institute of Technology
Department of Applied Physics and Instrumentation, 295
Cormack Planetarium, 864
Cornell University
Center for Radiophysics and Space Research (CRSR), 812
Department of Astronomy, 812
Icarus, 816
National Astronomy and Ionosphere Center (NAIC) Headquarters, 819
Press, 812
Cornell University Press, 812
Cornelly
(North_), 537, 546
Corning, 813
Corning Community College
Eileen Collins Observatory, 812
Corning Incorporated, 813
Cornwall, 541
Cornwall Astronomical Society, 541
Corona Borealis, 389
Coronado Filters, 541
Coronado Instrument Group (CIG) Inc., 594
Corona Park, 819
Corp. for Research Amateur Astronomy (CRAA), 617

Corpus Christi, 882
Corsicana, 878
Cortina d'Ampezzo, 308
Corvallis, 850
COSC
Chabot Observatory and Science Center, 616, 619
COSCQ
Central Organization for Standardization and Quality Control (Iraq), 294
Cosenza, 321, 337
COSI
Center of Science and Industry
Columbus, 836
Toledo, 836
Cosmo Astronomical Society (CAS), 680
Cosmodôme, 87
CosmoNorr, 488
Cosmonova, 488
COSPAR, 175
Committee on Space Research, 174
COSSA
CSIRO Office of Space Science and Applications, 22
Costa Mesa, 631, 632
COSTED
Committee on Science and Technology in Developing Countries, 282
Costitx, 480
Côte d'Or, 156
Cotswold Astronomical Society (CAS), 542
Cottbus, 243
Cotton Expressions Ltd., 699
Coulounieix-Chamiers, 178
Coulter Optical, 682
Council for Scientific and Industrial Research (CSIR), 465
Council for the Central Laboratory of the Research Councils (CCLRC), 542
Daresbury Laboratory, 542
Rutherford Appleton Laboratory (RAL), 542, 560
Chilbolton Observatory, 542
Space Science Department (SSD), 542
STARLINK Project, 543
Council of German Observatories, 242
County College of Morris (CCM)
Planetarium, 792
Courbevoie, 160, 181
COVENIN
Comisión Venezolana de Normas Industriales, 927
Coventry, 543, 582
Coventry and Warwickshire Astronomical Society (CAWAS), 543
Covina, 636
(West_), 636
Covington, 726
Covo, 332
Cowichan Valley StarFinders Astronomy Club (CVSF), 87
CPAC
Castle Point Astronomy Club, 540
CPC
Computer Physics Communications, 391
CPC Program Library, 543
CPEN
Centre Permanent d'Étude la Nature
Observatoire, 49
CPO
Central Pennsylvania Observers, 853
CPPM
Centre de Physique des Particules de Marseille, 165
CPT
Centre de Physique Théorique (CNRS), 167
CPTEC
Centro de Previsão de Tempo e Estudos Climáticos, 67
CRAA
Corp. for Research Amateur Astronomy, 617
CRAAE
Centro de Radio-Astronomia e Aplicações Espaciais, 68
CRAAG
Centre de Recherches en Astronomie, Astrophysique et

Géophysique, 4
Cracow, 428, 430
Crago Observatory, 15
Craigmont High School Planetarium, 871
CRAL
    Centre de Recherche Astronomique de Lyon, 165
Cranbury, 798
Crane Observatory (Zenas_), 724
Cranford, 791
CrAO
    Crimean Astrophysical Observatory, 525
        Katziveli, 525
        Simeiz, 525
CRAS
    Comet Rapid Announcement Service, 837
Craven E. Williams Observatory, 828
Crawford Hill, 793
Crawley, 543
Crawley Astronomical Society (CAS), 543
Crayford, 543
Crayford Manor House Astronomical Society (CMHAS), 543

Creare Inc., 788
CRESTech, 87
    Centre for Research in Earth and Space Technology, 83
Crestwood, 727, 777
Crete, 269
Crevillente, 477
CRI
    Cambridge Research and Instrumentation, 749
CRICYT
    Centro Regional de Investigaciones Científicas y Tecnológicas, 6
Crimea, 450, 525
Crimean Amateur Astronomical Society (CAAS), 525
Crimean Astrophysical Observatory (CrAO), 525
    Radioastronomical Department
        Katziveli, 525
        Simeiz, 525
Crimmitschau, 233
CRL
    Communications Research Laboratory
        Hiraiso Solar-Terrestrial Research Center, 342
        Inubo Radio Observatory, 343
        Kansai Advanced Research Center, 343
        Kashima Space Research Center, 343
        Okinawa Radio Observatory, 343
        Tokyo, 343, 362
        Wakkanai Radio Observatory, 344
        World Data Center C2 for Ionosphere, 362
        Yamagawa Radio Observatory, 344
Croatia, 121
Croatian Physical Society, 121
Cronyn Observatory (Hume_), 104
Crosby Ramsey Memorial Observatory, 739
Crosland Hill, 549
Cross Plains, 920
Cross River, 825
Crossroads Astronomy Club (CAC), 878
Crown Point, 521
Crown Space Center and OmniMax Theater (Henry_), 703
Crowthorne, 539
Croydon, 540
Croydon Astronomical Society (CAS), 544
CRPE
    Centre de Recherches en Physique de l'Environnement Terrestre et Planétaire, 166
CRPP
    Centre for Research in Particle Physics, 82
CRSR
    Center for Radiophysics and Space Research, 812
Cryomech Inc., 813
Cryophysics SA, 503
CSA
    Cambridge Scientific Abstracts, 735
    Canadian Space Agency
        David Florida Laboratory, 82

        Ottawa, 82
        Space Science Program, 82
CSAO
    Chinese Academy of Sciences Shaanxi Astronomical Observatory, 114
CSAS
    Colorado Springs Astronomical Society, 653
CSBTS
    China State Bureau of Technical Supervision, 113
CSC
    Computer Sciences Corp.
        Corporate Headquarters, 617
CSDO
    Canberra Space Dome and Observatory, 21
CSG
    Centre Spatial Guyanais, 168
CSI
    College of Staten Island
        Astrophysical Observatory, 811
CSIC
    Consejo Superior de Investigaciones Científicas, 475
        Instituto de Astrofísica de Andalucia, 476
        Instituto de Astronomía y Geodesia, 478
        Observatorio del Ebro, 476, 486
CSICOP
    Committee for the Scientific Investigations of Claims of the Paranormal, 812
CSIR
    Council for Scientific and Industrial Research, 465
CSIRO, Commonwealth Scientific and Industrial Research Organisation
    Australia Telescope National Facility (ATNF)
        Epping, 21
        Headquarters, 21
        Mopra Telescope, 21
        Narrabri, 21
        Parkes Radio Observatory, 22
    Earth Observation Centre, 22
    Editorial Services, 23
    Office of Space Science and Applications, 22
    Paul Wild Observatory, 22
    Publishing, 23
    Wild Observatory (Paul_), 22
CSIRO Office of Space Science and Applications (COSSA), 22
CSIRO Publishing, 22
CSK
    Committee for Standardization of the DPRK, 366
CSL
    Centre Spatial de Liège, 58
CSM
    Cumberland Science Museum
        Sudekum Planetarium, 874
CSN
    Centro di Scienze Naturali, 309
CSNSM
    Centre de Spectrométrie Nucléaire et de Spectrométrie de Masse, 204
CSO
    CalTech Submillimeter Observatory, 692
CSP
    Center for Space Physics, Boston University, 749
    Center for Space Power, 885
    Composants et Systèmes de Précision, 175
CSPAAR
    Center for Space Plasma, Aeronomy, and Astrophysics Research, 590
CSR
    Center for Space Research (MIT), 754
    Center for Space Research (U. Texas), 888
CSS
    Center for Space Science (UTD), 888
CSSC
    Chabot Space and Science Center, 616
CST
    Centre Spatial de Toulouse, 169
CSTG

Commission on International Coordination of Space Techniques for Geodesy and Geodynamics, 221
CSUN
   California State University, Northridge
      Department of Physics and Astronomy, 615
      Planetarium, 615
      San Fernando Observatory, 615
CSUS
   California State University, Sacramento
      Department of Physics and Astronomy, 615
CTAN
   Comprehensive TEX Archive Network
      England, 850
      Germany, 850
      USA-TX, 850
CTAS
   Central Texas Astronomical Society, 878
CTIO
   Cerro Tololo Interamerican Observatory, 109
CTS
   Centre for Theoretical Study (Charles University), 126
CTU
   Czech Technical University
      Astronomical Observatory, 127
CUA
   Catholic University of America
      Department of Physics, 672
      Institute for Astrophysics and Computational Sciences, 672
CUAS
   Cambridge University Astronomical Society, 540
   Champaign-Urbana Astronomical Society, 699
Cuba, 123
Cuba Astronomical Observatory (MCAO) Inc. (Mount.), 667
Cuba (Mount.), 667
Cubillas, 483
Cukurova University
   Department of Physics
      Center for Space Sciences, 522
Culcheth, 568
Culgoora Solar Observatory, 24
Culiacan, 382
Culture and Cosmos, 544
Culver City, 631
Cumberland Astronomical Society (CAS), 871
Cumberland Astronomy Club (CAC), 736
Cumberland Science Museum (CSM)
   Sudekum Planetarium, 874
Cumming, 687
Cuneo, 331
CUNY
   City University of New York
      Bernard Baruch College, Department of Physics and Astronomy, 810
      Herbert H. Lehman College, Department of Physics and Astronomy, 810
      Hunter College, Department of Physics and Astronomy, 810
CUP
   Cambridge University Press
      Cambridge, 540
      New York, 809
      United Kingdom, 540
      USA-NY, 809
Cupertino, 628
CUREA
   Consortium for Undergraduate Research and Education in Astronomy, 700
Current Science Association, 283
Custer Institute Inc.
   Observatory, 813
Custom Scientific, 594
Cuyahoga, 837
Cuyahoga Astronomical Association (CAA), 837
CVA
   Central Valley Astronomers, 616

CVAC
   Catawba Valley Astronomy Club, 827
   Comox Valley Astronomy Club, 86
CVAS
   Chagrin Valley Astronomical Society, 836
   Chippewa Valley Astronomical Society, 915
CVC Products Inc., 813
C-VII
   Company Seven, 736
CVSF
   Cowichan Valley StarFinders, 87
CWAS
   Central Wyoming Astronomical Society, 922
CWB
   Central Weather Bureau, 516
CWI
   Centrum voor Wiskunde en Informatica, 389
CWRU
   Case Western Reserve University
      Department of Astronomy, 835
      Department of Physics, 836
      Particle Astrophysics Group, 836
      Warner and Swasey Observatory, 835
CXC
   Chandra X-ray Center, 749
CXO
   Chandra X-ray Observatory, 750
Cyanogen Productions Inc., 87
CyberSphere Planetarium, 871
Cygnet, 840
Cygnus-Quasar Books, 837
Cypress, 650
Cyril and Methodius University
   Faculty of Natural Sciences and Mathematics, 377
Czech Astronomical Society (CAS), 126
Czech Hydrometeorological Institute (CHMI), 127
Czech Republic, 124
Czech Standards Institute, 127
Czech Technical University (CTU)
   Astronomical Observatory, 127
C.Z. Instruments India Pvt. Ltd., 283

D

DAA
   Devon Astronomical Association, 544
DAAS
   Department of Astronomy and Atmospheric Sciences (KNU), 369
Ďáblice, 130
Ďáblice Observatory, 127, 130
Daeduk Radio Astronomy Observatory (DRAO), 367, 368
Daeyang Observatory, 369
Dahlonega, 690
Daimler Benz Aerospace, 221
Dakota
   (North.), 834
   (South.), 869
Dakota Astronomical Society (DAS), 834
Dale Planetarium (Fred G..), 782
Dallas, 877, 879, 881–885
Damascus, 515
DAMTP
   Department of Applied Mathematics and Theoretical Physics, University of Cambridge, 575
Danbury, 664, 665
Daniel B. Milliken Planetarium, 617
Daniel Oberti - Architectural Ceramics, 617
Daniel Observatory (Charles E..), 868
Däniken, 500
Danish Astronautical Society, 136
Danish Astronomical Association, 136
Danish Meteorological Institute (DMI), 136
Danish Natural Science Research Council, 137
Danish Physical Society, 137
Danish Space Research Advisory Board, 137
Danish Space Research Institute (DSRI), 137

Danish Standards Association (DS), 137
Dansk Fysisk Selskab (DFS), 137
Dansk Selskab for Rumfartsforksning (DSR), 137
Danville, 615
DAO
    Dominion Astrophysical Observatory, 93, 94
    Canadian Astronomy Data Centre, 93
DAPNIA
    Département d'Astrophysique, de Physique des Particules, de Physique Nucléaire et de l'Instrumentation Associée, 174
D'Appolonia SpA, 317
Dar El Beïda, 5
Daresbury Laboratory, 542
Darling Hill Observatory, 824
Darlington, 541
Darlington Astronomical Society (CaDAS) (Cleveland and_), 541, 544
Darmstadt, 227, 228, 251, 260
Darnhall, 579
Dartmouth College
    Department of Physics and Astronomy, 788
    MDM Observatory, 598
    Michigan Dartmouth MIT Observatory, 598, 788
    Wilder Laboratory
    Department of Physics and Astronomy, 788
DAS
    Dakota Astronomical Society, 834
    Delaware Astronomical Society, 667
    Denver Astronomical Society, 654
    Dundee Astronomical Society, 544
DASA
    Deutsche Aerospace AG, 222
Data centres
    ADC, 740
    ADF, 740
    ARI, 214
    ASIS, 162
    Astronomisches Rechen-Institut, 214
    Astronomy and Space Information Service, 162
    Astronomy Data Center, 740
    Astrophysics Data Facility, 740
    BADC, 113
    BDL, 193
    Beijing Astronomical Data Center, 113
    CAD, 445
    CADC, 93
    Canadian Astronomy, 93
    Canadian Institute for Scientific and Technical Information, 93
    CAS, 836
    CDS, 165
    Center for Astronomical Data, 445
    Centre de Données astronomiques de Strasbourg, 165
    Chemical Abstracts Service, 836
    CISTI, 93
    CNRC, 93
    CNRS, 167
    CODATA, 174
    Committee on Data for Science and Technology, 174
    ERS, 318
    ESIS, 318
    ESRIN, 318
    FAGS, 138
    Fédération Internationale d'Information et de Documentation, 392
    Federation of Astronomical and Geophysical Data Analysis Services, 138
    FID, 392
    Goddard Space Flight Center (GSFC)
        National Space Science Data Center, 740
        World Data Center A for Rockets and Satellites, 745
    GSFC
        National Space Science Data Center, 740
        World Data Center A for Rockets and Satellites, 745
    ICIST, 93
    IEE, 550
    IERS, 234
    IMCCE, 193
    INIST, 167
    INSPEC, 550
    Institut Canadien de l'Information Scientifique et Technique, 93
    Institut de l'Information Scientifique et Technique, 167
    Institute for Scientific Information, 856
    Institution of Electrical Engineers, 550
    International Earth Rotation Service, 234
    International Federation for Information and Documentation, 392
    International Soil Reference and Information Centre, 394
    IRS, 318
    ISI, 856
    ISRIC, 394
    Japan Science and Technology Corporation, 348
    JST, 348
    LEDA, 318
    NAOJ, 352
    NASA, National Aeronautics and Space Administration
        National Space Science Data Center, 740
        World Data Center A for Rockets and Satellites, 745
    National Astronomical Observatory of Japan, 352
    National Climatic Data Center, 829, 832
    National Earth Rotation Service, 677
    National Institute of Standards and Technology
        Administrative Offices, 741
        Gaithersburg, 741
    NCDC, 829, 832
    NEOS, 677
    NIST
        Administrative Offices, 741
        Gaithersburg, 741
    NRCC, 93
    Scientific and Technical Information Network
        Columbus, 843
        Germany, 245
        Karlsruhe, 245
        Ohio, 843
        USA-OH, 843
    SDAC, 741
    SDR&I, 676
    Servicio Meteorológico Nacional, Mexico, 384
    SIDC, 57
    Solar Data Analysis Center, 741
    Solar Influences Data analysis Center, 57
    Solar Terrestrial Dispatch, 99
    Space Data Resources & Information, 676
    Space Science Data Operations Office, 741
    SSDOO, 741
    STD, 99
    STN
        Columbus, 843
        Germany, 245
        Karlsruhe, 245
        Ohio, 843
        USA-OH, 843
    WDC
        A for Meteorology, 832
        A for Rockets and Satellites, 745
        A for Solar-Terrestrial Physics, 661
        B for Solar-Terrestrial Physics, 451
        C2 for Ionosphere, 362
        for Solar Activity, 362
    World Data Center
        A for Meteorology, 832
        A for Rockets and Satellites, 745
        A for Solar-Terrestrial Physics, 661
        B for Solar-Terrestrial Physics, 451
        C2 for Ionosphere, 362
        D for Astronomy, 118
        for Solar Activity, 362
Daun, 253
Davenport, 718

David Chandler Co., 618
David Dunlap Observatory (DDO), 103
David Florida Laboratory (DFL), 82, 87
Davie, 679
Davis, 620, 647
Davis Astronomy Club, 618, 620
Davis Hall Observatory, 704
Davis Instruments, 618
Davis Planetarium (Alan C._), 739
Davis Planetarium (Russell C._), 774, 775
Davos-Dorf, 507
Dawson Springs, 728
Dax, 192
DayStar Filters, 618
Dayton, 839
Daytona Beach, 680, 682
DCESS
    Danish Center for Earth System Science, 137
DDO
    David Dunlap Observatory, 103
DEAA
    Downeast Amateur Astronomers, 732
Deakin West, 23
Deal, 534
Dealers and distributors
    Adirondack Video Astronomy, 805
    Adorama, 805
    Advanced Telescope Supplies, 14
    AIT, 212
    Alain Carion, 179
    Anacortes Telescopes, 907
    Andromeda, 207
    AOK, 502
    APD, 408
    Apollo-Foco, 652
    ASCPL, 456
    ASP, 610
    Astro, 134
    AstroArt, 533
    Astro Cards, 852
    Astrocom, 209
    Astromart, 609
    Astromedia, 488
    Astronomical Adventures, 593
    Astronomical Pocket Diary, 408
    Astronomical Society of the Pacific, 610
    astronomics, 846
    Astroptx, 662
    Astro Scientific Centre, 456
    astro-shop, 218
    AstroSystems, 652
    Astrotech, 273
    Astro-Versand, 218
    Astrovisuals, 17
    A Tech, 721
    Auriga, 312
    AVA, 805
    AWR Technology, 534
    Baader, 218
    Bethany Sciences, 665
    Binocular and Telescope Shop, 19
    Blaxall and Steven, 408
    Broadhurst Clarkson & Fuller, 539
    Carion (Alain_), 179
    Casper (Michael I._), 818
    Catalana de Telescopios, 475
    Celestaire, 721
    CKC Power, 811
    Company Seven, 736
    Composants et Systèmes de Précision, 175
    Cryophysics, 503
    CSP, 175
    C-VII, 736
    Cygnus-Quasar, 837
    C.Z. Instruments, 283
    Deutsch (Verlag Harri_), 259
    Digitec Optical, 813
    Dr. Vehrenberg, 257
    Eagle Optics, 915
    Eclipse Edge, 736
    Europtik, 855
    Everything in the Universe, 620
    Explorers Tours, 546
    F.C. Meichsner, 751
    Finley Holiday, 621
    Fuller (Broadhurst Clarkson &_), 539
    Galaxy Contact, 179
    Gemmary, 644
    Goodwill, 283
    Gordhandas Desai, 283
    Haag - Meteorites (Robert A._), 601
    Hardin Optical, 848
    Harri Deutsch (Verlag_), 259
    Infinity Infocus, 378
    Intercon Spacetec, 233
    Jameco, 624
    Jenson Tools, 598
    Joerger, 816
    Khan Scope, 89
    Knollwood, 916
    Kohler, 502
    Konus, 330
    Kowa, 624
    Land, Sea and Sky, 881
    LensPlus, 625
    Librairie Scientifique Uranie, 188
    Lichtenknecker, 55
    LightWedge, 753
    Lire la Nature, 90
    LLN, 90
    LNP, 753
    LO, 55
    Loch Ness Productions, 753
    LOT-Oriel, 555
    Maison de l'Astronomie, 90, 189
    Manasek, 893
    Meichsner (F.C._), 751
    Merlan Scientific, 92
    Michael I. Casper, 818
    Micromedia, 92
    Mineralogical Research, 627
    MMI, 739
    Mountain Instruments, 628
    Murmel, 506
    New England Meteoritical Services, 756
    Newport Industrial Glass, 631
    Nick Enterprise, 517
    Nikon USA, 820
    Nonius, 399
    Novaspace Galleries, 600
    OAM, 8
    Oberwerk, 840
    Observatorio Astronómico Móvil, 8
    Obsession Telescopes, 918
    Oceanside Photo and Telescope, 631
    OES, 240
    Optische und Elektronische Systeme, 240
    Oriel (LOT-_), 555
    Orion & Binocular Telescope Center, 632
    Owl Services, 859
    Photosolve, 633
    Pick Travel, 562
    Pocono Mountain, 860
    Quasar (Cygnus-_), 837
    Radio Research Instruments, 664
    Radio-Sky, 727
    Ransburg (Teleskop Service Wolfgang_), 252
    RICL, 564
    Rivers Camera Shop, 789
    Robert A. Haag - Meteorites, 601
    Roditi International, 564
    Rolyn Optics, 636
    RRI, 664
    Safelight, 566

SCI, 641
Science and Art Products, 639
Scientific Expeditions, 684
Scope City, 639
ScopeTronix, 684
Seiler, 778
Shibayama Scientific, 357
Siebert Optics, 831
Sirius Instruments, 706
Skylab, 411
Sky Laboratories, 411
Sky Optics, 98
Sovietski Collection, 641
Space Craft International, 641
Spacetec (Intercon_), 233
Spectra Astro Systems, 641
Speedibrews, 570
SPOT
   Reston, 904
   Toulouse, 202
   USA-VA, 904
   Virginia, 904
SPOT Image, 202, 904
Starsplitter Telescopes, 642
Stefan Thiele, 212
Swift, 759
Telescopes and Astronomy, 29
Teleskopschmiede, 252
Teleskop Service Wolfgang Ransburg, 252
Texas Nautical Repair, 881
The Universe Collection, 665
Thiele (Stefan_), 212
True Technology Ltd., 573
Twilight Tours, 646
Universe Collection (The_), 665
University Optics, 769
Unterlinden, 195
Vehrenberg (Dr._), 257
Velmex, 825
Verlag Harri Deutsch, 259
Warsash, 32
What in The World, 651
Wolfgang Ransburg (Teleskop Service_), 252
Woodstock Products, 651
Wyss, 508
Zarvan, 293
Dearborn, 763, 768
Dearborn Observatory, 700, 704
De Bilt, 395
Debrecen, 275
Debrecen Heliophysical Observatory, 274, 275
   Gyula Observing Station, 275
Decatur, 687, 703
Decimomannu, 306
Decker-Grebner-Van Zandt Observatory, 704
Deep River, 87
Deep River Astronomy Club, 87
Deep Space Exploration Society (DSES), 654
Deere Planetarium (John_), 700, 702, 705
Deer Lakes Park, 851
Deer Park, 880
Defence Evaluation and Research Agency (DERA), 544
Defford, 579
De Kalb, 704
Delaware, 667, 668, 841, 842
Delaware Astronomical Society (DAS), 667
Delaware Valley Amateur Astronomers (DVAA), 854
Délémont, 510
Delft, 390, 397–399, 403
Delft Institute for Earth-Oriented Space Research (DEOS), 390
Delft Instruments NV, 390
Delft University of Technology, 390
   Delft Institute for Earth-Oriented Space Research (D-EOS), 403
Delfzijl, 407
Delingha, 114

Delingha Radio Astronomy Station, 114, 116
Del Mar, 624
Delmarva Star Gazers, 667
Delta College Planetarium and Learning Center, 762
Deltona, 686
Demmin, 209, 231
Denekamp, 391
Den Haag, 392, 393, 397, 399
Denison University
   Department of Physics and Astronomy, 838
Denitron SnC, 317
Denmark, 134
Denton, 887
Denver, 653–655, 657, 659, 661
Denver Astronomical Society (DAS), 654
Denver Museum of Natural History (DMNH)
   Charles C. Gates Planetarium, 653
DEOS
   Delft Institute for Earth-Oriented Space Research, 403
Department for Standardization, Metrology and Certification (SARM), 12
Department of Meteorological Services, 929
Department of Meteorology
   Sri Lanka, 487
Department of Standards, Metrology and Technical Supervision, 387
De Pauw University
   Physics and Astronomy Department, 710
DERA
   Defence Evaluation and Research Agency, 544
DERA Astronomy Society, 544
Desert, 610
Des Moines, 716–718
Des Moines Astronomical Society (DMAS) Inc., 716
Dessau, 215
DESY
   Deutsches Elektronen-Synchrotron
      Hamburg, 223
      Zeuthen, 224
Detroit, 762, 769
Detroit Science Center, 762
Detwiler Planetarium, 858
Deutsche Aerospace AG (DASA), 222
Deutsche Agentur für Raumfahrtangelegenheiten (DARA) GmbH, 222
Deutsche Forschungsanstalt für Luft- und Raumfahrt (DLR) eV, 222
Deutsche Forschungsgemeinschaft (DFG), 222
Deutsche Geophysikalische Gesellschaft (DGG) eV, 222
Deutsche Gesellschaft für Chronometrie (DGC) eV, 222
Deutsche Gesellschaft für Luft- und Raumfahrt (DGLR) eV, 222
Deutsche Meteorologische Gesellschaft (DMG) eV, 223
Deutsche Physikalische Gesellschaft (DPG) eV, 202, 223
Deutscher Wetterdienst (DWD), 223
   Berlin, 223
   Offenbach/Main, 223
   Referat Hydrometeorologische Entwicklungen und Anwendungen, 223
Deutsches Elektronen-Synchrotron (DESY)
   Hamburg, 223
   Zeuthen, 224
Deutsches Fernerkundungsdatenzentrum (DFD), 224
Deutsches Geodätisches Forschungsinstitut (DGFI), 221, 224
Deutsches Hydrographisches Institut (DHI), 224
Deutsches Institut für Normung (DIN), 224
Deutsches Klimarechenzentrum (DKRZ) GmbH, 224
Deutsches Museum München
   Abteilung Astronomie, 224
Deutsches Zentrum für Luft- und Raumfahrt (DLR) eV, 225
   Berlin, 225
   Institut für Planetenerkundung, 225
   Köln, 225
   Oberpfaffenhofen, 225
   Standort Oberpfaffenhofen, 225

Deutsch (Verlag Harri_), 259
Devon Astronomical Association (DAA), 544
Devon Astronomical Observatory, 101
Dewan Standardisasi Nasional (DSN), 291
De Zonnewijzerkring, 390
DFG
    Deutsche Forschungsgemeinschaft, 222
DFL
    David Florida Laboratory, 82
DFM Engineering Inc., 654
DFNT
    Dipartimento di Fisica Nucleare e Teorica - Università di Pavia, 339
DFRC
    Dryden Flight Research Center, 630
DFRF
    Dryden Flight Research Facility, 630
DFS
    Dansk Fysisk Selskab, 137
DGC
    Deutsche Gesellschaft für Chronometrie, 222
DGFI
    Deutsches Geodätisches Forschungsinstitut, 221, 224
DGG
    Deutsche Geophysikalische Gesellschaft, 222
DGLR
    Deutsche Gesellschaft für Luft- und Raumfahrt, 222
DGN
    Dirección General de Normas de México, 383
DGPA
    Department of General Physics and Astronomy (RAS), 444
DGT
    Directorate General of Telecommunications
        Lunping Observatory, 516
Dhaka, 44
Dhaun, 212
DHDC
    Don Harrington Discovery Center, 878
DHM
    Dirección de Hidrología y Meteorología, 928
Diablo (Mount_), 629
Diadema, 71
DIAS
    Dublin Institute for Advanced Studies
        Astronomy Section, 295
        Cosmic-Ray Section, 295
        Dunsink Observatory, 295
Dickinson College
    Department of Physics and Astronomy, 854
Dickson, 17, 21, 871
Didcot, 542, 543, 560
Diedorf, 216
Die Sterne, 225
Diez, 234
Digger Pines, 645
Digger Pines Observatory, 627
Digistar, 890
Digitec Optical, 813
Dijon, 156, 199
DIN
    Deutsches Institut für Normung, 224
DINAC
    Dirección Nacional de Aeronáutica Civil, Paraguay, 422
Dirección de Hidrología y Meteorología (DHM), 928
Dirección General de Normas (DGN), 383
Dirección Meteorológica de Chile (DMC), 107
Dirección Nacional de Aeronáutica Civil (DINAC)
    Dirección de Meteorología e Hidrología (DMH), 422
Dirección Nacional de la Meteorología (DNM), 586
Directorate General of Telecommunications (DGT)
    Lunping Observatory, 516
Directorate of Standardization, 3
Discover Magazine, 813
Discovery Center Museum, 700
Discovery Museum
    Henry B. duPont III Planetarium, 662

Discovery Museum Science Center, 618
Discovery Park
    Gov Aker Observatory, 595
Discovery Telescopes Inc., 618
Diso, 318
District of Columbia, 669–677
Ditzingen, 222
Dixie County Radio Observatory, 686
DKRZ
    Deutsches Klimarechenzentrum, 224
DLR, Deutsches Zentrum für Luft- und Raumfahrt eV
    Berlin, 225
    Institut für Planetenerkundung, 225
    Köln, 225
    Oberpfaffenhofen, 225
    Standort Oberpfaffenhofen, 225
DMAS
    Des Moines Astronomical Society, 716
DMC
    Dirección Meteorológica de Chile, 107
DMG
    Deutsche Meteorologische Gesellschaft, 223
DMH
    Dirección de Meteorología e Hidrología, Paraguay, 422
DMI
    Danish Meteorological Institute, 137
DMNH
    Denver Museum of Natural History
        Charles C. Gates Planetarium, 653
DMS
    Dutch Meteor Society, 390
Dniepropetrovsk, 528, 529
DNM
    Dirección Nacional de la Meteorología (Uruguay), 586
DOA
    Dutch Occultation Association, 391
Doane Observatory, 698
Dodaira Observatory, 344
Dome Fuji Station, 355
Dominion Astrophysical Observatory (DAO), 87, 93, 94
Dominion Radio Astrophysical Observatory (DRAO), 87, 93, 94
Dompcevrin, 161
Donald A. Schaefer Planetarium, 716
Donald E. Bianchi Planetarium, 615
Donetsk, 525
Donetsk Planetarium, 525
Don Harrington Discovery Center (DHDC), 878
Donzdorf, 247
Dooley Planetarium, 866
Door Peninsula Astronomical Society (DPAS), 915
Doran Planetarium, 89
Dordrecht, 395, 406
Dorigny, 513
Dorking, 574
Dorrance Planetarium, 593
Dortmund, 214
Dorval, 83
Dourbies, 190
Dourouti Observatory, 269
Dover, 570, 606, 789, 792
Downeast Amateur Astronomers (DEAA), 732
Downing Planetarium, 614, 618
Downs
    (South_), 569
Downsview, 81, 104
Dow Planetarium, 96
Doylestown, 853
DPAS
    Door Peninsula Astronomical Society, 915
DPG
    Deutsche Physikalische Gesellschaft, 202, 223
DPRK
    Democratic People's Republic of Korea, 366
Dr. Adalbert Jeszenkowitsch Gesellschaft, 35
Dr. A.F. Philips Sterrenwacht, 390
Drake Planetarium, 838

Drake Planetarium (Marie.), 592
Drake University
    Department of Physics and Astronomy, 717
Drake University Municipal Observatory, 717
DRAO
    Daeduk Radio Astronomy Observatory, 368
    Dominion Radio Astrophysical Observatory, 93, 94
Drebach, 263
Drenthe
    (Noord.), 405
Drescher Planetarium (John.), 619, 638
Dresden, 225, 240, 251, 261
Dresdner Verein für Himmelskunde eV, 225
Drexel University
    Department of Physics, 854
Dreyfuss Planetarium (Alice and Leonard.), 791, 792
Droke Observatory (James Wesley.), 606
Dr. Petur Beron Observatory, 74, 76
Dr. Remeis-Sternwarte, 226, 254
Drummondville, 83
Druscìe (Col.), 308
Dr. Vehrenberg KG, 226, 257
Dr. W. Russell Blake Planetarium (WRBP), 750
Dryden Flight Research Center (DFRC), 619, 630
Dryden Flight Research Facility (DFRF), 619, 630
DS
    Danish Standards, 137
DSC
    Drejtoria e Standardizimit dhe Cilesise, 3
DSES
    Deep Space Exploration Society, 654
DSN
    Dewan Standardisasi Nasional, 291
DSR
    Dansk Selskab for Rumfartsforksning, 136
DSRI
    Danish Space Research Institute, 137
DSTU
    State Committee of Ukraine for Standardization, Metrology and Certification, 528
DTM
    Department of Terrestrial Magnetism (CIW), 672
Dublin, 295–298
Dublin Institute for Advanced Studies (DIAS)
    School of Cosmic Physics
        Astronomy Section, 295
        Astrophysics Section, 295
Dubuque, 717
Dudley Observatory, 814
Duisburg, 244
Duke University
    Department of Physics, 828
Duluth, 771
Dumand Observatory, 696
Dumfries, 544
Dumfries Museum and Camera Obscura, 544
Duncan, 87
Duncan Observatory, 103
Dundas, 88
Dundee, 544, 557
Dundee Astronomical Society (DAS), 544
Dunkerque, 195
Dunkirk, 195
Dunlap Observatory (DDO) (David.), 103
Dunmore, 855
Dunsink Observatory, 295, 296
duPont III Planetarium (Henry B..), 662
Dupont Planetarium, 866
Durazno, 587
Durban, 464, 465, 468
Durban Natural Science Museum
    Astronomy Interest Group, 465
Durham, 575, 790, 827, 828, 830, 831
Durmersheim, 247
Dushanbe, 518
Düsseldorf, 257
Dut'Astro, 175

Dutch Meteor Society (DMS), 390
Dutch Occultation Association (DOA), 391
Dutch Physical Society, 397
Dutch Youth Association for Space and Astronomy, 391, 397
Duttlenheim, 163, 175
DVAA
    Delaware Valley Amateur Astronomers, 854
DWD
    Deutscher Wetterdienst
        Berlin, 223
        Offenbach/Main, 223
        Referat Hydrometeorologische Entwicklungen und Anwendungen, 223
Dwingeloo, 394, 400, 402
Dyer Observatory (Arthur J..), 870, 872, 876
DynaVac, 750
DZNM
    Croatian State Office for Standardization and Metrology, 122

E

E&S
    Evans & Sutherland Computer Corp., 890
EAAA
    Escambia Amateur Astronomers Association, 680
EAAE
    European Association for Astronomy Education, 226
EAAS
    Euro-Asian Astronomical Society, 441
EAC
    European Astronauts Centre, 228
EADS
    European Aeronautic, Defense and Space Co., 154, 222
EAGE
    European Association of Geoscientists and Engineers, 391
Eagle Optics, 915
EAON
    European Asteroidal Occultation Network, 226
EARG
    Estación Astronómica Rio Grande, 7, 10
Earl and Hazel Faulkner Planetarium, 697
Earlham College
    Department of Physics and Astronomy, 711
Earlysville, 897
EARMA
    European Association for Research Managers and Administrators, 317
Earth & Sky Radio Series, 878
Earth and Space Science Center Planetarium, 776
Earth Observation Centre (EOC), 22
Earthwatch, 750
EARTO
    European Association of Research and Technology Organisations, 51
EAS
    Eastbay Astronomical Society, 619
    Eastbourne Astronomical Society, 544
    Eugene Astronomical Society, 848
    European Astronomical Society, 127
    Evansville Astronomical Society, 711
    Everglades Astronomical Society, 680
    Ewell Astronomical Society, 545
    Exeter Astronomical Society, 546
Eastbay Astronomical Society (EAS), 619
Eastbourne, 545
Eastbourne Astronomical Society (EAS), 544
East Central Wisconsin Astronomers (ECWA), 916
Eastern College
    Observatory and Planetarium, 855
Eastern Connecticut State University (ECSU)
    Robert K. Wickware Planetarium, 664
Eastern Jutland, 134
Eastern Kentucky University (EKU)
    Arnim D. Hummel Planetarium, 725

Eastern Mennonite University
    M.T. Brackbill Planetarium, 899
Eastern Michigan University (EMU)
    Astronomy Club, 762
    Department of Physics and Astronomy, 763
Eastern Missouri Dark Sky Observers (EMDSO), 777
Eastern Washington University (EWU)
    Department of Physics, 907
East Gothia, 492
East Greenwich, 865
East Hartford, 665
East Lansing, 761, 762, 766
East Meadow, 806
East Northport, 816
Easton, 857
East Peoria, 701
East Pittsburgh, 851
East Riding, 549
East Riding Astronomical Society (HERAS) (Hull and_), 545

East Tennessee Discovery Center (ETDC), 870, 874
East Tennessee State University (ETSU)
    Department of Physics, 871
East Valley Astronomy Club (EVAC), 595
Eastwood, 14
East York, 82
Eau Claire, 919
Eberhard-Karls-Universität Tübingen, 226
Ebingen (Albstadt-_), 216, 250
Ebro, 486
Eclipse Edge Expeditions, 736
ECMOG
    Erie County Mobile Observing Group, 855
ECMWF
    European Centre for Medium-Range Weather Forecasts, 545
ECO
    Etscorn Campus Observatory, 802
École Centrale de Paris (ECP)
    Club d'Astronomie, 175
École Normale Supérieure de Lyon (ENSL)
    Groupe d'Astrophysique, 165, 175
École Normale Supérieure (ENS)
    Département de Physique
        Laboratoire de Radioastronomie, 175
École Polytechnique
    Centre de Physique Théorique, 175
    Laboratoire de Météorologie Dynamique (LMD), 175
    Laboratoire de Physique Nucléaire des Hautes Énergies (LPNHE), 175, 187
École Polytechnique Fédérale, 503
École Royale Militaire (ERM)
    Département d'Astronomie, de Géodésie et de Topographie, 50
EcoTarium
    Norton Observatory, 750
ECP
    École Centrale de Paris
        Club d'Astronomie, 175
E.C. Schouweiler Planetarium, 711
ECSL
    European Centre for Space Law, 177
ECSO
    European Council of Skeptical Organizations, 226
ECSU
    Eastern Connecticut State University
        Robert K. Wickware Planetarium, 664
Ecuador, 141
ECWA
    East Central Wisconsin Astronomers, 916
Eddy Co., 619
Edgartown, 754
Edgware, 583
Edinboro, 855
Edinboro University Planetarium (EUP), 855
Edinburgh, 533, 534, 545, 552, 562, 566, 576

Edinburgh City Observatory, 533
Edinburgh Mathematical Society (EMS), 545
Edison High School Planetarium, 898
Edison Infrared Space Observatory (EISO), 923
Éditions Burillier, 176, 188
Éditions de Physique, 176
Éditions Diffusion Presse Sciences, 176
Éditions du CNRS, 173, 176
Edizioni Scientifiche Coelum, 317
EDL
    Equatorial Dust Lane, 150
Edmonton, 95, 101
Edmonton Space and Science Centre (ESSC), 87, 96
Edmund Scientific Co., 814
EDP Sciences, 176
Edward Naylor Observatory, 852
Edward R. Byers Co., 619
Edwards, 630
Edwin A. Link Planetarium, 814, 820
Edwin E. Aldrin Astronomical Center, 794
Edwin P. Hubble Planetarium, 814
Edwin Ritchie Observatory, 907
Eesti Füüsika Selts (EFS), 144
Effelsberg, 237
EFS
    Eesti Füüsika Selts, 144
EG&G Inc., 736
Ege University
    Department of Astronomy and Space Sciences, 522
    Observatory, 522
Eggenstein, 229, 245
Egham, 565
Egloffstein, 240, 247
EGO
    European Gravitational Observatory, 318
EGS
    European Geophysical Society, 227
Egypt, 142
Egyptian Meteorological Authority (EMA), 142
Egyptian Organization for Standardization and Quality Control (EOS), 142
EIA
    Engineering in Astronomy, 574
Eidgenössische Technische Hochschule (ETH)
    Institut für Astronomie, 503, 511
Eileen Collins Observatory, 813, 814
Eilenburg, 264
Eindhoven, 390
Einstein Planetarium (Albert_), 672, 676
EISCAT
    European Incoherent Scatter Facility
        Headquarters, 489
        Ionospheric Heating Facility, 415
        Sodankylä Station, 146
        Svalbard Radar, 415
        Tromsø Station, 415
    Scientific Association, 489
Eise Eisinga's Planetarium, 391
Eisenstadt, 35
Eisner-Sternwarte, 35
EISO
    Edison Infrared Space Observatory, 923
Ekar (Monte_), 327
Ekaterinburg, 451
EKU
    Eastern Kentucky University
        Arnim D. Hummel Planetarium, 725
El Cajon, 622
El Camino College
    Planetarium and Observatory, 619
Elche, 477
Electrim Corp., 792
Electro Optical Industries (EOI) Inc., 619
Elgin, 700
Elginfield Observatory, 104
Elgin Planetarium and Observatory, 700
Elizabeth, 19

Ellenwood, 691
El Leoncito, 7, 10, 666
Ellijay, 690
ELOT
    Hellenic Organization for Standardization, 268
Elp, 406
El Paso, 879
El Paso Astronomy Club, 879
El Paso Independent School District (EPISD)
    Planetarium, 879
El Salvador, 143
El Segundo, 608, 617
Elsevier Science BV
    Physics and Astronomy Department, 391
Elsevier Science Inc., 814
EMA
    Egyptian Meteorological Authority, 142
EMDSO
    Eastern Missouri Dark Sky Observers, 777
EM Electronics, 545
EMF
    Evaporated Metal Films, 814
EMF Corp., 814
EMI
    Ernst-Mach-Institut, 230
Emmenbrücke, 502
Emmendingen, 234, 247
Emmering, 259
Emporia, 721
Emporia State University (ESU)
    Division of Physical Sciences, 721
    Peterson Planetarium, 721
Emptinne, 52
EMS
    Edinburgh Mathematical Society, 545
EMU
    Eastern Michigan University
        Astronomy Club, 762
EMVIC (Observatorio Astronómico.), 928
Encinitas, 644
Enfield, 559
Engadine, 14
Engelhardt Astronomical Observatory, 440
Engineering and Physical Sciences Research Council (EPSRC), 545
Enid, 846
Ennepetal, 260
Enrico Fermi Institute (EFI), 700, 707
ENS
    École Normale Supérieure
        Département de Physique, 175
        Laboratoire de Radioastronomie, 175
Ensenada, 386
Ensisheim, 170
ENSL
    École Normale Supérieure de Lyon
        Groupe d'Astrophysique, 165
Ente Nazionale Italiano di Unificazione (UNI), 317
EOC
    Earth Observation Centre (CSIRO), 22
EOI
    Electro Optical Industries, 619
E.O. Lawrence Berkeley National Laboratory (LBNL)
    Advanced Light Source (ALS) Project, 608
    Institute for Nuclear and Particle Astrophysics (INPA), 620
EOS
    Egyptian Organization for Standardization and Quality Control, 142
    Estación de Observación Solar, 385
    European Optical Society, 227
EOST
    EOS Technologies, 595
EOS Technologies Inc. (EOST), 595
Eötvös Loránd Physical Society, 274
Eötvös Loránd University
    Department of Astronomy, 274

Gothard Astrophysical Observatory (GAO), 274
EPD
    ESA Publications Division, 392
Épendes, 504
EPISD
    El Paso Independent School District
        Planetarium, 879
Epner Technology Inc., 815
EPO
    European Patent Office, 227, 240
Epping, 14, 21
EPS
    Euregio Publiekscentrum voor Sterrenkunde, 391
    European Physical Society, 177
Epsom, 545
EPSRC, Engineering and Physical Sciences Research Council, 545
Equatorial Platforms, 620
Équinoxe-Astronomie, 176
Equipo Sirius, 476
ERAC
    European Radio Astronomy Club, 228
Erciyes University
    Department of Astronomy and Space Sciences, 522
Eretz Israel Museum, 300
Erhard Friedrich Verlag GmbH & Co. KG, 226
Erie, 855, 858
Erie County Mobile Observing Group (ECMOG), 855
Erindale, 103
Erkrath, 232, 249
Erlangen-Nürnberg, 254
ERM
    École Royale Militaire
        Département d'Astronomie, de Géodésie et de Topographie, 50
Erna and Victor Hasselblad Foundation, 488
Ernest B. Wright Observatory, 815
Ernst-Abbe-Stiftung, 226
    Zeiss-Planetarium, 266
Ernst-Mach-Institut (EMI), 226, 230
Erstes Physikalisches Institut, 255
Erwin W. Fick Observatory, 718
ESA, European Space Agency, 296
    Astrophysics Division, 392
    Darmstadt, 228
    ESA Publications Division, 392
    ESRIN, 318
    ESTEC, 392, 399
    European Astronauts Centre, 228
    European Centre for Space Law, 177
    European Space Operations Centre, 228
    France, 177
    Frascati, 318
    Garching, 245
    Germany, 228, 245
    Headquarters, 177
    Information Retrieval System, 318
    Infrared Space Observatory, 476
    ISO Observatory, 476
    Italy, 318
    Kiruna, 489
    Köln, 228
    Madrid, 476
    Netherlands, 392, 399
    Noordwijk, 392
    Noordwijk Space Expo, 399
    Paris, 177
    Satellite Station
        Kiruna, 489
        Sweden, 489
    Space Telescope - European Coordinating Facility, 245
    Spain, 476
    Sweden, 489
    Villafranca Satellite Tracking Station, 476
    VILSPA, 476
Esashi Earth Tides Station, 353
ESC

Euro Space Center, 51
Escambia Amateur Astronomers Association (EAAA), 680
Eschwege, 258
Escondido, 646
ESF
    European Science Foundation, 177
Esk Valley College Planetarium (Jewel and_), 545, 552
ESOC, 226
    European Space Operations Centre, 228
ESO, European Southern Observatory, 226
    Antofagasta Office, 108
    Chile, 108
    Headquarters, 228
    La Serena Office, 108
    La Silla Observatory, 108
    Paranal Observatory, 109
    Santiago, 108
Espace Mendès France, 176
España y América (Sociedad Astronómica de_), 482
Espoo, 146, 148, 149
ESR
    Earth Space Review, 856
Esrange, 488, 496
ESRF
    European Synchrotron Radiation Facility, 178
ESRIN, 317, 318
ESSC
    Edmonton Space and Science Centre, 96
Essen, 208, 257, 261
Estación Astronómica Jean Nicolini, 586
Estación Astronómica Rio Grande (EARG), 7, 10
Estación Carlos Cesco, 11
Estación de Observación Solar (EOS), 385
Estación Magnética de Trelew, 10
Estaimbourg, 49
ESTEC, 391
    Astrophysics Division, 392
    ESA Publications Division, 392
    European Space Research and Technology Centre, 392
    Noordwijk Space Expo, 399
Estenfeld, 263
Estonia, 144
Estonian Academy of Sciences, 144
Estonian Physical Society, 144
ESU
    Emporia State University
        Division of Physical Sciences, 721
        Peterson Planetarium, 721
ETCA
    Études Techniques et Constructions Aérospatiales, 47
ETDC
    East Tennessee Discovery Center, 870, 874
Etelä-Karjalan Nova, 146
Etel SA, 503
ETH
    Eidgenössische Technische Hochschule
        Institut für Astronomie, 503, 511
Ethyl Universe Planetarium - Space Theater, 898
Etna, 326
Etobicoke, 88
ETR
    Educational Technology Review, 896
Etscorn Campus Observatory (ECO), 802
ETSI
    European Telecommunications Standards Institute, 178
ETSU
    East Tennessee State University
        Department of Physics, 871
Ettlingen, 241
Ettore Majorana Centre for Scientific Culture (EMCSC), 317

Études Techniques et Constructions Aérospatiales (ETCA), 47
Eugene, 848, 850
Eugene Astronomical Society (EAS), 848
Eugenides Foundation
    Planetarium, 268

EUMETSAT, 226, 227
EUP
    Edinboro University Planetarium, 855
Eurastro, 176
Euregio Publiekscentrum voor Sterrenkunde (EPS), 391
Eureka-Plus, 176
Eureka Scientific Inc., 672
Euro-Asian Astronomical Society (EAAS), 440
Euroconsult, 177
Eurocopter, 188
European Aeronautic, Defense and Space Co. (EADS), 154, 222
European Association for Astronomy Education (EAAE), 226
European Association for Research Managers and Administrators (EARMA), 317
European Association of Geoscientists and Engineers (EAGE), 391
European Association of Research and Technology Organisations (EARTO), 51
European Asteroidal Occultation Network (EAON), 226
European Astronauts Centre (EAC), 226, 228
European Astronomical Society (EAS), 127, 503
European Centre for Medium-Range Weather Forecasts (ECMWF), 545
European Centre for Space Law (ECSL), 177
European Committee for Electrotechnical Standardization, 51
European Committee for Standardization, 51
European Council of Skeptical Organizations (ECSO), 226
European Geophysical Society (EGS), 227
European Gravitational Observatory (EGO), 318
European Incoherent Scatter Facility (EISCAT)
    Headquarters, 488
    Ionospheric Heating Facility, 415
    Scientific Association, 489
    Sodankylä Station, 146
    Svalbard Radar, 415
    Tromsø Station, 415
European Industrial Space Study Group, 177, 178
European Optical Society (EOS), 227
European Organisation for the Exploitation of Meteorological Satellites, 227
European Organization for Nuclear Research, 502
    Astronomy Club, 503
European Patent Office (EPO), 227
European Physical Society (EPS), 177
European Planetarium Network (EuroPlaNet), 227
European Radio Astronomy Club (ERAC), 228
European Science Foundation (ESF), 177
European Southern Observatory (ESO), 226
    Antofagasta Office, 108
    Chile, 108
    Headquarters, 228
    La Serena Office, 108
    La Silla Observatory, 108
    Paranal Observatory, 109
    Santiago, 108
    Space Telescope - European Coordinating Facility (ST-ECF), 245
European Space Agency (ESA), 296, 543
    Darmstadt, 228
    ESRIN, 318
    ESTEC, 392
        Astrophysics Division, 392
        ESA Publications Division (EPD), 392
        Noordwijk Space Expo, 399
    European Astronauts Centre (EAC), 228
    European Centre for Space Law (ECSL), 177
    European Space Operations Centre (ESOC), 228
    France, 177
    Frascati, 318
    Garching, 245
    Germany, 228, 245
    Headquarters, 177
    Information Retrieval System, 318
    ISO Observatory, 476

Italy, 318
Kiruna, 489
Köln, 228
Madrid, 476
Netherlands, 392
Noordwijk, 392
Noordwijk Space Expo, 399
Paris, 177
Satellite Station
Kiruna, 489
Sweden, 489
Space Telescope - European Coordinating Facility (ST-ECF), 245
Spain, 476
Sweden, 489
Villafranca Satellite Tracking Station (VILSPA)
Infrared Space Observatory (ISO), 476
VILSPA, 476
European Space Operations Centre (ESOC), 228
European Space Research and Technology Centre (ESTEC), 392
European Synchrotron Radiation Facility (ESRF), 178
European Telecommunications Standards Institute (ETSI), 178
European VLBI Network (EVN), 394
EuroPlaNet
European Planetarium Network, 227
Europlanetarium vzw, 51
Europtik Ltd., 855
Euroscience, 178
EUROSPACE, 178
European Industrial Space Study Group, 178
Euro Space Center (ESC), 51
Eurospace Technische Entwicklungen GmbH, 228
Euskirchen, 243
EVAC
East Valley Astronomy Club, 595
Evans & Associates (Charles_), 617
Evans & Sutherland Computer Corp. (E&S), 890
Evans (Mount_), 661
Evans Observatory (Mark_), 702
Evanston, 700, 703, 704
Evansville, 711
Evansville Astronomical Society (EAS), 711
Evansville Museum of Arts and Science, 711
Evaporated Metal Films (EMF) Corp., 814
Everett, 907
Everett Astronomical Society, 907
Everglades Astronomical Society (EAS), 680
Everton Park, 27
Everything in the Universe, 620
Evesham, 548
EVN
European VLBI Network, 394
Évora, 435
Évry, 155
EVS
National Standards Board of Estonia, 144
Ewell, 545
Ewell Astronomical Society (EAS), 545
EWU
Eastern Washington University
Department of Physics, 907
Exeter, 546, 576, 789
Exeter Astronomical Society (EAS), 546
Exploration Place, 722
Exploratorium, 620
Explorers Tours, 546
Explorit Science Center
Davis Astronomy Club, 620
Express Optics Inc., 620
Extrasolar Research Corp., 621
Eyes on the Skies, 645

F

FAA
Föreningen för Astronomi och Astronautik, 493
FAAC
Ford Amateur Astronomy Club, 763
FAAQ
Fédération des Astronomes Amateurs du Québec, 88
Fachhochschule Mannheim, Astronomische Arbeitsgemeinschaft, 228
Fachinformationszentrum (FIZ) Karlsruhe, 229, 245
Facultés Universitaires Notre-Dame de la Paix (FUNDP)
Département de Mathématique, 51
Faget Station, 438
FAGS
Federation of Astronomical and Geophysical Data Analysis Services, 138
Fairbanks Museum and Planetarium, 893
Fairfax, 898, 906
Fairy Meadow, 33
Fallbrook, 644
Fall Creek, 915
Falls Church, 898
Falls Church High School Planetarium, 898
Falmouth, 733
Falun, 493
Falun Science Center, 493
Fan Mountain Station, 904
FAPESP
Fundação de Amparo à Pesquisa do Estado de São Paulo, 66
Farmingdale, 809
Farnborough, 544, 557
Farra d'Isonzo, 314
Farran Technology Ltd. (FTL), 296
FAS
Federation of American Scientists, 672
Federation of Astronomical Societies, 546
Foothills Astronomical Society, 828
Forsyth Astronomical Society, 828
Fasano, 321
FASAS
Federation of Asian Scientific Academies and Societies, 283
FAST COMTEC GmbH, 229
FASTS
Federation of Australian Scientific and Technological Societies, 23
Faulkner Planetarium (Earl and Hazel_), 697
Fayette, 776
Fayetteville, 606
FBAC
Fort Bend Astronomy Club, 879
FBO
Foggy Bottom Observatory, 811
FCAG
Facultad de Ciencias Astronómicas y Geofísicas (La Plata), 10
F.C. Meichsner Co., 751
FCRAO
Five College Radio Astronomy Observatory, 751
FCT
Fundação para a Ciência e a Tecnologia, 433
FDSC
Flagstaff Dark Skies Coalition, 595
Fécamp, 161
Federal Hydrometeorological Institute (FMHI), 454
Federal Office of Meteorology and Climatology (MeteoSwiss), 504
Fédération des Astronomes Amateurs du Québec (FAAQ), 87
Fédération Internationale d'Information et de Documentation (FID), 392
Fédération Laïque d'Éducation Populaire (FLEP)
Section Astronomie, 178
Federation of American Scientists (FAS), 672
Federation of Asian Scientific Academies and Societies (FASAS), 283
Federation of Astronomical and Geophysical Data Analysis Services (FAGS), 138

Federation of Astronomical Societies (FAS), 546
Federation of Australian Scientific and Technological Societies (FASTS), 23
Feder Observatory (Paul P._), 770, 772
Fehrenbach-Planetarium (Richard-_), 229, 243
Feira de Santana, 68
Feldberg, 240
Felix Aguilar (Observatorio Astronómico_), 10
Felsina, 304
Fels Planetarium, 855
Feltre, 308
Fenton Hill Observatory (FHO), 800
Ferberts Associates, 602
Ferdinando Caliumi Observatory, 310
Ferguson Planetarium (Whitworth_), 815, 826
Ferme des Étoiles (La_), 179, 188
Ferme du Héron, 171
Fermi Institute (EFI) (Enrico_), 700, 707
Fermilab, 700
    Experimental Astrophysics Group, 700
    Theoretical Astrophysics Group, 700
Fermi National Accelerator Laboratory (FNAL)
    Experimental Astrophysics Group, 700
    Theoretical Astrophysics Group, 700
Fernbank Observatory, 688
Fernbank Science Center (FSC), 688
Ferndale, 911
Ferrara, 319
Ferrovia Monte Generoso (FMG) SA, 504
Fesenkov Astrophysical Institute (V.G._), 364
Festus, 776
FEVARO
    Friedrich Ebert Visual And Radiotelescope Observatory, 231
FHO
    Fenton Hill Observatory, 800
Fick Observatory (Erwin W._), 718
FID
    Fédération Internationale d'Information et de Documentation, 392
Figl-Observatorium für Astrophysik (Leopold-_), 41
Fikirtepe, 523
FIL
    Formation Information Liaison, 52
Findlay, 839
Finland, 146
Finley Holiday Film Corp., 621
Finnish Academy of Science and Letters, 146
    Sodankylä Geophysical Observatory (SGO), 146
Finnish Academy of Technology, 146
Finnish Astronautical Society, 146, 149
Finnish Astronomical Society, 147
Finnish Meteorological Institute (FMI), 147
Finnish Physical Society, 147
Finnish Standards Association (SFS), 147
Fireball Data Center (FIDAC), 229
Firenze, 304, 316, 325, 326, 334, 335, 338
FirstLight Astronomy Club, 621
First Magnitude Corp. (FMC), 922
Fish Lake, 763
Fiske Planetarium and Science Center (Wallace_), 654, 660, 661
FIT
    Florida Institute of Technology
        Department of Physics and Space Sciences, 680
Fitchburg, 746
FitzRandolph Observatory, 795
FIU
    Florida International University
        Department of Physics, 681
Fiumefreddo di Sicilia, 305
Five College Astronomy Department
    FCRAO, 751
    Radio Observatory (FCRAO), 751
    Smith College, 758
    University of Massachusetts, 759
Five College Radio Astronomy Observatory (FCRAO), 751

FIZ
    Fachinformationszentrum, 229, 245
FKS
    Franz-Kroller-Sternwarte, 35
FLAG
    Flemish Aerospace Group, 52
Flagstaff, 593–595, 598, 600, 603, 604
Flagstaff Dark Skies Coalition (FDSC), 595
Flandrau Science Center and Planetarium (Grace H._), 595, 596
Fleet Space Theater and Science Center (Reuben H._), 621, 634
Fleetwood, 564
Fleischmann Planetarium and Science Center, 786
Flemish Aerospace Group (FLAG), 52
Flensburg, 238
FLEP
    Fédération Laïque d'Éducation Populaire
        Section Astronomie, 178
Fleurance, 155, 188
Fliegenfelde, 208
Flint, 763, 767
Flint River Astronomy Club (FRAC), 688
Flöha, 228
Floirac, 203
Floreat Park, 14
Florence, 304, 316, 325, 326, 334, 335, 338, 591, 866
Florence and George Wise Observatory, 299, 300
Florianopolis, 67
Florida, 678–686, 688
    (South_), 681, 685
Florida Institute of Technology (FIT)
    Department of Physics and Space Sciences, 680
Florida International University (FIU)
    Department of Physics, 681
Florida State University (FSU)
    Pat Thomas Planetarium, 683
Florida Tech, 681
    Department of Physics and Space Sciences, 680
Flower and Cook Observatory, 862
Floyd Bennett Field, 806
Fluid Metering Inc. (FMI), 815
Flushing Meadows, 819
FLWO
    Fred Lawrence Whipple Observatory, 597
Flying M Ranch, 593
FMC
    First Magnitude Corp., 922
FMG
    Ferrovia Monte Generoso, 504
FMHI
    Federal Hydrometeorological Institute, 454
FMI
    Finnish Meteorological Institute, 147
FMU
    Francis Marion University
        John K. Cauthen Educational Media Center, 866
FNAL
    Fermi National Accelerator Laboratory
        Experimental Astrophysics Group, 700
        Theoretical Astrophysics Group, 700
FNRS
    Fonds National de la Recherche Scientifique, 52
Focus Software Inc. (FSI), 595
Foerster-Sternwarte (Wilhelm-_), 265
Foggy Bottom Observatory (FBO), 811
FOM
    Institute for Atomic and Molecular Physics (AMOLF), 393
Fomalhaut Astronomical Society, 441
Fondation de l'Observatoire de Vérossaz, 504
Fondation de l'Observatoire François-Xavier Bagnoud, 504
Fondation Européenne de la Science, 177, 179
Fondation Jungfraujoch-Gornergrat, 504
Fondation Louis de Broglie, 179
Fondation Robert-A. Naef, 504
Fondazione Internazionale Premio E. Balzan, 318

Fondo Nazionale Svizzero per la Ricerca Scientifica, 504, 509

Fonds National de la Recherche Scientifique (FNRS), 52
Fonds National Suisse de la Recherche Scientifique, 504, 509

Fonds voor Wetenschappelijk Onderzoek (FWO) - Vlaanderen, 52
Fonds zur Förderung der Wissenschaftlichen Forschung (FWF), 35
Foothill College
    Astronomy Department, 621
    Observatory, 621, 633
Foothills Astronomical Society (FAS), 828
Foran High School
    Planetarium (Joseph A._), 662
FORBAIRT, 296
Ford Amateur Astronomy Club (FAAC), 763
Förderkreis Planetarium Göttingen (FPG) eV, 229
Föreningen för Astronomi och Astronautik (FAA), 489, 493
Foresthill, 639
Forschungsstationen Jungfraujoch und Gornergrat (HFSJG)
    (Internationale Stiftung Hochalpine_), 504
Forst-Sternwarte, 263
Forsyth Astronomical Society (FAS), 828
Fortaleza, 69–71
Fort Bend Astronomy Club (FBAC), 879
Fort Collins, 652, 654, 657
Fort Davis, 751, 801, 880, 888
Fort Lauderdale, 685
Fort McMurray, 78
Fort Myers, 679
Fort Pickens, 680
Fort Skala Station, 428
Fort Smith, 606
Fort Wayne, 711
Fort Wayne Astronomical Society (FWAS), Inc., 711
Fort Worth, 879, 882, 885
Fort Worth Astronomical Society (FWAS), 879
Fort Worth Museum of Science and History
    Noble Planetarium, 882
Fort Wright, 726
Forum der Technik Betriebsgesellschaft mbH
    Planetarium, 229
Forum Weltraumforschung Aachen, 230, 251
Foundation for Research Development (FRD), 466
Fountainwood Observatory, 884
Foxdale, 552
Fox Island County Park, 711
Fox Observatory, 685
Fox Park Public Observatory (FPPO), 763
Fox Valley Astronomical Society (FVAS), 701
Fox Valley Planetarium, 915
Fox Valley Skywatchers (FVSW), 701
FPG
    Förderkreis Planetarium Göttingen, 229
FPOA
    Fremont Peak Observatory Association, 621
FPPO
    Fox Park Public Observatory, 763
FRAC
    Flint River Astronomy Club, 688
Framtidsmuseet, 489
France, 154
    (Société Astronomique de_), 199
Francfort, 210, 234, 240, 254, 259
Franche-Comté, 156
Francis Marion University (FMU)
    John K. Cauthen Educational Media Center, 866
Franeker, 391
Frankfurt/Main, 210, 234, 240, 254, 259
Franklin, 759, 921
Franklin and Marshall College
    Department of Physics and Astronomy, 855
Franklin Institute Science Museum, 855
Franklin Miller Jr. Observatory, 839
Frank P. Brackett Observatory, 634
Franz-Kroller-Sternwarte (FKS), 35

Frascati, 318, 329
Fraser Valley Astronomers Society (FVAS), 88
Frauenfeld, 501
Fraunhofer Gesellschaft zur Förderung der angewandten Forschung eV, 230
Fraunhofer-Institut für Atmosphärische Umweltforschung, 230

Fraunhofer-Institut für Kurzzeitdynamik
    Ernst-Mach-Institut (EMI), 230
Fraunhofer-Institut Physikalische Meßtechnik (IPM), 230
Fraunhofer USA Inc., 763
FRD
    Foundation for Research Development, 466
Frederick, 742, 743
Fredericksburg, 899, 904
Fredericton, 102
Fred G. Dale Planetarium, 782
Fred Lawrence Whipple Observatory (FLWO), 596, 597, 751

Freehold, 793
Freiburg-im-Breisgau, 208, 230, 235, 243
Freie Universität Berlin, 230
    Institut für Meteorologie, 223
Frejlev, 134
Fréjus, 171
Fremont, 617, 783
Fremont Peak Observatory Association (FPOA), 621
French Camp, 774, 775
French Camp Academy
    Rainwater Observatory and Planetarium, 774, 775
French Space Agency, 168
Frente Unido Pro Astronomía Zuliana (FUPAZ), 928
Fresnay/Sarthe, 188
Fresno, 614, 618
Freundeskreis der Himmelskunde Bad Salzschlirf eV, 230
Frewsburg, 818
Friedrich-Alexander-Universität Erlangen-Nürnberg, 231
Friedrich Ebert Visual And Radiotelescope Observatory (FEVARO), 231
Friedrich-Schiller-Universität Jena, 231
Friedrichs-Gymnasium der Stadt Herford, Sternwarte, 231
Friedrich Verlag GmbH & Co. KG (Erhard_), 226
Friuli, 309
Froburg, 500
Frombork, 427, 429
Frombork Planetarium and Observatory, 427
Frosinone, 308
Frösön, 488
Frostburg, 736, 737
Frostburg State University (FSU)
    Planetarium, 736
Frosty Drew Observatory, 864
FSC
    Fernbank Science Center, 688
FSI
    Focus Software Inc., 595
FSU
    Florida State University
        Pat Thomas Planetarium, 683
FTL
    Farran Technology Ltd., 296
Fuertes Observatory, 812
Fuerteventura, 471
Fuji, 351
Fujigane Station, 351
Fuji (Mount_), 360
Fujitok Corp., 344
Fukuoka, 344
Fukuoka University of Education
    Department of Earth Sciences and Astronomy, 344
Fulda, 233, 242
Fuldatal, 262
Fuller (Broadhurst Clarkson &_), 539
Fullerton, 614
Funchal, 435
Fundação de Amparo à Pesquisa do Estado de São Paulo (FAPESP), 66

Fundação para a Ciência e a Tecnologia (FCT), 433
Fundação Planetário da Cidade do Rio de Janeiro, 66
Fundación José María Aragón, 7
Fundación Mundo Juvenil
  Planetario, 141
Fundamentals of Cosmic Physics, 815
Funding institutions
  Aard- en Levenswetenschappen (Netherlands), 393
  ALW (Netherlands), 393
  Aragón (José María_) (Argentina), 7
  ASTRON (Netherlands), 402
  Astronomical Foundation of Nógrád County (Hungary), 273
  CCLRC (UK), 542
  CCM (Mexico), 382
  Centre National de la Recherche Scientifique (France), 166
  CNRS (France), 166
  Comisión Nacional de Actividades Espaciales (Argentina), 6
  CONACYT (El Salvador), 143
  CONACYT (Mexico), 383
  CONAE (Argentina), 6
  CONCYTEC (Peru), 423
  CONICIT (Venezuela), 927
  CONICYT (Uruguay), 586
  Consejo Cultural Mundial (Mexico), 382
  Consejo Nacional de Ciencia y Tecnología (El Salvador), 143
  Consejo Nacional de Ciencia y Tecnología (Mexico), 383
  Consejo Nacional de Ciencia y Tecnología (Peru), 423
  Consejo Nacional de Investigaciones Científicas y Técnicas (Uruguay), 586
  Consejo Nacional de Investigaciones Científicas y Técnologicas (Venezuela), 927
  Consejo Superior de Investigaciones Científicas (Spain), 475
  Council for Scientific and Industrial Research (South Africa), 465
  Council for the Central Laboratory of the Research Councils (UK), 542
  CSIC (Spain), 475
  CSIR (South Africa), 465
  Deutsche Forschungsgemeinschaft (Germany), 222
  DFG (Germany), 222
  Dudley Observatory (USA-NY), 814
  Erna and Victor Hasselblad Foundation (Sweden), 488
  FAPESP (Brazil), 66
  FCT (Portugal), 433
  FNRS (Belgium), 52
  Fondation Louis de Broglie (France), 179
  Fondazione Internazionale Premio E. Balzan (Italy), 318
  Fondo Nazionale Svizzero per la Ricerca Scientifica, 509
  Fonds National de la Recherche Scientifique (Belgium), 52
  Fonds National Suisse de la Recherche Scientifique, 509
  Fonds voor Wetenschappelijk Onderzoek - Vlaanderen (Belgium), 52
  Fonds zur Förderung der Wissenschaftlichen Forschung (Austria), 35
  FORBAIRT (Ireland), 296
  Fraunhofer Gesellschaft (Germany), 230
  Fundação de Amparo à Pesquisa do Estado de São Paulo, 66
  Fundação para a Ciência e a Tecnologia (Portugal), 433
  FWF (Austria), 35
  FWO (Belgium), 52
  Gruber Foundation (Peter_) (US Virgin Islands), 924
  Hasselblad Foundation (Erna and Victor _) (Sweden), 488
  Hermann-von-Helmholtz-Gemeinschaft Deutscher Forschungszentren, 233
  HFSJG (Switzerland), 505
  HGF (Germany), 233
  ICCTI (Portugal), 433
  Icelandic Research Council, 279
  Internationale Stiftung Hochalpine Forschungsstationen Jungfraujoch und Gornergrat (Switzerland), 505
  Irish Science and Technology Agency, 296
  José María Aragón (Argentina), 7
  KACST (Saudi Arabia), 453
  King Abdulaziz City for Science and Technology (Saudi Arabia), 453
  Max-Planck-Gesellschaft (Germany), 236
  MPG (Germany), 236
  National Research Foundation (South Africa), 466
  National Science Foundation (USA-VA), 900
  NATO, 55
  Natural Environment Research Council (UK), 558
  Naturvetenskapliga Forskningsrådet (Sweden), 496
  Nederlandse Organisatie voor Toegepast-Natuurwetenschappelijk Onderzoek, 397
  Nederlandse Organisatie voor Wetenschappelijk Onderzoek, 397
  NERC (UK), 558
  NFR (Sweden), 496
  Nobel (Sweden), 492
  North Atlantic Treaty Organization, 55
  NRF(South Africa), 466
  NSF (Sri Lanka), 487
  NSF (USA-VA), 900
  NWO (Netherlands), 397
  Organisation du Traité de l'Atlantique Nord, 55
  OTAN, 55
  Pachart Foundation (USA-AZ), 600
  Particle Physics and Astronomy Research Council (UK), 561
  Peter Gruber Foundation (US Virgin Islands), 924
  Planetary Studies Foundation (USA-IL), 704
  PPARC (UK), 561
  PSF (USA-IL), 704
  Rannsóknarráð Íslands, 279
  Research Council of Norway, 418
  Schweizerischer Nationalfonds zur Förderung der Wissenschaftlichen Forschung, 509
  Science Council of Japan, 356
  Scientific and Technical Research Council of Turkey, 523
  Stichting Astronomisch Onderzoek in Nederland, 402
  Swedish Natural Science Research Council, 496
  Swiss National Science Foundation, 509
  Technology Development Centre (Finland), 151
  TEKES (Finland), 151
  Third World Academy of Sciences, 336
  TNO (Netherlands), 397
  TÜBITAK (Turkey), 523
  TWAS, 336
  Victor Hasselblad Foundation (Erna and_) (Sweden), 488
  Virginia Space Grant Consortium (USA-VA), 906
  VSGC (USA-VA), 906
FUNDP
  Facultés Universitaires Notre-Dame de la Paix Département de Mathématique, 51
FUPAZ
  Frente Unido Pro Astronomía Zuliana, 928
Furness Astronomical Society, 546
Fürstenfeldbruck, 259
Future's Museum, 489
Futurestar Corp., 770
Futuroscope, 168
FVAS
  Fox Valley Astronomical Society, 701
  Fraser Valley Astronomers Society, 88
FVSW
  Fox Valley Skywatchers, 701
FWAS
  Fort Wayne Astronomical Society, 711
  Fort Worth Astronomical Society, 879
FWF
  Fonds zur Förderung der Wissenschaftlichen Forschung (Austria), 35

FWO
 Fonds voor Wetenschappelijk Onderzoek, 52
Fyfield, 531
Fyshwick, 20

G

GAA
 Gruppo Astrofili Astigiani, 319
GAAE
 Gruppo Astrofili "Arthur Eddington", 318
GAAG
 Groupement des Astronomes Amateurs de la Gâtine, 181
GAB
 Gesellschaft für astronomische Bildung, 231
Gablitz, 39
GAC
 Goddard Astronomy Club, 737
 Greenbelt Astronomy Club, 737
 Greensboro Astronomy Club, 828
 Gruppo Amici del Cielo, 318
GACB
 Gruppo Astrofili di Cinisello Balsamo, 319
GACE
 Grupo de Astronomía y Ciencias del Espacio, 486
GAD
 Gruppo Astronomia Digitale, 324
GAG
 Gruppo Astrofili Genovesi, 320
Gahberg, 34
GAHQ
 Groupe Astronomique Hague Querqueville, 180
Gainesville, 678, 686, 690
Gaithersburg, 736, 742, 745
GAK
 Göteborgs Astronomiska Klubb, 489
GAL
 Gruppo Astrofili Lariani, 321
Galaxis, Vereniging voor Amateur Astronomie, 393
Galaxy Contact, 179
Galaxy Optics, 654
Gale Observatory (Grant O._), 717
Galerie Alain Carion, 179
Galileo Society, 732
Galion, 842
Gallatin, 871
Galvoptics Ltd.
 Telescope Division, 546
Galway, 297
GAM
 Gruppo Astrofili Manfredonia, 321
 Gruppo Astrofili Massesi, 321
 Gruppo Astrofili Menkalinan, 321
GAMA
 Group of Amateur Astronomers, 127
Gambier, 839
Gand, 53, 59
Gandersheim (Bad._), 232
GAO
 Gothard Astrophysical Observatory, 274
GAP
 Gruppo Astrofili di Padova, 319
 Gruppo Astrofili Pavese, 322
 Gruppo Astrofili Pesarese, 322
 Gruppo Astrofili Piceni, 322
Gap Astronomie - Association Copernic, 179
GAPRA
 Groupement Astronomique Populaire de la Région d'Antibes, 180
GAPS
 Groupement d'Astronomie Populaire Sottevillais, 181
GAR
 Gruppo Astrofili di Rozzano, 320
 Gruppo Astrofili Romani, 323
Garbagnate Monastero, 333

Garching, 226, 228, 236, 245
Garden City, 805
Garden Route, 464
Gardner Memorial Planetarium, 763
Gardner-Webb University (GWU)
 Craven E. Williams Observatory, 828
Garmisch-Partenkirchen, 230
Garny Space Astronomy Institute (GSAI), 12
Garrett County Board of Education Planetarium, 737
GAS
 Genesee Astronomical Society, 763
 Gisborne Astronomical Society, 409
 Gislaved Astronomiska Sällskap, 489
 Groupe Astronomie de Spa, 52
 Grupo Astronómico Silos, 476
 Gruppo Astrofili di Schio, 320
 Gruppo Astrofili Savonesi, 323
 Gruppo Astrofili Soresinesi, 323
 Guildford Astronomical Society, 547
Gastonia, 831
GAT
 Gruppo Astronomico Tradatese, 324
Gates Planetarium (Charles C._), 653, 654
Gâtine, 181
Gauhati University
 Physics Department, 281
 Astrophysics Group, 283
Gauribidanur, 285, 288
GAV
 Gruppo Astrofili Vesuviano, 323
 Gruppo Astrofili Vicentini, 323
 Gruppo Astronomico Viareggio, 324
Gávea, 66
Gävle, 493
GAVRT
 Goldstone-Apple Valley Radio Telescope, 625
GAWH
 Gruppo Astrofili "William Herschel", 323
Gayle Planetarium (W.A._), 589, 591
Gazan Mountains, 293
G.B. Manasek Inc., 893
GCEV
 Grupo Canario de Estrellas Variables, 477
GCNS
 Glasgow College of Nautical Studies
  Planetarium, 547
GCO
 Grove Creek Observatory, 23
Gdansk, 427
Gdansk University
 Institute of Theoretical Physics and Astrophysics, 427
GDPL
 Gordhandas Desai Pvt. Ltd., 283
GEA
 Grup d'Estudis Astronòmics, 476
 Grupo de Estudios Astronómicos, 476
 Grupo de Estudos de Astronomia, 67
GEAI
 Groupe Électronique Astronomie Informatique, 180
GEAP
 Grupo de Estudios Astronómicos de Puertollano, 477
GEAQ
 Groupement d'Études Astronomiques de Queven, 161
Gebiedsbestuur Aard- en Levenswetenschappen (ALW), 393

Gebze, 523
G.E.D. Alcock (Gruppo_), 324
Geer Planetarium (Willard_), 907, 908
Gemenos, 180
Gemini Observatory, 671
 Northern Operations Center, 693
 Southern Operations Center, 109
 Tucson Project Office, 596
Gemmary Inc. (The_), 621, 644
General Nathan Twining Observatory (GNTO), 803
Generoso (Monte_), 504
Gênes, 309, 317, 318, 320, 340

Genesee Astronomical Society (GAS), 763
Geneseo, 806, 822
Geneva, 502, 503, 505, 507, 509, 510, 514
Genève, 502, 503, 505, 507, 509, 510, 514
Gengras Planetarium, 662
Genk, 51
Gennevilliers, 172
Genoa, 309, 317, 318, 320, 340
Genova, 309, 317, 318, 320, 340
Gent, 53, 59
GeoForschungsZentrum Potsdam, 222
Geological Society of America (GSA), 654
Georg-August-Universität Göttingen, 231
George, 464
George Carroll Observatory, 651
George C. Marshall Space Flight Center (MFSC), 589
    Space Science Laboratory, 590
George E. Coleman Sr. Planetarium, 690
George F. Beattie Planetarium, 622
George H. Clever Planetarium, 622
George Mason University (GMU)
    Astronomy and Space Sciences Group, 898
    Center for Earth Observing and Space Research (CEO-SR), 898
    Department of Physics and Astronomy, 898
George Observatory, 879
George R. Wallace Jr. Astrophysical Observatory, 754
Georgetown, 92, 884, 889
Georgia, 206, 687–691
    (North_), 690
Georgia Institute of Technology
    School of Physics, 688
Georgian Academy of Sciences
    Abastumani Astrophysical Observatory (AbAO), 206
        Town Department, 206
Georgia Southern University (GSU)
    Department of Physics, 688, 691
Georgia State University (GSU)
    Center for High Angular Resolution Astronomy (CHARA), 688
    Department of Physics and Astronomy, 689
GEOS
    Groupe Européen d'Observation Stellaire, 180
    Grupo Europeo de Observaciones Estelares, 477
Géospace Observatoire d'Aniane, 179
Gerald Hughes Phipps Observatory, 653
Geretsried, 264
Germany, 207
GERMEA
    Groupe d'Entraînement et de Recherche pour les Métho des d'Éducation Active, 180
Germering, 248
Gesellschaft der Freunde der Urania-Sternwarte (GFUS), 504
Gesellschaft für astronomische Bildung (GAB) eV, 231
Gesellschaft für astronomische Bildung in Mecklenburg-Vorpommern, 231
Gesellschaft für volkstümliche Astronomie eV (GvA), 231
Gesellschaft für Weltallkunde (GfW) eV, 232
Gesellschaft zur wissenschaftlichen Untersuchung von Parawissenschaften eV (GWUP), 232
Gettysburg, 854, 856
Gettysburg College, 854
    Observatory, 856
Geza, 142
GFOES
    Groupement Français pour l'Observation et l'Étude du Soleil, 181
GFUS
    Gesellschaft der Freunde der Urania-Sternwarte, 504
GfW
    Gesellschaft für Weltallkunde, 232
GHAAS
    Greater Hazleton Area Astronomical Society, 856
Gheens Science Hall and Rauch Planetarium, 725
Ghent, 53, 59
Giant Metrewave Radio Telescope (GMRT), 286

Gibbes Planetarium, 866
Gibbs Research Laboratory (J. Willard_), 666
Gibraltar, 267
Gibraltar Astronomical Society (GAS), 267
Gibson Observatory, 685
Gibson Observatory Astronomy Group (GOAG), 681
Giessen, 210
GIFAS
    Groupement des Industries Françaises Aéronautiques et Spatiales, 181
GIFO
    Groupement des Industries Françaises de l'Optique, 181
Gif-sur-Yvette, 174, 186
Gifu, 350, 354, 361
Gilbert, 602
Gillette, 922, 923
Gilliland Observatory, 750
Gisborne, 409
Gisborne Astronomical Society (GAS), 409
Gislaved, 489
Gislaved Astronomiska Sällskap (GAS) Orion, 489
GISS
    Goddard Institute for Space Studies, 819
Gistrup, 134
Giza, 142
Gladwin Planetarium, 638
Glarus, 501
Glasgow, 532, 534, 547, 567, 576
Glasgow College of Nautical Studies (GCNS)
    Planetarium, 546
Glastonbury, 663
Glavgidromet, 925
Glen D. Riley Observatory, 703
Glen Ellen, 650
Glenfield Planetarium, 792
Glengormley, 551
Glenlea Astronomical Observatory, 102
Glenn Research Center (GRC) (John H._), 838, 840
    Visitor Center, 840
Glen Oaks, 806
Glens Falls, 805
Glenshaw, 851
Global Network of Automatic Telescopes (GNAT), 596
Global Volcanism Program (GVP), 675
Gloucester, 797
GLPA
    Great Lakes Planetarium Association, 838
Glücksburg, 238
GMA
    Gruppo Marsicano Astrofili, 324
GMRT
    Giant Metrewave Radio Telescope, 286
GMU
    George Mason University
        Astronomy and Space Sciences Group, 898
        Center for Earth Observing and Space Research, 898
        Department of Physics and Astronomy, 898
Gmunden, 35
GNAT
    Global Network of Automatic Telescopes, 596
Gnomonicae Societas Austriaca (GSA), 36
GNTO
    General Nathan Twining Observatory, 803
GOA
    Stichting Geologisch, Oceanografisch en Atmosferisch Onderzoek, 403
GOAG
    Gibson Observatory Astronomy Group, 681
Goddard Astronomy Club (GAC), 737
Goddard Institute for Space Studies (GISS), 815, 819
Goddard Optical Research Facility, 737
Goddard Planetarium (Robert H._), 800, 803
Goddard Space Flight Center (GSFC), 737
    Astronomy Data Center (ADC), 740
    Astrophysics Data Facility (ADF), 740
    Goddard Astronomy Club (GAC), 737
    Laboratory for High-Energy Astrophysics (LHEA), 740

National Space Science Data Center (NSSDC), 740
Solar Data Analysis Center, 741
Solar Physics Branch, 741
Space Science Data Operations Office (SSDOO), 741
Visitor Center, 741
World Data Center A for Rockets and Satellites, 745
Godlee Observatory, 555
Goethe Link Observatory, 712
Goiânia, 72
Golden, 655, 658
Goldendale, 908
Goldendale Observatory State Park (GOSP), 908
Golden Pond, 725
Golden Pond Planetarium and Observatory, 725
Goldstone, 613
Goldstone-Apple Valley Radio Telescope (GAVRT), 625
Goleta, 625
Goloseev, 529
Golosiiv, 529
Gönnsdorf, 251
Goodsell Observatory, 770
Goodwill Cryogenics Enterprises, 283
Gordhandas Desai Pvt. Ltd. (GDPL), 283
Gordon and Breach Science Publishers, 856
Gordon Southam Observatory, 89
Görlitz, 264
Gornergrat, 255, 316, 317, 505
GOSP
    Goldendale Observatory State Park, 908
GOST
    State Committee for Standardization and Metrology (Russia), 449
Göteborg, 488–490, 493
Göteborg Astronomical Club, 489
Göteborgs Astronomiska Klubb (GAK), 489, 490
Göteborg University
    Chalmers Technical University
        Department of Astronomy and Astrophysics, 490
        Onsala Space Observatory (OSO), 490
Gotenburgsternwarte, 232
Gotha, 258
Gothard Astrophysical Observatory (GAO), 274
Gothenburg, 488–490, 493
Gothia
    (East_), 492
Gotoh Planetarium and Astronomical Museum, 344
GOTO Optical Manufacturing Co., 344
Göttingen, 207, 213, 229, 256
Goucher College
    Department of Physics and Astronomy, 737
Gov Aker Observatory, 595, 596
GPPA
    Great Plains Planetarium Association, 782
GPSP
    Great Plains Star Party, 722, 777
GRAAA
    Grand Rapids Amateur Astronomical Association, 763
GRAAL
    Groupe de Recherche en Astronomie et Astrophysique du Languedoc, 183
Grace H. Flandrau Science Center and Planetarium, 596
Grady Planetarium (Joseph_), 858
Gräfelfing, 209
Grafton, 918
Gragnola, 329
Graham (Mount_), 604, 926
Grahamstown, 466
Grainger Observatory, 789
Grakovo, 529
Granada, 475, 478, 481, 483, 485
GRAND
    Gamma-Ray Astrophysics at Notre Dame, 714
Grand Forks, 834
Grand Haven, 767
Grand Island, 784
Grand Junction, 652, 661
Grand-Mère, 85

Grand Rapids, 761, 767
Grand Rapids Amateur Astronomical Association (GRAAA) Inc., 763
Grand Traverse Astronomical Society (GTAS), 764, 766
Grange Observatory, 332
Granite Falls, 827
Gran Sasso, 328
Grant O. Gale Observatory, 717
Granville, 838
Grasse, 191
Grasslands Observatory, 596
Grass Valley, 620
Grasweg Sternwarte, 232
Graz, 37, 39, 40
Greater Hazleton Area Astronomical Society (GHAAS), 856
Great Island Science and Adventure Park, 88
Great Kills Park, 806
Great Lakes Planetarium Association (GLPA), 838
Great Plains Planetarium Association (GPPA), 782
Great Plains Star Party (GPSP), 722, 777
Grebner-Van Zandt Observatory (Decker-_), 704
Greece, 268
Green Bank, 913
Green Bay, 917
Greenbelt, 737, 738, 740, 741, 745
Greenbelt Astronomy Club (GAC), 737
Greencastle, 710
Greene, 732
Green Garden Observatory, 765
Green Mountain Observatory, 894
Greensboro, 828, 830, 832
Greensboro Astronomy Club (GAC), 828
Greenville, 667, 866–868
Greenwich, 565
    (East_), 865
Greifswald, 248
Grenoble, 178, 182, 203
Gresham, 848
Gretz-Armainvilliers, 205
Griffin, 688
Griffith, 710
Griffith Observatory, 622
    Los Angeles Astronomical Society, 626
Grimbergen, 60
Grinnell, 717
Gröbenzell, 217
Groenkloof, 467
Groningen, 396, 401, 404, 406
Grosse Pointe, 764
Grosse Pointe Schools Planetarium, 764
Grosseto, 310
Grossmont College
    Department of Physics, Astronomy and Physical Sciences, 622
Großröhrsdorf, 250
Großschwabhausen, 255
Groton, 753
Group 70, 622
Groupe Astronomie & CCD, 88
Groupe Astronomie de Spa (GAS), 52
Groupe Astronomique Hague Querqueville (GAHQ), 180
Groupe d'Astronomie Quasar, 52
Groupe d'Entraînement et de Recherche pour les Méthodes d'Éducation Active (GERMEA), 180
Groupe de Recherche en Astronomie et Astrophysique du Languedoc (GRAAL), 183
Groupe Électronique Astronomie Informatique (GEAI), 180
Groupe Européen d'Observation Stellaire (GEOS), 180
Groupement Astronomique Populaire de la Région d'Antibes (GAPRA), 180
Groupement d'Astronomie Populaire Sottevillais (GAPS), 180
Groupement des Astronomes Amateurs de la Gâtine (GAAG), 181
Groupement des Industries Françaises Aéronautiques et Spa-

tiales (GIFAS), 181
Groupement des Industries Françaises de l'Optique (GIFO), 181
Groupement Français pour l'Observation et l'Étude du Soleil (GFOES), 181
Groupe SFIM, 198
Group of Amateur Astronomers (GAMA), 127
Grout Museum of History and Science, 717
Grove Creek Observatory (GCO), 23
Gruber Foundation (Peter_), 924
Grudziądz, 429
Grundy Observatory (J.R._), 855
Grup d'Estudis Astronòmics (GEA), 476
Grupo Astronómico Silos (GAS), 476
Grupo Canario de Estrellas Variables (GCEV), 477
Grupo de Estudios Astronómicos de Puertollano (GEAP), 477
Grupo de Estudios Astronómicos (GEA), 476, 477
Grupo de Estudos de Astronomia (GEA), 67
Grupo Europeo de Observaciones Estelares (GEOS), 477
Grupo Joven de Astronomía Scorpius Elche, 477
Gruppo Amici del Cielo (GAC), 318
Gruppo Antares, 318
Gruppo Astrofili Aretini, 318
Gruppo Astrofili "Arthur Eddington" (GAAE), 318
Gruppo Astrofili Astigiani (GAA) "Beta Andromedae", 319

Gruppo Astrofili Catanesi "Guido Ruggieri", 319
Gruppo Astrofili Columbia, 319
Gruppo Astrofili del Dopolavoro Ferroviario di Rimini, 319
Gruppo Astrofili di Cinisello Balsamo (GACB), 319
Gruppo Astrofili di Padova (GAP), 319
Gruppo Astrofili di Palermo, 320
Gruppo Astrofili di Rozzano (GAR), 320
Gruppo Astrofili di Schio (GAS), 320
Gruppo Astrofili Frentani, 320
Gruppo Astrofili Genovesi (GAG), 320
Gruppo Astrofili "Giovanni e Angelo Bernasconi", 320
Gruppo Astrofili Hipparcos, 320
Gruppo Astrofili "Isaac Newton", 321
Gruppo Astrofili "La Nuova Selene", 321
Gruppo Astrofili Lariani (GAL), 321
Gruppo Astrofili Manfredonia (GAM), 321
Gruppo Astrofili Massesi (GAM), 321
Gruppo Astrofili Menkalinan (GAM), 321
Gruppo Astrofili "N. Copernico", 321
Gruppo Astrofili ORSA, 322, 330
Gruppo Astrofili Pavese (GAP), 322
Gruppo Astrofili Persicetani, 322
Gruppo Astrofili Pesarese (GAP), 322
Gruppo Astrofili Piceni (GAP), 322
Gruppo Astrofili Reggini M31 del Dopolavoro Ferroviario, 322
Gruppo Astrofili Rigel, 322
Gruppo Astrofili Romani (GAR), 323
Gruppo Astrofili Savonesi (GAS), 323
Gruppo Astrofili Soresinesi (GAS), 323
Gruppo Astrofili Vesuviano (GAV), 323
Gruppo Astrofili Vicentini (GAV) "Giorgio Abetti", 323
Gruppo Astrofili "William Herschel" (GAWH), 323
Gruppo Astronomia Digitale (GAD), 324
Gruppo Astronomico Tradatese (GAT), 324
Gruppo Astronomico Viareggio (GAV), 324
Gruppo "G.E.D. Alcock", 324
Gruppo Marsicano Astrofili (GMA) "F. Angelitti", 324
GSA
    Geological Society of America, 654
    Gnomonicae Societas Austriaca, 36
GSFC
    Goddard Space Flight Center
        Astronomy Data Center, 737, 740
        Astrophysics Data Facility, 740
        Laboratory for High-Energy Astrophysics, 740
        National Space Science Data Center, 740
        Solar Data Analysis Center, 741
        Solar Physics Branch, 741
        Space Science Data Operations Office, 741
        Visitor Center, 741
        World Data Center A for Rockets and Satellites, 745
GSU
    Georgia Southern University
        Department of Physics, 688, 691
        Planetarium, 688
    Georgia State University
        Center for High Angular Resolution Astronomy, 688
        Department of Physics and Astronomy, 689
GTAS
    Grand Traverse Astronomical Society, 764, 766
Guadalajara, 385, 481
Guadeloupe, 159
Guam, 271
Guanajuato, 385
Guelph, 102
Guéret, 194
Guernesey, 106
Guernsey, 106
Guido Ruggieri (Circolo Astrofili di Mestre_), 313
Guildford Astronomical Society (GAS), 547
Guilford, 547, 571
Gulbenkian (PCG) (Planetário Calouste_), 434
Guwahati, 281, 283
Guyancourt, 170
Guyane Française, 168
GvA
    Gesellschaft für volkstümliche Astronomie, 231
GVP
    Global Volcanism Program, 675
Gwent, 562
GWU
    Gardner-Webb University
        Craven E. Williams Observatory, 828
GWUP
    Gesellschaft zur wissenschaftlichen Untersuchung von Parawissenschaften eV, 232
Gwynedd, 547
Gwynedd Astronomical Society, 547
Gymnasium Philippinum Sternwarte, 232
Gyula Observing Station, 275

H

HAA
    Hamilton Amateur Astronomers, 88
    Hawkesbury Astronomical Association, 23
Haag - Meteorites (Robert A._), 596, 601
Haarlem, 407
Haastrecht, 407
HAC
    Hilltopper Astronomy Club, 725
Hadlow, 543
HAG
    Hampshire Astronomical Group, 547
Hagar Planetarium (Charles F._), 637
Hagen, 260
Hagerstown, 743, 745
Hahneberg, 219
Haifa, 299
Hail, 453
Hainaut, 58
Hainberg, 256
Halberstadt, 215
Haleakala Observatory, 696
Hale Pohaku, 693
Hale Solar Laboratory, 629
Halifax, 82, 95, 97, 100
Halifax Planetarium, 100
Halle, 231, 243
Halley Hill Observatory, 622
Hall in Tirol, 36
Halls Valley Astronomical Group (HVAG), 622
Halmstad, 490
Halmstad Astronomical Association, 490
Halmstads Astronomiska Sällskap (HAS), 490
Haltern, 207

Hamburg, 218, 220, 223, 224, 231, 241, 254
Hamburger Sternwarte, 232, 254
Hämeenlinna, 147
Hämeenlinnan Tähtitieteen Harrastajlen Yhdistys Vega ry, 147
Hamill Observatory, 849
Hamilton, 91, 409, 412, 811
Hamilton Amateur Astronomers (HAA), 88
Hamilton Astronomical Society (HAS), 409
Hamilton (Mount_), 649
H. Ammar, 453
Hampshire Astronomical Group (HAG), 547
Hampton, 688, 900, 903, 905, 906
Hancock, 801
Hanle, 285
Hanna City Robotic Observatory (HCRT), 701
Hannover, 219, 227, 254, 260
Hanover, 740, 788, 789
Hansen Planetarium, 890
Hans-Nüchter Sternwarte, 232
Hanson Center for Space Science (William B._), 879
HAO
    High Altitude Observatory, 656
HAPS
    Huddersfield Astronomical and Philosophical Society, 548
Haramiata, 74
Harare, 929
Hardin Optical, 848
Hardin Planetarium, 728
Hard Labor Creek Observatory (HLCO), 689
Harford, 737
Harford County Astronomical Society (HCAS), 737
Haringey, 534
Hariq, 453
Harper College Observatory (HCO), 701, 709
Harri Deutsch (Verlag_), 259
Harriman, 873
Harriman Observatory, 825
Harrington Discovery Center (DHDC) (Don_), 879
Harrisburg, 852, 853, 859, 861
Harrisonburg, 899
Harrogate, 547
Harrogate Astronomical Society (HAS), 547
Harry A. Burke High School
    Planetarium, 782
Harry Bayley Observatory, 45
HART
    Heartland Astronomical Research Team, 722
Hartebeesthoek Radio Astronomy Observatory (HartRAO), 466
Hartford, 662
    (East_), 665
    (West_), 662
Hartha, 220
HartRAO
    Hartebeesthoek Radio Astronomy Observatory, 466
Hartung Boothroyd Observatory, 812
Hartwick College
    Ernest B. Wright Observatory, 815
Harvard College Observatory (HCO), 751
    Harvard-Smithsonian Center for Astrophysics (CfA), 751
Harvard-Smithsonian Center for Astrophysics (CfA), 751, 752
    Chandra X-ray Center (CXC), 749
    Fred Lawrence Whipple Observatory (FLWO), 596
    IAU Central Bureau for Astronomical Telegrams (CBAT), 749
    Minor Planet Center (MPC), 755
    Multiple Mirror Telescope Observatory (MMTO), 597
    New England Light Pollution Advisory Group, 756
Harvard University
    Center for Astrophysics (CfA), 751
    Department of Astronomy, 751
    Department of Physics, 751
Harvard University Press (HUP), 752

Harvey Mudd College (HMC)
    Department of Physics, 623
Harwich, 749
Harzplanetarium, 233
HAS
    Halmstads Astronomiska Sällskap, 490
    Hamilton Astronomical Society, 409
    Harrogate Astronomical Society, 547
    Hawaiian Astronomical Society, 693
    Highlands Astronomical Society, 548
    Houston Astronomical Society, 880
Hasselblad Foundation (Erna and Victor _), 488, 490
Hasselt, 55
Hastings, 547, 782
Hastings and Battle Astronomical Society, 547
Hastings College
    Sachtleben Observatory, 782
Hastings Museum of Natural and Cultural History, 782
Hat Creek, 611
Hat Creek Observatory, 623, 647
Hat Creek Radio Observatory (HCRO), 611, 623
Hatfield, 577
Hatter Planetarium, 856
Hausen bei Brugg, 500
Haute Corréo, 179
Haute-Marne, 200
Hautes-Plates, 199
Haut-Léman, 509
Haut-Rhin, 200
Havaalani, 524
Haverford, 856
Haverford College
    Department of Physics and Astronomy, 856
Havering Astronomical Society, 547
Hawaii, 692–696
Hawaiian Astronomical Society (HAS), 693
Hawkes Bay Astronomical Society, 409
Hawkes Bay Holt Planetarium, 409
Hawkesbury Astronomical Association (HAA) Inc., 23
Hawksley Education Ltd. (PHEL) (Paton_), 562
Hawthorn, 28
Hayden Planetarium, 815
Hayden Planetarium (Charles_), 750, 752
Hayfield Secondary School Planetarium, 898
Haymarket, 20, 23, 28
Haynald Observatory, 274
Haystack Observatory, 746, 752, 756
Hayward, 616, 618, 644
HBOSC
    Howard B. Owens Science Center, 738
HCA
    Hill Country Astronomers, 880
HCAS
    Harford County Astronomical Society, 737
HCC
    Hibbing Community College
        Paulucci Space Theatre, 772
HCO
    Harper College Observatory, 709
    Harvard College Observatory, 751
HCRO
    Hat Creek Radio Observatory, 611
HCRT
    Hanna City Robotic Observatory, 701
Headwaters Science Center (HSC)
    Planetarium, 770
Heartland Astronomical Research Team (HART), 722
Heart of England Astronomical Society (HOEAS), 548
Hebden Bridge, 548
Hebden Bridge Literary and Scientific Society
    Astronomy Section, 548
Hebrew University
    Racah Institute of Physics, 299
Hedgesville, 912
Heerhugowaard, 403
Heerlen, 402
Heesch, 402

Hefei, 118
Heidelberg, 214, 235–237, 246, 254
Heilbronn, 243
Heinrich Hertz Submillimeter Telescope Observatory (SMTO), 603
Heitkamp Memorial Planetarium, 717
HELAS
    Hellenic Astronomical Society, 268
Helius Designs, 548
Hellenic Astronomical Society (HELAS), 268
Hellenic National Meteorological Service (HNMS), 268
Hellenic Organization for Standardization (ELOT), 268
Hellenic Physical Society, 268
Hellinikon, 268
Helsinki, 146, 147, 150–152
Helsinki University of Technology (HUT)
    Department of Electrical Engineering, 148
        Institute of Photogrammetry and Remote Sensing, 148
    Laboratory of Space Technology, 148
    Metsähovi Radio Observatory, 148
Helston, 541
Helwân Observatory, 142
Hempstead, 816
Hencoup Enterprises, 548
Henry B. duPont III Planetarium, 662, 663
Henry Buhl Jr. Planetarium and Observatory, 856, 858
Henry Crown Space Center and OmniMax Theater, 703
Henry Laboratories (Joseph_), 792
    Physics Department, 795
Henry R. Barber Research Observatory, 708
Heppenheim, 246, 258
Heraklion, 269
HERAS
    Hull and East Riding Astronomical Society, 549
Herbert H. Lehman College, 816
    Department of Physics and Astronomy, 810
Herbert Trackman Planetarium, 701
Herford, 231
Herforshire, 577
Hermann-von-Helmholtz-Gemeinschaft Deutscher Forschungszentren (HGF), 233
Hermanus, 466
Hermosillo, 385
Herndon, 895, 899, 902, 904
Herndon High School Planetarium, 898
Herne, 211
Héron (Ferme du_), 171
Herrett Center for Arts and Science, 697
    Earl and Hazel Faulkner Planetarium, 697
Herrsching, 219
Herschel Museum (William_), 584
Herschel Society (William_), 548
Hertford, 577
Hertogenbosch ('s-_), 393
Herzberg/Elster, 257, 266
HET
    Hobby-Eberly Telescope, 880
HETE
    High-Energy Transient Experiment, 754
Heuchelheim, 210
Heverlee, 59
Hextek Corp., 597
HFSJG
    Hochalpine Forschungsstationen Jungfraujoch und Gornergrat, 505
HGF
    Hermann-von-Helmholtz-Gemeinschaft Deutscher Forschungszentren, 233
H.H. Wills Physics Laboratory, 548, 574
HIA, Herzberg Institute of Astrophysics, 93
    Dominion Astrophysical Observatory, 94
        Canadian Astronomy Data Centre, 93
    Dominion Radio Astrophysical Observatory, 94
Hibbing, 772
Hibbing Community College (HCC)
    Paulucci Space Theatre (PST), 772
Hickory Hill, 684

Hida Observatory, 345, 350
HiDAS
    High Desert Astronomical Society, 623
High Altitude Observatory (HAO), 655, 656
High Bridge, 794
High Desert Astronomical Society (HiDAS), 623
High-Energy Transient Experiment (HETE), 754
Highland Park ISD Planetarium, 879, 884
Highlands Astronomical Society (HAS), 548
Highlands Stargazers Astronomy Group, 681
High Point, 832
High Res Technologies (HRT), 88
Highstown, 795
High Wycombe, 584
Hija de Dios, 472
Hill Country Astronomers (HCA), 880
Hill Observatory (Rosemary_), 686
Hillsboro, 789, 848
Hillside Observatory, 720
Hilltopper Astronomy Club (HAC), 725
Hilo, 692, 693, 695, 696
Hinds Instruments Inc., 848
Hingham, 750
Hiraiso Solar-Terrestrial Research Center, 342, 344, 345
Hiroshima, 345
Hiroshima Children's Museum, 345
Hiroshima Planetarium, 345
Hirrlingen, 218
Hisar Astronomical Observatory, 518
Hitchin, 554
Hitotsubashi University
    Laboratory of Astronomy and Geophysics, 345
HKAAS
    Hong Kong Amateur Astronomical Society, 116
HKAS
    Hong Kong Astronomical Society, 116
HKO
    Hong Kong Observatory, 116
HKU
    Hong Kong University
        Astronomy Club, 116
HLCO
    Hard Labor Creek Observatory, 689
HMC
    Harvey Mudd College
        Department of Physics, 623
HMNS
    Houston Museum of Natural Science
        Burke Baker Planetarium, 878
        George Observatory, 879
HMS
    Hungarian Meteorological Service, 276
HNMS
    Hellenic National Meteorological Service, 268
Hobart, 15, 16, 31
Hobby-Eberly Telescope (HET), 880
Hochstetten-Dhaun, 212
HOEAS
    Heart of England Astronomical Society, 548
Hoenheim, 170
Hoeven, 400
Hof, 248
Hofheim/Taunus, 261
Höfingen, 249
Hofstra University
    Department of Physics and Astronomy, 816
Hoher List, 253
Hohmann-Sternwarte Essen (Walter-_), 208
Hokkaido, 344
Holcomb Observatory and Planetarium (JIHOP) (J.I._), 710, 711
Holland, 767
Hollidaysburg, 861
Hollywood, 626, 640
Hollywood General Machining Inc., 623
Holt Planetarium (Hawkes Bay_), 410
Holt Planetarium (William Knox_), 623, 651

Holyport, 571
Homburg, 257
Honduras, 272
Honeywell
    Space Systems Headquarters, 681
Hong Kong, 116, 118
Hong Kong Amateur Astronomical Society (HKAAS), 116
Hong Kong Astronomical Society (HKAS), 116
Hong Kong Observatory (HKO), 116
Hong Kong Space Museum
    Planetarium, 116
Hong Kong University (HKU)
    Astronomy Club, 116
Honnef (Bad.), 223
Honolulu, 692–696
Hood (Mount.), 848
Hook Memorial Observatory (John C..), 711, 712
Hooper Planetarium (T.C..), 868
Hoover-Price Planetarium, 838
Hopkins, 770
Hopkins (Mount.), 597, 599, 787
Hopkins Observatory, 760
Hopkins Planetarium, 899, 903
Hopkins Technology, LLC, 770
Hopkins University (JHU) (Johns.), 737
Horn-Gesellschaft (Walter-.), 264
Horrocks (Jeremiah.) Observatory, 563, 575
Hoshinoko Yakata Observatory, 345
Hoskinstown, 31
Hostalets, 476
Hotel und Sternwarte Heiligkreuz, 36
Houghton, 766
House of Technology, 490, 497
Houston, 878–881, 884, 886, 887, 889
    (North.), 882
Houston Astronomical Society (HAS), 880
Houston Museum of Natural Science (HMNS)
    Burke Baker Planetarium, 878
    George Observatory, 879
Houston State University (SHSU) (Sam.), 880
    Department of Physics, 883
Houten, 391
Hove, 60, 537
Howard Astronomical League of Central Maryland, 737
Howard B. Owens Science Center (HBOSC), 737
    Planetarium, 738
Howell Memorial Planetarium, 866
Howell Observatory, 774
Hoyerswerda, 214, 241
HPO
    Huntsville Program Office (USRA), 590
Hradec Králové, 128
Hradec Králové Observatory and Planetarium, 128
HRM
    Hudson River Museum, 807
H.R. MacMillan Planetarium, 89
H.R. MacMillan Planetarium and Gordon Southam Observatory, 88
H.R. MacMillan Space Centre (HRMSC), 89
HRMSC
    H.R. MacMillan Space Centre, 89
HRT
    High Res Technologies, 88
HSC
    Headwaters Science Center
        Planetarium, 770
Hsinchu, 516
HSO
    Hungarian Space Office, 276
HST
    Hubble Space Telescope, 245
Huachuca Astronomy Club, 597
Huairou Station, 113
Huancayo, 423
Hubbard Scientific, 916
Hubble Planetarium (Edwin P..), 814, 816
Hubble Space Telescope (HST), 245

Huddersfield, 549
Huddersfield Astronomical and Philosophical Society (HAPS), 548
Hudnall Planetarium, 877, 880
Hudson River Museum (HRM), 807
Huesca, 471
Hughes Danbury Optical Systems (HDOS) Inc., 663, 664
Hull, 549
Hull and East Riding Astronomical Society (HERAS), 549
Humacao, 437
Humain Radio Astronomy Station, 55, 56
Hume Cronyn Observatory, 104
Hume Observatory, 608, 611
Hummel Planetarium (Arnim D..), 725, 726
Hummingbird Astronomical Observatory, 623
Hungarian Academy of Sciences, 274
    Debrecen Heliophysical Observatory, 274
        Gyula Observing Station, 275
    Institute for Particle and Nuclear Physics, 275
    Konkoly Observatory, 275, 276
Hungarian Amateur Astronomical Society, 275
Hungarian Astronautical Society, 275
Hungarian Astronomical Association, 276
Hungarian Meteorological Service (HMS), 276
Hungarian Space Office (HSO), 276
Hungarian Standards Institution, 276
Hungary, 273
Hunter, 16
Hunter College, 816
    Department of Physics and Astronomy, 810
Huntingdon Valley, 859
Huntington, 585, 913
Hunt Observatory (Colin.), 535
Huntsville, 590, 591, 883
HUP
    Harvard University Press, 752
Hurbanovo, 460
Hurky, 128
Huron Valley Astronomers Group, 838
Hurst, 885
HUT
    Helsinki University of Technology
        Department of Electrical Engineering, 148
        Institute of Photogrammetry and Remote Sensing, 148
        Laboratory of Space Technology, 148
        Metsähovi Radio Observatory, 148
Hutchinson, 722
Huuhanmäki, 149
HVAG
    Halls Valley Astronomical Group, 622
Hvar, 122
Hvar Observatory, 121
    Hvar, 122
    Zagreb, 122
Hvězdárna Úřadu Města Karlovy Vary, 128
H.V. McKay Planetarium, 23, 24
Hyde Memorial Observatory, 783, 784
Hyderabad, 282, 287
Hydrometeorological Department of Armenia, 12
Hydrometeorological Institute of Albania, 3
Hydrometeorological Institute of Macedonia, 377
Hydrometeorological Institute of Slovenia, 462
Hyogo, 343, 345, 355

I

IAA
    Institut d'Astronomie et d'Astrophysique (ULB), 59
    Instituto de Astrofísica de Andalucia, 476
    International Academy of Astronautics, 184
    Irish Astronomical Association, 551
    Ithaca Amateur Astronomers, 816
IAAA
    International Association of Astronomical Artists
        Arizona, 597
        Birmingham, 551

European Office, 551
Tucson, 597
United Kingdom, 551
USA-AZ, 597
IAAS
    International Association for Astronomical Studies, 655
IAAT
    Institut für Astronomie und Astrophysik Tübingen, 256
IAB
    Interessengemeinschaft Astrofotografie Bochum, 233
IAC
    Instituto de Astrofísica de Canarias, 477
        Observatorio del Roque de los Muchachos (ORM), 477
IACS
    Institute for Astrophysics and Computational Sciences, 672
IAE
    Instituto de Aeronautica e Espaço
        Observatório Young, 70
IAF
    International Astronautical Federation, 184
IAFE
    Instituto de Astronomía y Física del Espacio, 6, 8
IAG
    Instituto de Astronomía y Geodesia, 478
    International Association of Geodesy, 138, 186
IAGA
    International Association of Geomagnetism and Aeronomy, 186, 655
IAGL
    Institut d'Astrophysique et de Géophysique de Liège, 56, 58
IAHS
    International Association of Hydrological Sciences, 186
IAJ
    Irish Astronomical Journal, 552
IAMAP
    International Association of Meteorology and Atmospheric Physics, 186
IAO
    Indian Astronomical Observatory, 285
IAP
    Institut d'Astrophysique de Paris, 167, 202
IAPG
    Institut für Astronomische und Physikalische Geodäsie, 252
IAPPP
    International Amateur-Professional Photoelectric Photometry, 872
IAPSO
    International Association for the Physical Sciences of the Ocean, 186
IAR
    Institute for Astrophysical Research (Boston University), 749
    Instituto Argentino de Radioastronomía, 8
IAS
    Indiana Astronomical Society, 711
    Institut d'Astrophysique Spatiale, 167
    Institute for Advanced Study, 793
    Internationale Amateur-Sternwarte, 234
    Iowa Academy of Science, 717
    Irish Astronomical Society, 296
IASB
    Institut d'Aéronomie Spatiale de Belgique, 53
IASC
    International Association for Statistical Computing, 393
IASc
    Indian Academy of Sciences, 284
IASFC
    Istituto di Astrofisica Spaziale e Fisica Cosmica, 315
IASI
    Integral Applied Studies Institute
        Astrophysical Observatory, 387
IASL
    Institute of Air and Space Law, 91

IASPEI
    International Association of Seismology and Physics of the Earth's Interior, 186
IAU
    International Association of Universities, 184
    International Astronomical Union
        Central Bureau for Astronomical Telegrams, 749
        Information Bulletin on Variable Stars, 276
        Minor Planet Center (MPC), 755
        Secretariat, 184
        Symposiums, 610
IAU Observatory Codes
    007 (Paris), 192
    008 (Alger/Bouzareah), 4
    009 (Bern), 513
    010 (Caussols/Calern), 192
    012 (Brussels/Uccle), 55, 56
    014 (Leiden), 405
    014 (Marseille), 187
    015 (Sonnenborgh), 405
    016 (Besançon), 191
    017 (Hoher List), 253
    019 (Neuchâtel), 507
    020 (Nice), 191
    021 (Karlsruhe), 216
    022 (Torino/Pino Torinese), 328
    023 (Wiesbaden), 212
    024 (Heidelberg/Königstuhl), 235
    025 (Stuttgart), 245
    026 (Bern/Zimmerwald), 513
    027 (Milano/Brera), 327
    029 (Hamburg/Bergedorf), 254
    030 (Firenze/Arcetri), 326
    031 (Sonneberg), 250
    032 (Jena), 255
    033 (Tautenburg/KSO), 252
    034 (Roma/Monte Mario), 328
    035 (København), 136
    036 (Castel Gandolfo), 926
    037 (Teramo/Colluriana), 328
    038 (Trieste), 328
    039 (Lund), 491
    040 (Dresden/Lohrmann), 251
    041 (Innsbruck), 40
    042 (Potsdam/Babelsberg), 218
    043 (Padova), 339
    044 (Napoli/Capodimonte), 327
    045 (Wien), 41
    046 (České Budějovice/Klet), 129
    047 (Poznań), 431
    048 (Hradec Králové), 128
    049 (Uppsala/Bro/Kvistaberg), 499
    051 (Cape Town/SAAO), 467
    052 (Stockholm/Saltsjöbaden), 494
    053 (Budapest/Konkoly), 275
    055 (Kraków), 428
    056 (Skalnaté Pleso), 459
    057 (Beograd), 454
    059 (Lomnický Štít), 459
    060 (Warszawa/Ostrowik), 432
    063 (Turku/Tuorla), 152
    064 (Kevola), 151
    066 (Athinai), 269
    067 (L'viv/Observatory), 527
    068 (L'viv/Polytechnical), 527
    069 (Baldone), 372
    071 (Smolyan/Rozhen), 75
    073 (București), 438
    075 (Tartu/Tõravere), 144
    080 (Istanbul), 523
    083 (Kyiv/Golosiiv), 529
    084 (St. Petersburg/Pulkovo), 447
    085 (Kyiv), 526
    086 (Odessa), 528
    087 (Cairo/Helwân), 142
    089 (Mikolaiv), 527
    092 (Torun/Piwnice), 432

093 (Skibotn), 418, 420
094 (Crimea/Simeiz), 525
095 (Crimea/Nauchny), 525
096 (Merate/Brera), 326
097 (Tel-Aviv/Wise), 301
098 (Asiago/Monte Ekar), 327
101 (Kharkiv), 526
102 (Zvenigorod), 445
103 (Ljubljana), 463
104 (San Marcello Pistoiese), 334
107 (Cavezzo), 305
108 (Montelupo/San Giuseppe), 333
114 (Kazan/Zelenchukskaya), 442
118 (Modra), 458
119 (Abastumani), 206
120 (Visnjan), 122
121 (Kharkiv/Chuguev), 526
122 (Dourbies/Pises), 191
123 (Byurakan), 12
126 (Monte Vissegi), 307
130 (Lumezzane/Zani), 332
134 (Jena/Großschwabhausen), 255
136 (Kazan/Engelhardt), 442
151 (Eschenberg), 512
152 (Moletai), 374
154 (Cortina d'Ampezzo), 308
155 (Århus/Ole Rømer), 134, 139, 140
156 (Catania), 338
165 (Barcelona/Piera), 476
166 (Úpice), 132
167 (Bülach), 508
175 (St-Luc/Bagnoud), 504
179 (Capolago/Monte Generoso), 504
181 (La Réunion/Makes), 156
185 (Vicques/Jurassien), 510
186 (Kitab), 925
187 (Kórnik/Borowiec), 430
188 (Maidanak), 925
192 (Tashkent/Ulug Beg), 925
193 (Mt. Sanglok), 518
199 (Ris-Orangis/ANSTJ), 198
204 (Varese/Schiaparelli), 335
209 (Asiago/Monte Ekar), 327
210 (Almaty), 364
215 (Buchloe), 212
219 (Japal-Rangapur), 287
220 (Kavalur/VBO), 285
236 (Tomsk), 450
251 (Arecibo), 437
252 (Goldstone), 613
253 (Goldstone), 613
286 (Kunming/Yunnan), 115
290 (Mt. Graham), 926
299 (Lembang/Bosscha), 290
301 (Mont Mégantic), 100, 101
304 (Las Campanas), 103, 107
309 (Paranal), 108, 109
323 (Perth/Bickley), 26
324 (Beijing/Shahe), 113
327 (Beijing/Xinglong), 113
330 (Nanjing/Purple Mountain), 114
344 (Bohyunsan), 367
345 (Sobaeksan), 368
359 (Wakayama/Misato), 351
371 (Okayama/Mount Chikurinji), 354
377 (Kwasan), 350
381 (Kiso), 361
382 (Norikura), 353
388 (Tokyo/Mitaka), 352
413 (Siding Spring), 14, 18
414 (Mount Stromlo), 18
420 (Sydney), 28
463 (Boulder/Sommers-Bausch), 659
467 (Auckland), 408
468 (Campo Catino), 312
473 (Remanzacco), 309
474 (Lake Tekapo/Mount John), 412

476 (Bussoleno/Grange), 332
482 (St Andrews), 581
485 (Wellington/Carter), 409
488 (Newcastle), 580
491 (Yebes), 481
493 (Calar Alto), 475, 476, 481
495 (Altrincham), 531
503 (Cambridge), 575
511 (St.-Michel-l'Observatoire/Haute Provence), 191
513 (Lyon/St.-Genis-Laval), 165
520 (Bonn), 254
521 (Bamberg/Remeis), 254
522 (Strasbourg), 205
525 (Marburg), 262
528 (Göttingen), 256
532 (München), 256
533 (Asiago), 327
535 (Palermo/Vaiana), 327
540 (Linz), 37
544 (Berlin/Wilhelm Foerster), 265
547 (Wrocław), 432
549 (Uppsala), 498
551 (Hurbanovo/SCO), 460
552 (Bologna/San Vittore), 333
553 (Chorzów), 429
554 (Wetzlar/Burgsolms), 213
555 (Kraków/Fort Skala), 428
557 (Ondřejov), 124
559 (Catania/Serra la Nave), 326
561 (Piszkéstető), 275
562 (Wien/Figl), 41
563 (Seewalchen/Gahberg), 34
564 (Herrsching), 219
565 (Bassano Bresciano), 332
566 (Haleakala), 696
568 (Mauna Kea), 694
569 (Helsinki), 151
570 (Vilnius), 375
577 (Basel/Metzerlen), 512
579 (Novi Ligure/Urania), 335
580 (Graz/Lustbühel), 37, 40
581 (Sedgefield), 467
583 (Odessa/Mayaki), 528
584 (St. Petersburg), 449
586 (Pic du Midi), 194
587 (Sormano), 333
589 (Santa Lucia di Stroncone), 332
590 (Basel/Metzerlen), 512
592 (Solingen/WHG), 264
595 (Farra d'Isonzo), 314
598 (Bologna/Loiano), 326
599 (Campo Imperatore), 328
602 (Wien/Urania), 41
604 (Berlin/Archenhold), 209
606 (Norderstedt), 262
610 (Pianoro), 333
611 (Heppenheim/Starkenburg), 246
615 (Saint-Véran), 163
616 (Brno), 126
619 (Sabadell), 472
620 (Mallorca), 480
623 (Liège/Cointe), 58
625 (Kihei/Maui), 692
628 (Mülheim-Ruhr/TSO), 253
629 (Szeged/JATE), 277
630 (Osenbach/SAHR), 200
632 (San Polo a Mosciano), 305
636 (Essen/Walter Hohmann), 208
640 (Senftenberg), 242
643 (Costa Mesa/Anza), 632
645 (Sunspot/Apache Point), 799
649 (Powell), 776
651 (Tucson/Grasslands), 596
654 (Wrightwood/TMO), 613
656 (Victoria/DAO), 94
657 (Victoria/Climenhaga), 104
660 (Berkeley/Leuschner), 646

661 (Priddis/Rothney), 102
662 (Mt. Hamilton/Lick), 649
671 (Temecula/Stony Ridge), 643
672 (Pasadena/Mt. Wilson), 629
673 (Wrightwood/TMO), 613
677 (Lake Arrowhead), 629
679 (San Pedro Mártir), 386
681 (Calgary), 80
686 (Mt. Lemmon), 603
687 (Flagstaff/NAU), 600
688 (Flagstaff/NURO), 600
688 (Lowell/Anderson Mesa), 598
689 (Flagstaff/USNO), 677
690 (Lowell/Flagstaff), 598
691 (Tucson/Steward), 603
692 (Tucson/Steward), 603
694 (Tumanoc Hill), 603
695 (Kitt Peak), 599
696 (Mt. Hopkins/Whipple), 597, 751
697 (Sells/McGraw-Hill/MDM), 598
705 (Sunspot/Apache Point), 799
707 (Denver/Chamberlin), 654, 661
711 (Fort Davis/McDonald), 888
719 (Socorro/Etscorn), 802
729 (Winnipeg/Glenlea), 102
735 (Needville/George), 879
740 (Nacogdoches/SFASU), 885
741 (Northfield/Goodsell), 770
742 (Des Moines/Drake), 717
746 (Mount Pleasant/Brooks), 762
747 (Baton Rouge), 730
748 (Iowa City/Van Allen), 719
750 (Fall Creek/Hobbs), 915, 919
752 (Ellijay/Puckett), 690
753 (Madison/Washburn), 920
754 (Williams Bay/Yerkes), 919
755 (Lowell/Optec), 767
756 (Evanston/Dearborn), 704
758 (Cocoa/BCC-AMPO), 678
760 (Bloomington/Goethe Link), 712
764 (Ellijay/Puckett), 690
765 (Cincinnati), 837
766 (East Lansing/MSU), 766
767 (Ann Arbor), 768
773 (Cleveland/Warner & Swasey), 836
774 (Nassau/Warner & Swasey), 836
776 (Hamilton/Foggy Bottom), 811
778 (Pittsburgh/Allegheny), 862
779 (Toronto/DDO), 103
780 (Charlottesville/Leander McCormick), 904
781 (Quito), 141
784 (Alfred/Stull), 806
786 (Washington/USNO), 677
788 (Greenville/Mount Cuba), 667
791 (Malvern/Flower & Cook), 862
794 (Poughkeepsie/Vassar), 825
797 (New Haven/Yale), 666
798 (Bethany/Yale), 666
801 (Oak Ridge), 751
802 (Cambridge/CfA), 751
806 (Santiago/Cerro Calán), 111
807 (Cerro Tololo), 109
808 (San Juan/El Leoncito), 7, 666
809 (La Silla), 109
811 (Nantucket/MMO), 754
814 (North Scituate/Seagrave), 865
817 (Sudbury), 100
820 (Tarija), 63
821 (Córdoba/Bosque Alegre), 10
825 (Granville/Swasey), 838
829 (San Juan/El Leoncito), 7
831 (Gainesville/Rosemary Hill), 686
833 (Mercedes), 9
834 (Buenos Aires), 6
834 (Córdoba), 10
837 (Jupiter/Palermiti), 683
839 (La Plata), 10
841 (Blacksburg/Martin), 906
842 (Gettysburg), 856
844 (Montevideo/Los Molinos), 587
851 (Halifax/Burke-Gaffney), 98
854 (Tucson/Sabino Canyon), 601
860 (Valinhos), 72
867 (Tottiri/Saji), 356
870 (Campinas), 70
872 (Tokushima/Kainan), 359
874 (Itajubá/Pico dos Dias), 68
880 (Rio de Janeiro), 70
907 (Melbourne), 17
908 (Toyama), 359
911 (Corning/Collins), 813
913 (Montevideo/Kappa Crucis), 587
918 (Quinn/Badlands), 869
920 (Rochester/RIT), 821
923 (St. Davids), 855
927 (Madison/Yanna), 917
932 (New Milford/JJMO), 665
940 (Waterlooville/Clanfield), 547
954 (Teide/Izaña), 58, 235, 256, 257, 477
958 (Dax), 192
961 (Edinburgh/City), 533
965 (Portimão/Algarve), 433
971 (Lisboa), 434
974 (Genova), 340
981 (Armagh), 532
982 (Dublin/Dunsink), 295
983 (San Fernando/ROA), 482
987 (Isle of Man), 552
998 (London/Mill Hill Park), 574
999 (Bordeaux/Floirac), 203
IAVCEI
: International Association of Volcanology and Chemistry of the Earth's Interior, 186
IAYC
: International Astronomical Youth Camp, 53
Ibaraki, 342, 343
Ibaraki University
: Institute for Astrophysics and Planetary Science, 345
IBM, International Business Machines
: T.J. Watson Research Center, 815
IBN
: Institut Belge de Normalisation, 52
IBNORCA
: Instituto Boliviano de Normalización y Calidad, 62
IBVS
: Information Bulletin on Variable Stars, 276
ICAMER
: International Centre of Astronomical, Medical and Ecological Research, 526
Icarus, 816
ICAT
: Institut Canadien d'Astrophysique Théorique, 81
ICCR
: Institute for Cosmic Ray Research, 360
  Akeno Observatory, 360
  Kamioka Underground Observatory, 361
ICCTI
: Instituto para a Cooperação Científica e Tecnológica International (Portugal), 433
Iceland, 279
Icelandic Astronomical Society, 279
Icelandic Council for Standardization (STRÍ), 279
Icelandic Meteorological Office (IMO), 279
Icelandic Physical Society, 279
Icelandic Research Council, 279
Icelandic Standards (IST), 279
ICIST
: Institut Canadien de l'Information Scientifique et Technique, 93
Icon Isotopes, 792
ICONTEC
: Instituto Colombiano de Normas Técnicas, 119
ICP
: Imperial College Press, 549

ICQ
    International Comet Quarterly, 752
ICRA
    International Center for Relativistic Astrophysics, 324
ICSTI
    International Council for Scientific and Technical Information, 185
ICSTM
    Imperial College of Science, Technology and Medicine
        Astrophysics Group, 549
            Physics Department, 549
        Space and Atmospheric Physics Group, 549
ICSU
    International Council of Scientific Unions, 174, 185
        Panel on World Data Centres, 655, 672
ICTP
    International Centre for Theoretical Physics (Abdus Salam_), 302, 336
IDA
    International Dark-Sky Association, 597
Idaho, 697
    (Southern_), 697
Idaho Falls, 697
Idaho Falls Astronomical Society (IFAS), 697
Idaho State University (ISU)
    Department of Physics, 697
IDEM
    Instituto de Hidrología, Meteorología y Estudios Ambientales, 120
IDSC
    INTEGRAL Science Data Centre, 505
IEAUST
    Institution of Engineers, Australia, 23
IEC
    International Electrotechnical Commission, 505
IEE
    Institution of Electrical Engineers, 550
        INSPEC, 550
IERS
    International Earth Rotation Service, 234
IFAS
    Idaho Falls Astronomical Society, 697
IFCAI
    Istituto di Fisica Cosmica ed Applicazioni dell'Informatica, 315
IfE
    Institut für Erdmessung, 254
IFIAS
    International Federation of Institutes for Advanced Study, 89
IFLA
    International Federation of Library Associations and Institutions, 393
IfP
    Institut für Planetologie, Münster, 256
IFS
    Institute for Fusion Studies, 888
    Institute of Fundamental Studies, 487
IFSI
    Istituto di Fisica dello Spazio Interplanetario, 315
IGA
    Instituto de Geofísica y Astronomía, 123
IGAC
    Interessengemeinschaft Astronomie Crimmitschau, 233
IGAM
    Institut für Geophysik, Astrophysik und Meteorologie, 40
IGG Component Technology Ltd., 549
IGN
    Instituto Geográfico Nacional
        Institut de Radioastronomie Millimétrique (IRAM), 182
IGNS
    Institute of Geological and Nuclear Sciences, 410
IGP
    Instituto Geofísico del Perú, 423
IGPP
    Institute of Geophysics and Planetary Physics (LNLL), 625
    Institute of Geophysics and Planetary Physics (UCLA), 648
    Institute of Geophysics and Planetary Physics (UCR), 648
IGRAP
    Institut Gassendi pour la Recherche Astronomique en Provence, 183
IGY
    International Geophysical Year, 673
IHA, Institut Herzberg d'Astrophysique
    Observatoire Fédéral d'Astrophysique, 94
        Canadian Astronomy Data Centre, 93
    Observatoire Fédéral de Radioastrophysique, 94
IIA
    Indian Institute of Astrophysics, 281, 284
IIB
    Institut International de Bibliographie, 392
IISc
    Indian Institute of Science
        Joint Astronomy Programme, 285
IISL
    International Institute of Space Law, 185
IJET
    International Journal of Educational Telecommunications, 896
IJRS
    International Journal of Remote Sensing, 564
IKFIA
    Institute of Cosmophysical Research and Aeronomy of the Russian Academy of Sciences, 446
IKI
    Space Research Institute of the Russian Academy of Sciences, 448
IL
    International Light, 752
ILC Data Device Corp., 816
Île de France, 205
Ilfracombe, 559
Illawara Science Centre Planetarium and Observatory, 32
Illini Space Development Society (ISDS), 701
Illinois, 698–709, 852
    (Northern_), 704
    (Southern_), 698, 706
Illinois Mathematics and Science Academy (IMSA), 701
Illinois State University (ISU)
    Physics Department, 702
    Planetarium, 702, 707
Illinois Wesleyan University (IWU)
    Mark Evans Observatory, 702
    Physics Department, 702
Illkirch-Graffenstaden, 185
Ilminster, 537
ILOC
    International Lunar Occultation Center, 347
Ilsan Station, 370
IMA
    Institute for Mathematics and its Applications (Univ. Minnesota), 773
Image Reduction and Analysis Facility (IRAF), 597, 599
Images, Reflets, Initiation Scientifique (IRIS), 181
IMASTRO Rhône-Alpes, 181
IMAU
    Instituut voor Marien en Atmosferisch Onderzoek Utrecht, 405
IMCCE
    Institut de Mécanique Céleste et de Calcul des Éphémérides, 193
IMCO
    Inter-Governmental Maritime Consultative Organization, 551
IMD
    India Meteorological Department
    New Delhi, 284
    Weather Forecasting, 284
IMGC

Istituto di Metrologia G. Colonnetti, 316
IMO
    Icelandic Meteorological Office, 279
    International Maritime Organization, 551
    International Meteor Organization, 234
Imola, 305
Imperial College of Science, Technology and Medicine (ICSTM)
    Physics Department
        Astrophysics Group, 549
        Space and Atmospheric Physics Group, 549
Imperial College Press (ICP), 549
Impression 5 Science Museum
    Planetarium, 764
IMS
    Industrial Mathematics Society, 764
IMSA
    Illinois Mathematics and Science Academy, 701
IMSS
    Istituto e Museo di Storia della Scienza
        Planetario, 325
IMU
    International Mathematical Union, 68
IN2P3
    Institut National de Physique Nucléaire et de Physique des Particules
        Centre de Physique des Particules de Marseille, 165
        Laboratoire de l'Accélérateur Linéaire, 187
        Laboratoire de Physique Nucléaire des Hautes Énergies, 187
INAF
    Istituto Nazionale di Astrofisica, 325
        Osservatorio Astrofisico di Arcetri, 326
        Osservatorio Astrofisico di Catania, 326
        Osservatorio Astronomico di Bologna, 326
        Osservatorio Astronomico di Brera-Merate, 326
        Osservatorio Astronomico di Brera-Milano, 326
        Osservatorio Astronomico di Cagliari, 327
        Osservatorio Astronomico di Capodimonte, 327
        Osservatorio Astronomico di Padova, 327
        Osservatorio Astronomico di Palermo "Giuseppe S. Vaiana", 327
        Osservatorio Astronomico di Roma - Monte Mario, 328
        Osservatorio Astronomico di Roma - Monteporzio Catone, 328
        Osservatorio Astronomico di Teramo "Vincenzo Cerulli", 328
        Osservatorio Astronomico di Torino (OATo), 328
        Osservatorio Astronomico di Trieste (OAT), 328
        Telescopio Nazionale Galileo, 479
INAMHI
    Instituto Nacional de Meteorología e Hidrología, Ecuador, 141
INAOE
    Instituto Nacional de Astrofísica, Optica y Electronica
        Observatorio Astrofísico Guillermo Haro, 383
        Santa María Tonantzintla, 383
INAPI
    Institut National Algérien de la Propriété Industrielle, 4
INASAN
    Institute for Astronomy of the Russian Academy of Sciences, 445
Incline Village, 787
INDECOPI
    Instituto Nacional de Defensa de la Competencia y de la Protección de la Propiedad Intelectual, 423
Independence High School
    Planetarium, 623
India, 281
India Meteorological Department (IMD)
    New Delhi, 284
    Weather Forecasting, 284
Indiana, 710–715
    (Northern.), 713
Indiana Astronomical Society (IAS) Inc., 711

Indian Academy of Sciences (IASc), 283, 284
Indianapolis, 710, 712–714
Indiana State University
    Department of Geography and Geology
        John C. Hook Memorial Observatory, 712
Indian Astronomical Observatory, 285
Indian Astronomical Observatory (IAO), 285
Indiana University
    Department of Astronomy, 712
Indiana University - Purdue University at Indianapolis (IUPUI)
    Department of Physics, 712
Indian Hills, 659
Indian Institute of Astrophysics (IIA), 281, 284
Indian Institute of Science (IISc)
    Joint Astronomy Programme (JAP), 285
Indianola, 716
Indian River Astronomical Society, 681
Indian Skeptics, 285
Indian Space Research Organization (ISRO), 174, 285
    Bangalore, 285
    Satellite Centre (ISAC), 285
Indonesia, 290
Indonesian Institute of Sciences (LIPI), 290
Indre, 157
Industrial Mathematics Society (IMS), 764
INEN
    Instituto Ecuatoriano de Normalización, 141
Infinity Infocus, 378
INFN
    Istituto Nazionale di Fisica Nucleare
        Astrofisica Solare e Stellare, 329
        Laboratori Nazionali del Gran Sasso, 328
        Laboratori Nazionali di Frascati, 329
        Sezione di Napoli, 329
Information Bulletin on Variable Stars (IBVS), 276
Information Retrieval System (IRS), 318
Infrared Laboratories Inc., 597
Infrared Processing and Analysis Center (IPAC), 612, 623
Infrared Space Observatory (ISO), 476, 477
    US Support Center, 476
Infrared Telescope Facility (IRTF), 693, 695
ING
    Isaac Newton Group (ORM), 479
    Istituto Nazionale di Geofisica, 329
Inglemoor High School Planetarium, 908
Ingolstadt, 213
INIST
    Institut de l'Information Scientifique et Technique, 167
INM
    Instituto Nacional de Meteorología (Spain), 478
INMET
    Instituto Nacional de Meteorologia (Brazil), 67
INN
    Instituto Nacional de Normalización (Chile), 109
Innsbruck, 40
Inönü University
    Physics Department, 523
INPE
    Instituto Nacional de Pesquisas Espaciais
        Cachoeira Paulista, 67
        Centro de Previsão de Tempo e Estudos Climáticos, 67
        São José dos Campos, 67
Inrad Inc., 793
INSA
    Institut National des Sciences Appliquées, Club Astronomie, 183
INSMET
    Instituto de Meteorología (Cuba), 123
INSPEC, 549, 550
INSPIRE
    Interactive NASA Space Physics Ionospheric Research Experiment, 623
Institut Aéronautique et Spatial du Canada, 80, 89
Institut Belge de Normalisation (IBN), 52
Institut Canadien d'Astrophysique Théorique (ICAT), 81, 89

Institut Canadien de l'Information Scientifique et Technique (ICIST), 93
Institut d'Aéronomie Spatiale de Belgique (IASB), 53
Institut d'Astrophysique de Paris (IAP), 167, 182, 202
Institut d'Astrophysique Spatiale (IAS), 167, 182
Institut de France, 154
Institut de l'Information Scientifique et Technique (INIST), 167, 182
Institut de Mécanique Céleste et de Calcul des Éphémérides (IMCCE), 182, 193
Institut de Physique du Globe de Paris (IPGP), 182
Institut de Radioastronomie Millimétrique (IRAM)
    Grenoble, 182
    Observatoire du Plateau de Bure, 182
    Saint-Martin d'Hères, 182
Institut des Sciences de la Terre, de l'Eau et de l'Espace de Montpellier (ISTEEM), 182
    Groupe de Recherche en Astronomie et Astrophysique du Languedoc (GRAAL), 183
Institute Boris Kidric, 455
Institute for Advanced Study (IAS)
    School of Natural Sciences, 793
Institute for Astrophysics and Computational Sciences (IACS), 672
Institute for Cosmic Ray Research (ICCR), 360
    Akeno Observatory, 360
    Kamioka Underground Observatory, 361
Institute for Geodesy, Cartography and Remote Sensing
    Satellite Geodetic Observatory (SGO), 276
Institute for Physical-Technical and Radio-Technical Measurements (VNIIFTRI), 441
Institute for Scientific Information (ISI), 856
Institute for Space and Terrestrial Science (ISTS), 83
Institute for Terrestrial and Planetary Atmospheres (ITPA), 823
Institute of Air and Space Law (IASL), 91
Institute of Fundamental Studies (IFS), 487
Institute of Geodesy and Cartography, 427
Institute of Geological and Nuclear Sciences (IGNS) Inc., 410

Institute of Geophysics and Planetary Physics (IGPP), 625
    Los Angeles, 648
    Riverside, 648
Institute of Meteoritics (IOM), 800, 804
    Meteorite Museum, 801
Institute of Meteorology and Water Management, 427
Institute of Physics, 454
    Stellar Systems Department, 374
Institute of Physics (IOP), 202, 223, 550
Institute of Physics (IOP) Publishing Inc., 857
Institute of Physics (IOP) Publishing Ltd., 550
Institute of Space and Astronautical Science (ISAS), 345, 346
Institute of Standardization (SZS), 455
Institute of Standards and Industrial Research of Iran (ISIRI), 292
Institute of Terrestrial Magnetism, Ionosphere and Radio Wave Propagation (IZMIRAN), 446
Institute of Theoretical Physics and Astronomy, 374, 375
    Moletai Astronomical Observatory (MAO), 374
    Planetarium, 374
Institutet för Rymdfysik (IRF), 490, 495
    Kiruna, 495
    Lund, 495
    Umeå, 495
    Uppsala, 495
Institut für Astronomie und Astrophysik Tübingen (IAAT), 256
Institut für Hochenergiephysik, 224
Institut für Physikalische Weltraumforschung
    Freiburg-im-Breisgau, 233
Institut für Weltraumforschung (IWF), 37
Institut Gassendi pour la Recherche Astronomique en Provence (IGRAP), 183
Institut International de Bibliographie (IIB), 392
Institution of Electrical Engineers (IEE), 550

INSPEC, 550
Institution of Engineers, Australia (IEAUST), 23
Institut National Algérien de la Propriété Industrielle (INAPI), 4
Institut National de Physique Nucléaire et de Physique des Particules (IN2P3)
    Centre de Physique des Particules de Marseille (CPPM), 165, 183
    Laboratoire de l'Accélérateur Linéaire (LAL), 183, 187
    Laboratoire de Physique Nucléaire des Hautes Énergies (LPNHE), 183, 187
Institut National des Sciences Appliquées (INSA), Club Astronomie, 183
Institut National des Sciences de l'Univers (INSU), 183
    Centre d'Analyse des Images (CAI), 183
    Centre de Données astronomiques de Strasbourg (CDS), 165, 184
Instituto Argentino de Racionalización de Materiales (IRAM), 7
Instituto Argentino de Radioastronomía (IAR), 8
Instituto Boliviano de Normalización y Calidad (IBNORCA), 62
Instituto Colombiano de Normas Técnicas (ICONTEC), 119

Instituto Copérnico, 8
Instituto de Aeronautica e Espaço (IAE)
    Observatório Young, 70
Instituto de Astrofísica de Andalucia (IAA), 476, 477
Instituto de Astrofísica de Canarias (IAC), 477
    Observatorio del Roque de los Muchachos (ORM), 477
Instituto de Astronomía y Física del Espacio (IAFE), 6, 8
Instituto de Astronomía y Geodesia (IAG), 478
Instituto de Física de Cantabria, 478
Instituto de Geofísica y Astronomía (IGA), 123
Instituto de Hidrología, Meteorología y Estudios Ambientales (IDEAM), 120
Instituto de Meteorología (INSMET), 123
Instituto de Meteorologia do Portugal, 433
Instituto de Radioastronomía Milimétrica (IRAM), 478
    Observatorio Pico de Veleta, 478
Instituto Ecuatoriano de Normalización (INEN), 141
Instituto Geofísico del Perú (IGP), 423
Instituto Geográfico Nacional (IGN)
    Institut de Radioastronomie Millimétrique (IRAM), 182
Instituto Nacional de Astrofísica, Optica y Electronica (INAOE)
    Observatorio Astrofísico Guillermo Haro (OAGH), 383
    Santa María Tonantzintla, 383
Instituto Nacional de Defensa de la Competencia y de la Protección de la Propiedad Intelectual (INDECOPI), 423
Instituto Nacional de Meteorología e Hidrología (INAMHI), 141
Instituto Nacional de Meteorología (INM), 478
Instituto Nacional de Meteorologia (INMET), 67
Instituto Nacional de Normalización (INN), 109
Instituto Nacional de Pesquisas Espaciais (INPE)
    Cachoeira Paulista, 67
    Centro de Previsão de Tempo e Estudos Climáticos (CPTEC), 67
    São José dos Campos, 67
Instituto Nacional de Técnica Aeroespacial (INTA)
    División de Investigaciones Espaciales, 478
    Laboratorio de Astrofísica Espacial y Física Fundamental (LAEFF), 478
Instituto Nacional de Tecnología y Normalización (INTN), 422
Instituto para a Cooperação Científica e Tecnológica International (ICCTI), 433
Instituto Politécnico Maritimo Pesquero de Pasajes
    Planetario, 479
Instituto Politécnico Nacional (IPN)
    Centro de Investigación y de Estudios Avanzados (CINVESTAV), 383
Instituto Português da Qualidade (IPQ), 434
Instituto Presbiteriano Mackenzie
    Centro de Radio-Astronomia e Aplicações Espaciais (CR-

AAE), 67
Instituto Uruguayo de Normas Técnicas (UNIT), 587
Institut Royal Météorologique de Belgique (IRMB), 53
INSU
    Institut National des Sciences de l'Univers, 183
        Centre d'Analyse des Images, 183
        Centre de Données astronomiques de Strasbourg, 165
INT
    Isaac Newton Telescope, 479
INTA
    Instituto Nacional de Técnica Aeroespacial
        División de Investigaciones Espaciales, 478
        Laboratorio de Astrofísica Espacial y Física Fundamental, 479
Integral Applied Studies Institute (IASI)
    Astrophysical Observatory, 387
INTEGRAL Science Data Centre (ISDC), 505
Interactive NASA Space Physics Ionospheric Research Experiment (INSPIRE), 623
Intercon Spacetec GmbH, 233
Interessengemeinschaft Astrofotografie Bochum (IAB), 233
Interessengemeinschaft Astronomie Crimmitschau (IGAC) eV, 233
Interessengemeinschaft Astronomie Diez und Umgebung eV, 234
Interferometrics Inc., 899
International Academy of Astronautics (IAA), 184
International Amateur-Professional Photoelectric Photometry (IAPPP), 872
International Association for Astronomical Studies (IAAS), 655
International Association for Statistical Computing (IASC), 393
International Association for the Physical Sciences of the Ocean (IAPSO), 186
International Association of Astronomical Artists (IAAA), 597
    Arizona, 597
    Birmingham, 551
    European Office, 551
    Tucson, 597
    United Kingdom, 551
    USA-AZ, 597
International Association of Geodesy (IAG), 138, 186
International Association of Geomagnetism and Aeronomy (IAGA), 186, 655
International Association of Hydrological Sciences (IAHS), 186
International Association of Meteorology and Atmospheric Physics (IAMAP), 186
International Association of Seismology and Physics of the Earth's Interior (IASPEI), 186
International Association of Universities (IAU), 184
International Association of Volcanology and Chemistry of the Earth's Interior (IAVCEI), 186
International Astronautical Federation (IAF), 184
International Astronomical Union (IAU), 184
    Central Bureau for Astronomical Telegrams (CBAT), 749, 752
    Information Bulletin on Variable Stars, 276
    Minor Planet Center (MPC), 752, 755
    Secretariat, 184
    Symposiums, 610
International Astronomical Youth Camps (IAYC) Workshop for Astronomy (IWA), 53
International Business Machines (IBM)
    T.J. Watson Research Center, 815
International Center for Relativistic Astrophysics (ICRA), 324
International Centre for Theoretical Physics (ICTP) (Abdus Salam_), 302, 325
International Centre of Astronomical, Medical and Ecological Research (ICAMER), 526
International Comet Quarterly (ICQ), 752
International Council for Scientific and Technical Information (ICSTI), 185
International Council of Scientific Unions (ICSU), 174, 185

Panel on World Data Centres, 655, 672
International Dark-Sky Association (IDA), 597
Internationale Amateur-Sternwarte (IAS) eV, 234
International Earth Rotation Service (IERS), 234
International Electrotechnical Commission (IEC), 505
Internationale Stiftung Hochalpine Forschungsstationen Jungfraujoch und Gornergrat (HFSJG), 505
International Federation for Information and Documentation, 392, 393
International Federation of Institutes for Advanced Study (IFIAS), 89
International Federation of Library Associations and Institutions (IFLA), 393
International Geophysical Year (IGY), 673
International Information Services for the Physics and Engineering Communities, 550, 551
International Institute of Space Law (IISL), 185
International Light (IL) Inc., 752
International Lunar Occultation Center (ILOC), 347
International Maritime Organization (IMO), 551
International Mathematical Union (IMU), 68
International Meteor Organization (IMO), 234
International Occultation Timing Association (IOTA), 722, 738
    Greenbelt, 738
    Kansas, 722
    Maryland, 738
    Topeka, 722
    USA-KS, 722
    USA-MD, 738
International Organization for Standardization (ISO), 505
International Planetarium Society (IPS), 738
International Research Council (IRC), 185
International School for Advanced Studies (ISAS), 325, 334
International Science Writers Association (ISWA), 752
International Society for Optical Engineering, 908
International Soil Reference and Information Centre (ISRIC), 394
International Space Science Institute (ISSI), 505
International Space University (ISU), 185
International Statistical Institute (ISI), 394
International Supernovae Network (ISN), 325
International Telecommunication Union (ITU), 505
International Telegraph Union, 506
International Thermal Instrument (ITI) Co., 624
International Union of Geodesy and Geophysics (IUGG), 186
International Union of Geological Sciences (IUGS), 36
International Union of Pure and Applied Physics (IUPAP), 186
International Union of Radio Science (URSI), 53
International Union of the History and Philosophy of Science (IUHPS), 53, 186
INTERSPUTNIK, 441
Inter-University Centre for Astronomy and Astrophysics (IUCAA), 285
Intespace, 186
INTN
    Instituto Nacional de Tecnología y Normalización, 422
Inubo Radio Observatory, 343, 345
Invercargill, 411
Inverness, 548
IoA
    Institute of Astronomy
        X-Ray Astronomy Group, 575
Ioannina, 269
Ioffe Physical Technical Institute (A.F._), 441
    Division of Plasma Physics, Atomic Physics, and Theoretical Astrophysics, 444
IOM
    Institute of Meteoritics, 804
        Meteorite Museum, 801
Ionospheric Prediction Service (IPS), 23
IOP
    Institute of Physics, 202, 223, 550
IOP Publishing
    Institute of Physics Publishing, 550, 857

IOP Publishing Inc., 857
IOP Publishing Ltd., 551
IOTA
   International Occultation Timing Association
      Greenbelt, 738
      Kansas, 722
      Maryland, 738
      Topeka, 722
      USA-KS, 722
      USA-MD, 738
Iowa, 716–719
   (Central.), 716
   (Northern.), 717, 720
   (Southeastern.), 719
Iowa Academy of Science (IAS), 717
Iowa Automated Telescope Facilities (IRTF), 719
Iowa City, 719
Iowa State University (ISU)
   Department of Aerospace Engineering and Engineering Mechanics, 717
   Department of Physics and Astronomy, 718
IPAC
   Infrared Processing and Analysis Center, 612
IPG
   Institut für Physikalische Geodäsie, 251
IPGP
   Institut de Physique du Globe de Paris, 182
IPL
   Isotope Products Laboratories, 624
IPM
   Institut Physikalische Meßtechnik, 230
IPN
   Instituto Politécnico Nacional
      Centro de Investigación y de Estudios Avanzados (CINVESTAV), 383
IPQ
   Instituto Português da Qualidade, 434
IPS
   International Planetarium Society, 738
   Ionospheric Prediction Service, 23
   Israel Physical Society, 299
IPS Radio & Space Services, 23
Ipswich, 551, 561
Ipswich School Astronomical Society (ISAS), 551
IRA
   Istituto di Radioastronomia
      Bologna, 316
      Firenze, 316
      Stazione di Medicina, 316
      Stazione di Noto, 316
IRAF
   Image Reduction and Analysis Facility, 599
IRAF Project, 599
IRAM
   Institut de Radioastronomie Millimétrique
      Grenoble, 182
      Observatoire du Plateau de Bure, 182
      Saint-Étienne-en-Devoluy, 182
      Saint-Martin d'Hères, 182
   Instituto Argentino de Racionalización de Materiales, 7
   Instituto de Radioastronomía Milimétrica, 478
      Observatorio Pico de Veleta, 478
Iran, 292
Iraq, 294
Iraqi Meteorological Organization, 294
IRAS
   Istituto Spezzino Ricerche Astronomiche, 329
IRC
   International Research Council, 185
Ireland, 295
   (Northern.), 531–533, 543, 551, 563
IRF
   Institutet för Rymdfysik, 495
      Kiruna, 495
      Lund, 495
      Umeå, 495
      Uppsala, 495
IRFK
   Institutet för Rymdfysik, Kiruna, 495
IRFL
   Institutet för Rymdfysik, Lund, 495
IRFU
   Institutet för Rymdfysik, Uppsala, 495
IRFUm
   Institutet för Rymdfysik, Umeå, 495
IRIMO
   Islamic Republic of Iran Meteorological Organization, 292
IRIS
   Images, Reflets, Initiation Scientifique, 181
Irish Astronomical Association (IAA), 551
Irish Astronomical Journal (IAJ), 552
Irish Astronomical Society (IAS), 296
Irish Meteorological Service, 296, 297
Irish National Committee for Astronomy and Space Research, 296
Irish National Committee for Physics, 296
Irish Science and Technology Agency, 296, 297
Irkutsk, 441, 445, 448
Irkutsk State University
   Astronomical Observatory, 441
IRMB
   Institut Royal Météorologique de Belgique, 53
IRS
   Information Retrieval System, 318
IRSOL
   Istituto Ricerche Solari Locarno, 506
IRTF
   Infrared Telescope Facility, 695
   Iowa Automated Telescope Facilities, 719
Irvine, 627, 647
Isaac Newton Group (ING)
   Observatorio del Roque de los Muchachos (ORM), 479
Isaac Newton Institute for Mathematical Sciences, 552
Isaac Newton Telescope (INT), 479
ISAC
   ISRO Satellite Centre, 285
ISAS
   Institute of Space and Astronautical Science, 345, 346
   Institute of Space and Atmospheric Studies, 103
   International School for Advanced Studies, 334
   Ipswich School Astronomical Society, 551
ISDS
   Illini Space Development Society, 701
ISI
   Institute for Scientific Information, 856
   International Statistical Institute, 394
ISIRI
   Institute of Standards and Industrial Research of Iran, 292
Islamic Republic of Iran Meteorological Organization (IRIMO), 292
Island Planetarium (The.), 552, 572
Islands
   Anglo-Normandes, 106
   Bainbridge, 907
   Baleares, 483, 485
   Canary, 58, 235, 256, 257, 471, 474, 477, 479, 480, 482, 483
   Channel, 106
   Fuerteventura, 471
   Gran Canaria, 471, 474, 477
   Guernsey, 106
   La Palma, 471, 477, 479, 480, 482
   Long, 807
   Madeira, 435
   Mallorca, 480, 485
   Man, 542, 552
   Middle, 817
   Rhode, 864
   Rock, 702, 705
   Staten, 811
   Tenerife, 477, 479, 483

　　　　　Virgin, 801
　　　　　Wallops, 900
　　　　　Wight, 572, 582
Islas Baleares, 483
Islas Canarias, 58, 235, 256, 257, 471, 473, 477, 479, 480, 482, 483
Isle of Man, 542, 552
Isle of Man Astronomical Society, 552
Isle of Wight, 572, 582
ISN
　　International Supernovae Network, 325
ISO
　　Infrared Space Observatory, 476
　　　　US Support Center, 476
　　International Organization for Standardization, 505
Iso-Heikkilä Observatory, 151
Isonzo (Farra d'_), 314
Isotope Products Laboratories (IPL), 624
Ispra, 317
Israel, 299
Israel Academy of Sciences and Humanities, 299
Israel Institute of Technology
　　Asher Space Research Institute (ASRI), 299
Israel Meteorological Service, 299
Israel Physical Society (IPS), 299
ISRIC
　　International Soil Reference and Information Centre, 394
ISRO
　　Indian Space Research Organization, 174
　　　　Bangalore, 285
　　　　Satellite Centre, 285
ISSI
　　International Space Science Institute, 505
ISS Inc., 702
ISS Numbers
　　0000-037x (Journal of Chemical Research - Part M - Miniprint/RSC), 566
　　0001-0782 (Communications of the ACM), 807
　　0001-3765 (Anais da Academia Brasileira de Ciências), 65
　　0001-4141 (Bulletin Classe des Sciences Acad. Royale de Belgique), 47
　　0001-4338 (Izvestiia Russian Academy of Sciences: Atmospheric and Oceanic Physics/AGU), 670
　　0001-4354 (Izvestiia Russian Academy of Sciences: Physics of the Solid Earth/AGU), 670
　　0001-4370 (Oceanology/AGU), 670
　　0001-4842 (Accounts of Chemical Research/ACS), 670
　　0001-5237 (Acta Astronomica), 427
　　0001-5245 (Acta Astronomica Sinica), 112
　　0001-5245 (CAS Bulletin), 115
　　0001-7701 (General Relativity and Gravitation), 395
　　0002-3302 (Bulletin ITA), 445
　　0002-7537 (Bulletin of the American Astronomical Society), 670
　　0002-7863 (Journal of the American Chemical Society/ACS), 670
　　0002-8231 (Journal of the American Society for Information Science), 735
　　0002-9505 (American Journal of Physics), 734
　　0002-9920 (Notices of the American Mathematical Society), 864
　　0003-0503 (American Physical Society Bulletin/AIP), 734
　　0003-1305 (The American Statistician), 896
　　0003-2654 (The Analyst/RSC), 566
　　0003-2689 (Analytical Abstracts/RSC), 566
　　0003-2700 (Analytical Chemistry/ACS), 670
　　0003-6935 (Applied Optics/AIP), 734
　　0003-6951 (Applied Physics Letters/AIP), 734
　　0004-5411 (Journal of the ACM), 807
　　0004-6256 (Astronomical Journal), 670
　　0004-6302 (L'Astronomie/SAF), 199
　　0004-6337 (Astronomische Nachrichten), 212
　　0004-6345 (Astronomisk Tidskrift), 495
　　0004-6361 (Astronomy and Astrophysics), 176

0004-637x (Astrophysical Journal), 670
0004-640x (Astrophysics and Space Science), 395
0004-9506 (Australian Journal of Physics), 23
0004-9525 (Australian Journal of Chemistry), 23
0005-2175 (Aviation Week & Space Technology), 808
0005-6995 (Abhandlungen der Mathematische-naturwissenschaftliche Klasse/BAW), 218
0005-710x (Abhandlungen der Philosophische-historische Klasse/BAW), 218
0006-2960 (Biochemistry/ACS), 670
0007-0297 (Journal of the British Astronomical Association), 538
0007-084x (Journal of the British Interplanetary Society), 538
0007-8328 (CERN Reports), 503
0008-2821 (Canadian Aeronautics and Space Journal), 80
0008-2821 (Journal Aéronautique et Spatial du Canada), 80
0008-4204 (Canadian Journal of Physics), 86, 93
0008-8994 (Centaurus), 134
0009-0700 (Czechoslovak Journal of Physics), 132
0009-2347 (Chemical & Engineering News/ACS), 670
0009-2665 (Chemical Reviews/ACS), 670
0009-2703 (Chemtech/ACS), 670
0009-3068 (Chemistry & Industry/ACS), 670
0009-3106 (Chemistry in Britain/RSC), 566
0009-6709 (Ciel et Terre), 56
0010-4655 (Computer Physics Communications), 391
0010-4884 (Computing Reviews/ACM), 807
0010-9525 (Cosmic Research), 395
0011-3786 (Current Papers in Physics/INSPEC), 550
0011-3891 (Current Science), 283, 284
0011-5517 (SAWB Daily Weather Bulletin), 468
0012-821x (Earth and Planetary Science Letters), 391
0012-9820 (Universo), 63
0013-1350 (Education in Chemistry/RSC), 566
0013-5127 (IEE Review), 551
0013-5194 (Electronics Letters/IEE), 551
0013-936x (Environmental Science & Technology/ACS), 670
0016-7169 (Geofísica Internacional), 386
0016-7932 (Geomagnetism and Aeronomy/AGU), 670
0016-8017 (Geophysical Magazine/Japan Meteo Agency), 348
0016-8025 (Geophysical Prospecting/EAGE), 392
0016-8521 (Geotectonics/AGU), 670
0017-4645 (Großwetterlagen Europas/DWD), 223
0019-1035 (Icarus), 816
0019-8528 (Industrial Mathematics), 764
0020-1669 (Inorganic Chemistry/ACS), 670
0020-2525 (Geofysische Waarnemingen/KMIB), 54
0020-2525 (Observations Géophysiques/IRMB), 53
0020-2533 (Ionosferische Waarnemingen - Kosmische Straling/KMIB), 54
0020-2533 (Observations Ionosphériques - Rayonnement Cosmique/IRMB), 53
0020-255x (IRMB Publications Série A), 53
0020-255x (KMIB Publikaties Reeks A), 54
0020-3319 (Civil Engineering Transactions), 23
0020-6032 (IAU Newsletter), 184
0021-0382 (Ionospheric Data in Japan/CRL), 343
0021-1052 (Irish Astronomical Journal), 552
0021-8286 (Journal for the History of Astronomy), 567
0021-8561 (Journal of Agricultural and Food Chemistry/ACS), 670
0021-8979 (Journal of Applied Physics/AIP), 734
0021-9142 (Journal of the Astronautical Sciences), 895
0021-9568 (Journal of Chemical and Engineering Data/ACS), 670
0021-9584 (Journal of Chemical Education/ACS), 670
0022-1392 (Earth, Planets and Space), 357
0022-2488 (Journal of Mathematical Physics/AIP), 734
0022-2623 (Journal of Medicinal Chemistry/ACS), 670
0022-3262 (Journal of Organic Chemistry/ACS), 670
0022-3654 (Journal of Physical Chemistry/ACS), 670
0022-3670 (Journal of Physical Oceanography/AMS),

747
0022-3727 (Journal of Physics D: Applied Physics), 550
0022-4928 (Journal of Atmospheric Sciences/AMS), 747
0022-4936 (Chemical Communications/RSC), 566
0024-8266 (Monthly Notes of the Astronomical Society of Southern Africa), 464
0024-9297 (Macromolecules/ACS), 670
0025-4738 (Mass Spectrometry Bulletin/RSC), 566
0026-1114 (Meteoritics and Planetary Science), 695
0026-1130 (Meteorological and Geoastrophysical Abstracts/AMS), 747
0026-1173 (CHMI Meteorological Bulletin), 127
0027-0644 (Monthly Weather Review/AMS), 747
0028-0836 (Nature), 558
0028-1050 (Naturwissenschaftliche Rundschau), 239
0029-0181 (Butsuri/JPS), 358
0029-179x (Normalizacja/PKN), 431
0029-7682 (Klimatologische Waarnemingen/KMIB), 54
0029-7682 (Observations Climatologiques/IRMB), 53
0029-7704 (Observatory Magazine), 560
0030-3917 (Optica Pura y Aplicada/SEDO), 484
0030-400x (Optics and Spectroscopy), 675
0030-557x (Orion), 508
0031-9007 (Physical Review Letters/AIP), 734
0031-9015 (Journal of the Physical Society of Japan), 358
0031-9120 (Physics Education), 550
0031-9147 (La Physique au Canada), 80
0031-9147 (Physics in Canada), 80
0031-9155 (Physics in Medicine and Biology), 550
0031-921x (Physics Teacher/AIP), 734
0031-921x (The Physics Teacher), 734
0031-9228 (Physics Today/AIP), 734
0031-9279 (Physikalische Blätter/DPG), 223
0032-0633 (Planetary and Space Science), 391
0033-068x (Progress of Theoretical Physics/Kyoto Univ.), 350
0033-2135 (Review of Geophysics), 431
0034-1223 (La Recherche Aérospatiale/ONERA), 194
0034-2963 (The Reflector), 776
0034-4885 (Reports on Progress in Physics), 550
0035-2160 (Revue des Questions Scientifiques), 57
0035-8711 (Monthly Notices of the Royal Astronomical Society), 565
0035-872x (RASC Journal), 97
0036-8075 (Science Magazine/AAAS), 669
0036-8091 (Optoelectronics/IEE), 551
0036-8091 (Physics Abstracts/INSPEC), 550
0036-8105 (Electrical and Electronics Abstracts/IEE), 551
0036-8113 (Computer and Control Abstracts/IEE), 551
0036-8423 (Science News), 675
0037-6604 (Sky & Telescope), 757
0037-8720 (Memorie della Società Astronomica Italiana), 335
0037-9360 (Bulletin/SFP), 202
0038-0911 (Solar-Geophysical Data/NGDC), 657
0038-0938 (Solar Physics), 395
0038-0946 (Solar System Research), 395
0038-5514 (Journal of Optical Technology), 675
0038-6308 (Space Science Reviews), 395
0038-6340 (Spaceflight/BIS), 538
0039-1263 (Sterne und Weltraum), 246
0039-1271 (Der Sternenbote), 38
0039-2502 (The Journal of the ALPO), 609
0040-1676 (Technology Ireland/FORBAIRT), 296
0040-1706 (Technometrics), 896
0040-5787 (Theoretical Chemical Engineering Abstracts /RSC), 566
0042-0764 (Urania), 430
0042-9767 (World Meteorological Organization Bulletin), 514
0043-1397 (Water Resources Research/AGU), 670
0043-1656 (Weather/RMS), 565
0043-1680 (Webb Society Quarterly Journal), 583
0044-3948 (Zemlja y Vsselennaja), 440
0044-9253 (Revista Astronómica), 6

0044-9814 (Journal of the Astronomical Society of Victoria), 17
0047-2689 (Journal of Physical and Chemical Reference Data/ACS), 670
0047-6773 (Mercury), 610
0048-4024 (Physical Review Abstract/AIP), 734
0048-4024 (Physical Review Abstracts/AIP), 734
0048-6604 (Radio Science/AGU), 670
0049-1640 (Southern Stars), 411
0060-130x (Handbook of the British Astronomical Association), 538
0066-4146 (Annual Review of Astronomy and Astrophysics), 608
0066-4189 (Annual Review of Fluid Mechanics), 608
0066-4200 (Annual Review of Information Science and Technology), 735
0066-5495 (Physical Review Supplements/AIP), 734
0066-9997 (ASA Proceedings), 15
0067-0006 (ASV Astronomical Yearbook), 17
0067-0014 (Astronomische Grundlagen für den Kalender/ARI), 215
0067-0049 (Astrophysical Journal Supplement Series), 670
0067-0367 (SNOI Progress Reports), 100
0068-1725 (UBC Physics Annual Report), 101
0068-4236 (IERS Annual Report), 234
0068-8339 (CAP Annual Report), 80
0069-3030 (Annual Reports on the Progress of Chemistry - Section B: Organic Chemistry/RSC), 566
0072-4432 (Gent Sterrenkundig Observatorium Meddelinge), 59
0074-1809 (IAU Symposia Proceedings), 185
0074-6002 (IFLA Directory), 394
0075-3343 (JARE Data Reports/NIPR), 355
0075-9325 (Lick Observatory Bulletins), 649
0076-8421 (Univ. Michigan Dept. Astronomy Publications), 768
0077-6194 (The Nautical Almanac/USNO), 677
0077-6211 (Nautisches Jahrbuch/BSH), 220
0078-1193 (NGI Publications Series), 417
0078-2246 (Nova Kepleriana - Abhandlungen - Neue Folge/BAW), 218
0078-6322 (Monografías/OAS), 675
0078-6780 (Oslo Institute of Theoretical Astrophysics Reports), 419
0079-1067 (Communications of Perth Observatory), 26
0079-2735 (S+T News/FORBAIRT), 296
0080-1372 (Highlights of Astronomy/IAU), 185
0080-4193 (RASC Observer's Handbook), 97
0080-4614 (Philosophical Transactions A - Mathematical and Physical Sciences), 566
0080-5971 (Efemérides Astronómicas/ROA), 482
0081-0304 (IAU CBAT Circulars), 749
0081-1076 (Annuaire/SFP), 202
0082-5239 (UTIAS Annual Progress Report), 104
0082-5247 (UTIAS Reviews), 104
0082-5255 (UTIAS Reports), 104
0082-5263 (UTIAS Technical Notes), 104
0083-2421 (Astronomical Phenomena/USNO), 677
0083-2448 (Publications of the US Naval Observatory), 677
0084-6090 (Jahrbuch/BAW), 218
0084-6597 (Annual Review of Earth and Planetary Sciences), 608
0084-7518 (ANU RSES Research Papers), 18
0085-3968 (Institute of Meteoritics Special Publications), 804
0090-3213 (The Planetarian), 738
0091-0805 (CASI Special Publications), 740
0091-3286 (Optical Engineering), 908
0091-6338 (Astronomy), 916
0092-640x (Atomic Data and Nuclear Data Table), 808
0094-243x (AIP Conference Proceedings), 734
0094-5846 (Fundamentals of Cosmic Physics), 815
0094-8276 (Geophysical Research Letters/AGU), 670
0095-2338 (Journal of Chemical Information and Computer Sciences/ACS), 670

0095-4403 (Bulletin of the American Society for Information Science), 735
0096-3941 (EOS/AGU), 670
0098-3500 (ACM Trans. Mathematical Software), 807
0098-907x (Optics & Photonics News), 675
0098-9819 (Current Physics Index/AIP), 734
0099-0003 (Physical Review Index/AIP), 734
0101-3440 (Boletim da Sociedade Astronômica Brasileira), 71
0101-935x (Efemérides Astronômicas), 70
0102-9495 (Perspicillum), 68
0103-0019 (Climanálise), 67
0103-0795 (Espacial), 67
0110-4667 (Standards New Zealand), 411
0118-3648 (BPS Directions), 425
0126-0480 (Majalah Lapan), 290
0129-0908 (SISIR News), 456
0132-4624 (Vestnik Sankt-Peterburskogo Universiteta - Series of Mathematics, Mechanics, and Astronomy), 449
0133-249x (Meteor), 276
0135-0420 (Shemakha Circulars), 43
0135-129x (Zvaigžņotá Debess), 372
0135-1303 (Investigations of the Sun and Red Stars), 372
0135-2415 (Circular A/VNIIFTRI), 441
0136-8109 (Transactions of the Saint Petersburg Astronomical Institute), 449
0136-8141 (Transactions of the Saint Petersburg Astronomical Institute), 449
0138-6026 (Ciencias de la Tierra y del Espacio), 123
0138-8118 (Normalización), 123
0140-0118 (Medical and Biological Engineering and Computing/IEE), 551
0142-7253 (Archaeoastronomy), 567
0143-0807 (European Journal of Physics), 550
0145-0670 (NSF Bulletin), 900
0146-7662 (Eyepiece), 806
0146-9592 (Optics Letters), 675
0148-0227 (Journal of Geophysical Research/AGU), 670
0149-1989 (Collected Algorithms Supplements/ACM), 807
0151-0304 (Sciences), 159
0153-9949 (Bulletin de l'AFOEV), 159
0154-4101 (Pulsar/SAP), 201
0155-624x (ANU RSES Annual Report), 18
0160-2500 (Warner and Swasey Observatory Publications), 836
0162-1459 (Journal of the American Statistical Association), 896
0163-0946 (Odyssey), 788
0163-1824 (Physical Review B: Condensed Matter/AIP), 734
0163-2574 (Current Contents - Physical, Chemical and Earth Sciences), 857
0163-8998 (Annual Review of Nuclear and Particle Science), 608
0163-9617 (Amstat News), 896
0164-0925 (ACM Trans. Programming Languages and Systems), 807
0165-0211, 398
0165-0211 (Zenit), 402
0167-6792 (NWO Jaarboek), 397
0167-7764 (Journal of Atmospheric Chemistry/EGS), 227
0167-9295 (Earth, Moon and Planets), 395
0167-9473 (Computational Statistics & Data Analysis), 393
0169-3298 (Surveys in Geophysics/EGS), 227
0170-7434 (Physics Briefs/AIP), 734
0172-0570 (Agrarmeteorologischer Wochenhinweis/DWD), 223
0172-1518 (forschung - Mitteilungen der DFG), 222
0172-1526 (german research - Reports of the DFG), 222
0173-6264 (Luft- und Raumfahrt/DGLR), 223
0178-2770 (Distributed Computing/ACM), 807
0179-0730 (Duisburger Komet), 244
0180-3344 (Revue du Palais de la Découverte), 195
0181-3048 (La Connaissance des Temps), 193
0182-4295 (Annales de la Fondation Louis de Broglie), 179
0185-1101 (Revista Mexicana de Astronomía y Astrofísica), 385
0190-2717 (Bulletin/CrAO), 525
0194-6730 (The Skeptical Inquirer), 812
0195-3982 (Griffith Observer), 622
0199-5197 (Cobblestone), 788
0201-5099 (Astronomy and Geodesy), 450
0201-7342 (Circulars/Astron.-Geod. Soc.), 440
0202-2214 (Calendar of Tartu Observatory), 144
0208-4325 (Journal of the Institute of Meteorology and Water Management Observer), 428
0209-7567 (Publications of the Debrecen Observatory), 275
0210-4105 (Astrum), 472
0210-735x (Almanaque Náutico/ROA), 482
0210-8127 (Fenómenos Astronómicos ROA), 482
0211-4534 (Publicaciones del Observatorio del Ebro - Miscelánea), 486
0211-5166 (Boletín del Observatorio del Ebro - Ionosfera), 486
0212-6036 (ASTER), 470
0213-5892 (Tribuna de Astronomía), 476
0213-6198 (Publicaciones/IAG Madrid), 478
0213-621x (Astronomía, Astrofotografía y Astronautica/SADEYA), 483
0213-862x (Revista Española de Física), 482
0213-893x (Noticias del IAC), 477
0216-9754 (Warta Lapan), 290
0222-5123 (Met Mar), 189
0232-0401 (CHMI Transactions), 127
0234-1069 (Circular E/VNIIFTRI), 441
0235-3431 (Soobshcheniia/Astron.-Geod. Soc.), 440
0238-2091 (Communications/Konkoly), 275
0238-2423 (Publications of the Astronomy Department of the Eötvös Loránd University), 274
0239-622x (Bibliography of Water Management and Engineering/IMGW), 428
0239-6238 (Water Management and Water Protection/IMGW), 428
0239-6246 (Bibliography of Hydrology and Oceanology/IMGW), 428
0239-6254 (Water Engineering/IMGW), 428
0239-6262 (Meteorology/IMGW), 428
0239-6270 (Bibliography of Meteorology/IMGW), 428
0239-6297 (Hydrology and Oceanology/IMGW), 428
0240-8368 (Épiménides Nautiques/SHOM), 199
0240-8376 (Épidécides Lunaires/SHOM), 199
0246-1390 (Revue Montpelliéraine d'Astronomie Languedocienne), 164
0249-7522 (ADION Bulletin), 161
0250-0671 (SAAO Annual Report), 467
0250-1589 (ESA Brochures), 392
0250-4707 (Bulletin of Materials Science/IASc), 284
0250-6335 (Journal of Astrophysics and Astronomy), 284
0253-2379 (Acta Astrophysica Sinica), 112
0253-4126 (Proceedings - Earth and Planetary Sciences /IASc), 284
0253-4134 (Proceedings - Chemical Sciences/IASc), 284
0253-4142 (Proceedings - Earth and Planetary Sciences /IASc), 284
0254-3796 (Report of the Lunping Observatory - Sunspots), 516
0256-2499 (Sadhana - Engineering Science/IASc), 284
0256-596x (Earth Observation Quarterly/ESA), 392
0258-3836 (Vedráttan), 279
0258-7041 (Journal of Shanghai University), 118
0258-9710 (Sri Lanka Journal of Social Science), 487
0259-9163 (Space Horizons), 421
0260-1818 (Annual Reports on the Progress of Chemistry - Section A: Inorganic Chemistry/RSC), 566
0260-9428 (Learned Publishing/ALPSP), 533
0261-0892 (Popular Astronomy), 568

0261-2917 (Laboratory Hazards Bulletin/RSC), 566
0262-6438 (Chemical Engineering Abstracts/RSC), 566
0263-8355 (Rutherford Appleton Laboratory Report Series), 542
0264-3391 (Current Biotechnology Abstracts/RSC), 566
0264-4185 (Circulars of the British Astronomical Association), 538
0264-9381 (Classical and Quantum Gravity), 550
0265-0096 (Electronics Education/IEE), 551
0265-0568 (Natural Products Report/RSC), 566
0265-4245 (Methods in Organic Synthesis/RSC), 566
0265-5721 (Chemical Hazards in Industry/RSC), 566
0266-5611 (Inverse Problems), 550
0267-9272 (BAA Variable Star Section Circulars), 538
0268-1242 (Semiconductor Science and Technology), 550
0268-6961 (IEE Proceedings - Software Engineering), 551
0273-1177 (Advances in Space Research), 391
0273-1177 (Advances in Space Research/COSPAR), 174
0274-7529 (Discover Magazine), 814
0275-1062 (Chinese Astronomy and Astrophysics), 391
0276-1106 (The University of Texas Publications in Astronomy), 887
0276-7333 (Organometallics/ACS), 670
0277-788x (Proceedings of the SPIE), 908
0278-7407 (Tectonics/AGU), 670
0279-4631 (Higher Education Policy/IAU), 184
0280-7173 (Astro/SAAF), 494
0284-169x (Annual Report/IRF), 495
0284-6667 (MARS-Bulletinen), 491
0285-0818 (Bulletin of the Research Institute of Civilization), 358
0285-2853 (Uchuken Hokoku/ISAS), 346
0285-2861 (ISAS News), 346
0285-6808 (ISAS Report), 346
0287-2633 (Report of Hydrographic Observations - Series of Astronomy and Geodesy/JHOD), 347
0288-4232 (Bulletin of the Yamaguchi Museum), 362
0288-4240 (Yamaguchi-ken no shizen - Nature Study of Yamaguchi-ken), 362
0293-0072 (Latitude 5), 168
0293-082x (ESF Communications), 177
0295-5075 (Europhysics Letters), 177
0297-1038 (Procyon), 170
0300-922x (Perkin Transactions I/RSC), 566
0300-9246 (Dalton Transactions/RSC), 566
0300-9254 (Journal of the National Science Foundation of Sri Lanka), 487
0300-9580 (Perkin Transactions II/RSC), 566
0303-4070 (Physics Reports), 391
0303-4070 (Physics Reports/Kumamoto Univ.), 349
0303-805x (ISO Bulletin), 505
0304-2871 (Liste des Publications Scientifiques/CERN), 503
0304-288x (CERN Courier), 503
0304-2901 (CERN Annual Report), 503
0304-291x (Rapport Annuel du CERN), 503
0304-4289 (Pramana - Journal of Physics), 284
0304-9523 (Bulletin of the Astronomical Society of India), 281
0305-4470 (Journal of Physics A: Mathematical and General), 550
0306-0012 (Chemical Society Reviews/RSC), 566
0308-0684 (IEE News), 551
0308-2342 (Journal of Chemical Research - Part S - Synopsis/RSC), 566
0308-2350 (Journal of Chemical Research - Part M - Microfiche/RSC), 566
0309-0108 (ROAE Annual Report), 562
0320-930x (Astronomicheskij Vestnik), 440
0320-9318 (Bulletin of the Special Astrophysical Observatory), 448
0321-1762 (Izvestiia/Engelhardt Astron. Obs.), 442
0323-8180 (Mitteilungen des Lohrmann-Observatoriums), 251
0323-9918 (Bulgarian Geophysical Journal), 75
0324-1114 (Geodesy), 75

0324-6329 (Időjárás), 276
0332-5962 (Nytt om Romfart), 415
0332-6909 (Norwegian Academy of Science and Letters Yearbook), 416
0340-0352 (IFLA Journal), 394
0340-4366 (PTB Jahresbericht), 240
0340-4552 (Promet - meteorologische Fortbildung/DWD), 223
0340-4781 (Osnabrücker Naturwissenschaftliche Mitteilungen), 238
0340-7586 (Sitzungsberichte der Mathematische-naturwissenschaftliche Klasse/BAW), 218
0341-2970 (Europäischer Wetterbericht/DWD), 223
0341-4183 (Bibliothek Forschung und Praxis/IFLA), 394
0341-7727 (MPG Spiegel), 236
0341-7778 (MPG Berichte und Mitteilungen), 236
0342-5991 (Sitzungsberichte der Philosophische-historische Klasse/BAW), 218
0347-5719 (Documenta/Royal Swedish Academy of Sciences), 493
0349-2710 (EISCAT Annual Report), 489
0349-4217 (Lund Observatory Reports), 491
0350-3283 (Belgrade Univ. Dept. of Astronomy Publications), 454
0351-2657 (Hvar Observatory Bulletin), 122
0351-6652 (Presek), 462
0352-7859 (Sveske Fizičkih Nauka), 454
0354-9291 (Sveske Fizičkih Nauka Series A: Conferences), 454
0355-0826 (Sodankylä Geophysical Observatory Publications), 146
0355-9289 (Report/University of Helsinki Observatory), 151
0355-9459 (Tähdet 19xx), 152
0355-9467 (Tähdet ja Avaruus), 152
0356-021x (Avaruusluotain), 150
0356-1089 (SFS-tiedotus), 147
0360-0300 (Computing Surveys/ACM), 807
0362-5915 (ACM Trans. Database Systems), 807
0365-0936 (Mémoires Classe des Sciences Acad. Royale de Belgique in 8°), 47
0365-0952 (Mémoires Classe des Sciences Acad. Royale de Belgique in 4°), 47
0366-5690 (CERN-HERA Reports), 503
0366-757x (CODATA Bulletin), 174
0367-8466 (Izvestiia/CrAO), 525
0367-8466 (Izvestiia/Pulkovo Obs.), 447
0370-8691 (Soobshcheniia), 12
0371-6791 (Trudy/Sternberg State Astron. Inst.), 450
0371-6813 (Bulletin/Tokyo Gakugei University - Section IV: Mathematics and Natural Sciences), 359
0371-8247 (Trudy/Kazan), 442
0373-3602 (Report of Hydrographic Researches/JHOD), 347
0373-3629 (Annales Hydrographiques/SHOM), 199
0373-3696 (Japanese Ephemeris/JHOD), 347
0373-3734 (Belgrade Astronomical Observatory Bulletin), 454
0373-3742 (Belgrade Astronomical Observatory Publications), 454
0373-4854 (Journal of Magnetic Observations), 420
0373-7675 (Ulugh Beg Astronomical Institute Circulars), 925
0373-7683 (Peremennye Zvezdy), 445
0373-8280 (Hvězdářská Ročenka), 130
0373-9139 (Ciel et Espace), 159
0374-0676 (Information Bulletin on Variable Stars), 276
0374-1583 (Abhandlungen aus der Hamburger Sternwarte), 254
0374-1958 (Mitteilungen der Astronomischen Gesellschaft), 211
0374-2288 (Courrier CERN), 503
0374-2466 (Publications of the Astronomical Society of Japan), 342
0374-6402 (Bulletin de La Murithienne), 511
0375-6644 (Abastumanskaya Astrofizicheskaya Observatorija Byulleten), 206

0376-4265 (ESA Bulletin), 392
0379-6566 (ESA Special Publications), 392
0380-1314 (Consensus/SCC), 99
0380-1322 (Consensus/CCN), 99
0386-2194 (Proceedings of the Japan Academy A - Mathematical Sciences), 358
0386-2208 (Proceedings of the Japan Academy B - Physical and Biological Sciences), 358
0387-5857 (Standard Frequency and Time Service Bulletin/CRL), 343
0388-0036 (Transactions of the Japan Academy), 358
0388-0230 (Contributions/Kyoto University/Dept. Astronomy), 350
0388-2349 (Contributions of the Kwasan and Hida Observatories), 350
0388-5607 (Scientific Reports of the Tohoku University, Eighth Series: Physics and Astronomy), 358
0388-5852 (NMS Astronomical Circular), 355
0389-0341 (Star Friends/NMS), 355
0389-3057 (Memoirs of the Faculty of Education/Kagawa University - Part II), 348
0389-8229 (Catalog of Data in World Data Center C2 for Ionosphere), 362
0389-8237 (Ionospheric Data at Syowa Station-Antarctica /CRL), 343
0390-1106 (Il Giornale di Astronomia/SAIt), 335
0392-2308 (Astronomia UAI), 337
0392-3932 (Giornale dell'AAB), 304
0392-6737 (Il Nuovo Cimento/SIF), 335
0393-697x (La Rivista del Nuovo Cimento/SIF), 335
0399-4864 (GIFAS Letter), 181
0399-9939 (Pégase), 189
0400-8456 (The Air Almanac/USNO), 677
0405-5497 (BAV-Rundbrief), 220
0433-8251 (DWD Jahresbericht), 223
0435-7965 (Monatlicher Witterungsbericht/DWD), 223
0446-5059 (Seismological Bulletin/Japan Meteo Agency), 348
0453-9478 (Kiruna Geophysical Data), 495
0458-2128 (Communications from University of London Observatory), 574
0465-7926 (Journal of Nanjing Univ./Natural Sciences Edition), 117
0473-7466 (Obzornik za Matematiko in Fiziko), 462
0493-2285 (Tianwen Aihaozhe/Astronomical Amateur), 115
0497-137x (ITU News), 506
0499-9320 (CASI Technical Memorandums), 740
0499-9339 (CASI Technical Notes), 740
0506-4295 (Vasiona), 455
0515-6831 (Agrarmeteorologische Bibliographie/DWD), 223
0516-9518 (AAVSO Bulletin), 747
0524-7780 (Annuaire de l'ORB), 56
0524-7780 (KSB Jaarboek), 55
0524-7780 (Rayonnement Solaire/IRMB), 53
0524-7780 (Zonnestraling/KMIB), 54
0531-2728 (EAGE Yearbook), 392
0531-4283 (CERN School of Physics Proceedings), 503
0531-4496 (ESO Annual Report), 228
0531-7479 (Europhysics News), 177
0538-6918 (CODATA Newsletter), 174
0556-2791 (Physical Review A: General Physics/AIP), 734
0556-2813 (Physical Review C: Nuclear Physics/AIP), 734
0556-2821 (Physical Review D: Particle and Fields/AIP), 734
0557-1069 (HUT Annual Report), 148
0563-8038 (Contributions/Department of Astronomy/ University of Tokyo), 360
0568-6016 (Trudy ITA), 445
0568-6199 (Komety y Meteory), 518
0568-6865 (Bulletin of Tajik Academy of Sciences, Institute of Astrophysics), 518
0571-205x (Artificial Satellites), 430
0571-7191 (ASSA Handbook), 464

0571-7248 (Astrophysical Journal Letters), 670
0571-7256 (Astrophysics), 395
0588-6228 (Sessions des Comités Consultatifs/BIPM), 164
0670-1826 (Annual Reports on the Progress of Chemistry - Section C: Physical Chemistry/RSC), 566
0705-3797 (Episodes/IUGS), 36
0721-0094 (Sonne/VdS), 258
0721-8168 (Sternzeit), 213, 214, 216, 217, 238, 248, 249, 257–259, 262, 264
0722-2912 (Elektronorm/DIN), 224
0722-6691 (The Messenger/ESO), 228
0723-1121 (Telescopium), 259
0724-7125 (Deutsches Meteorologisches Jahrbuch/DWD), 223
0728-5833 (AAO Newsletter), 14
0728-6554 (AAO Annual Report), 14
0728-6554 (Annual Report of Anglo-Australian Telescope Board), 14
0730-0301 (ACM Trans. Graphics), 807
0733-8724 (Journal of Lightware Technology), 675
0734-2071 (ACM Trans. Multimedia Systems), 807
0734-2101 (Journal of Vacuum Science and Technology A), 808
0734-211x (Journal of Vacuum Science and Technology B), 808
0735-0015 (Journal of Business and Economic Statistics), 896
0736-3680 (Planetary Report), 634
0736-6884 (MPC Circulars), 755
0736-6922 (International Comet Quarterly), 752
0737-6421 (The Astronomical Almanac/USNO), 677
0737-6766 (Occultation Newsletter), 722, 738
0739-0572 (Journal of Atmospheric and Oceanic Technology/AMS), 747
0740-3224 (Journal of the Optical Society of America B: Optical Physics), 675
0740-3232 (Journal of the Optical Society of America A: Optics, Image Science and Vision), 675
0740-722x (Aerospace America), 895
0741-3335 (Plasma Physics and Controlled Fusion), 550
0742-450x (Fernbank Magazine), 688
0743-7463 (Langmuir/ACS), 670
0748-5492 (Issues in Science and Technology), 673
0749-1387 (Faces), 788
0758-234x (Les Cahiers Clairaut), 173
0763-8992 (Canadian Journal of Remote Sensing), 80
0764-048x (SPOT Magazine), 202
0764-4442 (Comptes-Rendus de l'Académie des Sciences - Série I), 154
0764-4450 (Comptes-Rendus de l'Académie des Sciences - Série II), 154
0764-9614 (Publications Spéciales du CDS), 165
0769-0878 (Observations et Travaux/SAF), 199
0769-1025 (Configurations des Huit Premiers Satellites de Saturne/BDL), 193
0769-1033 (Phénomènes et Configurations des Satellites Galiléens de Jupiter/BDL), 193
0769-1041 (Éphémérides des Satellites Faibles de Jupiter et de Saturne/BDL), 193
0770-0164 (Observations de l'Ozone/IRMB), 53
0770-0164 (Ozon Waarnemingen/KMIB), 54
0770-0261 (IRMB Miscellanea A), 53
0770-0261 (KMIB Miscellanea A), 54
0770-0318 (IRMB Miscellanea B), 53
0770-0318 (KMIB Miscellanea B), 54
0770-3244 (NATO Scientific Publications), 55
0770-4569 (Aardmagnetisme/KMIB), 54
0770-4569 (Magnétisme Terrestre/IRMB), 53
0770-4615 (IRMB Publications Série B), 53
0770-4615 (KMIB Publikaties Reeks B), 54
0770-7371 (IRMB Rapport Annuel), 53
0772-3288 (KMIB Jaarverslag), 54
0772-4330 (IRMB Hors Série), 53
0772-4330 (KMIB Buiten Reeks), 54
0772-4349 (Documentation Météorologique/IRMB), 53
0772-4349 (Meteorologische Documentatie/KMIB), 54

0772-4357 (IRMB Miscellanea C), 53
0772-4357 (KMIB Miscellanea C), 54
0772-6422 (Heelal), 60
0774-5834 (Nouvelles Brèves/Comité PARA), 50
0775-0315 (OBAFGKM), 49
0780-7945 (Ursa Minor), 152
0781-1466 (Valkoinen Kääpiö/White Dwarf), 148
0781-8062 (Astronomy & Space Magazine), 295
0785-5672 (Radiantti), 150
0788-5474 (Photogrammetric Journal of Finland), 148
0789-6719 (Informo/Tuorla Observatory), 152
0794-7976 (Nigerian Academy of Sciences Proceedings), 414
0797-0072 (Revista/CIDA-E), 586
0797-8812 (The Deep-Sky Observer/CONICYT), 586
0825-9984 (Annuaire Astronomique/SAM), 99
0843-8978 (Astro-Notes/SAM), 99
0858-2637 (Tang Chang Pueg), 519
0858-4648 (TISI Newsletter), 519
0861-1270 (Astronomical Calendar), 75
0862-173x (Contr. Brno N. Copernicus Observatory and Planetarium), 126
0864-0645 (Datos Astronómicos para Cuba), 123
0866-2851 (Hungarian Astronomical Association Yearbook), 276
0870-2856 (Anais Obs. Astron. Coimbra), 435
0871-1542 (Publicações/Obs. Astron. Prof. Manuel de Barros), 436
0871-1550 (Informações/Obs. Astron. Prof. Manuel de Barros), 436
0872-2884 (OPÇÃO/IPQ), 434
0882-8156 (Weather and Forecasting/AMS), 747
0883-8305 (Paleoceanography/AGU), 670
0884-2114 (Journal of Materials Research), 858
0886-6236 (Global Biochemical Cycles/AGU), 670
0886-6961 (IAPPP Communications), 872
0887-0624 (Energy & Fuels/ACS), 670
0888-2257 (Air & Space), 676
0888-5850 (SPOTLight), 904
0888-5885 (Industrial & Engineering Chemistry Research /ACS), 670
0890-216x (ALPO Solar System Ephemeris), 609
0893-228x (Chemical Research in Toxicology/ACS), 670
0893-7694 (MRS Bulletin), 858
0894-1866 (Computers in Physics), 736
0894-8755 (Journal of Climate/AMS), 747
0894-8763 (Journal of Applied Meteorology/AMS), 747
0896-3207 (TUGboat), 850
0897-2532 (The Shallow Sky Bulletin), 837
0897-4756 (Chemistry of Materials/ACS), 670
0898-8242 (SSI Update), 797
0905-2410 (Dansk Rumfart), 136
0905-8958 (Aktuel Astronomi), 140
0911-5501 (NRO Report), 353
0911-5870 (NRO Newsletter), 353
0914-5753 (Report of Hydrographic Observations - Series of Astronomy and Geodesy/JHOD), 347
0914-9260 (Journal of the Communications Research Laboratory), 343
0915-0021 (Reprint of the NAOJ), 352
0915-3640 (Publications of the NAOJ), 352
0915-3780 (Technical Reports/Mizusawa Kansoku Ctr Natl Astron. Obs.), 353
0915-5392 (Annual Report/Kiso Observatory), 361
0915-6321 (Report of the NAOJ), 352
0915-6410 (NAOJ Annual Report), 354
0916-6343 (Annual Report/Mizusawa Astrogeodynamics Obs. - Time Service and Geo physical Observations), 353
0917-1185 (NIFS Annual Report), 354
0917-6926 (Annual Report of the Nishi-Harima Astronomical Observatory), 355
0919-8296 (ICRR Report), 360
0922-6435 (Experimental Astronomy), 395
0923-2958 (Celestial Mechanics and Dynamical Astronomy), 395
0925-8566 (Radiant), 391

0927-6505 (Astroparticle Physics), 391
0927-880x (Research Reports from the Netherlands/NWO), 397
0928-6403 (Forschungsnachrichten aus den Niederlanden/NWO), 397
0928-6411 (Bulletin de Recherche des Pays-Bas/NWO), 397
0928-6640 (NWO Onderzoekberichten), 397
0930-4916 (Polaris), 208
0930-7575 (Climate Dynamics/EGS), 227
0934-4438 (Astronomische Gesellschaft Abstract Series), 211
0934-8220 (New Sunspot Indices Bulletin/VdS), 258
0935-4956 (Astronomy and Astrophysics Reviews), 246
0936-0252 (Plasma Sources Science and Technology), 550
0936-5818 (Wetterkarte/DWD), 223
0936-9244 (Skeptiker/GWUP), 232
0937-0420 (DLR-Nachrichten), 225
0938-0205 (Regiomontanusbote), 239
0938-846x (TUM/IAPG Jahresbericht), 252
0941-1445 (Reviews in Modern Astronomy/AG), 211
0941-2352 (Sirius), 258
0942-4962 (Multimedia Systems/ACM), 807
0944-1999 (Mitteilungen zur Astronomiegeschichte), 211
0947-8787 (MPE Jahresbericht), 237
0948-4418 (Astronomie & Philatelie), 238
0949-7714 (Journal of Geodesy), 138
0950-0340 (Journal of Modern Optics), 572, 862
0950-1711 (Natural Products Updates/RSC), 566
0950-3366 (Power Engineering Journa/IEE), 551
0951-7715 (Nonlinearity), 550
0952-326x (Gnomon), 532
0952-4746 (Journal of Radiological Protection), 550
0953-2048 (Superconductor Science and Technology, 550
0953-4075 (Journal of Physics B: Atomic, Molecular and Optical Physics), 550
0953-8100 (Observing Section Reports/Webb Society), 583
0953-8585 (Physics World), 550
0953-8984 (Journal of Physics C: Condensed Matter), 550
0954-0083 (High-Performance Polymers), 550
0954-027x (Journal of Hard Materials), 550
0954-0695 (Electronics and Communication Engineering Journal/IEE), 551
0954-3889 (Journal of Physics G: Nuclear Physics), 550
0954-898x (Network: Computation in Neural Systems), 550
0954-8998 (Quantum Optics), 550
0955-419x (Geophysical Journal International/RAS), 565
0956-3385 (Computing and Control Engineering Journal/IEE), 551
0956-5000 (Faraday Transactions/RSC), 566
0956-9944 (Manufacturing Engineer/IEE), 551
0957-0233 (Measurement Science and Technology), 550
0957-4484 (Nanotechnology), 550
0959-5228 (The Skeptic), 573
0959-7171 (Waves in Random Media), 550
0959-9428 (Journal of Materials Chemistry/RSC), 566
0960-1317 (Journal of Micromechanics and Microengineering), 550
0960-7919 (Engineering Management Journal/IEE), 551
0963-2700 (Spectrum), 562
0963-2700 (The UKIRT Newsletter), 693
0963-6625 (Public Understanding of Science), 550
0963-7346 (Engineering Science and Education Journal/IEE), 551
0963-9659 (Pure and Applied Optics), 550
0964-1726 (Smart Materials and Structures), 550
0965-0393 (Modelling and Simulation in Materials Science and Engineering), 550
0966-9809 (OLE - Opto and Laser Europe), 550
0967-1846 (Distributed Systems Engineering), 550
0967-3334 (Physiological Measurements), 550
0967-6139 (The Deep-Sky Observer/Webb Society), 583
0968-3356 (Academia Europaea Newsletter), 531

0970-2628 (Standard India/BIS), 282
0970-3985 (Standards Monthly Additions/BIS), 282
0971-1600 (Journal of Spacecraft Technology/ISRO), 285
0971-8044 (Resonance/IASc), 284
0981-6410 (Pégase), 156
0989-5876 (Rapport Annuel/SHOM), 199
0990-2862 (Astro News/Maison de l'Astronomie), 189
0992-7689 (Annales Geophysicae), 155
1000-3681 (Acta Astronomica Sinica), 114
1000-8349 (Progress in Astronomy), 117
1001-1544 (Publications of the Shaanxi Astronomical Observatory), 114
1001-1811 (Time and Frequency Bulletin), 114
1001-7526 (Publications of Yunnan Observatory), 115
1009-9271 (Chinese Journal of Astronomy and Astrophysics), 116
1011-6257 (Science International/ICSU), 185
1013-3410 (Image/EUMETSAT), 227
1013-9044 (Reaching for the Skies/ESA), 392
1015-5295 (Journal of Astronomy and Physics/Istanbul), 523
1015-8081 (IMU Bulletin), 68
1016-3115 (WGN/IMO), 234
1016-6114 (Annual Report of the Time Section/BIPM), 164
1017-9909 (Journal of Electronic Imaging), 908
1018-3051 (ESO-MIDAS Courier), 228
1018-9580 (ICSTI Forum), 185
1019-4568 (IERS Technical Notes), 234
1019-584x (Nojum Magazine), 293
1019-8695 (Report of the Lunping Observatory - Geomagnetism), 516
1020-7007 (ICTP Annual Report), 302
1021-8823 (Mitteilungen der Rudolf Wolf Gesellschaft), 508
1022-0038 (Wireless Networks/ACM), 807
1023-5809 (Nonlinear Processes in Geophysics/EGS), 227
1024-1787 (Fulcrum), 929
1024-4530 (Radio Science Bulletin), 53
1025-921x (Boletín de SOMETCUBA), 123
1031-508x (Nova Search), 30
1032-4739 (Space Frontier News), 29
1032-5999 (Earth Sciences at ANU), 18
1035-4751 (Univ. Tasmania Dept. Physics Research Report), 31
1035-932x (Sky & Space), 27
1036-3831 (The Physicist), 18
1037-5759 (CSIRO Space Industry News), 22
1039-3048 (The Sydney Sky Guide), 28
1041-102x (Ad Astra/NSS), 674
1042-0851 (Announcer/AAPT), 734
1043-1802 (Bioconjugate Chemistry/ACS), 670
1046-1914 (Solar Indices Bulletin/NGDC), 657
1046-8188 (ACM Trans. Information Systems), 807
1048-6879 (Optical Engineering Reports), 908
1048-8820 (Quantum), 901
1048-8841 (GPS World), 835
1049-331x (ACM Trans. Software Engineering and Methodology), 807
1050-7086 (Calliope), 788
1055-5269 (Surface Science Spectra), 808
1055-6796 (Astronomical and Astrophysical Transactions/EAAS), 441
1060-1848 (Earth Space Review), 856
1061-8600 (Journal of Computational and Graphical Statistics), 896
1062-7987 (European Review/Academia Europaea), 531
1063-6692 (ACM Trans. Networking), 807
1064-2358 (Reports of the National Center for Science Education), 630
1064-6965 (IRLIST Digest/ACM), 807
1064-9409 (access), 703
1065-3597 (EUVE Electronic Newsletter), 646
1067-9936 (Multimedia Systems/ACM), 807
1072-5520 (interactions/ACM), 807

1073-0516 (ACM Trans. Computer-Human Interactions), 807
1074-2697 (Amateur Telescope Making Journal), 907
1074-3197 (Compendium/NASS), 663
1074-8059 (Digital Compendium/NASS), 663
1075-6485 (EUVE Science Bulletin), 646
1079-3135 (Technical Support/NaSPA), 917
1084-4309 (ACM Trans. Design Automation of Electronic Systems), 807
1084-6654 (Journal of Experimental Algorithmics/ACM), 807
1085-7117 (Journal of Agricultural, Biological, and Environmental Statistics), 896
1086-7651 (Journal of Graphic Tools/ACM), 807
1091-3556 (netWorker/ACM), 807
1094-9224 (ACM Trans. Information and System Security), 807
1101-1718 (Aurora), 489
1120-6330 (Rendiconti Lincei - Matematica e Applicazioni), 302
1120-6349 (Rendiconti Lincei - Scienze Fisiche e Naturali), 302
1121-3094 (Rendiconti Lincei - Supplemento), 302
1130-0892 (Publicaciones del Observatorio Astronómico Ramón María Aller), 485
1130-491x (La Alternativa Racional), 474
1132-2306 (Boletín de Astronomía/Aranzadi), 483
1135-223x (Boletín Informativo del GAS), 476
1137-9111 (Meteors), 483
1149-168x (Infos GIFO), 181
1154-4317 (Annuaire/SFO), 202
1161-3289 (SPOT Flash), 202
1161-3297 (SPOT Flash), 202
1164-5873 (Sciences de la Terre), 154
1168-1195 (Le Point Astro), 156
1170-7372 (AANZ Journal), 408
1173-2245 (Meteorite), 410
1173-5392 (Carter Observatory Annual Report), 409
1173-8138 (RASNZ Newsletter), 411
1173-8812 (Carter Observatory Newsletter), 409
1183-5362 (Astronomie-Québec), 88
1210-0463 (Time and Latitude), 124
1210-5168 (Romanian Astronomical Journal), 438
1211-2453 (Kvart), 132
1217-7725 (Space Activities in Hungary), 276
1223-1118 (Romanian Journal of Meteorology), 438
1223-1126 (Romanian Journal of Hydrology and Water Resources), 438
1225-052x (Journal of Astronomy and Space Sciences), 369
1230-8242 (Biuletyn Informacyjny/PKN), 431
1235-1083 (Ceres), 151
1238-0091 (Tähtiharrastustietoa), 146
1238-9137 (Finnish Academy Year Book), 146
1243-4833 (Le Planétarium/Montpellier), 196
1243-8219 (Astronomie en Touraine et Centre-Ouest), 162
1244-9148 (Special'IST), 168
1256-7406 (Astrorama), 195
1266-7390 (Capella), 160
1270-9638 (Aerospace, Science and Technology), 225
1279-7782 (Mémoires de l'ISTEEM), 183
1283-3339 (Astronomie Passion), 199
1283-9817 (CNES Magazine), 168
1290-0958 (Air and Space Europe), 154
1310-3571 (Andromeda), 74
1311-3879 (Telescop), 74
1318-0614 (Astronomical Ephemeris/Ljubljana), 463
1323-3580 (PASA), 15, 23
1323-6326 (ATNF News), 21
1330-2620 (Acta Astronomica Visnianiensis), 122
1330-4410 (Nebeske Krijesnice), 122
1331-5765 (foton), 121
1341-7282 (Report of Lunar Occultation Observations/JHOD), 347
1343-4284 (Antarctic Meteorite Research/NIPR), 355
1344-3194 (Polar Geoscience/NIPR), 355

1344-3437 (Polar Meteorology and Glaciology/NIPR), 355
1344-6231 (Polar Bioscience/NIPR), 355
1345-1065 (Advances in Polar Upper Atmospheric Research/NIPR), 355
1350-2344 (IEE Proceedings - Science, Measurement and Technology), 551
1350-2352 (IEE Proceedings - Electric Power Applications), 551
1350-2360 (IEE Proceedings - Generation, Transmission and Distribution), 551
1350-2379 (IEE Proceedings - Control Theory and Applications), 551
1350-2387 (IEE Proceedings - Computers and Digital Techniques), 551
1350-2395 (Radar, Sonar and Navigation/IEE), 551
1350-2409 (IEE Proceedings - Circuits, Devices and Systems), 551
1350-2417 (Microwaves, Antennas and Propagation/IEE), 551
1350-2425 (IEE Proceedings - Communications), 551
1350-2433 (Optoelectronics/IEE), 551
1350-245x (IEE Proceedings - Vision, Image and Signal Processing), 551
1351-5497 (The JCMT Newsletter), 693
1352-3325 (Royal Society of Edinburgh News), 566
1359-5865 (CCLRC Annual Report), 542
1359-7337 (Analytical Communications/RSC), 566
1361-4126 (FAS Newsletter), 546
1366-8781 (Astronomy & Geophysics/RAS), 565
1368-6534 (Culture and Cosmos), 544
1381-5652 (Hypothese/NWO), 397
1383-469x (Mobile Networks and Applications/ACM), 807
1384-1076 (New Astronomy), 391
1387-6473 (New Astronomy Reviews), 391
1388-3828 (Space Debris), 395
1392-0049 (Baltic Astronomy), 374
1392-1932 (Lithuanian Journal of Physics), 375
1394-8636 (Semesta/BAKSA), 379
1403-3542 (carpe scientiam/NFR), 496
1406-0574 (Annual Report/Estonian Physical Society), 144
1431-8059 (Innovation/Zeiss), 221
1432-0754 (Astronomy and Astrophysics Reviews), 246
1433-8351 (Living Reviews in Relativity), 237
1435-0424 (Meteoros/AKM), 208
1440-2807 (Journal of Astronomical History and Heritage), 14
1455-9579 (Metsähovi Radio Observatory Report), 148
1455-9587 (Metsähovi Publications on Radio Science), 148
1462-3854 (Academia Europaea Directory), 531
1466-8017 (BUFORA Bulletin), 539
1510-091x (Canopus), 586
1531-1074 (Astrobiology), 818
1556-6558 (Qualité Espace/CNES), 168
1575-3476 (Boletín SEA), 484
1594-1299 (Coelum Astronomia), 317
1629-4475 (Photoniques/SFO), 202
1680-8096 (ISO Management Systems), 505
3025-0733 (Dinámica), 7
8750-0183 (Earthwatch Magazine), 750
8750-9350 (AAS Newsletter), 670
8755-1209 (Review of Geophysics/AGU), 670
8756-7938 (Biotechnology Progress/ACS), 670
Istanbul, 522–524
Istanbul University
    Astronomy and Space Sciences Department, 523
    Physics Department, 524
ISTEEM
    Institut des Sciences de la Terre, de l'Eau et de l'Espace de Montpellier, 182
        Groupe de Recherche en Astronomie et Astrophysique du Languedoc (GRAAL), 183
Istituto di Astrofisica Spaziale e Fisica Cosmica (IASFC), 315, 325
Istituto di Fisica Cosmica ed Applicazioni dell'Informatica (IFCAI), 315, 325
Istituto di Fisica dello Spazio Interplanetario (IFSI), 315, 325
Istituto di Metrologia G. Colonnetti (IMGC), 316, 325
Istituto di Radioastronomia (IRA), 316, 325
    Firenze, 316
Istituto di Tecnologie e Studio delle Radiazioni Extraterrestri (TESRE), 317, 325
Istituto e Museo di Storia della Scienza (IMSS)
    Planetario, 325
Istituto Nazionale di Astrofisica (INAF), 325
    Osservatorio Astrofisico di Arcetri (OAA), 325
    Osservatorio Astrofisico di Catania (OAC), 326
    Osservatorio Astronomico di Bologna (OAB), 326
    Osservatorio Astronomico di Brera (OAB)
        Merate, 326
        Milano, 326
    Osservatorio Astronomico di Cagliari (OAC), 327
    Osservatorio Astronomico di Capodimonte (OAC), 327
    Osservatorio Astronomico di Padova, 327
    Osservatorio Astronomico di Palermo (OAP) "Giuseppe S. Vaiana", 327
    Osservatorio Astronomico di Roma (OAR)
        Monte Mario, 327
        Monteporzio Catone, 328
    Osservatorio Astronomico di Teramo "Vincenzo Cerulli", 328
    Osservatorio Astronomico di Torino (OATo), 328
    Osservatorio Astronomico di Trieste (OAT), 328
    Telescopio Nazionale Galileo (TNG), 479
Istituto Nazionale di Fisica Nucleare (INFN)
    Laboratori Nazionali del Gran Sasso (LNGS), 328
    Laboratori Nazionali di Frascati (LNF), 329
    Sezione di Napoli
        Astrofisica Solare e Stellare, 329
Istituto Nazionale di Geofisica (ING), 329
Istituto Ricerche Solari Locarno (IRSOL), 506
Istituto Spezzino Ricerche Astronomiche (IRAS), 329
Istituto Tecnico Nautico "Artiglio"
    Planetario, 329
Istituto Tecnico Nautico Statale
    Planetario, 329
Istituto Universitario Navale
    Cattedra di Astronomia Nautica, 329
ISTS
    Institute for Space and Terrestrial Science, 83
ISU
    Idaho State University
        Department of Physics, 697
    Illinois State University
        Physics Department, 702
        Planetarium, 702
    International Space University, 185
    Iowa State University
        Department of Aerospace Engineering and Engineering Mechanics, 717
        Department of Physics and Astronomy, 718
ISWA
    International Science Writers Association, 752
ITA
    Institute for Theoretical Astronomy, 445
    Institut für Theoretische Astrophysik - Heidelberg, 254
Itabashi Science and Education Center Planetarium, 345
Itajubá, 68
Italy, 302
Itancourt, 155
Itapetinga, 68
Itchen, 569
Ithaca, 812, 814, 816, 818, 819, 821
Ithaca Amateur Astronomers (IAA), 816
ITI
    International Thermal Instrument, 624
ITPA
    Institute for Terrestrial and Planetary Atmospheres, 823
ITSS

Information Technology and Scientific Services (Raytheon), 742
ITU
International Telecommunication Union, 505
IUCAA
Inter-University Centre for Astronomy and Astrophysics, 285
IUGG
International Union of Geodesy and Geophysics, 186
IUGS
International Union of Geological Sciences, 36
IUHPS
International Union of the History and Philosophy of Science, 53, 186
IUPAP
International Union of Pure and Applied Physics, 186
IUPUI
Indiana University - Purdue University at Indianapolis, 712
IWA
IAYC Workshop for Astronomy, 53
Iwate, 352, 353
IWF
Institut für Weltraumforschung (ÖAW), 37
IWU
Illinois Wesleyan University
Mark Evans Observatory, 702
Izaña, 58, 235, 256
Izhevsk, 450
Izieux, 198
Izmir, 523
IZMIRAN, 441
Institute of Terrestrial Magnetism, Ionosphere and Radio Wave Propagation, 446

J

JAA
Jackson Astronomical Association, 774
Jablunkov, 125
JAC
Joint Astronomy Centre, 693
JACARA
Joint Australian Centre for Astrophysical Research in Antarctica, 30
Jackson, 761, 774, 775, 872, 922
Jackson Astronomical Association (JAA), 774
Jackson (Hole) Astronomy Club, 922
Jacksonville, 682
Jacobus Kapteyn Telescope (JKT), 479
Jadiriyah, 294
Jaén, 473
Jagiellonian University, 428
JAIPA
Japan Association for Information Processing in Astronomy, 346
Jakarta, 290, 291
Jakutsk, 446
Jameco Electronics, 624
James and Barbara Moore Observatory, 681
Jamesburg, 792
James Clerk Maxwell Telescope (JCMT), 693
James C. Veen Observatory, 764
James Lipp Observatory, 636
James Lockyer Planetarium, 552, 559
James Madison University (JMU)
John C. Wells Planetarium, 899
Physics Department, 899
James Schwartz Observatory, 631
Jamestown, 818, 840
James Wesley Droke Observatory, 606
Jamieson Science and Engineering Inc., 650
Janesville, 919
Janos Technology Inc., 893
Jan Paagman Sterrenwacht (JPS), 394
Janus (Club_), 172

JAP
Joint Astronomy Programme (IISc), 285
Japal-Rangapur Observatory, 287
Japan, 342
Japan Academy (The_), 346, 358
Japan Aerospace Exploration Agency (JAXA), 346
Japan Association for Information Processing in Astronomy (JAIPA), 346
Japanese Dark-Sky Association (JDA), 346
Japanese Industrial Standards Committee (JISC), 346
Japanese National Committee for Astronomy, 357
Japan Hydrographic and Oceanographic Department (JHOD)
Bisei Hydrographic Observatory, 347
Geodesy and Geophysics Division, 347
Shimosato Hydrographic Observatory (SHO), 347
Shirahama Hydrographic Observatory, 347
Japan Information Center of Science and Technology (JICST), 347, 348
Japan Meteorological Agency (JMA), 347
Japan Physical Society (JPS), 348, 358
Japan Planetarium Society, 348
Japan Science and Technology Corporation (JST), 348
Japon, 342
JAS
Junior Astronomical Society, 568
JASS
Journal of Astronomy and Space Sciences, 369
JATE University
Astronomical Observatory, 277
Department of Physics, 277
Jävan Station, 491
Javornik Astronomical Society, 462
JAXA
Japan Aerospace Exploration Agency, 346
JBO
Jodrell Bank Observatory, 579
JBSPO
John Bryan State Park Observatory, 839
JCA
Joint Center for Astrophysics, 744
JCCE
Journal of Computing in Childhood Education, 896
JCMS
Johnson County Middle School
Rocket and Astronomy Club, 726
JCMST
Journal of Computers in Mathematics and Science Teaching, 896
JCMT
James Clerk Maxwell Telescope, 693
JDA
Japanese Dark-Sky Association, 346
Jeddah, 453
Jefferson City, 776
Jefferson County, 776
Jefferson High School for Science and Technology Planetarium (Thomas_), 899, 904
Jefferson High School Planetarium, 712
JEMH
Journal of Educational Multimedia and Hypermedia, 896
Jena, 211, 220, 234, 255, 263, 266
Jenny Jump State Park, 791
Jenoptik AG, 234
Jensen Planetarium, 783
Jensen Tools Inc., 598
Jeremiah Horrocks Observatory, 563, 575
Jersey City, 793
Jerusalem, 299
Jeseník, 128
Jeseník School Observatory, 128
Jesse Besser Museum and Sky Theatre, 764
Jeszenkowitsch Gesellschaft (Dr. Adalbert_), 35, 36
Jet Propulsion Laboratory (JPL), 21, 612, 624
Deep Space Tracking Systems Group, 612
Table Mountain Observatory (TMO), 613

US Space VLBI Project, 613
Jewel and Esk Valley College Planetarium, 552
Jewett Observatory, 910
JHA
    Journal for the History of Astronomy, 567
JHOD
    Japan Hydrographic and Oceanographic Department
        Bisei Hydrographic Observatory, 347
        Geodesy and Geophysics Division, 347
        Shimosato Hydrographic Observatory, 347
        Shirahama Hydrographic Observatory, 347
JHU
    Johns Hopkins University
        Center for Astrophysical Sciences, 739
        Department of Earth and Planetary Sciences, 738
        Department of Physics and Astronomy, 738, 739
        Maryland Space Grant Consortium, 739
        Press, 739
JHU Press, 738
Jiang Su, 116
Jicamarca, 423
Jicamarca Radio Observatory (JRO), 423
JICST
    Japan Information Center of Science and Technology, 348
J.I. Holcomb Observatory and Planetarium (JIHOP), 710, 712
JIHOP
    J.I. Holcomb Observatory and Planetarium, 710
JILA
    Joint Institute for Laboratory Astrophysics, 655
Jilin, 113
JILR
    Journal of Interactive Learning Research, 896
Jim Cherry Memorial Planetarium, 688
Jim's Mobile Inc. (JMI), 655
Jindřichuv Hradec, 128
Jindřichuv Hradec Public Observatory, 128
JISC
    Japanese Industrial Standards Committee, 346
JISM
    Jordanian Institution for Standards and Metrology, 363
JIVE
    Joint Institute for VLBI in Europe, 394
Jizerou, 125
JJC
    Joliet Junior College
        Herbert Trackman Planetarium, 701
JJMO
    John J. McCarthy Observatory (JJMO), 665
JKT
    Jacobus Kapteyn Telescope, 479
JMA
    Japan Meteorological Agency, 347
J.M. McDonald Planetarium, 782
JMU
    James Madison University
        John C. Wells Planetarium, 899
        Physics Department, 899
Jodrell Bank Observatory (JBO), 552, 579
Jodrell Bank Science Centre
    Planetarium and Arboretum, 552
Joensuu, 150
Joerger Enterprises Inc., 816
Johannesburg, 464, 469
Johannes-Kepler-Sternwarte, 37
Johannes-Kepler-Volkssternwarte, 39
Johann Palisa Observatory and Planetarium, 128
Johann-Wolfgang-Goethe-Universität Frankfurt/Main, 234
John Bryan State Park Observatory (JBSPO), 839
John C. Hook Memorial Observatory, 712
John C. Wells Planetarium, 899
John Deere Planetarium, 702, 705
John Drescher Planetarium, 624, 638
John F. Kennedy Space Center (KSC), 681
    Visitor Complex, 682
John H. Glenn Research Center (GRC), 838, 840

    Visitor Center, 840
John H. Witte Observatory, 719
John J. McCarthy Observatory (JJMO), 665
John K. Cauthen Educational Media Center, 866, 867
John (Mount_), 412
Johns Hopkins University (JHU)
    Department of Earth and Planetary Sciences, 738
    Department of Physics and Astronomy, 738
        Center for Astrophysical Sciences (CAS), 738
    Maryland Space Grant Consortium (MSGC), 739
    Press, 739
Johns Hopkins University Press, 739
Johnson City, 872
Johnson County Middle School (JCMS)
    Rocket and Astronomy Club, 726
Johnson Space Center Astronomical Society (JSCAS), 880
Johnson Space Center (JSC) (Lyndon B._), 880
    Astronomical Society, 880
    Public Affairs Office, 881
Johnson-Weil Observatory, 805
John W.H. Simpson Observatory, 791
John Young Planetarium, 683
Joint Astronomy Centre (JAC), 693
Joint Australian Centre for Astrophysical Research in Antarctica (JACARA), 30
Joint Center for Astrophysics (JCA), 744
Joint Institute for Laboratory Astrophysics (JILA), 655
Joint Institute for VLBI in Europe (JIVE), 394
Joint Organization for Solar Observations (JOSO), 235
Joint Scientific and Educational Center (JSEC), 441
Jokioinen, 147
Joliet, 701
Joliet Junior College (JJC)
    Herbert Trackman Planetarium, 701
Jolimont, 201
Jones Observatory (Clarence T._), 875
Jones Planetarium (Clarence T._), 875
Jongerenvereniging voor Sterrenkunde (JvS), 54
Jonquière, 85
Jonsdorf, 261
Jordan, 363
Jordanian Amateur Astronomers Society, 363
Jordanian Astronomical Society (JAS), 363
Jordanian Institution for Standards and Metrology (JISM), 363
Jordan Meteorological Department, 363
Jordan Planetarium and Observatory (Maynard F._), 732, 733
Jorge Polman Observatory, 66
Joseph A. Foran High School
    Planetarium, 663
Joseph Grady Planetarium, 858
Joseph Henry Laboratories, 793
    Physics Department, 795
Joseph H. Rogers Observatory, 764, 766
JOSO
    Joint Organization for Solar Observations, 235
Joyce Memorial Observatory, 409
Józefoslaw University of Technology, 428
József Attila University
    Department of Physics
        Astronomical Observatory, 277
JPL, Jet Propulsion Laboratory, 612
    Deep Space Tracking Systems Group, 612
    Table Mountain Observatory, 613
    US Space VLBI Project, 613
JPS
    Jan Paagman Sterrenwacht, 394
    Japan Physical Society, 358
JPSJ
    Journal of the Physical Society of Japan, 358
JRA Aerospace & Technology Ltd., 552
J.R. Grundy Observatory, 855
JRO
    Jicamarca Radio Observatory, 423
JSC
    Lyndon B. Johnson Space Center

Astronomical Society, 880
Public Affairs Office, 881
JSCAS
Johnson Space Center Astronomical Society, 880
JSE
Jamieson Science and Engineering, 650
Journal of Scientific Exploration, 903
JSEC
Joint Scientific and Educational Center, 441
JST
Japan Science and Technology Corporation, 348
JTATE
Journal of Technology and Teacher Education, 896
Judson Broadman Coit Memorial Observatory, 749
Juneau, 592
Jungfraujoch, 505, 507
Junior Astronomical Society (JAS), 568
Junior Museum Planetarium, 816
Junnar, 286
Junnar Taluk, 286
Jupiter, 683
Jura, 510
Justice Planetarium, 722
Jutland
(Eastern.), 134
Jutz Volkssternwarte Berg eV (Christian.), 235
JvS
Jongeren Vereniging voor Sterrenkunde, 54
J. Willard Gibbs Research Laboratory, 666
JYP
John Young Planetarium, 683
Jyväskylä, 148
Jyväskylän Sirius ry, 148

K

Kabardino-Balkaria, 444
KAC
Krasnoyarsk Astronomical Club, 442
KACST
King Abdulaziz City for Science and Technology, 453
KAF
Karlskrona Astronomi Förening, 490
Københavns Astronomiske Forening, 135
Kagawa, 348
Kagawa University
Department of Astronomy and Earth Sciences, 348
Kagoshima, 344, 351
Kagoshima Space Center, 346
Kailua, 694
Kailua-Kona, 694
KAIST
Korea Advanced Institute of Science and Technology
Department of Physics, 367
Space Science Laboratory, 367
Kaivopuisto Observatory, 152
Kalamazoo, 764, 765
Kalamazoo Astronomical Society (KAS), 764
Kalamazoo Public Museum, 764
Kalamazoo Valley Museum (KVM)
Universe Theater and Planetarium, 765
Kalinenkov Astronomical Observatory, 526, 527
Kalmbach Publishing Co., 916
Kalocsa, 274
Kamioka Underground Observatory, 348, 360, 361
Kamuela, 692
Kanagawa, 358
Kanagawa University
Department of Physics, 348
Kanawha Valley Astronomical Society (KVAS), 912
Kandilli Observatory and Earthquake Research Institute, 522, 523
Kandy, 487
Kane, 861
Kanev Planetarium (Paul.), 854
Kankakee, 706

Kanobili (Mount.), 206
Kansai Advanced Research Center (KARC), 343, 348
Kansas, 721-724
(Northeast.), 723
Kansas Astronomical Observers (KAO), 722
Kansas Astrophotographers and Observers Society (KAOS), 722
Kansas City, 722, 776, 777
Kansas City Museum, 777
Kansas Cosmosphere and Space Center (KCSC), 722
Kansas University
Department of Physics and Astronomy, 721
KAO
Kansas Astronomical Observers, 722
Korea Astronomy Observatory, 367
Bohyunsan Optical Astronomy Observatory, 367
Daeduk Radio Astronomy Observatory, 368
Sobaeksan Optical Astronomy Observatory, 367
Taeduk Radio Astronomy Observatory, 368
KAOS
Kansas Astrophotographers and Observers Society, 722
KAP
Kluwer Academic Publishers
Boston, 752
Dordrecht, 395
Massachusetts, 752
Netherlands, 395
Norwell, 752
USA-MA, 752
Kapteyn Instituut, 394
Kapteyn Sterrenwacht - StarTel BV, 395
Karachai-Cherkesia, 448
Karachi, 421
Karadag, 528
Karaj, 292
Karaklidere, 523
KARC
Kansai Advanced Research Center, 343
Kardjali, 74
Kariya, 342
Karl-Franzens-Universität Graz, 36
Sonnenobservatorium Kanzelhöhe, 40
Karl Lambrecht Corp. (KLC), 702
Karl-Marx-Stadt, 244
Karlovy Vary, 128
Karlovy Vary City Observatory, 128
Karl-Schwarzschild-Observatorium (KSO), 235, 252
Karlskrona, 490
Karlskrona Astronomi Förening (KAF), 490
Karlskrona Astronomy Association, 490
Karlsruhe, 216, 229, 245
Kärnten, 34
KAS
Kalamazoo Astronomical Society, 764
Kern Astronomical Society, 624
Kopernik Astronomical Society, 817
Korean Astronomical Society, 368
Kashima Space Research Center (KSRC), 343, 344, 348
Kassel, 213, 238
Katholieke Universiteit Leuven, 54
Center for Plasma Astrophysics, 59
Instituut voor Sterrenkunde, 59
Katholieke Universiteit Nijmegen, 395
Katlenburg-Lindau, 227, 236
Katsushika City Museum
Planetarium, 349
Katziveli, 525
Kavalur, 285
Kayseri, 522
Kayser-Threde GmbH, 235
Kazakhstan, 364
Kazan, 442
Kazan State University
Department of Astronomy, 442
Engelhardt Astronomical Observatory, 442
KCSC
Kansas Cosmosphere and Space Center, 722

KDAAS
    Kettering and District Amateur Astronomers Society, 559
Kearney, 784
KEBS
    Kenya Bureau of Standards, 365
Keck Observatory (W.M._), 692, 693
Kecskemét, 277
Kecskemét Planetarium, 277
Keeble Observatory, 899, 902
Keele, 577
Keene Amateur Astronomers, Inc., 789
Kelce Planetarium (L. Russell_), 723
Keldysh Institute of Applied Mathematics, 442, 446
Kelly College
    Science Department, 572
Kelly Planetarium, 829
Kelowna, 92
Kendall Hyde Ltd., 553
Kenmore, 908
Kennebunk, 732
Kennedy Space Center (KSC) (John F._), 681
    Visitor Complex, 682
Kenner, 729, 730
Kenner Historical Museum
    Planetarium, 729
Kenneth Mees Observatory (C.E._), 825
Kenneth Mees Solar Observatory (C.E._), 693, 696
Kennon Observatory, 775
Kensington, 30, 88
Kent, 577, 838
Kent State University (KSU)
    Physics Department
        Planetarium, 838
Kentucky, 725–728
    (Eastern_), 725
    (Western_), 726, 728
Kentucky Space Grant Consortium (KSGC), 726
Kenya, 365
Kenya Bureau of Standards (KEBS), 365
Kenya Meteorological Department (KMD), 365
Kenya National Academy of Sciences (KNAS), 365
Kenyon College
    Department of Physics, 839
Kepler-Sternwarte (Johannes_), 37
Kepler-Volkssternwarte (Johannes-_), 39
Kermanshah, 292
Kern, 624
Kern Astronomical Society (KAS), 624
Kernen Farm, 103
Kettering and District Amateur Astronomers Society (KDAAS), 553, 559
Kevola Observatory, 151
Kew, 564
Kew Astronomical Society (RKAS) (Richmond and_), 564
Keynsham, 562
Keyport, 908
Keystone College
    Astronomy Observatory, 857
Keystone Oaks High School
    Planetarium, 857
Keyworth, 568
Khadjeh Nassir Addin Toussi Observatory, 293
Khan Scope Centre, 89
Kharkiv, 526, 528
Kharkiv State University
    Astronomical Observatory, 526
Kharkiv Y.A. Gagarin Planetarium, 526
Kharkov, 526, 528
Kharkov Amateur Astronomers' Society (KAAS), 526
Khurel Togót Observatory, 388
Kidderminster, 556, 584
Kiel, 241, 242, 255
Kieler Planetarium, 241
Kiepenheuer-Institut für Sonnenphysik (KIS), 235
Kiev, 526–529
Kihei, 692

Kilohani Planetarium, 692, 694
Kilwinning, 535
Kimberton, 854
King Abdulaziz City for Science and Technology (KACST), 453
King Abdulaziz University
    Astronomy and Geophysics Research Institute (AGRI), 453
King College
    Burke Observatory, 872
Kingman, 593
Kingman Museum of Natural History, 763
King Saud University (KSU)
    College of Science
        Astronomy Department, 453
Kings Bay Astronomy Club, 689
Kingsland, 689
Kingsport, 870, 871
Kingston, 97, 100
Kingswood, 32
Kingwood, 882
Kinohi Institute, 624
Kirchhain, 261, 262
Kirkland, 840
Kirkpatrick Planetarium, 846
Kirkpatrick Science and Air Space Museum at Omniplex
    Kirkpatrick Planetarium, 846
Kirkwood Observatory, 712
Kirov, 442
Kirov Planetarium, 442
Kirtland AFB, 799
Kiruna, 489, 495–497
KIS
    Kiepenheuer-Institut für Sonnenphysik, 235
Kishinev, 387
Kiskunhalas, 277
Kiskunhalas Popular Observatory, 277
Kislovodsk, 447
Kislovodsk Astronomical Station, 442, 447
Kiso, 351
Kiso Observatory, 349, 361
Kitab, 925
Kitab Latitude Station, 925
Kitt Peak, 801
Kitt Peak National Observatory (KPNO), 598, 599
Kitt Peak Station, Warner and Swasey Observatory, 836
KJS
    Kenya Journal of Sciences, 365
Klagenfurt, 34, 39
KLC
    Karl Lambrecht Corp., 702
Klet Observatory, 128
Kluwer Academic Publishers (KAP), 395
    Boston Office, 752
KMD
    Kenya Meteorological Department, 365
KMIB
    Koninklijk Meteorologisch Instituut van België, 54
Knaresborough, 547
KNAS
    Kenya National Academy of Sciences, 365
KNAW
    Koninklijke Nederlandse Akademie van Wetenschappen, 395
Knighton, 573
KNITQ
    Korean National Institute of Technology and Quality, 368
KNMI
    Koninklijke Nederlands Meteorologisch Instituut, 395
Knockin, 579
Knollwood Books, 916
Knoxville, 870, 872, 874–876
Knoxville Observers (KO), 872
KNU
    Kyungpook National University
        Department of Astronomy and Atmospheric Sciences,

369
KO
    Knoxville Observers, 872
Kobau (Mount-), 92
Kobe, 349
København, 135-140
Københavns Astronomiske Forening (KAF), 135, 138
Kobe University
    Department of Earth and Planetary Sciences, 349
Kocaeli, 523
Koch Planetarium, 711
Kodaikanal Observatory, 285
Kohl Observatory, 817
Kolkata, 282, 286, 287
Köln, 209, 225, 228, 238, 250, 255, 261
Kölner Observatorium für Submillimeter und Millimeter Astronomie (KOSMA), 255
Koninklijke Nederlandse Akademie van Wetenschappen (KNAW), 395
Koninklijke Nederlands Meteorologisch Instituut (KNMI), 395
Koninklijke Sterrenwacht van België (KSB), 54
Koninklijke Vlaamse Academie van België voor Wetenschappen en Kunsten, 54
Koninklijk Meteorologisch Instituut van België (KMIB), 54
Koninklijk Sterrenkundig Genootschap van Antwerpen (KSGA), 54
Konkoly Observatory, 275-277
Konus Italia Group Srl, 330
Kopernik Astronomical Society (KAS), 817
Kopernik Space Education Center (KSEC), 817
Korea
    (DPRK), 366
    (ROK), 367
Korea Advanced Institute of Science and Technology (KAIST)

    Department of Physics
        Space Science Laboratory, 367
Korea Astronomy Observatory (KAO), 367
    Bohyunsan Optical Astronomy Observatory (BOAO), 367
    Daeduk Radio Astronomy Observatory (DRAO), 368
    Sobaeksan Optical Astronomy Observatory (SOAO), 367
    Taeduk Radio Astronomy Observatory (TRAO), 368
Korean Astronomical Society (KAS), 368
Korean Meteorological Administration, 368
Korean National Institute of Technology and Quality (KNITQ), 368
Korean Physical Society (KPS), 368
Korean Space Science Society (KSSS), 368
Kórnik, 430
Koshka (Mount-), 525
KOSMA
    Kölner Observatorium für Submillimeter und Millimeter Astronomie, 255
Kosmorama Rymdteater, 489
Kot Astro, 55
Kötcse, 275
Kötzting, 220
Kountze Planetarium, 785
Kourou, 168
Kourov, 451
Kowa Optimed Inc., 624
Kowloon, 116
KPNO
    Kitt Peak National Observatory, 599
K. Price Bryan Planetarium, 739
KPS
    Korean Physical Society, 368
Kraków, 428, 430
Kraków Jagiellonian University
    Astronomical Observatory, 428
Kraków Pedagogical University
    Department of Physics, Mount Suhora Observatory, 428
Krasnoyarsk, 442

Krasnoyarsk Astronomical Club (KAC), 442
Kraví Hora, 129
Kredarica, 462
Krefeld, 258
Kreuzbergl, 34
Kreuzlingen, 502
Kroller-Sternwarte (Franz-_), 35, 36
Kronshagen, 249
Krugersdorp, 466
Kryonerion Station, 269
Kryzhanovska, 528
KSB
    Koninklijk Sterrenwacht van België, 54
KSC
    Kennedy Space Center
        Visitor Complex, 682
KSEC
    Kopernik Space Education Center, 817
KSGA
    Koninklijk Sterrenkundig Genootschap van Antwerpen, 54
KSGC
    Kentucky Space Grant Consortium, 726
KSO
    Karl-Schwarzschild-Observatorium, 252
KSRC
    Kashima Space Research Center, 343
KSSS
    Korean Space Science Society, 368
KSU
    Kent State University
        Physics Department, 838
        Planetarium, 838
    King Saud University
        Astronomy Department, 453
KSW
    Kuffner-Sternwarte, 36
Kuala Lumpur, 378
Kuffner-Sternwarte (KSW), 36
Kula, 695
Kula Maui, 696
Kumamoto, 349
Kumamoto University
    Department of Physics, 349
Kunming, 115
Kuopio, 149
Kuopion Tähtitieteellinen Seura Saturnus, 148, 149
Kurri Kurri, 16
Kutina, 121
KVAS
    Kanawha Valley Astronomical Society, 912
Kvissleby, 494
Kvistaberg Station, 490, 498
KVM
    Kalamazoo Valley Museum, 765
Kwa Dlangezwa, 467
Kwasan and Hida Observatories, 350
Kwasan Observatory, 349, 350
Kyiv, 526-529
Kyiv University
    Astronomical Observatory, 526
    Astronomy and Space Physics Department, 527
Kylmälä, 148
Kyoto, 349, 350, 357
Kyoto Sangyo University
    Department of Physics, 349
Kyoto University
    Data Analysis Center for Geomagnetism and Space Magnetism, 357
    Department of Astronomy, 349
        Ouda Station, 350
    Department of Physics, 350
    Graduate School of Science, 357
    Kwasan and Hida Observatories
        Hida Observatory, 350
        Kwasan Observatory, 350
    Yukawa Institute for Theoretical Physics (YITP), 350

Kyunggi-do, 368
Kyungpook, 367
Kyungpook National University (KNU)
    Department of Astronomy and Atmospheric Sciences (DAAS), 369

L

L-3 Communications ESSCO, 753
LAA
    Los Alamos Astrophysics, 800
    Louisville Amateur Astronomers, 726
LAAS
    Los Angeles Astronomical Society, 626
LAB
    Latvijas Astronomijas Biedrība, 371
Laben SpA, 330
Laboratoire d'Astronomie Spatiale (LAS), 186, 187
Laboratoire d'Astrophysique de Marseille (LAM)
    Site Le Verrier-Longchamp, 186
    Site Peiresc-les-Olives, 187
Laboratoire de Glaciologie et de Géophysique de l'Environnement (LGGE), 187, 203
Laboratoire de l'Accélérateur Linéaire (LAL), 187
Laboratoire de Physique Nucléaire des Hautes Énergies (LPN-HE), 187
Laboratoire Primaire du Temps et des Fréquences (LPTF), 164, 187
Laboratori Nazionali del Gran Sasso (LNGS), 328, 330
Laboratori Nazionali di Frascati (LNF), 329, 330
Laboratório Nacional de Astrofísica (LNA)
    Itajubá, 68
Laboratory for Astronomical Imaging (LAI), 708
Laboratory for Atmospheric and Space Physics (LASP), 660
Laboratory for Computational Astrophysics (LCA), 708
Laboratory for High-Energy Astrophysics (LHEA), 740
Labsphere Inc., 789
La Canada, 617
LACC
    Los Angeles City College
        Department of Physics and Astronomy, 626
La Ciotat, 181
Lackawanna Astronomical Society (LAS), 857
La Coruña, 474
La Crosse, 916, 919
La Crosse Area Astronomical Society (LCAAS), 916
Ladd Observatory, 864
LAEFF
    Laboratorio de Astrofísica Espacial y Física Fundamental (INTA), 479
Lafayette, 712, 729, 731
    (West_), 713, 714
Lafayette College
    Department of Physics, 857
Lafayette Natural History Museum (LNHM)
    Planetarium, 729
La Ferme des Étoiles, 188
La Ferté Saint-Aubin, 202
LAG
    Linzer Astronomische Gemeinschaft, 37
La Garandie, 158
Lagos, 414
La Grange Park, 698
Laguna (Mount_), 637
La Habana, 123
    Optical Station, 123
    Radioastronomical Station, 123
Lahall, 489
La Haye, 392, 393, 397, 399
Lahden URSA ry, 148
Lahore, 421
Lahore Astronomical Society Pakistan (LAST), 421
LAI
    Laboratory for Astronomical Imaging, 708
Laize-la-Ville, 160

La Jolla, 614, 648
Lake Afton Public Observatory (LAPO), 722
Lake Arrowhead, 629, 651
Lake County Astronomical Society (LCAS), 702
Lake Erie Nature and Science Center (LENSC), 843
Lake Huntington, 810
Lake Mills, 918
Lake Orion, 766
Lake Tekapo, 412
Lakeview Museum of Arts and Sciences
    Planetarium, 703
Lake Whitney Astronomy Association, 880
Lakewood, 655, 817, 841
LAL
    Laboratoire de l'Accélérateur Linéaire, 187
La Laguna, 477, 479
LAM
    Laboratoire d'Astrophysique de Marseille
        Site Le Verrier-Longchamp, 187
        Site Peiresc-les-Olives, 187
La Mariquita (Cerro de_), 383
Lambrecht Corp. (KLC) (Karl_), 702
Lambuth University
    M.D. Anderson Planetarium, 872
La Murithienne, 511
Lancaster, 553, 609, 855, 859
Lancaster and Morecambe Astronomical Society, 553
Lancaster University
    Environmental Science Department
        Planetary Science Research Group (PSRG), 553
Lancexport, 187
Lanciano, 320
Landelijk Samenwerkende Publiekssterrenwachten (LSPS), 395
Landelijk Samenwerkende Volkssterrenwachten (LSV), 396
Landers, 650
Landessternwarte Heidelberg, 235
Land, Sea and Sky Inc., 881
Lane Cove, 23
Lane Education Service District (ESD) Planetarium, 848
Langhorne, 856
Langley Research Center (LaRC), 899, 900
    Virginia Air and Space Center (VASC), 905
Langmuir Laboratory for Atmospheric Research, 800, 802
Langoiran, 226
Lanham, 737, 738, 742
LANL, Los Alamos National Laboratory
    Space Science and Technology Division
        Space Astronomy and Astrophysics Group, 800
        T-6 Division (Theoretical Astrophysics), 800
Lansbergen (Volkssterrenwacht Philippus_), 406
Lansing, 764
    (East_), 761, 762, 766
Lanusei, 310
Lanzenhäusern, 500
LAOG
    Laboratoire d'Astrophysique de l'Observatoire de Grenoble, 203
La Oliva, 471
La Palma, 471, 477, 479, 480, 482
LAPAN
    National Institute of Aeronautics and Space (Indonesia), 290
La Paz, 62, 63
La Plata, 10
Laplume, 857
LAPO
    Lake Afton Public Observatory, 722
Lappeenranta, 146
LAR
    La Alternativa Racional, 474
Laramie, 922, 923
Laramie Astronomical Society and Space Observers (LASSO), 923
La Rampisole, 179
LaRC
    Langley Research Center, 900

Virginia Air and Space Center, 905
Larchmont, 818
La Réunion, 156
Large Binocular Telescope (LBT), 326, 603
Lario, 321
La Rivière, 156
La Roche des Arnauds, 179
La Rochelle, 164
La Roche-sur-Yon, 164
LAS
    Laboratoire d'Astronomie Spatiale, 187
    Lackawanna Astronomical Society, 857
    Lebanese Astronomical Society, 373
    Leeds Astronomical Society, 553
    Leicester Astronomical Society, 553
    Lethbridge Astronomy Society, 90
    Liverpool Astronomical Society, 554
    Longmont Astronomical Society, 656
    Loughton Astronomical Society, 555
    Louisville Astronomical Society, 726
La Sainte Baume, 162
La Salle, 652
LaSalle, 90
La Sarre, 78
Las Brisas Observatory (LBO), 656
Las Campanas, 103
Las Campanas Observatory (LCO), 107, 109, 672
La Science par et pour l'Homme, 188
Las Cruces, 799, 802
La Serena, 107–111
Laser Fantasy International (LFI), 908
Laser Images Inc., 625
Laser Interferometer Gravitational-wave Observatory (LIGO) Project, 613, 625
Lasermax Inc., 817
La-Seyne-sur-Mer, 190
La Silla, 108, 507
La Silla Observatory, 109
Lasky Planetarium, 300
La Société Guernesiaise
    Section Astronomique, 106
LASP
    Laboratory for Atmospheric and Space Physics, 660
Las Palmas de Gran Canaria, 471, 474, 477
La Spezia, 306, 324, 329
LASSO
    Laramie Astronomical Society and Space Observers, 923
LAST
    Lahore Astronomical Society Pakistan, 421
L'astronomia, 330
Las Vegas, 787
    (North.), 786
Las Vegas Astronomical Society (LVAS), 786
LAT
    Large Amateur Telescope, 622
L'Atelier d'Helios, 188
Latina, 333
La Trinité, 195
La Trobe University
    Space Physics Group, 24
Lattrop, 391
Latvia, 371
Latvian Academy of Sciences
    Radioastrophysical Observatory, 371
    Ventspils International Radioastronomy Centre (VIRAC), 371
Latvian Astronomical Society, 371
Latvian Hydrometeorological Agency (HMP), 371
Latvian National Center for Standardization and Metrology (LVS), 371
Latvian Physical Society (LPS), 371
Latvijas Astronomijas Biedriba (LAB), 371
Lauenstein, 249
Laufen, 211, 248
Launceston, 16, 24
Launceston Planetarium, 24

Laupheim, 261
Laurel, 736
Laurentian University
    Department of Physics and Astronomy, 89
Laurier University (WLU) (Wilfrid.), 89, 105
Laurin Publishing Co. Inc., 753
Lausanne, 511, 513
Laval, 85, 87, 189, 194
La Villette, 159, 169
Lawrence, 721, 723
Lawrence Berkeley National Laboratory (LBNL) (E.O..), 625
    Institute for Nuclear and Particle Astrophysics (INPA), 620
Lawrence Hall of Science (LHS), 610
    William Knox Holt Planetarium, 651
Lawrence Livermore National Laboratory (LLNL)
    Institute of Geophysics and Planetary Physics (IGPP), 625
Lawrence Whipple Observatory (FLWO) (Fred.), 598
Laws Observatory, 778
Laxou, 202
Layton P. Ott Planetarium, 890, 891
LBNL
    Lawrence Berkeley National Laboratory (E.O..)
        Advanced Light Source, 608
        Institute for Nuclear and Particle Astrophysics, 620
LBO
    Las Brisas Observatory, 656
LBT
    Large Binocular Telescope, 326, 603
LCA
    Laboratory for Computational Astrophysics, 703, 708
LCAAS
    La Crosse Area Astronomical Society, 916
LCAS
    Lake County Astronomical Society, 702
LCC
    Leeward Community College
        Observatory, 694
LCER
    Lewis Center for Educational Research, 625
LCO
    Las Campanas Observatory, 107, 672
LDAS
    Letchworth and District Astronomical Society, 554
Leander McCormick Observatory, 904
Learmonth Solar Observatory, 24
Learning Technologies Inc., 753
Leatha House Park, 837
Leatherhead, 555
Lebanese Astronomical Society (LAS), 373
Lebanese Standards Institution (LIBNOR), 373
Lebanon, 373
Lèbe (Col de la.), 170
Lebedev Physics Institute (P.N..), 442, 447
    Nuclear Physics and Astrophysics, 447
Le Bourget, 189
Lebow Co., 625
Lecce, 310, 338
LEDA
    Lyon-Meudon Extragalactic Database, 165
LEDAS
    Leicester Database and Archive Service, 578
Leeds, 553, 578
Leeds Astronomical Society (LAS), 553
Leesport, 860
Leeward Community College (LCC)
    Observatory, 694
Legg Middle School (LMS)
    Planetarium, 765
Legnano, 318
Le Havre, 200
Lehigh Valley Amateur Astronomical Society (LVAAS) Inc., 857
Lehman College (Herbert H..), 817
    Department of Physics and Astronomy, 810

Leicester, 553, 554, 558, 562, 578
Leicester Astronomical Society (LAS), 553
Leicester Database and Archive Service (LEDAS), 578
Leicester University Astronomy Society, 553
Leiden, 390, 396, 398, 399, 401, 404, 405, 407
Leidse Sterrewacht (WLS) (Werkgroep_), 396
Leighton Buzzard, 535
Leinfelden-Echterdingen, 244
Leipzig, 220, 258
Lelekovice, 125
Léman, 509
Lembang, 290
Lemberg, 527
Lemmon (Mount_), 603
LEMSC
    Lake Erie Nature and Science Center, 843
Leningrad, 441, 444, 445, 447, 449, 451
LensPlus, 625
León, 382
Leonardo's Star Quest Astronomy Club, 846
Leonberg Höfingen, 249
Leoncito (El_), 7, 10, 666
Leopold-Figl-Observatorium für Astrophysik, 37, 41
Leopold-Franzens-Universität Innsbruck, 37
    Institut für Astronomie, 40
Leopoldshafen, 229, 245
L.E. Phillips Planetarium, 919
Lepton, 549
Les Marthes de Veyre, 162
Les Observateurs de la Magnitude Absolue, 89
Les Paroches, 161
Les Pléiades
    Beaumont-lès-Valences, 158
    Société d'Astronomie de Saint-Imier, 506
Les Sommets, 86
LeSueur Manufacturing Co. Inc., 589
Les Ulis, 176
Letchworth, 554
Letchworth and District Astronomical Society (LDAS), 554
Lethbridge, 90, 102
Lethbridge Astronomy Society (LAS), 90
Leuschner Observatory, 625, 646
Leuven, 59, 61
Lévis, 78
Lewisberry, 852
Lewisburg, 853, 862
Lewis Center for Educational Research (LCER), 625
Lewis Research Center (LeRC), 839
Lewis Science Center
    Observatory, 607
    Planetarium, 607
Lewiston, 732
Lexington, 725, 727, 748, 755, 905
LFD
    Lietuvos Fiziku Draugija, 375
LFI
    Laser Fantasy International, 908
LGGE
    Laboratoire de Glaciologie et de Géophysique de l'Environnement, 203
LGSCV
    Local Group of Santa Clarita Valley, 644
LHEA
    Laboratory for High-Energy Astrophysics, 740
LHMS
    Lithuanian Hydrometeorological Service, 374
LHS
    Lawrence Hall of Science (UCB), 610
        William Knox Holt Planetarium, 651
LIADA
    Liga Ibero-Americana de Astronomía, 62
Liberec, 129
Liberec Astronomy Club, 129
Liberty
    (North_), 801
Liberty Science Center (LSC), 793
Libertyville, 702

LIBNOR
    Lebanese Standards Institution, 373
Librairie Scientifique Uranie, 188
Lichtenknecker Optics (LO) sa, 55
Lick Observatory, 625, 649
    Headquarters, 649
Lido, 308
Lido di Ostia, 303
Liebert Inc. (Mary Ann_), 817, 818
Liège, 54, 56, 58
Lietuvos Fiziku Draugija (LFD), 375
Liga Ibero-Americana de Astronomía (LIADA), 62
Liga Venezolana de Astronomía (LIVA), 928
LightPath Technologies Inc., 800
LightWedge, 753
LIGO
    Laser Interferometer Gravitational-wave Observatory, 613
LIGO Project, 613, 625
Liguria, 309, 340
Lille, 170, 204
Lima, 423
Limagne, 162
Limburgse Volkssterrenwacht (LVS) vzw, 51, 55
Limerick, 854
Limoux, 170
Lincoln, 782–784
Lincoln Laboratory, 753, 755
Linda G. Sharpe Planetarium, 872, 873
Lindau, 227, 236
Lindon, 892
Linköping, 492
Link Planetarium (Edwin A._), 817, 820
Linz, 37
Linzer Astronomische Gemeinschaft (LAG), 37
LIPI
    Indonesian Institute of Sciences, 290
Lipp Observatory (James_), 636
Lire La Nature Inc. (LLN), 90
Lisboa, 433–435
Lisbon, 433–435
Lisnyky, 526
Lithuania, 374
Lithuanian Academy of Sciences, 374
Lithuanian Hydrometeorological Service (LHMS), 374
Lithuanian Physical Society (LPS), 375
Lithuanian Standards Board (LST), 375
Little Ferry, 796
Little Rock, 606
Little Thompson Observatory (LTO), 656
LIVA
    Liga Venezolana de Astronomía, 928
Livermore, 625, 627, 645, 732
Liverpool, 554, 578
Liverpool Astronomical Society (LAS), 554
Liverpool John Moores University
    Astrophysics Research Institute, 554
Liverpool Museum Planetarium, 554
Livingston, 627
Livonia, 765
Livorno, 334
Ljubljana, 462, 463
Llanfairpwll, 547
Llano del Hato Observatory, 927
LLN
    Lire La Nature, 90
LLNL
    Lawrence Livermore National Laboratory
        Institute of Geophysics and Planetary Physics, 625
LMD
    Laboratoire de Météorologie Dynamique, 175
LMO
    Lucile Miller Observatory, 827, 829
LMS
    Legg Middle School
        Planetarium, 765
LNA

Laboratório Nacional de Astrofísica
  Itajubá, 68
LNF
  Laboratori Nazionali di Frascati, 329
LNGS
  Laboratori Nazionali del Gran Sasso, 328
LNHM
  Lafayette Natural History Museum
    Planetarium, 729
LNP
  Loch Ness Productions, 753
LO
  Lichtenknecker Optics, 55
LOBO electronic GmbH, 235
Local Group of Deep Sky Observers, 681
Local Group of Santa Clarita Valley (The_), 625, 644
Locarno, 511
Locarno-Monti, 506, 509, 511
Lochem, 400
Loch Ness Productions (LNP), 753
Locke (Mount_), 888
Lockheed Martin Corp.
  Advanced Technology Center
    Solar and Astrophysics Laboratory, 626
Lockwood Park, 705
Lockyer Observatory (Norman_), 554, 559
Lockyer Planetarium (James_), 554, 559
LodeStar Project, 800
Lodi, 336
Logan, 891
Logroño, 472
Lohrmann-Observatorium, 236, 251
Loiano Observatory, 326
Loire, 201
Loisir Art Culture, Section Astronomie, 188
Lomnický Štít Coronal Observatory, 459
Lomonosov Moscow State University (M.V._), 442
  Institute of Nuclear Physics, 443
London, 90, 98, 104, 531, 532, 534, 538, 539, 541, 549–551,
    554, 555, 557–559, 561, 564–566, 569, 572–574, 578,
    579
  (North East_), 559
  (West_), 583
London Astronomical Society (NELAS) (North East_), 559
London Planetarium, 538, 554
London Regional Children's Museum, 90
Long Beach, 609, 614
Long Island, 807
Longmont, 654, 656
Longmont Astronomical Society (LAS), 656
Longueuil, 85, 90
Longway Planetarium (Robert T._), 763, 765, 767
Longwood Regional Planetarium, 817
Longyearbyen, 415, 419, 489
Loomis Laboratory of Physics, 703
Loosley Row, 548
Lorain, 835
Loral Space Systems, 626
Loránd University (Eötvös_), 277
Loras College
  Physics Department, 717
Lorraine, 202
Los Alamos, 800–803
Los Alamos Astrophysics (LAA), 800
Los Alamos National Laboratory (LANL)
  Space Science and Technology Division
    Space Astronomy and Astrophysics Group, 800
  T-6 Division (Theoretical Astrophysics), 800
Los Altos Hills, 621
Los Angeles, 608, 622, 623, 626, 648–650
Los Angeles Astronomical Society (LAAS), 626
Los Angeles City College (LACC)
  Department of Physics and Astronomy, 626
Los Angeles Sidewalk Astronomers, 626
Los Angeles Valley College (LAVC)
  Planetarium, 626
Los Dolores, 472

Los Gatos, 644
Losmandy Astronomical Products, 627
Los Mochis, 384
Los Ruices, 927
LOT-Oriel Ltd., 555
Loughton Astronomical Society (LAS), 555
Louisiana, 729–731
Louisiana Arts and Science Center, 729
Louisiana Nature Center and Planetarium, 729
Louisiana State University (LSU)
  Department of Physics and Astronomy, 729
  Press, 730
Louisiana State University Press, 730
Louisville, 725–727
Louisville Amateur Astronomers (LAA), 726
Louisville Astronomical Society (LAS) Inc., 726
Louisville Science Center (LSC), 726
Louvain, 59, 61
Louvain-la-Neuve, 55, 57
Lowbrows Astronomers, 765
Lowell, 763, 766
Lowell Observatory, 598
Lower Hutt, 410
LPGP
  Laboratoire de Physique des Gaz et des Plasmas, 204
LPI
  Lunar and Planetary Institute, 886
LPIB
  Lunar and Planetary Information Bulletin, 886
LPNHE
  Laboratoire de Physique Nucléaire des Hautes Énergies,
    187
LPS
  Latvian Physical Society, 371
  Lithuanian Physical Society, 375
LPTF
  Laboratoire Primaire du Temps et des Fréquences, 164
L. Russell Kelce Planetarium, 723
LSC
  Liberty Science Center, 793
L.S. Noblitt Planetarium, 712
LSPS
  Landelijk Samenwerkende Publiekssterrenwachten, 395
LST
  Lithuanian Standards Board, 375
LSU
  Louisiana State University
    Department of Physics and Astronomy, 729
    Press, 730
LSU Press, 730
LSV
  Landelijk Samenwerkende Volkssterrenwachten, 396
LTO
  Little Thompson Observatory, 656
Lubbock, 881, 883
Lübeck, 208
Lublin, 428
Lübz, 242
Lucent Technologies
  Bell Laboratories, 793
Lucerne, 514
Lucile Miller Observatory (LMO), 827, 829
Ludiver - Observatoire-Planétarium de la Hague, 188
Lueninghoener Planetarium, 783
Luleå, 497
Luling, 730
Lumberton, 831
Lumezzane, 303, 332
Lumicon, 627
Luminy, 167
Lunar and Planetary Institute (LPI), 881, 886
Lunas, 390
Lund, 491, 495, 498
Lund Observatory, 490, 491
Lund Planetarium, 490, 491
Lund University
  Department of Physics, 491

Lund Observatory, 491
    Planetarium, 491
Lunping Observatory, 516
Luqa, 380
Lure, 200
Lustenau, 41
Luxembourg, 376
Luzern, 514
LVAAS
    Lehigh Valley Amateur Astronomical Society, 857
LVAS
    Las Vegas Astronomical Society, 786
L'viv, 527
L'viv Astronomical Society, 527
L'viv Polytechnical State University
    Department of Geodesy and Astronomy, 527
L'viv University
    Astronomical Observatory, 527
    Chair of Astrophysics, 527
L'vov, 527
LVS
    Limburgse Volkssterrenwacht, 51
Lycksele, 495
Lycoming College
    Department of Physics and Astronomy, 858
Lynchburg, 897, 902
Lyndon B. Johnson Space Center (JSC), 881
    Astronomical Society, 880
    Public Affairs Office, 881
Lyon, 165, 171, 200, 203
Lyon-Meudon Extragalactic Database (LEDA), 165

M

M42 (Association Stéphanoise d'Astronomie_), 161
M53 Mayenne Astronomie, 188
MAA
    Martz Astronomical Association, 818
    Mathematical Association of America, 673
MAAA
    Mankato Area Astronomy Association, 771
MAC
    Midlands Astronomy Club, 867
    Millstream Astronomy Club, 839
Macclesfield, 552, 555, 579
Macclesfield Astronomical Society, 555
Mac Cormick Observatory (Leander_), 904
Macedonia, 377
Maceió, 65
Mac Farland Park Observatory, 716
Machine Automatique à Mesurer pour l'Astronomie (MAMA), 183
Mac Kim Observatory, 710
Mac Laughlin Planetarium, 91
Mac Lean Observatory, 787
Mac Master University
    Department of Physics and Astronomy, 91
MacMillan Planetarium (H.R._), 89, 90
Macon, 689
MACSIT, 275, 277
Madeira, 435
Madison, 917, 920, 921
Madison Astronomical Society (MAS) Inc., 916
Madison Metropolitan School District (MMSD)
    Planetarium and Observatory, 917
Madison University (JMU) (James_), 899
    Physics Department, 899
Madonna di Campiglio, 308
Madras, 282
Madrid, 472, 474–476, 478, 479, 481, 482, 484
Mafraq, 363
Magdalena Ridge Observatory, 802
Magdeburg, 216
Magic Valley Astronomical Society (MVAS), 697
Magyar Szabványügyi Testület (MSZT), 276
Mahoning Valley Astronomical Society (MVAS), 839

Mahoning Valley Cortese Observatory (MVCO), 839
Maidanak, 374
Maidanak Foundation (The_), 415, 418
Maidanak Observatory, 925
Maidanak Observatory (Mount_), 418
Maiden, 829
Maidenhead, 555, 571
Maidenhead Astronomical Society, 555
Main Astronomical Observatory (MAO), 527, 529
Maine, 732, 733
    (Southern_), 733
Maine Astronomical Society, 732
Mainz, 207, 210, 244
Maison de l'Astronomie, 189
Maison de l'Astronomie P.L. Inc., 90
MAK
    Mariestads Astronomiska Klubb, 491
Makati, 425
Makka, 453
Málaga, 484
Malatya, 523
Malaysia, 378
Malaysian Meteorological Service (MMS), 378
Malé Bielice, 458
Malibu, 639
Mali Lošinj, 121
Malling, 134
Mallorca, 480
Malmö, 491
Malmö Astronomi & Rymfarts Sällskap (MARS), 491
Malmö Astronomy and Space Society, 491
Måløv, 135
Malta, 380
Malta Astronomical Society (MAS), 380
Malta Standardisation Authority (MSA), 380
Malvern, 862
Malý Diel, 461
MAM
    Musée des Arts et Métiers, 189
MAMA
    Machine Automatique à Mesurer pour l'Astronomie, 183
Mamaroneck, 819
Mammendorf, 218
Manasek Inc. (G.B._), 893
Manchester, 555, 567, 573, 579, 789, 873
Manchester Astronomical Society (MAS), 555
Manching, 213
Manfred Olson Planetarium, 917, 920
Manfredonia, 321
Manila, 425, 426
Manila Observatory (MO), 425
Manitoba, 102
Manitoba Astronomy Club, 90
Manitoba Museum of Man and Nature
    Manitoba Planetarium, 90
Manitoba Planetarium, 90
Mankato, 771
Mankato Area Astronomy Association (MAAA), 771
Mankato State University (MSU), 771
    Department of Physics and Astronomy, 772
Mannheim, 228, 229, 242
Manora Peak, 288
Manresa, 472
Mansfield, 556, 747, 842, 845
Mansfield and Sutton Astronomical Society (MSAS), 556
Manufacturers
    ACE, 593
    Adirondack Video Astronomy, 805
    Advanced Mechanical and Optical Systems, 47
    AEDC, 870
    AeroAstro, 895
    Aerospace Industries, 669
    Aerospatiale, 154
    Aerovox
        Massachusetts, 746
        New Bedford, 746

Manufacturers – Manufacturers

United Kingdom, 535
USA-MA, 746
Weymouth, 535
AIA, 669
Alcatel
   Études Techniques et Constructions Aérospatiales, 47
Alenia, 303
American Educational Products, 652
AMOS, 47
Analogic, 747
Angénieux, 155
AOK, 502
APM Telescopes, 208, 747
Apogee Instruments, 609
Apollo-Foco, 652
Applied Geomechanics, 609
Arcane, 155
Arnold Engineering, 870
Ash, 698
   Bradenton, 678
   Florida, 678
   Richmond, 896
   USA-FL, 678
   USA-VA, 896
   Virginia, 896
ASL
   Massachusetts, 748
   Milton Keynes, 534
   United Kingdom, 534
   USA-MA, 748
   Wilmington, 748
Astrocom, 209
AstroDomes, 15
Astrolab, 791
Astromedia, 488
Astronomical Consultants and Equipment, 593
Astro-Physics, 698
AstroSystems, 652
Astrotech, 273
Astrovisuals, 17
Automatic Systems Laboratories
   Massachusetts, 748
   Milton Keynes, 534
   United Kingdom, 534
   USA-MA, 748
   Wilmington, 748
AVA, 805
AWR Technology, 534
Baader, 218
Barr Associates
   Massachusetts, 748
   Newbury, 535
   United Kingdom, 535
   USA-MA, 748
   Westford, 748
Beacon Hill Telescopes, 535
BHC Aerovox, 535
Bielser Observatorien, 502
Bowen, 710
Boyd, 761
Brailsford, 809
Brashear LP, 852
Brooklyn Thermometer, 809
Burleigh, 809
Bushnell, 721
Byers (Edward R._), 619
Cambridge Research and Instrumentation, 749
Canberra Semiconductor, 48
Carl Zeiss
   Chester, 897
   Jena, 220
   USA-VA, 897
   Virginia, 897
Catalana de Telescopios, 475
CCM, 389
Celestron, 616

Centorr, 788
Centronic, 540
Centrum voor Constructie en Mechatronica, 389
Ceramaseal, 810
Charles Evans, 617
CIG, 594
Clear Night, 762
Collins, 653
Company Seven, 736
Composants et Systèmes de Précision, 175
Comstock, 871
COMTEC (FAST_), 229
Corion, 759
Corning, 813
Coronado, 542, 594
Cotton Expressions, 699
Coulter Optical, 682
CRI, 749
Cryomech, 813
Cryophysics, 503
CSP, 175
Custom Scientific, 595
CVC Products, 813
C-VII, 736
Daimler Benz, 221
Daniel Oberti, 618
Davis Instruments, 618
DayStar, 618
Denitron, 317
DFM Engineering, 654
Digistar, 890
Discovery Telescopes, 618
DynaVac, 750
E&S, 890
Eddy, 619
Edmund Scientific, 814
Edward R. Byers, 619
EG&G, 736
Electrim, 792
Electro Optical Industries, 619
EM Electronics, 545
EMF, 814
EOI, 619
EOST, 595
EOS Technologies, 595
Epner, 815
Equatorial Platforms, 620
ETCA, 47
Etel, 503
Études Techniques et Constructions Aérospatiales, 47
Evans & Sutherland, 890
Evans (Charles_), 617
Evaporated Metal Films, 814
Express Optics, 620
Farran, 296
FAST COMTEC, 229
First Magnitude Corp., 922
FLAG, 52
Flemish Aerospace Group, 52
Fluid Metering, 815
FMC, 922
FMI, 815
FTL, 296
Fujitok, 344
Futurestar, 770
Galaxy Optics, 654
Galvoptics, 546
GIFAS, 181
GOTO, 344
Hawksley (Paton_), 562
Helius Designs, 548
Hextek, 597
High Res Technologies, 88
Hinds Instruments, 848
Hollywood General Machining, 623
Honeywell, 681
HRT, 88

Hubbard Scientific, 916
Icon Isotopes, 792
IGG Component Technology, 549
IL, 752
ILC Data Device Corp., 816
Infrared Laboratories, 597
Inrad, 793
Intercon Spacetec, 233
Interferometrics, 899
International Light, 752
International Thermal Instrument, 624
IPL, 624
Isotope Products, 624
ISS, 702
ITI, 624
Janos Technology, 893
Jenoptik, 234
Jim's Mobile Inc., 655
JMI, 655
Joerger, 816
Karl Lambrecht, 702
Kayser-Threde, 235
Kendall Hyde, 553
KLC, 702
Kohler, 502
L-3, 753
Laben, 330
Labsphere, 789
Lambrecht (Karl_), 702
Laser Fantasy International, 908
Laser Images, 625
Learning Technologies, 753
Lebow, 625
LeSueur, 589
LFI, 908
Lichtenknecker, 55
LightPath, 800
LNP, 753
LO, 55
LOBO, 235
Loch Ness Productions, 753
Lumicon, 627
Maxtek, 627
Meade, 627
MegaSystems, 858
Minolta, 350, 793
MMR Technologies, 628
Monoptec, 628
Mountain Instruments, 628
Murmel, 506
Murnaghan Instruments, 682
Nanjing Astronomical Instruments Co. Ltd., 116
National Photocolor, 819
Newport Industrial Glass, 631
Nippon Silica Glass, 794
Noctilume, 631
Nonius, 399
NPC, 819
Oberti (Daniel_), 618
Observa-Dome, 774
Obsession Telescopes, 918
OCLI, 631
OES, 240
OGS, 859
OLI, 682
Omega Engineering, 663
Omega Optical, 893
Optec, 766
Optical Coating Laboratory Inc., 631
Optical Guidance Systems, 859
Optische und Elektronische Systeme, 240
Optometrics, 757
Optronic, 682
Orion Telescope & Binocular Center, 632
Oxford Instruments, 561
OZ Optics, 96
Parallax Instruments, 894

Parks, 632
Pasco Scientific, 633
Paton Hawksley, 562
PCP, 683
Penny & Giles, 562
PerkinElmer Optoelectronics, 757
PHEL, 562
Physik Instrumente, 241
PI, 241
PIC Design, 663
Picosecond Pulse Labs, 658
Precision Cryogenic, 713
Precision Industrial Components, 663
PRI, 795
Princeton Research Instruments, 795
PSPL, 658
QSP, 634
Quadrant Engineering, 634
Quantum Technology, 909
Questar, 860
Radio Research Instruments, 664
Rayleigh Optical Corp., 742
Raytheon, 664
Remote Measurement Systems, 909
REOSC, 198
Resonance, 97
Reynard, 635
RGL, 820
Richardson Grating Laboratory, 820
Rigel Systems, 635
RMS, 909
ROC, 742
Roper Scientific
    Arizona, 601
    Marlow, 564
    Tucson, 601
    United Kingdom, 564
    USA-AZ, 601
RRI, 664
RSAI, 198
R.S. Automation Industrie, 198
S&Y, 790
Santa Barbara Instrument Group, 638
SBIG, 638
Schott Glas, 244
SCI, 641
Science and Art Products, 639
Sciencetech, 98
SciTech Astronomical Research, 639
Shibayama Scientific, 357
Sira, 568
SkyComm Engineering, 685
Sky-Skan, 790
Software Systems Consulting, 640
Sonoma Instruments, 640
SORL, 758
South Coast Telescopes, 569
Space Craft International, 641
Space Optics Research Labs, 758
Spacetec (Intercon_), 233
SPE, 403
Spectrum Sciences, 642
Speedibrews, 570
Spirus, 843
Spitz, 861
SRS, 642
SSC, 640
SSI, 642
SSTL, 571
Stanford Research Systems, 642
Stargazer Steve, 99
Starlight Xpress, 571
StarMaster, 723
Starsplitter Telescopes, 642
Stellarium, 643
Stellar Products, 643
Stocker & Yale, 790

Stork Product Engineering, 403
Surrey Satellite Technology, 571
Sutherland (Evans &_), 890
TBE, 590
Technical Innovations, 743
Teknofokus, 151
Tekton, 644
Teledyne Brown Engineering, 590
Telescopes and Astronomy, 29
Telescope Technologies Ltd., 572
Teleskoptechnik Halfmann, 252
Tele Vue, 824
TFT, 645
Thermionics
   California, 644
   Hayward, 644
   Port Townsend, 910
   USA-CA, 644
   USA-WA, 910
Thin Film Technology, 645
TLI, 644
TL Systems, 645
Torus Precision Optics, 719
Trango, 645
Triplett, 843
TTL, 572
UniMeasure, 850
Unitron, 824
Vacuum Industries, 788
Van Slyke, 661
Velmex, 825
Verkehr Raumfarht und Systemtechnik, 258
Vixen, 361
VRS, 258
What in The World, 651
Wilcoxon Research, 745
Woodstock Products, 651
XIA, 651
X-Ray Instrumentation Associates, 651
Zarvan, 293
Zeiss (Carl_)
   Chester, 897
   Jena, 220
   USA-VA, 897
   Virginia, 897
MAO
   Main Astronomical Observatory, 529
   Moerser Astronomische Organisation, 238
   Moletai Astronomical Observatory, 374
MAPS
   Middle Atlantic Planetarium Society, 858
Maracaibo, 928
Maracay, 928
Maranha, 71
Marburg, 232, 261
Marcinelle, 49
Margaret C. Woodson Planetarium, 829
Margate, 685
Marghera, 313
Maria Curie-Sklodowska University
   Institute of Physics, Astrophysics and Didactics
      Astronomy Laboratory, 428
Maria Mitchell Observatory (MMO), 754
Marian Blakemore Planetarium, 881, 889
Mariazell, 34
Marie Drake Planetarium, 592
Mariestad, 491
Mariestads Astronomiska Klubb (MAK), 491
Marietta, 687
Marignane, 188
Marina, 610, 628
Marinha Grande, 434
Mario (Monte_), 328
Mark Evans Observatory, 702
Mark Smith Planetarium, 689
Marlboro, 747, 893
Marlboro College
   Department of Physics and Astronomy, 893
Marlow, 553, 564
Marly-le-Roi, 176
Marmara Radio Telescope Observatory, 524
Marmara Research Center, 523
Marmara University
   Department of Physics, 523
Marne (Haute-_), 200
Marquardt, 234
Marquette, 765, 767
Marquette Astronomical Society (MAS), 765
MARS
   Malmö Astronomi & Rymfarts Sällskap, 491
Marseille, 154, 159, 165, 167, 180, 187
Marsfield, 21
Marshall, 714, 772
Marshall Space Flight Center (MFSC) (George C._), 590
   Space Science Laboratory, 590
Marshall W. Alworth Planetarium, 771
Marsh Rice University (William_), 881
Mars Society (The_), 656, 659
Martha (Mount_), 15
Marthasville, 777
Martha's Vineyard Amateur Astronomy Club (MVAAC), 754
Martz Astronomical Association (MAA) Inc., 818
Martz Observatory, 818
Mary Ann Liebert Inc., 818
Mary D. Pretlow Planetarium, 901
Maryland, 734–745
   (Central_), 738
   (Southern_), 742
Maryland Science Center (MSC), 739
   Alan C. Davis Planetarium, 739
   Baltimore Astronomical Society (BAS), 735
   Crosby Ramsey Memorial Observatory, 739
Maryland Space Grant Consortium (MSGC), 739
Mary Washington College (MWC)
   Department of Physics, 899
MAS
   Madison Astronomical Society, 916
   Malta Astronomical Society, 380
   Manchester Astronomical Society, 555
   Marquette Astronomical Society, 765
   Memphis Astronomical Society, 872
   Michiana Astronomical Society, 713
   Milwaukee Astronomical Society, 917
   Minnesota Astronomical Society, 771
   Mobile Astronomical Society, 590
   Murdoch Astronomical Society, 25
   Muskegon Astronomical Society, 766
Masaryk University
   Department of Theoretical Physics and Astrophysics, 129
Mason University (GMU) (George_), 899
MASP
   Mid-Atlantic Star Party, 829
Massa, 321
Massachusetts, 694, 736, 746–760
Massachusetts Institute of Technology (MIT), 758
   Atmospheric Sciences Group, 755
   Center for Space Research (CSR), 754
   Department of Earth, Atmospheric and Planetary Science, 754
   Department of Earth, Atmospheric, and Planetary Sciences
      Wallace Astrophysical Observatory (WAO), 754
   Department of Physics, 755
      Astrophysics Division, 755
   Laboratory for Computer Science, 760
   Lincoln Laboratory, 755
   Michigan Dartmouth MIT Observatory, 598, 755
   Millstone Hill Observatory (MHO), 755
   Press, 755, 756
   World Wide Web Consortium (W3C), 760
Massapequa, 808, 822
   (North_), 820

Masterton, 412
Matera, 302
Materials Research Society (MRS), 858
Mathematical Association of America (MAA), 673
Matthews, 828
Maui Optical Station, 692, 694
Mauna Kea, 109, 596, 693
Mauna Kea Astronomical Society (MKAS), 694
Mauna Kea Observatory (MKO), 692–695, 801
Mauna Loa Solar Observatory (MLSO), 656
Mauritius, 381
Mauritius Meteorological Services (MMS), 381
Mauritius Radio Telescope (MRT), 381
Mauritius Standards Bureau (MSB), 381
Mawson Lakes, 14, 30
Max-Koch-Sternwarte, 232
Max-Planck-Gesellschaft (MPG), 236
Max-Planck-Gesellschaft zur Förderung der Wissenschaften eV
    Institut de Radioastronomie Millimétrique (IRAM), 182
Max-Planck-Institut für Aeronomie (MPAe)
    Katlenburg-Lindau, 236
Max-Planck-Institut für Astronomie (MPIA)
    Centro Astronómico Hispano-Alemán
        Observatorio de Calar Alto, 475
    Heidelberg, 236
Max-Planck-Institut für Astrophysik (MPA), 236
Max-Planck-Institut für extraterrestrische Physik (MPE), 236

Max-Planck-Institut für Gravitationsphysik, 237
Max-Planck-Institut für Kernphysik (MPI-K)
    Bereich Astrophysik, 237
Max-Planck-Institut für Physik (MPP), 237
Max-Planck-Institut für Radioastronomie (MPIfR), 237
Maxtek Inc., 627
Mayaki, 528
Mayence, 207, 210
Mayer, 601
Mayer-Volks-und-Schulsternwarte Heilbronn (Robert-_), 243

Maynard, 750
Maynard F. Jordan Planetarium and Observatory, 732, 733
Maynooth, 297
MBO
    Modine-Benstead Observatory, 919
MCAO
    Mount Cuba Astronomical Observatory, 667
MCAS
    Medina County Astronomical Society, 839
McAuliffe Planetarium (CMP) (Christa_), 788
MCC
    Museo de la Ciencia y el Cosmos, 479
McCallion Planetarium (William J._), 91
McCarthy Observatory (JJMO)
    (John J._), 665
McCormick Observatory (Leander_), 904
McDonald Observatory, 881, 887, 888
    Hobby-Eberly Telescope (HET), 880
McDonald Planetarium (J.M._), 782
McDonnell Center for the Space Sciences, 779
McFarland Park Observatory, 716
McGill University
    Department of Earth and Planetary Sciences, 91
    Department of Physics, 91
    Institute and Centre of Air and Space Law (IASL), 91
McGraw Hill Observatory, 598, 754
McKay Planetarium (H.V._), 24
McKim Observatory, 710
McLaughlin Planetarium, 91
McLean, 901
McLean Observatory, 787
McMaster University
    Department of Physics and Astronomy, 91
McMath-Hulbert Astronomical Society, 765
MCP
    Merrillville Community Planetarium, 713
M.D. Anderson Planetarium, 872

MDAS
    Mount Diablo Astronomical Society, 629
MDM Observatory, 598
MDOA
    Mount Diablo Observatory Association, 629
Meade Instruments Corp., 627
Mead Observatory, 688
Meadow
    (East_), 806
Mead Valley, 643
Meadville, 851
Mecklenburg-Vorpommern, 231
Medea, 4
Medford, 759, 849
Medicina, 316
Medina, 824, 839
Medina County Astronomical Society (MCAS), 839
Meduza, 129
Meelick, 298
Meerbusch-Osterath, 257
Mees Observatory (C.E. Kenneth_), 825
Mees Solar Observatory (C.E. Kenneth_), 694, 696
Mégantic (Mont_), 100, 101
MegaSystems Inc., 858
Mehrabad, 292
Meichsner Co. (F.C._), 751, 755
Meinhard-Hitzelrode, 258
Melbourne, 15, 16, 21, 24, 28, 680
    (North_), 18
Melbourne Planetarium at Scienceworks, 24
Melilla, 587
Melle, 211
Melo, 587
Melton Memorial Observatory, 868
Melville, 805, 806, 820
Memorial University of Newfoundland (MUN)
    Department of Physics, 91
Memphis, 871–875
Memphis Astronomical Society (MAS), 872
Memphis Pink Palace Museum and Linda G. Sharpe Planetarium, 872
Menasha, 915, 918
Mendon, 756
Mendoza, 6
Menke Planetarium und Sternwarte, 237
Mentor, 840
MEPA
    Meteorology and Environmental Protection Administration, 453
Merate, 326
Merced, 627
Mercedes, 8, 587
Merced Society for Telescope and Astronomy Recreation (M-STAR), 627
Mercyhurst College
    Department of Physics, 858
Mergentheim (Bad._), 217
Mérida, 927, 928
Merlan Scientific Ltd., 92
MERLIN
    Multi-Element Radio Linked Interferometer Network, 579
Merrick
    (North_), 807
Merrillville, 713
Merrillville Community Planetarium (MCP), 713
Merseburg, 242
Mesa, 595
Mesquite, 882
Messelberg-Sternwarte, 247
Mestre, 313, 317
Metcalf Nature Center, 771
Met Éireann, 297
Meteo Consult BV, 396
Météo France, 189
Meteor Crater Enterprises Inc., 598
Meteor Group Hawaii, 694

Meteorite Museum, 801
Meteoritical Society, 694
Meteorological and Geophysical Agency, 290
Meteorological Department, 414
    Bangkok, 519
    Damascus, 515
    Lagos, 414
    Nigeria, 414
    Syria, 515
    Thailand, 519
Meteorological Office, 380, 556
    Bracknell, 556
    Crown Point, 521
    Luqa, 380
    Malta, 380
    Piarco, 521
    Tobago (Trinidad &_), 521
    Trinidad & Tobago, 521
    United Kingdom, 556
Meteorological offices
    Albania, 3
    Alger, 5
    Algeria, 5
    Ankara, 524
    Argentina, 10
    Armenia, 12
    Asunción, 422
    Australia, 21
    Austria, 42
    Azerbaijan, 43
    Baghdad, 294
    Baku, 43
    Bangkok, 519
    Bangladesh, 44
    Barbados, 45
    Beijing, 113
    Belgium, 53, 54
    Beograd, 454
    Berlin, 223
    Bet Dagan, 299
    Bilt (De_), 395
    BMD (Bangladesh), 44
    BMG (Indonesia), 290
    Bogotá, 120
    Bolivia, 63
    Bracknell, 556
    Brasília, 67
    Bratislava, 460
    Brazil, 67
    Bridgetown, 45
    Brussel, 53, 54
    Brussels, 53, 54
    Bruxelles, 53, 54
    Bucureşti, 438
    Budapest, 276
    Buenos Aires, 10
    Bulgaria, 75
    Cachoeira Paulista, 67
    Cairo, 142
    Canada, 81
    Caribbean, 45, 521
    Changi, 456
    Chile, 107
    China (PRC), 113
    CHMI (Czech Republic), 127
    CIMH (Barbados), 45
    CMA (China/PRC), 113
    CMO (Trinidad & Tobago), 521
    Colombia, 120
    Colombo, 487
    Croatia, 121
    Crown Point, 521
    Cuba, 123
    CWB (Taiwan), 516
    Czech Republic, 127
    Damascus, 515
    Dar El Beïda, 5
    Darmstadt, 227
    De Bilt, 395
    Denmark, 137
    Deutscher Wetterdienst
        Berlin, 223
        Offenbach, 223
    Deutsches Klimarechenzentrum, 224
    Dhaka, 44
    DHM (Venezuela), 928
    DINAC, 422
    DKRZ (Germany), 224
    DMC (Chile), 107
    DMH (Paraguay), 422
    DMI (Denmark), 137
    DNMI (Norway), 417
    DNM (Uruguay), 586
    Downsview, 81
    Dublin, 297
    DWD (Germany)
        Berlin, 223
        Offenbach/Main, 223
    Ecuador, 141
    Egypt, 142
    El Salvador, 143
    EMA (Egypt), 142
    EUMETSAT, 227
    Finland, 147
    FMHI (Serbia), 454
    FMI (Finland), 147
    Germany, 223
    Glavgidromet (Uzbekistan), 925
    Greece, 268
    Hamburg, 224
    Harare, 929
    Hellinikon, 268
    Helsinki, 147
    HKO (Hong Kong), 116
    HMP (Latvia), 371
    HMS (Hungary), 276
    HNMS (Greece), 268
    Honduras, 272
    Hong Kong, 116
    Hungary, 276
    Iceland, 279
    IDEAM (Colombia), 120
    IMD (India), 284
    IMO (Iceland), 279
    INAMHI (Ecuador), 141
    India, 284
    Indonesia, 290
    INMET (Brazil), 67
    INM (Spain), 478
    INPE (Brazil), 67
    INSMET (Cuba), 123
    Iran, 292
    Iraq, 294
    Ireland, 297
    IRIMO (Iran), 292
    IRMB (Belgium), 53
    Israel, 299
    Italy, 336
    Jakarta, 290
    Japan, 347
    Jeddah, 453
    JMA (Japan), 347
    Karachi, 421
    Kenya, 365
    KMD (Kenya), 365
    KMIB (Belgium), 54
    KNMI (Netherlands), 395
    København, 137
    Korea
        (ROK), 368
    Kowloon, 116
    Lagos, 414
    La Havana, 123
    La Paz, 63

Latvia, 371
LHMS (Lithuania), 374
Lima, 423
Lisboa, 433
Lithuania, 374
Ljubljana, 462
Luqa, 380
Luxembourg, 376
Macedonia, 377
Madrid, 478
Mahrabad, 292
Malaysia, 378
Malta, 380
Manila, 425
Maracay, 928
Mauritius, 381
Melbourne, 21
MEPA (Saudi Arabia), 453
Meteo Consult, 396
Météo France, 189
MeteoSchweiz, 506
MétéoSuisse, 506
MeteoSvizzera, 506
MeteoSwiss, 506
México, 384
Mexico, 384
MMS (Malaysia), 378
MMS (Mauritius), 381
Montevideo, 586, 587
Moscow, 448
MSS (Singapore), 456
Nairobi, 365
National Weather Service, 742
Netherlands, 395
New Delhi, 284
New Zealand, 410
Nigeria, 414
NIMH (Bulgaria), 75
Norrköping, 496
Norway, 417
NWS (USA), 742
Offenbach/Main, 223
Oslo, 417
PAGASA (Philippines), 425
Pakistan, 421
Peru, 423
Petaling Jaya, 378
Philippines, 425
Piarco, 521
PMD (Pakistan), 421
Poland, 427
Port of Spain, 521
Portugal, 433
Praha, 127
Pretoria, 468
Pune, 284
Quito, 141
Reykjavík, 279
Riga, 371
Roma, 336
Romania, 438
ROSHYDROMET (Russia), 448
Russia, 448
Saint-James, 45
San Salvador, 143
Santiago, 107
Saudi Arabia, 453
SAWB (South Africa), 468
SENAMHI (Bolivia), 63
SENAMHI (Peru), 423
Seoul, 368
Serbia, 454
Silver Spring, 742
Singapore, 456
Skopje, 377
Slovak Republic, 460
Slovenia, 462

SMHI (Sweden), 496
SMN (Argentina), 10
SMNH (El Salvador), 143
SMN (Honduras), 272
SMN (Mexico), 384
Sofia, 75
SOHMA (Uruguay), 587
South Africa, 468
Spain, 478
Sri Lanka, 487
St.-James, 45
Sweden, 496
Switzerland, 506
Syria, 515
Taipei, 516
Taiwan, 516
Tashkent, 925
Tegucigalpa, 272
Tehran, 292
Thailand, 519
Tirana, 3
Tobago (Trinidad &_), 521
Tokyo, 347
Trinidad & Tobago, 521
Turkey, 524
UGM (Italy), 336
United Kingdom, 556
Uruguay, 586, 587
USA-MD, 742
Uzbekistan, 925
Vacoas, 381
Venezuela, 928
Vilnius, 374
Wageningen, 396
Warszawa, 427
Wellington, 410
Wien, 42
Yerevan, 12
Zagreb, 121
ZAMG (Austria), 42
Zimbabwe, 929
Zürich, 506
Meteorological Service
    Dublin, 297
    Ireland, 297
Meteorological Service of New Zealand Ltd., 410
Meteorological Service Singapore (MSS), 456
Meteorology and Environmental Protection Administration (MEPA), 453
MeteoSchweiz, 502, 506
MétéoSuisse, 506, 507
MeteoSvizzera, 506, 512
MeteoSwiss, 504, 506
Metsähovi Radio Observatory, 149
Metzerlen, 512
Meudon, 167, 174, 192
Meuse, 49
Mexborough, 556
Mexborough and Swinton Astronomical Society (MSAS), 556

México, 382–386
Mexico, 382
    (New_), 799
Meyer Observatory (Paul J._), 878
MHAA
    Mid-Hudson Astronomy Association, 818
MHCC
    Mount Hood Community College, 848
MHO
    Millstone Hill Observatory, 755
MHOA
    Mount Hood Observatory Association, 848
Miami, 681, 682, 685
Miami Museum of Science and Space Transit Planetarium, 681
Miami Valley Astronomical Society (MVAS), 839
Michael Adrian Observatorium, 250

Michael I. Casper - Meteorites Inc., 818
Michael L. Britton Memorial Observatory, 854
Michalovce, 458
Michalovce Observatory, 458
Michiana Astronomical Society (MAS), 713
Michigan, 761-769
    (Central_), 762
    (Eastern_), 762, 763
Michigan-Dartmouth-MIT Observatory, 598
Michigan State University (MSU)
    Abrams Planetarium, 761, 762
    Department of Physics and Astronomy, 765
Michigan Technological University (MTU)
    Department of Physics, 766
Micromedia Ltd., 92
Mid-Atlantic Star Party (MASP), 829
Middelburg, 407
Middle Atlantic Planetarium Society (MAPS), 858
Middlebury, 663, 893
Middlebury College
    Department of Physics, 893
Middle East Technical University
    Physics Department
        Astrophysics Group, 523
Middle Georgia Astronomical Society, 689
Middle Island, 817
Middle Tennessee Astronomical Society (MTAS), 873
Middle Tennessee State University (MTSU)
    Department of Physics and Astronomy, 873
Middleton, 915
Middletown, 665
Mid Glamorgan, 537, 546
Mid-Hudson Astronomy Association (MHAA), 818
Mid-Kent Astronomical Society, 556
Midland, 881, 889
Midland Lutheran College (MLC)
    Lueninghoener Planetarium, 783
Midlands Astronomy Club (MAC), 867
Midlands Spaceflight Society (MSS), 556
Midlands Technical College Astronomy Club (MTCAC), 867

Mid-South Star Gaze, 775
Midtown Observatory, 666
Midwestern Astronomers (MWA), 726
Mijas, 473
Mijas Costa, 473
Milano, 309, 312-314, 317, 318, 320, 323, 325, 326, 330, 338
Milford, 663
Milham Planetarium, 760
Miller Jr. Observatory (Franklin_), 839
Miller Observatory (LMO) (Lucile_), 827, 829
Millikan Planetarium, 634
Milliken Planetarium (Daniel B._), 617, 627
Millikin University
    Department of Physics and Astronomy, 703
Mills Observatory, 556
Millstone Hill Observatory (MHO), 755
Millstream Astronomy Club (MAC), 839
Millville, 797
Milton, 755
Milton Academy
    Science Department, 755
Milton J. Rubenstein Museum of Science & Technology
    Silverman Planetarium, 821
Milton Keynes, 534, 557, 560
Milton Keynes Astronomical Society (MKAS), 557
Milwaukee, 917, 920
Milwaukee Astronomical Society (MAS), 917
Milwaukee Public Museum (MPM), 917
Mineralogical Research Co., 627
Minneapolis, 771, 773
Minneapolis Planetarium, 771
Minnesota, 770-773
Minnesota Astronomical Society (MAS), 771
Minnesota State University at Mankato
    Department of Physics and Astronomy, 771
Minolta Planetarium, 628

Minolta Planetarium Co. Ltd., 350
    US Representation, 793
Minor Planet Center (MPC), 755
Minot State University (MSU)
    Observatory, 834
Minsk, 46
Mira (Volkssterrenwacht_), 60
Misato Observatory, 351
Mission College
    Astronomy Department, 628
Mission Valley Astronomy Club (MVAC), 780
Mississauga, 103
Mississippi, 774, 775
Mississippi State, 774
Mississippi State University (MSU)
    Department of Physics and Astronomy, 774
Missoula, 781
Missouri, 776-779
    (Eastern_), 777
    (Southwest_), 778
Mitaka, 342, 352
Mitcham, 564
Mitchell, 17
Mitchell Observatory (MMO) (Maria_), 754, 755
MIT, Massachusetts Institute of Technology, 758
    Atmospheric Sciences Group, 755
    Center for Space Research, 754
    Department of Earth, Atmospheric and Planetary Science, 754
    Department of Earth, Atmospheric, and Planetary Sciences
        Wallace Astrophysical Observatory, 754
    Department of Physics, 755
        Astrophysics Division, 755
    Laboratory for Computer Science, 760
    Lincoln Laboratory, 755
    Millstone Hill Observatory, 755
    Press, 756
    World Wide Web Consortium, 760
Mito, 345
MIT Press, 755
Mitzpe Ramon, 300
Miyun Station, 113
Mizusawa Astrogeodynamics Observatory, 351-353
MJC
    Maison des Jeunes et de la Culture
        Carcassonne, 170
MJUO
    Mount John University Observatory, 412
MKAS
    Mauna Kea Astronomical Society, 694
    Milton Keynes Astronomical Society, 557
MKO
    Mauna Kea Observatory, 692-695, 801
MKSP
    Mount Kobau Star Party, 92
MLC
    Midland Lutheran College
        Lueninghoener Planetarium, 783
Mlodziezowe Obserwatorium Astronomiczne, 428
MLSO
    Mauna Loa Solar Observatory, 656
Mlynany, 460
MMI Corp., 739
MMO
    Maria Mitchell Observatory, 754
MMOG
    Monmouth Mobile Observer's Group, 793
MMR Technologies Inc., 628
MMS
    Malaysian Meteorological Service, 378
    Mauritius Meteorological Services, 381
MMSD
    Madison Metropolitan School District
        Planetarium and Observatory, 917
MMTO
    Multiple Mirror Telescope Observatory, 598

MNASSA
    Monthly Notes of the Astronomical Society of Southern Africa, 464
MNHN
    Muséum National d'Histoire Naturelle, 189
MNISM
    Mongolian National Institute for Standardisation and Metrology, 388
MNRAS
    Monthly Notices of the Royal Astronomical Society, 565
MO
    Manila Observatory, 425
Mobile, 590
Mobile Astronomical Society (MAS), 590
Modena, 333
Modine-Benstead Observatory (MBO), 919
Modra, 458
Moers, 238
Moerser Astronomische Organisation eV (MAO), 238
Moffett Field, 630
Mohave Valley, 593
Mohawk Valley Astronomical Society (MVAS), 818
Moka, 381
Moldova, 387
Moletai, 374
Moletai Astronomical Observatory (MAO), 374, 375
Mollet, 476
Mölndal, 494
Molonglo Observatory Synthesis Telescope (MOST), 31
Molonglo Radio Observatory, 24, 31
Monash University
    Department of Mathematics
        Astrophysics Group, 24
    Department of Physics, 24
Moncton, 100
Monegrillo, 476
Mongolia, 388
Mongolian Academy of Sciences
    Centre of Astronomy and Geophysics, 388
Mongolian National Institute for Standardisation and Metrology (MNISM), 388
Monmouth County, 793
Monmouth Mobile Observer's Group (MMOG), 793
Monoptec Corp., 628
Monroe Station (Morgan-_), 712
Monroeville, 838, 843
Mons, 49, 50, 58
Monsummano Terme, 307
Montagne de Lure, 200
Montana, 780, 781
    (Southwest_), 780
    (Western_), 781
Montana State University (MSU)
    Department of Physics, 780, 781
    Museum of the Rockies, 780
    Taylor Planetarium, 780
Mont Blanc, 190
Montclair, 792, 793
    (Upper_), 793
Montclair State University (MSU)
    College of Science and Mathematics, 793
Monte Armidda, 310
Monte da Caparica, 434
Monte Ekar, 327
Monte Generoso, 504
Montelupo, 332
Monte Mario, 328
Montenegro (Serbia-_), 454
Monte Novegno, 320
Monteporzio, 340
Monteporzio Catone, 328
Monterey Institute for Research in Astronomy (MIRA), 610, 628
Monterrey, 382
Monte San Martino, 324
Montevideo, 586–588
Monte Vissegi, 307

Montgomery, 591
Montgomery Center, 894
Montgomery College Planetarium, 740
Monthly Notices of the Royal Astronomical Society (MNRAS), 565
Monticello, 607, 725
Mont Mégantic, 100, 101
Montpelier, 736
Montpellier, 164, 179, 182, 183, 196, 200
Montréal, 80, 86, 88, 90, 91, 96, 98, 100
Montredon-Labessonnié, 197
Mont-sur-Marchienne, 47
Mont Tibarine, 4
Moody Planetarium, 881
Moore Observatory, 727
Moore Observatory (James and Barbara_), 681, 682
Mooresville, 712, 867
Moorhead, 770, 772
Moorhead State University (MSU)
    Department of Physics and Astronomy, 772
    Paul P. Feder Observatory, 772
    Regional Science Center, 772
Mopra Observatory, 25
Mopra Telescope, 21
Moran, 922
Morecambe, 553
Morecambe Astronomical Society (Lancaster and_), 553
Morehead Observatory, 832
Morehead Planetarium, 829
Morel Astrographics, 25
Morelia, 382, 386
Morgan-Monroe Station, 712
Morgantown, 913, 914
Morningside, 20
Morris Museum Astronomical Society, 794
Morrison Observatory, 776
Morrison Planetarium (Alexander_), 608, 611, 628, 637
Morro de São Domingo, 65
Morro Santana, 73
Morville, 49
Moscou, 440–451
Moscow, 440–451, 860
Moscow Astronomical Club, 442
Moscow Physical-Technical Institute, 443
Moscow Planetarium, 443
Moscow State University (M.V. Lomonosov_), 443
    Institute of Nuclear Physics, 443
Moshannon Observatory (Black_), 860
Moshiri, 351
Moskva, 440–451
MOST
    Molonglo Observatory Synthesis Telescope, 31
Most, 129
Most Planetarium, 129
Môtiers, 503
Motivgruppe Astronomie & Philatelie, 238
Mount Abu, 287
Mount Abu Infrared Observatory, 286
    Headquarters, 287
Mountain Instruments, 628
Mountain Observers, 818
Mountainside, 797
Mountain Skies Astronomical Society (MSAS), 628
Mountain Skies Observatory, 923
Mountain View, 628, 633, 639
Mount Allison University
    Department of Physics, Engineering and Geoscience, 92
Mount Burnett, 25
Mount Chikurinji, 354
Mount Coot-tha, 27
Mount Cuba Astronomical Observatory (MCAO) Inc., 667
Mount Cuba Astronomical Planetarium, 667
Mount Diablo, 629
Mount Diablo Astronomical Society (MDAS), 629
Mount Diablo Observatory Association (MDOA), 629
Mount Evans Meyer-Womble Observatory, 661

Mount Fuji Observatory, 360
Mount Graham, 604, 926
Mount Hamilton, 649
Mount Hood Community College (MHCC)
    Planetarium, 848
Mount Hood Observatory Association (MHOA), 848
Mount Hopkins, 597, 599, 787
Mount John University Observatory (MJUO), 412
Mount Kanobili, 206
Mount Kobau Star Party (MKSP), 92
Mount Koshka, 525
Mount Laguna Observatory, 637
Mount Lemmon Station, 603
Mount Locke, 888
Mount Maidanak Observatory, 418
Mount Martha, 15
Mount Norikura, 353
Mount Norikura Observatory, 360
Mount Pleasant, 19, 762, 867
Mount Pleasant Radio Astronomy Observatory, 31
Mount Sanglok Observatory, 518
Mount Stromlo, 18
Mount Stromlo and Siding Spring Observatories (MSSSO), 18, 25
Mount Suhora Observatory, 428
Mount Tamalpais Observers, 629
Mount Tamborine Observatory (MTO), 30
Mount Union College (MUC)
    Department of Physics and Astronomy, 839
Mount Wilson Institute (MWI), 629
Mount Wilson Observatory Association (MWOA), 629
Mount Wilson Observatory (MWO), 629, 650
Mowbray, 464
MPA
    Max-Planck-Institut für Astrophysik, 236
MPAe
    Max-Planck-Institut für Aeronomie, Katlenburg-Lindau, 236
M.P. Birla Planetarium, Kolkata, 286
MPC
    Minor Planet Center, 755
MPE
    Max-Planck-Institut für extraterrestrische Physik, 236
MPG
    Max-Planck-Gesellschaft, 236
MPIA, Max-Planck-Institut für Astronomie
    Centro Astronómico Hispano-Alemán
        Observatorio de Calar Alto, 475
    Heidelberg, 236
MPIfR
    Max-Planck-Institut für Radioastronomie, 237
MPI-K
    Max-Planck-Institut für Kernphysik
        Bereich Astrophysik, 237
MPM
    Milwaukee Public Museum, 917
MPP
    Max-Planck-Institut für Physik, 237
MRAO
    Mullard Radio Astronomy Observatory, 557
MRS
    Materials Research Society, 858
MRT
    Mauritius Radio Telescope, 381
MSA
    Malta Standardisation Authority, 380
MSAS
    Mansfield and Sutton Astronomical Society, 556
    Mexborough and Swinton Astronomical Society, 556
    Mountain Skies Astronomical Society, 629
MSB
    Mauritius Standards Bureau, 381
MSC
    Maryland Science Center
        Alan C. Davis Planetarium, 739
        Baltimore Astronomical Society, 735
        Crosby Ramsey Memorial Observatory, 739

MSD
    Museum of Scientific Discovery
        Planetarium, 859
MSFC
    George C. Marshall Space Flight Center
        Space Science Laboratory, 590
MSGC
    Maryland Space Grant Consortium, 739
MSI
    Museum of Science and Industry, 703
MSS
    Meteorological Service Singapore, 456
    Midlands Spaceflight Society, 556
MSSL
    Mullard Space Science Laboratory, 574
MSSSO
    Mount Stromlo and Siding Spring Observatories, 18
MST Aerospace GmbH, 238
M-STAR
    Merced Society for Telescope and Astronomy Recreation, 627
MSU
    Mankato State University
        Department of Physics and Astronomy, 772
    Michigan State University
        Abrams Planetarium, 761, 762
        Department of Physics and Astronomy, 766
    Minot State University
        Observatory, 834
    Mississippi State University, 774
    Montana State University
        Department of Physics, 780, 781
        Museum of the Rockies, 780
        Taylor Planetarium, 780
    Montclair State University
        College of Science and Mathematics, 793
    Moorhead State University
        Department of Physics and Astronomy, 772
        Paul P. Feder Observatory, 772
        Regional Science Center, 772
MSZT
    Magyar Szabványügyi Testület, 276
Mt. Abu, 287
MTAS
    Middle Tennessee Astronomical Society, 873
Mt. Blanc, 190
M.T. Brackbill Planetarium, 899
Mt. Burnett, 25
MTCAC
    Midlands Technical College Astronomy Club, 867
Mt. Chikurinji, 354
Mt. Coot-tha, 27
Mt. Cuba Astronomical Observatory (MCAO) Inc., 667
Mt. Cuba Astronomical Planetarium, 667
Mt. Diablo, 629
Mte. Armidda, 310
Mte. da Caparica, 434
Mte. Ekar, 327
Mte. Generoso, 504
Mte. Mario, 328
Mte. Novegno, 320
Mte. San Martino, 324
Mt. Evans Meyer-Womble Observatory, 661
Mte. Vissegi, 307
Mt. Fuji Observatory, 360
Mt. Graham, 604, 926
Mt. Hamilton, 649
Mt. Hood Community College (MHCC)
    Planetarium, 848
Mt. Hood Observatory Association (MHOA), 848
Mt. Hopkins, 597, 599, 787
Mt. John University Observatory (MJUO), 412
Mt. Kanobili, 206
Mt. Kobau Star Party (MKSP), 92
Mt. Koshka, 525
Mt. Laguna Observatory, 637
Mt. Lemmon Station, 603

Mt. Locke, 888
Mt. Maidanak Observatory, 418
Mt. Martha, 15
Mt. Mégantic, 100, 101
Mt. Norikura, 353
Mt. Norikura Observatory, 360
MTO
    Mount Tamborine Observatory, 30
Mt. Pleasant, 19, 762, 867
Mt. Pleasant Radio Astronomy Observatory, 31
Mt. Sanglok Observatory, 518
Mt. Stromlo, 18
Mt. Stromlo and Siding Spring Observatories (MSSSO), 18
MTSU
    Middle Tennessee State University
        Department of Physics and Astronomy, 873
Mt. Suhora Observatory, 428
Mt. Tamalpais Observers, 629
Mt. Tamborine Observatory (MTO), 30
Mt. Tibarine, 4
MTU
    Michigan Technological University
        Department of Physics, 766
Mt. Union College (MUC)
    Department of Physics and Astronomy, 839
Mt. Wilson Institute (MWI), 629
Mt. Wilson Observatory Association (MWOA), 629
Mt. Wilson Observatory (MWO), 629, 650
MUC
    Mount Union College
        Department of Physics and Astronomy, 839
Muchachos (Roque de los_), 477
Mueller Planetarium (Ralph_), 782, 784, 842
Muesmatt, 500
Muiderberg Observatory, 406
Mulbarton, 560
Mulgrave North, 27
Mülheim/Ruhr, 252
Mulhouse, 177
Mullard Radio Astronomy Observatory (MRAO), 557
Mullard Space Science Laboratory (MSSL), 557, 574
Multi-Element Radio Linked Interferometer Network (MERLIN), 579
Multiple Mirror Telescope Observatory (MMTO), 598
Mumbai, 281, 283, 288
    (New_), 283
MUN
    Memorial University of Newfoundland
        Department of Physics, 92
Munakata, 344
München, 209, 218, 219, 221, 224, 227, 229, 230, 235–237, 252, 255
    (West-_), 217
Muncie, 710
Munich, 209, 218, 219, 221, 224, 227, 229, 230, 235–237, 252, 255
    (West_), 217
Münster, 256, 265
Münstereifel (Bad_), 237
Murdoch, 25
Murdoch Astronomical Society (MAS), 25
Murdoch University
    Physics and Energy Studies, 25
Murdock Sky Theater, 849
Murfreesboro, 873
Murithienne (La_), 511
Murmel Spielwerkstatt und Verlag, 506
Murnaghan Instruments Corp., 682
Murray Hill, 793
Musée de l'Air et de l'Espace, 189
Musée de l'Instrumentation Optique, 189
Musée des Arts et Métiers (MAM), 189
Musée Suisse des Transports et des Communications, 506, 514
Museo Civico di Rovereto
    Planetario, 330
Museo de la Ciencia y el Cosmos (MCC), 479

Museu de Astronomia e Ciências Afins, 68
Museu de la Ciència de la Fundació "La Caixa", 479
Museum am Schölerberg, Natur und Umwelt
    Planetarium, 238
Museum Boerhaave, 396, 401
Muséum d'Histoire Naturelle, Genève, 510
Museum für Astronomie und Technikgeschichte
    Planetarium, 238
Muséum National d'Histoire Naturelle (MNHN), 189
Museum of Applied Arts and Sciences, 28
Museum of Arts and Sciences
    Daytona Beach, 682
    Lakeview, 703
    Macon, 689
    Peoria, 703
    Planetarium, 682
Museum of Man and Nature (Winnipeg), 90
Museum of Natural History
    American, 815
    Ann Arbor, 768
    Battle Creek, 763
    Denver, 653
    Gastonia, 831
    Kingman, 763
    Lafayette, 729
    Lancaster, 859
    New York, 815
    Providence, 864
    Santa Barbara, 610, 638
    Schiele, 831
    Washington, 675
Museum of Natural History and Cormack Planetarium, 864
Museum of Science and Industry (MSI), 703
Museum of Scientific Discovery (MSD)
    Planetarium, 859
Museum of Texas Tech University, 881
Museum of the Rockies
    Taylor Planetarium, 780
Museum of York County, 868
Museums
    Adler, 698
    Aerospace
        San Diego, 636
    Aiken, 866
    Air and Space
        Bourget (Le_), 189
        Hampton, 905
        Le Bourget, 189
        VASC, 905
        Virginia, 905
        Washington, 673, 675, 676
    Alamogordo, 803
    Albuquerque, 801
    Alcalá de Henares, 481
    Alpena, 764
    Ann Arbor, 768
    Archaeological
        Warszawa, 431
    Århus, 139
    Arizona, 593
    Armstrong (Neil_), 840
    Art
        Columbia, 866
    Arts and Sciences
        Daytona Beach, 682
        Evansville, 711
        Haymarket, 28
        Herrett, 697
        Lakeview, 703
        Macon, 689
        Peoria, 703
        Sydney, 28
        Twin Falls, 697
        Valencia, 475
    Arts et Métiers
        Paris, 189
    Astronaut Hall of Fame, 678

Astronomia e Ciências Afins (Rio), 68
Astronomie und Technikgeschichte, 238
Atlanta, 688
Aurora, 705
Avon, 664
Baltimore, 739
Barcelona, 480
Bath, 584
Baton Rouge, 729
Battle Creek, 763
Beijing Ancient Observatory, 112
Bemidji, 770
Berkeley, 651
Bernice P. Bishop, 692
Besser (Jesse_), 764
Biesheim, 189
Binghamton, 820
Birr Castle, 295
Bishop (Bernice P._), 692
Boerhaave, 401
Borlänge, 489
Boston, 750
Bourget (Le_), 189
Bozeman, 780
Bridgeport, 662
British Columbia, 98
Buffalo, 809
Calgary, 80
Calusa, 679
Canton, 838
Cape Town, 468
Carnegie, 856, 858
CCCS, 81
CCSC, 81
Centerport, 825
Cernan, 699
Chabot, 616
Cherokee, 718, 719
Chicago, 698, 703
Children's
  Hiroshima, 345
  Indianapolis, 714
  London Regional, 90
  Sendai, 357
Cleveland, 842
CMNH, 842
Coca-Cola, 688
Columbia, 776, 866
Columbus, 688
Connecticut, 662
Cook, 878
Copernicus
  (Nicolas_), 429
Corona Park, 819
Corsicana, 878
Cosmodôme, 87
CSM, 874
Culiacan, 382
Cumberland, 874
Dallas, 883
Davis, 620
Daytona Beach, 682
Denver, 653
Des Moines, 718
Detroit, 762
Deutsches, 224
Discovery
  Bridgeport, 662
  East Tennessee, 870
  ETDC, 870
  Harrisburg, 859
  Knoxville, 870
  MSD, 859
  Paris, 195
  Rockford, 700
  Sacramento, 618
  Safford, 595

DMNH, 653
Dumfries, 544
Durban, 465
Durham, 830
Earth and Space
  Columbia, 776
  River Grove, 699
East Tennessee, 870
EcoTarium, 750
Edmonton, 95
Eise Eisinga, 391
Eisinga (Eise_), 391
Eretz Israel, 300
ESC, 51
ESSC, 95
ETDC, 870
Euro Space, 51
Evansville, 711
Exhibit, 768
Experimental de Ciencias (Rosario), 9
Exploration Place, 722
Exploratorium, 620
Exploratory, 18
Explorit, 620
Fairbanks, 893
Fairy Meadow, 32
Falun, 493
Fernbank, 688
Firenze, 325
Fleet (Reuben H._), 634
Fleischmann, 786
Florida (South_), 681, 685
Fort Myers, 679
Fort Worth, 882
Framtidsmuseet, 489
Franeker, 391
Franklin Institute, 855
Frombork, 429
FSC, 688
Fulda, 242
Future
  Borlänge, 489
Gastonia, 831
Göteborg, 489, 493
Gotoh, 344
Granada, 481
Grand Rapids, 761, 767
Great Island, 88
Green Bay, 917
Greenbelt, 741
Greensboro, 828
Greenwich, 565
Griffith Observatory, 622
Grout, 717
GSFC, 741
Hagerstown, 743
Hampton, 905
Harrisburg, 859, 861
Hartford (West_), 662
Hastings, 782
Haymarket, 28
HBOSC, 737, 738
Headwaters, 770
Herrett, 697
Herschel (William_), 584
Heureka, 152
Hiroshima, 345
Histoire Naturelle
  Genève, 510
  Paris, 189
HMNS, 878, 879
Hong Kong, 116
Honolulu, 692
Houston, 878, 879, 881
Howard B. Owens, 737, 738
HRM, 807
H.R. MacMillan, 89

HSC, 770
Hudson River, 807
Huntsville, 591
Impression 5, 764
IMSS, 325
Indianapolis, 714
Innovation
    San Jose, 644
Iowa, 718
Ithaca, 821
Jersey City, 793
Jesse Besser, 764
JSC, 881
Junior
    Troy, 817
Kalamazoo, 764, 765
Kansas City, 777
Kassel, 238
Katsushika, 349
Kenner, 729
Kensington, 88
Kingman, 763
Kirkpatrick, 846
Knoxville, 870
Kowloon, 116
KSW, 36
Kuffner, 36
KVM, 765
La Coruña, 474
Lafayette, 729
Lakeview, 703
La Laguna, 479
Lancaster, 859
Lanham, 737, 738
Lansing, 764
Launceston, 16, 24
Laval, 87
La Villette, 169
Lawrence Hall, 651
Le Bourget, 189
Leicester, 558
Leiden, 401
LHS, 651
Liberty, 793
Life and Science, 830
Liverpool, 554
Livorno, 334
LNHM, 729
London, 90, 557, 558, 565
Los Angeles, 622
Louisiana, 729
Louisville, 726
LSC, 726
Lubbock, 881
Lucerne, 514
Luleå, 497
Luzern, 514
Macon, 689
MAM, 189
Man and Nature
    Winnipeg, 90
Manila, 425
Maryland, 739
MCC, 479
McKinley, 838
Memphis, 873
Merseyside, 554
Meteorite, 801
México, 386
Miami, 682
Midland, 881
Milwaukee, 917
Minnesota, 772
MNHN, 189
Morris, 794
Morristown, 794
Mount Stromlo, 18

MPM, 917
MSD, 859
MSI, 703
München, 224
Münster, 265
Mystic Seaport, 663
Nagoya, 351
NASA
    GSFC, 741
    JSC, 881
Nashville, 874
NASM, 673, 675, 676
Natural History
    American, 815
    Ann Arbor, 768
    Battle Creek, 763
    Cleveland, 842
    Denver, 653
    Exhibit, 768
    Gastonia, 831
    Geneva, 510
    Göteborg, 489, 493
    Hastings, 782
    Kingman, 763
    Lafayette, 729
    Lancaster, 859
    Livorno, 334
    London, 558
    Münster, 265
    New York, 815
    Paris, 189
    Providence, 864
    Santa Barbara, 610, 638
    Schiele, 831
    Washington, 675
Natural Science
    Boerhaave, 401
    Durban, 465
    Greensboro, 828
    HMNS, 878, 879
    Houston, 878, 879
    Leiden, 401
    Prato, 309
    Taichung, 517
Naturkunde
    Münster, 265
Natur und Umwelt
    Osnabrück, 238
    Schölerberg, 238
Natuurwetenschappen en Geneeskunde
    Boerhaave, 401
    Leiden, 401
NCMLS, 830
NCSM, 351
Neil Armstrong, 840
Neville, 917
Newark, 791
Newfoundland, 95
New Jersey, 794
Newport News, 905
New York, 815, 819
Nicolas Copernicus, 429
NMSI, 557
NMSTC, 93
Noordwijk, 399
North York, 96
NSE, 399
NSSC, 558
NTM, 416
NYHoS, 819
Oakland, 616
OAN, 481
Odyssium, 95
Oklahoma City, 846
Ole Rømer, 138
Omniplex, 846
OMSI, 849

Ontario, 96
Optics (Biesheim), 189
Oregon, 849
Orlando, 683
ORM, 138
OSC, 96, 683
Oslo, 416
Osnabrück, 238
Ottawa, 93
Owens (Howard B._), 737, 738
Pacific, 908
Palais de la Découverte, 195
Palm Beach
 (West._), 681, 685
Parc aux Étoiles, 195
Paris, 169, 189, 195
Patrick (Ruth._), 866
Pennsylvania, 861
Peoria, 703
Philadelphia, 855
Phoenix, 593
Pinellas County, 684
Pink Palace, 873
Pittsburgh, 856, 858
Poitiers, 176
Portland, 849
Prato, 309
Pretoria, 467
Providence, 864
Queen Victoria, 16, 24
Reading, 860
Regional Children's, 90
Reno, 786
Reuben H. Fleet, 634
Richmond, 898
Rio de Janeiro, 68
River Grove, 699
RMSC, 823
Roanoke, 903
Roberson, 820
Rochester, 823
Rockford, 700, 706
Rock Hill, 868
Rockies, 780
ROM, 91
Rømer (Ole._), 138
Rosario, 9
Roswell, 803
Rovereto, 330
Royal Ontario, 91
RPSEC, 866
Ruth Patrick, 866
SAASTEC, 467
Sacramento, 618
Safford, 595
Saint John's, 95
Saint Johnsbury, 893
Saint Louis, 778
Saint Paul, 772
Saint Petersburg, 684
San Bernardino, 636
San Diego, 634, 636
Sanford (Tiel._), 718, 719
San Francisco, 620
San Jose, 644
Santa Barbara, 610, 638
Schenectady, 821
Schiele, 831
Schölerberg, 238
Science
 Aiken, 866
 Arizona, 593
 Atlanta, 688
 Avon, 664
 Baltimore, 739
 Barcelona, 480
 Baton Rouge, 729

Bemidji, 770
Berkeley, 651
Binghamton, 820
Boston, 750
British Columbia, 98
Buffalo, 809
Calgary, 80
Carnegie, 856, 858
CCCS, 81
CCSC, 81
Coca-Cola, 688
Columbus, 688
Connecticut, 662
Corona Park, 819
CSM, 874
Culiacan, 382
Cumberland, 874
Dallas, 883
Davis, 620
Des Moines, 718
Detroit, 762
Durham, 830
Edmonton, 95
Fairy Meadow, 32
Falun, 493
Fernbank, 688
Finland, 152
Fleet (Reuben H._), 634
Fleischmann, 786
Florida (South._), 681, 685
Fort Worth, 882
Franklin Institute, 855
FSC, 688
Granada, 481
Great Island, 88
Grout, 717
Hartford (West._), 662
HBOSC, 737, 738
Headwaters, 770
Heureka, 152
Howard B. Owens, 737, 738
HSC, 770
Iowa, 718
Ithaca, 821
Jersey City, 793
Kensington, 88
Kirkpatrick, 846
La Coruña, 474
La Laguna, 479
Lancaster, 859
Lanham, 737, 738
Liberty, 793
Louisiana, 729
Louisville, 726
LSC, 726
Maryland, 739
México, 386
Miami, 682
Minnesota, 772
Nagoya, 351
Nashville, 874
NCSM, 351
New York, 819
North York, 96
NTM, 416
NYHoS, 819
Oklahoma City, 846
Omniplex, 846
Ontario, 96
Orlando, 683
Oslo, 416
Owens (Howard B._), 737, 738
Pacific, 908
Palm Beach (West._), 681, 685
Paris, 195
Patrick (Ruth._), 866
Philadelphia, 855

Phoenix, 593
Pinellas County, 684
Pittsburgh, 856, 858
Reno, 786
Reuben H. Fleet, 634
Richmond, 898
RMSC, 823
Roanoke, 903
Roberson, 820
Rochester, 823
Rosario, 9
RPSEC, 866
Ruth Patrick, 866
Sacramento, 618
Saint Louis, 778
Saint Paul, 772
Saint Petersburg, 684
San Diego, 634
Seattle, 908
Sinaloa, 382
Singapore, 456
SLSC, 778
SMM, 772
SMV, 898
SMWV, 903
South Florida, 681, 685
Springfield, 758
Stanley Bridge, 88
St. Louis, 778
St. Paul, 772
St. Petersburg, 684
Talcott Mountain, 664
Tenerife, 479
TMSC, 664
Tokyo, 355
Toyama, 359
Vancouver, 98
Vantaa, 152
Virginia, 898
Virginia (Western_), 903
Waterloo, 717
Western Virginia, 903
West Hartford, 662
West Palm Beach, 681, 685
Wollongong, 32
Yokohama, 362
York (North_), 96
Science and Industry
  Canton, 838
  Chicago, 703
  Columbus, 836
  COSI, 836
  La Villette, 169
  London, 557
  McKinley, 838
  OMSI, 849
  Oregon, 849
  Paris, 169
  Portland, 849
  Tampa, 684
  Toledo, 836
Science and Technology
  Aurora, 705
  Cook, 878
  Corsicana, 878
  Ottawa, 93
  Pretoria, 467
  SAASTEC, 467
Sciencenter, 821
Science World British Columbia, 98
Scienze Naturali
  Prato, 309
SciTech, 705
Seattle, 908
Sendai, 357
Shibuya, 344
Sinaloa, 382

Singapore, 456
SLSC, 778
SMM, 772
SMV, 898
SMWV, 903
Södertälje, 498
South African, 468
Southend-on-Sea, 570
South Florida, 681, 685
Southwest, 881
Space
  Alamogordo, 803
  Armstrong (Neil_), 840
  Cernan, 699
  Chabot, 616
  Columbia, 776
  Edmonton, 95
  Hagerstown, 743
  Hong Kong, 116
  H.R. MacMillan, 89
  Huntsville, 591
  Kirkpatrick, 846
  Kowloon, 116
  Leicester, 558
  MacMillan (H.R._), 89
  Miami, 682
  Neil Armstrong, 840
  Noordwijk, 399
  Oakland, 616
  Oklahoma City, 846
  Omniplex, 846
  Pacific, 81
  River Grove, 699
  San Diego, 636
  Sendai, 357
  Toulouse, 169
  Vancouver, 81, 89
  Wapakoneta, 840
  Washington County, 743
Springfield, 758
SSM, 758
Stankey Bridge, 88
Steno, 139
St. John's, 95
St. Johnsbury, 893
St. Louis, 778
Stockholm, 492
Storia della Scienza
  Firenze, 325
Storia Naturale
  Livorno, 334
St. Paul, 772
St. Petersburg, 684
Sydney, 28
Taichung, 517
Taipei, 517
Talcott Mountain, 664
TAM, 517
Tampa, 684
Tåstrup, 138
Technology
  Luleå, 497
  NTM, 416
  Oslo, 416
  San Jose, 644
Tel-Aviv, 300
Tenerife, 479
Texas Tech University, 881
Tiel Sanford, 718, 719
Time, 706
Titusville, 678
TMSC, 664
Tokyo, 349, 355
Tom Tits, 498
Toronto, 91
Toulouse, 169
Toyama, 359

Transinne, 51
Trenton, 794
Triel-sur-Seine, 195
Troy, 817
TSM, 359
Twin Falls, 697
Universum, 386
US Space and Rocket, 591
US Space Camp, 591
Valencia, 475
Vancouver, 81, 89, 98
Vanderbilt, 825
VASC, 905
Verkehrshaus, 514
Vienna, 36
Virginia, 898, 905
  (Western_), 903
Virginia Living, 905
VLM, 905
Vonderau, 242
Wapakoneta, 840
Warszawa, 431
Washington, 673, 675, 676
Washington County, 743
Waterloo, 717
Western Virginia, 903
West Hartford, 662
Weston Creek, 18
West Palm Beach, 681, 685
Wichita, 722
Wien, 36
William Herschel, 584
Winnipeg, 90
Wollongong, 32
Worcester, 750
Yamaguchi, 362
Yokohama, 362
Yonkers, 807
York
  (North_), 96
York County, 868
YSC, 362
Musicosmic, 190
Muskegon, 766
Muskegon Astronomical Society (MAS), 766
Muttenz, 502
MVAAC
  Martha's Vineyard Amateur Astronomy Club, 754
MVAC
  Mission Valley Astronomy Club, 780
MVAS
  Magic Valley Astronomical Society, 697
  Mahoning Valley Astronomical Society, 839
  Miami Valley Astronomical Society, 839
  Mohawk Valley Astronomical Society, 818
MVCO
  Mahoning Valley Cortese Observatory, 839
M.V. Lomonosov Moscow State University
  Institute of Nuclear Physics, 443
MWA
  Midwestern Astronomers, 726
M.W. Burke-Gaffney Observatory (BGO) (Rev._), 92, 97
MWC
  Mary Washington College
    Department of Physics, 899
MWI
  Mount Wilson Institute, 629
MWO
  Mount Wilson Observatory, 629
MWOA
  Mount Wilson Observatory Association, 629
Mykolaiv, 527
Mykolaiv Astronomical Observatory, 527
Mykolaiv State Pedagogical Institute
  Kalinenkov Astronomical Observatory, 527
Mystic, 663
Mystic Seaport Planetarium, 663

N

NAA
  Naperville Astronomical Association, 703
  Northants Amateur Astronomers, 559
  Nürnberger Astronomische Arbeitsgemeinschaft, 239
NAAA
  Northern Arizona Astronomy Association, 600
NAAPO
  North American Astrophysical Observatory, 841
NAAS
  Newbury Amateur Astronomical Society, 558
NAC
  Nederlandse Astronomenclub, 396
  Nepean Astronomy Centre (UWS), 32
Nacogdoches, 885
Nadir - Astronomie et Sciences, 190
NAF
  Norsk Astronautisk Forening, 415
Nagano, 353, 354, 361
Nagoya, 351, 352
Nagoya City Science Museum (NCSM)
  Astronomy Section, 351
Nagoya Radio Astronomy Laboratory, 351
Nagoya University
  Department of Physics and Astrophysics, 351
  Solar-Terrestrial Environment Laboratory, 351
    Cosmic-Ray Section, 351
  X-Ray Astronomy Group, 352
NAI
  NASA Astrobiology Institute
    Outreach Office, 630
NAIC
  National Astronomy and Ionosphere Center
    Arecibo Observatory, 437
    Headquarters, 819
Naini Tal, 288
Nairobi, 365
NAK
  Norrköpings Astronomiska Klubb, 492
Namass, 453
Namibia, 466
NAMN
  North American Meteor Network, 867
Namur, 51, 57
Nançay, 193
Nancy, 167, 176
Nandrin, 56
Nanjing, 112, 114–117
Nanjing Astronomical Instruments Co. Ltd., 116
Nanjing Normal University
  Department of Physics, 116
Nanjing University
  Astronomy Department, 117
Nanking, 112, 114–117
Nantes, 196, 201
Nanteuil le Haudouin, 172
Nantucket, 753, 754
NAO
  Nikolaev Astronomical Observatory, 528
NAOJ
  National Astronomical Observatory of Japan, 342, 346, 352
    Astrometry and Celestial Mechanics Division, 352
    Astronomical Data Analysis Center, 352
    Division of Earth Rotation, 352
    Division of Theoretical Astrophysics, 353
    Mizusawa Astrogeodynamics Observatory, 353
    Nobeyama Radio Observatory, 353
    Norikura Solar Observatory, 353
    Okayama Astrophysical Observatory, 354
    Optical and Infrared Astronomy Division, 354
    Radio Astronomy Division, 354
    Solar Physics Division, 354
    Subaru Telescope, 695
    World Data Center for Solar Activity, 362

NAP
    National Academy Press, 673
Naperville, 701, 703
Naperville Astronomical Association (NAA), 703
Napier, 409, 410
Naples, 309, 327, 329, 330, 337, 680
Napoli, 309, 327, 329, 330, 337
Nara, 350
Narberth, 778
N.A. Richardson Astronomical Observatory, 629, 636
Narrabri, 21, 22
NAS
    National Academy of Sciences, 673
    Newcastle Astronomical Society, 26
    Newcastle-upon-Tyne Astronomical Society, 558
    Nigerian Academy of Sciences, 414
    Norfolk Astronomical Society, 901
    Norsk Astronomisk Selskap, 415
    Norwegian Astronomical Society, 415
    Norwich Astronomical Society, 560
    Nottingham Astronomical Society, 560
NASA
    National Aeronautics and Space Administration, 543
NASA Astrobiology Institute (NAI)
    Outreach Office, 630
NASA Institute for Advanced Concepts (NIAC), 689
NASA/IPAC Extragalactic Database (NED), 612
NASA, National Aeronautics and Space Administration, 671

    Ames Research Center
        Astrophysics Branch, 630
        NASA Astrobiology Institute, 630
        SOFIA, 630
        Stratospheric Observatory for Infrared Astronomy, 630
    Astrophysics Data System, 751
    BioServe Space Technologies, 653
    Center for Aerospace Information, 740
    Center for the Commercial Development of Space
        BioServe Space Technologies, 653
    Dryden Flight Research Center, 630
    George C. Marshall Space Flight Center (MSFC)
        Space Science Laboratory, 590
    Goddard Institute for Space Studies, 819
    Goddard Space Flight Center (GSFC)
        Astronomy Data Center, 740
        Astrophysics Data Facility, 740
        Goddard Astronomy Club, 737
        Laboratory for High-Energy Astrophysics, 740
        National Space Science Data Center, 740
        Solar Data Analysis Center, 741
        Solar Physics Branch, 741
        Space Science Data Operations Office, 741
        Visitor Center, 741
        World Data Center A for Rockets and Satellites, 745
    Headquarters, 673
    Infrared Telescope Facility, 695
    John F. Kennedy Space Center
        Visitor Complex, 682
    John H. Glenn Research Center, 840
        Visitor Center, 840
    Langley Research Center, 900
        Virginia Air and Space Center, 905
    Lyndon B. Johnson Space Center
        Astronomical Society, 880
        Public Affairs Office, 881
    Office of Space Science, 673
    Stennis Space Center
        Public Affairs Office, 774
    Tidbindilla Space Tracking Station, 21
    Wallops Flight Facility
        Visitor Center, 900
NASDA
    National Space Development Agency of Japan, 346, 355
Nashua, 788, 790
Nashville, 870, 872, 874–876
NASM

National Air and Space Museum, 673, 675
    Albert Einstein Planetarium, 676
    Center for Earth and Planetary Studies, 676
    Space History Division, 676
NaSPA
    Network and Systems Professionals Association, 917
NASS
    National Air and Space Society, 673
    North American Sundial Society, 663
Nassau Station (Warner and Swasey Observatory), 836
NASSE
    National Association of Space Simulating Educators, 840
Nassir Addin Toussi Observatory (Khadjeh_), 293
Natal, 464
    Midlands, 464
Nathan M. Patterson Planetarium, 689
Nathan Twining Observatory (GNTO) (General_), 803
Nationaal Centrum voor Wetenschappelijke en Technische Dokumentatie (NCWTD), 55
Nationaal Instituut voor Kernfysica en Hoge-Energie Fysica (NIKHEF), 396
Nationaal Lucht- en Ruimtevaartlaboratorium (NLR), 396
National Academy of Sciences
    Space Research Institute (SRI), 364
    V.G. Fesenkov Astrophysical Institute, 364
    Tien-Shan Observatory, 364
National Academy of Sciences (NAS), 673
National Academy of Sciences of Belarus
    Institute of Molecular and Atomic Physics, 46
National Academy Press (NAP), 673
National Aeronautics and Space Administration (NASA), 543, 671
    Ames Research Center (ARC)
        Astrophysics Branch, 630
        NASA Astrobiology Institute (NAI), 630
        Stratospheric Observatory for Infrared Astronomy (SOFIA), 630
    Astrophysics Data System (ADS), 751
    BioServe Space Technologies (BST), 653
    Center for Aerospace Information (CASI), 740
    Center for the Commercial Development of Space (CCDS)
        BioServe Space Technologies (BST), 653
    Dryden Flight Research Center (DFRC), 630
    George C. Marshall Space Flight Center (MSFC)
        Space Science Laboratory, 590
    Goddard Institute for Space Studies (GISS), 818
    Goddard Space Flight Center (GSFC)
        Astronomical Data Center (ADC), 740
        Astrophysics Data Facility (ADF), 740
        Goddard Astronomy Club (GAC), 737
        Laboratory for High-Energy Astrophysics (LHEA), 740
        National Space Science Data Center (NSSDC), 740
        Solar Data Analysis Center, 741
        Solar Physics Branch, 741
        Space Science Data Operations Office (SSDOO), 741
        Visitor Center, 741
        World Data Center A for Rockets and Satellites, 745
    Headquarters, 673
    Infrared Telescope Facility (IRTF), 695
    John F. Kennedy Space Center (KSC)
        Visitor Complex, 682
    John H. Glenn Research Center (GRC), 839
        Visitor Center, 840
    Langley Research Center (LaRC), 900
        Virginia Air and Space Center (VASC), 905
    Lyndon B. Johnson Space Center (JSC)
        Astronomical Society, 880
        Public Affairs Office, 881
    Office of Space Science (OSS), 673
    Stennis Space Center (SSC)
        Public Affairs Office, 774
    Tidbindilla Space Tracking Station, 21
    Wallops Flight Facility
        Visitor Center, 900
National Aerospace Laboratory - Netherlands, 396

National Air and Space Museum (NASM), 673, 675
　　Albert Einstein Planetarium, 676
　　　Center for Earth and Planetary Studies, 676
　　　Space History Division, 676
National Air and Space Society (NASS), 673
National Association of Space Simulating Educators (NASSE), 840
National Astronomical Observatories of the Chinese Academy of Sciences, 114, 117
National Astronomical Observatory of China, 118
National Astronomical Observatory of Japan (NAOJ), 342, 346, 352
　　Astrometry and Celestial Mechanics Division, 352
　　Astronomical Data Analysis Center (ADAC), 352
　　Division of Earth Rotation, 352
　　Division of Theoretical Astrophysics, 353
　　Dodaira Observatory, 353
　　Mizusawa Astrogeodynamics Observatory, 353
　　Nobeyama Radio Observatory (NRO), 353
　　Norikura Solar Observatory, 353
　　Okayama Astrophysical Observatory (OAO), 354
　　Optical and Infrared Astronomy Division, 354
　　Radio Astronomy Division, 354
　　Solar Physics Division, 354
　　Subaru Telescope, 695
　　World Data Center for Solar Activity, 362
National Astronomical Observatory of the Bulgarian Academy of Sciences, 75
National Astronomy and Ionosphere Center (NAIC)
　　Arecibo Observatory, 437
　　Headquarters, 819
National Capital Astronomers (NCA), 674, 741
National Center for Atmospheric Research (NCAR), 656
　　High Altitude Observatory (HAO), 656
National Center for Science Education (NCSE), 630
National Center for Supercomputing Applications (NCSA), 703
National Central University (NCU)
　　Institute of Astronomy, 516
National Centre for Radio Astrophysics (NCRA)
　　Junnar Taluk, 286
　　Ooty, 286
　　Pune, 286
National Centre of the Ukrainian Youth Airspace Education, 527
National Climatic Data Center (NCDC), 829, 832
National Committee for Astronomy (NCA), Australia, 25
National Committee of Russian Physicists, 443
National Committee on Space Engineering, Australia, 23
National Deep Sky Observers Society (NDSOS), 727
National Earth Orientation Service (NEOS), 674, 677
National Geophysical Data Center (NGDC), 656
National Information Standards Organization (NISO), 741
National Institute for Fusion Science (NIFS), 354
National Institute for Research Advancement (NIRA), 355
National Institute of Aeronautics and Space (LAPAN), 290
National Institute of Meteorology and Hydrology, 438
National Institute of Polar Research (NIPR), 354
National Institute of Standards and Technology (NIST)
　　Administrative Offices, 741
　　Boulder, 655
　　Joint Institute for Laboratory Astrophysics (JILA), 655, 657
National Library of Canada (NLC), 92
National Museum of Natural Science, 517
National Museum of Science and Industry (NMSI)
　　Science Museum
　　　Astronomy Section, 557
National Museum of Science and Technology Corp. (NMSTC), 92
National Museum Planetarium, 425
National Museums and Galleries on Merseyside
　　Department of Earth and Physical Sciences, 554
　　Liverpool Museum Planetarium, 554
National Observatory of Athens (NOA)
　　Institute of Astronomy and Astrophysics, 269
National Oceanic and Atmospheric Administration (NOAA)
　　Environmental Research Laboratories (ERL)
　　　Cooperative Institute for Research in Environmental Sciences (CIRES), 657
　　ICSU Panel on World Data Centres, 655
　　National Geophysical Data Center (NGDC), 656
　　National Weather Service (NWS), 742
　　Scientific Committee on Solar-Terrestrial Physics (SCOSTEP), 658
　　Solar-Terrestrial Physics Division, 657
　　Space Environment Center (SEC), 657
　　World Data Center A for Solar-Terrestrial Physics, 661
National Optical Astronomy Observatories (NOAO), 599, 671
　　Cerro Tololo Interamerican Observatory (CTIO)
　　　La Serena, 109
　　Image Reduction and Analysis Facility (IRAF), 599
　　IRAF Project, 599
　　Kitt Peak National Observatory (KPNO), 599
　　National Solar Observatory (NSO)
　　　Sacramento Peak, 801
　　　Tucson, 599
National Photocolor Corp. (NPC), 819
National Physical Laboratory (NPL)
　　Radio and Atmospheric Sciences Division (RASD), 286
National Planetarium, Malaysia, 378
National Radio Astronomy Observatory (NRAO)
　　Arizona Operations, 599
　　Array Operations Center (AOC), 801
　　Associated Universities Inc., 896
　　Charlottesville, 900
　　Green Bank, 912
　　Tucson, 600
　　Very Large Array (VLA), 801
　　Very Long Baseline Array (VLBA), 801
National Remote Sensing Agency (NRSA), 287
National Remote Sensing Centre (NRSC) Ltd., 557
National Research Council (NRC)
　　Board on Physics and Astronomy (BPA), 674
National Research Council of Canada (NRCC), 93
　　Canadian Institute for Scientific and Technical Information (CISTI), 93
　　Herzberg Institute of Astrophysics (HIA), 93
　　　Canadian Astronomy Data Centre (CADC), 93
　　　Dominion Astrophysical Observatory (DAO), 94
　　　Dominion Radio Astrophysical Observatory (DRAO), 94
National Research Foundation (NRF), 466
　　Hartebeesthoek Radio Astronomy Observatory (HartRAO), 466
National Research Institute of Astronomy and Geophysics (NRIAG)
　　Helwân Observatory, 142
National Science Foundation (NSF), 487, 671, 900
National Science Museum
　　Department of Physical Sciences, 355
National Science Teachers Association (NSTA), 900
National Solar Observatory (NSO), 600, 671, 801
　　Sacramento Peak, 801
　　Tucson, 599
National Space Development Agency of Japan (NASDA), 346, 355
National Space Science Centre (NSSC), 558
National Space Science Data Center (NSSDC), 740, 742
National Space Society (NSS), 674
National Space Society of Australia Ltd. (NSSA), 25
National Standards Authority of Ireland (NSAI), 297
National Standards Board of Estonia (EVS), 144
National Technical Information Service (NTIS), 901
National Tsing Hua University
　　Department of Physics, 516
National Undergraduate Research Observatory (NURO), 600

National University of Athens
　　Department of Physics, 268
　　　Section of Astrophysics, Astronomy and Mechanics, 269

National University of Ireland, Galway
Department of Physics
Astrophysics and Applied Imaging Research Group, 297
National University of Ireland, Maynooth
Experimental Physics Department, 297
National University of Malaysia
Department of Physics, 378
National University of Singapore (NUS)
Department of Mathematics, 456
National Weather Service (NWS), 742
NATO
North Atlantic Treaty Organization, 55
Natrona Heights, 852
Natural Environment Research Council (NERC), 558
Natural History Museum
Meteoritic Research, 558
Natural Resources Canada
Geodetic Survey Division, 94
Natural Science Center, Greensboro, 828
Nature, 558
Naturhistoriska Museet, Göteborg, 489, 493
Naturvetenskapliga Forskningsrådet (NFR), 492, 496
Naturwissenschaftliche Rundschau, 239
Naturwissenschaftlicher Verein für Bielefeld und Umgegend eV
Arbeitsgemeinschaft Astronomie, 239
NAU
Northern Arizona University
Department of Physics and Astronomy, 600
Nauchny, 525
Nauheim (Bad.), 263
Naval Academy (US.)
Physics Department, 744
Naval Observatory (US.)
Flagstaff Station, 604
National Earth Rotation Service (NEOS), 677
Washington, 676
Naval Research Laboratory (NRL)
Space Sciences Division (SSD), 674
NBANB
New Brunswick Astronomy - Astronomie Nouveau Brunswick, 94
NBI
Niels Bohr Institute, 136
NBIfAFG
Niels Bohr Institute for Astronomy, Physics and Geophysics
Astronomical Observatory, 135, 136
Danish Center for Earth System Science, 137
Department of Geophysics, 138
Niels Bohr Institute, 136
Ørsted Laboratory, 136
NCA
National Capital Astronomers, 741
National Committee for Astronomy
Australia, 25
Nederlands Comité Astronomie, 396
NCAC
N. Copernicus Astronomical Center, 429
Department of Astrophysics I, 430
NCAR
National Center for Atmospheric Research, 656
High Altitude Observatory (HAO), 656
NCAS
Northern Colorado Astronomical Society, 657
NCC
Network Cybernetics Corp., 881
NCDC
National Climatic Data Center, 829, 832
NCMLS
North Carolina Museum of Life and Science, 830
N. Copernicus Astronomical Center (NCAC), 428, 429
Department of Astrophysics I, 430
NCRA, National Centre for Radio Astrophysics
Junnar Taluk, 286
Ooty, 286
Pune, 286
NCSA
National Center for Supercomputing Applications, 703
NCSE
National Center for Science Education, 630
NCSF
Northern Cross Science Foundation, 918
NCSM
Nagoya City Science Museum
Astronomy Section, 351
NCSU
North Carolina State University
Astrophysics Group, 830
Department of Physics, 830
NCU
National Central University
Institute of Astronomy, 516
NCWTD
Nationaal Centrum voor Wetenschappelijke en Technische Dokumentatie, 55
NDAG
Notre Dame Astrophysics Group, 714
NDAS
North Devon Astronomical Society, 559
NDSOS
National Deep Sky Observers Society, 727
Nebraska, 782–785
Nebraska Star Party (NSP), 783
Nebraska Wesleyan University
Physics Department
Jensen Planetarium, 783
NED
NASA/IPAC Extragalactic Database, 612
Nederlands Comité Astronomie (NCA), 396
Nederlandse Astronomenclub (NAC), 396
Nederlandse Jeugdvereniging voor Ruimtevaart en Sterrenkunde (NJRS), 397
Nederlandse Natuurkundige Vereniging (NNV), 397
Nederlandse Organisatie voor Toegepast-Natuurwetenschappelijk Onderzoek (TNO), 397
Nederlandse Organisatie voor Wetenschappelijk Onderzoek (NWO), 397
Nederlandse Vereniging voor Ruimtevaart, 397
Nederlandse Vereniging voor Weer- en Sterrenkunde (NVWS), 398
Nederlands Instituut voor Vliegtuigontwikkeling en Ruimtevaart (NIVR), 398
Nederlands Normalisatie-Instituut (NNI), 398
Nederlands Onderzoekschool voor Astronomie (NOVA), 398

Nedlands, 31
Needville, 879
NEFAS
Northeast Florida Astronomical Society, 682
Nehru Planetarium, 287
Neigem, 61
Neil Armstrong Air & Space Museum, 840
NEKAAL
Northeast Kansas Amateur Astronomers League, 723
NELAS
North East London Astronomical Society, 559
NELPAG
New England Light Pollution Advisory Group, 756
Nentershausen, 234
NEOS
National Earth Orientation Service, 677
Nepean, 87
Nepean Astronomy Centre (NAC), 25, 32
NERC
Natural Environment Research Council, 558
NEROC
Northeast Radio Observatory Corp.
Haystack Observatory, 756
Nessa, 215
Netherlands, 389
Netherlands Agency for Aerospace Programmes, 398
Netherlands Foundation for Radio Astronomy (NFRA), 398

Netherlands Foundation for Research in Astronomy (NFRA), 398
Netherlands Industrial Space Organisation (NISO), 398
Netherlands Remote Sensing Board, 399
Netherlands Research School for Astronomy, 398, 399
Network and Systems Professionals Association (NaSPA) Inc., 917
Network Cybernetics Corp. (NCC), 881
Neuchâtel, 507, 511
Neuenhaus, 214
Neuf-Brisach, 189
Neufchâteau, 47
Neukirch, 508
Neumarkt, 219
Neusäß-Vogelsang, 252
Nevada, 786, 787
    (Southern_), 786
Neville Public Museum Astronomical Society (NPMAS), 917

Newark, 651, 667, 668, 791, 794
Newark Museum, 791
New Bedford, 746
New Bombay, 283
New Brunswick Astronomy - Astronomie Nouveau Brunswick (NBANB) Inc., 94
Newbury, 535, 558
Newbury Amateur Astronomical Society (NAAS), 558
Newburyport, 752
New Castle, 667
Newcastle Astronomical Society (NAS), 25
Newcastle-upon-Tyne, 558, 579
Newcastle-upon-Tyne Astronomical Society (NAS), 558
New Delhi, 281–284, 286, 287
New England
    (Northern_), 732
    (Southern_), 748
New England Light Pollution Advisory Group (NELPAG), 756
New England Meteoritical Services, 756
Newfoundland, 92
Newfoundland Science Centre, 95
Newhall, 644
New Hampshire, 788–790
New Hampshire Astronomical Society (NHAS), 789
New Haven, 662, 664–666, 777
New Hope, 860
New Jersey, 791–798
New Jersey Astronomical Association (NJAA), 794
New Jersey Institute of Technology (NJIT)
    Big Bear Solar Observatory (BBSO), 630, 794
    Center for Solar Research (CFSR), 794
    Owens Valley Radio Observatory (OVRO), 794
New Jersey State Museum
    Planetarium, 794
New Lebanon, 810
New London, 662
Newmarket, 105
New Mexico, 799–804
New Mexico Institute of Mining and Technology (NMIMT)
    Astronomy Club, 802
    Department of Physics
        Astrophysics Research Center (ARC), 801
    Langmuir Laboratory for Atmospheric Research, 802
New Mexico State University (NMSU)
    Department of Astronomy, 802
New Mexico Tech, 802
    Astronomy Club, 802
New Milford, 662, 665
New Mumbai, 283
New Orleans, 729
New Paltz, 818
Newport, 714
Newport Industrial Glass Inc., 631
Newport News, 895, 905
New Salem, 751
New South Wales, 15, 18, 30

NEWSTAR
    Northeast Wisconsin Stargazers, 918
Newton Abbot, 573
Newton Falls, 839
Newton-in-Furness, 546
Newton Institute for Mathematical Sciences (Isaac_), 552, 558
Newton Observatory, 851
New York, 670, 796, 805–816, 819, 822
New York Academy of Sciences (NYAS), 819
New York Hall of Science (NYHoS), 819
New York State, 805–826
New York University
    Physics Department, 819
New Zealand, 408
New Zealand Institute of Physics (NZIP), 410
New Zealand National Committee for Astronomy, 411
New Zealand Spaceflight Association (NZSA) Inc., 410
New Zealand Standing Committee on the Astronomical Sciences, 411
NFR
    Naturvetenskapliga Forskningsrådet, 496
NFS
    Norsk Fysisk Selskap, 416
NGAO
    North Georgia Astronomical Observatory, 690
NGC
    North Georgia College
        George E. Coleman Sr. Planetarium, 690
NGC 69
    Nouvelle Gazette du Club 69, 171
NGCSU
    North Georgia College and State University
        Physics Department, 690
NGDC
    National Geophysical Data Center, 656
NGI
    Norwegian Geotechnical Institute, 416
NHAC
    North Houston Astronomy Club, 882
NHAO
    Nishi-Harima Astronomical Observatory, 355
NHAS
    New Hampshire Astronomical Society, 789
NIA
    Northwest Iowa Astronomers, 718
NIAG
    Northern Indiana Astronomical Group, 713
NIAGR
    National Research Institute of Astronomy and Geophysics
        Helwân Observatory, 142
Nice, 160, 173, 191, 192, 195, 204
Nicholas E. Wagman Observatory, 851
Nick Enterprise Co. Ltd., 517
Nicolas Copernicus Museum, 429
Nicolas Copernicus Planetarium and Observatory in Chorzów, 429
Nicolaus Copernicus Planetarium, 239
Niederglatt, 501
Niedernhausen, 212
Niederteufen, 502
Niels Bohr Institute for Astronomy, Physics and Geophysics (NBIfAFG), 135, 138
    Astronomical Observatory, 136
    Danish Center for Earth System Science (DCESS), 137
    Department of Geophysics, 138
    Niels Bohr Institute (NBI), 136
    Ørsted Laboratory, 136
Niels Bohr Institute (NBI), 136, 138
Niepolomice, 428
NIFS
    National Institute for Fusion Science, 354
Nigeria, 414
Nigerian Academy of Sciences (NAS), 414
Night Sky Observers (NSO), 727
Nijmegen, 403, 405

NIKHEF
    Nationaal Instituut voor Kernfysica en Hoge-Energie Fysica, 396
Nikolaev, 527
Nikolaev Astronomical Observatory (NAO), 528
Nikon USA, 820
NIMH
    National Institute of Meteorology and Hydrology - Bulgaria, 75
Nipomo, 642
Nippes, 250
Nippon Meteor Society (NMS), 355
Nippon Silica Glass USA Inc., 794
NIPR
    National Institute of Polar Research, 354
NIRA
    National Institute for Research Advancement, 355
Nishi-Harima Astronomical Observatory (NHAO), 355
Niskayuna, 805
NISO
    National Information Standards Organization, 741
    Netherlands Industrial Space Organisation, 399
NIST
    National Institute of Standards and Technology
        Administrative Offices, 741
        Boulder, 655
        Joint Institute for Laboratory Astrophysics, 655
Nitra, 458
Nitra Observatory, 458
NIU
    Northern Illinois University
        Physics Department, 704
NIVR
    Nederlands Instituut voor Vliegtuigontwikkeling en Ruimtevaart, 398
Niwot, 654
Nizamiah Observatory, 287
Nizhny Novgorod, 443, 446
Nizhny Novgorod Planetarium, 443
NJAA
    New Jersey Astronomical Association, 794
NJAG
    North Jersey Astronomical Group, 794
NJIT
    New Jersey Institute of Technology
        Big Bear Solar Observatory, 630, 794
        Center for Solar Research, 794
        Owens Valley Radio Observatory, 794
NJRS
    Nederlandse Jeugdvereniging voor Ruimtevaart en Sterrenkunde, 397
NLR
    Nationaal Lucht- en Ruimtevaartlaboratorium, 396
NMC
    Natal Midlands Centre (ASSA), 464
    Northwestern Michigan College
        Joseph H. Rogers Observatory, 764, 766
NMIMT
    New Mexico Institute of Mining and Technology
        Astronomy Club, 802
        Astrophysics Research Center, 801
        Department of Physics, 801
        Langmuir Laboratory for Atmospheric Research, 802
NMS
    Nippon Meteor Society, 355
NMSI
    National Museum of Science and Industry
        Astronomy Section, 557
        Science Museum, 557
NMSTC
    National Museum of Science and Technology Corp., 93
NMSU
    New Mexico State University
        Department of Astronomy, 802
NNAC
    Northeast Nebraska Astronomy Club, 783
NNHS
    Northamptonshire Natural History Society
        Astronomy Section, 559
NNI
    Nederlands Normalisatie-Instituut, 398
NNV
    Nederlandse Natuurkundige Vereniging, 397
NOA
    National Observatory of Athens
        Institute of Astronomy and Astrophysics, 269
NOAA
    National Oceanic and Atmospheric Administration
        ICSU Panel on World Data Centres, 655
        National Geophysical Data Center, 656
        National Weather Service, 742
        Scientific Committee on Solar-Terrestrial Physics, 658
        Solar-Terrestrial Physics Division, 657
        Space Environment Center (SEC), 657
        World Data Center A for Solar-Terrestrial Physics, 661
Noank, 664
NOAO, National Optical Astronomy Observatories, 599, 671

    Cerro Tololo Interamerican Observatory
        La Serena, 109
    Image Reduction and Analysis Facility, 599
    Kitt Peak National Observatory, 599
    National Solar Observatory (NSO)
        Sacramento Peak, 801
        Tucson, 599
Nobel Foundation, 492
Nobeyama Radio Observatory (NRO), 352–354, 356
Noble Planetarium, 881
Noblitt Planetarium (L.S._), 712, 713
Noctilume, 631
NOFS
    Naval Observatory Flagstaff Station, 604
Nogales, 602
Nokogiriyama Station, 360
Nonius BV, 399
Noordwijk, 392, 399
Noordwijk Space Expo (NSE), 399
Nordenham, 258
Norderstedt, 262
Nordic Institute for Theoretical Physics (NORDITA), 138
Nordic Optical Telescope (NOT), 480
Nordic Optical Telescope Scientific Association (NOTSA), 480
Nordic Planetarium Association (NPA), 492
NORDITA
    Nordic Institute for Theoretical Physics, 138
Nordlysobservatoriet, 415, 420
Norfolk, 783, 896, 901
Norfolk Astronomical Society (NAS), 901
Norges Standardiseringsforbund (NSF), 417
Norikura (Mount._), 353, 360
Norikura Solar Observatory, 352, 353, 356
Normal, 702, 707
Norman, 846, 847
Norman Lockyer Observatory and James Lockyer Planetarium, 558
Norms
    ABNT (Brazil), 65
    AENOR (Spain), 474
    AFNOR (France), 159
    Albania, 3
    Alger, 4
    Algeria, 4
    American National Standards Institute, 670
    Amman, 363
    Ankara, 524
    ANSI (USA), 670
    Arabia (Saudi_), 453
    Argentina, 7
    Armenia, 12
    Asociación Española de Normalización y Certificación, 474
    Asociatia de Standardizare din Romania, 438

ASRO (Romania), 438
Associação Brasileira de Normas Técnicas, 65
Association Française de Normalisation, 159
Astana, 364
Asunción, 422
Athinai, 268
Australia, 28
Austria, 38
Azerbaijan, 43
Baghdad, 294
Baku, 43
Bangkok, 519
Bangladesh Standards and Testing Institution, 44
Barbados National Standards Institution, 45
BDS (Bulgaria), 76
Beijing, 113
Beirut, 373
Belarus, 46
Belarusian Committee for Standardization, Metrology and Certification, 46
Belgium, 52
BELST (Belarus), 46
Beograd, 455
Berlin, 224
Bethesda, 741
Beyrouth, 373
BIPM, 163
BIS (India), 282
BNQ (Québec), 79
BNSI (Barbados), 45
Bogotá, 119
Bolivia, 62
BPS (Philippines), 425
Bratislava, 460
Braunschweig, 240
Brazil, 65
Brussel, 51
Brussels, 51
Bruxelles, 51, 52
BSI (United Kingdom), 539
BSTI (Bangladesh), 44
Bucureşti, 438
Budapest, 276
Buenos Aires, 7
Bulgaria, 76
Bureau de Normalisation du Québec, 79
Bureau International des Poids et Mesures, 163
Bureau of Indian Standards, 282
Bureau of Product Standards - Philippines, 425
Cairo, 142
Canada, 79, 99
Caracas, 927
CCN (Canada), 99
CENELEC (EU), 51
CEN (EU), 51
Central Organization for Standardization and Quality Control (Iraq), 294
Chile, 109
China
  (PRC), 113
China State Bureau of Technical Supervision, 113
CN (Cuba), 123
Colombia, 119
Colombo, 487
Comisión Panamericana de Normas Técnicas, 927
Comisión Venezolana de Normas Industriales, 927
Comité Estatal de Normalización (Cuba), 123
Comité Européen de Normalisation, 51
Comité Européen de Normalisation Électrotechnique, 51
Committee for Standardization and Metrology - Bulgaria, 76
Committee for Standardization of the DPRK, 366
CONACYT (El Salvador), 143
Conseil Canadien des Normes, 99
COPANT, 927
COSCQ (Iraq), 294

COVENIN (Venezuela), 927
Croatia, 122
CSBTS (China/PRC), 113
CSK (Korea/DPRK), 366
Cuba, 123
Czech Republic, 127
Czech Standards Institute, 127
Damascus, 515
Danish Standards Association, 137
Delft, 398
Denmark, 137
Department for Standardization, Metrology and Certification (Armenia), 12
Deutsches Institut für Normung, 224
Dewan Standardisasi Nasional, 291
DGN (Mexico), 383
Dhaka, 44
DIN (Germany), 224
Dirección General de Normas - Mexico, 383
Direction of Standards (Albania), 3
DSC (Albania), 3
DS (Denmark), 137
DSN (Indonesia), 291
DSTU (Ukraine), 528
Dublin, 297
DZNM (Croatia), 122
Ecuador, 141
Egypt, 142
Egyptian Organization for Standardization and Quality Control, 142
ELOT (Greece), 268
El Salvador, 143
Ente Nazionale Italiano di Unificazione, 317
EOS (Egypt), 142
Estonia, 144
ETSI, 178
European Committee for Electrotechnical Standardization, 51
European Committee for Standardization, 51
European Telecommunications Standards Institute, 178
EVS (Estonia), 144
Finland, 147
Finnish Standards Association, 147
France, 159
Gaithersburg, 741
Genève, 505
Georgia, 206
Germany, 224
GOST (Russia), 449
Greece, 268
Harare, 929
Hellenic Organization for Standardization, 268
Hellerup, 137
Helsinki, 147
Hungarian Standards Institution, 276
Hungary, 276
IBN (Belgium), 52
IBNORCA (Bolivia), 62
Iceland, 279
Icelandic Standards, 279
ICONTEC (Colombia), 119
IEC, 505
INAPI (Algeria), 4
INDECOPI (Peru), 423
India, 282
Indonesia, 291
INEN (Ecuador), 141
INN (Chile), 109
Institut Belge de Normalisation, 52
Institute of Standardization -Serbia, 455
Institute of Standards and Industrial Research of Iran, 292
Institut National Algérien de la Propriété Industrielle, 4
Instituto Argentino de Racionalización de Materiales, 7
Instituto Boliviano de Normalización y Calidad, 62

Instituto Colombiano de Normas Técnicas, 119
Instituto Ecuatoriano de Normalización, 141
Instituto Nacional de Defensa de la Competencia y de la Protección de la Propiedad Intelectual - Peru, 423
Instituto Nacional de Normalización (Chile), 109
Instituto Português da Qualidade, 434
Instituto Uruguayo de Normas Técnicas, 587
International Electrotechnical Commission, 505
International Organization for Standardization, 505
INTN (Paraguay), 422
IPQ (Portugal), 434
IRAM (Argentina), 7
Iran, 292
Iraq, 294
Ireland, 297
ISIRI (Iran), 292
ISO, 505
Israel, 300
IST (Iceland), 279
Italy, 317
Jakarta, 291
Japan, 346
Japanese Industrial Standards Committee, 346
JISC (Japan), 346
JISM (Jordania), 363
Karachi, 421
Kazakhstan, 364
KEBS (Kenya), 365
Kenya Bureau of Standards, 365
Kiev, 528
Kishinev, 387
KNITQ (Korea/ROK), 368
Korea
    (DPRK), 366
    (ROK), 368
Korean National Institute of Technology and Quality, 368
Kyiv, 528
Kyunggi-do, 368
Lagos, 414
La Habana, 123
La Paz, 62
Latvia, 371
Latvian National Center for Standardization and Metrology, 371
Lebanon, 373
LIBNOR (Lebanon), 373
Lima, 423
Lisboa, 434
Lithuania, 375
Lithuanian Standards Board, 375
Ljubljana, 462
LST (Lithuania), 375
Luxembourg, 376
LVS (Latvia), 371
Macedonia, 377
Madrid, 474
Magyar Szabványügyi Testület, 276
Makati, 425
Malaysia, 379
Malta, 380
Manila, 425
Maryland, 741
Mauritius, 381
México, 383
Mexico, 383
Milano, 317
Minsk, 46
MNISM (Mongolia), 388
Moldova, 387
Mongolia, 388
Mongolian National Institute for Standardisation and Metrology, 388
Montevideo, 587
Moscow, 449
MSA (Malta), 380
MSB (Mauritius), 381

MSZT (Hungary), 276
Nairobi, 365
National Information Standards Organization (USA), 741
National Institute of Standards and Technology
    Administrative Offices, 741
    Gaithersburg, 741
National Standards Authority of Ireland, 297
National Standards Board of Estonia, 144
Nederlands Normalisatie-Instituut, 398
Netherlands, 398
New Delhi, 282
New York, 670
New Zealand, 411
Nigeria, 414
NISO (USA), 741
NIST
    Administrative Offices, 741
    Gaithersburg, 741
NNI (Netherlands), 398
Norges Standardiseringsforbund, 417
Norway, 417
Norwegian Standards Association, 417
NSAI (Ireland), 297
NSF (Norway), 417
ÖN (Austria), 38
Oslo, 417
Österreichisches Normungsinstitut, 38
Ottawa, 99
Pakistan Standards Institution, 421
Paraguay, 422
Paris, 159
Peru, 423
Philippines, 425
Physikalisch-Technische Bundesanstalt, 240
PKN (Poland), 431
Poland, 431
Polish Committee for Standardization, 431
Polski Komitet Normalizacyjny, 431
Port of Spain, 521
Portugal, 434
Praha, 127
Pretoria, 467
PSB (Singapore), 456
PSI (Pakistan), 421
PTB (Germany), 240
Pyongyang, 366
Québec, 79
Quito, 141
Reduit, 381
Reykjavík, 279
Riga, 371
Rio de Janeiro, 65
Riyadh, 453
Romania, 438
Russia, 449
SABS (South Africa), 467
SAI (Australia), 28
Saint-Michael, 45
San Salvador, 143
Santiago, 109
SARM (Armenia), 12
SASMO (Syria), 515
SASO (Saudi Arabia), 453
Saudi Arabia, 453
Saudi Arabian Standards Organization, 453
SAZ (Zimbabwe), 929
SCC (Canada), 99
Schweizerische Normen-Vereinigung, 509
SEE (Luxembourg), 376
Serbia, 455
Sèvres, 163
SFS (Finland), 147
Shah Alam, 379
SII (Israel), 300
Singapore, 456
SIRIM (Malaysia), 379

SISIR (Singapore), 456
SIS (Sweden), 497
Skopje, 377
Slovak Office of Standards, Metrology and Testing, 460
Slovak Republic, 460
Slovenia, 462
SLSI (Sri Lanka), 487
SMIS (Slovenia), 462
SNV (Switzerland), 509
SNZ (New Zealand), 411
Sofia, 76
SON (Nigeria), 414
Sophia-Antipolis, 178
South Africa, 467
South African Bureau of Standards, 467
Spain, 474
Sri Lanka Standards Institution, 487
Standardiseringen i Sverige, 497
Standardization Council of Indonesia, 291
Standards and Industrial Research Institute of Malaysia, 379
Standards and Metrology Institute of Slovenia, 462
Standards Association of Zimbabwe, 929
Standards Australia International, 28
Standards Council of Canada, 99
Standards Institution of Israel, 300
Standards New Zealand, 411
State Committee for Standardization and Metrology (Russia), 449
State Committee of Ukraine for Standardization, Metrology and Certification, 528
State Office for Standardization and Metrology - Croatia, 122
St.-Michael, 45
Stockholm, 497
Sweden, 497
Swedish Standards Institution, 497
Swiss Association for Standardization, 509
Switzerland, 509
Sydney, 28
Syria, 515
Syrian Arab Organization for Standardization and Metrology, 515
SZS (Serbia), 455
Tallinn, 144
Tashkent, 925
Tbilisi, 206
Tehran, 292
Tel-Aviv, 300
Thai Industrial Standards Institute, 519
Thailand, 519
Tirana, 3
TISI (Thailand), 519
Tobago (Trinidad &.), 521
Tokyo, 346
Trinidad & Tobago Bureau of Standards, 521
TSE (Turkey), 524
TTBS (Trinida & Tobago), 521
Turkey, 524
Turkish Standards Institution, 524
Ukraine, 528
Ulaanbataar, 388
UNI (Italy), 317
United Kingdom, 539
UNIT (Uruguay), 587
UNMS (Slovakia), 460
Uruguay, 587
USA-MD, 741
USA-NY, 670
Uzbekistan, 925
Uzbek State Centre for Standardization, Metrology and Certification, 925
UZGOST (Uzbekistan), 925
Valletta, 380
Venezuela, 927
Vilnius, 375
Warszawa, 431

Wellington, 411
Wien, 38
Winterthur, 509
Yerevan, 12
Zagreb, 122
Zavod za Standardizacija i Metrologija - Macedonia, 377
Zimbabwe, 929
ZSM (Macedonia), 377
Norrköping, 492, 496
Norrköpings Astronomiska Klubb (NAK), 492
Norsk Astronautisk Forening (NAF), 415
Norsk Astronomisk Selskap (NAS), 415
Norske Videnskaps-Akademi, 416
Norsk Fysisk Selskap (NFS), 416
Norsk Teknisk Museum (NTM), 416
North Alabama, 591
North American Astrophysical Observatory (NAAPO), 841
North American Meteor Network (NAMN), 867
North American Skies, 657
North American Sundial Society (NASS), 663
Northampton, 559, 758
Northamptonshire Natural History Society (NNHS)
    Astronomy Section, 559
Northants Amateur Astronomers (NAA), 559
North Atlanta Planetarium, 689
North Atlantic Treaty Organization (NATO)
    Division of Scientific Affairs, 55
North Bay, 95
North Bay Astronomy Club, 95
North Branch, 796
North Carolina, 827–832
North Carolina A&T State University
    Department of Physics, 830
North Carolina Museum of Life and Science (NCMLS), 830
North Carolina State University (NCSU)
    Department of Physics, 830
        Astrophysics Group, 830
North Cornelly, 537, 546
North Dakota, 834
North Devon Astronomical Society (NDAS), 559
Northeast Bronx Planetarium, 820
Northeastern University
    Department of Physics, 756
Northeast Florida Astronomical Society (NEFAS), 682
Northeast Kansas Amateur Astronomers League (NEKAAL) Inc., 723
North East London Astronomical Society (NELAS), 559
Northeast Nebraska Astronomy Club (NNAC), 783
Northeast Radio Observatory Corp. (NEROC)
    Haystack Observatory, 756
Northeast Wisconsin Stargazers (NEWSTAR), 918
Northern Arizona Astronomy Association (NAAA), 600
Northern Arizona University (NAU)
    Department of Physics and Astronomy, 600
Northern Colorado Astronomical Society (NCAS), 657
Northern Cross Science Foundation (NCSF), 918
Northern Illinois Astronomers, 703
Northern Illinois University (NIU)
    Physics Department, 703
Northern Indiana Astronomical Group (NIAG), 713
Northern Iowa, 720
Northern Ireland, 531–533, 543, 551, 563
Northern Skies Astronomical Society (NSAS), 834
Northern Sydney Astronomical Society (NSAS), 26
Northern Virginia Astronomy Club (NOVAC), 901
Northfield, 770
North Georgia Astronomers, 689
North Georgia Astronomical Observatory (NGAO), 690
North Georgia College and State University (NGCSU)
    Physics Department, 690
North Georgia College (NGC)
    George E. Coleman Sr. Planetarium, 690
North Houston Astronomy Club (NHAC), 882
North Jersey Astronomical Group (NJAG), 794
North Las Vegas, 786
North Liberty, 801

North Massapequa, 820
North Melbourne, 18
North Merrick, 807
Northmoor Observatory, 704
North Museum of Natural History and Science
    Planetarium, 859
North Olmsted, 837
Northport
    (East-), 816
North Richard Hills, 881
Northridge, 615
North Scituate, 864
North Shore Amateur Astronomy Club (NSAAC), 756
North Sutton, 789
North Syracuse, 823
North Texas Skywatch, 882
North Turramurra, 26, 28
Northvale, 793
North Valley Astronomers (NOVA), 631
North West Astronomy Group (NWAG), 849
Northwestern Michigan College (NMC)
    Joseph H. Rogers Observatory, 764, 766
Northwestern University
    Department of Physics and Astronomy
        Dearborn Observatory, 700, 704
North West Group of Astronomical Societies (NWGAS), 559

Northwest Iowa Astronomers (NIA), 718
Northwest Ohio Visual Astronomers (NOVA), 840
Northwest Suburban Astronomers (NSA), 704
North York, 96, 98, 105
North York Astronomical Association (NYAA), 95
Norton, 759
Norton Observatory, 750, 757
Norway, 415, 418
Norwegian Academy of Science and Letters, 416
Norwegian Astronautical Association, 415, 416
Norwegian Astronomical Society (NAS), 416
Norwegian Geotechnical Institute (NGI), 416
Norwegian Meteorological Institute (DNMI), 417
Norwegian Museum of Science and Technology, 416
Norwegian Physical Society, 417
Norwegian Space Centre (NSC), 417
    Andøya Rocket Range (ARR), 417
    Tromsø Satellite Station (TSS), 417, 419
Norwegian Standards Association, 417
Norwegian University of Science and Technology
    Institute of Physics, 418
Norwell, 753, 758
Norwich, 537, 560, 893
Norwich Astronomical Society (NAS), 559
NOT
    Nordic Optical Telescope, 480
Noto, 316
Notre Dame, 714
Notre Dame Astrophysics Group (NDAG), 714
NOTSA
    Nordic Optical Telescope Scientific Association, 480
Nottingham, 538, 560, 563, 580
Nottingham Astronomical Society (NAS), 560
NOVA
    Nederlands Onderzoekschool voor Astronomie, 398
    North Valley Astronomers, 631
    Northwest Ohio Visual Astronomers, 840
Nova Astronomics, 95
NOVAC
    Northern Virginia Astronomy Club, 901
Novaspace Galleries, 600
Novegno (Monte-), 320
Novespace SA, 190
Novi Ligure, 335
Novins Planetarium (Robert J.-), 794, 796
    ASTRA, 791
Novokuzneck, 443
Novokuzneck Planetarium, 443
NPA
    Nordic Planetarium Association, 492

NPC
    National Photocolor Corp., 819
NPL
    National Physical Laboratory
        Radio and Atmospheric Sciences Division, 286
NPMAS
    Neville Public Museum Astronomical Society, 917
NRAL
    Nuffield Radio Astronomy Laboratories, 579
NRAO
    National Radio Astronomy Observatory
        Associated Universities Inc., 896
        Charlottesville, 900
        Green Bank, 913
        Tucson, 600
        Very Large Array, 801
        Very Long Baseline Array, 801
NRC
    National Research Council
        Board on Physics and Astronomy, 674
NRCC, National Research Council of Canada, 93
    Canadian Institute for Scientific and Technical Information, 93
    Herzberg Institute of Astrophysics, 93
        Canadian Astronomy Data Centre, 93
        Dominion Astrophysical Observatory, 94
        Dominion Radio Astrophysical Observatory, 94
NRF
    National Research Foundation, 466
        Hartebeesthoek Radio Astronomy Observatory, 466
NRL
    Naval Research Laboratory
        Space Sciences Division, 674
NRO
    Nobeyama Radio Observatory, 352-354
NRSA
    National Remote Sensing Agency (India), 287
NRSC
    National Remote Sensing Centre, 557
NSA
    Northwest Suburban Astronomers, 704
NSAAC
    North Shore Amateur Astronomy Club, 756
NSAI
    National Standards Authority of Ireland, 297
NSAS
    Northern Skies Astronomical Society, 834
NSC
    Norwegian Space Centre, 417
        Andøya Rocket Range, 417
        Tromsø Satellite Station, 417, 419
NSE
    Noordwijk Space Expo, 399
NSF
    National Science Foundation, 671, 900
    National Science Foundation (Sri Lanka), 487
    Norges Standardiseringsforbund, 417
NSO
    National Solar Observatory, 671
        Sacramento Peak, 801
        Tucson, 599
    Night Sky Observers, 727
NSP
    Nebraska Star Party, 783
NSS
    National Space Society, 674
NSSA
    National Space Society of Australia, 25
NSSC
    National Space Science Centre, 558
NSSDC
    National Space Science Data Center, 740
NSTA
    National Science Teachers Association, 900
Nsukka, 414
NTIS
    National Technical Information Service, 901

NTM
- Norsk Teknisk Museum, 416

NTO
- Nuevas Tecnologías Observacionales, 480

Nüchter Sternwarte (Hans-_), 232, 239
Nuenen, 389
Nuevas Tecnologías Observacionales (NTO), 480
Nuffield, 579
Nuffield Radio Astronomy Laboratories (NRAL), 560, 579
Nuovo Orione, 330
Nuremberg, 217, 238, 239, 265
- (Erlangen-_), 254

Nurmijärví, 147
Nurmijärví Geophysical Observatory, 149
Nürnberg, 217, 238, 239, 265
- (Erlangen-_), 254

Nürnberger Astronomische Arbeitsgemeinschaft (NAA) eV, 239

NURO
- National Undergraduate Research Observatory, 600

Nursery, 878
Nürtingen, 216

NUS
- National University of Singapore
  - Department of Mathematics, 456

Nutssterrenwacht Ommen, 399

NVWS
- Nederlandse Vereniging voor Weer- en Sterrenkunde, 398

NWAG
- North West Astronomy Group, 849

NWGAS
- North West Group of Astronomical Societies, 559

NWO
- Nederlandse Organisatie voor Wetenschappelijk Onderzoek, 397

NYAA
- North York Astronomical Association, 95

NYAS
- New York Academy of Sciences, 819

NYHoS
- New York Hall of Science, 819

NZIP
- New Zealand Institute of Physics, 410

NZSA
- New Zealand Spaceflight Association, 410

O

OAA
- Observatório Astronômico Antares, 68
- Oglethorpe Astronomical Association, 690
- Osservatorio Astrofisico di Arcetri, 326

ØAA
- Østjyske Amatør Astronomer, 134

OAB
- Osservatorio Astronomico di Bologna, 326, 331
- Osservatorio Astronomico di Brera
  - Merate, 326
  - Milano, 326

OAC
- Array Operations Center, NRAO, 801
- Oakland Astronomy Club, 766
- Observatoire de la Côte d'Azur, 160, 191
- Observatorio Astronómico de Córdoba, 6, 10
- Osservatorio Astrofisico di Catania, 326
- Osservatorio Astronomico di Cagliari, 327
- Osservatorio Astronomico di Capodimonte, 327

OACD
- Observatoire de l'Association Culturelle de Dax, 192

OAEM
- Observatório Astronômico da Escuela de Minas, 71

OAFA
- Observatorio Astronómico Felix Aguilar, 10

OAG
- Observatori Astronòmic del Garraf, 480
- Observatorio Astronómico del Garraf, 480

OAGH
- Observatorio Astrofísico Guillermo Haro (INAOE), 383

OAHE
- Observatório Astronômico Herschel-Einstein, 69

OAI
- Ohio Aerospace Institute, 841

Oak Creek, 917
Oakland, 610, 616, 619, 620, 672, 737
Oakland Astronomy Club (OAC), 766
Oak Ridge, 751, 871, 873
Oak Ridge Isochronous Observation Network (ORION), 873

Oakton, 901
Oakton High School Planetarium, 901

OAL
- Observatório Astronômico de Lisboa, 434, 435

OALM
- Observatorio Astronómico Los Molinos, 587

OAM
- Observatoire des Alpes Maritimes, 193
- Observatorio Astronómico de Mallorca, 480
- Observatório Astronômico Monoceros, 69
- Observatorio Astronómico Móvil, 8

OAMM
- Observatoire Astronomique du Mont Mégantic, 100
- Observatorio Astronómico Municipal de Mercedes, 8

OAN
- Observatorio Astronómico Nacional, 480
  - Bogotá, 120
  - Centro Astronómico de Yebes, 481
  - Ensenada, 386
  - Observatorio de Calar Alto, 481
  - San Diego, 646
  - San Ysidro, 646

OAO
- Okayama Astrophysical Observatory, 352, 354

OAOG
- Ottawa Valley Astronomy and Observers Group, 96

OAP
- Observatoire Astronomique des Pises, 190, 200
- Observatorio Astronómico de Patacamaya (UMSA), 63
- Observatório Astronômico de Piedade (UFMG), 72
- Osservatorio Astronomico di Palermo, 327

OAQ
- Observatorio Astronómico de Quito, 141

OAR
- Osservatorio Astronomico di Roma
  - Monteporzio Catone, 328
  - Roma, 328

OAS
- Ogden Astronomical Society, 890
- Olympic Astronomical Society, 908
- Omaha Astronomical Society, 783
- Organization of American States
  - Office of Science and Technology, 675
- Orpington Astronomical Society, 561

ÖAS
- Östergötland Astronomiska Sällskap, 492

OASI
- Orwell Astronomical Society Ipswich, 561

OAT
- Osservatorio Astronomico di Trieste, 328

OATo
- Osservatorio Astronomico di Torino, 328

OAU
- Observatório Astronômico de Uberlândia, 69

ÖAW, Österreichische Akademie der Wissenschaften, 37
- Institut für Weltraumforschung, 37

Oberhaching, 229
Oberherten, 501
Oberkochen, 210
Oberlin, 840
Oberlin College
- Department of Physics, 840

Oberpfaffenhofen, 225
Oberti (Daniel_), 618, 631

Oberwallis, 501
Oberwerk Corp., 840
Observa-Dome Inc., 774
Observateurs de la Magnitude Absolue (Les_), 89, 95
Observatoire Antarès, 190
Observatoire Astronomique d'Aniane, 179, 190
Observatoire Astronomique de la Brie, 205
Observatoire Astronomique de Saint-Jean-le-Blanc, 190
Observatoire Astronomique des Pises (OAP), 190, 200
Observatoire Astronomique de Strasbourg, 191, 200, 205
    Centre de Données astronomiques de Strasbourg, 165
Observatoire Astronomique du Mont Mégantic (OAMM), 95, 100
Observatoire Astronomique Jurassien, 510
Observatoire Astronomique Marseille-Provence
    Laboratoire d'Astrophysique de Marseille (LAM), 187, 191
    Observatoire de Haute Provence (OHP), 191
Observatoire Astronomique Mont-Soleil, 506
Observatoire Cantonal de Neuchâtel, 506
Observatoire d'Alger, 4, 5
Observatoire de Besançon, 191
Observatoire de Bordeaux, 191
Observatoire de Genève, 507
Observatoire de Grenoble, 191
Observatoire de Haute Provence (OHP), 168, 191
Observatoire de Jolimont, 201
Observatoire de la Côte d'Azur (OCA), 160, 191
    Centre d'Études et de Recherches en Géodynamique et Astrométrie (CERGA), 191
    Station de Calern, 192
Observatoire de l'Association Culturelle de Dax (OACD), 192
Observatoire de Lyon, 165, 192
Observatoire de Marseille, 154, 187, 192
Observatoire de Meudon, 192
Observatoire de Nice, 160, 192
    ADION, 160
Observatoire de Paris, 192
    Institut de Mécanique Céleste et de Calcul des Éphémérides (IMCCE), 193
    Laboratoire Primaire du Temps et des Fréquences (LPTF), 164
    Station de Radioastronomie de Nançay, 193
Observatoire de Puimichel, 193
Observatoire de Rouen (OR), 193
Observatoire des Alpes Maritimes (OAM), 193
Observatoire des Creusets, 500, 507
Observatoire des Makes, 156
Observatoire des Vallons, 163
Observatoire de Toulouse, 193
Observatoire de Vérossaz, 504
Observatoire de Vignola, 169
Observatoire du Mont Blanc, 190
Observatoire du Mont Mégantic, 101
Observatoire du Parc Montsouris, 158
Observatoire du Pic du Midi, 193, 194
Observatoire du Plateau de Bure, 182, 204
Observatoire François-Xavier Bagnoud (OFXB), 504, 507
Observatoire "L'Étoile d'Acadie", 84
Observatoire Midi-Pyrénées (OMP), 193
    BGI, 163
    Bureau Gravimétrique International (BGI), 163, 186
    Observatoire du Pic du Midi, 194
Observatoire Mont-Cosmos, 95
Observatoire Planétarium des Monts de Guéret, 194
Observatoire Populaire de Laval (OPL), 194
Observatoire Royal de Belgique (ORB), 55
    Sunspot Index Data Center, 57
Observatori Astronòmic del Garraf (OAG), 480
Observatoriemuseet, 492
Observatories
    Aachen, 247
    Aalen, 210
    AAO
        Coonabarabran, 14
        Eastwood, 14
        Epping, 14
    Aarau, 500
    Abastumani, 206
    Abbitt, 905
    A. Betti, 305
    Ablis, 162
    Abrahão de Moraes, 72
    Abu (Mount_), 287
    Acquaviva, 303
    Acton, 757
    Addin Toussi (Khadjeh Nassir_), 293
    Adelaide, 29
    Adolph Diesterweg
        Radebeul, 259
        Wernigerode, 214
    Adrian (Michael_), 250
    A.F. Philips (Dr._), 390
    African (South_)
        Cape Town, 467
        Sutherland, 467
    Afton (Lake_), 722
    Aguilar (Felix_), 11
    Ahmedabad, 287
    Aichi, 351, 355
    Airdrie, 532
    Ajaccio, 169
    Akeno, 360, 361
    Aker (Gov_), 595
    Akron, 835
    Alamogordo, 799
    Albany, 805
    Albert Einstein, 586
    Albi, 154
    Ålborg, 134
    Albstadt, 216, 250
    Albuquerque, 800, 803, 804
    Alcalá de Henares, 481
    Além-Paraíba, 69
    ALFA, 273
    Alfred, 805
    Algarve, 433
    Alger, 4
    Allegheny, 862
    Allen (Tamke-_), 873
    Allentown, 858
    Alleppo, 515
    Aller (Ramón María_), 485
    Allgäuer, 207
    Alloway, 20
    Alma-Ata, 364
    Almería, 475, 481
    Alpes Maritimes, 193
    Alston, 575
    Alta Val Trebbia, 341
    Alta Vista, 627
    Altenmünster, 248
    Altenrheine, 211
    Alushta, 528
    Amado, 597
    Americana, 69
    Amersfoort, 400
    Ames, 716, 718
    Amherst, 751
    Ammar (H._), 453
    Amsterdam, 406
    Ancón, 423
    Ancona, 310, 314
    Anderson Mesa, 598
    Andreas, 771
    Angelo Secchi (Padre_), 311
    Angers, 156
    Anglo-Australian
        Coonabarabran, 14
        Eastwood, 14
        Epping, 14
    Angwin, 632
    Aniane, 179

Ánimas (Las_), 385
Ankara, 522, 523
Annapolis, 744
Ann Arbor, 768
Antalya, 524
Antarctica, 915, 919
Antarès, 190
Antares, 68, 165
Antioch, 623
Antist, 483
Antofagasta, 108, 109
Antwerpen, 54, 60
Anza, 632
AOCA, 500
AOCC, 312
Apache Point, 799, 802
Apeldoorn, 390
APO, 799, 802
Apollo, 839
Apple Valley, 625
Arbaz, 500
Arçay, 201
Arcetri, 326, 335
Archenhold, 209
Arcturus, 791
Ardenne (Manfred von_), 261
Arecibo, 437, 819
Århus, 134, 139, 140
A. Righi, 322
Arlington, 896
Armagh, 532
Armainvilliers (Gretz-_), 205
Armidda (Monte_), 310
Arosa, 503, 514
Arthur J. Dyer, 870, 872, 876
Artigas, 587
Arunah Hill, 758
ASA, 161
Ashburton, 409
Ashland, 902
Ashton, 716
Asiago, 339
ASNSW, 15
Asten, 394
ASTER, 470
ASTRA, 310
Astrorama, 195
Asunción, 422
ATC, 534
Athens, 269, 691
Athinai, 269
Atibaia, 70
Atlanta, 687–689
ATNF
  Epping, 21
  Marsfield, 21
  Mopra, 21
  Narrabri, 21, 22
  Parkes, 22
Aubagne, 162
Aubière, 158
Auckland, 408
Aufkirchen, 221
Augsburg, 216
Augusta, 687
Austin, 887
Australia Telescope
  Epping, 21
  Marsfield, 21
  Mopra, 21
  Narrabri, 21, 22
  Parkes, 22
Avila, 472
Avren, 74
AVSO, 207
Aydat, 158
Ayer, 755

Aylesbury, 535
AZH, 216
Babelsberg, 218
Bacchus (Col de_), 158
Baden, 500
Bad Gandersheim, 232
Badlands, 869
Bad Mergentheim, 217
Bad Münstereifel, 237
Bad Nauheim, 263
Bad Salzschlirf, 231
Baikal, 448
Bainbridge Island, 907
Baja, 273
Baker, 778
Bakker's Farm, 103
Baldone, 372
Baleares, 483
Ballarat, 19
Ballreich, 843
Balmoral, 84
Baltimore, 739, 743
Baltus, 858
Bandung, 290
Bangalore, 288
Bangkok, 519
Banská Bystrica, 458
BAO, 342
Bappu (Vainu_), 285
Barbara Moore (James and_), 681
Barber (Henry R._), 708
Barber (Walter_), 687
Barcelona, 470, 476, 480, 483
Bär (Erich_), 260
Barros (Prof. Manuel de_), 436
Barrow-in-Furness, 546
Bartholomäus Scultetus, 264
Basel, 512
Bassano Bresciano, 332
Baton Rouge, 730
Battle Point, 907
Bauduen, 163
Bausch (Sommers-_), 659, 660
Bautzen, 249
Bayfordbury, 577
Bayley (Harry_), 45
Baylor Regional Park, 771
Bayrischzell, 255
Bays Mountain, 871
BBSO, 631
Beacon Hill, 541
Beaumont, 188
Beaumont-lès-Valence, 158
Beauvoir, 86
Beaver Meadow, 809
Beg (Ulugh_), 925
Behlen, 784
Beijing, 113, 114, 118
Beisbroek, 60
Bel Air, 737
Belgrade, 454
Bella Unión, 587
Belogradtchik, 75
Belo Horizonte, 73
Beloit, 915
Belvedere, 929
Beman, 766
Benátky nad Jizerou, 125
Benson, 604
Benstead (Modine-_), 919
Bentley, 748
Beograd, 454, 455
Bergamo, 313
Berg-Aufkirchen, 221
Berkeley, 646, 647, 651
Berlin, 209, 219, 265
Bern, 500, 513

Berthoud, 656
Besançon, 156, 191
Bethany, 662, 666
Betti (A._), 305
Beveridge, 24
BFO, 653
BGO, 97
BGSU, 835
Biak, 290
Bickley, 26
Bieselsberg, 213
Bieszczady Mountains, 428
Big Bear, 631
Big Bear City, 635
Big Ear, 841
Big Hill Pond, 874
Big Meadows, 736
Big Pine, 613
Big Sky, 79
BIMA, 611, 647
Birmingham, 574, 589
Biruni, 292
Bisei, 342, 347
Bishop, 679
Bishopstone, 567
Bismarck, 834
Black Forest, 653
Blackhill, 16
Black Moshannon, 860
Blacksburg, 906
Blanc (Mont._), 190
Bloomington, 702, 712
Blue Mountain, 781
Blue Springs, 776
BO, 594
BOAO, 367
Bochum, 233, 249, 253
Bogotá, 119, 120
Boliviano-Ruso, 63
Bologna, 304, 316, 326, 333
Bolton, 536
Bombay, 281, 288
Bonn, 237, 253, 254, 259
Boone, 827
Boothroyd (Hartung._), 812
Bordeaux, 203
Boring, 848
Born (Max._), 248
Bornova, 522
Borowa Gòra, 427
Borowiec, 430
Boskovice, 125
Bosque Alegre, 10
Bosscha, 290
Boston, 746, 749, 750
Boulder, 656, 659, 660
Bourg-en-Bresse, 156
Bourges, 201
Bouzareah, 4
Bowie, 736
Bowling Green, 726, 728, 835
Bracken County, 726
Brackett (Franck P._), 634
Bradenton, 679
Bradley, 687
Braeside, 594
Brahe (Tycho_)
　Malmö, 491
　Oxie, 498
　Rostock, 215
Braintree, 759
Bratislava, 458, 460, 461
Brazil, 70
Breezy Point, 806
Breisgau, 247
Brentwood, 870, 876
Brera, 326, 327

Brescia, 337
Brevard, 678
Brewster, 801
Briars (The_), 15
Bridgwater, 537
Brie, 205
Brisbane
　(Sir Thomas_), 27
Bristol, 872, 897
Britton (Michael L._), 854
Brno, 126, 127, 129
Bro, 499
Broadman Coit (Judson_), 749
Brooklyn, 806
Brooks, 762
Brugg, 500
Brugge, 60
Bruncu Spina, 327
Bruno H. Bürgel
　Berlin, 219
　Hartha, 220
　Sohland, 213, 264
Brussel, 53–56
Brussels, 53–56
Bruxelles, 49, 53–57
Bryan (John_), 839
BSO, 79
Buchloe, 212
Bucknell, 853
Bucureşti, 438
Budapest, 275–278
Buehler, 679
Buenos Aires, 6, 9
Buffalo, 809
Buhl Jr. (Henry_), 856, 858
Bülach, 501, 508
Bulgaria, 75
Bunbury, 16
Bundaberg, 20
Bundoora, 24
Bure (Plateau de_), 182, 204
Burg, 215
Burgdorf, 512
Bürgel (Bruno H._)
　Berlin, 219
　Hartha, 220
　Sohland, 213, 264
Burgenland, 35
Burgsolms, 213
Burgum, 395, 406
Burke, 872, 897
Burke-Gaffney (Rev. M.W._), 97
Burnett (Mount_), 25
Bussloo, 406
Bussoleno, 332
Butts-Clark, 774
Buxy, 201
Byron, 699
Byurakan, 12, 449
Cáceres, 474
Caen, 160
Cagliari, 327
Cahall, 759
Cairo, 142
Calán (Cerro_), 111
Calar Alto, 475, 476, 481
Calden, 213
Calern, 192
Calgary, 80, 102
Cali, 119
Caliumi (Ferdinando_), 310
Calixto (San._), 63
Callaway Gardens, 689
CalTech, 613, 631
CalTech Submillimeter, 692
Cambridge, 540, 557, 579, 754
Campinas, 69

Campo Catino, 308, 312
Campo Imperatore, 328
Canada-France-Hawaii, 693
Cananea, 383
Canberra, 21
Canterbury, 409
  (South_), 412
Canyon Camp, 847
CAO, 313
Cape Point, 230
Cape Town, 464, 465, 467
Capilla Peak, 804
Capodimonte, 327
Capolago, 504
Capoterra, 327
Cap-Rouge, 95
Caracas, 928
Carcassonne, 170
Carlisle, 536, 854
Carloforte, 327
Carlos Cesco, 11
Carl Schurz Park, 806
Carmen de Carupa, 119
Carolyn Wine, 683
Carrasco, 587
Carroll (George_), 651
Carshalton, 544
Carter, 409
Casey, 919
Casole d'Elsa, 314
Casper, 922
Castel Gandolfo, 926
Castelnuovo di Sotto, 311
Castle Eden, 541
Catalunya (Parc_), 472
Catania, 326, 338
Caussols, 192
CAV, 680
Cavezzo, 305
CAY, 481
CCSSC, 688
Cedar Falls, 720
Cedar Hill, 778
Cederberg, 465
Ceduna, 31
C.E. Kenneth Mees, 696, 825
CEMA, 477
Cengelköy, 522
CERGA, 192
Cerro Calán, 111
Cerro Las Ánimas, 385
Cerro Mamalluca, 109
Cerro Pachon, 109, 596, 693
Cerro Pochoco, 107
Cerro Tololo, 109
Cesco (Carlos_), 11
České Budějovice, 126, 128
CfA, 597
CFHT, 693
Chabot, 616, 619
Chalk River, 100
Chalon-sur-Saône, 201
Chamberlin, 654, 661
Chamblon, 510
Chamiers, 179
Chandler Mountain, 589
Changchun, 113
Chapel Hill, 832
Chappaqua, 825
Charles E. Daniel, 868
Charleston, 866
Charlestown, 864
Charlottesville, 896, 900, 904
Charterhouse, 537
Chateau du Lac, 651
Chatswood
  (West_), 24

Chattanooga, 875
Chavannes-des-Bois, 513
Cheisacker, 500
Chernigovka, 525
Cherokee, 718
Cherry Grove, 771
Cheyenne, 922
Chiang Mai, 519
Chicago, 698
Chico, 631
Chiefland, 680
Chieti, 307
Chikurinji (Mount_), 354
Chilbolton, 542
China, 114
Chinon, 162
Chita, 452
Chorzów, 429
Christchurch, 409
Chuguev, 526
Chungbuk, 368
Ciampino, 313
CiAO, 313
CIDA, 927
CIG, 594
Cimini, 313
Cincinnati, 837, 844
Cinisello Balsamo, 319
Città del Vaticano, 926
Civitavecchia, 306
Clanfield, 547
Claremont, 634
Clarence T. Jones, 875
Clark (Butts-_), 774
Clermont-Ferrand, 158
Cleveland, 541, 836, 871
Climenhaga, 104
Close House, 580
Clovis, 800
Cluj-Napoca, 438
Clyde W. Tombaugh, 723
COAA, 433
Coca-Cola, 688
Cochabamba, 62
Cocoa, 678
Cogeces del Monte, 483
Coimbra, 435
Cointe, 58
Coit (Judson Broadman_), 749
Col de Bacchus, 158
Col de la Lèbe, 170
Col Druscié, 308
Colin Hunt, 535
Collins (Eileen_), 813
Colluriana, 328
Cologne, 261
Colombara, 314
Colombia, 120
Colombo (G._), 320
Colonia, 587
Colorado Springs, 652, 653, 656, 660
Columbia, 778, 868
Columbus, 688, 841
Como, 321
Conestoga, 705
Constantine, 4
Conway, 607
Cook (Flower and_), 862
Coonabarabran, 14, 29
Copérnico, 69
Copernicus, 407, 837
  (N._), 74, 126, 127, 129
  (Nicolas_), 429
  (Nicolaus_), 432
Córdoba, 6, 10, 471
Corning, 813
Corona Borealis, 390

Coronado Solar, 594
COSC, 619
Costa Mesa, 632
Costitx, 480
Côte d'Azur, 160, 191, 192
Coulounieix-Chamiers, 179
Covo, 332
CPEN, 49
Crago, 15
Crane (Zenas_), 724
Cranford, 791
CrAO, 525
Crawford Hill, 793
Crayford, 543
Crestwood, 728
Creusets, 500
Crevillente, 477
Crimea, 450, 525
Crimmitschau, 233
Cronyn (Hume_), 104
Crosby Ramsey, 739
Crosland Hill, 549
Cross Plains, 920
Cross River, 825
Croydon, 544
CSAO, 114
CSDO, 21
CSO, 692
CSSC, 616
CTAS, 878
CTIO, 109
Cuba (Mount_), 667
Cubillas, 483
Culgoora, 24
Cuneo, 331
Curtomartino, 303
Custer, 813
Cuxhaven, 232
C-VII, 736
Cypress, 650
Ďáblice, 130
Daeduk, 368
Daeyang, 369
Dahlonega, 690
Damascus, 515
Danbury, 666
Daniel (Charles E._), 868
DAO, 94
Dark Sky, 827
Darling Hill, 824
Darmstadt, 260
Darnhall, 579
Daun, 253
Davenport, 718
David Dunlap, 103
Davie, 679
Davis Hall, 704
Davos, 507
Dax, 192
Dayton, 839
Dearborn, 704
de Barros (Prof. Manuel_), 436
Debrecen, 275
Decatur, 687
Decker-Grebner-Van Zandt, 704
Deer Lakes Park, 851
Defford, 579
De Kalb, 704
Delaware, 841, 842
Délémont, 510
Delingha, 114
Demmin, 209
Denver, 654, 661
Des Moines, 716, 717
Dessau, 215
De Tiendesprong, 401
Devon, 101

Diablo (Mount_), 629
Diadema, 71
Dickson, 21
Diedorf, 216
Diesterweg (Adolph_)
    Radebeul, 259
    Wernigerode, 214
Digger Pines, 627, 645
Dijon, 156, 199
Dinant, 49
Dixie, 686
Doane, 698
Dodaira, 344, 353
Dome Fuji, 355
Dominion
    Astrophysical, 93, 94
    Radio Astrophysical, 94
Dompcevrin, 161
Donzdorf, 247
Dortmund, 214
Dourbes, 53, 54
Dourbies, 190
Dourouti, 269
Dows (Palisades-_), 716
Dr. A.F. Philips, 390
Drake University Municipal, 717
DRAO, 94, 368
Drebach, 263
Drenthe
    (Noord-_), 406
Dresden, 251
Drew (Frosty_), 864
Droke (James Wesley_), 606
Dr. Remeis, 254
Drusciè (Col._), 308
Dublin, 295, 296
Dudley, 814
Duisburg, 244
Dumand, 696
Dumfries, 544
Duncan, 103
Dundee, 556
Dunlap (David_), 103
Dunsink, 295
Durazno, 587
Dushanbe, 518
Dwingeloo, 394, 400, 402
Dyer (Arthur J._), 870, 872, 876
EARG, 7
Eastbourne, 545
East Peoria, 701
East Sky, 732
Eastwood, 14
Eau Claire, 919
Ebingen, 216, 250
Ebro, 486
ECO, 802
EcoTarium, 750
Edinburgh, 533, 534, 562, 576
Edison Infrared Space, 923
Edmonton, 95, 101
Edward Naylor, 852
Edwin Ritchie, 907
Effelsberg, 237
Ege, 522
Egloffstein, 247
Eileen Collins, 813
Eilenburg, 264
Einstein (Albert_), 586
Einstein (Herschel-_), 69
Einsteinturm, 218
Eisenstadt, 35
Eisner, 35
EISO, 923
Ekar (Monte-_), 327
Ekaterinburg, 451
El Camino, 619

Elgin, 700
Elginfield, 104
El Leoncito, 7, 10, 666
Ellijay, 690
Elp, 406
Emmendingen, 234, 247
EMVIC, 928
Engelhardt, 442
Ennepetal, 260
Ensenada, 386
Épendes, 504
Epping, 14, 21
EPS, 391
Erich Bär, 260
Erich Scholz, 260
Erie, 858
Erkrath, 249
Erlangen-Nürnberg, 254
Erwin W. Fick, 718
Esashi, 353
Eschenberg, 511
Eskridge, 723
ESO
   Antofagasta, 108
   Garching, 226, 228, 245
   La Serena, 108
   La Silla, 108
   Paranal, 109
   Santiago de Chile, 108
Essen, 208
Etna, 326
Étoile d'Acadie (L'_), 84
Etscorn Campus, 802
ETSU, 872
Eugene, 850
Evans (Mark_), 702
Evans (Mount_), 661
Evanston, 704
Evansville, 711
Exeter, 789
Eyes on the Skies, 645
Faget, 438
Fall Creek, 915
Falmouth, 733
Fan Mountain, 904
Farra d'Isonzo, 314
Fasano, 321
Fayette, 776
Fayetteville, 606
FBO, 811
FCRAO, 751
Feder (Paul P._), 772
Feira de Santana, 68
Feldberg, 240
Felix Aguilar, 11
Felsina, 304
Feltre, 308
Fenton Hill, 800
Ferdinando Caliumi, 310
Ferme du Héran, 171
Fernbank, 688
FEVARO, 231
FHO, 800
Fick (Erwin W._), 718
Figl (Leopold_), 41
Firenze, 305, 317, 326, 335
Fish Lake, 763
FitzRandolph, 795
Five College radio, 751
FKS, 35
Flagstaff, 593, 594, 598, 600, 604, 677
Fleetwood, 564
Flensburg, 237
Fliegenfelde, 208
Floirac, 203
Florence, 305, 317, 326, 335, 591, 866
Florence and George Wise, 301

Flower and Cook, 862
Floyd Bennett Field, 806
FLWO, 597
Flying M Ranch, 593
FMG, 504
FMU, 866
Foerster (Wilhelm_), 265
Foggy Bottom, 811
Foothill, 633
Foothill College, 621
Forst, 263
Fortaleza, 68–70
Fort Davis, 801, 888
Fort Lauderdale, 685
Fort Pickens, 680
Fort Skala, 428
Fort Wayne, 711
Fort Wright, 726
Foster (M._), 110
Fountainwood, 884
Fox, 685
Foxdale, 552
Fox Island County Park, 711
Fox Park, 763
FPPO, 763
Franck P. Brackett, 634
Frankfurt/Main, 234, 240
Franklin Miller Jr., 839
Frankston, 15
Franz (Johannes_), 249
Franz Kroller, 35
Frauenfeld, 501
Fred Lawrence Whipple, 597, 751
Freiburg-im-Breisgau, 208, 235
Frejlev, 134
Fremont Peak, 621
French Camp, 774, 775
Fresno, 616
Frewsburg, 818
Friedrich Ebert, 231
Froburg, 500
Frombork, 429
Frosinone, 308
Frösön, 488
Frosty Drew, 864
Fuertes, 812
Fuji, 351
Fujigane, 351
Fuji (Mount_), 360
Fulda, 232
Fuldatal, 262
Funchal, 435
Furness, 546
Fürstenfeldbruck, 259
Gagarin (Juri_), 264
Gahberg, 34
Gainesville, 678, 686
Gales (Grant O._), 717
Galilei (Galileo_), 310
Galileo Galilei, 310
Gambier, 839
Gamla, 493
Gandersheim (Bad_), 232
Gandolfo (Castel_), 926
Gap, 179
Garbagnate Monastero, 333
Garching, 226, 228, 237
Garmisch-Partenkirchen, 230
Garraf, 480
Gâtine, 181
Gauribidanur, 285, 288
Gävle, 493
Gazan Mountains, 293
GCO, 23
G. Colombo, 320
Gebze, 524
Geising, 249

Gemenos, 180
Gemini
    North, 109, 596, 693
        Northern Operations Center, 693
    South, 109, 596, 693
        Southern Operations Center, 109
        Tucson Project Office, 596
General Nathan Twining, 803
Generoso (Monte_), 504
Geneva, 507
Genève, 507
Genk, 51
Gennevilliers, 172
Genova, 340, 341
Gent, 59
George, 879
George Carroll, 651
Georgetown, 884
GEOS, 477
Gerald Hughes Phipps, 653
Geretsried, 264
Germering, 248
Geschwister Herschel, 260
Giant Metrewave Radio Telescope, 286
Gibson, 681, 685
Gifu, 350, 361
Gilliland, 750
Giuseppe (San_), 332
Giuseppe S. Vaiana, 327
Glasgow, 532, 577
Glen D. Riley, 703
Glen Ellen, 650
Glenlea, 102
Glenshaw, 851
Glücksburg, 237
GMRT, 286
Gmunden, 35
GNTO, 803
Godlee, 555
Goethe Link, 712
Goldendale, 908
Golden Pond, 725
Goldstone, 613
Goloseev, 529
Golosiiv, 529
Gönnsdorf, 251
Goodsell, 770
Gora Kanobili, 206
Gordon Southam, 89
Görlitz, 264
Gornergrat, 255, 317, 505
GOSP, 908
Göteborg, 489, 493
Gotenburg, 232
Gothard, 274
Gothenburg, 489, 493
Göttingen, 213, 256
Gov Aker, 595
Gragnola, 329
Graham (Mount_), 604, 926
Grahamstown, 466
Grainger, 789
Grakovo, 529
Granada, 476, 478, 483
Grange, 306, 332
Granite Falls, 827
Grant O. Gale, 717
Granville, 838
Grasse, 192
Grasslands, 596
Grasweg, 232
Gravesend, 543
Graz, 37, 39, 40
Great Kills Park, 806
Grebner-Van Zandt (Decker-_), 704
Greece, 269
Green Bank, 913

Green Bay, 918
Greenbelt, 737
Greencastle, 710
Green Garden Hill, 765
Green Mountain, 894
Greensboro, 832
Greenville, 667, 868
Greenwich, 565
Greifswald, 248
Grenoble, 182, 203
Gretz-Armainvilliers, 205
Griffith, 622, 626
Grimbergen, 60
Grinnell, 717
Gröbenzell, 217
Grosseto, 310
Großröhrsdorf, 250
Großschwabhausen, 255
Grove Creek, 23
Grudziądz, 429
Grundy (J.R._), 855
GSFC, 737
Guadalajara, 385
Guarcino, 312
Guelph, 102
Guéret, 194
Guernsey, 106
Guillermo Haro, 383
G.V. Schiaparelli, 335
Gyula, 275
Haarlem, 407
Haastrecht, 407
Hagen, 260
Hague (La_), 188
Hahneberg, 219
Hail, 453
Hainberg, 256
Halberstadt, 216
Haleakala, 696
Hale Pohaku, 693
Halifax, 97
Halle, 231, 243
Halley, 402
Halley Hill, 622
Hamburg, 231, 232, 254
Hamill, 849
Hamilton, 409, 811
Hamilton (Mount_), 649
H. Ammar, 453
Hampton, 903
Hancock, 801
Hanle, 285
Hannover, 260
Hanover, 788
Hans Nüchter, 232
Haramiata, 74
Harare, 929
Hard Labor Creek, 689
Harford County, 737
Hariq, 453
Haro (Guillermo_), 383
Harriman, 825, 873
Harrisburg, 852
Harrogate, 547
Harry Bayley, 45
Hartebeesthoek, 466
Hartha, 220
Hartung Boothroyd, 812
Hastings, 782
Hat Creek, 611, 647
Hattingen, 260
Hausen bei Brugg, 500
Haute Corréo, 179
Haute-Marne, 200
Haute Provence, 168, 191
Hautes-Plates, 199
Haverford, 856

Haymarket, 20, 28
Haynald, 274
Haystack, 746, 756
HCRO, 611
HCRT, 701
Heerhugowaard, 403
Heerlen, 402
Heesch, 402
Heidelberg, 235, 258
Heilbronn, 243
Heiligkreuz, 36
Heinrich Hertz, 603
Heinrich S. Schwabe, 215
Helsinki, 147, 151, 152
Helwân, 142
Henry Buhl Jr., 856, 858
Henry R. Barber, 708
Heppenheim, 246
Héran (Ferme du_), 171
Herbuchenne, 49
Hercules, 402
Herford, 231
Heringsdorf (Seebad_), 261
Hermanus, 466
Herne, 211
Herrsching, 219
Herschel-Einstein, 69
Herschel (Geschwister_), 260
Herseaux, 47
Hertford, 577
Hertz (Heinrich_), 603
Hickory, 827
Hickory Hill, 684
Hida, 350
Hidden Valley, 869
High Altitude, 656
High Bridge, 794
Hija de Dios, 472
Hill (Rosemary_), 686
Hillsboro, 789
Hillside, 720
Hilo, 109, 596, 692, 693, 695
Hingham, 537
Hiraiso, 343, 344
Hiroshima, 345
Hisar, 518
HLCO, 689
Hobbs, 915, 919
Hochdahl (Neanderhöhe_), 249
Hoeven, 400
Hof, 248
Hofheim-Marxheim, 261
Hofheim/Taunus, 261
Höfingen, 249
Hoher List, 253
Hohmann (Walter_), 208
Holcomb (J.I._), 710
Hollola, 149
Homburg, 257
Hong Kong, 116
Honolulu, 694–696
Hood (Mount_), 848
Hook (John C._), 712
Hopkins, 760
Hopkins (Mount_), 597, 599, 787
Horrocks (Jeremiah_), 563, 575
Hoshinoko Yakata, 345
Hoskinstown, 31
Hostalets, 476
Houston, 879
Hove, 60
Howell, 774
Hradec Králové, 128
HST, 245, 743
Huairou, 113
Huancayo, 423
Hubble Space Telescope, 245, 743

Huddersfield, 549
Humacao, 437
Humain, 55, 56
Hume, 608, 611
Hume Cronyn, 104
Hummingbird, 623
Hunt (Colin_), 535
Huntsville, 883
Hurbanovo, 460
Huuhanmäki, 149
Hvar, 122
Hyde, 783, 784
Hyderabad, 287
Hyogo, 345, 355
IAA, 476
IAO, 285
IAS, 234
Ibaraki, 343
Illawara, 32
Ilsan, 370
Imola, 305
Incline Village, 787
Indianapolis, 710
Indianola, 716
Infrared Space, 476
Infrared Telescope Facility, 695
INT, 479
Ioannina, 269
Iowa City, 719
Ipswich, 561
IRA, 316
IRAM, 182, 478
Irkutsk, 441, 448
IRTF, 695
Isaac Newton Telescope, 479
Isle of Man, 552
ISO, 476
Iso-Heikkilä, 151
Isonzo (Farra d'_), 314
Istanbul, 522, 523
Itajubá, 68
Itapetinga, 68
Itchen, 569
Ithaca, 812
Iwate, 353
IWU, 702
Izaña, 58, 235, 256, 257
Jackson, 774
Jacobus Kapteyn Telescope, 479
Jakarta, 290
James and Barbara Moore, 681
James Clerk Maxwell Telescope, 693
James C. Veen, 764
James Lipp, 636
James Schwartz, 631
James Wesley Droke, 606
Jan Paagman, 394
Jansky, 202
Japal-Rangapur, 287
Japan, 342, 346, 352–354, 362
Jävan, 491
JBO, 579
JCMT, 693
Jean Nicolini, 586
Jeddah, 453
Jena, 255, 263
Jenny Jump State Park, 791
Jeremiah Horrocks, 563, 575
Jeseník, 128
Jewett, 910
Jicamarca, 423
J.I. Holcomb, 710
JIHOP, 710
Jilin, 113
Jindřichuv Hradec, 128
Jizerou, 125
JJMO, 665

JKT, 479
JNSPO, 839
Jodrell Bank, 579
Joensuu, 150
Johannesburg, 464
Johannes Franz, 249
Johannes Kepler
   Crimmitschau, 233
   Graz, 39
   Linz, 37
   Magdeburg, 216
   Steinberg, 39
Johann Palisa, 128
John Bryan, 839
John C. Hook, 712
John H. Witte, 719
John J. McCarthy, 665
John (Mount_), 412
Johnson City, 872
Johnson-Weil, 805
John W.H. Simpson, 791
Jokioinen, 147
Jolimont, 201
Jones (Clarence T._), 875
Jonsdorf, 261
Jordan (Maynard F._), 733
Jorge Polman, 66
Joseph H. Rogers, 764, 766
Joyce Memorial, 409
JPS, 394
J.R. Grundy, 855
Judson Broadman Coit, 749
Jungfraujoch, 505, 507
Junnar Taluk, 286
Juri Gagarin, 264
Kagoshima, 351
Kaivopuisto, 152
Kaleakala, 696
Kalinenkov, 527
Kalocsa, 274
Kamioka, 360, 361
Kamuela, 692, 693
Kandilli, 522
Kanobili (Gora_), 206
Kanzelhöhe, 40
KAO, 367
Kappa Crucis, 587
Kapteyn, 395
Karachaevo, 448
Karadag, 528
Kardjali, 74
Karlovy Vary, 128
Karl Schwarzschild, 252
Karlskrona, 490
Karlsruhe, 216
Kashima, 343, 344
Katziveli, 525
Kavalur, 285
Kazan, 442
KCO, 857
Kearney, 784
Keck (W.M._), 692
Keeble, 902
Keene, 723
Kenley, 544
Kennebunk, 732
Kenneth Mees (C.E._), 825
Kennon, 775
Kepler (Johannes_)
   Crimmitschau, 233
   Graz, 39
   Linz, 37
   Magdeburg, 216
   Steinberg, 39
Kernen Farm, 103
Kevola, 151
Keystone College, 857

K.E. Ziolkowski, 244
Khadjeh Nassir Addin Toussi, 293
Kharkiv, 526, 529
Kharkov, 526, 529
Khurel Togót, 388
Kiel, 255
Kiev, 526, 529
Killdeer Mountains, 834
Kingsport, 871
Kingston, 97
Kirchhain, 262
Kirkwood, 712
Kiruna, 489, 495
Kiskunhalas, 277
Kislovodsk, 448
Kiso, 351, 361
Kitab, 925
Kitt Peak, 599, 754, 801, 836
Klagenfurt, 39
Klet, 128, 130
Knaresborough, 547
Knighton, 573
Knockin, 579
Knottingley, 584
Knoxville, 873
København, 136
Koch (Max_), 232
Kodaikanal, 285
Kohl, 817
Köln, 250, 255, 261
Königsleiten, 39
Königstuhl, 235
Konkoly, 275
Kornik, 430
Koshka (Mount_), 525
Kötcse, 275
Kourov, 451
Kowloon, 116
KPNO, 599, 754, 801, 836
Kraków, 428
Kraví Hora, 129
Kredarica, 462
Kreuzbergl, 34, 39
Kreuzlingen, 502
Kroller (Franz_), 35
Kronshagen, 249
Krugersdorp, 466
Kryonerion, 269
Kryzhanovska, 528
KSW, 36
Kuffner, 36
Kula Maui, 696
Kunming, 115
Kuopio, 149
Kurky, 128
Kurri, 16
Kvistaberg, 498, 499
Kwasan, 350
Kyiv, 526, 529
Kylmälä, 148
Kyoto, 350
Kyungpook, 367
La Culla, 472
Ladd, 864
Lafayette, 729
La Garandie, 158
Laguna (Mount_), 637
La Habana, 123
La Hague, 188
Lahall, 489
Laize-la-Ville, 160
Lajos Terkán, 277
Lake Afton, 722
Lake Arrowhead, 629, 651
Lake Tekapo, 412
Lakewood, 817
La Mariquita, 383

Lancashire, 575
Lancaster, 855
Landers, 650
Lane Cove, 23
Langmuir, 802
Lansbergen (Philippus_), 407
Lanusei, 310
Lanzenhäusern, 500
La Palma, 472, 478–480, 482
La Paz, 62, 63
La Plata, 10
Laplume, 857
LAPO, 722
Laramie, 923
La Rampisole, 179
La Réunion, 156
La Rivière, 156
La Roche des Arnauds, 179
La Rochelle, 165
Las Acacias, 10
La Sainte Baume, 162
Las Ánimas (Cerro), 385
Las Brisas, 656
Las Campanas, 103, 107, 615, 672
Las Cruces, 802
La Serena, 107–110
La-Seyne-sur-Mer, 190
La Silla, 108, 228, 253, 490, 507
La Spezia, 307, 329
Las Vegas, 787
LAT, 622
Latina, 333
La Trinité, 195
Lattrop, 391
Lauenstein, 249
Laufen, 248
Laupheim, 261
Lausanne, 511, 513
Laval, 85, 194
Lawrence, 723
Lawrence Hall of Science, 651
Laws, 778
LBO, 656
LBT, 603
LCC, 694
LCO, 107, 672
Leander McCormick, 904
Learmonth, 24
Leatha House Park, 837
Lèbe (Col de la_), 170
Le Havre, 201
Leiden, 407
Leinfelden-Echterdingen, 244
Lelekovice, 125
Lembang, 290
Lemmon (Mount_), 603
Leonberg, 249
Leoncito (El_), 7, 10, 666
León (Luis G._), 385
Leopold Figl, 41
Le Pin, 181
Lepton, 549
Les Paroches, 161
Les Sommets, 86
Lethbridge, 90, 102
L'Étoile d'Acadie, 84
Leuschner, 646
Lewisburg, 853, 862
Lewis Science Center, 607
LHS, 651
Liberec, 129
Liberty (North_), 801
Lick, 649
Lido, 308
Liège, 56, 58
LIGO, 613
Liguria, 341

Lille, 171
Lima, 423, 424
Limburg, 51
Limerick, 298
Limoux, 170
Lincoln, 783, 784
Link (Goethe_), 712
Linz, 37
Lipp (James_), 636
Lisboa, 434
Lisnyky, 526
Little Thompson, 656
Livermore, 627, 645
Ljubljana, 462, 463
Llano del Hato, 927
LMO, 827, 829
Locarno-Monti, 506, 509, 511
Lochem, 400
Locke (Mount_), 888
Lockwood Park, 705
Lockyer (Norman_), 559
LodeStar, 800
Lodi, 336
Logan, 891
Lohrmann, 251
Loiano, 326
Lomnický Štít, 459
London, 104, 565, 574
Longyearbyen, 489
Los Alamos, 800, 801
Los Altos, 622
Los Angeles, 622, 626, 650
Lošinj, 121
Los Molinos, 587
Louisville, 727
Lowell, 598, 767
LSPS, 395
LSV, 51
LTO, 656
Lübeck, 208
Lucile Miller, 827, 829
Ludiver, 188
Luis G. León, 385
Lumezzane, 332
Lunas, 390
Lund, 491
Lunping, 516
Lustbühel, 37, 40
L'viv, 527
L'vov, 527
Lycksele, 495
Lynchburg, 902
Lyon, 165, 171, 200
Lysomice, 432
MAA, 818
Mac Arthur, 893
Macclesfield, 579
Mac Connell Hall, 758
Mac Cormick (Leander_), 904
Mac Farland Park, 716
Mac Kim, 710
Mac Lean, 787
Macon, 689
Madeira, 435
Madison, 917
Madrid, 472, 476
Magdalena Ridge, 802
Magdeburg, 216
Mahoning Valley Cortese, 839
Maidanak, 374, 925
Maidanak (Mount_), 418
Maiden, 829
Mainz, 210
Makes, 156
Makka, 453
Malé Bielice, 459
Mali Lošinj, 121

Mallorca, 480
Malmö, 491
Malvern, 862
Malý Diel, 461
Mamalluca (Cerro-), 109
Manchester, 555, 579, 789
Manfred von Ardenne, 261
Manila, 425, 426
Mankato, 771
Manora Peak, 289
Manresa, 472
Mansfield, 556, 845
Manuel de Barros (Prof.-), 436
MAO, 374, 529
Maracaibo, 928
Marburg, 232, 262
Marcinelle, 49
Maria Mitchell, 754
Mariazell, 35
Mariazellerland, 35
Marietta, 687
Marine-on-St.-Croix, 773
Mario (Monte-), 328
Mark Evans, 702
Marlboro, 893
Marmara, 524
Marquette, 765
Marseille, 154, 160, 165, 180, 187
Marsfield, 21
Martha (Mount-), 15
Martin, 906
Martz, 818
Masterton, 412
Mauna Kea, 109, 562, 596, 692–695, 801
Mauna Loa, 656
Mauritius, 381
Max Born, 248
Max Koch, 232
Mayaki, 528
Mayer (Robert-), 243
Maynard F. Jordan, 733
MBO, 919
MCAO, 667
McArthur, 893
McCarthy (John J.-), 665
McConnell Hall, 758
McCormick (Leander-), 904
McDonald, 887, 888
McFarland Park, 716
McGraw-Hill, 754
McKim, 710
McLean, 787
McMundo, 668
MDM, 598
Mead, 688
Meadville, 851
Medea, 4
Medicina, 316
Meelick, 298
Mees (C.E. Kenneth-), 696, 825
Mégantic (Mont-), 100, 101
Meinhard-Hitzelrode, 258
Melbourne, 16
Melilla, 587
Melo, 587
Melton, 868
Memphis, 873, 874
Menke, 237
Merate, 326
Mercedes, 8, 587
Mercurius, 406
Mergentheim (Bad-), 217
Mérida, 927
Messelberg, 247
Metcalf Nature Center, 771
Metsähovi, 148
Metzerlen, 512

Meudon, 192
México, 385
Meyer (Paul J.-), 878
Meyer-Womble, 661
M. Foster, 110
MHO, 755
MHOA, 848
Michael Adrian, 250
Michael L. Britton, 854
Michalovce, 458
Michigan-Dartmouth-MIT, 598
Middelburg, 407
Middeltown, 665
Midi-Pyrénées, 193, 194
Midtown, 666
Mike Palermiti, 683
Mikolaiv, 527
Milano, 326, 327
Milford, 663, 726
Miller Jr. (Franklin-), 839
Miller (Lucile-), 827, 829
Mills, 556
Milstone Hill, 755
Milton, 755
Minneapolis, 773
Minot, 834
Mira, 60
Misato, 351
Mississippi State, 774
Missoula, 781
MIT, 754
Mitaka, 346, 352–354, 362
Mitchell (Maria-), 754
Mitzpe Ramon, 301
Miyun, 113
Mizusawa, 352, 353
MKO, 692–695, 801
Mlodziezowe, 428
MLSO, 656
Mlynany, 460
MMSD, 917
MMTO, 599
Modine-Benstead, 919
Modra, 458
Moffett Field, 630
Moletai, 374, 375
Molinos (Los-), 587
Mollet, 476
Molonglo, 31
Moncton, 100
Monegrillo, 476
Monoceros, 69
Monroe (Morgan--), 712
Mons, 58
Mont Blanc, 190
Mont-Cosmos, 95
Monte Armidda, 310
Monte Ekar, 327
Monte Generoso, 504
Montelupo, 332
Monte Mario, 328
Monte Novegno, 320
Monteporzio, 328, 340
Monterrey, 382
Monte San Martino, 324
Montevideo, 586, 587
Monte Vissegi, 307
Mont Mégantic, 100, 101
Montpellier, 179, 200
Montréal, 99, 100
Montredon-Labessonnié, 197
Mont-Soleil, 506
Montsouris (Parc-), 158
Mont Tibarine, 4
Moore, 728
Moore (James and Barbara-), 681
Moorhead, 772

Mopra, 21
Moraes (Abrahão de_), 72
Morehead, 832
Morgan-Monroe, 712
Morrison, 776
Morro de São Domingo, 65
Morro Santana, 73
Morville, 49
Moscow, 450
Moshiri, 351
MOST, 31
Mount Abu, 287
Mountain Skies, 923
Mountain View, 633
Mount Burnett, 25
Mount Chikurinji, 354
Mount Cuba, 667
Mount Diablo, 629
Mount Evans, 661
Mount Fuji, 360
Mount Graham, 604, 926
Mount Hamilton, 649
Mount Hood, 848
Mount Hopkins, 597, 599, 787
Mount John, 412
Mount Koshka, 525
Mount Laguna, 637
Mount Lemmon, 603
Mount Locke, 888
Mount Maidanak, 418
Mount Martha, 15
Mount Norikura, 353, 360
Mount Pleasant, 31, 762
Mount Sanglok, 518
Mount Stromlo, 18
Mount Suhora, 428
Mount Tamborine, 30
Mount Wilson, 629, 650
Mowbray, 464
MPIfR, 237
MRT, 381
MSSSO, 18
Mt. Abu, 287
Mt. Blanc, 190
Mt. Burnett, 25
Mt. Chikurinji, 354
Mt. Cuba, 667
Mt. Diablo, 629
Mte. Armidda, 310
Mte. Ekar, 327
Mte. Generoso, 504
Mte. Mario, 328
Mte. Novegno, 320
Mte. San Martino, 324
Mt. Evans, 661
Mte. Vissegi, 307
Mt. Fuji, 360
Mt. Graham, 604, 926
Mt. Hamilton, 649
Mt. Hood, 848
Mt. Hopkins, 597, 599, 787
Mt. John, 412
Mt. Koshka, 525
Mt. Laguna, 637
Mt. Lemmon, 603
Mt. Locke, 888
Mt. Maidanak, 418
Mt. Martha, 15
Mt. Mégantic, 100, 101
Mt. Norikura, 353, 360
MTO, 30
Mt. Pleasant, 31, 762
Mt. Sanglok, 518
Mt. Stromlo, 18
Mt. Suhora, 428
Mt. Tamborine, 30
Mt. Tibarine, 4

Mt. Wilson, 629, 650
Muchachos (Roque de los_), 477–480, 482
Muesmatt, 500
Muiderberg, 406
Mulbarton, 560
Mülheim/Ruhr, 252
Mullard, 557
Multiple Mirror Telescope, 599
Mumbai, 281, 288
München, 219, 225, 226, 228, 256, 258
    (West-_), 217
Muncie, 710
Münstereifel (Bad_), 237
Murfreesboro, 873
Murray Hill, 793
MVCO, 839
M.W. Burke-Gaffney (Rev._), 97
MWI, 629
Mykolaiv, 527
Nacogdoches, 885
Naef (Robert-A._), 504
Nagano, 353, 354
Nagoya, 351
Naini Tal, 289
Namass, 453
Nançay, 192, 193
Nandrin, 56
Nanjing, 114, 115
Nantucket, 754
NAO, 528
NAOJ, 362, 695
    Dodaira, 353
    Iwate, 353
    Mitaka, 342, 346, 352–354
    Nagano, 353, 354
    Nobeyama, 353
    Okayama, 354
    Tokyo, 342, 346, 352–354
Naperville, 703
Naples, 680
Nara, 350
Narbonne, 160
N.A. Richardson, 636
Narrabri, 21, 22
NASA, 630, 695
Nashville, 872, 876
Nassau, 836
Nassir Addin Toussi (Khadjeh_), 293
Natal Midlands, 465
Nathan Twining (General_), 803
National
    Brazil, 70
    Bulgaria, 75
    China, 114
    Colombia, 120
    Greece, 269
    Japan, 342, 346, 352–354, 362
    New Zealand, 409
Nauchny, 525
Nauheim (Bad_), 263
Naval (US_), 604, 677
Naylor (Edward_), 852
N. Copernicus, 74, 126, 127, 129
Neanderhöhe Hochdahl, 249
Nedlands, 32
Needville, 879
NEROC, 756
Nessa, 215
Neuchâtel, 507
Neumarkt/Opf., 219
Neunkirchen, 257
Newark, 668
Newcastle-upon-Tyne, 558, 580
New Haven, 662, 666
New London, 662
Newport News, 905
New Salem, 751

Newton, 851
Newton Falls, 839
Newton-in-Furness, 546
New Zealand, 409
NGCAO, 690
NHAO, 355
Nice, 160, 191
Nicholas E. Wagman, 851
Nicolas Copernicus, 429
Nicolaus Copernicus, 432
Nicolini (Jean-), 586
Niederglatt, 501
Niepolomice, 428
Nijmegen, 403, 405
Nikolaev, 527, 528
Nippes, 250
Nishi-Harima, 355
Niskayuna, 805
Nitra, 458
NIU, 704
Nizamiah, 287
NMIMT, 802
NOAO
   Cerro Tololo, 109
   CTIO, 109
   Kitt Peak, 599
   La Serena, 109
   Sacramento Peak, 801
   Sunspot, 801
   Tucson, 599
Nobeyama, 352-354
Nogales, 602
Nokogiriyama, 360
Noord Drenthe, 406
Norderstedt, 262
Nordic, 480
Norikura (Mount-), 352, 353, 360
Normal, 707
Norman Lockyer, 559
Northampton, 758
Northfield, 770
North Liberty, 801
Northmoor, 704
North Olmsted, 837
North Scituate, 865
Norton, 750, 760
Norwich, 560
NOT, 480
Noto, 316
Novegno (Monte-), 320
Novi Ligure, 335
NRAL, 579
NRAO
   Charlottesville, 896, 900
   Green Bank, 913
   Socorro, 801
   Tucson, 600
NRO, 352-354
NSO
   Sacramento Peak, 801
   Sunspot, 801
   Tucson, 599
Nüchter (Hans-), 232
Nuffield, 579
Nurmijärvi, 147
Nürnberg, 239, 254
NURO, 600
Nursery, 878
Nützweid, 500
OAA, 68, 326
OAC, 6, 10, 327
OACD, 192
OAG, 480
OAGH, 383
OAHE, 69
Oakland, 616, 619
Oak Ridge, 751

OALM, 587
OAM, 69, 193
OAMM, 8, 100
OAN, 481
   Ensenada, 386
   San Diego, 646
   San Pedro Mártir, 386
   San Ysidro, 646
   Tonantzintla, 386
OAP, 63, 190, 327
OAQ, 141
OAT, 328
OATo, 328
OAU, 69
Oberherten, 501
Oberkochen, 210
OCA, 160, 191, 192
Odessa, 528
OFXB, 504
OHL, 253
OHP, 191
Ohya, 360
Okayama, 342, 352, 354
Oklahoma City, 847
Oldman River, 90
Old Orchard, 777
Ole Rømer, 134, 139, 140
Olin, 662
Olivet, 766
Olmsted (North-), 837
Olomouc, 129
Olsztyn, 429
Olten, 500
OMA, 69
Omaha, 785
Ommen, 399
OMP, 193, 194
Onan, 771
ONBA, 9
Ondřejov, 124
Onsala, 490
OOO, 777
Oostzaan, 406
Ootacamund, 286
Ooty, 286, 288
OPL, 194
Optec, 767
Orange, 632
Orchard (Old-), 777
Orlando, 680, 683, 686
Orléans, 158, 202
ORM, 478-480, 482
ORO, 134, 139, 140
Orono, 733
Orsay, 162
ORT, 286
Osaka, 356
OSC, 63
Osenbach, 200
Osnabrück, 238
Oss, 402
Ostrava-Poruba, 128
Ostrowik, 432
Otter Creek, 727
Ottobeuren, 207
Ouda, 350
Ouro Preto, 71
Owens Valley, 613, 801
Owl Ridge, 718
Oxie, 498
Ozersk, 440
Paagman (Jan-), 394
Pachon (Cerro-), 109, 596, 693
Paderborn, 262
Padova, 320, 327, 339
Padre Angelo Secchi, 311
PAGASA, 426

Palermiti (Mike.), 683
Palermo, 327, 331
Palisades-Dows, 716
Palisa (Johann.), 128
Palla (Spartaco.), 335
Palma de Mallorca, 483
Palm Beach
 (West.), 681, 685
Palmer, 638
Palmieri, 640
Palomar, 614
Palomar College, 632
Pameungpeuk, 290
Pamir, 518
Pantai Achech, 378, 379
Papenburg, 214
Papendrecht, 406
Paranal, 108, 109
Parc aux Étoiles, 195
Parc Catalunya, 472
Parc Montsouris, 158
Paris, 158, 183, 192, 193, 200
Parkes, 22
Park Site, 103
Parkway Stargaze, 827
Parthenay, 181
Partizánske, 459
Pasadena, 613, 614, 629
Paso de los Toros, 587
Passau, 262
Patacamaya, 63
Pattonville, 777
Pauligne, 170
Paul J. Meyer, 878
Paul P. Feder, 772
Paul Robinson, 794
Paul Wild, 22
Paysandú, 587
PBO, 920
Peach Mountain, 765
Pearce, 594
Pearl City, 694
Peist, 514
Penang, 378, 379
Penc, 277
Pensacola, 680
Penticton, 94
Peoria, 704
 (East.), 701
Périgueux, 179
Perkins, 842
Perth, 26, 32
Perugia, 339
Petaling Jaya, 378
Peterberg, 257
Petrín, 130
Pettinger-Guiley, 910
Pforzheim, 213
PGAO, 97
Philadelphia, 862
Philippus Lansbergen, 407
Philips (Dr. A.F..), 390
Phipps (Gerald Hughes.), 653
Phoenix, 400
Pianoro, 333
Pic du Midi, 194
Pickmere, 579
Pico dos Dias, 68
Pico Veleta, 182, 478
Piedade, 73
Piera, 476
Pieri, 333
Pietermaritzburg, 465
Pie Town, 801
Pietrasanta, 335
Piikkiö, 152
Pikes Peak, 660

Pin, 289
Pine Bluff, 920
Pine Mountain, 689, 850
Pino Torinese, 328
Piracicaba, 70
Pirttiharju, 149
Piscataway, 796
Pises, 190, 200
Piszkéstetö Mountain, 275
Pittsburgh, 851, 856, 858, 862
Piwnice, 432
Plana, 75
Plateau de Bure, 182, 204
Pleasant (Mount.), 31, 762
Plzeň, 130
PMO, 850
Pochoco (Cerro.), 107
Poços de Caldas, 65
Pole (South.), 668
Polman (Jorge.), 66
Poltava, 529
Pomona, 796
Pompaire, 181
Pontefract, 584
Portimão, 433
Porto, 436
Porto Alegre, 73
Poste-De Flacq, 381
Potchefstroom, 466
Potosí, 62
Potsdam, 215, 218
Potterville, 763
Powell, 776
Powys, 573
Poznań, 431
PPO, 660
Praha, 127, 130
Prenzlau, 262
Přerov, 131
Presolana, 319
Prešov, 459
Preston, 563, 575
Pretoria, 465, 469
Prices Fork, 906
Priddis, 102
Prince Albert, 468
Prince George, 97
Princeton, 791, 795
Prof. Manuel de Barros, 436
Profondeville, 50
Prostějov, 131
Providence, 864
Provo, 890
Puckett, 690
Puebla, 383
Pufferfish, 705
Puimichel, 193
Pulkovo, 447, 448
Pullman, 910
Pulpit Rock, 858
Pune, 286, 288
Punta Gorda, 681
Purple Mountain, 114, 115
Puyallup, 910
Pylypovychi, 526
Pyongyang, 366
Qinghai, 114
Quabbin, 751
Queen's, 97
Querqueville, 180
Queven, 161
Quezon City, 425, 426
Quinn, 869
Quito, 141
Rabastens, 154
Rabbit Lake, 103
Racine, 919

Radeberg, 260
Radebeul, 209, 232, 259
Raden, 684
Raiba, 35
Rainwater, 774, 775
Ráktanya, 276
RAL, 647
Ralph A. Worley, 730
Ramat-Aviv, 301
Ramfjordbotn, 415
Ramón María Aller, 485
Rampisole (La_), 179
Ramponio Verna, 321
Ramsey (Crosby_), 739
RAO, 102
Rapid City, 869
Ravenna, 310
Ravenstein, 239
RBO, 923
Recife, 66
Recklinghausen, 265
Red Bridge, 697
Red Buttes, 923
Reduit, 381
Reggio di Calabria, 322
Regina, 103
Regiomontanus, 239
Remanzacco, 309
Remeis (Dr._), 254
Remscheid, 214
Reno, 786, 859
Rensselaer, 820
Repsold, 232
Réunion (La_), 156
Reutlingen, 263
Rev. M.W. Burke-Gaffney, 97
Rezh, 440
RGO, 565
Rheine, 211
Riboux, 160
Richardson (N.A._), 636
Richard Stockton, 796
Richmond, 725
Richmond Hill, 103
Riga, 371, 372
Rigel, 313
Righi A., 322
Rihlaperä, 148
Riley (Glen D._), 703
Rimavská Sobota, 459, 461
Rinnegg, 39
Rio de Janeiro, 70, 73
Rio Grande, 7, 10
RIT, 821
Ritchie (Edwin_), 907
Ritter, 844
Rivera, 587
River Falls, 921
Rivière (La_), 156
Riyadh, 453
RMCC, 846
Robert-A. Naef, 504
Robert Mayer, 243
Robinson, 680, 686
Robinson (Paul_), 794
Robledo de Chavela, 613
Rocha, 587
Rochefort, 171
Rochester, 821, 825
Rockbottom, 757
Rockford, 705
Rock Island, 705
Roden, 395
Rodewisch, 244
ROEN, 68, 70
Rogers (Joseph H._), 764, 766
Rogersville, 778

ROI, 68, 70
Rokycany, 131
Rolnick, 666
Roma, 304, 328, 340
Romans-sur-Isère, 171
Romeo, 769
Rømer (Ole_), 134, 139, 140
Römer (Rudolf_), 244
Roque de los Muchachos, 472, 477–480, 482
Roquetes, 486
Roquevaire, 162
Rosario, 9
Rosemary Hill, 686
Rossall-Assheton, 564
Rosse, 584
Rostock, 215
Rothney, 102
Rothwesten, 262
Rottmayr, 248
Rouen, 181, 193
Rovegno, 341
Royal
   Brussels, 55–57
   Edinburgh, 534, 562, 576
   Greenwich, 565
Rozhen, 75
Roztoky, 459
RPI, 820
RSCC, 873
RSCNJ, 796
Rtyné v Podkrkonoší, 131
Rudolf Römer, 244
Rudy Park, 863
Rueil-Malmaison, 163
Rümlang, 513
Rupp (Warren_), 845
Rusnjak, 122
Ryerson Nature Center, 706
Sabadell, 472
Sabino Canyon, 601
Sachtleben, 782
Sacramento Peak, 801
Saint Andrews, 581
Saint Ann, 777
Saint Augustine, 678
Saint-Caprais, 154
Saint-Claude, 170
Saint Cloud, 772
Saint Croix, 801
Saint Davids, 855
Sainte Baume (La_), 162
Sainte-Foy, 101
Saint-Elzéar-de-Beauce, 95
Saint-Étienne, 161
Saint-Étienne-en-Devoluy, 182
Saint-Genis-Laval, 165, 200
Saint-Imier, 506
Saint-Jean-le-Blanc, 190
Saint-Laurent-du-Var, 160
Saint Louis, 777
Saint-Luc, 504
Saint-Martin d'Hères, 182, 203
Saint-Michael, 45
Saint-Michel-l'Observatoire, 164, 191, 507
Saint-Nérée-de-Bellechasse, 78
Saint Paul, 771
Saint Peter Port, 106
Saint Petersburg, 445, 447, 449, 684
Saint-Valérien de Milton, 99
Saint-Vallier-de-Thiey, 192
Saint-Véran, 163
Saji, 356
Sakashita, 351
Sakushima, 351
Salem, 757
Salgótarján, 273, 277
Salisbury, 567

Salt Lake City, 891
Salto, 587
Saltsjöbaden, 494
Saludecio, 322
Salve, 310
Salzburg, 34
Salzkammergut, 34
Salzschlirf (Bad-), 231
Sampsor, 476
Samson, 171
Sanae, 466
San Antonio, 883
San Bernardino, 636
San Calixto, 63
San Diego, 636, 637, 646
Sandvreten, 498
San Fernando, 482, 615
Sanford, 685
San Francisco, 608, 611, 637
San Giuseppe, 332
Sanglok (Mount-), 518
San José, 9
San Jose, 627
San Juan, 7, 11
San Juan Bautista, 621
Sankt Gallen, 502
Sankt Georgen, 39
Sankt Sebastian, 35
San Marcello Pistoiese, 334
San Marcos, 632
San Martino (Monte-), 324
San Miguel, 9
San Pedro Mártir, 386
San Pio X, 306
San Polo a Mosciano, 305
Santa Ana, 63
Santa Barbara, 638, 651
Santa Clara, 622
Santa Cruz, 62, 649
Santa Cruz de la Palma, 472, 478–480, 482
Santa Lucia di Stroncone, 332
Santa María, 383
Santa María de Trassierra, 471
Santiago de Cali, 119
Santiago de Chile, 107–110
Santiago de Compostela, 485
San Vittore, 333
San Ysidro, 646
SAO, 448
São João Nepomuceno, 69
São José dos Campos, 70
São Luís, 71
São Paulo, 66, 68, 72
Sapello, 803
Saran, 161
Saskatoon, 103
Saturnus, 403
Sauvenière, 52
Sauverny, 507
Savona, 323
Sayan, 448
SBO, 659, 660
SCC, 685
Schanfigg, 514
Schauinsland, 235, 247
Schenectady, 814
Schiaparelli (G.V.-), 335
Schio, 320
Schkeuditz, 216
Schneeberg, 250
Scholz (Erich-), 260
Schothorst, 400
Schrieversheide, 402
Schwabe (Heinrich S.-), 215
Schwäbische, 245
Schwartz (James-), 631
Schwarzschild (Karl-), 252

Scituate
 (North-), 865
SCMT, 109
SCO, 460
Scotia, 805
Scultetus (Bartholomäus-), 264
Seagrave Memorial, 865
Secchi (Padre Angelo-), 311
Sedalia, 659
Sedgefield, 467
Seebad Heringsdorf, 261
Seething, 560
Seewalchen, 34
Selah, 911
Sells, 598
Selvino, 313
Sencelles, 476
Senftenberg, 242
Senigalliesi, 310
Seoul, 369, 370
Serafino Zani, 332, 337
Serin, 796
Serra la Nave, 326
Setenil de las Bodegas, 473
Seven Hills, 784
SFASU, 885
SFO, 615
SFSU, 637
SGO, 146, 707
Shaanxi, 114
Shahe, 113
Shanghai, 114
Sharru, 332
Shattuck, 788
Shemakha, 43
Shenandoah, 736
Sherbrooke, 86
Sherman Park, 718
Sherwood, 556
Sherzer, 762, 763
Sheshan, 115
Shields (South-), 570
Shiga, 355
Shimosato, 347
Shirahama, 347
Shiraz, 292
Shreveport, 730
SHSU, 883
Sidak, 525
Siding Spring, 14, 18
Sidmouth, 559
Siena, 337
Sierra Nevada, 476
Sierre, 504
Sigulda, 371
Simeiz, 525
Simferopol, 525, 528
Simpson (John W.H.-), 791
Singapore, 457
Singen, 210
Sion, 500
Sir Thomas Brisbane, 27
Sivry, 49
Sizuoka, 347
SJA, 510
Skalnaté Pleso, 459, 460
Skibotn, 418, 420
Skokie Valley, 706
Sky Shack, 645
Slavey Zlatev, 74
Sliven, 74
Slottsskogen, 493
Slough, 542
Smith Park, 725
Smolyan, 75, 76
SMTO, 603
SNO, 100

SOAO, 368
Sobaeksan, 368
Socorro, 801, 802
Sodankylä, 146, 147, 489
SOE, 218
Soest, 259
SOFIA, 630
Sofia, 75
Sohland, 213, 264
SOHO, 642
Sojn Valley, 770
Solingen, 264
Solms/Lahn, 213
Sommers-Bausch, 659, 660
Sommets (Les.), 86
Sonneberg, 250
Sonnenborgh, 402, 405
Sonora, 385
Sontra-Breitau, 258
Sorbonne, 200
Soresina, 323, 332
Sormano, 333
Sotteville-lès-Rouen, 181
South African
   Cape Town, 467
   Sutherland, 467
Southam (Gordon.), 89
South Canterbury, 412
Southern Columbia Millimeter Telescope, 109
South Mountain, 858
Southold, 813
South Point, 875
South Pole, 668
South Shields, 570
South Tyneside, 570
Spa, 52
Spartaco Palla, 335
Speicher, 502
Sperry (William Miller.), 791
Spezia (La.), 329
Spreeufontein, 468
Springfield, 708, 758, 778, 894
Sproul, 862
SRO, 643
Sroborova, 460
SSW, 245
Standeford, 771
St. Andrews, 581
Stanford, 642
St. Ann, 777
Stansbury, 891
Stará Lesná, 459, 460
Star Chaser, 659
Starfield, 732
Stargate, 769
Star Hill Inn, 803
Starkenburg, 246
Staten Island, 811
St. Augustine, 678
Stavanger, 418
St.-Caprais, 154
St.-Claude, 170
St. Cloud, 772
St. Croix, 801
St. Davids, 855
Stefanik, 130
Ste.-Foy, 101
Steina, 250
Steinberg, 39
St.-Elzéar-de-Beauce, 95
St.-Étienne, 161
St.-Étienne-en-Devoluy, 182
Stevens Point, 921
Steward, 597, 599, 603, 604
Steyning, 584
St.-Genis-Laval, 165, 200
St.-Jean-le-Blanc, 190

St.-Laurent-du-Var, 160
St. Louis, 777
St.-Luc, 504
St.-Martin d'Hères, 182
St.-Michael, 45
St.-Michel-l'Observatoire, 164, 191, 507
St.-Nérée-de-Bellechasse, 78
Stockert, 253
Stockholm, 492–494
Stockton (Richard.), 796
Stonehenge, 567
Stony Ridge, 643
Store Merløse, 135
Storrs, 665
St. Paul, 771
St. Peter Port, 106
St. Petersburg, 445, 447, 449, 684
Strasbourg, 159, 165, 200, 205
Strawbridge, 856
Stromlo (Mount.), 18
STScI, 743
St. Sebastian, 35
Stull, 805
Stuttgart, 221, 245
St.-Valérien de Milton, 99
St.-Vallier-de-Thiey, 192
Subaru, 695
Sucre, 62
Sudak, 528
Sudbury, 98, 99
Sudbury Neutrino Observatory, 100
Suesca, 119
Sugadaira, 351
Sugar Grove, 707
Suhl, 244
Suhora (Mount.), 428
Summit, 718
Sundsvall, 494
Sunrise, 685
Sunspot, 799
Susa, 306
Sutherland, 28, 467
Sutrieu, 170
Sutton-in-Ashfield, 556
Svetloe, 445
Swanage, 584
Swarthmore, 862
Swasey, 838
Swasey (Warner and.), 836
Swaythling, 569
Swisher, 716, 718
Sydney, 15, 20, 28
Sylmar, 615
Sylvania, 837
Syowa, 355
Syracuse, 824
Székesfehérvár, 277
Szombathely, 274
Tabernes Blanques, 477
Table Mountain, 613, 634
Tabriz, 293
Tacoma, 910
Tacuarembo, 587
Taeduk, 368
Taejon, 367, 368
Taipei, 516, 517
Talmassons, 314
TAM, 517
Tamborine (Mount.), 30
Tamke-Allen, 873
Tampere, 150
Tanjung Sari, 290
TAO, 359
Taoyuan, 516
Tarbes, 194
Tarija, 62, 63
Tarn, 154

Tartu, 144
Tashkent, 925
Tatranská Lomnica, 459
Tautenburg, 252
TBO, 498
TCO, 832
Tegucigalpa, 272
Teide, 58, 235, 256, 257, 477
Tel-Aviv, 301
Telescopio Nazionale Galileo, 327, 479
Tenagra, 602
Tenerife, 477
Teplice, 131
Teramo, 328
Terkán (Lajos_), 277
Terre Haute, 712
Terskol, 526, 529
The Briars, 15
The Cape, 464
Themis, 187
Thomas Brisbane (Sir_), 27
Thompson, 915
Thorigné-sur-Dué, 162
Three College, 832
Thule, 668
Thüringer, 252
Tibarine (Mont_), 4
Ticino, 509, 511
Tidbindilla, 613
Tiendesprong (De_), 401
Tien-Shan, 364
Tierra del Sol, 636
Tiffin, 843
Tighes Hill, 16
Tilburg, 401
Timişoara, 438
Timmendorfer Strand, 208
TIRGO, 317
Tirschenreuth, 263
Tixie Bay, 446
TLS, 252
TMO, 613, 634
TNG, 327, 479
Todmorden, 562
Togót (Khurel_), 388
Tokushima, 359
Tokyo, 344–347, 352–355, 360–362
Toledo, 844
Tololito, 110
Tololo (Cerro_), 109
Tombaugh (Clyde W._), 723
Tomsk, 450
Tonantzintla, 383, 386
Tonneville, 188
Tony Dome, 878
Toothill, 569
Topeka, 723, 724
Töravere, 144
Torgelow, 250
Torino, 328
Toronto, 103
Torres Arbide, 483
Tortugas Mountain, 802
Torun, 432
Toulouse, 193, 201
Toussi (Khadjeh Nassir Addin_), 293
Toyama, 359
Toyokawa, 351
Tradate, 324
Traiskirchen, 35
TRAO, 368
Traverse City, 764, 766
Trebur, 250
Treffen, 40
Treinta y Tres, 587
Trelew, 10
Treptow, 209

Treviso, 307
Triel-sur-Seine, 195
Trieste, 328
Trinité (La_), 195
Troitsk, 446
Tromsø, 420, 489
Troy, 820
TSO, 252
Tsumeb, 466
TÜBITAK, 524
Tucson, 596, 597, 599–601, 603, 604, 773
TUG, 524
Tullahoma, 875
Tullamore, 298
Tulsa, 846
Tumanoc Hill, 603
Tuorla, 152
Turku, 152
Turtle Star, 252
Tuscaloosa, 591
Twin Falls, 697
Twining (General Nathan_), 803
Two Mile Run Park, 859
Tycho Brahe
    Malmö, 491
    Oxie, 498
    Rostock, 215
Tyneside (South_), 570
Tyngsboro, 756
UAO, 115, 448
Uberlândia, 69
Uccle, 53–56
Udaipur, 288
Udine, 309
UFOP, 71
Uherský Brod, 132
Uitikon, 512
UKIRT, 693
Ukkel, 53–56
UKST, 29
Ulaanbataar, 388
Ulugh Beg, 925
UNA, 591
Union Grove, 919
United Kingdom Infrared Telescope, 693
United Kingdom Telescopes, 693
University Park, USA-PA, 860
Úpice, 125, 132
Uppsala, 495, 498
Urania
    Budapest, 278
    Burgdorf, 512
    Jena, 263
    Salgótarján, 273
    Wien, 41
    Zürich, 504, 513
Uranoscope de l'Île de France, 205
Urbana, 705
URSA, 152
Urumqi, 115
US Naval, 604, 677
USNO, 604, 677
Ussurijsk, 448
Utrecht, 402, 405
Uttar Pradesh, 289
UWRF, 921
UWSP, 921
Vaiana (Giuseppe S._), 327
Vainu Bappu, 285
Valašské Meziříčí, 132
Valcourt, 200
Valcuvia, 324
Valdosta, 691
Valence (Beaumont-lès-_), 158
Valencia, 486
Valinhos, 72
Valla, 488

Valley of the Moon, 650
Vallons, 163
Valmorey, 170
Valongo, 73
Valparaiso, 714
Van Allen, 719
Vancouver, 89
Van Vleck, 665
Van Zandt (Decker-Grebner-_), 704
Varese, 335
Varna, 74
Varsovie, 427, 428, 432
Vartovka, 458
Vatican, 604, 926
Vayres, 171
VBO, 285
Veen (James C._), 764
Vega-Bray, 604
Veleta (Pico_), 182, 478
Veli Lošinj, 121
Velp, 390
Venezia, 308
Ventspils, 371
Verna (Ramponio_), 321
Vernonia Peak, 849
Verona, 314
Vérossaz, 504
Veselí nad Moravou, 132
Vesper, 824
Vesta, 406
Vevey, 510
Vicques, 510
Victoria, 16, 93, 94, 104
Vicuña, 109
Vignola, 169
Vignui, 308
Vila Nova de Gaia, 436
Vilanova i la Geltrú, 480
Villa de Leyva, 119
Villafranca del Castillo, 476
Villemur, 172
Villeneuve d'Ascq, 171
Villepreux, 158
Vilnius, 374, 375
VILSPA, 476
Vinissan, 160
Violau, 248
VIRAC, 371
Virgin Islands, 801
Visnjan, 122
Vissegi (Monte_), 307
Vittore (San_), 333
VLA, 801
Vlašim, 133
VLBA, 801
VMOA, 650
Voggenberg, 34
von Ardenne (Manfred_), 261
VORG, 604
Vries, 406
VSA, 406
VSD, 260
Vsetín, 133
VSN, 262
Vulcan, 79
VVO, 665
Vyhne, 460
Vyškov, 126
Waalre, 390
Waco, 878
Wagman (Nicholas E._), 851
Wahnsiedler, 711
Wajh, 453
Wakayama, 347, 351
Wald, 501
Walden, 99
Waldhügel, 211

Wald im Pinzgau, 39
Wallace, 754
Wallaroo, 16
Walter Barber, 687
Walter Hohmann, 208
Waltham, 748
Wanganui, 413
Wank, 230
Wanne-Eickel, 211
WAO, 754
Wardle, 579
Ward Pound Ridge Reservation, 825
Warner and Swasey, 836
Warren, 769
Warren Rupp, 845
Warrenville, 706
Warsaw, 432
Warszawa, 427, 428, 432
Washburn, 920
Washburn Valley, 547
Washington, 677
Washington Crossing, 791
Wast Hills, 574
Waterlooville, 547
Watukosek, 290
WCAC, 845
WCOA, 629
Webster Groves, 777
Weil (Johnson-_), 805
Weiskopf, 699
Weißenschirmbach, 231
Weitkamp, 842
Wellesley, 759
Wellington, 409
Welzheim, 221
Wendelstein, 255, 256
Wernigerode, 214
West Chatswood, 24
Westerbork, 401, 402
Westerville, 842
Westfälische, 265
Westford, 746, 755, 756
West-München, 217
West Palm Beach, 681, 685
Westport, 666
Westside, 666
Wetterau, 263
Wetzlar, 213
WFS, 265
Whately, 758
Wheaton, 709
Whipple (Fred Lawrence_), 597, 751
Whiting, 791
WHO, 584
WHT, 479
Wichita, 722
Wien, 36, 41
Wiesbaden, 212
Wiesendangen, 511
Wikswo, 873
Wilcox, 642
Wild (Paul_), 22
Wilhelm Foerster, 265
William Herschel Telescope, 479
William Miller Sperry, 791
Williams Bay, 915, 919
Williamsburg, 903
Williamstown, 760
Williston, 894
Wilmington, 667
Wilmot, 845
Wilson (Mount_), 629, 650
Windach, 217
Wine (Carolyn_), 683
Winfree, 902
Winnipeg, 102
Winterslow, 567

WIRO, 923
Wiruna, 16
Wisconsin-Indiana-Yale-NOAO, 920
Wise (Florence and George_), 301
Within, 759
Witte (John H._), 719
Witzenhausen, 258
WIYN, 920
WLS, 407
W.M. Keck, 692
Wollongong, 32
Womble (Meyer-_), 661
Wood-Cock Hill, 298
Worcester, 750
World's View, 465
Worley (Ralph A._), 730
Worth Hill, 584
Worthing, 584
Wrightwood, 613
WRO, 845
Wrocław, 432
WSO, 642
WSRT, 401
Würzburg, 257, 263
WWO, 629
Wylam, 558
Wyoming Infrared, 923
Xinglong, 113
Xinjiang, 115
Yakima, 911
Yakutsk, 446
Yale, 666
    Southern Station, 666
Yamaguchi, 362
Yanna, 917
Yaroslavl, 451
Yatsugakate, 345
Yebes, 481
Yeoville, 464
Yerkes, 915, 919
Yokohama, 362
York, 585, 863
Young, 70, 587, 632
Youreyeball, 733
Ypsilanti, 762, 763
Yunnan, 115
Yverdon-les-Bains, 510
Zagreb, 122
Zalaegerszeg, 273
Zani (Serafino_), 332, 337
Zaporozhye, 530
Zaragoza, 470, 476
Ždánice, 133
Žebrák, 133
Zelenchukskaya, 442
Zenas Crane, 724
Zermatt, 326
Žilina, 461
Zimmerwald, 513
Ziolkowski (K.E._), 244
Zittau, 260
Zlatev (Slavey_), 74
Zlín, 133
Zografos, 269
Zugspitze, 230
Zuidlaren, 406
Zürich, 501–504, 512, 513
Zvenigorod, 445
Zwiggelte, 401
Observatoriet i Slottsskogen, 492, 493
Observatório Abrahão de Moraes, 72
Observatorio Albert Einstein, 586
Observatorio Astrofísico Guillermo Haro (OAGH), 383, 384
Observatório Astronômico Antares (OAA), 68
Observatorio Astronómico Boliviano-Ruso, 63
Observatório Astronômico, Coimbra, 434, 435
Observatório Astronômico da Escuela de Minas (OAEM), 71

Observatorio Astronómico de Córdoba (OAC), 6, 8, 10
Observatorio Astronómico de La Plata, 8
Observatorio Astronómico de la Plata, 10
Observatorio Astronómico del Garraf (OAG), 480
Observatório Astronômico de Lisboa (OAL), 434, 435
Observatorio Astronómico de Mallorca (OAM), 480
Observatorio Astronómico de Patacamaya (OAP), 63
Observatório Astronômico de Piedade (OAP), 68, 72
Observatorio Astronómico de Quito (OAQ), 141
Observatório Astronômico de Uberlândia (OAU), 69
Observatorio Astronómico de Valencia, 480, 486
Observatorio Astronómico EMVIC, 928
Observatorio Astronómico Felix Aguilar (OAFA), 8, 10
Observatório Astronômico Herschel-Einstein (OAHE), 69
Observatorio Astronómico Kappa Crucis, 587
Observatorio Astronómico Los Molinos (OALM), 587
Observatório Astronômico Monoceros (OAM), 69
Observatorio Astronómico Móvil (OAM), 8
Observatorio Astronómico Municipal de Mercedes (OAMM), 8
Observatorio Astronómico Nacional (OAN)
    Bogotá, 120
    Centro Astronómico de Yebes (CAY), 481
    Ensenada, 386
    Mexico, 384, 631
    Observatorio de Calar Alto, 481
    San Diego, 646
    San Ysidro, 646
    Spain, 480
Observatorio Astronómico, Planetario y Museo Experimental de Ciencias de Rosario, 9
Observatório Astronômico Prof. Manuel de Barros, 434, 436

Observatorio Astronómico Ramón María Aller, 481, 485
Observatorio CEMA, 477
Observatorio Cerro Las Ánimas, 385
Observatorio Cerro Mamalluca, 109
Observatório Copérnico, 69
Observatório das Ciências e das Tecnologias (OCT), 434
Observatorio de Calar Alto, 475, 481
Observatorio del Ebro, 481, 486
Observatorio del Roque de los Muchachos (ORM), 477, 479, 481, 482
Observatorio del Teide, 477
Observatório do Pico dos Dias (OPD), 68
Observatorio Luis G. León, 385
Observatorio Magnético Las Acacias, 10
Observatorio M. Foster, 110
Observatório Municipal de Americana (OMA), 69
Observatório Municipal de Campinas (OMC), 69
Observatório Municipal de Piracicaba (OMP), 70
Observatorio Nacional de Física Cosmica
    San Miguel (ONFCSM), 9
Observatório Nacional (ON), 70
Observatorio Naval de Buenos Aires (ONBA), 9
Observatorio Pico de Veleta, 478
Observatorio San Calixto (OSC), 63
Observatorio San José (OSJ), 9
Observatorio Tololito, 110
Observatório Young, 70
Observatorium für Solare Radioastronomie (OSRA), 218, 239

Observatorium Hoher List (OHL), 239, 253
Observatorium Lustbühel, 37, 40
Observatorium Ravenstein, 239
Observatorium Wendelstein, 240
Observatory (Cape Town), 465, 467
Observatory Magazine, 560
Obsession Telescopes, 918
OBU
    Oklahoma Baptist University
    Planetarium, 846
OCA
    Observatoire de la Côte d'Azur
    Centre d'Études et de Recherches en Géodynamique et Astrométrie, 191

Station de Calern, 192
Orange County Astronomers, 632
Ocean County College, 791
Oceanside, 618, 631
Oceanside Photo and Telescope, 631
Ocean View, 694
OCIA
    Onizuka Center for International Astronomy, 696
OCLI
    Optical Coating Laboratory Inc., 631
OCT
    Observatório das Ciências e das Tecnologias, 434
Odense, 138
Odense University
    Physics Department, 138
Odessa, 528, 889
Odessa State University
    Astronomical Observatory, 528
ODU
    Old Dominion University
        Physics Department, 901
Odyssium, 95
OEB
    Office Européen des Brevets, 227, 240
OES
    Optische und Elektronische Systeme, 240
Oetwill-am-See, 502
OFA
    Observatoire Fédéral d'Astrophysique, 94
    Canadian Astronomy Data Centre, 93
Offenbach/Main, 223
Office Européen des Brevets (OEB), 227, 240
Office Fédéral de Météorologie et de Climatologie, 507
Office Fédéral de Météorologie et de Climatologie (MétéoSuisse), 506, 507
Office National de la Météorologie, 5
Office National d'Études et de Recherches Aérospatiales (ONERA), 194
    Centre d'Études et de Recherches de Toulouse (CERT), 194
Office of Space Science (NASA OSS), 673
Oficina de Protección de la Calidad del Cielo (OPCC), 110
OFR
    Observatoire Fédéral de Radioastrophysique, 94
OFXB
    Observatoire François-Xavier Bagnoud, 504
ÖGAA
    Österreichische Gesellschaft für Astronomie und Astrophysik, 37
Ogden, 890–892
Ogden Astronomical Society (OAS), 890
Ogden College, 728
Oglethorpe Astronomical Association (OAA), 690
Ogliastra, 310
OGS
    Optical Guidance Systems, 859
O'Halloran Hill, 29
Ohio, 835–845
Ohio Aerospace Institute (OAI), 840
Ohio State University (OSU)
    Department of Astronomy, 841
    MDM Observatory, 598
    Perkins Observatory, 842
    Radio Observatory (OSURO), 841
Ohio Turnpike Astronomers Association (OTAA), 841
Ohio University
    Department of Physics and Astronomy, 841
Ohio Valley Astronomical Society (OVAS), 913
Ohio Wesleyan University (OWU)
    Department of Physics and Astronomy, 841
    Perkins Observatory, 842
OHL
    Observatorium Hoher List, 253
OHP
    Observatoire de Haute Provence, 191
Ohya Observatory, 360
Oil City, 859

Oil Region Astronomical Society (ORAS), 859
Okanagan, 94
Okayama, 342, 347, 354
Okayama Astrophysical Observatory (OAO), 352, 354, 356
OKCAC
    Oklahoma City Astronomy Club, 847
Okinawa, 343
Okinawa Radio Observatory, 343, 356
Oklahoma, 846, 847
Oklahoma Astronomical Society (Arkansas-_), 846
Oklahoma Baptist University (OBU)
    Planetarium, 846
Oklahoma City, 846, 847
Oklahoma City Astronomy Club (OKCAC), 846
Oklahoma State University
    Department of Physics
        Astrophysics Group, 847
OKU
    Osaka Kyoiku University
        Astronomical Institute, 356
Olbers-Planetarium, 240
Old Dominion University (ODU)
    Physics Department, 901
Oldenburg, 209
Old Forge, 818
Oldman River Observatory, 90
Old Orchard Observatory (OOO), 777
Old Royal Observatory (ORO), 560
OLE
    Opto and Laser Europe, 550
Olen, 48
Ole Rømer Museet (ORM), 138
Ole Rømer Observatory (ORO), 134, 139, 140
OLI
    Optronic Laboratories Inc., 682
Olin Observatory, 662
Olivet, 766
Olivet College
    Beman Observatory, 766
    Olivet Planetarium, 766
Olivet Nazarene University
    Strickler Planetarium, 706
Olivet Planetarium and Beman Observatory, 766
Olivos, 6
Olmsted
    (North_), 837
Olomouc, 129, 130
Olomouc Observatory, 129
Olson Planetarium (Manfred_), 918, 920
Olsztyn, 429
Olten, 500
Olympia, 909
Olympic Astronomical Society (OAS), 908
Olympos Astronomical Association, 149, 150
OMA
    Observatório Municipal de Americana, 69
Omaha, 782, 783, 785
Omaha Astronomical Society (OAS), 783
OMC
    Observatório Municipal de Campinas, 69
Omega Engineering Inc., 663
Omega Optical Inc., 893
OMM
    Organisation Météorologique Mondiale, 514
Ommen, 399
Omniplex, 847
    Kirkpatrick Planetarium, 846
Omniversum Space Theater and Digital Planetarium, 399
OMP
    Observatoire Midi-Pyrénées
        Bureau Gravimétrique International, 163, 186
        Observatoire de Toulouse, 193
        Observatoire du Pic di Midi, 194
    Observatório Municipal de Piracicaba, 70
OMPI
    Organisation Mondiale de la Propriété Intellectuelle (O-NU), 514

OMSI
    Oregon Museum of Science and Industry, 849
        Murdock Sky Theater, 849
ON
    Observatório Nacional (Brazil), 70
ÖN
    Österreichisches Normungsinstitut, 38
Onan Observatory, 771
ONBA
    Observatorio Naval de Buenos Aires, 9
Ondřejov, 124
Ondřejov Observatory, 124, 130
Oneonta, 815
ONERA
    Office National d'Études et de Recherches Aérospatiales, 194
        Centre d'Études et de Recherches de Toulouse, 194
ONFCSM
    Observatorio Nacional de Física Cosmica - San Miguel, 9
Onizuka Center for International Astronomy (OCIA), 695, 696
Onsala, 490
Onsala Space Observatory (OSO), 490, 492
Ontario, 623
    (Western_), 104
Ontario Science Centre (OSC), 96
OOO
    Old Orchard Observatory, 777
OOSA
    Office for Outer Space Affairs (UNO), 40
Oostzaan, 406
Ootacamund, 286
Ooty, 286
Ooty Radio Telescope (ORT), 286
OPCC
    Oficina de Protección de la Calidad del Cielo, 110
OPD
    Observatório do Pico dos Dias, 68
Open University (OU)
    Department of Physics
        Astronomy Research Group, 560
    Planetary Sciences Research Institute (PSRI), 560
ÖPG
    Österreichische Physikalische Gesellschaft, 38
OPL
    Observatoire Populaire de Laval, 194
Opole, 429
Opole Pedagogical University
    Department of Astrophysics, 429
Oporto, 433, 436
Optec Inc., 766
Optec Observatory, 767
Optical Coating Laboratory Inc. (OCLI), 631
Optical Corp. of America (OCA), 757
Optical Data Corp., 690
Optical Guidance Systems (OGS), 859
Optical Research Associates (ORA), 631
Optical Sciences Center (OSC), 604
Optical Society of America (OSA), 674
Optical Society of India (OSI), 287
Optique Unterlinden, 195
Optische und Elektronische Systeme (OES) GmbH, 240
Opto and Laser Europe (OLE), 550
Optometrics USA Inc., 757
Optronic Laboratories Inc. (OLI), 682
OR
    Observatoire de Rouen, 193
ORA
    Optical Research Associates, 631
Oran, 4
Orangeburg, 811, 868
Orange County Astronomers (OCA), 632
ORAS
    Oil Region Astronomical Society, 859
ORB
    Observatoire Royal de Belgique, 55
    Sunspot Index Data Center, 57
Orbital Sciences Corp. (OSC), 901
Oregon, 848–850, 916
    (Southern_), 849
Oregon Museum of Science and Industry (OMSI), 849
    Murdock Sky Theater, 849
Oregon Star Party (OSP), 849
Organisation du Traité de l'Atlantique Nord (OTAN)
    Division des Affaires Scientifiques, 55, 56
Organisation Météorologique Mondiale (OMM), 507, 514
Organisation Mondiale de la Propriété Intellectuelle (OMPI), 507, 514
Organization of American States (OAS)
    Office of Science and Technology, 675
Organizzazione Ricerche e Studi d'Astronomia (ORSA), 330

Oriel Ltd. (LOT-_), 555
ORION
    Oak Ridge Isochronous Observation Network, 873
Orion Planetarium, 139
Orion Telescope & Binocular Center, 632
Orlando, 680–683, 686
Orlando Science Center (OSC), 683
Orléans, 158
ORM
    Observatorio del Roque de los Muchachos, 477, 479, 482
    Ole Rømer Museet, 138
ORO
    Old Royal Observatory, Greenwich, 560
    Ole Rømer Observatory, 134, 139, 140
Orono, 733
Oroville, 634
Orpington, 561
Orpington Astronomical Society (OAS), 561
ORSA
    Organizzazione Ricerche e Studi d'Astronomia, 330
Orsay, 157, 162, 167, 173, 176, 187, 202, 204
Ørsted Laboratory, 136, 139
ORT
    Ooty Radio Telescope, 286
Ortona, 329
Orwell Astronomical Society Ipswich (OASI), 561
OSA
    Optical Society of America, 674
Osaka, 351, 356
Osaka Kyoiku University (OKU)
    Astronomical Institute, 356
OSC
    Observatorio San Calixto, 63
    Ontario Science Centre, 96
    Optical Sciences Center (University of Arizona), 604
    Orbital Sciences Corp., 901
    Orlando Science Center, 683
Osenbach, 200
Oshkosh, 918, 921
OSI
    Optical Society of India, 287
OSJ
    Observatorio San José, 9
Oslo, 415–419
Osmania University
    Centre of Advanced Study in Astronomy (CASA), 287
Osnabrück, 238
OSO
    Onsala Space Observatory, 490
OSP
    Oregon Star Party, 849
OSRA
    Observatorium für Solare Radioastronomie, 218
OSS
    Office of Space Science (NASA), 673
Osservatorio A. Righi, 322
Osservatorio Astrofisico di Arcetri (OAA), 326, 331, 335
Osservatorio Astrofisico di Catania (OAC), 326, 331
Osservatorio Astronomico dei Monti Cimini, 313, 331
Osservatorio Astronomico di Bologna (OAB), 326, 331

Osservatorio Astronomico di Brera (OAB), 331
    Merate, 326
    Milano, 326
Osservatorio Astronomico di Cagliari (OAC), 327, 331
Osservatorio Astronomico di Capodimonte (OAC), 327, 331

Osservatorio Astronomico di Cuneo, 331
Osservatorio Astronomico di Genova, 331
Osservatorio Astronomico di Merate, 331
Osservatorio Astronomico di Milano, 331
Osservatorio Astronomico di Padova, 327, 331
Osservatorio Astronomico di Palermo (OAP) "Giuseppe S. Vaiana", 327, 331
Osservatorio Astronomico di Perugia, 331
Osservatorio Astronomico di Roma (OAR), 331
    Monte Mario, 328
    Monteporzio Catone, 328
Osservatorio Astronomico di Sormano, 333
Osservatorio Astronomico di Teramo "Vincenzo Cerulli", 328, 331
Osservatorio Astronomico di Torino (OATo), 328, 331
Osservatorio Astronomico di Trieste (OAT), 328, 331
Osservatorio Astronomico Grange, 331
Osservatorio Astronomico "G.V.Schiaparelli", 335
Osservatorio Astronomico "Padre Angelo Secchi", 311
Osservatorio Astronomico Pubblico, Soresina, 323, 332
Osservatorio Astronomico Santa Lucia di Stroncone, 332
Osservatorio Astronomico "Serafino Zani", 332
Osservatorio Astronomico Sharru, 332
Osservatorio Bassano Bresciano, 332
Osservatorio di Rovegno, 340
Osservatorio G. Colombo, 320
Osservatorio San Giuseppe, 332
Osservatorio San Vittore, 333
Osservatorio Spartaco Palla, 335
Osservatorio Ticinese di Locarno-Monti, 507
Östergötland Astronomical Society, 492
Östergötlands Astronomiska Sällskap (ÖAS), 492, 493
Österreichische Akademie der Wissenschaften (ÖAW), 37
    Institut für Weltraumforschung (IWF), 37
Österreichische Gesellschaft für Astronomie und Astrophysik (ÖGAA), 37
Österreichische Gesellschaft für Weltraumfragen GesmbH, 37

Österreichische Physikalische Gesellschaft (ÖPG), 37
Österreichischer Astronomischer Verein, 38
    Arbeitsgruppe Sonnenuhren, 36, 38
Österreichisches Normungsinstitut (ÖN), 38
Ostia Lido, 303
Østjyske Amatør Astronomer (ØAA), 139
Ostrava Amateur Society, 130
Ostrava-Hrabuvka, 130
Ostrava-Poruba, 128
Ostrowik, 432
OSU
    Ohio State University
        Department of Astronomy, 841
        MDM Observatory, 598
        Perkins Observatory, 842
        Radio Observatory, 841
OSURO
    Ohio State University Radio Observatory, 841
OTAA
    Ohio Turnpike Astronomers Association, 841
OTAN
    Organisation du Traité de l'Atlantique Nord, 55
Otsu, 357
Ottawa, 78, 80, 82, 83, 86, 87, 92–94, 96, 99
Ottawa Valley Astronomy and Observers Group (OAOG), 96
Otterbein College
    Department of Physics and Astronomy, 841
Otter Creek Observatory, 727
Ottignies, 50
Ottobeuren, 207
Ott Planetarium (Layton P._), 890, 891
OU
    Open University
        Astronomy Research Group, 560
        Department of Physics, 560
        Planetary Sciences Research Institute (PSRI), 560
Ouda Station, 350, 356
Oulu, 151
OUP
    Oxford University Press, 561
Ouro Preto, 71
OUSAS
    Oxford University Space and Astronomical Society, 561
Outremont, 86
OVAS
    Ohio Valley Astronomical Society, 913
Overland Park, 721
OVRO
    Owens Valley Radio Observatory, 613, 794
        Headquarters, 613
Owatonna, 773
Owens Science Center (HBOSC) (Howard B._), 737
    Planetarium, 738, 742
Owens Valley, 801
Owens Valley Radio Observatory (OVRO), 613, 632, 794
    Headquarters, 613
Owl Ridge Observatory, 718
Owl Services, 859
OWU
    Ohio Wesleyan University
        Department of Physics and Astronomy, 841
        Perkins Observatory, 842
Oxford, 561, 580
Oxford Instruments
    Research Instruments, 561
Oxford University Press (OUP), 561
Oxford University Space and Astronomical Society (OUSAS), 561
Oxie, 498
Ozersk, 440
OZ Optics Ltd., 96

P

PAA
    Piedmont Amateur Astronomers, 867
Paagman Sterrenwacht (JPS) (Jan._), 394, 400
PAC
    Planetarium Association of Canada, 96
    Popular Astronomy Club Inc., 705
    Prairie Astronomy Club, 784
    Prescott Astronomy Club, 601
Pachart Foundation, 600
Pachart Publishing House, 600
Pachón (Cerro_), 109, 596, 693
Pacific, 610
Pacific Planetarium Association (PPA), 786
Pacific Science Center
    Willard W. Smith Planetarium, 908
Pacific Space Centre, 81
Pacific Union College (PUC)
    Physics Department, 632
    Young Observatory, 632
PADAS
    Preston and District Astronomical Society, 563
Paderborn, 262
Padova, 315, 319, 327, 337, 339
Padua, 319, 327, 337, 339
PAG
    Poole Astronomical Society, 583
PAGASA
    Philippine Atmospheric, Geophysical and Astronomical Services Administration
        Astronomy Research and Development Section, 425
Paintsville, 726
Pajarito Astronomers, 802
Pakistan, 421
Pakistan International Airlines (PIA)

Planetarium
  Lahore, 421
Pakistan Meteorological Department (PMD), 421
Pakistan Space & Upper Atmosphere Research Commission (SUPARCO), 421
Pakistan Standards Institution (PSI), 421
Palace of Pupils
  Technical Activities
    Astronomical Laboratory, 375
Palacky University
  Department of Theoretical Physics, 129, 130
Palais de la Découverte, 173
  Planétarium, 195
Palais de l'Univers, 195
Palaiseau, 175, 187
Palatine, 709
Palermiti Observatory, 683
Palermo, 315, 320, 327, 330
Palisades-Dows Observatory, 716
Palitzsch-Gesellschaft eV, 240
Pallasite Press, 410
Palma de Mallorca, 483, 485
Palma (La_), 471
Palm Beach
  (West_), 679, 681–683, 685
Palmieri Observatory, 640
Palo Alto, 608, 611, 626
Palomar College
  Observatory, 632
  Planetarium, 632
Palomar Mountain, 614
Palomar Observatory, 614, 632
Palos Verdes, 635
Pameungpeuk, 290
Pamir, 518
Pamplona, 482
Pannekoek (Sterrenkundig Instituut Anton_), 404
PAO
  Pyongyang Astronomical Observatory, 366
Papenburg, 214
Papendrecht, 406
Paraguay, 422
Parallax Instruments, 894
Parc aux Étoiles, 195
PARI
  Pisgah Astronomical Research Institute, 830
Paris, 154, 155, 158–160, 164, 166–169, 172–175, 177–179, 181–185, 189, 190, 192, 193, 195, 198, 199, 201–203, 205
Parkes, 22
Parkes Radio Observatory, 22, 26
Park Site, 103
Parks Optical Co., 632
Parkville, 29
Parma, 843
Parque de las Ciencias, 481
Parsec, 195
Parthenay, 181, 196
Particle Physics and Astronomy Research Council (PPARC), 561
  Royal Observatory Edinburgh (ROE), 562, 576
Partizánske, 458
Partizánske Observatory, 458
PAS
  Peninsula Astronomical Society, 621, 633
  Peoria Astronomical Society, 704
  Philippine Astronomical Society, 425
  Phoenix Astronomical Society, 600
  Physics Academic Software, 830
  Plymouth Astronomical Society, 563
  Pontchartrain Astronomy Society, 730
PASA
  Publications of the Astronomical Society of Australia, 15
Pasadena, 611–615, 621, 624, 629, 631, 633, 641
Pasajes, 479
Pasco Scientific, 633

PASJ
  Publications of the Astronomical Society of Japan, 342
Paso de los Toros, 587
PASP
  Publications of the Astronomical Society of the Pacific, 610
PASS
  Physics and Astronomy Student Society, 96
Passaic County Astronomical Association (PCAA), 794
Passau, 262
Patacamaya, 63
Paterson, 794
Paterswolde, 398
Patiala, 288
Paton Hawksley Education Ltd. (PHEL), 562
Patras, 269
Patterson Planetarium (Nathan M._), 689, 690
Pat Thomas Planetarium, 683
Pattonville Observatory and Planetarium, 777
Pau, 180
Pauligne, 170
Paul J. Meyer Observatory, 878
Paul Kanev Planetarium, 854
Paul P. Feder Observatory, 772
Paul Robinson Observatory, 794
Paulucci Space Theatre (PST), 772
Paul Wild Observatory, 22, 26
  Mopra Telescope, 21
Pavia, 322, 339
Paysandú, 587
PBO
  Pine Bluff Observatory, 920
PCAA
  Passaic County Astronomical Association, 794
PCG
  Planetário Calouste Gulbenkian, 434
PCP Inc., 683
PEA
  Phillips Exeter Academy
    Grainger Observatory, 789
Peabody, 747
Peach Mountain Observatory, 765
Pearce, 594
Pearl City, 694
Peddie School
  Planetarium, 795
Pegasus Productions, 713
Pégomas, 198
Peist, 514
Peking, 112–115, 118
Pembroke, 732
Penang, 378, 379
Penango, 319
Penc, 277
Peninsula Astronomical Society (PAS), 621, 633
Penkridge, 571
Penn State Astronomy Club, 859
Penn State University (PSU), 859
  Department of Astronomy and Astrophysics, 860
  Department of Statistics
    Statistical Consulting Center for Astronomy (SCCA), 860
Pennsylvania, 851–863
Pennsylvania State University (PSU), 859
  Department of Astronomy and Astrophysics, 860
  Department of Statistics
    Statistical Consulting Center for Astronomy (SCCA), 860
Penn Valley, 634
Penny & Giles Controls Ltd., 562
Penobscot Valley Star Gazers (PVSG), 733
Pensacola, 680, 683
Pensacola Junior College (PJC)
  Department of Physical Sciences, 683
Penticton, 94
People's Republic of China (PRC), 112
Peoria, 699, 703, 704

(East.), 701
Peoria Astronomical Society (PAS), 704
Périgueux, 179
Periodicals
  2002, 534
  A&R-bladet, 499
  AAC Actividades, 470
  AANZ Journal, 408
  AAO Annual Report, 14
  AAO Newsletter, 14
  Aardmagnetisme, 54
  AAS Job Register, 670
  AAS Newsletter, 670
  AAVSO Bulletin, 747
  AAVSO Circular, 747
  AAVSO Monographs, 747
  AAVSO Newsletter, 747
  AAVSO Reports, 747
  Abastumanskaya Astrofizicheskaya Observatorija Byulleten, 206
  Abhandlungen aus der Hamburger Sternwarte, 254
  Abhandlungen der Mathematische-naturwissenschaftliche Klasse, 218
  Abhandlungen der Philosophische-historische Klasse, 218
  Above the Fog, 637
  Academiae Analecta, 54
  Academia Europaea Directory, 531
  Academia Europaea Newsletter, 531
  Academy News, 395
  Academy Newsletter, 18
  Academy Update, 819
  Accent on Mathematics Engineering and Science Newsletter, 596
  access, 703
  Accounts of Chemical Research, 670
  ACMembernet, 807
  ACM Guide to Computing Literature, 807
  ACM No-Nonsense Guide to Computing Careers, 807
  ACM Transactions on
    Computer-Human Interactions, 807
    Computer Systems, 807
    Database Systems, 807
    Design Automation of Electronic Systems, 807
    Graphics, 807
    Information and System Security, 807
    Information Systems, 807
    Mathematical Software, 807
    Modeling and Computer Simulation, 807
    Networking, 807
    Programming Languages and Systems, 807
    Software Engineering and Methodology, 807
  Acrux, 9
  A Csillagvizsgáló, 273
  Acta Astronautica, 184
  Acta Astronomica, 427
  Acta Astronomica Sinica, 112, 114, 115
  Acta Astronomica Visnianiensis, 122
  Acta Astrophysica Sinica, 112, 113, 115, 116
  Acta Historica Astronomiae, 211
  Acta Humanistica et Scientifica Universitatis Sangio Kyotiensis, 349
  Acta Mathematica, 493
  Acta Polytechnica Scandinavica, 146
  Acta Universitatis Latviensis (Astronomy), 372
  Acta Zoologica, 493
  Ad Astra, 25, 531, 674
  ADION Bulletin, 161
  Adler Star, 698
  ADNDT, 808
  Advances in Polar Upper Atmospheric Research, 355
  Advances in Space Research, 174, 391
  Adventures, 830
  Aeronomica Acta, 48, 53
  Aerospace America, 895
  Aerospace Investigations in Bulgaria, 76
  Aerospace, Science and Technology, 225
  Aether (The.), 887
  Afascope, 159
  After Dark, 827
  Afterglow, 687
  Agenda Astronômica, 73
  AGO-Mitteilungen, 501
  Agrarmeteorologische Bibliographie, 223
  Agrarmeteorologischer Wochenhinweis, 223
  Agromet Bulletin of Pakistan, 421
  AIAA Journal, 895
  AIAA Student Journal, 895
  AIP Conference Proceedings, 734
  Air & Space, 676
  Air Almanac (The.), 677
  Air and Space Europe, 154
  Airborne Astronomy Flyer, 630
  AJP, 734
  Aktuel Astronomi, 140
  Alajpacha, 62
  Albedo, 557
  Albiréo, 164
  Albireo, 273, 307
  Alcor, 474
  Alert Notices, 747
  Algol, 585
  Allegheny Observer (The.), 851
  All India Weather Summary, 284
  Almanacco Astronomico, 334
  Almanacco Astronomico (AAI), 305
  Almanacco UAI, 337
  Almanac for Geodetic Engineers, 426
  Almanaque Náutico/ROA, 482
  Almanaque Nautico y Aeronautico (ONBA), 9
  ALPO Solar System Ephemeris, 609
  Alrukaba, 35
  Alsace Astronomie, 200
  Altair, 304, 544
  Amateur Astronomer, 112
  Amateur Astronomer (The.), 839
  Amateur Telescope Making Journal, 907
  Amatörcsillagászati Courier, 275
  Ambio, 493
  American Journal of Physics, 734
  American Physical Society Bulletin, 734
  American Scientist, 831
  American Statistician (The.), 896
  AMES Newsletter, 596
  AMS Annual Report, 806
  AMS Bulletins, 806
  Amstat News, 896
  AN, 212
  Anais da Academia Brasileira de Ciências, 65
  Anais Obs. Astron. Coimbra, 435
  Anales de Física, 482
  Anales ROA, 482
  Analyst (The.), 566
  Analytical Abstracts, 566
  Analytical Chemistry, 670
  Analytical Communications, 566
  A Naso in Sù, 334
  Andromeda, 74
  Annales Academiae Scientiarum Fennicae, 146
  Annales de Droit Aérien et Spatial, 91
  Annales de la Fondation Louis de Broglie, 179
  Annales de Physique, 176
  Annales Geophysicae, 155
  Annales Hydrographiques, 199
  Annales UMCS, 428
  Annali di Geofisica, 329
  Annals of Air and Space Law, 91
  Annals of Shanghai Observatory, 115
  Announcer, 734
  Annuaire Astronomique/SAM, 99
  Annuaire de la SFO, 202
  Annuaire de l'ORB, 56
  Annuaire de l'Union Internationale des Associations et Organismes Techniques, 203
  Annual Prediction of Maxima and Minima of Long Pe-

riod Variables, 747
Annual Report of Anglo-Australian Telescope Board, 14
Annual Report of the Estonian Physical Society, 144
Annual Report of the Kiso Observatory, 361
Annual Report of the Mizusawa Astrogeodynamics Observatory - Time Service and Geophysical Observations, 353
Annual Report of the Nantucket Maria Mitchell Association, 754
Annual Report of the National Astronomical Observatory of Japan, 352
Annual Report of the Nishi-Harima Astronomical Observatory, 355
Annual Report of the Time Section, 164
Annual Report of the Yamaguchi Museum, 362
Annual Reports on the Progress of Chemistry
    Section A: Inorganic Chemistry, 566
    Section B: Organic Chemistry, 566
    Section C: Physical Chemistry, 566
Annual Review of Astronomy and Astrophysics, 608
Annual Review of Earth and Planetary Sciences, 608
Annual Review of Fluid Mechanics, 608
Annual Review of Information Science and Technology, 735
Annual Review of Nuclear and Particle Science, 608
Annuario della Accademia delle Scienze di Torino, 302
Annuario Nautico (Unione Astrofili Bresciani), 337
ANSTJ Bonjour, 198
Antarctic Astrophysics, 915
Antarctic Meteorite Research, 355
Antares, 422
Anuário Astronômico, 72
Anuario del Observatorio Astronómico Nacional, 120, 481
Anuarios Meteorológicos, 63
Anuarul Astronomic, 438
Aperture (The_), 829
ApJ, 708
    Supplement, 708
APL, 734
Apolo, 65
Apparent Places of Fundamental Stars, 215
Applied Optics, 675, 734
Applied Physics Letters, 734
Applied Spectroscopy Journal, 742
Appulse (The_), 425
Appunti di Astronomia, 307
Arab Journal of Astronomy and Space Sciences, 363
Arcetri Astrophysical Observatory Annual Report, 326
Arcetri Astrophysical Preprints, 326
Arcetri Technical Reports, 326
Archives des Sciences, 510
Arkiv för Matematik, 493
Arlington Astroletter (The_), 896
Armagh Observatory Preprints, 532
ARSEC Journal, 747
Artificial Satellites, 430
Arts & Sciences, 682
ASA-Information Service, 37
ASA Newsletter, 15
ASA Proceedings, 15
ASGARD, 532
Asian & Pacific Rim Directory, 753
ASLI News, 808
As Plêiades, 65
ASRI News, 19
ASSA Handbook, 464
ASSA Newsletter (The_), 282
AST, 225
ASTC Directory, 671
ASTC Informs, 671
ASTC Newsletter, 671
AST Ephemeris, 16
ASTER, 470
Asterisken, 492
Asterism, 791

Asteroid Belt (The_), 841
Astra, 54
Astral Projections, 792
Astrea, 483
Astro, 143, 494
Astro 41, 157
Astro-Amateur, 207
Astrobiology, 818
Astroblick, 231
Astro Calendar, 116
Astrocosas, 422
Astrocourier, 441
Astrofax, 217
Astro Fax Zirkular, 258
Astrofilo Lariano (L'_), 321
Astro-Info, 34
Astro-Kurier, 238
Astrolog, 680
Astromag, 161
Astromundo, 62
Astro-Nachrichten, 214
Astro News, 189, 547
Astro-News, 213
Astronews, 324, 392, 889
Astronews (The_), 693
ASTRON/NFRA Newsletter, 402
Astronom, 462
Astronomía, Astrofotografía y Astronautica, 483
Astronomers' Bulletin (The_), 20
Astronomer (The_), 572, 821
Astronomi, 416
astronomia (L'_), 330
Astronomia Novae, 71
Astronomia UAI, 337
Astronomica, 306
Astronomical Almanac, 529
Astronomical Almanac (The_), 677
Astronomical and Astrophysical Transactions, 441
Astronomical Calendar, 75
Astronomical Circular, 355
Astronomical Circulars, 115
Astronomical Contributions from the University of Manchester, Series III, 579
Astronomical Events for the Liverpool Area, 554
Astronomical Handbook, 26
Astronomical Herald (The_), 342
Astronomical Journal, 670
Astronomical Phenomena, 677
Astronomical Society of the Pacific Conference Series, 610
Astronomical Society of Victoria Newsletter, 17
Astronomical Yearbook, 17
Astronomicheskij Vestnik, 440
Astronomie & Philatelie, 238
Astronomie en Touraine et Centre-Ouest, 162
Astronomie in Bielefeld, 239
Astronomie (L'_), 199
Astronomie Magazine, 162
Astronomie Passion, 199
Astronomie pour Tous, 171
Astronomie-Québec, 88
Astronomie + Raumfahrt im Unterricht, 226
Astronomische Gazet, 54
Astronomische Gesellschaft Abstract Series, 211
Astronomische Grundlagen für den Kalender, 215
Astronomische Kurzinformationen und Eilnachrichten, 34
Astronomische Nachrichten, 212
Astronomisches Informationsblatt der Volkssternwarte Freiburg, 208
Astronomisk Tidskrift, 495
Astronomy, 916
Astronomy & Geophysics, 565
Astronomy & Space Magazine, 295
Astronomy Almanac, 747
Astronomy and Astrophysics, 176
Astronomy and Geodesy, 450

Astronomy News, 898
Astronomy Newsletter, 517
Astronomy Now, 534
Astronomy Resource Manual, 740
Astronotes, 794
Astro-Notes/SAM, 99
Astroparticle Physics, 391
Astro Passion, 179
Astrophysical Journal, 670, 708
    Letters, 670
    Supplement Series, 670, 708
Astrophysics, 395
Astrophysics and Space Science, 395
Astrophysics and Space Science Library, 395
Astrorama, 195
Astrosplai i Natura, 480
AstroTent, 95
Astrotrotter, 265
Astrovisie, 406
Astrum, 397, 472
Asztronautikai Tájékoztató, 275
Atlas of Tracks of Storms and Depressions in the Bay of Bengal and Arabian Sea, 284
Atmosphère et Climat, 189
Atomic Data and Nuclear Data Tables, 808
Atti della Accademia delle Scienze di Torino
    Classe di Scienze Fisiche, 302
    Classe di Scienze Morali, 302
AU AstroNews, 610
Aurora, 489
Aus Astronomie und Raumfahrt, 257
Australian Journal of Chemistry, 23
Australian Journal of Physics, 23
Australian Meteorological Magazine, 21
AVA Aktuell, 216
Avance y Perspectiva, 384
Avaruusluotain, 150
Aviation Week & Space Technology, 808
AVZ-Mitteilung, 502
Aylesbury Astronomical Society Bulletin, 535
BAAS, 670
BAA Variable Star Section Circulars, 538
Baltic Astronomy, 374
Barnard Star (The-), 870
BAV-Circular, 220
BAV-Mitteilungen, 220
BAV-Rundbrief, 220
Becas y Cursos, 7
Bedford Astronomical Society Newsletter, 535
Beiträge zur Astronomiegeschichte, 211
Belgrade Astronomical Observatory Bulletin, 454
Belgrade Astronomical Observatory Publications, 454
Belgrade University Department of Astronomy Publications, 454
Bibliographie Informatik für Schule, Hochschule und Weiterbildung, 229
Bibliography of Hydrology and Oceanology, 428
Bibliography of Meteorology, 428
Bibliography of Water Management and Engineering, 428
Biblioteka Uranii, 430
Bibliothek Forschung und Praxis, 394
Big Bang, 380
Big Bear Solar Observatory Preprints, 631
Bilimve Teknik, 523
Bioastronomy News, 634
Biochemistry, 670
Bioconjugate Chemistry, 670
Biographical Memoirs of Fellows of the Royal Society, 566
Biotechnology Progress, 670
Bishop Planetarium Sky Reporter, 679
Biuletyn Informacyjny, 431
Blick ins All, 219
BNSI Standards News, 45
Boletim Agrometeorológico, 67
Boletim Astronômico, 66, 70

Boletim da Sociedade Astronômica Brasileira, 71
Boletim de Radiação Solar, 67
Boletim de SAAD, 71
Boletines Meteorológicos, 63
Boletín Bibliográfico en Ciencia y Tecnología, 586
Boletín Ciencia en América Latina, 927
Boletín Climatológico Mensual, 10
Boletín CONICYT, 586
Boletín de Actividad Solar, 64
Boletín de Astronomía, 483
Boletín de la Academia, 382
Boletín de la Asociación Argentina de Astronomía, 6
Boletín del Observatorio Astronómico de Quito, 141
Boletín del Observatorio del Ebro - Ionosfera, 486
Boletín de los Amigos de ASO, 62
Boletín de Observación (ASAFI), 119
Boletín de SOMETCUBA, 123
Boletín Electrónico Tecnología en América Latina, 927
Boletín Informativo de la AAC, 471
Boletín Informativo del GAS, 476
Boletín Informativo SAA, 424
Boletín Meteorológico Diario, 10
Boletín SEA, 484
Bollettino AMSA, 310
Bologna Astrophysical Preprints, 326, 338
Bologna Technical Reports, 338
Borowiec Laser Station Operational Report, 430
Botime Hidrometeorologike, 3
Boundless Universe, 248
BPAA Newsletter, 907
BPA News, 674
BPS Directions, 425
Braintree Astronomical Society Newsletter, 537
Breakthrough: A Journal On Science and Society, 282
Breckland Astronomical Society Newsletter, 537
Bridgend Astronomical Society Newsletter, 537
Brisbane Astronomical Society Newsletter, 20
Buchloer Astronomisches Zirkular, 212
BUFORA Bulletin, 539
Bulgarian Geophysical Journal, 75
Bulletin Climatique, 189
Bulletin Climatologique de la Station de Louvain-la-Neuve, 58
Bulletin de la Classe des Sciences de l'Académie Royale de Belgique, 47
Bulletin de l'AFOEV, 159
Bulletin de La Murithienne, 511
Bulletin de la Société d'Astronomie du Valais Romand, 510
Bulletin de la Société Vaudoise d'Astronomie, 511
Bulletin de l'IERS, 192
Bulletin de l'Observatoire de Rouen, 193
Bulletin de l'Union Internationale des Associations et Organismes Techniques, 203
Bulletin de Recherche des Pays-Bas, 397
Bulletin of Air Pollution, 460
Bulletin of Association of Planetaria of Russia, 441
Bulletin of Astronomical Observations, 426
Bulletin Officiel de la Propriété Industrielle, 5
Bulletin of Materials Science, 284
Bulletin of Meteorological Data, 292
Bulletin of Solar Photospheric Observations in Slovakia, 459
Bulletin of Tajik Academy of Sciences, Institute of Astrophysics, 518
Bulletin of the American Astronomical Society, 670
Bulletin of the American Physical Society, 734
Bulletin of the American Society for Information Science, 735
Bulletin of the Astronomical Observatory of Lisbon, 434
Bulletin of the Astronomical Society of India, 281
Bulletin of the Astronomical Society of South Australia, 16
Bulletin of the Astronomical Society of Tasmania Inc., 16
Bulletin of the Fukuoka University of Education - Part

III: Mathematics, Natural Sciences and Technology, 344
Bulletin of the Global Volcanism Network, 675
Bulletin of the Institute for Theoretical Astronomy, 445
Bulletin of the International Statistical Institute, 394
Bulletin of the Korean Space Science Society, 369
Bulletin of the National Research Institute of Astronomy ad Geophysics, 142
Bulletin of the National Science Museum, Series E (Physical Sciences and Engineering), 355
Bulletin of the Physics Section, 132
Bulletin of the Research Institute of Civilization, 358
Bulletin of the Slovak Office of Standards, Metrology and Testing, 460
Bulletin of the Slovak Seismological Stations, 460
Bulletin of the Special Astrophysical Observatory, 448
Bulletin of the Tokyo Gakugei University - Section IV: Mathematics and Natural Sciences, 359
Bulletin of the Yamaguchi Museum, 362
Bulletin of Toyama Science Museum, 359
Butsuri, 358
Cadernos de Astronomia, 68, 73
Cahiers Clairaut (Les_), 173
Calendar Data, 426
Calendario Astronómico, 64
Calendar of Tartu Observatory, 144
Calendar of the Night Sky, 562
Calendars of Sliven Astronomical Observatory, 74
Calendrier des Manifestations Spatiales, 168
Calliope, 788
CalTech Astrophysics Abstracts, 612
Canadian Aeronautics and Space Journal, 80
Canadian Journal of Earth Sciences, 86, 93
Canadian Journal of Physics, 86, 93
Canadian Journal of Remote Sensing, 80
Canberra Observer, 21
Canopus, 586
CAP Annual Report, 80
CapCom, 556
Cape Fear Skies, 827
Capella, 160, 540
Cape Observer (The_), 464
Capitol Skies, 917
CARJ Circular Astronômica, 66
CARJ Guia do Amador de Astronomia, 66
Carnegie Institution of Washington Academic Catalog, 672
Carnegie Institution of Washington Year Book, 615, 672
Carpe Noctem, 878
carpe scientiam, 496
Carter Observatory Annual Report, 409
Carter Observatory Newsletter, 409
Cartes Synoptiques de la Chromosphère Solaire, 192
CAS Bulletin, 115
CASI Special Publications, 740
CASI Technical Memorandums, 740
CASI Technical Notes, 740
CASMAG, 409
CAS Newsletter, 541
CAS Observer, 699
Cassiopeia, 80
Catalogo del Instituto Português da Qualidade, 434
Catalog of Cometary Orbits, 749
Catalog of Data in World Data Center C2 for Ionosphere, 362
Catalog of Hellenic Standards, 268
Catalog of Turkish Standards, 524
CBST, 348
CCD & Telescope, 158
CCLRC Annual Report, 542
C.E. Kenneth Mees Observatory Reprints, 825
Celestial Horizons, 650
Celestial Log, 704
Celestial Mechanic, 616
Celestial Mechanics and Dynamical Astronomy, 395
Celestial Observer (The_), 757, 806, 870

Celestial Onlooker (The_), 16
Celestial Times, 805
Centaurus, 134
Center Line, 905
Ceres, 151
CERN Annual Report, 503
CERN Courier, 503
CERN-HERA Reports, 503
CERN Reports, 503
CERN School of Physics Proceedings, 503
CFA Preprint Series, 751
CFHT Information Bulletin, 693
Chance, 896
CHARA Contributions, 689
Chemical & Engineering News, 670
Chemical Business Bulletins, 566
Chemical Business Update, 566
Chemical Communications
    Dalton Transactions, 566
    Faraday Transactions, 566
    Perkin Transactions I, 566
    Perkin Transactions II, 566
Chemical Engineering Abstracts, 566
Chemical Hazards in Industry, 566
Chemical Research in Toxicology, 670
Chemical Reviews, 670
Chemical Society Reviews, 566
Chemistry & Industry, 670
Chemistry in Britain, 566
Chemistry of Materials, 670
Chemtech, 670
Chinese Astronomy Abstracts, 113
Chinese Astronomy and Astrophysics, 391
Chinese Journal of Astronomy and Astrophysics, 115
Chinese Journal of Lasers B, 675
Chinese Physics, 734
Chinese Solar Geophysical Data, 113
CHMI Meteorological Bulletin, 127
CHMI Transactions, 127
Ciel de Nuit, 86
Ciel de vos Vacances (Le_), 181
Ciel du Mois (Le_), 181
Ciel et Espace, 159
Ciel et Terre, 56
Ciel Info, 93
Ciel (Le_), 56
Cielo, 311
Cielosservare, 323
Ciencia, 382
Ciencias de la Tierra y del Espacio, 123
CIP, 736
Circolare (Unione Astrofili Bresciani), 337
Circolo Astrofili Veronesi - Notiziario, 314
Circular GEA, 476
Circular of ADJ, 462
Circulars of the British Astronomical Association, 538
Circular T, 164
CISTI News, 93
Civil Engineering Transactions, 23
CJAA, 115
Classical and Quantum Gravity, 550
CLEA Newsletter, 854
Climanálise, 67
Climate Dynamics, 227
Climate Summary/Annual (Meteorological Office, Piarco), 521
Climate Summary of Southern Africa, 468
Climate Summary/Quarterly (Meteorological Office, Piarco), 521
Climatological Data, 515
Cloudy Nights, 929
Cluster (De_), 406
CNES Magazine, 168
COAA Annual Report, 433
COAA News, 433
Coal Sack, 685
Cobblestone, 788

CODATA Bulletin, 174
CODATA Newsletter, 174
Coelum Astronomia, 317
Collected Algorithms Supplements, 807
Comet Circular, 526
Communications from University of London Observatory, 574
Communications of Perth Observatory, 26
Communications of the ACM, 807
Communications of the Institute of Applied Astronomy, 445
Communicator, 53
Compendium, 663
Comptes-Rendus de l'Académie des Sciences
    Série I, 154
    Série II, 154
Comptes-Rendus des Séances de la Conférence Générale des Poids et Mesures, 164
Computational Statistics & Data Analysis, 393
Computer and Control Abstracts, 550, 551
Computer Physics Communications, 391
Computers in Physics, 734, 736
Computer Theoretikum und Praktikum für Physiker, 229
Computing and Control Engineering Journal, 551
Computing Reviews, 807
Computing Surveys, 807
Comunicando, 306
Concordia Constellation, 792
Configurations des Huit Premiers Satellites de Saturne, 193
Connaissance des Temps (La_), 193
Consensus, 99
Constellation, 298, 853
Constellation (The_), 858
Contacten, 407
Contributions/Department of Astronomy/ University of Tokyo, 360
Contributions from KPNO, 599
Contributions from the Nizamiah and Japal-Rangapur Observatories, 287
Contributions of Brno N. Copernicus Observatory and Planetarium, 126
Contributions of the Astronomical Observatory Skalnaté Pleso, 459
Contributions of the Geophysical Institute of the Slovak Academy of Sciences, 460
Contributions of the Kwasan and Hida Observatories, Kyoto University, 350
Contributions of the University of Waterloo, 104
Copia, 711
Corona Borealis, 390
Cortina Astronomica, 308
Cosmic Babbler, 922
Cosmic Echoes, 678
Cosmic Messenger, 776
Cosmic Research, 395
Cosmos Express, 205
COSPAR Colloquia Series, 174
COSPAR Directory of Organizations and Associates, 174
COSPAR Information Bulletin, 174, 391
Courrier CERN, 503
CPC, 391
CRAS Notices, 837
Crawley Astronomical Society Newsletter, 543
Crimmitschauer Astronomische Nachrichten, 233
Crossroads Astronomer (The_), 878
Cryophysics Newsletter, 503
Csillagvizsgáló (A_), 273
CSIRO Space Industry News, 22
CSTG Bulletin, 221
Culture and Cosmos, 544
Curierul de Fizicá, 439
Current Bibliography on Science and Technology, 348
Current Biotechnology Abstracts, 566
Current Contents, 857

Current Index to Statistics, 896
Current Papers in Physics, 550
Current Papers on Computers and Control, 550
Current Papers on Electrical and Electronics Engineering, 550
Current Physics Index, 734
Current Published Information on Standardization, 282
Current Science, 283, 284
Current (The_), 875
CVA Observer (The_), 616
C-VII Journal (The_), 736
CWI Quarterly, 389
CXC Newsletter, 750
Cygnus, 492, 560, 585
Czechoslovak Journal of Physics, 132
Dados Astronômicos para os Almanaques, 434
Dansk Rumfart, 136
Dansk Standard, 137
Dark-Sky Observer (The_), 777
Datos Astronómicos para Cuba, 123
DDC News, 816
De Cluster, 406
Deep-Sky Objects of the Month, 650
Deep-Sky Observer (The_), 583
Deep Sky (The_), 727
Delaware Valley Amateur Astronomer (The_), 854
de Lier, 389
Denver Observer, 654
Der Sternenbote, 38
Der Sternfreund, 209, 213, 225, 249, 260, 264
Desert Skies, 603, 879
De Sterrenkijker, 390
De Sterrenwachter, 60
Deutsche Geophysikalische Gesellschaft Mitteilungen, 222
Deutsches Meteorologisches Jahrbuch, 223
De Vangspiegel, 407
Die Sterne, 225
Die Sternenrundschau, 34
Digital Compendium, 663
Dinámica, 7
DIN Mitteilungen, 224
Directorio de Instituciones Científicas de América Latina, 927
Discover, 813
Discovery Center Newsletter, 878
Discovery Circulars, 572
Discovery Park Explorer, 595
Distant Targets, 60
Distributed Computing, 807
Distributed Systems Engineering, 550
DLR-Nachrichten, 225
Documenta, 493
Dome & Sky, 702
Duisburger Komet, 244
DWD Jahresbericht, 223
EAAS Bulletin, 441
EAGE Yearbook, 392
EAON Informations and Asteroidal Results, 226
Earth and Planetary Science Letters, 391
Earth, Moon and Planets, 395
Earth Observation Quarterly, 392
Earth, Planets and Space, 357
Earth Sciences at ANU, 18
Earth Space Review, 856
Earthwatch Magazine, 750
EAS Newsletter, 127
EAS Publication Series, 176
East Valley Astronomy Club Newsletter, 595
Eclipse, 570
Ecliptic, 573
Ecliptic (The_), 857
Eddy Currents, 619
Edinburgh Astronomy Preprints, 562
EDL, 150
Educational Technology Review, 896
Education in Chemistry, 566

EEC Norm Scan, 282
Efemérides Astronômicas, 70
Efemérides Astronómicas/ROA, 482
EGS Newsletter, 227
Electrical and Electronics Abstracts, 550, 551
Electrical Engineering Transactions, 23
Electronic Newsletter for the History of Astronomy, 211
Electronics and Communication Engineering Journal, 551
Electronics Education, 551
Electronics Letters, 551
Elektronische Mitteilungen zur Astronomiegeschichte, 211
Elektronorm, 224
Elementi Astronomici, 327
El-Fellek, 4
El-Marsad, 4
Encyclopedia of World Problems and Human Potential, 57
Energy & Fuels, 670
Energy Data, 229
Engineering Management Journal, 551
Engineering Science and Education Journal, 551
Enjeux, 159
Environmental Science & Technology, 670
EOS, 670
EOS Newsletter, 227
Éphémérides, 201
Éphémérides Astronomiques - Annuaire du Bureau des Longitudes, 193
Éphémérides des Satellites de Jupiter, Saturne et Uranus, 193
Éphémérides des Satellites Faibles de Jupiter et de Saturne, 193
Éphémérides Nautiques, 193
Ephemeris of Minor Planets, 445
Épidécides Lunaires, 199
Épiménides Nautiques, 199
Episodes, 36
Equatorial Dust Lane, 150
Equuleus, 571
ERAC Newsletter (The_), 228
ESA Annual Report, 392
ESA Brochures, 392
ESA Bulletin, 392
ESA Special Publications, 392
ESF Communications, 177
ESO Annual Report, 228
ESO Conference and Workshop Proceedings, 228
ESO-MIDAS Courier, 228
Espace Information, 168
Espacial, 67
ESR, 856
Estela, 470
ETR, 896
Europäischer Wetterbericht, 223
European Centre for Space Law News, 177
European Directory, 753
European Journal of Physics, 177, 550
European Physical Journal, 176
European Review, 531
European Space Directory, 178
Europhysics Conference Abstracts, 177
Europhysics Letters, 176, 177
Europhysics News, 177
Euroscience News, 178
EUVE Electronic Newsletter, 646
EUVE Science Bulletin, 646
EUVE Technical Bulletin, 646
Event Horizon, 88, 754
Ever Wonder, 719
EVS Teataja, 144
Exchange, 671
Experimental Astronomy, 395
Exploratorium, 620
Explorer (The_), 903
Eye on the Sky, 725

Eye on the Sky (Big Sky AS), 79
Eyepiece, 806
Eyepiece (The_), 711
Faces, 788
Faraday Discussions, 566
FAS Newsletter, 546
FAS Public Interest Report, 672
Faszinierendes Universum, 212
Fellowship Guide, 7
Fenix, 470
Fenómenos Astronómicos/ROA, 482
Fernbank Magazine, 688
FID Review, 393
Field of View, 666
FIL, 52
Finnish Academy Year Book, 146
Firmamento, 62
First Break, 392
FirstLight, 678
FIZ-KA Berichte, 229
FIZ-KA Referenzserie, 229
Fleischmann Flyer (The_), 786
Florida Skies, 679
Flyer, 674
FoCAAL, 85
Focal Plane (The_), 703
Focal Point, 678
Focal Point (The_), 687
Focus, 295, 683
Focused, 650
Focus (The_), 667
Forest Preserve Newsletter, 699
Formation Information Liaison, 52
forschung - Mitteilungen der DFG, 222
Forschungsnachrichten aus den Niederlanden, 397
Forschungsthemen, 233
foton, 121
Fraunhofer Gesellschaft Annual Report, 230
Fraunhofer Magazine, 230
Friends of Cincinnati Observatory Newsletter, 837
Frontiers, 562
Fulcrum, 929
Fundamentals of Cosmic Physics, 815
FWF-Info, 35
FWF-Statistics, 35
Galactée, 49, 50
Galaxie, 511
Galaxis, 247, 393
Ganymed, 217, 233
Garandie (La_), 158
GAR News, 323
Gazer's Gazette (The_), 918
Gazeta de Física, 435
Gemini, 771
Gemini Newsletter (The_), 596, 693
General Catalog, 9
General Physics Advance Abstracts, 734
General Relativity and Gravitation, 395
General Report, 186
Gent Sterrenkundig Observatorium Mededelingen, 59
Geochimica et Cosmochimica Acta, 695
Geodesy, 75
Geofísica Internacional, 386
Geofysische Waarnemingen, 54
Geological Maps IGNS, 410
Geology and Earth Resource Manual, 740
Geomagnetic Indices Bulletin, 657
Geomagnetism and Aeronomy, 670
Geophysical Journal International, 222, 227, 565
Geophysical Magazine, 348
Geophysical Prospecting, 392
Geophysical Research Letters, 670
GEOS Circulares, 477
GEOS Circulars, 180
Geotectonics, 670
german research - Reports of the DFG, 222
Giornale dell'AAB, 304

Giornale di Fisica, 335
Global Biochemical Cycles, 670
GLPA Newsletter, 838
Gluon (Le_), 173
Gnomon, 532
Gnomoniste (Le_), 86
GOAG News, 681
GPPA Newsletter, 782
GPS World, 835
Gradient, 461
Griffith Observer, 622
Grinnell Magazine, 717
Großwetterlagen Europas, 223
Guía de Licenciatorios, 7
Guidestar, 880
Guide Star (The_), 851
Guide to the World Data Center System, 655
Halley, 483
Halley Periodiek, 402
Halo, 60
Hamilton Astronomical Society Bulletin, 409
Hampshire Observer (The_), 547
Handbook of the British Astronomical Association, 538
Handbuch der Helmholtz-Zentren, 233
Harthaer Beobachtungszirkulare, 220
Harvard-Smithsonian Preprint Series, 597
Heelal, 60
Helios, 181, 546
Hellenic Astronomical Society Newsletter, 268
Hercules, 402
Hermes, 568
HGF-Mitteilungen, 233
High-Energy Physics Index, 229
Higher Education Policy, 184
Highlights in Space: Progress in Space Science, Technology and Applications, International Cooperation and Space Law, 40
Highlights of Astronomy, 185, 395
High-Performance Polymers, 550
Himmelsbeobachter, 501
Himmelskundliches aus Bad Salzschlirf, 231
Home-Dome Star (The_), 743
Horizons, 509
Horizonte, 509
Hungarian Astronomical Association Yearbook, 276
HUT Annual Report, 148
Hvar Observatory Bulletin, 122
Hvězdářská Ročenka, 130
Hydrology and Earth System Sciences, 227
Hydrology and Oceanology, 428
Hydrometeorology and Environment Monitoring, 43
Hypothese, 397
Hypoxic Observer, 653
I & Heavens, 292
IAGA Bulletin, 655
IAGA News, 655
IAJ, 552
IAPPP Communications, 872
IAU CBAT Circulars, 749
IAU Information Bulletin, 185, 395
IAU Newsletter, 184
IAU Symposia Proceedings, 185, 395
IAU Transactions, 185, 395
IAYC Report, 53
IBVS, 276
Icarus, 816
Içindekiler, 523
ICQ, 752
ICRR Report, 360
ICSTI Forum, 185
ICTP Annual Report, 302
IDA Newsletter, 598
Időjárás, 276
IEE News, 551
IEE Proceedings
    Circuits, Devices and Systems, 551
    Communications, 551
    Computers and Digital Techniques, 551
    Control Theory and Applications, 551
    Electric Power Applications, 551
    Generation, Transmission and Distribution, 551
    Microwaves, Antennas and Propagation, 551
    Optoelectronics, 551
    Radar, Sonar and Navigation, 551
    Science, Measurement and Technology, 551
    Software Engineering, 551
    Vision, Image and Signal Processing, 551
IEE Review, 551
IERS Annual report, 234
IERS Bulletin A, 234, 677
IERS Bulletin B, 234
IERS Bulletin C, 234
IERS Bulletin D, 234
IERS Messages, 234
IERS Technical Notes, 234
IFLA Directory, 394
IFLA Journal, 394
IISL Proceedings, 185
IJET, 896
IJRS, 564
Il Giornale di Astronomia, 335
Ilhuícatl, 384
Il Nuovo Cimento, 335
Il Nuovo Saggiatore, 335
Image, 227
Imaginings, 685
IMO News, 551
IMO Proceedings, 234
Impact (Spaceguard UK), 573
IMU Bulletin, 68
India Daily Weather Report, 284
Indian Skeptics, 285
India Weekly Weather Report, 284
Industrial & Engineering Chemistry Research, 670
Industrial Mathematics, 764
Info, 51
Info Astro, 50
In Focus, 559
Infoheft, 214
Informações do Observatório Astronômico Prof. Manuel de Barros, 436
Information Bulletin on Variable Stars, 275, 276
Informo, 152
Infos GIFO, 181
Infosheet: Crimmitschauer Astronomische Nachrichten, 233
ING Newsletter, 479
Innovation, 221
Inorganic Chemistry, 670
INPE Space News, 67
inside, 600
Inside Orbit, 767
Inside PEM, 848
INSPIRE Journal (The_), 623
Institute of Meteoritics Special Publications, 804
intelligence, 807
interactions, 807
Interface, 287
Interkomeet, 394
International Astronomical Youth Camps Report, 53
International Comet Quarterly, 752
International Congress Calendar, 57
International Geophysical Calendar, 657
International Handbook of Universities, 184
International Journal of Climatology, 565
International Journal of Educational Telecommunications, 896
International Journal of Remote Sensing, 564
International Reviews on Mathematical Education, 229
International Statistical Review, 394
INTER-SOL Reports, 262
Inter-Stellar Newsletter, 780
Intijiwaña, 63
Inverse Problems, 550

Investigations of the Sun and Red Stars, 372
Io, 848
Ionosferische Waarnemingen - Kosmische Straling, 54
Ionospheric Bulletin, 75
Ionospheric Data, 425
Ionospheric Data at Syowa Station (Antarctica), 343
Ionospheric Data in Japan, 343
Ionospheric Data Sweden, 495
Iowa Science Teachers Journal, 717
IPS Monthly Solar and Geophysical Summary, 24
IRAM Newsletter, 182
IRF Scientific Reports, 495
Irish Astronomical Journal, 532, 552
IRLIST Digest, 807
IRTF News, 695
ISAS News, 346
ISAS Report, 346
ISI Directories, 394
ISI Newsletter, 394
ISO Bulletin, 505
ISO Catalog, 505
ISO Info, 392, 476
ISO Management Systems, 505
ISO Memento, 505
ISSI Scientific Reports Series, 505
Issledovania po Geomagnetismu, Aeronomii i Fisike Solntsa, 448
Issues in Higher Education, 184
Issues in Science and Technology, 673
ITU News, 506
Izarra, 474
Izvestiia Russian Academy of Sciences
    Atmospheric and Oceanic Physics, 670
    Physics of the Solid Earth, 670
Jaarverslag, 51
Jackson (Hole) Astronomy Club Newsletter, 923
Jahrbuch der Bayerische Akademie der Wissenschaften, 218
Jahresberichte des Bundesministers für Forschung und Entwicklung in der Biologie, Ökologie, Energie, 229
Janus, 546
Japanese Ephemeris, 347
JARE Data Reports, 355
JASS, 369
JCCE, 896
JCMST, 896
JCMT Newsletter (The-), 693
JCMT-UKIRT Newsletter (The-), 562
JEMH, 896
JETP Letters, 734
JHA, 567
JILR, 896
Journal Aéronautique et Spatial du Canada (Le-), 80
Journal for the History of Astronomy, 567
Journal of AAVSO, 747
Journal of Agricultural and Food Chemistry, 670
Journal of Agricultural, Biological, and Environmental Statistics, 896
Journal of Aircraft, 895
Journal of Analytical Atomic Spectrometry, 566
Journal of Applied Meteorology, 113, 747
Journal of Applied Physics, 734
Journal of Astronomical History and Heritage, 14
Journal of Astronomy and Physics, 523
Journal of Astronomy and Space Sciences, 369
Journal of Astrophysics and Astronomy, 284
Journal of Atmospheric and Oceanic Technology, 747
Journal of Atmospheric Chemistry, 227
Journal of Atmospheric Sciences, 747
Journal of Birla Planetarium (The-), 286
Journal of Business and Economic Statistics, 896
Journal of Chemical and Engineering Data, 670
Journal of Chemical Education, 670
Journal of Chemical Information and Computer Sciences, 670
Journal of Chemical Physics, 734
Journal of Chemical Research
    Part M (Microfiche), 566
    Part M (Miniprint), 566
    Part S (Synopsis), 566
    Vouchers, 566
Journal of Climate, 747
Journal of Computational and Graphical Statistics, 896
Journal of Computers in Mathematics and Science Teaching, 896
Journal of Computing in Childhood Education, 896
Journal of Educational Multimedia and Hypermedia, 896
Journal of Electronic Imaging, 908
Journal of Experimental Algorithmics, 807
Journal of Geodesy, 138
Journal of Geodynamics, 227
Journal of Geophysical Research, 670
Journal of Graphic Tools, 807
Journal of Guidance, Control and Dynamics, 895
Journal of Hard Materials, 550
Journal of Information Processing and Management, 348
Journal of Interactive Learning Research, 896
Journal of Lightware Technology, 675
Journal of Magnetic Observations, 420
Journal of Materials Chemistry, 566
Journal of Materials Research, 858
Journal of Mathematical Physics, 734
Journal of Medicinal Chemistry, 670
Journal of Meteor Observations, 355
Journal of Meteorology, 113
Journal of Micromechanics and Microengineering, 550
Journal of Nanjing University (Natural Sciences Edition), 117
Journal of NIVAR, 292
Journal of Optical Technology, 675
Journal of Optics, 287
Journal of Organic Chemistry, 670
Journal of Physical and Chemical Reference Data, 670, 734
Journal of Physical Chemistry, 670
Journal of Physical Oceanography, 747
Journal of Physics
    A: Mathematical and General, 550
    B: Atomic, Molecular and Optical Physics, 550
    C: Condensed Matter, 550
    D: Applied Physics, 550
    G: Nuclear Physics, 550
Journal of Propulsion and Power, 895
Journal of Radiological Protection, 550
Journal of Scientific Exploration, 903
Journal of Shanghai University, 118
Journal of Space and Astronomy Research, 294
Journal of Spacecraft and Rockets, 895
Journal of Spacecraft Technology, 285
Journal of Technology and Teacher Education, 896
Journal of the ACM, 807
Journal of the ALPO (The-), 609
Journal of the American Chemical Society, 670
Journal of the American Society for Information Science, 735
Journal of the American Statistical Association, 896
Journal of the Antique Telescope Society, 687
Journal of the Astronautical Sciences, 895
Journal of the Astronomical Society of Victoria, 17
Journal of the Auckland Astronomical Society, 408
Journal of the British Astronomical Association, 538
Journal of the British Interplanetary Society, 538
Journal of the Chemical Society, 566
Journal of the Communications Research Laboratory, 343
Journal of the European Optical Society
    Part A: Pure and Applied Optics, 227
    Part B: Quantum Optics, 227
Journal of the Institute of Meteorology and Water Management Observer, 428
Journal of the Iowa Academy of Science, 717
Journal of the Korean Astronomical Society, 368

Journal of the National Science Foundation of Sri Lanka, 487
Journal of the Nottingham Astronomical Society, 560
Journal of the Optical Society of America
  A: Optics, Image Science and Vision, 675
  B: Optical Physics, 675
Journal of the Physical Society of Japan, 358
Journal of Thermophysics and Heat Transfer, 895
Journal of The Royal Society of New Zealand, 411
Journal of the Royal Statistical Society - Series A, B, C and D, 566
Journal of the Zeiss Historica Society, 798
Journal of Tropical Meteorology, 113
Journal of Vacuum Science and Technology A, 808
Journal of Vacuum Science and Technology B, 808
JPSJ, 358
JSE, 903
JTATE, 896
KAF-NYTT, 490
Kajuyali, 119
Kenneth Mees Observatory Reprints (C.E._), 825
Kenya Journal of Sciences (KJS)
  Series A: Physical Sciences, 365
  Series B: Biological Sciences, 365
  Series C: Humanities and Social Sciences, 365
Khagol: The IUCAA Bulletin, 286
Kinematics and Physics of Celestial Bodies, 529
Kiruna Geophysical Data, 495
KJS
  Series A: Physical Sciences, 365
  Series B: Biological Sciences, 365
  Series C: Humanities and Social Sciences, 365
Klimatologische Waarnemingen, 54
KNMG/ALW Nieuwsbrief, 393
KNMI Daily Weather Bulletin, 395
KNMI Monthly Rain Bulletin, 395
KNMI Monthly Weather Bulletin, 395
KNMI Seismological Bulletin, 395
Knudepunktet, 136
Komentarze Fromborskie, 429
Komety y Meteory, 518
Konkoly Observatory Communications, 275
Korona, 213
Kosmické Rozhledy, 127
Kougai Bushin, 346
Kozmos, 460, 461
Kristian Birkeland Lecture, 416
KSB Jaarboek, 55
KSC Countdown, 682
Kultur und Technik, 225
Kvart, 132
Kyoto University - Department of Astronomy, 350
La Alternativa Racional, 474
Laboratory Hazards Bulletin, 566
La Connaissance des Temps, 193
La Garandie, 158
Lakeviews, 703
La Lettre d'Aphélie, 162
La Lettre de l'OHP, 191
La Lettre du Club Éclipse, 173
Langmuir, 670
La Petite Ourse, 157
La Physique au Canada, 80
LAR, 474
La Recherche Aérospatiale, 194
La Revue de l'Observatoire Populaire de Laval, 194
La Rivista del Nuovo Cimento, 335
L'Astrofilo Lariano, 321
L'astronomia, 330
L'Astronomie, 199
Latitude 5, 168
Latitude Circulars, 432
LCAS Newsletter, 702
Learned Publishing, 533
Le Ciel, 56
Le Ciel de vos Vacances, 181
Le Ciel du Mois, 181

Le Courrier de l'Observatoire, 510
Lecturas de la Estación de la Paz, 63
Le Gluon, 173
Le Gnomoniste, 86
Le Journal Aéronautique et Spatial du Canada, 80
Le Micro Bulletin, 167
Le Planétarium, 196
Le Point Astro, 156
Le Point de Lagrange, 158
Les Cahiers Clairaut, 173
Le Trait d'Union, 50
Letters in Astronomy, 206
Letters on Programming Languages and Systems, 807
Lettre AstroQueyras, 163
Lettre d'Aphélie (La_), 162
Lettre de l'OHP (La_), 191
Lezioni di Astronomia, 307
LHS Programs for Schools, 651
LHS Quarterly (The_), 651
Lick Observatory Bulletins, 649
Lier (de_), 389
Liftoff, 410
Lists of Standards, Draft Standards and Technical Regulations del Instituto Português da Qualidade, 434
Lithuanian Journal of Physics, 375
Liverpool's Monthly Sky Diary, 554
Living Reviews in Relativity, 237
LMB Actu, 167
L'Observateur, 509
L'Observation du Ciel, 199
L'Obs'session, 190
Longmont Astronomical Society Journal, 656
L'Orionide, 85
L'Osservatorio, 309
Louisiana State University Observatory Contributions, 730
Louisville Amateur Astronomers Newsletter, 726
Lowell Observer (The_), 598
LPIB, 886
LST Bulletin, 375
Luft- und Raumfahrt, 223
Lunar and Planetary Information Bulletin, 886
Lunario, 314
Lund Observatory Reports, 491
L'Univers du Namurois, 49
LVAS Newsletter, 786
Lyra, 584
Lyre, 158
M-111, 842
M51, 281
Macromolecules, 670
Madison Skies, 917
Magnetic Bulletin, 75
Magnétisme Terrestre, 53
Majalah Lapan, 290
Månadens Standard, 497
Manakdoot, 282
Manual Astronómico, 9
Manufacturing Engineer, 551
Marine Observer, 556
MARS-Bulletinen, 491
Mars Underground News, 634
Mason-Dixon Astronomer, 744
Mass Spectrometry Bulletin, 566
Matematisk-Fysiske Meddelelser, 139
Materials of the World Data Center B, 451
Mathematics Abstracts, 229
MAUSAM, 284
McDonald Observatory Preprints, 888
Measurement Science and Technology, 550
Mechanical Engineering Transactions, 23
Meddelelser fra Ole Rømer Venner, 139
Mededelingen, 54
Medical and Biological Engineering and Computing, 551
Meduza Circular, 129
Mees Observatory Reprints (C.E. Kenneth_), 825

Mémoires de la Classe des Sciences de l'Académie Royale de Belgique in 4°, 47
Mémoires de la Classe des Sciences de l'Académie Royale de Belgique in 8°, 47
Mémoires de la Société de Physique et d'Histoire Naturelle, 510
Mémoires de l'ISTEEM, 183
Memoirs of Osaka Kyoiku University, 356
Memoirs of the Faculty of Education - Kagawa University (Part II), 348
Memoirs of the Faculty of Education, Shiga University, 357
Memoirs of the National Observatory of Athens, Series I, Astronomy, 269
Memoria del Instituto de Astrofísica de Canarias, 477
Memorie della Accademia delle Scienze di Torino
   Classe di Scienze Fisiche, 302
   Classe di Scienze Morali, 302
Memorie della Società Astronomica Italiana, 335
Mercury, 542, 610
Mercury (The_), 690
Meridian, 718
Meridiana, 509
Meridian (The_), 685
Messenger (The_), 228
Met Éireann Technical Note Series, 297
meteo.fr, 189
Météo-Hebdo, 189
Meteor, 276
Meteorite, 410
Meteorite (The_), 839
Meteoritics, 695
Meteoritika, 444
Meteor News, 806
Meteorological and Geoastrophysical Abstracts, 747
Meteorological Applications, 565
Meteorological Knowledge, 113
Meteorological Report, 133
Meteorologische Documentatie, 54
Meteorology, 113, 428
Meteoros, 62, 208
Meteors, 483
Meteor (The_), 680, 832
Meteor Write, 872
Methods in Organic Synthesis, 566
Met Mar, 189
Metrologia, 164
Metrology and Testing, 460
Metsähovi Publications on Radio Science, 148
Metsähovi Radio Observatory Report, 148
Micro Bulletin (Le_), 167
Microgravity News from ESA, 392
Mira, 180
Mira Ceti, 60
MIRA Newsletter (The_), 628
Mirror (The_), 710
Mitteilungen astronomischer Vereinigungen Rhein-Main-Nahe, 210, 212, 240, 261
Mitteilungen aus der DGLR, 223
Mitteilungen der Astronomischen Gesellschaft, 211
Mitteilungen der Bruno-H. Bürgel-Sternwarte Hartha, 220
Mitteilungen der Greifswalder Sternwarte, 248
Mitteilungen der Rudolf Wolf Gesellschaft, 508
Mitteilungen der Satellitenbeobachtungsstation Zimmerwald, 513
Mitteilungen der Universitäts-Sternwarte zu Jena, 255
Mitteilungen der Volkssternwarte Darmstadt, 260
Mitteilungen des Lohrmann-Observatoriums, 251
Mitteilungen des Sonnenobservatoriums Kanzelhöhe, 40
Mitteilungen zur Astronomiegeschichte, 211
MMTO Technical Reports, 599
MNASSA, 464
MNRAS, 565
Mobile Networks and Applications, 807
Modelling and Simulation in Materials Science and Engineering, 550

Modern Radio Science, 53
Monatlicher Witterungsbericht, 223
Monographs IGNS, 410
Mons Astrophysical Papers, 58
Monthly Bulletin on Solar Phenomena, 362
Monthly Climatic Summary of Pakistan, 421
Monthly Notes of the Astronomical Society of Southern Africa, 464
Monthly Notices of the Royal Astronomical Society, 565
Monthly Report on Solar Radio Emission, 428
Monthly Survey of Selected Events in the Peaceful Exploration and Use of Outer Space, 40
Monthly Weather Review, 747
Moonrise, Moonset, Sunrise, Sunset and Twilight Tables, 426
Moonstarer, 457
Morning Star, 894
MOST News, 821
Movement & Positioning, 241
MPC Circulars, 755
MPE Jahresbericht, 237
MPG Berichte und Mitteilungen, 236
MPG Spiegel, 236
MRS Bulletin, 858
M-STAR Newsletter, 627
MSZT Bulletin, 276
Multi-Disciplinary Transactions, 23
Multimedia Systems, 807
Murdoch Astronomical Society Newsletter, 25
MV-Astroblick, 248
Nachtschicht, 207
NAIC/Arecibo Observatory Newsletter, 437
NAIC Report Series, 819
Nanotechnology, 550
NAOJ Annual Report, 354
NASDA Report, 346
National Astronomical Observatory Reprint, 352
National Library News, 92
National Long-Term Plan for Space Activities, 417
National Standards and Codes of Practice, 45
NATO Science & Society Newsletter, 55
NATO Scientific Publications, 55
NATO Yearbook, 55
Natural Products Report, 566
Natural Products Updates, 566
Nature, 558
Nature Study of Yamaguchi-ken, 362
Naturwissenschaftliche Rundschau, 239
Nautical Almanac (The_), 677
Nautisches Jahrbuch, 220
Navigator, 679
NCAR Technical Notes, 656
Nebeske Krijesnice, 122
Nebula, 553
Nebula (The_), 90, 714
Nederlandse Tijdschrift voor Natuurkunde, 397
NEKAAL Newsletter, 723
NELPAG Circulars, 756
Neptune, 540
Network: Computation in Neural Systems, 550
netWorker, 807
New Astronomy, 391
New Astronomy Reviews, 391
News Bulletin, 186
New Science, 778
Newscope, 589
News from Prospace, 198
Newsletter of the AAAA, 716
Newsletter of the Astronomy Interest Group, Durban Natural Science Museum, 465
Newsletter of the Baltimore Astronomical Society (The_), 735
Newsletter of the British Sundial Society, 539
Newsletter of the Sydney Outdoor Lighting Improvement Society, 28
Newsline, 545
News of the American Physical Society, 734

New Sunspot Indices Bulletin, 258
Newtonian (The.), 861
New Zealand Journal of Agricultural Research, 411
New Zealand Journal of Botany, 411
New Zealand Journal of Crop and Horticultural Science, 411
New Zealand Journal of Geology and Geophysics, 411
New Zealand Journal of Marine and Freshwater Research, 411
New Zealand Journal of Zoology, 411
New Zenith (The.), 582
NFR Yearbook, 496
NGC 69, 171
NGI Publications Series, 417
NIFS Annual Report, 354
Nigerian Academy of Sciences Proceedings, 414
Nightlife, 570
Night Sights, 685
Night Sky (The.), 20, 835
Night Times, 842
Nihil Sub Astris Novum, 320
Nite Skies, 633
NLR News, 396
NOAO Annual Report, 109, 599, 801
NOAO Newsletter, 109, 599, 801
NOAO Quarterly Report, 109, 801
Nojum Magazine, 293
Nonlinearity, 550
Nonlinear Processes in Geophysics, 227
Normalización, 123
Normalizacja, 431
Normas y Calidad, 120
Norwegian Academy of Science and Letters Yearbook, 416
Notes Scientifiques et Techniques du BDL, 193
Notices of the American Mathematical Society, 864
Noticias del Instituto de Astrofísica de Canarias, 477
Notiziario AAT, 307
Notiziario di Astronomia, 322
Notiziario GAD, 324
Notiziario IRAS, 329
Notiziario Rheyta, 310
Notizie di Metrologia, 316
Nouvelle Gazette du Club 69, 171
Nouvelles Brèves/Comité PARA, 50
Nouvelles de la Bibliothèque Nationale, 79
Nova, 39, 208, 891
Nova 87-a, 473
NOVAC Newsletter, 901
Nova Kepleriana - Abhandlungen - Neue Folge, 218
Nova Search, 30
NPG-News, 48
NP Newsletter, 492
NRAO Newsletter, 801, 900
NRO Newsletter, 353
NRO Report, 353
NSF Bulletin, 900
NSO Update, 801
NSSDC News, 741
Nuncius, 325
Nuovo Orione, 330
Nuovo Saggiatore (Il.), 335
NWO Facts, 397
NWO Jaarboek, 397
NWO Onderzoekberichten, 397
Nytt om Romfart, 415
OBAFGKM, 49
OBAFGKM Informations, 49
Oberonnieuws, 60
Objectif Univers, 195
Objective View (The.), 657
Observaciones Solares, 9
Observateur (L'.), 509
Observation du Ciel (L'.), 199
Observations, 16, 854
Observations Climatologiques, 53
Observations de l'Ozone, 53
Observations et Travaux, 199
Observations Géophysiques, 53
Observations Ionosphériques - Rayonnement Cosmique, 53
Observations Solaires, 438
Observator et Emergo, 407
Observatory and Planetarium Reporter, 130
Observatory Magazine, 560
Observer, 743, 846, 847
Observer II (The.), 774
Observer's Handbook, 97
Observer (The.), 621, 638, 711, 743, 789, 837, 858
Observing Guides, 747
Obs'session (L'.), 190
Obzornik za Matematiko in Fiziko, 462
Occultation Express, 130
Occultation Newsletter, 722, 738
Oceanology, 670
Octantis, 413
Oddie-Baker Bulletin, 19
Odessa Astronomical Publications, 528
Odyssey, 788
Odyssium Annual Report, 96
Odyssium Observer, 96
OLE, 550
Olympic Scope (The.), 908
Omega, 549
onOrbit Magazine, 786
On Station, 392
OPÇÃO, 434
ÖPG-Mitteilungsblatt, 38
Optica Acta, 572, 862
Optical Engineering, 908
Optical Engineering Reports, 908
Optica Pura y Aplicada, 484
Optics & Photonics News, 675
Optics and Spectroscopy, 675
Optics in Education, 908
Optics Letters, 675
Opto & Laser Europe, 227
Opto and Laser Europe, 550
Orbit, 296, 545
Orbiter, 903
Orbiter (The.), 682, 849
Organometallics, 670
Orion, 508
Orionide (L'.), 85
Oslo Institute of Theoretical Astrophysics Reports, 419
Osnabrücker Naturwissenschaftliche Mitteilungen, 238
Osservatorio (L'.), 309
Osservazioni Solari, 328
Österreichischer Himmelskalender, 38
Our Friendly Skies, 745
Our Parks, 797
Owens Valley Radio Observatory Preprints, 613
Ozon Waarnemingen, 54
Paleoceanography, 670
Palomar Observatory Preprints, 614
Panorama, 787
PASA, 15
PASJ, 342
PAStimes, 601, 730
Paws and Reflect, 905
Pégase, 156
Pégase (Musée de l'Air et de l'Espace), 189
PeGASus, 97
Pegasus, 571, 852
Peremennye Zvezdy, 445
Perseus, 126
Perspicillum, 68
Petite Ourse (La.), 157
Petroleum Geoscience, 392
Phénomènes et Configurations des Satellites Galiléens de Jupiter, 193
Philippines Astronomical Handbook, 426
Philosophical Transactions of the Royal Society A - Mathematical and Physical Sciences, 566

B - Biological Sciences, 566
Phobos, 584
Photogrammetric Journal of Finland (The_), 148
Photonics Directory Set: Corporate Guide - Buyers Guide - Handbook & Dictionary, 753
Photonics Literature Showcase, 753
Photonics Postcard Deck, 753
Photonics Product Portfolio, 753
Photonics Spectra Magazine, 753
Photoniques, 202
Physicalia, 48
Physical Review
   Abstracts, 734
   A: General Physics, 734
   B: Condensed Matter, 734
   C: Nuclear Physics, 734
   D: Particle and Fields, 734
   Index, 734
   Letters, 734
   Supplements, 734
Physica Scripta, 493, 734
Physicist (The_), 18
Physics Abstracts, 550, 551
Physics and Chemistry of the Earth, 227
Physics Briefs, 734
Physics Data, 229
Physics Education, 550
Physics in Canada, 80
Physics in Medicine and Biology, 550
Physics of Fluids A, 734
Physics of Fluids B, 734
Physics Reports, 349, 391
Physics Teacher, 734
Physics Teacher (The_), 734
Physics Today, 734
Physics World, 550
Physikalische Blätter, 223
Physiological Measurements, 550
Physique au Canada (La_), 80
Planetarian (The_), 738
Planetárium, 128
Planétarium (Le_), 196
Planetarium Show, 345
Planetarium Stuttgart, Program, 221
Planetary and Space Science, 227, 391
Planetary Report, 634
Planetary Studies Foundation News, 705
Plants and Planets, 729
Plasma Physics and Controlled Fusion, 550
Plasma Sources Science and Technology, 550
Pleiades, 94
Plêiades (As_), 65
Pleyades, 69
Point Astro (Le_), 156
Point de Lagrange (Le_), 158
Polar Bioscience, 355
Polar Geoscience, 355
Polaris, 208, 877
Polar Meteorology and Glaciology, 355
Popular Astronomy, 568
Popular Science, 658
Postepy Astronomii, 431
Power Engineering Journal, 551
Practical Observer (The_), 725, 727
Pragna, 487
Prairie Astronomer (The_), 784
Prakriti: Vigyan O Samaj Vishayak Patrika, 282
Pramana - Journal of Physics, 284
Preliminary Report and Forecast of Solar-Geophysical Activity, 657
Preparing for the Future, 392
Preprint Series, 94
Presek, 462
Prime Focus, 645, 837, 879
Prime Focus (The_), 716
Princeton Observatory Preprints, 795
Probe (The_), 818

Proceedings of Annual Meeting of Italian Planetariums, 303
Proceedings of Research Works, 131
Proceedings of the Finnish Astronomy Days, 147
Proceedings of the Indian Academy of Sciences
   Chemical Sciences, 284
   Earth and Planetary Sciences, 284
   Mathematical Sciences, 284
Proceedings of the Israel Academy of Sciences and Humanities, 299
Proceedings of the Japan Academy
   A: Mathematical Sciences, 358
   B: Physical and Biological Sciences, 358
Proceedings of the LPS Conference, 371
Proceedings of the Mathematical and Physical Society of Egypt, 142
Proceedings of the Royal Society
   A - Mathematical and Physical Sciences, 566
   B - Biological Sciences, 566
Proceedings of the SPIE, 908
Proceedings of URSI General Assemblies, 53
Proceedings. Sect. A (Mathematical, Astronomical and Physical Sciences), 297
Procès-Verbaux des Séances du Comité International des Poids et Mesures, 164
Procyon, 170
Progress in Astronomy, 115, 117
Progress of Theoretical Physics, 350
Pro-ISSI Spatium Series, 505
Projections, 778
Promet - meteorologische Fortbildung, 223
Provisional Sunspot Numbers, 258
PTB Jahresbericht, 240
Publicaciones del Instituto de Astrofísica de Canarias, 477
Publicaciones del Observatorio Astronómico Nacional, 120
Publicaciones del Observatorio Astronómico Ramón María Aller, 485
Publicaciones del Observatorio del Ebro
   Memoria, 486
   Miscelánea, 486
Publicações do Observatório Astronômico Prof. Manuel de Barros, 436
Publicações do Observatório Nacional, 70
Publication of the Astronomical Society Rudjer Boskovitch, 455
Publications of the Astronomical Society of Australia, 15, 23
Publications of the Astronomical Society of Japan, 342
Publications of the Astronomical Society of the Pacific, 610
Publications of the Astronomy Department of the Eötvös Loránd University, 274
Publications of the Beijing Astronomical Observatory, 113
Publications of the Debrecen Observatory, 275
Publications of the Department of Astronomy of the University of Cape Town, 469
Publications of the Ege University Observatory, 522
Publications of the Istanbul University Observatory, 523
Publications of the Korea Astronomy Observatory, 367, 368
Publications of the National Astronomical Observatory, 352
Publications of the Purple Mountain Observatory, 114
Publications of the Rothney Astrophysical Observatory, 102
Publications of the Shaanxi Astronomical Observatory, 114
Publications of the US Naval Observatory, 677
Publications of the Yerkes Observatory, 919
Publications of Yunnan Observatory, 115
Publications Scientifiques et Techniques de l'IRMB, 53
Publications Spéciales du CDS, 165
Public Understanding of Science, 550

Pulsar, 201
Pure and Applied Optics, 550
QBSA, 362
Qualirama, 434
Qualité Espace, 168
Quantum, 901
Quantum Optics, 550
Quarterly Bulletin on Solar Activity, 362
Quasar, 119
Quasar/NOUT, 47
Radiant, 391
Radiant Rising, 867
Radiant (The.), 867
Radiantti, 150
Radio Astronomy, 822
Radio Astronomy Laboratory
   Preprints, 647
   Technical Reports, 647
Radio Physics and Radio Astronomy, 529
Radio Science, 670
Radio Science Bulletin (The.), 53
Rainwater Observer (The.), 774, 775
Rapport Annuel du CERN, 503
Rapport au COSPAR, 168
Rapporto Interno IAS, 315
RASNZ Newsletter, 411
RASTAR, 563
Raum und Zeit, 214
Rayonnement Solaire, 53
Reaching for the Skies, 392
re-actions, 698
Recherche Aérospatiale (La.), 194
Reflections, 26, 705
Reflections - Refractions, 765
Reflector (The.), 776
Refractor (The.), 874
Regard de l'Astronome, 172
Regiomontanusbote, 239
Relatório de Atividades, 67
Remote Sensing Society Newsletter, 564
Rendiconti Lincei
   Matematica e Applicazioni, 302
   Scienze Fisiche e Naturali, 302
   Supplemento, 302
Report of Hydrographic Observations
   Series of Astronomy and Geodesy, 347
   Series of Satellite Geodesy, 347
Report of Hydrographic Researches, 347
Report of International Lunar Occultation Center, 347
Report of the Lunping Observatory
   Geomagnetism, 516
   Sunspots, 516
Report of the National Astronomical Observatory, 352
Report of the University of Helsinki Observatory, 151
Report on Sunspot Observations, 516
Reports in the Fields of Science and Technology, 229
Reports of the Institute of Meteorology and Water Management, 428
Reports of the National Center for Science Education, 630
Reports on Progress in Physics, 550
Research Matters, 561
Research Reports from the Netherlands, 397
Resonance, 284
Results of Geomagnetic Observations at the Hurbanovo Geomagnetic Observatory, 460
Review of Geophysics, 431, 670
Review of Radio Science, 53
Review of Scientific Instruments, 734
Reviews in Modern Astronomy, 211
Reviews of Modern Physics, 734
Revista Astronómica, 6
Revista Española de Física, 482
Revista Mexicana de Astronomía y Astrofísica, 385
   Serie de Conferencias, 385
Revista Peruana de Astronomía y Astrofísica, 424
Revue Aérospatiale, 154

Revue de l'Observatoire Populaire de Laval (La.), 194
Revue des Questions Scientifiques, 57
Revue du Palais de la Découverte, 195
Revue Montpelliéraine d'Astronomie Languedocienne, 164
Rheticus, 308
Richmond Astronomical Society News, 902
Rivista del Nuovo Cimento (La.), 335
Rocznik Astronomiczny Obserwatorium Krakowskiego (Ephemeris of eclipsing variable stars), 428
ROE Annual Report, 562
Romanian Astronomical Journal, 438
Romanian Journal of Hydrology and Water Resources, 438
Romanian Journal of Meteorology, 438
Rosette Gazette, 849
Rowan Sky Line, 827
Royal Meteorological Society Quarterly Journal, 565
Royal Observatory Edinburgh Computing Newsletter, 562
Royal Society of Edinburgh News, 566
Royal Statistical Society News, 566
Ruimtevaart, 397
Rutherford Appleton Laboratory Report Series, 542
Ryusei-sokuho (Meteor Advance Report), 355
SAAO Annual Report, 467
SAAO Circulars, 467
SAAO Newsletter, 467
Sacramento Peak Observatory Contributions, 801
Sacramento Peak Observatory Research Notes, 801
SADEYA, 483
Sadhana (Engineering Science), 284
Sagittario, 332
Saint Petersburg Astronomy Club Examiner, 684
San Diego Aerospace Museum Newsletter, 636
SAWB Daily Weather Bulletin, 468
SAWB Technical Papers, 468
SAWB Technical Reports, 468
SAWB Yearly Weather Report, 468
Science and Public Affairs, 566
Science and Technology, 453
Science Center News, 616
Science International, 185
Science Magazine, 669
Science News, 675
Science Place Pages (The.), 883
Sciences, 159
Science Scope, 635
Sciences de la Terre, 154
Sciences (The.), 819
Scientific Publications of Józefoslaw Warsaw University of Technology - Series: Geodesy", "Reports on Geodesy, 432
Scientific Reports of the Tohoku University, Eighth Series: Physics and Astronomy (The.), 358
Scientifics, 814
Scienza & Paranormale, 315
Scoop, 198
Scoopium, 406
Scopeville Informer (The.), 722
Scorpius, 15
Scottish Astronomers Group Newsletter, 567
SDAA News & Notes, 636
SEA Report, 133
Seismic Bulletin, 75
Seismological Bulletin, 348
Selenology, 851
Semesta, 379
Semiconductor Science and Technology, 550
Seminars of the United Nations Programme on Space Applications, 40
Sendai Astronomiaj Raportoj, 358
SETI News, 639
Shallow Sky Bulletin (The.), 837
Shemakha Circulars, 43
Shoreline Observer (The.), 767
Short Book Reviews, 394

SIAM Journal on
  Applied Mathematics, 861
  Computing, 861
  Control and Optimization, 861
  Discrete Mathematics, 861
  Mathematical Analysis, 861
  Matrix Analysis and Applications, 861
  Numerical Analysis, 861
  Optimization, 861
  Scientific Computing, 861
SIAM News, 861
SIAM Review, 861
SIDC News, 57
Sidereal Messenger, 708, 837
Sidereal Times, 17, 719, 791, 803, 877
Sidewalk Astronomers (The_), 626
Signals, 842
Sirius, 246, 258, 380, 570
Sirius Astronomer, 632
Sirius Observer (The_), 713
Sirius Voice, 4
SISIR News, 456
Sitzungsberichte der Mathematische-naturwissenschaftliche Klasse, 218
Sitzungsberichte der Philosophische-historische Klasse, 218
SJAA Ephemeris, 637
Skeptical Briefs (The_), 812
Skeptical Inquirer (The_), 812
Skeptiker, 232
Skokie Valley Astronomers Newsletter, 706
Sky & Space, 27
Sky & Telescope, 757
Sky Almanac for Maltese Islands, 380
Sky Calendar, 761
Skye Stuff, 896
Sky-High, 296
Skylight, 680
Skylights, 723, 732, 871
Skyline News, 903
Sky Maps, 629
Sky News, 93
Skynotes, 585
Sky Observers (The_), 118
Sky Scanner, 679
Skytrack, 457
SkyWatch, 757
Skywatch, 568
Skywatcher Meeting Notice, 701
Sky Watchers, 337
Smånytt om Romfart, 415
Smart Materials and Structures, 550
Smithsonian Contributions to the Earth Sciences, 675
SNOI Progress Reports, 100
SNV Bulletin, 509
Société Astronomique de Lyon, 200
Sodankylä Geophysical Observatory Publications, 146
Solar Activity and its Influence on the Earth, 448
Solar Bulletin, 747
Solar Data, 441
Solar Data Bulletin, 447
Solar-Geophysical Data, 657
Solar Indices Bulletin, 657
Solar Maps and Activity, 425
Solar Observer (The_), 594
Solar Patrol, 425
Solar Physics, 395
Solar Radio Data, 124
Solar Radio Emission Hiraiso, 343
Solar System Research, 395
Solnechnye Dannye, 448
Sonne, 258
Sonne - Data Report, 260
Sonne Datenblatt, 258
Sonne Tageskarten, 258
Sonne Zirkular, 258
Sonoma Skies, 640

Sopročila/Messages, 462
South Downs Astronomical Society Mercury, 569
South Downs Planetarium News, 569
Southern Cross (The_), 20
Southern Observer, 28
Southern Skies, 685
Southern Stars, 411
Soviet Astronomy, 734
  Letters, 734
Soviet Journal of
  Low-Temperature Physics, 734
  Nuclear Physics, 734
  Particles and Nuclei, 734
  Plasma Physics, 734
  Quantum Electronics, 734
Soviet Physics
  Acoustics, 734
  Crystallography, 734
  Doklady, 734
  JETP, 734
  Semiconductors, 734
  Solid State, 734
  Technical Physics, 734
  Uspekhi, 734
Soviet Technical Physics Letters, 734
SPACE, 684
Space Activities in Hungary, 276
Space Business in Europe - Prospects to 2005, 177
Space Debris, 395
Spaceflight, 538
Space Frontier News, 29
Space Horizons, 421
Space India, 285
Space Information Systems Newsletter, 318
Space Log, 803
Space News, 539
Spaceport News, 682
Spacereport, 532
Space Research in Norway, 417
Space Science and Technology, 529
Space Science Reviews, 395
Space Sciences Series of ISSI, 505
SpaceTalk, 399
Space Technology and Industries in Norway, 417
Space Times, 895
SPACEWARN Bulletin, 745
SPACEX 90 - Nuevos Horizontes, 471
Spark (The_), 80
SPARRSO Newsletter, 44
Special'IST, 168
Special Publication of Toyama Science Museum, 359
Spectra: The Newsletter of the Carnegie Institution of Washington, 672
Spectrum, 562
Spectrum (The_), 809, 867
Sperry Observations, 791
Spica, 253
SPOT Flash, 202
SPOTLight, 904
Spotlight, 568
SPOT Magazine, 202
Sri Lanka Journal of Social Science, 487
SSI Update, 797
SSSI, 505
Stadlatídindi, 280
Standard, 524
Standard Frequency and Time Service Bulletin, 343
Standard India, 282
Standardization, 460
Standards and Metrology, 388
Standards Bulletin, 297
Standards Monthly Additions, 282
Standards New Zealand, 411
Standards Worldover, 282
Standard (The_), 521
Standard View, 807
Stand van Zaken, 399

Starcharts, 875
StarDate, 888
Star Diagonal (The_), 891
Stardust, 465, 551, 839, 852
Star Fields, 746
Star Friends, 355
Stargazer, 691, 859
Stargazer's Discovery Updates, 629
Stargazer (The_), 567, 700, 908
StarGazing, 93
Starlight Journal, 716
Starlight (Sternf. Durmersheim), 247
Starlink Bulletin, 543
Starlink Documentation Series, 543
Starlink Software Collection, 543
Star News, 540
Star Observer, 39
Starry Messenger (The_), 89, 878
STARSCAN, 880
Star Stuff, 763
Star (The_), 800
StarTimes, 765
Star Trails, 662
Starword, 726
Star Wrangler (The_), 780
Statistical Software Newsletter, 393
Statistical Theory and Method Abstracts, 394
Statistik, 222
Stats, 896
ST-ECF Newsletter, 245
Stellae, 473
Stella Online, 784
Stellar Data, 25
Stellar Matters, 629
Stellar Messenger, 834
Stellar Sentinel, 764
Sterne (Die_), 225
Sternenbote, 258
Sternenbote (Der_), 38
Sternenrundschau (Die_), 34
Sternenwelt, 34, 39
Sterne und Weltraum, 246
Sternfreund (Der_), 209, 213, 225, 249, 260, 264
Sterngucker, 221
Sternkieker, 232
Sternwarte Stuttgart, 245
Sternwarte Tübingen, 217
Sternzeit, 213, 214, 216, 217, 238, 248, 249, 257–259, 262, 264
Sterrengids, 398, 402
Sterrenkijker (De_), 390
Sterrenwachter (De_), 60
Steward Observatory Preprints, 603
Stjerneskuddet, 134
S+T News, 296
STNews, 229, 245, 843
STN International: Databases in Science and Technology, 229
St. Petersburg Astronomy Club Examiner, 684
STScI Newsletter, 743
Studii si Cercetari de Hidrologie, 438
Studii si Cercetari de Meteorologie, 438
Sundial, 829
Sunspot Bulletin (SIDC), 57
Sunspot Data, 9
Sunspot Observations, 517
Sunspot Report, 133
Sunspots Report, 62
Superba, 320
Superconductivity, 734
Superconductor Science and Technology, 550
Supernova, 569
Suplemento al Almanaque (ONBA), 9
Supplement to the Shallow Sky Bulletin, 837
Surface Science Spectra, 808
Surveys in Geophysics, 227
SuW, 246

Suzirya, 529
Sveske Fizičkih Nauka, 454
    Series A: Conferences, 454
Switec Information, 509
Sydney Sky Guide (The_), 28
Syzygy, 884
Syzygy (The_), 624
Tähdet 200x, 152
Tähdet ja Avaruus, 152
Tähtiharrastustietoa, 146
Taipei Skylight, 517
Tamix Journal, 473
Tang Chang Pueg, 519
Technical Reports of the Mizusawa Kansoku Center National Astronomical Observatory, 353
Technical Support, 917
Technologist Magazine, 821
Technology Ireland, 296
Technology Today, 884
Technometrics, 896
Tectonics, 227, 670
Teknik & Standard, 497
Telapo, 278
TeleScoop, 488
Telescop, 74
Telescopes and Astronomy Newsletter, 29
Telescopic Topics, 818
Telescopium, 259
The Aether, 887
The Air Almanac, 677
The Allegheny Observer, 851
The Amateur Astronomer, 839
The American Statistician, 896
The Analyst, 566
The Aperture, 829
The Appulse, 425
The Arlington Astroletter, 896
The ASSA Newsletter, 282
The Asteroid Belt, 841
The Astronews, 693
The Astronomer, 572, 821
The Astronomers' Bulletin, 20
The Astronomical Almanac, 677
The Astronomical Herald, 342
The Barnard Star", 870
The Cape Observer, 464
The Celestial Observer, 757, 806, 870
The Celestial Onlooker, 16
The Constellation, 858
The Crossroads Astronomer, 878
The Current, 875
The CVA Observer, 616
The Dark-Sky Observer, 777
The Deep Sky, 727
The Deep-Sky Observer, 583
The Delaware Valley Amateur Astronomer, 854
The Ecliptic, 857
The ERAC Newsletter, 228
The Explorer, 903
The Eyepiece, 711
The Fleischmann Flyer, 786
The Focal Plane, 703
The Focal Point, 687
The Focus, 667
The Gazer's Gazette, 918
The Gemini Newsletter, 109, 596, 693
The Guide Star, 851
The Hampshire Observer, 547
The Home-Dome Star, 743
The INSPIRE Journal, 623
The JCMT Newsletter, 693
The JCMT-UKIRT Newsletter, 562
The Journal of Birla Planetarium, 286
The Journal of the ALPO, 609
The LHS Quarterly, 651
The Lowell Observer, 598
The Mercury, 690

The Meridian, 685
The Messenger, 228
The Meteor, 680, 832
The Meteorite, 839
The Mills Observatory Night Sky Guide, 557
The MIRA Newsletter, 628
The Mirror, 710
The Nautical Almanac, 677
The Nebula, 90, 714
The Newsletter of the Baltimore Astronomical Society, 735
The Newtonian, 861
The New Zenith, 582
The Night Sky, 20, 835
The Objective View, 657
The Observer, 621, 638, 711, 743, 789, 837, 858
The Observer II, 774
The Olympic Scope, 908
The Orbiter, 682, 849
Theoretical Chemical Engineering Abstracts, 566
The Orpington Astronomical Society Times, 561
Theory of Probability and its Application, 861
The Photogrammetric Journal of Finland, 148
The Physicist, 18
The Physics Teacher, 734
The Planetarian, 738
The Practical Observer, 725, 727
The Prairie Astronomer, 784
The Prime Focus, 716
The Probe, 818
The Radiant, 867
The Radio Science Bulletin, 53
The Rainwater Observer, 774, 775
The Reflector, 776
The Refractor, 874
The Science Place Pages, 883
The Sciences, 819
The Scientific Reports of the Tohoku University, Eighth Series: Physics and Astronomy, 358
The Scopeville Informer, 722
The Shallow Sky Bulletin, 837
The Shoreline Observer, 767
The Sidewalk Astronomers, 626
The Sirius Observer, 713
The Skeptical Briefs, 812
The Skeptical Inquirer, 812
The Sky Observers, 118
The Solar Observer, 594
The Southern Cross, 20
The Spark, 80
The Spectrum, 809, 867
The Standard, 521
The Star, 800
The Star Diagonal, 891
The Stargazer, 567, 700, 908
The Starry Messenger, 89, 878
The Star Wrangler, 780
The Sydney Sky Guide, 28
The Syzygy, 624
The UKIRT Newsletter, 693
The Universe, 186
The University of Texas Publications in Astronomy, 887
The Western Observer, 650
Thumbprint, 90
Tianwen Aihaozhe (Astronomical Amateur), 115
Time and Frequency Bulletin, 114
Time and Latitude, 124
TISI Newsletter, 519
Titan, 403
TOAST, 561
Trait d'Union (Le_), 50
Transactions of the Japan Academy, 358
Transactions of the Saint Petersburg Astronomical Institute, 449
Transnational Associations, 57
Travaux de l'Association Internationale de Géodésie, 138
Tribuna de Astronomía, 476
Trudy of the Institute for Theoretical Astronomy, 445
TUGboat, 850
TUM/IAPG Jahresbericht, 252
TUM/IAPG Mitteilungen, 252
Tuorla Observatory Reports, 152
TWAS Newsletter, 336
Twinkles, 871
UAG Report Series, 657
UANotizie, 337
UBC Physics Annual Report, 101
UCAR Corporate Report, 659
UCAR Newsletter, 659
Uchuken Hokoku, 346
UKIRT Newsletter (The_), 693
Ukrainian Astronomical Association Information Bulletin, 528
UK Space Activities, 539
UK Space Index, 539
Ulugh Beg Astronomical Institute Circulars, 925
Universal Time and Pole Coordinates, 441
Universal Times, 27
Univers du Namurois (L'_), 49
Universe, 15
Universe and Ourselves, 441
Universe in the Classroom Newsletter, 610
Universe (The_), 186
Universidad Nacional de La Plata Series (Astronomy, Geophysics, Circulars and Special), 10
University of Michigan Department of Astronomy Publications, 768
University of Texas Publications in Astronomy (The_), 887
Universo, 63
Univ. Tasmania Dept. Physics Research Report, 31
Update, 890
Uppsala Astronomical Observatory Reports, 498
Uppsala Preprints in Astronomy, 498
Urania, 430
Urania in de Kijker, 60
Uranus, 216
Ürkaleidoszkóp, 275
Ursa Minor, 152
UTIAS Annual Progress Report, 104
UTIAS Reports, 104
UTIAS Reviews, 104
UTIAS Technical Notes, 104
Valkoinen Kääpiö (White Dwarf), 148
Valley Skies, 643
Vangspiegel (De_), 407
Variable Star Memorandum, 25
Vasiona, 455
VdS-Nachrichten, 258
Vedráttan, 279
Verhandlungen der Deutsche Physikalische Gesellschaft, 223
Veröffentlichungen der Astronomischen Institute der Universität Bonn, 254
Veröffentlichungen der Deutschen Geodätischen Kommission, 224
Veröffentlichungen des Astronomisches Rechen-Institut, 215
Vesta, 406
Vestnik Sankt-Peterburskogo Universiteta - Series of Mathematics, Mechanics, and Astronomy, 449
Vestnik Udmurtskogo Universiteta, 450
Via Stellaris, 591
ViewFinder, 715
Views from Astronomy and Astrophysics, 860
Views on Finnish Technology, 151
Visnyk Kyivskoho Universitetu. Astronomia, 526
VSOLJ Variable Star Bulletin, 361
VSRR-Infoblatt, 513
WACO News, 911
WAITRO News, 140
Warner and Swasey Observatory Publications, 836

Warren Astronomical Society Paper, 769
Warta Lapan, 290
Warta Standardisasi, 291
WAS News, 584
WASP, 769
Water Engineering, 428
Water Management and Water Protection, 428
Water Resources Research, 670
Waves in Random Media, 550
Weather, 565
Weather and Forecasting, 747
Webb Society Observing Section Reports, 583
Webb Society Quarterly Journal, 583
WEGA, 37
Wellington Astronomical Society Newsletter, 413
Weltallkunde, 232
Wessex Astronomical Society News & Notes, 583
Western Observer (The_), 650
Wetenschappelijke en Technische Publikaties van KMIB, 54
Wetterkarte, 223
WGN, 234
What's up in Yakima?, 911
White paper on Science and Technology, 348
Wiadomości Pulsara, 427
Window on Science, 890
WIP, 61
Wireless Networks, 807
Wisconsin Astrophysics Preprint Series, 920
Wollongong Observer, 32
Woodland Star, 622
World Directory of Mathematicians, 68
World List of Universities and Other Institutions of Higher Education, 184
World Meteorological Organization Bulletin, 514
World Space Communications and Broadcasting Markets Survey - Ten Year Outlook, 177
World Space Markets Survey - Ten Year Outlook, 177
Yamaguchi-ken no shizen, 362
Yearbook of International Organizations, 57
Your Universe Awaits, 633
Zeitschrift für Meteorologie, 223
Zemlja y Vsselennaja, 440
Zenit, 402
Zenit (Zenit), 398
Zonnestraling, 54
Zoologica Scripta, 493
Zpravodaj, 131
Zpravodaj Zorné Pole, 133
Zvaigžņotá Debess, 372
Zvezdociot, 441
PerkinElmer Optoelectronics, 757
Perkins Observatory, 842
Perris, 643
Perth, 15, 17, 25, 26
(West_), 26
Perth Observatory, 26, 32
Perth Omni Theatre and Planetarium, 26
Peru, 423
Perugia, 339
Pesaro, 322
Petaling Jaya, 378
Peter-Anich-Planetarium Kufstein, 38
Peterborough, 788
Peter Gruber Foundation, 924
Peterson Planetarium, 721
Petit-Bourg, 159
Petit-Lancy, 504
Pettinger-Guiley Observatory (PGO), 910
Petur Beron Observatory (Dr._), 74, 76
Peverell, 544
Pforzheim, 213
PGAO
    Prince George Astronomical Observatory, 97
PGAS
    Prince George Astronomical Society, 96
PGO

Pettinger-Guiley Observatory, 910
PHEL
    Paton Hawksley Education Ltd., 562
Philadelphia, 778, 854–857, 861, 862
Philatelie (Motivgruppe Astronomie &_), 238
Philippine Astronomical Society (PAS), 425
Philippine Atmospheric, Geophysical and Astronomical Services Administration (PAGASA)
    Astronomy Research and Development Section, 425
Philippines, 425
Philippinum Sternwarte (Gymnasium_), 232
Philippus Lansbergen (Volkssterrenwacht_), 406
Philips Sterrenwacht (Dr. A.F._), 390, 400
Phillips Exeter Academy (PEA)
    Grainger Observatory, 789
Phillips Planetarium (L.E._), 919
Phipps Observatory (Gerald Hughes_), 653
Phoenix, 593, 595, 598, 600, 602
Phoenix Astronomical Society (PAS), 600
Phoenix Public Observatory, 400
Photosolve, 633
Physical Research Laboratory (PRL)
    Mount Abu Infrared Observatory Headquarters, 287
    Udaipur Solar Observatory (USO), 288
Physical Sciences Resource Center (PSRC), 734
Physical Society of Japan (The_), 356, 358
Physics Academic Software (PAS), 830
Physics and Astronomy Student Society (PASS), 96
Physikalischer Verein Frankfurt
    Volkssternwarte, 240
Physikalisch-Meteorologisches Observatorium Davos (PMOD)
    Weltstrahlungszentrum, 507
Physikalisch-Technische Bundesanstalt (PTB), 240
Physik Instrumente (PI) GmbH & Co., 241
PI
    Physik Instrumente, 241
PIA
    Pakistan International Airlines (PIA), Planetarium Lahore, 421
Pianoro, 333
Pianoro Observatory, 333
Piarco, 521
PIC
    Precision Industrial Components, 663
Picardie, 157
PIC Design, 663
Pickmere, 579
Pick Travel Ltd., 562
Pico dos Dias, 68
Picosecond Pulse Labs (PSPL) Inc., 658
Pico Veleta, 182, 478
Piedmont Amateur Astronomers (PAA), 867
Piera, 476
Pieri Astronomy Observatory, 333
Pietermaritzburg, 464
Pieterse Planetarium, 394
Pie Town, 801
Pietrasanta, 335
Piikkiö, 152
Pikes Peak Observatory (PPO), 660
Pine Bluff Observatory (PBO), 918, 920
Pinedale, 616
Pine Hills, 680
Pine Knot, 725
Pinellas County, 684
Pine Mountain, 689
Pine Mountain Observatory (PMO), 850
Pino Torinese, 328
Piombino, 304
Piracicaba, 70
Pirttiharju, 149
Pisa, 305, 321, 333, 334, 339
Piscataway, 796
Pisgah Astronomical Research Institute (PARI), 830
Piszkéstető Mountain Station, 275

Pitomača, 121
Pittsburg, 723
Pittsburgh, 851-853, 856-858, 862, 863
    (East._), 851
Pittsburg State University
    Physics Department, 723
Pittsfield, 753, 757
Pittsfield Astronomy Club, 757
Pittsylvania County Schools' Planetarium, 902
Piwnice Observatory, 432
PJC
    Pensacola Junior College
        Science and Space Theatre, 683
PKN
    Polski Komitet Normalizacyjny, 431
Plainedge Planetarium, 820
Plainfield, 698
Plana, 75
Planetário Calouste Gulbenkian (PCG), 434
Planetario Comunale di Modena, 333
Planetario Comunale di Pisa, 333
Planetário da Cidade do Rio de Janeiro, 66, 70
Planetario de Cuidad Victoria, 384
Planetario de la Ciudad de Buenos Aires Galileo Galilei, 9
Planetario del Collegio Pio X, 307
Planetario del Museo Provinciale di Storia Naturale, 334
Planetario de Madrid, 481
Planetario de Pamplona, 482
Planetario di Ravenna, 310, 334
Planetario di Venezia Lido, 308
Planetario Dr. Max Schreier, 63
Planetario Humboldt, 928
Planetario Municipal de Montevideo, 587
Planetário Prof. José Baptista Pereira, 70, 73
Planetario Ulrico Hoepli, 313, 314
Planetarium and Astronomical Observatory
    Grudziądz, 429
    Olsztyn, 429
Planetarium Aschersleben, 241
Planetarium Association of Canada (PAC), 96
Planétarium d'Aix-en-Provence
    (Amis du._), 154
Planétarium de Bourbon-Lancy, 196
Planétarium de Bretagne, 196
Planétarium de Cholet, 196
Planétarium de La Villette, 169
Planétarium de Montpellier, 196
Planétarium de Montréal, 96
Planétarium de Nantes, 196
Planétarium de Nîmes, 196
Planétarium de Parthenay, 196
Planétarium de Poitiers, 196
Planétarium de Reims, 197
Planetarium der Fachhochschule Kiel, 241
Planetarium der Fachhochschule Stralsund, 241
Planetarium der Stadt Wien, 38
Planetarium der Stadt Wolfsburg GmbH, 241
Planétarium de Saint-Étienne, 197
Planétarium de Vaulx-en-Velin, 197
Planétarium Dow, 96
Planétarium du Palais de la Découverte, 195
Planétarium du Trégor, 197
Planetarium Europapark, 34
Planetarium Hamburg, 231, 241
Planetarium Hoyerswerda, 241
Planetarium im Vonderau Museum, 242
Planetarium Institute, 633
Planetarium Klagenfurt, 34
Planetarium Lübz, 242
Planetarium Mannheim GmbH, 242
Planetarium Merseburg, 242
Planétarium Observatoire de Montredon-Labessonnié, 197
Planétarium Peiresc, 197
Planetarium Praha, 130
Planetariums
    Abbe (Ernst-_), 266
    Aberdeen, 531

Abrams, 761, 762
Adelaide, 14
Adler, 698
Adolf Diesterweg
    Radebeul, 259
Aiken, 866
Aix-en-Provence, 154, 197
Ajaccio, 169
Akima, 870
Alamogordo, 803
Alan C. Davis, 739
Albany, 691
Albert Einstein, 676
Albstadt, 216, 250
Albuquerque, 800
Alden, 746
Aldrin, 685
Alexander Morrison, 608, 611, 637
Alexandria, 897, 898, 904
Algonquin, 705
Alice and Leonard Dreyfuss, 791
Alice G. Wallace, 746
Allentown, 851, 858
Alpena, 764
Altenmünster, 248
Alworth (Marshall W._), 771
Amarillo, 878
Amsterdam, 389
Ancona, 310
Anderson (M.D._), 872
Andrus, 807
Angered, 488
Annapolis, 744
Ann Arbor, 761, 768
Antares, 68
Antwerpen, 60
Antwerp Zoo, 60
APAP, 154
Argus, 761
Århus, 139
Arlington, 887, 896
Armagh, 532
Armainvilliers (Gretz-_), 205
Arnim D. Hummel, 725
Arthur Storer, 735, 738
Artis, 389
ASA, 161
Aschersleben, 241
Asten, 394
Astrodome, 533
Astronomy Roadshow, 572
Athens, 268
Athinai, 268
Atlanta, 688
    (North._), 689
Auckland, 408
Augsburg, 216, 245
Aurora, 709
Avon, 664
AZH, 216
Baader, 218
Baker (Burke._), 878
Baltimore, 739
Bangkok, 519
Banneker (Benjamin._), 735
BAP, 538
Baptista Pereira (Prof. José._), 73
Barcelona, 480
Barlow, 915
Bartholomäus Scultetus, 264
Battelle, 836
Battle Creek, 763
Bautzen, 249
Bay City, 762
Bays Mountain, 871
Bay Village, 843
Beattie (George F._), 622

Beaumont, 188
Beecher (Ward.), 844
Beijing, 112
Bel Air, 742
Bellingham, 911
Bemidji, 770
Benedum, 912
Benjamin Banneker, 735
Benson, 604
Beograd, 455
Berea, 728
Bergerac, 188
Berkeley, 651, 912
Berlin, 265
Bernice P. Bishop, 692
Besançon, 156
Besser (Jesse.), 764
Bettendorf, 716
Beverly Hills, 611
BGSU, 835
BHHS, 611
Bianchi (Donald E.), 615
Binghamton, 820
Birla (B.M.)
   Chennai, 282
   Hyderabad, 282
   Madras, 282
Birla (M.P.), 286
Bishop, 679, 685
Bishop (Bernice P.), 692
Blake (Dr. W. Russell.), 750
Blakemore (Marian.), 881, 889
Bloomington, 712
Bloomsburg, 853
B.M. Birla
   Chennai, 282
   Hyderabad, 282
   Madras, 282
Bochum, 249
Bognor Regis, 569
Bogotá, 119
Bondar (Roberta.), 98
Borlänge, 489, 492
Boston, 750
Bothell, 908
Boulder, 660, 661
Bourbon-Lancy, 196
Bourg-en-Bresse, 182
Bowling Green, 728, 835
Boynton Beach, 683
Bozeman, 780
Brackbill (M.T.), 899
Bradenton, 679, 685
Bradley, 687
Brahe (Tycho.), 135, 140
Bremen, 240
Brescia, 337
Brevard, 678
Brisbane
   (Sir Thomas.), 27
Brish (William M.), 745
Brno, 126, 127, 129
Broman, 488
Brønshøj, 140
Bronx, 820
Brooklyn, 814
Brugge, 60
Brussels, 56
Bryan (K. Price.), 739
Buckingham, 853
Buckstaff, 921
Budapest, 274
Buehler, 679
Buenos Aires, 9
Buffalo, 826
Buhl Jr. (Henry.), 856, 858
Burg, 215

Burke Baker, 878
Bussloo, 406
Caird, 565
Calcutta, 286
Calgary, 80
Calouste Gulbenkian, 434
Calusa, 679
Campbell County, 922, 923
Canberra, 21
Canton, 838
Cape Town, 468
Cappelle-la-Grande, 195
Caracas, 927, 928
Carlisle, 854
Carl Sandburg, 897
Carl Zeiss, 221
Carter, 409
Casper, 922
Catonsville, 735
Cauthen (John K..), 866
CCM, 792
CCNY, 810
CCSSC, 688
Cedar Rapids, 719
Centennial, 80
Centerport, 825
Central Bucks, 853
České Budějovice, 126
Chabot, 616
Chaffee (Roger B..), 761, 767
Champaign, 709
Chapel Hill, 829, 832
Charles C. Gates, 653
Charles F. Clise, 911
Charles F. Hagar, 637
Charles Hayden, 750
Charleston, 913
Charlotte, 829
Chatham, 902
Chattanooga, 875
Chemnitz, 244
Cheney, 907
Chennai, 282
Cherokee, 718, 719
Cherry (Jim.), 688
Chesapeake, 897
Chicago, 698, 703
Chico, 614
Cholet, 196
Chorzów, 429
Christa McAuliffe, 788
Cincinnati, 838
Ciudad Victoria, 384
Clarence T. Jones, 875
CLEA, 173
Cleveland, 842
Clever (George H..), 622
Clise (Charles F..), 911
Clyde W. Tombaugh, 803
CMP, 788
Coca-Cola, 688
Cocoa, 678
Coldwater, 765
Coleman Sr. (George E..), 690
Colorado Springs, 652
Columbia, 777, 853, 866
Columbus, 688, 689, 712, 836
Concord, 788
Constellarium E4, 181
Conway, 607
Cook, 878
Copernicus, 837
   (N..), 74, 126, 127, 129
   (Nicolas.), 429
   (Nicolaus.), 239
Cormack, 864
Corona Park, 819

Corpus Christi, 882
Corsicana, 878
Cosmonova, 488
Cottbus, 243
Craigmont, 871
Crown (Henry_), 703
CSDO, 21
CSSC, 616
CSUN, 615
Cuba (Mount_), 667
Culiacan, 382
Cupertino, 628
CyberSphere, 871
Dahlonega, 690
Dale (Fred G._), 782
Dallas, 879, 882–884
Danbury, 666
Daniel B. Milliken, 617
Davie, 679
Davis (Alan C._), 739
Davis (Russell C._), 775
Dax, 192
Daytona Beach, 682
Decatur, 687
Deere (John_), 702, 705
Delhi, 281, 287
Delta, 762
Den Haag, 399
Denton, 887
Denver, 653
Des Moines, 718
Dessau, 215
Detroit, 769
Detwiler, 858
DHDC, 878
Dickson, 21, 871
Diedorf, 216
Diesterweg (Adolf_)
 Radebeul, 259
Digistar, 890
Dompcevrin, 161
Donald A. Schaefer, 716
Donald E. Bianchi, 615
Donetsk, 525
Dooley, 866
Doran, 89
Dorrance, 593
Dow, 96
Downing, 614, 618
Drake, 838
Drake (Marie_), 592
Drebach, 263
Drescher (John_), 638
Dreyfuss (Alice and Leonard_), 791
Dr. Max Schreier, 63, 64
Dr. W. Russell Blake, 750
Dubuque, 717
Duluth, 771
Dundee, 557
Dupont, 866
duPont III (Henry B._), 662
Dwingeloo, 400
E&S, 890
Earl and Hazel Faulkner, 697
East Lansing, 761, 762
Eau Claire, 919
Ebingen, 216, 250
E.C. Schouweiler, 711
Edinboro, 855
Edinburgh, 552
Edison, 898
Edmonton, 95
Edwin A. Link, 820
Edwin P. Hubble, 814
Eilenburg, 264
Einstein (Albert_), 676
Eise Eisinga, 391

Eisinga (Eise_), 391
El Camino, 619
Elgin, 700
El Paso, 879
Emporia, 721
EPISD, 879
Erkrath, 249
Ernst-Abbe, 266
Esk Valley (Jewel and_), 552
ESU, 721
Ethyl, 898
ETSU, 872
Eugene, 848
Eugenides, 268
EUP, 855
Europapark, 34
EuroPlaNet, 227
Europlanetarium, 51
Evans & Sutherland, 890
Evansville, 711
Exploration Place, 722
Fairbanks, 893
Fairfax, 906
Fairy Meadow, 32, 33
Falls Church, 898
Falun, 493
Faulkner (Earl and Hazel_), 697
Fehrenbach (Richard_), 243
Feira de Santana, 68
Fels, 855
Ferguson (Whitworth_), 826
Firenze, 325
Fiske (Wallace_), 660, 661
Fitchburg, 746
Flandrau (Grace H._), 596
Fleet (Reuben H._), 634
Fleischmann, 786
Flensburg, 237
Fleurance, 155
Flint, 763, 767
Florence, 591, 866
FMU, 866
Foran (Joseph A._), 663
Fort Myers, 679
Fort Victoria, 572
Fort Wayne, 711
Fort Worth, 882
Fox Valley, 915
Framtidsmuseet, 489
Franeker, 391
Fred G. Dale, 782
Freiburg-im-Breisgau, 243
Fremont, 783
French Camp, 774, 775
Fresno, 614, 618
Frombork, 429
Fulda, 233, 242
Gagarin (Y.A._), 526
Galilei (Galileo_), 9, 320
Galileo Galilei, 9, 320
Gardner, 763
Garrett County, 737
Gastonia, 831
Gates (Charles C._), 653
Gávea, 66
Gayle (W.A._), 591
Geer (Willard_), 907
Gengras, 662
Genk, 51
George E. Coleman Sr., 690
George F. Beattie, 622
George H. Clever, 622
George Stahl, 907
Gettysburg, 856
Gibbes, 866
Gillette, 922, 923
Gillingham, 533

Gladwin, 638
Glasgow, 577
Glenfield, 792
Glücksburg, 237
Goddard (Robert H._), 803
Goiás, 72
Golden Pond, 725
Görlitz, 264
Göteborg, 493
Gothenburg, 493
Gotoh, 344
Göttingen, 229
GPPA, 782
Grace H. Flandrau, 596
Grady (Joseph_), 858
Granada, 481
Grand Rapids, 761, 767
Great Plains, 782
Greensboro, 830
Greenville, 667, 866, 868
Greenwich, 565
Gresham, 848
Gretz-Armainvilliers, 205
Grimbergen, 60
Grosse Pointe, 764
Grout, 717
Grudziądz, 429
GSU, 688
Guam, 271
Guéret, 194
Gulbenkian (Calouste_), 434
Haag (Den_), 399
Hagar (Charles F._), 637
Hagerstown, 743, 745
Hague (La_), 188
Hague (The_), 399
Halberstadt, 216
Halifax, 100
Halle, 243
Hamburg, 231, 241
Hamilton, 91
Hannover, 219
Hansen, 890
Hardin, 728
Harrisburg, 859, 861
Harrisonburg, 899
Hartford
  (West_), 662
Harz, 233
Hastings, 782
Hatter, 856
Hawkes Bay, 410
Hayden, 815
Hayden (Charles_), 750
Hayfield, 898
HBOSC, 738
Headwaters, 770
Hedgesville, 912
Heinrich S. Schwabe, 215
Heitkamp, 717
Helsinki, 152
Henry B. duPont III, 662
Henry Buhl Jr., 856, 858
Henry Crown, 703
Herbert Trackman, 701
Herndon, 899
Herzberg, 257, 266
Hibbing, 772
Highland Park, 879, 884
Highstown, 795
Hiroshima, 345
HMNS, 878
Hochdahl (Neanderhöhe_), 249
Hoepli (Ulrico_), 313, 314
Hof, 248
Holcomb (J.I._), 710
Holt, 410

Holt (William Knox_), 651
Hong Kong, 116
Honolulu, 692
Hooper (T.C._), 868
Hoover-Price, 838
Hopkins, 903
Houston, 878
Hove, 60
Howard B. Owens, 738
Hoyerswerda, 241
Hradec Králové, 128
H.R. MacMillan, 89
HSC, 770
Hubble (Edwin P._), 814
Huddersfield, 549
Hudnall, 877, 880
Humboldt, 928
Hummel (Arnim D._), 725
Huntsville, 591, 883
Hurbanovo, 460
Hutchinson, 722
Hyderabad, 282
Illawara, 32
Impression 5, 764
Indianapolis, 714
International Planetarium Society, 738
IPS, 738
Isle of Wight, 572
ISU, 702
Itabashi, 345
Jackson, 775, 872
James Lockyer, 559
Jefferson (Thomas_), 904
Jena, 266
Jensen, 783
Jesse Besser, 764
Jewel and Esk Valley, 552
J.I. Holcomb, 710
JIHOP, 710
Jim Cherry, 688
JJC, 701
J.M. McDonald, 782
JMU, 899
Jodrell Bank, 579
Johann Palisa, 128
John C. Wells, 899
John Deere, 702, 705
John Drescher, 638
John K. Cauthen, 866
Johnson City, 872
John Young, 683
Joliet, 701
Jones (Clarence T._), 875
Jordan (Maynard F._), 733
José Baptista Pereira (Prof._), 73
Joseph A. Foran, 663
Joseph Grady, 858
Juneau, 592
Junior, 817
Justice, 722
JYP, 683
Kalamazoo, 765
Kanec (Paul_), 854
Kankakee, 706
Kansas City, 777
Kassel, 238
Katsushika, 349
KCSC, 722
Kecskemét, 277
Kelce (L. Russell_), 723
Kelly, 829
Kenner, 729
Kensington, 88
Kent, 838
Keystone Oaks, 857
K.E. Ziolkowski, 244
Kharkiv, 526

Kharkov, 526
Kiel, 241
Kilohani, 692
Kingsport, 871
Kirkpatrick, 846
Kirov, 442
Klagenfurt, 34
Knighton, 573
Knoxville, 870
København, 135, 140
Koch, 711
Kolkata, 286
Köln, 250
Königsleiten, 39
Kosmorama, 489
Kountze, 785
Kowloon, 116
K. Price Bryan, 739
Kronshagen, 249
KSU, 838
Kuala Lumpur, 378
Kunming, 115
KVM, 765
La Coruña, 474
La Crosse, 919
Lafayette, 712, 729
Laguna (Mount.), 637
La Hague, 188
Lahore, 421
Lakeview, 703
La Laguna, 479
Lancaster, 859
Lane ESD, 848
Lanham, 738
Lansing, 764
 (East.), 761, 762
La Paz, 63, 64
Laramie, 923
La Rochelle, 165
Lasky, 300
Las Vegas, 786
Launceston, 24
Laupheim, 261
Laval, 194
LAVC, 626
La Villette, 169
Layton P. Ott, 890, 891
Le Bourget, 189
L.E. Phillips, 919
Lepton, 549
Lewis Science Center, 607
Lewiston, 732
Lido, 308
Liège, 56
Lille, 171
Lincoln, 782–784
Linda G. Sharpe, 873
Link (Edwin A..), 820
Lisboa, 434
Little Rock, 607
Liverpool, 554
Livorno, 334
LMS, 765
LNHM, 729
LNP, 753
Loch Ness Productions, 753
Lockyer (James.), 559
LodeStar, 800
London, 90, 538, 554, 565
Longway (Robert T..), 763, 767
Longwood, 817
Los Angeles, 622, 648
Louisville, 725
L. Russell Kelce, 723
L.S. Noblitt, 712
Lubbock, 881
Lübz, 242

Lucerne, 514
Ludiver, 188
Lueninghoener, 783
Luleå, 497
Luling, 730
Lumberton, 831
Lumezzane, 303, 332
Lund, 491
Luzern, 514
Lyon, 171
Macclesfield, 552, 579
Mac Laughlin, 91
MacMillan (H.R..), 89
Macon, 689
Madison, 917, 920
Madras, 282
Madrid, 481
Magdeburg, 216
Malaysia, 378
Mammendorf, 218
Manfred Olson, 920
Manila, 425, 426
Manitoba, 90
Mannheim, 242
Marburg, 232
Margaret C. Woodson, 829
Marian Blakemore, 881, 889
Marie Drake, 592
Mark Smith, 689
Marquette, 765, 767
Marseille, 155
Marshall, 714, 772
Marshall W. Alworth, 771
Mary D. Pretlow, 901
Massapequa
 (North.), 820
Mawson Lakes, 30
Max Schreier (Dr..), 63, 64
Maynard F. Jordan, 733
MCAO, 667
McAuliffe (Christa.), 788
MCC, 479
McCallion (William J..), 91
McDonald (J.M..), 782
McLaughlin, 91
MCP, 713
M.D. Anderson, 872
Meadville, 851
Melbourne, 16, 24
Memphis, 871, 873
Menasha, 915
Menke, 237
Merrillville, 713
Merseburg, 242
Mesquite, 882
MHCC, 848
Miami, 682
Middle Island, 817
Midland, 881, 889
Milano, 313, 314
Milford, 663
Milham, 760
Millikan, 634
Milliken (Daniel B..), 617
Mills, 557
Milwaukee, 920
Minneapolis, 771
Minolta, 350, 628
 USA, 793
Mira, 60
MLC, 783
MMSD, 917
Modena, 333
Moers, 238
Montclair, 792
Montevideo, 587
Montgomery, 591

Montgomery College, 740
Monticello, 607
Montpellier, 196
Montréal, 96
Montredon-Labessonnié, 197
Moody, 881
Moorhead, 772
Morehead, 829, 832
Morelia, 382
Morgantown, 914
Morrison (Alexander_), 608, 611, 637
Moscow, 443
Most, 129
Mountainside, 797
Mount Cuba, 667
Mount Laguna, 637
M.P. Birla, 286
MSD, 859
M.T. Brackbill, 899
Mt. Cuba, 667
Mt. Laguna, 637
Mueller (Ralph_)
  Cleveland, 842
  Lincoln, 782, 784
München, 219, 225
Muncie, 710
Mundo Juvenil, 141
Münster, 265
Murdock, 849
Mystic Seaport, 663
Nacogdoches, 885
Nagoya, 351
Nantes, 196
Nashville, 874
Nathan M. Patterson, 689
N. Copernicus, 74, 126, 127, 129
NCSM, 351
Neanderhöhe Hochdahl, 249
Nehru, 281, 287
Neumarkt/Opf., 219
Nevada
  (Southern_), 786
Newark, 791
New Castle, 667
New Haven, 664
New Orleans, 729
Newport News, 905
New York, 810, 815, 819
NGC, 690
Nice, 173, 195
Nicolas Copernicus, 429
Nicolaus Copernicus, 239
Nippes, 250
Nizhny Novgorod, 443
Noble, 882
Noblitt (L.S._), 712
Nordic, 492
Norfolk, 901
Normal, 702
North Atlanta, 689
North Branch, 796
North Las Vegas, 786
North Massapequa, 820
Northridge, 615
Novins (Robert J._), 791, 796
Novokuzneck, 443
NPA, 492
NTM, 416
Nürnberg, 239
NYHoS, 819
OAA, 68
Oakland, 616, 737
Oakton, 901
OBU, 846
Ogden, 890, 891
Oklahoma City, 846
Olbers, 240

Olivet, 766
Olson (Manfred_), 920
Olsztyn, 429
Omaha, 785
Omniplex, 846
Omniversum, 399
Orangeburg, 868
Orion, 139
Orlando, 683
Orono, 733
Orsay, 173
Ortona, 329
Oshkosh, 921
Oslo, 416
Osnabrück, 238
Ostrava-Poruba, 128
Ott (Layton P._), 890, 891
Owens (Howard B._), 738
Padova, 320
PAGASA, 426
Palisa (Johann_), 128
Palm Beach
  (West_), 685
Palomar College, 632
Pamplona, 482
PAO, 366
Paris, 169, 195
Parthenay, 181, 196
Pasajes de San Pedro, 479
Patterson (Nathan M._), 689
Pattonville, 777
Paul Kanev, 854
Paulucci, 772
PCG, 434
Pegasus, 713
Peiresc, 154, 197
Peißnitz, 243
Peninsula, 905
Pensacola, 683
Peoria, 703
Pereira (Prof. José Baptista_), 73
Perth
  (West_), 26, 27
Peterson, 721
Philadelphia, 855
Phillips (L.E._), 919
Phoenix, 593
Pieterse, 394
Pinellas County, 684
Pisa, 333
Pittsburg, 723
Pittsburgh, 856-858
Pittsylvania, 902
Plainedge, 820
Plains (Great_), 782
Planetron, 400
Pleumeur-Bodou, 172, 196
Plymouth, 789
Plzeň, 130
Poinciana, 683
Poitiers, 176
Pomeroy, 607
Pomona, 634
Portland, 733, 849
Porto, 436
Porto Alegre, 73
Potsdam, 215
Powys, 573
Praha, 130
Prato, 309
Prešov, 459
Pretlow (Mary D._), 901
Price (Hoover-_), 838
Prince Frederick, 735, 738
Prof. José Baptista Pereira, 73
Providence, 864
Provo, 890

PSF, 705
Pullman, 910
Puyricard, 154
Pyongyang, 366
Quezon City, 426
Quito, 141
Radebeul, 259
Radford, 902
Rainwater, 774, 775
Ralph Mueller
   Cleveland, 842
   Lincoln, 782, 784
Rancho Cucamonga, 617
Randolph, 792
Raritan Valley, 796
Rauch, 725
Rauma, 149
Ravenna, 310, 334
Reading, 852, 860
Recklinghausen, 265
Regensburg, 227
Reims, 197
Reno, 786
Reuben H. Fleet, 634
Reutlingen, 263
RFP, 243
RHS, 683
Richard Fehrenbach, 243
Richard R. Russell, 882
Richardson, 882
Richmond, 725, 898
Rio de Janeiro, 66
RISD, 882
Ritter, 844
River Falls, 921
Riverside, 635
Riverview, 683
Roanoke, 903
Roberson, 820
Roberta Bondar, 98
Robert H. Goddard, 803
Robert J. Novins, 791, 796
Robert K. Wickware, 664
Robert T. Longway, 763, 767
Robeson, 831
Rochester, 823
Rock Hill, 868
Rock Island, 702, 705
Rødding, 139
Rodewisch, 244
Roger B. Chaffee, 761, 767
Rollins, 691
Rosario, 9
Roswell, 803
Roth, 614
Rouen, 181, 193
Rovereto, 330
RPSEC, 866
Russell Blake (Dr. W._), 750
Russell C. Davis, 775
Russell (Richard R._), 882
Sacramento, 618
Saint Ann, 777
Saint Charles Parish, 730
Saint Cloud, 772
Saint Davids, 855
Saint-Étienne, 197
Saint-Exupéry, 416
Saint Johnsbury, 893
Saint Louis, 778
Saint-Michel-l'Observatoire, 164
Saint Petersburg, 684
Salisbury, 829
Salt Lake City, 890
San Angelo, 877
San Antonio, 883
San Bernardino, 622

Sandburg (Carl._), 897
San Diego, 634, 637
Sanford, 684, 718, 719
San Francisco, 608, 633, 637
San Giovanni in Persiceto, 322
San Marcos, 632
San Pedro (Pasajes de_), 479
Santa Ana, 644
Santa Barbara, 638
Santa Fe, 803
Santa Monica, 638
Santa Rosa, 638
Sarah B. Summerhays, 890
Sarasota, 683
Sargent, 718
Saunders, 684
Schaefer (Donald A._), 716
Schenectady, 821
Schiele, 831
Schkeuditz, 216
Schneeberg, 250
Schouweiler (E.C._), 711
Schreder, 638
Schreier (Dr. Max_), 63, 64
Schuele, 843
Schull, 298
Schwabe (Heinrich S._), 215
Schwaz, 38
Science Place, 883
Scienceworks, 24
SciWorks, 831
Scobee, 883
Scultetus (Bartholomäus_), 264
Seattle, 908
Seiler, 778
Seminole, 684
Sendai, 357
Senftenberg, 242
SEPA, 685
Serafino Zani, 332
Settlemyre, 868
Seymour, 758
SFASU, 885
SFCC, 803
SFSU, 637
Sharpe (Linda G._), 873
Shawnee, 846
Shibuya, 344
Shields (South_), 570
Shiras, 765, 767
Shreveport, 730
SHSU, 883
Sidmouth, 559
Silverman, 821
Sinaloa, 382
Singapore, 457
Sir Thomas Brisbane, 27
Sittingbourne, 572
Sivry, 49
Slottsskogen, 493
SMC, 638
Smith (Mark_), 689
Smith (Willard W._), 908
Smolyan, 76
SMWV, 903
Södertälje, 498
Sotteville-lès-Rouen, 181
South African Museum, 468
Southampton, 742
South Bend, 713
South Downs, 569
Southeastern Association, 685
Southend-on-Sea, 570
Southern Nevada, 786
South Shields, 570
South Tyneside, 570
Southwest Association, 884

Southworth, 733
SpaceQuest, 714
SPAR, 730
SPJC, 684
Springfield, 758
    (West.), 906
SRJC, 638
SSM, 758
SSU, 772
Staerkel (William M..), 709
Stahl (George.), 907
Stanback, 868
St. Ann, 777
Starlab, 753
Starwalk, 667
Statesboro, 688
STBP, 27
St. Charles Parish, 730
St. Cloud, 772
St. Davids, 855
Stella Nova, 493
Steno, 139
St.-Étienne, 197
Stevens Point, 921
Stjernekammeret, 140
St. Johnsbury, 893
St. Louis, 778
St.-Michel-l'Observatoire, 164
Stockholm, 488
Stockton, 622
Storer (Arthur.), 735, 738
Storrs, 665
St. Petersburg, 684
Stralsund, 241
Strasbourg, 205
Strasenburgh, 823
Strickler, 706
Struve, 719
Stuttgart, 221
Sudbury, 89, 98
Sudekum, 874
Suhl, 244
Summerhays (Sarah B..), 890
Sunshine, 358
Sutherland (Evans &.), 890
SWAP, 884
Sydney, 81
Sylvania, 837
Syracuse, 821
Taichung, 517
Takoma Park, 740
Talcott Mountain, 664
Tampa, 684
Tampere, 150
TASSF, 100
Taylor, 780
T.C. Hooper, 868
Teknikens Hus, 497
Tel-Aviv, 300
Tempe, 593
Tenerife, 479
Teplice, 131
Tessmann, 644
The Hague, 399
Thomas Brisbane (Sir.), 27
Thomas Jefferson, 904
Tipton, 714
TMSC, 664
TNSC, 797
Todmorden, 562
Tokyo, 345, 349
Toledo, 844
Tombaugh (Clyde W..), 803
Tomchin, 914
Tomsk, 450
Toms River, 791, 796
Tom Tits, 498

Tonneville, 188
Topeka, 724
Toronto, 91
Torun, 432
Toulouse, 169
Toyama, 359
Trackman (Herbert.), 701
Trailside Nature and Science Center, 797
Trégor, 172, 196
Trenton, 794
Treviso, 307
Troy, 817
TSM, 359
Tucson, 596, 604
Turkey Run, 714
Twin Falls, 697
Tycho Brahe, 135, 140
Tyler, 877, 880
Tyneside (South.), 570
UCLA, 648
Ufa, 450
UFG, 72
UFRGS, 73
Ulrico Hoepli, 313, 314
UNA, 591
United Kingdom, 538
University Park, USA-PA, 860
UNT, 887
Uranoscope de l'Île de France, 205
UTA, 887
UWRF, 921
UWSP, 921
Valdosta, 691
Valencia, 475
Valley City, 834
Vancouver, 89
Vanderbilt, 825
Van Nuys, 626
Vantaa, 152
Varna, 74
Vaulx-en-Velin, 197
VCSU, 834
Venezia, 308
Verne, 152
Viareggio, 329
Victoria, 16
Vilnius, 374
Violau, 248
Virginia Beach, 905
VLM, 905
Vonderau Museum, 242
Voorst, 406
W.A. Gayle, 591
Wald im Pinzgau, 39
Wallace (Alice G..), 746
Wallace Fiske, 660, 661
Walla Walla, 911
Ward Beecher, 844
Washburn, 724
Washington, 676
Washington County, 743
Waterloo, 717
Waubonsie Valley, 709
Wayne, 782
WCAC, 845
Weatherford, 728
Weitkamp, 842
Wellington, 409
Wells (John C..), 899
Wernigerode, 233
Westerville, 842
West Hartford, 662
West Palm Beach, 685
West Perth, 26, 27
West Springfield, 906
Wetherbee, 691
Whitworth Ferguson, 826

Wible, 851
Wichita, 722
Wickware (Robert K.-), 664
Wien, 38
Willard Geer, 907
Willard W. Smith, 908
William J. McCallion, 91
William Knox Holt, 651
William M. Brish, 745
William M. Staerkel, 709
Williamsport, 858
Williamstown, 760
Willimantic, 664
Willowdale, 98
Wilmington, 667
Wilmot, 845
Winnipeg, 90
Winston-Salem, 831
Wittenberg, 265
Wolfsburg, 241
Wollongong, 32, 33
Woodson, 906
Woodson (Margaret C.-), 829
Worcester, 746
WRBP, 750
W. Russell Blake (Dr.-), 750
WSU, 769, 910
WWU, 911
Y.A. Gagarin, 526
Yarmouth, 572
Yokohama, 362
Yonkers, 807
Young Harris, 691
Young (John-), 683
Youngstown, 844
YSU, 844
Yunnan, 115
Zani (Serafino-), 332
Zeiss, 265, 266
Zeiss (Carl-), 221
ZGP, 265
Ziolkowski (K.E.-), 244
Zürich, 507
Planetarium Schwaz GmbH, 38
Planetarium Senftenberg, 242
Planetarium Zürich, 507
Planetary Science Institute (PSI)
    San Juan Capistrano Division, 633
    Tucson Division, 601
Planetary Sciences Research Institute (PSRI), 560
Planetary Society, 633
Planetary Studies Foundation (PSF), 704
Planet:Earth Centre, 562
Planète Sciences, 197
Planetron, 400
Plasma Science Committee, 674
Plateau de Bure (Observatoire du-), 182, 204
Platte Valley Astronomical Observers (PVAO), 784
Pleasant (Mount-), 19, 31, 762, 867
Plenum Publishing Corp., 820
Pleumeur-Bodou, 172, 196
Plymouth, 544, 563, 748, 750, 763, 789
Plymouth Astronomical Society (PAS), 563
Plymouth State College
    Planetarium, 789
Plzeň, 130
Plzeň Observatory and Planetarium, 130
PMD
    Pakistan Meteorological Department, 421
PMO
    Pine Mountain Observatory, 850
PMOD
    Physikalisch-Meteorologisches Observatorium Davos
        Weltstrahlungszentrum, 507
P.N. Lebedev Physics Institute, 443, 447
    Astro Space Center (ASC), 447
    Nuclear Physics and Astrophysics, 447

PNTPM
    Physique Nucléaire Théorique et Physique Mathématique (ULB), 59
POA
    Polaris Observatory Association, 634
Pocatello, 697
Pochoco (Cerro-), 107
Pocono Mountain Optics Inc., 860
Poços de Caldas, 65
Podanur, 285
Poinciana Planetarium, 683
Point
    (West-), 687
Pointe-Claire, 83
Poitiers, 168, 176
Poland, 427
Polaris Observatory Association (POA), 634
Polish Academy of Sciences
    N. Copernicus Astronomical Center (NCAC), 429
        Department of Astrophysics I, 430
    Space Research Center, 430
        Astrogeodynamic Observatory, 430
        Solar Physics Department, 430
Polish Amateur Astronomical Society, 430
Polish Astronomical Society, 430
Polish Committee for Standardization, 431
Polish Geophysical Society, 431
Polish Physical Society, 431
Pollux, 149
Polman Observatory (Jorge-), 66
Polskie Towarzystwo Astronomiczne (PTA), 430
Poltava, 529
Poltava Gravimetrical Observatory, 528, 529
Pomeroy Planetarium, 606, 607
Pomona, 796
Pomona College
    Department of Physics and Astronomy, 634
Pomona Valley Amateur Astronomers (PVAA), 634
Pompaire, 181
Pontchartrain Astronomy Society (PAS), 730
Pontefract, 583
Ponte in Valtellina, 307
Pontificia Universidad Católica de Chile, 110
Poole Astronomical Society (PAG), 583
Poona, 284, 286
Popular Astronomy Club (PAC) Inc., 705
Popular Science Magazine, 658
Port Elgin, 79
Porterville, 645
Portimão, 433
Portland, 733, 849, 850
Port Macquarie, 26
Port Macquarie Astronomical Association, 26
Porto, 433, 436
Porto Alegre, 73
Port of Spain, 521
Porto Torres, 313
Portsmouth, 549, 570, 580
Port Townsend, 910
Portugal, 433
Poste-De Flacq, 381
Potchefstroom, 466
Potchefstroom University for Christian Higher Education
    Department of Physics
        Space Research Unit (SRU), 466
Potosí, 62
Potsdam, 208, 211, 212, 215, 217, 218, 222, 237, 256, 811, 822
Potterville, 763
Poughkeepsie, 825
Powell, 837
Powell Observatory, 776
Powys, 573
Poznań, 431
Poznań University A. Mickiewicza
    Astronomical Observatory, 431
Pozuelo de Alarcón, 480

PPA
    Pacific Planetarium Association, 786
PPARC, Particle Physics and Astronomy Research Council, 561
    Royal Observatory Edinburgh, 562, 576
PPO
    Pikes Peak Observatory, 660
PPPL
    Princeton Plasma Physics Laboratory, 795
Prague, 124, 126, 127, 130, 132
Praha, 124–127, 130, 132
Praha Observatory and Planetarium, 130
    Ďáblice Observatory, 130
    Planetarium Praha, 130
Prairie Astronomy Club (PAC), 784
PRAO
    Pushchino Radioastronomy Observatory, 447
Prato, 309
PRC
    People's Republic of China, 112
Precision Cryogenic Systems Inc., 713
Precision Industrial Components (PIC) Corp., 663, 664
Prenzlau, 262
Přerov, 131
Přerov Astronomical Club, 131
Prescott Astronomy Club (PAC), 601
Preserving Starry-Sky Association (PSA), 356
Presolana, 319
Prešov, 459
Prešov Observatory and Planetarium, 459
Preston, 563, 575
Preston and District Astronomical Society (PADAS), 563
Pretlow Planetarium (Mary D..), 901
Pretoria, 465–469
PRI
    Princeton Research Instruments, 795
Priddis, 102
Prince Albert, 468
Prince Frederick, 735, 738
Prince George, 96
Prince George Astronomical Observatory (PGAO), 97
Prince George Astronomical Society (PGAS), 96
Princeton, 791–793, 795, 797
Princeton Plasma Physics Laboratory (PPPL), 795
Princeton Research Instruments (PRI), Inc., 795
Princeton University
    Department of Astrophysical Sciences, 795
    Joseph Henry Laboratories
        Physics Department, 795
    Princeton Plasma Physics Laboratory (PPPL), 795
Princeton University Press, 795
PRL
    Physical Research Laboratory
        Mount Abu Infrared Observatory, Headquarters, 287
        Udaipur Solar Observatory, 288
    Plasma Research Laboratory, 19
Professional societies
    AAA (Argentina), 6
    AAPT (USA), 734
    AAS (USA), 669
    ACM (USA), 807
    Acoustical Society of America, 805
    ACP (Canada), 80
    ACS (USA), 670
    Aerospace Industries Association, 669
    AG (Germany), 211
        Arbeitskreis Astronomiegeschichte, 211
    AGU (USA), 670
    AIAA (USA), 895
    AIA (USA), 669
    AIP (USA), 734, 806
    Albanian Physical Society, 3
    ALPSP (UK), 532
    American Association of Physics Teachers, 734
    American Astronomical Society, 669
    American Chemical Society, 670
    American Geophysical Union, 670
    American Institute of Aeronautics and Astronautics, 895
    American Institute of Physics, 734, 806
    American Mathematical Society, 864
    American Meteorological Society, 747
    American Meteor Society, 806
    American Nuclear Society, 698
    American Physical Society, 734, 806
    American Society for Engineering Education, 671
    American Society for Information Science, 734
    American Statistical Association, 895
    American Vacuum Society, 808
    AMS (USA), 747, 806, 864
    ANS (USA), 698
    APS (USA), 734, 806
    Armenian Physical Society, 12
    ASA (Australia), 15
    ASA (USA), 805, 895
    ASEE (USA), 671
    ASGI (Ireland), 533
    ASGRG (Australia), 17
    ASI (India), 281
    ASIS (USA), 734
    ASJ (Japan), 342
    ASLIB (UK), 532
    ASM (Austria), 35
    Asociación Argentina de Astronomía, 6
    Association Canadienne des Physiciens, 80
    Association for Computing Machinery, 807
    Association for Information Management, 532
    Association of Learned and Professional Society Publishers, 532
    Astronomical Science Group of Ireland, 533
    Astronomical Society of Australia, 15
    Astronomical Society of India, 281
    Astronomical Society of Japan, 342
    Astronomische Gesellschaft, 211
        Arbeitskreis Astronomiegeschichte, 211
    Astronomisk Selskab i Danmark, 136
    Australasian Society for General Relativity and Gravitation, 17
    Austrian Society for Aerospace Medicine, 35
    AVS, 808
    Belarusian Physical Society, 46
    Belgian Physical Society, 48
    BPS (Belarus), 46
    BPS (Belgium), 48
    British Aerospace Companies, 569
    Canadian Association of Physicists, 80
    Canadian Astronomical Society, 80
    CAP (Canada), 80
    CASCA (Canada), 80
    CAS (China/PRC), 115
    Chinese Astronomical Society, 115
    CLEA (France), 173
    Comité de Liaison Enseignants Astronomes, 173
    Croatian Physical Society, 121
    Danish Astronomical Association, 136
    Danish Physical Society, 137
    Deutsche Geophysikalische Gesellschaft, 222
    Deutsche Gesellschaft für Luft- und Raumfahrt, 222
    Deutsche Meteorologische Gesellschaft, 223
    Deutsche Physikalische Gesellschaft, 223
    DGG (Germany), 222
    DGLR (Germany), 222
    DMG (Germany), 223
    DPG (Germany), 223
    Dutch Physical Society, 397
    EAAE, 226
    EAAS, 441
    EAGE, 391
    EARMA, 317
    EARTO, 51
    EAS
        Czech Republic, 127
        Petit-Lancy, 504
        Praha, 127

Switzerland, 504
Edinburgh Mathematical Society, 545
Eesti Füüsika Selts, 144
EFS (Estonia), 144
EGS, 227
EMS (Scotland), 545
Eötvös Loránd Physical Society, 274
EPS, 177
Estonian Physical Society, 144
Euro-Asian Astronomical Society, 441
European Association for Astronomy Education, 226
European Association for Research Managers and Administrators, 317
European Association of Geoscientists and Engineers, 391
European Association of Research and Technology Organisations, 51
European Astronomical Society
   Czech Republic, 127
   Petit-Lancy, 504
   Praha, 127
   Switzerland, 504
European Geophysical Society, 227
European Physical Society, 177
FASTS, 23
FAS (USA), 672
Federation of American Scientists, 672
Federation of Australian Scientific and Technological Societies, 23
Finnish Astronomical Society, 147
Finnish Physical Society, 147
FLAG (Belgium), 52
Flemish Aerospace Group, 52
Geological Society of America, 654
GSA (USA), 654
HELAS (Greece), 268
Hellenic Astronomical Society, 268
Hellenic Physical Society, 268
Hermann-von-Helmholtz-Gemeinschaft Deutscher Forschungszentren, 233
HGF (Germany), 233
IAGA, 655
IASC, 393
IAU, 184
Icelandic Astronomical Society, 279
Icelandic Physical Society, 279
IEAUST (Australia), 23
IEE (UK), 550
IMS (USA), 764
IMU, 68
Industrial Mathematics Society, 764
Institution of Electrical Engineers, 550
Institution of Engineers (Australia), 23
International Association for Statistical Computing, 393
International Association of Geomagnetism and Aeronomy, 655
International Astronomical Union, 184
International Mathematical Union, 68
International Society for Optical Engineering, 908
International Statistical Institute, 394
International Union of Geodesy and Geophysics, 186
International Union of Radio Science, 53
IPS (Israel), 299
ISI, 394
Israel Physical Society, 299
IUGG, 186
Japan Physical Society, 358
JPS (Japan), 358
KAS (Korea/ROK), 368
Korean Astronomical Society, 368
Korean Physical Society, 368
KPS (Korea/ROK), 368
Latvian Physical Society, 371
Lithuanian Physical Society, 375
LPS (Latvia), 371
LPS (Lithuania), 375
MAA (USA), 673

Materials Research Society, 858
Mathematical Association of America, 673
Meteoritical Society, 694
MRS (USA), 858
NAC (Netherlands), 396
NaSPA (USA), 917
National Committee of Russian Physicists, 443
National Science Teachers Association, 900
Nederlandse Astronomenclub, 396
Nederlandse Natuurkundige Vereniging, 397
Network and Systems Professionals Association, 917
NNV (Netherlands), 397
Norsk Fysisk Selskap, 416
Norwegian Physical Society, 416
NSTA (USA), 900
ÖGAA (Austria), 37
ÖPG (Austria), 38
Optical Society of America, 674
Optical Society of India, 287
OSA (USA), 674
OSI (India), 287
Österreichische Gesellschaft für Astronomie und Astrophysik, 37
Österreichische Physikalische Gesellschaft, 38
Physical Society of Japan (The_), 358
Polish Astronomical Society, 430
Polish Geophysical Society, 431
Polish Physical Society, 431
Polskie Towarzystwo Astronomiczne, 430
PTA (Poland), 430
RAS (UK), 564
Real Sociedad Española de Física, 482
Remote Sensing Society, 563
RMS (UK), 565
Romanian Physical Society, 439
Royal Astronomical Society, 564
Royal Meteorological Society, 565
Royal Society, 565
Royal Society of Chemistry, 566
Royal Society of Edinburgh, 566
Royal Society of New Zealand, 411
Royal Statistical Society, 566
RSC (UK), 566
RSE, 566
RSEF (Spain), 482
RSNZ (New Zealand), 411
RSS (UK), 563, 566
RS (UK), 565
SAB (Brazil), 70
SAIt (Italy), 335
SAS (Slovakia), 460
SAS (USA), 742
SBAC (UK), 569
Schweizerische Gesellschaft für Astrophysik und Astronomie, 511
SEDO (Spain), 484
SF2A (France), 202
SFO (France), 202
SFP (France), 201
SGAA (Switzerland), 511
SGEPSS (Japan), 357
SIAM (USA), 861
Slovak Astronomical Society, 460
Slovak Physical Society, 460
Sociedade Astronômica Brasileira, 70
Sociedade Portuguesa de Física, 435
Sociedad Española de Física (Real_), 482
Sociedad Española de Optica, 484
Società Astronomica Italiana, 335
Società Italiana de Fisica, 335
Societatea Română de Fizică, 439
Société Canadienne d'Astronomie, 80
Société Française d'Astronomie et d'Astrophysique, 202
Société Française de Physique, 201
Société Française d'Optique, 202
Société Royale des Sciences de Liège, 56
Société Scientifique de Bruxelles, 57

Société Suisse d'Astrophysique et d'Astronomie, 511
Société Suisse de Physique, 511
Society for Applied Spectroscopy, 742
Society for Industrial and Applied Mathematics, 861
Society for Scientific Exploration, 903
Society of Geomagnetism and Earth, Planetary and Space Sciences, 357
Society of Mathematicians, Physicists and Astronomers of Slovenia, 462
Society of Photo-Optical Instrumentation Engineers, 908
Society of Physicists of the Republic of Macedonia, 377
Society of Satellite Professionals International, 903
Space Transportation Association, 904
SPF (Portugal), 435
SPIE (USA), 908
SRSL (Belgium), 56
SSAA (Switzerland), 511
SSE (USA), 903
SSPI (USA), 903
SSP (Switzerland), 511
STA (USA), 904
Suomen Tähtitieteilijäseura, 147
Swedish Physical Society, 496
The Physical Society of Japan, 358
Turkish Physical Society, 524
UAA (Ukraine), 528
UAI, 184
UATI, 203
UGGI, 186
UKISC (UK), 573
Ukrainian Astronomical Association, 528
Ukrainian Physical Society, 529
Union Astronomique Internationale, 184
Union Géodésique et Géophysique Internationale, 186
Union Internationale des Associations et Organismes Techniques, 203
Union of Czech Mathematicians and Physicists
   Physics Section, 132
Union of Physicists in Bulgaria, 76
Union of Societies of Mathematicians, Physicists and Astronomers
   Physics Section, 455
United Kingdom Industrial Space Committee, 573
UPS (Ukraine), 529
URSI, 53
WAITRO, 140
World Association of Industrial and Technological Research Organisations, 140
Profondeville, 50
Progress in Astronomy, 117
Project Pluto, 733
Prospace, 198
Prostějov, 131
Prostějov Popular Observatory, 131
ProtoStar, 842
Provence
   (Haute-), 191
Provence Sciences Techniques Jeunesse (PSTJ), 198
Providence, 864
Provo, 890
PSA
   Preserving Starry-Sky Association, 356
PSB
   Productivity and Standards Board, Singapore, 456
PSF
   Planetary Studies Foundation, 704
PSI
   Pakistan Standards Institution, 421
   Planetary Science Institute
      San Juan Capistrano, 633
      Tucson Division, 601
PSPL
   Picosecond Pulse Labs, 658
PSRC
   Physical Sciences Resource Center, 734
PSRG
   Planetary Science Research Group, 553
PSRI
   Planetary Sciences Research Institute, 560
PST
   Paulucci Space Theatre, 772
PSTJ
   Provence Sciences Techniques Jeunesse, 198
PSU, Pennsylvania State University, 859
   Department of Astronomy and Astrophysics, 860
   Department of Statistics
      Statistical Consulting Center for Astronomy, 860
PTA
   Polskie Towarzystwo Astronomiczne, 430
PTB
   Physikalisch-Technische Bundesanstalt, 240
Publications of the Astronomical Society of Australia (PASA), 15
Publications of the Astronomical Society of Japan (PASJ), 342
Publications of the Astronomical Society of the Pacific (PASP), 610
Public Museum of Grand Rapids
   Astronomy Day Headquarters, 761
   Roger B. Chaffee Planetarium, 767
Publiekssterrenwacht Schothorst, 400
Publishers
   Advanstar Communications, 835
   Agence Spatiale Européenne
      Division des Publications, 392
   AIP, 806
   ALPSP, 532
   American Institute of Physics, 806
   American Physical Society, 806
   Andromeda, 207
   Annual Reviews, 608
   APD, 408
   APS, 806
   ARI, 608
   ASE
      Division des Publications, 392
   ASP, 610
   Association of Learned and Professional Society Publishers, 532
   Astral Press, 14
   Astronomical Pocket Diary, 408
   Astronomical Society of the Pacific, 610
   Astronomisk Forlag, 135
   Burillier, 188
   Cambridge Scientific Abstracts, 735
   Cambridge University Press
      Cambridge, 540
      New York, 809
      United Kingdom, 540
      USA-NY, 809
   CAS, 836
   Chandler (David-), 618
   Chemical Abstracts Service, 836
   CNRS Éditions, 173
   Cobblestone, 788
   Commonwealth Scientific and Industrial Research Organisation, 23
   Cornell University Press, 812
   CSA, 735
   CSIRO, 23
   CUP
      Cambridge, 540
      New York, 809
      United Kingdom, 540
      USA-NY, 809
   Cygnus-Quasar, 837
   David Chandler, 618
   Deutsch (Verlag Harri-), 259
   Dr. Vehrenberg, 257
   Earth & Sky, 878
   Éditions Burillier, 188
   Éditions de Physique (Les-), 176
   Éditions Diffusion Presse Sciences, 176

Éditions du CNRS, 173
Edizioni Scientifiche Coelum, 317
EDP Sciences, 176
Elsevier
   Amsterdam, 391
   Netherlands, 391
   New York, 814
   USA-NY, 814
EPD, 392
Equipo Sirius, 476
Erhard Friedrich, 226
ESA Publications Division, 392
European Space Agency
   ESA Publications Division, 392
Friedrich (Erhard_), 226
Galaxy Contact, 179
Gordon and Breach, 856
Harri Deutsch (Verlag_), 259
Harvard University Press, 752
HUP, 752
ICP, 549
Imperial College Press, 549
Institute of Physics
   Bristol, 550
   Pennsylvania, 857
   Philadelphia, 857
   United Kingdom, 550
   USA-PA, 857
IOP
   Bristol, 550
   Pennsylvania, 857
   Philadelphia, 857
   United Kingdom, 550
   USA-PA, 857
JHU Press, 739
Johns Hopkins University Press, 739
Kalmbach, 916
KAP
   Boston, 752
   Dordrecht, 395
   Massachusetts, 752
   Netherlands, 395
   Norwell, 752
   USA-MA, 752
Kluwer
   Boston, 752
   Dordrecht, 395
   Massachusetts, 752
   Netherlands, 395
   Norwell, 752
   USA-MA, 752
Laurin, 753
Les Éditions de Physique, 176
Liebert (Mary Ann_), 818
Louisiana State University Press, 730
LSU Press, 730
Mary Ann Liebert, 818
MIT Press, 756
Murmel, 506
NAP, 673
National Academy Press, 673
National Technical Information Service, 901
NCC, 881
Network Cybernetics, 881
North American Skies, 657
Novaspace Galleries, 600
NTIS, 901
OUP, 561
Oxford University Press, 561
Pallasite Press, 410
Princeton University Press, 795
Quasar (Cygnus-_), 837
Radio-Sky, 727
Science History Publications, 567
Science Service, 675
Sirius (Equipo_), 476
Sky & Space, 27

Sky Planner, 891
Sky Publishing, 757
Springer-Verlag
   Germany, 246
   Heidelberg, 246
   New York, 822
   USA-NY, 822
Star Farm (The_), 686
Star Observer, 39
Sunstone, 823
Taylor & Francis
   Bristol, 862
   London, 572
   Pennsylvania, 862
   United Kingdom, 572
   USA-PA, 862
The Star Farm, 686
UCP, 708
UMI, 767
Univelt, 646
University Microfilms International, 767
University of Arizona, 604
University of Chicago, 708
University of Illinois, 709
University of Iowa, 719
University of Michigan, 768
University of South Carolina Press, 868
University of Tennessee, 876
University of Texas Press, 889
University of Toronto, 104
University of Wisconsin, 921
University Science Books, 904
USB, 904
USC, 868
UTK, 876
Vanderbilt University Press, 876
Vehrenberg (Dr._), 257
Verlag Harri Deutsch, 259
Yale University Press, 666
Zarvan, 293
PUC
   Pacific Union College
      Physics Department, 632
      Young Observatory, 632
Puckett Observatory, 690
Puertollano, 477
Pufferfish Observatory, 705
Puimichel, 193
Pulkovo Observatory, 443, 447
   Kislovodsk Astronomical Station, 447
Pullman, 910
Pulpit Rock, 858
Pune, 284–286
Punjabi University
   Department of Physics, 288
Punta Gorda, 681
Purdue University
   Indianapolis, 713
   West Lafayette
      Department of Physics, 713
Purple Mountain Observatory, 114, 115, 117
   Qinghai Radio Observatory, 114
Pushchino, 447
Pushchino Radioastronomy Observatory (PRAO), 443, 447
Putney, 893
Putzbrunn/Solalinden, 252
Puyallup, 909, 910
Puyricard, 154
PVAA
   Pomona Valley Amateur Astronomers, 634
PVAO
   Platte Valley Astronomical Observers, 784
PVSG
   Penobscot Valley Star Gazers, 733
P. Wyss Photo-Video, 508
Pylypovychi, 526
Pyongyang, 366

Pyongyang Astronomical Observatory (PAO), 366
Pythagorica, 776, 777

Q

QBSA
    Quarterly Bulletin on Solar Activity, 362
QCAS
    Quad Cities Astronomical Society, 718
Qinghai Radio Observatory, 114, 117
QMWC
    Queen Mary and Westfield College
        Astronomy Unit, 578
        Astrophysics Group, 579
QSP Optical Technology Inc., 634
Quad Cities Astronomical Society (QCAS) Inc., 718
Quadrant Engineering Inc., 634
Quantum Technology Corp., 909
QUARG
    Queen's University Astronomy Research Group, 97
Quasar, 400
Quasar Books (Cygnus-_), 837
QUB, Queen's University of Belfast
    Department of Applied Mathematics and Theoretical
        Physics, 543, 563
    Department of Pure and Applied Physics
        Astrophysics and Planetary Science Division, 563
Québec, 86, 101
Queen Mary and Westfield College (QMWC), 563
    Astronomy Unit, 578
    Astrophysics Group, 579
Queensbury, 805
Queensland University of Technology (QUT)
    Space Centre for Satellite Navigation (SCSN), 27
Queen's Observatory, 97
Queen's University
    Department of Physics, 100
        Astronomy Research Group (QUARG), 97
Queen's University of Belfast (QUB)
    Department of Applied Mathematics and Theoretical
        Physics, 543, 563
    Department of Pure and Applied Physics
        Astrophysics and Planetary Science Division, 563
Queen Victoria Museum and Art Gallery, 16, 24
Queerswood, 465
Querqueville, 180, 188
Questar Corp., 860
Queven, 161
Quezon City, 425, 426
Quinn, 869
Quito, 141
QUT
    Queensland University of Technology
        Space Centre for Satellite Navigation, 27

R

RAA
    Rainwater Astronomical Association, 775
    Rockford Amateur Astronomers, 705
RAAGE
    Regionale Astronomische Arbeitsgemeinschaft Euskirchen, 243
Rabastens, 154
Rabbit Lake, 103
RAC
    Raleigh Astronomy Club, 830
    Rockland Astronomy Club, 796
Racah Institute of Physics, 299, 300
Racine, 919
Racine Astronomical Society (RAS), 918
Radeberg, 260
Radebeul, 209, 225, 232, 259
Raden Memorial Observatory, 684
Radford, 902

Radford University
    Planetarium, 902
RadioAstron, 613
Rádio Observatório do Itapetinga (ROI), 68, 70
Rádio Observátorio Espacial do Nordeste (ROEN), 68, 70
Radioobservatorium Stockert, 253
Radio Research Instrument (RRI) Co. Inc., 664
Radio-Sky Publishing, 727
Radiosterrenwacht Dwingeloo, 400, 402
Radiosterrenwacht Westerbork, 400, 402
Radolfzell, 210
Raiba Volkssternwarte Mariazellerland, 35
Rainwater Astronomical Association (RAA), 775
Rainwater Observatory and Planetarium, 775
RAIUB
    Radioastronomisches Institut der Universität Bonn, 253
RAL
    Radio Astronomy Laboratory
        Hat Creek Observatory, 647
    Rutherford Appleton Laboratory, 542, 560
        Chilbolton Observatory, 542
        Space Science Department, 542
        STARLINK Project, 543
Raleigh, 830
Raleigh Astronomy Club (RAC), 830
Ralph A. Worley Observatory, 730
Ralph Mueller Planetarium, 782, 784, 842
Raman Research Institute (RRI), 288
Ramat-Aviv, 300
Ramfjordbotn, 415
Ramón María Aller (Observatorio Astronómico_), 485
Ramonville-Saint-Agne, 199
Rampisole (La_), 179
Ramponio Verna, 321
Ramsey, 793
Ramsey Memorial Observatory (Crosby_), 739
Rancho Cucamonga, 617, 635
Rancho Mirage, 610
Randolph, 792
Randolph-Macon College (RMC)
    Keeble Observatory, 902
Randolph-Macon Woman's College (RMWC)
    Physics Department, 902
Random Factory (The_), 601
Rannsóknarráð Íslands, 279
RAO
    Rothney Astrophysical Observatory, 101
RARC
    Ritter Astrophysical Research Center, 844
Raritan Valley Community College
    Planetarium, 796
RAS
    Racine Astronomical Society, 918
    Reading Astronomical Society, 563
    Richland Astronomical Society, 842, 845
    Richmond Astronomical Society, 902
    Riverside Astronomical Society, 635
    Royal Astronomical Society, 532, 564
    Ryerson Astronomical Society, 705
RASC
    Royal Astronomical Society of Canada, 97
RASD
    Radio and Atmospheric Sciences Division (NPL), 286
RASNZ
    Royal Astronomical Society of New Zealand, 411
Rat Deutscher Sternwarten (RDS), 242
Rauch Planetarium, 725, 727
Rauma, 149
Rauma Vocational College
    Maritime Department
        Planetarium, 149
Raumflugplanetarium Cottbus, 243
Raumflugplanetarium Halle (RFP), 243
Ravalli, 780
Ravenna, 310, 334
Ravenstein, 239
Rayleigh Optical Corp. (ROC), 742

Raymond and Beverly Sackler Institute of Astronomy, 300
Raymond and Beverly Sackler Laboratorium voor Astrofysica, 401, 404
Raytheon Optical Systems Inc., 664
Raytheon Technical Services Co.
    Information Technology and Scientific Services (ITSS), 742
Ray Township, 761
RBO
    Red Buttes Observatory, 923
RCA
    Rose City Astronomers, 849
RCfTA
    Research Centre for Theoretical Astronomy, 31
RDS
    Rat Deutscher Sternwarten, 242
Reading, 545, 555, 563, 565, 860, 894
Reading Astronomical Society (RAS), 563
Reading Public Museum and Planetarium, 860
Real Instituto y Observatorio de la Armada (ROA), 482
Real Sociedad Española de Física (RSEF), 482
Recife, 66
Recklinghausen, 265
Red Bank, 797
Red Buttes Observatory (RBO), 923
Redding, 625, 638
Rede Nacional de Observação Astronómica (RNOA), 434
Redfern, 32
Redlands, 636
Redmond, 908
Reduit, 381
Regensburg, 227
Reggio di Calabria, 322
Regina, 102
Regionale Astronomische Arbeitsgemeinschaft Euskirchen (R-AAGE), 243
Rehoboth, 748
Reims, 197
Reisterstown, 744
Remanzacco, 309
Remeis-Sternwarte (Dr._), 243, 254
Remote Measurement Systems (RMS) Inc., 909
Remote Sensing Society (RSS), 563
Remscheid, 214
Reno, 786, 859
Rensselaer Astrophysical Society, 820
Rensselaer Polytechnic Institute (RPI), 820
    Department of Physics, Applied Physics and Astronomy, 820
REOSC
    Recherche et Études d'Optique et de Sciences Connexes, 198
    Recherche Étude Optique, 198
Repsold-Sternwarte, 232
Republican Palace of School Children Observatory, 364
République Populaire de Chine (RPC), 112
RESCEU
    Research Center for the Early Universe, 361
Research Council of Norway, 418
Research Institute of Geodesy, Topography and Cartography, 131
Research School of Astronomy and Astrophysics (RSAA)
    Mount Stromlo and Siding Spring Observatories (MSSSO), 18
Research Triangle Park, 827
Reseda, 642
Resonance Ltd., 97
Ressources Naturelles Canada
    Division des Levés Géodésiques, 94, 97
Reston, 895, 904, 905
Reuben H. Fleet Space Theater and Science Center, 634
Réunion (La_), 156
Reutlingen, 263
Revista Mexicana de Astronomía y Astrofísica, 385
Rev. M.W. Burke-Gaffney Observatory (BGO), 97
Reykjavík, 279, 280
Reynard Corp., 635
Reynella, 29
Rezh, 440
RFP
    Raumflugplanetarium Halle, 243
RGL
    Richardson Grating Laboratory, 820
RGO
    Royal Greenwich Observatory, 565
RHBNC
    Royal Holloway and Bedford New College
        Department of Physics, 565
Rheine, 210
Rheinhausen, 244
Rheinische Friedrich-Wilhelms-Universität Bonn, 243
Rheinische-Westfälische Technische Hochschule Aachen, 243
Rheita, 310
Rhin (Haut-_), 200
RHKO
    Royal Hong Kong Observatory, 116
Rhode Island, 864, 865
Rhodes College
    Department of Physics, 873
Rhodes University
    Astronomy Society, 466
    Department of Physics and Electronics, 466
Rhône, 182
Rhône-Alpes, 203
RHS
    Riverview High School
        Planetarium, 683
Ribemont, 155
Riboux, 160
Rice University (William Marsh_), 882
Richard-Fehrenbach-Planetarium, 243
Richard King High School
    Planetarium, 882
Richard R. Russell Planetarium, 882
Richardson, 888
Richardson Astronomical Observatory (N.A._), 635, 636
Richardson Grating Laboratory (RGL), 820
Richardson Independent School District (RISD)
    Planetarium, 882
Richard Stockton College of New Jersey (RSCNJ)
    Observatory, 796
Richland, 910
Richland Astronomical Society (RAS), 842, 845
Richland College
    Planetarium, 882
Richmond, 564, 711, 725, 896, 898, 902
Richmond and Kew Astronomical Society (RKAS), 564
Richmond Astronomical Society (RAS), 902
Richmond Hill, 103
RICL
    Roditi International Corp. Ltd., 564
Ridge, 806
Ridgecrest, 617
Ridgway, 851
Riding
    (East_), 549
Riding Astronomical Society (Hull and East_), 564
Rieti, 311
Riga, 371
Rigel Systems, 635
Riglione, 305
Rihlaperä, 148
Rijksmuseum voor de Geschiedenis van de Natuurwetenschappen en van de Geneeskunde, 401
Rijksuniversitair Centrum Antwerpen (RUCA), 56
    Onderzoeksgroep Astrofysica, 58
Rijksuniversiteit Gent (RUG), 56
Rijksuniversiteit Groningen, 401
Rijksuniversiteit Leiden, 401
Rijksuniversiteit Utrecht, 401
Riley Observatory (Glen D._), 703
Rimavská Sobota, 459, 461
Rimavská Sobota Observatory, 459, 461

Rimini, 319
Rimouski, 84, 88
Rinnegg, 39
Rio de Janeiro, 65, 66, 68, 70, 73
Rio Grande, 7, 10
Rio Hondo Astronomical Society, 635
Rioja, 472
Rion, 269
Rio Rico, 602
RISD
    Richardson Independent School District
        Planetarium, 882
Risø National Laboratory, 139
Ris-Orangis, 197, 198
RIT
    Rochester Institute of Technology, 821
        Department of Physics, 821, 824
Ritchie Observatory (Edwin_), 907
Ritter Astrophysical Research Center (RARC), 842, 844
Ritter Observatory, 844
Ritter Planetarium, 844
Rivanazzano, 307
Rivera, 587
River Falls, 921
River Grove, 699
Riverhead, 809
Rivers Camera Shop Inc., 789
Riverside, 635, 648
Riverside Astronomical Society (RAS), 635
Riverside Community College Planetarium, 635
Riverside Telescope Makers' Conference (RTMC) Inc., 635
River Valley Stargazers, 606
Riverview High School (RHS)
    Planetarium, 683
Rivière (La_), 156
Riyadh, 453
RKAS
    Richmond and Kew Astronomical Society, 564
RMA
    Roper Mountain Astronomers, 867
RMC
    Randolph-Macon College
        Keeble Observatory, 902
RMCC
    Ronald McDonald's Children's Charities, 846
RMCC Observatory, 846
RMPA
    Rocky Mountain Planetarium Association, 923
RMS
    Remote Measurement Systems, 909
    Royal Meteorological Society, 565
RMSC
    Rochester Museum and Science Center
        Strasenburgh Planetarium, 823
    Roper Mountain Science Center, 867
RMWC
    Randolph-Macon Woman's College
        Physics Department, 902
RNOA
    Rede Nacional de Observação Astronómica, 434
ROA
    Real Instituto y Observatorio de la Armada, 482
Roane State Community College (RSCC)
    Tamke-Allen Observatory, 873
Roanoke, 798, 902, 903
Roanoke Valley Astronomical Society (RVAS), 902
Roberson Museum and Science Center
    Edwin A. Link Planetarium, 820
Roberta Bondar Planetarium, 97
Robert A. Haag - Meteorites, 601
Robert H. Goddard Planetarium, 802, 803
Robert J. Novins Planetarium, 796
    ASTRA, 791
Robert K. Wickware Planetarium, 664
Robert-Mayer-Volks-und-Schulsternwarte Heilbronn eV, 243

Robert T. Longway Planetarium, 763, 767

Robeson Planetarium, 831
Robinson Astrophysics Laboratory, 611, 613
Robinson Observatory, 680, 686
Robinson Observatory (Paul_), 794
Robledo de Chavela, 613
ROC
    Rayleigh Optical Corp., 742
    Republic of China, 516
Rocha, 587
Rochefort, 50
Rochefort-Samson, 171
Rochester, 533, 813, 817, 820, 821, 823–825
Rochester Academy of Science
    Astronomy Section (ASRAS), 821
Rochester Institute of Technology (RIT)
    Department of Physics, 821, 824
Rochester Museum and Science Center (RMSC)
    Strasenburgh Planetarium, 823
Rochford, 540
Rockbottom Observatory, 757
Rock Creek Park Nature Center
    Planetarium, 675
Rockford, 698, 700, 705, 706
Rockford Amateur Astronomers (RAA) Inc., 705
Rock Hill, 868
Rock Island, 702, 705
Rockland Astronomy Club (RAC) of New York, 796
Rockledge, 679
Rock Valley Astronomical Society (RVAC), 919
Rockville, 741, 743
Rocky Mountain Planetarium Association (RMPA), 923
Rødding, 139
Roden, 395
Rodewisch, 244
Roditi International Corp. Ltd. (RICL), 564
Rodovre, 135
ROE
    Royal Observatory, Edinburgh, 562, 576
        Astronomy Technology Centre, 534
ROEN
    Rádio Observatório Espacial do Nordeste, 68, 70
Roepersdorf, 262
ROG
    Royal Observatory Greenwich, 565
Roger B. Chaffee Planetarium, 767
    Astronomy Day, 761
Rogers Observatory (Joseph H._), 764, 766, 767
Rogersville, 778
Rohnert Park, 640
ROI
    Rádio Observatório do Itapetinga, 68, 70
ROK
    Republic of Korea, 367
Rokycany, 131
Rokycany Astronomical Observatory, 131
Rollins Planetarium, 690, 691
Rolnick Observatory, 666
Rolyn Optics Co., 636
ROM
    Royal Ontario Museum, 91
Roma, 302–304, 311, 312, 315, 321, 323–325, 327–329, 336,
    339, 340, 926
Romanian Academy
    Astronomical Institute, 438
        Cluj-Napoca Astronomical Observatory, 438
        Timişoara Astronomical Observatory, 438
Romanian Physical Society, 439
Romanian Space Agency (ROSA), 438
Romans-sur-Isère, 171
Romedenne, 49
Romeo, 769
Rømer Museet (ORM) (Ole_), 138, 139
Rømer Observatory (ORO) (Ole_), 134, 139, 140
Römer-Sternwarte Rheinhausen (RRS) eV (Rudolf-_), 243,
    244
Ronald McDonald's Children's Charities (RMCC) Observatory, 846

Rondebosch, 468, 469
Roper Mountain Astronomers (RMA), 867
Roper Mountain Science Center (RMSC), 867
Roper Scientific, 564, 601
Roque de los Muchachos, 472, 477, 479, 480, 482
Roquetes, 486
Roquevaire, 161
ROSA
    Romanian Space Agency, 438
Rosario, 9
Roseburg, 850
Rose Center for Earth and Space, 815
Rose City Astronomers (RCA), 849
Rosemary Hill Observatory, 686
Roseville, 633, 764
ROSHYDROMET
    Russian Federal Service for Hydrometeorology and Environmental Monitoring, 448
Roskilde, 139
Rosman, 830
Rossall-Assheton Astronomical Observatory, 564
Rossdorf, 226, 232
Rosse Observatory, 584
Rostock, 215, 220
Rostov, 444
Rostov State University (RSU)
    Department of Space Research, 444
    Institute of Physics, 444
Roswell, 802, 803
Roswell Astronomy Club, 802
Roswell Museum and Art Center
    Robert H. Goddard Planetarium, 802
Rothney Astrophysical Observatory (RAO), 97, 101
Roth Planetarium, 614, 636
Rouen, 181, 193
Round Lake Park, 702
Rouyn-Noranda, 84
Rovegno, 340
Rovereto, 330
Rowan County, 827
Roy, 890, 891
Royal Astronomical Society of Canada (RASC), 97
Royal Astronomical Society of New Zealand (RASNZ) Inc., 411
Royal Astronomical Society (RAS), 532, 564
Royal Danish Academy of Sciences and Letters, 139
Royal Greenwich Observatory (RGO), 565
Royal Holloway and Bedford New College (RHBNC)
    Department of Physics, 565
Royal Hong Kong Observatory (RHKO), 116
Royal Irish Academy, 297
    Irish Irish National Committee for Astronomy and Space Research, 296
    Irish National Committee for Physics, 296
Royal Meteorological Society (RMS), 565
Royal Observatory
    Edinburgh, 562, 576
Royal Observatory Edinburgh (ROE), 562, 565, 576
    Astronomy Technology Centre (ATC), 534
Royal Observatory Greenwich (ROG), 565
Royal Observatory of Belgium, 55, 56
    Planetarium, 56
Royal Ontario Museum (ROM), 91
Royal Society of Chemistry (RSC), 566
Royal Society of Edinburgh (RSE), 566
Royal Society of New Zealand (RSNZ), 411
Royal Society (RS), 53, 174, 565
Royal Statistical Society (RSS), 566
Royal Swedish Academy of Sciences, 493
    Research Station for Astrophysics, 482, 493
Rozhen, 75
Rozhen Observatory, 75, 76
Roztoky, 459
Roztoky Observatory, 459
Rozzano, 320
RPC
    République Populaire de Chine, 112

RPI
    Rensselaer Polytechnic Institute, 820
        Department of Physics, Applied Physics, and Astronomy, 820
        Observatory, 820
RPSEC
    Ruth Patrick Science Education Center, 866
RRI
    Radio Research Instrument, 664
    Raman Research Institute, 288
RS
    Royal Society, 53, 174, 565
RSAA
    Research School of Astronomy and Astrophysics
        Mount Stromlo and Siding Spring Observatories, 18
RSAI
    R.S. Automation Industrie, 198
R.S. Automation Industrie (RSAI), 198
RSC
    Royal Society of Chemistry, 566
RSCC
    Roane State Community College
        Tamke-Allen Observatory, 873
RSCNJ
    Richard Stockton College of New Jersey
        Observatory, 796
RSE
    Royal Society of Edinburgh, 566
RSEF
    Real Sociedad Española de Física, 482
RSES
    Research School of Earth Sciences, 18
RSNZ
    Royal Society of New Zealand, 411
RSS
    Remote Sensing Society, 563
    Royal Statistical Society, 566
RSU
    Rostov State University
        Department of Space Research, 444
        Institute of Physics, 444
RTMC
    Riverside Telescope Makers' Conference, 635
Rtyně v Podkrkonoší, 131
Rtyně v Podkrkonoší Observatory, 131
Rubempré, 157
Rubenstein Museum of Science & Technology (Milton J._)
    Silverman Planetarium, 821
RUCA
    Rijksuniversitair Centrum Antwerpen, 56
        Onderzoeksgroep Astrofysica, 58
Rudjer Boskovitch Astronomical Society, 455
Rudolf-Römer-Sternwarte Rheinhausen (RRS) eV, 243
Rudolf Wolf Gesellschaft, 508
RUG
    Rijksuniversiteit Gent, 56
        Sterrenkundig Observatorium, 59
Ruggieri (Circolo Astrofili di Mestre Guido_), 313
Ruhr-Universität Bochum, 244
Ruimtevaart-Sterrenkunde Centrum Saturnus, 403
Rümlang, 513
Ruprecht-Karls-Universität Heidelberg, 244
Russell Blake Planetarium (WRBP) (Dr. W._), 750, 757
Russell C. Davis Planetarium, 775
Russell Planetarium (Richard R._), 882
Russellville, 606
Rüsselsheim, 250
Russia, 440
Russian Academy of Sciences, 174
    A.F. Ioffe Physical Technical Institute
        Division of Plasma Physics, Atomic Physics, and Astrophysics, 444
    Committee on Meteorites, 444
    Department of General Physics and Astronomy (DGPA), 444
    Geophysical Institute, 451
    Institute for Astronomy

Center for Astronomical Data (CAD), 445
Institute for Astronomy (INASAN), 445
Institute for Nuclear Research
  Baksan Neutrino Observatory (BNO), 444
Institute for Theoretical Astronomy (ITA), 445
Institute of Applied Astronomy, 445
  Irkutsk Department, 445
  Zelenchuk Department, 445
Institute of Applied Physics
  Millimeter Astronomy Group, 446
Institute of Cosmophysical Research and Aeronomy (IKFIA), 446
Institute of Spectroscopy, 446
Institute of Terrestrial Magnetism, Ionosphere and Radio Wave Propagation (IZMIRAN), 446
Institute of Theoretical and Experimental Physics, 446
Keldysh Institute of Applied Mathematics, 446
Nuclear Physics Institute, 446
P.N. Lebedev Physics Institute, 447
  Astro Space Center (ASC), 447
  Nuclear Physics and Astrophysics, 447
  Pushchino Radioastronomy Observatory (PRAO), 447
Pulkovo Observatory, 447
  Kislovodsk Astronomical Station, 447
Siberian Division
  Institute of Solar-Terrestrial Physics, 448
Space Research Institute (IKI), 448
Special Astrophysical Observatory (SAO), 448
Ussurijsk Astrophysical Observatory (UAO), 448
Russian Federal Service for Hydrometeorology and Environmental Monitoring, 448
Russian Space Agency, 449
Russie, 440
Rutgers University
  Department of Physics and Astronomy, 796
Rutherford Appleton Laboratory (RAL), 542, 560, 566
  Chilbolton Observatory, 542
  Space Science Department, 542
  STARLINK Project, 543
Ruth Patrick Science Education Center (RPSEC), 866
RVAC
  Rock Valley Astronomical Society, 919
RVAS
  Roanoke Valley Astronomical Society, 902
Rye, 809, 825
Ryerson Astronomical Society (RAS), 705

S

S*T*A*R Astronomy Society Inc., 797
SAA
  Space Association of Australia, 27
SAAA
  San Antonio Astronomical Association, 883
  Shoreline Amateur Astronomical Association, 767
SAAAA
  San Angelo Amateur Astronomy Association, 883
SAAD
  Sociedade de Astronomia e Astrofísica de Diadema, 71
SAAF
  Svensk Amatör Astronomisk Förening, 494
SAAO
  South African Astronomical Observatory
    Cape Town, 465, 467
    Sutherland Observing Station, 467
Saarbrücken, 208
Saarland, 257
SAAS
  Solent Amateur Astronomers Astronomical Society, 569
SAASIN
  Sociedad Astronómica Amateur de Sinaloa, 384
SAASTEC
  South African Association of Science and Technology Centres, 467
SAB
  Sociedad de Astronomía Balear, 483

Sociedade Astronômica Brasileira, 70
Société Astronomique de Bordeaux, 199
Société Astronomique de Bourgogne, 199
Sabadell, 472
Sabino Canyon Observatory (SCO), 601
SABS
  South African Bureau of Standards, 467
SAC
  Shannonside Astronomy Club, 298
SACA
  Société d'Astronomie de Cannes, 201
Sachtleben Observatory, 782, 784
Sackler Institute of Astronomy (Raymond and Beverly_), 300
Sackville, 92
Saclay, 174, 186
Sacramento, 615, 618, 636
Sacramento Peak, 801
Sacramento Valley Astronomical Society (SVAS), 636
SADAS
  Stevenage & District Astronomical Society, 554
SADEYA
  Sociedad Astronómica de España y America, 482
SAF
  Société Astronomique de France, 199
    Groupe d'Alsace, 200
  Stavanger Astronomiske Forening, 418
Safelight, 566
SAFGA
  Société Astronomique de France - Groupe d'Alsace, 200
SAG
  Schweizerische Astronomische Gesellschaft, 508
  Scottish Astronomers Group, 567
  Sociedad Astronómica Granadina, 483
SAGAS
  Southern Area Group of Astronomical Societies, 570
Saguaro Astronomy Club (SAC), 601
SAHL
  Société d'Astronomie du Haut-Léman, 509
SAHM
  Société Astronomique de Haute-Marne, 200
SAHR
  Société Astronomique du Haut-Rhin, 200
SAI
  Standards Australia International, 28
  Sternberg State Astronomical Institute, 441, 450
Saint Albans, 912
Saint Andrews, 580, 581
Saint Ann, 777
Saint-Apollinaire, 156
Saint-Aubin, 202
Saint Augustine, 521, 678
Saint-Brieuc, 169
Saint-Caprais, 154
Saint-Chamond, 172
Saint Charles Parish Library
  Planetarium, 730
Saint-Claude, 170
Saint Cloud, 772
Saint Cloud State University
  Department of Physics, Astronomy and Engineering Science, 772
Saint Croix, 801
Saint-Cyr, 158
Saint Davids, 855
Saint-Denis-la-Plaine, 159
Sainte Baume (La_), 162
Sainte-Feyre, 194
Sainte-Foy, 79, 101
Saint-Elzéar-de-Beauce, 95
Saint-Étienne, 161, 197
Saint-Étienne-des-Orgues, 200
Saint-Étienne du Bois, 182
Saint-Étienne-en-Devoluy, 182
Saint-Exupéry Planetarium, 416
Saint-Genis-Laval, 165, 200
Saint-Georges-le-Gaultier, 188

Saint-Héand, 155
Saint-Hubert, 82
Saint-Hyppolyte, 85
Saint-Imier, 506
Saint-James, 45
Saint-Jean-le-Blanc, 190
Saint John, 94
Saint John's, 92, 95
Saint Johnsbury, 893
Saint-Lambert, 85
Saint-Laurent-du-Var, 160
Saint Lawrence University
    Department of Physics, 821
Saint Louis, 777–779
Saint Louis Astronomical Society (SLAS), 777
Saint Louis Science Center (SLSC), 778
Saint-Luc, 504
Saint Lucia, 30
Saint-Lunaire, 159
Saint-Martin d'Hères, 182, 203
Saint Mary's University (SMU)
    Department of Astronomy and Physics
        Rev. M.W. Burke-Gaffney Observatory (BGO), 97
Saint-Maur-des-Fossés, 166
Saint-Maurice, 504
Saint-Michael, 45
Saint-Michel-l'Observatoire, 164, 191, 507
Saint-Nérée-de-Bellechasse, 78
Saint Patrick's College
    Experimental Physics Department, 297
Saint Paul, 770–772
Saint Peter Port, 106
Saint Petersburg, 441, 444, 445, 447, 449, 451, 684
Saint Petersburg Astronomy Club (SPAC), Inc., 683
Saint Petersburg Junior College (SPJC)
    Planetarium, 684
Saint Petersburg University
    Astronomical Institute, 449
    Department of Applied Mathematics and Control Processes, 449
Saint Pierre du Perray, 198
Saint-Quentin, 162
Saint Thomas, 924
Saint-Valérien de Milton, 99
Saint-Vallier-de-Thiey, 192
Saint-Véran, 163
Saint Xavier's College
    Physics Department, 281
SAIP
    South African Institute of Physics, 467
SAIt
    Società Astronomica Italiana, 335
Saitama, 348, 361
SAJ
    Station Astronomique Jansky, 202
Saji Astronomical Observatory, 356
Sakashita, 351
Sakushima, 351
SAL
    Société Astronomique de Liège, 56
    Société Astronomique de Lyon, 200
Salem, 757, 790
Salerno, 312
Salford, 567
Salford Astronomical Society (SAS), 567
Salgótarján, 273, 277
Salina, 723
Salina Astronomy Club, 723
Salisbury, 567, 827, 829
Salisbury Astronomical Society (SAS), 567
Salons Internationaux de l'Aéronautique et de l'Espace, 198

SALT
    Southern African Large Telescope, 467
Salt Lake Astronomical Society (SLAS), 891
Salt Lake City, 890, 891
Salto, 587

Saltsjöbaden, 494
Saludecio, 322
Salve, 310
Salzburg, 34, 38
Salzburger Sterngucker, 38
Salzkammergut, 34
Salzschlirf (Bad_), 230
SAM
    Sociedad Astronómica de México, 384
    Société Astronomique de Montpellier, 200
    Société d'Astronomie de Montréal, 98
SAMA
    Sociedade Astronômica Maranhense de Amadores, 71
Sam Houston State University (SHSU)
    Department of Physics, 882
Sampsor, 476
Samson, 171
SAN
    Société d'Astronomie de Nantes, 201
Sanae, 466
San Angelo, 877, 883
San Angelo Amateur Astronomy Association (SAAAA), 883

San Antonio, 883, 884
San Antonio Astronomical Association (SAAA), 883
San Antonio Space Society (SASS), 883
San Benedetto del Tronto, 322
San Benedetto Po, 305
San Bernardino, 622, 636
San Bernardino County Museum, 636
San Bernardino Valley Amateur Astronomers (SBVAA), 636

San Bernardino Valley College
    George F. Beattie College, 622
    N.A. Richardson Astronomical Observatory, 636
San Clemente, 635, 640
Sandburg Middle School Planetarium (Carl_), 897, 903
San Diego, 634, 636, 637, 641, 643, 645, 646
San Diego Aerospace Museum, 636
San Diego Astronomy Association (SDAA), 636
San Diego State University (SDSU)
    Astronomy Department, 637
Sandusky Valley Amateur Astronomy Club (SVAAC), 843
Sandvreten, 498
San Fernando, 482
San Fernando Observatory (SFO), 615, 637
Sanford, 684
Sanford Museum and Planetarium (Tiel_), 718, 719
San Francisco, 608–611, 617, 620, 628, 633, 637, 639
San Francisco Amateur Astronomers (SFAA), 637
San Francisco Sidewalk Astronomers, 637
San Francisco State University (SFSU)
    Department of Physics and Astronomy, 637
        Planetarium Institute, 633
Sangamon Astronomical Society, 705
Sangamon State University, 705
San Gersolè Planetary Group, 334
San Giovanni in Persiceto, 322
San Giuseppe Vesuviano, 323
Sanglok (Mount_), 518
Sanglok Observatory (Mount_), 518
San Joaquin Delta College, 622
San Jose, 622, 623, 627, 637, 644
San Jose Astronomical Association (SJAA), 637
San Jose State University (SJSU)
    Department of Physics, 637
San Juan, 7, 11
San Juan Bautista, 621
San Juan Capistrano, 633
Sankt Gallen, 502
Sankt Georgen, 39
Sankt Pölten, 41
San Luis Obispo, 616
San Marcello Pistoiese, 334
San Marcello Pistoiese Astronomical Observatory, 334
San Marcos, 632, 880
San Martino (Monte_), 324

San Mateo, 638
San Mateo County Astronomical Society, 638
San Miguel, 9
San Pedro Mártir, 386
San Polo a Mosciano, 305
San Rafael, 8, 629
San Salvador, 143
San Sebastian, 483
San Stefano a Macerata, 318
Santa Ana, 63, 634, 644
Santa Barbara, 610, 619, 638, 649, 651
Santa Barbara Instrument Group (SBIG), 638
Santa Barbara Museum of Natural History, 610, 638
Santa Clara, 622, 642
Santa Cruz, 62, 609, 632, 638, 649
Santa Cruz Astronomy Club (SCAC), 638
Santa Cruz de la Palma, 471, 477, 479, 480, 482
Santa Cruz de Tenerife, 473, 483
Santa Fe, 803
Santa Fe Community College (SFCC)
    Planetarium, 803
Santafé de Bogotá, 119
Santa Fe Springs, 627
Santa Lucia di Stroncone, 332
Santa María de Trassierra, 471
Santa María Tonantzintla, 383
Santa Maria a Monte, 321
Santa Maria Maddalena, 319
Santa Monica, 638
Santa Monica College (SMC)
    John Drescher Planetarium, 638
Santander, 470, 478
Santa Pola, 472
Santa Rosa, 631, 638, 640, 643
Santa Rosa Junior College (SRJC)
    Planetarium, 638
Sant Cugat del Vallès, 471
Santiago de Cali, 119
Santiago de Chile, 107–111
Santiago de Compostela, 485
SANV
    Société d'Astronomie du Nord Vaudois, 510
SANW
    Schweizerische Akademie der Naturwissenschaften, 508
San Ysidro, 646
SAO
    Smithsonian Astrophysical Observatory, 752
        Harvard-Smithsonian Center for Astrophysics, 751
        Smithsonian Submillimeter Array, 695
    Sociedad Astronómica Octante, 587
    Special Astrophysical Observatory, 448
São Carlos, 65
São João Nepomuceno, 69
São José dos Campos, 67, 70
São Luís, 71
Saône et Loire, 201
São Paulo, 66, 68, 70–72
SAP
    Société d'Astronomie Populaire, 201
SAPC
    Société Astronomique Populaire du Centre, 201
Sapello, 803
SARA
    Society of Amateur Radio Astronomers, 822
    Southeastern Association for Research in Astronomy, 690
Sarah B. Summerhays Planetarium, 890
Saran, 161
Sarasota, 683
Saratoga, 628, 633, 641
SARC
    Space and Astronomy Research Center (Baghdad), 294
Sardegna, 313
Sardinia, 313
Sargent Space Theater, 718
SARM
    Department for Standardization, Metrology and Certification (Armenia), 12
Saronno, 320
Sart Tilman, 47, 57, 58
SAS
    Salford Astronomical Society, 567
    Salisbury Astronomical Society, 567
    Seattle Astronomical Society, 909
    Sheboygan Astronomical Society, 919
    Sheffield Astronomical Society, 567
    Shenandoah Astronomical Society, 903
    Shropshire Astronomical Society, 568
    Skylight Astronomical Society, 757
    Slovak Astronomical Society, 460
    Société Astronomique de Suisse, 508
    Society for Amateur Scientists, 865
    Society for Applied Spectroscopy, 742
    Southampton Astronomical Society, 569
    Southern Astronomical Society, 27
    Southland Astronomical Society, 411
    Spokane Astronomical Society, 909
    Springfield Astronomical Society, 778
    Stockton Astronomy Society, 643
    Svenska Astronomiska Sällskapet, 494
    Swiss Academy of Sciences, 508
    Syracuse Astronomical Society, 823
SASI
    Sutherland Astronomical Society Inc., 28
Saskatchewan, 103
Saskatoon, 99, 103, 105
SASL
    Société d'Astronomie de Saône et Loire, 201
SASMO
    Syrian Arab Organization for Standardization and Metrology, 515
SASO
    Saudi Arabian Standards Organization, 453
SASS
    San Antonio Space Society, 883
SAT, 199
Satellitbild, 493, 497
Saturnus
    (Stichting Volkssterrenwacht_), 403
Saturnus Astronomical Association of Kuopio, 149
SAU
    Società Astronomica Urania, 335
Saudi Arabia, 453
Saudi Arabian Standards Organization (SASO), 453
Sauerbrunn (Bad_), 35
Saunders Planetarium, 684
Sauverny, 507, 511
SAV
    Società Astronomica Versiliese, 335
Savannah, 690
SAVAR
    Société d'Astronomie du Valais Romand, 510
Savona, 323
SAWB
    South African Weather Bureau, 468
SAX Scientific Data Center (SDC), 312, 334
Sayan Mountains Radiophysical Observatory, 448
Sayan Solar Observatory, 448
SAZ
    Standards Association of Zimbabwe, 929
SAZU
    Slovenian Academy of Sciences and Arts, 462
SBAA
    Sociedade Brasileira dos Amigos da Astronomia, 71
SBAC
    Society of British Aerospace Companies, 569
SBAS
    Shreveport-Bossier Astronomical Society, 730
    South Bay Astronomical Society, 640
SBEA
    Sociedade Brasileira para o Ensino da Astronomia, 71
SBIG
    Santa Barbara Instrument Group, 638
SBO

Sommers-Bausch Observatory, 659, 660
SBVAA
    San Bernardino Valley Amateur Astronomers, 636
SCAC
    Santa Cruz Astronomy Club, 638
Scandinavian Twin Auroral Radar Experiment (STARE), 236
Scarborough, 567
Scarborough and District Astronomical Society, 567
Scarisbrick, 570
SCAS
    Sonoma County Astronomical Society, 640
    Southern Cross Astronomical Society, 685
    Steele County Astronomical Society, 772
SCC
    Seminole Community College
        Planetarium, 684
    Standards Council of Canada, 99
SCCA
    Statistical Consulting Center for Astronomy (PSU), 860
Schaefer Planetarium (Donald A._), 716, 719
Schauinsland, 235
Schauinslandsternwarte, 247
Schaumburg, 704
Schenectady, 814, 821, 824
Schenectady Museum and Planetarium, 821
Schiele Museum of Natural History, 831
Schio, 320
Schkeuditz, 216
Schneeberg, 250
Schongau, 210
Schott Glas
    Optics Division, 244
Schouweiler Planetarium (E.C._), 711, 714
Schreder Planetarium, 638
Schuele Planetarium, 843
Schull, 298
Schull Planetarium, 298
Schulplanetarium Chemnitz, 244
Schulsternwarte Bartholomäus Scultetus, 264
Schulsternwarte Juri Gagarin, 264
Schulsternwarte und Planetarium Rodewisch, 244
Schul- und Volkssternwarte Bülach, 501, 508
Schul- und Volkssternwarte K.E. Ziolkowski, 244
Schul- und Volkssternwarte Leinfelden-Echterdingen eV, 244

Schuylkill Valley Astronomy Club, 860
Schwäbische Sternwarte (SSW) eV, 245
Schwarzschild-Observatorium (KSO) (Karl-_), 245, 252
Schwaz, 38
Schweizerische Akademie der Naturwissenschaften (SANW), 508
Schweizerische Astronomische Gesellschaft (SAG), 508
Schweizerische Gesellschaft für Astrophysik und Astronomie (SGAA), 509, 511
Schweizerische Normen-Vereinigung (SNV), 509
Schweizerische Physikalische Gesellschaft (SPG), 509
Schweizerischer Nationalfonds zur Förderung der Wissenschaftlichen Forschung, 509
SCI
    Space Craft International, 641
Science Abstracts, 550
Science and Art Products, 639
Science and Technology Interactive Center (SciTech), 705
Science Center of Connecticut, 662
Science Center of Iowa, 718
Science Center of Pinellas County Inc., 684
Science Council of Japan, 356
Science History Publications Ltd., 567
Science Museum, 567
Science Museum of Minnesota (SMM), 772
Science Museum of Virginia (SMV)
    Ethyl Universe Planetarium, 898
Science Museum of Western Virginia (SMWV)
    Hopkins Planetarium, 903
Science North Solar Observatory, 98
Sciencenter, 821

Science Place, 883
Science Service Inc., 675
Sciencetech Inc., 98
Science World British Columbia, 98
Scientific and Research Centre on Space Hydrometeorology "Planeta", 449
Scientific and Technical Information Network (STN) International, 245, 843
    Columbus, 843
    Germany, 245
    Karlsruhe, 245
    Ohio, 843
    USA-OH, 843
Scientific and Technical Research Council of Turkey (TÜBITAK), 523
    Marmara Research Center
        Space Sciences Department, 523
Scientific Committee on Solar-Terrestrial Physics (SCOSTEP), 658
Scientific Educational Society of Nógrádmegye, 277
Scientific Expeditions Inc., 684
Scientific Research Council
    Space and Astronomy Research Center (SARC), 294
SciTech
    Science and Technology Interactive Center, 705
SciTech Astronomical Research, 639
Scitech Discovery Centre, 26
Scituate
    (North_), 864
SciWorks Planetarium, 831
SCMT
    Southern Columbia Millimeter Telescope, 109
SCO
    Sabino Canyon Observatory, 601
    Slovak Central Observatory, 460
Scobee Planetarium, 883
Scope City, Inc., 639
ScopeTronix, 684
SCOSTEP
    Scientific Committee on Solar-Terrestrial Physics, 658
Scot, 199
Scotia, 805
Scotland, 531–535, 544, 545, 547, 552, 557, 562, 566, 567, 576, 581
Scottish Astronomers Group (SAG), 567
Scotts Valley, 650
Scripps Institution of Oceanography
    California Space Institute, 614
SCS
    Syrian Cosmological Society, 515
SCSN
    Space Centre for Satellite Navigation (QUT), 27
SCSU
    South Carolina State University
        Stanback Planetarium, 868
    Southern Connecticut State University
        Planetarium, 664
Scuola Internazionale Superiore di Studi Avanzati (SISSA), 334
Scuola Normale Superiore
    Astrophysics Group, 334
SDAA
    San Diego Astronomy Association, 636
SDAC
    Solar Data Analysis Center, 741
SDAS
    South Downs Astronomical Society, 569
    Stafford and District Astronomical Society, 571
SDC
    Scientific Data Center, 312
SDR&I
    Space Data Resources & Information, 676
SDSS
    Sloan Digital Sky Survey, 707
SDSU
    San Diego State University
        Astronomy Department, 637

SEA
   Sociedad Española de Astronomía, 484
SEAC
   Société Européenne pour l'Astronomie dans la Culture, 431
Seagoat Consulting, LLC, 602
Seagrave Memorial Observatory, 865
SEAOP
   Sociedade de Estudos Astronômicos de Ouro Preto, 71
Search for Extraterrestrial Intelligence
   SETI League Inc., 796
Search for Extraterrestrial Intelligence (SETI) Institute, 639

Seattle, 907–910
Seattle Astronomical Society (SAS), 909
Sebastopol, 618
SEC
   Space Environment Center, 657
Secção Portuguesa das Uniões Internacionais Astronômica e Geodesica e Geofisica (SPUIAGG), 435
SEDA
   Sociedad Einstein de Astronomía, 483
Sedalia, 659
Sedgefield, 467
Sedgefield Observatory, 467
SEDO
   Sociedad Española de Optica, 484
SEDS
   Students for the Exploration and Development of Space, 758
SEE
   Service de l'Énergie de l'État (Luxembourg)
   Département de Normalisation, 376
Seebad Heringsdorf, 261
Seelze, 226
Seething Astronomical Observatory, 560
Seewalchen-am-Attersee, 34
SEIAC
   Southeastern Iowa Astronomy Club Inc., 719
Seibersdorf, 39
Seiler Instrument & Manufacturing Company Inc.
   Planetarium Division, 778
Seinäjoen Ursa Astronomical Association, 149
Seinäjoki, 149
Sejong University
   Daeyang Observatory, 369
SEKAS
   South East Kent Astronomical Society, 569
Selah, 911
Selatan, 290
Sells, 598
Seltjarnarnes, 279
Selvino, 313
Seminole Community College (SCC)
   Planetarium, 684
SENAMHI
   Servicio Nacional de Meteorología e Hidrología (Bolivia), 63
   Servicio Nacional de Meteorología e Hidrología (Peru), 423
Sencelles, 476
Sendai, 357, 358
Sendai Children's Space Center, 357
Seneca College
   Roberta Bondar Planetarium, 98
Senftenberg, 242
Senigalliesi, 310
Seoul, 368, 369
Seoul National University (SNU)
   Department of Astronomy, 368, 369
SEPA
   Southeastern Planetarium Association, 685
Sept-Îles, 84
SEQAS
   South East Queensland Astronomical Society, 27
Serbia, 454, 455
Serbia-Montenegro, 454

Serin Observatory, 796
Serra la Nave, 326
Service de la Météorologie et de l'Hydrologie (GD Luxembourg), 376
Service de l'Énergie de l'État (SEE)
   Département de Normalisation, 376
Service Hydrographique et Océanographique de la Marine (SHOM), 199
Servicio de Meteorología e Hidrología Nacional (SMHN), 143
Servicio de Oceanografía, Hidrografía y Meteorología de la Armada (SOHMA), 587
Servicio Meteorológico Nacional (SMN), 10, 272, 384
Servicio Nacional de Meteorología e Hidrología (SENAMHI), 63, 423
Sessa Aurunca, 304
Setenil de las Bodegas, 473
SETI
   Search for Extraterrestrial Intelligence, 639
SETI Institute, 639
SETI League Inc., 796
Settimo di Pescantina, 330
Settlemyre Planetarium, 868
Seven-0-Five Astronomers (SOFA), 639
Seven Hills Observatory, 784
Sèvres, 163
Seymour Planetarium, 757, 758
SF2A
   Société Française d'Astronomie et d'Astrophysique, 202
SFAA
   San Francisco Amateur Astronomers, 637
SFAAA
   South Florida Amateur Astronomers Association, 685
SFASU
   Stephen F. Austin State University
   Department of Physics and Astronomy, 884
SFB
   Sternfreunde Breisgau, 247
SFCC
   Santa Fe Community College
   Planetarium, 803
SFF
   Sternfreunde Franken, 247
SFO
   San Fernando Observatory, 615
   Société Française d'Optique, 202
SFP
   Société Française de Physique, 201, 223
SFS
   Finnish Standards Association, 147
SFSM
   South Florida Science Museum, 681, 685
SFSU
   San Francisco State University
   Department of Physics and Astronomy, 633, 637
   Planetarium Institute, 633
SGAA
   Schweizerische Gesellschaft für Astrophysik und Astronomie, 511
SGEPSS
   Society of Geomagnetism and Earth, Planetary and Space Sciences, 357
SGO
   Satellite Geodetic Observatory, 277
   Sodankylä Geophysical Observatory, 146
   Sugar Grove Observatory, 707
SHA
   Society for the History of Astronomy, 568
Shaanxi, 114
Shaanxi Astronomical Observatory, 114, 117
Shah Alam, 379
Shahe Station, 113
Shaker Heights, 838
Shamshuipo, 118
Shanghai, 115, 117
Shanghai Astronomical Observatory (SHAO), 114, 117
Shanghai Institute of Electron Physics, 117

Shanghai Jiao Tong University (SJTU)
    Department of Applied Physics, 117
    Institute for Space Astrophysics, 117
Shanghai Teachers University
    Department of Physics, 117
Shanghai University
    Shanghai Institute of Electron Physics, 117
Shannonside Astronomy Club (SAC), 298
SHAO
    Shanghai Astronomical Observatory, 114
Sharpe Planetarium (Linda G._), 873, 874
Shattuck Observatory, 788
Shawnee, 846
Sheboygan Astronomical Society (SAS), 919
Sheboygan Falls, 919
Sheffield, 552, 568, 581
Sheffield Astronomical Society (SAS), 567
Sheikh Tahir Astronomical Centre, 378, 379
Shemakha, 43
Shemakha Astrophysical Observatory, 43
Shenandoah Astronomical Society (SAS), 903
Shepherd College
    Institute for Environmental Studies, 913
Shepherdstown, 913
Sherbrooke, 86
Sherman Park, 718
's-Hertogenbosch, 393
Sherwood Observatory, 556
Sherzer Observatory, 762, 763
Sheshan Station, 115
Shibayama Scientific Co. Ltd., 357
Shields
    (South_), 570
Shiga University
    Department of Earth Science, 357
Shimosato Hydrographic Observatory (SHO), 347, 357
Shippagan, 84
Shirahama Hydrographic Observatory, 347, 357
Shiras Planetarium, 765, 767
Shiraz, 292
Shiraz University
    Physics Department
        Biruni Observatory, 292
Shirley, 569
SHO
    Shimosato Hydrographic Observatory, 347
SHOM
    Service Hydrographique et Océanographique de la Marine, 199
Shoreline Amateur Astronomical Association (SAAA), 767
Shoumen, 76
Shoumen University
    Department of Physics
        Astronomical Group, 76
Shreveport, 730
Shreveport-Bossier Astronomical Society (SBAS) Inc., 730
Shreveport Parks and Recreation Planetarium, 730
Shrewsbury, 568
Shropshire Astronomical Society (SAS), 568
SHSU
    Sam Houston State University
        Department of Physics, 883
Shumen, 76
SIAM
    Society for Industrial and Applied Mathematics, 861
Siberia, 448
Siberian Solar Radio Telescope (SSRT), 448
Sidak, 525
SIDC
    Solar Influences Data analysis Center, 57
Sidewalk Astronomers, 639
Siding Spring, 18
Siding Spring Observatory, 27
Sidmouth, 559
Siebert Optics, 831
Siena, 337
Sierra Nevada, 476

Sierra Nevada College
    Department of Astronomy, 787
Sierra Vista, 597
Sierre, 504
SIF
    Società Italiana di Fisica, 335
SiFEZ
    Sternfreunde im FEZ, 247
Sigma Octante (Astronomía_), Centro de Investigación y Estudio en Astronomía, 62
Sigma Xi - The Scientific Research Society, 831
Signal Analytics Corp., 903
SIGRAV
    Società Italiana di Relatività Generale e Fisica della Gravitazione, 336
Sigulda, 371
SII
    Standards Institution of Israel, 300
Siloam Springs, 606
Silverman Planetarium, 821
Silver Spring, 734, 742
Simeiz, 525
Simferopol, 525, 528
Similkameen, 94
Simi Valley, 632, 639, 650
Simon Stevin Public Astronomical Observatory, 401
Simpson Observatory (John W.H._), 791
SINAS
    Sir Isaac Newton Astronomical Society, 861
Singapore, 456, 457
Singapore Institute of Standards and Industrial Research (SISIR), 456
Singapore Productivity and Standards Board, 456
Singapore Science Centre, 456
    Omnitheatre and Observatory, 457
Singapour, 456, 457
Singen, 210
Sion, 500, 510, 511
Sioux Empire Astronomy Club, 869
Sioux Falls, 869
Sira Ltd.
    Electro-Optics Division, 568
SIRIM
    Standards and Industrial Research Institute of Malaysia, 379
Sir Isaac Newton Astronomical Society (SINAS), 861
Sirius Astronomical Society of Kermanshah, 292
Sirius Instruments, 706
Sir Thomas Brisbane Planetarium (STBP), 27
SIS
    Standardiseringen i Sverige, 497
SISIR
    Singapore Institute of Standards and Industrial Research, 456
SISSA
    Scuola Internazionale Superiore di Studi Avanzati, 334
Sistrans, 36
Sittingbourne, 556, 572
SIU
    Southern Illinois University
        Department of Physics, 706
    Sterrekundig Instituut Utrecht, 405
Sivry, 49
Sizuoka, 347
SJA
    Société Jurassienne d'Astronomie, 510
SJAA
    San Jose Astronomical Association, 637
SJAC
    South Jersey Astronomical Club, 797
SJSP
    Star Jersey Star Party, 797
SJSU
    San Jose State University
        Department of Physics, 637
SJTU
    Shanghai Jiao Tong University

Department of Applied Physics, 117
Institute for Space Astrophysics, 117
Skalnaté Pleso, 460
Skalnaté Pleso Observatory, 459
Skeptic (The_), 568, 572
Skibotn, 418
Skibotn Astronomical Observatory, 418, 420
Skokie, 704
Skokie Valley Astronomers (SVA), 706
Skopje, 377
SKVFZ
    Sport- und Kulturvereins Forschungszentrum Seibersdorf
        Sektion Amateurastronomie, 39
Sky & Space Publishing, 27
Sky & Telescope, 757
SkyComm Engineering, 685
Sky Laboratories International Ltd., 411
Skylight Astronomical Society (SAS) Inc., 757
Skymap Software, 568
Sky Observers' Association (SOA), 118
Sky Optics, 98
Sky Planner Enterprises, 891
Sky Publishing Corp., 757
Skyscrapers, Inc., 864
Sky Shack, 645
Sky-Skan Inc., 790
Skywatchers Astronomy Club, 903
Skywatchers Inn, 602
SLA
    Société Lorraine d'Astronomie, 202
SLAS
    Saint Louis Astronomical Society, 777
    Salt Lake Astronomical Society, 891
Slavonski Brod, 121
Sleepy Hollow, 701
Slidell, 774
Sliven, 74
SLO
    Society of Low-energy Observers, 874
Sloan Digital Sky Survey (SDSS), 707
Slottsskogen Observatory, 493
Slovak Academy of Sciences
    Astronomical Institute, 459
    Geophysical Institute, 459
    Institute of Physics, 460
Slovak Astronomical Society (SAS), 460
Slovak Central Observatory (SCO), 460
Slovak Centre of Amateur Astronomy, 460
Slovak Hydrometeorological Institute, 460
Slovakia, 458–461
Slovak Office of Standards, Metrology and Testing (UNMS), 460
Slovak Physical Society, 460
Slovak Republic, 458
Slovak Union of Amateur Astronomers, 461
Slovak University of Technology
    Observatory, 461
Slovenia, 462
Slovenian Academy of Sciences and Arts (SAZU), 462
SLSC
    Saint Louis Science Center, 778
SLSI
    Sri Lanka Standards Institution, 487
SMA
    Smithsonian Submillimeter Array, 695
    Sociedad Malagueña de Astronomía, 484
SMAS
    Smoky Mountain Astronomical Society, 874
    Southern Maryland Astronomical Society, 742
    Southwest Montana Astronomical Society, 780
SMC
    Santa Monica College
        John Drescher Planetarium, 638
Smethwick, 556
SMHI
    Swedish Meteorological and Hydrological Institute, 496

SMHN
    Servicio de Meteorología e Hidrología Nacional, 143
SMIS
    Standards and Metrology Institute of Slovenia, 462
Smith College
    Clark Science Center, 758
    Five College Astronomy Department, 757
Smith Planetarium (Mark_), 689, 690
Smith Planetarium (Willard W._), 908, 909
Smithsonian Astrophysical Observatory (SAO), 752, 758
    Harvard-Smithsonian Center for Astrophysics (CfA), 751
    Smithsonian Submillimeter Array (SMA), 695
Smithsonian Institution
    Center for Astrophysics (CfA), 751
    Division of Meteorites, 675
    Fred Lawrence Whipple Observatory, 597
    Global Volcanism Program (GVP), 675
    Multiple Mirror Telescope Observatory (MMTO), 598, 602
    National Air and Space Museum (NASM), 673, 675
        Albert Einstein Planetarium, 676
        Center for Earth and Planetary Studies (CEPS), 676
        Space History Division, 676
    National Museum of Natural History, 675
Smithsonian Submillimeter Array (SMA), 695
SMN
    Servicio Meteorológico Nacional, 10, 272, 384
Smoky Mountain Astronomical Society (SMAS), 874
Smolyan, 75, 76
Smolyan Planetarium, 76
SMSU
    Southwest Missouri State University
        Department of Physics, Astronomy, and Materials Science, 778
SMTO
    Submillimeter Telescope Observatory, 603
SMU, Saint Mary's University
    Department of Astronomy and Physics
        Rev. M.W. Burke-Gaffney Observatory, 97
SMV
    Science Museum of Virginia
        Ethyl Universe Planetarium, 898
SMWV
    Science Museum of Western Virginia
        Hopkins Planetarium, 903
Smyrna, 667
SNA
    Société Neuchâteloise d'Astronomie, 510
SNO
    Sudbury Neutrino Observatory, 100
SNOI
    Sudbury Neutrino Observatory Institute, 100
SNSB
    Swedish National Space Board, 496
SNU
    Seoul National University
        Department of Astronomy, 368, 369
SNV
    Schweizerische Normen-Vereinigung, 509
SNZ
    Standards New Zealand, 411
SOA
    Sky Observers' Association, 118
SOAO
    Sobaeksan Optical Astronomy Observatory, 367
SOAR
    Southern Observatory for Astronomical Research, 832
Sobaeksan Optical Astronomy Observatory (SOAO), 367, 369
Sociedad Astronómica Amateur de Sinaloa (SAASIN), 384
Sociedad Astronómica de Aragón Ilhuícatl, 384
Sociedad Astronómica de España y América (SADEYA), 482

Sociedad Astronómica del Paraguay, 422
Sociedad Astronómica de México (SAM) AC, 384
Sociedad Astronómica Granadina (SAG), 483

Sociedad Astronómica Octante (SAO), 587
Sociedad Astronómica SYRMA, 483
Sociedad de Astronomía Balear (SAB), 483
Sociedad de Astronomia de Puerto Rico, 437
Sociedad de Ciencias Aranzadi
    Sección Astronomía, 483
Sociedad de Estudios Astronómicos, 422
Sociedad de Observadores de Meteoros y Cometas de España
    (SOMYCE), 483
Sociedade Astronômica Brasileira (SAB), 70
Sociedade Astronômica Maranhense de Amadores (SAMA), 71
Sociedade Brasileira dos Amigos da Astronomia (SBAA), 71

Sociedade Brasileira para o Ensino da Astronomia (SBEA), 71
Sociedade de Astronomia e Astrofísica de Diadema (SAAD), 71
Sociedade de Estudos Astronômicos de Ouro Preto (SEAOP), 71
Sociedad Einstein de Astronomía (SEDA), 483
Sociedade Portuguesa de Física (SPF), 435
Sociedad Española de Astronomía (SEA), 484
Sociedad Española de Física (Real.), 482
Sociedad Española de Optica (SEDO), 484
Sociedad Malagueña de Astronomía (SMA), 484
Sociedad Meteorológica de Cuba, 123
Sociedad Universitaria de Astronomía (SUNA), 928
Sociedad Uruguaya de Astronomía (SUA), 587
Società Astronomica "G.V. Schiaparelli", 335
Società Astronomica Italiana (SAIt), 335
Società Astronomica Ticinese, 509
Società Astronomica Urania (SAU), 335
Società Astronomica Versiliese (SAV), 335
Società Italiana di Fisica (SIF), 335
Società Italiana di Relatività Generale e Fisica della Gravitazione (SIGRAV), 336
Societatea Română de Fizicá, 439
Société Astronomique de Bordeaux (SAB), 199
Société Astronomique de Bourgogne (SAB), 199
Société Astronomique de France (SAF), 199
    Groupe d'Alsace (SAFGA), 200
Société Astronomique de Genève, 509
Société Astronomique de Haute-Marne (SAHM), 200
Société Astronomique de la Montagne de Lure, 200
Société Astronomique de Liège (SAL), 56
Société Astronomique de Lyon (SAL), 200
Société Astronomique de Montpellier (SAM) "Pierre Vauriot", 200
Société Astronomique de Suisse (SAS), 508, 509
Société Astronomique du Haut-Rhin (SAHR), 200
Société Astronomique du Havre, 200
Société Astronomique Populaire du Centre (SAPC), 201
Société Canadienne d'Astronomie (CASCA), 80, 98
Société d'Astronomie de Cannes (SACA), 201
Société d'Astronomie de Montréal (SAM), 98
Société d'Astronomie de Nantes (SAN), 201
Société d'Astronomie de Saint-Imier, 506, 509
Société d'Astronomie de Saône et Loire (SASL), 201
Société d'Astronomie du Haut-Léman (SAHL), 509
Société d'Astronomie du Nord Vaudois (SANV), 510
Société d'Astronomie du Valais Romand (SAVAR), 510
Société d'Astronomie Populaire (SAP), 201
Société de Physique et d'Histoire Naturelle (SPHN), 510
Société du Télescope Canada-France-Hawaii, 692
Société Européenne pour l'Astronomie dans la Culture (SEAC), 431
Société Française d'Astronomie et d'Astrophysique (SF2A), 202
Société Française de Physique (SFP), 201, 223
Société Française des Spécialistes d'Astronomie (SFSA), 202

Société Française d'Optique (SFO), 202
Société Jurassienne d'Astronomie (SJA), 510
Société Lorraine d'Astronomie (SLA), 202
Société Neuchâteloise d'Astronomie (SNA), 510
Société Royale Belge d'Astronomie, de Météorologie et de Physique du Globe (SRBA), 56
Société Royale d'Astronomie du Canada (SRAC), 99
Société Royale des Sciences de Liège (SRSL), 56
Société Scientifique de Bruxelles, 57
Société Suisse d'Astrophysique et d'Astronomie (SSAA), 511

Société Suisse de Physique (SSP), 511
Société Valaisanne des Sciences Naturelles "La Murithienne", 511
Société Vaudoise d'Astronomie (SVA), 511
Society for Amateur Scientists (SAS), 865
Society for Applied Spectroscopy (SAS), 742
Society for Industrial and Applied Mathematics (SIAM), 861

Society for Popular Astronomy (SPA), 568
Society for Scientific Exploration (SSE), 903
Society for the History of Astronomy (SHA), 568
Society of Amateur Radio Astronomers (SARA) Inc., 822
Society of Astronomy and Astronautics, 493
Society of British Aerospace Companies (SBAC) Ltd., 569
Society of Geomagnetism and Earth, Planetary and Space Sciences (SGEPSS), 357
Society of Low-energy Observers (SLO), 874
Society of Mathematicians, Physicists and Astronomers of Slovenia, 462
Society of Photo-Optical Instrumentation Engineers (SPIE), 909
Society of Physicists of the Republic of Macedonia, 377
Society of Satellite Professionals International (SSPI), 903
Socorro, 801, 802
Sodankylä, 146, 147, 489
Sodankylä Geophysical Observatory (SGO), 146, 149
Södertälje, 498
SOE
    Sonnenobservatorium Einsteinturm, 218
Soeborg, 136
Soest, 259
SOFA
    Seven-0-Five Astronomers, 639
SOFIA
    Stratospheric Observatory for Infrared Astronomy, 630
Sofia, 74–76
Sofia University
    Faculty of Physics, 76
Software Bisque Inc., 658
Software distributors and producers
    AIT, 212
    Andromeda, 807
    Applied Geomechanics, 609
    Astrotech, 273
    Atari Public Domain Service, 209
    CapellaSoft, 799
    Carina Software, 615
    Chandler (David.), 618
    Cyanogen Productions, 87
    David Chandler, 618
    Deutsch (Verlag Harri.), 259
    Focus Software, 595
    FSI, 595
    Harri Deutsch (Verlag.), 259
    Hopkins Technology, 770
    Intercon Spacetec, 233
    Interferometrics, 899
    MMI, 739
    National Technical Information Service, 901
    Nova Astronomics, 95
    NTIS, 901
    OES, 240
    Optical Research Associates, 631
    Optische und Elektronische Systeme, 240
    ORA, 631
    PAS, 830
    Physics Academic Software, 830
    Project Pluto, 733
    Radio-Sky, 727
    RSAI, 198
    R.S. Automation Industrie, 198

Santa Barbara Instrument Group, 638
SBIG, 638
Skymap Software, 568
Software Bisque, 658
Southern Stars, 641
Spacetec (Intercon.), 233
StarTel, 395
Stefan Thiele, 212
Stellar Software, 643
Thiele (Stefan.), 212
TL Systems, 645
Verlag Harri Deutsch, 259
Software Systems Consulting (SSC), 640
Sohland, 213, 264
SOHMA
    Servicio de Oceanografía, Hidrografía y Meteorología de la Armada (Uruguay), 587
Soissons, 157
Sojn Valley Radio Astronomy Observatory, 770
Solar Data Analysis Center (SDAC), 741
Solar Influences Data analysis Center (SIDC), 57
Solaris (Astroclub.), 500
Solar Terrestrial Dispatch (STD), 99
Solent Amateur Astronomers Astronomical Society (SAAS), 569
Solihull, 548
Solingen, 264
SOLIS
    Sydney Outdoor Lighting Improvement Society, 28
Sollid Optics Inc., 803
Solms/Lahn, 213
Solna, 496, 497
Solothurn, 501
Somerville, 753
Somis, 634
Sommers-Bausch Observatory (SBO), 659, 660
Sommets (Les.), 86
SOMYCE
    Sociedad de Observadores de Meteoros y Cometas de España, 483
SON
    Standards Organization of Nigeria, 414
SONAAS
    Stratford-upon-Avon Astronomical Society, 571
Sondrio, 307
Sonneberg, 250
Sonnenborgh, 402
Sonnenobservatorium Einsteinturm (SOE), 218, 245
Sonnenobservatorium Kanzelhöhe, 38, 40
Sonoma, 619
Sonoma County Astronomical Society (SCAS), 640
Sonoma Instrument Co., 640
Sonoma State University (SSU)
    Department of Physics and Astronomy, 640
Sontra, 258
Sontra-Breitau, 258
Sontra-Ulfen, 258
Sophia-Antipolis, 178, 204
Sophia University
    Department of Physics, 357
Sorbiers, 198
Soresina, 323, 332
SORL
    Space Optics Research Labs, 758
Sossano, 304
Sotteville-lès-Rouen, 181
South Africa, 464
South African Association of Science and Technology Centres (SAASTEC), 467
South African Astronomical Observatory (SAAO), 465, 467
    Sutherland Observing Station, 467
South African Bureau of Standards (SABS), 467
South African Institute of Physics (SAIP), 467
South African Museum
    Planetarium, 468
South African Weather Bureau (SAWB), 468
Southam Observatory (Gordon.), 89

Southampton, 569, 581
Southampton Astronomical Society (SAS), 569
Southampton Planetarium, 742
South Bay Astronomical Society (SBAS), 640
South Bend, 713
South Carelian Amateur Astronomical Association Nova, 146, 149
South Carolina, 866–868
South Carolina State University (SCSU)
    Stanback Planetarium, 868
South Coast Telescopes, 569
South Dakota, 869
South Downs Astronomical Society (SDAS), 569
South Downs Planetarium, 569
Southeastern Association for Research in Astronomy (SARA), 690
Southeastern Iowa Astronomy Club (SEIAC) Inc., 719
Southeastern Planetarium Association (SEPA), 685
South East Kent Astronomical Society (SEKAS), 569
South East Queensland Astronomical Society (SEQAS), 27
Southend-on-Sea, 570
Southend Planetarium, 570
Southern Adventist University
    Department of Physics, 874
Southern Africa, 464
Southern African Large Telescope (SALT), 467
Southern Area Group of Astronomical Societies (SAGAS), 570
Southern Astronomical Society (SAS) Inc., 27
Southern California, 649
Southern Columbia Millimeter Telescope (SCMT), 109
Southern Connecticut State University (SCSU)
    Planetarium, 664
Southern Cross Astronomical Society (SCAS) Inc., 685
Southern Illinois University (SIU)
    Carbondale
        Department of Physics, 706
Southern Maryland Astronomical Society (SMAS), 742
Southern Nevada, 786
Southern New England, 748
Southern Observatory for Astronomical Research (SOAR), 832
Southern Oregon Skywatchers, 849
Southern Stars Systems, 640
South Florida Amateur Astronomers Association (SFAAA), 685
South Florida Science Museum (SFSM), 685
South Jersey Astronomical Club (SJAC), 797
Southland Astronomical Society (SAS), 411
South Mountain, 858
Southold, 813
South Plains Astronomy Club (SPAC), 883
Southport Astronomical Society, 570
Southsea, 583
South Shields, 570
South Shore Astronomical Society, 758
South Texas Astro Retreat, 883
South Tyneside College
    Planetarium and Observatory, 570
Southwest Association of Planetariums (SWAP), 884
Southwestern University
    Department of Physics, 884
Southwest Missouri State University (SMSU)
    Department of Physics, Astronomy, and Materials Science, 778
Southwest Montana Astronomical Society (SMAS), 780
Southwest Research Institute (SwRI)
    Department of Space Studies, 658
    Instrumentation and Space Research Division, 884
Southwest State University (SSU)
    Planetarium, 772
Southwest Texas State University (SWTSU)
    Physics Department, 880
Southwest Washington Astronomical Society (SWWAS) Inc., 909
Southwood, 557
Southworth Planetarium, 733

Sovietski Collection, 641
SPA
    Society for Popular Astronomy, 568
Spa, 52
SPAC
    Saint Petersburg Astronomy Club, 684
    South Plains Astronomy Club, 883
SPACE
    Saint Petersburg Astronomy Club Examiner, 684
    Seminario de Astronomía y Ciencias Espaciales, 423
    Space, Physics and the Advancement of Cosmic Exploration, 570
Space and Astronomy Research Center (SARC), 294
Space Association of Australia (SAA) Inc., 27
Spacebel Instrumentation SA, 57
Space Center, 803
Space Centre for Satellite Navigation (SCSN), 27
Space Craft International (SCI), 641
Space Data Resources & Information (SDR&I), 676
SpaceDev, 641
Space Frontier Foundation, 641
Spaceguard Centre (The_), 570, 573
SpaceLaunch Foundation, 691
Space Optical Communications Research Center, 344
Space Optics Research Labs (SORL), 758
Space, Physics and the Advancement of Cosmic Exploration (SPACE), 570
Space Policy project (SPP), 672
Space Power Institute (SPI), 589
SpaceQuest Planetarium, 714
Space Radiation Laboratory (SRL), 612
Space Research and Remote Sensing Organization (SPARRSO), 44
Space Research Center (Polish Academy of Sciences), 430
Space Research Organization Netherlands (SRON)
    Groningen, 401
    Utrecht, 401
Space Science and Engineering Center (SSEC), 919, 920
Space Science Data Operations Office (SSDOO), 741, 743
Space Science Institute (SSI), 658
Space Sciences Series of ISSI (SSSI), 505
Space Science Studies Division, 378
Space Studies Institute (SSI), 797
Spacetec GmbH (Intercon_), 233, 245
Space Telescope - European Coordinating Facility (ST-ECF), 245
Space Telescope Science Institute (STScI), 245, 671, 743
Space Theater, 517
Space: The Future Frontier (S:TFF), 849
Space Transportation Association (STA), 903
Space Vacuum Epitaxy Center (SVEC), 884, 887
Spaceweek International Association, 884
Space West Science Activities (SWSA) Inc., 99
Spain, 470
SPAR
    Shreveport Parks and Recreation, 730
Sparkassen-Planetarium Augsburg, 245
SPAR Planetarium, 730
SPARRSO
    Space Research and Remote Sensing Organization, 44
SPE
    Stork Product Engineering, 403
Special Astrophysical Observatory (SAO), 448, 449
Specola Astronomica Lodi, 336
Specola Solare Ticinese, 509, 511
Specola Vaticana, 926
    Arizona, 604
    Città del Vaticano, 926
    Tucson, 604
    USA-AZ, 604
    Vatican, 926
    Vatican Observatory Research Group, 604
Spectra Astro Systems, 641
Spectrum Astro Inc., 602
Spectrum Coatings, 685
Spectrum Sciences Inc. (SSI), 642
Speedibrews, 570

Sperry Observatory (William Miller_), 791
SPF
    Sociedade Portuguesa de Física, 435
SPG
    Schweizerische Physikalische Gesellschaft, 511
SPHN
    Société de Physique et d'Histoire Naturelle, 510
SPI
    Space Power Institute, 589
SPIE
    International Society for Optical Engineering, 908
    Society of Photo-Optical Instrumentation Engineers, 908
Spirus Co., 843
Spitz Inc., 861
SPJC
    Saint Petersburg Junior College Planetarium, 684
Spokane, 909
Spokane Astronomical Society (SAS), 909
Sport- und Kulturvereins Forschungszentrum Seibersdorf (SK-VFZ)
    Sektion Amateurastronomie, 39
SPOT
    France, 202
    Reston, 904
    USA-VA, 904
    Virginia, 904
SPOT Image, 202
SPOT Image Corp., 904
Spotsylvania, 904
SPP
    Space Policy Project, 672
Spreeufontein Observatory, 468
Springer-Verlag, 246
    Germany, 246
    Heidelberg, 246
    New York, 822
    USA-NY, 822
Springfield, 705, 708, 758, 778, 894, 895, 901, 906
    (West_), 906
Springfield Astronomical Society (SAS), 778
Springfield Science Museum (SSM)
    Seymour Planetarium, 758
Springfield STARS Club, 758
Springfield Telescope Makers (STM) Inc., 894
Springville, 618, 645
Sproul Observatory, 861, 862
SPS
    Swiss Physical Society, 511
SPSP
    South Pacific Star Party, 15
SPUIAGG
    Secção Portuguesa das Uniões Internacionais Astronômica e Geodesica e Geofisica, 435
SRBA
    Société Royale Belge d'Astronomie, de Météorologie et de Physique du Globe, 56
SRI
    Space Research Institute of the National Academy of Sciences of Kazakhstan, 364
Sri Lanka, 487
Sri Lanka Standards Institution (SLSI), 487
SRJC
    Santa Rosa Junior College Planetarium, 638
SRL
    Space Radiation Laboratory, 612
SRML
    Solar Radiation Monitoring Laboratory, 850
SRO
    Stony Ridge Observatory, 643
Sroborova, 460
SRON, Space Research Organization Netherlands
    Groningen, 401
    Utrecht, 401
SRS

Stanford Research Systems, 642
SRSL
    Société Royale des Sciences de Liège, 56
SRU
    Space Research Unit, 466
SRZM
    Stichting Radiostraling van Zon en Melkweg, 403
SSAA
    Société Suisse d'Astrophysique et d'Astronomie, 511
SSC
    Software Systems Consulting, 640
    Stennis Space Center
        Public Affairs Office, 774
    Swedish Space Corp.
        Earth Observation Division, 419
        Esrange, 496
        Kiruna, 496, 497
        Satellitbild, 497
        Solna, 497
        Tromsø Satellite Station, 419
SSD
    Space Science Department
        (RAL), 542
    Space Sciences Division
        Naval Research Laboratory, 674
SSDOO
    Space Science Data Operations Office, 741
SSE
    Society for Scientific Exploration, 903
SSEC
    Space Science and Engineering Center, 920
SSFS
    Sydney Space Frontier Society, 29
SSI
    Space Science Institute, 658
    Space Studies Institute, 797
    Spectrum Sciences Inc., 642
SSL
    Space Sciences Laboratory (UCB), 647
SSM
    Springfield Science Museum
        Seymour Planetarium, 758
SSP
    Société Suisse de Physique, 511
SSRT
    Siberian Solar Radio Telescope, 448
SSSI
    Space Sciences Series of ISSI, 505
SSTL
    Surrey Satellite Technology Ltd., 571
SSU
    Sonoma State University
        Department of Physics and Astronomy, 640
    Southwest State University
        Planetarium, 772
SSW
    Schwäbische Sternwarte, 245
STA
    Space Transportation Association, 904
Staerkel Planetarium (William M..), 706, 709
Stafford, 879
Stafford and District Astronomical Society (SDAS), 571
St. Albans, 912
Stamford, 663
Stanback Planetarium, 868
Standardiseringen i Sverige (SIS), 493, 497
Standardization Council of Indonesia, 290
Standards and Industrial Research Institute of Malaysia (SIRIM), 379
Standards and Metrology Institute of Slovenia (SMIS), 462
Standards Association of Zimbabwe (SAZ), 929
Standards Australia International (SAI) Ltd., 28
Standards Council of Canada (SCC), 99
Standards Institution of Israel (SII), 300
Standards New Zealand (SNZ), 411
Standards Organization of Nigeria (SON), 414
St. Andrews, 580, 581

Stanford, 642
Stanford Research Systems (SRS) Inc., 642
Stanford University
    Astronomy Program, 642
    Solar Observatories Group
        Wilcox Solar Observatory (WSO), 642
St. Ann, 777
Stansbury Park Observatory Complex, 891
Stanton, 621
STAR
    Stockholms Amatörastronomer, 493
Stará Lesná, 460
Stará Lesná Observatory, 459
Star Chaser Observatory, 659
STARE
    Scandinavian Twin Auroral Radar Experiment, 236
Star Farm (The.), 686
Starfield, 732
Stargate Observatory, 769
Stargazer Steve, 99
Star Hill Inn, 803
Star Jersey Star Party (SJSP), 797
Starkenburg-Sternwarte eV Heppenheim, 246
Starkville, 775
Starlight Astronomy Club, 861
Starlight Xpress, 571
    Holyport, 571
    New-York, 805
    Queensbury, 805
    United Kingdom, 571
    USA-NY, 805
STARLINK Project, 543, 571
StarMaster Portable Telescopes, 723
Star Observer Verlag, 39
Starsplitter Telescopes, 642
StarTel BV, 395, 401
Starwalk Planetarium, 667
State Archaeological Museum, 431
State College, 853, 860
State Committee for Standardization and Metrology (GOST), 449
State Committee of Ukraine for Standardization, Metrology and Certification (DSTU), 528
State Department for Standardization, Metrology and Certification of Georgia, 206
State Hydrometeorological Committee of Azerbaijan, 43
State Hydrometeorological Institute of Croatia, 121
State Information Center for Standardization, 364
State Museum of Pennsylvania
    Planetarium, 861
Staten Island, 808, 811
State Office for Standardization and Metrology (DZNM), 122

Statesboro, 688, 691
Statesboro Area Astronomy Club, 691
State University of New York (SUNY)
    Albany, 808
    Binghamton
        Department of Physics and Astronomy, 822
    Buffalo
        Department of Physics, 822
    Geneseo
        Department of Physics and Astronomy, 822
    New Paltz, 818
    Potsdam
        Planetarium, 822
    Stony Brook
        Department of Physics and Astronomy, 823
        Institute for Terrestrial and Planetary Atmospheres (ITPA), 823
Station Astronomique Jansky (SAJ), 202
Station de Radioastronomie de Nançay, 193, 202
St.-Aubin, 202
St. Augustine, 521, 678
StAV
    Steirischer Astronomen Verein, 39
Stavanger, 418

Stavanger Astronomical Association, 418
Stavanger Astronomiske Forening (SAF), 418
Stazione Astronomica Rigel, 313
STBP
    Sir Thomas Brisbane Planetarium, 27
St.-Caprais, 154
St. Charles Parish Library
    Planetarium, 730
St.-Claude, 170
St. Cloud, 772
St. Croix, 801
St.-Cyr, 158
STD
    Solar Terrestrial Dispatch, 99
St. Davids, 855
Steacie Institute, 80
ST-ECF
    Space Telescope - European Coordinating Facility, 245
Steele County Astronomical Society (SCAS), 772
Stefanik Observatory, 130
Stefan Thiele (Astronomische Instrumente-), 212, 246
Ste.-Foy, 79, 101
Steinberg, 39
Steirischer Astronomen Verein (StAV), 39
Stellafane, 894
Stella Nova Planetarium, 493
Stellarium, 643
Stellar Products, 643
Stellar Software, 643
St.-Elzéar-de-Beauce, 95
Stennis Space Center (SSC), 774
    Public Affairs Office, 774
Steno Museet, 139
Steno Museum, 139
Stephen F. Austin State University (SFASU)
    Department of Physics and Astronomy, 884
Sternberg State Astronomical Institute (SAI), 441, 450
Sterne (Die_), 225, 246
Sterne und Weltraum (SuW), 246
Sternfreunde Braunschweig-Hondelage eV, 246
Sternfreunde Breisgau (SFB) eV, 247
Sternfreunde Donzdorf eV, 247
Sternfreunde Durmersheim und Umgebung eV, 247
Sternfreunde Franken (SFF) eV, 247
Sternfreunde im FEZ (SiFEZ), 247
Sternwarte Aalen, 210
Sternwarte Bautzen, 247, 249
Sternwarte Bochum, 249
Sternwarte Burgsolms, 213
Sternwarte Calden, 213
Sternwarte der Volkshochschule Aachen, 247
Sternwarte des Bruder-Klaus-Heim, 248
Sternwarte des Friedrichs-Gymnasiums der Stadt Herford, 231
Sternwarte des Max-Born-Gymnasiums, 248
Sternwarte des Rottmayr-Gymnasiums, 248
Sternwarte Eschenberg, 511
Sternwarte Greifswald, 248
Sternwarte Heiligkreuz, 36, 39
Sternwarte Hof, 248
Sternwarte Höfingen, 248
Sternwarte in Bad Salzschlirf, 231
Sternwarte Johannes Franz Bautzen, 249
Sternwarte Johannes Kepler, 233
Sternwarte Karlsruhe, 216
Sternwarte Köln-Nippes, 250
Sternwarte Königsleiten, 39
Sternwarte Kreuzbergl, 39
Sternwarte Kronshagen, 249
Sternwarte Lauenstein, 249
Sternwarte Neanderhöhe Hochdahl eV, 249
Sternwarte Nessa, 215
Sternwarte Nippes, 250
Sternwarte Nürnberg, 239
Sternwarte Peterberg, 257
Sternwarte-Planetarium Bochum, 249
Sternwarte Rümlang, 512

Sternwarte Sankt Sebastian, 35
Sternwarte Solingen, 249, 264
Sternwarte Sonneberg, 249
Sternwarte Steina, 250
Sternwarte Stuttgart, 245
Sternwarte Sursee, 512
Sternwarte Torgelow, 250
Sternwarte und Planetarium Albstadt-Ebingen, 250
Sternwarte und Planetarium Köln-Nippes, 250
Sternwarte und Planetarium Schneeberg, 250
Sternwarte Urania Burgdorf, 512
Sternwarte Wiesbaden, 212
Sterrekundig Instituut Utrecht (SIU), 405
Sterrenkundig Instituut Anton Pannekoek, 401, 404
Sterrenwacht De Tiendesprong, 401
Sterrenwacht Halley, 402
Sterrenwacht Schrieversheide, 402
Sterrewacht Leiden, 396, 402, 404, 405
Sterrewacht Sonnenborgh, 405
St.-Étienne, 161, 197
St.-Étienne-des-Orgues, 200
St.-Étienne-en-Devoluy, 182
Stevenage, 550
Stevenage & District Astronomical Society (SADAS), 554
Stevens Point, 921
Steward Observatory, 602, 603
    Vatican Observatory Research Group (VORG), 604
Steyning, 584
S:TFF
    Space: The Future Frontier, 849
St.-Genis-Laval, 165, 200
St.-Georges-le-Gaultier, 188
St.-Héand, 155
St.-Hyppolyte, 85
Stichting Astronomisch Onderzoek in Nederland (ASTRON), 402
Stichting De Koepel, 397, 402
Stichting Geologisch, Oceanografisch en Atmosferisch Onderzoek (GOA), 402
Stichting Radiostraling van Zon en Melkweg (SRZM), 403
Stichting Skepsis, 403
Stichting Volkssterrenwacht Saturnus, 403
Stichting Volkssterrewacht Hercules, 402, 403
Stichting Volkssterrewacht Nijmegen, 403
Stiftung Sternwarte Uitikon, 512
Stiftung Volkssternwarte Trebur, 250
Stillwater, 847
Stillwater Stargazers, 843
St-Imier, 506
Stirling, 99
St.-James, 45
St.-Jean-le-Blanc, 190
Stjernekammeret, 140
St. John's, 92, 95
St. Johnsbury, 893
St.-Lambert, 85
St.-Laurent-du-Var, 160
St. Lawrence University
    Department of Physics, 821, 822
St. Louis, 777–779
St. Louis Astronomical Society (SLAS), 777
St.-Luc, 504
St. Lucia, 30
St-Lunaire, 159
STM
    Springfield Telescope Makers, 894
St.-Martin d'Hères, 182, 203
St. Mary's University
    Department of Astronomy and Physics
        Rev. M.W. Burke-Gaffney Observatory (BGO), 97
St.-Maur-des-Fossés, 166
St.-Maurice, 504
St.-Michael, 45
St.-Michel-l'Observatoire, 164, 191, 507
STN
    Scientific and Technical Information Network
        Columbus, 843

Germany, 245
Karlsruhe, 245
Ohio, 843
USA-OH, 843
St.-Nérée-de-Bellechasse, 78
STN International, 250, 843
Stockbridge, 542
Stocker & Yale Inc. (S&Y), 790
Stockert, 253
Stockholm, 488, 492–494, 496, 497
Stockholm Amateur Astronomers, 493
Stockholm Observatory, 494
Stockholms Amatörastronomer (STAR), 493, 494
Stockholm University
    Manne Siegbahn Laboratory, 496
    Stockholm Observatory, 494
Stockton, 622, 643
Stockton Astronomical Society (SAS), 643
Stockton College of New Jersey (RSCNJ) (Richard_), 797
    Observatory, 796
Stockton-on-Tees, 541
Stonehenge, 567
Stony Brook, 823
Stony Ridge Observatory (SRO), 643
Store Merløse, 134
Storer Planetarium (Arthur_), 735, 738, 743
Stork Product Engineering (SPE) BV, 403
Storrs, 665
St. Patrick's College
    Experimental Physics Department, 297
    Physics Department, 297
St. Paul, 770–772
St. Peter Port, 106
St. Petersburg, 441, 444, 445, 447, 449, 451, 684
St. Petersburg University
    Astronomical Institute, 449
    Department of Applied Mathematics and Control Processes, 449
Stralsund, 241
Strasbourg, 159, 165, 170, 177, 178, 186, 190, 200, 205
Strasenburgh Planetarium, 823
Stratford-upon-Avon, 571
Stratford-upon-Avon Astronomical Society (SONAAS), 571

Strathmore, 17
Strathpine, 27
Stratospheric Observatory for Infrared Astronomy (SOFIA), 630, 643
Straubenhardt-Feldrennach, 213
Strawberry Hills, 32
Strawbridge Observatory, Haverford College, 856
Streetsboro, 835
STRÍ
    Icelandic Council for Standardization, 279
Strickler Planetarium, 706
Stromlo (Mount_), 18
Strongsville, 837
Stroud, 542
Struve Planetarium, 719
STScI
    Space Telescope Science Institute, 245, 671, 743
Students for the Exploration and Development of Space (SEDS), 758
Studio City, 641
Stull Observatory, 805, 823
Sturgeon Bay, 915
Stuttgart, 221, 239, 245
St.-Valérien de Milton, 99
St.-Vallier-de-Thiey, 192
St. Xavier's College
    Physics Department, 281
SUA
    Sociedad Uruguaya de Astronomía, 587
Subaru Telescope, 695
Subiaco, 17
Submillimeter Telescope Observatory (SMTO), 603
Sucre, 62, 63

Sudak, 528
Sudak Astronomical Club, 528
Sudbury, 89, 98, 99
Sudbury Astronomy Club, 99
Sudbury Neutrino Observatory Institute (SNOI), 99
Sudekum Planetarium, 874
Suesca, 119
Sugadaira, 351
Sugadaira Station, 351
Sugar Grove, 701
Sugar Grove Observatory (SGO), 707
Suginami Science Education Center, 348
Suhl, 244
Suhora (Mount_), 428
Summerhays Planetarium (Sarah B._), 890
Summit, 792
Summit Observatory, 718
SUNA
    Sociedad Universitaria de Astronomía, 928
Sun City, 594
    West, 594
Sundsvall, 494
Sundsvalls Astronomiska Förening, 494
Sunnyvale, 617, 642
Sunrise, 685
Sunrise Museum, 913
Sunsearch, 643
Sunshine Planetarium, 358
Sunspot, 799, 801
Sunspot Index Data Center (SIDC), 57
Sunstone Inc., 823
SUNY, State University of New York
    Albany, 808
    Binghamton
        Department of Physics and Astronomy, 822
    Buffalo
        Department of Physics, 822
    Geneseo
        Department of Physics and Astronomy, 822
    New Paltz, 818
    Potsdam
        Planetarium, 822
    Stony Brook
        Department of Physics and Astronomy, 823
        Institute for Terrestrial and Planetary Atmospheres, 823
Suomen Avaruustutkimusseura (ATS), 149
Suomen Tähtitieteilijäseura, 147
SUPARCO
    Pakistan Space & Upper Atmosphere Research Commission, 421
Surrey, 571
Surrey Satellite Technology Ltd. (SSTL), 571
Surrey Space Centre, 571
Sursee/Lu, 512
Susa, 306
SUSI
    Sydney University Stellar Interferometer, 31
Susquehanna Valley Amateur Astronomers (SVAA), 862
Sussex, 581
Suter Science Center
    M.T. Brackbill Planetarium, 899
Sutherland, 28
Sutherland Astronomical Society Inc. (SASI), 28
Sutherland Computer Corp. (Evans &_), 890
Sutherland Observing Station, 467, 468
Sutrieu, 170
Sutton
    (North_), 789
Sutton Astronomical Society (MSAS) (Mansfield and_), 571

Sutton-in-Ashfield, 556
SuW
    Sterne und Weltraum, 246
SVA
    Skokie Valley Astronomers, 706
    Société Vaudoise d'Astronomie, 511

SVAA
    Susquehanna Valley Amateur Astronomers, 862
SVAAC
    Sandusky Valley Amateur Astronomy Club, 843
Svalsat, 418, 419
SVAS
    Sacramento Valley Astronomical Society, 636
SVEC
    Space Vacuum Epitaxy Center, 887
Svenska Astronomiska Sällskapet (SAS), 494
Svensk Amatör Astronomisk Förening (SAAF), 494
Svetloe, 445
Swanage, 584
Swansea, 571
Swansea Astronomical Society, 571
SWAP
    Southwest Association of Planetariums, 884
Swarthmore, 862
Swarthmore College
    Department of Physics and Astronomy
        Sproul Observatory, 862
Swasey Observatory, 835, 838, 843
Swaythling, 569
Sweden, 488
Swedish Amateur Astronomical Society, 494
Swedish Astronomical Society, 494
Swedish Institute of Space Physics, 495
    Kiruna Division, 495
    Solar-Terrestrial Physics Division, 495
    Umeå Division, 495
    Uppsala Division, 495
Swedish Meteorological and Hydrological Institute (SMHI), 496
Swedish National Space Board (SNSB), 496
Swedish Natural Science Research Council, 496
Swedish Physical Society, 496
Swedish Space Corp. (SSC)
    Earth Observation Division, 419
    Esrange, 496
    Kiruna, 496, 497
    Satellitbild, 497
    Solna, 497
    Tromsø Satellite Station (TSS), 419
Swedish Standards Institution (SIS), 497
Swift Instruments Inc., 758
Swinburne University of Technology
    Astrophysics and Supercomputing, 28
Swindon, 545, 558, 562
Swinton, 556
Swisher, 716, 718
Swiss Academy of Sciences (SAS), 508
Swiss Association for Standardization, 509, 512
Swiss Museum of Transport and Communications, 512, 514
Swiss National Science Foundation, 509, 512
Swiss Physical Society (SPS), 511
Switzerland, 500
SwRI
    Southwest Research Institute
        Boulder, 658
        Colorado, 658
        Department of Space Studies, 658
        Instrumentation and Space Research Division, 884
        San Antonio, 884
        Texas, 884
        USA-CO, 658
        USA-TX, 884
SWSA
    Space West Science Activities, 99
SWTSU
    Southwest Texas State University
        Physics Department, 880
SWWAS
    Southwest Washington Astronomical Society, 909
Sydney, 15, 19, 20, 23–32, 81, 82
    (Western.), 32
Sydney Observatory, 20, 28
Sydney Outdoor Lighting Improvement Society (SOLIS) Inc., 28
Sydney Space Frontier Society (SSFS), 28
Sydney University Stellar Interferometer (SUSI), 31
Sylmar, 615
Sylvania, 837
Syosset, 815
Syowa Station, 343, 355
Syracuse, 813, 821, 823, 824
    (North.), 823
Syracuse Astronomical Society (SAS) Inc., 823
Syracuse University
    Department of Physics, 824
Syria, 515
Syrian Arab Organization for Standardization and Metrology (SASMO), 515
Syrian Cosmological Society (SCS), 515
Syrian Meteorological Department, 515
Szeged, 277
Székesfehérvár, 278
Szombathely, 274
SZS
    Serbia Institute of Standardization, 455

T

TAA
    Tulare Astronomical Association, 645
TAAA
    Tucson Amateur Astronomy Association, 602
TAAC
    Triad Action Astronomy Club, 832
TAAS
    Thames Amateur Astronomical Society, 664
    The Albuquerque Astronomical Society, 803
Tabernes Blanques, 477
Table Mountain Observatory (TMO), 613, 634, 644
Table Mountain Star Party Association (TMSPA), 909
Table Mountain Star Party (TMSP), 910
Tabriz, 292
Tabriz University
    Center for Applied Physics and Astronomical Research, 292
TAC
    Theoretical Astrophysics Center, 140
Tacoma Astronomical Society (TAS), 910
Tacuarembo, 587
Taeduk Radio Astronomy Observatory (TRAO), 368, 369
Taegu, 369
Taejon, 367, 368
Tag der Astronomie, 231
Tähtitieteellinen Yhdistys Olympos ry, 150
Tähtitieteellinen Yhdistys Tampereen Ursa ry, 150
Tähtitieteen Harrastajan Yhdistys Seulaset ry, 150
Taichung, 517
Taipei, 516, 517
Taipei Astronomical Museum (TAM), 517
Taipei Observatory, 517
Taiwan, 516
Tajik Academy of Sciences
    Institute of Astrophysics, 518
Tajikistan, 518
TAK
    Tierps Astronomiska Klubb, 497
Takoma Park, 740
Talcott Mountain Science Center (TMSC), 664
Tallahassee, 683, 686
Tallahassee Astronomical Society (TAS), 686
Tallinn, 144
Talmassons, 314
TAM
    Taipei Astronomical Museum, 517
Tama, 716
Tamalpais (Mount.), 629
Tamborine (Mount.), 30
Tamke-Allen Observatory, 873, 875
Tampa, 684
Tampere, 150

Tampereen Särkänniemi Oy, Planetarium, 150
Tandogan, 522
Tanjung Sari, 290
TAO
    Toyama Astronomical Observatory, 359
Taoyuan, 516
Tarbes, 194
Taree, 29
Taree Astronomical Society (TAS), 29
Tarija, 62, 63
Tarn, 154
Tarrant County College
    Astronomy Club, 885
Tar River Astronomy Club (TRAC), 831
Tartu, 144
Tartu Observatory, 144
Tartu Old Observatory, 144
Tartu University
    Institute of Physics, 144
TAS
    Tacoma Astronomical Society, 910
    Tallahassee Astronomical Society, 686
    Taree Astronomical Society, 29
    Texas Astronomical Society of Dallas, 885
    Thai Astronomical Society, 519
    Torbay Astronomical Society, 573
    Triangulum Astronomical Society, 904
    Tullamore Astronomical Society, 298
Tashkent, 925
Tashkent Observatory, 925
Tasmania, 15, 16, 31
TASOS
    The Astronomical Society of Singapore, 457
TASS
    The Amateur Sky Survey, 824
TASSF
    The Atlantic Space Sciences Foundation, 100
Tåstrup, 138, 140
Tata Institute of Fundamental Research (TIFR), 288
Tatranská Lomnica, 459, 460
Tautenburg, 252
Tavernerio, 321
Tavistock, 572
Tavistock Astronomical Society, 572
Taylor & Francis
    Bristol, 862
    London, 572
    Pennsylvania, 862
    United Kingdom, 572
    USA-PA, 862
Taylor & Francis Inc., 862
Taylor & Francis Ltd., 572
Taylor Planetarium, 780
TBE
    Teledyne Brown Engineering, 590
Tbilisi, 206
TBO
    Tycho Brahe Observatory, 498
TCAA
    Twin City Amateur Astronomers, 707
TCAC
    Tri-City Astronomy Club, 910
TCD
    Trinity College Dublin, 298
T.C. Hooper Planetarium, 868
TCO
    Three College Observatory (UNCG), 832
TCU
    Texas Christian University
        Department of Physics and Astronomy, 885
Tech 2000, 843
Tech Museum of Innovation (The_), 644
Technical Innovations Inc., 743
Technical Research Centre of Finland, 150, 152
Technische Hochschule Aachen
    Forum Weltraumforschung, 251
Technische Universität Berlin
    Institut für Astronomie und Astrophysik, 251
Technische Universität Braunschweig
    Institut für Flugmechanik und Raumfahrttechnik, 251
Technische Universität Darmstadt (TUD)
    Institut für Physikalische Geodäsie (IPG), 251
Technische Universität Dresden
    Institut für Planetare Geodäsie
        Lohrmann-Observatorium, 251
Technische Universität Graz
    Institut für Angewandte Geodäsie
        Abteilung für Positionierung und Navigation, 39
Technische Universität München
    Institut für Astronomische und Physikalische Geodäsie (IAPG), 252
Technische Universiteit Delft
    Delft Institute for Earth-Oriented Space Research (DEOS), 403
Technology Development Centre, 150, 151
Teesside, 582
Tegucigalpa, 272
Tehran, 292, 293
Teide, 58, 235, 256, 477
TEKES, 151
Teknikens Hus, 497
Teknofokus, 151
Teknoteket, 416, 418
Tekton Industries, 644
TELAPO
    Terkán Lajos Public Observatory, 277
Tel-Aviv, 300
Tel Aviv University
    Department of Geophysics and Planetary Science, 300
    Florence and George Wise Observatory, 300
    School of Physics and Astronomy, 300
Teledyne Brown Engineering (TBE), 590
Telescopes and Astronomy, 29
Telescope Technologies Ltd. (TTL), 572
Telescopio Infrarosso del Gornergrat (TIRGO), 317
Telescopio Nazionale Galileo (TNG), 327, 479, 484
Teleskopschmiede, 252
Teleskop Service Wolfgang Ransburg GmbH, 252
Teleskoptechnik Halfmann, 252
Tele Vue Optics, Inc., 824
Temecula, 621, 643
Tempe, 593
Tenagra Observatories Ltd., 602
Tenerife, 477, 479, 483
    (Santa Cruz de_), 473
Tennessee, 870–876
    (East_), 870, 871, 874
    (Middle_), 873
Tennessee State University (TSU)
    Center of Excellence in Information Systems, 875
Teplice, 131
Teplice Observatory and Planetarium, 131
Teramo, 328
Terkán Lajos Public Observatory (TELAPO), 277
Terni, 332
Terre Haute, 712
Terskol, 526, 529
TESRE
    Istituto di Tecnologie e Studio delle Radiazioni Extraterrestri, 317
Tessin, 509
Tessmann Planetarium, 644
Teufen, 502
Texas, 877–889
    (North_), 882, 887
    (South_), 883
    (West_), 889
Texas A & M University, 877
    Center for Space Power (CSP), 885
    Department of Physics, 877, 885
Texas Astronomical Society (TAS) of Dallas, 885
Texas Christian University (TCU)
    Department of Physics and Astronomy, 885
Texas Nautical Repair Co., 885

Texas Star Party (TSP) Inc., 885
TeX Users Group (TUG), 849
Thai Astronomical Society (TAS), 519
Thai Industrial Standards Institute (TISI), 519
Thailand, 519
Thai Meteorological Department, 519
Thames Amateur Astronomical Society (TAAS), 664
Thayer Academy, 759
The Albuquerque Astronomical Society (TAAS), 803
The Amateur Sky Survey (TASS), 824
The Astronomer, 572
The Astronomical Society of Singapore (TASOS), 457
The Astronomy Connection, 644
The Astronomy Roadshow, 572
The Atlantic Space Sciences Foundation (TASSF) Inc., 100
The Briars, 15
The Children's Museum of Indianapolis, 714
The Citadel
    Department of Physics, 868
The Gemmary Inc., 644
The Hague, 392, 393, 397, 399
The Island Planetarium, 572
The Japan Academy, 358
The Local Group of Santa Clarita Valley (LGSCV), 644
The Maidanak Foundation, 418
The Mars Society, 659
Themis, 187
Theoretical Astrophysics Center (TAC), 140
The Phoenix Astronomical Society (TPAS), 412
The Physical Society of Japan, 358
The Physics Teacher (TPT), 734
Thermionics Laboratory Inc. (TLI), 644
Thermionics Vacuum Products, 910
Thermo Corion Optical Filters, 759
The Skeptic, 572
The Spaceguard Centre, 573
Thessaloniki, 270
The Star Farm, 686
The Tech Museum of Innovation, 644
The Universe Collection, 665
Thiele (Astronomische Instrumente Stefan_), 212, 252
Thin Film Technology (TFT), 645
Third World Academy of Sciences (TWAS), 336
Third World Network of Scientific Organizations (TWNSO), 336
Thomas Brisbane Planetarium (STBP) (Sir_), 27, 29
Thomas Jefferson High School for Science and Technology Planetarium, 904
Thomas Planetarium (Pat_), 683, 686
Thompson Observatory, 915, 919
Thorigné-sur-Dué, 162
Thornton, 25
Thousand Oaks, 633, 634, 639, 643, 650
Three College Observatory (TCO), 831, 832
Thüringer Landessternwarte (TLS) - Tautenburg
    Karl-Schwarzschild-Observatorium (KSO), 252
Tibarine (Mont_), 4
Ticino, 509
Tidbindilla, 20, 613
Tidbindilla Space Tracking Station, 20
Tiel Sanford Museum and Planetarium, 718, 719
Tien-Shan Observatory, 364
Tierp, 497
Tierps Astronomiska Klubb (TAK), 497
Tierra del Fuego, 7
Tierra del Sol, 636
Tiffin, 843
TIFR
    Tata Institute of Fundamental Research, 288
Tilburg, 389, 401
Time Museum, 706
Timişoara, 438
Timişoara Astronomical Observatory, 438
Tipton, 714
Tipton Planetarium, 714
Tirana, 3
TIRGO

Telescopio Infrarosso del Gornergrat, 317
Tirschenreuth, 263
TISI
    Thai Industrial Standards Institute, 519
Titusville, 678
Tixie Bay, 446
TJC
    Tyler Junior College
        Hudnall Planetarium, 877, 880
T.J. Watson Research Center, 815
TLI
    Thermionics Laboratory Inc., 644
TLS
    Thüringer Landessternwarte, 252
TLSP
    Twin Lakes Star Party, 728
TL Systems, 645
TMO
    Table Mountain Observatory, 613, 634
TMSP
    Table Mountain Star Party, 910
TMSPA
    Table Mountain Star Party Association, 909
TMU
    Tokyo Metropolitan University
        Department of Physics, 359
TNG
    Telescopio Nazionale Galileo, 327, 479
TNO
    Nederlandse Organisatie voor Toegepast-Natuurwetenschappelijk Onderzoek, 397
TNSC
    Trailside Nature and Science Center, 797
TOAST
    The Orpington Astronomical Society Times, 561
Tobago, 521
    (Trinidad &_), 521
Toccoa, 691
Todmorden, 562
Togót Observatory (Khurel_), 388
Tohoku University
    Astronomical Institute, 358
Tokai University
    Research Institute of Civilization, 358
Tokushima, 359
Tokushima Kainan Observatory, 359
Tokyo, 342–349, 352–362
Tokyo Astronomical Observatory, 359
Tokyo Gakugei University
    Department of Astronomy and Earth Sciences, 359
Tokyo Metropolitan University (TMU)
    Department of Physics, 359
Toledo, 836, 844
Tololo (Cerro_), 109
Tombaugh Observatory (Clyde W._), 723
Tombaugh Space Theater (Clyde W._), 803
Tomchin Planetarium, 914
Tomsk, 440, 450
Tomsk Planetarium, 450
Tomsk State University
    Astronomical Observatory, 450
Toms River, 792, 796
Tom Tits Experiment, 498
Tonantzintla, 383, 386
Tonawanda, 814
Tonbridge, 534
Tonneville, 188
Toothill, 569
Topeka, 721–723
Topsfield, 756
Töravere, 144
Torbay Astronomical Society (TAS), 573
Torgelow, 250
Torino, 302, 315, 316, 323, 328, 336
Toronto, 81, 83, 89, 91, 92, 95, 97, 103, 104
Toronto Sidewalk Astronomers (TSA), 100
Torrance, 616, 619, 624, 640

Torrejón de Ardóz, 478
Torres Arbide, 483
Tortugas Mountain Station, 802
Torun, 430, 432
Torun Planetarium, 432
Torun University Nicolaus Copernicus
    Centre for Astronomy, 432
Torus Precision Optics Inc., 719
Tottiri, 356
Toulouse, 155, 163, 166, 169, 186, 193, 194, 199, 201, 202
Touraine, 162
Toussaint, 161
Toussi Observatory (Khadjeh Nassir Addin_), 293
Townshend, 893
Towson, 743
Towson State University (TSU)
    Department of Physics, Astronomy, and Geosciences, 743
Toyama, 359
Toyama Astronomical Observatory (TAO), 359
Toyama Science Museum (TSM), 359
Toyama University
    Department of Physics, 359
Toyokawa, 351
TPAS
    The Phoenix Astronomical Society, 412
TPT
    The Physics Teacher, 734
TRAA
    Tule River Amateur Astronomers, 645
TRAC
    Tar River Astronomy Club, 831
Trackman Planetarium (Herbert_), 706
Tradate, 324
Trailside Nature and Science Center (TNSC)
    Planetarium, 797
Traiskirchen, 35
Trango Systems Inc., 645
Transinne, 51
TRAO
    Taeduk Radio Astronomy Observatory, 368
Trapani, 302
Traverse City, 764, 766
Trebur, 250
Treffen, 40
Trégor, 172, 196
Treinta y Tres, 587
Trek Inc., 824
Trelew, 10
Tremsdorf, 218
Trento, 307
Trenton, 794
Treviso, 307, 312
Triad Action Astronomy Club (TAAC), 832
Triangulum Astronomical Society (TAS), 904
Tri-City Astronomy Club (TCAC), 910
Triel-sur-Seine, 195
Trier, 207
Trieste, 302, 328, 334, 336, 340
Trinidad, 521
Trinidad & Tobago, 521
Trinidad & Tobago Bureau of Standards (TTBS), 521
Trinité (La_), 195
Trinity College Dublin (TCD)
    Department of Physics, 298
Trinity School Observatory, 536
Triplett Corp., 843
Tri-State Astronomers, 743
Tri-Valley Stargazers (TVS), 645
Trivandrum, 289
Trois-Rivières, 84, 101
Troitsk, 446
Tromsø, 418–420, 489
Tromsø Satellite Station (TSS), 417, 418
    SvalSat, 419
Trondheim, 418
Troutdale, 848

Troutman, 867
Troy, 817, 820, 843
Troy State University Montgomery (TSUM)
    W.A. Gayle Planetarium, 591
True Technology Ltd., 573
TSA
    Toronto Sidewalk Astronomers, 100
TSE
    Turkish Standards Institution, 524
Tsiolkovskiy Astronomical Club, 450
TSM
    Toyama Science Museum, 359
TSO
    Turtle Star Observatory, 252
TSP
    Texas Star Party, 886
TSS
    Tromsø Satellite Station, 417, 418
        SvalSat, 419
TSU
    Tennessee State University
        Center of Excellence in Information Systems, 875
    Towson State University
        Department of Physics, Astronomy, and Geosciences, 743
        Watson-King Planetarium, 743
TSUM
    Troy State University Montgomery
        W.A. Gayle Planetarium, 591
Tsumeb, 466
TTBS
    Trinidad & Tobago Bureau of Standards, 521
TTL
    Telescope Technologies Ltd., 572
Tübingen, 217, 218, 256, 257
TÜBITAK, 524
    Scientific and Technical Research Council of Turkey, 523
    Marmara Research Center, 523
TÜBITAK Ulusal Gozlemevi (TUG), 524
Tucson, 594–597, 599–604
Tucson Amateur Astronomy Association (TAAA), 602
TUD
    Technische Universität Darmstadt
        Institut für Physikalische Geodäsie, 251
Tufts University
    Department of Physics and Astronomy, 759
TUG
    TEX Users Group, 850
    TÜBITAK Ulusal Gozlemevi, 524
Tulare, 645
Tulare Astronomical Association (TAA), 645
Tule River Amateur Astronomers (TRAA), 645
Tullahoma, 875
Tullamore, 298
Tullamore Astronomical Society (TAS), 298
Tulsa, 846
Tumanoc Hill Station, 603
Tunapuna, 521
Tuorla Observatory, 151, 152
Turin, 302, 315, 316, 323, 328, 336
Turkey, 522
Turkey Run State Park Nature Center Planetarium, 714
Turkish National Observatory, 524
Turkish Physical Society, 524
Turkish Standards Institution (TSE), 524
Turkish State Meteorological Service, 524
Turku, 151, 152
Turnov, 124
Turramurra (North_), 26, 28
Turtle Star Observatory (TSO), 252
Turun Ursa Astronomical Association, 151
Tuscaloosa, 591
TVS
    Tri-Valley Stargazers, 645
TWAS
    Third World Academy of Sciences, 336

Twilight Tours Inc., 645
Twin City Amateur Astronomers (TCAA) Inc., 707
Twin Falls, 697
Twining Observatory (GNTO) (General Nathan_), 803
Twin Lakes Star Party (TLSP), 728
TWNSO
    Third World Network of Scientific Organizations, 336
Two Mile Run Park, 859
Tycho Brahe Astronomical Society, 498
Tycho Brahe Observatory (TBO), 498
Tycho Brahe Planetarium, 135, 140
Tyler, 877, 880
Tyler Junior College (TJC)
    Hudnall Planetarium, 877, 880
Tyneside
    (South_), 570
Tyngsboro, 756

U

UAA
    Ukrainian Astronomical Association, 528
    Umpqua Amateur Astronomers, 850
    University of Alaska in Anchorage
        Department of Physics and Astronomy, 592
    Uppsala Amatörastronomer, 498
UAB
    Unione Astrofili Bresciani, 336
UAI
    Union Astronomique Internationale, 184
    Unione Astrofili Italiani, 337
    Universitair Instelling Antwerpen
        Department of Physics, 58
UALR
    University of Arkansas, Little Rock
        Department of Physics and Astronomy, 606
UAM
    Universidad Autónoma de Madrid
        Departamento de Física Teórica, 484
        Grupo de Astrofísica, 484
    University of Arkansas, Monticello
        Pomeroy Planetarium, 607
UAM-I, Universidad Autónoma Metropolitana-Iztapalapa
    Departamento de Física
        Area Gravitación y Astrofísica, 385
UAN
    Unione Astrofili Napoletani, 337
UAO
    Urumqi Astronomical Observatory, 115
    Ussurijsk Astronomical Observatory, 448
UAS
    Unione Astrofili Senesi, 337
UATI
    Union Internationale des Associations et Organismes Techniques, 203
UBC
    University of British Columbia
        Department of Physics and Astronomy, 101
Úbeda, 470
Uberlândia, 69
UCA
    University of Central Arkansas
        Department of Physics and Astronomy, 607
        Lewis Science Center Observatory, 607
        Lewis Science Center Planetarium, 607
UCAR
    University Corp. for Atmospheric Research, 656, 659
UCB
    University of California at Berkeley, 639
        Advanced Light Source Project, 608
        Berkeley-Illinois-Maryland Association, 647
        Center for Extreme Ultraviolet Astrophysics, 646
        Department of Astronomy, 646
        Department of Physics, 646
        E.O. Lawrence Berkeley National Laboratory, 620
        Hat Creek Observatory, 647
        Institute for Nuclear and Particle Astrophysics, 620
        Lawrence Hall of Science, 610, 651
        Space Sciences Laboratory, 647
        William Knox Holt Planetarium, 651
UCCS
    University of Colorado at Colorado Springs
        Department of Physics, 660
UCD
    University College Dublin
        Physics Department, 298
    University of California at Davis
        Astronomy Club, 647
        Department of Chemistry, 647
        Department of Physics, 647
UCF
    University of Central Florida
        Department of Physics, 686
UCI
    University of California at Irvine
        Department of Physics and Astronomy, 647
UCL
    Université Catholique de Louvain
        Institut d'Astronomie et de Géophysique G. Lemaître, 57
    University College London
        Department of Physics and Astronomy, 574
        Mullard Space Science Laboratory, 574
        University of London Observatory, 574
UCLA, University of California at Los Angeles
    Department of Earth and Space Sciences, 648
    Department of Physics and Astronomy
        Division of Astronomy, 648
    Institute of Geophysics and Planetary Physics, 648
UCP
    University of Chicago Press, 708
UCR
    University of California at Riverside
        Institute of Geophysics and Planetary Physics (IGPP), 648
UCSB
    University of California at Santa Barbara
        Institute for Theoretical Physics, 649
        Physics Department, 649
UCSC
    University of California at Santa Cruz
        Department of Astronomy and Astrophysics, 649
        Lick Observatory, 649
UCSD
    University of California at San Diego
        Center for Astrophysics and Space Sciences (CASS), 648
        Physics Department, 648
        Scripps Institution of Oceanography, 614
UCT
    University of Cape Town
        Department of Applied Mathematics, 468
        Department of Astronomy, 469
        Department of Physics, 469
UCV
    Universidad Central de Venezuela
        Escuela de Geologia, Minas y Geofísica, 928
Udaipur, 288
Udaipur Solar Observatory (USO), 288
Udine, 309
Udmurt State University
    Department of Astronomy and Mechanics, 450
Ufa, 450
Ufa Planetarium, 450
UFES
    Universidade Federal do Espírito Santo
        Observatório Astronômico, 72
Ufficio Federale di Meteorologia e Climatologia, 512
Ufficio Federale di Meteorologia e Climatologia (MeteoSvizzera), 506, 512
Ufficio Generale per la Meteorologia (UGM), 336
UFG
    Universidade Federal de Goiás

Planetário, 72
UFMG
   Universidade Federal de Minas Gerais
      Departamento de Física, 72
      Observatório Astronômico de Piedade, 72
UFOP
   Universidade Federal de Ouro Preto
      Observatório Astronômico, 71
UFRGS, Universidade Federal do Rio Grande do Sul
   Instituto de Física
      Departamento de Astronomia, 73
   Planetário Prof. José Baptista Pereira, 73
UFRJ
   Universidade Federal de Rio de Janeiro
      Observatório do Valongo, 73
UFSC
   Universidade Federal de Santa Catarina, 67
UGA
   University of Georgia at Athens
      Department of Physics and Astronomy, 691
UGGI
   Union Géodésique et Géophysique Internationale, 186
UGM
   Ufficio Generale per la Meteorologia, 336
Uherský Brod, 132
Uherský Brod Cultural House Observatory, 132
UIA
   Union of International Associations, 57
   Universitaire Instelling Antwerpen, 57
UIAS
   University of Illinois Astronomical Society, 708
UIB
   Universidad de las Islas Baleares
      Departamento de Física, 485
UIT
   Union Internationale des Télécommunications, 505
UK
   United Kingdom, 531
UKAFF
   United Kingdom Astrophysical Fluids Facility, 578
UK Astronomy Technology Centre, 534
UKIRT
   United Kingdom Infrared Telescope, 693
UKISC
   United Kingdom Industrial Space Committee, 573
Ukraine, 525
Ukrainian Astronomical Association (UAA), 528
Ukrainian National Academy of Sciences
   Institute of Radio Astronomy, 528
   Main Astronomical Observatory (MAO), 529
   Poltava Gravimetrical Observatory, 529
Ukrainian Physical Society (UPS), 529
Ukrainian Youth Aerospace Association (UYAA) "Suzirya", 529
UKSEDS
   United Kingdom Students for the Exploration and Development of Space, 573
UKST
   United Kingdom Schmidt Telescope, 14, 29
ULA
   Universidad de los Andes
      Departamento de Física, 928
Ulaanbataar, 388
ULB
   Université Libre de Bruxelles
      Institut d'Astronomie et d'Astrophysique, 59
      Physique Nucléaire Théorique et Physique Mathématique, 59
ULg
   Université de Liège
      Centre Spatial de Liège, 58
      Institut d'Astrophysique et de Géophysique, 56, 58
      Institut de Mathématique, 56
Ulm, 780
ULO
   University of London Observatory, 574
ULP
   Université Louis Pasteur, 205
Ulugh Beg Astronomical Institute, 925
   Space Geodynamics Department, 925
UMBC
   University of Maryland Baltimore County
      Joint Center for Astrophysics, 744
      Physics Department, 744
Umeå, 495
UMI
   University Microfilms International, 767
UMI Inc., 767
UMIST
   University of Manchester Institute of Science and Technology
      Astrophysics Group, 579
      Department of Physics, 579
Umpqua Amateur Astronomers (UAA), 850
UMSA
   Universidad Mayor de San Andrés
      Observatorio Astronómico de Patacamaya, 63
UMSL
   University of Missouri in Saint Louis, 778
UNA
   University of North Alabama
      Observatory, 591
      Planetarium, 591
UNAH
   Universidad Nacional Autónoma de Honduras
      Sección de Astrofísica, 272
UNAM, Universidad Nacional Autónoma de México
   Instituto de Astronomía, 385
      Unidad Morelia, 386
   Instituto de Geofísica, 386
   Observatorio Astronómico Nacional (OAN)
      Ensenada, 386
      San Diego, 646
      San Ysidro, 646
Unanderra, 32
Uncasville, 664
UNCG
   University of North Carolina in Greensboro
      Department of Physics and Astronomy, 832
      Three College Observatory, 832
UND
   University of North Dakota
      Department of Space Studies, 834
Undercliffe, 536
UNESCO, 202
UNESP
   Universidade Estadual Paulista
      Instituto de Física Teorica, 72
UNI
   Ente Nazionale Italiano di Unificazione, 317
   University of Northern Iowa
      Department of Earth Science, 720
      Hillside Observatory, 720
UniMeasure Inc., 850
Union Astronomique Internationale (UAI), 184, 203
Union College
   Department of Physics, 824
Unione Astrofili Bresciani (UAB), 336
Unione Astrofili Cosentini, 337
Unione Astrofili Italiani (UAI), 337
Unione Astrofili Napoletani (UAN), 337
Unione Astrofili Senesi (UAS), 337
Union Géodésique et Géophysique Internationale (UGGI), 186
Union Grove, 918
Union Internationale des Associations et Organismes Techniques (UATI), 203
Union Internationale des Télécommunications (UIT), 505, 512
Union (Mount_), 839
Union of Czech Mathematicians and Physicists
   Physics Section, 132
Union of International Associations (UIA), 57
Union of Physicists in Bulgaria, 76

Union of Societies of Mathematicians, Physicists and Astronomers
    Physics Section, 455
Union Rhône-Alpes des Clubs d'Astronomie (URACA), 203

UNISA
    University of South Africa
        Department of Mathematics, Applied Mathematics and Astronomy, 469
        Observatory, 469
UNIT
    Instituto Uruguayo de Normas Técnicas, 587
United Astronomy Clubs of New Jersey Inc., 797
United Kingdom Astrophysical Fluids Facility (UKAFF), 578
United Kingdom Industrial Space Committee (UKISC), 573

United Kingdom Infrared Telescope (UKIRT), 693
United Kingdom Schmidt Telescope (UKST), 29
United Kingdom Students for the Exploration and Development of Space (UKSEDS), 573
United Kingdom (UK), 531
United Nations Organization (UNO)
    Office for Outer Space Affairs (OOSA), 40
    World Intellectual Property Organization (WIPO), 514
United States Geological Survey (USGS)
    Flagstaff Field Center, 603
United States of America
    Alabama, 589
    Alaska, 592
    Arizona, 593
    Arkansas, 606
    California, 608
    Carolina
        (North.), 827
        (South.), 866
    Colorado, 652
    Connecticut, 662
    Dakota
        (North.), 834
        (South.), 869
    Delaware, 667
    District of Columbia, 669
    Florida, 678
    Georgia, 687
    Hawaii, 692
    Idaho, 697
    Illinois, 698
    Indiana, 710
    Iowa, 716
    Kansas, 721
    Kentucky, 725
    Louisiana, 729
    Maine, 732
    Maryland, 734
    Massachusetts, 746
    Michigan, 761
    Minnesota, 770
    Mississippi, 774
    Missouri, 776
    Montana, 780
    Nebraska, 782
    Nevada, 786
    New Hampshire, 788
    New Jersey, 791
    New Mexico, 799
    New York, 805
    North Carolina, 827
    North Dakota, 834
    Ohio, 835
    Oklahoma, 846
    Oregon, 848
    Pennsylvania, 851
    Rhode Island, 864
    South Carolina, 866
    South Dakota, 869
    Tennessee, 870
    Texas, 877
    Utah, 890
    Vermont, 893
    Virginia, 895
        (West.), 912
    Washington, 907
    West Virginia, 912
    Wisconsin, 915
    Wyoming, 922
United States Space Foundation, 659
United Technologies Research Center (UTRC), 665
Unitron Inc., 824
Univelt Inc., 646
Universe Collection (The.), 665
Universidad Autónoma de Madrid (UAM)
    Departamento de Física Teórica, 484
    Grupo de Astrofísica, 484
Universidad Autónoma Metropolitana-Iztapalapa (UAM-I)
    Departamento de Física
        Area Gravitación y Astrofísica, 385
Universidad Católica de Chile
    Departamento de Astrofísica y Astronomía, 110
Universidad Católica del Norte (UCN)
    Instituto de Astronomía, 110
Universidad Central de Venezuela (UCV)
    Escuela de Geologia, Minas y Geofísica (EGMG), 928
Universidad Complutense
    Departamento de Astrofísica, 484
    Facultad de Ciencias Físicas, 482
    Instituto de Astronomía y Geodesia (IAG), 478, 485
Universidad de Barcelona
    Departamento de Astronomía y Meteorología, 485
    Facultad de Física, 484
Universidad de Cantabria
    Instituto de Física de Cantabria, 478, 485
Universidad de Chile
    Departamento de Astronomía, 110
Universidad de Concepción
    Departamento de Física
        Grupo de Astronomía, 111
Universidad de Granada
    Departamento de Física Teórica y del Cosmos, 485
Universidad de Guadalajara
    Instituto de Astronomía y Meteorología, 385
Universidad de Guanajuato
    Departamento de Astronomía, 385
Universidad de la República
    Departamento de Astronomía, 588
Universidad de la Rioja
    Departamento de Matemática y Computación, 472
Universidad de La Serena
    Departamento de Física, 111
Universidad de las Islas Baleares (UIB)
    Departamento de Física, 485
Universidad de los Andes (ULA)
    Departamento de Física, 928
Universidad de Santiago de Compostela
    Observatorio Astronómico Ramón María Aller, 485
Universidad de Sonora
    Centro de Investigación en Física
        Área de Astronomía, 385
        Estación de Observación Solar (EOS), 385
Universidad de Tarapacá
    Departamento de Física, 111
Universidad de Valencia
    Departamento de Astronomía y Astrofísica, 485
    Facultad de Matemáticas
        Grupo de Astronomía y Ciencias del Espacio (GACE), 486
    Observatorio Astronómico, 486
Universidad de Valparaíso
    Instituto de Matemáticas y Física, 111
Universidad de Zaragoza
    Grupo de Mecanica Espacial, 486
Universidade da Madeira
    Grupo de Astronomia, 435
Universidade de Coimbra

Observatório Astronômico, 435
Universidade de Évora
   Departamento de Física
      Grupo de Astronomia, 435
Universidade de Lisboa
   Departamento de Física
      Centro de Astronomia e Astrofísica, 435
Universidade de São Paulo (USP)
   Instituto de Astronomia, Geofísica e Ciências Atmosféricas
      Departamento de Astronomia, 71
Universidade do Porto
   Centro de Astrofísica, 436
   Laboratório da Física, 435
   Observatório Astronômico Prof. Manuel de Barros, 436
Universidade Estadual de Campinas
   Instituto de Física
      Departamento Raios Cosmicos e Cronologia, 72
Universidade Estadual Paulista (UNESP)
   Instituto de Física Teorica, 72
Universidade Federal de Goiás (UFG)
   Planetário, 72
Universidade Federal de Minas Gerais (UFMG)
   Departamento de Física
      Observatório Astronômico de Piedade (OAP), 72
Universidade Federal de Ouro Preto (UFOP)
   Observatório Astronômico, 71
Universidade Federal de Rio de Janeiro (UFRJ)
   Observatório do Valongo, 73
Universidade Federal de Santa Catarina (UFSC), 67
Universidade Federal do Espírito Santo (UFES)
   Observatório Astronômico, 72
Universidade Federal do Rio Grande do Sul (UFRGS)
   Instituto de Física
      Departamento de Astronomia, 73
   Planetário Prof. José Baptista Pereira, 73
Universidad Mayor de San Andrés (UMSA)
   Observatorio Astronómico de Patacamaya (OAP), 63
Universidad Nacional Autónoma de Honduras (UNAH)
   Sección de Astrofísica, 272
Universidad Nacional Autónoma de México (UNAM)
   Instituto de Astronomía, 385
      Unidad Morelia, 386
   Instituto de Geofísica, 386
   Observatorio Astronómico Nacional (OAN), 386, 646
      Ensenada, 386
      San Diego, 646
      San Ysidro, 646
Universidad Nacional de Córdoba
   Observatorio Astronómico de Córdoba (OAC), 10
Universidad Nacional de La Plata (UNLP)
   Facultad de Ciencias Astronómicas y Geofísicas (FCAG), 10
   Observatorio Astronómico, 10
Universidad Nacional de San Juan
   Observatorio Astronómico Felix Aguilar (OAFA), 10
Universidad Nacional Mayor de San Marcos (UNMSM)
   Seminario Permanente de Astronomía y Ciencias Espaciales (SPACE), 423
Universidad Politécnica de Cataluña
   Departamento de Física Aplicada, 486
Universidad Ramón Llull (URL)
   Observatorio del Ebro, 486
Università degli Studi di Bologna, 337
Università degli Studi di Catania, 337
Università degli Studi di Padova, 337
   Dipartimento di Astronomia, 337
Università degli Studi di Roma
   Dipartimento di Fisica
      Consorzio Internazionale per l'Astrofisica Relativistica, 324
      International Center for Relativistic Astrophysics, 324
Università degli Studi di Trieste, 338
Università di Bologna
   Dipartimento di Astronomia, 338
Università di Catania
   Dipartimento di Astronomia, 338

Università di Firenze
   Dipartimento di Astronomia e Scienza dello Spazio, 338
Università di Lecce
   Dipartimento di Fisica
      Gruppo di Astrofisica, 338
Università di Milano
   Dipartimento di Fisica, 338
Università di Padova
   Dipartimento di Astronomia, 339
Università di Pavia
   Dipartimento di Fisica Nucleare e Teorica (DFNT), 339
Università di Perugia
   Osservatorio Astronomico, 339
Università di Pisa
   Dipartimento di Fisica
      Sezione di Astronomia e Astrofisica, 339
   Dipartimento di Matematica, Gruppo di Meccanica Spaziale, 339
Università di Roma La Sapienza
   Istituto Astronomico, 339
Università di Roma Tor Vergata
   Dipartimento di Fisica
      Astrofisica, 340
Università di Roma Tre
   Dipartimento di Fisica, 340
Università di Trieste
   Dipartimento di Astronomia, 340
Universitaire Instelling Antwerpen (UIA), 57
   Department of Physics, 58
Università Popolare Sestrese
   Osservatorio Astronomico di Genova, 340
Universität Basel
   Astronomisches Institut, 512
   Institut für Physik, 511, 512
   Theoretische Kern/Teilchen- und Astrophysik, 512
Universität Bern
   Astronomisches Institut, 513
   Institut für Angewandte Physik, 513
Universität Bochum
   Astronomisches Institut, 253
Universität Bonn
   Astronomische Institute
      Institut für Astrophysik und Extraterrestrische Forschung, 253
      Observatorium Hoher List (OHL), 253
      Radioastronomisches Institut (RAIUB), 253
      Sternwarte, 253
Universität Erlangen-Nürnberg
   Astronomisches Institut
      Dr. Remeis Sternwarte, 254
Universität Frankfurt/Main
   Institut für Theoretische Physik/Astrophysik, 254
Universität Graz
   Institut für Geophysik, Astrophysik und Meteorologie (IGAM), 40
   Sonnenobservatorium Kanzelhöhe, 40
Universität Hamburg
   Hamburger Sternwarte, 254
Universität Hannover
   Institut für Erdmessung (IfE)
      Astronomische Station, 254
Universität Heidelberg
   Institut für Theoretische Astrophysik (ITA), 254
Universität Innsbruck
   Institut für Astronomie, 40
Universität Jena
   Astrophysikalisches Institut und Universitäts-Sternwarte, 211, 255
Universität Kiel
   Institut für Astronomie und Astrophysik, 255
   Institut für Theoretische Physik und Astrophysik, 242, 255
Universität Köln
   Erstes Physikalisches Institut, 255
Universität München
   Institut für Astronomie und Astrophysik
      Observatorium Wendelstein, 255

Universitäts-Sternwarte München (USM), 255
Universität Münster
    Institut für Planetologie (IfP), 256
Universität Potsdam
    Lehrstuhl Astrophysik, 256
Universitäts-Sternwarte Bonn (USB), 253, 256
Universitäts-Sternwarte Göttingen (USG), 229, 256
Universitäts-Sternwarte Jena (USJ), 211, 255, 256
Universitäts-Sternwarte München (USM), 255, 256
Universität Tübingen
    Institut für Astronomie und Astrophysik
        Abteilung Astronomie, 256
    Lehr- und Forschungsbereich Theoretische Astrophysik, 257
Universität Wien
    Institut für Astronomie, 37, 40
    Institut für Geochemie, 41
    Institut für Mathematik, 41
Universität Würzburg
    Astronomisches Institut, 257
Université Catholique de Louvain (UCL)
    Institut d'Astronomie et de Géophysique G. Lemaître, 57
Université d'Aix-Marseille I
    Laboratoire d'Astrophysique de Marseille (LAM), 203
Université d'Aix-Marseille II
    Centre de Physique des Particules de Marseille (CPPM), 165
Université de Bordeaux 1
    Observatoire, 203
Université de Grenoble I
    Laboratoire de Glaciologie et de Géophysique de l'Environnement (LGGE), 203
    Observatoire de Grenoble
        Laboratoire d'Astrophysique (LAOG), 203
Université de la Méditerranée
    Centre de Physique des Particules de Marseille (CPPM), 165
Université de Lausanne
    Institut d'Astronomie, 513
        Observatoire, 513
Université de Liège
    Centre d'Histoire des Sciences, 53
Université de Liège (ULg)
    Centre Spatial de Liège (CSL), 58
    Institut d'Astrophysique et de Géophysique (IAGL), 56, 58
    Institut de Mathématique, 56
Université de Lille
    Laboratoire d'Astronomie, 204
Université de Moncton
    Département de Physique, 100
Université de Mons-Hainaut
    Département d'Astrophysique et de Spectroscopie, 58
Université de Montpellier II Sciences et des Techniques du Languedoc
    Institut des Sciences de la Terre, de l'Eau et de l'Espace de Montpellier (ISTEEM), 183, 204
Université de Montréal
    Département de Physique, 80
        Groupe d'Astronomie, 100
    Observatoire Astronomique du Mont Mégantic (OAMM), 100
Université de Nice-Sophia-Antipolis
    Département d'Astrophysique, 204
Université Denis Diderot, 205
Université de Paris-Sud, 204
Université de Paris XI
    Centre de Spectrométrie Nucléaire et de Spectrométrie de Masse (CSNSM), 204
    Laboratoire d'Astronomie, 173, 204
    Laboratoire de Physique des Gaz et des Plasmas (LPGP), 204
Université de Picardie Jules Verne (UPJV)
    Laboratoire de Physique Théorique et d'Astrophysique, 205
Université des Sciences et Technologies de Lille, 205

Université de Strasbourg I
    Centre de Données astronomiques de Strasbourg (CDS), 165, 205
    Institut d'Histoire des Sciences, 186
    Observatoire Astronomique, 165, 205
Université du Québec à Trois-Rivières (UQTR)
    Département de Physique, 101
Universiteit Amsterdam
    Faculteit der Natuurkunde en Sterrenkunde, 404
    Instituut voor Hoge-Energie Astrofysica, 404
    Sterrenkundig Instituut Anton Pannekoek, 404
Universiteit Antwerpen
    Rijksuniversitair Centrum Antwerpen (RUCA)
        Onderzoeksgroep Astrofysica, 58
    Universitaire Instelling Antwerpen (UIA)
        Department of Physics, 58
Universiteit Brussel
    Astronomy Group, 59
Universiteit Gent
    Sterrenkundig Observatorium, 59
Universiteit Groningen
    Kapteyn Instituut, 404
Universiteit Leiden
    Raymond and Beverly Sackler Laboratorium voor Astrofysica, 404
    Sterrewacht Leiden, 396, 405
Universiteit Leuven
    Center for Plasma Astrophysics, 59
    Instituut voor Sterrenkunde, 59
Universiteit Nijmegen
    Sterrenkunde, 405
Universiteit Utrecht
    Instituut voor Marien en Atmosferisch Onderzoek Utrecht (IMAU), 405
    Sterrekundig Instituut, 405
Université Joseph Fourier, 205
Université Laval
    Département de Physique
        Groupe de Recherche en Astrophysique, 101
Université Libre de Bruxelles (ULB)
    Institut d'Astronomie et d'Astrophysique (IAA), 59
    Physique Nucléaire Théorique et Physique Mathématique (PNTPM), 59
Université Louis Pasteur (ULP), 205
Université Pierre et Marie Curie, 205
Universities Space Research Association (USRA)
    Center for Advanced Space Studies (CASS), 886
    Corporate Office, 744
    Division of Educational Programs (DEP), 886
    Division of Space Life Sciences (DSLS), 886
    Huntsville Program Office (HPO), 590
    Lunar and Planetary Institute (LPI), 886
University College Dublin (UCD)
    Experimental Physics Department, 298
University College London (UCL)
    Department of Physics and Astronomy, 574
        University of London Observatory (ULO), 574
    Mullard Space Science Laboratory (MSSL), 574
University Corp. for Atmospheric Research (UCAR), 656, 659
University Microfilms International (UMI), 767
University of Adelaide
    Department of Physics
        High-Energy Astrophysics Research Group, 29
University of Akron
    Department of Physics, 844
University of Alabama
    Birmingham
        Department of Physics, 590
    Huntsville
        Center for Space Plasma, Aeronomy, and Astrophysics Research (CSPAAR), 590
    Tuscaloosa
        Department of Physics and Astronomy, Astronomy Program, 591
University of Alaska in Anchorage (UAA)
    Department of Physics and Astronomy, 592

University of Alberta
  Department of Physics, 101
  Space Physics Group, 101
University of Arizona
  Astronomy Club, 603
  Center for Astronomical Adaptive Optics (CAAO), 603
  Department of Astronomy
    Steward Observatory, 603
  Department of Planetary Sciences
    Lunar and Planetary Laboratory, 603
  Lunar and Planetary Laboratory, 603
  Multiple Mirror Telescope Observatory (MMTO), 598, 603
  Optical Sciences Center (OSC), 604
  Press, 604
  Steward Observatory, 603, 604
  Vatican Observatory Research Group (VORG), 604
University of Arizona Press, 604
University of Arkansas, Fayetteville
  Department of Physics, 606
University of Arkansas, Little Rock (UALR)
  Department of Physics and Astronomy, 606
University of Arkansas, Monticello (UAM)
  Pomeroy Planetarium, 607
University of Bergen
  Department of Physics, 419
University of Birmingham
  School of Physics and Space Research
    Astrophysics and Space Research Group, 574
University of Bradford
  Faculty of Engineering
    Engineering in Astronomy (EIA) Group, 574
University of Bristol
  H.H. Wills Physics Laboratory, 574
University of British Columbia (UBC)
  Department of Physics and Astronomy, 101
University of Budapest, 278
University of Calgary
  Department of Physics and Astronomy
    Rothney Astrophysical Observatory (RAO), 101
  Institute for Space Research, 102
University of California at Berkeley (UCB), 639
  Berkeley-Illinois-Maryland Association (BIMA), 647
  Center for Extreme Ultraviolet Astrophysics (CEA), 646
  Department of Astronomy, 646
  Department of Physics, 646
  E.O. Lawrence Berkeley National Laboratory (LBNL)
    Advanced Light Source Project (ALS), 608
    Institute for Nuclear and Particle Astrophysics (INPA), 620
  Lawrence Hall of Science (LHS), 610, 651
  Radio Astronomy Laboratory (RAL)
    Hat Creek Observatory, 646
  Space Sciences Laboratory (SSL), 647
  William Knox Holt Planetarium, 651
University of California at Davis (UCD)
  Astronomy Club, 647
  Department of Chemistry, 647
  Department of Physics, 647
University of California at Irvine (UCI)
  Department of Physics and Astronomy, 647
University of California at Los Angeles (UCLA)
  Department of Earth and Space Sciences, 647
  Department of Physics and Astronomy
    Division of Astronomy, 648
    Institute of Geophysics and Planetary Physics (IGPP), 648
University of California at Riverside (UCR)
  Institute of Geophysics and Planetary Physics (IGPP), 648
University of California at San Diego (UCSD)
  Center for Astrophysics and Space Sciences (CASS), 648
  Physics Department, 648
  Scripps Institution of Oceanography, 614
University of California at Santa Barbara (UCSB)
  Institute for Theoretical Physics, 649
  Physics Department, 648
University of California at Santa Cruz (UCSC)
  Department of Astronomy and Astrophysics, 649
  Lick Observatory, 649
    Headquarters, 649
University of Cambridge
  Department of Applied Mathematics and Theoretical Physics (DAMTP), 575
  Institute of Astronomy (IoA), 540, 575
    X-Ray Astronomy Group, 575
  Isaac Newton Institute for Mathematical Sciences, 552, 575
University of Canterbury
  Department of Physics and Astronomy, 412
    Mount John University Observatory (MJUO), 412
University of Cape Town (UCT)
  Department of Applied Mathematics, 468
  Department of Astronomy, 469
  Department of Physics, 469
University of Central Arkansas (UCA)
  Department of Physics and Astronomy, 607
  Lewis Science Center
    Observatory, 607
    Planetarium, 607
University of Central Florida (UCF)
  Department of Physics, 686
University of Central Lancashire
  Department of Physics, Astronomy and Mathematics, 575
University of Chicago
  Center for Astrophysical Thermonuclear Flashes, 707
  Department of Astronomy and Astrophysics, 707
  Department of Mathematics, 707
  Department of Physics, 707
  Enrico Fermi Institute (EFI), 707
  Yerkes Observatory, 915, 919
University of Chicago Press (UCP), 708
University of Cincinnati
  Department of Physics, 844
University of Colorado
  BioServe Space Technologies (BST), 653
  Center for Astrophysics and Space Astronomy (CASA), 659
    Astrophysics Research Laboratory (ARL), 659
  Cooperative Institute for Research in Environmental Sciences (CIRES), 660
  Department of Aerospace Engineering Sciences, 660
  Department of Astrophysical and Planetary Sciences (APS), 660
  Joint Institute for Laboratory Astrophysics (JILA), 655, 660
  Laboratory for Atmospheric and Space Physics (LASP), 660
University of Colorado at Colorado Springs (UCCS)
  Department of Physics, 660
University of Connecticut
  Department of Physics, 665
University of Crete
  Department of Physics
    Astrophysics and Space Physics, 269
University of Delaware
  Bartol Research Institute (BRI), 667
  Department of Physics and Astronomy, 668
University of Denver
  Department of Physics and Astronomy, 660
University of Durham
  Department of Physics, 575
University of Edinburgh
  Institute for Astronomy, 576
  School of Mathematics, 576
University of Exeter
  School of Physics
    Astrophysics Group, 576
University of Florida
  Department of Astronomy, 686
University of Georgia at Athens (UGA)

Department of Physics and Astronomy, 691
University of Glasgow
   Department of Aerospace Engineering, 576
   Department of Physics and Astronomy
      Astronomy and Astrophysics Group, 576
      Observatory, 576
University of Guam (UoG)
   Planetarium, 271
University of Guelph
   Department of Physics, 102
University of Hawaii
   Hawaii Institute of Geophysics and Planetary Physics, 694
University of Hawaii, Hilo
   Department of Physics and Astronomy, 695
University of Hawaii, Honolulu
   C.E. Kenneth Mees Solar Observatory and Lunar Ranging Facility, 695
   Department of Physics and Astronomy, 696
   Institute for Astronomy, 696
      Infrared Telescope Facility (IRTF), 695
   Institute of Geophysics and Planetology, 696
   Onizuka Center for International Astronomy (OCIA), 696
University of Helsinki
   Observatory, 151
   Observatory and Astrophysical Laboratory, 147
University of Hertfordshire
   Department of Physical Sciences
      Division of Physics and Astronomy, 577
      Observatory, 577
University of Houston
   Department of Physics, 886
   Space Physics Group, 886
   Space Vacuum Epitaxy Center (SVEC), 887
University of Iceland
   Science Institute, 279, 280
University of Illinois
   Astronomical Society (UIAS), 708
   Berkeley-Illinois-Maryland Association (BIMA), 647
   National Center for Supercomputing Applications (NCSA), 703
University of Illinois at Springfield
   Department of Astronomy-Physics, 708
University of Illinois at Urbana
   Department of Aero-Astro Engineering, 701
   Department of Astronomy, 708
   Laboratory for Astronomical Imaging (LAI), 708
   Laboratory for Computational Astrophysics (LCA), 708
University of Illinois Press, 709
University of Ionnina
   Department of Physics
      Section of Astro-Geophysics, 269
University of Iowa
   Department of Physics and Astronomy, 719
University of Iowa Press, 719
University of Istanbul, 523
University of Joensuu
   Department of Physics, 150
University of Kansas
   Department of Physics and Astronomy, 723
University of Keele
   Department of Physics, 577
University of Kent
   Electronic Engineering Laboratories
      Radio Astronomy Group, 577
   Unit for Space Sciences and Astrophysics, 577
University of Kentucky
   Department of Physics and Astronomy, 727
University of Kolkata
   Department of Applied Physics, 287
University of Lagos
   Nigerian Academy of Sciences, 414
University of Latvia
   Institute of Astronomy, 371
University of Leeds
   Department of Physics and Astronomy, 577

University of Leicester
   Astronomy Society, 578
   Department of Physics and Astronomy, 554, 578
University of Lethbridge
   Astrophysical Observatory, 102
   Department of Physics, 102
University of Liverpool
   Department of Physics, 578
University of Ljubljana
   Department of Physics
      Astronomical and Geophysical Observatory, 463
University of London
   Goldsmith's College
      Department of Mathematical and Computer Sciences, 578
   Queen Mary and Westfield College (QMWC)
      Astronomy Unit, 578
      Astrophysics Group, 579
University of London Observatory (ULO), 574, 579
University of Louisiana at Lafayette
   Department of Physics, 731
University of Louisville
   Department of Physics, 727
   Gheens Science Hall and Rauch Planetarium, 725
   Moore Observatory, 727
University of Maine
   Department of Physics and Astronomy, 733
   Maynard F. Jordan Planetarium and Observatory, 733
University of Manchester
   Department of Physics and Astronomy, 579
      Jodrell Bank Observatory (JBO), 579
   Institute of Science and Technology
      Goddlee Observatory, 555
University of Manchester Institute of Science and Technology (UMIST)
   Department of Physics
      Astrophysics Group, 579
University of Manitoba
   Department of Physics and Astronomy, 102
University of Maryland
   Berkeley-Illinois-Maryland Association (BIMA), 647
   Department of Astronomy, 744
   Laboratory for Millimeter-Wave Astronomy, 744
University of Maryland Baltimore County (UMBC)
   Joint Center for Astrophysics (JCA), 744
   Physics Department, 744
University of Massachusetts
   Astronomy Department, 759
   Five College Radioastronomy Observatory, 751
University of Mauritius (UoM)
   Faculty of Science, 381
University of Melbourne
   School of Physics
      Astrophysics Group, 29
University of Memphis
   Department of Physics, 875
University of Michigan
   Department of Aerospace Engineering, 767
   Department of Astronomy, 768
   Department of Atmospheric, Oceanic and Space Sciences, 768
   Department of Natural Sciences, 768
   Exhibit Museum of Natural History, 768
   MDM Observatory, 598
   Michigan Dartmouth MIT Observatory, 598, 768
   Solar and Heliospheric Research Group, 768
   Student Astronomical Society, 768
University of Michigan Press, 768
University of Minnesota
   Institute for Mathematics and its Applications (IMA), 773
   School of Physics and Astronomy, 773
University of Mississippi
   Department of Physics and Astronomy, 775
University of Missouri
   Columbia
      Department of Physics and Astronomy, 778

Saint Louis (UMSL)
   Department of Physics and Astronomy, 778
University of Montana
   Department of Physics and Astronomy, 780
University of Nebraska
   Department of Physics, 784
   Mueller Planetarium, 782
University of Nebraska, Kearney (UNK)
   Department of Physics and Physical Science, 784
University of Nebraska, Lincoln (UNL)
   Behlen Observatory, 784
   Department of Physics and Astronomy, 784
   Ralph Mueller Planetarium, 784
University of Nevada
   Department of Physics, 787
   Fleischmann Planetarium and Science Center, 786
University of New Brunswick
   Department of Physics, 102
University of Newcastle
   Department of Physics, 26
University of Newcastle-upon-Tyne
   Department of Physics, 579
University of New Hampshire
   Physics Department
      Observatory, 790
   Space Science Center, 790
University of New Mexico (UNM)
   Department of Physics and Astronomy, 803
   Institute for Astrophysics, 804
   Institute of Meteoritics (IOM), 804
      Meteorite Museum, 801
   Meteorite Museum, 801
University of New South Wales (UNSW)
   Australian Defence Force Academy (AFDA)
      School of Physics, 18
   Centre for Remote Sensing, 30
   Department of Astrophysics and Optics, 30
   University College
      Department of Physics, 30
University of Nigeria
   Department of Physics and Astronomy, 414
University of North Alabama (UNA)
   Planetarium and Observatory, 591
University of North Carolina
   Department of Physics and Astronomy, 832
   Morehead Planetarium, 829
University of North Carolina in Greensboro (UNCG)
   Department of Physics and Astronomy
      Three College Observatory (TCO), 832
University of North Dakota (UND)
   Department of Space Studies, 834
University of Northern Iowa
   Chemistry Department, 717
   Iowa Academy of Science, 717
University of Northern Iowa (UNI)
   Department of Earth Science, 720
   Hillside Observatory, 720
University of North Texas (UNT)
   Department of Physics, 887
   Sky Theater, 887
University of Notre Dame
   Department of Physics
      Astrophysics Group, 714
University of Nottingham
   School of Geography, 563
   School of Physics and Astronomy, 580
University of Oklahoma
   Department of Physics and Astronomy, 847
University of Oregon
   Department of Physics, 850
   Solar Radiation Monitoring Laboratory (SRML), 850
University of Oslo
   Department of Physics, 416, 419
   Institute of Theoretical Astrophysics, 419
University of Oulu
   Department of Physical Sciences
      Astronomy Division, 151

University of Oxford
   Clarendon Laboratory, 580
   Department of Physics
      Astrophysics, 580
      Atmospheric, Oceanic and Planetary Physics, 580
      Theoretical Physics, 580
University of Patras
   Department of Physics, 269
University of Pennsylvania
   Department of Physics and Astronomy, 862
University of Pittsburgh
   Allegheny Observatory, 862
   Department of Geology and Planetary Science, 862
   Department of Physics and Astronomy, 862
University of Portsmouth
   School of Computer Science and Mathematics
      Relativity and Cosmology Group, 580
University of Pretoria
   Faculty of Science, 467
University of Puerto Rico (UPR)
   Humacao
      Department of Physics, 437
      Observatory, 437
University of Queensland
   Department of Physics, 30
      Mount Tamborine Observatory (MTO), 30
University of Regina
   Department of Physics, 102
University of Rochester
   Department of Physics and Astronomy, 825
University of Saint Andrews
   Mathematics and Statistics Department
      Solar Theory Group, 580
   School of Physics and Astronomy, 581
University of Saint Francis
   E.C. Schouweiler Planetarium, 711
University of Saskatchewan
   Department of Physics and Engineering Physics, 103
   Institute of Space and Atmospheric Studies (ISAS), 103
University of Science and Technology of China (USTC)
   Center for Astrophysics, 118
University of Science Malaysia
   Astronomy and Atmospheric Research Unit (AARU), 379
   Sheikh Tahir Astronomical Centre, 378
University of Sheffield
   Department of Physics and Astronomy, 581
University of Sofia
   Department of Astronomy, 76
University of South Africa (UNISA)
   Department of Mathematics, Applied Mathematics and Astronomy, 469
University of Southampton
   Physics and Astronomy Department, 581
University of South Australia
   School of Environmental and Recreation Management
      Planetarium, 30
University of South Carolina Press, 868
University of South Carolina (USC)
   Department of Physics and Astronomy, 868
   Melton Memorial Observatory, 868
   Press, 868
University of Southern California (USC)
   Department of Physics and Astronomy, 649
   Space Sciences Center, 650
University of Southern Maine
   Southworth Planetarium, 733
University of Southwestern Louisiana
   Department of Physics, 731
University of St. Andrews
   Mathematical and Statistics Department
      Solar Theory Group, 580
   School of Physics and Astronomy, 581
University of Surrey
   Surrey Satellite Technology Ltd. (SSTL), 571
   Surrey Space Centre, 571
University of Sussex

Astronomy Centre, 581
School of Chemistry, Physics and Environmental Science, 581
SPRU - Science and Technology Policy Research, 581
University of Sydney
Chatterton Astronomy Department, 30
Molonglo Radio Observatory, 31
Research Centre for Theoretical Astrophysics (RCfTA), 31
School of Mathematics and Statistics, 31
School of Physics, 31
University of Szeged, 278
University of Tasmania
Department of Physics, 15, 31
Mount Pleasant Radio Astronomy Observatory, 31
University of Teesside
School of Computing and Mathematics, 582
University of Tennessee at Chattanooga (UTC)
Clarence T. Jones Observatory, 875
Clarence T. Jones Planetarium, 875
Department of Physics, Geology and Astronomy, 875
University of Tennessee at Knoxville (UTK)
Department of Physics and Astronomy, 875
Press, 876
University of Tennessee at Tullahoma (UTT)
Space Institute (UTSI), 875
Astronomy Club, 875
University of Tennessee Press, 875
University of Texas at Arlington (UTA)
Astronomical Society, 887
Department of Physics, 887
Planetarium, 887
University of Texas at Austin
Astronomy Department, 887
Center for Space Research (CSR), 887
Department of Physics, 888
Institute for Fusion Studies (IFS), 888
McDonald Observatory, 888
Administrative and Research Offices, 888
Hobby-Eberly Telescope (HET), 880
University of Texas at Dallas (UTD)
William B. Hanson Center for Space Science (CSS), 888
University of Texas Press, 889
University of Thessaloniki
Department of Physics
Section of Astrophysics, Astronomy and Mechanics, 270
University of the Witwatersrand
Department of Computational and Applied Mathematics, 469
Department of Physics, 469
University of Tokyo
Department of Astronomy, 360
Department of Earth and Planetary Physics, 360
Department of Earth Science and Astronomy, 360
Department of Physics
Theoretical Astrophysics Group, 360
Institute for Cosmic Ray Research (ICCR), 360
Akeno Observatory, 360
Kamioka Underground Observatory, 361
Institute of Astronomy
Kiso Observatory, 361
Research Center for the Early Universe (RESCEU), 361
Tokyo Astronomical Observatory, 361
University of Toledo
Department of Physics and Astronomy
Ritter Astrophysical Research Center (RARC), 844
University of Toronto
Canadian Institute of Theoretical Astrophysics (CITA), 81
Department of Astronomy and Astrophysics, 103
David Dunlap Observatory (DDO), 103
Erindale Campus, 103
Institut Canadien d'Astrophysique Théorique (ICAT), 81
Institute for Aerospace Studies (UTIAS), 104
University of Toronto Press Inc., 104

University of Toronto Southern Observatory (UTSO), 103
University of Tromsø
Department of Physics
Astrophysics Group, 420
Auroral Observatory, 419
University of Turku
Space Research Laboratory, 152
Tuorla Observatory, 152
University of Utah
Physics Department
Observatory, 891
University of Vermont
Physics Department, 894
University of Victoria
Department of Physics and Astronomy, 96, 104
Physics and Astronomy Student Society (PASS), 96
University of Virginia
Department of Astronomy, 904
University of Waikato
Mathematics Department, 412
University of Wales, Aberystwyth
Department of Physics, 582
University of Wales, College of Cardiff (UWCC)
Physics and Astronomy Department, 582
University of Warwick
Physics Department
Space and Astrophysics Group, 582
University of Washington
Department of Astronomy, 910
Department of Physics, 910
University of Waterloo
Department of Physics
Astronomy/Gravitation Group, 104
University of Western Australia (UWA)
Department of Physics, 31
University of Western Ontario (UWO)
Physics and Astronomy Department, 104
University of Western Sydney (UWS)
Nepean Astronomy Centre (NAC), 32
University of West Indies (UWI)
Saint Augustine
Department of Physics, 521
Trinidad
Department of Physics, 521
University of Wisconsin, Eau Claire
Department of Physics and Astronomy, 919
University of Wisconsin, Fox Valley
Barlow Planetarium, 915
University of Wisconsin, La Crosse
Department of Physics, 916, 919
Planetarium, 919
University of Wisconsin, Madison
Department of Astronomy, 920
Department of Physics, 920
Pine Bluff Observatory (PBO), 920
Press, 921
Space Science and Engineering Center (SSEC), 920
University of Wisconsin, Milwaukee (UWM)
Department of Physics
Manfred Olson Planetarium, 920
University of Wisconsin, Oshkosh
Department of Physics and Astronomy, 921
University of Wisconsin Press, 921
University of Wisconsin, River Falls (UWRF)
Department of Physics, 921
University of Wisconsin, Stevens Point (UWSP)
Department of Physics and Astronomy, 921
University of Wollongong (UoW)
Department of Physics
Astronomy and Astrophysics Group, 32
University of Wyoming
Department of Physics and Astronomy, 923
Edison Infrared Space Observatory (EISO), 923
Planetarium, 923
Red Buttes Observatory (RBO), 923
Wyoming Infrared Observatory (WIRO), 923
University of Zagreb

Faculty of Geodesy
    Hvar Observatory, 122
    Institute of Physics, 122
University of Zululand
    Physics Department
        South African Institute of Physics (SAIP), 467
University Optics, 769
University Park, 859, 860
University Science Books (USB), 904
Universum - Museo de las Ciencias, 386
UNK
    University of Nebraska, Kearney
        Department of Physics, 784
UNL
    University of Nebraska, Lincoln
        Behlen Observatory, 784
        Department of Physics and Astronomy, 784
        Ralph Mueller Planetarium, 784
UNLP
    Universidad Nacional de La Plata
        Facultad de Ciencias Astronómicas y Geofísicas, 10
        Observatorio Astronómico, 10
UNM
    University of New Mexico
        Department of Physics and Astronomy, 803
        Institute for Astrophysics, 804
        Institute of Meteoritics, 801, 804
        IOM, 804
        Meteorite Museum, 801
UNMS
    Slovak Office of Standards, Metrology and Testing, 460
UNMSM
    Universidad Nacional Mayor de San Marcos
        Seminario Permanente de Astronomía y Ciencias Espaciales, 423
        SPACE, 423
UNO
    United Nations Organization
        Office for Outer Space Affairs, 40
        World Intellectual Property Organization, 514
UNSW
    University of New South Wales
        Australian Defence Force Academy, 18
        Centre for Remote Sensing, 30
        Department of Astrophysics and Optics, 30
        Department of Physics, 30
UNT
    University of North Texas
        Department of Physics, 887
        Sky Theater, 887
UoM
    University of Mauritius
        Faculty of Science, 381
UoW
    University of Wollongong
        Astronomy and Astrophysics, 32
        Department of Physics, 32
Úpice, 125, 131, 132
Úpice Observatory, 125, 132
UPJV
    Université de Picardie Jules Verne
        Laboratoire de Physique Théorique et d'Astrophysique, 205
Upland, 634
Upminster, 548
Upper Montclair, 793
Uppsala, 495, 498
Uppsala Amatörastronomer (UAA), 498
Uppsala Southern Station, 498
Uppsala University
    Astronomical Observatory, 498
    Kvistaberg Station, 498
UPR
    University of Puerto Rico
        Department of Physics, 437
        Humacao, 437
        Observatory, 437

UPS
    Ukrainian Physical Society, 529
UPSO
    Uttar Pradesh State Observatory, 288
Upton, 809
UQAR
    Université du Québec à Rimouski, 84
UQTR
    Université du Québec à Trois-Rivières, 101
URACA
    Union Rhône-Alpes des Clubs d'Astronomie, 203
Ural State University
    Department of Astronomy and Geodesy, 451
Urania, 340
    Budapest, 278
    Volkssterrenwacht van Antwerpen, 60
    Wien, 41
    Zürich, 504
Urania Observatory, 278
    Salgótarján, 273
Urania-Sternwarte, 513
    (Gesellschaft der Freunde der_), 504
    Wien, 41
Uranoscope de France, 205
Uranoscope de l'Île de France, 205
Urbana, 699, 701, 705, 708
URL
    Universidad Ramón Llull
        Observatorio del Ebro, 486
Ursa Astronomical Association, 152
URSA Astronomical Society, 148, 152
URSA Observatory, 152
URSI
    International Union of Radio Science, 53
Uruguay, 586
Urumqi Astronomical Observatory (UAO), 115, 118
USA
    Alabama, 589
    Alaska, 592
    Arizona, 593
    Arkansas, 606
    California, 608
    Carolina
        (North_), 827
        (South_), 866
    Colorado, 652
    Connecticut, 662
    Dakota
        (North_), 834
        (South_), 869
    Delaware, 667
    District of Columbia, 669
    Florida, 678
    Georgia, 687
    Hawaii, 692
    Idaho, 697
    Illinois, 698
    Indiana, 710
    Iowa, 716
    Kansas, 721
    Kentucky, 725
    Louisiana, 729
    Maine, 732
    Maryland, 734
    Massachusetts, 746
    Michigan, 761
    Minnesota, 770
    Mississippi, 774
    Missouri, 776
    Montana, 780
    Nebraska, 782
    Nevada, 786
    New Hampshire, 788
    New Jersey, 791
    New Mexico, 799
    New York State, 805
    North Carolina, 827

North Dakota, 834
Ohio, 835
Oklahoma, 846
Oregon, 848
Pennsylvania, 851
Rhode Island, 864
South Carolina, 866
South Dakota, 869
Tennessee, 870
Texas, 877
United States of America, 589
Utah, 890
Vermont, 893
Virginia, 895
(West.), 912
Washington, 907
West Virginia, 912
Wisconsin, 915
Wyoming, 922
US Air Force Academy Planetarium, 652
US Air Force Office of Scientific Research (AFOSR), 669
USB
    Universitäts-Sternwarte Bonn, 253
    University Science Books, 904
USC
    University of South Carolina
        Department of Physics and Astronomy, 868
        Melton Memorial Observatory, 868
        Press, 868
    University of Southern California
        Department of Physics and Astronomy, 649
        Space Sciences Center, 650
USG
    Universitäts-Sternwarte Göttingen, 229, 256
US Geological Survey (USGS), 905
USGS
    United States Geological Survey, 905
        Flagstaff Field Center, 603
USJ
    Universitäts-Sternwarte Jena, 211, 255
USM
    Universitäts-Sternwarte München, 255
USNA
    US Naval Academy
        Physics Department, 744
US Naval Academy (USNA)
    Physics Department, 744
US Naval Observatory (USNO), 676
    Flagstaff Station (NOFS), 604
    National Earth Orientation Service (NEOS), 677
    Washington, 676
USNO
    US Naval Observatory
        Flagstaff Station, 604
        National Earth Orientation Service, 677
        Washington, 676
USO
    Udaipur Solar Observatory, 288
USP
    Universidade de São Paulo
        Departamento de Astronomia, 72
        Instituto de Astronomia, Geofísica e Ciências Atmosféricas, 72
USRA
    Universities Space Research Association
        Center for Advanced Space Studies, 886
        Corporate Office, 744
        Division of Educational Programs, 886
        Division of Space Life Sciences, 886
        Huntsville Program Office, 590
        Lunar and Planetary Institute, 886
US Space Camp/US Space and Rocket Center, 591
US Space VLBI Project, 613
Ussurijsk, 448
Ussurijsk Astronomical Observatory (UAO), 448
Ussurijsk Astrophysical Observatory (UAO), 451
USTC
    University of Science and Technology of China
        Center for Astrophysics, 118
USU
    Utah State University
        Center for Atmospheric and Space Sciences, 891
US Virgin Islands, 924
UTA
    University of Texas at Arlington
        Astronomical Society, 887
        Department of Physics, 887
        Planetarium, 887
Utah, 890–892
Utah State University (USU)
    Center for Atmospheric and Space Sciences (CASS), 891
Utah Valley Astronomy Association (UVAA), 891
UTC
    University of Tennessee at Chattanooga
        Clarence T. Jones Observatory, 875
        Clarence T. Jones Planetarium, 875
        Department of Physics, Geology and Astronomy, 875
UTD
    University of Texas at Dallas
        William B. Hanson Center for Space Science, 888
UTIAS
    University of Toronto Institute for Aerospace Studies, 104
UTK
    University of Tennessee at Knoxville
        Department of Physics and Astronomy, 875
        Press, 876
UTRC
    United Technologies Research Center, 665
Utrecht, 397, 401–403, 405
UTSI
    University of Tennessee Space Institute, 875
UTSO
    University of Toronto Southern Observatory, 103
UTT
    University of Tennessee at Tullahoma, 875
        Space Institute, 875
Uttar Pradesh State Observatory (UPSO), 288
UVAA
    Utah Valley Astronomy Association, 891
UWA
    University of Western Australia
        Department of Physics, 31
UWCC
    University of Wales, College of Cardiff
        Physics and Astronomy Department, 582
UWI, University of West Indies
    Saint Augustine
        Department of Physics, 521
    Trinidad
        Department of Physics, 521
UWM
    University of Wisconsin, Milwaukee
        Manfred Olson Planetarium, 920
UWO
    University of Western Ontario
        Elginfield Observatory, 104
        Physics and Astronomy Department, 104
UWRF
    University of Wisconsin, River Falls
        Department of Physics, 921
        Observatory, 921
        Planetarium, 921
UWS
    University of Western Sydney
        Nepean Astronomy Centre, 32
UWSP
    University of Wisconsin, Stevens Point
        Department of Physics and Astronomy, 921
UYAA
    Ukrainian Youth Aerospace Association "Suzirya", 529
Uzbek Academy of Sciences
    Maidanak Observatory, 925

Ulugh Beg Astronomical Institute, 925
   Space Geodynamics Department, 925
Uzbekistan, 374, 925
Uzbek Main Administration of Hydrometeorology, 925
Uzbek State Centre for Standardization, Metrology and Certification (UZGOST), 925
UZGOST
   Uzbek State Centre for Standardization, Metrology and Certification, 925
Uzhgorod, 529
Uzhgorod State University
   Laboratory for Space Research, 529

V

VAA
   Vorarlberger Amateur Astronomen, 41
VAAE
   Vereinigte Amateur-Astronomen Eschwege 1975 eV, 258
Vacoas, 381
Vadodara, 281
Vainu Bappu Observatory (VBO), 285
Väisälä Institute for Space Physics and Astronomy (VISPA), 152
Valais, 510
Valašské Meziříčí, 132
Valašské Meziříčí Astronomical Observatory, 132
Valcourt, 200
Valcuvia, 324
Valdosta, 690, 691
Valdosta State University (VSU)
   Department of Physics, Astronomy and Geosciences, 691
      Astronomy Program, 690
Valence, 158
Valencia, 475, 485, 486, 624
Valinhos, 72
Valladolid, 483
Valletta, 380
Valley City, 834
Valley City State University (VCSU)
   Planetarium, 834
Valleyfield, 85
Valley of the Moon Observatory Association (VMOA), 650
Valongo, 73
Valparaiso, 714
Valparaiso University
   Department of Physics and Astronomy, 714
Valparaíso, 111
Valromey, 170
Vancouver, 81, 89, 98, 101
Vanderbilt Museum
   Planetarium, 825
Vanderbilt University
   Arthur J. Dyer Observatory, 870, 876
   Department of Physics and Astronomy, 876
   Dyer Observatory (Arthur J._), 870, 876
   Press, 876
Vanderbilt University Press, 876
Vandoeuvre-lès-Nancy, 167
Vanguard Research Inc.
   Astronomy and Astrophysics, 650
Vannes, 188
Van Nuys, 625, 626
Van Slyke Engineering, 661
Vantaa, 152
Van Vleck Observatory (VVO), 665
Van Zandt Observatory (Decker-Grebner-_), 704
Varese, 335
Variable Star Observers League in Japan (VSOLJ), 361
Varna, 74
Várzea, 66
VAS
   Vectis Astronomical Society, 582
   Verein der Amateurastronomen des Saarlandes, 257
   Vermont Astronomical Society, 894

Villanova Astronomical Society, 863
Vlašim Astronomical Society, 133
VASC
   Virginia Air and Space Center, 905
Vassar College
   Department of Physics and Astronomy, 825
Västerås, 499
Västerås Astronomi och Rymdforsknings Förening, 499
Vastra Frolunda, 488
Vatican, 926
Vatican Observatory, 926
Vatican Observatory Research Group (VORG), 604
Vaticano (Città del_), 926
Vaucresson, 187
Vaud, 511
Vaulx-en-Velin, 197
Vaure, 198
Vayres, 171
VBAS
   Von Braun Astronomical Society, 591
VBO
   Vainu Bappu Observatory, 285
VCAS
   Ventura County Astronomical Society, 650
VCSU
   Valley City State University
      Planetarium, 834
VdS
   Vereinigung der Sternfreunde, 258
VEB
   Volkssternwarte Erich Bär, 260
Vectis Astronomical Society (VAS), 582
Veen Observatory (James C._), 764
Vega-Bray Observatory, 604
Vehrenberg KG, 257
Veleta (Pico_), 182, 478
Vélizy, 166
Velmex Inc., 825
Velp, 390
Venango County Astronomy Club, 859
Vendée, 164
Venedig, 308
Venezia, 308
Venezuela, 927
Venice, 308, 684
Ventnor, 582
Ventspils, 371
Ventspils International Radioastronomy Centre (VIRAC), 371, 372
Ventura County Astronomical Society (VCAS), 650
Venus, 686
Verein Antares, 41
Verein der Amateurastronomen des Saarlandes (VAS) eV, 257
Verein für Volkstümliche Astronomie Essen (VVA) eV, 257
Verein Herzberger Sternfreunde eV, 257
Verein Historische Sternwarten Gotha eV, 258
Vereinigte Amateur-Astronomen Eschwege 1975 eV (VAAE), 258
Vereinigung der Nordenhamer Sternfreunde eV, 258
Vereinigung der Sternfreunde eV (VdS), 258
Vereinigung Krefelder Sternfreunde eV, 258
Verein Sternwarte Rotgrueb Rümlang (VSRR), 513
Vereniging voor Sterrenkunde, Meteorologie, Geofysica en Aanverwante Wetenschappen (VVS), 60
Vereniging voor Weer- en Sterrenkunde Noord Drenthe, 405

Verkehr Raumfarht und Systemtechnik (VRS) GmbH, 258
Verkehrshaus der Schweiz, 514
Verlag Harri Deutsch, 259
Vermont, 893, 894
Vermont Astronomical Society (VAS), 894
Vernadsky Institute of Geochemistry and Analytical Chemistry, 444
Verna (Ramponio_), 321
Verne Theatre, 152
Vernonia, 849

Vernonia Peak, 849
Vero Beach, 681
Verona, 314
Vérossaz, 504
Verrières-le-Buisson, 168
Versailles, 170, 180
Versoix, 505
Very Large Array (VLA), 801
Very Long Baseline Array (VLBA), 801
Veselí nad Moravou, 132
Veselí nad Moravou Observatory, 132
Vestal, 817
Vesta Volkssterrenwacht, 406
Vevey, 510
V.G. Fesenkov Astrophysical Institute, 364
Viareggio, 324, 329
Vic, 473
Vicenza, 323
Vich, 473
Victor, 809
Victor Hasselblad Foundation (Erna and_), 488, 499
Victoria, 16, 93, 94, 96, 104
Victor Valley Community College (VVCC)
    Planetarium, 650
Victorville, 650
Vicuña, 110
Vienna, 34–42
Vigevano, 312
Vignui, 308
Vikram Sarabhai Space Centre (VSSC), 289
Vila Nova de Gaia, 436
Vilanova i la Geltrú, 480
Villa de Leyva, 119
Villa Elisa, 8
Villafontana, 316
Villafranca del Castillo, 476
Villafranca Satellite Tracking Station (VILSPA), 476, 479, 486
Villanova, 852, 863
Villanova Astronomical Society (VAS), 863
Villanova University
    Department of Astronomy and Astrophysics, 852, 863
Villanueva de la Cañada, 479
Villa Park, 706
Villemur, 171
Villeneuve d'Ascq, 169, 171
Villeneuve-Saint-Georges, 164
Villepreux, 157
Villers Bocage, 157
Villers-lès-Nancy, 176
Villeurbanne, 183
Vilnius, 374, 375
Vilnius Planetarium, 374, 375
Vilnius University
    Astronomical Observatory, 375
VILSPA, 486
    Villafranca Satellite Tracking Station, 476, 479
Vimodrone, 330
Vinassan, 160
Vineuil, 157
Violau, 248
VIRAC
    Ventspils International Radioastronomy Centre, 371
Virginia, 895–906
    (Northern_), 901
    (Southeast_), 895
    (West_), 912, 913
    (Western_), 903
Virginia Air and Space Center (VASC), 905
Virginia Beach, 896, 905
Virginia Beach City Public Schools Planetarium, 905
Virginia Living Museum (VLM), 905
Virginia Military Institute (VMI)
    Department of Physics and Astronomy, 905
Virginia Polytechnic Institute and State University
    Astronomy Club, 896
    Department of Physics, 905

Virginia Space Grant Consortium (VSGC), 906
Virginia Tech, 906
    Department of Physics, 905
Virgin Islands, 801
Virgin Islands (US_), 924
Visnjan, 122
Visnjan Observatory, 122
VISPA
    Väisälä Institute for Space Physics and Astronomy, 152
Vissegi (Monte_), 307
Vista, 645
Viterbo, 313
Vitória, 72
Vixen Optical Industries Ltd., 361
VLA
    Very Large Array, 801
Vlašim, 133
Vlašim Astronomical Society (VAS), 133
VLBA
    Very Long Baseline Array, 801
VLBI Space Observatory Program (VSOP), 613
VLM
    Virginia Living Museum, 905
VMI
    Virginia Military Institute
    Department of Physics and Astronomy, 905
VMOA
    Valley of the Moon Observatory Association, 650
VNIIFTRI, 451
    Institute for Physical-Technical and Radio-Technical Measurements, 441
Vogelsang, 252
Voggenberg, 34
Volkshochschule der Stadt Fürstenfeldbruck eV, Arbeitsgemeinschaft Astronomie, 259
Volkshochschule der Stadt Soest, Astronomische Arbeitsgemeinschaft, 259
Volkssternwarte Adolph Diesterweg Radebeul, 259
Volkssternwarte Augsburg, 216
Volkssternwarte Bartholomäus Scultetus, 264
Volkssternwarte Bonn (VSB)
    Astronomische Vereinigung eV, 259
Volkssternwarte Darmstadt (VSD) eV, 260
Volkssternwarte Diedorf bei Augsburg, 216
Volkssternwarte Dortmund, 214
Volkssternwarte Ennepetal eV, 260
Volkssternwarte Erich Bär (VEB), 260
Volkssternwarte Erich Scholz, 260
Volkssternwarte Freiburg, 208
Volkssternwarte Geschwister Herschel Hannover eV, 260
Volkssternwarte Hagen eV, 260
Volkssternwarte Hattingen, 260
Volkssternwarte Heilbronn, 243
Volkssternwarte Herschel, 260
Volkssternwarte im Volksbildungswerk Hofheim-Marxheim, 261
Volkssternwarte Jonsdorf, 261
Volkssternwarte Juri Gagarin, 264
Volkssternwarte Köln, 261
Volkssternwarte Laupheim eV und Planetarium, 261
Volkssternwarte Manfred von Ardenne, 261
Volkssternwarte Marburg eV, 261
Volkssternwarte München, 219
Volkssternwarte Neumarkt/Opf., 219
Volkssternwarte Norderstedt (VSN) eV, 262
Volkssternwarte Nützweid, 500
Volkssternwarte Paderborn eV, 262
Volkssternwarte Papenburg, 214
Volkssternwarte Passau, 262
Volkssternwarte Prenzlau, 262
Volkssternwarte Radeberg, 260
Volkssternwarte Rothwesten, 262
Volkssternwarte Schanfigg Arosa (VSA), 514
Volkssternwarte Singen, 210
Volkssternwarte Soest, 259
Volkssternwarte Tirschenreuth, 263
Volkssternwarte und Planetarium Drebach, 263

Volkssternwarte und Planetarium Reutlingen, 263
Volkssternwarte Urania Jena eV, 263
Volkssternwarte Voggenberg, 34
Volkssternwarte von Ardenne, 261
Volkssternwarte Wetterau eV, 263
Volkssternwarte Würzburg eV, 263
Volkssterrenwacht Amsterdam (VSA), 406
Volkssterrenwacht Beisbroek vzw, 60
Volkssterrenwacht Burgum, 406
Volkssterrenwacht Bussloo (VSB), 406
Volkssterrenwacht Mercurius, 406
Volkssterrenwacht Mira vzw, 60
Volkssterrenwacht Nysa, 60
Volkssterrenwacht Philippus Lansbergen, 406
Volkssterrenwacht Stichting Copernicus, 407
Volks- und Schulsternwarte Bartholomäus Scultetus, 264
Volks- und Schulsternwarte Bruno H. Bürgel eV, 213, 264
Volks- und Schulsternwarte Geretsried eV, 264
Volks- und Schulsternwarte Juri Gagarin, 264
Von Braun Astronomical Society (VBAS), 591
Voorburg, 393, 394
Voorst, 406
Vorarlberg, 41
Vorarlberger Amateur Astronomen (VAA), 41
VORG
    Vatican Observatory Research Group, 604
Vorpommern (Mecklenburg-_), 231
Vries, 406
Vrije Universiteit Amsterdam, 407
Vrije Universiteit Brussel (VUB), 61
VRS
    Verkehr Raumfarht und Systemtechnik, 258
VRS GmbH, 258, 264
VSA
    Volkssternwarte Schanfigg Arosa, 514
    Volkssterrenwacht Amsterdam, 406
VSB
    Volkssternwarte Bonn, 259
    Volkssterrewacht Bussloo, 406
VSD
    Volkssternwarte Darmstadt, 260
Vsetín, 133
Vsetín Observatory, 133
VSGC
    Virginia Space Grant Consortium, 906
VSN
    Volkssternwarte Norderstedt, 262
VSOLJ
    Variable Star Observers League in Japan, 361
VSOP
    VLBI Space Observatory Program, 613
VSRR
    Verein Sternwarte Rotgrueb Rümlang, 513
VSSC
    Vikram Sarabhai Space Centre, 289
VSU
    Valdosta State University
        Astronomy Program, 690
        Department of Physics, Astronomy and Geosciences, 690, 691
VTT, 152
VUB
    Vrije Universiteit Brussel, 61
    Astronomy Group, 59
Vulcan, 79
VVA
    Verein für Volkstümliche Astronomie Essen, 257
VVCC
    Victor Valley Community College
        Planetarium, 650
VVO
    Van Vleck Observatory, 665
VVS
    Vereniging voor Sterrenkunde, Meteorologie, Geofysica en Aanverwante Wetenschappen, 60
Vyhne, 460
Vyškov, 126

W

W3C
    World Wide Web Consortium, 760
WAA
    Westchester Amateur Astronomers, 825
WAAC
    Wollongong Amateur Astronomy Club, 32
Waalre, 390
Wabash Valley Astronomical Society (WVAS), 714
WACO
    Whatcom Association of Celestial Observers, 911
Waco, 878
Wadhurst, 582
Wadhurst Astronomical Society (WAS), 582
W.A. Gayle Planetarium, 591
Wageningen, 394, 396
Wagman Observatory (Nicholas E._), 851
Wahnsiedler Observatory, 711
Waikato, 412
Waimea, 693
Wairarapa Astronomical Society, 412
WAITRO
    World Association of Industrial and Technological Research Organisations, 140
Wajh, 453
Wakayama, 347, 351
Wake Forest University (WFU)
    Department of Physics, 832
Wakkanai, 344
Wakkanai Radio Observatory, 344, 362
Wald, 501
Waldbronn, 241
Walden Observatory, 99
Waldhügel, 210
Wald im Pinzgau, 39
Waldorf, 743
Wales, 537, 540, 546, 562, 568, 572, 573, 582
    (New South_), 15, 18, 30
Wallace Astrophysical Observatory (WAO), 754
Wallace Fiske Planetarium and Science Center, 660, 661
Wallace Planetarium (Alice G._), 746, 759
Wallasey, 554
Walla Walla, 911
Wallops Flight Facility, 906
    Visitor Center, 900
Wallops Island, 900
Wallsend, 16
Walnut Creek, 609, 629
Wals, 38
Walsall, 583
Walsall Astronomical Society, 583
Walter Barber Observatory, 687
Walter-Hohmann-Sternwarte Essen, 208
Walter-Horn-Gesellschaft eV (WHG), 264
Waltham, 748, 749
Wanganui, 412
Wanganui Astronomical Society, 412
Wanne-Eickel, 211
Wantagh, 813
WAO
    Wallace Astrophysical Observatory, 754
Wapakoneta, 840
Ward Beecher Planetarium, 844
Wardle, 579
Ward Pound Ridge Reservation, 825
Warner and Swasey Observatory, 835, 844
Warner Robbins, 689
Warren, 761, 769, 791, 797
Warren Astronomical Society (WAS), 769
Warrendale, 858
Warren Rupp Observatory (WRO), 844
Warrenville, 706
Warrington, 542
Warsash Pty. Ltd., 32
Warsaw, 427, 429–432

Warsaw Astronomical Society (WAS), 715
Warsaw Józefoslaw University of Technology
    Institute of Geodesy and Geodetic Astronomy, 432
Warsaw University
    Astronomical Observatory, 432
Warsaw University of Technology, 432
Warszawa, 427, 429–432
WAS
    Wadhurst Astronomical Society, 582
    Warren Astronomical Society, 769
    Wehr Astronomical Society, 921
    Wellington Astronomical Society, 413
    Weltastronomie Studienreisen, 264
    Wessex Astronomical Society, 583, 584
    Westminster Astronomical Society, 744
    Westport Astronomical Society, 666
    Willingboro Astronomical Society, 797
    Worthing Astronomical Society, 584
    Wycombe Astronomical Society, 584
Washburn Observatory, 920
Washburn University of Topeka
    Department of Physics and Astronomy, 723
    Zenas Crane Observatory, 723
Washburn Valley, 547
Washington, 669–677, 741
Washington College (MWC) (Mary-), 906
    Department of Physics, 899
Washington County Planetarium and Space Science Center, 743, 744
Washington State, 907–911
    (Eastern-), 907
    (Southwest-), 909
    (Western-), 911
Washington State University (WSU)
    Astronomy Club, 910
    Planetarium, 910
    Program in Astronomy, 910
Washington University in Saint Louis (WUStL)
    Department of Earth and Planetary Sciences, 779
    Department of Physics, 779
    Laboratory for Space Sciences, 779
    McDonnell Center for the Space Sciences, 779
Wast Hills Observatory, 574
Waterbury, 664
Waterloo, 50, 104, 105, 717
Waterloo Center for Groundwater Research (WCGR), 83
Waterlooville, 547
Watson-King Planetarium, 743
Watson Research Center (T.J.-), 815
Watukosek, 290
Waubonsie Valley High School
    Planetarium, 709
Waukesha, 916
Waunakee, 917
Wauwatosa, 917
Wayne, 782, 858
Wayne State College (WSC)
    Fred G. Dale Planetarium, 782
Wayne State University (WSU)
    Department of Physics and Astronomy, 769
WBA
    Welsh Border Astronomers, 583
WCAC
    Western Colorado Astronomy Club, 661
    Wilderness Center Astronomy Club, 845
    Williamson County Astronomy Club, 889
WCCSAS
    Western Connecticut Chapter of the Society for Amateur Scientists, 665
WCGR
    Waterloo Center for Groundwater Research, 83
WCSU
    Western Connecticut State University
        Department of Physics, Astronomy and Meteorology, 666
WDC
    World Data Center
        A for Meteorology, 832
        A for Rockets and Satellites, 745
        A for Solar-Terrestrial Physics, 661
        B for Solar-Terrestrial Physics, 451
        C2 for Ionosphere, 362
        for Solar Activity, 362
Weatherford Planetarium, 728
Webb Society, 583
Weber State University (WSU)
    Department of Physics, 892
    Layton P. Ott Planetarium, 890, 891
Webster Groves, 777
Weer- en Sterrenkundige Kring Eemsmond, 407
Wega vzw, 61
Wehr Astronomical Society (WAS) Inc., 921
Weikersheim, 217
Weil Observatory (Johnson-_), 805
Weiskopf Observatory, 699
Weißenschirmbach, 231
Weitkamp Observatory, 842
Welden, 248
Wellesley, 759
Wellesley College
    Department of Astronomy
        Whitin Observatory, 759
Wellingborough, 559
Wellington, 408–413, 835
Wellington Astronomical Society (WAS), 413
Wells Planetarium (John C.-), 899
Welsh Border Astronomers (WBA), 583
Weltastronomie Studienreisen (WAS), 264
Welzheim, 221
Wembley, 14
Wendelstein, 255, 256
Werkgroep Leidse Sterrewacht (WLS), 407
Werner-Heisenberg-Institut (WHI), 237, 265
Wernigerode, 214, 233
Wesleyan University
    Department of Astronomy
        Van Vleck Observatory (VVO), 665
Wesley Droke Observatory (James-), 606
Wessex Astronomical Society (WAS), 583, 584
Weßling, 225
West Bend, 918
West Burlington, 719
Westchester Amateur Astronomers (WAA), 825
West Covina, 636
Westerbork Synthesis Radio Telescope (WSRT), 401
Western Australia, 17, 31
Western Colorado Astronomy Club (WCAC), 661
Western Connecticut Chapter of the Society for Amateur Scientists (WCCSAS), 665
Western Connecticut State University (WCSU)
    Department of Physics, Astronomy and Meteorology, 665
Western Kentucky Amateur Astronomers (WKAA), 728
Western Kentucky University (WKU)
    Department of Physics and Astronomy, 725, 726, 728
    Kentucky Space Grant Consortium (KSGC), 726
Western Montana Astronomical Association (WMAA), 781
Western Observatorium Inc., 650
Western Ontario, 104
Western Space Education Network (WSEN) Inc., 104
Western Sydney Amateur Astronomy Group (WSAAG), 32
Western Washington University (WWU)
    Physics/Astronomy Department, 910
    Planetarium, 911
Westerville, 841
Westerwälder Tag der Astronomie (WTA), 234
Westfälisches Museum für Naturkunde
    Planetarium, 265
Westfälische Volkssternwarte und Planetarium, 265
Westfälische Wilhelms-Universität Münster, 265
Westford, 746, 748, 755, 756
West Hartford, 662
West Indies, 521
West Lafayette, 713, 714

Westminster, 744
Westminster Astronomical Society (WAS), 744
Westmont College
    Department of Physics, 650
        George Carroll Observatory, 651
West-München, 217
West of London Astronomical Society (WOLAS), 583
Weston Creek, 18
West Palm Beach, 679, 681-683, 685
West Perth, 26
West Point, 687
Westport, 666
Westport Astronomical Society Inc. (WAS), 666
Westside Observatory, 666
West Springfield High School Planetarium, 906
West Texas Astronomers (WTA), 889
West Virginia, 912, 913
West Virginia Space Grant Consortium (WVSGC), 913
West Virginia University (WVU)
    Astronomy Club, 913
    Department of Physics, 913
Westwego, 730
West Yorkshire Astronomical Society (WYAS), 583
Wetherbee Planetarium, 691
Wetterau, 263
Wetzlar, 213
Weymouth, 536
WFS
    Wilhelm-Foerster-Sternwarte, 265
WFU
    Wake Forest University
        Department of Physics, 832
Whangarei, 413
Whangarei Astronomical Society Inc., 413
Whatcom Association of Celestial Observers (WACO), 911
Whately Observatory, 758
What in The World, 651
Wheaton, 709
Wheaton College
    Astronomy Program, 759
    Department of Physics and Astronomy, 709
Wheeling, 912
WHG
    Walter-Horn-Gesellschaft, 264
WHI
    Werner-Heisenberg-Institut, 237
Whipple Observatory (FLWO) (Fred Lawrence_), 597, 604, 751
Whitchurch, 538
White Plains, 743
White River Junction, 893
Whiting, 791
Whitin Observatory, 759, 760
Whitman College
    Department of Astronomy, 911
Whittier, 621, 635
Whitworth Ferguson Planetarium, 825
WHO
    Worth Hill Observatory, 584
WHT
    William Herschel Telescope, 479
Wible Planetarium, 851
Wichita, 721, 722, 724
Wichita State University
    Physics Department, 724
Wickware Planetarium (Robert K._), 666
Widener University
    Department of Physics, 863
Wien, 34-42
Wiener Arbeitsgemeinschaft für Astronomie (WAA), 41
Wiesbaden, 212
Wiesendangen, 511
Wikswo Observatory, 873
Wilcoxon Research Inc., 745
Wilcox Solar Observatory (WSO), 642, 651
Wilder Laboratory, 790
Wilderness Center Astronomy Club (WCAC), 845

Wild Observatory (Paul_), 22, 32
Wilfrid Laurier University (WLU)
    Department of Physics and Computing, 105
Wilhelm-Foerster-Sternwarte (WFS) eV, 265
Willard Geer Planetarium, 907, 911
Willard Gibbs Research Laboratory (J._), 666
Willard W. Smith Planetarium, 908, 911
William B. Hanson Center for Space Science, 888, 889
William Herschel Museum, 584
William Herschel Society, 584
William Herschel Telescope (WHT), 479
William J. McCallion Planetarium, 91
William Knox Holt Planetarium, 651
William Marsh Rice University
    Physics and Astronomy Department, 889
William M. Brish Planetarium, 745
William Miller Sperry Observatory, 791
William M. Staerkel Planetarium, 709
William Rainey Harper College
    Observatory, 709
Williams Bay, 915, 919
Williamsburg, 903
Williams College
    Astronomy Department, 760
        Hopkins Observatory, 760
        Milham Planetarium, 760
Williams Observatory (Craven E._), 828, 832
Williamson County Astronomy Club (WCAC), 889
Williamsport, 858
Williamstown, 760
Willimantic, 664
Willingboro Astronomical Society (WAS), 797
Williston, 894
Wills Physics Laboratory (H.H._), 574, 584
Wilmington, 667, 748, 827
Wilmot, 845
Wilrijk, 58
Wilson, 831
Wilson (Mount_), 629, 650
Winchester, 873, 903
Windach, 217
Windber, 851
Windsor, 23
Wine Observatory (Carolyn_), 683
Winfree Observatory, 902
Wingham, 539
Winnipeg, 90, 91, 102
Winona Lake, 713, 715
Winston-Salem, 828, 831, 832
Winston-Salem Astronomical League (WSAL), 832
Winterslow, 567
Winter Star Party (WSP), 685
Winterthur, 509
WIPO
    World Intellectual Property Organization (UNO), 514
WIRO
    Wyoming Infrared Observatory, 923
Wiruna, 16
Wisbech, 573
Wisconsin, 915-921
    (East Central_), 916
Wisconsin-Indiana-Yale-NOAO (WIYN) Observatory, 920
Wise Observatory (Florence and George_), 300, 301
Witham, 537
Wittenberg, 265
Wittenberger Planetarium, 265
Wittenheim, 170
Witte Observatory (John H._), 719
Witzenhausen, 258
WIYN (Wisconsin-Indiana-Yale-NOAO) Observatory, 920
WKAA
    Western Kentucky Amateur Astronomers, 728
WKU
    Western Kentucky University
        Department of Physics and Astronomy, 725, 726, 728
        Kentucky Space Grant Consortium, 726
WLU

Wilfrid Laurier University
    Department of Physics and Computing, 105
WMAA
    Western Montana Astronomical Association, 781
W.M. Keck Observatory, 692, 696
WMO
    World Meteorological Organization, 514
Woking, 571
WOLAS
    West of London Astronomical Society, 583
Wolf Gesellschaft (Rudolf_), 508, 514
Wolfsburg, 241
Wollongong, 32
Wollongong Amateur Astronomy Club (WAAC), 32
Wollongong Science Centre and Planetarium, 32
Wolverhampton Astronomical Society, 584
Wood-Cock Hill, 298
Woodson High School Planetarium, 906
Woodson Planetarium (Margaret C._), 829, 832
Woodstock Products Inc., 651
Worcester, 746, 750
World Association of Industrial and Technological Research Organisations (WAITRO), 140
World Data Center A for Meteorology, 832
World Data Center A for Rockets and Satellites, 745
World Data Center A for Solar-Terrestrial Physics, 661
World Data Center B for Solar-Terrestrial Physics, 451
World Data Center C2 for Ionosphere, 362
World Data Center D for Astronomy, 118
World Data Center for Solar Activity, 362
World Intellectual Property Organization (WIPO), 514
World Meteorological Organization (WMO), 514
World Radiation Center, 514
World Radiation Center (WRC), 507
World Wide Web Consortium (W3C), 760
Worley Observatory (Ralph A._), 730
Worth Hill Observatory (WHO), 584
Worthing, 532
Worthing Astronomical Society (WAS), 584
Worthington, 842
Wraysbury, 546
WRBP
    W. Russell Blake Planetarium (WRBP), 750
WRC
    World Radiation Center, 507
Wright Nuclear Structure Laboratory (WNSL) (A.W._), 666

Wright Observatory (Ernest B._), 815, 826
Wrightwood, 613
WRO
    Warren Rupp Observatory, 845
Wrocław, 430, 432
Wrocław University
    Astronomical Institute, 432
W. Russell Blake Planetarium (WRBP) (Dr._), 750, 760
WSAAG
    Western Sydney Amateur Astronomy Group, 32
WSAL
    Winston-Salem Astronomical League, 832
WSC
    Wayne State College
        Fred G. Dale Planetarium, 782
WSEN
    Western Space Education Network, 105
WSO
    Wilcox Solar Observatory, 642
WSP
    Winter Star Party, 685
WSRT
    Westerbork Synthesis Radio Telescope, 401
WSU
    Washington State University
        Astronomy Club, 910
        Planetarium, 910
        Program in Astronomy, 910
    Wayne State University
        Department of Physics and Astronomy, 769

    Planetarium, 769
    Weber State University
        Layton P. Ott Planetarium, 890, 891
WTA
    Westerwälder Tag der Astronomie, 234
    West Texas Astronomers, 889
Würzburg, 257, 263
WUStL
    Washington University in Saint Louis
        Department of Earth and Planetary Sciences, 779
        Department of Physics, 779
        Laboratory for Space Sciences, 779
        McDonnell Center for the Space Sciences, 779
WVAS
    Wabash Valley Astronomical Society, 714
WVSGC
    West Virginia Space Grant Consortium, 913
WVU
    West Virginia University
        Astronomy Club, 913
        Department of Physics, 913
WWU
    Western Washington University
        Physics/Astronomy Department, 911
        Planetarium, 911
WYAS
    West Yorkshire Astronomical Society, 583
Wyboston, 535
Wycombe Astronomical Society (WAS), 584
Wylam, 558
Wyoming, 922, 923
Wyoming Infrared Observatory (WIRO), 923
Wyomissing, 852
Wyss Photo-Video (P._), 508, 514

X

Xinglong Station, 113
Xinjiang, 115
X-Ray Instrumentation Associates (XIA), 651
X-Ray Timing Explorer (XTE), 754
XTE
    X-Ray Timing Explorer, 754

Y

Yakima, 911
Yakima Valley Astronomy Club (YVAC), 911
Yakutsk, 446
Yale Southern Station, 666
Yale University
    A.W. Wright Nuclear Structure Laboratory (WNSL), 666
    Department of Astronomy, 666
    Department of Physics, 666
    Press, 666
Yale University Press, 666
Yamagawa Radio Observatory, 344, 362
Yamaguchi, 362
Yamaguchi Prefectural Museum, 362
Yamanashi, 356, 361
Yandina, 15
Yanna Research Station, 917
Yarmouth, 572
Yaroslavl, 451
Yaroslavl Astronomical and Geodetical Society "Meridian", 451
Yaroslavl State Pedagogical University (YSPU)
    Astronomical Observatory, 451
YAS
    York Astronomical Society, 585
Yatsugakate, 345
YCAS
    York County Astronomical Society, 863
Yebes, 481

Yellowstone Valley Amateur Astronomers (YVAA), 781
Yeoville, 464
Yerevan, 12, 13
Yerevan Physics Institute (YerPhi)
    Department of Theoretical Physics, 12
Yerevan State University (YSU)
    Department of Astrophysics, 13
Yerkes Observatory, 915, 919, 921
YerPhi
    Yerevan Physics Institute
        Department of Theoretical Physics, 12
YITP
    Yukawa Institute for Theoretical Physics, 350
Ylistaro, 149
Yohkoh Public Outreach Project (YPOP), 781
Yokohama, 342, 346, 348, 356, 362
Yokohama Science Center (YSC)
    Astronomy Section, 362
Yonkers, 807
Yonsei University
    Center for Space Astrophysics, 369
    Department of Astronomy, 368, 369
York, 585, 863
    (East.), 82
    (North.), 96, 98, 105
York Astronomical Society (YAS), 585
York County Astronomical Society (YCAS), 863
York Observatory, 585
Yorkshire
    (West.), 583
York-Simcoe Amateur Astronomers, 105
Yorktown, 815
York University, 105
    Department of Earth and Atmospheric Sciences, 105
    Department of Physics and Astronomy, 105
York University Astronomy Club (YUAC), 105
York University Astronomy Physics Club, 105
Young, 587
Young Harris, 691
Young Harris College (YHC)
    Rollins Planetarium, 691
Young Observatory, 632
Young Planetarium (JYP) (John.), 683
Youngstown, 844, 845
Youngstown State University (YSU)
    Department of Physics and Astronomy, 845
    Ward Beecher Planetarium, 844
Young University (Brigham.), 890, 892
    Department of Physics and Astronomy, 890
    Sarah B. Summerhays Planetarium, 890
Youreyeball Observatory, 733
Youth Astronomical School, 451
YPOP
    Yohkoh Public Outreach Project, 781
Ypsilanti, 762, 763
YSC
    Yokohama Science Center
        Astronomy Section, 362
    Young Harris College
        Rollins Planetarium, 691
YSPU
    Yaroslavl State Pedagogical University
        Astronomical Observatory, 451
YSU
    Yerevan State University
        Department of Astrophysics, 13
    Youngstown State University
        Department of Physics and Astronomy, 845
        Ward Beecher Planetarium, 844
YUAC
    York University Astronomy Club, 105
Yukawa Institute for Theoretical Physics (YITP), 350, 362
Yunnan Observatory, 115, 118
Yuzhno-Sakhalinsk, 441
YVAA
    Yellowstone Valley Amateur Astronomers, 781
YVAC
    Yakima Valley Astronomy Club, 911
Yverdon-les-Bains, 510

Z

Zabaikalsky State Pedagogical University, 451
Zadar, 121
Zagreb, 121, 122
Zalaegerszeg, 273
ZAMG
    Zentralanstalt für Meteorologie und Geodynamik, 42
Zaporozhye, 530
Zaporozhye Astronomical Club Altair, 530
Zaragoza, 470, 473, 476, 486
Zarvan Co. Ltd., 293
Zavod za Standardizacija i Metrologija (ZSM), 377
Ždánice, 133
Ždánice Public Observatory, 133
Zdiby, 131
Žebrák, 133
Žebrák Popular Observatory, 133
Zeiss (Carl.), 234
    Chester, 897
    Germany, 220
    Jena, 220
    USA-VA, 897
    Virginia, 897
Zeiss-Grossplanetarium Berlin (ZGP), 265
Zeiss Historica Society of America (ZHSA), 797
Zeiss Jena GmbH (Carl.), 265
Zeiss-Kleinplanetarium Herzberg, 266
Zeiss Optical (Carl.), 897, 906
Zeiss-Planetarium der Ernst-Abbe-Stiftung, 266
Zelenchukskaya, 445
Zemun, 454
Zenas Crane Observatory, 724
Zentralanstalt für Meteorologie und Geodynamik (ZAMG), 41
Zermatt, 326
Zeuthen, 224
ZGP
    Zeiss-Grossplanetarium Berlin, 265
ZHSA
    Zeiss Historica Society of America, 798
Žilina, 461
Žilina Observatory, 461
Zimbabwe, 929
Zimmerwald Station, 513
Zittau, 260
Zlín, 133
Zlín Astronomical Observatory, 133
Zografos, 268, 269
Zonnewijzerkring (De.), 390, 407
Zuidlaren, 406
Zúlia, 928
Zürich, 501–504, 506–508, 511–513
Zvenigorod, 445
Zwiggelte, 400

Updating Form

Copy the following form for adding entries and for updating or completing the data given in the directory. Please list on a separate sheet the different departments, divisions, committees, commissions or sections of the organization, as well as the names **and addresses** of collaborating organizations. If the organization has several branches with different geographical locations, a questionnaire should be filled in for each of them.

Please return the form, together with **a sample copy of the most recent annual report or periodical** published by the organization to:

> **Dr. Harry (J.J.) Blom**
> Kluwer Academic Publishers
> P.O. Box 17
> NL- 3300 AA Dordrecht
> The Netherlands
> Telephone: +31 78 657 6315
> Telefax: +31 78 657 6388
> Electronic mail: harry.blom@wkap.nl
> WWW: http://www.wkap.nl/

Please note that all questionnaires returned become our property and we reserve the right to edit the answers for concision and homogeneity (cross-references or pointers are provided within the directory and the databases whenever necessary).

We reserve also the right to conduct independent checks and to reject entries if the information provided seems suspicious or if it is not accompanied by any kind of documentation. In spite of all the care devoted to this service and repeated verifications, inaccuracies may exist, for which we disclaim responsibility. It is always advisable to fill in the questionnaires with a typewriter and to identify properly a contact point for possible further queries.

Thanks in advance for your assistance!

Data to be published in the next releases of the **Star*s Family of Astronomy and Related Resources** (explanations of the notes are given on the back of this sheet):

- Full name (1) ....:

- Abbreviation (2) ....:

- English translation (3) .:

- Address ....:

- Postal address (4) ......:

- Country ....:

Indicate to which category/categories your organization belongs:

o academy	o journal	o public observatory
o advisory or expert committee	o manufacturer	o publisher
o association, club or society	o meteorological office	o research inst. (astronomy)
o data or documentation centre	o museum	o research inst. (Earth)
o dealer or distributor	o norms and standards office	o research inst. (space)
o educational institution	o planetarium	o research inst. (other – specify)
o funding agency or institution	o private consultant	o software producer/distributor (specify)
o other (specify)		

- Telephone number (5) ..:   - Telefax number (5):
- E-mail address(es) (6) ..:
- WWW (7) ....:

- Foundation year (8) ....:
- Members or staff (9) ...:
- Major activities (10) ...:

- Periodicals (11) ....:

- Awards (12) ....:

- Observing sites (13) ....:

- Planetariums (14) ......:

Person who filled in the questionnaire:

Full name and position: ....

E-mail: ....

Signature: ....Date: ....

NOTES:

(1) Full name of the organization.

(2) If used.

(3) If applicable.

(4) If different from previous item.

(5) Including the area code (not the country code).

(6) Electronic mail addresses (including names of networks).

(7) Uniform Resource Locators for World-Wide Web access.

(8) Specific to the organization, department, unit, ...

(9) Number of members or of staff.

(10) Maximum 20 (key)words.

(11) If applicable, titles, ISS-Numbers, frequencies and circulations of periodicals PUBLISHED BY the organization.

(12) If applicable, prizes, distinctions, etc. AWARDED BY the organization. Please indicate also their frequencies.

(12) If applicable, names and geographical coordinates of observing sites BELONGING TO the organization. Please indicate the longitude and latitude in degrees, minutes, seconds, and the altitude in meters.

(14) If applicable, names of planetariums BELONGING TO the organization. Please indicate also their addresses if different.

Please return this form, together with **a sample copy of the most recent annual report or periodical** published by the organization to:

Dr. Harry (J.J.) Blom
Kluwer Academic Publishers
P.O. Box 17
NL- 3300 AA Dordrecht
The Netherlands
Telephone: +31 78 657 6315
Telefax: +31 78 657 6388
Electronic mail: harry.blom@wkap.nl
WWW: http://www.wkap.nl/